Westcott's
Plant Disease
Handbook

Westcott's Plant Disease Handbook

Fifth Edition

REVISED BY

R. Kenneth Horst, Ph.D.

An avi Book
Published by Van Nostrand Reinhold
New York

An AVI Book
(AVI is an imprint of Van Nostrand Reinhold)

Copyright © 1990 by Van Nostrand Reinhold

Library of Congress Catalog Card Number 89–27795
ISBN 0–442–31853–7

Printed in the United States of America

Van Nostrand Reinhold
115 Fifth Avenue
New York, New York 10003

Van Nostrand Reinhold International Company Limited
11 New Fetter Lane
London EC4P 4EE, England

Van Nostrand Reinhold
480 La Trobe Street
Melbourne, Victoria 3000, Australia

Nelson Canada
1120 Birchmount Road
Scarborough, Ontario M1K 5G4, Canada

16 15 14 13 12 11 10 9 8 7 6 5 4 3 2 1

Library of Congress Cataloging-in-Publication Data
Westcott, Cynthia, 1898–
   [Plant disease handbook]
   Westcott's plant disease handbook. — 5th ed. / revised by R.
Kenneth Horst.
      p.   cm.
   Includes bibliographical references.
   ISBN 0–442–31853–7
   1. Plant diseases—Handbooks, manuals, etc.   2. Phytopathogenic
microorganisms—Control—Handbooks, manuals, etc.   3. Plant
diseases—United States—Handbooks, manuals, etc.   4. Phytopatho-
genic microorganisms—Control—United States—Handbooks, manuals,
etc.   I. Horst, R. Kenneth (Ralph Kenneth), 1935-   .   II. Title.
III. Title: plant disease handbook.
SB731.W47   1990
632'.3—dc20   89–27795
                                                              CIP

To my wife, Hope, who is always there
to give me encouragement and support.

Contents

Preface to the Fifth Edition, ix
Preface to the First Edition, xi
How to Use This Book, xv

Introduction   1

What is Plant Disease? 2   Plant Pathology in the United
States, 4   Principals of Control, 5

1. Garden Chemicals and Their Application   7

Fungicides, 10   Bactericides, 40   Nematicides, 41
Virocides, 47   Sources of Pesticides, 47   Applying
the Chemicals, 51   Mixing the Chemicals, 56   All-
Purpose Sprays and Dusts, 58   Integrated Pest
Management, 58

2. Classification of Plant Pathogens   60

Fungi, 62   Bacteria, 78   Viruses, 81   Nematodes, 84

3. Plant Diseases and Their Pathogens   86

Anthracnose, 87   Bacterial Diseases, 98   Black
Knot, 124   Blackleg, 126   Black Mildew, 127   Black-
spot, 130   Blights, 135   Blotch Diseases, 187
Broomrapes, 191   Cankers and Diebacks, 192   Club
Root, 226   Damping-off, 227   Dodder, 228   Downy
Mildews, 231   Fairy Rings, 237   Fruit Spots, 238
Galls, 239   Leaf Blister and Leaf Curl Diseases, 242   Leaf
Scorch, 246   Leaf Spots, 248   Lichens, 295
Mistletoe, 296   Molds, 299   Needle Casts, 301
Nematodes, 306   Nonparasitic Diseases, 333   Powdery

Mildews, 349   Rots, 362   Rusts, 421   Scab, 451
Surf, 458   Slime Molds, 458   Smuts, 459   Snow-
mold, 467   Sooty Mold, 468   Spot Anthracnose, 470
Virus Diseases, 475   White Rusts, 500   Wilt
Diseases, 502   Witchweed, 515

4. Host Plants and Their Diseases 516

*List of Land-Grant Institutions and
Agricultural Experiment Stations in
The United States, 875*

Glossary, 879

Selected Bibliography, 889

Index, 899

Preface to the Fifth Edition

It was a compliment to me to be asked to prepare the fourth edition of *Westcott's Plant Disease Handbook*, and the decision to accept the responsibility for the fourth edition and now the fifth edition was not taken lightly. The task has been a formidable one. I have always had a great respect professionally for Dr. Cynthia Westcott. That respect has grown considerably with the completion of the two editions. I now fully realize the tremendous amount of effort expended by Dr. Westcott in developing the *Handbook*. A book such as this is never finished, since one is never sure that everything has been included that should be. I would quote and endorse the words of Dr. Westcott in her preface to the first edition: "It is easy enough to start a book on plant disease. It is impossible to finish it."

This revision of the *Handbook* retains the same general format contained in the previous editions. The chemicals and pesticides regulations have been updated; a few taxonomic changes have been made in the bacteria, fungi, and mistletoes; the changing picture in diseases caused by viruses and/or viruslike agents has been described. A few new host plants have been added, and many recently reported diseases as well as previously known diseases listed now on new hosts have been included. In addition, photographs have been replaced where possible, and the color photograph section has been retained. For the photography work I am grateful for the help and expertise of Kent E. Loeffler. I also had access to the Cornell Plant Pathology Herbarium, which contains a wealth of photographic work on plant diseases that has been supplied by numerous scientists over many years.

This book should be useful to gardeners, botanical gardens, landscape architects, florists, nurserymen, seed and fungicide dealers, pesticide applicators, arborists, cooperative extension agents and specialists, plant pathologists, and consultants. The book should also be a useful reference book for plant pathology classrooms.

Acknowledgments

I am indebted to many people for advice and suggestions for the present revision. A few who have been particularly helpful are T. J. Burr, R. A. Dunn, R. P. Korf, K. E. Loeffler, W. F. Mai, and M. S. Szyndel. I want to recognize the help of my technician, S. O. Kawamoto, in my research program at Cornell University while this revision was being made. Finally, I recognize and appreciate the professional and efficient job of typing the manuscript by Helene J. Croft, and the help I received from Helene Croft, Jeff Hogue, Stanley Kawamoto, Cristi Palmer, Leah Porter, M. Schollenberger, and my wife, Hope, in proofreading the manuscript.

Preface to the First Edition

The *Plant Disease Handbook* was designed as a companion volume to *The Gardener's Bug Book*, a reference book for professional and amateur gardeners and those who advise them. It turns out to be a formidable tome, and a hybrid to boot, composed of purely technical information crossed with admonitions to the layman. The result is neither a comforting bedside volume for the first-year gardener nor a treatise for the specialist in any one field. It is a compendium (and that word means inclusion within *small* compass of a *large* subject) of available information on diseases of plants grown in gardens or in the home in continental United States. It includes some references to Alaska and the subtropical region of southern Florida, but excludes the purely tropical problems of Hawaii, the Canal Zone, and Puerto Rico. It includes florists' crops grown for home decoration and native plants sometimes grown in wild gardens, but excludes cotton, wheat, and other field crops.

This information is filtered through, and somewhat colored by, my own experience. Once upon a time I was a normal plant pathologist. Since 1933 I have been a practicing plant physician called upon to minister to private garden patients and expected to act as a liaison agent between the university and the gardening public. When requests for free information get so numerous I cannot salvage enough time to take care of the paying patients I write a book—first, to save my time in finding the answers, second, to encourage a few gardeners to look up the answers for themselves.

The Plant Doctor, published in 1937, was quickly written, for it was based entirely on the doctor's casebook and limited to diseases and pests found most commonly in northeastern gardens. The day that galley proofs went back to the printer I set out to discover how many of my observations were true for the rest of the country. This handbook has been in the making ever since. While New Jersey gardens sleep in the winter the Ford and I have wandered thousands of miles. I have visited tiny backyard gardens and large estates, public gardens and parks, commercial nurseries and greenhouses, universities and experiment stations, from New England to Florida, from New York to California. To cite individually the people who have opened their homes and gardens, answered questions, provided bulletins and reprints, and shown experiments in progress, would fill many pages. I can give here only a collective thank-you. Garden visiting is by no means a sunny weather proposition and I am particularly grateful to the indefatigable souls who trudged around with me in the cloudbursts of "sunny" Florida and California, in Texas snowstorms, Louisiana icestorms, Iowa windstorms, and Virginia high tides. We who garden in the heat and humidity of New Jersey summers have no monopoly on bad weather.

Superimposed on such a background is the literature summary presented here, arranged for finger-tip reference. I hope that it will prove useful not only to gardeners but to those who serve them—the landscape architects who design their plantings, the florists and nurserymen who grow their plants and cut-flowers, the dealers who supply their seeds and fungicides, the arborists who care for their trees and shrubs, the county agents who answer their questions, and the plant pathologists who diagnose their diseased specimens.

The charge has been made that plant pathologists can tell you what a disease is but seldom what to do about it except to "Remove diseased plants or parts; rake up and burn fallen leaves." The charge has been made against this handbook, reviewed in manuscript, that too many possible control measures are offered without indication of the best. I plead guilty on both counts. We really know very little about disease control in home gardens and, while sanitary measures do not always lessen the incidence of disease, they work fairly well in some instances. In no instance is it possible for me to recommend a single chemical that will control a given disease in all parts of the country in all seasons without plant injury. I can make no such recommendation for rose diseases, and I have tested fungicides on roses for a quarter of a century. The most I can say is that a certain spray is expedient in the conditions under which I practice. I can cite numerous instances where the same chemical applied in the same dilution on the same dates controlled black spot in one rose garden and not in the one across the street.

The use of chemicals by amateurs is hazardous in any event. Many of the control measures mentioned have been developed for professional growers and in some cases should be left to them. It is far better for the grower to sell a healthy plant than for the gardener to try to cure a sick one. Without implying that any reputable nurseryman would deliberately sell a diseased plant, I think that there is room for a good deal of education. For that reason disease symptoms which remain constant are given more prominence than control measures which are constantly changing. The grower must learn not to sell, and the gardener to reject, dangerously diseased plants. While scouting for azalea petal blight in Georgia some years ago I met a conscientious nurseryman who said he would certainly like to see that disease and learn to recognize it. I asked him what he thought was the matter with the row of azaleas he was standing by. "That? Oh, that's just weather! We had rain a couple of days ago and the blossoms all collapsed." The blossoms had collapsed from the petal blight, encouraged by the weather.

On the other hand, I found it equally difficult to persuade a park superintendent that his azaleas were actually suffering from drought and not petal blight. Because few gardeners realize the importance of distinguishing between a disease caused by a living organism and that caused by an unfavorable environmental influence I have stressed the latter in the section on Physiogenic Diseases and have included spray injuries under the same heading.

The backbone of the handbook is the list of diseases under host plants in Chapter 5 [Chapter 4 in the fifth edition]. Credit for this section goes almost entirely to Freeman Weiss whose *Check List Revision, Diseases of Economic Plants of the United States* has been published in installments, from 1940 to 1949, in the *Plant Disease Reporter*, issued by The Plant Disease Survey, Division of Mycology and Disease Survey, Bureau of Plant Industry, Soils, and Agricultural Engineering, Agricultural Research Administration, United States Department of Agriculture. A very particular thank-you goes to Dr. Weiss, not only for permission to adapt the check list for my purposes but for painstakingly going over Chapter 5 and making many suggestions as to nomenclature. The record of states where diseases are found is compiled by The Plant Disease Survey from reports by collaborators all over the country. My grateful appreciation goes to these collaborators and to Paul R. Miller, head of the Survey. Dr. Miller has read the entire manuscript and made helpful suggestions.

The bibliography must stand as my acknowledgment of debt to many authors. A great many others, uncited, have also provided information.

The decision to sandwich in classification of the pathogen along with the usual description of the disease (for the benefit of people with a microscope and most especially myself) has added many unexpected months to the preparation of the manuscript. During the years of working in gardens I had forgotten that names could be so baffling and that classifications changed almost as rapidly as control measures. Authorities used in naming bacteria, fungi, and viruses are given in the text.

Finding no consistency in present-day usage of host names — e.g. horsechestnut, horse-chestnut and horse chestnut, mountain ash, mountain-ash and mountainash — I have taken the middle-of-the-road policy and followed *Hortus Second*. The hyphen is used to designate plants which are not what the name implies. That is, mountain-ash (Sorbus) is not a true ash (Fraxinus) nor horse-chestnut (Aesculus) a chestnut (Castanea). This distinction seems useful in dealing with diseases, for they are often confined to a single genus or family.

The photographs are of material collected in doctoring gardens or on cross-country jaunts and are intended to emphasize the more common home garden problems. Many were taken by my former assistant, Lacelle Stites, who deserves special thanks. The drawings are diagrams adapted from various sources.

It is easy enough to start a book on plant disease. It is impossible to finish it. Every garden visited, every meeting attended, every journal read, means additions, deletions, and corrections. So I chop it off, unfinished, while the river of knowledge keeps rolling along.

Glen Ridge, New Jersey Cynthia Westcott
October, 1949

How to Use This Book

This book is a reference manual. You will certainly not read it through from cover to cover, but I hope you will read the Introduction and the first and last section of Chapter 1 on garden chemicals. The chemicals themselves are listed in alphabetical order, by common names where possible, by trade names where these are used in lieu of approved common names. A few materials still in the experimental stage but very promising are included. A few uses are suggested, but many more, with correct dosages, will be found on the labels or in recent publications.

Chapter 2, on the classification of plant pathogens, can be taken or not as desired. It provides a mycological background for students and a review for professional workers. The bibliography gives some of the taxonomic references consulted in preparing this very condensed treatment.

The rest of the book is in two main sections. Chapter 3 describes specific diseases and gives remedies when known. The diseases are grouped according to their common names into forty types treated in alphabetical order.

Chapter 4 gives 1,206 host plants in alphabetical order, from Abelia to Zizia, according to common names except where the Latin name may mean less confusion. Under the hosts the disease are sorted out according to types and you can quickly thumb back to the corresponding section — Anthracnose, Blight, Wilt, and so on — in Chapter 3 by means of the running head at the top of each page.

The book works like a dictionary. In both the disease and host section the Latin name of the pathogen causing the disease is given in **boldface** type and provides the key word, for the individual diseases in each group are listed in alphabetical order according to the names of their pathogens. A specific example in the use of this book is given on page 517.

You may be able to find the information you are seeking directly from the index, which includes common and Latin names of hosts plants, Latin names of pathogens, and common names of the diseases described in Chapter 3. More than 2,400 diseases are included in that chapter and some additional pathogens are listed under Host Plants without a corresponding description of disease.

Addresses of state universities and agricultural experiment stations, sources of help for every gardener, are given on pages 875 through 877, followed by a glossary and a bibliography.

The very best way to use this book is to take it in small doses as needed. Do not let the hundreds of diseases you will never meet worry you too much. And remember that most plants survive, despite their troubles!

Westcott's
Plant Disease
Handbook

Introduction

The chief hazard any garden plant has to endure is its owner or gardener. Moreover, many plants will suffer undue hardship from the publication of this handbook. It is human nature to read symptoms of an ailment and immediately assume it is your own affliction. Jumping to conclusions is as dangerous to plants as it is to humans. A sore throat does not necessarily mean diphtheria. Only a trained physician can diagnose probable diphtheria, and for positive identification a laboratory culture is necessary.

A spotted or yellowed rose leaf does not necessarily mean rose blackspot. Mite injury, spray injury, or reaction to weather conditions may also cause spotted or yellow rose leaves; yet gardeners blithely continue increasing the spray dosage, confident that more and stronger chemicals will control the "disease" and they seldom notice that they are nearly killing the patient in the process.

A browning azalea flower does not necessarily mean the dreaded petal blight. Some years ago a Westcott article on possible azalea troubles appeared in print about the time azalea blooms in a northern region were turning brown from a combination of unusual weather conditions. Some gardeners immediately assumed the worst, thought that the southern blight had arrived in the North, and started spraying. The poor plants, suffering from drought and a heat wave, suffered additional injury from the additional stress of sprays.

All chemicals used as sprays or dusts are injurious to plants under some conditions, the injury varying with the chemical and the dosage, with the species and even the variety of plant, with temperature, soil moisture, and many other factors. Plants suffering from drought are commonly injured by sprays.

So please don't jump to conclusions. Don't do anything in a hurry because the plants are getting sick fast and there is no time for a proper diagnosis. Don't rush to the seed store to buy some chemical you vaguely remember reading about. Relax! You have all the time in the world for proper identification, because, by the time the disease is serious enough for you to notice, it is probably too late for protective spraying this season anyway.

Browning of an azalea flower means nothing as a diagnostic symptom. It could just as well come from frost, heat, or old age as from a pathogen. If the flowers are limp and collapsed with a *slimy* feel, these are good symptoms, but signs of the fungus are needed as well. Thin, slightly curved black bodies (sclerotia) formed at the base of petals are distinctive, but even more conclusive are spores taken from the inside of the petals and examined under a microscope. If these are one-celled, with a little boxlike appendage, then you may reasonably conclude that you have the true azalea petal blight.

This book is about garden diseases, but it is not expected that anyone, amateur or professional, can read a brief description, look at an unfamiliar disease in the garden, and make a very reliable diagnosis. I certainly cannot, and after compiling this tome I am less likely to try than ever before. I have written "water-soaked" or "reddish brown" too many hundreds of times for different diseases to make such symptoms seem very distinctive.

However, if you are a gardener, you can narrow the field down considerably by consulting Chapter 4, where host plants are listed in alphabetical order, and under each the type of disease—blight, canker, leaf spot, and so forth—and then the organisms causing these diseases by their scientific names and the states where they have been reported. By eliminating the types of disease that are obviously different from yours, and by eliminating diseases that are reported only on the West Coast when you live in New York, you may find only two or three possibilities to look up in Chapter 3, which lists, under the different disease groups, the pathogens in alphabetical order, followed by a discussion of each disease. Don't let all the scientific names worry you. It is the only way to make this book a quick and easy reference, for there are very few common names of plant diseases that can be used without confusion. It works just like the telephone book. While thumbing your way down to Smith, John, you do not worry about spelling Smiecinski, C., which you pass on the way.

If you are a quasi-professional, with little or no formal mycology but trying to keep abreast of a flood of miscellaneous specimens, there is a brief review for you of the salient microscopic characteristics of each genus, together with its classification. These reviews are in small type and may be readily passed over by those interested solely in macroscopic characteristics.

WHAT IS PLANT DISEASE?

There are many definitions of plant disease, the simplest being any deviation from the normal. The concept of the late professor H. H. Whetzel, a great teacher of plant pathology who influenced many students including Dr. Cynthia Westcott, is valid and appropriate even today. "Disease in plants is an injurious physiological process, caused by the *continued* irritation of a primary causal factor, exhibited through abnormal cellular activity and expressed in characteristic pathological conditions called symptoms." The causal factor may be a living organism or an environmental condition. Injury differs from disease in being due to the *transient* irritation of a causal factor, as the wound of an insect, sudden freezing or burning, or application of a poison.

Plant diseases may be *necrotic*, with dying or death of cells, tissues, or organs; *hypoplastic*, resulting in dwarfing or stunting; or *hyperplastic*, with an overgrowth of plant tissue, as in crown gall or club root.

All species of plants, wild and cultivated, are subject to disease. Fossil remains suggest that plant diseases were present on Earth before humans. Certainly humanity has been punished by them ever since the garden of Eden. "I smote you with blasting and with mildew and hail in all the labors of your hands yet ye turned not to me, saith the Lord" (Haggai 2:17).

Attempts at controlling plant disease go back at least to 700 B.C. when the Romans instituted the Robigalia to propitiate the rust gods with prayer and sacrifice. About 470 B.C., Pliny reported that amurca of olives should be sprinkled on plants to prevent attacks of blight, this being our earliest known reference to a fungicide, although Homer, 1000 B.C., wrote of "pest-averting sulfur."

In 1660, at Rouen, France, a law was passed calling for eradication of the barberry as a means of fighting wheat rust, two centuries before anyone knew the true nature of rust or how barberry affected wheat. In the latter part of the eighteenth century, the Englishman Forsyth discoursed on tree surgery and treatment of wounds and cankers. His seemingly fantastic recommendation of a paste of cow dung to promote healing of tree wounds has modern corroboration in research showing that urea speeds up healing of such wounds.

Much of our progress in dealing with plant disease has followed spectacular catastrophes. Modern plant pathology had its start with the blight that swept the potato fields of Europe in 1844 and 1845, resulting in the Irish famine. This lesson in the importance of plant disease to the economic welfare of humankind marked the beginning of public support for investigations into the cause of disease. Two men, both German, laid the firm foundations of our present knowledge. Mycologist Anton de Bary, 1867 to 1888, first proved beyond doubt that fungi associated with plant diseases were pathogenic. Julius Kuhn, a farmer with a doctor's degree in science, first showed the relation between science and practice in the problems of plant disease control. His textbook on *Diseases of Cultivated Plants*, published in 1858, is still useful.

The accidental discovery of bordeaux mixture in France in 1882 marks the beginning of protective spraying for disease control, but the use of drugs goes back to 1824, when sulfur was recommended as an eradicant for powdery mildew. The development of synthetic organic fungicides was sparked by World War II, partly as a result of a search for chemicals to mildew-proof fabrics used by the armed forces. Antibiotics for plant disease control followed their use in medical practice, with a great deal of research in this field since 1949.

Since the establishment of the Environmental Protection Agency in 1972, there has been increased concern about the use of toxic chemicals for controlling plant disease. Moreover, this concern has generated renewed interest in integrated pest management (IPM) and biological control strategies in the 1980s. IPM utilizes all available pertinent information regarding the crop or plant, its pathogens, the environmental conditions expected to prevail, locality, availability of materials, and costs in developing the control program. Biological control is the total or partial destruction of pathogen populations by other organisms. This phenomenon occurs routinely in nature. There are several diseases in which the pathogen cannot develop because the soil, called suppressive soils, contain microorganisms antagonistic to the pathogen, or because the plant that is attacked has been naturally inoculated before or after the pathogen attack, with antagonistic microorganisms. Even higher plants may reduce the amount of pathogen inoculum by trapping available pathogens (trap plants) or by releasing substances toxic to the pathogen into the soil. Although biological antagonisms are subject to numerous ecological limitations, they can be expected to become an important part of control measures employed against many more diseases in future years.

PLANT PATHOLOGY IN THE
UNITED STATES

Organized plant pathology in the United States started in 1885 with a section of mycology in the U.S. Department of Agriculture. In 1904, the start of the great epiphytotic of chestnut blight, which was to wipe out our native trees, stimulated more public interest and support for plant pathology. In 1907, the first university department of plant pathology was established at Cornell University.

The United States Quarantine Act of 1912 officially recognized the possibility of introducing pests and diseases on imported plants, after low-priced nursery seedlings from Europe had brought in the white pine blister rust. This attempt was our first at control by exclusion.

In 1917, during World War I, the Plant Disease Survey was organized as an office of the Bureau of Plant Industry "to collect information on plant diseases in the United States, covering such topics as prevalence, geographical distribution, severity, etc., and to make this information immediately available to all persons interested, especially those concerned with disease control." During World War II, the Plant Disease Survey was in charge of the emergency project "to protect the country's food, feed, fiber and oil supplies by ensuring immediate detection of enemy attempts at crop destruction through the use of plant diseases and providing production specialists and extension workers with prompt and accurate information regarding outbreaks of plant diseases whether introduced inadvertently or by design while still in incipient stages." As a by-product of these wartime surveys we accumulated a good deal of evidence on the prevalence of new and established diseases across the country, in home gardens as well as on farms.

In 1946, a century after *Phytophthora infestans* had made history with the potato blight, a strain of the same fungus started an unprecedented epiphytotic of tomato blight. This disaster led to the forecasting service that warns dealers and growers when certain diseases are imminent.

The Plant Disease Survey has now become the Epidemiology Investigations Section of the Agricultural Research Service of the U.S. Department of Agriculture. The Agricultural Research Service became a part of the Science and Education Administration in 1978. It issues a monthly bulletin, *The Plant Disease Reporter*, based on reports from qualified volunteer collaborators all over the country. The American Phytopathological Society assumed the responsibility for publishing this journal in 1980 and the journal was renamed *Plant Disease*. Much of the material in this handbook is taken from these reports.

PRINCIPLES OF CONTROL

Control of a plant disease means reduction in the amount of damage caused. Our present annual toll from disease is nearly $4 billion. Perfect control is rare, but profitable control, when the increased yield more than covers the cost of chemicals and labor, is quite possible. Commercial growers now average a return of $4 for each $1 so invested. Keeping home plantings ornamental yields a large return in satisfaction and increased property value.

The five fundamental principles of control are exclusion, eradication, protection, resistance, and therapy.

1. *Exclusion* means preventing the entrance and establishment of pathogens in uninfested gardens, states, or countries. For home gardeners it means using certified seed or plants, sorting bulbs before planting, discarding any that are doubtful, possibly treating seeds or tubers or corms before they are planted, and, most especially, refusing obviously diseased specimens from nursery people or dealers. For states and countries, exclusion means quarantines, prohibition by law. Sometimes restricted entry of nursery stock is allowed; the plants are to be grown in isolation and inspected for one or two years before distribution is permitted.

2. *Eradication* means the elimination of a pathogen once it has become established on a plant or in a garden. It can be accomplished by *removal* of diseased specimens, or parts, as in roguing to control virus diseases or cutting off cankered tree limbs; by *cultivating* to keep down weed hosts and deep ploughing or spading to bury diseased plant debris; by *rotation* of susceptible crops with nonsusceptible crops to starve out the pathogen; and by *disinfection*,

usually by chemicals, sometimes by heat treatment. Spraying or dusting foliage with sulfur after mildew mycelium is present is eradication, and so is treating the soil with chloropicrin to kill nematodes and fungi.

3. *Protection* is the interposition of some protective barrier between the susceptible part of the suscept or host and the pathogen. In most instances this barrier is a protective spray or dust applied to the plant in advance of the arrival of the fungus spore; sometimes it means killing insects or other inoculating agents; sometimes it means the erection of a windbreak or other mechanical barrier.

Chapter 1 gives an alphabetical list of chemicals used in present-day protective spraying and dusting, along with eradicant chemicals, and includes notes on compatibility and possibilities of injury. It is here that home gardeners, sometimes commercial growers, can do their plants irreparable harm instead of the good they intend. Spraying is never to be undertaken lightly or thoughtlessly. Stop and think! Read all of the fine print on the label; be sure of your dosage and the safety of that particular chemical on the plant you want to protect, to say nothing of precautions necessary for your own safety.

4. *Resistance* is control by the development of resistant varieties. Resistant varieties are as old as time. Nature has always eliminated the unfit, but since about 1890 humans have been speeding up the process by deliberately breeding, selecting, and propagating plants resistant to the more important diseases. Resistant ornamental plants have lagged behind food plants, but we do have wilt-resistant asters, rust-resistant snapdragons, wilt-resistant mimosas. Here is the ideal way for home gardeners to control their plant diseases—in the winter when the seed order and the nursery list is made out—so easy, and so safe!

5. *Therapy* is control by inoculating or treating the plant with something that will inactivate the pathogen. Chemotherapy is the use of chemicals to inactivate the pathogen, whereas heat is sometimes used to inactivate or inhibit virus development in infected plant tissues so that newly developing tissue may be obtained that is free of the pathogen. The use of this procedure is discussed in Chapter 3.

1

Garden Chemicals and Their Application

A *fungicide* is a toxicant or poison for fungi, a chemical or physical agent that kills or inhibits the development of fungus spores or mycelium. It may be an *eradicant*, applied to a plant, plant part, or the environment as a curative treatment to destroy fungi established within a given area or plant; or preferably it may be a *protectant*, applied to protect a plant or plant part from infection by killing, or inhibiting the development of, fungus spores or mycelium that may arrive at the infection court. A *bactericide* is a toxicant or poison for bacteria. Among the more recent bactericides are antibiotics, products of other living organisms. They also have value against certain fungi. There are also a few *virocides*, which are toxic or poisonous to viruses.

A *disinfectant* is an agent that frees a plant or plant part from infection by destroying the pathogen established within it. A disinfestant kills or inactivates organisms present on the surface of the plant or plant part or in the immediate environment. Chemicals for seed treatment can be either eradicants or protectants, but most of them are disinfestants, in that they kill organisms on the surface of the seed rather than those within. In common usage, however, they are called disinfectants.

A *nematicide* is, of course, a chemical that kills nematodes in the soil or in the plant. Most nematicides are *fumigants*, chemical toxicants that act in volatile form.

Not so long ago the chemicals on the garden medicine shelf consisted of copper and sulfur for protectants, lime sulfur as an eradicant, mercuric chloride as a disinfectant, and formalin and carbon bisulfide as fumigants. You sometimes got plant injury; you did not always get the best possible control, but at least you did not have to be an organic chemist. Now we have the following classes of fungicides:

Inorganic

1. Sulfur
2. Copper
3. Mercury (Discontinued)

Organic

1. Dithiocarbamates	7. Fumigants
2. Thiazoles	8. Antibiotics
3. Triazines	9. Dinitrophenols
4. Substituted aromatics	10. Quinones
5. Dicarboximides	11. Organotins
6. Systemics	12. Aliphatic nitrogens

The search for new fungicides goes on, with hundreds of synthetic organic compounds being screened each year. This screening is often a cooperative venture between manufacturers, state experiment stations, and the U.S. Department of Agriculture. After safety precautions for the operator, and the effectiveness of a compound for certain diseases have been determined, the chief question is whether the material is *phytotoxic*, that is, injurious to plants, at concentrations required for control.

Phytotoxicity is an elusive factor, not to be pinned down in a few tests. It varies not only with the kind of plant but with the particular variety, the amount of moisture in the soil when the spray is applied, the temperature, whether or not the application is followed by rain or high humidity, the section of the country, and the compatibility of the chemical with spreaders or wetting agents, as well as with other fungicides or insecticides. Coordinated tests with new materials in many different states are extremely valuable. Some compounds give rather uniform results over the country; others vary widely with climatic conditions.

The 1947 Federal Insecticide, Fungicide and Rodenticide Act (FIFRA) provides that all fungicides must be registered with the U.S. Department of Agriculture before being marketed. Materials highly toxic to humans must be prominently marked, instructions given for avoiding injury to plants or animals, the toxicant chemical named, and the percentage of active or inactive ingredients given. All labels submitted for registration must be accompanied by some sort of proof that the claims for performance are valid.

In 1954, Public Law 518, known as the Miller Bill, was passed, providing for tolerances. A tolerance is the legal limit of a poisonous residue, expressed in parts per million (ppm), that may remain on an edible product at the time it is distributed for consumption. In 1958, the Food Additives Amendment was passed, which also controls pesticide residues in processed foods. It included the Delaney clause, which states that any chemical found to be a carcinogen in laboratory animals may not appear in a human food.

In 1959, the FIFRA was amended to include nematicides, plant growth regulators, defoliants, and dessicants as pesticides. Since that time, poisons and repellents used against all classes of animals (from invertebrates to mammals) have been brought under control.

FIFRA was further amended in 1972 as the Federal Environmental Pesticide Control Act (FEPCA), making violations by growers, applicators, or dealers subject to heavy fines and/or imprisonment. All pesticides had been classified into either general-use or restricted-use categories by October 1977, with anyone applying restricted pesticides required to be state-certified. Pesticide manufacturing plants are to be registered and government-inspected. All pesticide products must be registered whether shipped in interstate or intrastate commerce. Other provisions are of various degrees of importance to concerned persons or companies.

Additional modifications were made in FIFRA in 1989. The modifications specifically will (1) accelerate reregistration of older pesticides (those registered prior to November 1984) and impose fees on chemical manufacturers for reregistration; (2) essentially eliminate indemnification payments to those holding inventories of suspended or canceled pesticides, except farmers and "certain end users"; and (3) shift part of the burden for storage and disposal of banned pesticides from the government to the manufacturer. The 1989 FIFRA also empowers the Environmental Protection Agency (EPA) to change regulations on how applicators handle, rinse, and dispose of pesticide containers. In the next five years EPA will be determining ways to maximize reuse and refill of containers, minimize disposal of containers, and maximize containment of rinsing. EPA is expected to report to Congress on the progress made on these issues by December 1990.

The federal government considers these to be minimum pesticide regulations. Any state may choose to establish more rigid pesticide regulations within its boundaries than those legislated by the federal government, and some have done so. Some states require notification to be posted prior to commercial pesticide application including chemicals used. Thus, pesticide applicators must be familiar with individual state pesticide regulations as well as federal pesticide regulations.

Consumers, therefore, are well protected against fraud, but they must be willing to read the fine print on labels if they are to choose intelligently from the bewildering array of proprietary compounds on dealers' shelves. And they must also read the fine print and follow directions exactly if their homegrown vegetables are to be as safe for consumption as those from commercial growers who have to comply with the law in the matter of residues.

Even if you follow exactly the directions for dosage given on the label, you may have some plant injury under your particular combination of soil, weather, and kinds of plants. Keep a notebook. Put down the date you sprayed, the dosage used, the approximate temperature and humidity, whether it was cloudy or sunny, in a period of drought or prolonged wet weather. Go around later and check for burning; for leaf spotting and defoliation from the spray or from failure to control the disease; for leaf curling or stunting; for too much unsightly residue. Note which varieties can take the spray and which cannot.

The following alphabetical list includes chemicals now commercially available, a few that are rather outmoded but still found in textbooks, a few that were marketed in the past but have now been discontinued, and a few that will probably be marketed before this text is published. By that time there will be many more that should have been included, for the search for better chemicals is unending. There will also be more that will be discontinued. The list presented herein must be considered only as a guide. Exclusive reliance must be placed on directions and information supplied by the manufacturer or by agricultural specialists, agents, or advisors. *Be sure to read the label.*

Because so many of the new compounds have long, complex chemical names, they have been given short common names by the American Standards Association. Such common names are not capitalized. Frequently, however, no official common name has yet been given to a product, and so the trade name is used as a common name, in which case it should be capitalized. For a comprehensive list of proprietary chemicals and mixtures of pesticides see *Pesticide Handbook*, compiled annually by Dr. E. E. H. Frear and available by writing to College Science Publishers, State College, PA. 16801. The *Farm Chemicals Handbook*, which is published each year by Meister Publishing Co., 37841 Euclid Ave., Willoughby, OH 44094, also gives an up-to-date listing of pesticides.

FUNGICIDES

AATack. See Thiram.

Aaterra. See Koban.

AC 5223. See Dodine.

Acarelte. See Dinobuton.

Acarelte forte. See Dinobuton.

Acetic acid. Present in vinegar, one of the oldest preservatives, formerly used to some extent for damping-off of evergreen seedlings.

Acrex. See Dinobuton.

Acti-dione. See Actidione.

Actidione (discontinued 1987 by Nor-Am Chemical Co.). Cycloheximide, from *Streptomyces griseus*, the first antibiotic introduced (1949) for control of plant disease. Effective in control of powdery mildews, cherry leaf spot, cedar-apple rusts (applied to juniper host to prevent spore horns), white pine blister rust (for treating cankers), turf diseases. Available in different formulations: Actidione Ferrated, with ferrous sulfate for dollar spot, fading out, melting out of turf; Actidione RZ, with pentachloronitrobenzene for azalea petal blight, brown patch, and other turf diseases; Actidione PM, with a safener to prevent phytotoxicity when used for powdery mildew of roses and other ornamentals; in combination with Thiram as Actidione Thiram; also Actispray, Acti-dione TGF and Hizarocin.

Acquinite. See Chloropicrin.

Actispray. See Actidione.

Afugan (Hoechst AG, West Germany). 0,6-ethoxycarbonyl-5-methylpyrazolo [1,5, -a] pyrimidi-n-2-yl 0,0-diethyl phosphorothioate), for control of powdery mildew on cucumber, melon, pumpkin, squash, watermelon, ornamentals, strawberry, apple, grape, and so on. Other names: Curamil and Hoe 002873.

Agri-Mycin 17. See Streptomycin.

Agri-Strep. See Streptomycin.

Agrosol S (discontinued 1984 by Chipman Chemical, Inc.). Combination of maneb and captan. Protective seed treatment for corn, peas, beans, including damping-off, seed decay organisms, and seed blight.

Agrox Strep (discontinued by Chipman Chemical, Inc.). Combination bactericide and fungicide, seed treatment of beans where halo blight is a problem. Registered only in Michigan.

Agrox 2-Way (Chipman Chemical, Inc.). Combination of captan and diazinon. Protective seed treatment for corn, peas, and beans.

Agrox 3-Way (discontinued 1985 by Chipman Chemical, Inc.). Combination of captan, lindane, and diazinon. Protective seed treatment for corn, peas, and beans.

Airone. See Antracol.

Akzo Chemie Maneb. See Maneb.

Allisan. See DCNA.

Ammonium Sulfate. Used around trees and shrubs affected with Phymatotrichum root rot.

Amobam (Artel Chemical Corp.). Diammonium ethylene bisdithiocarbamate. Used for damping-off and some vegetable blights; forms zineb in a tank mix with zinc sulfate. Other name: Chemo-O-Bam.

Anisomycin (Chas. Pfizer). Antibiotic from *Streptomyces griseolus*. Has controlled apple powdery mildew but is phytotoxic; effective for bean rust and anthracnose. See also Streptomycin.

Antracol (Bayer AG, West Germany; Agrimont S.p.A., Italy; Visplant Chimiren S.r.l., Italy). Zinc-N,N'-propylene-1,2-bis-(dithiocarba-mate), used for downy mildew and red fire disease of grape vines, early and late blights of potatoes and tomatoes, blue mold of tobacco, and Sigatoka disease of bananas. Other names: Airone; Bay 46131; LH 30/Z; Taifer.

Apadodine. See Dodine

APL-Luster. See Thiabendazole.

Arasan (discontinued by DuPont). Tetra-methyl thiuram disulfide (thiram). Seed treatment for peanut, corn, bean, pea, beet, carrot, and other vegetables and flowers; used mixed with fertilizer for onion smut control.

Arbotect. See Thiabendazole.

Aspor. See Zineb.

Aules. See Thiram.

Avicol. See PCNB.

Award. See Penconazole.

Banol Turf Fungicide. See Propamocarb hydrochloride.

Banrot (Sierra Chemical Co.). 5-ethoxy-trichloromethyl-1,2,4-thiodiazole, 15%; dimethyl 4,4-0-phenylenebis (3-thioallophanate), in a 3 to 5 ratio. Used as a soil drench for control of damping-off, root and seed rot diseases.

Bardac (Lonza, Inc.). Quaternary ammonium compounds.

Barquat Compounds (Lonza, Inc.). Quaternary ammonium compounds. Germicidal effectiveness against a wide variety of microorganisms in antiseptics, germicides, algicides, deodorants, and detergent-sanitizers.

BASF-Maneb Spritzpulver. See Maneb.

Basfungin (discontinued by BASF Aktiengesellschaft, Federal Republic of Germany). Coprecipitation of zinc ammonia propylene-bio-dithiocarbamate + polypropylene-bis-thiocar-bamoyldisulfide. Used for control of fungus diseases of grapes and hops.

Bavistin. See Carbendazin.

Bay 4631. See Antracol.

Bay 47531. See Euparen.

Bay 49854. See Euparen M.

Bay 572. See Euparen M.

Baycor (Bayer AG, West Germany; Mobay Corp.). *B*-([+1, 1'-biphenyl]-4 xloxy)-*α* (1, 1 dimethylethyl) 1H-1, 2, 4-triazole-1-ethanol. Used for scab and *Monilinia* on pome and stone fruit; leafspot diseases and powdery mildew on stone fruits, bananas, vegetables, sugar beets, peanuts, and ornamentals.

Bayleton (Bayer AG, Federal Republic Germany; Mobay Corp.). 1-(4-chlorophenoxy)-3,3-dimethyl-1-(1,2, 4-triazol-1-yl)-butan-2-one. Used on mildew and rust affecting vegetables, grapes, and ornamentals.

Benlate, benomyl (Du Pont). Methyl 1-(butylcarbamoyl)-2-benzimidazolecarbamate. Highly active fungicide with systemic, residual, and curative properties, also ovacidal for mites, of low order of toxicity to plants and animals although phytotoxicity has been reported. Especially effective for powdery mildew, brown rot, Cercospora and cherry leaf spots, rose black spot, Botrytis blights, and club root. Other names: Benex, Benor, Tersan 1991.

Benomyl. See Benlate.

Bentonite (Applied Industrial Materials Corp.; Lowe's Industrial Products). Hydrated aluminum silicate, a clay mineral used as a diluent in dusts.

Benzimidazoles. A group of systemic fungicides including benomyl (Benlate), thiabendazol (Mertect), thiophanatemethyl, and OM 2424 (Terrazole).

Binapacryl. Sold as Morocide (discontinued 1987 by Hoechst AG, F.R.G.). 2-*sec* butyl-4,6-dinitrophenyl-3-methyl-2-butenoate. Miticide also used for powdery mildew, registered for apples and other fruits.

Bioguard. See Thiabendazole (discontinued by Merck Chemical Div.).

Bioquin 1 (Monsanto). Copper oxinate or copper 8-quinolinolate organic fungicide; has been used for sycamore anthracnose, avocado fruit rot.

Biotrol-Plus (Nutrilite Products, Inc.). Contains *Bacillus thuringiensis* 1.2% with pyrethrins 2.0% in petroleum distillate for insecticide. Used for control of cabbage loopers, cabbage worms, leaf rollers, leaf hoppers, thrips, aphids, and certain insects on ornamental flowers, cole crops, celery, tomatoes, and potatoes.

Biotrol VHZ (discontinued 1985 by Zoecon Corp.). Contains Heliothis Polyhedrosis Virus. Used for bollworm and corn earworm.

Biotrol XK (discontinued by Nutrilite Products, Inc.). *Bacillus thuringiensis*.

Borax (Kerr-McGee Chemical Corp.). Used as a wash for citrus fruit molds, a dip for sweetpotato sprouts, in boron-deficient soils for black heart of beets, cracked heart of celery, internal cork of apples.

Bordeaux mixture (Leffingwell, Uniroyal Chemical Co.). Mixture of hydrated lime and copper sulfate; or as produced, for example, a fixed copper fungicide with 12.75% copper (elemental) content. Prepared from copper sulfate and lime to form a membranous coating over plant parts, the first protective spray and still widely used. About 1878, French vineyards were threatened with downy mildew, which had been introduced from the United States. Millardet, one of the workers assigned to the problem, noticed that where grapes near the highways to Bordeaux had been treated with a poisonous-looking mixture of copper and lime to prevent stealing, there was little or no downy mildew. A description of the preparation of bordeaux mixture was published in 1885, and it remains a most efficient fungicide. It does, however, have a most conspicuous residue, and is injurious to some plants.

Bordeaux mixture is made in varying concentrations. The most usual formula is 8-8-100 (often stated as 4-4-50), which means 8 pounds copper sulfate, 8 pounds of hydrated lime to 100 gallons of water. Stock solutions are made up for each chemical (1 pound per gallon of water), the lime solution placed first in the sprayer, diluted to nearly the full amount, and the copper sulfate solution added. Or, for power sprayers, finely divided copper sulfate (snow) can be washed through the strainer into the spray tank, and when the spray tank is two-thirds full the weighed amount of hydrated lime can also be washed through the strainer while the agitator is running. Casein or other spreader is added toward the end.

For ornamentals, a 4-4-100 is usually strong enough and can be made in small amounts by dissolving 2 ounces of copper sulfate in 1 gallon of water, 2 ounces of hydrated lime in 2 gallons of water, pouring the copper sulfate solution into the lime water, and straining into the spray tank through fine cheesecloth.

For some plants, as stone fruits, the proportion of lime is increased; for others, as azaleas, a low-lime bordeaux is used. Once the two solutions have been mixed, the preparation must be used immediately. Fresh lime is essential, not some left over from a previous season. Somewhat less effective than homemade bordeaux but easier for the home gardener are the various powders and pastes available under trade names; to these add only water at the time of use.

Phytotoxicity comes from both the lime and the copper. Plants are often stunted, with yield reduced; fruit-setting of tomatoes may be delayed. Bordeaux is not safe on peaches during the growing season, may burn and russet apples (both foliage and fruits), may cause red spotting, yellowing, and dropping of rose leaves (often confused with blackspot by amateur and sometimes professional gardeners), and may cause defoliation of Japanese plums. Injury is most prominent early in the season when temperature is below 50° F and in dull, cloudy weather when light rain or high humidity prevents rapid drying of the spray. Late summer use of bordeaux is credited with making some plants more susceptible to early fall frosts. Other name: Bordocop.

Bordeaux-oil emulsion. For citrus fruits. Enough oil emulsion is added to the dilute bordeaux mixture to make 1% actual oil. The copper kills the beneficial fungi, keeping down scale insects, so the oil is added to kill the scales.

Bordeaux paint. Raw linseed oil is stirred into dry bordeaux powder until thin enough to apply with a brush to pruning wounds, especially those on apple trees after pruning for fire blight.

Bordeaux wash. Water is added to dry bordeaux powder and half as much lime until the mixture is like thin paint. It is applied to the lower trunks of citrus trees to prevent brown rot gummosis.

Bordocop. See Bordeaux mixture.

Botran (Nor-Am Chemical Co.). See DCNA.

Botrilex (Uniroyal Chemical Co.). See PCNB.

Brassicol (discontinued 1985 by Hoechst A G). See PCNB.

Bravo (Fermenta Plant Protection Co.). Tetrachloroisophthalonitrile. Used as a broad-spectrum fungicide on snap beans, cole crops, carrot, celery, sweet corn, cucumber, onion, cantaloupe, muskmelon, honeydew, watermelon, squash, pumpkin, peanut, potato, tomato, turf, and ornamentals. Other names: Clorto Caffaro, Clorto Caffaro Flow, Clortocaf Romato, Clortosip, Daconil 2787, Exotherm Termil, Turfcide.

Brestan (Hoechst AG, F. R. G.). Triphenyltin acetate. Effective for pecan scab, potato late blight, downy mildew of cucumbers; slightly phytotoxic.

Brestanid. See Du-Ter.

Bromofume. See Ethylene dibromide.

Brom-O-Gas. See Methyl bromide.

Brom-O-Gaz. See Methyl bromide.

Bromoethane. See Methyl bromide.

Brom-O-Sol (Great Lakes Chemical Corp.). Contains methyl bromide 68.6%, chloropicrin 1.4%, organic solvent 30%. Used to control soil-borne diseases, nematodes, and insects. Growth difficulty may be experienced on carnations, holly, snapdragons, and multiflora rose. Apply at least 18 inches from plants to be maintained.

Bupirimate (ICI Agrochemicals). 5-butyl-2-ethylamino-6-methyl-pyrimidin-4-yl-dimethylsulfamate. Systemic fungicide for control of powdery mildew of fruit and ornamentals. Other name: Nimrod.

Burcop. See Burgundy mixture.

Burgundy mixture (La Cornubia S.A.). A soda bordeaux formerly used, prepared with washing soda instead of lime. Other names: Comac, Burcop.

Busan 72 A (Discontinued by Buckman Laboratories, Inc.). 2-(thiocyano-methylthio) benzothiazole. Used for treatment of pine, fir, and cotton seed and as a soaking treatment for iris, narcissus, tulip bulbs, and gladiolus corms.

Caddy (W. A. Cleary Chemical Corp.). Contains 20.1%, equivalent to 12.3% elemental cadmium. Used for control of dollar spot, copper spot, Helminthosporium, and Curvularia leaf spots on turf. Other names: Cadmium chloride, Vi-Cad.

Cadminate (Mallinckrodt, Inc.). Cadmium succinate, used for dollar spot, copper spot, red thread, and snow mold of turf.

Cadmium chloride. See Caddy.

Cadmium sulfate. 14% solution. Used for painting bark surface in infected areas of apple, pear trees.

Cad-Trete (W. A. Cleary Chemical Corp.). Thiram and cadmium chloride hydrate. Used as multipurpose turf fungicide.

Calcium sulfide. Gives good control of apple scab.

Calixin (BASF AG, F. R. G.). Reaction mixture of C_{11}-C_{14} 4 alkyl-2,6 dimethylmorpholine homologues containing 60% to 70% of 4-tridicyl isomers. Systemic fungicide with protective, curative properties. Controls wide range of diseases (black sigatoka, pink disease, powdery mildew, rust), of economic importance to fruit (bananas, mango), vegetables (cucurbits).

Caocobre. See Copper oxides.

Captafol. See Difolatan.

Captan. N-trichloromethylthiotetrahydrophthalimide, sold as Orthocide Garden Fungicide (Chevron Chemical Co.). Used extensively for control of many fruit and vegetable diseases — apple scab, peach brown rot, cherry leaf spot, black rot of grapes; for bluegrass melting-out and other turf diseases; for rose blackspot. Also used as a seed protectant and for conifer and other seedlings to prevent damping-off. It is of questionable compatibility with oils, dinitros, alkaline materials.

Capthion (ICI Australia Ltd.). Mixture of captan, sulfur, and malathion. Used for control of home garden diseases and insects. Not available in the United States.

Carbam. See Metam-sodium.

Carbamate. See Ferbam.

Carbendazin (Hoechst AG, F. R. G.; BASF AG, F. R. G.; DuPont). 2-(methoxycarbonylamino)-benzimidazole. Controls wide range of Ascomycetes and Fungi Imperfecti and numerous Basidiomycetes. Used on vegetables, fruits, ornamentals, and turf. Other names: Bavistin, Delsene, Derosal, Equitdazin, Hoe 017411.

Carbon disulfide (ICI Americas). Soil fumigant used for control of Armillaria root rot and sometimes for nematodes. This chemical is highly inflammable, also dangerous to living plants. Use only on fallow soil.

Carboxin. See Vitavax.

Carpene. See Dodine.

Casein. One of the proteins in dried skim milk, used as a spreader for sprays, to reduce surface tension.

Cela W 524. See Triforine.

Celfume. See Methyl bromide.

Cercobin. See Thiophanate.

Cercobin M. See Thiophanate-methyl.

CGA 38140. See Fongarid.

CGA 48988. See Metalaxyl.

CGA 71818. See Penconazole.

Chem Bam. See Nabam.

Chem Neb. See Maneb.

Chem-O-Bam. See Amobam.

Chemsect. See Dinitro compounds.

Chem Zineb. See Zineb.

Chinosol (Probelte, S. A.). 8-Hydroxyquinoline sulfate. Soil control of certain vascular wilts on citrus, fruit trees, vegetables, vine. Bacterial diseases of plants, fungus on cutting flowers.

Chipco 26019. See Iprodione.

Chipco 26019 Flo. See Iprodione.

Chipco Spot Kleen (discontinued by Rhone-Poulenc Inc.). 4,4'-0-phenylenebis (3-thio) allophanate. Used to control major turf diseases on golf course greens, tees, and fairways.

Chipco Thiram 75. See Thiram.

Chloranil (Uniroyal). Tetrachloro-para-benzoquinone, sold as Spergon. Seed protectant to prevent damping-off of peas, beans, soybeans, corn, peanuts; root and sprout dip for sweet potatoes; bulb dip; fungicide for turf brown patch and dollar spot. Other name: Spergon.

Chloroneb (Kincaid Enterprises, Inc.). 1,4-dichloro-2,5-dimethoxybenzene. Used for treatment of turfgrass to control snow mold (*Typhula*) and Pythium blight.

Cloro-O-Pic. See Chloropicrin.

Cloro-O-Pic 70. See Chloropicrin.

Chloropicrin (Great Lakes Chemical, Corp.). Tear gas, sold as Larvacide, fumigant for nematodes and some fungi, particularly *Sclerotium rolfsii*, not to be used around living plants. The soil should be prepared as for planting and be moderately loose. Spot injections, 10 inches apart, are made with a hand applicator resembling a huge hypodermic needle, 6 inches deep. It helps to mark the area like a checkerboard with crosswise and lengthwise lines 10 inches apart. On the first row injections are made at the intersections of lines, on the next row they are staggered halfway between; they are again at the intersections on the third row, and so on. Tear gas is disagreeable to handle and must be applied with the special applicator, which can be rented. A gas mask is required for application. The gas works much better at high temperatures, with most rapid killing of root-knot nematodes at 98° F. Chloropicrin should not be used in early spring before the ground has warmed to at least 60° F. The soil should be moist. Apply a water seal immediately after smoothing the soil, wetting the surface for an inch or more. A delay of 2 hours after injection before adding water means a great reduction in effectiveness. Other names: Acquinite, Chlor-O-Pic, Chlor-O-Pic 70, Dojyopicrin, Dolochlor, Larvacide, Pic-Clor, Tri-clor.

Chlorothalonil. See Bravo.

Clorto Caffaro. See Bravo.

Clorto Caffaro Flow. See Bravo.

Clortocof Ramato. See Bravo.

Clortosip. See Bravo.

Comac. See Burgundy mixture.

Cop-O-Cide (discontinued by Tower Chemical Co.). An emulsion of copper salts of fatty and rosin acids.

Copper Acetate. First developed by 1889; became the first factory-made basic copper fungicide.

Copper ammonium carbonate (Mineral Research and Development Corp.). An aqueous blue liquid containing 8% copper as active fungicide. Available as Copper-Count-N.

Copper Carbonate, Basic (Uniroyal Chemical Co.; Tennessee Chemical Co.). Basic cupric carbonates. Used as a seed treatment. Controls smut diseases of grasses. Applied as early-season sprays for tree fruits and nuts. Used to impregnate pear wraps.

Copper chloride, basic. See Copper oxychloride.

Coppercide. See Copper Sulfate, Basic.

Copper compounds. So-called fixed coppers, more stable than bordeaux mixture, less phytotoxic, easier to use, and with less objectionable residue. They include basic sulfates (Tennessee Tribasic, Basicop), basic chlorides (Copper-A Compound, C−O−C−S), copper oxides (Cuprocide), copper ammonium silicate (Coposil), copper zeolite, and copper phosphate. Copper sprays control many blights, leaf spots, downy and powdery mildews.

They are incompatible with lime sulfur, questionable with cryolite, benzene hexachloride, tetraethyl pyrophosphate, organic mercuries, and dithiocarbamates. They may injure plants in cool, cloudy, or moist weather. Injury to apple and rose foliage varies from reddish spots to yellowing and defoliation.

Copper-Count-N. See Copper ammonium carbonate.

Copper-Count-NS. See Copper ammonium carbonate.

Copper-fixed (Agtrol Chemical Products; Tennessee Chemical Co.). Includes basic sulfates, oxychlorides, and oxides. Developed to replace bordeaux mixture.

Copper hydroxide (Agtrol Chemical Products; Griffin Corp.). Cupric hydroxide. Used as general fungicide on many fruit, vegetable, and ornamental plants. Available in various formulations: Kocide 101, Kocide 404, Kocide SD, Kocide 404S, Kocide 101S.

Copper-lime dust. Usually prepared of 20% monohydrated copper sulfate and 80% hydrated lime, sometimes used on potatoes and other vegetables as a substitute for bordeaux mixture. It is more effective applied when foliage is wet.

Copper naphthenate (Agtrol Chemical Products; Troy Chemical Corp.). Chiefly a preservative for wood and fabrics but tested as a fungicide in aerosol sprays.

Copper Nordox. See Copper oxides.

Copper oxides (Agtrol Chemical Products; CP Chemical Corp.). Cuprous oxide; cupric oxide. Used as a seed treatment to control damping-off and as a dust to control vegetable and fruit diseases. Other names: Caocobre, Copper Nordox, Copper-Sandoz, Fungi-Rhap, Nordox SD-45, Nordox SD-50, Oleocuivre, Oleo Nordox, Yellow Cuprocide.

Copper oxychloride. Solution used on potatoes and tomatoes.

Copper Power. See Copper Sulfate, Basic.

Copper Pride. See Copper Sulfate, Basic.

Copper-Sandoz. See Copper oxides.

Copper Sulfate, Basic (Agtrol Chemical Products; CP Chemical, Inc.; Old Bridge Chemical; Tennessee Chemical Co.). Contains a minimum 53% copper (metallic basis). Used widely to control citrus diseases, bacterial and fungus diseases of tomatoes and peppers as well as fruit, vegetables, and ornamentals. Other names: Coppercide, Copper Pride, Copper Power, CP Basic Copper TS-53 WP, KOP 300, Phyto-Bordeaux, Super CU, Tennessee Brand Tri-Basic Copper Sulfate 53 WP, TNCS 53, Tricop.

Corozate. See Ziram.

CP Basic Copper TS-53WP. See Copper Sulfate, Basic.

Crisfolatan. See Difolatan.

Crotothane. See Karathane.

Cryptonol (discontinued by Midox Ltd.). Potassium hydroxyquinoline sulfate. Used as a soil drench, dip, or spray for control of fungal and bacterial diseases or as a glass-house disinfestant.

Cufram Z (Universal Crop Protection Ltd.). Dithiocarbamate complex containing copper, manganese, iron, and zinc. Used to control potato blight, mildew of hops, apple scab, blackcurrant leafspot, downy mildew of vines.

Cufraneb. See Cufram Z.

Cuman. See Ziram.

Cumene. See Zineb.

Curamil. See Afugan.

Curitan. See Dodine.

Curzate M. See Mancozeb; Maneb.

Curzate M8. See Mancozeb.

Cycloheximide. See Actidione.

Cyclomorph. See Dodemorph.

Cyprex. See Dodine.

Daconil 2787. (Fermenta Plant Protection Co.). Tetrachloroisophthalonitrile. Broad-spectrum foliage and fruit protectant, registered for turf and certain ornamentals, controlling various vegetable and fruit diseases. Used as a thermal dust in greenhouses under name of Termil. See also Bravo.

Dazomet (BASF AG; Hopkins Agri. Chem. Co.; ICI Americas, Inc.; UCB Chem. Corp.). Tetrahydro-3,5-dimethyl-2H-thiadiazine-2-thione. Used to control soil fungi, nematodes, weeds, and soil insects; preplant treatment for turf and ornamentals.

DCNA (Nor-Am Chem. Co.). 2,6-dichloro-4-nitroaniline. Active against *Botrytis, Monilinia, Rhizopus, Sclerotinia,* and *Sclerotium* species. Used on apricots, blackberries, boysenberries, carrots, cucumbers, garlic, grapes, lettuce, nectarines, onions, peaches, plum, prune, potatoes, red raspberries, rhubarb, snap beans, sweet cherries, tomatoes, sweet potatoes, and ornamentals. Other names: Allisan, Botran, Kiwi Lustr 277, Resisan.

Dehydroacetic acid. 3-acetyl-6-methyl-2,4 pyrandione. Mold preventative used for processed fruits and vegetables; formerly used as a dip or wrapper impregnant.

Deksonal. See Lesan.

Delan. See Dithianon.

Delan-Col. See Dithianon.

Delsene. See Carbendazin.

Demosan. See Chloroneb. Discontinued by DuPont.

Denarin. See Triforine.

Desain. See Dinobuton.

Devizeb. See Zineb.

Dexon (discontinued by Chemagro, Bayer AG). *p*-dimethylaminobenzene diazo sodium sulfonate. Nonmercurial seed and soil fungicide available as wettable powders, as granules, and mixed with PCNB. Used to control damping-off and Pythium and Phytophthora root rots. A practical soil drench for container-grown plants.

Diazoben. Discontinued.

Dicamate. See Karamate; Mancozeb.

Dichlofluanide. See Euparen.

Dichlone (discontinued by Uniroyal). Common name for 2,3-dichloro-1,4-naphthoquinone, sold as Phygon (Uniroyal). Used for early-season control of fruit diseases, sometimes for celery and tomato blights, and for seed treatment. It is effective in control of rose blackspot and azalea petal blight but slightly phytotoxic. Other name: Phygon.

Dichloran. Common name for 2,6-dichloro-4-nitroaniline, DCNA, and sold as Botran (Nor-Am Chem. Co.). Effective for Botrytis, Sclerotina, and related blights on fruits, vegetables, and ornamentals. Can be used as a spray or dust in the field and as a postharvest dip.

Dichloropropene. See Telone C.

Diethofencarb (Sumitomo Chem. Co.). Isopropyl 3, 4-diethoxyphenylcarbamcate. Use for benzimidasole-resistant strains of various fungi. Strong activity against strains of *Botrytis cinerea*.

Difolatan (ICI Agrochem.; Chrystal Chem. Inter-America; Sunko Chem. Co.). Folcid, *cis*-N-[(1,1,2,2 -tetrachloroethyl)thio]-4-cyclohexene-1,2,dicarboximide. Useful for control of foliage, soil, and seed diseases; registered for early and late blight on potatoes; good for downy mildews and various other fruit and vegetable diseases and for turf. Discontinued names: Difolatan, Folcid. Other names: Captafol, Foltaf, Haipen, Mycodifol, Sanspor.

Dikar (Rohm & Haas). Coordination product of zinc ion, dinitro (1-methyl heptyl) phenyl crotonate, and manganese ethylene bisdithiocarbamate. Especially recommended for apples, to control scab, mildew, rust, and brown rot.

Dimethirimol. See Milcurb.

Dinitro compounds (A. H. Marks & Co., Ltd.; Pennwalt Holland B.V.; Tifa Ltd.). Derivatives of cresol and phenol used as dormant sprays for some insects, as herbicides, and occasionally as eradicant fungicides. See also Elgetol. Other names: Chemsect, DNOC, DNC, Elgetol 30, Nitrador, Selinon, Sinox, Trifocide, Trifrina.

Dinobuton (Keno Gard AB, Sweden). 2-(1 -methyl-2-propyl)-4,6-dinitrophenyl isopropylcarbonate. Used to control powdery mildew on apples, cucumbers, and hops. Other names: Acarelte, Acarelte Forte, Acrex, Desain, Dinofen, Dravinol, Talan.

Dinocap. See Karathane.

Dinofen. See Dinobuton.

Dipher. See Zineb.

Direx. See Dyrene.

Dithane D-14. See Nabam (Discontinued 1987 by Rohm and Haas Co.).

Dithane M-22. See Maneb.

Dithane M-22 Special. See Maneb.

Dithane M-45 (Rohm & Haas). Coordination product of zinc ion and manganese ethylene bisdithiocarbamate, related to both maneb and zineb. Used as a protectant against a wide spectrum of diseases of many fruit and vegetables. Other names: Mancozeb, Manzate 200, Nemispor, Penncozeb, Policar MZ, Policar S, Riozeb, Vondozeb Plus, Ziman-Dithane.

Dithane Z-78 (Name discontinued). See Zineb.

Dithianon (Shell Agrar GMBH & Co. KG). 5,10-dihydro-5,10-dioxonapths-(2,3b)-*p*-dithiin-2,3-dicarbonitrile. Effective against many diseases of pome fruit, stone fruit, small fruit, grapes, and ornamentals. Other names: Delan, Delan-Col.

Dithiocarbamates. Organic sulfur fungicide derivatives of dithiocarbamic acid. See also Nabam, Thiram, Maneb, Zineb, Ziram.

Ditranil. See DCNA.

DNC. See Dinitro compounds.

DNOC. See Dinitro compounds.

DMTT, 3,5-dimethyl tetrahydro-1,3,4,2H-thiadiazine-2-thione, sold as Mylone (Union Carbide). A fumigant for soil fungi and nematodes, to be applied as a drench or as granules. See also Dazomet.

Dodemorfe. See Dodemorph (discontinued).

Dodemorph (BASF Aktiengesellschaft) (discontinued). N-cyclododecyl-2,6-dimethylmorpholinium acetate. Used for control of powdery mildew in ornamentals.

Dodine. Common name for *n*-dodecylguanidine acetate, sold as Cyprex (American Cyanamid; Keno Garb AB; Rhone-Poulenc Agrochimie; Shell Agrar Gmb H & Co. KG). Effective for foliage diseases of fruits, vegetables, and ornamentals, controlling scab on apple, pear, pecan, and cherry leaf spot. It also controls rose blackspot, but may be somewhat phytotoxic at normal dilution. Other names: AC 5223, Apadodine, Carpene, Curitan, Melprex, Syllit, Venturol.

Doguadine. See Dodine.

Dojyopicrin. See Chloropicrin.

Dolochlor. See Chloropicrin.

Dowco 186. See Du-Ter (discontinued by Dow Chemical Co.).

Dofume MC-33 (discontinued by Dow Chemical). Methyl bromide 67%, chloropicrin 33%.

Drawinol. See Dinobuton.

Drazoxolon (ICI Arochemicals). 4-(2-chloro-phenylhydrazono)-3-methyl-5-isoxazolone. Effective against powdery mildews and rusts. Used for treatment against some seedling diseases. Other names: Ganocide, Mil-Col, SA Isan.

DSE. See Nabam.

Duter. See Du-Ter.

Du-Ter (Agtrol Chem. Products; Duphar B.V.; Hoechst AG; Wesley Industries, Inc.), triphenytin hydroxide. Used to control early and late blight on potatoes, leaf spot on sugar beets and peanuts, scab and several diseases on pecans, leaf spot and Alternaria blight on carrots. Other names: Brestanid, Duter, Haitin, Phenostat-H, Suzu H, TPTH, TPTOH, Triple Tin, Tubotin.

Dwell. See Koban.

Dynone. See Previcur.

Dyrene (Bayer AG; Mobay Corp., Agri. Chem. Div.). 2,4-dichloro-60-chloroanilino-s-triazine. Foliar fungicide, for control of anthracnose, *Botrytis,* early and late blights of potato and tomato, and other vegetable diseases, strawberry leaf spot and scorch, dollar spot, melting-out and rust of turf, and leaf spot of gladiolus. Other names: Direx, Kemate, Triasyn.

Earthcide. See PCNB.

EL-273. See Triarimol (discontinued by Elanco Products Co.).

Elgetol 30 (A. H. Marks & Co., Ltd.; Pennwalt Holland B.V.; Tifa Ltd.). 4,6-dinitro-o-cresol. Used for apple scab control. Other names: Chemsect, DNOC, DNC, Nitrador, Selinon, Sinox, Trifocide, Trifrina.

Elvaron. See Euparen.

Endosan. See Morocide (discontinued in 1987 by Hoechst AG).

Equitdazin. See Carbendazin.

Eradex (Discontinued by Bayer AG). 2,3-quinoxalinedithiol cyclic trithiocarbonate. Used on ornamentals to kill spider mites and their eggs and also effective against powdery mildew.

Eraditon. See Eradex (discontinued).

Erazidon. See Eradex (discontinued).

ETCMTD. See Koban.

Ethazol. See Koban.

Ethylene dibromide. See under nematicides.

Etridiazole (Mallinckrodt, Inc.; Uniroyal Chem. Co.; United Agri-Products). 5-ethoxy-3-trichloromethyl-1,2,4-thiadiazole. Controls *Pythium* and some *Phytophthora* spp. Terra-Coat, Terraclor Super-X for seedling disease complex (*Fusarium, Rhizoctonia, Pythium*) of beans, sugar beets, and wheat; as a seed treatment or as in-furrow soil applications. Controls root rot (*Phytophthora cinnamoni*) of avocado; *Pythium* on strawberries; used on ornamentals, turf, vegetables. Other names: Aaterra, Dwell, ETCMTD, Ethazol, Koban, Pansoil, Phorate TSX, Truban. Combinations: Terra-Coat, Terraclor Super-X (Terraclor+Terrazole), Truban+methyl thiophanate (Banrot).

Etrimix. See Mancozeb.

Euparen (Bayer AG). N'-dichlorofluoromethylthio-N,N-dimethyl-N'-phenylsulfamide. Used for controlling *Botrytis* on strawberries, raspberries, currants, and grapes; rose mildew and other fungal diseases on orchard fruits, garden crops, and ornamentals. Other names: Bay 47531, Elvaron, Euparence, KUE 13032c.

Euparen M (Bayer AG). N'-dichlorofluoromethylthio-N,N-dimethyl-N'-(4-tolyl) sulfamide. Used for controlling scab on apples, *Botrytis* on strawberries, currants, and ornamentals. Other names: Bay 5712, Bay 49854, KUE 13183b.

Euparene. See Euparen.

Exotherm Termil (Diamond Shamrock). Tetrachloroisophthalonitrile. Used on greenhouse tomatoes and ornamentals for control of *Botrytis*. When heated it vaporizes to form a gas, which condenses, forming ultrafine particles deposited on plant surfaces. See also Bravo.

Fenaminosulf. See Lesan.

Fenarimol. See Rubigan.

Fenolovo acetate. See Brestan.

Fentin acetate. See Brestan.

Fentin hydroxide. See Du-Ter.

Ferbam (Pennwalt Holland B.C.; FMC Corp.; UCB Chem. Corp.). Common name for ferric dimethyl dithiocarbamate, sold in many pesticide combinations, the first organic fungicide to come into wide use. Used as a spray or as a 10% dust, often with sulfur, for damping-off of flower cuttings, apple rust and scab, brown rot of stone fruits, black rot of grapes, some Botrytis blights, anthracnose, downy mildews, leaf spots, including rose blackspot. It does not control powdery mildews. The black color is objectionable on some flowers, but there is little visible residue on foliage. Ferbam is of questionable compatibility with Paris green, TEPP, lime sulfur, lime, bordeaux mixture, and some fixed coppers. It is not highly toxic, with a tolerance of 7 ppm set for food crops, but may cause irritation if inhaled. Other names: Carbamate, Ferberk, Hexaferb, Knockmate, Trifungol.

Ferberk. See Ferbam.

Fermasan. See Thiram.

Fermate (discontinued by DuPont). See Ferbam.

Fermid 850. See Thiram.

Ferrous sulfate. Used to correct chlorosis from iron deficiency.

Filipin. Antibiotic that controls some seed-rot fungi and partially protects against some diseases of beans and tomatoes.

Flotation sulfur. See Sulfur.

FMC 9102. See Polyram-Combi.

Folcid. See Difolatan.

Folicur (Bayer AG; Mobay Corp.). α-[2-(4-chlorophenyl)ethyl]-α-(1, 1-di-methyl-ethyl)-1 H-1,2,4--triazole-1-ethanol. Effective against powdery mildew, rusts, leaf spots on fruits, vegetables, peanuts, and grasses grown for seed. Other name: Raxil.

Folosan. See PCNB.

Folpan. See Folpet.

Folpet. Common name for N-trichloromethyl-thiophthalimide, sold as Phaltan (Chevron). A protectant and eradicant fungicide for fruits, vegetables, and ornamentals controlling apple scab, cherry leaf spot, rose blackspot, with some effect on rose mildew, and other diseases. Compatible with most common pesticides but cannot be used with strong alkalies. Other names: Fungitrol II, Folpan, Thiophal.

Foltaf. See Difolatan.

Fonganil. See Fongarid.

Fongarid (Ciba-Geigy Ltd.). Methyl N-2,6-dimethylphenyl-N-furoyl (2)-alaninate. For soil-borne diseases caused by *Phytophthora* and *Pythium* spp. on ornamentals. Systemic properties. Other name: Fonganil, CGA 38140.

Fore (Rohm & Haas). A special formulation of Dithane M-45 developed for control of diseases of turf and certain ornamentals. Protects against dollar spot, red thread, copper spot, *Helminthosporium*, brown patch, rust, slime mold, and Fusarium and Pythium blights of lawn grasses, also rose blackspot, chrysanthemum petal blight, gladiolus diseases.

Formaldehyde. Soil fumigant for damping-off and other diseases, not very efficient for nematodes; sold as Formalin, a 35–40% solution of a colorless gas in water and methanol. In preparing soil for flats, mix 3 tablespoons formalin with 1 cup of water and sprinkle over 1 bushel of soil; mix well. Fill flats; plant seed 24 hours later; then water. As a drench for fallow soil dilute 1 part formalin to 50 parts water and apply ½ to 1 gallon to each square foot of soil. As a treatment for potato scab, soak tubers for 2 hours in 1 pint formalin to 30 gallons water. Disinfest tools in a 5% solution. Do not use near living plants.

Formalin. See Formaldehyde.

Forturf. See Bravo.

Frucote, 2-aminobutane, or *sec*-butylamine. Used to control green or blue mold in lemons, oranges, and grapefruits and stem-end rot of oranges. Applied as dip, drench, or spray.

Fuberidazol. See Voronit.

Fuklasin. (discontinued name). See Ziram.

Fungiclor. See PCNB.

Funginex. See Triforine.

Fungi-Rhap. See Copper oxides.

Fungi-Rhap CU6; Liquid Copper Fungicide (CP Chem., Inc.), copper salts of fatty and rosin acids. Used to control leaf spots in carrots, peanuts, and sugar beets; bacterial spot in peppers and tomatoes; and melanose in citrus.

Fungitrol II. See Folpet.

Fungo 50 (Mallinckrodt, Inc.). Dimethyl 4,4-0-phenylenebis (3-thioallophanate). Used to control brown patch, Fusarium blight, dollar spot, red thread, stripe smut, and powdery mildews on turf.

Fusarex (ICI). 2,3,5,6-tetrachloronitrobenzene. Used to control dry rot in seed and ware potatoes.

Galben (Agrimont S.p.A). DL-alanine, N-(2,6-dimethylphenyl)-N (phenylacetyl) -methylester. Systemic fungicide controls Oomycetes including blue mold, late blight and downy mildew of potatoes, tomatoes, tobacco, hops, grapes, lettuce, peppers, onions, strawberries, sunflowers, soybeans, turf, flowers, ornamentals. Other names: Tairel, M9834.

Ganocide. See Drazoxolon.

Glyodex (discontinued by Agway, Inc.). glyodin 37.5% and dodine 22.5%. Used to control apple scab on apples and leaf spot on sour cherries.

Glyodin (discontinued by Agway, Inc.). 2-heptadecyl-2-imidazoline acetate, sold as Crag Fruit Fungicide 341 and Crag Glyodin. Protectant fungicide for control of scab and other apple diseases, cherry leaf spot, rose blackspot, and some other diseases of ornamentals. It may be slightly phytotoxic in certain mixtures and should not be used on solanaceous plants. Glyodin acts as a wetting agent, and the residue is invisible on foliage or fruit. The tolerance is 5 ppm. It is of questionable compatibility with benzene hexachloride, cryolite, oils, rotenone, pyrethrum, and some dinitro compounds.

Glyoxide. 2-heptadecyl imidazoline. Used on apples, sour cherries, and pears.

Glyrophene. See Iprodione.

Granox PFM (Chipman Chem., Inc.). Combination of maneb, captan, and molybdenum. Used as a protective seed treatment for peanuts to control seed-borne pathogens, including damping-off, seed decay organisms, and seedling blight.

Guanidine. See Dodine.

Haipen. See Difolatan.

Haitin. See Du-Ter.

Harven. See Dehydroacetic acid.

Hexachlorophene (discontinued in 1984 by Kalo Laboratories, Inc.). 2,7-methylene bis (3,4,6-trichlorophenol). Used on tomatoes, peppers, and cucumbers for fungus and bacterial diseases and as a soil fungicide against *Rhizoctonia*.

Hexaferb. See Ferbam.

Hexasul. See Sulfur.

Hexathane. See Zineb.

Hexathir. See Thiram.

Hexazir. See Ziram.

Hizarocin. See Actidione.

Hoe 002873. See Afugan.

Hoe 017411. See Carbendazin.

Hoe 2784 (discontinued in 1987 by Hoechst AG). See Morocide.

Hoe 2873 (discontinued in 1984 by Hoechst AG). See Afugan.

Hoe 2989 (Hoechst AG). See Sicarol.

Hoe 6052 (Hoechst AG). See Sicarol.

Hoe 6053 (Hoechst AG). See Sicarol.

Hoe 13764 (Hoechst AG). See Sicarol.

Hoe 17411 (Hoechst AG). See Carbendazin.

Homai (Nippon Soda Co., Ltd.). Mixture of thiophanate-methyl and thiram. Used as a seed protectant on vegetables.

Hot water, used in disinfection of seeds, bulbs, and sometimes living plants to kill internal bacteria, fungi, or nematodes, the temperature and time of treatment varying with the plant.

Hydroxydiphenyl. See Ortho-Phenylphenol.

Hydroxyisoxazole. See Tachigaren.

Hymexazol. See Tachigaren.

Iprodione (Rhone-Poulenc Inc. Ag. Co.). 3-(3,5-dichlorophenyl)-N-(1-methylethyl) 2,4-dioxo-1-imidazolidinecarboxamide. Active on a broad spectrum of diseases on vines, grapes, fruits, berries, vegetables, ornamentals, flowers, turf, potatoes. Other names: 26019RP, Chipco 26019, Chipco 26019 Flo, Glycophene, LFA 2043, NRC 910, ROP500F, Rovral.

Iscothane. See Karathane.

Karamate. See Dithane M-45.

Karabation. See Metam-Sodium.

Karathane (Rohm & Haas). Common name dinocap, 2,4-dinitro-6-octyl-phenyl crotonate, and other nitrophenols. Of some value as a miticide and excellent for control of powdery mildews of apple and other fruits, vegetables, especially cucurbits that are sensitive to sulfur, roses, and other ornamentals. Karathane is also included in many dust mixtures sold for roses. It may be slightly phytotoxic above 85° F. Other names: Crotothane, Dinocap.

Kasugamycin (Hokko Chemical Industry Co., Ltd.). Kasugamycin hydro-chloride. Used to control leaf mold of tomatoes, halo blight of beans, and apple scab.

Kasumin. See Kasugamycin.

Kayafume. See Methyl bromide.

K-Cop Liquid Agricultural Fungicide (Griffin Corp.). Aqueous solution containing 8% copper. Used on diseases of beans, cantaloupe, honeydews, muskmelon, watermelon, celery, cucumber, peanuts, peppers, potatoes, squash, and tomatoes.

Kemate. See Dyrene.

Kiwi Luster 277. See DCNA.

Knockmate. See Ferbam.

Koban (Mallinckrodt, Inc.; Uniroyal Chem. Co.; United Agri. Products). 5 ethoxy-3-trichloromethyl-1,2,3-thiadiazole. Used to control Pythium blight, cottony blight, grease spot, spot blight, and damping-off on turf. Other names: Aaterra, Dwell, ETCMTD, Ethazol, Pansoil, Phorate TSX, Truban.

Kobu. See PCNB.

Kobutol. See PCNB.

Kocide. See Copper hydroxide.

KOP 300. See Copper Sulfate, Basic.

Kroma-Clor. See Cadminate.

Kromad (Mallinckrodt, Inc.). Cadmium sebacate 5%, potassium chromate 5%, malachite green 1%, thiram 16%. Used to control brown patch, dollar spot, pink patch (red thread), copper spot, and leaf spot diseases of turf.

K-Tea Algaecide (Griffin Corp.). Copper (8%) as copper-triethanolamine complex. Controls planktonic and filamentous algae, hydrilla verticillata in golf course ornamental, fish and fire ponds, potable water reservoirs, freshwater lakes, and fish hatcheries.

Kue 13032c. See Euparen.

Kue 13183b. See Euparen M.

Kumulan (BASF AG). Sulphur and 5-nitro-benzene-1, 3-dicarboxylic acid bis (1-methylethyl) ester. Controls powdery mildew on apples and hops.

Kumulus S. See Sulfur.

Kypman. See Maneb.

Kypzin. See Zineb.

Labilite. See Maneb.

Larvacide. See Chloropicrin.

Lesan (Bayer AG). Sodium [4-(dimethylamino) phenyl]diazene sulfonate. Protects germinating seeds and seedlings in corn, beans, peas, spinach, cucumbers, and ornamentals.

LFA 910. See Iprodione.

LH 3012. See Antracol.

Lime, hydrated. Calcium hydroxide, used in preparing bordeaux mixture, and as a filler in pesticide dusts. Until recently considered relatively inert but may cause some of the dwarfing and hardening of bordeaux-sprayed plants.

Lime sulfur. Polysulfides formed by boiling together sulfur and milk of lime. The standard liquid has a specific gravity of 32 Baumé and the commercial product is far superior to the homemade. Lime sulfur dates back to 1851, when the head gardener, Grison, at Versailles, France, boiled together sulfur and lime for a vegetable fungicide called "Eau Grison." In 1886, this fungicide was used in California as a dormant spray for San Jose scale and later for peach leaf curl. A self-boiled lime sulfur made without heat was produced in 1908 as a summer spray for sensitive plants, but it was later replaced by wettable sulfurs for most fruit-spray programs. A dry form of lime sulfur was marketed about 1908.

Lime sulfur is still used as a dormant spray for fruits, roses, and some other plants for mildews, Volutella blight of boxwood, and other diseases. It should not be used at temperatures above 85° F.

Lime sulfur is compatible with nicotine sulfate and glyodin. It is incompatible with soaps, Paris green, cryolite, rotenone, pyrethrum, oils, dinitro compounds, benzene hexachloride, TEPP, bordeaux mixture, and fixed coppers. It is questionable with toxaphene, parathion, and dithiocarbamates.

Lonacol. See Zineb.

Lysol. Sometimes used for treatment of gladiolus corms. A 5% solution is used for dipping the cutting knife to prevent potato ring rot.

M 9834. See Galben.

Magnesium sulfate. Epsom salts, sometimes used as a safener and to correct nutrient deficiencies.

Magnetic 70. See Sulfur.

Malachite. See Copper carbonate.

Mancozeb. See Dithane M-45.

Maneb (BASF AG; Rhone Poulenc; Rohm & Haas). Common name for manganese ethylene bisdithiocarbamate. Used to control early and late blights of potato, tomato, celery, tomato anthracnose, leaf spots, downy mildews, purple blotch of onions, shothole of almond and peaches. Nontoxic to most plants, relatively safe to use, and useful for ornamentals because it leaves little visible residue. It controls anthracnose of violet and pansy, spot anthracnose of dogwood, rose blackspot, and Cercospora leafspot (but not powdery mildew), and some Botrytis blights. Other names; Akzo Chemie Maneb, BASF-Maneb Spritz-pulver, Dithane M-22,

Dithane M-22 Special, Kypman 80, Manex 80, Maneba, Manesam, Manex, M-Diphar, Polyram M, Remasan Chloroble M, Rhodianebe, Sopranebe, Trimangol, Tubothane, Unicrop.

Maneba. See Maneb.

Manebgan. See Maneb.

Manesan. See Maneb.

Manex. See Maneb.

Manex 80. See Maneb.

Manzate. See Maneb.

Manzate 200 Fungicide (DuPont). Coordination product of zinc ion and manganese ethylene bisdithiocarbamate. Used as a protectant against a wide spectrum of diseases of fruits, vegetables, and nuts. Also used as a seed treatment on potatoes and peanuts.

Manzeb. See Dithane M-45.

Maposol. See Matham-Sodium.

MBC. Methyl 2-benzimidazolecarbamate. Common breakdown product of several fungicides, including benomyl, thiophanate-methyl, and other ethyl and methyl thiophanates. These contain the common characteristic that once a fungus has developed resistance to one, the fungus also possesses resistance to other fungicides of the same group (cross-resistance).

MC 1053 (discontinued name). See Dinobuton.

M-Diphar. See Maneb.

M-Dipher. See Maneb.

MEB 6447. See Bayleton.

MeBr. See Methyl bromide.

Melprex. See Dodine.

Meltatox. See Dodemorph.

Mepronil (Kumiai Chem. Industry Co., Ltd.). 3'-isopropoxy-2-methyl-benzanilide. Controls rusts of pear and chrysanthemum, and damping-off and southern blight of vegetables.

Mercuran. See Thiram.

Merpan. See Captan.

Mertect (Merck). See Thiabendazole.

Metalaxyl (Ciba Geigy Corp.). N-(2,6-Dimethylphenyl)-N-(methoxyacetyl)-alanine methyl ester. Controls soil-borne diseases caused by *Pythium* and *Phytophthora*, and foliar diseases caused by Phycomycetes (downy mildews). Other names: CGA 48988, Ridomil 2E, Ridamil 5G, Subdue 2E.

Metam-Sodium (BASF Aktiengesellschaft; ICI Plant Protection Division; Procida; Stauffer Chemical Co.). Sodium N-methyldithiocarbamate. General-purpose soil fumigant that is highly effective in control of weeds, weed seeds, nematodes, and soil fungi.

Methanal. See Formaldehyde.

Metho-O-Gas. See Methyl bromide.

Methyl bromide (Great Lakes Chem. Corp.). Soil fumigant, supplied under that name by a number of manufacturers. It is used in greenhouses because it is somewhat less toxic to plants than most other fumigants and for balled or potted nursery stock in special fumigating chambers. As Dowfume MC-2, methyl bromide with 2% chloropicrin, it comes in cans with a special dispenser. The area to be treated is covered with plastic film, under which are evaporating pans with hoses leading to the edge of the cover. The dispenser is attached to each hose in turn to fill the pans, and then the edges of the cover are held down with soil. For small areas injections are made 10 inches apart. This material is very poisonous; follow all safety precautions. Other names: Brom-O-Gas, Brom-O-Gaz, Brom-O-Sol, Celfume, Kayafume, MeBr, Meth-O-Gas, Terr-O-Cide II, Terr-O-Gas.

Methylmetiram. See Basfungin.

Methyl thiophanate. See Fungo 50.

Metiram. See Polyram-Combi.

Metiram-Complex. See Polyram-Combi.

Mezene. See Ziram.

Mezineb. See Antracol.

MF-344. See Koban.

Micofume. See Dazomet.

Mil-Col. See Drazoxolon.

Milcurb (ICI Agrochemicals). 5-*n*-butyl-2-dimethylamino-4-hydroxy-6-methylpyrimidine. A systemic eradicant fungicide used for control of powdery mildew affecting cucumbers, melons, and certain ornamentals. One soil application may give protection for 6 weeks or more.

Mildex. (discontinued name). See Karathane.

Mildothane. See Thiophanate-Methyl.

Miltox (Sandoz, Ltd.). Zineb, copper oxychloride. Controls downy mildew (*Plasmopara*) and Brenner disease (*Pseudopeziza*) in grapes and other diseases in most crops.

Monceren (Bayer AG; Mobay Corp.). N-[(4-chlorophenyl)-methyl]-N-cyclopentyl-N'-phenylurea. For *Rhizoctonia solani* caused diseases in potatoes, rice, sugar beets, and ornamentals.

Monox. See Polynox.

Morestan (Bayer AG; Mobay Corp.). 6-methyl-2,3-quinoxalinedithio cyclic carbonate. Insecticide, miticide also effective for powdery mildew on apple and other crops; may cause some fruit spotting.

Morocide (Hoechst AG). 2-*sec*-butyl-4,6-dinitrophenyl-3-methyl-2-butenoate. A contact miticide with ovicidal action and fungicide for control of powdery mildews on apples, pears, plums, prunes, almonds, and walnuts.

Morrocid (discontinued name by Hoechst AG in 1987). See Morocide.

Mycodifol. See Difolatan.

Mylone. See Dazomet.

Nabac 25 EC (discontinued in 1984 by Kalo Laboratories, Inc.). 2,2-methylenebis (3,4,6-trichlorophenol). Used as a broad-spectrum foliar fungicide and bactericide.

Nabam (Rhone-Poulenc Agrochimie). Common name for disodium ethylene-1,2-bisdithiocarbamate, sold as Dithane D-14, a liquid, very useful for some vegetable and flower diseases, especially tomato and potato blights, azalea flower blight. It is usually used with zinc sulfate to form zineb in the spray tank. It controls some root-rotting fungi and enhances the effect of a nematicide by stimulating hatching of eggs of root-knot and some cyst nematodes. Other names: Chem Bam, DSE, Nabasan, Parzate, Spring-Bak.

Nabasam. See Nabam.

Naramycin. See Actidione.

Natriphene (Natriphene Co.). Sodium salt of o-hydroxyphenyl. Use for damping-off and other diseases of ornamentals, especially orchids, and of some merit for powdery mildew on rose.

Nemacur. See Nematicides section.

Nemispor. See Dithane M-45.

Nia 9044 (FMC Corp.). See Morocide.

Nia 9102 (name discontinued by FMC Corp.). See Polyram-Combi.

Niacide (discontinued by FMC Corp.). Mixture of manganous dimethyl dithiocarbamate and mercaptobenzothiazole. Apple and pear foliage spray for scab and summer diseases.

Nimrod. See Bupirimate.

Nitrador. See Dinitro compounds.

Nomersan. See Thiram.

Nordox SD-45. See Copper oxides.

Nordox SD-50. See Copper oxides.

Nu-Z. See Zinc sulfate.

Ofurace (Chevron Chem. Co, Ortho Agri. Chem. Div.). 2-chloro-N-(2,6, dimethylphenyl)-N-(tetrahydro-2-oxo-3-furanyl) acetamide. Systemic action both acropetal and basipetal. For Phycomycete plant pathogens, notably downy mildew of grapes, hops and lettuce, late blight of potato, tomato, and Phytophthora crown and root rots of safflower and tobacco.

Oleocuivre. See Copper oxides.

Oleo Nordox. See Copper oxides.

OM-2424 (discontinued name). See Etridiazole.

Ornalin. See Vinclozolin.

Orthocide. See Captan.

Ortho-phenylphenol (Dow Chemical Co.). 2-phenylphenol; or ortho-phenylphenol. Used as postharvest treatment in wax to retard spoilage of fruits and vegetables in transit to market.

Orthoxenol (discontinued name). See Ortho-phenylphenol.

Oxadixyl (Sandoz Ltd., Agro Div.). 2-methoxy-N-(2-oxo-1,3 oxazolidinyl) acet-2',6'-xylidide. Preventive and curative activity against many Oomycetes on grape vines, potatoes, vegetables, ornamentals, and seed treatments. Other names: Pulsan, Recoil, Ripost, Sandofan, Wakil.

Oxycarboxin. See Plantvax.

Oxyquinoline sulfate (Probelte, S.A.). 8-hydrozyquinoline sulfate. Used as a drench to control root-rotting fungi such as *Rhizoctonia solani*. Also used for treating citrus fruits for stem end rot and green mold, orchids for black mold, and carnations for wilt. Other name: Chinosol.

Oxythioquinox. See Morestan.

Pallinal (BASF AG). tris[ammine[ethylene-bis (dithocarbamato)] zinc (2+)] [tetrahydro-1,2,4,7-dithia-diazocine-3,8 dithione], polymer, 5-nitro-benzene-1,3-dicarboxylic acid bis (1-methylethyl) ester. Controls powdery mildew on apples and apple scab.

Pansoil. See Koban.

Parinol. See Parnon.

Parnon (discontinued by Elanco). a,a-bis (4-chlorophenyl)-3-pyridylmethanol, for powdery mildews of rose, fruits, and vegetables; nontoxic to ornamental plants.

Parzate. See Nabam.

Parzate C. See Zineb.

PCNB (Uniroyal Chem. Co.). Pentachloronitrobenzene, excellent soil fungicide that may be used around living plants. It controls various root, stem, and crown rots of vegetables and ornamentals, is available as a dust, 75% wettable powder, and an emulsifiable concentrate. For club root of crucifers, work dust into soil before planting and use in liquid form for transplants. Mix into soil before planting for southern blight of peanuts but apply at transplanting for peppers and tomatoes. For camellia blight, apply to soil under bushes in early winter. Use as a soil dust or drench for stem rot of carnation, poinsettia, African violet, snapdragon, and other ornamentals. Use the manufacturer's dosage chart for different formulations and situations. Other names: Avicol, Botrilex, Earthcide, Folosan, Kobu, Kobutol, Pentagen, Quintox, Terraclor, Tilcarex, Tri-PCNB.

Penconazole (Ciba-Geigy Corp.). 1-[2-(2,4-dichlorophenyl)-n-pentyl]-1H-1,2,4-triazole. Systemic fungicide for protective, curative, and eradicative use against powdery mildews, pome fruit scab, and other pathogenic ascomycetes, basidiomycetes, and deuteromycetes. For use in grapes, deciduous fruits, vegetables and ornamentals. Other names: Award, Topas, Topaz, Topaze, CGA-71818.

Penncozeb. See Dithane M-45.

Pentagen. See PCNB.

Perecot (discontinued name). See Copper oxides.

Perenox (discontinued by ICI Plant Protection Division). Cuprono oxide. Used to control many common leaf and fruit diseases of tomato, potato, celery, peach, banana, cocoa, tea, and citrus.

PETD (discontinued name). See Polyram-Combi.

Phaltan. See Folpet.

Phenamiphos (discontinued name). See Nemacur.

Phenostat-H. See Du-Ter.

Phentinacetate. See Brestan.

Phenylphenol. See Ortho-phenylphenol.

Phleomycin. Isolated from *Streptomyces* by Japanese workers. Used to control rust on snap beans.

Phorate TSK. See Koban.

Phygon (discontinued by Uniroyal). See Dichlone.

Phyto-Bordeaux Super CU. See Copper Sulfate, Basic.

Phytomycin. See Streptomycin.

Phyton-27 (Source Technology Biologicals, Inc.). Tannate complex of picro cupric ammonium formate. Broad spectrum for tree injection of Dutch elm disease prevention and control. Other applications by foliar spray and soil drench.

Pic-Clor. See Chloropicrin.

Picfume. See Chloropicrin.

Piperalin (Elanco Products Co., Div. Eli Lilly and Co.). Common name for 3-(2-methylpiperidino) propyl 3,4-dichlorobenzoate. Sold as Pipron. For powdery mildew on rose and other ornamentals. The liquid form leaves less visible residue than the wettable powder.

Pipron. See Piperalin.

PKhNB. See PCNB.

Plantomycin. See Streptomycin.

Plantvax (Uniroyal). 2,3-dihydro-5-carboxanilido-6-methyl-1,4-oxathiin-4, 4-dioxide, systemic fungicide, effective in controlling rust diseases.

Polycar MZ. See Dithane M-45.

Polycar S. See Dithane M-45.

Polynox (discontinued by Nihon Nohyaku Co., Ltd.). Mixture of Polyoxin and Monox, zinc dimethyl dithiocarbamate bis (dimethyl) dithiocarbamoyl ethylenediamine. Used against diseases of apple and pear.

Polyoxin (Kaken Pharmaceutical Co., Ltd.). Antibiotic fungicide. Used to control Alternaria leafspot of apple, black spot of pear, gray mold and leaf mold of tomato, Sclerotinia rot, stem rot, leaf blight and leaf spot of cucumber, Alternaria leaf spot of welsh onion, gray mold and powdery mildew of strawberry, and leaf blight of carrot. Other names: Polyoxin AB, Polyoxin B.

Polyoxin AB. See Polyoxin.

Polyoxin B. See Polyoxin.

Polyram-Combi (BASF Aktiengesellschaft; FMC Corp.). Coprecipitation of zinc ammonia ethylene-bis-dithiocarbamate and polyethylene-bis-thio-carbamoyl disulfide. Used to control diseases of apples, asparagus, peanuts, pecans, potatoes, sweet corn, and tobacco; also scab and cedar apple rust on apple and rose blackspot. Other names: Metiram, Metiram-Complex.

Polyram M. See Maneb.

Polyram-Ultra. See Thiram.

Polyram Z. See Zineb.

Pomarsol Forte. See Thiram.

Pomarsol Z Forte. See Ziram.

Potassium permanganate. Occasionally used as a disinfectant for bulbs and rhizomes and for dipping grafting knives and other tools (1 ounce to 2 gallons of water). Applied to citrus trunks, 1 teaspoon to 1 pint of water, after cleaning scaly bark wounds.

PP-588. See Bupirimate.

PP-675. See Milcurb.

PP-781. See Drazoxolon.

Previcur (discontinued by NOR-AM Agricultural Products, Inc.; Schering AG). ethyl-N-(3-dimethyl-amino-propyl)-thiol-carbamate hydrochloride. Used to control *Pythium, Peronospora, Bremia, Phytophthora,* and other Peronosporales; emergence promoting, and is partially systemic.

Prezervit. See Dazomet.

Propineb. See Antracol.

Prothiocarb. See Previcur.

PTF (discontinued name). See Polyram.

Pyracarbolid. See Sicarol.

Pyrazophos. See Afugan.

Quinomethionate. See Morestan.

Quintox. See PCNB.

Quintozene. See PCNB.

Raxil. See Folicur.

Readex (discontinued by Bayer AG). See Eradex.

Remasan Chloroble M. See Maneb.

Resisan. See DCNA.

Rhodianebe. See Maneb.

Ridomil. See Metalaxyl.

Ridomil MZ. See Metalaxyl.

Ridomil MZ58. See Metalaxyl.

Ridomil MZ72. See Metalaxyl.

Ridomil Plus. See Metalaxyl.

Rizolex (Sumitomo Chem. Co.). 0-2, 6-dichloro-4-methylphenyl 0,0-dimethyl phosphorothioate. For control of soil-borne diseases caused by *Rhizoctonia, Sclerotium* and *Typhula* on potatoes, sugar beets, cotton, peanuts, vegetables, cereals, ornamentals, turf by soil and seed treatment. Other name: S-3349.

Ronilan. See Vinclozolin.

Rop 500F. See Iprodione.

Rovral. See Iprodione.

RPH. See Thiabendazole.

Rubigan (Elanco Products Co., Div. of Eli Lilly and Co.). 3-(2-chlorophenyl)-3-(4-chlorophenyl)-5-pyrimidine-methanol. Provides protectant, curative, and eradicant activity against certain diseases such as powdery mildew, scab and rust of apple, dollar spot, large brown patch, fusarium blight and snow mold of turf, powdery mildew of roses.

S-3349. See Rizolex.

SAI San. See Drazoxolon.

Salsan. See Drazoxolon.

Sanspor (ICI Agrochemicals; Crystal Chem. Inter-America). Used to control potato blight, especially tuber blight. Other names: Criafolatan, Foltaf, Haipsen, Mycodifol.

Saprol. See Triforine.

Selinon. See Dinitro compounds.

SF-6505. See Tachigaren.

Sicarol (discontinued in 1984 by Hoechst AG). 2-methyl-5,6-dihydro-4-H-pyran-3-carboxylic acid anilide. Used for seed and foliar treatment to control rusts and smuts on ornamentals, coffee, tea, and vegetables.

Sinox. See Dinitro compounds.

SMDC. See Metam-Sodium.

SN 41703. See Previcur.

Sodium dehydroacetate. See Dehydroacetic acid.

Sodium hypochlorite. Recommended as a disinfestant for pruning tools for fire blight control to replace dangerous mercurials. Dipping for 2 seconds in 10% solution kills bacteria. Can be purchased as Chlorox.

Sodium methyldithiocarbamate. See Metam-Sodium.

Sofril. See Sulfur.

Sopranebe. See Maneb.

Spergon. See Chloranil.

Spotrete. See Thiram.

Spotrete-F. See Thiram.

Spotrete-WP 75. See Thiram.

Spring-Bak. See Nabam.

SR-406. See Captan.

SS 1451. See Eradex.

SS 2074. See Morestan.

Streptomycin (MSD Agvet, Div. of Merck Co., Inc.; Pfizer Inc. Chem. Div.). Antibiotic formulated as a sulfate or a nitrate, effective for many bacterial and fungus diseases. As Agrimycin 17 (Pfizer) it is rather widely used, applied at blossom stage, for control of fire blight of apple and pear. Streptomycin is also used for walnut blight; bacterial leaf spots of tomato, pepper, philodendron; chrysanthemum bacterial blight; stem rot of stock; bean and celery blights; downy mildew, wilt, and angular leaf spot of cucurbits; crown gall on rose and cherry. Dusts are formulated with pyrophyllites, hydrated lime, sulfur, or calcium or magnesium carbonates as carriers. Phytotoxicity shows as a chlorotic flecking, sometimes stunting. Recent research indicates that spraying at night, when humidity is high, increases the absorption of streptomycin for fire blight control. Other names: Agri-Mycin 17, Agri-Strep, Plantomycin, Anisomycin, Phytomycin.

Subdue 2E. See Metalaxyl.

Sul-Cide. See Sulfur.

Sulfacop. See Copper sulfate.

Sulfur (Agtrol Chem. Products; BASF AG; Chem. Enterprises, Inc.; FMC Agri Chem.; Hoechst AG; ICI Americas Inc.). The oldest known fungicide, antedating written history, and somewhat effective as a miticide. In dust form, the particles should be fine enough to go through a 325-mesh screen.

Flowers of sulfur, small crystals produced by sublimation, are not fine enough.

Wettable sulfurs have wetting agents added for ready mixing with water for sprays. Flotation sulfurs are by-products of the manufacture of fuel gas from coal, so finely divided that they are almost colloidal. Micronized sulfurs also have particles approaching colloidal size. Sulfur sprays and dusts are effective in control of powdery mildews, rusts, apple scab, brown rot of stone fruits, rose blackspot, and other diseases. They are compatible with many other fungicides (having a synergistic effect with copper) and most insecticides, but should never be used with, or within a month of, oil sprays. Sulfur is of questionable compatibility with dinitro compounds and parathion. It is exempt from tolerance.

Wettable sulfurs and sulfur dusts are safer than lime sulfur at high temperatures, but should be used with caution above 85° F. They are not safe on many varieties of cucurbits, decreasing yield of squash and melon except in sulfur-resistant varieties. Other names: Brimstone, Sul-Cide.

Sulkol. See Sulfur.

Sultricop. See Copper Sulfate, Basic.

Sup'r Flo (discontinued by Rhone-Poulenc). See Maneb.

Super X Macclesfield. See Bordeaux mixture.

Suzu. See Brestan.

Suzu H. See Du-Ter.

Syllit. See Dodine.

Tachigaren (Sankyo Co., Ltd.). 3-hydroxy-5-methylisoxazole. Used to control fungi causing damping-off, such as *Fusarium, Aphanomyces, Corticium,* and *Pythium*.

Taifen. See Antracol.

Tairel. See Galben.

Talan. See Dinobuton.

TBCS-53. See Copper Sulfate, Basic.

TBZ. See Thiabendazole.

Tecto. See Thiabendazole.

Tecto RPH. See Thiabendazole.

Telone C. See Chloropicrin.

Tennessee Brand Tri-Basic Copper Sulfate 53WP. See Copper Sulfate, Basic.

Termil. Formulation of Daconil. A thermal dust effective for Botrytis flower spotting on orchids, geraniums, and other ornamentals. See also Daconil.

Terraclor. See PCNB.

Terraclor Super X. See Terrazole.

Terra-Coat. See Terrazole.

Terrazole (Mallinkroott, Inc.; Uniroyal Chem. Co., Inc.; United Agri Products). 5-ethoxy-3-trichloromethyl-1,2,4 thiadiazole. Used to control *Pythium* at low rates. Other names: Aaterra, Dwell, ETC MTD, Ethazol, Koban, Pansoil, Phorate TSX, Truban.

Terr-O-Cide (discontinued by Great Lakes Chemical Corp.). Combinations of ethylene dibromide and chloropicrin, or 1,3-dichloropropene, 1,2-dichloropropane, and related chlorinated hydrocarbons with chloropicrin. Used to control nematodes and fungi in soils.

Terr-O-Gas (Great Lakes Chemical Corp.). Various percentage proportions of methyl bromide and chloropicrin. Used to control nematodes and fungi in soils.

Tersan 75 (discontinued by DuPont Co.). See Thiram.

Tersan 1991. See Benomyl.

Tersan SP (discontinued by DuPont Co.). See Chloroneb.

Tetrapom. See Thiram.

Thiabendazole (Merck & Co., Inc.). 2-(4'-thiazolyl)-benzimidazole. Used to control green mold, blue mold, and stem end rot of citrus fruits; Cercospora leaf spot on sugar beets; brown patch, Fusarium patch, and dollar spot on turf; Fusarium basal rot, and Penicillium blue mold on ornamental bulbs and corms; brown rot on bananas; blue mold rot, bull's eye rot, and gray mold on apples and pears; black rot, scurf, and foot rot on sweet potatoes; Fusarium storage rot on Hubbard squash. Other names: Apl-Luster, Arbotect, Mertect, TBZ, Tecto, Tecto RPH, Thibenzole.

Thibenzole. See Thiabendazole.

Thimer. See Thiram.

Thioknock. See Thiram.

Thiolux. See Sulfur.

Thioneb. See Polyram-Combi.

Thion 80. See Sulfur.

Thion 95. See Sulfur.

Thiophal. See Phaltan.

Thiophan. See Thiophanate-Methyl.

Thiophanate (W. A. Cleary Chem. Corp.; Nippon Soda Co., Ltd.). 1,2-bis (3-ethoxycarbonyl-2-thioureido) benzene, or diethyl (1,2-phenylene) bis-(iminocarbonothioyl) (carbamate), active ingredient Topsin. Used to control diseases of turf as a systemic fungicide. Other names: Cercobin, Topsin E.

Thiophanate-Methyl (Nippon Soda Co., Ltd.; Pennwalt Corp.). Dimethyl (1,2-phenylene) bis-(iminocarboxo-thioyl) bis-(carbamate), also known as dimethyl 4,4'—o-phenylenebis (3-thioallophanate). Technical at least 96% active ingredient. Used as a systemic fungicide on a broad spectrum of diseases in vegetables, fruit, and turf. Other names: Cercobin M, Mildothane, Thiophan, Topsin M 70W, Topsin M 4.5F, Topsin Turf and Ornamentals.

Thioquinox (discontinued by Bayer AG). See Eradex.

Thiotex. See Thiram.

Thiovit. See Sulfur.

Thiram (W. A. Clearly; Pennwalt Corp.; UCB Chemicals Corp.; R.T. Vanderbilt Co., Inc.). bis-(dimethylthio-carbamoyl) disulfide; or tetramethylthiurani disulfide. Used as a seed protectant against seed decays and damping-off and also seedling blights. Controls certain fungus diseases of apples, peaches, strawberries, celery, and tomatoes. Used also as a fungicide on turf to control large brown patch and dollar spot. Other names: AAtack, Aules, Chipco Thiram 75, Fermide 850, Fernasan, Hexathir, Mercuram, Nomersam, Polyram Ultra, Pomarsol Forte, Spotrete-F, Spotrete WP75, Tetrapom, Thimer, Thioknock, Thiotex, Thiram Tech, Thiramad, Thirasan, Thiuramin, Tirampa, Trametan, Tripomol, Tuads.

Thirama D. See Thiram.

Thiram Tech. See Thiram.

Thirasan. See Thiram.

Thiuramin. See Thiram.

Thylate (discontinued by DuPont Co.). See Thiram.

Thynon (discontinued). See Dithane.

Tiazin. See Zineb.

Tiezene. See Zineb.

Tilcarex. See PCNB.

Tineston. See Triphenyltin acetate.

Tinnate (Nihon Nohyaku Co., Ltd.). Triphenyltin chloride 10%. Used to control late blight of potato.

Tirampa. See Thiram.

TMTDS. See Thiram.

TNCS 53. See Copper Sulfate, Basic.

Tobaz (discontinued by Merck Chemical Div.). See Thiabendazole.

Tolyfluanid. See Euparen M.

Topas. See Penconazole.

Topaz. See Penconazole.

Topaze. See Penconazole.

Topsin E. See Thiophanate.

Topsin Turf and Ornamentals. See Thiophanate-Methyl.

Topsin wettable powder. Thiophanate 50%. A broad-spectrum fungicide with preventive, curative, and systemic properties.

Topsin M. See Thiophanate-Methyl.

TPTA. See Brestan.

TPTH. See Du-Ter.

TPTOH. See Du-Ter.

Trametan. See Thiram.

Triadimefon. See Bayleton.

Triarimol (discontinued by Elanco Products Co.). Mildew fungicide.

Triasyn. See Dyrene.

Tribasic copper sulfate. See Copper Sulfate, Basic.

Tricarbamix. See Ziram.

Tricarbasul (discontinued by Pennwalt Holland B.V.). Co-manufactured ethylene bis-dithiocarbamate containing zinc and manganese ions and wettable sulfur. Used for control of downy mildew and powdery mildew of apples and cucurbits.

Tri-Clor. See Chloropicrin.

Tri-Con. See Chloropicrin.

Tricop. See Copper Sulfate, Basic.

Trifocide. See Dinitro compounds.

Triforine (Em Industries; Shell Agrar GMBH & Co. KG). N,N'-(1,4-pipera-zinediyl-(2,2,2-trichloroethylidene))-bis (formamide). Used as a systemic fungicide to control powdery mildew, scab, rust, and other diseases of ornamentals, fruits, and vegetables.

Trifrina. See Dinitro compounds.

Trifuncit. See Dithiocarbamates.

Trifungol. See Ferbam.

Trimangol. See Maneb.

Trimastan (Pennwalt Holland B.V.). Mixture of maneb and triphenyltin acetate. Used to control potato blight and Cercospora diseases of sugar beet and celery.

Trimaton. See Metam-Sodium.

Tri-Miltox (Sandoz Ltd.). Mancozeb + 3 copper salts (copper oxychloride, copper sulfate, copper carbonate). For control of downy mildews (*Phytophthora, Plasmopara*) and other diseases in grapes, potatoes, tomatoes, and most other crops.

Triofterol. See Zineb.

Trioneb. See Polyram-Combi.

Tri-PCNB. See PCNB.

Triphenyltin acetate. See Brestan.

Triphenyltin chloride. See Tinnate.

Triphenyltin hydroxide. See Du-Ter.

Triple Tin. See Du-Ter.

Tripomol. See Thiram.

Triquintam (discontinued by Pennwalt Holland B.V.). Mixture of PCNB and thiram. Used for soil disinfestation of sclerotiae-forming fungi.

Triscabol. See Ziram.

Tritisan. See PCNB.

Tritoftorol. See Zineb.

Trizone (Dow Chemical). Methyl bromide, chloropicrin, 3-bromoproparyl bromide, and related compounds. A nematicide to be used with caution. See also Methyl bromide.

Truban (Sierra Chem. Co.). 5 ethoxy-3-trichloromethyl-1,2,4-thiadiazole. Used to control *Pythium* and *Phytophthora*.

Tsitrex. See Dodine.

Tuads. See Thiram.

Tubothane. See Maneb.

Tubotin. See Du-Ter.

Turfcide. See Bravo.

Tuzet. See Urbacid.

UC 19786. See Dinobuton.

Unicrop Maneb. See Maneb.

Urbacid (discontinued by Bayer AG). bis(dimethylthio-carbamoylthio) methyl arsine. Used to control coffee diseases and apple scab.

Validacin (Takeda Chemical Industries, Ltd.). N-(1 S)-(1,46/5)-3-hydroxymethyl-4,5,6-trihydroxy-2-cyclohexenyl) (O-beta-D-glucopyranosyl-(-3-(15)-(1,2,4/3,5)-2,3,4-trihydroxy-5-hydroxy-methylcyclohexyl) amine. Used to control damping-off of vegetables caused by *Rhizoctonia* and black scurf of potatoes.

Validamycin A. See Validacin.

Vancide-TM Flowable (discontinued by R. T. Vanderbilt Co.). See Thiram.

Vancide TM-95 (discontinued by R. T. Vanderbilt Co.). See Thiram.

Vapam. See Metam-Sodium.

Vencedor. See Copper sulfate; Copper Sulfate, Basic.

Venturol. See Dodine.

Vi-Cad. See Caddy.

Vinclozolin (BASF AG; Sierra Chem. Co.). 3-(3,5 dichlorophenyl)-5-ethenyl-5-methyl-2,4-oxazolidinedione. For control of *Botrytis* spp., *Sclerotinia* spp., *Monilia* spp. in grapes, strawberry, rape, soft fruits, hops, vegetables and ornamentals, also turf diseases. Other names: Ronilan, Ornalin, Vorlan.

Vitavax (Uniroyal). 2,3-dihydro-5-carboxanilido-6-methyl 1,4-oxathiin, systemic fungicide used for damping-off rusts and seed treatment.

Vondcaptan. See Captan.

Vondodine. See Dodine.

Vondozeb Plus. See Dithane M-45.

Vorlan. See Vinclozolin.

Voronit (discontinued). Mixture of 2 (2'-furyl)-benzimidazole (I) and hexachlorobenzene (II). Used as a seed dressing with special action against *Fusarium.*

VPM. See Metam-Sodium.

Yellow Cuprocide. See Copper oxides.

Z-C Spray (discontinued by FMC Corp.). See Ziram.

Zebtox. See Zineb.

Zerlate. See Ziram.

Zidan (discontinued by Makhteshim-Agan). See Zineb.

Ziman-Dithane. See Dithane M-45.

Zincmate. See Ziram.

Zinc Metiram. See Polyram-Combi.

Zinc sulfate (discontinued by Cities Service Co.). Used with Dithane D-14 to form zineb and also to control zinc deficiency diseases such as little-leaf or mottle-leaf.

Zineb (Bayer AG; Drexel Chem. Co.; Pennwalt Holan B.V.; Rhone-Poulenc Agrochemie), common name for zinc ethylene bisdithiocarbamate, available as Dithane Z-78 and as Parzate, also in many pesticide mixtures. Effective for azalea petal blight, Botrytis blight of peony, tulip, rose blackspot, snapdragon rust, potato and tomato blights, downy (but not powdery) mildews, anthracnose diseases, citrus fruit russet, strawberry root rot. It is used some as a blossom spray for fire blight and for other fruit diseases. The tolerance is 7 ppm for most food crops. Other names: Aspor, Chem Zineb, Devizeb, Dipher, Hexathane, Kypzin, Lonccol, Parzate, Parzate C, Polyram Z, Tiezene, Tritoftorol, Zebtox, Zineb 75, Zineb 75%, Zineb 75WP, Zinosan.

Zineb 75%. See Zineb.

Zineb 75WP. See Zineb.

Zinosan. See Zineb.

Ziram (FMC Agri. Chem. Group; Pennwalt Corp.; Rhone-Poulenc Agrochemie; UCB Chem. Corp.; R.T. Vanderbilt Co., Inc.). Common name for zinc dimethyl dithiocarbamate, sold as Zerlate or Karbam White. Effective for some vegetable blights and leaf spots, some fruit diseases.

Zirbeck. See Ziram.

Ziride. See Ziram.

Zitox. See Ziram.

BACTERICIDES

Agri.-Mycin 17 (Pfizer Inc., Chem. Div.). Streptomycin sulfate, for control of bacterial plant diseases.

Agri.-Strep. See Strystomycin in Fungicides section.

Agritol (discontinued by Merck & Co.). *Bacillus thuringiensis.*

Agrox Strep (discontinued by Chipman Chem., Inc.). Combination bactericide, insecticide, and fungicide, seed treatment of beans where halo blight is a problem. Registered only in Michigan.

Bacticin (discontinued by TUCO Div., Upjohn). Antibiotic bactericide for eradicating crown gall on fruits, rose, and other ornamentals, and for treating olive knot.

Barquat Compounds (Lonza, Inc.). Quarternary ammonium compounds. Germicidal effectiveness against a wide variety of microorganisms in antiseptics, germicides, algicides, deodorants, and detergent-sanitizers.

Barquat MB-50. See Benzalkonium chloride.

Barquat MB-80. See Benzalkonium chloride.

Bayclean. See Dimanin A.

Benzalkonium Chloride (Lonza Inc.). Alkyl dimethyl benzylammonium chloride. Bactericide, fungicide.

Bromofume. See Ethylene dibromide in Nematicides section.

Brom-O-Gas. See Methyl bromide in Nematicides section.

Bromoethane. See Methyl bromide in Nematicides section.

Brom-O-Sol (Great Lakes Chemical Corp.). Contains methyl bromide 68.6%, chloropicrin 1.4%, organic solvent 30%. Used to control soil-borne diseases, nematodes, and insects. Growth difficulty may be experienced on carnations, holly, snapdragons, and multiflora rose. Apply at least 18 inches from plants to be maintained.

Bronopol (Schering Ag.). Used as a bactericide and bacteriostat against *Xanthomonas malvacearum* and *Erwinia amylovora.*

Chinosol. See Fungicides.

Copper hydroxide. See Fungicides.

Copper Sulfate, Basic. Contains a minimum 53% copper (metallic basis). Used widely to control citrus diseases, bacterial and fungus diseases of tomatoes and peppers as well as fruit, vegetables, and ornamentals. See also Fungicides.

Cryptonol (discontinued by Duphar-Midox Ltd., England). Potassium hydroxyquinoline sulfate. Used as a soil drench, dip, or spray for control of fungal and bacterial diseases or a glass-house disinfestant.

Dimanin A (Bayer AG). Alkyldimethylbenzylammonium chloride. Algicide, bactericide. Other name: Bayclean.

Dimanin C (Bayer AG). Sodium dichloroisocyanurate. Algicide, bactericide, viricide.

Gallex (Ag Bio Chem, Inc.). 2,4-Xylenol, meta-Cresol, and penetrants. Crown gall eradicant.

Galltrol-A (Ag Bio Chem, Inc.). *Agrobacterium radiobacter* (strain 84). Biological control; ecological preventative.

Hexachlorophene (discontinued by Kalo Laboratories, Inc.). 2,7-methylene bis-(3,4,6-trichloro-phenol). Used on tomatoes, peppers, and cucumbers for fungus and bacterial diseases and as a soil fungicide against *Rhizoctonia.*

Kasugamycin (Hokko Chemical Industry Co., Ltd.). Kasugamycin hydrochloride. Used to control leaf mold of tomatoes, halo blight of beans, and apple scab.

Mycoshield (Pfizer Inc.). Contains 17% oxytetracycline. Antibacterial, antibiotic.

Nabac 25 EC (discontinued by Kalo Laboratories, Inc.). 2,2-methylenebis (3,4,6-trichlorophenol). Used as a broad-spectrum foliar fungicide and bactericide.

Phytomycin (Olin Mathieson). Streptomycin nitrate, promising antibiotic for lima bean downy mildew, tomato blight. See Streptomycin in Fungicides section.

Phyton 27 (Technology Biologicals, Inc.). Tannate complex of pirro cupric ammonium formate. Systemic fungicide, bactericide. Other name: Copper complex.

TBCS-53. See Copper Sulfate, Basic in Fungicides section.

TNCS-53. See Copper Sulfate, Basic in Fungicides section.

Tribasic Copper Sulfate. See Copper Sulfate, Basic in Fungicides section.

Zinc Sulfate, Basic (discontinued in 1985 by Woolfolk Chemical Works, Inc.). $ZnSO_4 - H_2O$ (the monohydrate is used in agriculture). Used as bactericide and nutritional spray on peaches.

NEMATICIDES

A7 Vapam. See Metam-Sodium

Basamid Granular. See Dazomet.

Bay 25141. See Dasanit.

Bay 68138. See Nemacur.

Bay 70143. See Furadan.

Bay SRA 3886. See Nemacur.

Brifur. See Furadan.

Bromofume. See Ethylene dibromide.

Brom-O-Gas. See Methyl bromide.

Brom-O-Gaz. See Methyl bromide.

Bromoethane. See Ethylene dibromide.

Bromomethane. See Methyl bromide.

Brom-O-Sol. See Methyl bromide.

Brozone (discontinued by Dow Chemical Co.). Formulation of methyl bromide and chloropicrin in a petroleum solvent.

Busan 1020. See Metam-Sodium.

Celfume. See Methyl bromide.

Celmide. See Ethylene dibromide.

Chloropicrin. See Fungicides.

Crag Fungicide 974. See Dazamet.

Crag Nematicide. See Dazamet.

Crisfuran. See Furadan.

Curaterr. See Furadan.

D 1221. See Furadan.

Dasanit (Bayer AG; Mobay Corp.). 0,0-diethyl 0-[P-(methylsulfinyl) phenyl] phosphorothioate. General nematicide, also used for plant dip but extremely toxic to mammals. (Use ceased in 1989.) Other names: Bay 25141, S767, Terracur P.

Dazomet (BASF Aktiengesellschaft; Hopkins Agri. Chem. Co.; ICI Americas, Inc.; UCB Chem Corp.). Tetrahydro-3,5-dimethyl-2H-thiadiazine-2-thione. Used to control soil fungi, nematodes, weeds, and soil insects; preplant treatment for turf and ornamentals. Other names: Basamid Granular, Crag Fungicide 974, Crag nematicide, Dazomet-Powder BASF, DMTT, Micofume, Mylone, N-521, Prezervit.

Dazomet-Powder BASF. See Dazamet.

DBCP. 1,2-dibromo-3-chloropropane, nematicide safe around living plants. Sold as Nemagon Soil Fumigant (Shell), Fumazone (Dow), and under other trade names. Can be used as preplant treatment or as side dressing. May be phytotoxic to carnation, chrysanthemum, and dwarf palms. See also Dibromochloropropane.

D-D 92. See Dichloropropene.

D-D Soil Fumigant (discontinued by Shell). Dichloropropene-dichloropropane, excellent nematicide, well-suited for large-scale operation, since the fumes need not be confined. Injections are made 12 inches apart. There is little or no control of soil fungi. D-D is not safe around living plants; fields should be treated well in advance of planting.

Diamidfos. See Nellite.

Dibromochloropropane (Dow; Occidental; Shell). 1,2-dibromo-3-chloropropane. Used on nuts, vegetables, and ornamentals. Other names: Nemafume, Nemanox, Nemaset.

Dichlofenthion. (discontinued name). See Dichlorofenthion; Mobilawn.

Dichlorofenthion (Pennwalt; Sintesul S.A.). 0-2,4-dichlorophenyl 0,0 diethylphosphorothioate. Used to control non-cyst-forming nematodes on ornamentals and turf. Other names: Tri-VC13, VC13 nematicide. See also Mobilawn (discontinued name).

Dichloropropene (The Dow Chemical). 1, 3-dichloropropene. Preplant for nematodes, disease, insect, and weed control on a variety of crops such as vegetables, field crops, citrus, deciduous fruits and nuts, bush and vines, and nursery crops. Apply only as preplant to control nematodes since the chemical is phytotoxic. Applications should not be made in glass-houses containing plants or within 1 meter of the root zone of growing crops in the field. Other names: D-D 92, Telone II Soil Fumigant.

Di-Trapex. See Vorlex.

DMTT. See Dazomet.

Dorlone. See Telone.

Dow-Fume 75 (discontinued name by The Dow Chemical). Ethylene dichloride 70%, carbon tetrachloride 30%.

Dowfume C (discontinued name by The Dow Chemical). Carbon disulfide 12.1%, carbon tetrachloride 81.3%, ethylene dibromide 6.6%.

Dowfume EB-5 (discontinued name by The Dow Chemical). Ethylene dichloride 29.2%, carbon tetrachloride 63.6%, ethylene dibromide 7.2%.

Dowfume-59 (discontinued name by The Dow Chemical). Ethylene dibromide 59%, carbon tetrachloride 32%, ethylene dichloride 9%.

Dowfume F (discontinued name by The Dow Chemical). Ethylene dichloride 65%, carbon tetrachloride 27%, ethylene dibromide 5%.

Dowfume MC-2 (discontinued name by The Dow Chemical). Methyl bromide 98%, chloropicrin 2%.

Dowfume MC-33 (discontinued name by The Dow Chemical). Methyl bromide 67%, chloropicrin 33%.

Dowfume N (discontinued name by The Dow Chemical). 1,3-dichloropropene 50%, other chloropropenes.

Dowfume V (discontinued name by The Dow Chemical). Carbon tetrachloride 85.1%, ethylene dichloride 12.1%, ethylene dibromide 2.8%.

Dowfume W-85 (discontinued name by The Dow Chemical). Ethylene dibromide 83%.

Du Nema, 4-chloropyridine-N-oxide. Used on turf.

EDB. See Ethylene dibromide. (Uses canceled by EPA.)

EDB-85. See Ethylene dibromide. (Uses canceled by EPA.)

E-D-Bee. See Ethylene dibromide. (Uses canceled by EPA.)

EDC. See Ethylene dichloride (Uses canceled by EPA.)

ED/CT. See Ethylene dichloride (Uses canceled by EPA.)

ENT 27164. See Furadan.

Ethoprop. See Mocap.

Ethylene Dibromide (Excel Industries Ltd.; United Phosphorus Ltd.). 1,2-Dibromoethane. Soil fumigant for nematodes and other pests. Sold as Dowfume, Soil-fume, and Bromofume. Treat in late summer or early fall. Do not use around living plants. Other names: Bromofume, Celmide, E-D-Bee, EDB, EDB-85, Kop Fume, Nephis. (Uses canceled by EPA.)

Ethylene Dichloride (All India Medical Corp.). 1,2-Dichloroethane. Soil fumigant for nematodes and other pests. Treat in late summer or early fall. Do not use around living plants. Other name: EDC. (Uses canceled by EPA.)

Fensulfothion. See Dasanit. (Uses canceled by EPA.)

FMC 10242. See Furadan.

Fumazone (discontinued name). See Dibromochloropropane.

Fumigant-1 (Great Lakes Chem. Corp.) (discontinued name). See Methyl bromide.

Furadan (FMC Agri-Chem Group). 2,3-dihydro-2,2 dimethyl-7-benzofuranyl methylcarbamate. Other names: Bay 70143, Brifur, Crisfuran, Curaterr, D1221, ENT 27164, FMC10242, Furadan, NIA 10242, Yaltox.

Hexa-Nema (discontinued). See Mobilawn.

Hoe 002960. (Hoechst AG). See Hostathion.

Hostathion (Hoechst AG). 1-phenyl-1,2,4-triazolyl-3-(0,0-diethyl-thionophosphoryl). Used on free-living nematodes in vegetable and fruit crops. Other name: HOE 002960.

Jolt (discontinued name). See Mocap.

Karbation. See Metam-Sodium.

Kayafume. See Methyl bromide.

Kop-Fume. See Ethylene dibromide.

Lannate. See Methomyl. (Not registered for use as nematicide.)

Maposol. See Metam-Sodium.

MeBr. See Methyl bromide.

Metam 32.7. See Metam-Sodium.

Metam 42. See Metam-Sodium.

Metam-Fluid BASF. See Metam-Sodium.

Metam-Sodium (BASF Aktiengesellschaft; Buckman Laboratories, Inc.; ICI Plant Protection Division; Pennwalt Holland B.V.; United Agri Products, Inc.). Sodium N-methyldithiocarbamate. General-purpose soil fumigant that is highly effective in control of weeds, weed seeds, nematodes, and soil fungi. Other names: A7 Vapam, Busan 1020, Karbation, Maposol, Metam 32.7, Metam 42, Metam Fluide BASF, Nemasol, Solasan 500, Sometam, Trimaton, Vapam, VPM.

Metham. See Metam-Sodium.

Metham-Sodium. See Metam-Sodium.

Meth-O-Gas. See Methyl bromide.

Methomyl (DuPont; Chrystal Chem. Inter-America). S-methyl-N-((methylcarbamoyl) oxy) thioacetimidate. An insecticide-nematicide used in vegetables, fruits, and ornamentals. Other names: Lannate, Lanox 90, Lanox 216, Methomex, Metox-900, Nudrin. (Not registered for use as nematicide.)

Methyl bromide (Great Lakes Chemical Corp.). Bromomethane. Soil fumigant, supplied under that name by a number of manufacturers. It is used in greenhouses because it is somewhat less toxic to plants than most other fumigants and for bailed or potted nursery stock in special fumigating chambers. As Dowfume MC-2, methyl bromide with 2% chloropicrin, it comes in cans with a special dispenser. The area to be treated is covered with plastic film, under which are evaporating pans with hoses leading to the edge of the cover. The dispenser is attached to each hose in turn to fill the pans, and then the edges of the cover are held down with soil. For small areas injections are made 10 inches apart. This material is very poisonous; follow all safety precautions. Other names: Brom-O-Gas, Brom-O-Gaz, Brom-O-Sol, Celfume, Kayafume, MeBr, Meth-O-Gas, Terr-O-Cide II, Terr-O-Gas.

Methyl isothiocyanate. See Vorlex.

Micofume. See Dazomet.

Mobilawn (discontinued name by Mobil Chemical Co.). 0-2,4-dichlorophenyl 0,0 diethylphosphorothioate. Used to control non-cyst-forming nematodes on ornamentals and turf.

Mocap (Rhone-Poulenc Ag. Co.). 0-ethyl, S-S-dipropyl phosphorothioate. Used to control nematodes on sweet potato, banana, cabbage, corn, pineapple, peanuts, cucumber, snap and lima beans, white potato, and selected ornamentals and turf.

Mylone. See Dazomet.

N 521. See Dazomet.

Nellite (discontinued by The Dow Chemical), phenyl N,N'-dimethyl phosphorodiamidate. Used on tobacco to control rootknot nematodes.

Nemacur (Bayer AG; Mobay Agri. Chem. Div.). Ethyl 3-methyl-4-(methylthio) phenyl (1-methylethyl) phosphoramidate. Used to control major genera of nematodes attacking peanuts, cabbage, brussel sprouts, and turf. Other names: Bay 68138, Bay SRA 3886.

Nemafene. See D-D Soil Fumigant.

Nemafume. See Dibromochloropropane.

Nemagon (discontinued by Shell). Nematicide safe around living plants. See also Dibromochloropropane (DBCP).

Nemanex. See Dibromochloropropane.

Nemaset. See Dibromochloropropane.

Nemasol. See Metam-Sodium.

Nephis. See Ethylene dibromide.

NIA 10242 (FMC Corp.). See Furadan.

Nudrin. See Methomyl. (Not registered for use as nematicide.)

OMS 771. See Temik.

Pestmaster (discontinued by Velsicol Corp.). See Methyl bromide.

Prezervit. See Dazomet.

Profume (discontinued by Dow Chem.). See Methyl bromide.

Prophos (discontinued name). See Mocap.

Rotox (discontinued by Ferguson Fumigants). See Methyl bromide.

S 767. See Dasanit.

Sarolex (Geigy). Nematicide-insecticide, a special formulation of Diazinon and formerly for some nematodes in southern turf grasses.

SMDC. See Metam-Sodium.

Sodium methyldithiocarbamate. See Metam-Sodium.

Solasan 500. See Metam-Sodium.

Sometam. See Metam-Sodium.

Telone. See Dichloropropene.

Telone II Soil Fumigant. See Dichloropropene.

Temik (Rhone-Poulenc Ag. Co.; Union Carbide). 2-methyl-2-(methylthio) propionaldehyde O-(methylcarbamol) oxime. Insecticide-acaricide with nematocidal activity. Other names: OMS 771, UC 21149.

Temik Brand. See Temik.

Terracur. See Dasanit.

Terr-O-Cide (discontinued by Great Lakes Chemical Corp.). Combinations of ethylene dibromide and chloropicrin, or 1,3-dichloropropene, 1,2-dichloropropane, and related chlorinated hydrocarbons with chloropicrin. Used to control nematodes and fungi in soils.

Terr-O-Gas (Great Lakes Chemical Corp.). Various percentage proportions of methyl bromide and chloropicrin. Used to control nematodes and fungi in soils.

Tiazon. See Dazomet.

Tirpate (discontinued by 3M Co.). 2,4-dimethyl-1,3-dithiolane-2-carboxaldehyde 0-(methylcarbamoyl)oxime. Used to control nematodes.

Triazophos. See Hostathion.

Trimaton. See Metam-Sodium.

Tri-VC 13. See Dichlofenthion.

Trizone (Dow Chemical). Methyl bromide, chloropicrin, 3-bromoproparyl bromide, and related compounds. A nematicide to be used with caution. See also Methyl bromide.

UC 21149. See Temik.

Vapam. See Metam-Sodium.

VC-13 Nemacide. See Dichlofenthion.

Vidden D (Dow Chemical Co.). Mixture of dichloropropane-dichloropropene. Used to control nematodes in soils.

Vorlex (Nor-Am Chem. Co.; Schering AG). Methyl isothiocyanate, 1,3-dichloropropene and other chlorinated hydrocarbons). Preplant soil fumigant to control weeds, fungi, insects, and nematodes in potatoes, tobacco, vegetables, and ornamentals. Other name: Di-Trapex.

VPM. See Metam-Sodium.

Yaltox. See Furadan.

VIROCIDES

Cytovirin. Antiviral antibiotic, inhibiting mosaic in bean and tomato.
Dimanin C. See Bactericides.

SOURCES OF PESTICIDES

Applied Industrial Materials Corp. One Parkway North, Suite 400, Deerfield, IL 60015.

AgBiochem Inc., 3 Fleetwood Ct., Orinda, CA 94563.

Agrimont S.p.A, Piazza della Repubblica 14/16, 20124 Milano, Italy.

Agtrol Chemical Products, 7324 Southwest Freeway, New Orleans, LA 70112.

Agway, Inc., Box 4741, Syracuse, NY 13221.

All India Medical Corp., Akhand Jyoti 8th Road, P.O. Box 16806, Santacruz East, Bombay 400 055, India.

American Cyanamid Company, One Cyanamid Plaza, Wayne, NJ 07470.

American Hoechst Corporation, Agricultural Division, 11312 Hartland Street, North Hollywood, CA 91605.

Artel Chemical Corp., 91 Carolyn Blvd., Farmingdale, NY 11735-1527.

BASF Aktiengesellschaft, Carl-Bosch-Str. 38, D-6700/Ludwigshafen, Federal Republic of Germany.

Bayer AG, Sektor Landwirtschaft, Pflanzenschutzzentrum Monheim, 5090 Leverkusen Bayerwerk, Federal Republic of Germany.

The Boots Co., Ltd., Agro Chemical Division, 1 Thane Rd. West, Nottingham, England.

Buckman Laboratories, Inc., 1256 N. McLean Blvd., Memphis, TN 38108.

Celamerck GMBH & Co. KG, P.O. Box 202, 6507 Ingelheim/Rhein, West Germany.

Chemagro Agricultural Division, Mobay Chemical Corporation, P.O. Box 4913, Hawthorn Road, Kansas City, MO 64120.

Chemical Formulators, Inc., Box 26, Nitro, WV 25143.

Chemical Interprises, Inc., 8582 Katy Freeway, Suite 202, Houston, TX 77024-1854.

C.P. Chemicals Inc., P.O. Box 21, 7B Terminal Way, Avenel, NJ 07001.

Chevron Chemical Company, 6001 Bollinger Canyon Road, Building T, P.O. Box 5047, San Ramon, CA 94583-0947.

Chipman Chemicals, Inc., Box 718, River Rouge, MI 48218.

Ciba-Geigy Corp, P.O. Box 18300, Greensboro, NC 27419-8300.

Cities Service Co., Industrial Chemicals Marketing Dept., Box 50360, Atlanta, GA 30302.

W. A. Cleary Corporation, Box 10, 1049 Somerset St., Somerset, NJ 08873.

Crystal Chemical Co.-Inter America, 1525 N. Post Oak Rd., Houston, TX 77055.

Davison Chemical Division, W. R. Grace & Co., Charles and Baltimore Streets, Box 247, Baltimore, MD 21203.

Diamond Shamrock Chemical Co., 300 Union Commerce Building, 1100 Superior Ave., Cleveland, OH 44114.

Dow Chemical Company, 2020 Willard H. Dow Center, Midland, MI 48674.

Drexal Chemical Co., 2487 Pennsylvania St., Memphis, TN 38109.

Duphar B.V., Crop Protection Div., P.O. Box 4, 1243 2G-s-Graveland, The Netherlands.

Duphar-Midox Ltd., Smarden, Kent, England TN 278QL.

E. I. DuPont de Nemours & Company, Industrial & Biochemical Department, 1007 Market St., Wilmington, DE 19898.

Elanco Products Company, Division of Eli Lilly & Co., Lily Corporate Center, Indianapolis, IN 46285.

Excel Industries, Ltd., 184-87 Swani Vivekanand Road, P.O. Box 7474, Jogeshwari, Bombay 400102, India.

Fabriek van Chemische Producten Vondelingenplaat B.V., P.O. Box 7120, Rotterdam 3031, Netherlands.

Ferguson Industries, 1900 W. Northwest Highway, Dallas, TX 75220.

Fermenta Plant Protection Co., 5966 Heisley Rd., P.O. Box 8000, Mentor, OH 44060-8000.

FMC Corporation, Agricultural Chemicals Group, 2000 Market St., Philadelphia, PA 19103.

Great Lakes Chemical Corp., P.O. Box 2200, West Lafayette, IN 47906.

Griffin Ag. Products Co., Inc., P.O. Box 1847, Valdosta, GA 31603.

Harshaw Chemical Co., 1945 E. 97th St., Cleveland, OH 44106.

Hoechst AG, Agricultural Div., Postfach 80 03 20, D-6230 Frankfurt (80) F.R.G., Germany.

Hokko Chemical Industry Co., Ltd., Mitsui Bldg. No. 2, 4-4-20, Nihonbashi Hongokucho-Cho, Chuo-ku, Tokyo, Japan.

Hopkins Agricultural Chemical Co., P.O. Box 17532, Madison, WI 53707.

ICI Agrochemicals, Fernhurst, Haslemere, Surrey GU27 3JE, United Kingdom.

ICI Americas, Inc., Agricultural Chemicals Div., Wilmington, DE 19897.

Kaken Pharmaceutical Co., Ltd., 3-4-10, Nihonbashi-Honcho, Chuo-Ku, Tokyo 103, Japan.

Kalo Inc., 4550 West 109 Street, Overland Park, KS 66211-1351.

Keno Gard AB, P.O. Box 11555, S-10061 Stockholm, Sweden.

Kerr-McGee Chemical Corp., Kerr-McGee Center, Oklahoma City, OK 73125.

Kincaid Enterprises, Inc., P.O. Box 549, Nitro, WV 25143.

Kocide Chemical Corp., 12701 Almeda Rd., Houston, TX 77045.

Kumiai Chemical Industry Co., Ltd., 4-26 Ikenohata, 1-chome Taitoh-ku, Tokyo 110, Japan.

LaCorunbia S.A., 85 Quai de Brazza, Bordeaux 33100, France.

Leffingwell, A Business of Uniroyal Chemical Co., Inc., 111 South Berry St., P.O. Box 1880, Brea, CA 92621.

Lonza, Inc. 22-10 Rt. 208, Fairlawn, NJ 07410.

Lowes Industrial Products Div., Edward Lowe Industries, 348 South Columbia St., P.O. Box 16, South Bend, IN 46624.

Makhteshim-Agan, P.O. Box 60, 84100 Beer-sheva, Israel.

Mallinckrodt Chemical Works, Mallinckrodt and Second St., P.O. Box 5439, St. Louis, MO 63147.

A. H. Marks & Co., Ltd., Wyke Lane, Wyke Bradford, W. Yorkshire, England.

Merck & Company, Inc., 126 Lincoln Ave., P.O. Box M, Rahway, NJ 07065.

Michigan Chemical Corp., 2 N. Riverside Plaza, Chicago, IL 60606.

Miller Chemical & Fertilizer Corporation, P.O. Box 333, Hanover, PA 17331.

Mineral Research and Development Corporation, One Woodlawn Green, Charlotte, NC 28217.

Mobay Corp. Agricultural Chemicals Div., P.O. Box 4913, 8400 Hawthorn Rd., Kansas City, MO 64120.

Mobil Chemical Company, 150 East 42nd St., New York, NY 10017.

Monsanto Chemical Company, 800 N. Lindbergh Boulevard, St. Louis, MO 63167.

Montedison DIPA, Agricultural Products Div., Piazza Della Republica 14/16, 20100 Milano, Italy.

Nationwide Chemical Company, 2209 Fowler St., P.O. Box 775, Fort Myers, FL 33902.

Nihon Nohyaku Co., Ltd., 2-5 Nihonbashi 1-chome Chuo-ku, Tokyo, Japan.

Nippon Soda Co., Ltd. Agro-Pharm Div., 2-2-1 Ohtemachi, Chiyoda-ku, Tokyo 100, Japan.

Nor-Am Agricultural Products, Inc., 20 N. Wacker Dr., Chicago, IL 60606.

Nor-Am Chemical Co., 3509 Silverside Rd., P.O. Box 7495, Wilmington, DE 19803.

Nutrilite Products, Inc., 5600 Beach Blvd., Buena Park, CA 90620.

Nysjon Soda Co., Ltd., Fine Chemical Division, Shin-Oktemachi Bldg. No. 2-2-1 Ohtemachi, Chiyoda-Ku, Tokyo, Japan.

Occidental Chemical Co., P.O. Box 198, Lanthrop, CA 95330.

Old Bridge Chemicals, P.O. Box 194, Old Bridge, NJ 08857.

Olin Corporation, P.O. Box 991, Little Rock, AR 72203.

Pennwalt Corp., Agchem. Div., 3 Parkway, Philadelphia, PA 19102.

Pennwalt Holland B.V., Production Dept., P.O. Box 7120, 3000 H.C. Rotterdam, Holland.

Pfizer Inc., MPM Division, 235 East 42nd St., New York, NY 10017.

Phelps Dodge Refining Corp., P.O. Box 20001, El Paso, TX 79998.

Probelte, S.A., Ctra. de Madrid, Km 384'6 Apartado 579, Murcia, Spain.

Procida S. A., 5 rue Bellini, 92 806 Puteaux, France.

Rhodia, Inc., Agricultural Division, P.O. Box 125, Monmouth Junction, NJ 08852.

Rhone-Poulenc Agrochimie S.A., 14-20 rue Pierre Baizet, 69009 France.

Roberts Chemicals, Inc., Box 446, Nitro, WV 25143.

Rohm & Haas Company, Independence Mall West, Philadelphia, PA 19105.

Sandoz, Ltd., Agro Div., Lichstrasse 35, CH4002, Basle, Switzerland.

Sankyo Company, Ltd., No. 7-12, Ginza 2-chome, Chuo-Ku, Tokyo 104, Japan.

Schering AG, Agrochemical Div., P.O. Box 65 03 11, D-100 Berlin 65, West Germany.

Shell Agrar Gmbtt & Co., KG, P.O. Box 202, 6507 Ingleheim/Rhein, F.R.G.

Shell Development Company, Agricultural Research Division, 2401 Crow Canyon Rd., San Ramon, CA 94583.

Shell International Chemical Co., Ltd., Shell Centre, London SE1 7PG England.

Sierra Chemical Co., 1001 Yosemite Drive, Milpitas, CA 95035.

Sintesul, Rua Joas Thomaz Munhoz 218, Caixa Postal No. 263, Pelotas, Rio Grande do Sul, Brazil 96.100.

Stauffer Chemical Company, Agricultural Chemical Div., Westport, CT 06880.

Sumitomo Chemical America, Inc., 345 Park Ave., New York, NY 10154.

Sunko Chemical Co., Ltd., No. 12, Lane 42, Jen-Hua Rd., Ta-li Hsiang, Taichung Hsien, Taiwan, R.O.C.

Takeda Chemical Industries, Ltd., 12-10 Nihonbashi 2-chome, Chou-ku Tokyo 103. Japan.

Tennessee Chemical Co., 3475 Lenox Rd., N.E., Suite 670, Atlanta, GA 30326.

Thompson-Hayward Chemical Co., 5200 Speaker Rd., Kansas City, KS 66106.

Tifa Ltd., Tifa Square, 50 Division Ave., Millington, NJ 07946.

Tower Chemical Co., P.O. Box 585, Clermont, FL 32711.

Transvaal, Inc., P.O. Box 69, Marshall Rd., Jacksonville, AR 72076.

Troy Chemical Corp., One Avenue L, Newark, NJ 07105.

UCB Chemicals Corp., 5365-A Robin Hood Rd., Norfolk, VA 23513.

Union Carbide Corporation, Agricultural Products Co., P.O. Box 12014, Research Triangle Park, NC 27709.

Uniroyal Chemical Co., Inc., Crop Protection Div., World Headquarters, Middlebury, CT 06749.

Uniroyal, U.S. Rubber Company, Emic Bldg., Naugatuck, CT 06770.

United Agri-Products, P.O. Box 1286, Greeley, CO 80632.

United Phosphorus Ltd., 167 Dr. Annie Besant Road, Worli, Bombay 400 018, India.

Universal Crop Protection, Ltd., Park House, Maidenhead Rd., Cookham, Berkshire, SL6 9DS United Kingdom.

Upjohn Company, TUCO Products Co. Div., 7000 Portage Rd., Kalamazoo, MI 49001.

R. T. Vanderbilt Co., Inc., 30 Winfield St., Norwalk, CT 06855.

Velsicol Chemical Corporation, 341 E. Ohio Street, Chicago, IL 60611.

Vineland Chemical Co., Inc., 1611 West Wheat Rd., Vineland, NJ 08360.

Visplant Chemiren S.R.L., Via Salviz, 44045 Renazzo de Cento Ferrara, Italy.

Wesley Industries, Inc., P.O. Box 490, Montrose, AL 36559.

Woolfolk Chemical Works, Inc., P.O. Box 938, Fort Valley, GA 31030.

APPLYING THE CHEMICALS

Spraying is the application of a chemical to a plant in liquid form; *dusting* is the application of a fine dry powder. The difference between spraying and dusting was very clear-cut before aerosol bombs, mist blowers, and fog machines were developed to apply liquids in such concentrated form that the particles are practically dry before they reach the plant and before spray-dusters were made to deliver wetted dusts.

Sprayers vary from a flit gun or pint atomizer that takes an hour to discharge a gallon, to power apparatus that discharges 60 gallons a minute at 800 pounds pressure from a 600-gallon spray tank. Dusters vary from the small cardboard or plastic carton in which the dust is purchased to helicopters. Applicators for pressurize sprays or aerosols vary from the one-pound "bomb" to truck-mounted fog generators or air blast machines. See Figure 1 for various applicators.

Mist Sprayers

In orchards and in shade tree work there has been increasing use of mist blowers, air blast machines that carry droplets of concentrated pesticides to plants in air rather than water. They are speedier than hydraulic sprayers, use far less water, which may be scarce in times of drought, and do not leave puddles or poisonous runoff that may be dangerous to pets and birds. They cannot, however, be operated in much wind; for that reason, and also in order to see the distribution of the concentrates, they often have to be used at night. They are not too efficient for very tall trees, and the droplet size has to be rather carefully regulated. Too large drops may fall out before they reach a tree, and too small drops may not settle down.

Although we usually think of mist blowers on trucks for large-scale operations, there are now some about the size of knapsack sprayers that, engine and all, are worn on the back around the garden. They weigh around 35 pounds and will cover foliage up to 30 feet. They cost somewhat more than the hydraulic power sprayers of small estate size.

Hydraulic Sprayers

Mist blowers will probably never entirely outmode hydraulic sprayers, which can place the spray more accurately, at a greater height, and can operate under more unfavorable weather conditions. For trees, high gallonage per minute and enough pressure to drive sprays high in the air have advantages, but for garden plants the emphasis should be on cutting down gallonage and pressure.

Shaker Can	Pressure Can	Trigger Sprayer	Continuous Sprayer	Hose End Sprayer
Bucket Pump Sprayer	Slide Pump Sprayer	Compressed Air Sprayer	Knapsack Sprayer	Wheel-barrow Sprayer
Plunger Duster	Crank Duster	Drop Spreader	Centrifugal Spreader	Root Irrigator

Figure 1. Pesticide application equipment. (Modified from the National Sprayer and Duster Association.)

Power sprayers for home gardens are available in almost any size, from 5-gallon capacity on up, and may have gasoline or electric motors (see Fig. 2). For the orchard a spray gun is satisfactory, but for flowering shrubs—azaleas, roses, and so on—a spray rod, curved at the end, or with an angle nozzle, is easier on the plants and more effective, allowing for better coverage on the underside of foliage.

The size of the hole in the nozzle disc and the pressure determine the amount of spray used. The volume of spray ejected per minute doubles or triples with each small increase in the hole size or pressure used. Thus, in a home garden where the objective is to cover a few rose bushes effectively, a

Figure 2. Spray application techniques.

large amount of spray can be wasted at too high pressure, an expensive item with many pesticide mixtures costing 20 to 30 cents a diluted gallon. Most chemicals are corrosive, and even if you start with a mist nozzle with a small hole at the beginning of the season, you will soon be delivering more spray per minute because the hole is enlarging. Therefore, there is more conspicuous residue left on the plant as well as more expense.

Hand-Operated Sprayers

"Aerosol bombs" are pressurized sprays in push-button containers. A gas propellent reduced to liquid form is added to a pesticide concentrate and a fine mist is released when the button is pushed. Unless the container is held 12 to 18 inches away from plants, to allow the gas to evaporate, there will be some burning (more literally a freezing) when the liquid gas hits foliage. Such cans are good for house plants and for spot treatment of insects outdoors, but air currents make it difficult to place fungicides effectively. Aerosols are also used for the application of wound dressings to trees (see Fig. 3).

Household sprayers of the atomizer type are intermittent, discharging spray material with each forward stroke of the pump; or continuous, maintaining constant pressure. They are too small and too tiresome to operate for

Figure 3. Aerosol pressurized spray.

more than a few plants, and it is hard to get adequate coverage on the underside of foliage.

Compressed air sprayers are adequate for small gardens and are relatively inexpensive. Capacity varies from 1 to 6 gallons. They are meant to be carried slung over one shoulder, but some come mounted on a cart. Air is compressed into the tank above the spray liquid by a hand-operated pump. A short hose, extension rod, and adjustable nozzle make it possible to cover undersurfaces. Such sprayers are a bit hard to pump up, and some models have carbon dioxide cylinders to provide operating pressure.

Knapsack sprayers, of 2 to 6 gallons capacity, are carried on the back of the operator and are pumped by moving a lever up and down with the right hand as you spray with the left. These are more expensive than compressed air sprayers, but deliver a fine continuous mist and are excellent for larger gardens.

Slide or trombone sprayers have a telescoping plunger, operated with two hands. They draw material from an attached jar or separate pail and discharge it as a continuous spray. They develop good pressure and can be used for small trees, but are tiring to use.

Wheelbarrow sprayers are manually operated hydraulic sprayers, holding 7 to 18 gallons, that are mounted on a frame with wheelbarrow-type handles and one or two wheels. Pressures up to 250 pounds may be developed, providing excellent coverage for shrubs and small trees. This type works best with two people: one to control the pump, the other to operate the spray rod.

Hose-end sprayers are attached to the garden hose so that water supplies the pressure. The action is that of a siphon. The concentrated pesticide is placed in a jar, and as water under pressure is passed over the metering jet a small amount of chemical is drawn into it. This way to spray is very easy and some models are relatively accurate in materials discharged. Be sure to purchase a type with an extension tube and deflector, so that spray can be directed to underside of the foliage, with a shutoff at the jar, not just back at the hose, and with a device to prevent back-siphonage. Hose-end sprayers can be used for roses and other shrubs and for low trees. The droplets may be somewhat larger than those from a wheelbarrow or knapsack sprayer, and slightly more chemical may be used.

Dusters

Pesticide dusts are most often made with talcs, pyrophyllite, clays, calcium carbonate, precipitated hydrated silicates and silicon dioxides, synthetic calcium silicate, and diatomaceous earth as the diluents, although finely ground plant material such as tobacco dust or walnutshell flour is sometimes used. In some cases, a solution of the toxicant in a volatile organic solvent such as acetone or benzene is mixed with the dust diluents, the solvent is allowed to evaporate, and the mixture is then ground. A solution of toxicant may be sprayed on the dust diluent during mixing and grinding or the toxicant may be dissolved in a nonvolatile solvent and mixed with the diluent. Care must be taken to avoid an excess of solvent that might impair dusting qualities of the finished product. Many technical pesticides in solid form lend themselves to direct grinding with a sorptive clay carrier in adequate milling equipment. Field-strength dusts may be produced by diluting or cutting down dust concentrates that contain from 10% to 50% a.i. (dust bases). Because of their good dusting properties, attapulgites, diatomite, talc, pyrophyllite, kaolins, and treated calcium carbonate are used as diluents to provide the volume per acre needed to facilitate metering of the dust through the duster mechanism. Since many formulations contain more than one a.i., dry concentrates must have the proper qualities to make a good formulation with relatively little or no diluent. From a toxicity standpoint, it is desirable to have a very small particle size, since immediate toxicity is generally inversely proportional to particle size. There are several important disadvantages to extremely small particle size: high wind losses, more or less rapid volatilization, and the prohibitive cost of extremely fine grinding. Also, to obtain better toxicant exposure of technical concentrates absorbed on a carrier, it is desirable to have the extender or diluent in as large a particle size as possible and still give good dusting characteristics. In a 5% dust-effective toxicant exposure is obtained with the extender averaging 10 times the size of the toxicant particles. At present, particle size specifications are usually 10 to 30 microns for ground dusters and 20 to 40 microns for aircraft units. For use in fertilizer mixtures, granulated powders of 20 to 80 mesh are prepared by impregnation of Fuller's earth and bentonite fractions with the desired toxicants.

Some dusts are sold in a can with a shaker top, meant to be applied like salt, which is certainly not going to place a fungicide where it will do the most good. Some dusts are sold in small cardboard cylinders to be used as dusters, which work for a little while if the cardboard is well paraffined to slide easily; but the dust soon gets damp and clogs. Many more dusts are sold in plastic containers, with the dust supposedly coming out in clouds as you squeeze, but more often it doesn't after the first few days. Dusts are tricky to use because of these disadvantages.

Spraying Versus Dusting

There is really no answer to the question of whether it is better to dust or to spray. In most gardens you will do both, depending on the weather, the plant, the fungicide you want to use, and how much time and help you have.

Some orchardists prefer dusting because they can get around the trees quickly in a rain, whereas to apply a spray they must wait until the foliage is dry. But for ornamentals exactly the opposite is true! You cannot dust a shrub even slightly wet with rain or dew without having a hideous splotchy effect that persists for a long time. If absolutely necessary you can spray while the plants are still slightly wet, though the spray may not stick quite as well, and you may want to make the next application a bit sooner. It is easier to spray than to dust on a windy day. Also, in dusting you are somewhat more likely to get possibly toxic materials into your lungs than in spraying. The chief points in favor of dusting are the ease and speed of application and the fact that you do not have to clean out the duster after each dusting.

Sprayers have to be cleaned, often between different sprays, and they must be rinsed with at least two changes of water pumped through the system at the end of every day. Occasionally they must be taken apart, the tank soaked in trisodium phosphate or washing soda, the strainers and nozzles in kerosene, wire run through the spray rods, then all put together and rinsed with water. Details of this cleaning chore are given in the *Gardener's Bug Book*.

MIXING THE CHEMICALS

It still seems incredible that so many gardeners continue to treat their plants in a haphazard fashion. Buy a set of measuring spoons and a measuring cup, marked in ounces. Buy a large pail and mark it off in gallons. Then measure, exactly!

Dosage directions are usually given in pounds per 100 gallons of water, with or without translation on the label into small amounts. Not much arithmetic is required to figure a smaller dosage, if you remember a few measurements:

Conversion Table

3 teaspoons	=	1 tablespoon		
2 tablespoons	=	1 fluid ounce		
16 tablespoons,				
8 fluid ounces	=	1 cup		
16 ounces, 2 cups	=	1 pint		
2 pints, 4 cups	=	1 quart		
16 cups, 8 pints	=	4 quarts	=	1 gallon
1 acre	=	43,560 square feet		

Suppose 3 gallons of a 2 to 100 dilution of lime sulfur is desired. That is the same as a 1 to 50 dilution. Three gallons constitute 48 cups; so if 1 cup of liquid lime sulfur is added to 3 gallons, you will have a 1 to 49 dilution, and that is close enough.

Or suppose you want to make 4 gallons of Zineb at the rate of 1½ pounds per 100 gallons. That is 24 ounces per 100 gallons, or .24 ounce for 1 gallon and .96 ounce for 4 gallons. That is approximately 1 ounce to weigh on your small scales. It also works out at about 1 level tablespoon of the Dithane powder per gallon, and it is easier to measure than to weigh. There is, however, a good deal of volume variation, depending on how fluffed up the material is at the time you measure it, so weighing is preferable.

When you buy chemicals in small packages designed for the home garden, the dosages given on the labels will probably be in terms of tablespoons per gallon, and you need only follow directions. When, to save a good deal of expense, you buy the larger sizes intended for farmers, the directions may be given only in terms of pounds per 100 gallons. As a very rough rule of thumb, you can figure 1 tablespoon per gallon where directions call for 1 pound per 100 gallons, but the different mixtures have different weights, so this is not very accurate.

At the rate of 1 pound to 100 you would use, accurately, ¾ tablespoon captan 50%, 1 tablespoon chloranil (Spergon), ⅓ tablespoon copper sulfate, ⅔ tablespoon dichlone 50% (Phygon), 1¼ tablespoon ferbam, ½ tablespoon maneb, 1 tablespoon spray lime, ¾ tablespoon thiram, ½ tablespoon sulfur, ⅔ tablespoon zineb (Dithane Z-78 or Parzate), 1¼ tablespoon ziram to 1 gallon of water.

Sometimes materials for soil treatment are given in pounds per acre. Knowing that one acre contains 43,560 square feet, you can make a proportion to find out how many pounds are required per 1000 square feet.

ALL-PURPOSE SPRAYS AND DUSTS

The practicability of combination insecticide-fungicide mixtures is sometimes argued. The proprietary compounds are more expensive, but they are more properly prepared than can be done at home and certainly save a lot of time. Nobody today could put on separate applications of all the materials needed. The trouble is that the mixtures follow fads, as in human medicine. Just as penicillin was given for most human ills some years ago, DDT was put in almost all pesticide mixtures, followed a little later by malathion. Both are excellent insecticides. The trouble is they are somewhat too efficient, killing the parasites and predators that keep mites and some other pests in check and they also may damage the environment. Some of the new fungicides leave a rather conspicuous residue; some are somewhat phytotoxic under certain conditions. Some of these pesticides are no longer available owing to new federal pesticide legislation; however, new materials are available that will replace those whose use is illegal. Every mixture must be evaluated for particular climatic situations and kinds of plants. There are hundreds and hundreds of combinations on the market under brand names. In order not to be out of date before this text is printed, I have used as few brand names as possible.

INTEGRATED PEST MANAGEMENT

In recent years pesticides have been constantly scrutinized. Registrations of many pesticides have been canceled. The diminished availability of pesticides may limit choices to more costly materials. In addition, there is growing concern about groundwater contamination by pesticides and fertilizers, pesticide resistance in plant pathogens, insects and weeds, destruction of beneficial organisms, atmospheric contamination by pollutants, and concern for endangered species, all of which combine to make the problem of pest control more serious.

For the past 20 years integrated pest management (IPM) has received increased interest. Investigations have concentrated on enhancement of a broad arsenal of integrated strategies for control of pests and diseases on selected commodities. A key goal of IPM strategies is the reduction of pesticide use to the absolute minimum and the reliance on other strategies to assist in controlling pests. One of those strategies is biological control.

Intense research in biological control of root diseases has been proceeding in the United States and in Europe. Some microbial agents, although sometimes sensitive to environmental variation, can be effective in controlling soil-borne plant pathogens. Although there are many promising fungal and bacterial biocontrol agents, and experiments demonstrate successful biocontrol in the greenhouse and field, there are few commercially available biocontrol products. The reasons may be due to an insufficient understanding of the mode of action of most biocontrol agents, to a need to develop mass production and delivery systems, to little methodology for integrating biocontrol with other control strategies and crop production methods, and to competition of the biocontrol agent with other microorganisms.

The most widely known biocontrol of plant disease is Galltrol-A, which contains *Agrobacterium radiobacter* (strain 84) and is used as a preventative control of crown gall. Moreover, *Trichoderma* species, fungi associated with debris in many soils, are known to be antagonistic against several fungal pathogens that cause root and stem diseases. A strain of *Trichoderma* (T-1-R9) has been developed that is tolerant to benomyl. When T-1-R9 is added to soil in pots, Fusarium wilt disease of chrysanthemum and carnation is reduced more than 50%. In this case, benomyl can be applied to the foliage to control foliar diseases without endangering the biocontrol of Fusarium wilt. Another promising biocontrol fungus is *Sporidermium sclerotivorum*. A single application of this fungus to soil reduces lettuce drop by 65%–85%. Field experiments revealed efficacy through consecutive lettuce plantings over a 2-year period. Some *Trichoderma* strains are being investigated as seed coatings.

Mass production and good delivery systems are required for biocontrol agents. Biocontrol fungi are grown in industrial fermentation systems on cheap feed sources such as molasses and brewer's yeast; then they are mixed with inert powders or converted into sprays or pellets. Gelling agents used in food processing are added to make pellets that resemble some plant fertilizers. The pellets are uniform in size, biodegradable, and nontoxic.

2

Classification of
Plant Pathogens

The plant diseases described in this handbook are caused by bacteria, fungi, nematodes, a few seed plants (such as dodder, mistletoe, and witchweed), viruses, physiological disturbances, and air and water pollutants. The classification of bacteria, fungi, and viruses is somewhat involved and is given here as a background for the specific descriptions in Chapter 3. There are many classifications of bacteria, fungi, and viruses, with numerous disagreements among mycologists, bacteriologists, virologists, and plant pathologists. Names and groups have been chosen that are widely accepted and most readily adapted to the alphabetical requirements of a reference that works like a dictionary.

Classification of the bacteria is based on that given in volume 1 (1984) and volume 2 (1986) of *Bergey's Manual of Systematic Bacteriology* and in *Laboratory Guide for Identification of Plant Pathogenic Bacteria* (1980) edited by N. W. Schaad. Classification of the viruses is based on that given by R. E. F. Matthew's book entitled *Classification and Nomenclature of Viruses* (1979) by Van Regenmortel and in *Descriptions of Plant Viruses*, published by the Commonwealth Mycological Institute and Association of Applied Biologists. Other helpful sources were *European Handbook of Plant Diseases* (1988) by I. M. Smith, J. Dunez, R. A. Lelliott, D. W. Phillips, and S. A. Archer and *A Textbook of Plant Virus Diseases*, 1972 edition, by Kenneth M. Smith.

So far as possible, the genera, orders, and families of fungi agree with those given in *Plant Pathogenic Fungi* (1987) by J. A. von Arx. Helpful sources included *A Dictionary of the Fungi*, 1961 edition, by G. C. Ainsworth and G. R. Bisby, which includes G. W. Martin's *Key to the Families of Fungi; The*

Genera of Fungi, by F. E. Clements and C. L. Shear; *The Fungi*, by Frederick A. Wolf and Frederick R. Wolf; *The Lower Fungi: Phycomycetes*, by H. M. Fitzpatrick; *Morphology and Taxonomy of Fungi*, by E. A. Bessey; and *Illustrated Genera of Imperfect Fungi*, by H. L. Barnett and B. B. Hunter.

Classification in accordance with convention or law is called *taxonomy*. Common names vary from locality to locality and from country to country. Scientific names are international and are based on the binomial system except for viral taxonomy, which is based on English vernacular names. Each kind of bacterium, fungus, nematode, or higher plant is a species, and has two Latin words for its name. The first name indicates the genus to which the species belongs, and the second indicates the species itself. The latter name is usually descriptive. *Diplocarpon rosae* means that *Diplocarpon*, the blackspot fungus, is found on rose. Sometimes the species name honors a person, as *Coniothyrium wernsdorffiae*, for the fungus causing brand canker of rose. Such a species name, derived from a proper name, has sometimes been written with a capital, but present custom is to decapitalize all species names. The names of genera should always be written with a capital.

Correctly, the author of the name should be written after the species. Then, if someone else places the species in a new genus, the name of the first author is put in parentheses followed by the name of the second author. When a number of taxonomists have worked on a group, the list of authors gets quite unwieldy. For simplicity all authors have been omitted from the scientific names in this text. The correct name for a fungus with more than one stage is that first given, with a valid description, for the perfect or sexual stage. That rule is followed here with a few exceptions—as when a fungus is almost universally recognized by another name.

Species are grouped into genera, related genera into families, designated with the suffix *aceae*, as Erysiphaceae, and families into orders with the suffix *ales*, as Erysiphales. Groups of related orders form classes.

Strange as it may seem, scientists are not yet agreed on what constitutes a plant or even a living organism. The old definition of bacteria as unicellular plants is disputed, and some question exists if fungi are truly plants. Bacteria are prokaryotes. Prokaryotes are generally single-celled microorganisms that have a cell membrane or a cell membrane and a cell wall surrounding the cytoplasm and no organized nucleus. Eukaryotes contain membrane-bound nuclei, mitochondria, and—in plants only—chloroplasts. Although viruses are known to multiply inside their hosts or vectors, the question of their being a living entity has not been resolved. The arguments continue. Meanwhile, entities have to be grouped into some sort of order. Whittaker in 1969 introduced the five (5) kingdom classification for all living organisms: Monera (or Procaryota), Protista, Animalia, Plantae, and Fungi (or Mycota). The Monera are organisms with small cells lacking nuclei, mitochondria, and plastids, namely, the bacteria. The Protista include microorganisms with one-celled, often motile thalli (cells contain nuclei). The plants, animals, and fungi are believed to have evolved from Protista. The Fungi are characterized as heterotrophic organisms, dependent on organic food, which they absorb. The following scheme, adapted from the *Plant Pathogenic Fungi* (Nova Hedwigia **87**:288pp. by J. A. von Arx) is an attempt to show the position of fungal plant pathogens in the Kingdom Mycota. The listing of families is restricted to those containing such pathogens.

FUNGI

Fungi are organisms having no chlorophyll, reproducing by sexual and asexual spores, not by fission like bacteria, and typically possessing a mycelium or mass of interwoven threads (hyphae) containing well-marked nuclei. According to Ainsworth and Bisby (1961), there are about 4,300 valid genera, and many more that are synonyms, and about 50,000 species living as parasites or saprophytes on other organisms or their residues. More than 8,000 species cause plant disease. Fungi are divided into six phyla.

Kingdom: Mycota—Fungi

Phylum: Myxomycota
Phylum: Oomycota
Phylum: Chytridiomycota
Phylum: Zygomycota
Phylum: Ascomycota
Phylum: Basidiomycota

Oomycetes, Zygomycetes and Chytridiomycetes were formerly listed as subclasses within the class Phycomycetes. Oomycota, Zygomycota, and Chytridiomycota are now generally accepted as separate phyla of fungi. The mycelium of these three phyla has many nuclei that are not marked off by cross-wells (or nonseptate mycelium) except where reproductive structures arise, a condition known as coenocytic. Asexual reproduction is by means of spores borne in sacs called sporangia. The Zygomycota have sexual spores called zygospores that are formed by the union of two similar sex cells or gametes; the Oomycota have sexual spores called oospores formed from dissimilar gametes; the Chytridiomycota have neither type of sexual spore; the Ascomycota have septate mycelium and sexual spores in asci; the Basidiomycota have septate mycelium, frequently with clamp connections, and sexual spores; the Myxomycota have thalli as a motile mass of protoplasm (a plasmodium or myxamoeba—no mycelium) that is transformed into a mass of small, aseptate resting spores that on germination form motile cells with or without flagella.

The Myxomycota include protists with amoeboid thalli and their status as fungi often has been questioned. The thalli of the Myxomycota are naked, amoeboid, plasmotic masses without cell walls and are termed plasmodia or pseudoplasmodia. They are also able to move by the formation of pseudopodia and by plasma-streaming. The Plasmodiophoromycetes is the only class of the Myxomycota that includes parasites of vascular plants. The best-known species is *Plasmodiophora brassicae*, which causes "club root" of cabbage.

Chytridiomycota

The thalli are usually vesicular, occasionally filamentous, and are transformed to sporangia, gametangia, or resting spores. Motile cells may function as zoospores, or as gametes, and are radially symmetrical, with an anterior, whiplash-type flagellum. The Chytridiomycetes are the only class in this phylum.

Chytridiales

Simple aquatic fungus have almost no mycelium, the thallus at maturity acting as a single sporangium, or dividing to become a sorus of sporangia; zoospores posteriorly uniflagellate. The families are sometimes separated into two series: inoperculate, with the sporangia opening by the rupturing of one or more papillae; and operculate, opening by a lid. The families Chytridiaceae and Megachytriaceae are operculate.

Olpidiaceae. Inside living organisms, holocarpic, the whole thallus functioning as a sporangium.

Achlyogetonaceae. Thallus converted into a linear series of sporangia.

Synchytriaceae. Thallus converted into a prosorus or sorus surrounded by a common membrane.

Phylctidiaceae. Thallus divided into reproductive and vegetative portions, inside plant tissue or with reproductive bodies on surface of host.

Rhizidiaceae. Thallus between host cells; sporangium, prosporangium, or resting spore formed from enlarged body of encysted zoospore.

Cladochytriaceae. Mycelium widespreading with terminal and intercalary enlargements transformed wholly or in part into sporangia or resting spores.

Physodermataceae. Sporangia and resting spores formed on separate thalli, the former on the surface, the latter abundant inside host; plant parasites.

Chytridiaceae. Monocentric, in algae.

Megachytriaceae. Thallus with many centers, in living tissue or saprophytic.

Blastocladiales

Saprophytes in water or soil; thallus anchored in substratum by tapering rhizoids.

Monoblepharidales

Saprophytes in water; thallus of much-branched delicate hyphae.

Plasmodiophorales

The placement of this order has always been uncertain. Some put it with the Myxomycetes, the slime molds; others put it between the Myxomycetes and the true fungi. Some have considered it a family in the Chytridiales. Parasitic, assimilative phase a multinucleate thallus within host cells, chiefly of vascular plants, often causing hypertrophy; germinating in place by amoeboid, occasionally uniciliate, zoospores.

Plasmodiophoraceae. The only family in this order but with two important genera: *Plasmodiophora*, causing club root, and *Spongospora*, causing potato scab.

Oomycota

The thalli may be vesicular, often irregular, but are usually filamentous. Sporangia on germination release biflagellate zoospores. One flagellum is the whiplash type and the other the tinsel type. Motile sex cells are absent. Sessile gametangial cells conjugate and form an oogonium containing one or several egg cells.

The Oomycota are related to autotrophic algae with similar characters. The Oomycetes are the only class in this phylum; however, the small classes Hyphochytriomycetes and Labrinthulomycetes may also be included.

Hyphochytriales

Zoospores anteriorly uniflagellate, usually formed outside the sporangium.

Saprolegniales

Marine forms, parasites of diatoms and algae, or in fresh water and soil, the water molds, with abundant mycelium; hyphae without constrictions; oogonium with several oospores.

Leptomitales

Water forms; hyphae constricted, with cellulin plugs; oogonium with a single oospore.

Lagenidiales

Primarily aquatic, mostly parasitic on algae and water molds; thallus simple; zoospores formed by cleavage within sporangium or partly or wholly in an evanescent external vesicle.

Peronosporales

Downy mildews and white rusts. Primarily terrestrial, living in soil or parasitic on vascular plants; in the latter case, zoosporangia function as conidia. (see Fig. 4).

Albuginaceae. The white rusts. Conidia (sporangia) in chains on club-shaped conidiophores borne in dense sori beneath epidermis of host, the sori forming white blisters; intercellular mycelium with globose haustoria.

Pythiaceae. Conidiophores differing little from assimilative hyphae; mycelium saprophytic or parasitic, but if latter within cells and without haustoria. Two genera, *Phytophthora*, which includes the potato blight and other pathogens, and *Pythium*, causing damping-off, are especially important.

Peronosporaceae. Downy mildews. Conidia are borne singly or in clusters at tips of usually branched, rarely clavate, conidiophores emerging through stomata; haustoria various.

Zygomycota

The thalli are vesicular, or more often represent a coenocytic, multinucleate mycelium (with aseptate hyphae). The gametangial cells conjugate and form a thick-walled, persistent resting spore, called a zygospore (see Fig. 5). Motile sex cells are absent, but sporangiospores and conidia are usually formed and

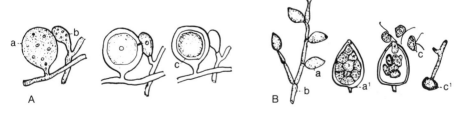

Figure 4. Reproduction of an Oomycete (*Phytophthora*, order Peronosporales). *A*, multinucleate oogonium *(a)* and male antheridium *(b)* in contact; fertilization tube formed between gametes after all nuclei except one have disintegrated; thick-walled oospore *(c)* formed inside oogonium. *B*, asexual reproduction by sporangium *(a)* formed on sporangiophore *(b)*; a¹ sporangium germinating by formation of ciliate zoospores; c¹ zoospores germinating with germ tube.

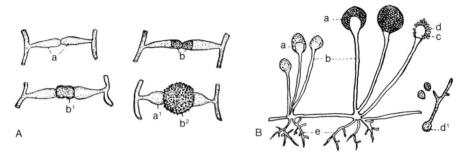

Figure 5. Reproduction of a Zygomycete (*Rhyzopus*, order Mucorales). *A*, suspensors *(a)* from different hyphae cut off gametes *(b)* of equal size that fuse *(b¹)* to form a spiny zygospore *(b²)*. B, asexual sporangiospores *(d)* formed inside a sporangium *(a)* formed on a sporangiophore *(b)* around a columella *(c)*. Hyphae are attached to substratum by rhizoids *(e)*. Sporangiospore germinates by a germ tube *(d¹)*.

dispersed by air. There are three classes, the Zygomycetes, the Endogonomycetes (fungi that form mycorrhizae in roots of vascular plants), and the Entomophthoromycetes (mainly parasitic on insects). These classes differ by morphological and chemical characteristics.

Zygomycetes

Mucorales

Profuse mycelium, much branched; asexual reproduction by sporangia or conidia; sexual reproduction by zygospores from union of two branches of the same mycelium or from different mycelia. Some species damage fruits and vegetables in storage. Only two families are of much interest to plant pathologists.

Mucoraceae. Sporangiophores liberated by breaking up of thin sporangial wall; zygospores rough. *Mucor* and *Rhizopus* cause storage molds.

Choanephoraceae. Both sporangia and conidia present, the latter borne on swollen tips; zygospores naked. *Choanephora* is a weak parasite causing blossom blight or blossom-end rot of young fruits.

Entomophthoromycetes

Entomophthorales

Profuse mycelium, species frequently parasitic on insects or other animals, rarely on plants; imperfect spores modified sporangia functioning as conidia; zoospores free within a gametangial vesicle.

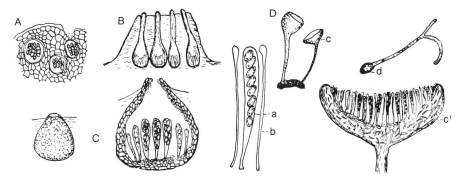

Figure 6. Sexual reproduction in the Ascomycetes. *A,* asci borne singly in locules in stroma (Myriangiales). *B,* perithecia with long necks or beaks immersed in stroma (Sphaeriales). *C,* papillate perithecium in host tissue, opening with a mouth or ostioles (Sphaeriales). *D,* Discomycetes (Heliotiales), ascus *(a)* and paraphyses *(b)* formed in an hymenial layer in a cuplike apothecium *(c)* and *(c¹)*; ascospore *(d)* germinates by germ tube.

Ascomycota

The thalli may consist of aseptate yeast cells or septate hyphae. Following meiosis, endogenous spores (ascospores) form within a cell called an ascus. There are two classes: the Endomycetes (usually yeastlike) and the Ascomycetes (form septate hyphae). The asci in Ascomycetes are aggregated in fructifications called ascomata (apothecia, cleistothecia, perithecia). The asexual states (anamorphs) of the Ascomycetes usually are classified in a separate class called Deuteromycetes.

Ascomycetes

The diagnostic characteristics of this class are a septate mycelium (hyphae with cross walls) and the ascus, a sac, typically club-shaped or cylindrical, bearing the sexual spores, ascospores, usually eight in number (see Fig. 6). Asci may be formed on or in hyphae or cells but are usually grouped in structures, ascocarps, either in locules in a stroma or lining a cup-shaped fruiting body called an apothecium or the walls of an enclosed round or flask-shaped perithecium. The young ascus has two nuclei, which fuse and then undergo generally three divisions to give the eight spores. In many genera paraphyses, thin sterile clubs, are formed between the asci.

Many ascomycetes have both a parasitic and a saprophytic stage. In their parasitic stage they usually produce conidia or imperfect spores, sometimes on groups of conidiophores growing out of the mycelium, sometimes in a special pycnidium. Similar structures sometimes found are spermagonia containing spermatia, which are small sex cells.

Subclass **Hemiascomycetidae**. Asci or ascuslike spore sacs formed singly; no ascocarp formed.

Protomycetales

Spore sac compound (a synascus) regarded as equivalent of numerous asci. Parasitic.

Protomycetaceae. Chlamydospores thick-walled, germinating after a rest period, the exospore splitting and the endospore emerging to form a large multispored spore-sac. Parasitic on vascular plants.

Endomycetales

Zygote or single cell transformed directly into an ascus; mycelium sometimes lacking. Mostly saprobic.

Ascoideaceae. Asci multispored, borne on hyphal tips or arising from gametangia; mycelium abundant.

Saccharomycetaceae. Asci 1- to 16-spored; mycelium scanty or lacking; reproduction mainly by budding. The yeast fungi.

Endomycetaceae. Asci on well-developed mycelium. Gametangia, if present, uninucleate, not dividing internally to form gametes.

Spermophthoraceae. Mycelium abundant; gametangia multinucleate, forming numerous nonmotile gametes, which fuse in pairs and produce asci on short hyphae.

Taphrinales

Hyphae bearing terminal chlamydospores or ascogenous cells, each of which produces a single ascus, usually forming a continuous hymenium-like layer on often modified tissues of host. Parasitic on vascular plants.

Taphrinaceae. Chlamydospores thin-walled; asci eight-spored but may become multispored by budding. Genera *Exoascus* and *Taphrina* cause leaf curl and leaf blisters and now *Exoascus* is usually considered a synonym of *Taphrina*.

Subclass **Euascomycetidae**. Asci borne in ascocarps.

Eurotiales

Asci borne in tufts or singly at various levels in interior of ascocarp or stroma, but extensive stroma lacking. Perithecia without ostioles (cleistothecia).

Gymnoascaceae. Perithecia around asci of loosely interwoven hyphae.

Eurotiaceae. Ascocarp sessile, small; peridium weak, tardily and irregularly dehiscent.

Onygenaceae. Ascocarp stalked and capitate, small to medium; peridium tough, opening above.

Erysiphales (Perisporiales)

Parasites of higher plants; mycelium generally on surface of host; perithecia without true ostioles.

Erysiphaceae. The powdery mildews. White mycelium, with conidia in chains; perithecia rupturing with an apical tear or slit.

Meliolaceae. Dark or black mildews. Mycelium dark; stroma unilocular, resembling a perithecium.

Englerulaceae. Mycelium dark; asci exposed by gelatinization of upper portion of ascocarp.

Martin places the last two families in a separate order, Meliolales.

Myriangiales

Stroma well developed, often gelatinous; asci borne singly in locules. Nearly all are parasites on higher plants.

Atichiaceae. Tropical fungi on insect secretions. Thallus gelatinous, superficial on leaves, typically of yeastlike cells; asci arising at various levels.

Myriangiaceae. Stroma pulvinate, often with lobes, nearly homogeneous.

Elsinoaceae. Stroma effused, with gelatinous interior and crustose rind.

Dothideales

Mycelium immersed in substratum; stroma with hard, dark rind, soft and pale within; locules more or less spherical, resembling perithecial cavities.

Capnodiaceae. Sooty molds. Often on living plants associated with insect secretions. Stroma massive, carbonaceous, often excessively branched; fruiting bodies borne singly at tips of branches, resembling perithecia.

Coryneliaceae. Stroma lobed, each lobe with a single locule that is finally wide open. Martin places this in the Coryneliales.

Dothideaceae. Stroma not markedly lobed, locules immersed in groups; at maturity stroma is erumpent and superficial.

Acrospermaceae. Stroma typically uniloculate, clavate, erect; dehiscence by a fimbriate, often spreading, tip.

Martin places this family in the Coryneliales and adds, under Dothideales, Pseudosphaeriaceae, with asci more or less separated by stromatic tissue.

Microthyriales

Mycelium largely superficial; stroma flattened; dimidiate; opening by a pore or tear, simulating the upper half of a perithecium.

Polystomellaceae (including Stigmateae). Mycelium largely internal, forming a hypostroma; fruiting stroma subcuticular or superficial.

Micropeltaceae (Hemisphaeriaceae). Internal mycelium scanty; stromatic cover not of radially arranged hyphae; chiefly tropical species.

Microthyriaceae (including Asterineae and Trichopelteae). Stromatic cover of radial or parallel hyphae.

Trichothyriaceae. Superficial mycelium irregular or lacking; base of stroma well developed; parasitic on other fungi.

Hysteriales

Ostiole an elongated slit on a usually flattened, elongate apothecium, bearing asci in a flat, basal layer.

Hysteriaceae. Ascocarps superficial from the first; black, carbonaceous, round, or elongate.

Hypocreales

Perithecia, and stromata if present, bright colored, soft, and fleshy. Martin gives two families.

Nectriaceae. Asci elliptical to cylindrical; inoperculate; ascospores various but never long-filiform.

Clavicipitaceae. Asci long-cylindrical, with a thickened tip, ascospores long-filiform.

Sphaeriales (Pyrenomycetes)

Mycelium well developed; perithecia dark, more or less hard, carbonaceous, with an ostiole typically circular in section; with or without stromata; asci inoperculate (without a lid) but spores discharged with force; paraphyses and periphyses usually present.

Chaetomiaceae. Perithecia superficial, hairy, walls membranous; asci deliquescent; ascospores dark; paraphyses wanting.

Sordariaceae (Fimetariaceae). Perithecia superficial, walls membranous, naked or sparsely setose; asci discharging spores forcibly.

Sphaeriaceae. Perithecia superficial, walls carbonaceous, mouths papillate.

Ceratostomataceae. Perithecia superficial, carbonaceous, with long, hair-like beaks.

Cucurbitariaceae. Stroma present but perithecia completely emergent at maturity; formed in groups.

Amphisphaeriaceae. Bases of perithecia persistently immersed in stroma; mouths circular.

Lophiostomataceae. Bases of perithecia persistently immersed in stroma; mouths compressed, elongate.

Sphaerellaceae (Mycosphaerellaceae). Perithecia immersed in substratum; stroma lacking or poorly developed; asci not thickened at tips; mouths of perithecia papillate.

Gnomoniaceae. Perithecia immersed in substratum; usually beaked; asci thickened at tips.

Clypeosphaeriaceae. Stroma a shieldlike crust (clypeus) over perithecia, through which necks protrude.

Valsaceae. Stroma composed of mixed host and fungal elements; perithecia immersed, with long necks; conidia borne in cavities in stroma.

Melanconidiaceae. Like Valsaceae but conidia borne superficially on the stroma.

Diatrypaceae. Stroma composed wholly of fungus elements; in some genera present only in conidial stage; perithecia develop under bark; ascospores small, allantoid, hyaline to yellow-brown.

Melogrammataceae. Conidia typically borne in hollow chambers in stroma composed of fungal elements; ascospores one- to many-celled, hyaline or brown.

Xylariaceae. Conidia borne in superficial layer on surface of stroma; ascospores one- to two-celled, blackish brown.

Martin does not use the order Sphaeriales. He places some of the above families in separate orders.

Laboulbeniales

Minute parasites on insects or spiders; mycelium represented by a small number of basal cells functioning as haustorium and stalk.

Phacidiales (= Rhytismatales)

Discomycetes in which the hymenium is covered by a membrane until ascospores are mature, then splitting stellately or irregularly.

Phacidiaceae. Ascocarps leathery or carbonaceous, black, remaining embedded in host tissue or in stroma; hypothecium thin.

Martin includes Tryblidiaceae, ascocarps leathery, immersed, hypothecium thick; but Ainsworth and Bisby place members of this family in the Helotiales.

Helotiales

Discomycetes without a membrane; asci inoperculate, opening with a definite pore. Cup fungi.

Geoglossaceae. Ascocarps clavate or caplike, hymenium covering convex upper portion.

Ascocorticiaceae. Fructification effused, indeterminate, without excipulum; paraphyses lacking.

Stictidiaceae. Ascocarps first immersed in substratum, then erumpent; asci long-cylindrical with thickened apex; ascospores filiform, breaking up into segments at maturity.

Cyttariaceae. Ascocarps compound, in form of subglobose stromata bearing numerous apothecial pits.

Patellariaceae. Apothecia leathery, horny, cartilaginous, or gelatinous; tips of paraphyses united to form an epithecium; asci thick-walled.

Mollisiaceae. Apothecia waxy or fleshy; peridium of rounded or angular, mostly thin-walled, dark cells forming a pseudoparenchyma.

Helotiaceae. Apothecia soft, fleshy, stalked; peridium of elongate, thin-walled, bright-colored hyphae, arranged in parallel strands.

Sclerotiniaceae. Apothecia arising from a definite sclerotium or stromatized portion of the substratum; stalked, cup-shaped, funnel-form, or saucer-shaped; usually brown; asci inoperculate, usually 8-spored; spores ellipsoidal, often flattened on one side, usually hyaline; spermatia globose to slightly ovate; conidial forms lacking in many genera.

These families are from Martin's 1954 *Key to Families*. His 1961 list puts Ostropaceae in the Ostropales and Patellariaceae in the Hysteriales. Ainsworth and Bisby list Geoglossaceae and put all other genera under "other Helotiales."

Pezizales

Asci operculate, opening by a lid; hymenium exposed before maturity of spores; apothecia often brightly colored; most forms saprophytic.

Pezizaceae. Apothecia cup-shaped or discoid; sessile or stalked.

Helvellaceae. Fruit bodies upright, columnar or with a stalk and cap; sometimes edible.

Tuberales

Ascocarp hypogeic, remaining closed; hymenium covered with a pseudotissue or hymenium-lacking and asci-filling cavities; mostly subterranean; includes edible truffles.

Tuberaceae. Interior waxy at maturity; asci persistent.

Elaphomycetaceae. Interior powdery at maturity; asci disappearing early, leaving interior filled with spores.

Basidiomycota

The thalli may contain budding cells that are formed successively by new inner layers bursting through the outer layers. After meiosis, the haploid cells are formed exogenously by budding and are called basidiospores or sporidia. Endogenous spores (sporangiospores or ascospores) are absent in Basidiomycota.

The structures on which haploid spores resulting from meiosis are formed are termed basidia and usually bear a constant number of spores, two or four, occasionally more. The basidia are differentiated on dikaryotic hyphae usually in or on fruiting bodies called basidiomata. The basidia may also be formed on resting spores called teliospores. Dikaryotic resting spores may also germinate with a shorter or longer tube, which is termed promycelium.

The three classes now distinguished are the Ustomycetes, the Urediniomycetes and the Basidiomycetes. The Ustomycetes propagate mainly by budding cells; septate hyphae may be present, but are rare. After meiosis, resting spores form short, often septate, promycelia, that produce budding cells laterally or terminally. Characteristic basidia or basidiospores are absent.

The Urediniomycetes form basidia, which after meiosis form uninucleate cells by transverse septation. Each cell forms a single, stalked basidiospore. Nearly all Urediniomycetes are obligate parasites of vascular plants and are known as rust fungi.

The Basidiomycetes form basidia, which usually remain aseptate after meiosis; the basidiospores are arranged in an apical whorl and are sessile or stalked (see Fig. 7). The septa of the hypha have characteristic central pores termed dolipores, with thickened walls and caps. Dolipores are not present in the Ustomycetes and the Urediniomycetes.

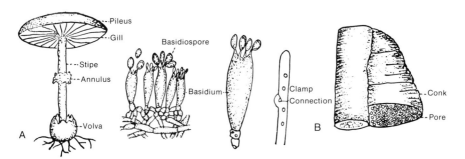

Figure 7. Reproduction in Basidiomycetes. *A*, mushroom (Agaricaceae) with cap of pileus lined with gills bearing basidia germinating by basidiospores. *B*, sporophore, or conk, in Polyporaceae where basidia line pores instead of gills. Mycelium in basidiomycetes sometimes has a structure around a septum called a clamp connection.

Ustomycetes

Ustomycetes include about 500 species belonging to two orders: the plant parasitic Ustilaginales (smut fungi) and the Sporidiales (red yeasts).

Ustilaginales

The smuts. Spore masses are usually black; spores are heavy-walled chlamydospores, germinating by a promycelium (basidium) and four or more sporidia (basidiospores).

Ustilaginaceae. Smuts. Basidiospores are produced on sides of a four-celled promycelium.

Urediniomycetes

Urediniomycetes have cylindrical, often slightly curved, transversely septate basidia. Each cell forms a sterigma with a basidiospore, which is forcibly discharged when mature. Usually basidia develop on resting spores called teliospores. The Urediniomycetes contain two orders: the Uredinales (rust fungi, obligate parasites on vascular plants) and the Auriculariales.

Uredinales

The rusts. More than 5,000 species have been described in about 300 genera. Always parasitic in vascular plants, teliospores or probasidia germinate with a promycelium divided transversely into four cells, each producing a single basidiospore on a sterigma; spore masses are yellowish or orange, and there are several spore forms.

Melampsoraceae. Teliospores sessile, in crusts, cushions, or cylindrical masses, or solitary, or in clusters, in mesophyll or epidermis of host.

Pucciniaceae. Teliospores usually stalked, separate, or held together in gelatinous masses; sometimes several on common stalks; less frequently sessile, catenulate, breaking apart.

Auriculariaceae. Basidia with transverse septa; typically gelatinous. The genus *Helicobasidium* causes violet root rot and the genus *Herpobasidium* causes blight of lilac.

Septobasidiaceae (Felt fungus). Arid, lichenoid, parasitic on scale insects; probasidia often with thickened walls.

There are six other families, of no particular interest from the standpoint of plant disease.

Basidiomycetes

About 10,000 species have been described and include the mushrooms and the bracket fungi formed on trees. Most grow in the soil and many form mycorrhiza with roots of forest trees. The hyphae in general are septate and dikaryotic. The septa of the hyphae often have clamp connections, hyphal outgrowths formed during cell division that form a connection between two cells. The basidia are formed in or on basidiomata on dikaryotic hyphae or on dikaryotic resting spores (teliospores). At maturity they are arranged either in a free, open layer termed *hymenium* or enclosed in fungal structures termed *gleba*. The basidiospores are sessile or more often develop on sterigmata. Young basidia are dikaryotic, until the nuclei fuse and meiosis follows. The two, four, or more haploid nuclei migrate into the basidiospores, which usually are uninucleate but occasionally binucleate. Those orders containing plant parasitic species are included below.

Tilletiales

Tilletiaceae. Smuts. Elongated basidiospores produced in a cluster at tip of a non-septate promycelium or basidium.

Graphioliales

Graphiolaceae. False smuts. Black, erumpent sori and spores in chains; on palms in warmer regions.

Tremellales

Trembling fungi. Basidiocarp usually well developed, often gelatinous varying to waxy or leathery hornlike when dry; mostly saprophytic, sometimes parasitic on mosses, vascular plants, insects, or other fungi.

Agaricales

Hymenium (fruiting layer) present, exposed from beginning or before spores are matured.
Exobasidiaceae. Hymenium on galls or hypertrophied tissues of hosts, which are vascular plants. Martin places this family in a separate order, Exobasidiales.

Thelephoraceae. Hymenium smooth or somewhat roughened or corrugated; basidiocarp weblike or membranous, leathery or woody; hymenium on lower side.

Clavariaceae. Hymenium smooth, pileus more or less clavate or club-shaped, erect, simple or branched, fleshy or rarely gelatinous; hymenium on all surfaces.

Hydnaceae. Hymenium covering downward-directed spines, warts, or teeth.

Polyporaceae. Hymenium lining pores (pits or tubes); hymenophore woody, tough or membranous, rarely subfleshy but never soft. Martin places this family and the preceding three in another order, Polyporales.

Boletaceae. Fruiting surface poroid or occasionally pitted; basidiocarp fleshy to tough or membranous.

Agaricaceae. The mushrooms. Fruiting bodies usually fleshy, sometimes tough or membranous, often with a stipe and cap; hymenophore lamellate, with gills.

Hymenogastrales

Hymenium present in early stages, lining chambers of the gleba, closed fruiting body that is fleshy or waxy, sometimes slimy and fetid at maturity.

Phallales

Gleba slimy and fetid; exposed at maturity on an elongated or enlarged receptacle.

Lycoperdales

The puffballs. Gleba powdery and dry at maturity; spores usually small, pale.

Sclerodermatales

Gleba powdery at maturity; chambers not separating from peridium or each other; spores usually large, dark.

Nidulariales

Bird's nest fungi. Gleba waxy; chamber with distinct walls forming peridioles (the eggs in the nest) that serve as propagules of dissemination.

Deuteromycetes—Fungi Imperfecti

Imperfect fungi are those for which a perfect stage is not yet known or does not exist. Most of them are in the Ascomycetes. The groupings are based on conidia: hyaline or colored; with one, two, or several cells; formed in pycnidia, on acervuli (little cushions of hyphae breaking through the host epidermis), or free on the surface of the host (see Figs. 8 and 9).

Sphaeropsidales

Conidia borne in pycnidia or chambered cavities.

Sphaerioidaceae (Sphaeropsidaceae, Phyllostictaceae). Pycnidia more or less globose, ostiolate or closed; walls dark, tough, leathery or carbonaceous.

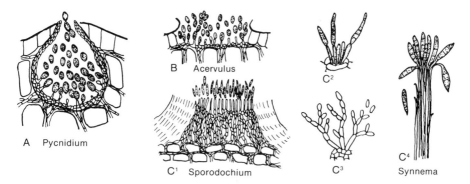

Figure 8. Spore formation in the Deuteromycetes. *A,* Sphaeropsidales, conidia in pycnidium. *B,* Melanconiales, conidia in acervulus. *C,* Moniliales—*C*[1], sporodochium of Tuberculariaceae; *C*[2], dark conidiophores and conidia of Dematiaceae; *C*[3], hyaline conidia in chains, Moniliaceae; *C*[4], conidiophores grouped into a synnema, Stilbaceae.

Figure 9. Spore forms in the Deuteromycetes, commonly designated by letters and figures. *A,* Amerosporae, one-celled; *A*[1], Hyalosporae, spores hyaline; *A*[2], Phaeosporae, spores dark. *B,* Didymosporae, two-celled; *B*[1], Hyalodidymae, hyaline; *B*[2], Phaeodidymae, dark. *C,* Phragmosporae, spores with two or more cross septa; *C*[1], Hyalophragmiae, hyaline or light; *C*[2], Phaeophragmiae, dark. *D,* Dictyosporae, muriform spores. *E,* Scolecosporae, fifliform spores. *F,* Helicosporae, spirally coiled spores. *G,* Staurosporae, starlike spores.

Nectrioidaceae. As above but walls or stroma bright-colored, fleshy or waxy.

Leptostromataceae. Pycnidia dimidate (having the outer wall covering only the top half); usually radiate, sometimes long and cleft.

Excipulaceae. Pycnidia discoid or cupulate.

Melanconiales

Conidia borne in definitely circumscribed acervuli; erumpent (breaking through the substratum).

Melanconiaceae. Conidia are slime-spores; cause anthracnose diseases.

Moniliales

Conidiophores (specialized hyphae bearing conidia) superficial, entirely free or bound in tufts or in cushionlike masses (sporodochia).

Pseudosaccharomycetaceae (Cryptococcaceae). False yeasts. Hyphae scanty or nearly lacking; reproduction by budding but not germinating by repetition.

Sporobolomycetaceae. False yeasts. Reproduction by budding and germination by repetition; probably imperfect species of the Tremellales, in the Basidiomycetes.

Moniliaceae. Hyphae and spores hyaline or brightly colored; conidiophores not grouped together.

Dematiaceae. Same as Moniliaceae but hyphae or conidia, or both, brownish to black.

Stilbaceae (Stilbellaceae). Conidiophores united into a coremium or synnema, an upright group of hyphae.

Tuberculariaceae. Hyphae and conidiophores combined in a sporodochium, a tight, spore-bearing mass.

BACTERIA

The fact that bacteria can cause plant diseases was discovered almost simultaneously in four different countries, with the United States claiming first honors. In 1878, Professor T. J. Burrill of the University of Illinois advanced the theory that fire blight of apple and pear was due to the bacteria that he found constantly associated with blighted tissues. In 1879, the French scientist Prillieux published a paper on bacteria as the cause of rose-red disease of wheat; in 1880, the Italian Comes recognized bacteria as pathogenic to plants; in 1882, Burrill named his fire-blight organism *Micrococcus amylovorus*; and in 1883, Walker in Holland reported the bacterial nature of yellows disease of hyacinth. It remained, however, for Erwin F. Smith, of the U. S. Department of Agriculture, to do most of the pioneer work in this field and to convince the world that bacteria were to blame for so many diseases. He spent a lifetime in the process, starting with peach yellows, and going on to a study of crown gall and its relation to human cancer. In 1905, the first volume of his monumental work *Bacteria in Relation to Plant Diseases* was published.

There are about 80 species of bacteria that cause plant disease and many of them consist of numerous pathovars. Bacterial diseases fall into three categories: (1) a wilting, as in cucumber wilt, due to invasion of the vascular system, or water-conducting vessels; (2) necrotic blights, rots, and leaf spots, where the parenchyma tissue is killed, as in fire blight, delphinium black spot, soft rot of iris and other plants with rhizomes or fleshy roots; (3) an overgrowth or hyperplasia, as in crown gall or hairy root.

Pathogenic bacteria apparently cannot enter plants directly through unbroken cuticle but get in through insect or other wounds, through stomata, through water-pores, possibly through lenticels, and often through flower nectaries. They can survive for some months in an inactive state in plant tissue, as in holdover cankers of fire blight, and perhaps years in the soil, although claims for extreme longevity of the crown-gall organism in soil are discounted.

Most of these plant disease bacteria have had their genus names changed several times since they were first described, and some species have been combined. Classification of bacteria will probably change further in future years. Where genus and/or species names have been changed, the old name is given in parentheses. The genera and species used in this text agree with those given in volume 1 (1984) and volume 2 (1986) of *Bergey's Manual of Systematic Bacteriology*. Walter H. Burkholder, of Cornell University, who revised the portions of the manual dealing with plant pathogens, followed in the footsteps of Erwin F. Smith by spending his life with bacterial diseases of plants, as did Charlotte Elliott of the U.S. Department of Agriculture, from whose *Manual of Bacterial Plant Pathogens* much information on disease symptoms has been taken.

Two kinds of prokaryotes cause disease in plants. Bacteria have a cell membrane, a rigid cell wall, and often one or more flagella. The mollicutes, or mycoplasma-like organisms (MLO) lack a cell wall and have only a single-unit membrane. A general classification of plant pathogenic prokaryotes is shown below.

Kingdom: Prokaryotae. Organisms with genetic material not organized into a nucleus, that is, not surrounded by a membrane.

Bacteria: Have a cell membrane and cell wall.

Part I: Gram-negative aerobic rods and cocci.

Family: Pseudomonadaceae

Genus: Pseudomonas, rod-shaped, one or several polar flagella, colonies white or yellow.

Xanthomonas, rod-shaped, one polar flagellum, colonies yellow.

Xylella, rod-shaped, under some cultural conditions filamentous; nonmotile, aflagellate, nonpigmented.

                    Family: Rhizobiaceae

                        Genus: Agrobacterium, rod-shaped sparse lateral flagella, colonies white, rarely yellow.

Part II: Gram-negative facultative anaerobic rods.

                    Family: Enterobacteriaceae

                        Genus: Erwinia, peritrichous flagella, colonies white or yellow.

Part III: Irregular, Gram-positive, nonsporing rods.

                    Family: Mycobacteriaceae

                        Genus: Clavibacter. Contains important phytopathogenic bacteria formerly classified as Corynebacterium. Nonmotile, pleomorphic (no defined shape) rods, often arranged in V-formations; obligate aerobic. A few phytopathogenic Corynebacterium species are still listed as Corynebacterium but they also are expected to be transferred to other genera.

Part IV: Actinomycetes, bacteria forming branching filaments.

                    Family: Streptomycetaceae

                        Genus: Streptomyces. Gram-positive, aerial mycelium with chains of nonmotile conidia.

Part V: Mollicutes, prokaryotes that have a cell membrane but no cell wall.

                    Family: Mycoplasmataceae, the plant mycoplasma-like organisms. Spiroplasmataceae.

                    Genus: Spiroplasma, helical, motile but lacking flagella.

The taxonomy of the plant pathogenic fastidious phloem-limited bacteria is still unknown, and the taxonomy of the plant pathogenic MLOs, and of the spiroplasmas is tentative. Furthermore, Rickettsia-like organisms (RLOs) have been reported to be associated with a number of plant diseases. RLOs are also cultured with difficulty, which is a characteristic similar to the MLOs. On this basis, both MLOs and RLOs are referred to as fastidious prokaryotes. There are more than 200 distinct plant diseases affecting several hundred genera of plants that have been shown to be caused by the Mollicutes.

The taxonomic scheme for mollicutes and MLO is difficult to present in this handbook since morphological criteria are limited and both the criteria used in bacteriology and the serological methods used in virology are difficult to apply because MLOs (except for spiroplasma) have not been cultured. Thus, the true nature of MLO and RLOs, and their taxonomic position among microorganisms, is uncertain. In practice, the diseases caused by mollicutes have been taxonomically treated individually. The elucidation of true relatedness among these organisms awaits further research.

The general nature of the symptoms and the name of the host plant will, in many cases, leave little doubt as to the identity of a bacterial disease. In the case of the soft rot due to *Erwinia carotovora* the nose alone is a reliable guide. In other cases identification must be left to the technically trained bacteriologist. It involves special staining technique, for examination of form and motility under the microscope, and to see whether it is Gram-negative or Gram-positive, and special culture technique to determine shape, color, and texture of colonies on agar and gelatin, production of gases, fermentation of sugar, coagulation of milk, etc. If you are in doubt about a plant disease, and the absence of fungus fruiting bodies leads you to believe that bacteria may be at work, send a specimen to your State Experiment Station for expert diagnosis.

VIRUSES

The word *virus* means poison or venom. When it is used in connection with a plant disease, it means a filterable virus, an infective principle or etiological agent so small it passes through filters that will retain bacteria. Virus diseases in humans range from infantile paralysis to the common cold and in plants from "breaking" of tulip flowers to the deadly raspberry ringspot disease on the Malling Jewel variety of raspberry.

Viruses are obligate parasites in that they are capable of increasing only in living cells. They are not organisms in the usual sense because they do not multiply by growth and fission, and they are too complex to be chemical molecules. F. C. Bawden, in the 1964 edition of his *Plant Viruses and Virus Diseases*, defines viruses as "submicroscopic infective entities that multiply only intracellularly and are potentially pathogenic."

Virus diseases are old; our knowledge of them is relatively recent. Tulip mosaic, shown as breaking of flower color, was described in a book published in 1576. In 1892, it was shown that the cause of tobacco mosaic could pass through a bacteria-proof filter, and in 1935, a crystalline protein was prepared from tobacco mosaic virus juice. At present we believe that virus particles contain only two major components, nucleic acid embedded in a protein structure, and that they are built of uniform-sized subunits arranged in a fixed and regular manner. Many plant viruses contain ribonucleic acid (RNA). Some plant viruses and many animal viruses contain deoxyribonucleic acid (DNA) instead of RNA. X-ray diffraction and electron microscopy have shown something of the morphology of virus particles. Some are rods, some are filiform, and some are isometric, but polyhedral rather than spherical. They apparently act not as organisms but as disturbances in the host metabolism of nucleic acid.

There are 500 or more plant pathogens thought to be viruses; however, more than two-thirds of these have not been conclusively proved to be viruses and could be agents of other kinds. For example, aster yellows and elm phloem necrosis were thought for some time to be caused by viruses, but have now been determined to be caused by mycoplasma-like organisms. In addition, potato spindle tuber and chrysanthemum stunt disease were long thought to be caused by viruses, but have now been determined to be caused by viroids. Viroids consist solely of small RNAs with no protein coat. There are now about 15 plant diseases that have been identified as having viroid causal agents including potato spindle tuber, chrysanthemum stunt, citrus exocortis, chrysanthemum chlorotic mottle, cucumber pale fruit, and cadang-cadang of palm. More diseases caused by viroids will probably be identified in future years. Finally, some plant diseases formerly thought to be caused by viruses have now been determined to be caused by spiroplasma, such as citrus stubborn disease. Thus, the field of virology has changed somewhat in recent years. In order to simplify the discussion of these viruses and viruslike agents and the diseases they cause, these agents may be grouped under virus diseases, since the symptoms that they cause in plants are similar.

Some viruses attack a large number of different plants and are of great economic importance; others are confined to a single host. Virus symptoms fall into several categories, but commonly there is loss of color due to the suppression of chlorophyll development. Foliage may be mottled green and yellow, mosaic, or have yellow rings (ring spot); or there may be a rather uniform yellowing (yellows). Stunting is common. The reduction in manufactured food from the chlorophyll loss leads to smaller size, shorter internodes, smaller leaves and blossoms, and reduced yield. There may be various distortions of leaves and flowers, witches' brooms, or rosettes. There may be necrotic symptoms with death as the end result, and sometimes symptoms are "masked," not showing up under certain conditions such as hot weather, or they are latent, not appearing until another virus is also present.

Viruses are transmitted from plant to plant by insects, mites, fungi, and nematodes; rubbing, abrasion, or other mechanical means (sometimes handling tobacco and merely touching a healthy plant spreads mosaic); grafting or propagation by cuttings and bulbs; occasionally seeds; sometimes soil (probably by means of nematodes); and dodder, parasitic vines whose tendrils link one plant to another. About half of the insect vectors are aphids; a third are leafhoppers. Mealybugs and whiteflies transmit some viruses, and three, including tomato spotted wilt, are transmitted by thrips. In some cases the virus multiplies within the insect as well as in the plant. Some viruses have many different vectors, 50 being recorded for onion yellow dwarf, and some have but a single known vector.

Control of virus diseases starts with obtaining healthy seed, cuttings, or plants. "Certified" means that plants have been inspected during the growing season and found free of certain diseases. Virus-free foundation stock can be built up from heat treatment—growing plants at high temperatures for weeks or even months—and/or meristem tip cultured plants. Virus-free stock is tested by "indexing," bioassays and/or serological assays, before using stock for propagating. Controlling insect vectors (by spraying plants or treating soil with systemic insecticides), eliminating weed hosts, roguing diseased plants before insects can transmit the virus, and using resistant varieties are all ways of combating virus diseases.

This handbook does not deal predominantly with the characteristics of the causal viral agent, but with the disease caused by the virus. For those plant diseases that have been identified and characterized to be caused by virus, the classification and nomenclature are generally by groups according to particle shape and size and several criteria unique to plant viruses. The groups are named after a type virus in the group; the groups and type virus are listed below. In addition, acronyms for virus names have been given in parentheses following the list reported by Van Regenmortel (1982).

Carlavirus Group: Elongated, slightly flexuous particles (600–700 × 12–15 nm). Carnation latent virus (CLV).

Closterovirus Group: Elongated, very flexuous particles (600–2,000 × 12 nm). Beet yellows virus (BYV).

Cucumovirus Group: Isometric particles (29 nm in diameter). Cucumber mosaic virus (CMV).

Ilarvirus Group: Quasi-isometric particles different in size (26–35 nm in diameter). Tobacco streak virus (TSV).

Luteovirus Group: Isometric particles (25–30 nm in diameter). Barley yellow dwarf virus (BYDV).

Nepovirus Group: Three types of isometric particles of the same size (28 nm) that are distinguished by sedimentation constants. Tobacco ringspot virus (TRSV).

Potexvirus Group: Elongated slightly flexuous particles (470–580 × 13 nm). Potato virus X (PVX).

Potyvirus Group: Elongated flexuous particles (680–900 × 11 nm). Potato virus Y (PVY).

Tobamovirus Group: Elongated rigid particles (300 × 18 nm). Tobacco mosaic virus (TMV).

Viroid Group: A naked molecule of single stranded RNA. Potato spindle tuber viroid (PSTV).

There are 17 very diverse and minor virus groups listed below:

Alfalfa Mosaic Virus Group: Bacilliform particles of different lengths. Alfalfa mosaic virus (AMV).

Bromovirus Group: Isometric particles (ca. 29 nm in diameter). Brome mosaic virus (BMV).

Carnation Mottle Virus Group: Isometric particles (ca. 28 nm in diameter). Carnation mottle virus (CarMV).

Caulimovirus Group: Isometric particles (ca. 50 nm in diameter). Cauliflower mosaic virus (CaMV).

Comovirus Group: Two types of isometric particles (ca. 28 nm in diameter). Cowpea mosaic virus (CPMV).

Dianthovirus Group: Isometric particles (ca. 32 nm in diameter). Carnation ringspot virus (CRSV).

Geminivirus Group: Geminate consisting of two incomplete icosahedra (18 × 20 nm). Maize streak virus (MSMV).

Hordeivirus Group: Rod-shaped particles (100–150 × 20 nm). Barley stripe mosaic virus (BSMV).

Penamovirus Group: Isometric, polyhedral particles (28 nm in diameter). Pea enation mosaic virus (PEMV).

Reovirus Group: Isometric particles (60–80 nm in diameter). Wound tumor virus (WTV).

Rhabdovirus Group: Enveloped bacilliform or bullet-shaped particles (135–380 × 45–95 nm). Lettuce necrotic yellow virus (LNYV).

Sobemovirus Group: Isometric virus particles (ca. 30 nm in diameter). Southern bean mosaic virus (SBMV).

Tobacco Necrosis Virus Group: Isometric particles (ca. 28 nm in diameter). Tobacco necrosis virus (TNV).

Tobravirus Group: Two types of elongated rigid particles (180–215 × 22 nm and 46–114 × 22 nm). Tobacco rattle virus (TRV).

Tomato Spotted Wilt Virus Group: Special particles (ca. 85 nm in diameter) with a lipoprotein coat. Tomato spotted wilt virus (TSWV).

Tombusvirus Group: Isometric particles (ca. 30 nm in diameter). Tomato bushy stunt virus (TBSV).

Tymovirus Group: Isometric particles (30 nm in diameter). Turnip yellow mosaic virus (TYMV).

NEMATODES

In the four decades since the first edition of this book was prepared, nematodes have become of major importance in plant pathology. Several hundred species are known to cause plant disease. All plant parasitic nematodes are in the animal kingdom and belong to the phylum Nematoda. Some examples are given after each genus.

Phylum: Nematoda
  Order: Tylenchida
    Suborder: Tylenchina
      Superfamily: Tylenchoidea
        Family: Tylenchidae
          Genus: *Ditylenchus*, stem or bulb nematode of onion, narcissus
        Family: Tylenchorhynchidae
          Genus: *Tylenchorhynchus*, stunt nematode tobacco, corn
        Family: Pratylenchidae
          Genus: *Pratylenchus*, lesion nematode of nearly all plants
                Radopholus, burrowing nematode of citrus

Family: Hoplolaimidae
Genus: *Hoplolaimus*, lance nematode of corn, turf grass, carnation
Rotylenchus, spiral nematode of turf grass, tomato, gardenia
Helicotylenchus, spiral nematode of turf grass, gardenia, azalea, apple, grape
Family: Belonolaimidae
Genus: *Belonolaimus*, sting nematode on wide variety of plants
Superfamily: Heteroderoidea
Family: Heteroderidae
Genus: *Globodera*, cyst nematode of potato
Heterodera, cyst nematode on wide variety of plants
Meloidogyne, root-knot nematode on wide variety of plants
Family: Nacobbidae
Genus: *Naccobus*, false root-knot nematode of garden beets, cacti, crucifers, lettuce
Rotylenchus, reniform nematode of turf grass, tomato, gardenia
Superfamily: Criconematoidea
Family: Criconematidae
Genus: *Criconemella*, ring nematode of citrus, fig, zoysia
Hemicycliophora, sheath nematode of beet, bean, blueberry, dracaena
Family: Paratylenchidae
Genus: *Paratylenchus*, pin nematode of carnation, celery, fig
Family: Tylenchulidae
Genus: *Tylenchulus*, citrus nematode of citrus, grapes, lilac
Suborder: Aphelenchina
Superfamily: Aphelenchoidea
Family: Aphelenchoididae
Genus: *Aphelenchoides*, foliar nematode of chrysanthemum, strawberry, lily, begonia
Bursaphelenchus, pine wood nematode
Rhadinaphelenchus, coconut red ring nematode
Order: Dorylaimida
Family: Longidoridae
Genus: *Longidorus*, needle nematode of grape, celery, leek, lettuce, parsley
Xiphinema, dagger nematode of rose, trees, many annuals
Family: Trichodoridae
Genus: *Paratrichodorus*, stubby root nematode of apple, vegetables
Trichodorus, stubby root nematode of vegetables, turf grass, dahlia, azalea

3

Plant Diseases and Their Pathogens

Because this book is for reference, and not one to be read for pleasure or continuity, most of you will come to the material you need in this section by way of the index or the lists of diseases given under the different hosts in Chapter 4. On page 518 you will find a list of headings under which diseases are grouped and described, from Anthracnose to Wilts. In the Host section the key word, for example, rot or blight, is given in capital and small capitals, followed by the name of the pathogen (agent causing disease) in boldface. In this Diseases section the pathogens are listed in boldface in alphabetical order under each heading such as Rots or Blights and so on, followed by the common name of the disease. This system was adopted for quick and easy reference because trying to alphabetize hundreds of similar common names would lead to endless confusion. Also, it allows a very brief summary of the classification and diagnostic characters of each genus before going on to a consideration of diseases caused by the various species. This brief summary is in small type, so that it can be readily skipped by readers uninterested in the technical details. Perhaps I am the only one who feels the need for this quick review, to be used in conjunction with the classification given in Chapter 2; perhaps others who have to answer questions over a broad field instead of their own specialty can make use of these capsules sandwiched in between nontechnical descriptions.

An alphabetical arrangement has the great disadvantage of being thrown out of alignment every time the name of a fungus is changed, as it so frequently is. In some such cases the old name is retained to avoid change in order, but the present accepted name is also given. Sometimes names have been changed under several hosts and the old name inadvertently retained under others. And sometimes the old name is purposely retained because it is so familiar to everyone, which is particularly true of a few fungi far better known by their imperfect stages than by the correct name of the perfect stage.

A fungus not only can have several names; it also can cause more than one type of disease. For instance, *Pellicularia filamentosa* is the present name of the fungus formerly known as *Corticium vagum* when causing Rhizoctonia rot of potatoes and *Corticium microsclerotia* when causing web blight of beans. As *Rhizoctonia solani*, the name given to the sclerotial stage, the same fungus causes damping-off of seedlings, root rots of many plants, and brown patch of lawn grasses. There are lots of plant diseases, and there are lots of fungi causing them, but there are not nearly as many separate pathogenic organisms as all the names would indicate.

I cannot think of anything more deadly than ploughing straight through this section from Anthracnose to Wilts. By doctor's orders, take it in small doses, as needed. But do read the few introductory remarks as you look up each group, and please, before starting any control measures, read the opening remarks in Chapter 1 on Garden Chemicals, and look up, in the list of chemicals, any material you propose to use, noting precautions to be taken along the lines of compatibility, weather relations, and phytotoxicity.

Although the disease descriptions, fungus life cycles, and general principles of control given here will remain fairly valid, it must be stressed that chemicals suggested for control are constantly changing. Today's discovery may be obsolete tomorrow. Therefore, this handbook should be used in conjunction with the latest advice from your own county agent or experiment station. Addresses of the state agricultural experiment stations are given on page 875.

ANTHRACNOSE

The term *anthracnose* has been used for two distinct types of disease, one characterized by a typical necrotic spot, a lesion of dead tissue, and the other by some hyperplastic symptom, such as a raised border around a more or less depressed center. The word was coined in France for the latter type, to differentiate a grape disease from a smut of cereals, both of which were called *charbon*. The new word was taken from the Greek *Anthrax* (carbuncle) and *nosos* (disease), and was first used for the grape disease, caused by *Sphaceloma ampelina*, the chief symptom of which was a bird's-eye spot with a raised border.

A disease of brambles, raspberry and blackberry, was then named anthracnose because it looked like the grape disease. The fungus, however, instead of being correctly placed in the genus *Sphaceloma*, was mistakenly named *Gloeosporium venetum*. The next disease entering the picture was a bean trouble, and, because the fungus was identified as *Gloeosporium* (though later transferred to the genus *Colletotrichum*), this common bean disease with typical necrotic symptoms was also called anthracnose and came to typify diseases so designated.

Recently the term *spot anthracnose* has been given to those diseases similar to the original hyperplastic grape disease. Those with slight hyperplastic symptoms are still commonly called anthracnose, and those with pronounced overgrowth of tissue are commonly called scab. Both types are caused by the genus *Elsinöe*, imperfect state *Sphaceloma*, and are treated in this text as a separate group. See under Spot Anthracnose.

Anthracnose in the modern sense is a disease characterized by distinctive limited lesions on stem, leaf, or fruit, often accompanied by dieback and usually caused by a *Gloeosporium* or a *Colletotrichum*, imperfect fungi producing slime spores oozing out of fruiting bodies (acervuli) in wet, pinkish pustules. These spores (conidia) on germinating form an appressorium (organ of attachment) before entering the host plant. The perfect state of the fungus, when known, is *Gnomonia* or *Glomerella*.

Colletotrichum

Fungi Imperfecti, Melanconiales, Melanconiaceae

Spores are formed in acervuli, erumpent, cushionlike masses of hyphae bearing conidiophores and one-celled, hyaline, oblong to fusoid conidia. Acervuli have stiff marginal bristles (setae) that are sometimes hard to see. Conidia (slime-spores), held together by a gelantinous coating, appear pinkish in mass. They are not wind-borne but can be disseminated by wind-splashed rain. On landing on a suitable host, the conidium sends out a short germ tube that on contact with the epidermis enlarges at the tip into a brown thick-walled appressorium. From this, a peglike infection hypha penetrates the cuticle.

Colletotrichum antirrhini. Snapdragon anthracnose; on snapdragon, chiefly in greenhouses, sometimes outdoors in late summer. Stems have oval, sunken spots, grayish white with narrow brown or reddish borders, fruiting bodies showing as minute black dots in center. Spots on leaves are circular, yellow green turning dirty white, with narrow brown borders. Stem cankers may coalesce to girdle plant at base, causing collapse of upper portions, with leaves hanging limp along the stem.

Control. Take cuttings from healthy plants; provide air circulation; keep foliage dry; destroy infected outdoor plants in autumn. Spray every 7 to 10 days.

Colletotrichum atramentarium (or *C. coccodes*). **Potato anthracnose;** black dot disease; on potato stems and stolons following wilt and other stem diseases, occasionally on tomato, eggplant, and pepper; general distribution but minor importance. Starting below the soil surface, brown dead areas extend up and down the stem. The partial girdling causes vines to lose their fresh color and lower leaves to fall. Infection may extend to stolons and roots. The black dots embedded in epidermal cells, inside hollow stems and on tubers, are sclerotia to carry the fungus over winter and to produce conidia the following spring.

The fungus is a wound parasite ordinarily not serious enough to call for control measures other than cleaning up old refuse and using healthy seed potatoes.

Colletotrichum bletiae and other species. Orchid anthracnose; leaf spot; on orchids coming in from the tropics. Lemon-colored acervuli are formed in soft, blackish spots in ragged leaves. Burn diseased plants or parts. Spray with a copper fungicide.

Colletotrichum dematium f.sp. **truncata.** Anthracnose; fruit on tomato. Found in Georgia on *Dolichos.*

Colletotrichum erumpens.* Rhubarb anthracnose; stalk rot. Oval, soft watery spots on petioles increase until whole stalks are included; leaves wilt and die. Small dark fruiting bodies with setae survive winter in stems, produce conidia in spring. Clean up all rhubarb remains in fall.

Colletotrichum fragariae. Strawberry anthracnose; found in Florida and Louisiana. Runners are girdled and killed before young runner plants are rooted. Rhizomes or leaf petioles of rooted plants may be infected. Spots are light brown, oval, sunken, small at first but gradually increasing to length of runner, which turns brown to black and is covered with bristlelike tufts of setae. Spores are produced in great abundance. The disease is more severe after August 1, during warm, moist weather. Spray with bordeaux mixture at 10-day intervals in late July, August, and September. May also cause anthracnose of blue lupine.

Colletotrichum fuscum. Foxglove anthracnose; small spots to ⅛ inch, circular to angular, brown to purple brown, on leaves; sunken, fusiform lesions on petioles and veins; minute black acervuli, with bristles, in center of spots. Seedlings damp-off, older plants are killed or stunted in warm moist weather. Use clean seed or treat with hot water (131° F for 15 minutes).

Colletotrichum gloeosporioides. See *Glomerella cingulata.*

Colletotrichum graminicolum. Cereal anthracnose; widely distributed on barley, oats, rye, wheat, sorghum, and also on cultivated lawn grasses, causing a root decay and stem rot. Leaf spots are small, circular to elliptical, reddish purple, enlarging and fading with age; centers have black acervuli. The fungus winters on seed and plant refuse in or on soil. Improved soil fertility reduces damage from this disease. This pathogen also causes fruit anthracnose of tomato.

Colletotrichum higginsianum. Turnip anthracnose; also on rutabaga, mustard greens, radish, and Chinese cabbage in southeastern states. Very small, circular gray spots on leaves and elongate brown or gray spots on midrib, petiole, and stem show pink pustules in centers of dead tissue. Heavily

---

*Chupp and Sherf believe this is wrong identification and give *Colletotrichum* sp.

infected leaves turn yellow and die; young seeds in diseased pods may be killed. Mustard variety Southern Curled Giant is highly resistant.

Colletotrichum lagenarium. Melon anthracnose; on muskmelon, watermelon, cucumber, and other cucurbits. This disease is the most destructive of watermelons, found everywhere that melons are grown and particularly destructive in the South. There are at least three races of the fungus differing in ability to infect different cucurbits. One race is virulent on cucumber, slight on watermelon, moderate on Butternut squash; another is virulent on both watermelon and cucumber; Butternut squash is immune to a third.

Leaf symptoms are small yellow or water-soaked areas that enlarge and turn black on watermelon, brown on muskmelon and cucumber. The dead tissue shatters; leaves shrivel and die. Elongated, narrow, sunken lesions appear on stems and petioles; vines may die. Young fruit darkens, shrivels, and dies if pedicels are infected; older fruit shows circular, black, sunken cankers or depressions, from ¼ inch to 2 inches across and ⅓ inch deep on watermelon. In moist weather the centers of such spots are covered with gelatinous masses of salmon-colored spores. Infected fruit has a bitter taste or the flesh is tough and insipid. Soft rots often follow the anthracnose. Epiphytotics occur only in periods of high rainfall and temperature, near 75° F.

Control. Treating seed before planting is essential. Use a 3-year crop rotation with noncucurbits; destroy plant refuse. Watermelon varieties Charleston Gray, Congo, Fairfax, and Black Kleckly are resistant but not to all races of the fungus.

Colletotrichum liliacearum. Found on dead stems of daylilies and many other plants and perhaps weakly parasitic.

Colletotrichum lindemuthianum. Bean anthracnose; a major bean disease, sometimes mistakenly called "rust," generally present in eastern central states, rare from the Rocky Mountains to the Pacific Coast. It may also affect lima bean, Scarlet runner, tepary, mung, kudzu, broad beans, and cowpea. It is worldwide in distribution, known in the United States since 1880. There are at least 34 strains of the fungus, in three different groups, but the disease has decreased in importance with the use of western-grown, anthracnose-free seed.

The most conspicuous symptoms are on the pods, small, brown specks enlarging to black, circular, sunken spots, in moist weather showing the typical pinkish ooze of the slime-spores (see Fig. 10). Older spots often have narrow reddish borders. After the spores are washed away, the acervuli look like dark pimples. If pods are infected when young, the disease extends through to the seed, which turns yellow, then rusty brown or black under the pod lesion. The infection may extend deep enough to reach the cotyledons. Leaf lesions are dark areas along veins on underside of the blade and on petioles. Seedlings may show stem spotting below diseased cotyledons. The fungus is spread by splashing rain, tools, and gardeners working with beans when they are wet. Optimum temperature is between 63° F and 75° F, with maximum around 85° F.

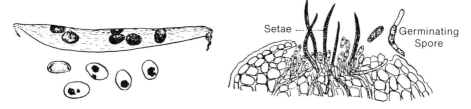

Figure 10. Bean anthracnose. Pod and seeds with dark, sunken areas; section through bean seed showing spores formed in an acervulus marked with prominent black setae.

Control. Use western-grown seed. Saving home-grown seed is dangerous unless you can be sure of selecting from perfectly healthy plants and pods. Clean up, or spade under, old bean tops; rotate crops. Never pick or cultivate beans when vines are wet. There are some resistant varieties, but more reliance should be placed on obtaining seed grown where the disease is not present.

Colletotrichum malvarum. Hollyhock anthracnose; seedling blight; on hollyhock, mallow, and abutilon, particularly destructive to greenhouse seedlings. Black blotches are formed on veins, leaf blades, petioles, and stems. Remove and burn all old plant parts in autumn.

Colletotrichum omnivorum. Anthracnose; on aspidistra and hosta. Large, whitish spots with brown margins are formed on leaves and stalks. Remove and burn infected plant parts.

Colletotrichum pisi. Pea anthracnose; leaf and pod spot; commonly associated with Ascochyta blight and often a secondary parasite. Spots on pods, stems, and leaves are sunken, gray, circular, with dark borders. Crop rotation is the best control.

Colletotrichum schizanthi. Anthracnose; on butterfly flower. Symptoms are small brown spots on leaves and water-soaked areas on young stems. Cankers on stems and branches of older plants may cause leaves to turn yellow, branches to die back from the tip, and finally death of all parts above the canker.

Colletotrichum spinaciae. Spinach anthracnose; known on spinach since 1880 but unimportant in most years. Leaves have few to many circular spots, water-soaked, turning gray or brown, with setae prominent in spore pustules. The fungus is seed-borne.

Colletotrichum sublineolum. Anthracnose; on wild rice (*Zizania*).

Colletotrichum trichellum. Fruit anthracnose; of tomato.

Colletotrichum truncatum. Stem anthracnose; prevalent in the South on bean, lima bean, and soybean, also on clovers. Brick-red spots appear on veins on underside of leaves and on pods. Plants are chlorotic, stunted, may die prematurely; blossoms or pods may drop. Use healthy seed grown in arid states; clean up plant refuse; rotate with nonlegumes.

Colletotrichum violae-tricoloris. Anthracnose of violet, pansy; circular dead spots with black margins, sometimes zonate, appear on leaves; flowers have petals spotted or not fully developed and producing no seed; entire plants are sometimes killed. Remove and burn infected plants or parts; clean up old leaves in fall. Copper sprays may be injurious.

Colletotrichum sp. Azalea anthracnose; new disease serious on Indian and Kurume azaleas in Louisiana since 1954. Very small rusty brown spots appear on both surfaces of young leaves, followed by defoliation. Spores appear on fallen leaves, which serve as source of inoculum for the next season. Copper and organic fungicides are effective in control.

Gloeosporium

Fungi Imperfecti, Melanconiales, Melanconiaceae

Genus characters are about the same as for *Colletotrichum* except that there are no setae around the acervuli. Conidia are hyaline, one-celled, appearing in masses or pustules on leaves or fruit. Leaf spots are usually light brown, with foliage appearing scorched.

Gloeosporium allantosporum. Anthracnose; dieback; on raspberry in Oregon, Washington. See *Elsinöe veneta* under Spot Anthracnose for the common raspberry disease called anthracnose.

Gloeosporium apocryptum. Maple anthracnose; leaf blight; an important leaf disease of silver maple, common also on other maples and boxelder, appearing from late May to August. The leaf spots are light brown, often merging with the leaves, appearing scorched. The effect may be confused with the physiological scorch caused by hot weather. On Norway maples the leaf lesions are confined to purple to brown lines along the veins. In rainy seasons there may be severe defoliation.

Control. If trees have been affected more than a year or so, feed to stimulate vigorous growth. Spray with a copper fungicide two or three times at 14-day intervals, starting when buds break open.

Gloeosporium limetticolum. Lime anthracnose; withertip; only on lime in southern Florida. Shoots, leaves, and fruits are infected when young; mature tissues are immune. Twigs wither and shrivel from one inch to several inches back from the tip; young leaves have dead areas or are distorted; buds fail to open and may drop; fruits drop, or are misshapen, or have shallow spots or depressed cankers.

Control. Spray with bordeaux-oil emulsion as fruit is setting, with two or three applications of 1 to 40 lime sulfur at 7- to 14-day intervals.

Gloeosporium melongenae (possibly identical with *G. piperatum*). Eggplant anthracnose; ripe rot; an occasional trouble. Yellow to brown spots on leaves and small to medium depressed spots on fruit show pink spore masses following rain or heavy dew. Spores are splashed by rain and spread by tools, insects, and workmen. Rotation of crops and sanitary measures may be sufficient control.

Gloeosporium piperatum. Pepper anthracnose; fruit spot; sometimes a leaf and stem spot but more often a disease of green or ripe fruit. Spots are dark, sunken, with concentric rings of acervuli and pink masses of spores, which are washed to other fruit. Seed is infected internally and contaminated externally. Harvest seed only from healthy fruit.

Gloeosporium quercinum. Oak anthracnose; see *Gnomonia quercina* and Figure 11.

Gloeosporium thuemenii f. **tulipi. Tulip anthracnose**; found in California in 1939. Lesions on peduncles and leaf blades of Darwin tulips are small to

Figure 11. Oak anthracnose.

large, elliptical, first water-soaked then dry with black margins and numerous black acervuli in center of spots.

Gloeosporium sp. **Peony anthracnose**; on stems, leaves, flowers, petals of peony. Stem lesions are sunken, with pink spore pustules, and may completely girdle the stalks, causing death of plants. Also a destructive anthracnose on strawberry.

Glomerella

Ascomycetes, Sphaeriales, Gnomoniaceae

Perithecia are dark, hard, carbonaceous, usually beaked, immersed in substratum so only the neck protudes. Ascospores are hyaline, one-celled; asci are thickened at tips, inoperculate but spores sometimes discharged with force; paraphyses present.

Glomerella cingulata (*Colletotrichum gloeosporioides*). **Anthracnose**; canker, dieback, withertip, fruit rot; of a great many plants, generally distributed except on the Pacific Coast, more common in the South. Infection is often secondary, in tissues weakened from other causes. See also under Canker and under Rots.

On citrus, orange, lemon, grapefruit there is a dying back or withertip of twigs. Leaf spots are light green turning brown, with pinkish spore pustules prominent in wet weather. Decayed spots are produced on ripening fruits in storage. Similar withertip symptoms may also appear on avocado, aucuba, cherimoya, fig, loquat, roselle, rosemallow, royal palm, dieffenbachia, rubber-plant, strawberry, and other ornamentals and fruits. The disease has also been reported on European white birch in Virginia. Lack of water and nutrient deficiency predispose plants to infection by this weak parasite.

The fungus attacks blue lupine and statice or sea lavender; peach anthracnose became important in Georgia when lupine was used as a ground cover in orchards. Sweet pea anthracnose is often more severe near apple orchards where the fungus winters on cankered apple limbs and in bitter rot mummies. Whitish lesions disfigure sweet pea leaves, shoots, and flower stalks. Leaves wither and fall; stalks dry up before blossoming; seed pods shrivel. There may be general wilting and shoot dieback.

Anthracnose and twig blight are widespread on privet. Leaves dry and cling to the stem; cankers at the base of stems are dotted with pink pustules. Bark turns brown and splits; death follows complete girdling of stems. European privet is highly susceptible; California, Amur, Ibota, and Regal privets are fairly resistant.

Control. Remove infected twigs and branches from trees and shrubs, taking care to make smooth cuts at base of limbs and painting surfaces with a wound dressing. Plant sweet peas, from healthy pods, at a distance from apple and privet, in clean soil; rake and burn plant refuse at the end of the season.

Glomerella glycines. **Fruit anthracnose**; of tomato.

Glomerella gossypii. Fruit anthracnose; of tomato.

Glomerella nephrolepis. Fern anthracnose; tip blight; of Boston and sword ferns. The soft growing tips of fronds turn brown and dry. Keep foliage dry; remove and burn diseased leaves.

Glomerella phomoides (*Colletotrichum phomoides*). Tomato anthracnose; common rot; of ripe tomatoes, most frequent in Northeast and North Central districts. Symptoms appear late in the season, causing more loss to canning crops. Small, circular, sunken spots, increasing to an inch in diameter, penetrate deeply into the flesh. At first water-soaked, the spots turn dark, with pinkish, cream, or brown spore masses in the depressed centers, often arranged in concentric rings. The disease is worse in warm, moist weather. The fungus winters in tomato refuse, sometimes in cucumber and melon debris.

Control. Clean up trash and rotting fruit.

Gnomonia

Ascomycetes, Sphaeriales, Gnomoniaceae

Perithecia innate, beaked, separate; paraphyses absent; ascospores two-celled, hyaline; imperfect state *Gloeosporium* or *Marssonina*. Diseases caused by *Gnomonia* are classified as anthracnose, scorch, or leaf spot.

Gnomonia caryae. Hickory anthracnose; leaf spot; widespread. The disease is common in eastern states, causing defoliation in wet seasons. Large, roundish spots are reddish brown on upper leaf surface, dull brown underneath. The fruiting bodies are minute brown specks, and the fungus winters in dead leaves on the ground.

Gnomonia leptostyla (*Marssonina juglandis*). Walnut anthracnose; leaf spot; general on butternut, hickory, and walnut. Spring infection comes from ascospores shot from dead leaves on the ground, secondary infection from conidia. Irregular dark brown spots appear on leaflets in early summer; if these are numerous, there is defoliation. An unthrifty condition of black walnuts and butternuts is often due to anthracnose.

Gnomonia platani (*G. veneta*). Sycamore anthracnose; twig blight; general on American and Oriental planes (London plane is rather resistant) and on California and Arizona sycamores.

The fungus winters as mycelium in fallen leaves, producing perithecia that discharge ascospores when young foliage is breaking out. Mycelium also winters in twig cankers. Young sycamore leaves turn brown and die, looking as if hit by late frost. Leaves infected later in the season have irregular brown areas along the veins. Conidia ooze out from acervuli on underside of veins in flesh-colored masses, in rainy weather, and are splashed to other leaves. Twigs and branches have sunken cankers with more acervuli. Native sycamores may be nearly defoliated, with smaller twigs killed. Larger branches die with several successive wet springs. The trees usually put out a second crop of leaves after defoliation, but this process is devitalizing. Dead twigs and branches give a witches' broom effect to the trees.

On white oaks anthracnose appears as brown areas adjacent to midribs and lateral veins.

Control. Although raking and burning all fallen leaves has been stressed for years, the overwintering of the fungus on twigs makes this measure rather ineffective. The spray schedule has called for three applications of bordeaux mixture; a dormant spray, one when the buds swell, and another 7 days later. Trees should be fertilized to stimulate vigorous growth.

Gnomonia quercina (*Gloeosporium quercinum*). Oak anthracnose; the fungus is closely related to *Gnomonia platani*, usually reported as *G. veneta*, but is now considered a separate species. The anthracnose appears as brown areas adjacent to midribs and lateral veins (see Fig. 11).

Gnomonia tiliae (*Gloeosporium tiliae*). Linden anthracnose; leaf spot, leaf blotch, scorch on American and European linden. Small, circular to irregular brown spots with dark margins form blotches along main veins in leaves, leaf stalks, and young twigs, with rose-colored pustules. In wet seasons, defoliation in early summer may be followed by wilting and death of branches. Cut out and burn such branches.

Marssonina

Fungi Imperfecti, Melanconiales, Melanconiacae

Hyaline, two-celled spores are formed in acervuli without setae. Spores are rounded at ends and are formed in pale to black masses on leaves.

Marssonina panattoniana. Lettuce anthracnose; small, dead, brown spots appear on blades and petioles, centers often falling out leaving black-margined shot holes. Spots progress from older to young inner leaves; outer leaves are broken off and blown around by wind. The disease is important only during prolonged periods of wet weather, when it may cause heavy losses. Sanitary measures and treating seed before planting suffice for control.

Mycosphaerella

Ascomycetes, Sphaeriales, Mycosphaerellaceae

Perithecia immersed in substratum, not beaked, not setose, paraphyses lacking; spores hyaline, two-celled. The genus contains more than 1,000 species, many destructive to plants, with conidial stages in many genera.

Mycosphaerella opuntiae. Cactus anthracnose; on Cereus, Echinocactus, Mammillaria, and Opuntia. The curved spores of the imperfect stage

(*Gloeosporium cactorum*) form light pink pustules on the surface of moist, light brown rotten areas. Cut out and destroy diseased segments.

Neofabraea

Ascomycetes, Helotiales, Mollisiaceae

This is one of the discomycetes, cup fungi. The apothecia, formed on living plants, are fleshy, bright-colored with a peridium of dark cells forming a pseudoparenchyma. Spores are hyaline, fusoid.

Neofabraea malicorticis. Northwestern apple anthracnose; on apple, crabapple, pear, quince, chiefly in the Pacific Northwest, where it is a native disease, serious in regions with heavy rainfall.

Cankers are formed on younger branches—elliptical, dark, sunken, up to 3 or 4 inches wide and 10 to 12 inches long, delimited when mature by a crack in the bark. Conidia of the imperfect state (*Gloeosporium malicorticis*) are formed in cream-colored cushions, which turn black with age, in slits in the bark. Young cankers, reddish brown, circular spots appear on the bark in late fall. Fruit is infected, usually through lenticels from either ascospores or conidia in pustules on bark, but the disease may not show up until the apples are in storage.

Control. Cut out diseased limbs or excise cankers, burning all prunings and dead bark. Spray with bordeaux mixture before fruit is picked and fall rains start; repeat after harvest, and again about 2 weeks later.

Pseudopeziza

Ascomycetes, Helotiales, Mollisiaceae

Apothecia brown, cup-shaped, arising from leaves on short stalks, not setose, paraphyses present; spores one-celled, hyaline, ovoid.

Pseudopeziza ribis. Currant anthracnose; leaf, stem, and fruit spot; generally distributed on currant, flowering currant, and gooseberry, first reported on black currants in Connecticut in 1873. Very small, brown, circular spots appear first on lower, older leaves, which turn yellow if spots are numerous. Hyaline, crescent-shaped conidia are formed in moist, flesh-colored masses in center of spots. In severe infections there is progressive defoliation from below upward.

Other occasional symptoms are black, sunken spots on leaf stalks, light brown to pale yellow lesions on canes, and black flyspeck spots on green berries, with considerable reduction in yield. Apothecia are formed on fallen leaves; ascospores are forcibly discharged in spring and carried by wind to young leaves.

Control. Clean up and burn old leaves under the bushes. Spray with bordeaux mixture (preferred to the newer organics) shortly after leaves appear (about 3 weeks after blossoming) and immediately after picking. Include a good spreader and cover both leaf surfaces thoroughly.

BACTERIAL DISEASES
Agrobacterium

Rhizobiaceae

Small, motile, short rods, with two to six peritrichous flagella or a polar or subpolar flagellum, ordinarily Gram-negative, not producing visible gas or detectable acid in ordinary culture media; growth on carbohydrate media usually accompanied by copious entracellular, polysaccharide slime; gelatin liquefied slowly or not at all; optimum temperatures 77° F to 86° F (25° C to 30° C). Found in soil, or plant roots in soil, or in hypertrophies or galls on roots or stems of plants.

Agrobacterium rhizogenes. Hairy root of apple; also recorded on cotoneaster, hollyhock, honey locust, honeysuckle, mulberry, peavine, peach, quince, Russian olive, rose, and spirea. "Woolly root" and "woolly knot" are other names given to this disease, which was long considered merely a form of crown gall. Both diseases may appear on the same plant and in early stages be confused. In hairy root a great number of small roots protrude either directly from stems or roots or from localized hard swellings that frequently occur at the graft union. The disease is common on grafted nursery apple trees, 1 to 3 years old, and the root development may be as profuse as witches' brooms. Control measures are the same as for crown gall.

Agrobacterium rubi. Cane gall of brambles; on blackberry, black and purple raspberries, and, very rarely, red raspberry. Symptoms appear on fruiting canes in late May or June as small, spherical protuberances or elongated ridges of white gall tissue, turning brown after several weeks. Canes often split open and dry out; produce small seedy berries. Cane gall is not as important as crown gall, but one should use the same preventive measures. Avoid runner plants from infected mother plants.

Agrobacterium tumefaciens. Crown gall; on a great variety of plants in more than 40 families, general on blackberry, raspberry, and other brambles, on grapes and on rose (see Fig. 12); on fruit trees—apple, apricot, cherry, fig, peach and nectarine, pear (rarely), plum—on nuts, almond very susceptible, walnut fairly susceptible, pecan occasionally; on shade trees, willow and other hard woods; rare on conifers but reported on incense cedar and juniper; on many shrubs and vines, particularly honeysuckle and euonymus; on perennials such as asters, daisies, and chrysanthemums; and on beets, turnips, and a few other vegetables, with tomato widely used in experiments.

Crown gall was first noticed on grape in Europe in 1853, and the organism was first isolated in 1904 in the United States from galls on Paris daisy. It is of first importance as a disease of nursery stock, but may cause losses of large

Figure 12. Crown gall on rose.

productive trees in neglected orchards, especially almonds and peaches in California and other warm climates. It is very important to rose growers and to the amateur gardeners who sometimes receive infected bushes.

The galls are usually rounded, with an irregular rough surface, ranging up to several inches, usually occurring near the soil line, commonly at the graft union, but sometimes on roots or aerial parts. On euonymus, galls are formed anywhere along the vine. This disease is primarily of the parenchyma, starting with a rapid proliferation of cells in the meristematic tissue and the formation of more or less convoluted soft or hard overgrowths or tumors. The close analogy of the unorganized cell growth of plant galls to wild cell proliferation in human cancer has intrigued scientists for many years. In some fashion bacteria provide stimulus for this overdevelopment, but similar galls have been produced on plants experimentally by injecting a virus or growth-promoting substances. Again experimentally, the injection of penicillin and other antibiotics has inhibited the development of bacterial crown galls.

Entrance of bacteria into plants for natural infection is through wounds. In nurseries and orchards nematodes, the plow, the disc, or the hoe may be responsible; on the propagating bench grafting tools are indicted. Many claims have been made for the longevity of crown gall bacteria in soil, but it now seems to be established that they do not live in the absence of host plants more than a couple of years, and that sudden outbreaks of crown gall on land not previously growing susceptible crops are due to irrigation water bringing in viable bacteria from other infected orchards. The addition of lime to the soil may encourage crown gall, for bacteria do not live in an acid medium. The period of greatest activity is during the warm months.

Control. For home gardens rigid exclusion of all suspected planting stock is the very best control. Do not accept from your nurseryman blackberries, raspberries, roses, or fruit trees showing suspicious bumps. If you have had previous trouble, choose a different location for new, healthy plants. Be careful not to wound stems in cultivating.

For nurserymen, sanitary propagating practices are a must. Stock should be healthy. Grafting knives should be sterilized by frequent dipping in potassium permanganate, 1 ounce in 2 gallons of water, or in denatured alcohol. If nursery soil is infested, 2 years' growth of cowpeas, oats, or crotalaria between crops will minimize crown gall.

Fruit and nut growers can perhaps plant less susceptible varieties, although fruit that is resistant in one locality may be diseased in another. American grape varieties are considered more resistant than European. Apples may be better on mahaleb root-stock, nut trees on black walnut understock. Budding rather than grafting reduces the chance of infection.

Painting galls with a solution of Elgetol-methanol has given control of crown gall on peaches and almonds in California. One part Elgetol (sodium dinitrocresol) is shaken with four parts synthetic wood alcohol and applied with a brush, covering the surface of the gall and extending ½ inch to 1 inch beyond the margin into healthy bark.

Some years ago it was found that colchicine inhibited some plant tumors. Since 1945, there have been encouraging reports on the effects of penicillin, streptomycin, and other antibiotics, administered by hypodermic injection, by immersing galled roots, by applying a cotton pad soaked in an antibiotic to a gall, and by dipping nursery seedlings. Dipping young cherry trees in 200 ppm of terramycin for 1 hour has reduced the incidence of crown gall by half. Antibiotics, such as Bacticin, offer much promise, but the easy practical control of crown gall in home gardens is not yet at hand.

Bacillus

Bacillus polymyxa. **Seedling blight**; of tomato; gram-variable; rod-shaped bacterial cells; motile by means of peritrichous flagella; cells do not form chains.

Clavibacter

Coryneform Group Mycobacteriaceae

Slender, straight to slightly curved rods, with irregularly stained segments or granules, often with pointed or club-shaped swellings at ends; nonmotile with a few exceptions (*C. flaccumfaciens* and *C. poinsettiae*). Gram-positive.

Corynebacterium agropyri. Yellow gum disease; on western wheat grass. Enormous masses of surface bacteria form yellow slime between stem and upper sheath and glumes of flower head; plants dwarfed or bent; normal seeds rare.

Rhodococcus fasciens. Fasciation; widespread on sweet pea, also on carnation, chrysanthemum, gypsophila, geranium, petunia, impatiens, *Hebe* sp. and pyrethrum. Sweet pea symptoms are masses of short, thick, and aborted stems with misshapen leaves developing near the soil line at first or second stem nodes. The fasciated growth on old plants may have a diameter of 3 inches but does not extend more than an inch or two above ground. The portion exposed to light develops normal green color. Plants are not killed, but stems are dwarfed and blossom production is curtailed.

Control. Sterilize soil or use fresh.

Curtobacterium flaccumfaciens pv. flaccumfaciens. Bacterial wilt of bean; widespread on kidney and lima beans and soybean, causing considerable loss. Plants wilt at any stage from seedling to pod production, with leaves turning dry, brown, and ragged after rains. Plants are often stunted. Bacteria winter on or in seed, which appear yellow or wrinkled and varnished. When infected seed is planted, bacteria pass from cotyledons into stems and xylem vessels. Other plants are infected by mechanical injury and perhaps by insects, but there is not much danger from splashed rain. Plants girdled at nodes may break over.

Control. Use seed grown in Idaho or California.

Corynebacterium humiferum subsp. **michiganense.** Reported from wetwood of poplar, in Colorado.

Clavibacter michiganense. Bacterial canker of tomato; widespread, formerly causing serious losses of tomato canning crops. The disease has now been reported on browallia, brunfelsia, cestrum, *Datura* sp., eggplant, Jerusalem cherry, bittersweet, pepper, painted tongue, potato, ground cherry, and butterfly flower in Wyoming. This is a vascular wilt disease, seedlings remaining stunted. Symptoms on older plants start with wilting of margins of lower leaflets, often only on one side of a leaf. Leaflets curl upward, brown, and wither, but remain attached to stem. One-sided infection may extend up through the plant and open cankers from pith to outer surface of stem. Fruit infection starts with small, raised, snow-white spots, centers later browned and roughened but the white color persisting as a halo to give a bird's-eye spot. Fruits can be distorted, stunted, yellow inside. In the field, bacteria are spread by splashed rain and can persist in soil for 2 years or more. Seeds carry the bacteria internally as well as externally.

Control. Use certified seed, a 2- or 3-year rotation; clean up tomato refuse at end of season and diseased plants throughout season. Fermenting tomato pulp for 4 days at a temperature near 70° F will destroy bacteria on surface of seed; hot-water treatment, 25 minutes at 122° F will kill some, perhaps not all, of internal bacteria. Start seedlings in soil that has not previously grown tomatoes.

Clavibacter poinsettiae. Stem canker and leaf spot of poinsettia; a relatively new disease, first noted in greenhouses in 1941. Longitudinal water-soaked streaks appear on one side of green stems, sometimes continuing through leaf petioles to cause spotting or blotching of leaves and complete defoliation. The cortex of stems turns yellow, the vascular system brown. Stems may crack open and bend down, with glistening, golden brown masses of bacteria oozing from stem ruptures and leaf lesions.

Control. Discard diseased stock plants; place cuttings from healthy mother plants in sterilized media; avoid overhead watering and syringing; rogue suspicious plants promptly.

Clavibacter sepedonicum. Bacterial ring rot of potato; widespread since 1931, when it probably was introduced from Europe. All commercial varieties are susceptible, with losses formerly in millions of dollars in decay of tubers in field and storage. Now a single infected plant in a potato field disqualifies the whole field for certification. Symptoms appear when plants are nearly full grown, with one or more stems in a hill wilted and stunted while the rest seem healthy. Lower leaves have pale yellow areas between veins; these turn deeper yellow, and margins roll upward. A creamy exudate is expelled when the stem is cut across. This bacterium may also occur in sugar beet which are symptomless.

Tuber infection takes place at the stem end, and the most prominent symptoms appear some time after storage. The vascular ring turns creamy yellow to light brown, with a crumbly or cheesy odorless decay followed by decay from secondary organisms. Bacteria are not spread from plant to plant in the field, but by cutting knife and fingers at planting. A knife used to cut one infected tuber may contaminate the next 20 seed pieces.

Control. Use certified seed potatoes. Use several knives and rotate them in disinfestant. Commercial growers use a rotating knife passed through a chemical or hot-water bath between cuts. Disinfest tools, grader, digger, and bags; sweep storage house clean and spray with copper sulfate, 1 pound to 5 gallons of water.

Clavibacter xyli sub. sp. **cynodontis.** Stunting disease; of Bermuda grass.

Erwinia

Enterobacteriaceae

Motile rods (usually) with peritrichous flagella; Gram-negative; producing acid with or without visible gas from a variety of sugars; invading tissues of living plants producing dry necroses, galls, wilts, and soft rots. The genus is named for Erwin F. Smith, pioneer in plant diseases caused by bacteria.

Erwinia amylovora. Fire blight; general on many species in several tribes of the Rosaceae, particularly serious on apple, pear, and quince. Other hosts include almond, amelanchier, apricot, aronia, blackberry, cherry, chokecherry, cotoneaster, crabapple, exochorda, geum, hawthorn, holodiscus, kerria, Japanese quince, loquat, medlar, mountain ash, plum, photinia, pyracantha, raspberry, rose, spirea, and strawberry.

Apparently a native disease, first noticed near the Hudson River in 1780, fire blight spread south and west with increased cultivation of pears and apples. By 1880, it had practically wrecked pear orchards in Illinois, Iowa, and other states in the Northern Mississippi Valley. Then it devastated pears on the Texas Gulf. Reaching California by 1910, it played havoc up the coast to Washington.

Blossoms and leaves of infected twigs suddenly wilt, turn dark brown to black, shrivel and die, but remain attached to twigs (see Fig. 13). The bark is shrunken, dark brown to purplish, sometimes blistered with gum oozing out. Brown or black blighted branches with dead persistent leaves look as if scorched by fire. The bacteria survive the winter in living tissue at the edge of "holdover cankers" on limbs. These are dead, slightly sunken areas with a definite margin or slight crack where dead tissue has shrunk away from living. In moist weather bacteria appear on the surface of cankers in pearly viscid drops of ooze that is carried by wind-blown rain or insects to blossoms. Infection spreads from the blighted bloom to the young fruit, then down the pedicel to adjacent leaves, which turn brown, remaining hanging around the blighted blossom cluster. Leaf and fruit blight is also possible by direct

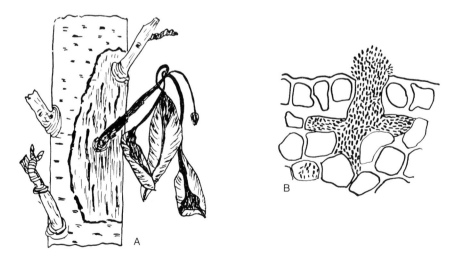

Figure 13. Fire blight. *A,* hold over canker developed on apple limb at base of blighted twig. *B,* bacteria swarming through tissue.

invasion, a secondary infection via bacteria carried from primary blossom blight by ants, aphids, flies, wasps, fruit-tree bark beetles, and honeybees, sometimes tarnished plant bugs, and pear psyllids.

The tissue first appears water-soaked, then reddish, then brown to black as the bacteria swarm between the dying parenchyma cells. Division may take place every half hour, so they multiply rapidly and are usually well in advance of discolored external tissue. A collar rot may develop when cankers are formed near the base of a tree. Water sprouts are common sources of infection.

As spring changes to summer, the bacteria gradually become less active and remain dormant at the edge of a woody canker until the next spring at sap flow. Ordinarily they do not winter on branches smaller than ½ inch in diameter.

Control. Spraying during bloom is now a standard means of preventing blossom blight. Use bordeaux mixture or a fixed copper or streptomycin at 60 ppm to 100 ppm. The latter is very effective at relatively high temperatures; at 65° F and below, copper is more satisfactory. Start spraying when about 10% of the blossoms are open and repeat at 5- to 7-day intervals until late bloom is over. A dormant spray for aphid control helps in preventing fire blight. One or more sprays may be needed for leafhoppers, starting at petal fall.

Inspect trees through the season and cut or break out infected twigs 12 inches below the portion visibly blighted. If lesions appear on large limbs they may be painted with one of the following mixtures:

I. 1 quart denatured alcohol, ¼ pint distilled water, ¾ ounce muriatic acid, 1½ pounds zinc chloride.

II. 100 grams cobalt nitrate, 50 cubic centimeters glycerine, 100 cubic centimeters oil of wintergreen, 50 cubic centimeters acetic acid, 80 cubic centimeters denatured alcohol.

III. 5 parts cadmium sulfate stock solution (1 pound stirred into 2 pints warm water), 2 parts glycerine, 2 parts muriatic acid, 5 parts denatured alcohol.

Formulas I and II were developed for use on the West Coast; III was developed for use in New York. The paint is brushed on the unbroken bark over the lesions and for several inches above and below the canker; it may injure if there are wounds or cuts.

In cutting out cankered limbs during the dormant season, take the branch off at least 4 inches back from edge of the canker, and disinfect the cut. The home gardener may want to use Clorox for tools and bordeaux paint for cut surfaces. Dry bordeaux powder is stirred into raw linseed oil until a workable paste is formed.

Almost all desirable pear varieties are susceptible to fire blight, particularly Bartlett, Flemish Beauty, Howell, Clapps Favorite. Varieties Old Home, Orient, and the common Kieffer are more or less resistant. Jonathan apples are very susceptible. Less apt to be severely blighted are Baldwin, Ben Davis, Delicious, Duchess, McIntosh, Northern Spy, Stayman, and Winter Banana.

At the University of California some work has been done on susceptibility of ornamentals to fire blight. *Pyracantha angustifolia* is quite susceptible, but *P. coccinea* and *P. crenulata* are rather resistant. *Cotoneaster salicifolia* is susceptible; *C. dammeri*, *C. pannosa*, and *C. horizontalis* are more resistant; and *C. adpressa* and *C. microphylla* show marked resistance.

Cultural methods influence the degree of fire blight, which is worse on fast-growing succulent tissue. Avoid heavy applications of nitrogen fertilizers; apply such nitrogen as is required in autumn or in spring in foliar sprays after danger of blossom blight is over.

Erwinia carnigieana. Bacterial necrosis of giant cactus; in the entire habitat of *Carnegia gigantea*. Long present in southern Arizona, this disease was not described until 1942, after it had encroached on cactus parks and private estates. Many giant cacti in the Saguaro National Monument have been killed, with heaviest mortality in trees 150 to 200 years old.

Symptoms start with a small, circular, light spot, usually with a water-soaked margin. The tissues underneath turn nearly black; the spot enlarges and has a purplish hue with the center cracking and bleeding a brown liquid. The rotten tissues dry, break up into granular or lumpy pieces, and fall to the ground. Rotting on one side means leaning to that side; when the trunk is girdled near the base, the giant is likely to fall in a wind storm. If it does not break, it stands as a bare, woody skeleton, with all parenchyma tissue disintegrated. An insect, *Cactobrosis fernaldialis*, is largely responsible for the rapid spread of the disease. The larvae tunnel inside the stems most of the year, emerging from May to August to pupate for a month or so before the adult, a tan and brown nocturnal moth, lays eggs.

Control. A phosphate dust, applied monthly from April to September, has effectively controlled the insect vector. Incipient infections can be cut out and the cavity allowed to dry out and cork over. Before the insect vector was known, fallen trees were cut into short lengths, dragged to a burial pit, covered with a disinfectant, and then with soil.

Erwinia carotovora subsp. **aroideae.** Soft rot of calla; originally described from common calla, found on golden calla, and also on beet, cactus, cabbage, cauliflower, celery, cucumber, carrot, eggplant, geranium (*Pelargonium*), hyacinth, iris, onion, parsnip, pepper, potato, salsify, sansevieria, tobacco, tomato, and turnip.

On calla lily the soft rot starts in upper portion of the corm and progresses upward into leaf and flower stalks or down into roots, with the corm becoming soft, brown, and watery. Sometimes infection starts at edge of petiole, which turns slimy. Leaves with brown spots and margins die or rot off at the base before losing color. Flowers turn brown; stalks fall over; roots are soft and slimy inside the epidermis. Corms may rot so fast that the plant falls over without other symptoms, or the diseased portion may dry down to sunken dark spots, in which the bacteria stay dormant to the next season.

On tomatoes, infection takes place through growth cracks, insect wounds, or sunscald areas. The tissue is at first water-soaked, then opaque, and in 3 to 10 days the whole fruit is soft, watery, colorless, with an offensive odor.

Control. Scrub calla corms, cut out rotted spots, and let cork over for a day or two. Plant in fresh or sterilized soil in sterilized containers and keep pots on clean gravel or wood racks, never on beds where diseased callas have grown previously. Grow at cool temperatures and avoid overwatering.

Erwinia carotovora subsp. **atroseptica. Potato blackleg**; basal stem rot, tuber rot, general on potato. This disease is systemic and is perpetuated by naturally infected tubers. Lower leaves turn yellow; upper leaves curl upward; stems and leaves tend to grow up rather than spread out; stem is black-spotted, more or less softened at base and up to 3 or 4 inches from ground, and may be covered with bacterial slime; shoots may wilt and fall over. Tubers are infected through the stem end. The disease is most rapid in warm, moist weather, and may continue in storage. The bacteria are spread on the cutting knife, as with ring rot, and by seed-corn maggots, and may persist for a time in soil.

Control. Use certified seed potatoes and plant whole tubers; if cut seed must be used, allow to cork over to prevent infection from soil. Practice long rotation; disinfest cutting knife. Late varieties seem to be more resistant.

Erwinia carotovora subsp. **atroseptica. Delphinium blackleg**; foot rot; bacterial crown rot of perennial delphinium; stem and bud rot of rocket larkspur. In delphinium there is a soft black discoloration at the base of the stem, with bacteria oozing out from cracks. In larkspur there is a black rot of buds as well as yellowing of leaves, blackening of stem, stunting of plants. The bacteria are apparently carried in seed; hot-water treatment is helpful. Drenching delphinium crowns with bordeaux mixture has been recommended in the past. Soil treatment with Agrimycin, 1 ounce in 10 gallons of water, is suggested for trial. Avoid excessive watering or irrigation.

Erwinia carotovora subsp. **carotovora. Wilt**; of sunflower, kalanchoë; zucchini squash, and dracaena.

Erwinia carotovora. Soft rot; general on many vegetables, in field, storage, and transit, and many ornamentals, especially iris. The bacteria were first isolated from rotten carrots, whence the name, but they are equally at home in asparagus, cabbage, turnips and other crucifers, celery, cucumber, eggplant, endive, garlic, horseradish, melon, parsnip, pepper, spinach, sunflower (stalk rot), sweet potato, and tomato. Besides wide distribution on iris, soft rot has been reported, among ornamentals, on chrysanthemum, dahlia, Easter lily, geranium, orchid, sansevieria, poinsettia, and yellow calla.

The bacteria enter through wounds, causing a rapid, wet rot with a most offensive odor. The middle lamella is dissolved, and roots become soft and pulpy. Soft rot in iris often follows borer infestation. Tips of leaves are withered, the basal portions wet and practically shredded. The entire interior of a rhizome may disintegrate into a vile yellow mess while the epidermis remains firm. The rot is more serious in shaded locations, when iris is too crowded or planted too deeply.

Control. Borer control, starting when fans are 6 inches high, has greatly reduced the incidence of rot. If it appears, dig up the clumps, cut away all rotted portions, cut leaves back to short fans. Allow to dry in the sun for a day or two, then replant in well-drained soil, in full sun with upper portion of the rhizome slightly exposed. Many good iris growers do not agree with this "sitting duck" method, preferring to cover with an inch of soil; but the sun is an excellent bactericide, and shallow planting is one method of disease control. Clean off all old leaves in late fall after frost.

Prevent rot on stored vegetables by saving only sound, dry tubers, in straw or sand, in a well-ventilated room with temperature not too much above freezing. In the garden, rotate vegetables with fleshy roots with leafy varieties. Avoid bruising at harvest time.

Erwinia chrysanthemi pv. chrysanthemi. Bacterial blight of chrysanthemum; a florists' disease, first noted in 1950. First evidence of blight is a gray water-soaked area mid-point on the stem, followed by rot and falling over. The diseased tissue is brown or reddish brown; the rot progresses downward to the base of the stem or, under unfavorable conditions, may be checked with axillary buds below the diseased area producing normal shoots. Cuttings rot at the base. Sometimes affected plants do not show external symptoms, and cuttings taken from them spread the disease. Bacteria can be spread via cutting knife, or fingernails in pinching, and can live several months in soil. A form of this species causes a leaf blight of philodendron and may also infect banana, carnation, corn, and sorghum.

Control. Snap off cuttings; sterilize soil and tools.

Erwinia cypripedii. Reported from California, causing brown rot of Cypripedium orchids. Small, circular to oval, water-soaked, greasy light brown spots become sunken, dark brown to chestnut. Affected crowns shrivel; leaves drop.

Erwinia dissolvens. Corn rot; corn leaves show light or dark brown rotting at base; husks and leaf blades have dark brown spots; lower portion of stalk is rotten, soft, brown, with strong odor of decay; plants may break over and die, with little left but a mass of shredded remnants of fibrovascular bundles. Bacteria enter through hydathodes (water pores), stomata, and wounds.

Erwinia herbicola. Leaf spot; of dracaena.

Erwinia herbicola f.sp. gypsophilae. On Dracaena sanderana, gypsophila, and related plants. Galls are formed at crown and roots of grafted plants from ¼ to 1 inch in diameter, but with a flat nodular growth rather than the usual globose crown gall.

Erwinia lathyri (=*E. herbicola*). A saprophyte, formerly credited with causing streak of sweet pea, now known to be a virus disease.

Erwinia nimipressuralis. Wetwood of elm; slime flux, due to bacteria pathogenic in elm trunk wood, especially Asiatic elms, but possibly occurring in many other trees, including maple, oak, mulberry, poplar, and willow. A water-soaked dark discoloration of the heartwood is correlated with chronic bleeding at crotches and wounds and abnormally high sap pressure in trunk, with wilting a secondary symptom. The pressure in diseased trees increases from April to August or September, reaching 5 to 30 pounds per square inch (as much as 60 pounds in one record). The bacteria inhabit ray cells mostly and do not cause a general clogging of water-conducting tissues. This pressure, caused by fermentation of tissues by bacteria, causes fluxing, a forcing of sap out of trunks through cracks, branch crotches, and wounds. The flux flows down the trunk, wetting large areas of bark and drying to a grayish white incrustation. Bacteria and yeasts working in the flux cause an offensive odor that attracts insects.

Control. Bore drain holes through the wood below the fluxing wound, slightly slanted to facilitate drainage. Install ½-inch copper pipe to carry the dripping sap away from the trunk and buttress roots. Screw the pipe in only far enough to be firm; if it penetrates the water-soaked wood, it interferes with drainage.

Erwinia rhapontica. **Rhubarb crown rot**; similar to soft rot.

Erwinia stewartii (*Xanthomonas stewartii*). **Bacterial wilt of corn**; Stewart's disease; on sweet corn, sometimes field corn, in the middle regions of the United States, from New York to California. This disease is vascular with yellow slime formed in the water-conducting system, resulting in browning of nodes, and dwarfing of plants; or long pale green streaks on leaf blades, followed by wilting and death of whole plant. Tassels may be formed prematurely and die before the rest of the plant. The bacteria are chiefly disseminated by corn flea beetles, and winter either in the beetles or in seed. Primary infections come from flea beetles feeding in spring, from infected seed, and occasionally from soil; but secondary spread is mostly by insects.

Corn grown in rich soil is more susceptible to wilt, and so are early varieties, especially Golden Bantam. Winter temperatures influence the amount of wilt the following summer. If the winter index, which is the sum of mean temperatures for December, January, and February, is above 100, bacterial wilt will be present in destructive amounts on susceptible varieties. If the index is below 90, the disease will be very sparse in northeastern states; if the index is between 90 and 100, there will be a moderate amount of wilt. Disease surveys over a period of years testify to the reliability of such forecasts (based on the amount of cold the flea beetle vectors can survive); but with the increasing use of hybrid sweet corn resistant to wilt, the importance of winter temperatures is reduced.

Control. Use insecticides to control flea beetles; substitute commercial fertilizer for manure; destroy infected refuse; try treating seed with terramycin or streptomycin. Use resistant varieties such as Golden Cross Bantam, Carmelcross, Ioana, Marcross, and Iochief.

Erwinia tracheiphila. **Bacterial wilt of cucurbits**; cucumber wilt; on cucumber, pumpkin, squash, and muskmelon but not watermelon. The disease is generally east of the Rocky Mountains and is also present in parts of the West; is most serious north of Tennessee. Total loss of vines is rare, but a 10% to 20% loss is common.

This disease is a vascular wound disease transmitted by striped and 12-spotted cucumber beetles. Dull green flabby patches on leaves are followed by sudden wilting and shriveling of foliage, and drying of stems. Bacteria ooze from cut stems in viscid masses. Partially resistant plants may be dwarfed, with excessive blooming and branching, wilting during the day but partially recovering at night. The bacteria winter solely in the digestive tract of the insects and are deposited on leaves in spring with excrement, entering through wounds or stomata.

Control is directed chiefly at the insects. Start vines under Hotkaps and spray or dust with rotenone or other insecticide when the mechanical protection is removed. Experimental spraying with antibiotics—streptomycin, terramycin, and neomycin—has reduced wilt and increased yield.

Pseudomonas

Pseudomonadaceae

Motile with polar flagella; straight or curved rods; Gram-negative. Many species produce a greenish, water-soluble pigment. Many species are found in soil and water; many are plant pathogens causing leaf spots or blights. Many bacterial plant pathogens causing leaf, stem, and fruit spots, and necroses are pathovars of *Pseudomonas syringae*. The only reliable test for differentiating pathovars is the host test. For example, *Pseudomonas aceris* may now be called *P. syringae* pv. *aceris*.

Pseudomonas aceris. Maple leaf spot; found in California on big leaf maple. Small, water-soaked spots, surrounded by yellow zones, turn brown or black; cankers develop on petioles and bracts in serious cases; leaves may drop; disease present in cool, damp weather of early spring.

Pseudomonas albopreciptans. Bacterial spot; of cereals, grasses, and corn. Light or dark brown spots or streaks on grass blades. Bacteria enter through stomata or water pores.

Pseudomonas alcaligenes. Leaf spot; on gloxinia.

Pseudomonas andropogonis. Bacterial stripe; of sorghum and corn. Also causes blight of chickpea, and bacterial leaf spot on white clover. Red streaks and blotches appear on leaves and sheaths, with abundant exudate drying down to red crusts or scales, readily washed off in rains. Bacteria enter through stomata.

Pseudomonas angulata. Blackfire; of tobacco.

Pseudomonas asplenii. Bacterial leaf blight of bird's-nest fern; first reported from greenhouses in California. Small translucent spots enlarge to cover whole frond; bacteria may invade crown and kill whole plant. Control depends on strict sanitation—sterilizing flats, pots, media, and foreceps used in transplanting. Avoid excessive watering and too high humidity.

Pseudomonas adzukicola. Stem rot; of adzuki bean.

Pseudomonas berberidis. Bacterial leaf spot of barberry; small, irregular, dark green water-soaked areas on leaves turn purple-brown with age; occasional spotting occurs on leaf stalks and young shoots. If twigs are infected, buds do not develop in the next season; if they are girdled, the entire twig is blighted. Cut out infected twigs and spray with bordeaux mixture or an antibiotic.

Pseudomonas caryophylli. Bacterial wilt of carnation; usually under glass.

Plants wilt, turn dry, colorless, with roots disintegrating. Grayish-green foliage is the first symptom, but leaves rapidly turn yellow and die. Yellow streaks of frayed tissue in vascular areas extend a foot or two up the stem. It takes a month for disease to show up after inoculation, but it can be transmitted on cuttings taken from plants before appearance of symptoms. The sticky character of diseased tissue distinguishes this wilt from Fusarium wilt. Varieties Cardinal Sim, Laddie, Mamie, Portrait, and others may have severe cankers at base of stems, orange-yellow when young, very sticky. Bacteria are spread by hands, tools, splashing water. Also causes crown and leaf rot of statice.

Control. Remove and burn diseased plants and all within 1½-foot radius. After handling wash with hot water and soap, sterilize tools (10% Clorox for 5 minutes). Obtain rooted cuttings from propagators of cultured, disease-free material; keep in shipping bags until ready for benching and then place in raised, steam-pasteurized benches. Never place cuttings in water or a liquid fungicide (use dust if a fungicide is required for other diseases); never place temporarily on an unsterilized table; never cut or trim with hands or knives; never plant in outdoor "nurse beds"; never use overhead watering.

Pseudomonas cattleyae. Brown spot of orchids; *Phalaenopsis* and *Cattleya*, common in greenhouses. Infection is through stomata of young plants, wounds of older plants. Dark green, circular water-soaked spots change to brown and finally black. On *Cattleya* the disease is limited to older leaves.

Pseudomonas cepacia. Sour skin rot of onion; slimy yellow rot of outer fleshy scales, with a vinegar odor. Let crop mature well before harvesting, tops dry before topping; cure bulbs thoroughly before storage.

Pseudomonas cichorii. Bacterial rot; of chicory, Belgium endive, French endive, iris, soft rot of potato, and bacterial leaf spot of hibiscus, geranium, magnolia. May also cause a leaf spot and stem necrosis on chrysanthemum (see Fig. 14) and bacterial leaf blight on dwarf *Schefflera*. A yellowish olive center rot, affecting young inner leaves.

Pseudomonas corrugata. Stem rot; of tomato, also pith necrosis.

Pseudomonas fluorescens (*marginalis*). Marginal blight of lettuce; Kansas lettuce disease; also on witloof chicory, soft rot of potato tubers. Leaf margins are dark brown to almost black, first soft, then like parchment. Yellowish red spots, turning dark, are scattered over leaves. Infected tissue disintegrates into an odorous mass. Bacteria live in the soil, which should not be splashed on plants by careless watering.

Pseudomonas gladioli. Leaf spot and blight; on bird's nest fern.

Pseudomonas jaggeri (*apii*). Bacterial blight of celery; small, irregularly circular rusty leaf spots, with a yellow halo, are occasionally numerous enough to cause death of foliage, but commonly are only disfiguring. Spray plants in seedbed with bordeaux mixture, or dust with copper lime dust; clean up old refuse.

Pseudomonas maculicola. Bacterial leaf spot of crucifers; pepper spot of cabbage, cauliflower, chinese cabbage, and turnip, mostly in northeastern and Middle Atlantic states. Numerous brown or purple spots range from pinpoint to ⅛ inch in diameter. If spots are very numerous, leaves yellow and

Figure 14. Bacterial black spot on chrysanthemum.

drop off. Cauliflower is more commonly affected than cabbage. Bacteria, disseminated on seed or in diseased plant parts, enter through stomata, and visible symptoms appear in 3 to 6 days. Disease is most severe in seedbeds.

Control. Change location of hotbed starting seedlings; use 2-year rotation in field; have seed hot-water treated.

Pseudomonas marginata (*alliicola*). Onion bulb rot; a storage disease; inner scales of bulb water-soaked and soft, sometimes entire bulb rotting.

Pseudomonas marginata. Gladiolus scab; stem rot, neck rot; widespread on gladiolus, also on iris, bell peppers and tigridia. Lesions on corms are pale yellow, watersoaked circular spots deepening to brown or nearly black, eventually sunken with raised, horny, or brittle margins that are scablike and exude a gummy substance. Bacteria overwinter on corms. First symptoms after planting are tiny reddish raised specks on leaves, mostly near the base, enlarging to dark sunken spots, which grow together into large areas with a firm or soft rot. Sometimes plants fall over, but the disease is not ordinarily very damaging in the garden. The chief loss is to the grower in disfigured, unsalable corms. Brown streaks in husks sometimes disintegrate, leaving holes.

Gladiolus scab is increased by bulb mites, may be related to grub and wireworm injury.

Pseudomonas melophthora (=*Acetobacter aceti*). Apple rot; probably widespread. This disease is a decay of ripe apples following after apple maggots and eventually rotting whole fruit.

Pseudomonas primulae. Bacterial leaf spot of primrose; in ornamental and commercial plantings in California. Infection is confined to older leaves — irregularly circular brown lesions surrounded by conspicuous yellow halos.

Spots may coalesce to kill all or part of leaf. Spraying with bordeaux mixture has prevented infection.

Pseudomonas pseudoalcaligenes subsp. **citrulli**. Angular leaf spot; of muskmelon and watermelon.

Pseudomonas ribicola. On golden currant in Wyoming.

Pseudomonas sesami. Bacterial leaf spot on sesame; brown spots on leaves and stems. Can be controlled by treating seed with streptomycin.

Pseudomonas solanacearum. Southern bacterial wilt; also called brown rot, bacterial ring disease, slime disease, Granville wilt (of tobacco), present in many states but particularly prevalent in the South, from Maryland around the coast to Texas. Southern wilt is common on potatoes in Florida but also appears on many other vegetables—bean, lima bean, castor bean, soybean, velvet bean, beet, carrot, cowpea, peanut, sweet potato, tomato, eggplant, pepper, and rhubarb. Ornamentals sometimes infected include ageratum, dwarf banana, garden balsam, geranium, canna, cosmos, croton, chrysanthemum, dahlia, hollyhock, lead tree, marigold, nasturtium, Spanish needle, sunflower, and zinnia. The symptoms are those of a vascular disease, with dwarfing or sudden wilting, a brown stain in vascular bundles, and dark patches or streaks in stems. Often the first symptom is a slight wilting of leaves at end of branches in the heat of the day, followed by recovery at night, but each day the wilting is more pronounced and recovery less until the plant dies. Young plants are more susceptible than older ones. In potatoes and tomatoes there may be a brown mushy decay of stems, with bacterial ooze present. Potato tubers often have a browning of vascular ring, followed by general decay.

Bacteria live in fallow soil for 6 years or more and may persist indefinitely in the presence of susceptible plants. They are spread by irrigation water, in crop debris, or soil fragments on tools and tractors, or by farm animals. Optimum temperatures are high, ranging from 77° F to 97° F, with inhibition of disease below 55° F.

Control. Use northern-grown seed potatoes and Sebago and Katahdin varieties, more resistant than Triumph and Cobbler. Use a long rotation for tomatoes. Soil can be acidified with sulfur to kill bacteria, followed by liming in the fall before planting.

Pseudomonas stizolobii. Bacterial leaf spot of velvet bean; clovers. Translucent angular brown leaf spots have lighter centers and chlorotic surrounding tissue; there is no exudate. Bacteria enter through stomata and fill intercellular spaces of parenchyma.

Pseudomonas syringae. Canker; on kiwifruit. Blight; on mock orange. Leaf spot; on English and American elm.

Pseudomonas syringae pv. **aptata**. Bacterial spot; on beets and nasturtium. Spots on nasturtium leaves are water-soaked, brownish, ⅛ to ¼ inch across. On beets they are dark brown or black, irregular, and in addition there are narrow streaks on petioles, midribs, and larger veins. Petiole tissue may be softened as with soft rot. Infection is only through wounds.

Pseudomonas syringae pv. coronafaciens. Halo blight; on grasses, such as *Poa* spp. and *Calamagrostis* spp.

Pseudomonas syringae pv. delphinii. Delphinium black spot; on delphinium and aconite (monkshood). Irregular tarry black spots on leaves, flower buds, petioles, and stems may coalesce in late stages to form large black areas. The bacteria enter through stomata or water pores. Occasionally this bacterial leaf spot results in some distortion, but most abnormal growth and blackening of buds is due to the cyclamen mite, a much more important problem than black spot.

Control. Remove diseased leaves as noticed; cut and burn all old stalks at end of season; avoid overhead watering. In a wet season spraying with bordeaux mixture may have some value.

Pseudomonas syringae pv. glycinea. Bacterial blight of soybean; perhaps the most common and conspicuous disease of soybean, appearing in fields when plants are half-grown and remaining active until maturity, with defoliation during periods of high humidity or heavy dews. Small, angular, translucent leaf spots, yellow to light brown, turn dark reddish brown to nearly black with age. There is often a white exudate drying to a glistening film on under leaf surfaces. Black lesions appear on stems and petioles, and on pods water-soaked spots enlarge to cover a wide area, darken, and produce an exudate drying to brownish scales; seeds are often infected. Seedlings from infected seed have brown spots on cotyledons and often die. Flambeau and Hawkeye varieties are somewhat less susceptible. Use seed taken from disease-free pods.

Pseudomonas syringae pv. helianthi. Bacterial leaf spot of sunflower; leaves show brown, necrotic spots, first water-soaked, then dark and oily.

Pseudomonas syringae pv. hibisci. Bacterial leaf spot; on *Hibiscus*.

Pseudomonas syringae pv. lachyrmans. Angular leaf spot of cucurbits; general on cucumber, muskmelon, summer squash, occasional on other cucurbits. Leaves or stems have irregular, angular, water-soaked spots with bacteria oozing out in tearlike droplets that dry down to a white residue. Eventually the spots turn gray, die, and shrink, leaving holes in foliage. Fruit spots are small, nearly round, with the tissue turning white, sometimes cracking. The bacteria overwinter in diseased plant tissue and in the seed coat. They are spread from soil to stems and later to fruit in rainy weather, also transferred from plant to plant on hands and clothing. Infection is most severe in plants gone over by pickers early in the morning before dew has dried off.

Control. Plow under or remove vines immediately after harvest.

Pseudomonas syringae pv. mori. Bacterial blight of mulberry; general on black and white mulberry. Numerous water-soaked leaf spots join to form brown or black areas with surrounding yellow tissue. Young leaves may be distorted, with dark sunken spots on midribs and veins. Dark stripes with translucent borders on young shoots exude white or yellow ooze from lenticels. Dead twigs and brown leaves resemble fire blight; trees are stunted but seldom killed. Remove and burn blighted branches; do not plant young mulberry trees near infected specimens.

Pseudomonas syringae pv. **mors-prunorum.** Bacterial canker of stone fruits; citrus blast; lilac blight; on many unrelated plants, including apple, plum, peach, cherry, pear, almond, avocado, citrus fruits, lilacs, flowering stock, rose, beans, cowpeas, oleander, and leaf spot on peas.

On stone fruits all plant parts are subject to attack, but most destructive are elongated water-soaked lesions or gummy cankers on trunks and branches, usually sour-smelling. Dormant buds of cherry and apricot are likely to be blighted, pear blossoms blasted. Small purple spots appear on leaves of plum and apricot, black lesions on fruit of cherry and apricot. All varieties of apricot are very susceptible to the disease. Plums on Myrobalan rootstock are more resistant, and varieties California, Duarte, and President are tolerant.

On citrus, and particularly lemons, dark sunken spots, called black pit, are formed on fruit rind, but there is no decay. The blast form of the disease is most often on oranges and grapefruit—water-soaked areas in leaves that may drop or hang on, twigs blackened and shriveled. The disease is most serious in seasons with cold, driving rainstorms.

On lilac, brown water-soaked spots on leaves and internodes on young shoots blacken and rapidly enlarge. Young leaves are killed; older leaves have large portions of the blade affected. Infection starts in early spring in rainy weather. The bacteria are primarily in the parenchyma, spreading through intercellular spaces, blackening and killing cells, forming cavities. The vascular system may also be affected, followed by wilting of upper leaves.

Control. Prune out infected twigs and branches. In California spray fruits in fall with bordeaux mixture, at the time first leaves are dropping. Grow bushy, compact citrus trees less liable to wind injury; use windbreaks for orchards.

Pseudomonas syringae pv. **papulans.** Blister spot of apple; small, dark brown blisters on fruit and rough bark cankers on limbs start at lenticels. Bark may have rough scaly patches from a few inches to a yard long, bordered with a pimpled edge, and with outer bark sloughing off in spring.

Pseudomonas syringae pv. **phaseolicola.** Bean halo blight; halo spot on common, lima, and scarlet runner beans. The symptoms are those of other bean blights except that there are wide green or yellowish green halos around water-soaked leaf spots, such spots later turning brown and dry. Leaves wilt and turn brown; young pods wither and produce no seed; sometimes plants are dwarfed with top leaves crinkled and mottled. In hot weather, spots are often angular, reddish brown, and without halo. Stem streaks are reddish, with gray ooze; pod spots are red to brown with silver crusts; seeds are small, wrinkled, with cream-colored spots. All snap beans are susceptible;

many dry beans—Pinto, Great Northern, Red Mexican, Michelite—are rather resistant.

Control. Use seed from blight-free areas. Blight is rare in California, occasional in Idaho. Plan a 3-year rotation. Do not pick beans when foliage is wet.

Pseudomonas syringae pv. **pisi.** Bacterial blight of pea; general on field and garden peas, especially in the East and South, and causing a leaf spot of sweet peas. Dark green water-soaked dots on leaves enlarge and dry to russet brown; stems have dark green to brown streaks. Flowers are killed or young pods shriveled, with seed covered with bacterial slime. Bacteria enter through stomata or wounds, and if they reach the vascular system, either leaflets or whole plants wilt. Vines infected when young usually die. Alaska and Telephone varieties are particularly susceptible.

Control. Avoid wounding vines during cultivation. Sow peas in early spring in well-drained soil. Use disease-free seed and plan a 4-year rotation.

Pseudomonas syringae pv. **savastanoi.** Olive knot; bacterial knot of olive. Irregular, spongy, more or less hard, knotty galls on roots, trunk, branches, leaf, or fruit pedicels start as small swellings and increase to several inches with irregular fissures. Terminal shoots are dwarfed or killed; whole trees may die. Bacteria enter through wounds, often leaf scars or frost cracks. Variety Manzanilla is most susceptible of the olives commonly grown in California. Another form of this species causes similar galls on ash.

Control. Cut out galls carefully, disinfesting tools; paint larger cuts with bordeaux paste and spray trees with bordeaux mixture in early November, repeating in December and March if infection has been abundant. Do not plant infected nursery trees or bring equipment from an infected orchard into a healthy one.

Pseudomonas syringae pv. **syringa.** Brown spot, foliar; on wild rice (*Zizania*).

Pseudomonas syringae pv. **tabaci.** Tobacco wildfire; on tobacco, tomato, eggplant, soybean, cowpea, pokeberry, and ground-cherry, in all tobacco districts sporadically. Leaf spots have tan to brown dead centers with chlorotic halos. The disease appears first on lower leaves and spreads rapidly in wet weather. The bacteria persist a few months in crop refuse and on seed and enter through stomatal cavities. In buried soybean leaves the bacteria have lived less than 4 months; so fall plowing may be beneficial. Seed stored for 18 months produces plants free from wildfire.

Pseudomonas syringae pv. **tomato.** Bacterial speck of tomato; numerous, dark brown raised spots on fruit are very small, less than $1/16$ inch; they do not extend into flesh and are more disfiguring than harmful.

Pseudomonas syringae pv. **tonelliana.** Oleander bacterial gall; galls or tumors are formed on branches, herbaceous shoots, leaves, and flowers but not on underground parts. Small swellings develop on leaf veins, surrounded by yellow tissue, with bacterial ooze coming from veins in large quantity. Young shoots have longitudinal swellings with small secondary tubercles;

young leaves and seedpods may be distorted and curled. On older branches tumors are soft or spongy and roughened with projecting tubercles; they slowly turn dark. Prune out infected portions, sterilizing shears between cuts; propagate only from healthy plants.

Pseudomonas syringae pv. **tagetes. Bacterial leaf spot;** circular necrotic lesions on leaves and petioles. The lesions have dark purple margins. This disease occurs on marigold, sunflower, Jerusalem artichoke, and common ragweed. Apical chlorosis is also caused by this pathogen on sunflower and sunflower seed may be a source of inoculum.

Pseudomonas syringae pv. **zizaniae. Leaf spot and stem spot;** of wild rice.

Pseudomonas tabaci. Blackfire; of tobacco.

Pseudomonas viburni. Bacterial leaf spot of viburnum; widespread. Circular water-soaked spots appear on leaves, and irregular sunken brown cankers on young stems, and the bacteria overwinter in leaves, stems or buds. Remove and burn infected leaves. Spray with bordeaux mixture or an antibiotic such as Agrimycin two or three times at weekly intervals.

Pseudomonas viridilivida. Louisiana lettuce disease; on lettuce, bell pepper, and tomatoes. Numerous water-soaked leaf spots fuse to infect large areas, first with a soft rot, then a dry shriveling. Sometimes outer leaves are rotted and the heart sound. This bacterium also causes greasy canker of poinsettia.

Pseudomonas washingtoniae. This bacterium causes spots on leaves of Washington palm.

Pseudomonas woodsii. Bacterial spot of carnation; leaf lesions are small, elongated, brown with water-soaked borders, withering to brown sunken areas, with masses of bacteria oozing out of stomata. They are spread in greenhouses by syringing, and outdoors by rain. Follow cultural practices suggested under *P. caryophylli* for carnation wilt.

Pseudomonas sp. **Blueberry canker;** reported from Oregon. Reddish brown to black cankers appear on canes of the previous season; all buds in the cankered areas are killed; stems are sometimes girdled. Varieties Weymouth, June, and Rancocas are resistant, but Jersey, Atlantic, Scammel, Coville, and Evelyn are highly susceptible.

Xanthomonas

Pseudomonadaceae

Small rods, motile with a single polar flagellum; form abundant slimy yellow growth. Most species are plant pathogens causing necroses. Many bacteria plant pathogens causing necroses are pathovars of *Xanthomonas campestris*. The only reliable test for differentiating pathovars is the host test. For example, *Xanthomonas barbareae* may now be called *X. campestris* pv. *barbareae*.

Xanthomonas barbareae. Black rot of winter-cress (*Barbarea vulgaris*); similar to black rot of cabbage; small greenish spots turn black.

Xanthomonas begoniae. Begonia bacteriosis; leaf spot of fibrous and tuberous begonias. Blisterlike, roundish dead spots are scattered over surface of leaves. Spots are brown with yellow translucent margins. Leaves fall prematurely, and in severe cases the main stem is invaded, with gradual softening of all tissues and death of plants. Bacteria remain viable for at least 3 months in yellow ooze on surface of dried leaves. Leaves are infected through upper surfaces during watering, with rapid spread of disease when plants are crowded together under conditions of high humidity.

Control. Keep top of leaves dry, avoiding syringing or overhead watering; keep pots widely spaced; spray with bordeaux mixture and dip cuttings in it.

Xanthomonas campestris. Black rot of crucifers; bacterial blight, wilt, stump rot of cabbage, cauliflower, broccoli, brussels sprouts, kale, mustard, radish, rutabaga, sunflower, stock, and turnip. Black rot was first observed in Kentucky and Wisconsin about 1890 and is generally distributed in the country, with losses often 40% to 50% of the total crop. It is one of the most serious crucifer diseases, present each season but epidemic in warm, wet seasons.

The bacteria invade leaves through water pores or wounds and progress to the vascular system. Veins are blackened, with leaf tissue browning in a V-shape. With early infection plants either die or are dwarfed, with a one-sided growth. Late infection results in defoliation, long bare stalks with a tuft of leaves on top. When stems are cut across, they show a black ring, result of the vascular invasion, and sometimes yellow bacterial ooze. Black rot is a hard odorless rot, but it may be followed by soft, odorous decays. Primary infection comes from bacteria carried on seed, or in refuse in soil, but drainage water, rain, farm implements, and animals aid in secondary infection.

Control. Use seed grown in disease-free areas in the West or treat with hot water, 122° F, 25 minutes for cabbage, 18 minutes for broccoli, cauliflower, and collards. Plan a 3-year rotation with plants other than crucifers, and clean up all crop refuse.

Xanthomonas campestris. Horse-radish leaf spot; leaves are spotted but there is no vascular infection. Also causes leaf spot of *Pilea* sp., *Pellionia* sp. and leaf spot and blight of bird of paradise, white butterfly.

Xanthomonas campestris pv. cyamopsidis. Rot; of *Lithops* spp.

Xanthomonas campestris pv. dieffenbachiae. Blight; of *Anthurium*; also leaf spot of cocoyam.

Xanthomonas campestris pv. malvacearum. Leaf spot; on *Hibiscus.*

Xanthomonas campestris pv. zinnae. Leaf and flower spot of zinnia.

Xanthomonas campestris pv. citri. Citrus Canker; on all citrus fruits, but not apparently eradicated from the United States. It came from the Orient and appeared in Texas in 1910, becoming of major importance in Florida and the Gulf states by 1914, ranking with chestnut blight and white pine blister rust as a national calamity. But here is one of the few cases on record where humans have won the fight, where a disease has been nearly eradicated by spending enough money and having enough cooperation early in the game. Several million dollars, together with concerted intelligent effort by growers, quarantine measures, destruction of every infected tree, sanitary precautions so rigid they included walking the mules through disinfestant, and sterilization of clothes worn by workers, all saved us from untold later losses.

Symptoms of citrus canker are rough, brown corky eruptions on both sides of leaves and fruit. On foliage the lesions are surrounded by oily or yellow halos. Old lesions become brown and corky.

Xanthomonas carotae. Bacterial blight of carrot; the chief damage is to flower heads grown for seed, which may be entirely killed. Symptoms include irregular dead spots on leaves, dark brown lines on petioles and stems, blighting of floral parts, which may be one-sided. Use clean seed, or treat with hot water; rotate crops.

Xanthomonas corylina. Filbert blight; bacteriosis; the most serious disease of filberts in the Pacific Northwest, known since 1913 from the Cascade Mountains west in Oregon and Washington. The disease is similar to walnut blight (see *X. juglandis*) with infection on buds, leaves, and stems of current growth; on branches; and on trunks 1 to 4 years old. The bacteria are weakly pathogenic to the nuts. Copper-lime dusts are effective, with four to six weekly applications, starting at the early prebloom stage.

Xanthomonas cucurbitae. Bacterial spot; on winter squash and pumpkin. Leaf spots are first small and round, then angular between veins, with bright yellow halos; sometimes translucent and thin but not dropping out; often coalescing to involve whole leaf. Bacterial exudate is present.

Xanthomonas dieffenbachiae. Dieffenbachia leaf spot; spots are formed on all parts of leaf blade except midrib, but not on petioles and stems. They range from minute, translucent specks to lesions ⅜ inch in diameter, circular to elongated, yellow to orange-yellow with a dull green center. Spots may grow together to cover large areas, which turn yellow, wilt, and dry. Dead leaves are dull tan to light brown, thin and tough but not brittle. The exudate on lower surface of spots dries to a waxy, silver-white layer.

Control. Separate infected from healthy plants; keep temperature low; avoid syringing; try protective spraying with streptomycin.

Xanthomonas fragariae. Angular leaf spot; on strawberry.

Xanthomonas geranii (= *X. pelargonii?*). Geranium leaf spot; on *Pelargonium* spp. Leaf spots are small, brown, necrotic, sometimes with reddish tinge on upper surface and a slightly water-soaked condition on underside. Young leaves may die and drop. Petioles are occasionally spotted. Bacteria winter in old leaves or under mulch.

Xanthomonas glycines (*phaseoli* var. *sojense*). Bacterial pustule of soybean; similar to regular bean blight but chiefly a foliage disease, present in most soybean areas, more severe in the South. Small, yellow-green spots with reddish brown centers appear on upper surface of leaves with a small raised pustule at the center of the spot on the under leaf surface. Spots run together to large irregular brown areas, portions of which drop out, giving a ragged appearance. Bacteria overwinter in diseased leaves and on seed. Variety CNS is highly resistant; Ogden has some resistance.

Xanthomonas gummisudans. Bacterial blight of gladiolus; narrow, horizontal, water-soaked, dark green spots turn into brown squares or rectangles between veins, covering entire leaf, particularly a young leaf, or middle section of the blade. Bacteria ooze out in slender, twisted, white columns or in a gummy film, in which soil and insects get stuck. Disease is spread by planting infected corms or by bacteria splashed in rain from infected to healthy leaves. The small dark brown corm lesions are almost unnoticeable. Soak corms unhusked for 2 hours before planting.

Xanthomonas hederae. Bacterial leaf spot of English ivy; small water-soaked area on leaves develop dark brown to black centers as they increase in size, sometimes cracking, with reddish purple margins. Spots are sometimes formed on petioles and stems, with plants dwarfed and foliage yellow-green. Spray with bordeaux mixture or an antibiotic. Keep plants well spaced; avoid overhead watering and high humidity.

Xanthomonas hyacinthi. Hyacinth yellows; yellow rot of Dutch hyacinth, occasionally entering the country in imported bulbs. The disease was first noted in Holland in 1881 and named for the yellow slime or bacterial ooze seen when a bulb is cut. The bulbs rot either before or after planting, producing no plants above ground or badly infected specimens, which do not flower and have yellow to brown stripes on leaves or flower stalks. Bacteria are transmitted by wind, rain, tools, and clothes, with rapid infection in wet or humid weather, particularly among luxuriantly growing plants. The disease is usually minor in our Pacific Northwest but worse in warm, wet weather on rapidly growing plants. Innocence is more susceptible than King of the Blues.

Control. Cover infected plants with a jar or can until the end of the season; then dig after the others. Never work or walk in fields when plants are wet; avoid bruising; discard rotten bulbs; rotate plantings; avoid fertilizer high in nitrogen.

Xanthomonas incanae. Bacterial blight of garden stocks; causing, since 1933, serious losses on flower-seed ranches in California; also present in home gardens. This disease is a vascular disease of main stem and lateral branches, often extending into leaf petioles and seed peduncles. Seedlings suddenly wilt when 2 to 4 inches high, with stem tissues yellowish, soft and mushy, and sometimes a yellow exudate along stem. On older plants, dark water-soaked areas appear around leaf scars near ground, stem is girdled, and lower leaves turn yellow and drop; or entire plants wilt or are broken by wind at ground level. Bacteria persist in soil and on or in seed; they are also spread in irrigation water.

Control. Use a 2- to 3-year rotation. Treat seed with hot water, 127.5° F to 131° F for 10 minutes, followed by rapid cooling.

Xanthomonas juglandis. Walnut blight; on English or Persian walnut, black walnut, butternut, Siebold walnut. Black, dead spots appear on young nuts, green shoots, and leaves. Many nuts fall prematurely, but others reach full size with husk, shell, and kernel more or less blackened and destroyed. Bacteria winter in old nuts or in buds, and may be carried by the walnut erinose mite.

Control. Spray with a fixed copper, as copper oxalate, or with streptomycin. Apply when 10% of the blossoms are open, repeat when 20% are open, and again after bloom.

Xanthomonas nigromaculans f.sp. **zinniae. Leaf spot of zinnia.**

Xanthomonas oryzae. Carnation pimple; reported from Colorado as caused by a new form of the rice blight organism. Very small, 1 millimeter, pimples are formed near base and tips of leaves, which may shrivel.

Xanthomonas papavericola. Bacterial blight of poppy; on corn poppy and on Oriental, opium, and California poppies. Minute, water-soaked areas darken to intense black spots bounded by a colorless ring. Spots are scattered, circular, small, often zonate, with tissue between yellow and then brown. There is a noticeable, slimy exudate. Infection is through stomata and often into veins. Stem lesions are long, very black, sometimes girdling and causing young plants to fall over. Flower sepals are blackened, petals stop developing; pods show conspicuous black spots.

Control. Remove and destroy infected plants; do not replant poppies in the same location. Try Agrimycin as a preventive spray.

Xanthomonas pelargonii. (*X. campestris* pv. *pelargonii*). **Bacterial leaf spot of geranium** (*Pelargonium*); irregular to circular brown leaf spots start as water-soaked dots on undersurface, becoming sunken as they enlarge and with tissue collapsing. If spots are numerous, the entire leaf turns yellow, brown, and shriveled, then drops. The leaves sometimes wilt and droop but hang on the plant for a week or so. Exterior of stem is gray and dull, the pith and cortex black, later disintegrating into a dry rot. The roots are blackened but not decayed. Cuttings fail to root, and rot from the base upward. Bacteria can live 3 months in moist soil and are spread by handling, splashing water, cutting knives, and whiteflies.

Control. Remove diseased plants. Take cuttings from plants known to be healthy; place in sterilized media and pots. Commercial growers should purchase culture-indexed cuttings. Be sure to sterilize cutting knives. Use 1-year rotation. Try Agrimycin as a preventive spray, or copper.

Xanthomonas phaseoli. Bacterial bean blight; general and serious on beans but rare in some western states. Leaf spots are at first very small, water-soaked or light green wilted areas, which enlarge, turn brown, are dry and brittle, and have a yellow border around edge of lesions and often a narrow, pale green zone outside the border. Leaves become ragged in wind and rainstorms. Reddish brown horizontal streaks appear in stem, which may be girdled and break over at cotyledons or first leaf node.

Pod lesions are first dark green and water-soaked, then dry, sunken and brick red, sometimes with a yellowish encrustation of bacterial ooze. White seeds turn yellow, are wrinkled with a varnished look.

Control. Use disease-free western-grown seed. Keep away from beans when plants are wet.

Xanthomonas pruni. Bacterial spot of stone fruits; also called canker, shot hole, black spot; general on plum, Japanese plum, prune, peach, and nectarine east of the Rocky Mountains; one of the more destructive stone fruit diseases, causing heavy losses in some states.

Symptoms on leaves are numerous, round or angular, small reddish spots with centers turning brown and dead, dropping out to leave shot holes. Spots may run together to give a burned, blighted, or ragged appearance, followed by defoliation, with losses running high in devitalized trees. On twigs dark blisters dry out to sunken cankers. Fruit spots turn into brown to black, saucer-shaped depressions with small masses of gummy, yellow exudate, often with cracking through the spot.

Control. Plant new orchards from nurseries free from the disease. Prune to allow air in the interior of trees. Feed properly; trees with sufficient nitrogen do not defoliate so readily. Zinc sulfate-lime sprays have been somewhat effective.

Xanthomonas vesicatoria. Bacterial spot; of tomato and pepper, common in wet seasons. Small, black, scabby fruit spots, sometimes with a translucent border, provide entrance points for secondary decay organisms. Small, dark greasy spots appear on leaflets and elongated black spots on stems and petioles. Bacteria are carried on seed.

Control. Rotate crops; destroy diseased vines. Spraying or dusting with copper may reduce infection. These may be combined with streptomycin.

Xanthomonas vesicatoria var. raphani. Leaf spot; of radish, turnip, and other crucifers, similar to bacterial spot on tomato.

Xanthomonas vignicola. Cowpea canker; on cowpeas and red kidney beans, a destructive disease, first described in 1944. Beans are blighted; cowpea stems have swollen, cankerlike lesions, with the cortex cracked open and a white bacterial exudate. The plants tend to break over. Leaves, stems, pods, and seeds are liable to infection. Chinese Red cowpeas seem particularly susceptible, but the disease appears on other varieties.

Xanthomonas vitians. Bacterial wilt of lettuce; South Carolina lettuce disease; wilting and rotting of lettuce leaves and stems. In early stages plants are lighter green than normal. Leaves may have definite brown spots coalescing to large areas or may wilt following stem infection. Use windbreaks to prevent injuries affording entrance to bacteria.

Mycoplasma-like Organisms (MLOs)

Aster Yellows. Throughout the United States, also called lettuce Rio Grande disease, lettuce white heart, potato purple top.

Bean Phyllody. Perhaps caused by a strain of aster-yellows MLO.

California Aster Yellows. In the West, also known as celery yellows, western aster yellows, potato late break, strawberry green petal. Aster yellows may appear in more than 170 species of 38 families of dicotyledons. The disease is serious on China aster, may also affect anemone, calendula, coreopsis, cosmos, delphinium, daisies, golden-glow, hydrangea, marigold, petunia, phlox, scabiosa, strawflower, and other flowers. It is serious on lettuce, endive, carrot, parsley, New Zealand spinach, and some other vegetables, but not on peas, beans or other legumes. This disease is now known to be caused by a mycoplasma-like organism.

In most plants vein clearing is followed by chlorosis of newly formed tissues, adventitious growth, erect habit, virescence of flowers. Asters have a stiff yellow growth with many secondary shoots; are stunted, with short internodes; flowers are greenish, dwarfed, or none. The chief vector is the six-spotted leafhopper (*Macrosteles fascifrons*). The virus multiplies in the insect, and there is a delay of 10 days or more after the insect feeds on a diseased plant before it can infect a healthy specimen. There is no transmission through insect eggs or aster seeds.

Celery petioles are upright, somewhat elongated, with inner petioles short, chlorotic, twisted, brittle, often cracked, yellow. The celery strain of the virus causes yellowing and stunting of cucumber, squash, pumpkin; infects gladiolus and zinnia.

Control of aster yellows is directed against the leafhoppers. Asters are grown commercially under frames of cheesecloth, 22 threads to the inch, or wire screening, 18 threads to the inch. In home gardens all diseased plants should be rogued immediately and overwintering weeds, which harbor leafhopper eggs, destroyed. Spraying or dusting ornamentals and vegetables with pyrethrum will reduce the number of vectors but will not entirely eliminate the disease.

Recent work raises the probability that the etiological agent of aster yellows is a mycoplasma rather than a virus. Therefore, treatment with antibiotics, such as chlortetracycline, has suppressed the development of yellows symptoms. Mycoplasma-like bodies have been seen in microscopic study of diseased plants and in transmitting leafhopper vectors, but not in healthy plants or nontransmitting vectors.

Corn Stunt. A dwarfing disease present primarily in the South; transmitted by leafhoppers. Mycoplasma-like bodies present; see Aster Yellows.

Elm Phloem Necrosis. On American elm from West Virginia and Georgia to northern Mississippi, eastern Oklahoma, Kansas, and Nebraska. Origin unknown but apparently present since 1882; the disease reached epidemic proportions in Ohio in 1944, killing 20,000 trees that year near Dayton and 10,000 at Columbus. The most reliable diagnostic character is a buttercup yellow discoloration of the phloem, often flecked with brown or black and an odor of wintergreen. Destruction of phloem causes the bark to loosen and fall away. Roots die first, then the phloem in lower portions of tree, followed by wilting and defoliation. American elms may be attacked at any age; they wilt and die suddenly within 3 or 4 weeks or gradually decline for 12 to 18 months. This disease is now thought to be caused by a mycoplasma-like agent.

Transmission is by the white-banded elm leafhopper (*Scaphoideus luteolus*) and possibly other species. Nymphs hatch about May 1 from eggs wintered on elm bark and feed on leaf veins. Adults move from diseased to healthy trees.

There is hope of propagating elms resistant to phloem necrosis. Communities should interplant existing elms with Asiatic or European varieties or with some other type of tree to provide shade if and when present elms die.

Peach Western X-Disease. Perhaps same as X-disease but usually treated separately; also known as cherry buckskin and western-X little cherry. The pathogen is transmitted by leafhoppers (*Colladonus germinatus, Fieberella florii, Osbornellus borealis,* and others) to peach, nectarine, and cherry in western states. Symptoms vary according to rootstock, but cherry fruit is smaller than normal. Sour cherries are puttylike, pinkish; sweet cherries are small, conical, hang on trees late, fail to develop normal color. Symptoms on peach are similar to those of X-disease.

Peach X-Disease. On peach and chokecherry, sometimes cherry in the northern United States and of major importance in Connecticut, Massachusetts, and New York. Peach trees appear normal in spring for 6 or 7 weeks after growth starts; then foliage shows a diffused yellow and red discoloration with a longitudinal upward curling of leaf edges; spots may drop out, leaving a tattered effect. Defoliation starts by midsummer. Fruits shrivel and drop or ripen prematurely. Seed do not develop. Weakened trees are killed by low temperatures or remain unproductive.

Chokecherry has conspicuous premature reddening of foliage, dead embryos in fruit. The second and third seasons after infection foliage colors are duller, there are rosettes of small leaves on terminals, and death may follow. Natural infection is apparently from chokecherry to peach (not peach to peach or peach to chokecherry) by a leafhopper (*Colladonus clitellarius*). Elimination of chokecherries within 500 feet of peach trees provides the best control.

Peach Yellow Leaf Roll. Form of western X-disease; perhaps caused by a more severe strain of the MLO.

Peach Yellows. Little Peach. First noted near Philadelphia in 1791 and so serious that in 1796 the American Philosophical Society offered a $60 prize for the best method of preventing premature decay of peach trees. Present in eastern states on peach, almond, nectarine, apricot, and plum. Not found west of the Mississippi or in the South. In peach, clearing of veins, production of thin erect shoots with small chlorotic leaves, premature ripening of fruit (with reddish streaks in flesh and insipid taste) is followed by death of the tree in a year or so. The little peach strain of the MLO causes distortion of young leaves at tips of branches, small fruit, delayed ripening. Plum is systemically infected, with few obvious symptoms. Transmission is by the plum leafhopper or budding.

Control. Budsticks and dormant nursery trees can be safely treated with heat sufficient to kill the MLO (122° F for 5 to 10 minutes), but cured trees are susceptible to reinfection. Most effective control is removal of wild plum trees around peach orchard and spraying to control leafhoppers.

Pierce's Grape Disease. First described as California vine disease by Pierce in 1892, now known as cause of grape degeneration in Gulf states; reported from Rhode Island. First symptoms are scalding and browning of leaf tissues, often with veins remaining green; canes die back from tips in late summer;

growth is dwarfed, fruit shriveled; roots die. The bacterium invades the xylem and turns it brown. Alfalfa plants are stunted with short stems and small leaves. Many species of sharpshooter leafhoppers transmit the bacterium to grape from alfalfa, clovers, grasses, also from ivy, acacia, fuchsia, rosemary, zinnia, and other ornamentals that are symptomless carriers. There is no adequate control; roguing of diseased vines and spraying for leafhoppers has proved ineffective. Propagate by cuttings from disease-free vineyards.

Potato Apical Leaf Roll and **Arizona Purple Top Wilt.** Caused by aster yellows.

Strawberry Green Petal. Perhaps due to a strain of aster yellows MLO, as is chlorotic phyllody reported from Louisiana. Flowers have enlarged sepals, small green petals.

Bud Proliferation and **Delayed Maturity.** On soybean.

Decline. Of ash.

Lethal Yellowing. On palms.

Phloem Necrosis. Of chrysanthemum.

Rickettsialike bacteria. Bacterial wilt; on Toronto creeping bentgrass; bacteria found in xylem of roots, crown, and leaves. Initially, leaf blades wilt from tip down and within several days entire leaf wilts, becomes dark green, shriveled, and twisted; also leaf scorch of mulberry.

Rickettsialike bacteria. Pierce's disease of grape; a serologically similar bacterium infects sycamore.

Spiroplasma citri. Corn stunt; has been reported on corn, onions, horseradish, shepherd's purse, yellow rocket, and wild mustard.

Stunt. Of blueberry.

Virescence. On horseradish.

Witches Broom. On pigeon pea (*Cajanus cajan*), and black raspberry.

Witches Broom. On Japanese persimmon, and lilac.

Witches Broom and **Yellowing.** On annual statice.

Yellows. Of elm.

Xylella fastidiosa. Bacterial leaf scorch; on northern red oak; this is a xylem-limited bacterium.

BLACK KNOT

The term *black knot* is used to designate a disease with black knotty excrescences.

Dibotryon

Ascomycetes, Dothideales, Dothideaceae

Asci are in locules, without well-marked perithecial walls, immersed in a massive, carbonaceous stroma, erumpent and superficial at maturity. Spores are hyaline, unequally two-celled.

Dibotryon morbosum. Black knot of plum and cherry; prunus black knot; plum wart; widespread and serious on garden plums, also present on sweet and sour cherries, chokecherry, and apricot. Apparently a native disease, destructive in Massachusetts by 1811 and the pathogen described from Pennsylvania in 1821, black knot has been reported on peach, long thought to be immune.

The chief symptoms are black, rough, cylindrical or spindle-shaped enlargements of twigs into knots two to four times their thickness and several inches long (see Fig. 15). Infection takes place in spring, but swelling is not evident until growth starts the following spring, at which time the bark ruptures, and

Figure 15. Black knot on *Prunus* sp.

a light yellowish growth fills the crevices. In late spring this growth is covered with an olive green, velvety layer made up of brownish conidiophores and one-celled hyaline conidia of the imperfect *Hormodendron* stage. Conidia are spread by wind.

In late summer black stromata cover the affected tissues and the galls become hard. Asci are formed during the winter in cavities in the stroma; ascospores are discharged and germinate in early spring, completing the 2-year cycle. Knots are produced from primary infection by ascospores or from secondary infection from mycelium formed in old knots and growing out to invade new tissue. Limbs may be girdled and killed; trees are stunted and dwarfed, nearly worthless after a few years. Old knots may be riddled with insects or covered with a pink fungus growing on the *Dibotryon* mycelium.

Control. Cut out infected twigs and branches, 3 or 4 inches beyond the knot, to include advancing perennial mycelium. Do this in winter or before April 1. Eradicate or thoroughly clean up wild plums and cherries in the vicinity. Spray at delayed dormant stage in spring (just as buds break) with bordeaux mixture or with liquid lime sulfur. The latter is preferable unless oil is combined in the spray as an insecticide. Spray with lime sulfur at full bloom.

Gibberidea

Ascomycetes, Sphaeriales, Sphaeriaceae

Perithecia in clusters on wood; spores dark, with several cells.

Gibberidea heliopsidis. Black knot, black patch; on goldenrod and sunflower.

BLACKLEG

The term *blackleg* is used to describe darkening at the base of a stem or plant. Blackleg of potatoes and delphinium are described under Bacterial Diseases; blackleg of geraniums is under Rots.

Phoma

Fungi Imperfecti, Sphaeropsidales, Sphaerioidaceae

Pycnidia dark, ostiolate, lenticular to globose, immersed in host tissue, erumpent or with short beak piercing the epidermis; conidiophores short or obsolete, conidia small, one-celled, hyaline, ovate to elongate; parasitic on seed plants, chiefly on stems and fruits, rarely on leaves.

Phoma lingam. Blackleg of crucifers; foot rot; phoma wilt of plants of the mustard family, including cabbage, cauliflower, Chinese cabbage, brussels sprouts, charlock, garden cress, pepper grass, kale, kohlrabi, mustard, rape, radish, rutabaga, turnip, stock, and sweet alyssum. The perfect stage, *Lystosphaeria maculans*, has been found on cabbage. The fungus was first noticed in Germany in 1791; the disease was reported in France in 1849, and in the United States in 1910. It is generally distributed east of the Rocky Mountains and formerly caused from 50% to 90% loss. With improved seed and seed treatment it has become less important.

The first symptom is a sunken area in the stem near the ground, which extends until the stem is girdled and the area turns black. Leaves, seed stalks, and seed pods have circular, light brown spots. Small black pycnidia appearing on the lesions distinguish blackleg from other cabbage diseases. The leaves sometimes turn purple and wilt, but there is no defoliation, as in black rot.

The fungus reaches the soil via infected plant debris, remaining alive 2 or more years. Spores are spread by splashing rain, or manure, on tools, and perhaps by insects, with new lesions resulting in 10 to 14 days. But the chief spread is by mycelium wintering in infected seed. When such seed is planted, fruiting bodies are formed on cotyledons as they are pushed above ground, and these serve as a source of inoculum for nearby plants. A few diseased seed can start an epiphytotic in wet weather.

Control. Use seed grown on the Pacific Coast, which is usually, although not always, disease-free. If the seed is infected, tie loosely in cheesecloth bags and immerse in hot water, held at 122°F for 30 minutes. It is sometimes possible to buy seed already treated. Sterilize soil for the seedbed; use a 3-year rotation; do not splash seedlings when watering; do not transplant any seedlings if the disease shows up in the seedbed; do not feed cabbage refuse to cattle; do not transfer cultivators and other tools from a diseased to a healthy field without using a disinfestant.

BLACK MILDEW

The terms *black mildew, sooty mold,* and *black spot* have been used to some extent interchangeably. In this text sooty mold is restricted to those fungi living on insect exudate and hence not true parasites. Included in this section are parasitic fungi that have a superficial dark mycelium. They are members of the Erysiphales (Meliolales according to some classifications) and hence similar to powdery mildews except for the dark color, or they belong to the Hemisphaeriales, characterized by a dark stroma simulating the upper portion of a perithecium. In a few cases the diseases are called black spot rather than mildew.

Apiosporina

Ascomycetes, Erysiphales (or Meliolales), Meliolaceae

Perithecia and mycelium superficial; mycelium with setae and perithecia usually hairy; paraphysoids present; spores two-celled; dark.

Apiosporina collinsii. Witches broom of serviceberry (*Amelanchier*); widespread. Perennial mycelium stimulates the development of numerous stout branches into a broom. A sooty growth on underside of leaves is first olive brown, then black. Numerous globose, beadlike, black perithecia appear in late summer. The damage to the host is not serious.

Asterina

Ascomycetes, Hemisphaeriales, Microthyriaceae

Asterina species are parasites on the surface of leaves and are usually found in warm climates. In some cases the disease is called black mildew; in others, black spot. The perithecia are dimidiate, having the top half covered with a shield, a small, round stroma composed of radially arranged dark hyphae. Underneath this stromatic cover, called scutellum, there is a single layer of fruiting cells; paraphyses are lacking; spores are dark, two-celled. The mycelium, which is free over the surface, has lobed appendages, hyphopodia, acting as haustoria in penetrating the cuticle and obtaining nourishment from the host.

Asterina anomala. Black mildew; on California-laurel, California.
Asterina delitescens. Black spot; on redbay.
Asterina diplopoides. Black spot; on leucothoë.
Asterina gaultheriae. Black mildew; on bearberry, Wisconsin.
Asterina lepidigena. Black mildew; on lyonia, Florida.
Asterina orbicularis. Black spot; on American holly and *Ilex* spp.

Asterinella

Ascomycetes, Hemisphaeriales, Microthyriaceae

Like *Asterina* but lacking hyphopodia; with or without paraphyses; spores dark, two-celled.

Asterinella puiggarii. Black spot; on eugenia.

Dimerium

Ascomycetes, Erysiphales (or Meliolales), Meliolaceae

Perithecia smooth; spores two-celled, dark; paraphyses lacking.

Dimerium juniperi. Black mildew; on Rocky Mountain juniper, California.

Dimerosporium

According to some authorities this fungus is the same as *Asterina* but the name *Dimerosporium* is in common use.

Dimerosporium abietis. Black mildew; on Pacific silver and lowland white firs. Black patches are formed on older needles, usually on under surface. There is no apparent injury to trees.

Dimerosporium hispidulum. Black mildew; on boxelder.

Dimerosporium pulchrum. Black mildew; on ash.

Dimerosporium robiniae. Black mildew; on ailanthus.

Dimerosporium tropicale. Black mildew; on bignonia, Mississippi.

Irene (Asteridiella)

Ascomycetes, Erysiphales (or Meliolales), Meliolaceae

Mycelium with capitate hyphopodia but no bristles; perithecia with larviform appendages; spores dark, with several cells.

Irene araliae. Black mildew; on magnolia, Mississippi.

Irene calastroma. Black mildew; on wax-myrtle, Gulf States.

Irene perseae. Black mildew; on avocado, Florida.

Irenina

Like *Irene* except that perithecia have no appendages.

Irenina manca. Black mildew; on wax-myrtle, Mississippi.

Irenopsis

Like *Irene* except that mycelium has setae (stiff bristles) and perithecia lack larviform appendages.

Irenopsis martiniana. Black mildew; on redbay, swampbay, Alabama, Florida, Mississippi.

Lembosia (Morenoella)

Ascomycetes, Hemisphaeriales, Microthyriaceae

Brown vegetative mycelium with hyphopodia on surface of host; linear stroma, scutellum, over single layer of fruiting cells; paraphyses present; spores dark, two-celled.

Lembosia cactorum. Black mildew; on cactus, Florida.

Lembosia coccolobae. Black mildew; on sea-grape, Florida; also *L. portoricensis* and *L. tenella.*

Lembosia illiciicola. Black mildew; on anise-trees, Alabama, Mississippi.

Lembosia rugispora. Black mildew; on redbay, swampbay, Mississippi, North Carolina.

Meliola

Ascomycetes, Erysiphales (or Meliolales), Meliolaceae

Most abundant in tropics. Superficial dark mycelium with hyphopodia and setae; perithecia globose, coal black without ostiole or appendages but often with setae; spores several-celled, dark; paraphyses lacking. Conidia are lacking in most species, of *Helminthosporium* type in others.

Meliola amphitricha. Black mildew; on boxelder, magnolia, redbay, swampbay.

Meliola bidentata. Black mildew; on bignonia.

Meliola camelliae. Black mildew of camellia; abundant black growth may cover camellia leaves and twigs. Spraying with a light summer oil is sometimes effective.

Meliola cookeana. Black mildew; on callicarpa, lantana.

Meliola cryptocarpa. Black mildew; on gordonia.

Meliola lippiae. Black mildew; on lippia.

Meliola magnoliae. Black mildew; on magnolia.

Meliola nidulans. Black mildew; on blueberry, wintergreen.

Meliola palmicola. Black mildew; on palmetto.

Meliola tenuis. Black mildew; on bamboo.

Meliola wrightii. Black mildew; on chinaberry.

Morenoella

Considered by many authorities a synonym for *Lembosia*, but the name is still used.

Morenoella angustiformis; M. ilicis; M. orinoides. Black mildew; on holly (*Ilex* spp.), Mississippi.

BLACKSPOT

In common use the term *black spot* means rose black spot, with the two words currently written as one. This section is limited to the rose disease. Delphinium black spot will be found under Bacterial Diseases, elm black spot under Leaf Spots, and other black spots under Black Mildew.

Diplocarpon

Ascomycetes, Helotiales, Dermateaceae (Mollisiaceae)

Apothecia innate, formed in dead leaves, but at maturity rupturing overlying tissues; horny to leathery with a thick margin or outer wall (excipulum) of dark, thick-walled cells; spores two-celled, hyaline; paraphyses present. Imperfect state is a *Marssonina* with two-celled hyaline spores in an acervulus.

Diplocarpon rosae. Rose blackspot; general on rose but less serious in the semiarid Southwest; reported from all states except Arizona, Nevada, and Wyoming.

For nearly 100 years the fungus was known only by its imperfect stage, which has had about 25 different names. The first definite record is by Fries in Sweden in 1815, under the name *Erysiphe radiosum*, but the first valid description was by Libert in 1827 as *Asteroma rosae*. Later Fries called it *Actinonema rosae*, and that term was widely used until *Actinonema* species were transferred to *Marssonina*. The blackspot fungus was first reported in the United States in 1831, from Philadelphia, and in 1912 Wolf made the connection with the perfect stage, so that the correct name became *Diplocarpon rosae.*

Blackspot is probably the most widely distributed and best known rose disease. It is confined to roses, garden and greenhouse, and may affect practically all varieties, although not all are equally susceptible. There has been some progress made in breeding resistant varieties, but recent investigation disclosing many physiological races of the fungus explains why roses that are almost immune to blackspot in one location may succumb in another. *Rosa bracteata* is the only species thus far shown to be reasonably resistant to all the different races tested. Roses with the Pernetiana parentage, which has given us the lovely yellows, coppers, and blends, are especially prone to blackspot. Some roses, like Radiance, are tolerant of blackspot, usually holding their leaves, even though they cannot be considered resistant.

Symptoms are primarily more or less circular black spots, up to ½ inch in diameter, with radiating fimbriate or fringed margins (see Fig. 16). This fimbriate margin is a special diagnostic character, differentiating blackspot from other leaf spots and from discolorations due to cold or chemicals. The spots vary from one or two to a dozen or more on a leaf, usually on the upper surface. With close examination you can see small black dots or pimples in the center of the spots. These are the acervuli, bearing conidia, and they glisten when wet (see Fig. 17).

In susceptible varieties the appearance of black spots is soon followed by yellowing of a portion or all of leaflets and then by defoliation. The leaf fall is apparently correlated with increased production of ethylene gas in diseased tissue and perhaps by a difference in auxin gradient between leaf and stem. Some roses lose almost all their leaves, put out another set and lose those, and often are trying to leaf out for the third time by late summer. The process is so devitalizing that some bushes may die during the following winter. On

Figure 16. Rose blackspot. Note fimbriate margin to spot.

tolerant varieties leaf spots are present, though usually in smaller numbers, but there is much less yellowing and defoliation. Cane lesions are small indistinct black areas, slightly blistered, without fimbriate margins.

Infection occurs through either leaf surface, the fungus sending its germ tube directly through the cuticle by mechanical pressure. The hyphae form a network under the cuticle, joining together into several parallel filaments radiating from the point of infection. The hyphae are actually colorless, the black color of the spot coming from the death and disorganization of host cells. The mycelial growth is between cells, with haustoria (suckers) invading epidermal and palisade cells for nourishment.

Acervuli, summer fruiting bodies, formed just under the cuticle, bear two-celled hyaline conidia on short conidiophores on a thin, basal stroma. Splashed by rain or overhead watering, or spread by gardeners working among wet plants, the conidia germinate and enter a leaf if there is continued moisture for at least 6 hours. Rain, heavy dew, fog, and sprinklers used late in the day so foliage does not dry off before night provide the requisite moisture.

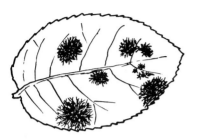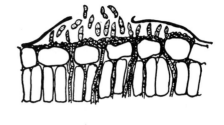

Figure 17. Rose blackspot. Two-celled conidia formed in Acervulus under cuticle.

New spots show up within a week and new spores within 10 days. Secondary cycles are repeated all summer—from late May to late October—around New York City.

In my personal experience, the spread of disease is most rapid where large numbers of susceptible varieties are massed together. If all the yellows, for instance, are planted together, the disease gets such a head start and builds up so much inoculum to spread to the more tolerant red and pink varieties nearby that these varieties also are more heavily infected than usual. When roses are mixed in beds so that one or two particularly susceptible bushes are surrounded by more resistant types, the infective material cannot increase as rapidly and the net result is less disease in the garden as a whole. Protected corners in the garden where air circulation is poor also increase the disease potentiality. Spores are apt to be splashed farther when water hits hard-packed soil without a mulch.

When old leaves drop to the ground, the mycelium continues a saprophytic existence, growing through dead tissue with hyphae that are now dark in color. In spring three types of fruiting bodies may be formed: microacervuli or spermagonia containing very small cells that perhaps act as male cells; apothecia, the sexual fruiting bodies formed on a stroma between the epidermis and palisade cells and covered with a circular shield of radiating strands; and winter acervuli, formed internally and producing new conidia in spring. The *Diplocarpon* or apothecial stage is apparently not essential; it is known only in the northeastern United States and south-central Canada. The shield over the apothecium ruptures, and the two-celled ascospores are forcibly discharged into the air to infect lowest leaves.

Where the sexual stage is not formed, primary spring infection comes from conidia splashed by rain to foliage overhead, from acervuli either in overwintered leaves on the ground or in cane lesions. New roses from a nursery sometimes bring blackspot via these cane lesions to a garden previously free of disease.

Control. The importance of sanitation may have been somewhat overstressed; it cannot replace routine spraying or dusting. It is certainly a good idea to pick off for burning the first spotted leaves, if this is done when bushes are dry so that the act of removal does not further spread the fungus. Raking up old leaves from the ground at the end of the season makes the garden neater and may reduce the amount of inoculum in spring, but, because the fungus winters also on canes in most sections of the country, removal of leaves cannot be expected to provide a disease-free garden the next season. Comparative tests have shown that fall cleanup is ineffectual. A good mulch, applied after uncovering and the first feeding in spring, serves as a mechanical barrier between inoculum from overwintered leaves on the ground and developing leaves overhead. A mulch also reduces disease by reducing the distance spores can be splashed from one bush to another during the season. Drastic spring pruning, far lower than normal, reduces the amount of inoculum from infected canes.

The importance of a dormant spray is debatable. Experiments have shown that as a true eradicant, applied in winter, it has little value in reducing the amount of blackspot the next summer. Use liquid lime sulfur after pruning, provided the buds have not broken far enough to show the leaflets.

Summer spraying or dusting, weekly throughout the season (from late April to early November in New Jersey) is essential if you want to keep enough foliage on bushes for continuous production of fine flowers (it takes food manufactured in several leaves to produce one bloom) and for winter survival. Some strong varieties will, however, live for years without chemical treatment; they are usually scraggly bushes with erratic bloom. The idea that floribunda varieties do not require as much spraying as hybrid teas is a misconception. Some floribundas are quite resistant; others are very susceptible. The same holds true for old-fashioned shrub roses. All too often blackspot gets a head start in a garden from shrub roses we thought it unnecessary to spray.

Roses can be defoliated as readily by chemicals as by the blackspot fungus, so the fungicide chosen must be safe under the conditions of applications as well as effective. There are many chemicals that will control blackspot if they are applied regularly and thoroughly. Choice depends somewhat on climate. Some copper sprays and dusts cause red spotting and defoliation in cool, cloudy weather. Bordeaux mixture is both unsightly and harmful, unless used in very weak dilution. At strengths recommended for vegetables it will quickly turn rose leaves yellow and make them drop off. Dusts containing more than 3% to 4% metallic copper are injurious under some weather conditions. Dusting sulfur fine enough to pass through a 325-mesh screen has been successfully used for years for blackspot control, but in hot weather it burns margins of leaves. Copper and sulfur have a synergistic effect; a mixture of the two is more effective than either used alone, but such a mixture also combines injurious effects.

There are literally hundreds of combination rose sprays and dusts on the market under brand names, and it seems to me easier, and even cheaper, considering the time saved, for home gardeners to make use of them to control blackspot and other rose diseases as well as insects in one operation. You will have to determine by trial and error the best combination for your area, and you may not find one that combines remedies for all the pests you may have to fight through the season. Choose one that contains ingredients required every week all summer, and then add other chemicals if and when necessary.

Whatever mixture is chosen, coverage should be complete on upper- and lower-leaf surfaces, and applications must be repeated at approximately weekly intervals, which may mean every 5 or 6 days when plants are growing rapidly in a rainy spring and perhaps every 7 to 9 days in dry weather when growth is slow. Intervals of 10 to 14 days between sprays seldom give adequate control. Most directions call for application ahead of rain so that the foliage will be protected when spores germinate during the rain; but if sprays are applied every 7 days, there will always be enough residue left on the foliage to give protection during the next rain. It is not necessary to make an additional application immediately after a rain if your spraying is on a regular basis.

BLIGHTS

According to *Webster's*, blight is "any disease or injury of plants resulting in withering, cessation of growth and death of parts, as in leaves, without rotting." The term is somewhat loosely used by pathologists and gardeners to cover a wide variety of diseases, some of which may have rotting as a secondary symptom. In general, the chief characteristic of a blight is sudden and conspicuous leaf and fruit damage, in contradistinction to leaf spotting, where dead areas are definitely delimited, or to wilt due to a toxin or other disturbance in the vascular system. Fire blight, discussed under Bacterial Diseases, is a typical blight, with twigs and branches dying back but holding withered, dead foliage.

Alternaria

Fungi Imperfecti, Moniliales, Dematiaceae

Dark, muriform conidia formed in chains, simple or branches, or sometimes singly, on dark, simple conidiophores growing from dark hyphae (See Fig. 18). The apical portion of each conidium is narrowed and often elongated, bearing at its tip the next ovoid, tapering conidium. Species with this characteristic formerly placed in *Macrosporium* are now in *Alternaria*; those with spores rounded at both ends have been transferred to *Stemphyllium*.

There are many saprophytic species in *Alternaria*, the spores of which are windborne for many miles and are a common cause of hayfever. There are also parasitic forms causing blights and leaf spots. Sometimes the disease starts as a leaf spot, but the lesions, typically formed in concentric circles, run together to form a blight, the dark conidia making the surface appear dark and velvety.

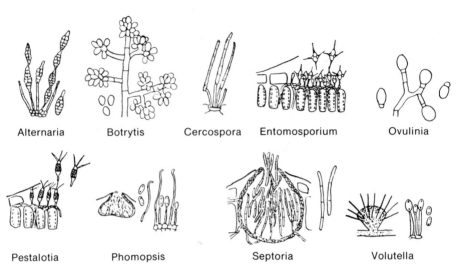

Alternaria   Botrytis   Cercospora   Entomosporium   Ovulinia

Pestalotia   Phomopsis   Septoria   Volutella

Figure 18. Conidial production among some fungi causing blights. *Alternaria,* dark muriform spores in chains; *Botrytis,* hyaline spores in clusters; *Cercospora,* pale to dark septate spores on dark conidia protruding from stomata; *Entomosporium,* peculiarly appendaged spores in acervulus; *Ovulinia,* hyaline spore with basal disjunctor cell, borne free of mycelium; *Pestalotia,* in acervulus, mediam cells colored, end cells hyaline, apical cell with appendages; *Phomopsis,* oval and filiform hyaline spores in pycnidium; *Septoria,* septate hyaline spores in Pycnidium; *Volutella,* hyaline spores formed on a hairy sporodochium.

Alternaria alternata. Blight, foliage and **pod;** of pea.

Alternaria cassiae. Seedling blight; of *Cassia* (sicklepod, and coffee senna), and showy crotalaria.

Alternaria cucumerina. Alternaria blight of cucurbits; cucumber blight, black mold, general on cucumbers, muskmelon, watermelon, and winter and summer squash. Symptoms appear in the middle of the season, first nearest the center of the hill. Circular brown spots with concentric rings are visible only on upper surface of leaves, but a black, moldy growth, made up of conidiophores and large brown spores, can be seen on both leaf surfaces. Leaves curl and dry up, cantaloupe foliage being more sensitive than that of other cucurbits. The disease spreads rapidly in warm, humid weather, and, with the vines drying, the fruit is exposed to sunburn. Sunken spots develop on the fruit, covered with an olive green mass of conidia. Other species of *Alternaria* cause a decay of melons in transit and storage.

Control. Purdue 44 and some other varieties of muskmelon are rather resistant.

Alternaria dauci. Alternaria blight of carrot; carrot leaf blight, general on carrot and parsley. Affected leaves and petioles are spotted, then turn yellow and brown; entire tops are killed in severe infections. In California the disease is known as late blight, with the peak coming in November. The fungus apparently winters in discarded tops and on seed.

Control. Clean up refuse. Spray with a fixed copper spray or dust, starting soon after seedlings emerge and repeating at 7- to 10-day intervals.

Alternaria dianthi. Carnation collar blight; leaf spot, stem and branch rot, general on carnation, widespread on garden pinks and sweet william. The chief symptom is a blight or rot at leaf bases and around nodes, which are girdled. Spots on leaves are ashy white but centers of old spots are covered with dark brown to black fungus growth. Leaves may be constricted and twisted, the tip killed. Branches die back to the girdled area, and black crusts of spores are formed on the cankers. Conidia are spread during watering in the greenhouse or in rains, outdoors. Entrance is through wounds, stomata, or directly through the cuticle. The spores are carried on cuttings.

Control. Commercial growers can often avoid Alternaria blight by keeping plants growing continuously in the greenhouse. Cuttings should be disease-free, taken from midway up the stem, broken at the joint rather than cut, and started in sterilized soil. Ordinarily the foliage should be kept dry, but under mist propagation chemicals introduced into the mist system have reduced blight.

Alternaria helianthi. Blight and stem lesion; of sunflower.

Alternaria panax. Alternaria blight; root rot, leaf spot of ginseng, ming aralia, and goldenseal, generally distributed. In Ohio the disease appears each year in semiepidemic form and has been controlled with bordeaux mixture or a fixed copper spray plus a wetting agent, starting when plants emerge in early May and repeating every 2 weeks until 3 weeks after bloom.

Alternaria solani. Early blight of potato and tomato; general on these hosts, occasional on eggplant and pepper. The pathogen was first described from New Jersey, in 1882.

Leaf symptoms are dark brown, circular to oval spots, marked with concentric rings in a target effect, appearing first on lower, shaded foliage, with the spots growing together to blight large portions or all of leaves, exposing fruits. There may be a collar rot of young tomato seedlings, sunken spots or cankers on older stems, blossom-drop with loss of young fruits, or dark leathery spots near the stem end of older fruits. Alternaria blight is the most common leaf spot disease of tomatoes in the central and Atlantic states but is somewhat less important elsewhere.

Foliage symptoms on potato are similar to those on tomato. Small round spots on tubers afford entrance to secondary rot organisms. Each leaf spot may produce three or four crops of dark spores, which remain viable more than a year. They are blown by wind, splashed by rain, sometimes transmitted by flea beetles. The fungus is a weak parasite, entering through wounds and thriving in warm, moist weather, with 85°F as optimum temperature. It can survive in soil as long as the host refuse is not completely rotted; it also winters on seed and on weed hosts.

Control. Plan, if possible, a 3-year rotation with crops not in the potato family; dig under diseased refuse immediately after harvest. Use seed from healthy tomatoes, or purchase plants free from collar rot.

Alternaria tagetica. Blight; of marigold.

Alternaria violae. Alternaria blight; leaf spot of violet, and pansy. Spots vary from greenish yellow to light buff with burnt amber margins. Brown patches run together to form large, blighted areas. Clean up and burn old leaves in fall.

Alternaria zinniae. Zinnia blight; Alternariosis, on zinnia. Small reddish brown spots with grayish white centers increase to irregular, large, brown, dry areas. Similar spots on stem internodes or at nodes may girdle the stem, with dying back of upper portions. Dark brown to black basal cankers with sunken lesions are common. Roots may turn dark gray, rot, and slough off. Small brown flower spots enlarge to include whole petals, causing conspicuous blighting. The fungus apparently winters on seed and in soil.

Control. Clean up refuse; use a long rotation if growing plants commercially.

Ascochyta

Fungi Imperfecti, Sphaeropsidales, Sphaerioidaceae

Pycnidia dark, globose, separate, immersed in host tissue, ostiolate; spores two-celled, hyaline ovoid to oblong.

Ascochyta asparagina. Stem blight of asparagus fern; small branchlets dry and drop prematurely; small branches are killed if attacked at crown.

Ascochyta chrysanthemi (*Mycosphaerella ligulicola*). Ascochyta ray blight of chrysanthemum; a conspicuous and rapid disease of ray flowers. If young buds are infected, the head does not open; if the attack is later, there may be one-sided development of flowers. A tan or brown discoloration proceeds from the base toward the tip of each individual flower, followed by withering. Upper portions of stems and receptacles may turn black. Keep plants well spaced; avoid overhead watering and excessive humidity.

Ascochyta piniperda. Spruce twig blight; on young shoots of red, Norway, and blue spruce; apparently a minor disease.

Ascochyta pisi, A. pinodes, A. pinodella. Ascochyta blight or mycosphaerella blight of peas; all three fungi may be connected with the disease complex known as Ascochyta blight, are carried in infected seed, and overwinter in plant debris. *A. pinodes* has *Mycosphaerella pinodes* as its ascospore stage so that the life cycle can start from either pycnidia or perithecia produced on plants or stubble.

Lesions begin as small purplish specks on leaves and pods. When infection is caused by *M. pinodes* or *A. pinodella*, the specks enlarge to round, targetlike spots, which join together to form irregular, brownish purple blotches. *M. pinodes* often withers and distorts young pods; *A. pinodella* causes a severe foot rot, a dark region at the soil line. Elongated, purplish black stem lesions are common. *A. pisi* causes leaf spots with dark brown margins, stem and pod spots, but no foot rot.

Control. Use western-grown seed, usually free from the disease; clean up all pea refuse and use a 3- or 4-year rotation.

The host range now includes many plants such as carrot, banana, and foliage plants.

Balansia

Ascomycetes, Hypocreales, Clavicipitaccae

Balansia cyperi. Diseased inflorescence, blight; of purple nutsedge; fungus is systemic and transmitted through tubers.

Beniowskia

Fungi Imperfecti, Moniliales, Moniliaceae

Hyphae are coiled at the periphery of mature sporodochia; spherical spores are borne on short denticles.

Beniowskia sphaeroidea. Blight; of knotroot bristlegrass.

Botryodiplodia

Fungi Imperfecti, Sphaeropsidales, Sphaeropsidaceae

Picnidia black, ostiolate, erumpent, stromatic, confluent; conidiophores simple, short; conidia dark and two-celled, ovoid to elongate.

Botryosphaeria

Ascomycetes, Sphaeriales, Sphaeriaceae

Asci in locules in a stroma; spores one-celled, hyaline, eight in an ascus. There is a good deal of variation in the genus. The locules may be scattered throughout the stromatic tissue, or seated on the surface, or like perithecia, as in *Botryosphaeria ribis*. In *B. ribis* there are two pycnidial forms: a *Dothiorella* stage containing very small spores that may function as male cells and a *Macrophoma* stage containing larger spores, one-celled, hyaline, functioning as other conidia.

Botryosphaeria ribis var. **chromogena.** Currant cane blight; canker, dieback, of currant, flowering currant, gooseberry, apple, rose, and many other plants (see also under Cankers). There are two forms of this species, one being a saprophyte developing on already dying tissue. The parasitic form *chromogena* is so named from its developing a purple-pink color when grown on starch paste. There are also a number of pathogenic strains, varying from high to low in virulence. Some currant varieties are quite resistant, but the widely grown Wilder and Red Lake are rather susceptible.

Dieback and death of fruiting branches occur as the berries are coloring, with leaves wilting and fruit shriveling. Later in the season small, dark, wartlike fruiting bodies appear in rather definite parallel rows on the diseased canes. Rose canes show a similar dying back and wilting above a canker. The fungus winters in the canes; ascospores infect new shoots; secondary infection is by spores oozing from pycnidia. The mycelium grows downward through bark and wood to the main stem, which it encircles and kills.

Control. Cut out and burn diseased canes as soon as noticed. Take cuttings from healthy bushes.

Botryotinia

Ascomycetes, Helotiales, Sclerotiniaceae

Stroma a typical black sclerotium, loaf-shaped or hemispherical, just on or beneath cuticle or epidermis of plant and firmly attached to it; apothecia cupulate, stalked, brown; ascospores hyaline, one-celled; conidiophores and conidia of the *Botrytis cinerea* type.

Botryotinia fuckeliana. The apothecial stage of *Botrytis cinerea*, the connection having been made with isolates from grape, apple, celery, and potato. The name of the conidial stage is still widely used for the pathogen causing gray mold blights.

Botryotinia ricini. Gray mold blight of castor bean; soft rot of caladium. A pale to olive gray mold develops on castor-bean inflorescence, and when fading flowers drop onto stem and leaves, they are infected in turn.

Botrytis

Fungi Imperfecti, Moniliales, Moniliaceae

Egglike conidia hyaline, one-celled, are formed on branched conidiophores over the surface, not in special fruiting bodies (see Fig. 18). The arrangement of the spores gives the genus its name, from the Greek *botrys*, meaning a cluster of grapes. Flattened, loaf-shaped, or hemispherical black sclerotia are formed on or just underneath cuticle or epidermis of the host and are firmly attached to it. These sclerotia, with a dark rind and light interior made up of firmly interwoven hyphae, serve as resting bodies to carry the fungus over winter. Microconidia, very minute spores that are spermatia or male cells, function in the formation of apothecia in the few cases where a definite connection has been made between the *Botrytis* stage and the ascospore form, *Botryotinia*.

Botrytis species are the common gray molds, only too familiar to every gardener. Some are saprophytic or weakly parasitic on senescent plant parts on a wide variety of hosts; others are true parasites and cause such important diseases as peony blight, lily blight, tulip fire.

Botrytis cinerea. Gray mold blight; bud and flower blight (see Fig. 19), blossom blight, gray mold rot, Botrytis blight, of general distribution on a great many flowers, fruits, and vegetables. There are undoubtedly many strains of this fungus and perhaps more than one species involved, but they have not been definitely separated.

This gray-mold disease is common on soft ripe fruits after picking, as any cook knows after throwing out half a box of strawberries or raspberries. But in continued humid weather the blight appears on fruits before harvest. Blackberries in the Northwest are subject to gray mold. The fungus winters in blighted blueberry twigs, and spores infect blossom clusters.

Vegetables are commonly afflicted as seedlings grown in greenhouses and in storage after harvest. If lettuce plants are set in the garden too close

Figure 19. Botrytis petal spot on Magnolia.

together, they may blight at the base in moist weather, as will endive and escarole. Gray mold is common on lima beans, and is sometimes found on snap and kidney beans. In rainy or foggy periods globe artichoke may be covered with a brownish gray, dusty mold, with bud scales rotten. Asparagus shoots are sometimes blighted, tomato stems rotted.

Some of the ornamentals on which *Botrytis cinerea* is troublesome are given in the following annotated list:

African violet—leaf and stem rot, cosmopolitan in greenhouses.
Amaryllis—gray mold, mostly in the South, on outdoor plants after chilling.
Anemone—occasional severe rotting of crowns.
Arborvitae—twig blight.
Aster—brown patches in flower heads of perennial aster; gray mold on flowers of China aster grown for seed in California.
Begonia—dead areas in leaves and flowers rapidly enlarging and turning black in a moist atmosphere; profuse brownish gray mold.
Calendula—gray-mold blight.
Camellia—flower and bud blight, common after frost.
Carnation—flower rot or brown spotting, worse in a cool greenhouse.
Century plant—gray mold after overwatering and chilling.
Chrysanthemum—cosmopolitan on flowers, buds, leaf tips, and cuttings. Ray blight on flowers starts as small, water-soaked spots, which rapidly enlarge with characteristic gray mold.
Dahlia—bud and flower blight.
Dogwood—flower and leaf blight. In wet springs anthers and bracts of aging flowers are covered with gray mold,˚ and when these rot down on top of young leaves, there is a striking leaf blight.
Eupatorium—stem blight, common in crowded plantings. A tan area girdles stem near ground with tops wilting or drying to that point.
Geranium (Pelargonium)—blossom blight and leaf spot, most common in cool, moist greenhouses where plants are syringed frequently. Petals are discolored, flowers drop, gray mold forms on leaves.
Lily—*Botrytis cinerea* is common on lilies, but see also *B. elliptica*.
Marigold—gray mold prevalent on fading flowers.
Peony—late blight, distinguished from early blight (see *B. paeoniae*) by the sparse mold, usually standing far out from affected tissues, rather than a thick, short velvety mold, and by much larger, flatter sclerotia formed near base of the stalk. Late flowers are infected, and when they drop down onto wet foliage, irregular brown areas are formed in leaves.
Pine—seedling blight.
Pistachio—shoot blight.
Poinsettia—tip blight and stem canker.
Primrose—crown rot and decay of basal leaves, with prominent gray mold, very common in greenhouses where plants are heavily watered.

Rhododendron—flower, twig, and seedling blight.

Rose—bud or flower blight, cane canker. When half-open buds ball, the cause is often an infestation of thrips; but if gray mold is present, *Botrytis* is indicated. Canes kept too wet by a manure mulch, or wet leaves, or injured in some way, are often moldy.

Snapdragon—flower spikes wilt; tan cankers girdle stems.

Sunflower—bud rot and mold.

Sweet pea—blossom blight.

Viola spp.—gray mold and basal rot of violet and pansy.

Zinnia—petal blight, head blight, moldy seed.

Botrytis cinerea may also infect arabis, cineraria, eucharis, euphorbia, fuchsia, gerbera, gypsophila, heliotrope, hydrangea, iris, lilac, lupine, Mayapple, pyrethrum, periwinkle, rose-of-Sharon, stokesia, viburnum, and wallflower.

Control. Sanitation is more important than anything else. Carry around a paper bag as you inspect the garden; put into it all fading flowers and blighted foliage; if infection is near the base, take the whole plant up for burning. Keep greenhouse plants widely spaced, with good ventilation; avoid syringing, overhead watering, and too cool temperature. Propagate cuttings from healthy plants in a sterilized medium.

Botrytis douglasii. Seedling blight; of giant sequoia and redwood, perhaps a form of *B. cinerea.*

Botrytis elliptica. Lily botrytis blight; general on lilies, also reported on tuberose and stephanotis in California. Lily species vary in susceptibility to the disease, but there are several strains of the fungus, and few lilies are resistant to all strains. Madonna lily, *L. candidum,* is particularly susceptible, with infection starting in autumn on the rosette of leaves developed at that time.

If the blight strikes early, the entire apical growth may be killed with no further development. More often the disease starts as a leaf spot when stems are a good height. Spots are orange to reddish brown, usually oval. In some species there is a definite red to purple margin around a light center; in others the dark margin is replaced by an indefinite water-soaked zone. If spots are numerous, they grow together to blight the whole leaf. Infection often starts with the lowest leaves and works up the stem until all leaves are blackened and hanging limp. This spreading is the result of many spot infections and not from an invasion of the vascular system.

Buds rot or open to distorted flowers with irregular brown flecks. There are sometimes severe stem lesions, but the rot rarely progresses into the bulbs. Spores formed in the usual gray-mold masses in blighted portions are spread by rain, air currents, and gardeners. Optimum spore germination is in cool weather, around 60° F, but once infection has started 70° F promotes most rapid blighting. With sufficient moisture the cycle may repeat every few days through the season. The fungus winters as very small black sclerotia, irregular or elliptical in shape, in fallen flowers or blighted dead stems and leaves, or as mycelium in the basal rosette of Madonna lilies.

Control. Avoid too dense planting, and shady or low spots with little air circulation and subject to heavy dews. Clean up infected plant parts before sclerotia can be formed. Copper sprays are more effective for the lily *Botrytis* than the newer organics. Spray with bordeaux mixture; start when lilies are 5 or 6 inches high and continue at 10- to 14-day intervals until flowering.

Botrytis galanthina. Botrytis blight of snowdrop; sometimes found in the sclerotial state on imported bulbs. If the black dots of sclerotia are present only on outer scales, remove scales before planting; otherwise discard bulbs.

Botrytis gladiolorum. Gladiolus botrytis blight; corm rot, first reported in Oregon in 1939 and now serious in all important gladiolus-growing areas — the Pacific Coast, the Midwest, Florida — in cool, rainy weather. In northern areas the disease is a corm-rotting problem, in the South a flower blight, damaging in transit, and in all areas it is a leaf spot or blight.

In dry weather and in more resistant varieties the leaf spots are very small, rusty brown, appearing only on the exposed side of the leaf. In more humid weather the spots are large, brown, round to oval or smaller, pale brown with reddish margins. Flower stems have pale brown spots that turn dark. There may be a soft rotting at the base of florets. The disease starts on petals as pinpoint, water-soaked spots, but in moist weather the whole flower turns brown and slimy. Flowers with no visible spotting when packed often arrive ruined. After the flowers are cut, infection spreads down the stalk and into the corm, producing dark brown spots, irregular in shape and size, most numerous on the upper surface. Corms may become soft and spongy with a whitish mold. Oval, flat, black sclerotia, ⅛ to ¼ inch long, are formed on corms in storage and in rotting tissue in the field or in refuse piles. They may persist in the soil several years.

Control. Cure corms rapidly after digging; bury or burn all plant refuse.

Botrytis hyacinthi. Hyacinth botrytis blight; recently found in Washington on plants grown from imported bulbs. Leaves have brown tips with gray mold or brown spots on lower surface. Leaves may be killed, with small black sclerotia formed in rotting tissue. Flowers rot and are covered with powdery gray spores. Do not work with plants when they are wet; remove infected parts or whole plants.

Botrytis narcissicola. See *Sclerotinia narcissicola*, under Rots.

Botrytis paeoniae. Peony botrytis blight; early blight, bud blast, gray mold, probably present wherever peonies are grown. It is also recorded on lily of the valley, but that may be a form of *Botrytis cinerea*. Peony blight was first noticed in epiphytotic form in this country in 1897 and has been important in wet springs ever since.

Young shoots may rot off at the base as they come through the ground or when a few inches high, with a dense velvety gray mold on the rotting portions. This early shoot blight is far more common when the young stems are kept moist by having to emerge through a mulch of manure or wet leaves. Flowers are attacked at any stage. Buds turn black when they are very tiny, never developing, or they may be blasted when they are half open. If it is dry in early spring, infection may be delayed until flowers are in full bloom, at which time they turn brown. Infection proceeds from the flower down the stem for a few inches, giving it a brown and tan zoned appearance. Leaf spots develop when infected petals fall on foliage. Continued blighting of leaves through the summer and late blasting of flowers may be due to *Botrytis cinerea*, which produces a sparser mold and conidiophores projecting farther from the petal or leaf surface.

Conidia are blown by wind, splashed by rain, carried on gardeners' tools, and sometimes transported by ants. Secondary infection is abundant in cool moist weather. In late summer small, shiny black, slightly loaf-shaped sclerotia are formed near the base of stalks, just under the epidermis. They are quite different from the large, flat, black sclerotia often formed by *B. cinerea* on the same stalks.

Control. Sanitation is the most important step. Cut down all tops in autumn at ground level, or just below, to get rid of sclerotia wintering near base of stems. Burn this debris; never use it for a mulch. Avoid any moisture-retentive covering. If you insist on manure, apply it in a wide ring around the plant, well outside the area of emerging shoots. Go around with a paper bag periodically, cutting off for burning all blighted parts; never carry these parts loose through the garden for fear of shedding spores to healthy plants.

Botrytis polyblastis. See *Sclerotinia polyblastis.*

Botrytis streptothrix (perfect stage *Streptotinia arisaemae*). **Leaf and stalk blight**; of Jack-in-the-pulpit and golden club. This species has conidiophores with strikingly twisted branches, producing a reddish brown mat of conidia. Sclerotia are very small, seldom over $1/32$ inch, black, shiny, and somewhat hemispherical.

Botrytis tulipae. Tulip fire; botrytis blight of tulips; general wherever tulips are grown, causing much damage in rainy springs. The first indication of disease is the appearance of a few malformed leaves and shoots among healthy tulips or large light patches resembling frost injury on leaves. Gray mold forming on such blighted areas of plants grown from infected bulbs provides an enormous number of conidia to be splashed by rain to nearby tulips. Secondary infection appears as minute, slightly sunken, yellowish leaf spots, surrounded with a water-soaked area, and gray to brown spots on stems, often zonate, and resulting in collapse. Small white spots appear on colored flowers, brown spots on white petals; but with continued moisture the spots grow together, and in a day or so the fuzzy gray mold has covered rotten blooms and large portions of blighted leaves (see Fig. 20).

Very small, shiny black sclerotia are formed in leaves and petals rotting into the ground, or on old flower stems or bulbs. Sometimes the latter have

Figure 20. Botrytis blight on tulip.

yellow to brown, slightly sunken, circular lesions on outermost fleshy scales without the formation of sclerotia. Spring infection comes from spores produced on such bulbs or from sclerotia on bulbs or sclerotia left loose in the soil after infected tissues have rotted.

Control. Inspect all bulbs carefully before planting; discard those harboring sclerotia or suspicious brown lesions. It is wise, though seldom possible in a small garden, to plant new bulbs where tulips have not grown for 3 years. Plant where there is good air circulation. Make periodic inspections, starting early, removing into a paper bag plants with serious primary infection and blighted leaves. Cut off all fading flowers before petals fall; cut off all foliage at ground level when it turns yellow. Burn all debris.

Briosia

Fungi Imperfecti, Moniliales, Stilbaceae

Conidia on synnemata or coremia, erect fascicles of hyphae ending in a small head; spores globose, dark, one-celled, catenulate (formed in chains).

Briosia azaleae (*Pycnostysanus azaleae*). **Bud and twig blight of azalea and rhododendron**; widespread but occasional. The disease was reported from New York in 1874 and, as a rhododendron bud rot, from California in 1920. It was particularly serious on Massachusetts azaleas in 1931 and 1939. Flower buds are dwarfed, turn brown and dry; scales are silvery gray. Twigs die when lateral leaf buds are infected. Successive crops of coremia are produced on old dead buds for as long as three years, the first crop appearing the spring after summer infection. The coremia heads are dark, and the buds look as if stuck with tiny, round-headed pins. Prune out and burn infected buds and twigs in late autumn and early spring. Spraying with bordeaux mixture before blossoming and at monthly intervals after bloom may be wise in severe cases.

Cenangium

Ascomycetes, Helotiales, Patellariaceae

Apothecia small, brown to black, sessile or substipitate on bark; spores hyaline, elliptical, one-celled; paraphyses filiform.

Cenangium ferruginosum. Pine twig blight; pruning disease; cenangium dieback, of fir and pine. The fungus is ordinarily saprophytic on native pines but may become parasitic when their vigor is reduced by drought. The disease is considered beneficial to ponderosa pine in the Southwest because it prunes off the lower branches; on exotic pines it can be damaging.

Infection starts near a terminal bud in late summer and progresses down a twig into a node, sometimes beyond into 2-year wood. The needles redden and die; they are conspicuous in spring but drop in late summer. Then brown to black apothecia with a greenish surface to the cup appear on twigs. Cut off and destroy infected twigs.

Cercospora

Fungi Imperfecti, Moniliales, Dematiaceae

Conidia hyaline to pale to medium green or brown; long, usually with more than three cross walls; straight or curved, with the base obconate or truncate, tip acute to obtuse; thin-walled; not formed in a fruiting body but successively on slender conidiophores, which emerge in fascicles or groups from stomata and usually show joints or scars where conidia have fallen off successively. The conidiophores are always colored, olivaceous to brown, pale to very dark (see Fig. 18).

This group is the largest one of the Dematiaceae, with about 400 species, all parasitic, causing leaf spots or blights. The perfect state, when known, is *Mycosphaerella*.

Cercospora apii. Early blight of celery; general on celery and celeriac, first noted in Missouri in 1884 and since found in varying abundance wherever celery is grown. The disease is most severe from New Jersey southward. The name is somewhat misleading; in Florida early blight rarely appears before the Septoria disease known as late blight. Foliage spots appear when plants are about 6 weeks old. Minute yellow areas change to large, irregular, ash gray lesions, covered in moist weather with velvety groups of conidiophores and spores on both sides of leaves. Sunken, tan, elongated spots appear on stalks just before harvest. The disease spreads rapidly in warm, moist weather, the spores being splashed by rain, carried with manure or cultivators, or blown by wind. The life cycle is completed in 2 weeks.

Control. Seed more than 2 years old is probably free from viable spores; other seed should be treated with hot water, 30 minutes at 118°F to 120°F. Bordeaux mixture and other copper sprays have been recommended. Spray applications should start soon after plants are set and be repeated weekly, or more often. Emerson Pascal is blight-resistant.

Cercospora carotae. Early blight of carrot; lesions on leaves and stems are subcircular to elliptic, pale tan to gray or brown or almost black; lobes or entire leaflets are killed. The disease is more severe on young leaves and builds up as the plant grows. Spores, produced on both leaf surfaces, are spread by wind.

Control. Rotate crops and clean up refuse.

Cercospora microsora. Linden leaf blight; general on American and European linden. Small circular brown spots with darker borders coalesce to form large, blighted areas, often followed by defoliation; most serious on young trees.

Cercospora sordida (*Mycosphaerella tecomae*). Trumpetvine leaf blight; from New Jersey to Iowa and southward. Small, angular, sordid brown patches run together; edge of leaflets may be purplish; the fungus fruits on underside of leaves. The blight is seldom important enough to warrant control measures.

Cercospora thujina. Arborvitae blight; fire, on oriental arborvitae and Italian cypress in the South; destructive in ornamental plants. First reported from Louisiana in 1943, the fungus was named as a new species of *Cercospora* in 1945, but it is nearer *Heterosporium* in spore character. Affected leaves and branchlets are killed, turn brown, and gradually fall off, leaving shrubs thin and ragged. The lower two-thirds of the bush is affected most severely, with a tuft of healthy growth at the top. When close to a house, the side away from the wall shows most symptoms. Plants crowded in nurseries are killed in 1 year to 3 years, but in home gardens they may persist for years in an unsightly condition. Conidiophores in fascicles produce conidia after girdling cankers have killed the twigs. There is often a swelling above the girdle that resembles an insect gall.

Choanephora

Phycomycetes, Mucorales, Choanephoraceae

Mycelium profuse; sporangia and conidia present; sporangiola lacking. Sporangium pendent on recurved end of an erect, unbranched sporangiophore with a columella, containing spores provided at both ends and sometimes at the side with a cluster of fine, radiating appendages. Conidia formed in heads on a few short branches or an erect conidiophore enlarged at the tip; conidia longitudinally striate, without appendages.

Choanephora cucurbitarum. Blossom blight; fruit rot, common on summer squash and pumpkin, occasional on amaranth, cowpea, cucumber, okra, and pepper; on sweet potato foliage, on fading hibiscus and other flowers. This blight is often found in home gardens in seasons of high humidity and rainfall. Flowers and young fruits are covered with a luxuriant fungus growth, first white, then brown to purple with a definite metallic luster. The fruiting bodies look like little pins stuck through this growth. Both staminate and pistillate flowers are infected, and from the latter the fungus advances into young fruits, producing a soft wet rot at the blossom end. In severe cases all flowers are blighted or fruits rotted.

Control. Grow plants on well-drained land; rotate crops. Remove infected flowers and fruits as noticed.

Choanephora infundibulifera. Blossom blight; on hibiscus and jasmine.

Ciboria

Ascomycetes, Helotiales, Sclerotiniaceae

Stroma a dark brown to black sclerotium in catkins or seed, simulating in shape the stromatized organ and not resembling a sclerotium externally. Apothecia cupulate to shallow saucer-shaped; brown.

Ciboria acerina. Maple inflorescence blight; on red and silver maple. Apothecia, developed in great numbers from stromatized inflorescences on ground beneath trees, start discharging spores when maple flowers appear overhead. Mycelium spreads through stamens, calyx, and bud scales until flower cluster drops.

Ciboria carunculoides. Popcorn disease of mulberry; a southern disease, not very important. Sclerotia are formed in carpels of fruit, which swells to resemble popcorn but remains green.

Ciborinia

Ascomycetes, Helotiales, Sclerotiniaceae

Stroma a thin, flat, black sclerotium of discoid type in leaves; one to several stalked apothecia arise from sclerotia; apothecia small, brown, cupulate to flat when expanded.

Ciborinia erythronii and **C. gracilis.** Leaf blight of erythronium; flat black sclerotia are prominent in leaves.

Corticium

Basidiomycetes, Agaricales, Thelephoraceae

Hymenium or fruiting surface of basidia consisting of a single resupinate or horizontal layer. This genus has contained a rather heterogeneous collection of species; some of the more important have been transferred to the genus *Pellicularia*.

Corticium koleroga. Thread blight; see *Pellicularia koleroga.*

Corticium microsclerotia. Web blight; see *Pellicularia filamentosa.*

Corticium salmonicolor. Limb blight; of fig, pear, apple in Gulf states. The spore surface is pinkish.

Corticium stevensii. Thread blight; see *Pellicularia koleroga.*

Corticium vagum, now *Pellicularia filamentosa.* Perfect stage of *Rhizoctonia solani,* causing black scurf of potatoes and damping-off and root rot of many plants. See both *Pellicularia* and *Rhizoctonia* under Rots.

Coryneum

Fungi Imperfecti, Melanconiales, Melanconiaceae

Acervuli subcutaneous or subcortical, black, cushion-shaped or disc-shaped; conidiophores slender, simple; spores dark with several cross walls, oblong to fusoid; parasitic or saprophytic.

Coryneum berckmanii. Coryneum blight of oriental arborvitae; also on Italian cypress, causing serious losses in nurseries and home gardens in the Pacific Northwest. Small twigs or branches are blighted, turn gray-green then reddish brown; many small branchlets drop, leaving a tangle of dead gray stems; larger limbs may be girdled. Twigs are dotted with black pustules bearing five-septate spores. As new growth develops in blighted areas, the spores spread the disease to young contiguous foliage. Reinfection continues until the plant is so devitalized that it dies. The fungus fruits only on scale leaves or young stems.

Control. Remove and destroy blighted twigs. Apply a copper spray in September to healthy bushes as a preventive spray; apply in September and repeat in late October to infected bushes.

Coryneum carpophilum (*Cladosporium beijerinckii*). **Peach shoot blight, coryneum blight of stone fruits;** shot hole, fruit spot, winter blight, pustular spot, general on peach in the West, also on almond, apricot, nectarine, and cherry. Twig lesions are formed on 1-year shoots, reddish spots developing into sunken cankers; fruit buds are invaded, and there is copious gum formation. Small spots are formed on foliage, dropping out to leave typical shot holes, followed by considerable defoliation.

Apricot buds are blackened and killed during winter; fruiting wood in peaches is killed before growth starts. In late rains leaves and fruit are peppered with small, round, dead spots. Fruit lesions are raised, roughened, scabby. The fungus winters in twigs, diseased buds, and spurs.

Control. In California, the standard spray for peach is bordeaux mixture applied in autumn immediately after leaf fall and before the rainy season. On apricots additional sprays are suggested for late January and at early bloom. On almonds at least two spring sprays are recommended, one at the popcorn stage of bloom, the other at petal fall.

Coryneum microstictum. Twig blight; of American bladdernut. Young twigs are killed; the fungus winters in acervuli on this dead tissue, and spores are disseminated in spring. Prune out and burn diseased twigs during the winter.

Cryptocline

Fungi Imperfecti, Melanconiales, Melanconiaceae

Cryptocline cinerescens. Twig blight; of oaks.

Cryptospora

Ascomycetes, Sphaeriales, Sphaeriaceae

Perithecia immersed in a stroma, with long necks converging into a disc; ascospores long, filiform, hyaline; conidia on a stroma.

Cryptospora longispora. Araucaria branch blight; lower branches are attacked first, with disease spreading upward; tip ends are bent and then broken off; plants several years old may be killed. Prune off and burn infected branches.

Cryptopstictis

Fungi Imperfecti, Melanconiales, Melanconiaceae

Spores dark, with several cross walls, formed in acervuli.

Cryptostictis sp. Twig blight; of dogwood.

Curvularia

Fungi Imperfecti, Moniliales, Dematiaceae

Conidiophores brown, simple or sometimes branched, bearing conidia successively on new growing tips; conidia dark, three- to five-celled, with end cells lighter, more or less fusiform, typically bent or curved with central cells enlarged; parasitic or saprophytic.

Curvularia cymbopogonis. Blight and leaf spot; of itchgrass. Leafspots coalesce after 3 or 4 days to form larger lesions and final blighting symptoms.
Curvularia lunata (*C. trifolii* f.sp. *gladioli*). **Gladiolus flower blight and leaf spot**; Curvularia disease. Suddenly, in 1947, a blight showed up in

Florida as a serious threat to the gladiolus cut-flower industry, ruining hundreds of acres there and in Alabama in a few months. The disease is now recorded as far north as New York and Wisconsin and on the Pacific Coast. The pathogen is usually identified as *Curvularia lunata*, known as a crop pest for many years, especially in the tropics, but studies indicate it is a special form of *C. trifolii*, cause of a leaf spot of clover.

Curvularia spots on leaf or stem are oval, tan to dark brown, showing on both sides of the leaf, bordered with a brown ring, slightly depressed and with a narrow yellowish region between the spot and normal green of the leaf. Tan centers of spots are covered with black spores resembling powder. Premature death comes when stems of young plants are girdled; florets fail to open when petioles are girdled.

Under favorable weather conditions tan spots on petals turn into a smudgy flower blight. Brown to black irregular lesions appear on corms of blooming stock and develop further in storage; the fungus survives in corms from one season to the next. This is a high-temperature disease, with optimum for fungus growth 75° F to 85° F and no infection under 55° F. A 13-hour dew period is sufficient moisture. Leaf spots show up in 4 to 5 days, spots on florets and stems in only 2 to 3 days. The complete life cycle is as short as a week in warm rainy weather, and the fungus can survive in the soil for 3 years. Many gladiolus varieties are more or less resistant; Picardy and some others are very susceptible.

Cylindrocladium

Fungi Imperfecti, Moniliales, Moniliaceae

Conidiophores dichotomously branched; spores hyaline, two- or several-celled.

Cylindrocladium scoparium. Cylindrocladium blight; damping-off of seedlings and cuttings—conifers, azalea, magnolia, hydrangea, holly, pyracantha, bottle brush, and poinsettia—in greenhouses under very moist conditions. Infected azalea leaves turn black, with petiole bases softened, and drop in a few days; the bark turns brown. Leaves and stems are covered with brownish mycelial strands and white powdery masses of conidia. Control by proper humidity and aeration.

Cylindrocladium avesiculatum. Blight and leaf spot; of *Leucothoë axillaris.*

Cylindrosporium

Fungi Imperfecti, Melanconiales, Melanconiaceae

Acervuli subepidermal, white or pale, discoid or spread out; conidiophores short, simple; conidia hyaline, filiform, straight or curved, one-celled or becoming septate; parasitic on leaves. Many species have *Higginsia* or *Coccomyces* as a perfect state.

Cylindrosporium defoliatum. Leaf blight of hackberry; may cause defoliation but usually unimportant.

Cylindrosporium griseum. On western soapberry.

Cylindrosporium juglandis. On walnut.

Dendrophoma

Fungi Imperfecti, Sphaeropsidales, Sphaerioidaceae

Pycnidia dark or light brown, superficial or submerged and erumpent; globose or elongate, ostiolate; conidiophores elongated, branched; conidia hyaline, one-celled, elongate to ellipsoid; parasitic or saprophytic.

Dendrophoma obscurans. Strawberry leaf blight; angular leaf spot. The lesions are large, circular to angular, reddish purple, zonate with age, having a dark brown center, a light brown zone, and a purple border. Spots may extend in a V-shaped area from a large vein to the edge of the leaf, with black fruiting bodies appearing in the central portion. Not serious before midsummer, the disease may be destructive late in the season. The fungus winters on old leaves.

Diaporthe

Ascomycetes, Sphaeriales, Valsaceae

Perithecia in a hard black stroma made up of host and fungal elements, first immersed, then erumpent; ascospores fusoid or ellipsoid, two-celled, hyaline. Imperfect stage a *Phomopsis* with two types of spores; alpha conidia, hyaline, one-celled ovate to fusoid, and beta conidia, curved or bent stylospores.

Diaporthe arctii. Diaporthe blight of larkspur; stem canker, on annual larkspur and delphinium. Lower leaves turn brown and dry but remain attached; brown lesions at base of stems extend several inches upward and down into roots. Scattered dark pycnidia are present in stems, petioles, leaf blades, and seed capsules, the latter probably spreading the blight. Crowns are sometimes developed in a cottony weft of mycelium; perithecia develop on decaying stems. Remove and destroy diseased plants; use seed from healthy plants.

Diaporthe phaseolorum. Lima bean pod blight; leaf spot, apparently native in New Jersey, where it was first noticed in 1891, more abundant on pole than on bush beans. Leaf spots are large, irregular, brown, often with discolored borders and large black pycnidia formed in concentric circles in dead tissue. Necrotic portions may drop out, making leaves ragged. Pod lesions spread; pods turn black and wilted, with prominent black pycnidia. Seeds are shriveled or lacking. Spores are produced in great numbers, are disseminated by wind and pickers, and enter through stomata or wounds. The disease is most severe along the coast; optimum temperature is around 80° F. The fungus is seed-borne, but most lima bean seed is produced where the disease does not occur. Use healthy seed; clean up refuse; rotate crops.

Diaporthe phaseolorum var. **sojae.**   Soybean pod and stem blight; widespread. This disease was formerly confused with the more acute stem canker caused by *D. phaseolorum* var. *caulivora* (see under Rots). The pod blight is a slower disease, killing plants in later stages of development. It can be identified by the numerous small black pycnidia scattered over the pods and arranged in rows on stems. The blight is more serious in wet seasons. The fungus winters on the seed and on diseased stems in the field. Use clean seed; clean up plant refuse; rotate crops.

Diaporthe phaseolorum var. **caulivora.**   Stem blight; of soybean; also causes pod and seedling blight, stem canker, and seed decay of soybean.

Diaporthe vaccinii.   Blueberry twig blight; the same fungus that causes cranberry rot blights new shoots of cultivated blueberries, entering at tips, progressing toward the base, and ultimately girdling old branches. Pycnidia develop on leaves and dead twigs. The disease is seldom serious enough for control measures.

Diaporthe vexans (*Phomopsis vexans*).   Phomopsis blight of eggplant; fruit rot, general in field and market, especially in the South. Destruction is often complete, with every above-ground part affected. Seedlings rot at ground level. The first leaf spots are near the ground, definite, circular, gray to brown areas with light centers and numerous black pycnidia. The leaves turn yellow and die. Stem cankers are constrictions or light gray lesions. Fruit lesions are pale brown, sunken, marked by many black pycnidia arranged more or less concentrically. Eventually the whole fruit is involved in a soft rot or shriveling. Spores winter on seed and in contaminated soil. There is no fungicidal control. Use resistant varieties Florida Market and Florida Beauty.

Dichotomophora

See under Cankers.

Dichotomophora indica.   Stem blight; of common parsley.

Didymascella (*Keithia*)

Ascomycetes, Phacidiales, Stictidiaceae

Apothecia brown, erumpent on leaves of conifers; spores dark, two-celled, ovoid; paraphyses filiform; asci two- to four-spored.

Didymascella thujina.   Arborvitae leaf blight; seedling blight of arborvitae in eastern states and of giant arborvitae, sometimes called western red cedar.

The fungus is a native of North America and occurs abundantly in the West, damaging seedlings and saplings, often killing trees up to 4 years old, if they are in dense stands in humid regions. Older trees do not die, but foliage appears scorched, particularly on lower branches, and young leaf twigs may drop. Cushionlike, olive brown apothecia embedded in leaf tissue, usually upper, are exposed by rupture of the epidermis. After summer discharge of spores (round, brown, unequally two-celled) the apothecia drop out of the needles, leaving deep pits.

Control. Spray small trees and nursery stock several times during summer and fall with bordeaux mixture.

Didymascella tsugae. Hemlock needle blight; needles of Canada hemlock turn brown and drop in late summer. Spores are matured in apothecia on fallen needles with new infection in spring. The damage is not heavy.

Didymella

Ascomycetes, Sphaeriales, Mycosphaerellaceae

Perithecia (or perithecia-like stromata) membranous, not carbonaceous; innate; not beaked; paraphyses present; spores two-celled, hyaline.

Didymella applanata. Raspberry spur blight; purple cane spot, gray bark, general on raspberries, also on dewberry, blackberry. Named because it partially or completely destroys spurs or laterals on canes. The disease, known in North America since 1891, may cause losses of up to 75% of the crop of individual plants of red raspberries. Dark reddish or purple spots on canes at point of attachment of leaves enlarge to surround leaf and bud and may darken lower portion of cane. Affected areas turn brown, then gray. If buds are not killed outright during the winter, they are so weakened that the next season's spurs are weak, chlorotic, seldom blossoming. Pycnidia of the imperfect *Phoma* stage and perithecia are numerous on the gray bark; ascospores are discharged during spring and early summer; on germination they can penetrate unwounded tissue.

Control. Keep plants well-spaced, allowing plenty of sunlight for quick drying of foliage and canes. Remove infected canes and old fruiting canes after harvest. A delayed dormant spray of lime sulfur or Elgetol may be advisable, followed by two sprays of ferbam or bordeaux mixture, applied when new shoots are 6 to 10 inches high and 2 weeks later.

Didymosphaeria

Ascomycetes, Sphaeriales, Sphaeriaceae

Perithecia innate or finally erumpent; not beaked; smooth; paraphyses present; spores dark, two-celled.

Didymosphaeria populina (*Venturia populina, V. tremulae*). Shoot blight of poplar; leaf and twig blight. Young shoots are blackened and wilted. In moist weather dark olive green masses of spores are formed on leaves.

Diplodia

Fungi Imperfecti, Sphaeropsidales, Sphaerioidaceae

Pycnidia innate or finally erumpent; black, single, globose, smooth; ostiole present; conidiophores slender, simple; conidia dark, two-celled, ellipsoid or ovoid. Parasitic or saprophytic.

Some species cause twig blights that are not too important: **Diplodia coluteae** on bladder senna; **D. longispora** on white oak; **D. pinea** on pine; **D. sarmentorum** on pyracantha. **Diplodia natalensis** (imperfect stage of *Physalospora rhodina*) causes blight, stem gumming, or stem-end rot of melons, as well as twig blight of peach and citrus. See further under Rots.

Diplodia gossypina. Blight; of slash pine and loblolly pine seedlings.

Discula

Fungi Imperfecti, Melanconiales, Melanconiaceae.

Discula quercina, Twig blight; of oaks.

Dothistroma

Fungi Imperfecti, Sphaeropsidales, Sphaerioidaceae

Stroma dark, elongate, innate, becoming erumpent and swollen, with a stalk extending into the substratum, composed internally of dense, vertical hyphae; locules separate, one to several in the upper part of the stroma; conidiophores simple, slender; conidia several-celled, hyaline, long-cylindrical to filiform.

Dothistroma pini. Needle blight; on Austrian pine and red pine.

Endothia

Ascomycetes, Sphaeriales, Melogrammataceae

Perithecia deeply embedded in a reddish to yellow stroma, with long necks opening to the surface but not beaked; paraphyses lacking; spores two-celled, hyaline. Conidia borne in hollow chambers or pycnidia in a stroma and expelled in cirrhi.

Endothia parasitica. Chestnut blight; endothia canker, general on chestnut. To most gardeners this disease is of only historical importance, for practically all of our native chestnuts are gone. The disease, however, persists in sprouts starting from old stumps and in the chinquapin. One of the most destructive tree diseases ever known, chestnut blight at least served to awaken people to the importance of plant disease and to the need for research in this field.

First noticed in the New York Zoological Park in 1904, the blight rapidly wiped out the chestnut stands in New England and along the Allegheny and Blue Ridge mountains, leaving not a single undamaged tree. In 1925, the disease eliminated chestnuts in Illinois, and by 1929, it had reached the Pacific Northwest.

Conspicuous reddish bark cankers are formed on trunk and limbs, often swollen and splitting longitudinally. As the limbs are girdled, the foliage blights, so that brown, dried leaves are seen from a distance. The fungus fruits abundantly in crevices of broken bark, first producing conidia extruded in yellow tendrils from reddish pycnidia and later ascospores from perithecia embedded in orange stromata. Fans of buff-colored mycelium are found under affected bark.

Ascospores can be spread many miles by the wind, landing in open wounds, but the sticky conidia are carried by birds and insects. The fungus can live indefinitely as a saprophyte, and new sprouts developing from old stumps may grow for several years before they are killed.

Control. All eradication and protective measures have proved futile. Hope for the future lies in cross-breeding resistant Asiatic species with the American chestnut (and there has been some success in this line) or in substituting Chinese and Japanese chestnuts for our own.

Fabraea

Ascomycetes, Helotiales, Dermateaceae

Apothecia develop on fallen leaves; small, disclike, leathery when dry, gelatinous when wet; asci extend above the surface of the disc; ascospores two-celled, hyaline. Conidial stage an *Entomosporium* with distinctive cruciate four-celled conidia, each cell with an appendage, formed in acervuli (see Fig. 18).

Fabraea maculata (*Entomosporium maculatum*). Pear leaf blight; Entomosporium leaf spot; fruit spot, generally distributed on pear and quince, widespread on amelanchier, sometimes found on apple, Japanese quince, medlar, mountain-ash, Siberian crab, cotoneaster, loquat, photinia.

Pears may be affected as seedlings in nurseries or in bearing orchards. Very small purple spots appear on leaves, later extending to a brownish circular lesion, ¼ inch or less in diameter, with the raised black dot of a fruiting body in the center of each spot. If spots are numerous, there is extensive defoliation. Fruit spots are red at first, then black and slightly sunken; the skin is roughened, sometimes cracked. Quince has similar symptoms.

Twig lesions appear on the current season's growth about midsummer, indefinite purple or black areas coalescing to form a canker. Primary spring infection comes more from conidia produced in these twig lesions than from ascospores shot from fallen leaves on the ground. Most commercial varieties of pear and quince are susceptible, although some are moderately resistant.

Fabraea thuemenii (*Entomosporium thuemenii*). Hawthorn leaf blight; widespread on *Crataegus* species. Symptoms are similar to those of pear leaf blight and for a long time the pathogen was considered identical. Small dark brown or reddish brown spots, with raised black dots, are numerous over leaves, which drop prematurely in August. In wet seasons trees may be naked by late August.

Control. Because the fungus winters in twig cankers as well as in fallen leaves, sanitation has little effect. Standard recommendation has been to spray three times with bordeaux mixture, starting when leaves are half out and repeating at 2-week intervals. The copper may be somewhat phytotoxic, causing small reddish spots similar to those of blight, but it does prevent defoliation.

Furcaspora

Fungi Imperfecti, Melanconiales, Melanconiaceae

Starlike botryoblastospores; acervuli become erumpent at maturity and grade into sporodochia and pycnidia.

Furcaspora sp. Needle cast, pine.

Fusarium

See under Rots.

Fusarium moniliforme var. **subglutinans.** Blight; of slash pine and loblolly pine seedlings.

Fusarium solani. Stem and leaf blight; on Spanish moss.

Gibberella

Ascomycetes, Hypocreales, Nectriaceae

Perithecia superficial, blue, violet, or greenish; spores hyaline with several cells. Conidial stage in genus *Fusarium* with fusoid curved spores, several-septate. The species causing stalk rots of corn and producing gibberellic acid are more important than those causing blights.

Gibberella baccata (*Fusarium lateritium*). **Twig blight**; of ailanthus, citrus, cotoneaster, fig, hibiscus, hornbeam, peach, and other plants in warm climates, sometimes associated with other diseases.

Glomerella

See under Anthracnose.

Glomerella cingulata. Cyclamen leaf and bud blight.

Gnomonia

See under Anthracnose.

Gnomonia rubi. Cane blight; of blackberry, dewberry, raspberry.

Hadrotrichum

Fungi Imperfecti, Moniliales, Tuberculariaceae

Sporodochia cushion-shaped, dark; conidiophores dark, simple, forming a palisade and arising from a stroma-like layer; conidia dark, nearly spherical, one-celled, borne singly; parasitic on leaves.

Hadrotrichum globiferum. Leaf blight; of lupine.

Helminthosporium

Fungi Imperfecti, Moniliales, Dematiaceae

Mycelium light to dark; conidiophores short or long; septate, simple or branched, often protruding from stomata of host; more or less irregular or bent, bearing conidia successively on new growing tips; conidia dark typically with more than three cells, cylindrical or ellipsoid, sometimes slightly curved or bent, ends rounded. Parasitic, often causing leaf spots or blights of cereals and grasses.

Helminthosporium catenarium (*Drechslera catenaria*). **Leaf blight or crown rot**; on creeping bentgrass; red leaf lesions and leaf tip dieback; eventually entire plant becomes blighted to crown.

Helminthosporium gigantea (*Drechslera gigantea*). Blight or zonate leafspot on wild rice and grasses.

Helminthosporium maydis (*Cochliobolus heterostrophus*). **Southern corn leaf blight**; easily confused with southern corn leaf spot due to *H. carbonum*.

The leaf blight occurs throughout the corn areas of the South and north to Illinois, more important on field than on sweet corn. Grayish tan to straw-colored spots with parallel sides unite to blight most of the leaf tissue. The fields appear burned by fire. Resistant varieties offer the only control.

Helminthosporium turcicum (*Trichometasphaeria turcica*). Northern corn leaf blight; on field and sweet corn and on grasses; found from Wisconsin and Minnesota to Florida but more severe in states with heavy dews, abundant rainfall, and warm summers, losses running from a trace to 50%. The disease starts on the lower leaves and progresses upward. Small, elliptical, dark grayish green, water-soaked spots turn greenish tan and enlarge to spindle-shape, ½ inch to 2 inches wide, 2 inches to 6 inches long. Spores developing on both leaf surfaces after rain or heavy dew give a velvety dark green appearance to the center of the lesions. Whole leaves may be killed; entire fields turn dry. The fungus winters in corn residue in the field and produces spores the next spring; these are spread by wind.

Control. Use a 3-year or longer rotation.

Herpotrichia

Ascomycetes, Sphaeriales, Sphaeriaceae

Mycelium dark, perithecia superficial; spores with several crosswalls, olivaceous when mature.

Herpotrichia nigra. Brown felt blight; of conifers at high elevations; on fir, juniper, incense cedar, spruce, pine, yew when under snow. When the snow melts, lower branches are seen covered with a dense felty growth of brown to nearly black mycelium, which kills foliage by excluding light and air as well as by invading hyphae. Small, black perithecia are scattered over the felt. This pathogen also found on dwarf mistletoe.

Heterosporium

Fungi Imperfecti, Moniliales, Dematiaceae

Conidiophores dark, simple, bearing conidia successively on new growing tips; conidia dark, with three or more cells, cylindrical, with rough walls (echinulate to verrucose); parasitic, causing leaf spots, or saprophytic.

Heterosporium syringae. Lilac leaf blight; a velvety, olive green bloom of spores if formed in blighted, gray-brown leaf areas, which may crack and fall away. Infection is on mature leaves and the fungus is often associated with *Cladosporium*. If necessary, spray after mid-June with bordeaux mixture.

Higginisia

See *Coccomyces* under Leaf Spots.

Higginisia hiemalis. Cherry leaf blight; see *Coccomyces hiemalis*, cherry leaf spot.

Higginisia kerriae. Kerria leaf and twig blight; see *Coccomyces kerriae* under Leaf Spots.

Hypoderma

Ascomycetes, Phacidiales, Phacidiaceae

Ascospores formed in hysterothecia (elongated perithecia or apothecia) extending along evergreen needles; asci long-stalked; ascospores one-celled hyaline, fusiform, surrounded by a gelantinous sheath.

Hypoderma lethale. Gray blight; of hard pines, from New England to the Gulf states. Hysterothecia are short, narrow, black; often seen on pitch pine.

Hypodermella

Ascomycetes, Phacidiales, Phacidiaceae

Like *Hypoderma* but one-celled spores are club-shaped at upper end, tapering toward base.

Hypodermella abietis-concoloris. Fir needle blight; on firs and southern balsam.

Hypodermella laricis. Larch needle and shoot blight; yellow spots are formed on needles, which turn reddish brown but stay attached, giving a scorched appearance to trees. Hysterothecia are very small, oblong to elliptical, dull black, on upper surface of needles.

Hypomyces

Ascomycetes, Hypocreales, Nectriaceae

Perithecia bright colored with a subicle (crustlike mycelial growth underneath); spores two-celled, light, with a short projection at one end.

Hypomyces ipomoeae. Twig blight; of bladdernut.

Hyponectria

Ascomycetes, Hypocreales, Nectriaceae

Perithecia bright colored, soft; innate or finally erumpent; paraphyses lacking; spores one-celled, light-colored, oblong.

Hyponectria buxi. Leaf blight; leaf cast of boxwood.

Itersonilia

Fungi Imperfecti, Moniliales, Sporobolomycetaceae

Cells reproducing by budding and germinating by repetition; clamp connections as in Basidiomycetes and probably imperfect species of *Tremellales*. The genus is not well understood.

Itersonilia perplexans. Petal blight of chrysanthemum; the fungus was isolated from greenhouse chrysanthemums in Minnesota in 1951 but apparently has been present, as a parasite or saprophyte, on many other plants. On pompom chrysanthemums the tip half of outer petals turns brown and dries; the diseased tissue is filled with broad hyphae and clamp connections. Inoculated snapdragons show similar symptoms. Adequate greenhouse ventilation seems to prevent trouble.

Itersonilia sp. Leaf blight, canker of parsnip; seasonal in New York and neighboring states. Plants are defoliated in cool, moist weather. Spores from leaves produce a chocolate brown dry rot on shoulder or crown of the root. Good drainage and long rotation aid in control.

Kellermannia

Fungi Imperfecti, Sphaeropsidales, Sphaerioidaceae

Pycnidia black, globose, separate; immersed in host tissue; ostiolate; conidiophores short, simple; conidia hyaline, mostly two-celled, cylindrical with an awl-shaped appendage at the tip; parasitic or saprophytic.

Kellermannia anomala (*K. yuccaegena*). **Yucca leaf blight**; general on nonarborescent forms of yucca; in Florida and California on arborescent forms.

Kellermannia sisyrinchi. Leaf blight; of blue-eyed grass.

Labrella

Fungi Imperfecti, Sphaeropsidales, Leptostromataceae

Pycnidia with a radiate shield, rounded; innate or erumpent; spores hyaline, one-celled.

Labrella aspidistrae. Leaf blight; of aspidistra.

Leptosphaeria

Ascomycetes, Sphaeriales, Sphaeriaceae

Perithecia membranous, not beaked, opening with an ostiole; innate or finally erumpent; paraphyses present; spores dark, with several cells. Imperfect stage a *Coniothryium* with black, globose pycnidia and very small, dark, one-celled conidia, extruded in a black cirrhus.

Leptosphaeria (*Melanomma*) **conithyrium** (*Coniothyrium fuckelii*). Raspberry cane blight; general on raspberry, dewberry, blackberry. The same fungus causes cankers of apple and rose (see under Cankers). On raspberry, brown dead areas extend into wood; whole canes or single branches wilt and die; often between blossoming and fruiting. The fungus enters the bark at any time during the season, through an insect wound or mechanical injury. Smutty patches on the bark come from small olive conidia of the *Coniothyrium* stage and larger, dark, four-celled ascospores. Ascospores are spread by rain; conidia by rain and insects.

Control. Sanitation is very important; cut out and burn all diseased canes. A control program for spur blight should suffice for cane blight.

Leptosphaeria korrae. Blight; on turfgrass (associated with Fusarium blight syndrome); disease is also called spring dead spot.

Leptosphaeria thomasiana. Cane blight; of dewberry, raspberry, in Pacific Northwest.

Lophodermella

Ascomycetes, Phacidiales, Cryptomycetaceae

Hymenium on a fleshy, gelatinous stroma under the bark of woody plants; ascospores aseptate. **Lophodermella** sp. Needle cast; of pine.

Macrophomina

See under Rots.

Macrophomina phaseoli. Ashy stem blight; charcoal rot of soybeans, sweet potatoes, many other plants. See under Rots.

Micropeltis

Ascomycetes, Hemisphaeriales, Hemisphaeriaceae (Micropeltaceae)

A single hymenium, fruiting layer, covered with an open, reticulate scutellum; paraphyses present; spores hyaline, with several cells.

Micropeltis viburni. Leaf blight; of viburnum.

Monilinia (Sclerotinia)

Ascomycetes, Helotiales, Sclerotiniaceae

Stroma is a sclerotium formed in fruit by the fungus digesting fleshy tissues and replacing them with a layer of broad, thick-walled, interwoven hyphae forming a hollow sphere enclosing core or seed of fruit, which has become a dark, wrinkled, hard mummy. Apothecia funnel-form or cupulate, rarely flat-expanded, some shade of brown; asci eight-spored; ascospores one-celled, ellipsoidal, often slightly flattened on one side, hyaline. Conidia hyaline, one-celled, formed in chains in grayish masses called sporodochia.

Monilinia azaleae. Shoot blight; of native or pinxter azalea (*Rhododendron roseum*). Apothecia are formed on overwintered mummied fruits (capsules) in leaf mold under shrubs in moist places. Ascospores infect leaves and succulent shoots when the azalea is in full bloom. The conidial stage is common on young developing fruits in late June and July (New York).

Monilinia fructicola. Leaf and shoot blight; of peach.

Monilinia johnsonii. Leaf blight; fruit rot, of hawthorn.

Monilinia laxa. Blossom blight, brown rot; of apricot, almond, cherry, plum, and prune on Pacific Coast. Blossoms and twigs are blighted with a good deal of gum formation. *Monilinia laxa* is sometimes coincident with, and confused with, *M. fructicola*, which causes a more general rot of stone fruits. Both are discussed more fully under Rots.

Monilinia rhododendri (*Sclerotinia seaveri*). Twig blight; seedling blight, of sweet cherry.

Mycosphaerella

See under Anthracnose.

Mycosphaerella citrullina (*M. melonis*) conidial stage *Phyllosticta citrullina*. Gummy stem blight; stem end rot, leaf spot of watermelon, muskmelon, summer squash, pumpkin, and cucumber. Gray to brown dead areas in leaves are marked with black pycnidia; leaves may turn yellow and shrivel. Stem infection starts with a water-soaked oily green area at nodes. The stem is girdled, covered with a dark exuded gum, and the vine wilts back to that point. Fruit rot starts gray, darkens to nearly jet black, with gummy exudate.

Control. Clean up crop refuse; practice rotation. Some varieties are more resistant than others.

Mycosphaerella melonis. Gummy stem blight; of cucumbers.

Mycosphaerella pinodes. Pea blight; see *Ascochyta pinodes*.

Mycosphaerella rabiei (Imperfect stage, *Ascochyta rabiei*). Blight; of chickpea

Mycosphaerella sequoiae. Needle blight; of redwood.

Myriogenospora

Ascomycetes, Clavicipitales, Clavicipitaceae.

Ascomata superficial or in a stroma, fleshy, bright-colored; ascus with a thick cap traversed by a slender pore; ascospores filiform, multiseptate, often fragmenting.

Myrogenospora atramentosa. Blight; on turf grass, centipedegrass.

Mystrosporium

Fungi Imperfecti, Moniliales, Dematiaceae

Conidia dark, muriform; hyphae long.

Mystrosporium adustum. Leaf blight, ink spot; of bulbous iris; also on montbretia and lachenalia. Irregular black patches or blotches appear soon after leaves push through the ground; under moist conditions the foliage withers and dies prematurely. Inky black stains appear on husks of bulbs (usually *Iris reticulata*), and yellow dots or elongated sunken black craters show on fleshy scales. The bulbs may rot, leaving only the husk and a mass of black powder. The fungus spreads through the soil, invading adjacent healthy bulbs.

Control. Dig bulbs every year; discard all diseased bulbs and debris; plant in a new location. Spray with bordeaux mixture.

Myxosporium

Fungi Imperfecti, Melanconiales, Melanconiaceae

Conidia hyaline, one-celled, in discoid to pulvinate acervuli on branches.

Myxosporium diedickii. Twig blight; of mulberry.
Myxosporium everhartii. Twig blight; of dogwood. **M. nitidum.** Twig blight and dieback of native dogwood. Prune twigs back to sound wood; feed and water trees.

Neopeckia

Ascomycetes, Sphaeriales, Sphaeriaceae

Perithecia hairy, not beaked, formed on a mycelial mass; paraphyses present; spores two-celled, dark.

Neopeckia coulteri. Brown felt blight; on pines only; otherwise similar to brown felt blight caused by *Herpotrichia*, a disease of high altitudes on foliage under snow.

Ovulinia

Ascomycetes, Helotiales, Sclerotiniaceae

Stroma a sclerotium, thin, circular to oval, shallowly cupulate, formed in petal tissue but falling away; minute globose spermatia; apothecia of *Sclerotinia* type, small; asci eight-spored; paraphyses septate with swollen tips; conidia large, obovoid, one-celled except for basal appendage or disjunctor cell; borne singly at tips of short branches of mycelium forming a mat over surface of petal tissue (see Fig. 18).

Ovulinia azaleae. Azalea flower spot, petal blight; very destructive to southern azaleas in humid coastal regions, occasional on mountain laurel and rhododendron. Starting as a sudden outbreak near Charleston, South Carolina, 1931, the disease spread rapidly north of Wilmington, North Carolina, down the coast to Florida, and around the Gulf. It reached Texas by 1938 and was in California by 1940; it was reported in Maryland in 1945, in Virginia in 1947, and in Philadelphia in 1959. Petal blight was reported from a Long Island, New York, greenhouse in 1956, apparently present there since 1952, and in 1959 infected all the azaleas in one New Jersey greenhouse. In both cases the blight started on plants purchased from the South. This disease is the most spectacular one that I have ever witnessed, with most of the bloom on all the azaleas in a town blighting simultaneously and seemingly overnight under special weather conditions. The blight does not injure stem or foliage; it is confined to the flowers. The loss is aesthetic and economic from the standpoint of tourist trade. For many years, before a control program was worked out, the great azalea gardens of the South had to close their gates to visitors far too early in the season.

Primary infection comes from very small apothecia produced from sclerotia on the ground under shrubs, usually in January or February, occasionally as early as December. Spores shot into the air are carried by wind drift to flowers near the ground of early varieties, initial spots being whitish. If you put your finger on such a spot, the tissue melts away. With continued high humidity, heavy fog, dew, or rain, conidia are produced over the inner surfaces of petals and are widely disseminated to other petals by wind, insects, and splashed rain. Within a few hours colored petals are peppered with small white spots, and white flowers have numerous brown spots. By the next day flowers have collapsed into a slimy mush, bushes looking as if scalding water was poured over them. If the weather stays wet, small black sclerotia are formed in the petals in another 2 or 3 days. Infected blooms seldom drop normally but remain hanging on the bushes in an unsightly condition for weeks and months, some even to the next season. Many of the sclerotia, however, drop out and remain in the litter on the ground ready to send up apothecia the next winter.

Both Indian and Kurume varieties are attacked, the peak of infection coming with midseason varieties such as Pride of Mobile or Formosa. In some seasons dry weather during early spring allows a good showing of azaleas; in other years blight starts early and there is little color unless azaleas are sprayed. On Belgian azaleas in greenhouses blight may start in December.

Control. Some mulches and soil treatments will inhibit apothecial production. Secondary infection is bound to come from some untreated azalea in the neighborhood. Spraying gives very effective, even spectacular, control if started on time, when early varieties are in bloom and midseason azaleas are showing color. Sprays must be repeated three times a week as long as petal surface is expanding, about 3 or 4 weeks. After that, weekly spraying is sufficient. Spraying is mandatory now for the big azalea gardens, and the admission fees from the lengthened season pay for the program many times over.

The original successful formula was Dithane D-14 (nabam) $1\frac{1}{3}$ quarts to 100 gallons water, plus 1 pound 25% zinc sulfate, $\frac{1}{2}$ pound hydrated lime, and 1 ounce of spreader Triton B 1956. Later work showed that the lime could be omitted, Dithane reduced to 1 quart, and zinc sulfate to $\frac{2}{3}$ pound to prevent injury in periods of drought. The spray should be a fine mist, applied from several directions to get adequate coverage.

Commercial growers should beware of ordering azaleas from the South unless they are bare-rooted and all flower buds showing color are removed.

As a matter of fact, any potted or balled and burlapped plant grown in a nursery near azaleas could very easily bring along some of the tiny sclerotia in the soil, and they might remain viable for more than 1 year. All traces of soil should be washed off roots, and the plants wrapped in polyethylene for shipping.

Pellicularia

Basidiomycetes, Agaricales, Thelephoracae

Includes some species formerly assigned to *Corticium*, *Hypochnus*, and *Peniophora*. Hyphae stout, very short-celled; mycelium branching at right angles; basidia very stout, formed on a resupinate, cottony or membranous layer of mycelium. Imperfect state a *Rhizoctonia*, with sclerotia made up of brown, thin-walled, rather angular cells, or *Sclerotium*, with sclerotia having a definite brown rind and light interior.

Pellicularia filamentosa, perfect stage of *Rhizoctonia solani*. This is a variable fungus with some strains or forms causing leaf blights but best known as cause of Rhizoctonia rot of potatoes and damping-off of many plants. See under Rots.

Pellicularia filamentosa f. sp. **microsclerotia** (*Corticium microsclerotia*). Web blight; of snap bean, lima bean, also reported on fig, elder, hibiscus, hollyhock, tung oil, and phoenix tree, from Florida to Texas. Many small brown sclerotia and abundant weblike mycelium are found on bean stems, pods, and foliage. Infection starts with small circular spots that appear water-soaked or scalded. They enlarge to an inch or more, become tan with a darker border, are sometimes zonate. The whitish mycelium grows rapidly over the leaf blade, killing it, and spreads a web from leaf to leaf, over petioles, flowers, and fruit, in wet weather and at temperatures of 70° F to 90° F; in dry weather growth is inconspicuous except on fallen leaves. The fungus is spread by wind, rain, irrigation water, cultivating tools, and bean pickers; it survives in sclerotial form from season to season.

Control. Destroy infected plants; clean up refuse. In Florida, do not plant beans between June and September if web blight has been present. Use a copper spray or dust.

Pellicularia filamentosa f. sp. **sasakii.** Leaf blight; of grasses, clover, and so on.

Pellicularia filamentosa f.sp. **timsii.** Leaf blight of fig.

Pellicularia koleroga (*Corticium stevensii*). Thread blight; a southern disease, from North Carolina to Texas, important on fig and tung, sometimes defoliating pittosporum, crape myrtle, roses, and other ornamentals, and some fruits. The disease is recorded on apple, azalea, banana shrub, blackberry, boxwood, camphor, cherry laurel, chinaberry, columbine, crabapple, crape myrtle, casuarina, currant, dewberry, dogwood, elderberry, elm, erythrina, euonymus, fig, flowering almond, flowering quince, goldenrod, gooseberry, guava, honeysuckle, hibiscus, morning glory, pear, pecan, pepper vine, persimmon, pittosporum, plum, pomegranate, quince, rose, satsuma orange, soapberry, silver maple, sweetpotato, tievine (*Jacquemontia*), tung, Virginia creeper, and viburnum.

The fungus winters as sclerotia on twigs and leaf petioles, and in May and June produces threadlike mycelia that grow over lower surface of leaves, killing them and causing premature defoliation, although often dead leaves hang on the tree in groups, matted together by threadlike spider webs. Fruiting patches on leaves are first white, then buff. The fungus flourishes in moist weather, temperatures 75° F to 90° F.

Control. On figs, one or two applications of tribasic copper sulfate, or bordeaux mixture, are satisfactory until the fruit ripens in July. Pruning out infected branches may be sufficient on tung and pecan, but at least one spray of bordeaux mixture may be required.

Pellicularia rolfsii (*Sclerotium rolfsii*). Southern blight; crown rot; the disease has been known, in its sclerotium stage, for many years on hundreds of plants. The connection with *Pellicularia* is recent, and the name does not have universal agreement. One strain of the fungus has been called *Sclerotium delphinii* in the North, where the disease is usually designated crown rot. However, this fungus is variable with single-spore cultures from the *Pellicularia* stage producing sclerotia typical of *Sclerotium delphinii* and *S. rolfsii*, with intermediate forms. Sclerotia of the southern blight strain are very small, round, tan, about the size, shape, and color of mustard seed, the pathogen being frequently called the mustard-seed fungus.

Southern blight affects almost all plants except field crops like wheat, oats, corn, and sorghum. Fruits and vegetables include Jerusalem artichoke, avocado, bean, beet, carrot, cabbage, cucumber, eggplant, endive, lettuce, melon, okra, onion, garlic and shallot, pea, peanut, pepper, potato, rhubarb, strawberry, sweet potato, tomato, turnip, and watermelon. Ornamentals, too numerous to list in entirety, include ajuga, ageratum, amaryllis, azalea, caladium, calendula, campanula, canna, carnation, cosmos, China aster, chrysanthemum, dahlia, delphinium, daphne, duranta, gladiolus, hollyhock, hydrangea, iris, jasmine, lemon verbena, lily, lupine, marigold, morning-glory, myrtle, narcissus, orchids, phlox, pittosporum, rose, rose-mallow, rudbeckia, scabiosa, sedum, sweet pea, star of bethlehem, tulip, violet, and zinnia.

The first sign of blight is the formation of white wefts of mycelium at the base of the stem, spreading up in somewhat fan-shaped fashion and some-times spreading over the ground in wet weather. The sclerotia formed in the wefts are first white, later reddish tan or light brown. They may be numerous enough to form a crust over the soil for several inches around a stem, or they may be somewhat sparse and scattered.

In the white stage, droplets of liquid often form on the sclerotia, and the oxalic acid in this liquid is assumed to kill plant cells in advance of the fungus hyphae. Thus, the pathogen never has to penetrate living tissue, which explains why so many different kinds of plants succumb so readily to southern blight. Fruits touching the ground, as well as vegetables with fleshy roots, like carrots and beets, or plants with bulbs or rhizomes, like onions, narcissus, and iris, seem particularly subject to this disease. Low ornamentals such as ajuga blight quickly, the whole plant turning black; tall plants like delphinium rot at the crown and then die back or topple over; bulbs have a cheesy interior, with sclerotia forming on or between the scales.

Control. Remove diseased plants as soon as they are noticed. Take out surrounding soil, for 6 inches beyond the diseased area, wrapping it carefully so that none of the sclerotia drop back. Increasing the organic content of the soil reduces southern blight, as does the addition of nitrogenous fertilizers, such as ammonium nitrate. Treating narcissus bulbs in hot water for 3 hours, as for nematodes, kills the fungus in all except the very largest bulbs.

Pestalotia

Fungi Imperfecti, Melanconiales, Melanconiaceae

Acervuli dark, discoid or cushion-shaped, subcutaneous; conidiophores short, simple; conidia fusiform, several-celled with median cells colored, end cells hyaline, a short stalk at the basal cells and a crest of two or more hyaline appendages, setae, from the apical cell (see Fig. 18). Weak parasites or saprophytes; some are treated under Leaf Spots.

Pestalotia funerea. Tip blight of conifers; needle blight, twig blight of chamaecyparis, retinospora, cypress, bald cypress, arborvitae, juniper, yew, and giant sequoia. The fungus is saprophytic on dead and dying tissue and also weakly parasitic, infecting living tissue through wounds under moist conditions. It appears in sooty pustules on leaves, bark, and cones.

Pestalotia hartigii. Associated with a basal stem girdle of young conifers but parasitism not proven. The stem has a swelling above the girdling lesions, and the tree gradually turns yellow and dies. The effect may be more from high temperature than the fungus; shading transplants is helpful.

Pestalotia sp. and **Penicillium** sp. **Flower blight**; on camellia.

Phacidium

Ascomycetes, Phacidiales, Phacidiaceae

Apothecia innate, concrete above with the epidermis and slitting with it into lobes; spores one-celled, hyaline.

Phacidium abietinellum. Needle blight; of balsam fir.
Phacidium balsameae. Needle blight; of balsam fir in New England, of white and alpine fir in the Northwest.

Phacidium infestans. Snow blight of conifer seedlings; on fir and young pines in the Northeast, also on arborvitae and spruce; on white and alpine fir in the Northwest. This native fungus is most damaging in nurseries, attacking foliage under the snow. The needles turn brown, with a covering of white mycelium, just as the snow melts. In late summer and fall, brown to nearly black apothecia appear on underside of browned needles. Ascospores are spread by wind, primary infection being in autumn. Additional infection occurs in late winter, when mycelium grows out under the snow from diseased to dormant, healthy needles.

Control. Spray nursery beds with dormant-strength lime sulfur in late fall; remove infected seedlings; dip new stock in lime sulfur before planting.

Phialophora

See under Rots.

Phialophora graminicola. Blight; on turfgrasses (associated with Fusarium blight syndrome).

Phleospora

Fungi Imperfecti, Sphaeropsidales, Sphaerioidaceae

Pycnidia dark, imperfectly formed, globose, innate in tissue, not in distinct spots; conidia hyaline or subhyaline, several-celled, elongate fusoid to filiform; parasitic or saprophytic. One of the conidial forms linked with *Mycosphaerella* as a perfect stage.

Phleospora adusta. Leaf blight; of clematis.

Phoma

See under Blackleg.

Phoma conidiogena. Boxwood tip blight; ashy gray necrotic areas at leaf tips, with pycnidia on both leaf surfaces.

Phoma fumosa. Twig blight; occasional on maple.

Phoma macdonaldii. Blight, premature ripening; of sunflower.

Phoma mariae. Twig blight; on Japanese honeysuckle.

Phoma piceina. Twig and needle blight; of Norway spruce. May cause defoliation and sometimes death of forest trees.

Phoma strobiligena. On cone scales of Norway spruce.

Phomopsis

Fungi Imperfecti, Sphaeropsidales, Sphaerioidaceae

Pycnidia dark, ostiolate, immersed, erumpent, nearly globose; conidiophores simple; conidia hyaline, one-celled, of two types: ovate or ellipsoidal and long, filamentous, sickle-shaped or hooked at upper end (see Fig. 18). Imperfect stage of *Diaporthe*; parasitic causing spots on various plant parts.

Phomopsis ambigua (*Diaporthe eres*). **Twig blight of pear;** widespread.
Phomopsis diospyri. Twig blight; of native persimmon.
Phomopsis japonica. Twig blight; of kerria.
Phomopsis juniperovora. Nursery blight; juniper blight; cedar blight, canker, on red cedar and other junipers, cypress, chamaecyparis, Japanese yew (*Cephalotaxus*), arborvitae, giant sequoia, and redwood. This disease occurs in virulent form from New England to Florida and through the Middle West; it may also occur on the Pacific Coast.

Tips of branches turn brown with progressive dying back until a whole branch or even a young tree is killed. Trees over 5 years old are less seriously injured. Spores produced in quantity in pycnidia on diseased twigs ooze out in little tendrils in moist weather, to be spread by splashing water, insects, and workers. Entrance is through unbroken tissue as well as wounds; the stem is killed above and below the point of entrance. Small, sunken lesions give a flattened appearance to some seedlings. Overhead irrigation in a nursery is a predisposing factor, and a large amount of stock can be blighted in a very short time. Older trees in home plantings suffer from twig blight. The fungus winters on infected plant parts and remains viable at least 2 years.

Control. Have seedbeds well drained; water by ditch irrigation; remove and burn diseased seedlings early in the season; keep seedbeds away from older cedar trees; do not use cedar branches or needles for mulching. Spray with fixed copper or bordeaux mixture plus a wetting agent, starting when growth begins and repeating to keep new foliage covered. Spiny Greek and Hill junipers and Keteller red cedars are somewhat resistant.

Phomopsis kalmiae. Mountain-laurel leaf blight; blotch. Circular, brown, often zonate areas on leaves, frequently starting near margin or tip, gradually enlarge and coalesce until most of the blade is involved. The fungus often works down the petiole to cause a twig blight. The disease is more prominent on bushes in the shade or under drip of trees. Remove blighted leaves or clean up fallen leaves.

Phomopsis oblonga. Twig blight; on Chinese elm.
Phomopsi occulta. Shoot blight; of Colorado blue spruce.
Phomopsis vexans. Phomopsis blight of eggplant; see *Diaporthe vexans.*
Phomopsis vaccinii. Twig blight; of blueberry.

Phyllosticta

Fungi Imperfecti, Sphaeropsidales, Sphaerioidaceae

Pycnidia dark, with ostiole, in spots in leaves; spores one-celled, hyaline. The characteristics are the same as *Phoma* except that leaves rather than stems are infected. Other species are listed under Leaf Spots.

Phyllosticta batatas. Sweet potato leaf blight; occasional from New Jersey to Florida, more prevalent in the South but seldom important enough for control measures. Numerous white spots on leaves are bordered with narrow reddish zones; pycnidia are numerous; spores are extruded in tendrils.

Phyllosticta cryptomeriae. Needle blight; found on *Cryptomeria*.

Phyllosticta lagerstroemia. Tip blight; of crape myrtle.

Phyllosticta pteridis. Tip blight; of fern. Leaves lose green color; spots are ash gray with purple brown margins and numerous black pycnidia in center. A very weak bordeaux mixture has been suggested for control; if overhead watering is avoided, spraying may not be necessary.

Physalospora

Ascomycetes, Sphaeriales, Mycosphaerellaceae

Perithecia with papillate mouths, immerse in substratum but without well-defined stromata; paraphyses present; spores one-celled, hyaline. A few species cause blights; many cause rots.

Physalospora dracaenae. Dracaena tip blight; leaf spot. Disease starts at the tips of lower leaves and spreads down toward the base. Infected areas are sunken and straw-colored, dotted with black specks of pycnidia. All leaves on the plant may die except a few at the top. Remove infected leaves as soon as noticed. Spray with a copper fungicide.

Physalospora gregaria. Twig blight; of yew.

Physalospora obtusa. Cane blight; of rose, also black rot of apple, canker and dieback of many plants. See under Cankers and Diebacks and also under Rots.

Phytophthora

Phycomycetes, Peronosporales, Pythiaceae

This most important genus contains many species causing destructive blights, cankers, and rots. The name, which means "plant destroyer," was given in 1876 for the potato blight fungus. Sporangia, formed successively on sporangiophores, slender, sparsely branched hyphae emerging from stomata, germinate either by a germ tube or by zoospores. The sexual spore is an oospore.

Phytophthora cactorum. Lilac shoot blight; blossoms and succulent growing tips are blighted and turn brown; suckers are killed back 4 or 5 feet. Blight is most severe in wet springs when shrubs are crowded, shaded, and improperly pruned. The same fungus causes a canker, foot rot, and dieback of rhododendron and other plants and is considered again under Cankers and Diebacks. Avoid planting lilacs and rhododendrons close together. Prune each year for air circulation and to remove dead twigs.

Phytophthora capsici. Phytophthora blight of pepper; leaf and stem blight of squash, fruit rot of pepper, eggplant, tomato, cucumber, and melon. The disease was first found in New Mexico in 1918 injuring chili peppers; it occurs chiefly in southwestern and Gulf states. In 1953, however, it was reported that for some years it had been causing a leaf blight of squash in North Carolina.

Pepper plants are girdled at the soil line with a dark green water-soaked band, which dries and turns brown, followed by wilting and death of the entire plant. Leaf spots are dark green and small at first, later large bleached or scalded areas. Dark, water-soaked patches on fruits are covered with white mycelium. The fruit withers but remains attached; 60% of green fruit may be infected in southwestern commercial plantings. Seed are infected from the fruit. Symptoms on squash are somewhat similar: green leaf lesions spreading over the blade, a basal stem rot, and wilting. Wet soil and high temperatures encourage blight.

Control. Place seedbeds on land that has not previously grown peppers; rotate crops. Avoid overirrigation.

Phytophthora citrophthora (also *P. citricola* and *P. nicotianae* var. *parasitica*). Shoot and stem blight; on azalea.

Phytophthora erythroseptica. Leaf blight of pink and golden calla; leaves are wilted and distorted; petioles are black and soft.

Phytophthora ilicis. Holly blight; Phytophthora leaf and twig blight, the most serious disease of English holly, particularly serious in the Northwest. For many years the trouble was ascribed to *Boydia insculpta* and called Boydia canker, but this fungus merely invades tissue killed by *Phytophthora*. Leaf spots are dark, developing on lower leaves in cool rainy weather and progressing upward in late fall and winter. Young twigs die back; black stem cankers kill older twigs. Young plants in nurseries are defoliated and sometimes killed.

Control. Choose a planting site with moderate air movement; space trees well apart. Prune out all cankered and blighted twigs; prune also for air movement through trees. Spray with tribasic copper sulfate, starting the middle of October.

Phytophthora infestans. Late blight of potato and tomato; general on potato in the Northeast, in Middle Atlantic and North Central states, sometimes in Gulf and western states; on tomato in humid regions and seasons.

Here is a pathogen that has not lost its destructive virulence with passage of time. In 1946, a whole century after potato blight caused the famous Irish famine, tomato blight devastated tomatoes along the eastern seaboard, both in home gardens and canning fields.

The potato went to Europe from South America shortly before 1600, seemingly leaving its pathogens at home. For 200 years, potatoes thrived in Europe as the main source of carbohydrate food, but in August 1845, the *Gardener's Chronicle* reported: "A fatal malady has broken out amongst the potato crop. On all sides we hear of destruction. In Belgium the fields are said to have been completely desolated. There is hardly a sound sample in Covent Garden Market." The editor went on to describe the decay and to say "As to cure for this distemper there is none. One of our correspondents is today angry with us for not telling the public how to stop it; but he ought to consider that Man has no power to arrest the dispensations of Providence. We are visited by a great calamity which we must bear." And in 1946, American gardeners were again blaming the editor, for lack of information on tomato blight.

In 1845, the weather was continued gloom and fog, with below-average temperatures. The *Gardener's Chronicle* editor was sure blight was due to potatoes being overladen with water. The Rev. M. J. Berkeley disagreed. He insisted blight was due to a fungus, with the weather contributing to spread of a moisture-loving parasite. The argument raged, for this was long before Pasteur and his germ theory, and the first time anyone believed a fungus could be the cause and not the consequence of plant disease. A French scientist, Montagne, named the fungus *Botrytis infestans*, but the first really good description of it was published by Berkeley, and it remained for the German de Bary, in 1876, actually to prove the pathogenic nature of the fungus and to erect the new genus *Phytophthora* to include it.

Meanwhile the disease was making history. The loss of the potato crop in 1845 and 1846 killed off a million people and caused another million and a half to emigrate; the first Government Relief program was instigated; and the English Corn Laws were repealed with a change to a policy of free trade and unbounded expansion of commerce.

Late Blight of Potato

After blossoming, large, dark green, water-soaked spots appear on leaves in wet weather, first on lower leaves. As a spot enlarges the center is shriveled, dry, dark brown to black, and a downy, whitish growth appears on the

Figure 21. Late blight on potato.

underside of leaves. Similar lesions are formed on stems and petioles, and there is a characteristic strong odor as tops are blighted. On tubers, first symptoms are small brown to purple discolorations of skin on upper side, changing to depressed pits when tubers are removed from soil and put in storage (see Fig. 21). On cutting through the potato, a reddish brown dry rot is seen.

The primary cycle starts with infected tubers, which have harbored mycelia in the dry rot patches over winter. If infected seed pieces are planted, the fungus grows systemically into the shoots and finally fruits by sending sporangiophores through the stomata on lower-leaf surfaces (see Fig. 22). These swell at the tips into ovoid bodies, sporangia, then branch and produce successively more sporangia. The latter may function as conidia, putting out a germ tube, but more often are differentiated into a number of swarmspores (zoospores), which have cilia enabling them to swim about after they are splashed by rain to another leaf. Eventually they stop swimming and send a germ tube in through the leaf cuticle or enter through a stoma. Initial infection in the field also comes from conidia blown from sprouts produced on infected tubers in cull piles. Blighting follows rapidly, with first symptoms 5 days or less from the time of infection and with the fungus fruiting again in a whitish layer on the underside of leaves.

Tubers with only a thin covering of soil may be infected by swarmspores washing onto them from blighted leaves overhead; they are also infected

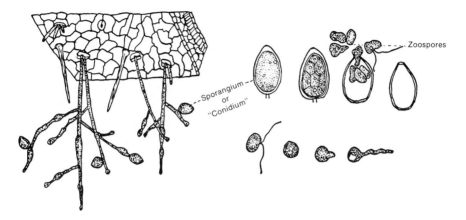

Figure 22. Late blight of potatoes. Sporangiophores of *Phytophthora infestans* emerging from leaf, bearing sporangia, sometimes called conidia, which germinate by zoospores.

during digging if it is done in moist weather while tops are still green. Swarmspores remain viable in the soil several weeks while awaiting favorable conditions. Oospores, the sexual spores, are apparently not required in the life cycle for they are not found with potatoes grown in the field. They have been produced in culture.

This is a disease entirely dependent on weather conditions. Temperature and moisture conditions are right for an epiphytotic about 2 years out of 5. Zoospores are produced only in cool weather, 60° F and under, but they invade leaves most rapidly at higher temperatures. Because they are swimming spores, rain is required. A cool, wet July is usually followed by blight in August and September.

Control. Some varieties, such as Kennebec, Essex, Pungo, and Cherokee, are resistant to the common strain of the fungus but not to some of the newer strains. Treat potato dumps and cull piles with a weed spray to control sprouts. Delay digging crop until 2 weeks after tops die, or kill the tops with a weed killer to prevent infection at early digging.

Late Blight of Tomato

Although there are potato and tomato strains of *Phytophthora infestans*, each is capable of infecting the other host. Ordinarily blight starts with potatoes in midsummer; when the fungus moves to tomatoes, it has to go through several cycles to build up a strain virulent enough to produce general blighting, and by that time the tomato season is nearly over. Now we know that it is possible for the tomato strain to winter in potato tubers and be ready

Figure 23. Late blight of tomato.

to inflict damage on tomatoes with the first crop of zoospores produced on potato sprouts. Conversely, tomato seedlings brought up from the South and planted near potato fields can start an epiphytotic of late blight on potatoes.

The 1946 tomato blight saga—the one that awakened eastern gardeners to the fact that plant disease could be as important to home gardeners as to farmers—started in Florida late in November 1945. By January the disease was extremely destructive in tomato seedbeds, and it continued so intermittently whenever temperatures ranged from 60° F to 70° F and relative humidity was nearly 100% for more than 15 hours. Evidence indicated spores could be wind-borne for as far as 30 miles. The wave of late blight went west to Alabama, taking 75% of the early crop, and rolled up the Atlantic Coast, reaching the Carolinas in May and Virginia and Maryland in June, again taking 75% of the early crop. It rolled into Delaware and New Jersey in July, but did not reach peak epidemic form until after an extended rainy period in August, and ended in Massachusetts in August and September.

In 1947, a blight-forecasting service was started, based on weekly graphs prepared by plotting daily the cumulative rainfall and mean temperatures and aided by reports from key pathologists in various states. If conditions are unfavorable for blight, we can save time and money by eliminating useless spraying.

On seedlings small, dark spots on stems or leaves are followed by death within 2 or 3 days. On mature plants blight starts with dark, water-soaked leaf spots and large, dark brown spots on fruit, with most of the leaves soon hanging lifeless and fruit rotting on the ground (see Fig. 23).

Control. Bordeaux mixture applied to young tomato plants will either prevent fruit setting or cause stunting. It can be used after blossoming, or a fixed copper can be substituted.

Phytophthora parasitica. Leaf, stem, and bud blight; on bougainvillea, dogwood, hibiscus, artillery plant, and aluminum plant.

Phytophthora syringae. Citrus blight; also on lilac, but the more common lilac blight is due to *P. cactorum*. On citrus trees leaves have semitransparent spots similar to frost damage. Other *Phytophthora* species may be present with *P. syringae* to cause brown rot of fruits. See under Rots. On lilacs large irregular leaf patches have a lighter zone at margin. There may be some defoliation.

Piricularia

Fungi Imperfecti, Moniliales, Moniliaceae

Conidiophores long, slender, simple or rarely branched, septate, single or in tufts; conidia pyriform to nearly ellipsoid, borne singly and attached at broader end; spores hyaline, two- to three-celled; parasitic, chiefly on grasses.

Piricularia grisea. Leaf blight; on creeping bent grass.

Pyrenochaeta

Fungi Imperfecti, Sphaeropsidales, Sphaerioidaceae

Pycnidia dark, ostiolate, nearly globose, erumpent with a few bristles near ostiole; conidiophores simple or branched; conidia small, one-celled, hyaline, ovate to elongate; parasitic or saprophytic. See also under Rots.

Pyrenochaeta phlogis. Stem blight of phlox.

Rehmiellopsis

Ascomycetes, Sphaeriales, Sphaeriaceae

Perithecia single, globose, rupturing irregularly; asci in fascicles, no paraphyses; spores hyaline, two-celled.

Rehmiellopsis balsameae. Tip blight; needle blight of balsam fir; on native balsam fir in northern New England and on ornamental firs in southern New England and New York. Infection is in spring with needles of current season shriveled, curled, and killed, often with a dieback of terminal or lateral shoots and sometimes cankers at base of infected needles. Satisfactory control on ornamental firs has been obtained by three sprays, at 10-day intervals, of bordeaux mixture, the first application made as new growth starts.

Rhizoctonia

Fungi Imperfecti, Mycelia Sterilia

Sclerotial form of some species of *Pellicularia*, *Corticium*, *Macrophomina*, and *Helicobasidium*. Young mycelium colorless, with branches constricted at points of origin from main axis, but soon colored, a weft of brownish yellow to brown strands, organized into dense groups, sclerotia made up of short, irregular, angular or somewhat barrel-shaped cells.

Rhizoctonia ramicola. Silky thread blight; a southern disease similar to web blight caused by *Pellicularia koleroga*. Perennial ornamental hosts in Florida include elaeagnus, erythrina, crape myrtle, holly, guava, pittosporum, pyracantha, Carolina jessamine, feijoa, and rhododendron. Tan spots with purple-brown margins appear on leaf blades, dead lesions on petioles, and young twigs. When leaves are abscissed, they are often held dangling and matted together by brown fungus threads. Infection recurs annually in moist weather with high daytime temperatures. The fungus winters as mycelium in leaf lesions and diseased twigs. Sclerotia are apparently lacking in this species.

Rhizoctonia sp. (Perfect Stage, *Aquathanatephorus pendulus*). Blight; on water hyacinth.

Rhizoctonia solani. Blight; of *Cynodon* spp., and foliar blight of soybean.

Rhizopus

See under Rots.

Rhizopus stolonifer. Seedling blight; on lupine; also caused by *Pleiochaeta setosa*, *Alternaria* sp., *Aspergillus flavus*, *Aspergillus niger*, and *Curvularia* sp.

Rosellinia

Ascomycetes, Sphaeriales, Sphaeriaceae

Perithecia separate, superficial from the first, carbonaceous, not beaked, ostioles papillate; spores dark, one-celled with a small groove.

Rosellinia herpotrichioides. Hemlock needle blight; needle-bearing portions of twigs become covered on underside with a grayish brown mycelial mat; black perithecia are produced in this mat in great abundance. Ovoid, hyaline conidia are formed on *Botrytis*-like conidiophores.

Schirrhia

Ascomycetes, Dothideales, Dothideaceae

Asci usually short, cylindrical, and relatively numerous in spherical, ostiolate locules.

Scleropycnium

Fungi Imperfecti, Sphaeropsidales, Excipulaceae

Pycnidia open out to a deep cupulate or discoid structure, tough, dark or black, subepidermal or subcortical, then erumpent; spores hyaline, one-celled. Largely saprophytic on twigs, sometimes parasitic on leaves.

Scleropycnium aureum. Leaf blight; of mesquite.

Sclerotinia (Whetzelinia)

Ascomycetes, Helotiales, Sclerotiniaceae

Apothecia arising from a tuberoid sclerotium that is sometimes enclosed in natural cavities of suscept or host, as in hollow stem of perennials, though formed free on aerial mycelium. Interior (medulla) of sclerotium white, completely enveloped by a dark rind; gelatinous matrix lacking. Conidia wanting but spermatia (very small microconidia) formed on sporodochia borne free or enclosed in cavities. Apothecia some shade of brown; cupulate to funnel-form; usually at maturity saucer-shaped to flat-expanded; ascospores hyaline, one-celled, ovoid. Species formerly included in *Sclerotinia* but possessing monilioid conidia are now in *Monilinia*.

Sclerotinia camelliae. Camellia flower blight; long known in Japan, first noted in California in 1938, confirmed in Georgia in 1948, although probably there several years previously, reported in Oregon in 1949, Louisiana and North Carolina in 1950, South Carolina in 1954. The blight is now widespread in Virginia, confined to certain counties in other states. It was not officially recorded from Texas until 1957 but must have been there earlier. The 1950 outbreak at Shreveport, Louisiana, is said to have started on plants brought in from Texas that probably originated in California.

Floral parts only are affected, infection taking place any time after tips of petals are visible in opening buds. Few to many brownish specks on expanding petals enlarge until the whole flower turns brown and drops. In early stages darkened veins are prominent diagnostic symptoms. When the flowers rest on moist earth, spermatia are produced on petals in shiny black masses. Hard, dark brown to black sclerotia formed at the base of petals frequently unite into a compound structure simulating petal arrangement. This compound sclerotium may be an inch or more in diameter. Although the petals do not melt when touched as do azaleas with petal blight, there is a distinctive moist feeling that helps to differentiate flower blight from frost injury. Rarely, a flower blight of camellias is caused by another *Sclerotinia* (*S. sclerotiorum*).

Sclerotia lie dormant on ground or in mulching materials until the next winter when, from January on (possibly earlier), after wet periods with rising temperature, they produce one to several apothecia on long or short stipes with brown, saucerlike discs ¼ to ¾ inch across, rarely up to 1 inch. Spores, discharged forcibly, are carried by wind currents to flowers, thus completing the cycle. Spores may be wind-borne at least ⅓ mile, but presumably a large proportion of them land on opening petals of the bush overhead. The sclerotia remain viable in the soil at least 2 or 3 years, sending up more apothecia each season. No conidia are known, so there is no secondary infection from flower to flower as with azalea blight. The amount of primary inoculum is very large, however. One afternoon in New Orleans I collected nearly a thousand sclerotia that were producing apothecia from under a single camellia.

Control. The first line of defense is exclusion. Most southern states have quarantines against known infected areas; they require that plants be shipped bare-rooted, with all flower buds showing color removed. Northern gardeners ordering plants for greenhouses should insist on the same precautions even without specific quarantines. Practically all outbreaks of camellia flower blight have been traced to plants shipped in cans, presumably carrying sclerotia in the soil. The disease has also appeared on flowers shipped in by air for camellia shows. Schedules should state that all specimens become the property of the show committee, to be destroyed at the end of the show; no blooms should be taken home for propagation.

Theoretically, because there is no conidial stage to spread the fungus, this disease should be easy to eradicate, but it has not proved so in practice. Camellias have thousands of flowers produced over a period of months. They drop into various ground covers, and it is almost impossible to find and destroy all infected blooms before rotting tissues release sclerotia into the litter. Some cities have quarantined infected properties and provided a host-free period of 2 years, during which all flower buds are removed from all camellias in the area, but this approach has been only partially successful. Various chemicals have been tried as ground treatment to inhibit formation of apothecia.

Sclerotinia minor. Blight; of soybean.

Sclerotinia (*Botryotinia*) **polyblastis.** Narcissus fire; a serious flower blight in England, known in the United States on the Pacific Coast. In England overwintering sclerotia produce apothecia when *Narcissus tazetta* comes into flower, the ascospores infecting the perianth and causing flower spotting. From withered flowers numerous large conidia, germinating with several germ tubes, infect foliage, on which large sclerotia are formed late in the season. Remove infected parts immediately; spray early in the season.

Sclerotinia sclerotiorum. Shoot and twig blight; of lilac, grape, soybean, and malaviscus; flower blight of camellia resembling that caused by *S. camelliae* but far less serious. This ubiquitous fungus more often causes stem rots on its many different hosts. See under Rots.

Sclerotium

Fungi Imperfecti, Mycelia Sterilia

Asexual fruit bodies and spores lacking; there is merely a resting body, sclerotium, made up of a compact, rounded mass of light-colored hyphae with a brown to black rind; parasitic, often on underground plant parts. *Pellicularia* has proved to be the perfect stage for some forms.

Sclerotium bataticola. Ashy stem blight; see *Macrophomina phaseoli* under Rots.

Sclerotium hydrophilum. Of wild rice.

Sclerotium oryzae. Blight; of wild rice.

Sclerotium rhizodes. White tip blight of grass; see under Snowmold.

Sclerotium rolfsii. Southern blight; see *Pellicularia rolfsii.*

Septoria

Fungi Imperfecti, Sphaeropsidales, Sphaerioidaceae

Pycnidia dark, separate, globose, ostiolate; produce in spots, erumpent; conidiophores short, conidia hyaline, narrowly elongate to filiform, several septate; parasitic, typically causing leaf spots, but also blights and blotches (see Fig. 18). There are about a thousand species.

Septoria apii and **S. apii-graveolentis.*** Celery late blight; general on celery, also on celeriac. The two species, singly or together, produce the disease known as late blight, first reported in Delaware in 1891, and causing much crop destruction since then, one California county reporting half a million dollars loss from celery blight in 1908 and Michigan a million in 1915. It was not known until 1932 that two distinct species were involved.

Early symptoms are similar. Large leaf spot, due to *S. apii*, starts as a light yellow area, which soon turns brown and dies. Spots are up to ¼ inch in

*Recent study indicates that these species are one species and that the name should be *S. apiicola.*

diameter, with small black pycnidia. In small leaf spot, due to *S. apii-graveolentis*, the more common and destructive pathogen, pycnidia appear at the first sign of chlorotic spotting and are often outside of the indefinite margins of the spots, which are not over 2 millimeters. If infection is severe, the spots fuse, and the leaves turn brownish black and rot. Leaf stalks may also be infected. Pycnidia winter on seed and in plant refuse in garden and compost.

A single pycnidium of the small-spot fungus has an average of 3,675 spores, extruded in gelatinous tendrils. A single leaf spot may average 56 pycnidia, and a single plant may have 2,000 spots. Thus, there are enormous amounts of inoculum to be spread by rain, insects, people, and tools. Some years ago on Long Island, when celery was intercropped with spinach, it was found that workers spread blight spores on their sleeves as they cut the spinach in early-morning dew. And there is a case on record where a man walked through his own blighted celery before taking a diagonal path across his neighbor's healthy field. In a few days blight showed up all along that diagonal path.

Control. The fungus usually dies in the seed coat while the seed is still viable. Using celery seed more than 2 years old obviates the necessity for treatment. Fresh seed can be soaked in hot water for 30 minutes at 118° F to 120° F. Use crop rotation; do not plant near where celery was grown the year before. Spray with bordeaux mixture or a fixed copper, starting in the seedbed when plants are just out of the ground.

Septoria leucanthemi.* Leaf blight; blotch on chrysanthemum, shasta daisy, and oxeye daisy. The generally destructive *Septoria* on chrysanthemum is *S. chrysanthemi*. See under Leaf Spots.

Septoria petrosellini. Leaf blight of parsley; similar to late blight of celery but confined to parsley.

Septotinia

Ascomycetes, Helotiales, Sclerotiniaceae

Stroma a definite, small, thin, elongate to angular black sclerotium maturing in host tissue after it has fallen to ground. Apothecia shallow, cup-shaped, stipitate; spores hyaline, ovoid, one-celled. Conidial stage a *Septogloeum*, with hyaline spores, two or more cells, formed on sporodochia.

Septotinia podophyllina. Leaf blight of May-apple; found on leaves and stalks of this plant only.

---

*Recent study indicates that S. *leucanthemi* and S. *petrosellini* are one species and the name should be S. *apiicola.*

Sirococcus

Fungi Imperfecti, Sphaeropsidales, Sphaerioidaceae

Small, rounded, black, semi-immersed pycnidia with wide ostioles; conidia hyaline, fusiform, slightly constricted, 1-septate.

Sirococcus strobilinus. Shoot blight; of *Picea, Abies, Pinus,* and *Tsuga* spp.

Sphaerulina

Ascomycetes, Sphaeriales, Sphaeriaceae

Perithecia innate or finally erumpent, not beaked; paraphyses and paraphysoids lacking; spores hyaline, several-celled.

Sphaerulina polyspora. Twig blight; of sourwood, and oxydendron.
Sphaerulina taxi. Needle blight; of yew.

Sporodesmium

Fungi Imperfecti, Moniliales, Dematiaceae

Conidiophores clustered, dark, short, simple, each bearing a terminal conidium; conidia dark, quite large, muriform with many cells, oblong to ovoid; usually saprophytic, sometimes parasitic.

Sporodesmium maclurae. Leaf blight; of osage-orange.
Sporodesmium scorzonerae. Salsify leaf blight; leaves have many circular spots, varying from pinpoint to ¼ inch, brown with red borders. Leaves or whole tops die; roots are small and unsalable. The fungus winters as mycelium and spores in plant refuse. May be the same as *Alternaria tenuis.*

Stemphylium

See Leaf Spots.

Stemphylium vesicarium. Stemphylium blight of onions; lesions are non-delineated, light yellow to brown, water-soaked and range in length from one centimeter to the entire leaf.

Systremma

Ascomycetes, Dothideales, Dothideaceae

Asci in locules in an elongated stroma, which is erumpent and superficial at maturity; spores light brown, two-celled. Conidial state *Lecanosticta* with brown conidia, two to four cells, formed on a conidial stroma resembling an acervulus.

Systremma acicola. Pine brown spot needle blight; on southern pines, most serious on longleaf. The name and classification of the fungus has been in dispute. The conidial stage, known since 1876, was first listed as *Septoria*, later placed in *Lecanosticta*. The perfect stage was named *Scirrhia acicola* in 1939 but later transferred to *Systremma* because of its colored spores.

Most injurious on seedlings, needle blight may also injure large trees. Small, gray-green spots on needles turn brown and form a narrow brown band, the needle tips dying. Three successive seasons of brown spot kill longleaf seedlings. The fungus is more severe on trees in unburned areas because of accumulation of inoculum. Spray seedlings in plantations with bordeaux mixture every 2 weeks from May to October or November.

Thelephora

Basidiomycetes, Agaricales, Thelephoraceae

Fruiting body leathery, upright, stalked; pileate or fan-shaped or much lobed, or in an overlapping series; hymenium on the underside, smooth or slightly warty; spores one-celled.

Thelephora spiculosa. Stem blight; found on azalea, fern, and other ornamentals in a Maryland garden. The fungus formed a dense weft of mycelium on surface of the soil and on plants.

Thelephora terrestris. Seedling blight; smother; the mycelium ramifies in the soil, and the leathery fruiting body grows up around the stem of a seedling conifer or deciduous tree, smothering it or strangling it without being actually parasitic on living tissue. The disease occurs most often in crowded stands in nurseries. The damage is seldom important.

Tryblidiella

Ascomycetes, Helotiales, Tryblidiaceae

Apothecia opening by a wide cleft; spores dark, cylindrical, with several cells.

Tryblidiella rufula. Twig blight; on citrus.

Volutella

Fungi Imperfecti, Moniliales, Tuberculariaceae

Sporodochia discoid, with marginal dark setae; conidiophores usually simple, in a compact palisade; conidia hyaline, one-celled, ovoid to oblong; parasitic or saprophytic (see Fig. 18).

Volutella buxi. Boxwood leaf blight; Nectria canker. Pinkish spores occur as pustules on leaves and twigs. Leaves often turn straw-colored. See further under Cankers.

Volutella pachysandrae (*Pseudonectria pachysandricola*). **Pachysandra leaf and stem blight**; large areas of leaves turn brown to black, along with portions of stems, and in wet weather numerous pinkish spore pustules appear along stems. The blight is most serious when pachysandra has been injured or is too crowded or is kept too moist by tree leaves falling into the bed. Spraying once or twice with bordeaux mixture gives excellent control if severely blighted plants have been removed before treatment. Keep pachysandra thinned and sheared back periodically.

BLOTCH DISEASES

Diseases designated as blotch have symptoms that are intermediate between blights, where the entire leaf or shoot dies, and leaf spots, where the necrotic lesions are definitely delimited. Blotches are irregular or indefinite large or small necrotic areas on leaves or fruit.

Alternaria

See under Blights.

Alternaria porri. Purple blotch of onion; also on garlic, and shallot, a problem in southern and irrigated areas. Small, white, circular to irregular spots increase to large purplish blotches, sometimes surrounded by orange and yellow bands, on leaves and flower stalks. Leaves often turn yellow and die beyond the spots; girdled stalks die before seeds mature. Brown muriform spores form a dusky layer on the blotches. Varieties with a waxy foliage are more resistant than those with glossy leaves. The fungus winters as mycelia and spores in crop refuse. Rotation, cleaning up plant debris, and seed treatment are recommended.

Two other species of *Alternaria*, *A. tenuis* and *A. tenuissima*, may cause purple or brown blotches on onion, and there are physiological races as well.

Cercospora

See under Blights.

Cercospora concors. Potato leaf blotch; an unimportant disease; leaflets turn yellow with small blackened dead areas or larger, irregular brown areas.

Cercospora purpurea. Avocado blotch; Cercospora spot, considered the most important avocado disease in Florida with no commercial variety entirely resistant. Leaf spots are angular, brown to chocolate brown, scattered and distinct, less than $\frac{1}{16}$ inch or coalescing to larger patches. With a hand lens, grayish spore groups can be seen on both sides of the leaf. Successive crops of spores are produced in moist periods throughout the year. Fruit spots are $\frac{1}{4}$ inch or less in diameter, brown to dark brown, irregular, sunken, with cracked surfaces and grayish spore tufts. Lesions are confined to the rind so that the flesh is not affected, but the cracks furnish entrance to anthracnose and other decay organisms. The fungus winters in leaves, and appears to be progressively more abundant.

Cladosporium

Fungi Imperfecti, Moniliales, Dematiaceae

Conidiophores dark, branched variously near upper or middle portion, clustered or single; conidia dark, one- or two-celled, variable in size and shape, ovoid to cylindrical, borne singly or in chains of two or three; parasitic or saprophytic.

Cladosporium herbarum. Leaf blotch of lilac; the fungus is usually secondary, saprophytic, following blights.

Cladosporium paeoniae. Peony leaf blotch; red stem spot, measles. Leaf and stem spots are purplish or brownish red. On stems the spots are raised, up to 4 millimeters long; on leaves the lesions are small specks. Small reddish spots are also present on floral bracts and petals. The disease is widely distributed in commercial plantings and may sometimes destroy the value of flowers for cutting. Cut down tops in fall as for Botrytis blight. Spraying the ground with Elgetol in spring before new growth starts has given good control in some fields.

Gloeodes

Fungi Imperfecti, Sphaeropsidales, Leptostromataceae

Pycnidia dimidiate, having a radiate cover over the top half only, on a dark subicle or mycelial crust; pseudoparaphyses present; conidia hyaline, one-celled.

Gloeodes pomigena. Sooty blotch of fruit; apple, crabapple, blackberry, pear, and citrus, in eastern and central states down to the Gulf, rare in the West. Fruit may be infected by heavy spore dissemination from pycnidia on twigs of various wild trees, including persimmon, prickly ash, white ash, bladdernut, hawthorn, red elm, sassafras, maple, sycamore, and willow. On apples, clusters of short dark hyphae make a superficial thallus on the cuticle, which appears as a sooty brown or black blotch, ¼ inch in diameter. Numerous spots may coalesce to cover the apple, a condition known as cloudy fruit. Because the lesion is superficial the fruit flesh is little affected, but the grade and market value are reduced. On citrus the fungus does not penetrate the rind, and spots can be removed by gentle hand rubbing. The disease develops in cool rainy weather during the summer. To control open up the trees in the orchards to facilitate quick drying.

Guignardia

Ascomycetes, Sphaeriales, Mycosphaerellaceae

Perithecia immersed in substratum, stroma lacking, mouths papillate; spores hyaline unequally two-celled, with lower cell cut off just before maturity.

Guignardia aesculi. Horse-chestnut leaf blotch; buckeye leaf blotch, general on horse-chestnut and Ohio buckeye, sometimes on red and yellow buckeye. Large, reddish brown blotches in foliage are usually surrounded by a yellowish area. Numerous pinpoint black dots, pycnidia, distinguish blotch from scorch due to drought. Petioles often have reddish oval spots. In a rainy season there is a good deal of secondary infection from spores spread by wind and rain. Blotches appear on nearly every leaflet with extensive defoliation. Primary infection in spring comes from ascospores developed in fallen overwinter leaves.

Control. Rake up and burn leaves in fall. Feed trees that have been defoliated for successive years.

Mycosphaerella

See under Blights.

Mycosphaerella dendroides (*Cercospora halstedii*). Pecan leaf blotch; on pecan in the South, on hickory in East and South, a foliage disease of nursery and orchard trees. Olive green velvety tufts of conidiophores and spores appear on undersurface of mature leaves in June and July (in Florida), and yellow spots appear in corresponding areas on upper leaf surfaces. Black pimplelike perithecia are produced in the tufts about midsummer, united in groups to give the leaf a shiny black, blotched appearance after the spores are washed away. In nursery trees, defoliation, starting with basal leaves and progressing upward, may be serious. The disease is of little consequence to orchard trees unless they have been weakened by overcrowding, borer attack, or other cause. The fungus winters in fallen leaves. To control clean up fallen leaves.

Mycosphaerella diospyri. Leaf blotch; of Japanese persimmon.

Mycosphaerella lythracearum (*Cercospora punicae*). **Leaf blotch**; fruit spot of pomegranate. The imperfect stage has been thought the same as that on crape myrtle (*Cercospora lythracearum*), but is now considered distinct. Leaf spots are circular, small, dark reddish brown to almost black, sometimes grayish brown.

Phoma

See under Blackleg.

Phoma arachidicola. Web blotch; of peanut.

Phyllosticta

See under Blights.

Phyllosticta congesta. Leaf blotch; of garden plum.

Phyllosticta solitaria. Apple blotch; widespread on apple and crabapple in eastern states, serious in the South and in the Ozark section of Missouri, Arkansas, Oklahoma, and Texas. The disease is also called fruit blotch, dry rot, black scab, late scab, cancer, and tar blotch. From Kansas eastward it is second in importance to apple scab. Leaf spots are very small, round, white, with a single black pycnidium in the center of each. Larger elongate lesions are formed on veins, midribs, and petioles. Leaves do not turn yellow, but they drop prematurely if spots are numerous. Cankers on twigs and branches are located at leaf nodes or base of spurs. The first season they are small, purple to olive in color; the next season this portion is tan and the new area is dark purple, often slightly raised. Pycnidia formed in twig lesions wash to leaves, fruit, and new shoots, discharged only after heavy rains and in warm weather. Heavily fertilized trees are more susceptible.

Fruit blotches are brown, irregular, feathery at the margin, studded with numerous pycnidia. They afford entrance to secondary decay organisms and may develop deep cracks, but the blotch fungus itself is superficial. It winters in infected twigs and bark cankers.

Control. Secure healthy nursery stock. Some varieties, including Grimes Golden, Jonathan, Stayman Winesap, and Winesap, are rather resistant.

Septoria

See under Blights.

Septoria agropyrina. Brown leaf blotch; on wheat grasses.
Septoria elymi. Speckled leaf blotch; on wheat grasses. A salt and pepper effect with numerous pycnidia in pale gray, tan, or fuscous lesions.
Septoria macropoda. Purple leaf blotch; general on blue grasses. Irregular blotches on blades are mottled greenish, then gray, tan or brown, finally bleached nearly white. Pycnidia are round, flattened, and light brown.

Zygophiala

Fungi Imperfecti, Moniliales

A genus described from banana leaves in Jamaica.

Zygophiala jamaicensis. Greasy blotch of carnation; a tropical fungus found causing serious losses in California greenhouses in 1953 and reported from Pennsylvania in 1957. Small, radiate patterns, resembling spider webs, appear as if dipped in oil. Leaves become brittle, turn yellow, and die prematurely. The same fungus is present as a flyspeck on apple.

BROOMRAPES

Broomrapes are parasitic seed plants like dodder and mistletoe. They are leafless herbs, of the family Orobanchaceae, living on roots of other plants and arising from them in clumps of whitish, yellowish, brownish, or purplish stems. There are 130 or more species, mostly from North Temperate regions, but few have any garden importance. The seed germinates in soil and produces a filiform plant body that grows into the ground, penetrating crown or root of the host plant and forming a more or less tuberous enlargement from which the flowering shoots arise. Such shoots may be nearly naked, clothed only with a few scattered rudimentary leaves, or they may be covered with conspicuous, overlapping scalelike leaves. The seed may remain viable in the soil several years but probably not as long as has been believed, for they can live on some weeds between crops.

Orobanche ludoviciana. Louisiana broomrape; on tomato and other plants, including Spanish needle and coldenia, becoming a problem in California. Tomatoes are stunted and do not produce a full crop of fruit.

Orobanche racemosa. Branched broomrape; hemp broomrape, most serious on hemp but parasitizing tomatoes, lettuce, tobacco, eggplant, *Ganra, Melitlotus, Silene,* poppy mallow, cranesbil, *Chaerophyllum, Verbena, Coreopsis,* fleabank, engelmann daisy, and other hosts in California. In small infections destroy the aerial stems before they set seed; practice crop rotation. Deep plowing gives some control.

CANKERS AND DIEBACKS

A canker is a localized lesion or diseased area often resulting in an open wound and usually on a woody structure. Starting as a definite necrotic spot, it may girdle cane, stem, or tree trunk, killing the water-conducting tissues so that the most prominent symptom becomes a dieback. When twigs and branches die back from the tip, the condition may be a blight, with the pathogen directly invading the dying area, or it may be a secondary effect from a canker some distance below.

Aleurodiscus

Basidiomycetes, Agaricales, Thelephoraceae

Hymenium resupinate, of one layer, with projecting spinose or short-branching cystidia (swollen sterile cells); spores hyaline. Facultative parasite on trees.

Aleurodiscus acerina. Bark patch; widespread on maple.

Aleurodiscus amorphus. Balsam fir canker; cankers are formed on main stems of saplings, which are sometimes killed, but the fungus is also widespread as a saprophyte on dead bark of firs and other conifers. Cankers center around a dead branch, are narrowly elliptical with a raised border; the dead bark is covered with a light-colored layer of the fungus.

Aleurodiscus oakesii. Oak bark patch; smooth patch of white oak. Irregularly circular, smooth, light gray sunken areas in bark vary from several inches to a foot across. The fungus is confined to dead bark; trees are not injured.

Amphobotrys

Fungi Imperfecti, Moniliales, Moniliaceae

Conidiophores are long, slender, pigmented, and highly branched; clusters of conidia at apex of each branch; conidia ovoid, one-celled, hyaline.

Amphobotrys ricini. Stem canker; on texasweed, and castorbean.

Apioporthe

Ascomycetes, Sphaeriales, Valsaceae

Perithecia in a black, carbonaceous stroma; spores two-celled, hyaline; conidia in cavities in a stroma.

Apioporthe anomala. Canker; twig blight of hazelnut.
Apioporthe apiospora. Twig canker; Dieback; of elm.

Ascospora

Ascomycetes, Sphaeriales, Sphaeriaceae

Perithecia with a subicle; paraphyses lacking; spores two-celled, hyaline.

Ascospora ruborum (*Hendersonia rubi*). **Cane spot;** dieback of red and black raspberry, dewberry.

Atropellis

Ascomycetes, Helotiales, Tryblidiaceae

Apothecia black, sessile or with short stalk; asci clavate, with longer, hairlike paraphyses; spores needlelike to slightly club-shaped, hyaline, one-celled.

Atropellis apiculata. Twig canker; on southern pines.
Atropellis arizonica. Branch and trunk canker; on western yellow pine.
Atropellis pinicola. Pine branch and trunk canker; on western white, sugar, and lodgepole pines in Pacific Northwest and California. Branches are girdled and killed, but not the trees. Perennial cankers are smooth, elongated, flattened depressions covered with bark, in which appear very small black apothecia, 2 millimeters to 4 millimeters in diameter.
Atropellis piniphila (*Cenangium piniphilum*). **Branch and trunk canker;** on lodgepole and ponderosa pines on Pacific Coast, on cultivated pines in the South. Trees 5 years to 25 years old are damaged by deformation of main stem and branches. Infection is at branch whorls. Cankers are elongated, flattened depressions covered with bark and copious resin. Apothecia have short stalks, are black with brownish discs, 2 millimeters to 5 millimeters across.
Atropellis tingens. Branch and trunk canker; of native and exotic hard pines from New England and lake states to Gulf states. Slash pine saplings are most susceptible. Smaller branches are girdled; perennial target cankers are

formed on larger branches and main stems. Cankers persist for many years, but extension stops after about 10 years.

Botryodiplodia

See under Blights.

Botryodiplodia gallae. Canker; of oak.
Botryodiplodia theobromae. Canker; of rose, and citrus.

Botryosphaeria

See under Blights.

Botryosphaeria dothidea. Canker; gummosis; dieback; on peach, Bradford pear, thornless blackberry, sequoiadendron and sequoia.

Botryosphaeria obtusa. Canker; on thornless blackberry.

Botryosphaeria ribis. Saprophytic on dying tissue, and var. **chromogena**, parasitic. Canker; dieback; of at least 50 woody plants, including apple, avocado, eucalyptus, fig, forsythia, hickory, pecan, pyracantha, quince, rhododendron, sequoia, sequoiadendron, sweet gum, and willow. See under Blights for the disease caused on currant and rose, under Rots for apple and avocado diseases.

On redbud, sunken oval cankers nearly girdle branches, the fungus entering through wounds, and dead and dying twigs. On rhododendron there is a leaf spot and dieback similar to that caused by *Phytophthora* except that the surface is roughened by protruding fruit bodies. Cankers on twigs, larger branches, and trunks of willow may kill trees in a few years. Trunk lesions are very small, ¼ inch to ½ inch, and numerous or else large, from the union of several small cankers, with fissured bark. Apples have watery blisters on bark and decline in vigor. Forsythia has affected canes girdled and killed with conspicuous brown dead leaves above the canker.

Control. Prune and burn dead twigs and heavily infected branches; paint wounds with a disinfectant followed by tree paint; avoid injuries. Copper sprays may help.

Botrytis

See under Blights.

Botrytis cinerea. Canker; of rose.

Caliciopsis

Ascomycetes, Dothideales, Coryneliaceae

Stroma lobed, each lobe containing a single locule, which is finally wide open; perithecia stalked; asci on long slender stalks; spores dark, one-celled.

Caliciopsis pinea. Pine canker; on eastern white pine and other species, also on Douglas fir. Cankers are sharply depressed areas in bark, reddish brown and smoother than rest of bark, up to several inches in diameter. Small, globose, clustered black pycnidia, and stalked perithecia looking like slender black bristles, arise from stroma in cankered bark. The disease is most serious on suppressed saplings.

Cenangium

See under Blights.

Cenangium singulare. Sooty-bark canker of aspen; on *Populus tremuloides* in Rocky Mountain area. Cankers on older trees, at any point on trunk up to 60 feet to 70 feet may extend 10 feet to 15 feet before they girdle the tree. The bark is sooty black with a thin white outer layer.

Ceratocystis (*Ceratostomella*)

Ascomycetes, Sphaeriales, Ceratostomataceae

Perithecia with very long beaks, carbonaceous or leathery; ascospores hyaline, one-celled; brown, ovoid conidia and one-celled rodlike endospores formed inside tubelike conidiophores and extruded endwise. Some species are important tree pathogens; see Oak Wilt and Dutch Elm Disease under Wilts.

Ceratocystis platani (*Endoconidiophora fimbriata* f. *platani*). Canker stain of London plane; plane blight; on London plane and also on American plane or sycamore. This serious disease started as a killing epidemic in the Philadelphia area about 1935, destroying city shade trees by the thousands there and in Baltimore during the next few years. The disease now extends from New Jersey to North Carolina and Mississippi. Trees show sparse foliage, smaller leaves, and elongated sunken cankers on trunks and larger branches. Cross sections through cankers reveal blue black or reddish brown discoloration of wood, usually in wedge-shaped sectors. First-year cankers may not be more than 2 inches wide and a yard or so long, but they widen annually, girdling and killing trees in 3 years to 5 years. Several cankers coalescing around the trunk kill more quickly. Once infection starts, the tree is doomed.

Ascospores and the two types of conidia are produced in moist spring weather (see Fig. 24). They may be spread by rain a short distance, but most dissemination is by humans in pruning operations, and ordinary tree paint carry viable spores. Some beetles may be vectors. Infection is solely through wounds.

Control. Do not try to save trees where trunk has been invaded; diseased branches may sometimes be removed, cutting at least 3 feet from infected area. Do not prune unless absolutely necessary and then only in winter when trees are less susceptible. Use tree wound dressing fortified with a disinfectant.

Ceratocystis sp. Canker; Dieback; on poplar.

Chondropodium

Fungi Imperfecti, Sphaeropsidales, Sphaerioidaceae

Pycnidia stromatic, stalked, columnar, externally black, hard, internally gelatinous; conidiophores simple; conidia hyaline, with several cells, crescent- or sickle-shaped; weakly parasitic or saprophytic.

Chondropodium pseudotsugae. Bark canker of Douglas fir; this is a superficial canker with outer layers of bark killed over small, circular to

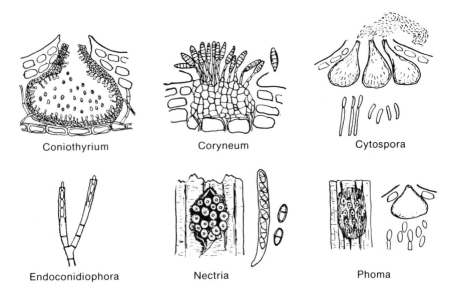

Coniothyrium Coryneum Cytospora

Endoconidiophora Nectria Phoma

Figure 24. Spore formation of some canker fungi. *Coniothyrium,* small dark spores on short conidiophores in pycnidium; *Coryneum,* dark, septate spores in acervulus; *Cytospora,* sausage-shaped spores in valsoid pycnidia expelled in cirrhi; *Endoconidiophora,* spores formed on inside of conidiophores; *Nectria,* two-celled bright ascospores in reddish perithecia clustered on bark; *Phoma,* hyaline spores in pycnidia formed in spots on bark.

elliptical areas, in which pycnidia project as short, blunt, black spines. Trees are not noticeably injured.

Coniothyrium

Fungi Imperfecti, Sphaeropsidales, Sphaerioidaceae

Pycnidia black, globose, separate, erumpent, ostiolate; conidiophores short, simple; conidia small, dark, one-celled, ovoid or ellipsoid; parasitic or saprophytic (see Fig. 24).

Coniothyrium fuckelii (imperfect stage of *Leptosphaeria coniothyrium*). Rose common canker; stem canker; widespread on rose, also causing raspberry cane blight (see *Leptosphaeria* under Blights), sometimes associated with apple rots, peach cankers, and stem canker of Virginia creeper. Of the three species of *Coniothyrium* that cause rose cankers, *C. fuckelii* is by far the most common. Any plant part may be affected. Pycnidia have even been found within blackspot lesions on leaves, but this disease is primarily a cane disease, starting as a red or yellow spot on bark, drying out and turning brown as it increases in size, with the epidermis somewhat wrinkled and perhaps rupturing irregularly over sooty masses of very small, olive brown spores. The stem may be girdled with dieback to that point.

Stem cankers are found around insect punctures, thorn pricks, leaf or thorn scars, or abrasions caused by tying, but the majority of cankers are formed at the cut end of a cane when a stub has been left in pruning above a leaf axil or bud. Roses cut properly close to a bud seldom develop this canker. A rose stub usually dies back to the first node, and since this fungus is a weak parasite, it starts most readily in such dead or dying tissue. When a cut is made close to the node, it is quickly callused over, and the callus is a good defense against wound fungi.

Control. Prune out cankered and dying stems as soon as noticed. Make all cuts just above a bud or leaf axil, not only at spring pruning but in cutting flowers for the house or cutting off dead blooms during the season.

Coniothyrium rosarum. Rose graft canker; this is a disease of roses under glass, starting at the union of stock and scion in the warm moist propagating frame and continuing in a large amount of dead wood when plants are removed to the greenhouse bench. Some consider the pathogen a form of *C. fuckelii.* Having measured spores of the type specimen, in the Kew Herbarium, I think they are distinct species, but that some cases of graft canker are due to the common canker fungus.

Coniothyrium wernsdorffiae. Rose brand canker; a rather rare but very serious disease. The pathogen was named in Germany in 1905 and was not reported in the United States until 1925, although it was subsequently shown to have been collected in Canada in 1912 and in Pennsylvania and Minnesota in 1914 and 1916. In 1926, a severe epiphytotic appeared at Ithaca, New York, in the Cornell rose garden, infecting about 90% of the climbers so seriously that the canes had to be cut to the ground. Since then it has been reported from a few other states, but in several instances it has been confused with common canker.

Small, dark reddish spots on canes enlarge and acquire a more or less definite reddish brown or purple margin, contrasting sharply with the green of the cane. The center of the spot turns light brown as the cells die, and little longitudinal slits appear over the developing pycnidia. Spores are olive brown, nearly twice the size of *C. fuckelii*, and released through epidermal slits instead of being spread in a sooty mass under the epidermis. Cankers formed under the winter protection of soil are black when roses are first uncovered in spring, which explains the name *Brandfleckenkrankheit*, meaning firespot disease.

C. wernsdorffiae is a cold-temperature fungus, infecting rose canes under the winter covering, entering through insect wounds, thorn scars, scratches, and occasionally through dormant buds. During a 4-year investigation at Ithaca, I found no infection on canes not hilled with earth or other moist cover over winter and no natural infection during the summer.

Control. Omit the usual winter protection of soil or other materials that keep canes moist. If brand canker is a problem, just fasten canes of climbers down near the ground, uncovered, and hope for the best. Loss from winter injury will be less than from the canker. Cut out diseased canes carefully.

Coryneum

See under Blights.

Coryneum cardinale (*Leptosphaeria* sp.). **Coryneum canker of Cypress;** bark canker, of cypress, incense cedar, common juniper, and oriental arborvitae. This disease, since its discovery in 1927, has been gradually exterminating Monterey cypress in most parts of California and is also serious on Italian cypress. Twigs, branches, and whole trees turn sickly, lose their leaves, and finally die.

The fungus attacks living bark and cambium, girdling twig and branch. Cankers appear first at base of lateral twigs; they are slightly sunken, dark, resinous, rough, with black spore pustules. Conidia have dark median cells, five cross-walls (see Fig. 24). They are spread by tools, in nursery stock, by wind and rain, and perhaps by birds and insects. Infection appears first in upper parts of trees, usually in spring during moist weather. Yellowing and browning of foliage together with gummy ooze at the cankers form conspicuous symptoms.

Control. Drastic surgery, removing wood well below infected parts, and spraying foliage heavily with bordeaux mixture help some, but with heavy infection the price of saving healthy trees is the removal and destruction of all diseased specimens. California citizens, threatened with the loss of the famous native stands of Monterey cypress at Point Lobos and Cypress Point, voluntarily destroyed their own plantings by the thousands.

Coryneum foliicola. Twig canker; fruit rot, widespread on apple, affecting twigs, foliage, and fruit.

Cryptodiaporthe

Ascomycetes, Sphaeriales, Valsaceae

Like *Diaporthe* but without blackened zones in substratum; spores hyaline, two-celled.

Cryptodiaporthe aculeans (*Sporocybe rhois*) Dieback; canker; of sumac.

Cryptodiaporthe castanea. Dieback; canker; of Asiatic chestnut, widespread, chiefly on seedlings or on larger trees in poor sites. Canker starts as a brown discoloration of bark of the trunk, limb, or twig, often girdling twig and then invading larger branch. Leaves on girdled branches wilt without yellowing, turn brown, and die. Bark splitting over callus formation at edge of diseased area forms pronounced canker. Conidia, two-celled, fusoid, are formed in pustules in bark; beaked perithecia are formed in groups by midsummer.

Control. Maintain vigor; plant on well-drained, fertile soil. Prune out diseased portions several inches below affected area.

Cryptodiaporthe salicina. Twig and branch canker; of willow.

Cryptomyces

Ascomycetes, Phacidiales, Phacidiaceae

Apothecia effuse, splitting irregularly; paraphyses present; spores hyaline, one-celled.

Cryptomyces maximum. Blister canker; on common and purple osier.

Cryptosporella

Ascomycetes, Sphaeriales, Melanconidiaceae

Perithecia in a circle in a stroma, with long necks converging in a common canal; spores one-celled, hyaline; conidia borne on surface of stroma.

Cryptosporella umbrina. Rose brown canker; a widespread and serious rose disease, first reported in Virginia in 1917 but known from herbarium specimens to have been present since 1903. The fungus was first placed in *Diaporthe* because of occasional two-celled spores.

Symptoms are most noticeable on canes, starting with very small purplish spots, the center soon turning white with a reddish purple margin (see Fig. 25).

Figure 25. Brown canker on rose.

Many small spots may be grouped on a single cane. During the winter, and especially on portions of canes covered with earth, cankers or girdling lesions are formed, often several inches long, with tan centers and purplish borders. In moist weather the surface of these large cankers is covered with yellow spore tendrils from pycnidia just under the bark; asci are also extruded in tendrils from perithecia.

Leaf spots are small purplish specks or larger dead areas, cinnamon buff to white, bordered with purple and with black pycnidia in the center. Marginal spots are subcircular. Buds are sometimes blighted; exposed petals of flowers have cinnamon-buff spots without the purple border. Infection is through wounds and also uninjured tissue.

Control. The best time to take care of brown canker is at spring pruning. Cut out every diseased cane possible. A dormant lime sulfur spray, immediately after pruning, kills spores that may have been spread in the process and may inhibit the fungus in initial lesions. Copper or sulfur sprays largely prevent summer infections. Brown canker is more likely to be serious where roses are overprotected for winter with salt, hay, leaves, or other material added to the mound of soil. I have no trouble with brown canker when roses are left unhilled over winter.

Cryptosporella viticola. Dead-arm disease of grapes; branch necrosis, widespread, especially in the Northeast, serious in Illinois, important in California. Small, angular spots with yellowish margins and dark centers are formed on leaves, stems of flower clusters and canes. The latter may split to diamond-shaped cankers, and by the next season the arm is dead or producing yellowed, dwarfed and crimped foliage. Lesions on cluster stems advance into fruit late in the season causing rotting. Pycnidia are developed on old wood; infection is often through pruning wounds.

Control. Make pruning cuts at least 6 inches below the lower margin of the infected part. Spray with bordeaux mixture when spores are extruded.

Cryptosporium

Fungi Imperfecti, Melanconiales, Melanconiaceae

Acervuli erumpent, becoming cup-shaped or disclike; stroma brownish; conidiophores simple or branched; conidia hyaline or subhyaline, one-celled filiform.

Cryptosporium minimum. Canker; on rose, not common.
Cryptosporium pinicola. Canker; branch mortality of *Abies* spp.

Cylindrocarpon

See under Rots.

Cylindrocarpon cylindroides. Canker; branch mortality of Abies spp.

Cylindrocladium

See under Blights.

Cylindrocladium scoparium. Crown canker of rose; the cane is attacked at or just below the union of stock and scion, the bark darkening into a black, water-soaked punky region. The cankers girdle but do not kill the canes; there are fewer and more inferior blooms. The disease was long thought confined to greenhouse roses but has appeared once or twice in outdoor fields. The fungus lives in the soil and enters through wounds in the presence of sufficient moisture. Before planting of fresh stock, greenhouse benches should be washed with boiling water and soil sterilized or changed.

The same fungus injures seedling conifers in nursery rows, causing damping-off, root rot, stem canker, and needle blight to white pine and Douglas fir. See under Blights for a discussion of the pathogen on cuttings of azaleas and other ornamentals.

Cytospora

Fungi Imperfecti, Sphaeropsidales, Sphaerioidaceae

Cosmopolitan species, imperfect stage of *Valsa*. Pycnidia in a valsoid stroma with irregular cavities, incompletely separated; conidia hyaline, one-celled, allantoid, expelled in cirrhi (see Fig. 24).

Cytospora abietis. Canker; branch mortality, of *Abies* spp.

Cytospora annularis. Canker, dieback; of ash, on twigs and branches.

Cytospora chrysosperma (*Valsa sordida*). Cytospora Canker of poplar; aspen, cottonwood, willow, occasional on mountain ash, maple, cherry, and elder. Cankers form on trunks and large branches, most often on trees of low vigor. Bark is discolored in more or less circular areas; sapwood is reddish brown. In old cankers exposed wood is surrounded by layers of callus tissue. In moist weather spring spore tendrils are extruded from pycnidia in dead bark. Perithecia are found infrequently in aspen, arranged circularly around a grayish disc; they are flask-shaped with long necks pushing through the bark. Twigs and small branches may die back without a definite canker. The fungus is often present on healthy trees, not becoming pathogenic until the trees are weakened by neglect, drought, pollarding, or other causes. Entrance is through wounds. Lombardy and Simon poplars are frequently killed.

Control. Remove dead and dying branches and trees with extensive cankers. Avoid wounds; feed and water as necessary. Plant poplars that are less susceptible than Lombardy. Rio Grande cottonwood is resistant to twig blight.

Cytospora kunzei (perfect stage *Valsa kunzei* var. *piceae*). Cytospora canker of spruce; twig blight, common and serious New England to the Midwest. Cankers start around bases of small twigs or on trunks. Browning and death of Colorado blue spruce branches starts near the ground and progresses upward, a large flow of resin on affected limbs. Needles drop immediately or persist for a time. Cankers are formed near resin spots and yellow tendrils extruded. Spores are splashed by rain and wind to other branches; infection is mostly through wounds.

Another form of the pathogen, *Valsa kunzei* var. *superficialis*, occurs on pine and variety *kunzei* on balsam fir, Douglas fir, larch, and hemlock.

Control. There are no satisfactory control measures except removal of diseased branches and perhaps carefully excising cankered bark. Spraying with bordeaux mixture has been recommended but is seldom very effective. Avoid wounding ornamental trees with lawn mowers; sterilize pruning tools between cuts; feed to renew vigor.

Cytospora leucostoma. Canker; of black cherry.

Cytospora nivea. Canker; dieback; of poplar and willow, similar to that caused by *C. chrysosperma*; occasional.

Cytospora sambucicola. Branch canker; of elder.

Cytospora sp. Canker; on pecan.

Cytospora spp. Cytospora canker of Italian prune; causing severe injury to prune and apricot in Idaho orchards since 1951, also present on cherries, peach, apple, and willows. Some orchards have been lost, others hard hit. Symptoms are yellow to brown flags of dead leaves and erumpent, gummy cankers or elongated necrotic streaks in the bark. All suspicious wood should be cut out, hauled out of the orchard, and burned.

See *Valsa cincta* for further discussion of cankers on stone fruits.

Dasyscypha

Ascomycetes, Helotiales, Helotiaceae

Apothecia stalked, white and hairy on the outside with a bright disc; paraphyses filiform; asci inoperculate; spores elliptical to fusoid.

Dasyscypha agassizi. Common on blister-rust lesions of white pine; saprophytic on dead branches.

Dasyscypha calycina (*Trichoscyphella hahniana*). On larch and fir, ordinarily a saprophyte but can be a weak parasite; occasional on blister-rust cankers.

Dasyscypha ellisiana. Canker of Douglas fir and pine; in eastern United States. This is a native fungus on twigs and branches of native and introduced pines and on basal trunk and branches of Douglas fir. Bark on trunk may be infected for 10 feet to 15 feet, with copious resin flow and numerous swellings, but trees are not killed. Apothecia are short-stalked, covered with white hairs, with an orange to yellow disc, 2 millimeters to 4 millimeters across. Remove trees with trunk cankers.

Dasyscypha pseudotsugae. Canker; on Douglas fir. Swollen open cankers, 2 inches to 3 inches long, are formed on suppressed saplings.

Dasyscypha resinaria. Canker; on balsam fir. Swollen cankers at base of branches; younger stems girdled and killed.

Dasyscypha willkommii (*Trichoscyphella willkommii* syn. *Lachnellula* *willkommii*). **European larch canker;** found in Massachusetts in 1927 on nursery stock from Great Britain. Infected trees were removed and the fungus was not seen again until 1935, near the original location. Perennial branch or trunk cankers are flattened depressions, swollen on the flanks and on the opposite side of the stem. Neighboring bark is somewhat cracked and dark with heavy exudation of resin. Cup-shaped apothecia are 3 millimeters to 6 millimeters across with white hairs and orange to buff discs, very short stalks. Young trees may be killed; older trees usually survive. Frost wounds are a contributing but not an essential factor. Promptly remove all trees showing cankers; continue periodic inspection.

Dermatea

Ascomycetes, Helotiales, Dermatiaceae

Apothecia small, brownish to black with a circular opening; innate at first, on a stromoid base, rupturing host at maturity; spores one-celled, hyaline, globose to oblong.

Dermatea acerina. Bark canker; of maple, occasional.
Dermatea balsamea. Twig canker; of hemlock.
Dermatea livida. Bark canker; of redwood.

Dermea

Ascomycetes, Helotiales, Dermatiaceae

Cup fungi (ascocarp cup-shaped); excipulum of subglobose cells; sclerotia absent.

Dermea pseudotsugae. Branch canker; on fir.

Diaporthe

See under Blights.

Diaporthe cubensis. Canker; of *Eucalyptus* spp.
Diaporthe eres. Canker, dieback; of English holly in the Northwest. The fungus name is a species complex that may include a *Diaporthe* on rose petals and one causing a peach constriction disease.
Diaporthe helianthi. Canker; of sunflower; also leaf spot of sunflower.
Diaporthe oncostoma. Canker, dieback; of black locust.
Diaporthe phaseolorum var.**caulivora. Canker;** of soybean.
Diaporthe pruni. Twig canker; on black cherry. *D. prunicola* on American plum.

Dichotomophthora

Fungi Imperfecti, Moniliales, Dematiaceae

Conidiophores brown, branching dichotomous to subdichotomous, elongated, terminal branches four- to eight-lobed, each lobe bearing single conidium; conidia dark, ovoid to elongate—ovoid, one- to six-celled.

Dichotomophthora portulacae. Stem canker; root rot; on common purslane.

Didymella

See under Blights.

Didymella sepincoliformis. Dieback; of rose.

Diplodia

See under Blights.

Diplodia sp. Rose dieback; sometimes after drought and other contributing factors. In Texas the disease is most evident in autumn, progressing on roses in storage or overwintering in the ground. Canes die from tip downward, often starting in the flower stem. Diseased wood turns brown or black, and is somewhat shriveled. Pycnidia are produced in dead canes. Improve general rose vigor; use fungicides as for blackspot. May also cause canker of Russian olive.

Diplodia camphorae. Canker, dieback; of camphor-tree.

Diplodia infuscans. Ash canker; dieback, northeastern states.

Diplodia juglandis. Dieback; widespread on branches of walnut.

Diplodia natalensis. Stem canker; of prickly-ash. Dieback; of citrus twigs, also causing citrus stem-end rot. See under Rots.

Diplodia quercina. Canker; blight; of oaks.

Diplodia sophorae. Dieback; of pagoda tree.

Diplodia sycina. Canker, dieback; of fig.

Discella

Fungi Imperfecti, Sphaeropsidales, Excipulaceae

Pycnidia cupulate or discoid; spores two-celled, hyaline.

Discella carbonacea. Twig canker; of willow.

Dothichiza

Fungi Imperfecti, Sphaeropsidales, Sphaerioidaceae

Pycnidia innate, finally erumpent; conidiophores lacking; conidia hyaline, one-celled.

Dothichiza populea. Dothichiza canker of poplar; European poplar canker, widespread but sporadic as a branch and trunk canker. Lombardy poplars are most susceptible, but hosts include black and eastern cottonwoods, balsam, black and Norway poplars. Japanese poplars are rather resistant. Young trees in nurseries are most injured, cankers often starting around wounds. They start as slightly darker, sunken areas, often at base of twigs and limbs, and become elongated. The bark is killed to the cambium; sapwood is brown. If a stem is completely girdled, it dies; otherwise, callus formation goes on through the summer, over the canker. In time diseased bark turns brown and cracks. Spores are extruded in amber tendrils, drying to brown, and are washed to wounds in the wood.

Control. Destroy infected stock in nurseries and plantations; do not move stock from a nursery where the disease is known. Avoid pruning and other wounds as much as possible; sterilize tools between cuts. Spraying nursery trees with bordeaux mixture in spring may be helpful.

Dothiora

Ascomycetes, Pseudosphaeriales, Pseudosphaeriaceae

Ascocarps hairy and phragonosporous or muriform ascospores are colored.

Dothiora polyspora. Canker; of aspen.

Dothiorella

Fungi Imperfecti, Sphaeropsidales, Sphaerioidaceae

Pycnidia dark, globose, grouped in a subcortical stroma; conidiophores simple, short; conidia hyaline, one-celled, ovoid to ellipsoid; parasitic or saprophytic on wood.

Dothiorella fraxinicola. Branch canker; of ash.

Dothiorella quercina. Dothiorella canker of oak; very destructive to red and white oaks in Illinois, affecting twigs, branches, and occasionally trunks. Cankers are dark brown, elongated, sunken, often with cracks at the margin. Pustules of pycnidia develop in bark and erupt through cracks, spores oozing on the surface. Sapwood has dark streaks.

Dothiorella ulmi. Dieback; wilt of elm. See under Wilts.

Dothiorella sp. London plane canker; first noted in New York City in

1947. Infected trees have sparse, undersized foliage and narrow, longitudinal cankers on trunk and branches, varying from 1 inch to 4 inches wide and often extending from ground level to branch top. The bark is rough, deeply fissured; inner bark is brown, dry; sapwood is only superficially discolored. Branches wilt and die back.

Endothia

See under Blights.

Endothia gyrosa. Branch canker; on oak.

Epicoccum

See under Leaf Spots.

Epicoccum purpurescens. Canker; on thornless blackberry.

Eutypa

Ascomycetes, Sphaeriales

Stroma effuse; perithecia with necks at right angles to surface.

Eutypa armeniaca. Cytosporina dieback of apricot; imperfect state reported from California in 1962, perithecia in 1965. Bark cankers with gum are formed at pruning wounds.

Fusarium

See under Rots.

Fusarium moniliforme var. **subglutinans.** Pitch (branch) cankers; shoot dieback; on southern pine species, loblolly and pond pines.
Fusarium solani. Stem canker; of sweet potato, black walnut, and poinsettia.

Fusicoccum

Fungi Imperfecti, Sphaeropsidales, Sphaerioidaceae

Pycnidia one to several in a stroma, spherical or flattened, subepidermal, erumpent; opening separately or with a common pore; conidiophores simple, short; conidia hyaline, one-celled, fusoid; parasitic or saprophytic.

Fusicoccum amygdali. Twig canker of peach; increasingly important on peaches in North Atlantic coastal area. Leaf spots are large, irregular or circular, often zonate, brown with scattered pycnidia near center. Cankers at buds and bases of young twigs result in death of the distal portions; trunks of young trees may be girdled. Infections occur throughout the season at bud scales, stipules, fruit and leaf scars. Prune only in winter.

Fusicoccum elaeagni. Canker; on Russian-olive.

Gibberella

See under Blights.

Gibberella baccata. Twig canker; of acacia, ailanthus, apple, boxwood, mimosa, mulberry, and also on other plants where twig blight is the most important symptom. See under Blights.

Gloeosporium

See under Anthracnose.

Gloeosporium sp. Canker; on holly.

Gloeosporium sp. (*Gnomonia rubi*, perfect stage). Canker; on thornless blackberry.

Glomerella

See under Anthracnose.

Glomerella cingulata. Camellia dieback; canker; widespread; sometimes on azalea, blackberry, bittersweet, rose, raspberry, soapberry, mountain ash, and English ivy; also causing bitter rot of apple (see under Rots) and anthracnose of various hosts (see under Anthracnose). Camellia tips die back; leaves wilt, turn dull green, and finally brown. The stem dries out, turns brown, and there is a girdle of dead bark. Elliptical cankers are present on older wood. Infection is solely through wounds, principally leaf scars in early spring but also through bark wounded by cultivating tools or lawn mowers, frost cracks, or the graft union.

Governor Moulton, Professor Sargent, and some other varieties are rather resistant; Flora Plena, Prince Eugene Napoleon, and many others are highly susceptible. Spraying with bordeaux mixture to prevent infection through leaf and bud scars gives fair control.

Glutinium

Fungi Imperfecti, Sphaeropsidales, Sphaerioidaceae

Pycnidia innate, without a stroma; spores borne at tip and sides of conidiophores, hyaline, one-celled.

Glutinium macrosporum. Canker; fruit rot, of apple.

Griphosphaeria

Ascomycetes, Sphaeriales, Sphaeriaceae

Perithecial wall carbonaceous, mouths papillate; spores dark, with several cells.

Griphosphaeria corticola (imperfect state *Coryneum microstictum*). Rose canker; dieback; cankers are formed near base of canes, often showing dark glistening pustules of conidia. Occasionally when the canker has girdled the cane, a large gall forms above the lesion (see Fig. 26). It resembles crown gall but is apparently due to interference with downward transfer of food. Cut out infected canes.

Hendersonula

Fungi Imperfecti, Sphaeropsidales, Sphaeropsidaceae

Pycnidia black, stromata, one to several per stroma, locules occurring at different levels in stroma; conidophores long, flexuous; conidia often extruded in cirrhi; at first one-celled, hyaline to yellowish, later becoming three- to four-celled and dark.

Hendersonula toruloidea. Canker; on *Arbutus menziesii*.

Hymenochaete

Basidiomycetes, Agaricales, Thelephoraceae

Pileus, fruiting structure, resupinate, of several layers, with long, stiff, usually brown setae (cystidia).

Hymenochaete agglutinans. Hymenochaete canker; on apple, birch, hazelnut, sweetgum, mistletoe, and various young hardwoods. When an infected dead stem comes in contact with a live one, the mycelium forms a thin leathery fruiting body around the living stem, holding it to the dead stem.

Figure 26. *Coryneum* canker on rose.

This resupinate structure is deep brown in the center, with a yellow margin. The stem is constricted at the point of encirclement, and the sapling usually dies in 2 or 3 years. If the dead stem is removed before girdling, a sunken canker appears on one side, but this may be overgrown with callus and disappear. Do not leave severed stems in contact with living seedlings or saplings in nursery stands.

Hypoxylon

Ascomycetes, Sphaeriales, Xylariaceae

Perithecia in a pulvinate stroma, often confluent and crustose; ascospores with one cell, rarely two, blackish brown; conidia in superficial layer on surface of young stroma.

Hypoxylon pruinatum. Hypoxylon canker of poplar; aspen and large-tooth aspen are most commonly attacked, balsam poplar less frequently. This disease is usually a forest, rather than a home garden, disease. Trees less than 30 years old, growing on poor sites, are most susceptible. Trunk cankers start as small, yellow to reddish brown, slightly sunken areas, centering around a wound, then grow together to form a canker marked off by vertical cracks. The bark is mottled, gray, with black patches where the blackened cortex is exposed. Conidia appear in blisterlike stromata on first- and second-year cankers, whereas perithecia are formed on third-year cankers in hard, black stromata covered with a white pruinose coat. Ascospores are ejected in winter. Eliminate infected trees when thinning stands.

Kabatina

Fungi Imperfecti, Melanconiales

Kabatina juniperi. Blight; on eastern red cedar; conidia produced in black acervuli on discolored foliage.

Lachnellula

Ascomycetes, Helotiales, Helotiaceae

Apothecia mostly cup-shaped.

Lachnellula willkommii (= *Trichoscyphello willkommii*). Canker; of European larch (See Dasyscypha).

Leptosphaeria

See under Blights.

Leptosphaeria coniothyrium. Canker; on thornless blackberry.

Macrophoma

Fungi Imperfecti, Sphaeropsidales, Sphaerioidaceae

Like *Phoma*, with discrete pycnidia arising innately, but with much larger spores; conidia hyaline; one-celled.

Macrophoma candollei. Dieback; of boxwood but apparently saprophytic only. The large black pycnidia are, however, quite striking on straw-colored leaves.

Macrophoma cupressi. Dieback; of Italian cypress.

Macrophoma phoradendron. Defoliates Mistletoe, but it grows back.

Macrophoma tumefaciens. Branch gall canker; of poplar. Nearly spherical round galls, not over 1½ inches in diameter, at base of twigs, which usually die; not serious.

Massaria

Ascomycetes, Sphaeriales, Sphaeriaceae

Spores dark, with several cells, oblong-fusiform, with mucous sheath.

Massaria platani. Canker; widespread on branches of American, London, and California plane trees.

Melanconis

Ascomycetes, Sphaeriales, Melanconidiaceae

Perithecia in an immersed black stroma; paraphyses present; spores two-celled, light; conidia superficial on a stroma.

Melanconis juglandis. Walnut canker; butternut dieback; widespread on butternut, also on black, Japanese, and English walnut. The disease was first described from Connecticut in 1923, but evidently was responsible for slow dying of butternuts long before that. If trees have been previously weakened, the fungus proceeds rapidly; otherwise there is the slow advance of a weak parasite. Dead limbs are sprinkled with small, black acervuli, looking like drops of ink and occasionally, in wet weather, developing spore horns of olive gray conidia. In the perfect stage, which is rare, perithecia are embedded in the bark singly or in groups. Mycelium invades bark and wood, with a dark discoloration, and grows slowly down a branch to the trunk. When the latter is reached, the tree is doomed. In final stages trees have a stag-headed effect from loss of leaves.

Control. Remove diseased branches promptly, cutting some distance below infection; remove trees developing trunk cankers; keep the rest growing well with food and water.

Meria

Fungi Imperfecti, Moniliales, Moniliaceae.

Hyaline mycelium, branched; conidiophores simple, septate; conidia hyaline, one-celled, produced singly or in clusters.

Meria laricis. Dieback and blight; on western larch seedlings.

Monochaetia

Fungi Imperfecti, Melanconiales, Melanconiaceae

Acervuli dark, discoid or cushion-shaped, subcutaneous; conidia several-celled, dark median cells, hyaline end cells, and a single apical appendage; parasitic.

Monochaetia mali. Canker, leaf spot; of apple. Fungus enters through deep wounds and grows into wood, then attacks resulting wound callus and produces numerous fruiting bodies on exposed wood and callus layer. Killing of successive callus layers results in a canker similar to European apple canker. The disease is not common enough to be serious.

Nectria

Ascomycetes, Hypocreales, Nectriaceae

Perithecia bright, more or less soft and fleshy, in groups, basal portion seated on a stroma; spores two-celled, hyaline or subhyaline (see Fig. 24).

Nectria cinnabarina. Dieback; twig canker; coral spot; cosmopolitan on hardwoods, most common on maples but also found on ailanthus, amelanchier, apple, crabapple, apricot, ash, blackberry, chokecherry, beech, birch, elm, hickory, horsechestnut, mimosa, linden, paper mulberry, pear, peach, sophora, locust, and honey locust. It may also appear in stem cankers on vines and shrubs—ampelopsis, barberry, boxwood, callicarpa, cotoneaster, currant, gooseberry, fig, honeysuckle, kerria, California laurel, rose, and syringa. The fungus is widespread as a saprophyte. On ornamental trees and shrubs it is weakly parasitic, producing cankers around wounds and at base of dead branches or causing a dieback of twigs and branches.

On maple, the fungus is more pathogenic, killing twigs, small branches, young trees, and girdling larger branches. It is more frequent on Norway maple and boxelder; it may also invade red, sycamore, Japanese, and other maples. First symptoms are small, depressed, dead areas in bark near wounds or branch stubs. Conspicuous flesh-colored or coral pink sporodochia, formed in dead bark, bear conidia. Later the pustules turn chocolate brown and form pockets, in which perithecia are produced. The canker is most common in severely wounded or recently pruned trees. Sapwood has a greenish discoloration. Open cankers are eventually formed with successive rolls of callus. Remove diseased wood and bark, cutting beyond the greenish discoloration.

Nectria coccinea var. **faginata. Nectria beech bark canker**; on beech in the Northeast. The disease occurs solely in connection with the woolly beech scale insects (*Cryptococcus fagi* and *C. fagisuga*), but it has caused high mortality in Canada, killing 50% of beech stands; it is epidemic in Maine on American beech and is now present in much of New England and New York.

The scale nymphs, covered with a woolly white down, cluster thickly around cracks and wounds in bark, often making trunk and branches appear to be coated with snow. The small yellow larvae establish themselves on the bark in autumn, each inserting its sucking organ, stylet, into the living bark, which shrinks and cracks. *Nectria* enters through these cracks and kills surrounding tissue in bark and cambium. When the cells are dead, the insects can no longer obtain food; therefore, they disappear.

White pustules of sporodochia are pushed out through dead bark, bearing elongate, three- to nine-celled, slightly curved macroconidia. Red perithecia, slightly lemon-shaped, appear in clusters on the bark, often so abundant that the bark appears red. After ascospores are discharged, the upper half of the perithecium collapses and sinks into the lower. The eventual canker is a deeply depressed cavity surrounded by callus. After the cambium dies, the leaves wilt; the twigs, branches, and roots finally die.

Control. Ornamental trees can be sprayed or scrubbed to kill the insects. A dormant lime sulfur spray is very effective. Oil sprays will kill the scale but may injure beech. Late summer spraying for crawlers can supplement the dormant spray.

Nectria desmazierii (*Fusarium buxicola*). **Canker; dieback;** of boxwood (see Fig. 27).

Nectria ditissima. Sometimes reported but not confirmed in the United States; reports probably refer to *Nectria galligena*.

Nectria galligena (*Cylindrosporium mali*). **European nectria canker;** trunk canker, widespread on apple, pear, quince, aspen, beech, birch, maple, hickory, Pacific dogwood, and various other hardwoods. This disease is one of the more important ones of apple and pear in Europe but is less serious in the United States. In eastern United States it is primarily an apple disease; on the Pacific Coast it is more common on pear.

Young cankers are small, depressed or flattened areas of bark near small wounds or at base of dead twigs or branches, darker than the rest of the bark and water-soaked. Older cankers are conspicuous and somewhat like a target,

Figure 27. Volutella blight or "Nectria" canker on boxwood.

with bark sloughed off to expose concentric rings of callus. Cankers on elm, sugar maple, and birch are usually circular; those on oak irregular; on basswood elongate, pointed at ends. If the canker is nearly covered with a callus roll, it indicates that the infection is being overcome.

Small red perithecia are formed singly or in clusters on bark or on wood at margin of cankers. Ascospores discharged during moist weather are disseminated by wind and rain. Creamy white sporodochia protruding through recently killed bark of young cankers produce cylindrical macroconidia and ellipsoidal microconidia. Invasion is through bark cracks or other wounds in living or dying, but not dead, wood. Infection is slow, with annual callus formation; only the smallest branches are likely to be girdled. Younger, more vigorous apple trees receiving nitrogenous fertilizer appear to be more susceptible.

Control. Remove and destroy small branches with cankers. Clean out trunk cankers and cut back to sound bark; treat with bordeaux paste. On the West Coast spray pome fruits immediately after leaf fall in autumn with bordeaux mixture to prevent infection through leaf scars.

Nectria magnoliae. Nectria canker; similar to the preceding but found on magnolia and tuliptree.

Neofabraea

See under Anthracnose.

Neofabraea perennans (*Gloeosporium perennans*). **Perennial canker**; of apple, also bull's-eye rot of fruit. The disease is much like northwestern anthracnose. It often follows after winter injury or starts at pruning cuts where aphids congregate, or may appear after an application of wound dressing.

Nummularia

Ascomycetes, Sphaeriales, Xylariaceae

Stroma superficial, composed entirely of fungus elements, covered with a conidial layer when young. Perithecia flask-shaped, embedded in stroma; spores one-celled, dark.

Nummularia discreta. Blister canker; of apple, crabapple, pear, mountain ash; also reported on serviceberry, birch, elm, magnolia, and honey locust. This is a major apple disease east of the Rocky Mountains, especially in Upper Mississippi and Lower Missouri river valleys, where millions of apple trees have been killed. Large and small limbs are affected. Cankers are dead areas, up to 3 feet long, mottled with living wood and dotted with numerous round cushions of stromata, looking like nailheads. Perithecia, with dark ascospores, are buried in the stromata; hyphae bearing small, light-colored conidia grow over the surface. The fungus enters through branch stubs, bark injuries, and other wounds.

Control. Avoid especially susceptible varieties like Ben Davis. Shape trees early to prevent large pruning wounds on older trees; the canker seldom appears on trees less than 10 years old. Shellac pruning cuts immediately; sterilize tools between cuts.

Ophionectria (*Scoleconectria*)

Ascomycetes, Hypocreales, Hypocreaceae

Perithecia red to white, globoid, with a round ostiole, superficial, paraphyses lacking; spores needle-shaped to filiform, light colored.

Ophionectria balsamea. Bark canker; of balsam fir.
Ophionectria scolecospora. Bark canker; of balsam and alpine firs.

Penicillium

Fungi Imperfecti, Moniliales, Moniliaceae

Conidia in heads; conidiophores unequally verticillate at tip in whorls; globose conidia formed in chains, one-celled, hyaline or brightly colored in mass; parasitic or saprophytic.

Penicillium vermoeseni. Penicillium disease of ornamental palms; serious in southern California with symptoms varying according to type of palm.

On queen palm (*Arecastrum* or *Cocos plumosa*) the disease is a trunk canker, which may remain inconspicuous for several years but leads to weakening and breaking of trunk. Infected trees should be removed at an early stage. On Canary date palm the disease is a leaf-base rot, and on Washington a bud rot. See under Rots.

Pezicula

Ascomycetes, Helotiales, Dermateaceae

Apothecia similar to *Dermatea* but lighter.

Pezicula carpinea. Bark canker; of hornbeam.
Pezicula corticola. Superficial bark canker; fruit rot; rather common on apple and pears. Hyaline, one-celled conidia of the *Myxosporium* stage are formed in acervuli.
Pezicula pruinosa. Canker; on branches of amelanchier.

Phacidiella

Ascomycetes, Phacidiales, Phacidiaceae

Asci borne in hymenial layers, covered with a membrane until mature, then splitting; apothecia remain embedded in a stroma; paraphyses present; asci clavate.

Phacidiella coniferarum [imperfect state *Phacidiopycnis* (*Phomopsis*) *pseudotsugae*]. **Phomopsis disease of conifers**; the fungus is usually saprophytic, but it is parasitic on Douglas fir and larch in Europe and on living pine in Maine.

Phomopsis

See under Blights.

Phomopsis alnea. Canker; of European black alder.
Phomopsis boycei. Phomopsis canker; of lowland white fir. Branches or main stem of saplings may be girdled and killed; there is often swelling at base of canker where dead tissues join living. The reddish brown needles of dead branches are prominent against living foliage.
Phomopsis elaeagni (= *Phomopsis arnoldia*). **Canker**; of Russian-olive.
Phomopsis gardeniae. (perfect state *Diaporthe gardeniae*). **Gardenia canker**; stem gall, widespread in greenhouses. Although not reported until about 1933, this seems to be the most common gardenia disease. Symptoms start with brown dead areas on stem, usually near the soil line. The canker is first

sunken, then, as the stem enlarges, swollen with a rough, cracked outer cork. The stem is bright yellow for a short distance above the canker, a contrast to its normal greenish white. When stems are completely girdled, the foliage wilts and dies; the plant may live a few weeks in a stunted condition. Flower buds fall before opening. When humidity is high, black pycnidia on cankers exude yellowish spore masses. Entrance is through wounds; spores may be spread on propagating knives. Infection often starts at leaf joints at the base of cuttings after they have been placed in a rooting medium. Because the cankers may be only slightly visible on rooted cuttings, the disease may be widely distributed by the sale of such cuttings.

Control. Use sterilized rooting medium. Use steam for a sand and peat mixture. Destroy infected plants; sometimes it is possible to wait until blooms are marketed.

Phomopsis livella (perfect state *Diaporthe vincae*). **Canker, dieback**; of vinca, and periwinkle.

Phomopsis lokoyae. Phomopsis canker; of Douglas fir mostly on saplings in poor sites in California and Oregon. Long, narrow cankers, somewhat pointed at ends, develop during the dormant season after young shoots are infected. If the tree is not girdled during the first season, the canker heals over.

Phomopsis mali. Bark canker; of pear, and apple. The bark is rough.

Phomopsis padina (perfect state *Diaporthe decorticans*). **Canker**; twig blight of sour cherry.

Phragmodothella

Ascomycetes, Dothideales, Dothideaceae

Asci in locules immersed in groups in a cushionlike stroma; spores hyaline, many-celled.

Phragmodothella ribesia. Dieback, black pustule; on currant, flowering currant, and gooseberry.

Physalospora

See under Blights.

Physalospora corticis. Blueberry cane canker; in Southeast on cultivated blueberries. The fungus enters through unbroken bark, probably through lenticels, with cankers starting as reddish, broadly conical swellings, enlarging the next year to rough, black, deeply fissured cankers that girdle the shoots. The portions above cankers are unfruitful and finally die. Avoid very susceptible varieties like Cabot and Pioneer.

Physalospora glandicola (*Sphaeropsis quercina*). **Sphaeropsis canker; dieback**; of red, chestnut, and other oaks. Shade and ornamental trees of all ages may be killed. Infection may start anywhere through wounds but more often on small twigs and branches, passing to larger branches and trunk. Twigs and branches die; leaves wither and turn brown; infected bark is sunken and wrinkled, with small black pycnidia breaking through. On larger stems the bark has a ridge of callus around the canker, the sapwood in this area turning dark with black streaks extending longitudinally for several inches. Numerous water sprouts grow from below the dead crown. The fungus winters on dead twigs, producing a new crop of conidia in spring, readily infecting most trees weakened by unfavorable environmental conditions.

Control. Prune out diseased portions at least 6 inches below cankers. Fertilize and water to improve vigor. Remove seriously diseased trees.

Physalospora miyabeana. Willow black canker; accompanying scab to form the disease complex known as willow blight in New England and New York. Starting in leaf blades, the fungus proceeds through petioles into twigs; it also causes cankers on larger stems, followed by defoliation. Pinkish spore masses of the imperfect *Gloeosporium* stage are formed on dead twigs and branch cankers and then short-necked perithecia, which overwinter. Remove and destroy dead twigs and branches during the dormant period. Spray three times with bordeaux mixture, starting just after leaves emerge in spring.

Physalospora obtusa (*Sphaeropsis malorum*). Dieback; canker; of hardwoods, New York apple-tree canker, black rot of apple. The fungus attacks leaves, twigs, and fruits, is more important east of the Rocky Mountains, and is found on many plants, including alder, ampelopsis, birch, bignonia, bittersweet, callicarpa, catalpa, ceanothus, chestnut, currant, cotoneaster, hawthorn, Japanese quince, maple, peach, pear, and persimmon. On hardwoods the canker is similar to that caused by *P. glandicola* on oaks. Limbs are girdled with large areas of rough bark with numerous protruding black pycnidia. For the fruit rot phase of this disease see under Rots.

Physalospora rhodina. Black rot canker of tung; in Mississippi and Louisiana. Black, sunken cankers on trunks, limbs, twigs, and shoots, may girdle and kill trees. Rogue and burn diseased specimens.

Phytophthora

See under Blights.

Phytophthora cactorum. Bleeding canker; of maple, beech, birch, elm, horsechestnut, linden, oak, sweetgum, and willow; Crown canker; of dogwood. Dieback; of rhododendron. Trunk canker; of apple, almond, apricot, cherry, and peach.

Bleeding Canker, first noticed in Rhode Island on maple about 1939 and found in New Jersey the next year, is now present on many trees in the Northeast. The most characteristic symptom is the oozing of a watery light brown or thick reddish brown liquid from fissures in bark at the root collar and at intervals in trunk and branches. When dry, this sap resembles dried blood, hence the name, bleeding canker. Sunken, furrowed cankers are more definite on young trees than on older trees with rough bark. Symptoms are most prominent in late spring and early fall, with trees in moist situations most often affected. The fungus lives in the soil and advances upward from a primary root infection. Wilting of leaves and blighting of branches is evidently from a toxin. Mature trees have fewer, smaller, yellow-green leaves, and there is an acute dieback of branches. Reddish-brown areas with intense olive-green margins are found in wood extending vertically from roots to dying branches, marked at irregular intervals with cavities containing the watery fluid.

Control. Although there is no real cure, injecting trees with Carosel, a mixture of helione orange dye and malachite green, has inhibited the fungus and neutralized the toxin. In some cases trees recover without treatment. Avoid heavy feeding, which seems to encourage the spread of disease and causes chronic cases to become acute.

Crown Canker, collar rot, is the most serious disease of dogwood reported in New York, New Jersey, and Massachusetts. The first symptom is a general unhealthy appearance, with leaves smaller and lighter green than normal, turning prematurely red in late summer. Leaves may shrivel and curl during dry spells (normal leaves often do likewise). Twigs and large branches die, frequently on one side of the tree. The canker develops slowly on the lower trunk near the soil level. Inner bark, cambium and sapwood are discolored; the cankered area is sunken; the bark dries and falls away, leaving wood exposed. Trees die when the canker extends completely around the trunk base or root collar. The fungus lives in the soil in partially decayed organic matter, and spores are washed to nearby trees. Entrance is through wounds. The disease affects transplanted dogwoods, seldom natives growing in woods.

Control. Transplant carefully, avoiding all unnecessary wounds; avoid hitting base with lawn mower, by using a wire guard around the tree. It is difficult to save trees already infected, but cutting out small cankers and painting the wound with bordeaux paste is worth trying. If trees have died from crown canker, do not replant with dogwoods in the same location for several years.

Rhododendron dieback is a disease in which terminal buds and leaves turn brown, roll up, and droop as in winter cold. A canker encircles the twigs, which shrivel with the terminal portion wilting and dying. In shady locations leaves have water-soaked areas, changing to brown, zonate spots. Do not plant rhododendrons near lilacs, for they are blighted by the same fungus. Prune diseased tips well below the shriveled part, and spray after blooming with bordeaux mixture, two applications 14 days apart.

Trunk canker of apple is an irregular canker often involving the entire trunk and base of scaffold branches, the first outward symptom a wet area on bark. Trees must be 5 years old or older for infection. Grimes Golden and Tomkins King are especially susceptible, often being completely girdled.

Phytophthora cinnamomi. Basal canker of maple; particularly Norway maple. Trees have a thin crown, fewer and smaller leaves, and die a year or two after cankers are formed at the base of the trunk. Sapwood is reddish brown; the roots decay. Remove diseased trees. Plant new Norway maples in good soil, well drained, rich in organic matter; treat injuries at base of trunk promptly. See under Rots and Wilts for other manifestations of this pathogen.

Phytophthora syringae. Pruning wound canker; of almond.

Plenodomus

Fungi Imperfecti, Sphaeropsidales, Sphaeropsidaceae

Pycnidia dark, immersed, irregular in shape and opening irregularly; conidia hyaline, one-celled, oblong; parasitic.

Plenodomus fuscomaculans. Canker; on apple.

Pseudonectria

Ascomycetes, Hypocreales, Nectriaceae

Perithecia superficial, blight-colored, smooth; spores one-celled, hyaline.

Pseudonectria rouselliana. Nectria canker of boxwood; leaf cast, twig blight. The perithecia are formed on dead leaves, but the fungus is thought to be the perfect stage of *Volutella buxi*, which see.

Pseudovalsa

Ascomycetes, Sphaeriales

Perithecia in a stroma; spores dark, with several cells.

Pseudovalsa longipes. Twig canker; on coast live oak and white oak.

Rhabdospora

Fungi Imperfecti, Sphaeropsidales, Sphaerioidaceae

Pycnidia separate, not produced in spots, erumpent, ostiolate; conidiophores short, simple conidia hyaline, filiform to needle-shaped, with several cells; parasitic or saprophytic.

Rhabdospora rubi. Cane spot; canker of raspberry.

Scleroderris

Ascomycetes, Helotiales (or Phacidiales), Tryblidiaceae

Apothecia black, opening with lobes, crowded together or with a stroma, short-stalked; spores hyaline, elongate, with several cells.

Scleroderris abieticola. Canker of balsam fir; on Pacific Coast. An annual canker, starting in autumn and ceasing when cambium is active in spring, is formed on twigs, branches, and trunks of saplings. Only twigs and small branches are girdled, and if this does not happen before spring, the wound heals over. Small black apothecia with short stalks appear on dead bark. Ascospore infection is through uninjured bark or leaf scars.

Scleroderris lagerbergii (= *Gremmeniella abietina*). **Canker;** on pine.
Scleroderris lateritium. Canker; on pine.

Sclerotinia

See under Blights.

Sclerotinia (= *Whetzelinia*) **sclerotiorum. Basal canker;** on Euonymus.

Septobasidium

Basidiomycetes, Tremellales, Septobasidiaceae

All species are on living plants in association with scale insects; the combination causes damage to trees. Fungus body variable, usually resupinate, dry, crustaceous or spongy, in most species composed of subiculum growing over bark; a middle region of upright slender or thick pillars of hyphae supports the top layer, in which hymenium is formed. Basidium transversely septate into two, three, or four cells, rarely one-celled; basidiospores elliptical, colorless, divided into two to many cells soon after formation, budding with numerous sporidia if kept moist. Some species with conidia.

The fungus lives by parasitizing scales, obtaining food via haustoria. The insects pierce the bark to the cambium, sometimes killing young trees. The fungus kills a few scales but protects many more in its enveloping felty or leathery covering, a symbiotic relationship. Spores are spread by scale crawlers and by birds. Most felt fungi are found in the South, abundant on neglected fruit, nut, or ornamental trees, rare on those well kept.

Septobasidium burtii. Felt fungus; on southern hackberry, beech, pear, apple, and peach. This growth is perennial, with a new ring added to the patch each summer. Probasidia are formed during the winter, and four-celled basidia in spring.

Septobasidium castaneum. Felt fungus; abundant on willow and water oaks, and holly; may injure azaleas. The surface is smooth, shiny, chocolate brown to nearly black.

Septobasidium curtisii. Felt fungus; widespread on many trees in the Southeast, commonly on sour gum (tupelo) and American ash, also on hickory, hawthorn, Japanese quince, and others. The felt, purple black throughout, is mounded over the insects.

Septobasidium pseudopedicellatum. Felt fungus; on citrus twigs, sometimes on main stem or branches of hornbeam. Surface is smooth, buff-colored over dark brown pillars.

Solenia (*Henningsomyces*)

Basidiomycetes, Agaricales, Thelephoraceae

Fruiting layers erect, cylindrical, formed in groups, membranous.

Solenia (*Henningsomyces*) **anomala.** Bark patch; canker, widespread on alder.
Solenia ochracea. Bark patch; of birch, hornbeam, hickory, and alder.

Sphaeropsis

Fungi Imperfecti, Sphaeropsidales, Sphaerioideaceae

Pycnidia black, separate or grouped, globose, erumpent, ostiolate; conidiophores short; conidia large, dark, one-celled, ovate to elongate, on filiform conidiophores. Some species have *Physalospora* as the perfect stage.

Sphaeropsis ellisii. Bleeding canker; on pine.
Sphaeropsis tumefaciens. Canker; gall; on *Carissa*.
Sphaeropsis ulmicola. Sphaeropsis canker; of American elm. The disease spreads downward from small twigs to larger branches with a brown discoloration of wood just under the bark. Secondary shoots sometimes develop below the cankers. Trees weakened by drought or poor growing conditions are particularly susceptible. Prune out infected wood, cutting well below cankers.

Steganosporium

Fungi Imperfecti, Melanconiales, Melanconiaceae

Steganosporium sp. Maple canker; dieback; reported from New Jersey. Large branches die back with conspicuous flagging. Black tarlike fruiting bodies are formed in cankers.

Strumella

Fungi Imperfecti, Moniliales, Tuberculariaceae

Sporodochia wartlike, gray to black, of interwoven hyphae; conidiophores dark, branches; conidia dark, one-celled, ovoid to irregular.

Strumella coryneoidea. Strumella canker of oak; especially the red oak group, also on American beech and chestnut, occasional on pignut and hickories, red maple, and tupelo. Primarily a forest disease, this canker may become important on red and scarlet ornamental oaks. Starting as a yellowish discoloration of bark around a dead branch or other point of infection, the canker develops into a diffuse lesion or into a target canker with concentric rings of callus. Whitish mycelium is present near outer corky bark, and the infected portion of the trunk may be flattened or distorted. Target cankers may be up to 2 feet wide and 5 feet long. The small black nodules bear no spores while trees are living, but after death dark brown spore pustules are formed, which blacken with age. New pustules are formed yearly. Canker eradication has been unsuccessful in forest stands. The diseased trees should be removed and utilized before spores can spread infection.

Sydowia

Ascomycetes, Dothidiales, Dothideaceae

Asci usually short, cylindrical, and relatively numerous, in spherical, ostiolate locules.

Sydowia polyspora. Twig dieback; on fir.

Thyronectria

Ascomycetes, Hypocreales, Hypocreaceae

Stroma valsoid with several perithecia, bright-colored; spores muriform, hyaline to subhyaline.

Thyronectria austro-americana. Canker; wilt of honeylocust. Slightly depressed cankers ranging from pinhead size to ½ inch grow together and enlarge to girdle a branch. Underlying wood is streaked reddish brown for several inches from the canker, and there is often a gummy exudate. Some trees die, but many survive.

Thyronectria berolinensis. Cane-knot canker; of fruiting and flowering currants.

Trichothecium

See under Rots.

Trichothecium roseum. Canker; of rose.

Tubercularia

Fungi Imperfecti, Moniliales, Moniliaceae.

Forms bright colored cushions, mostly on wood or bark; fine branching conidiophores bearing small, elipsoidal hyaline conidia.

Tubercularia ulmea. Canker; on Russian olive, and honeylocust.

Tympanis

Ascomycetes, Helotiales, Helotiaceae

Ascocarp cup-shaped; sclerotia absent; expiculum usually, if parallel hyphae.

Tympanis confusa. Canker; on pine.

Valsa

Ascomycetes, Sphaeriales, Valsaceae

Many perithecia in a circle in a stroma in bark; flask-shaped with long necks opening to the surface; spores hyaline, one-celled, curved, slender.

Valsa cincta. Perennial canker of peach; dieback; also on nectarine. The fungus is apparently infective during the dormant season, entering through wounds, dead buds, leaf scars, and fruit spurs. It forms a canker complex with *V. leucostoma* and sometimes the brown-rot fungus. It is more common in northern latitudes than in southern, but is not important in well-cared-for orchards.

Valsa kunzei. See *Cytospora kunzei*.

Valsa leucostoma. Apple canker; Dieback; twig blight on apple, apricot, peach, pear, quince, plum, cherry, willow, and mountain-ash. The fungus is a weak parasite entering through wounds or twigs killed by frost.

Valsa salicina (*Cytospora salicis*). Twig and branch canker; of willow.

Valsa sordida. See *Cytospora chrysosperma*.

Vermicularia

Fungi Imperfecti, Melanconiales, Melanconiaceae

Like *Colletotrichum* but setae are scattered throughout the acervuli, not just marginal; spores hyaline, globose to fusoid.

Vermicularia ipomoearum. Stem canker; of morning glory.

Volutella

See under Blights.

Volutella buxi. Boxwood "nectria" canker; Volutella blight. The perfect stage of the fungus is supposed to be *Pseudonectria rouselliana*, which see. As a canker the disease often follows after winter injury, with salmon-pink spore pustules on dying twigs, branches, and main stems. As a blight, the fungus spreads rapidly in moist weather in summer, attacking healthy twigs when humidity is high and often discernible at a distance by a straw yellow "flag." On such yellowing branches the backs of leaves and the bark of twigs are both covered with the pinkish spore pustules (see Fig. 27).

Control. Cut out branches where the bark has been loosened by winter ice and snow. Have a yearly "housecleaning," brushing out accumulated leaves and other debris from interior of bushes and cutting out all twigs with pink pustules. If there are signs of disease, follow cleaning with thorough spraying, from ground up through interior of bushes, with lime sulfur.

CLUB ROOT

Plasmodiophora

Phycomycetes, Plasmodiophorales, Plasmodiophoraceae

This genus, founded on the club root organism, has a somewhat doubtful taxonomic position. Formerly considered a slime mold, one of the Myxomycetes, then placed in the Chytridiales, lowest order of true fungi, it is now placed in a separate order, Plasmodiophorales.

Thallus amoeboid, multinucleate in host cell; spores lying free in host cell at maturity; frequently causing hypertrophy; parasitic on vascular plants.

Plasmodiophora brassicae. Club root of cabbage; and other crucifers; finger-and-toe disease, on alyssum, brussels sprouts, cabbage, Chinese cabbage, candytuft, cauliflower, hesperis, honesty, peppergrass, garden cress, mustard, radish, rutabaga, stock, turnip, and western wallflower.

Club root was present in western Europe as early as the thirteenth century, but the true cause was not known until the classic paper of the Russian Woronin in 1878. The disease was important in the United States by the middle of the nineteenth century, and is now present in at least 37 states. Losses come from death of the plants and also from soil infestation, for susceptible crucifers cannot be grown again on the same land for several years, unless it is treated.

The first symptom is wilting of tops on hot days, followed by partial recovery at night; affected plants may be stunted and not dead; outer leaves turn yellow and drop. The root system becomes a distorted mass of large and small swellings, sometimes several roots swollen like sweet potatoes, and sometimes joined in one massive gall. Lateral and tap roots are scabby and fissured, with rot starting from secondary fungi.

When diseased roots decompose, small spherical spores are liberated in the soil; they are capable of surviving there many years between crops. In spring, with suitable temperature and moisture, the resting spores germinate, each becoming a motile swarm spore with a flagellum. This whiplike appendage is soon lost, and the organism becomes amoebalike, moving by protoplasmic streaming until it reaches a root hair or other root tissue. The plasmodium continues to grow and divide until it reaches the cambial cells, in which it develops up and down the root. The swelling is produced by division of plasmodia and of the infected cells. Eventually the multinucleate plasmodium breaks into many small resting spores, each rounded around a single nucleus. They are set free by the millions when the root rots, and are spread in soil clinging to shoes or tools and in drainage water, manure, and plant refuse. Spores are not seed-borne. Long-distance spread is probably by infected seedlings. Infection takes place chiefly in a neutral to acid soil, pH 5.0 to 7.0, at temperatures below 80° F, and when moisture of soil is above 50% of its water-holding capacity.

Control. Inspect seedlings carefully before planting. Dispose of infested crops with caution; resting spores passed through animals are still viable. A long rotation of crops has been recommended, combined with adding lime to soil, which must be applied in large amounts, about 6 weeks before the cabbage crop is set. The pH will be too high to use potatoes as a following crop. Most turnip and rutabaga varieties are relatively resistant to strains of the club root organism present in the United States.

DAMPING-OFF

Damping-off is the destruction of young seedlings by soil organisms. There are two types. Preemergence damping-off rots the sprouting seed before it breaks through the soil; it is recognized by bare spaces in what should be uniform rows. Such a poor stand may be due to poor viability of seed, but more often it is due to soil fungi functioning in cold, wet soils when germination is slow. Postemergence damping-off is the rotting or wilting of seedlings soon after they emerge from the soil. Succulent stems have a water-soaked, then necrotic and sunken, zone at ground level; the little herbaceous plants fall on the ground, or in woody seedlings, wilt and remain upright. Root decay follows. This type of damping-off is most common in greenhouses or outdoors in warm humid weather and where seedlings are too crowded. Tree seedlings in nursery rows are subject to this type of damping-off, and so are perennial flowers started in late summer for the next year.

Many fungi living saprophytically in the upper layers of soil can cause damping-off. *Pythium debaryanum* and *Rhizoctonia solani* are probably most common, but other species of these two genera and *Aphanomyces, Botrytis, Cylindrocladium, Diplodia, Fusarium, Macrophomina, Helminthosporium, Sclerotium rolfsii, Fusarium equiseti*, and *Phytophthora* may be important on occasion. A synergistic interaction of *Pythium myriotylum, Fusarium solani*, and *Meloidogyne arenaria* causes damping-off of peanut, which has been reported in Florida. See under Rots for details. Also, *Caloscypha fulgens* (Imperfect Stage, *Geniculodendron pyriforme*) causes damping-off of spruce seed.

Damping-off is prevented by starting seed in a sterile medium, such as vermiculite, perlite, or sphagnum moss, or by treating the soil or the seed before planting. Commercial operators treat soil with steam or electricity.

Seed treatment, the coating of seed with a protectant dust, is crop insurance. In some seasons, good stands can be obtained without it, but it scarcely pays to take a chance. Seed disinfection is used to kill organisms of anthracnose and other specific diseases carried on seed. The damping-off organisms are in the soil, not on the seed, and coating the seed with a chemical is intended to kill or inhibit fungi in the soil immediately surrounding the seed and so provide temporary protection during germination.

DODDER

Dodders are seed plants parasitic on stems and other parts of cultivated or wild plants. They are leafless, orange to yellow twining vines, without chlorophyll and hence incapable of manufacturing their own food. They are called love vine, strangle weed, gold thread, hairweed, devil's hair, devil's ringlet, pull down, clover silk, and hell-bind, the last being most appropriate. There are about 40 species in the United States, causing serious agricultural losses in clovers, alfalfa, and flax, and becoming more and more important in gardens on ornamentals and sometimes vegetables. Dodders belong to the single genus *Cuscuta*, family Cuscutaceae, close to the morning glory family.

Dodder seed is grayish to reddish brown, resembling small legume seed but roughened with three flattened sides. It germinates as ordinary seed but is synchronized to start a little later than its host seedlings. The parasite is a slender, yellowish, unbranched thread with the growing tip circling around in search of support. When it touches the host it twines like a morning glory and

Figure 28. Dodder on oleander.

puts out little suckers, haustoria, into the stem of the victim, after which its
original connection with the soil dries up (see Fig. 28). Although seedlings can
live for a few weeks without a susceptible host, they finally die if a connection
is not established. Successful parasites continue to twine and to spread orange
tendrils from one plant to the next, often making a tangle of matted orange
hairs many feet across, with a black region in the center where plants have
died. Such tangles are conspicuous in weeds along roadsides.

In ornamental plantings host plants are not often killed but exhibit stunting and pallor, symptoms of starvation. Minute scales or rudimentary leaves form on the dodder tendrils followed by dense clusters of beautiful white blossoms (sometimes pale pink or yellow), which ripen seed in late summer, with as many as 3,000 seed being produced on a single plant.

Cuscuta spp. Much of the dodder infesting ornamentals is not readily identified as to species, but it is widespread on a great many shrubs, perennials and annuals. It is found very commonly on chrysanthemum, also strangling any other plant in the vicinity. Many hours may be spent cleaning up ivy and trumpet vine, petunias, and asters. Dodder is reported on camellias in the South. It is even a pest of house plants, if field soil has been used for the potting mixture. Dodder has, however, one virtue for plant pathologists. It is used as a bridge between plants to carry viruses and mycoplasma-like organisms in testing their host range.

Cuscuta americana. On citrus. **C. californica.** On beet.

Cuscuta coryli. Hazel dodder. **C. epithymum.** Clover dodder; on legumes.

Cuscuta exaltata. On redbud, ilex, and sumac.

Cuscuta gronovii. Common dodder; on buttonbush, cucumber, raspberry, members of the potato family, and many garden ornamentals, including hedge plants.

Cuscuta indecora. Bigseed alfalfa dodder; on alfalfa from Colorado westward, also on sweet pea and tomato.

Cuscuta paradoxa on rose, Texas and Florida.

Cuscuta pentagona (*C. arvensis*). Field dodder; widely distributed, most common and serious east of Mississippi on many cultivated and wild herbaceous plants.

Cuscuta planifera. Littleseed alfalfa dodder; on some legumes in the West.

Control. Avoid dodder-infested seed. Commercial seed containing one or more dodder seed per 5-gram sample is prohibited entry into the United States. Many states have laws regulating sale of infested seed, but it may still be included inadvertently in a seed packet. If any contamination with rough, flat-sided seed is found, do not use any of the lot. Commercial dealers sometimes clean infested seed by screening or treating with an iron powder, which sticks to the rough dodder seed so it can be drawn out by magnets.

Before breaking new ground for a garden on native sod, examine it carefully. If dodder is found, burn over the area, then hoe lightly but repeatedly for several weeks to allow buried seed to germinate and die. When dodder is present on cultivated plants, the only thing to do is to remove and burn infested parts before seed is formed. Pulling off the orange tendrils is not sufficient. All parts of the plant attacked must be cut off and burned, for even a small fraction of a tendril left twined around a stem will start growing again.

A fungus, *Colletotrichum destructivum*, has been found to parasitize dodder and offers a slight possibility of biological control.

DOWNY MILDEWS

Downy mildews, sometimes called false mildews, are Phycomycetes, in the order Peronosporales and all in the family Peronosporaceae except *Phytophthora* in the Pythiaceae. They form mycelium in higher plants and produce sporangiophores that protrude through stomata in great numbers, their sporangia making white, gray, or violet patches on the leaves. The downy effect distinguishes these mildews from the true or powdery mildews that form white felty or powdery patches.

The sporangiophores are often branched; they bear a single sporangium at the tip of each branch simultaneously, or successively in *Phytophthora*. Sporangia germinate by swarm spores or with a germ tube as a conidium. An oospore, resting spores with external ridges or knobs, is formed in an oogonium, large globular multinucleate female cell, after it is fertilized by the antheridium, a smaller male cell. The oospores are set free by weathering and decay of host parts.

Basidiophora

Sporangiophore a single trunk with a swollen apex from which short branches grow out, each bearing a nearly globose sporangium; germination by swarm spores; oospore wall not confluent with that of oogonium. Mycelium is intercellular, haustoria small, knoblike (see Fig. 29).

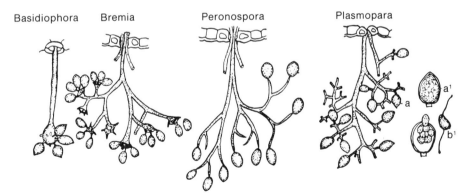

Figure 29. Downy mildews fruiting from stomata on underside on leaves. *Basidiophora*, sporangiophore with swollen apex; *Bremia*, sporangiophore tip enlarged to a disc, dichotomous branching; *Peronospora*, sporangia on sharply pointed terminal branches; Plasmopara, on obtuse tips; *a* and *a*[1], sporangium; *b*[1], zoospore.

Basidiophora entospora. Downy mildew; of aster, China aster, goldenrod, and erigeron. Aster losses are reported by commercial growers in the South, but apparently this is not an important garden problem.

Bremia

Dichotomous branching of sporangiophores; tips enlarged into discs bordered with sterigmata bearing sporangia; swarm spores rare; germination usually by a germ tube protruded through an apical papilla (see Fig. 29).

Bremia lactucae. Downy mildew of lettuce; and other composites, endive, cornflower, centaurea, celtuce, escarole, romaine, and various weeds. First noticed around Boston in 1875, the disease is serious in greenhouses and in states where outdoor winter crops are grown. Light green or yellowish areas on upper surface of leaves are matched by downy patches on the under surface. Affected portions turn brown, and leaves die, the older ones first. Entrance is through stomata. The disease is worse in damp, foggy, cool weather (43° F to 53° F).

Control. The pathogen has numerous physiological races so that lettuce varieties like Imperial 44 and Great Lakes that are resistant in some localities may not be so in others. Avoid excessive irrigation; eliminate crop residue and weeds.

Peronospora

Mycelium intercellular; haustoria in a few species short and knoblike, but in most filamentous and more or less branched. Sporangiophore with erect trunk two to ten times dichotomously branched, with branches somewhat reflexed and terminal branches sharp-pointed; sporangia colored, lacking an apical papilla, germinating from a indeterminate point on the side. Oospores smooth or variously marked, germinating by germ tubes (see Fig. 29).

Peronospora arborescens. Downy mildew of prickly-poppy; on leaves, buds, and capsules. Yellow or light brown blotches on upper leaf surface turn dark, with light gray mold on the underside. The fungus winters in old plant debris in soil. Remove and burn infected plants. Use clean seed.

Peronospora antirrhini. Snapdragon downy mildew; reported from California, Oregon, Oklahoma, Pennsylvania, and Maryland.

Peronospora arthuri. Downy mildew; of godetia, clarkia, gaura, and evening primrose.

Peronospora destructor. Onion downy mildew; blight, general on onion, shallot, and garlic. One of the more serious diseases of onion, reported in the United States in 1884. All varieties are susceptible, but red onions have some resistance. Reduction in yield may be as high as 75%.

The first sign of onion mildew is the production of conidiophores with a purplish tinge a short distance back from tips of older leaves. Leaves turn yellow, wither, and break over; seed stalks may be infected. Onion mildew is sporadic, abundant in years of heavy rainfall. Spores, produced in great numbers in rain or when plants are wet with dew, lose vitality quickly when exposed to sun. Low temperature, optimum 50° F, favors infection. The fungus winters as mycelium in bulbs, in overwintering plants in mild areas, or as oospores in soil. Perennial onions in home gardens are considered an important source of primary inoculum, but oospores have been known to survive 25 years in soil.

Control. Calred is a resistant variety adapted to California. More onion seed is being produced in Idaho, where dry summers preclude mildew.

Peronospora dianthicola. Carnation downy mildew; common in California on seedlings. Leaves turn pale, curl downward; terminal growth is checked, and plants may die. There is a white growth on lower leaf surfaces.

Peronospora effusa. Spinach downy mildew; chard blue mold; found wherever spinach and swiss chard are grown, absent some seasons, nearly destroying the crop in others. Large pale yellow spots grow together to cover all or part of the leaf; lower leaves are infected first, and then the blight is scattered through the plant. Gray to violet mold forms on underside of leaves; sometimes the whole plant decays and dries. Initial infection comes from oospores in the soil; it requires humidity above 85% and a mean temperature between 45° F and 65° F for a week. Secondary infection is from conidia. The fungus is an obligate parasite and does not live on hosts other than spinach.

Control. Plant on well-drained, fertile ground; do not crowd; if overhead irrigation is used, water early on sunny days; practice a 2- to 3-year crop rotation. Resistant varieties such as Califlay and Texas Early Hybrid 7 are being introduced.

Peronospora fragariae. Strawberry downy mildew.

Peronospora grisea. On veronica, a grayish mildew on underside of leaves.

Peronospora lepidii. On garden cress. **P. leptosperma.** On artemisia. **P. linariae.** On linaria. **P. lophanti.** On agastache.

Peronospora manshurica. Soybean downy mildew; general. Yellow-green foliage spots turn brown, with a grayish mold underneath; there may be premature defoliation. The pathogen winters as mycelium in seed and oospores in soil. There are at least three races.

Peronospora myosotidis. Forget-me-not downy mildew; also on lappula. Pale spots on upper surface of leaves, with downy growth underneath.

Peronospora oxybaphi. On sand verbena and four-o'clock.

Peronospora parasitica. Downy mildew of crucifers; general on cabbage, Chinese cabbage, broccoli, cauliflower, horseradish, radish, turnip, cress, peppergrass, also on sweet alyssum, arabis, stock, and hesperis. Chief damage is to cabbage seedlings or plants grown for seed. Leaf lesions are light green, then yellow, with downy mold on both sides of the leaf in the widening yellow zone but not in the dead, shrunken, gray or tan central portion. Secondary fungi often cover dead parts with a black sooty mold. Fleshy roots or turnips and radishes may be discolored internally. Warm days and cool nights favor the disease. The pathogen lives between crops in perennial plants or winter annuals. There are several strains of *P. parasitica*; one, often reported as *P. matthiolae*, blights stock in greenhouse and nursery. Leaves wilt; tender stems and flower parts are stunted and dwarfed.

Control. Avoid crowding plants; keep foliage dry. Spray cabbage seedlings; repeat two or three times a week until plants are set in field. Treat heading cabbage every 6 or 7 days beginning 1 to 3 weeks before harvest.

Peronospora pisi. Pea downy mildew; water-soaked tissue and white growth appear on any aerial plant part. The mycelium winters in vetch stems, fruiting there in spring, and spores are disseminated back to peas. The disease is not important enough for control measures.

Peronospora potentiallae. Downy mildew; of agrimony and mock strawberry.

Peronospora rubi. Downy mildew; of blackberry, dewberry, and black raspberry.

Peronospora rumicis. Rhubarb downy mildew; a European disease reported from California on garden rhubarb. Fungus winters in root stalks and grows up into new leaves.

Peronospora schactii. Beet downy mildew; on beet, sugar beet, and swiss chard. Inner leaves and seed stalks are stunted and killed, covered with violet down. The disease appears on the Pacific Coast during the fall rainy season. Oospores can survive in the soil several years.

Peronospora sparsa. Rose downy mildew; chiefly on roses under glass, rarely outdoors. Young foliage is spotted, leaves drop; flowers are delayed or unmarketable. Abundant spores are produced on undersurface of leaves. To control, keep humidity below 85% and daytime temperature relatively high.

Peronospora tabacina. Blue mold of tobacco; downy mildew; also on eggplant, pepper, and tomato. This is a seedling disease that can be controlled by sprays on eggplant and pepper; it is unimportant on tomato.

Peronospora trifoliorum. Downy mildew; of lupine, and alfalfa.

Phytophthora

See under Blights.

Phytophthora phaseoli. Downy mildew of lima bean; most important in Middle and North Atlantic states, in periods of cool nights, heavy dews, and fairly warm days. Some seasons it takes 50% to 90% of the crop; in other years it is of little consequence. The white downy mold is conspicuous on the pod, either in patches or covering it completely. The fungus grows through the pod wall into the bean, then the pod dries, turns black. On leaves the white mycelial weft appears sparingly, but veins are often twisted, purplish, or otherwise distorted. Young shoots and flowers are also attacked, bees and other insects carrying spores from diseased to healthy blossoms. The fungus fruits abundantly on pods, stems, and leaves; spores are splashed by rain.

Control. Use seed grown in the West where mildew is not present; plan a 2- to 3-year rotation. Copper dusts are satisfactory.

Plasmopara

Sporangiophores with monopodial branches, with obtuse tips, arising more or less at right angles; haustoria unbranched and knoblike; sporangia (conidia) small, hyaline, papillate, germinating sometimes by germ tubes but usually by swarm spores; oospores yellowish brown, outer wall wrinkled, sometimes reticulate, oogonial wall persistent but not fused with oospore wall (see Fig. 29).

Plasmopara acalyphae. Acalypha downy mildew.

Plasmopara geranii. On geranium. **P. gonolobi.** On gonolobus.

Plasmopara halstedii. Downy mildew; of bur marigold, centaurea, erigeron, eupatorium, gnaphalium, goldenrod, hymenopappus, Jerusalem artichoke, ratibida, rudbeckia, senecio, silphium, verbesina, and vernonia. Zoospores germinate in soil moisture and invade seedlings via root hairs; mycelium moving into stem and leaves causes early wilting and death. Older plants may not die but exhibit a light yellow mottling. Sporangiophores project through stomata on underside of leaves. The fungus winters in seed and as oospores in soil.

Plasmopara nivea. Downy mildew of carrot; parsley, parsnip, and chervil. Yellow spots on upper surface of foliage and white mycelial wefts on under surface turn dark brown with age. The disease is relatively infrequent, important when plants are so crowded they cannot dry off quickly after rain or heavy dew. Control by spacing rows properly.

Plasmopara pygmaea. On anemone, and hepatica. Fine white mildew covers underside of leaves; plants are distorted, stems aborted.

Plasmopara viburni. Viburnum downy mildew.

Plasmopara viticola. Grape downy mildew; general on grape, also on Virginia creeper and Boston ivy. This is a native disease, endemic in eastern United States, first observed in 1834 on wild grapes. It appeared in France after 1870, imported with American stock resistant to the Phylloxera aphid, and in a few years had become as ruinous to the wine industry of Europe as the potato blight had been to Ireland. The efficacy of bordeaux was first discovered in connection with this mildew.

In this country downy mildew is most destructive on European varieties of grape. Pale yellow spots, varying in form but often nearly circular and somewhat transparent, appear on upper leaf surfaces, and a conspicuous white coating appears on lower surfaces. The spots turn brown with age; in dry weather the downy growth is scanty. Young canes, leafstocks, and tendrils may be infected; flowers may blight or rot; young fruits stop growing, turn dark, and dry with a copious grayish growth. Older fruits have a brown rot but lack the mildew effect. Fruits from diseased vines have less juice; bunches are very poorly filled.

Initial infection comes from a swarm spore stopping on the lower side of a leaf, putting out a germ tube and entering through a stoma. In 5 to 20 days the mycelium has spread through the leaf between cells, obtaining food through thin-walled, globular haustoria. The hyphae mass in compact cushions just beneath the stomata; under humid conditions a few grow through the openings and develop into branched conidiophores (sporangiophores). Each has three to six main branches, and they branch again. The terminal branches end in two to four short, slender sterigmata, each of which produces a single multinucleate spore. With moisture, each nucleus with adjacent protoplasm is organized into a swarm spore, motile with two cilia. They swim around for a while, then settle down, absorb their cilia, and put out a germ tube. If they happen to be on the upper side of a leaf, nothing happens; if on the lower surface, the germ tube may reach a stoma and start an infection.

Toward the end of the growing season thick-walled resting spores, oospores, are produced in intercellular spaces of the infected leaves. These are set free in spring by disintegration of host tissue, are rain-splashed to other vines, and germinate by production of a short, unbranched hypha bearing a single large sporangium, to start the cycle anew.

Control. Copper sprays are effective. Apply bordeaux mixture immediately before and just after blooming; repeat 7 to 10 days later and possibly when fruit is half grown. Destroy fallen leaves by burning.

Pseudoperonospora

Like *Plasmopara* but with branches of sporangiophores forming more or less acute angles; tips more acute.

Pseudoperonospora celtidis. Downy mildew; of hackberry.

Pseudoperonospora cubensis. Downy mildew of cucurbits; destructive to cucumber, muskmelon, and watermelon, particularly along the Atlantic seaboard and the Gulf Coast, occasional on gourd, pumpkin, and squash. The disease was first noted in 1889 in New Jersey, and in 1896 destroyed most of the cucumbers on Long Island. Irregular yellow spots appear on upper leaf surfaces, often on leaves nearest the center of the hill. The lesion is brown on the opposite side, covered with a purple growth in rain or dew. The whole leaf may wither and die, with the fruit dwarfed to nubbins and of poor flavor. The fungus does not live in the soil and is not prevalent in the North until July or August. It winters in greenhouses or comes from the South by degrees. Sporangia are spread by wind and cucumber beetles. The disease is favored by high humidity, but temperatures need not be as cool as for other downy mildews.

Control. Resistant cucumbers are of rather poor quality. Cantaloupe varieties Texas Resistant No. 1 and Georgia 47 combine resistance to aphids with resistance to downy mildew.

Sclerospora

Oospore wall confluent with that of oogonium; sporangiophore typically stout with heavy branches clustered at apex; mycelium intercellular, with small, knoblike, unbranched haustoria; germination by germ tube or swarmspores. Common in moist tropic regions on corn, millet, sorghum, and sugar cane.

Sclerospora farlowii. Downy mildew of Bermuda grass; in the Southwest. Short, black, dead areas prune off tips of leaves without serious damage to grass. Tissues are filled with thick-walled, hard oospores.

Sclerospora graminicola. On cereals.

Sclerospora (*Sclerophthora*) **macrospora. Downy mildew of oats**; crazy top of corn, wheat, barley, St. Augustine grass, and wild grasses. Plants bunch owing to shortening of internodes.

FAIRY RINGS

Several species of mushrooms growing in circles in lawns and golf greens cause a condition known as fairy ring, rather common when the soil is quite moist and contains a superabundance of organic matter. Less commonly, some of these mushrooms are responsible for a poor condition of other herbaceous plants and of roses. The chief symptom in turf is the appearance of continuous or interrupted bands of darker green, due to the fungus mycelium breaking down organic matter into products easily assimilated by grass roots. Following the zone of stimulated growth there may be a zone of dying grass due to temporary exhaustion of nutrients, or to toxic substances from the mushroom mycelium, or because a layer has developed that is rather impervious to water. The green rings are more conspicuous on underfertilized lawns, and their presence can sometimes be masked by adequate fertilization. Breaking off the mushrooms, possibly spiking the sod, is all the control ordinarily recommended.

The following species are merely representative of the Basidiomycetes found in fairy rings. They are in the order Agaricales, family Agaricaceae.

Lepiota morgani. On turf and also in rose greenhouses, causing poor growth. The caps are 2 inches to 12 inches across, white with scattered brown scales; flesh white; gills green when mature, spores green turning yellow, stem bulbous at base with a large ring (annulus). Poisonous, though other members of this genus, also causing fairy rings, are edible.

Marasmius oreades. Cap 2 inches or less, convex to plane, thin, tough, withering but not decaying; gills free from stem; spores white. Edible.

Psalliota (*Agaricus*) **campestris.** Cap 1½ inches to 3 inches; white, silky, nearly flat; flesh white to pinkish; gills pink, then brown; spores brownish purple; stem white, with a ring when young. Edible.

Other Basidiomycetes found on lawns in moist weather include puffballs, which are very good for eating when white and firm inside, and bird's nest fungi, which are tiny cups filled with "seed," resembling a nest of eggs.

Trechispora

Basidiomycetes, Corticiaceae

Trechispora alnicola, Blight, fairy ring; of Kentucky bluegrass.

FRUIT SPOTS

Many fruit blemishes are symptoms of rot diseases and are treated under Rots; others are due to physiological disturbances; a few others, limited to fruits and known primarily as fruit spots or specks, are included here.

Cribropeltis

Fungi Imperfecti, Sphaeropsidales, Leptostromataceae

Brown mycellium, branches profusely; black, irregularly circular pycnidia; simple, hyaline, clavate conidiophores; pale, oblong, straight or slightly curved conidia.

Cribopeltis citrullina. Fly speck; of watermelon fruits.

Zygophiala

See under Blotches.

Zygophiala jamaicensis (= *Schizothyrium pomi*). **Fly speck;** on apple.

Helminthosporium

See under Blights.

Helminthosporium papulosum. Black pox; on apples and pears in eastern states. Fruit spots are small, sunken, dark, scattered in profusion over the surface. Blackish papules on bark are followed by a pitted or scaly condition. Spray with sulfur (except at high temperatures).

Microthyriella

Ascomycetes, Hemisphaeriales (Microthyriales),
Microthyriaceae

Vegetative mycelium lacking; stromata with radial structure appearing as black superficial dots on leaves or stems.

Microthyriella rubi. Fly speck of pome fruits; general on apple, also on
pear, quince, citrus fruits, banana, Japanese persimmon, plum, blackberry,
raspberry, and grape. The pathogen has long been recorded as *Leptothyrium
pomi*, but this is apparently a misconception. The imperfect state is *Zygophiala jamaicensis*, originally isolated from banana and recently reported as
causing a greasy blotch of carnations. Flyspeck is often associated with sooty
blotch on apples, but the two diseases are distinct. Flyspeck looks like its
name, groups of 6 to 50 very small, slightly elevated, superficial black dots
connected with very fine threads. Spots may extend entirely around blackberry canes and shoots.

Mycosphaerella

See under Anthracnose.

Mycosphaerella pomi. Brooks fruit spot; phoma fruit spot; quince blotch,
of apple and quince, most prevalent in northeastern states. Spots appear on
fruits in July or early August, deeper red on the colored face of apples, darker
green on the lighter surface. They are irregular, slightly sunken, more abundant near the calyx end of the fruit, usually with centers flecked with black.
The symptoms on quince are more of a blotch than a definite spot.

GALLS

Galls are local swellings, hyperplastic enlargements of plant tissue due to
stimulation from insects, bacteria, fungi, viruses, and occasionally physiological factors. Crown gall, a common and serious problem, is discussed under
Bacterial Diseases. Cedar galls are treated under Rusts. See Black Knot for
hypertrophy of plum branches.

Exobasidium

Basidiomycetes, Agaricales, Exobasidiaceae

Mycelium intercellular with branched haustoria entering host cells; basidia extend
above the layer of epidermal cells much like the layer of asci in *Taphrina*; each

basidium bears two to eight basidiospores. Species cause marked hypertrophy in the Ericaceae.

Exobasidium azaleae. Leaf gall; widespread on flame azalea.

Exobasidium burtii. Leaf gall; yellow leaf spot, on azalea and rhododendron.

Exobasidium camelliae. Camellia leaf gall; on camellia in the Southeast, more common on sasanqua than on japonica. Symptoms are a striking enlargement and thickening of leaves and a thickening of stems of new shoots. Diseased leaves are four or more times as wide and long as normal leaves, very thick and succulent. Color of the upper surface is nearly normal, but the underside is white with a thin membrane that cracks and peels back in strips or patches exposing the spore-bearing layer. There is seldom more than one diseased shoot on a stem, and not many on the whole bush; so the disease does not cause serious damage.

Control. Handpicking of affected parts, searching carefully for diseased leaves at base of new growth, and removing them before spores are formed keeps sporadic infection at a minimum. Spraying with a low-lime bordeaux may be effective but is seldom necessary.

Exobasidium oxycocci. Cranberry rose bloom; shoot hypertrophy; on cranberry, and manzanita. The disease appears in cranberry bogs soon after water is removed in spring. Bud infection results in abnormal lateral shoots with enlarged, swollen, pink or light rose distorted leaves that somewhat resemble flowers. Excessive water supply promotes the disease. Remove water early in spring. If necessary, spray with bordeaux mixture.

Exobasidium rhododendri. Rhododendron leaf gall; large vesicular galls, especially on *Rhododendron catawbiense* and *R. maximum.*

Exobasidium symploci. Bud gall; on sweetleaf.

Exobasidium uvae-ursi. Shoot hypertrophy; of bearberry.

Exobasidium vaccinii. Azalea leaf gall; red leaf spot, shoot hypertrophy, of andromeda, arbutus (*A. menziesii*), bearberry, blueberry, box sandmyrtle, chamaedaphne, cranberry, farkleberry, huckleberry, ledum, leucothoë, manzanita, and rhododendron. On azaleas and other ornamentals the galls are bladder-shaped enlargements of all or part of a leaf, sometimes a flower bud (see Fig. 30). They are white or pink, soft and succulent when young, brown and hard with age. This disease is seldom serious, but in wet seasons, particularly in the South, and in shaded gardens, the number of galls may become rather alarming. On cranberries and blueberries the gall is a small, round, red blister in the leaf, with spores packed in a dense layer on the underside. The fungus is systemic in blueberries, fruiting on the leaves in June and July.

Control. Handpick and destroy galls as they appear. Spraying is seldom required for cranberries and other fruits.

Exobasidium vaccinii-uliginosi. Shoot and leaf gall; witches' broom, of rhododendron, manzanita, and mountain heath. An excessive number of twigs is formed on infected branches. Leaves are yellowish white covered with

Figure 30. Azalea leaf gall.

a dense mealy fungus growth. The mycelium penetrates the whole plant so that it is wiser to remove the shrub than to attempt remedial measures.

Kutilakesa

Fungi Imperfecti, Moniliales, Tuberculariaceae

Sporodochia erumpent, pale olive-green, cushion-shaped; similar to Kutilakesopsis but differs by having larger two-celled conidia; perfect state is *Nectriella*.

Kutilakesa pironii. Stem and leaf galls, cankers; on croton, zebra plant, and *Clorodendron.*

Nocardia

Actinomycetales, Actinomycetaceae

Related to bacteria with mycelial filaments breaking up into rod forms.

Nocardia vaccinii. Blueberry bud-proliferating gall; first observed in Maryland in 1944, described as a new species in 1952. Galls, similar to crown gall, are formed at the soil line. Abnormal buds abort at an early stage or grow into weak shoots, 1 inch to 6 inches high, forming a witches' broom effect.

Phoma, Phomopsis

See under Blights.

Phoma sp. or **Phomopsis** sp. Stem gall; on winter jasmine, privet, forsythia, and rose, at scattered locations. Both pathogens have been reported causing roundish, rather rough stem enlargements on ornamentals. It has not been determined whether more than one fungus is involved.

Plasmopora

See under Downy Mildew.

Plasmopora halstedii. Basal gall; on sunflower.

Protomyces

Phycomycetes, Protomycetales, Protomycetaceae

Protomyces macrosporus. Leaf gall; on hedge parsley (*Torilis* sp.).

Sphaeropsis

See under Cankers.

Sphaeropsis tumefaciens. Canker and gall; on *Carissa.*

Synchytrium

Phycomycetes, Chytridiales, Synchytriaceae

Mycelium lacking; thallus converted into a soros with a membrane, at maturity functioning in entirety as a resting sporangium or divided to form many sporangia in a common membrane; zoospores with one cilium at posterior end. Various species cause excrescences on leaves, and fruit; potato wart.

Synchytrium anemones. Leaf gall; flower spot, of anemone and thalictrum. Flowers are spotted, distorted, dwarfed, and may fall. Red spots are formed on leaves and stems.

Synchytrium aureum. Red leaf gall; false rust, on many plants, 130 species in widely separated genera, including calypha, artemisia, clintonia, delphinium, geum, golden-glow, marsh-marigold, and viola. Pick off and burn affected parts.

Synchytrium endobioticum. Potato wart; black wart of potatoes, a warty hypertrophy of tubers. A European disease wart was found in 1918 in backyard gardens in mining towns of Pennsylvania, Maryland, and West Virginia. Diseased tubers had apparently been brought in by immigrants. A strict quarantine was placed on infested districts, and there has been no spread to commercial potato fields. The disease shows as prominent outgrowths or warts originating in the eyes, varying from the size of a pea to that of the tuber itself. Numerous yellow sporangia are released into the soil by decay of the malformed tissue. The disease, which may affect other species of *Solanum*, is spread by contaminated soil or infected tubers. Buds and adventitious shoots of tomato are infected below the soil line.

Control. By 1953, potato wart had been eradicated from more than half of the 1,112 infested gardens in Pennsylvania. The plan called for applying copper sulfate the first year, keeping the land clean and cultivated, applying lime the next year, growing vegetables the third year, and going back to potatoes the fourth year to test results.

Synchytrium vaccinii. Red leaf gall; on cranberry, azalea, chamaedaphne, gaultheria, and ledum, from New Jersey northward. On cranberry the disease appears just before blossoms open. Buds, flowers, and young leaves are covered with small, red, somewhat globular galls about the size of birdshot; affected shoots bear no fruit. The disease is erratic in appearance but is most frequent in bogs that have excessive or uneven water supply.

Synchytrium sp. Stem gall; on castor bean, in Texas. Small red galls on stems, petioles, and leaves of seedlings.

LEAF BLISTER AND LEAF CURL DISEASES

A single genus, *Taphrina*, is responsible for most of the hyperplastic (overgrowth) deformities known as leaf blister, leaf curl, or, occasionally, as pockets.

Taphrina

Ascomycetes, Taphrinales, Taphrinaceae

Parasitic on vascular plants, causing hypertrophy. Asci in a single palisade layer, not formed in a fruiting body; hyphal cells become thin-walled chlamydospores; on germination the inner spore protrudes from the host and is cut off by a septum to form an eight-spored ascus, which may become many-spored by budding or the ascospores.

Taphrina spp. Maple leaf blister; leaves after expanding in spring show dark spots, shrivel, and fall. The disease may be locally epidemic; it is more common in shaded locations.

Taphrina aesculi. Leaf blister; of California buckeye; yellow turning to dull red; witches' brooms formed.

Taphrina aurea. Poplar yellow leaf blister; conspicuous blisters, small to large, an inch or more in diameter, are brilliant yellow on the concave side when the asci are fully developed; later the color changes to brown.

Taphrina australis. American hornbeam leaf curl.

Taphrina bartholomaei. Western maple leaf blister.

Taphrina caerulescens. Oak leaf blister; on various oak species, with red oak particularly susceptible but often defoliating and sometimes killing water, willow, laurel, and live oaks in the South. Blisters start on young partially grown leaves as gray depressed areas on the undersurface, convex and yellow on the upper surface. Individual blisters are ¼ inch to ½ inch across but often become confluent, causing the leaf to curl. Ascospores are borne on the surface of the blistered area. The disease is most serious in a cool wet spring.

Control. A single dormant eradicant spray, before the buds swell, controls the disease; later sprays are ineffective.

Taphrina carnea. Birch red leaf blister.

Taphrina castanopsidis. California chinquapin leaf blister.

Taphrina cerasi. Cherry witches' broom; leaf curl, on wild and cultivated cherries. **T. flavorubra.** On sand cherry. **T. flectans.** On western wild cherry. **T. farlowii.** Leaf curl and fruit pockets on eastern wild cherry. **T. confusa.** On chokecherry. **T. thomasii.** Witches' broom of cherry laurel in California.

Taphrina communis. Plum pockets; common on American plums. **T. pruni.** On European species, not in United States. **T. prunisubcordata.** In western United States. Leaves, shoots, and fruits become puffy and enlarged into reddish or white swollen bladders. Fruits are sometimes ten times the size of normal plums. Most garden plums are of foreign origin and not susceptible to the American species of *Taphrina*. Bordeaux mixture applied in spring before flower buds open gives satisfactory control.

Taphrina coryli. Hazelnut leaf blister.

Taphrina deformans. Peach leaf curl; general on peach, also on nectarine and almond but not on apricot. This disease is old, known in the United States for well over a century but not quite as important since 1900, when a control was worked out. Young leaves are arched and reddened, or paler than normal

Figure 31. Peach leaf curl; deformed leaf; palisade layer of asci formed on curled portion; germinating spore.

as they emerge from the bud, then much curled, puckered, and distorted, greatly increased in thickness (see Fig. 31). Any portion or the entire leaf may be curled, and one or all leaves from a bud. The leaves often look as if a gathering string had been run along the midvein and pulled tight. Leaves may drop, lowering vitality of tree, with partial or total failure to set fruit, and increasing chances of winter injury. Young fruits may be distorted or cracked. Defoliation for several seasons kills tree outright.

The fungus has no summer stage, and the asci are formed not in a fruiting body but in a layer over infected surfaces, giving them a silvery sheen. Before leaves fall, ascospores are discharged from this layer, and they land on bark or twigs and bud scales, there to germinate by budding into yeastlike spores, which remain viable over winter, sometimes for 2 years. In spring they are washed by rain to opening leaf buds.

Control. One spray during the dormant season gives effective control. This is best applied just before the buds swell, but can be done any time after leaf fall in autumn when the temperature is above freezing. Applications after the buds swell have little effect.

Taphrina faulliana. Leaf blister; of Christmas fern. **T. filicina.** On sensitive fern. **T. struthiopteridis.** On ostrich fern.

Taphrina flava. Yellow leaf blister; of gray and paper birches in northeastern states.

Taphrina japonica (*T. macrophylla*). Leaf curl; on red alder. Young leaves are enlarged to several times normal size and curled. They dry up after ascospore discharge, and a new crop of healthy leaves is formed.

Taphrina populina. Leaf blister; yellow, on poplar.

Taphrina rugosa, T. robinsoniana, T. occidentalis, T. amentorum. Catkin hypertrophy of alder; scales of catkins enlarge and project as reddish curled tongues covered with a white glistening layer. Infection can be reduced with a lime sulfur spray.

Taphrina sacchari. Maple brown leaf blister.

Taphrina ulmi. Elm leaf blister; very small blisters on elm leaves. Dusting nursery trees with sulfur has helped.

LEAF SCORCH

According to the dictionary scorching means to heat so as to change color and texture without consuming. Sometimes leaves are literally scorched in summer heat, and sometimes symptoms caused by fungi resemble those of a heat scorch. This section includes some of the latter.

Ceratocystis

See under Cankers.

Ceratocystis (*Ceratostomella*) **paradoxa.** Black scorch; bud scorch; heart rot of coconut, Canary, Washington, and Guadaloupe palms, also causing a pineapple disease in the tropics. The most striking symptom is a black, irregular, necrotic condition of the leaf stalk. The tissues look as if they had been burned, whence the name black scorch. Furled pinnae of leaf fronds show pale yellow spots with broad margins that later converge and turn black; infection spreads rapidly, and in severe cases the heart leaves dry up. The heart rot discolors trunk tissues and rots the pithy material between cells. Infection is through wounds during periods of relatively high humidity, or through roots, or sometimes through uninjured fruit strands, petioles, or pinnae. Palms with vitality lowered, as when the normal crown of leaves has been reduced but the water supply to the leaves is not reduced, are most susceptible.

Control. Destruction of infected parts seems to be the chief control measure. It is easier to bury than to burn palm trunks.

Diplocarpon

See under Blackspot.

Diplocarpon earliana. Strawberry leaf scorch; general where strawberries are grown but more prevalent in the South. Dark purplish spots about ¼ inch in diameter are scattered profusely over upper surface of leaves in all stages of development. Later the spots enlarge to scorch wide areas of the leaf, and black fruiting bodies give a tar-spot appearance. Scorch spots always lack the white centers so characteristic of Mycosphaerella leaf spot on strawberry. Lesions are found on petioles, stolons, and fruit stalks as well as leaves. If the fruit stems are girdled, flowers or young fruits die. Rarely the disease appears on green berries as a superficial red or brown discoloration and flecking. Spores, produced in quantity in acervuli on lesions, are distributed by birds, insects, and pickers on tools and clothing. The fungus winters in old leaves. Perfect and imperfect stages are both produced in spring, and repeated infections occur throughout the summer in moist weather.

Control. Remove all old leaves when setting plants in spring. Spray with bordeaux mixture at 10-day intervals, starting in January in Louisiana, late February in North Carolina. Fairly resistant varieties include Catskill, Midland, Fairfax, Howard 17, Blakemore, Southland.

Hendersonia

Fungi Imperfecti, Sphaeropsidales, Sphaerioidaceae

Pycnidia dark, separate, globose, ostiolate, immersed then usually erumpent; conidia dark, several-celled, elongate to fusoid; saprophytic or parasitic.

Hendersonia opuntiae. Scorch; sunscald; common and serious on prickly pear cactus (*Opuntia*). Segments turn reddish brown and die; centers are grayish brown and cracked.

Pseudopezicula

Ascomycetes, Helotiales, Peziculoidaceae

Hyaline, gelatinous apothecia containing paraphyses and 20–80 asci; asci contain four reniform, binucleate ascospores; five-spored asci rarely observed.

Pseudopezicula tetraspora. Leaf scorch; of grapevines.

Septoria

See under Blights.

Septoria azalea. Azalea leaf scorch; leaf spot. Small, yellowish, round spots enlarge irregularly, turn reddish brown, with dark brown centers. Leaves fall prematurely; black fruiting bodies are produced in fallen leaves. The disease is most severe in greenhouses in fall and winter and under high humidity.

Stagonospora

Fungi Imperfecti, Sphaeropsidales, Sphaerioidaceae

Pycnidia dark, separate, superficial, or erumpent, globose, ostiolate; conidiophores short; conidia hyaline, typically with three or more cells, cylindrical to elliptical; parasitic or saprophytic.

Stagonospora curtisii. Narcissus leaf scorch; red blotch of amaryllis, red leaf spot, red fire disease, also on crinum, eucharis, hymenocallis, leucojum, nerine, sternbergia, vallota, and zephyranthes.

Leaf tips of narcissus are blighted for 2 or 3 inches as in frost injury and separated from healthy basal portions of leaves by a definite margin or yellow area. Spores formed in pycnidia in the dead area furnish inoculum for secondary infection, which consists of lesions in lower portions of leaves, minute water-soaked or yellowish spots becoming raised, scabby, and reddish brown. Flower stalks may be spotted; brown spots appear on petals. Bulbs suffer loss in weight due to killing of foliage a month or two before normal dying down. All types may be infected but the most susceptible varieties are in the Leedsii and Polyanthus groups. The fungus was described on narcissus in 1878 but was not considered a threat to it, nor was it known to be connected with amaryllis red blotch before 1929.

On amaryllis or hippeastrum red spots are formed on leaves, flower stems, and petals. On foliage the spots are bright red to purplish, small at first but often increasing to 2 inches. Leaf or flower stalks are bent or deformed at the point of attack. This disease should not be confused with "red disease" caused by mites. The spores are variable in size and number of cells, one to six. They are embedded in a gelatinous matrix and are disseminated in rain. The fungus apparently winters in or on bulbs, infecting new leaves as they grow out in spring.

Control. Treat suspected narcissus bulbs before planting. Control secondary infection in the field with bordeaux mixture. Discard seriously diseased amaryllis bulbs; remove infected leaves and bulb scales; avoid syringing and heavy watering.

LEAF SPOTS

Leaf spots are the most prevalent of plant diseases, so common we seldom notice them, and rightly so, for if we should attempt to control all the miscellaneous leaf spots that appear in a small suburban garden in a single season, we would quickly go mad. A typical leaf spot is a rather definitely delimited necrotic lesion, often with a brown, sometimes white, center and a darker margin. When the spots are so numerous they grow together to form large dead areas, the disease becomes a blight, or perhaps a blotch, or scorch. Certain types of lesions are called anthracnose, spot anthracnose, blackspot. All of these have been segregated in their different sections. What is left is a very large collection of names.

The genus *Septoria*, for instance, has about 1000 species, *Mycosphaerella* 500, *Cercospora* 400, chiefly identified by the hosts on which they appear. *Cercospora beticola* is so named because it causes a leaf spot of beet, *C. apii* for its celery host. Species recorded in this country as causing a definite disease are listed under their respective hosts. They are not repeated here unless the leaf spot is of some importance or there is some useful information that can be added to the name.

Most leaf spot diseases flourish in wet seasons. A comparative few may be important enough to call for control measures other than general sanitation. Adequate protection usually means several applications of fungicides, and the cost of spraying trees and shrubs must be balanced against the expected damage. Calling in a tree expert with high-pressure apparatus is often an expensive proposition. If the budget is limited, it is more important to have an elm sprayed for elm leaf beetles, which cause d⌐foliation every season, than for elm black spot, which may be serious in only one year out of three or four. When it comes to rose blackspot (no relation to elm black spot), weekly protection with a fungicide is necessary, but to save labor it can be combined with insecticides.

Actinopelte

Fungi Imperfecti, Sphaeropsidales

Actinopelte dryina (= *Tubakia dryina*). On oak. Very small dark spots between veins. Conspicuous in midsummer but not serious.

Actinothyrium

Fungi Imperfecti, Sphaeropsidales, Leptostromataceae

Pycnidia superficial, globose, with a more or less fimbriate shield; spores filiform, hyaline.

Actinothyrium gloeosporioides. Leaf spot; on sassafras.

Alternaria

See under Blights.

Alternaria alternata. Leaf spot; of *Calathea* spp.
Alternaria brassicae (with large spores) and **A. brassicicola** (with small spores). Black leaf spot of crucifers; cabbage, Chinese cabbage, collards, turnip, garden cress, radish, and horseradish. Head browning; leaf and pod spot of cauliflower; Damping-off; wire-stem of seedlings.

Seedlings are subject to pre- or postemergence damping-off, with dark brown to black sunken spots on cotyledons, narrow dark spots on stems, followed by wire-stem, a blackening toward the base. Leaf spots are small, circular, yellowish, enlarging in concentric circles with a sooty black color from the spores. In storage the spots unite to form a moldy growth over the entire leaf. On seed pods, spots are purplish at first, later brown; in moist weather entire pods may be infected. Cauliflower infection is a browning of the head, starting at the margin of an individual flower or cluster. Spores are blown, splashed by tools, spread on feet of people and animals. Seed bears spores externally, mycelium internally. Wounds are not necessary for infection.

Control. Hot-water treatment of seed, 122° F for 30 minutes, is fairly effective. Use long rotation for cauliflower, avoiding all other crucifers in intermediate years.

Alternaria catalpae. Catalpa leaf spot; widespread in rainy seasons. Small, water-soaked spots, up to ¼ inch, appear over the leaf; they turn brown and sometimes drop out leaving shot holes; there is more or less defoliation. The fungus is sometimes secondary following bacterial infection or midge infestation. Rake up and burn fallen leaves.

Alternaria chrysanthemi. Leaf spot; on shasta daisy, and Canada thistle.

Alternaria citri. Cherry leaf spot; occasional, more often a rot of citrus fruits. See under Rots.

Alternaria fasciculata. Leaf spot; on rose acacia and asclepiodora.

Alternaria longipes. Brown spot of tobacco; including ornamental flowering tobacco. Small spots on lower leaves rapidly enlarge and turn brown. The fungus winters on old stalks, which should be removed and burned.

Alternaria oleracea. Cabbage leaf spot; occasional on crucifers. Has been confused with *A. brassicicola*.

Alternaria panax. Leaf spot; of schefflera, *Dizygotheca*, and *Tupidanthurs*.

Alternaria passiflorae. Brown spot of passion flower; minute brown leaf spots, enlarging to an inch across are concentrically zoned with various shades of brown. Dark green water-soaked spots on fruit turn brown; the fruit shrivels, but the spots stay firm.

Alternaria polypodii. Fern leaf spot; brown, circular to ovate, concentrically zonate spots are formed along margins of fronds. Chains of spores are spread by syringing or air currents. Keep foliage dry; remove and burn diseased leaves.

Alternaria raphani. Radish leaf spot; yellow spots with black sporulation, often with centers dropping out. Also occurs on turnip.

Alternaria sonchi. Leaf spot; of lettuce, escarole, endive, and chicory.

Alternaria tagetica. Leaf spot; of marigold.

Alternaria tenuis. Leaf spot; of magnolia, hibiscus, clarkia, and many ornamental and other hosts. The fungus is a general saprophyte and an occasional weak parasite. It discolors beet, chard, and spinach seed.

Alternaria tenuissima. Leaf spot; on blueberry.

Alternaria tomato. Nailhead spot of tomato; a leaf, stem, and fruit spot. On leaves and stems the disease is much like early blight (see *A. solani* under Blights) with small dark brown spots with yellow margins. But on fruit the disease is quite different. Very small tan spots, ¹⁄₁₆ inch to ⅛ inch in diameter, become slightly sunken, with grayish brown centers and darker margins.

Spores produced abundantly on fruit and foliage are spread by winds and splashing rain. Treat seed and spray as for early blight. Varieties Marglobe, Pritchard, Glovel, and Break O'Day are quite resistant to nailhead spot. The same fungus causes ghost spot of apple.

Alternaria sp. Leaf spot; of schefflera, and umbrella tree.

Amerosporium

Fungi Imperfecti, Sphaeropsidales, Excipulaceae

Pycnidia superficial, discoid to cupulate, hairy; spores one-celled, hyaline.

Amerosporium trichellum. Leaf and stem spot; on English ivy. In some cases stems are girdled, causing collapse and death.

Annellophora

Fungi Imperfecti, Moniliales, Dematiaceae

Conidiophores brown, simple, slender, elongating by successive proliferations through conidial scars; conidia brown, multiseptate, obclavate to fusoid.

Annellophora phoenicis. Leaf spot; of date palm.

Aristastoma

Fungi Imperfecti, Sphaeropsidales, Sphaerioidaceae

Pycnidia brown, globose, erumpent, separate, with dark brown setae near ostiole; conidiophores short, simple; conida hyaline, several-celled.

Aristastoma oeconomicum. Zonate leaf spot; of cowpea, kidney bean.

Ascochyta

See under Blights.

Ascochyta abelmoschi (possibly identical with *A. phaseolorum*). Leaf spot, pod spot; stem spot of okra. Dark, small, water-soaked spots slowly enlarge, turn brown, with many large black pycnidia in concentric rings in dead tissue (see Fig. 32). Young okra pods are severely infected, and the mycelium grows into the seed.

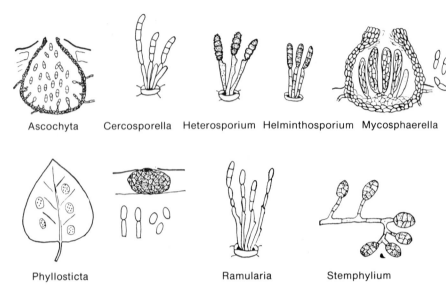

Ascochyta   Cercosporella   Heterosporium   Helminthosporium   Mycosphaerella

Phyllosticta   Ramularia   Stemphylium

Figure 32. Some leaf-spot fungi. *Ascochyta,* hyaline, two-celled conidia in pycnidium; *Cercosporella,* hyaline, septate spores on conidiophores emerging from a stoma; *Heterosporium,* spiny, dark, septate spores; *Helminthosporium,* smooth, dark, septate spores; *Mycosphaerella,* two-celled hyaline ascospores in a perithecium; *Phyllosticta,* hyaline, one-celled conidia in pycnida formed in spots on leaves; *Ramularia,* hyaline spores, becoming septate, formed successively on conidiophores; *Stemphylium,* colored muriform spores borne free on mycelium.

Ascochyta althaeina. Leaf spot; of hollyhock, rose mallow.

Ascochyta armoraciae. Leaf spot; of horse-radish.

Ascochyta aspidistrae. Aspidistra leaf spot; large, irregular pale spots on leaves.

Ascochyta asteris. Leaf spot of china aster; spray foliage with bordeaux mixture.

Ascochyta boltschauseri. Leaf spot; pod spot of beans; on snap, kidney, lima, and scarlet runner beans, reported in Oregon. Spots on leaves and pods are dark to drab, zonate; light to dark brown pycnidia are numerous.

Aschochyta brachypodii. Leaf spot; on big bluestem, little bluestem (both species of *Andropogon*) and on Indian grass.

Ascochyta cheiranthi. Leaf and stem spot of wallflower; grayish spots up to ½ inch long, may girdle stems. Leaf spots are circular to elongate, brown with darker brown margins. Dark pycnidia contain hyaline, two-celled spores. Leaves wilt and fall; potted plants may be infected. Keep greenhouse on the dry side.

Ascochyta clematidina. Clematis leaf and stem spot; widespread. On outdoor plants stems are infected near the ground and are often girdled, upper portions dying back. Spores for initial infection probably come from pycnidia on stumps of old stems. Leaf spots are more common in greenhouses, small, water-soaked, then buff with reddish margins. Remove and destroy infected leaves and stems.

Ascochyta compositarum. Leaf spot; on aster, eupatorium, silphium, and sunflower.

Ascochyta cornicola. Dogwood leaf spot.

Ascochyta cypripedii. Cypripedium leaf spot; reported on orchid from Wisconsin. Leaf lesions are narrow, brownish, with a dark brown border.

Ascochyta juglandis. Walnut ring spot; very small, round, brown leaf spots between veins, ringed with targetlike ridges. The disease is unimportant in trees sprayed for walnut blight.

Ascochyta lycopersici (*Didymella lycopersici*). Leaf spot; Ascochyta blight, of tomato, eggplant, and potato. Brown spots with concentric rings are formed on leaves and stems, sometimes cankers at base of young stems. Black pustules in center of spots discharge spore tendrils in wet weather. The fungus winters in old plant refuse, is a weak parasite, and is ordinarily too unimportant for control measures.

Ascochyta phaseolorum. Leaf spot of snap beans; recent isolation and inoculation studies indicate that the Ascochyta leaf blights of hollyhock, okra, pepper, eggplant, and tomato are all caused by strains of the bean pathogen.

Ascochyta pisi. Leaf spot, pod spot of pea; general, but rare in the Northwest. One of three species causing the disease complex known as Ascochyta blight (see also under Blights). Foliage spots are circular to irregular, pinhead size to ½ inch. Stem lesions, at nodes or base, are brown to purplish black. Brown pycnidia exude spore tendrils in wet weather.

Asteroma

Fungi Imperfecti, Sphaeropsidales, Sphaerioidaceae

Pycnidia globose with a radiate subicle, a compact, crustlike growth of mycelium underneath; without an ostiole; spores hyaline, one-celled.

Asteroma garretianum. Black spot; on primrose.
Asteroma solidaginis. Black spot; black scurf, on goldenrod.
Asteroma tenerrimum. Black spot; on erythronium.

Asteromella (*Stictochlorella*)

Fungi Imperfecti, Sphaeropsidales, Sphaerioidaceae

Pycnidia smooth, with ostiole, densely gregarious in asteroma-like spots; spores hyaline, one-celled.

Asteromello lupini. Leaf spot; on lupine.

Cephaleuros

One of the green algae, possessing chlorophyll but not differentiated into root, stem, and leaves; forming motile spores in sporangia.

Cephaleuros virescens. Algal spot; red leaf spot; green scurf; in the far South or in greenhouses on acacia, albizzia, ardisia, avocado, bischofia, bixa, camellia, camphor-tree, cinnamon-tree, citrus, grevillea, guava, jasmine, jujube, loquat, magnolia, mango, pecan, Japanese persimmon, privet, rhododendron, viburnum.

On some hosts this is a disease of twigs and branches, which may be girdled and stunted, covered with reddish brown hairlike fruiting bodies. On magnolia leaves velvety, reddish brown to orange, cushiony patches are formed, but in the absence of sporangia (tiny globular heads on fine, dense reddish hairs) the leaf spots remain greenish brown. Occasionally citrus fruits as well as leaves are attacked.

The sporangia formed on the fine hairs germinate in moist weather, producing zoospores that enter through stomata and form mycelium-like chains of algal cells in host tissue. On twigs the alga invades outer cortical tissue, which may swell abnormally, crack, and afford entrance to injurious fungi. Weakened trees are most susceptible, and disease spread is most rapid in periods of frequent and abundant rains. Twigs may die, and there may be reduced yield of citrus fruit.

Control. Improve draining and other growing conditions. Citrus trees sprayed regularly with copper seldom have algal trouble. If it gets started, follow cleanup pruning with a bordeaux mixture spray in December or January. Repeat with bordeaux at start of rainy season or when red stage of the alga is first seen, and spray again 1 month later. A neutral copper may substitute for bordeaux for the last two applications. The copper kills beneficial insects parasitic on scales, but the oil controls the scale insects.

Cephalosporium

Fungi Imperfecti, Moniliales, Moniliaceae

Conidiophores slender or swollen, simple; conidia hyaline, one-celled, produced successively at the tip and collecting in a slime drop, produced endogenously in some species; saprophytic or parasitic, some species causing vascular wilts of trees.

Cephalosporium apii. Celery brown spot; a new disease first reported from Colorado in 1943, later from New York and Ohio. Irregular light tan or reddish brown shallow lesions are formed on celery leaf stalks, petioles, and leaflets. They may unite to make a scurfy brown streak up the inside of the stalk and may develop transverse cracks. Utah and Pascal varieties are most susceptible.

Cephalosporium cinnamomeum. Leaf spot of nephthytis and syngonium; small circular to irregular spots, reddish brown with pale yellow borders enlarge, with centers becoming gray and papery. In severe cases leaves turn yellow and die. Pick off infected leaves. Maintain low temperature and humidity.

Cephalosporium dieffenbachiae. Dieffenbachia leaf spot; small red lesions with dark borders appear on young leaves. Spots sometimes run together, and the whole leaf turns yellow and dies. Infection is often through mealybug wounds. Avoid promiscuous syringing; keep temperature and humidity low; control mealybugs, and ants that transport them.

Cercospora

See under Blights.

Cercospora abeliae. Abelia leaf spot; reported from Louisiana. Irregular purple to brown spots; defoliation.

Cercospora abelmoschi. Leaf spot; on okra, hibiscus. Spots indistinct, but a sooty fruiting of spores on under leaf surface.

Cercospora albo-maculans (*Cercosporella brassicae*). White spot; of turnip, Chinese cabbage, mustard, and other crucifers, common in the Southeast. Small, pale, circular slightly sunken spots; may coalesce.

Cercospora althaeina. Leaf spot; of hollyhock and abutilon. Spots circular, angular or irregular, 1.5 millimeters, olivaceous to grayish brown, with the dead tissue falling out. The fungus winters in old plant parts.

Cercospora angulata. Leaf spot; on philadelphus, currant, flowering currant, and gooseberry. Circular to angular spots, dingy gray centers, dark purple to nearly black margins.

Cercospora aquilegiae. Columbine leaf spot; reported from Kansas, Wisconsin, Oregon. Spots circular to elliptical, reddish brown to nearly black; fruiting is on both sides of the leaf.

Cercospora arachidicola (*Mycosphaerella arachidicola*). Peanut early leaf spot; spots light tan, aging to reddish or dark brown with a yellow halo, often confluent. Conidiophores on both sides of the leaf, emerging from stomata or breaking through epidermal cells. Conidia colorless to pale yellow or olive, with 5 to 12 cells. Control with sulfur-copper dust.

Cercospora armoraciae. Horse-radish leaf spot; tan to dingy gray lesions with yellow-brown margin; often slightly zonate.

Cercospora beticola. Cercospora leaf spot of beet; general on garden and sugar beets, also on swiss chard, spinach. Brown flecks with reddish purple borders become conspicuous spots with ash-gray centers and purple margins. The brittle central tissue often drops out, leaving ragged holes. The spots usually remain small but are often so numerous that foliage is killed. If successive crops of leaves are lost, the crown of the beet root is elongated and roughened. Leaf spotting is of little direct importance except in chard, where foliage is used for greens. The beet root yield is reduced.

The grayish color of the spots is due to long, thin, septate conidia produced on coniophores protruded through stomata in fascicles or groups, coming from a knotted mass of mycelium resembling a sclerotium. Conidia are spread by rain, wind, tools, and insects. Infection is through stomata; disease spread is most rapid under conditions of high humidity that keep stomata open. Hot weather favors the disease.

Control. Crop rotation is highly important. In a small garden pick off the first spotted leaves.

Cercospora bougainvilleae. Leaf spot first seen in Florida in 1962 and now the most important pathogen of this host. Lesions are 1 millimeter to 5 millimeters, circular, depressed, with brown or tan centers, reddish brown margins, and a diffuse chlorotic area.

Cercospora brunkii. Geranium leaf spot; mostly in the South. Spots are circular, light reddish brown with dark brown borders, sometimes coalescing to kill entire leaf.

Cercospora calendulae. Calendula leaf spot; spots run together to blight and kill leaves; plants may be destroyed early in the season. Spores enter through stomata of plants more than a month old.

Cercospora cannabina. Leaf curl and wilt; on hemp.

Cercospora cannabis. Leaf spot; on hemp.

Cercospora capsici. Pepper leaf spot; stem-end rot, common in the Southeast, serious in rainy seasons. Spots $1/7$ inch to 1 inch in diameter are first water-soaked then white with dark brown margins. Leaves turn yellow and drop. The fungus grows through the pedicel into fruit, causing a rot of the stem end. Loss of foliage exposes the fruit to sunscald. Spray or dust with copper.

Cercospora circumscissa (*Mycosphaerella cerasella*). **Leaf spot, shot hole**; of apricot, plum, cherry, cherry-laurel, oriental cherry, and chokecherry. Dead spots are somewhat larger than those caused by other shot-hole fungi, but the damage is not serious.

Cercospora citrullina. Leaf spot; of watermelon, muskmelon, and other cucurbits. Spots are small, circular, black with grayish centers, occurring first on leaves in center of watermelon hills. On cucumber, muskmelon, and squash the spots are large and ochre-gray. Defoliation of vines causes reduction in fruit size, but the disease is not considered important. Clean up diseased vines; use a 2- or 3-year rotation; spray or dust as for bacterial wilt.

Cercospora concors. Potato leaf spot; leaf blotch. Spots none to large irregular brown areas. Fruiting on undersurface; conidiophores very pale; conidia almost hyaline.

Cercospora cornicola. Dogwood leaf spot; in the Gulf states, often with *Septoria florida*. Spots irregular without definite borders.

Cercospora fusca. Pecan brown leaf spot; prevalent throughout the pecan belt but minor, serious only with high rainfall and in neglected orchards where trees lack vigor. Spots are circular to irregular, reddish brown, often with grayish concentric zones. The fungus winters in old spots on leaves. In Florida the disease appears first in June or July on mature leaves and may cause premature defoliation in October. Stuart variety is particularly susceptible; others are more resistant. Control with one application of bordeaux mixture between May 15 and June 15.

Cercospora lathyrina. Leaf spot; on pea and sweet pea, in southern states and north to New Jersey and Missouri. Angular to elongate spots have dirty gray centers with a black line border.

Cercospora lythracearum. Leaf spot; on crape myrtle, in Texas. Spots circular, pale brown to gray with a greenish fringe or yellow halo.

Cercospora magnoliae (*Mycosphaerella milleri*). On magnolia in South. Leaf spots are small, angular, dark, with narrow yellow halo.

Cercospora melongenae. Eggplant leaf spot; more common in tropical areas. Yellow lesions change to large brown areas with concentric rings.

Cercospora nandinae. Nandina leaf spot; one of the few diseases of this usually healthy shrub. Red blotches appear on upper leaf surface with centers of older spots almost black. There is a scant sooty fruiting layer on the undersurface. Reported from Alabama and North Carolina.

Cercospora personata (*Mycosphaerella berkeleyii*). Peanut leaf spot; general on peanut. Spots are circular, 1 millimeter to 7 millimeters, but may coalesce; dark brown to black, often with a yellow halo. Conidiophores on both sides of the leaf, more numerous on the lower, are arranged concentrically in tufts; the epidermis is ruptured. Spores are pale brown to olivaceous, one- to eight-septate. In wet seasons vines may be nearly defoliated. Primary infections come from ascospores on overwintered peanut leaves. Sulfur dust with 3.5% copper is recommended; apply every 10 to 14 days.

Cercospora piaropi. Leaf spot; on water hyacinths.

Cercospora pittospori. Pittosporum leaf spot; reported from Mississippi, Florida, Louisiana, and Texas. Spots small, angular, yellow to dull brown, fruiting in fawn-colored effuse patches on lower surface.

Cercospora puderi. Leaf spot; on rose, reported from Georgia and Texas. Spots are circular, to 5 millimeters, with dingy gray centers, brown or reddish brown margins. Fruiting is chiefly on the upper surface in dense fascicles of short conidia.

Cercospora resedae. Leaf spot; blight of mignonette, a rapid disease killing much of the foliage. Numerous small circular spots, pale yellow with reddish brown borders, run together, discoloring the entire leaf. Spores are spread by wind and rain; lower leaves are most affected.

Cercospora rhododendri. Rhododendron leaf spot; angular dark brown spots with grayish down in center. Control seldom necessary.

Cercospora richardiaecola. Leaf spot; on calla lily, sometimes injurious. Spots circular, brown, tan, or gray. Avoid syringing; keep plants well spaced; ventilate greenhouse.

Cercospora rosicola (*Mycosphaerella rosicola*). Cercospora spot of rose; wherever roses are grown but more important in the South. Spots are circular,

1 millimeter to 4 millimeters, but coalescing to irregular areas, purplish or reddish brown with pale brown, tan, or gray centers. Perithecia are formed in fallen leaves.

Cercospora smilacis. Smilax leaf spot; spots are more or less circular up to ¼ inch, dark purplish red, centers fading with age but margins remaining definite and dark.

Cercospora sojina. Frog-eye disease of soybean; typical frog-eye spots are formed on leaves and elongated reddish lesions on stems, changing to brown, gray, or nearly black with age. Pods of late varieties may be infected. The fungus winters on diseased leaves and stems. Seed treatment is not effective; crop rotation is necessary. Early varieties often escape injury. There is a wide difference in varietal susceptibility.

Cercospora symphoricarpi. Leaf spot; on snowberry, coralberry, and wolfberry. Very small circular to angular spots, uniformly brown or with tan centers and brown margins.

Cercospora sp. Leaf spot; on kalanchoë.

Cercosporella

Fungi Imperfecti, Moniliales, Moniliaceae

Conidiophores hyaline, bearing conidia apically or on short branches; conidia hyaline, cylindrical to filiform with several cells (see Fig. 32); like *Cercospora* except for light conidiophores; parasitic.

Cercosporella brassicae. Leaf spot; of cabbage, turnip, mustard, on West Coast. Lesions on cabbage are black; those on turnip and mustard are gray with tan margins.

Ciborinia

See under Blights.

Ciborinia bifrons (*Sclerotinia whetzelii*). Black leaf spot of poplar; ink spot, from New England states to the Rocky Mountains on aspen, black poplar, and other species. Saucerlike, thin black sclerotia are formed in leaves, fall to the ground, and produce apothecia in spring. There is often considerable defoliation, and small trees may be killed.

Ciborinia confundens (*Sclerotinia bifrons*). Ink spot; in western states, producing apothecia on ground under cottonwoods and poplars but pathogenic state confused.

Cladosporium

See under Blotch Diseases.

Cladosporium epiphyllum. Leaf spot; on locust.

Coccomyces (Higginsia)

Ascomycetes, Phacidiales, Phacidaceae

Apothecia with black leathery hypothecium and thin epithecium embedded in stroma, breaking through a starlike slit; asci broadly clavate; spores hyaline, one-celled. Conidia of *Cylindrosporium* stage hyaline, threadlike; spermatia formed in conidial acervuli apparently function in fertilization.

Coccomyces hiemalis. Cherry leaf spot; blight; shot hole; general on sweet and sour cherries, the most common and destructive leaf disease of cherries. Leaf spots are circular, first purplish, then brown, falling out to give the shot-hole effect (see Fig. 33). If lesions are numerous, the leaves turn yellow and fall by midsummer, this premature defoliation reducing next season's harvest. The fungus winters in fallen leaves, producing disc-shaped apothecia for primary infection. Secondary infection comes from conidia, formed in whitish masses on the spots in moist weather, more numerous on the undersurface. New infection continues through the summer after harvest.

Figure 33. Shot hole on *Prunus* sp.

Defoliation prior to ripening reduces size and quality of fruit and exposes it to sunscald. Some seasons' shoots, spurs, and branches are killed, followed by a light crop the next year. Thousands of sour cherry trees have been killed.

Control. An eradicant spray of a dinitro compound, such as Elgetol, applied to the ground in early spring, reduces the amount of primary inoculum, but summer sprays are also necessary. On sour cherry this application may mean a spray at petal fall, another 10 days later, two sprays in June, and another just after fruit is picked, with more applications, especially on nursery trees, needed in some seasons. Consult your state experiment station for suitable materials and schedule for your area.

Coccomyces kerriae. Kerria leaf spot; twig blight, widespread on kerris from eastern states to Texas. Leaves have small, round to angular, light brown or reddish brown spots with darker borders. When spots are numerous, leaves turn yellow and die. Similar lesions on young stems may run together into extended cankers, the bark splitting to show black pycnidia, from which ooze out masses of long, white, curved spores. The fungus winters in old dead leaves. Spraying with bordeaux mixture may help.

Coccomyces lutescens. Leaf spot, shot hole; on cherry-laurel, black cherry, and chokecherry. Similar to the disease caused by *C. hiemalis.*

Coccomyces prunophorae. Leaf spot, shot hole; on garden plum and apricot. Reddish to brown spots, dark blue initially, produce pinkish spore masses on underside of leaves in wet weather. The shot-hole effect from dropping out of dead tissue may be very prominent and accompanied by heavy fruit drop. Spray when shucks are off young fruit, 2 or 3 weeks later, and before fruit ripens, with lime sulfur, or with wettable sulfur.

Colletotrichum

See under Anthracnose.

Colletotrichum acutatum. Fruit spot; crown and petiole spot; on strawberry.

Colletotrichum gloeosporioides. Leaf spot; of jasmine, passion flower, leaf and stem spot of calendula; on many other hosts as anthracnose.

Colletotrichum coccodes. Leaf spot; slight blight; of velvet-leaf.

Colletrotrichum dematium f. sp. **truncata.** Leaf spot; stem canker; of *Stylosanthes* spp.

Colletotrichum elastica. Leaf spot; on fig (*Fiscus carica*).

Coniothyrium

See under Cankers.

Coniothyrium concentricum. Leaf spot; of century plant and yucca. Spots are zoned, light grayish brown, an inch or more in diameter, with concentric rings of tiny black pycnidia. Large portions of leaves may be destroyed. Remove and burn diseased leaves.

Coniothyrium hellebori. Black spot of Christmas-rose; large, irregular, dark brown to black spots on both sides of leaves, often running together with concentric zonation; many leaves turn yellow prematurely and die; plants are weakened and fail to mature the normal number of leaves. Stems may be cankered, shrivel, and fall over, with wilting of unopened flower buds. Open petals sometimes have black spots. In wet weather in spring and fall the disease can spread through an entire planting in 2 or 3 days, but continuous moisture is necessary for infection. Spray with bordeaux mixture.

Coniothyrium pyrina. Leaf spot; fruit spot, of apple, pear.

Corynespora

Fungi Imperfecti, Moniliales, Dematiaceae

Hyphae and conidia both dark.

Corynespora cassiicola (*Helminthosporium vignicola*). Soybean target spot; also on cowpea, tomato, poinsettia, and privet; general in South. Circular to irregular, reddish brown leaf spots, pinpoint to ¼ inch, often zonate and surrounded by yellow-green halos. Dark brown spots on petioles, pods, and seed. Variety Ogden is moderately resistant. The same fungus causes reddish purple spots on azalea, hydrangea and leaf spots on lipstick vine, and on weeping fig.

Cristulariella

Fungi Imperfecti, Moniliales, Moniliaceae

Sterile hyphae decumbent; fertile hyphae hyaline; ascending in a branched head with conidia at tips of intermediate branches; spores globose, hyaline, one-celled.

Cristulariella depraedans. Leaf spot; on sugar and other maples. Spots gray, definite or confluent.

Cristulariella pyramidalis (= *Grovesinia pyramidalis*). Leaf spot; on maple, tree-of-heaven, apple, bean, blueberry, cherry, dogwood, hibiscus, sycamore, tung tree, viburnum, walnut, black walnut, beggar-ticks, trumpet vine, Mexican tea, dayflower, blue waxweed, tick clover, mistflower, white snakeroot, morning glory, Indian tobacco, blue cardinal-flower, beefsteak plant, poke, smart weed, false buckwheat, yellow dock, prickly mallow, goldenrod, catbird grape, nectarine, grape, maple, serviceberry, and boxelder. Spots yellow-gray with definite margins.

Cryptomycina

Ascomycetes, Phacidiales, Phacidiaceae

Apothecium splitting irregularly into lobes, hyphal layer thin; spores hyaline, one-celled.

Cryptomycina pteridis. Tar spot of fern, bracken; spots are usually on lower surface and between veins; leaves may roll.

Cryptostictis

See under Blights.

Cryptostictis arbuti. Leaf spot; on *Arbutus menziesii, Manzanita,* ledum.

Cycloconium

Fungi Imperfecti, Moniliales, Dematiaceae

Mycelium coiled, spores small, dark, two-celled; scarcely different from short hyphae.

Cycloconium oleaginum. Olive leaf spot; peacock spot, ring spot. Blackish, more or less concentric rings on leaves, especially those weakened or old.

Cylindrocladium

Fungi Imperfecti, Moniliales, Moniliaceae

Conidiophores repeatedly dichotomously or trichotomously branched, each terminating in two or three phialides (cells developing spores); conidia hyaline, with two or more cells, cylindrical, borne singly; parasitic or saprophytic.

Cylindrocladium avesiculatum. Leaf spot; twig dieback; on holly, and *Leucothoë* sp.
Cylindrocladium macrosporium. Leaf spot; leaf blight, of Washington palm. Numerous small dark brown spots with light margins are somewhat disfiguring.
Cylindrocladium pteridis. Fern leaf spot; leaf blotch. Reddish brown lesions run together to cover large areas. Pick off and burn infected fronds.

Cylindrosporium

Fungi Imperfecti, Melanconiales, Melanconiaceae

Acervuli subepidermal, white or pale; conidiophores short, simple; conidia hyaline, filiform, straight or curved, one-celled or septate; parasitic on leaves.

Cylindrosporium betulae. Brown leaf spot of birch; sometimes serious enough to defoliate but not often present on ornamental trees.

Cylindrosporium chrysanthemi. Chrysanthemum leaf spot; spots are dark brown with yellowish margins, increasing to take in the whole leaf, which hangs down. Similar to more common Septoria leaf spot.

Cylindrosporium clematidinis. Clematis leaf spot; reddish brown spots on lower leaves, which may drop. Dusting with sulfur has been suggested.

Cylindrosporium salicinum. Willow leaf spot; sometimes causing defoliation; can be controlled with bordeaux mixture if necessary.

Cylindrosporium sp. Leaf spot; on spirea, recorded from a Kansas nursery. Light yellow lesions turn dark brown, with masses of yellow conidia on underside.

Cytospora

See under Cankers.

Cytospora sp. Leaf spot; on mulberry.

Dichotomophthoropsis

Fungi Imperfecti, Hyphomycetes?

Dichotomophthoropsis nymphaerum. Leaf spot; on water lily, and water shield.

Didymaria

Fungi Imperfecti, Moniliales, Moniliaceae

Conidiophores simple, arising from leaf surface in loose groups; conidia hyaline, two-celled, ovate-oblong, borne singly; parasitic on leaves.

Didymaria didyma. Leaf spot; on anemone. Angular brown spots.

Didymellina

Acomycetes, Sphaeriales, Mycosphaerellaceae

Perithecia separate, innate or finally erumpent, not beaked; spores two-celled, hyaline.

Didymellina macrospora (*Heterosporium iridis*, *H. gracilis*). Iris leaf spot; blotch; fire, on both bulbous and rhizomatous iris. The spotting is conspicuous toward the end of the season but is not too serious in a normally dry season. Usually the spots are confined to the upper half of leaves, but if plants are crowded and shaded and the summer is wet, the spotting appears earlier, covers more of the leaf, and is more damaging.

Spots are dark brown at first, surrounded by a water-soaked and then yellowing region; they enlarge into rather oval lesions, up to ½ inch long, with a red-brown border (see Fig. 34). Flower buds and stems of bulbous iris may be attacked. Tufts of olive conidia turn the centers grayish, the spores being produced in abundance and splashed by rain to neighboring leaves. Infection is through stomata or directly through the epidermis. The fungus winters as mycelium in old leaves, and in spring produces a fresh crop of conidia or perithecia of the *Didymellina* stage. Soils deficient in lime apparently favor the disease. Repeated spotting reduces bloom and, after a number of years, may kill plants.

Control. It is often sufficient to remove and burn all old leaves at the end of the season; shearing back spotted leaves in midsummer is helpful. If the

Figure 34. Iris leaf spot.

disease is regularly a problem, spray with bordeaux mixture, starting when fans are 6 inches to 8 inches high and repeating at 10- to 14-day intervals.

Didymellina ornithogali (*Heterosporium ornithogali*). Leaf spot; on star of bethlehem. Occasional sooty spots on leaves, with foliage blackened and killed in severe infections.

Didymellina poecilospora. A weak parasite sometimes causing black discoloration of iris foliage.

Didymosporium

Fungi Imperfecti, Melanconiales, Melanconiaceae

Conidia are slime spores in acervuli; dark, two-celled.

Didymosporium arbuticola. Leaf spot; on *Arbutus menziesii*.

Dilophospora

Fungi Imperfecti, Sphaeropsidales, Sphaerioidaceae

Pycnidia distinct in a stroma; conidia very long, filiform, with bristlelike hairs at each end. Usually found on cereals and sometimes with the wheat nematode, causing a disease called twist.

Dilophospora geranii. Leaf spot; on native geranium.

Diplodina

Fungi Imperfecti, Sphaeropsidales, Sphaerioidaceae

Pycnidia black, separate, immersed or erumpent, globose or flattened, ostiolate; conidiophores simple, slender; conidia hyaline, two-celled, ovoid or ellipsoid; parasitic or saprophytic. Similar to *Ascochyta* but not produced in spots.

Diplodia eurhododendri. Leaf spot on rhododendron.

Diplotheca (*Stevensea*)

Ascomycetes, Myriangiales, Myriangiaceae

Asci born singly in locules at various levels in a massive stroma; spores dark, several-celled.

Diplotheca wrightii. Black spot; charcoal spot of Opuntia cacti in Florida and Texas uncommon in the North. Dark spots, ¼ inch or more in diameter, are surrounded by a ring of fruiting bodies.

Dothichiza

See under Cankers.

Dothichiza caroliniana. Leaf spot, double spot; of blueberry, found only on *Vaccinium australis* in North Carolina, but causing extensive defoliation. Leaf spots are small, circular, with brown centers and a dark brown ring, but in late summer infection spreads to a secondary necrotic area around the original spot, giving the common name of double spot. Black pycnidia are formed sparsely in the spots. All varieties of high bush blueberries are somewhat susceptible, but Cabot, Dixie, Pioneer, and Rancocas are most damaged.

Ectostroma

Fungi Imperfecti, Mycelia Sterilia

Black stromata formed in leaves and stems.

Ectostroma liriodendri. Tar spot; widespread in tulip trees but perhaps secondary after insect injury.

Epicoccum

Fungi Imperfecti, Moniliales, Tuberculariaceae

Sporodochia dark, rather cushion-shaped; conidiophores compact or loose, rather short; conidia dark, with one or more cells, globose; mostly saprophytic.

Epicoccum asterinum. Leaf spot; of yucca. **E. neglectum.** On royal palm. **E. nigrum.** On *Magnolia grandiflora*. **E. purpurascens.** On amaryllis. All of these may be secondary infections.

Exosporium

Fungi Imperfecti, Moniliales, Tuberculariaceae

Conidia on subglobose to convex sporodochia; spores dark, with two to several cells, somewhat club-shaped.

Exosporium concentricum. Leaf spot; on euonymus and ligustrum (privet) in the South.

Exosporium liquidambaris. Leaf spot; on sweet gum.

Exosporium palmivorum. Leaf spot of palms; in greenhouses and in the South. Small, round, yellowish transparent spots run together to form large, irregular, gray-brown blotches; leaves may die. The disease is more serious with insufficient light. Spores are long, club-shaped, brown, with many cells. Remove and burn infected leaves. Spray with bordeaux mixture.

Fusicladium

Fungi Imperfecti, Moniliales, Dematiaceae

Mycelium forming a stroma under cuticle of host; conidiophores dark, short; conidia dark, two-celled, produced successively as pushed-out ends of new growing tips. Parasitic on higher plants, causing scab as well as leaf spots.

Fusicladium pisicola. Black leaf; of peas, first reported in Utah in 1921, causing trouble with canning peas. Spots start as small, irregular, whitish areas on undersurface of leaflets and stipules, but they darken to gray or black from the closely packed layer of dark conidia. The disease is not very important.

Fusicladium robiniae. Leaf spot; seedling leaf blight, of black locust. Spots are small, with light centers and dark margins. There is frequently defoliation of seedlings, sometimes stunting and death.

Gibbago

Fungi Imperfecti

Gibbago trianthemae. Leaf spot; of horse purslane; a new genus and species, recently described (1986), with potential for bioherbicide activity.

Gloeocercospora

Fungi Imperfecti, Moniliales, Tuberculariaceae

Sporodochia formed on surface of host above stomata from hyphae emerging through openings; conidiophores hyaline, simple or branched; conidia hyaline, elongate to filiform, one- to many-septate, straight or curved, in a slimy matrix.

Gloeocercospora inconspicua. Leaf spot; of highbush and rabbit-eye blueberry. Circular to angular brownish spots on leaves, with sporodochia more frequent on upper surface. These are flat discs when dry, glistening globules when wet, containing curved, septate conidia.

Gloeocercospora sorghi. Copper spot of turf; see *Ramulispora sorghi.*

Gloeosporium

See under Anthracnose.

Gloeosporium betularum. Leaf spot; anthracnose of river birch. Spots are more or less circular, ⅛ inch across, brownish with pale centers and yellow margins.

Gloeosporium inconspicuum. Elm leaf spot; twig blight, anthracnose, on American and English elms. Subcircular brown spots with darker margins and centers are visible on upper and lower leaf surfaces.

Gloeosporium mezerei. Leaf spot; on daphne. Small brown spots on both sides of leaves.

Gloeosporium rhododendri. Leaf spot; on rhododendron, tulip tree.

Gloeosporium ulmicolum. Elm leaf spot; elongated spots on midribs, veins, and margins, visible on both leaf surfaces.

Glomerella

See under Anthracnose.

Glomerella cincta. Leaf spot; widespread on queen palm, dracaena, and maranta. Sobralia blight of orchids. Dark discoloration starts at tip of leaves and advances toward base.

Glomerella cingulata. Leaf spot; on aucuba, wampi, and croton. See under Anthracnose for this fungus on many other hosts.

Glomerella sp. Black spot; of Vanda orchids.

Gnomonia

See under Anthracnose.

Gnomonia caryae var. **pecanae.** Pecan liver spot; dark brown circular spots, mostly along midribs on underside of leaves, appear in May and June. In autumn the color changes to cinnamon brown, and dark fruiting bodies appear; there may be premature defoliation. Spray in May with bordeaux mixture.

Gnomonia fragariae. Leaf spot; leaf blotch of strawberry. Often associated with *Dendrophoma* causing leaf blight, but not connected.

Gnomonia nerviseda. Pecan vein spot; lesions resemble pecan scab on veins or leaf stems; sometimes a narrow brown lesion extends nearly the length of a midrib. Defoliation may be moderate or severe. Stuart variety is especially susceptible. Spray with bordeaux mixture just before and just after pollination; repeat 3 to 4 weeks later.

Gnomonia ulmea (*Gloeosporium ulmeum*). Elm black spot; black leaf spot of elm, general on American, English, and Chinese elms. Spots on leaves are small but conspicuous, shining coal black, and slightly raised. Leaves may turn yellow and drop, with severe defoliation in a wet season, especially on Siberian elm. Defoliation in spring means death of twigs, but the disease is more common and less important toward fall. Ascospores are formed in spring in perithecia on fallen dead leaves; conidia are produced as a creamy exudate of spores in summer. The fungus also winters as mycelium in dormant buds.

Control. Rake and burn fallen leaves. Chemical control is required only in a wet spring, difficult to determine in advance.

Gnomoniella

Ascomycetes, Sphaeriales, Gnomoniaceae

Perithecia in substratum, beaked, membranous, separate; spores hyaline, one-celled.

Gnomoniella coryli. Leaf spot; on hazel, frequent in northern states. Controlled with bordeaux mixture aided by cleaning up fallen leaves.

Gnomoniella fimbriata. Leaf spot; of hornbeam.

Gonatobotryum

Fungi Imperfecti, Moniliales, Dematiaceae

Conidiophores dark, with spiny inflations at intervals, around which are borne ovoid, dark, one-celled conidia.

Gonatobotryum maculicolum. Leaf spot; on witchhazel.

Graphium

Fungi Imperfecti, Moniliales, Stilbaceae

Synnema or coremium tall, dark, with a rounded terminal mass of conidia embedded in mucus; simple, hyaline conidiophores; oblong conidia reproducing by budding; parasitic.

Graphium sorbi. Leaf spot; of mountain ash.

Guignardia

See under Blotch Diseases.

Guignardia biddwellii f. **parthenocissi.** Leaf spot; on Boston ivy, pepper vine, and Virginia creeper. Spots are numerous, angular, reddish brown, usually dark brown at margins, with black dots in center, minute pycnidia of the imperfect *Phyllostictina* stage. Leaves are quite unsightly and there may be defoliation. Bordeaux mixture applied two or three times, starting as leaves are expanding, gives some control, but the cure looks about as bad as the disease. This fungus is a form of the species causing black rot of grapes.

Helminthosporium

See under Blights.

Helminthosporium catenarium. Leaf spot; on ribbon-grass.

Helminthosporium cynodontis. Bermuda grass leaf blotch; general in South. Olive brown indefinite lesions on dry leaves.

Helminthosporium dictyoides. Fescue netblotch; general on fescue. Dark streaks across green leaves with darker lengthwise streaks form a net pattern. Leaves turn yellow and die back from tips.

Helminthosporium erythrospilum. Red leaf spot; on redtop and bent grasses, widespread in eastern and midwestern states. Under wet conditions lesions have small, pale centers with russet borders; in dry weather leaves wither as in drought but with less evident spotting. Conidia are typically cylindrical, rounded at both ends, yellowish, and germinate from any or all cells (see Fig. 32).

Helminthosporium giganteum. Zonate leaf spot; eye spot; on bent grasses, Canada and Kentucky bluegrass, and Bermuda grass. The disease is present in turf and in nursery rows. Spots are small, $\frac{1}{16}$ inch to $\frac{1}{8}$ inch, bleached straw color in centers. In presence of moisture (dew is sufficient) the fungus grows periodically into new areas, giving the zoned appearance. In continued wet weather leaves are killed and grass turns brown. Metropolitan and velvet bent grasses are less susceptible. Most injury is in July and August. The fungus overwinters as dormant mycelium in old leaves.

Helminthosporium rostratum. Leaf spot; on bromelia, areca palm, fishtail palm, rhapis palm, sweet sorghum, and *Chamaedorea* spp.

Helminthosporium sativum. Melting-out; prevalent on bent grass in warm weather. Leaf spot on Russian wildrye (*Elymus*); spot blotch on switchgrass (*Panicum*).

Helminthosporium setariae (*Drechslera setariae*). Leaf and petal or greasy spot; on geranium, areca palm, fishtail palm, rhapis palm, *Calathea* spp. *Maranta* spp., and *Chamaedorea* spp.

Helminthosporium siccans (*Pyrenophora lolii*). Brown blight; on fescue, and ryegrass. Leaves die back with numerous dark chocolate-brown spots, oval to elongate and often coalescing. The disease appears in early spring in cool, moist weather.

Helminthosporium (*Bipolaris*) **sorokiniana.** Leaf spot; stem spot; of wild rice.

Helminthosporium stenacrum. Leaf mold; on redtop and bent grasses. Indefinite spots; leaves dry, withered, in fall.

Helminthosporium triseptatum. Leaf spot; gray leaf mold, on redtop, spike, and bent grasses in Oregon, Washington, and New York. Leaf tips are killed with vague lesions; gray mold appears on dying tissue.

Helminthosporium tritici-repentis. Leaf spot; on Russian wildrye (*Elymus*).

Helminthosporium vagans. Bluegrass leaf spot; going-out, melting-out, foot rot, general but most injurious in northeastern states, on bluegrass only. Scattered circular to enlongate leaf spots, 0.5 to 3 millimeters by 1 to 8 millimeters, have prominent reddish brown to black borders; centers are brown changing to straw-colored or white with age. The disease, favored by cool rainy weather, usually appears in early spring, sometimes in late fall, and is most severe on close-clippped turf. Grass thins out in large areas; roots rot; weeds invade exposed soil.

Control. Merion bluegrass is quite resistant to leaf spot and will stand close clipping. For other bluegrasses cut high and fertilize well to help turf withstand the disease.

Hendersonia

Fungi Imperfecti, Sphaeropsidales, Sphaerioidaceae

Pycnidia smooth, innate or finally erumpent, ostiolate; conidia dark, several-celled, elongate to fusoid; saprophytic or parasitic.

Hendersonia concentrica. Leaf spot; on rhododendron.

Hendersonia crataegicola. Leaf spot; on hawthorn. Spots irregular, dark brown.

Heterosporium

Fungi Imperfecti, Moniliales, Dematiaceae

Conidiophores dark, simple; conidia dark, spiny, cylindrical, with three or more cells (see Fig. 32); parasitic, causing leaf spots, or saprophytic.

Heterosporium allii. Leaf spot; on onion, leek, shallot, chive, and garlic; rare in North America. Leaves have elliptical, depressed, pale brown spots, and yellow and wither from tip downward.

Heterosporium echinulatum. Fairy ring spot; leaf mold on carnation, occasional in greenhouses. Bleached spots on leaves have black spore groups in ring formation. Syringe as little as possible and on bright days; control ventilation.

Heterosporium eschscholtziae. Capsule spot; leaf spot, stem spot of California poppy. Lesions faint purplish brown; seed capsules may shrivel. Treat seed with hot water, 125° F, for 30 minutes.

Heterosporium gracile. Leaf spot; on chlorogalum, daylily, confused with **H. iridis** on iris.

Heterosporium iridis (conidial state of *Didymellina macrospora*, which see). Leaf spot; on iris, blackberry-lily, freesia, and gladiolus.

Heterosporium variable. Leaf spot; pinhead "rust" of spinach, cabbage mold, sometimes severe in cold, wet weather. Circular, chlorotic spots with brown or purple margins enlarge and multiply until they cover most of the leaf, which turns yellow, withers, dies. There is a greenish-black mold on both leaf surfaces, made up of large olive conidia, one- to six-celled, covered with warts. Keep plants growing vigorously in well-drained soil.

Illosporium

Fungi Imperfecti, Moniliales, Tuberculariaceae

Sporodochia cushionlike, light-colored; conidiophores hyaline, branched with phialides bearing conidia apically; spores hyaline, one-celled; parasitic or saprophytic, often secondary.

Illosporium malifoliorum. Leaf spot; of apple and crabapple.

Isariopsis

Fungi Imperfecti, Moniliales, Stilbaceae

Dark, synnemata composed of loose conidiophores with spores at or near tips; conidia dark or pale, with two or more cells, cylindrical to obclavate, often curved; parasitic.

Isariopsis griseola. Angular leaf spot; pod spot of beans, also sweet pea. Small, angular brown spots are so numerous that they give a checkerboard appearance to leaves. The fungus forms a gray moldy covering over dead areas on underside of leaves. Pod spots are conspicuous when present, black with red or brown centers, varying from a speck to the width of the pod. Small, dark synnemata scattered over the surface bear large conidia, with two to four cells, at top of stalks. They are probably wind-disseminated. Control measures are seldom practical.

Isariopsis laxa. Leaf spot; on kidney bean.

Kabatia

Fungi Imperfecti, Sphaeropsidales, Leptostromataceae

Pycnidia with a radiate shield or scutellum, with an ostiole; spores two-celled, hyaline, like a tooth at the apex.

Kabatia lonicerae. Leaf spot; on honeysuckle.

Lasiobotrys

Ascomycetes, Sphaeriales

Perithecia in a ring around a sclerotial stroma; spores dark, two-celled.

Lasiobotrys affinis. Leaf spot; on honeysuckle. Spot is well-marked with small, dark, wartlike stromas.

Leptostromella

Fungi Imperfecti, Sphaeropsidales, Leptostromataceae

Pycnidia elongate, with a cleft; separate; spores filiform, with rounded ends, hyaline, continuous to septate on simple conidiophores.

Leptostromella elastica. Leaf spot; of rubber plant. The symptoms appear in spots and streaks, but infection spreads until the entire leaf is involved. Black lines outline spots in which small black pycnidia produce long, colorless spores. Remove and burn infected leaves.

Leptothyrella

Fungi Imperfecti, Sphaeropsidales, Leptostromataceae

Pycnidia with a radiate shield, separate; spores two-celled, hyaline.

Leptothyrella liquidambaris. Leaf spot; red on sweetgum.

Leptothyrium

Fungi Imperfecti, Sphaeropsidales, Leptostromataceae

Pycnidium flattened with a more or less radiate shield, opening with an ostiole; spores one-celled, hyaline, on simple conidiophores.

Leptothyrium californicum. Leaf spot; on coast live oak.
Leptothyrium dryinum. Leaf spot; on white oak.
Leptothyrium periclymeni. Leaf spot; on honeysuckle, widespread.

Linospora

Ascomycetes, Sphaeriales, Clypeosphaeriaceae

Perithecia innate, beak often lateral, with a shield; paraphyses lacking; spores spindle-shaped to filiform, hyaline.

Linospora gleditsiae. Leaf spot; tar spot, on honey locust in the South. Numerous black fruiting bodies are formed on undersurface of leaves.

Lophodermium

Ascomycetes, Phacidiales, Phacidiaceae

Fruiting body a hysterothecium, midway between an elongated perithecium and a compressed apothecium, hard, black, opening with a long narrow slit; paraphyses present; hooked at tip; spores filiform, septate or continuous. Most species cause needle casts.

Lophodermium rhododendri. Rhododendron leaf spot; large silvery white spots with red, raised margins have very prominent oval, black fruiting bodies on the upper surface. Lower side of spots is a light chocolate brown. Infected portions may fall out, leaving irregular holes. The disease is more common on native than on hybrid varieties.

Macrophoma

See under Cankers.

Macrophoma candollei. Leaf spot of boxwood; Conspicuous black pycnidia on dead leaves, usually straw-colored, sometimes brown or tan. The fungus is a weak parasite coming in secondarily after winter injury or other predisposing factors.

Marssonina

See under Anthracnose.

Marssonina brunnea. Leaf spot; on poplar.

Marssonina daphnes. Daphne leaf spot; small, thick brown spots on both sides of leaf, which turns yellow, dies.

Marssonina delastrei. Leaf spot; on corncockle and campion.

Marssonina fraxini. Ash leaf spot; sometimes serious in nursery stock, controlled by spraying with bordeaux mixture.

Marssonina juglandis. See *Gnomonia leptostyla* under Anthracnose.

Marssonina ochroleuca. Leaf spot; on oak, American chestnut. Spots are circular, yellow to brown with concentric markings, small on chestnut, up to an inch on oak.

Marssonina populi. Poplar leaf spot; brown spots with darker margins. There may be premature defoliation and killing of twigs.

Marssonina rhabdospora (perfect stage *Pleuroceras populi*). Poplar leaf spot; brown spots on living leaves; beaked pyriform perithecia formed in fallen leaves over winter.

Marssonina rosae. Imperfect stage of the rose blackspot fungus, *Diplocarpon rosae.*

Marssonina truncalata (= *Didymosporina aceris*). Leaf spot; leaf blight; of Norway maple.

Mastigosporium

Fungi Imperfecti, Moniliales, Moniliaceae

Conidiophores hyaline, very short, simple; conidia with four or more cells, with or without apical appendages, broadly cylindrical with rounded or pointed ends; parasitic on grasses.

Mastigosporium rubricosum. Leaf fleck; on redtop and bent grasses. Spores with rounded ends, without appendages.

Melanconium

Fungi Imperfecti, Melanconiales, Melanconiaceae

Acervuli subcutaneous or subcortical, conic or discoid, black; with setae; conidiophores simple; conidia dark, one-celled, ovoid to ellipsoid; parasitic or saprophytic.

Melanconium pandani. Leaf spot; on pandanus.

Melasmia

Fungi Imperfecti, Sphaeropsidales, Leptostromataceae

Pycnidia in a broad, black, flattened stroma that is superficial or nearly so, dimidiate; conidiophores simple or branched; spores hyaline or subhyaline, one-celled, allantoid or fusoid; parasitic on leaves; imperfect stages of *Rhytisma*.

Melasmia menziesii. Leaf spot; tar spot of azalea.
Melasmia falcata. Tar spot; of persimmon.

Micropeltis

See under Blights.

Micropeltis alabamensis. Black leaf spot; on magnolia.

Microstroma

Fungi Imperfecti, Melanconiales, Melanconiaceae

Sporodochia small, white, breaking through epidermis; conidiophores hyaline, one-celled, somewhat clavate, bearing conidia on short sterigmata; spores hyaline, one-celled, small, oblong; parasitic.

Microstroma juglandis. Leaf spot; white mold, downy spot, witches' broom of pecan, walnut, and hickory. Yellow blotching of upper side of leaves and a glistening white coating on underside, due to pustules with enormous numbers of spores, may be accompanied by premature defoliation. On shagbark hickory the fungus also invades the stems, causing witches' brooms up to 3 feet across. Leaves formed on them in spring are yellow-green, with white powder on underside. Leaflets are small, curled, and soon drop. Prune out witches' brooms; spray with bordeaux mixture.

Microthyriella

See under Fruit Spots.

Microthyriella cuticulosa. Black spot of holly; dark spots on leaves of American holly, Georgia.

Monochaetia

See under Cankers.

Monochaetia desmazierii. Leaf spot; on chestnut, white, red, and coast live oaks, winged elm, hickories, especially destructive in the Southeast. Spots are large, 1 inch to 2 inches in diameter, with pale green or yellow centers with a red and brown border or concentric zones of gray, yellow, and brown. Symptoms appear most often in late summer when loss of green tissue is not as important.

Morenoella

See *Lembosia* under Black Mildew.

Morenoella quercina. Leaf spot; black mildew, of red and black oaks; twig blight of white oak, common in Southeast. Spots are purplish black, roughly circular, up to ⅓ inch across, on upper surface and irregular brown areas on underside. Mycelium is superficial in early summer, but by late summer there are subcuticular hyphae and a black shield formed over a flat cushion of fertile cells. Asci are mature and shield is fissured by spring.

Mycosphaerella

See under Blights.

Mycosphaerella angulata. Angular leaf spot; of muscadine grapes. Many small, angular black spots, more conspicuous on lower surface of leaves, which may turn yellow and die.

Mycosphaerella arachidicola. Peanut leaf spot; see *Cercospora arachidicola*.

Mycosphaerella aurea. Leaf spot; of flowering currant.

Mycosphaerella berkeleyi. Peanut leaf spot; see *Cercospora personata*.

Mycosphaerella (*Cercospora*) **bolleana. Leaf spot**; of fig, and rubber tree.

Mycosphaerella (*Phyllosticta*) **brassicicola. Ring spot of crucifers**; chiefly cabbage and cauliflower, sometimes brussels sprouts, broccoli, and turnip. Dead spots in leaves, small to ½ inch, are surrounded by a green zone that keeps its color even if the rest of the leaf turns yellow. Small black pycnidia are deeply embedded in the dead tissue, often in concentric rings. In moist weather conidia ooze from pycnidia in pink tendrils. The fungus winters in old plant refuse, and ascospores are forcibly ejected from perithecia in spring (see Fig. 32). The disease is confined to the Pacific Coast and, as black blight, is serious on the seed crop in the Puget Sound area. Sanitary measures and crop rotation keep it in check.

Mycosphaerella caroliniana. Leaf spot; purple blotch, on oxydendron (sourwood). Reddish or purple spots on foliage in midsummer have dry, brown centers. Pycnidia embedded in tissue break through lower surface, spores being formed in great numbers.

Mycosphaerella caryigena. Pecan downy spot; conidial stage has been listed as a *Cercosporella* and as *Cylindrosporium carigenum*. Leaf spots are pale yellow when young, turning yellow-brown, brown, or black. Conidia produced in minute acervuli on underside of leaves form a white downy or frosty coating; leaves may drop early. Spores are spread in rain, fog, and dew. The fungus overwinters in leaves, liberating ascospores in spring to infect new foliage. Moneymaker and Stuart varieties are especially susceptible.

Control. Turn under old leaves before spring (plowing under winter cover in spring takes care of this). Spray as for scab, bordeaux mixture when leaves are half-grown and bordeaux plus 4 pounds of zinc sulfate when tips of small nuts have turned brown.

Mycosphaerella cerasella. See *Cercospora circumscissa.*

Mycosphaerella (*Cercospora*) **cercidicola.** Redbud leaf spot; general. Spots are circular to angular or irregular with raised dark brown borders. With age, lesions become grayish above and rusty brown on the undersurface, with the leaf tissue yellow-green outside the borders. Spores are formed on fascicles of conidiophores projecting through stomata. The fungus winters on fallen leaves, producing perithecia in spring. Twigs may be attacked as well as foliage.

Mycosphaerella citri. Leaf spot or greasy spot; on citrus.

Mycosphaerella colorata. Mountain laurel leaf spot; see *Phyllosticta kalmicola.*

Mycosphaerella (*Cercospora*) **cruenta.** Leaf spot; leaf blotch of soybean, and kidney bean. Leaf spots distinct to indistinct, circular to irregular, greenish to yellowish to rusty brown to almost red, sometimes with gray centers.

Mycosphaerella effigurata (*Piggotia fraxini*). Ash leaf spot; general east of the Great Plains. Spots small, purple to brown with yellow borders.

Mycosphaerella fragariae. Strawberry leaf spot; black-seed disease, general on strawberries. Leaf spots are first purple then reddish with light brown or white centers, ⅛ inch to ¼ inch across. Spots are also present on petioles and fruit stems, and occasionally there are black spots on fruit, with blackened achenes prominent against the white of unripe berries. Fruit is poor; total yield is reduced; runner plants are weakened. Conidia of the *Ramularia* stage are produced in clusters of short conidiophores on underside of diseased areas; perithecia are formed in autumn at the edge of the leaf spots where the fungus winters. New conidia are produced in spring with most infection taking place through stomata. There is a difference in varietal susceptibility.

Control. Set healthy plants in well-drained soil; remove diseased leaves before planting; spray with bordeaux mixture before planting and follow with two or three more applications. The conidia are very sensitive to copper, which prevents sporulation and kills nongerminated spores.

Mycosphaerella fraxinicola (*Phyllosticta viridis*). Ash leaf spot; east of the Rocky Mountains.

Mycosphaerella juglandis. Leaf spot; of black walnut.

Mycosphaerella liriodendri (*Phyllosticta liriodendrica*). Tulip-tree leaf spot.

Mycosphaerella louisianae. Purple leaf spot of strawberry; in the South. Large, irregular, reddish purple areas.

Mycosphaerella mori. Mulberry leaf spot; widespread, with the conidial stage reported variously as *Cercosporella, Cylindrosporium, Phleospora, Septogloeum,* and *Septoria.* Yellow areas on upper leaf surface are matched by whitish patches underneath, the fungus forming a white downy or powdery coating. The disease is most serious in shady locations.

Mycosphaerella nigromaculans. Black stem spot of cranberry; reported from all cranberry areas, often associated with red leaf spot. The fungus enters through leaves, grows down the petioles, and forms elongated black spots on the stems, which may be completely girdled, followed by defoliation. Fruiting bodies are produced in autumn on dead stems and ascospores are discharged in rainy periods in spring. The imperfect stage of the fungus is a *Ramularia.*

Mycosphaerella nyssaecola (*Phyllosticta nyssae*). Tupelo leaf spot; on sour gum and water tupelo. Purplish irregular blotches, an inch or more across, are scattered on upper leaf surface with lower surface dark brown. There may be heavy defoliation. Perithecia mature in spring on fallen leaves.

Mycosphaerella personata (*Isariopsis clavispora*). Leaf spot; widespread on muscadine and other grapes after midseason. Spots are dark brown, ¼ inch to ½ inch, surrounded by a yellow circle but with a narrow band of normal green between spot and circle.

Mycosphaerella populicola (*Septoria populicola*); **M. populorum** (*S. musiva*). Leaf spot; of native poplar. Canker; on twigs and branches of hybrid poplars.

Mycosphaerella psilospora (*Septoria querceti*). Oak leaf spot; on red and other oaks, common in Iowa. Spots very small, circular, with straw-colored centers and dark margins.

Mycosphaerella ribis (*M. grossulariae, Septoria ribis*). Leaf spot of gooseberry, currant; Numerous small brown spots with grayish centers are formed on both sides of leaves; there may be premature defoliation. The fungus winters in leaves, producing ascospores in late spring. Two sprays of bordeaux mixture plus 1 pint of self-emulsifying cottonseed oil per 100 gallons have given good control of leaf spot on gooseberries in New York. The first application is about June 1, the second in July right after fruit is picked.

Mycosphaerella rosicola. See *Cercospora rosicola.*

Mycosphaerella rubi. See *Septoria rubi.*

Mycosphaerella sentina (*Septoria pyricola*). Pear leaf spot; also on quince, occasional on apple. Spots are small, ⅛ inch to ¼ inch, grayish in center, dotted with black fruiting bodies, with a well-defined dark brown margin. There are marked differences in susceptibility in pear varieties. Flemish Beauty, Duchess, and Winter Nellis are moderately resistant and Kieffer is very resistant. Sprays applied for leaf blight or scab control leaf spot.

Myocentrospora

Mycentrospora sp. Leaf spot; on euonymus.

Myrothecium

Fungi Imperfecti, Moniliales, Tuberculariaceae

Sporodochia cushionlike, light or dark; conidiophores subhyaline to colored, repeatedly branched, bearing conidia terminally; conidia subhyaline to dark, one-celled, ovoid to elongate; weakly parasitic or saprophytic.

Myrothecium roridum. Leaf spot; on snapdragon, stock, eremurus, gardenia, hollyhock, aeschynanthus, aglaonema, aphelandra, dieffenbachia, episcia, fittonia, nematanthus, hoya, peperomia, pilea, and spathiphyllum. Tissues are dry, brittle, with black sporodochia. Snapdragon leaves and flowering stems wilt, with sunken cracked cankers. Avoid excessive moisture; sterilize soil.

Neottiospora

Fungi Imperfecti, Sphaeropsidales, Sphaerioidaceae

Pycnidia dark, smooth, innate; spores hyaline, one-celled with two to several appendages at the apex.

Neottiospora yuccifolia. Yucca leaf spot.

Ophiodothella

Ascomycetes, Dothideales, Phyllachoraceae

Asci in locules immersed in groups in a stroma, covered by host tissue at maturity; paraphyses lacking; spores filiform.

Ophiodothella vaccinii. Leaf spot; on huckleberry, and farkleberry.

Ovularia

Fungi Imperfecti, Moniliales, Moniliaceae

Conidiophores emerging from leaves in clusters, simple or branched; conidia hyaline, one-celled, ovoid or globose, apical or lateral, single or sometimes catenulate; parasitic.

Ovularia aristolochiae. Leaf spot; on Dutchmans pipe.
Ovularia pulchella. Tan leaf spot; on creeping bent grass.

Pestalotia

See under Blights.

Pestalotia aquatica. Leaf spot; of arrow arum. Irregular, chestnut-brown spots, up to 1 inch in diameter, have purplish or dark borders and are wrinkled concentrically. Acervuli are sparse, black, erumpent on upper side of leaf. Spores are five-celled with three widely divergent setae.

Pestalotia aucubae. Aucuba leaf spot; the fungus appears as a weak parasite in sunscald spots or after other fungi.

Pestalotia cliftoniae. Leaf spot; on buckwheat tree. Ashy or pale brown spots. Spores usually curved, constricted at septa, three setae at crest.

Pestalotia funerea. Leaf spot; bark and cone spot on conifers. Pathogenicity of the fungus is questionable. Median spore cells are dark brown; apical hyaline cell has four or five erect setae.

Pestalotia guepini. Camellia leaf spot; widespread. Numerous, punctiform, black fruiting bodies are scattered over papery gray spots. The spores are five-celled, bright olivaceous, with one to four divergent, sometimes branched, setae, and a straight, short pedicel. This species seems to be a true parasite.

Pestalotia leucothoës. Leucothoë leaf spot; apparently following winter injury or other disease.

Pestalotia macrotricha. Rhododendron leaf spot; gray blight, twig blight, widespread on azalea and rhododendron after winter injury. Dark or pale spots with black raised pustules are scattered over stems and leaves. Spots are often silvery gray on upper surface and dark brown underneath, with densely gregarious acervuli sooty from dark spores.

Pestalotia palmarum. Palm leaf spot; gray leaf; black pustules are sparsely produced on both surfaces of pale, dead areas with definite, reddish brown borders. Spores are five-celled, with two or three setae, usually knobbed. The fungus is a wound parasite.

Pestalotia rhododendri. Rhododendron leaf spot; black pustules are scattered without order over dried brown areas of living leaves. Spores are broader than those of *P. macrotricha* and have shorter setae.

Pestalozziella

Fungi Imperfecti, Melanconiales, Melanconiaceae

Conidia hyaline, one-celled, with a branched appendage at apex; acervuli subcutaneous; conidiophores slender, simple or branched.

Pestalozziella subsessilis. Leaf spot; on geranium.

Pezizella (*Allophylaria*)

Ascomycetes, Helotiales, Helotiaceae

Apothecia sessile, bright-colored, smooth; paraphyses filiform, blunt; spores elliptical to fusoid, hyaline, one-celled.

Pezizella (*Discohainesia*) **oenotherae.** Leaf spot; fruit rot of blackberry, raspberry, and strawberry; leaf spot of evening primrose, eugenia, galax, loosestrife, ludwigia, mock strawberry, May apple, peony, and sumac. Spots are irregular, gray in center with a dark brown border. Fruiting bodies are light amber discs; spores are amber in masse.

Phacidium

See under Blights.

Phacidium curtisii. Tar spot; leaf spot of American holly, more serious in southern commercial plantings. Small yellow spots appearing in early summer age to reddish brown with narrow yellow borders. At end of season flat, black, cushion-shaped stromata develop beneath the epidermis. Leaves seldom drop prematurely, but infected areas may fall out leaving holes. In years of heavy rainfall berries as well as leaves are spotted. Remove lower branches; clean up and burn or turn under fallen leaves. Spray with bordeaux mixture.

Phaeotrichoconis

Phaeotrichoconis crotolariae. Leaf spot; on areca palm; leaf spots on palms that are similar in appearance are caused more often by *Bipolaris, Helminthosporium setariae,* and *Helminthosporium (Exserohilum) rostratum.*

Phleospora

See under Blights.

Phleospora aceris. Leaf spot of maple; including vine and dwarf maples. The spot is small, rather angular, common but not important.

Phylctaena*

Fungi Imperfecti, Sphaeropsidales, Excipulaceae

Pycnidia dark, separate or sometimes confluent, in or under epidermis or bark; closed or ostiolate; conidiophores simple or forked; conidia hyaline, one-celled, cylindrical or long spindle-shaped, mostly bent, sickle-shaped; saprophytic usually.

Phylctaena ficuum. Leaf spot; on strangler fig.

* The 1961 edition of Ainsworth and Bisby places *Phylctaena* in the Melanconiales and *Phyllachora* in the Sphaeriales.

Phyllachora

Ascomycetes, Dothideales, Phyllachoraceae

Asci in locules, immersed in groups in a dark stroma covered by host tissue at maturity; spores one-celled, hyaline; paraphyses present; asci cylindrical with short pedicels.

Phyllachora graminis. Tar spot; black leaf spot, general on wheatgrass, ryegrass, fescues, redtop, and bent grass. Elongated grayish violet to dark olive green spots, on both leaf surfaces, turn glossy black. The disease is seldom serious.

Phyllachora sylvatica. Tar spot; on fescues in Northwest.

Phyllosticta

See under Blights.

Phyllosticta althaeina. Leaf spot; stem canker on abutilon and hollyhock. Ashy spots have black dots of pycnidia. The tissue sometimes becomes brittle and falls away, leaving jagged holes.

Phyllosticta antirrhini. Snapdragon leaf spot; stem rot, blight. Large circular, dark brown or black spots, with concentric ridges, are located most often near tips and margins of leaves; centers may be cream to pale brown, dotted with dark pycnidia (see Fig. 32). Young leaves may be curled; older leaves shrivel and hang down along the stem. Petioles are girdled with brown elongated lesions. Stems have firm brown rot with shoots or branches wilting or have ashy white spots with dark brown or purplish margins and stems cracking in area of spots. Young seedlings may damp off. Spray with bordeaux mixture; keep greenhouse cool; avoid wetting foliage in watering; clean up diseased plants.

Phyllosticta aucubae. Aucuba leaf spot; brown or black zonate spots are mostly along margins of leaves, sometimes with much defoliation. Spores are exuded from leaves in yellow tendrils, then spread by rain, or syringing in the greenhouse.

Phyllosticta camelliae and **P. camelliaecola.** Camellia leaf spot; lesions are irregular brown spots.

Phyllosticta catalpae. Catalpa leaf spot; dark brown or black spots ⅛ inch to ¼ inch in diameter, may run together to give a blotched appearance. Minute black fruiting bodies pepper the spots, which are often associated with injury by the catalpa midge. Heavy infection may mean defoliation.

Phyllosticta circumscissa. Leaf spot; widespread on apricot, peach, sour cherry, chokecherry, and garden plum.

Phyllosticta concentrica. English ivy leaf spot; also a twig blight, widespread. Plants look ragged. Fruiting bodies are arranged in spots in concentric circles.

Phyllosticta cookei. Magnolia leaf spot; spots are grayish without definite margins.

Phyllosticta decidua. Leaf spot; of agrimony, aralia, basil weed, betony, cynoglossum, eupatorium, germander, hierachia, hoarhound, motherwort, lycopus, mint, and monarda.

Phyllosticta ilicis (*Physalospora ilicis*). Holly leaf spot; on American and English holly and on winterberry.

Phyllosticta hamamelidis. Witch hazel leaf spot; small spots enlarge to reddish brown blotches, causing some defoliation.

Phyllosticta hydrangeae. Hydrangea leaf spot; widespread. Brown spots usually near leaf margins; in severe cases both leaves and blossoms are killed. Spray with bordeaux mixture.

Phyllosticta kalmicola (*Mycosphaerella colorata*). Mountain laurel leaf spot; kalmia leaf spot. Circular, grayish white to silvery spots with red or

Figure 35. Phyllosticta leaf spot on mountain laurel.

purple borders, up to ¼ inch across, are sparsely or thickly covered with black pycnidia (see Fig. 35). Heavy infection means disfigured foliage and some defoliation. The disease is worse in shady locations where shrubs are under drip of trees.

Phyllosticta maculicola. Dracaena leaf spot; irregular small brown spots have yellowish margins and long coils of spores from black pycnidia.

Phyllosticta maxima. Rhododendron leaf spot; widespread. Spots are marginal or terminal, large, dark brown, and zonate.

Phyllosticta minima. Maple leaf spot; gray spot, also on boxelder, widespread. Spots are irregular, ¼ inch or more across, with brownish centers, containing black pycnidia, and purple-brown margins. The disease is seldom serious enough for control measures.

Phyllosticta richardiae. Calla leaf spot; small, round, ash-gray spots run together, producing irregular decayed areas.

Phyllosticta saccardoi. Rhododendron leaf spot; similar to that caused by *P. maxima*.

Phyllosticta sanguinariae. Bloodroot leaf spot; spots reddish brown with a darker border, then a zone of Indian red.

Phyllosticta sojaecola. Leaf spot and pod spot; of soybean; lesions have purplish red borders surrounding lighter brownish centers that contain numerous dark pycnidia.

Phyllosticta vaccinii. Leaf spot; of farkleberry and highbush blueberry.

Phyllosticta wistariae. Wisteria leaf spot; more important in the South.

Phyllostictina

Fungi Imperfecti, Sphaeropsidales, Sphaerioidaceae

Pycnidia membranous, with an ostiole; conidiophores obsolete; spores hyaline, one-celled, ovoid.

Phyllostictina vaccinii. Blueberry leaf spot; fruit rot. Small, circular gray spots, with one to six pycnidia in center, have brown margins. The disease is unimportant as a leaf spot; fruits have a hard, dry rot.

Physoderma

Phycomycetes, Chytridiales, Cladochytriaceae

Definite mycelium with terminal and intercalary enlargements that are transformed wholly or in part into sporangia and resting spores; sporangia rare, oospores abundant, globose or ellipsoidal. Affected plant parts are discolored or slightly thickened.

Physoderma maydis. Brown spot of corn; corn measles, corn pox, dropsy; most prevalent in the South. Very small, bleached or yellowish spots darken to brown or reddish brown with a light margin. Adjacent spots may coalesce to give the whole blade a rusty appearance. Spots on midrib and leaf sheath are larger, up to ¼ inch, irregular to square, darker than leaf lesions. The entire sheath may turn brown on death of host cells; the epidermis ruptures, exposing brown spore dust. In severe infections low nodes are girdled so stalks break over. The resting spores remain in soil or plant refuse over winter, germinating by swarmspores the next spring. A fairly high temperature and low, wet land favor the disease. Remove plant refuse early; rotate crops.

Pirostoma

Fungi Imperfecti, Sphaeropsidales, Leptostromataceae

Pycnidia superficial, with a shield; spores one-celled, dark.

Pirostoma nyssae. Tupelo leaf spot.

Placosphaeria

Fungi Imperfecti, Sphaeropsidales, Sphaerioidaceae

Pycnidia globose, dark, in a discoid stroma; spores hyaline, one-celled; perfect stage in *Dithideales.*

Placosphaeria graminis. Tar spot; on redtop grass.
Placosphaeria haydeni. Black spot, tar spot; on goldenrod and aster, stems and leaves.

Plagiostoma (*Laestadia*)

Ascomycetes, Sphaeriales, Valsaceae

Spores two-celled, hyaline.

Plagiostoma asarifolia. Leaf spot; on wild ginger.
Plagiostoma prenanthis. Leaf spot; on prenanthis.

Pleospora

Ascomycetes, Sphaeriales, Sphaeriaceae

Perithecia membranous, paraphyses present; spores muriform, dark; some species have *Alternaria*, some *Stemphylium* as imperfect stage; wide saprophytic and pathogenic relationships.

Pleospora herbarum (*Stemphylium botryosum; S. sarcinaeforme*). Leaf spot of clovers; leaf blight of lilac; seed mold; of China aster and other plants. Spots on legumes are small, irregular, dark brown, sunken, changing to concentric zonated light and dark brown areas. In final stages leaves are wrinkled, dark brown, and sooty. Conidia, like ascospores, are muriform, olivaceous. Annual phlox has tan lesions. Asparagus has purple spots.

Pringsheimia (*Pleosphaerulina, Saccothecium*)

Ascomycetes, Sphaeriales, Dothioraceae

Perithecia innate, not beaked, paraphyses and paraphysoids lacking; spores muriform, hyaline.

Pringsheimia sojaecola. Leaf spot; of soybean.

Ramularia

Fungi Imperfecti, Moniliales, Moniliaceae

Conidiophores growing out from host through stoma, clustered, short, dark to hyaline; conidia hyaline, cylindrical, mostly two-celled, often in chains; found on living leaves causing leaf spots or white mold (see Fig. 32).

Ramularia armoraciae. Pale leaf spot of horse-radish; few to numerous light green to yellowish spots appear on leaves in early summer, the invaded areas quickly turning thin and papery with dead portions dropping out, leaving ragged holes late in the season. Innumerable small sclerotium-like bodies in the dead tissue carry the fungus over winter, producing short knobby conidiophores in spring, which either push out through stomata or break through either epidermis. There is no special control.

Ramularia (*Cercosporella*) **pastinaeae.** Leaf spot of parsnip; lesions are circular, very small, at first brown, then with a white center and brown border. Long, slender, septate, hyaline conidia are produced on exposed conidiophores. No control is necessary.

Ramularia primula. Primrose leaf spot; yellow blotches have ash-colored centers.

Ramularia vallisumbrosae. Narcissus white mold; sometimes destructive on Pacific Coast. Small, sunken, grayish or yellow spots appear on leaves, especially near tips, increasing to dark green to yellow-brown patches, on which, in moist weather, spores are formed in white powdery masses. The disease may become epidemic with the foliage killed several weeks before normal ripening. Flower stalks of late varieties may be attacked. Black "sclerotia" winter in leaf fragments on ground, producing spores in spring to infect young shoots.

Control. Spray with bordeaux mixture, starting when leaves are 4 inches to 6 inches high. Clean bulbs thoroughly after digging and replant in a new location.

Ramularia variabilis. Foxglove leaf spot; irregular spots, up to ¼ inch in diameter, brown with a reddish border, are formed most often on lower leaves. Spores in tufts give a white, moldy appearance.

Ramulispora

Fungi Imperfecti, Moniliales, Tuberculariaceae

Conidia on sporodochia, two- to many-septate, hyaline to subhyaline, oblong to fusoid, irregularly united or branched at base; produced in gelatinous masses.

Ramulispora (*Gloeocercospora*) **sorghi.** Copper spot; of turf grasses, sooty stripe of sorghum, Sudan grass, and Johnson grass. Black superficial sclerotia are formed on both leaf surfaces, with conidia in pinkish gelatinous masses. Spots on leaves are straw-colored with purple borders. Dead areas in turf are small, 1 inch to 3 inches, copper-red to orange. Velvet bent grass in acid soil is very susceptible. Liming the soil may help.

Rhizoctonia

See under Rots.

Rhizoctonia solani. Leaf spot; of tobacco.

Rhytisma

Ascomycetes, Phacidiales, Phacidiaceae

Apothecia concrete with epidermis and in black, stroma-like spots, tar spots, on leaves; spores filiform, typically hyaline.

Rhytisma acerinum. Tar spot of maple; especially on cut-leaf varieties. Black, thickened, raised, tarlike spots, up to ½ inch in diameter, are formed on upper leaf surface. They may be numerous enough to cause some defoliation but ordinarily are more disfiguring than destructive. Red and silver maples are commonly affected in the East. The lesions are light yellow-green at first, forming black stomata in summer along with the conidial stage (*Melasmia acerina*) (see Fig. 36). Ascospores are developed in spring in tar spots on fallen overwintered leaves and are forcibly ejected, to be carried by air currents to young leaves overhead.

Control. Collect and burn fallen leaves. Spray in early May with copper, repeating in 3 weeks in an unusually wet season.

Figure 36. Tar spot of maple: black tarry spot on leaf; section through spot; ascus, paraphyses, and filiform ascospores.

Rhytisma andromedae. Tar spot; on bog rosemary and lyonia.

Rhytisma bistorti. Tar spot; on polygonum. Black tarry spots similar to those on maple.

Rhytisma liriodendri. Leaf spot; on tulip tree.

Rhytisma punctatum. Speckeled tar spot of maple; a black speckled leaf spot on all species but especially on silver, striped, and bigleaf maple in Pacific Coast states, rare in the East. Black, raised specks, pinhead size, are formed in groups on upper leaf surface, in yellow-green areas about ½ inch in diameter. Such areas retain their color even after leaves have faded in the fall.

Rhytisma salicinum. Tar spot of willow; on pussy willow and other varieties. Spots are very thick, jet black, definitely bounded, ¼ inch in diameter. The fungus winters in old leaves, which should be raked and burned.

Schizothyrium

Ascomycetes, Hemisphaeriales, Micropeltaceae

Brown scutellum or shield, radiate at margin, with a single hymenium underneath; apothecia round to linear, opening with a cleft or lobes; spores hyaline, two-celled.

Schizothyrium gaultheriae. Leaf spot; on wintergreen.

Scolecotrichum

Fungi Imperfecti, Moniliales, Dematiaceae

Conidiophores in loose clusters, simple, bearing conidia on pushed-out ends of successive new growing points; spores dark, two-celled, ovoid or oblong, often pointed; parasitic.

Scolecotrichum graminis. Brown stripe of lawn grasses; streak, of blue-grass and redtop. Grayish brown to dark linear streaks on leaf blade may extend into leaf sheath and cause defoliation. Dark gray masses of conidiophores emerge in rows through stomata of upper leaf surface.

Selenophoma

Fungi Imperfecti, Sphaeropsidales, Sphaerioidaceae

Pycnidia brown, globose, immersed, erumpent, ostiolate; conidia hyaline, one-celled, bent or curved, typically crescent-shaped, parasitic.

Selenophoma donacis, S. everhartii, S. obtusa. Speckle; leaf blotch on bluegrass and other grasses. Brown flecks and frog-eye spots on blades in early spring enlarge to straw-colored blotches scattered with minute pycnidia. Spots may drop out, leaving holes.

Septocylindrium

Fungi Imperfecti, Moniliales, Moniliaceae

Conidiophores hyaline, short and simple or longer and branched, with irregular somewhat inflated cells; conidia hyaline, two- to several-celled, in chains that are sometimes branched; parasitic or saprophytic.

Septocylindrium hydrophyllis. Hydrophyllium leaf spot.

Septogloeum

Fungi Imperfecti, Melanconiales, Melanconiaceae

Acervuli subepidermal, erumpent, pale; conidiophores short, simple; conidia hyaline, several-celled, oblong to fusoid; parasitic.

Septogloeum acerinum. Maple leaf spot; a small leaf spot occasionally defoliating Norway and Schwedler maples in the Midwest.

Septogloeum oxysporum. Char spot of lawn grasses; lesions are tawny with yellow margins, circular becoming elliptical, pointed at each end, covered with dull black to brown stromatic tissue.

Septogloeum parasiticum. Elm leaf spot; twig blight.

Septogloeum rhopaloideum (*Guignardia populi*). Grayish brown, circular or irregular spots on poplar.

Septoria

See under Blights.

Septoria agropyrina. Brown leaf blotch; on wheatgrasses.

Septoria bataticola. Sweetpotato leaf spot; occasional, most common in northern tier of sweet-potato states. Minute white spots on leaves are bordered with a narrow reddish zone. Older lesions have one or more pycnidia barely visible to the naked eye. The spores, oozing out in tendrils when water is present on the leaf, are spread by rain and insects. No control is needed except cleaning up crop refuse.

Septoria clamagrostidis. Leaf spot; on bent grasses. Scattered gray to straw-colored lesions at tip of blades, appearing in Northwest in late winter and early spring. Seaside creeping bent is especially susceptible.

Septoria callistephi. Leaf spot; damping-off, stem rot of China aster.

Septoria chrysanthemella and **S. obesa. Chrysanthemum leaf spot**; also on oxeye daisy, general through eastern and central states to Florida; also reported in the West. This disease is sometimes confused with nematode injury, but the leaf nematode browns the leaves in wedge-shaped areas between veins, and the fungi cause definite spots. These are first small and yellowish, then dark brown to nearly black. Sometimes the spots coalesce into blotches; minute black fruiting bodies are faintly visible. Leaves may turn yellow and drop prematurely or dry and hang down along the stems. Spores are splashed from plant to plant in watering or rain, and are spread on cultivating tools.

Control. Avoid syringing greenhouse plants; do not cultivate outdoor plants when they are wet.

Septoria citri. Citrus septoria spot; on leaves but more serious on fruits. Small, shallow, light brown depressions on green immature fruit retain a green marginal ring as the fruit colors. Usually a minor trouble, sometimes important in California.

Septoria citrulli. Watermelon leaf spot; the pathogen is like *S. cucurbitacearum* except that spores are shorter.

Septoria cornicola. Dogwood leaf spot; angular lesions between veins are grayish with dark purple margins.

Septoria cucurbitacearum. Septoria leaf spot of cucurbits; on cucumber, winter squash, muskmelon, and watermelon. Foliage spots are small, gray, circular, rather conspicuous, often bordered with a zone of yellow tissue. The fungus fruits abundantly on upper side of leaves, with long thin septate spores in black pycnidia. It winters in old plant parts; clean up all refuse at end of the season.

Septoria cyclaminis. Leaf spot; on cyclamen.

Septoria dianthi. Septoria leaf spot of dianthus; on carnation and sweet william. Spots are more or less circular, light brown with purplish brown borders, scattered over leaves and stems, particularly on lower leaves. The spots may enlarge, and the leaves die. Take cuttings from disease-free plants; avoid syringing, or do it early in the day.

Septoria divaricata. Septoria leaf spot of phlox; dark brown circular spots, up to ¼ inch in diameter, have light gray to white centers and often run together in blotches.

Septoria gladioli. Leaf spot; more important as a hard rot of gladiolus corms.

Septoria glycines. Brown spot of soybean; primarily a foliage disease that may also appear on stems and pods. It starts with irregular brown patches on cotyledons, then reddish brown zones on both sides of leaves, often with pale green or chlorotic zones surrounding the lesions. Spots may cover the whole leaf, defoliation starting from lowest leaves. Brown discolorations with indistinct margins extend an inch or more along stems. The pathogen winters in diseased leaves and in seed. Some varieties are quite resistant. Use healthy seed; treatment is unsatisfactory; rotate crops.

Septoria lactucae. Septoria leaf spot of lettuce; occasionally destructive to some varieties. Lesions are more common on lower leaves—irregular reddish marks, dotted sparsely with black pycnidia. The fungus is disseminated with seed.

Septoria loligena. Leaf spot; on ryegrass, in California. Chocolate brown spots, paler in the center, surrounded by lighter areas.

Septoria lycopersici. Septoria leaf spot of tomato; leaf blight, quite destructive in Atlantic and central states, less important in the South and West. In seasons with moderate temperature and abundant rainfall enough foliage is destroyed so that fruits do not mature properly and are subject to sunscald. The disease appears at any age but more often after fruit is set. Infection starts on older leaves near the ground, with small, thickly scattered, water-soaked spots, which become roughly circular with gray centers and prominent dark margins. The spots are smaller, $\frac{1}{16}$ to $\frac{1}{8}$ inch, and more numerous than those of early blight. Leaflets may die with progressive loss of foliage from the bottom up. The pathogen winters on tomato refuse and solanaceous weed hosts; spores are washed from pycnidia by rain or spread by brushing against moist leaves. Optimum temperature is 60° F to 80° F.

Control. Bury plant remains deep in soil or burn; control weeds; use long rotations.

Septoria oudemansii. Leaf spot of bluegrass; in northern states. Dark brown, purple spots turning straw-colored appear on leaf sheaths and spread to blades, with turf turning yellowish brown. Plants may be defoliated in cold wet seasons, but they are rarely killed.

Septoria paeoniae. Septoria leaf spot of peony; stem canker. Round gray spots with reddish borders are found on stems and leaves. Control with sanitary measures.

Septoria pistaciarum. Leaf spot; on pistachio.

Septoria rubi. (*Mycosphaerella rubi*). Blackberry leaf spot; on blackberry, and dewberry, perhaps with more than one strain. See *Sphaerulina rubi* for forms reported on red raspberry. Leaf spots are light brown, sometimes with a purple border. Infection is usually so late in the season that it is of minor importance, but it may cause some defoliation.

Septoria secalis var. **stipae.** Leaf spot; on bent grass. White spots turn straw-colored, with scattered pycnidia.

Septoria spraguei. Leaf spot; on Russian wildrye (*Elysum*).

Septoria tageticola. Marigold leaf spot; reported in 1958 from Florida. Spots are oval to irregular, smoky gray to black, speckled with minute black pycnidia. The disease advances upward from the lower leaves and also infects younger branches, peduncles, bracts, and seed. African marigolds are very susceptible, French almost immune.

Septoria tenella. Leaf spot; on fescue grasses. Small, vague, greasy brown spots.

Septoria tritici var. **lilicola.** Leaf spot; on ryegrass. Indefinite green to yellow mottled or blotched spots becoming fuscous to deep brown.

Sphaerulina

Ascomycetes, Sphaeriales, Mycosphaerellaceae

Perithecia separate, innate to erumpent, not beaked, lacking paraphyses and paraphysoids; hyaline, with several cells; clavate-cylindrical.

Sphaerulina rubi (*Cylindrosporium rubi*). Raspberry leaf spot; on red and black raspberry only, common east of the Rocky Mountains. This disease and a similar one on blackberry and dewberry were for many years considered due to *Septoria rubi* and then attributed to *Mycosphaerella* as the perfect stage. Later it was shown that two species were involved, with *Sphaerulina* the ascomycete on raspberry, *Septoria rubi* the pathogen commonly found on blackberry and dewberry.

Spots are small, circular to angular, first greenish black, then grayish; pycnidia produce elongate, three- to nine-septate spores. Perithecia, formed in fallen leaves, are black, subepidermal, later erumpent; ascospores are cylindrical, curved, pointed at both ends, usually four septate.

Sporonema

Fungi Imperfecti, Sphaeropsidales, Excipulaceae

Pycnidia dark, membranous or carbonaceous, innate, opening with torn lobes; spores hyaline, one-celled.

Sporonema camelliae. Camellia leaf spot.

Stemphylium

Fungi Imperfecti, Moniliales, Dematiaceae

Conidiophores dark, mostly simple, bearing a single terminal conidium or successive conidia on new growing tips; conidia dark, muriform, smooth or spiny; parasitic or saprophytic (see Fig. 32).

Stemphylium sp. (*?Pleospora herbarum*). Red leaf spot of gladiolus; widely distributed, causing an annual loss in Florida since 1938. Spots are small, round, translucent, pale yellow with reddish brown centers. Leaves may be killed before flowering or after spikes are cut, resulting in smaller corms. Infection takes place with 10 hours of dew or fog; rain is unnecessary; optimum temperature is 75° F. Leaves may be killed within 2 weeks of inoculation. Picardy variety is moderately susceptible; it is damaged more severely when grown near very susceptible Stoplight and Casablanca. The disease, starting on particularly susceptible varieties, spreads radially to less susceptible plants, decreasing in severity with distance from focal point. The leaf spot disappears in summer and autumn, reappears in winter 3 weeks after a cold period.

Control. Use resistant varieties to separate very susceptible types from those partly susceptible.

Stemphylium bolicki. Leaf spot; of echeveria, kalanchoë, and sedum. On some species lesions are small, raised, irregular to circular, brown to purplish black. On other species spots are larger, with tan centers, purplish margins.

Stemphylium botryosum (*Pleospora herbarum*). Leaf spot; black seed rot, seed mold on kidney beans, pea, onion, garlic, shallot, salsify, asparagus, pepper, and tomato.

Stemphylium callistepha. Leaf spot of China aster; brown, nearly circular, concentrically zonate spots with dark margins on leaves, bracts, petals, and stems.

Stemphylium ?cucurbitacearum. Leaf spot of cucurbits; on cucumbers, muskmelon, and winter and summer squash. The pathogen is possibly secondary, perhaps confused with *S. botryosum*. Small brown spots with lighter centers have mycelium growing over the lesion, producing globose, multiseptate spores.

Stemphylium floridanum. Tomato leaf spot; similar to gray leaf spot but the conidia and conidiophores are longer.

Stemphylium solani. Gray leaf spot; stemphylium leaf spot; in pepper, tomato, groundcherry, eggplant, and other *Solanum* species, mostly in the South, occasionally a problem elsewhere. In warm, humid weather, plants are defoliated in seedbed or field. First infection is on older leaves, which exhibit numerous small, dark brown spots extending through to the undersurface. Centers are often a glazed gray-brown with cracking and tearing. Leaves turn yellow and wither; all leaves may be killed except those at the top; seedbeds are often destroyed.

Control. Use clean soil for seedbed; spray seedlings at weekly intervals.
Stemphylium vesicarium. Purple spot; of asparagus.

Stigmea (*Stigmatae*)

Ascomycetes, Hemisphaeriales, Stigmateaceae

Fruiting structure subcuticular, hymenium a single disclike layer covered with a scutellum; spores dark, two-celled; mycelium scanty.

Stigmea geranii. Black leaf speck of geranium (cranesbill).
Stigmea rubicola. Black spot of raspberry; spot formed in late summer with a membranous layer under the cuticle; fruiting bodies produced in spring.

Stigmella

Fungi Imperfecti, Moniliales, Dematiaceae

Conidiophores short, dark, with a single terminal spore; conidia dark, muriform but with few cells, ovoid to oblong to nearly spherical; parasitic on leaves.

Stigmella platani-racemosae. Leaf spot of California sycamore; sometimes causing premature defoliation.

LICHENS

A lichen is a fungus body, usually one of the Ascomycetes with apothecia, enclosing a green or blue-green alga. The fungus receives some food from the alga and the alga some food and protection from the fungus, a relationship termed symbiotic. Lichens frequently grow on living trees and shrubs, but their injury is indirect, an interference with light or gas exchange to stems or foliage, rather than from penetration of living cells of the suscept plant. There are three types associated with plants: crustose, a crust closely appressed to bark of main trunk or larger limbs; foliose, leaflike, prostrate but not so firmly attached to the substratum; and fructicose, bushlike, erect or hanging.

Lichens are more abundant on garden shrubs—boxwood, camellias, azaleas, and so on—and on citrus in the South. They flourish in neglected gardens and orchards, and in shady damp locations, and may sometimes kill twigs and branches of weak trees growing on poor sites.

In most gardens control is unnecessary. If lichens become too disfiguring or too abundant for plant health, they may be killed by spraying affected parts with bordeaux mixture or other copper spray; spray when the lichens are dry.

They may be removed from main trunks by rubbing the bark with a steel brush after they are softened by rain.

MISTLETOE

Mistletoes are seed plants belonging to the family Viscaceae. They are semiparasites, manufacturing food but depending on a host plant for water and mineral salts. There are three genera in North America: *Phoradendron* and *Viscum*, which are true mistletoes, and *Arceuthobium*, dwarf mistletoe.

The mistletoe seed is naked embryo and endosperm invested with a fibrous coat and borne in white, straw-colored, pink, or red fruits—"berries"—embedded in a sticky gelatinous pulp enabling them to cling to bark of trees or stick to feet and beaks of birds, which disseminate them.

The seeds can germinate almost anywhere but penetrate only young thin bark, by means of a haustorium sent out from a flattened disc. Branches of the haustorium extend up and down and around the tree and occasionally produce secondary haustoria. The number of annual rings on a tree between the tip of the primary haustorium and the bark tells the age of the mistletoe. Many are 60 to 70 years old, and one has been reported as living 419 years.

The aerial portions of mistletoes are leafy, evergreen tufts of shoots on the stems of host plants, most conspicuous on hardwoods after leaf fall (see Fig. 37). The stems and leaves contain chlorophyll and are green but often with a yellowish, brown, or olive cast, depending on the season. All species have opposite leaves and round, jointed stems, and are dioecious with inconspicuous petal-less flowers. They occasionally become so large or numerous that the weight of the parasite breaks branches of the host. Growth is slow at first, but in 6 to 8 years the tufts may be 3 feet across. The aerial part does not live much longer than that, but the haustoria live as long as the tree, producing new bunches from adventitious buds.

Because they manufacture their own food, mistletoes require a lot of sun, which may be one reason why they flourish in the Southwest. Leafy mistletoes are relatively harmless in some situations; in others they handicap shade and forest trees, and occasionally kill hackberries and oaks. There are a few leaf spots and other fungus diseases that keep mistletoes from getting too abundant. They are harvested for Christmas greens with a curved mistletoe hook, which can be used to keep aerial portions cut off valuable trees. Breaking off or cutting off the bunches, however, may lead to more shoots in an ever-widening area.

Dwarf mistletoes are far more injurious, especially to forest trees, and much less conspicuous. In western coniferous forests they rank next to heart rots in importance, reducing the quality and quantity of timber and paving

Figure 37. Mistletoe, common in southern trees.

the way for bark beetle infestations. Infected branches should be pruned out; if the trunk is infected, the tree should be felled and removed.

Phoradendron (True Mistletoe)

Phoradendron means tree thief. The genus is restricted to the Americas, ranging from southern New Jersey and Oregon southward. Most are on hardwoods.

Phoradendron californicum. California mistletoe; ranging from southern California to Arizona, chiefly on Leguminosae—mesquite, carob, squaw bush, creosote bush, parkinsonia. This species is leafless, generally pendent, with long stems and reddish pink berries.

Phoradendron juniperinum. Juniper mistletoe; a leafless species with straw- or wine-colored berries, ranging from Colorado and Utah through New Mexico and Arizona.

Phoradendron libocedri. Incense cedar mistletoe; confined to incense cedar and occurring throughout its range in Oregon, California, and Nevada. The pendent plants are leafless with straw-colored berries. It may injure plants severely, causing spindle-shaped swellings in limbs at point of attack and living in the trunk as a parasite for hundreds of years after external portions have disappeared.

Phoradendron serotinum (*flavescens*). Eastern mistletoe; from southern New Jersey west to Ohio and Missouri and south to the Gulf, on many hardwoods—oaks, elm, maple, sycamore, gums, hickory, pecan, hackberry, hawthorn, persimmon, black locust, western soapberry, sassafras, and trumpet vine. This species has white berries and is the common Christmas mistletoe.

Phoradendron tomentosum. Texas mistletoe; abundant in Texas on elms, oaks, mesquite, osage-orange, and sugarberry; has white berries.

Phoradendron villosum. Hairy mistletoe; ranging from Oregon through California, usually on oaks, also on Oregon myrtle, California buckeye, chestnut, and manzanitas. It has pinkish white berries and may cause large hypertrophies on oaks.

Viscum (True Mistletoe)

The genus is restricted to California. It is now known that Luther Burbank introduced the parasite into the state in about 1900. Burbank's notes indicate that seed was supplied to him by J. C. Vaughan of Chicago, Illinois. This mistletoe has spread about 3.5 miles in 75 years.

Viscum album. European mistletoe, on alder, ash, birch, hawthorn, hickory, buckeye, maple, mountain ash, pear, persimmon, plum, poplar, pyracantha, willow, crabapple, and elm.

Arceuthobium (Dwarf Mistletoe)

The genus is restricted to conifers, and most species are found in the Northwest. Trees of any age may be deformed or killed, but the greatest mortality is among seedlings and saplings, with lodgepole and ponderosa pines most susceptible. The most striking symptom is the formation of witches' brooms, with sometimes the whole crown transformed into a huge broom. In other cases fusiform swellings in trunks turn into cankers. Foliage of affected trees is reduced.

The mistletoes themselves are small, rarely attaining a maximum of 8 inches, sometimes less than an inch. They are perennial shoots, simple or branched, jointed, with leaves reduced to opposite pairs of scales at the top of each segment. Stems range in color from yellow to brown to olive green. Berries are olive green to dark blue; each contains a single seed, rarely two. The seed is ejected with force and is spread horizontally for some feet. Animals and birds account for infection at a distance.

Arceuthobium americanum. Lodgepole pine dwarf mistletoe; common on the Rocky Mountain form but not the Pacific lodgepole pine; rare on other pines. The flowers bloom in spring, accessory branches forming a whorl.

Arceuthobium campylopodum. Western dwarf mistletoe; it forms witches' brooms and flowers late in summer. Widespread in Northwest principally on coastal ponderosa pine; species that were formerly called *A. campylopodum* are *A. abietinum* on white and grand firs, *A. divaricatum* on pinon pines, *A. laricis* on western larch, *A. microcarpum* on blue and Englemann spruce, *A. tsugense* on western hemlock, *A. cyanocarpum* on limber pine. Found also on exotic pines in California.

Arceuthobium cyanocarpum. Dwarf mistletoe; on pine, timber pine, and hemlock.

Arceuthobium douglasii. Douglas fir dwarf mistletoe; confined to this host. Plants are small, only 1½ inches high, greenish, slender.

Arceuthobium occidentale. Dwarf mistletoe; on exotic pines in California.

Arceuthobium pusillum. Eastern dwarf mistletoe; the only species in the East, from Minnesota to New Jersey and north to Canada, common on spruce, also on tamarack, and pines. The fruit matures in autumn; shoots are very short, less than an inch.

Arceuthobium tsugense. Hemlock dwarf mistletoe; on western and mountain hemlock.

Arceuthobium vaginatum subsp. **cryptopodum.** Southwestern ponderosa pine dwarf mistletoe; plant yellowish, robust.

MOLDS

The word *mold*, or *mould*, has many meanings. The first one given in *Webster's* is "a growth, often woolly, produced on various forms of organic matter, especially when damp and decaying, by saprophytic fungi." Leaf mold is organic matter reduced to friable earth by these saprophytic fungi. When rhododendrons are fed with a fertilizer having a cottonseed meal base, one can often see a moldy growth, showing that beneficial organisms are at work breaking down the material for plant use.

Some of these saprophytic fungi have a harmful, parasitic phase. The common black bread mold, *Rhizopus nigricans*, causes soft rot of sweet potatoes, "leak" of strawberries and grapes. *Penicillium* spp., the common blue molds on jellies, cause a decay of citrus and other fruits. Such diseases are discussed under Rots.

The word *mold* is used loosely to cover any profuse fungus growth on the surface of plant tissue. See Blights for a discussion of botrytis gray mold, so common on many plants; see Leaf Spots for alternaria brown molds and ramularia white molds, and for moldy leaf spots due to *Heterosporium* and *Pleospora*; see Sooty Molds for the black growths on insect exudate; and see Snowmold for turf diseases.

Botryosporium

Fungi Imperfecti, Moniliales, Moniliaceae

Conidiophores, tall, slender, hyaline producing numerous lateral branches of nearly equal length, each producing two or more secondary branches that are enlarged at the tip and bear heads of conidia; spores one-celled, hyaline; saprophytic.

Botryosporium pulchrum. Leaf mold; on tomato, also geranium (pelargonium), occasional in greenhouses.

Chalaropsis

Fungi Imperfecti, Moniliales, Dematiaceae

Mycelium at first hyaline, then greenish; two types of conidia-macroconidia or chlamydospores, olive green, thick-walled when mature, sessile or borne in short conidiophores in compact groups; endoconidia, hyaline, formed inside end cells of a dark endoconidiophore and extruded in chains.

Chalaropsis thielavioides. Black mold of rose grafts; Manetti mold, usually on grafted roses, sometimes on budded roses in nursery fields. The fungus grows over and blackens cut surfaces of stock and scion, preventing union and resulting in death of scions. When outdoor roses are budded on Manetti understock, the bud often turns black and dies. Infection is only through wounds. *Rosa odorata* and *R. chinensis* var. *Manetti* are both very susceptible understocks; *R. multiflora* is moderately susceptible; Ragged Robin is immune.

Control. Use healthy understock. Spray greenhouse benches, tools, etc., with copper sulfate; prevent spread of spores by workers on hands, clothing, and budding knife.

Cladosporium

See under Blotch Diseases.

Cladosporium fulvum. Leaf mold of tomatoes; general on greenhouse crops, occasionally serious in gardens in wet seasons in the Southeast and sometimes present in other states. Diffuse, whitish spots on upper surface of older leaves enlarge, turn yellow; the undersurface of the patches has a velvety olive brown coating of spores that are spread by air currents and in watering. Spores remain viable about the greenhouse for several months after plants are removed, and are sometimes carried on seed. Infection occurs only when humidity is high.

Control. Resistant varieties such as Globelle, Bay State, and Vetomold have been developed, but the fungus has mutated to more virulent forms. Regulating ventilators in greenhouses to reduce humidity seems to be the most practical control, sometimes providing heat on cool nights, even in summer.

Cladosporium herbarum. Leaf mold; pod and seed spot; the fungus is a weak parasite causing black mold of peanut, pod spot and seed mold of lima and kidney beans, glume spot of bluegrass, leaf mold of pepper and tomato, sometimes a fruit mold.

Cladosporium macrocarpum. Black mold of spinach; on old leaves or secondary after other leaf spots.

Erostrotheca (*Melanospora*)

Ascomycetes, Hypocreales

Perithecia bright, more or less soft, without beak, paraphyses lacking; spores ellipsoid, yellow to olivaceous. Conidial stage has many spore forms.

Erostrotheca multiformis (*Cladosporium album*). White mold of sweet pea; white blight, also on perennial pea, observed on greenhouse crops. Leaflets are covered with tan or buff, circular to irregular, small to large spots with cinnamon brown pustules giving a granular appearance. White tufts of mold represent the *Cladosporium* stage. Pseudosclerotia are also formed in the leaves, which may die and drop. The fungus enters through stomata under conditions of high humidity. Dusting with sulfur has been suggested.

Torula

Fungi Imperfecti, Moniliales, Dematiaceae

Conidiophores lacking; entire branches of mycelium develop into simple or branched chains of dark conidia, which separate readily; saprophytic.

Torula maculans. Leaf mold; on yucca.

NEEDLE CASTS

Certain diseases of conifers that result in conspicuous shedding of needles are termed needle casts, sometimes needle blights. Most of the fungi causing such symptoms are members of the Phacidiales.

Adelopus (*Phaeocryptopus*)

Ascomycetes, Erysiphales, Meliolaceae

One of the black mildews, with superficial, dark mycelium; perithecia innate with a central foot, without ostiole; spores two-celled, hyaline.

Adelopus gäumannii. Adelopus needle cast of Douglas fir; Swiss needle cast. Although first noted in Switzerland in 1925, this seems to be a native American disease occurring in relatively harmless fashion on the Pacific Coast, somewhat injurious to native Douglas fir in the Southwest and to trees in New England and New York. Needles fall prematurely, leaving only the current season's growth. If this disease occurs for several consecutive years, trees have thin foliage, appear yellow or brown, and finally die. Needles are yellow-green to brown, often mottled, and on undersurface tiny black perithecia, issuing from stomata, appear as sooty streaks, one on each side of the middle nerve.

Bifusella

Ascomycetes, Phacidiales, Phacidiaceae

Apothecia elongate, slitting with a cleft; paraphyses lacking; spores hyaline, one-celled, club-shaped at both ends with halves joined by a narrow neck (see Fig. 38).

Bifusella abietis. Needle cast of fir; on alpine and corkbark fir from New Mexico to Idaho. Dark brown to black hysterothecia extend the entire length of the middle nerve on undersurface of needle. Pycnidia are in two rows on upper surface.

Bifusella faullii. Needle cast of balsam fir; the most common and destructive of the needle casts of this host. Ascospores are discharged in July, but infected young needles do not change color until spring, then turn light brown to buff. Effused pycnidia in the same color appear in the groove on upper surface of the needle, followed by dusky brown hysterothecia (apothecia with a covering), with maturing ascospores the second summer.

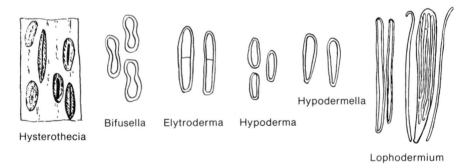

Hysterothecia   Bifusella   Elytroderma   Hypoderma   Hypodermella   Lophodermium

Figure 38. Needle cast fungi, which form ascospores in hysterothecia, elongate apothecia opening with a cleft. *Bifusella*, spores constricted in middle; *Elytroderma*, fusiform spores; *Hypoderma*, short fusiform spores; *Hypodermella*, spores tapering at base; *Lophodermium*, ascus with filiform spores, and paraphyses.

Bifusella linearis. Needle cast of pine; tar spot, on various pine species. Hysterothecia are variable in length, shining black, on two-year needles.
Bifusella saccata. Needle cast of pine.

Elytroderma

Ascomycetes, Phacidiales, Phacidiaceae

Ascospores two-celled, broadly fusiform (see Fig. 38).

Elytroderma deformans. Needle cast; witches' broom, on Coulter, ponderosa, lodgepole, Jeffrey, pinon, and Jack pines. Elongated dull, dark hysterothecia are on both leaf surfaces. The tissues of ponderosa and Jeffrey pines may be penetrated and loose witches' brooms formed. Saplings may have entire crown converted; they die or make little growth.

Fusarium

See under Blights.

Fusarium lateritium. Needle cast; on *Torreya taxifolia.*

Hypoderma

Ascomycetes, Phacidiales, Phacidiaceae

Hysterothecia elliptical to oblong, opening by a cleft; asci long-stalked, spores hyaline, fusiform, surrounded by a gelatinous sheath (see Fig. 38).

Hypoderma desmazierii. Needle cast; tar spot of pines, most frequent on eastern white pine. Infected needles are at first yellow, then reddish brown, and finally deep brown with a grayish cast. The tips are infected first, the fungus being a weak parasite, completing its cycle in a year. Hysterothecia are shining black, elliptical.
Hypoderma hedgecockii. Needle cast of hard pines; in southeastern states. Elliptical shining black hysterothecia are present in discolored areas on green needles. Each ascus contains four normal and four aborted spores.
Hypoderma lethali. Gray blight; needle cast of hard pines; from New England to Gulf states. Hysterothecia are short, narrow, black, often found on pitch pine.
Hypoderma robustum. Needle cast of firs; in West, usually white fir. Concolorous pycnidia, which form two rows, one in each needle wing, often turn black after spore discharge.

Hypodermella

Ascomycetes, Phacidiales, Phacidiaceae

Like *Bifusella* with elongate apothecia, with a cleft, but paraphyses present; spores hyaline, one-celled, club-shaped at upper end, tapering toward base (see Fig. 38).

Hypodermella abietis-concoloris. On firs and southern balsam.

Hypodermella ampla. Needle cast of Jack pine; all needles may drop except those of the current season. Short, elliptical, dull black hysterothecia are scattered over light buff-colored areas.

Hypodermella concolor. Needle cast of Jack and lodgepole pines; virulent fungus infects young needles, in summer, which turn brown the next season. Short hysterothecia are concolorous with the leaf and appear as shallow depressions.

Hypodermella laricis. Larch needle and shoot blight of eastern and western larches; yellow spots are formed on needles, which turn reddish brown but stay attached, giving a scorched appearance to trees that are normally deciduous. Hysterothecia are very small, oblong to elliptical, dull black, on upper surface of needles.

Hypodermella nervata. Needle cast of balsam fir; pycnidia are in a groove along upper surface of needle in continuous or occasionally interrupted row, turning nearly black after spores are discharged.

Lirula

Fungi Imperfecti, Rhytismatales, Phacidiaceae

Lirula macrospora. Needle cast or blight; on spruce.

Lophodermium

See under Leaf Spots.

Lophodermium durilabrum. Needle cast; on pine.

Lophodermium filiforme. Spruce needle cast; sometimes causing serious defoliation of red and black spruce. Hysterothecia are long or short, shining black (see Fig. 38).

Lophodermium juniperinum. Widespread and abundant on common juniper and red cedar but apparently not parasitic. Hysterothecia are elliptical, shining black, on both leaf surfaces.

Lophodermium nitens. Frequent but apparently saprophytic on five-needle pines. Hysterothecia short, black, shining.

Lophodermium piceae. Needle cast; needle blight of fir; tar spot, on fir and spruce, most severe on young specimens. Needles turn yellow, reddish, or brown, and drop. Short, shining black hysterothecia are formed on all needle surfaces.

Lophodermium pinastri. Pine needle cast; widespread. Pycnidia appear in spring or early summer as tiny black spots on browned needles, followed by dull, occasionally shining, black, short, elliptical hysterothecia. The fungus is a weak parasite but can be epidemic in nurseries. Bordeaux mixture will control it.

Lophodermium seditiosum. Needle cast; of scotch pine.

Mycosphaerella

See under Blights.

Mycosphaerella laricina. Needle cast of European larch and western larch.

Naemacyclus

Ascomycetes, Phacidiales, Stictidiaceae

Apothecia bright-colored, soft, opening with a cleft; paraphyses much branched; spores worm-shaped.

Naemacyclus niveus. Needle cast; occasional on various pines. Fruiting bodies tiny, elliptical, first waxy, dark brown, later concolorous with leaf surface.

Phoma

See under Blackleg.

Phoma eupyrena. Needle cast and blight; of red fir and Douglas fir.

Rhabdocline

Ascomycetes, Phacidiales, Stictidiaceae

Apothecia innate, brown, exposed by irregular rupture of epidermis; paraphyses present; spores one-celled, becoming septate after discharge from ascus, rounded at ends and constricted in the middle.

Rhabdocline pseudotsugae. Needle cast of Douglas fir; needle blight, common on Pacific Coast and in Rocky Mountain states on native Douglas fir and in northeastern states on ornamental forms. The disease has reached Europe on trees from western North America and is causing much concern there.

Needles are infected in spring or early summer, with first symptoms showing as slightly yellow spots, usually at ends of needles, in autumn or winter. By the next spring the color is reddish brown, and leaves have a mottled appearance. In severe infection needles turn a more uniform brown, and the entire tree appears scorched. Apothecia are usually on underside of needles, sometimes on upper. They are at first round cushions; then the epidermis ruptures to expose a brown, elongated disc. Infected needles drop after ascospore discharge, thereby living only 1 year instead of the normal 8 years or so.

Control. Spraying with bordeaux mixture when new needles develop, repeating twice at 10- to 14-day intervals has been suggested; also, spraying with lime sulfur at time of ascospore discharge in early summer. In forests, control will probably depend on early elimination of susceptible trees.

Rhabdocline weirii. Needlecast; of Douglas fir.

Rhizosphaera

Fungi Imperfecti, Sphaeropsidales, Sphaerioidaceae

Pycnidia brown, on a stalk; spores ovoid, one-celled, hyaline.

Rhizosphaera kalkoffi. Needle cast of blue spruce; lowest needles are affected first, becoming mottled yellow, and the disease progresses up the tree. It has been controlled in ornamentals with three sprays of bordeaux mixture.

NEMATODES

In the four decades since the first edition of this book was prepared nematodes have become of major importance in plant pathology. It used to be stated that plant pests, insects, and diseases took a toll of one-tenth of all our crops. Now we believe that nematodes alone may cause a 10% crop loss, and some place the figure as high as 25%. The monetary loss is not easy to figure. Guesses range from $500,000 to $3 billion a year. Nematodes may be as damaging in home gardens as on farms.

Nematodes used to be considered primarily a southern problem, with the root-knot nematode the major culprit. Now we know that nematodes can be as serious in Maine or Minnesota as in Florida or Texas, and that root-knot species are responsible for only a fraction of total losses.

A 1957 report from Maryland states that samples were taken from around the roots of crop plants on 1,210 different farms and gardens, and that every sample included at least one species of nematode known to be a plant parasite, with root-knot nematodes making only 3.2% of the total. A 1959 report from New Jersey states that, on the basis of 2,500 soil and root samples taken since 1954, a very conservative estimate of annual loss in the state is $15 million. The root-knot nematodes reduced by cold winters were in third place because of their importance as pests of greenhouse crops, including African violets, roses, and other ornamentals, as well as vegetable seedlings.

Nematodes (eelworms or roundworms) are threadlike animals in the group Nematoda (or Nemata). Some taxonomists consider this group a separate phylum; others call it a class, in the phylum Nemathelminthes or Aschelminthes. Nematodes live in moist soil, water, decaying organic matter, and tissues of other living organisms. Some cause diseases of humans or animals; others cause plant diseases. The animal parasites include hookworms, pinworms, and the worms in pork causing trichinosis, and they range in length from less than an inch to nearly a yard. Most plant parasites are practically microscopic in size, sometimes just barely visible to the naked eye. They mostly range in length from $1/50$ inch to $1/10$ inch (0.5 mm to 2 mm).

Nematode diseases of plants are not new. The wheat eelworm was recorded more than two centuries ago; root knot has been a recognized problem since 1855. Our systematic investigation of plant parasitic nematodes is very new. Only in the past few years have we made surveys to find out how widespread nematodes are and how many cases of "decline" in plants are due to them. Nematodes injure plants directly by their feeding, causing root loss and general stunting, and indirectly by wounding the tissues and affording entrance to bacteria and fungi causing rots and wilts. Some nematodes also are vectors of ring spots and other virus diseases.

Many nematodes may merely live in the soil close to the plant and cause no damage, and a few are actually beneficial, feeding on such harmful pests as Japanese beetle grubs. Only an expert nematologist can determine species and decide which are responsible for a plant's ill health. In submitting samples to your experiment station for diagnosis, dig up roots and some surrounding soil, place immediately in a plastic bag to prevent drying out, and mail as soon as possible.

Plant parasitic nematodes may be sedentary or migratory, though even the migratory forms do not move through soil to any great distance. Major dispersal is by shipment of infested nursery stock and soil; locally nematodes are spread on tools, and feet, in irrigation water, in plant parts, and sometimes as dry cysts by the wind. Plant nematodes are facultative or obligate parasites. They may be endoparasitic, living inside roots or other tissues, or ectoparasitic, living outside the plant, inserting only the head for feeding; and some forms are intermediate between the two types. Most plant nematodes are root parasites, but some live in stems, bulbs, leaves, or buds. Some cause galls or other distinctive symptoms; others produce a general yellowing, stunting, or dieback that is often ascribed to other causes.

Nematodes are usually long and cylindrical, tapering at both ends, round in cross section. In some genera the female is pear-shaped or saclike, the male fusiform. Nematodes in general lack coloration, being transparent or with a whitish or yellowish tint. They are covered with a cuticle, made up of three main layers, largely protein, under which is a cellular layer called the hypodermis. The body cavity, pseudocoel, is filled with fluid. The body wall musculature, directly beneath the epidermis, consists of longitudinal fibers only. Thus, nematodes cannot contract transversely. They move through moist soil with a threshing motion, or a series of undulations.

Nematodes have a complete digestive tract with a mouth at the anterior end surrounded by lips bearing the sensory organs, but there is no true head, and nematodes lack eyes and nose. Basically there are six lips, but they may be fused in pairs. The sense organs, amphids, are important diagnostic characters, one class of nematodes having amphids with conspicuous openings, the other having amphids with minute pores. Most plant-parasitic nematodes belong to the latter group.

Behind the mouth there is a cavity (stoma), then the esophagus, the intestine, and the rectum. The latter terminates in a ventral terminal or subterminal anus in females, in a cloacal opening in males. The sexes are usually separate, but sometimes males are missing and females are hermaphroditic. The body region behind the anus or cloacal opening is called the tail.

Near the posterior end of many nematodes there is a pair of cuticular pouches called phasmids, believed to be sense organs like the amphids. They are used to divide nematodes into two main groups, the Secernentea, or Phasmidia, with phasmids, and the Adenophora, without phasmids.

Almost all of the plant-parasitic nematodes feed by means of a stylet, which works something like a hypodermic needle. It is a conspicuous protrusible spear used to puncture tissue. In most families this spear is a stomatostylet, a hollow spear derived from the sclerotized walls of the buccal cavity or stoma. Commonly the nematode punctures plant tissue with its stylet, then injects a secretion from its salivary gland that predigests the food before it is sucked in through the stylet. In the family Dorylaimidae the spear is an enlarged tooth, odontostylet (onchiostyle), originating in the esophagus wall. It is usually hollow, but in the genus *Trichodorus* the tooth is solid but grooved.

The structure of the esophagus varies in different groups and is an important diagnostic character. The esophagus commonly has one or two swellings, known as bulbs. Those provided with a glandular apparatus are true bulbs; those lacking such apparatus are pseudobulbs. True bulbs are the chief pumping and sucking structures. They may be median, situated at midlength, or posterior, at the end of the esophagus.

Control measures for nematodes include crop rotation and other cultural practices and soil treatment with chemicals. Most chemicals are meant for fallow soil; a few are safe around living plants. Details of nematicides and their application are given in Chapter 1. Greenhouse soils are often steam-sterilized, and plants are sometimes dipped in hot water, the duration of the soak and the temperature depending on the tolerance of the plant and the kind of nematode to be eradicated. Some plants are antagonistic to nematodes. Asparagus roots produce a chemical that is toxic to many species, and marigolds grown with or in advance of some flower crops reduce the numbers of *Pratylenchus*, lesion nematodes. Some soil fungi trap nematodes but do not provide a practical control.

Anguina

Tylenchidae. Endoparasitic nematodes feeding in aboveground plant tissue and transforming seeds or leaves into galls. Males and females both elongate (wormlike), but females are obese. Cuticle finely striated; stylet short with well-developed basal knobs; tail cone-shaped; single ovary.

Anguina agrostis. Grass nematode; serious on bent grass and chewings fescue in the Pacific Northwest. Second-stage larvae remain in sheaths near growing tips most of the year, entering embryonic flowers in late spring. There the larvae mature, and the females lay large quantities of eggs. The quickly hatching young larvae transform developing seed into elongated dark purple galls. When the galls fall to the ground, nematodes are released to reinfect grass in the vicinity. There is only one generation a year, and larvae cannot exist in moist soil more than a year without access to a host plant with developing inflorescence. The disease is important only on grass grown for seed; it is not a problem on clipped turf. When seed is threshed, galls can be carried 300 feet or more from the machines by air currents, and still further in heavy winds.

Control. Rotate with a crop other than bent grass or fescue or plow under and prevent inflorescence for 1 year. Soak seed for 2 hours in tepid water with a wetting agent; then hold for 15 minutes at 126°F.

Anguina balsamophila. On balsam-root; galls on underside of leaves.

Anguina graminis. Galls on leaves of fescue grasses.

Anguina tritici. Wheat nematode; on wheat and rye, a field crop pest forming galls in place of grain. The disease was recognized in 1745, the first to be attributed to nematodes. The species is long-lived, viable nematodes having been found in seed stored 28 years. Brine flotation was the old method of eliminating galled seed.

Aphelenchoides

Aphelenchoididae. Bud and leaf nematodes, foliar nematodes. Ecto- and endoparasites; males and females wormlike, very slender; cuticle finely annulated; stylet with small basal knobs; tail with acute tip.

Aphelenchoides besseyi (including *A. oryzae*). Summer dwarf nematode; of strawberry, present from Maryland to Louisiana, also reported from Oklahoma, Missouri, southern Illinois, California, and Washington. The nematodes live in the soil and are washed into buds by rains and irrigation water, affecting young leaves as they develop. Leaflets are crimped or crinkled, cupped, narrow, with a reddish cast to veins and petioles. Older leaves are darker green, more brittle than normal. This disease is major in Florida, commonly noted from July to October. Cold weather stops its progress, often masking symptoms, but plants do not recover; runner plants from infested mother plants are diseased. In spring the nematode population may be low, allowing nearly normal formation of early leaves, but in summer a single bud may harbor up to 1,300 individuals, causing center leaves to be deformed and dwarfed. The same species causes a serious disease of rice in Arkansas and Louisiana.

Control. Buy certified plants; rogue and burn diseased plants as soon as noticed. Treat dormant infested plants with hot water, 2 minutes at 127°F.

Aphelenchoides fragariae (including *A. olesistus*). Spring dwarf nematode; of strawberry. Fern nematode; a leaf nematode. A bud parasite of strawberry from Cape Cod to Maryland and found in scattered localities along the Pacific Coast. A cold-weather species, it persists through the winter with several thousand nematodes present in a single bud as leaves unfold in spring. The foliage is small, twisted, thickened, glossy, with swollen petioles; blossom buds are killed or poor, and no fruit is set. Some plants are killed; others recover.

As the fern nematode, or begonia leaf blight nematode, this species is recorded on anemone, aquatic plants (*Cabomba* sp., *Limnophila* sp., *Peplis* sp., and *Potamogeton* sp.), begonia, bouvardia, calceolaria, chrysanthemum, clematis, coleus, crassula, dianthus, doronicum, fern, geranium, hydrangea, lily, peony, primrose, saintpaulia, scabiosa, zinnia, and other ornamentals. Fern leaves have a patchy or blotched appearance with dark brown to black areas on the fronds. In some species these are rather narrow dark bands from midrib to border, limited by parallel side veins; in bird's nest fern there is a profuse brown discoloration from the base halfway up the leaf.

On begonias the disease is most serious on semituberous varieties grown in greenhouses. Small brown spots with water-soaked margins, on underside of leaves, enlarge, coalesce, turn dark brown, and become visible on the upper surface. Whole leaves may turn dark; plants may be stunted. On fibrous-rooted begonias spots stay small, and leaves become shiny with a tendency to curl, lose color, and drop. Nematodes are spattered from plant to plant by syringing or careless watering; there is no disease spread when foliage is kept dry.

Dieback of Easter lilies grown in the Northwest is also attributed to this bud and leaf nematode. Leaves are first blotched with yellow, then turn brownish, drooping and curling against the stem (see Fig. 39). The nematodes live over winter in the bulbs and are splashed from leaves of one plant to

Figure 39. Foliar nematode on lily.

another in the field. Lilies from diseased bulbs develop "bunchy-top" symptoms, with thick, twisted foliage and dieback.

Control. Strawberry plants in nurseries should be inspected and certified in spring. Mother plants, near the end of the dormant period, can be treated with hot water, 2 minutes at 127°F. Crop rotation helps.

Bulbs may be treated with hot water, for 1 hour at 111°F. Potted begonias can be submerged, pot and all, for 1 minute at 120°F, or for 3 minutes at 116°F. African violets may be treated for 30 minutes at 110°F, ferns for 10 to 15 minutes at the same temperature.

Aphelenchoides parietinus. Causing root plate and scale necrosis of bulbous iris.

Aphelenchoides ribes (= *A. ritzema-bosi?*). Currant nematode; a bud parasite on black currants and gooseberries in England; reported from California on gooseberries. Treat cuttings for 30 minutes in hot water, 110°F.

Aphelenchoides ritzema-bosi. Chrysanthemum foliar nematode; common and serious on this host in home gardens and greenhouses, first reported in New Jersey in 1890. It is also recorded on dahlia, zinnia, and some other ornamentals but possibly confused with *A. fragariae.* A morphologically similar species produces a yellow bud blight of Vanda orchids. The first symptoms are dark spots on areas on underside of leaves, but by the fifth day after infestation discolored veins stand out sharply on upper leaf surface, and diseased leaves turn brown or black, starting in distinctive wedge-shaped areas between veins (see Fig. 40). Later the leaves dry, wither, and hang along the stems. The nematodes swim from the soil up the stem in a film of water, the disease going from lowest leaves progressively upward. Almost any variety may be attacked, but Koreans are particularly susceptible. The nematodes may not survive the winter in old dead leaves but they do survive in living leaves in old crowns.

Control. Keep foliage dry; avoid overhead watering. Use a mulch to avoid splashing. Avoid crown divisions; make tip cuttings, which are usually free from nematodes. Dormant plants can be treated with hot water, 5 minutes at 122°F or 30 minutes at 112°F.

Aphelenchoides subtenuis. Bud and leaf nematode; on narcissus, causing scale necrosis. Reported from the Southeast and Pacific Coast states.

Belonolaimus

Criconematidae. Sting nematodes, migratory obligate ectoparasites, usually found free in soil near growing tips; both sexes long, slender, with blunt ends; body strongly annulate; about 2 mm long, stylet long, with well-developed knobs; two ovaries.

Figure 40. Leaf nematode of chrysanthemum. Wormlike male and female nematodes cause wedge-shaped browning between veins, followed by general blighting of leaf.

Belonolaimus gracilis. Sting nematode; on a wide variety of hosts from Virginia southward, also reported from New Jersey and from a rose greenhouse in Connecticut. This pest is major in strawberries, celery, and sweet corn in Florida. It injures Bermuda, centipede, and St. Augustine grasses and seedlings of slash and long-leaf pines, being first recorded from pine. Other plants damaged by *Belonolaimus* species include peanut, pea, lupine, Austrian winter pea, cowpea, bean, lima bean, soybean, beets, cabbage, cauliflower, lettuce, endive, onion, potato, and sweet potato. The slender worms feed at root tips and along the sides. Soil fungi enter roots through feeding punctures. Roots develop short stubby branches with necrotic lesions; plants are stunted. On woody plants decline symptoms include chlorosis, twig dieback, premature dropping of fruit (such as grapefruit), and rapid wilting under moisture stress. The nematodes seem to be limited to light, sandy soils.

Control. Rotate crops; cultivate to remove weed hosts.

Belonolaimus longicaudatus. This species may be responsible for some of the injury ascribed to *B. gracilis*. It occurs in the same southeastern states and may injure roots of celery, peanut, grasses, cabbage, bean, and other vegetables. Potato and soybean are considered especially susceptible.

Bursaphelenchus

Aphelenchoididae. Ecto- and endoparasites; females (adult) have a vulval flap.

Bursaphelenchus lignicolus. Causes wilt of pine and the nematode is vectored by cerambycid beetle (pine sawyer beetle).

Cacopaurus

Criconematidae. Cuticle finely annulate; female small but very obese; eggs large; male lacks stylet.

Cacopaurus pestis. Reported from roots of Persian (English) walnut in California, causing typical decline with reduction in size and number of leaves, fewer nuts, eventually complete defoliation and death.

Criconemella

Criconemella xenoplax. Ring nematode; on peach cover crops including curly dock, perennial ryegrass, vetch, crimson clover, hairy vetch, and cowpea; also tall fescue and white clover.

Criconema

Criconematidae. Ring nematodes, short, thick, sedentary ectoparasites; cuticle thick with spines or scales; usually found in woodlands, in damp areas, seldom in cultivated soil.

Criconema civellae. Reported on citrus roots in a Maryland greenhouse.
Criconema decalineatum. Fig spine nematode; on figs.
Criconema spinalineatum. Zoysia spine nematode; on Zoysia.

Criconemoides

Criconematidae. Ring nematodes; short, thick-bodied; cuticle thick with retrose (projecting backward) annules; ectoparasites with a wide host range.

Criconemoides annulatum. On holly oak, Montana; beans and citrus, Louisiana.
Criconemoides citri. Citrus ring nematode; on citrus in Florida. The broadly annulated head is often buried deep in root tissue, which dies near the feeding puncture.
Criconemoides curvatum. Reported in large numbers on carnations but apparently not very injurious; also on grasses, in Ohio.
Criconemoides crotaloides. On Douglas fir and poplar, in Utah.
Criconemoides cylindricum. On peanut, in Georgia.
Criconemoides komabaensis. On camellia, in Florida.
Criconemoides lobatum. On pines, Florida; potato, in New York; also grasses.
Criconemoides mutabile. On marigold, in Washington, D.C.
Criconemoides ornatum. On grasses, in Ohio.
Criconemoides parvum. On grasses, in Ohio.
Criconemoides rusticum. On grasses, in Ohio.
Criconemoides similis. Cobb's ring nematode; apparently an important factor in decline of peaches in Maryland and North Carolina, reported on pine in Florida.
Criconemoides teres. On oak, in California.
Criconemoides xenoplax. On carnation, causing reduced root system, stunting, reduced flower yield; also reported on grape and grasses.

Crossonema

Crossonema sp. Decline; of Alaska cedar.

Ditylenchus

Tylenchidae. Bulb and stem nematodes, slender, of moderate length, conelike tail, finely striated cuticle, mostly endoparasites.

Ditylenchus destructor. Potato rot nematode; feeding on underground stem structures of a large number of plants but important on potato, especially in Idaho and Wisconsin. Discolored spots on tubers progress to a gray or brown decay. The tissues have a granular appearance; they dry and shrink and the skin may crack. Invasions continue in storage, sometimes with complete destruction of tubers.

Ditylenchus dipsaci. Stem and bulb nematode; an internal parasite of bulbs, stems, leaves, rarely roots, causing eelworm disease of narcissus, ring disease of hyacinth, onion bloat, stem disease of phlox. The name *dipsaci* covers many strains and probably more than one species. The type was found in 1857 on Fuller's teasel. The nematodes are thought to release a pectinase during feeding, which results in a dissolution of the middle lamella and the production of large intercellular spaces. Besides hyacinth and narcissus, they injure grape hyacinth, tulip, galtonia, garlic, shallot, and onion, and cause a stem disease of alfalfa and many flowers besides phlox.

The strains of hyacinth and narcissus are not reciprocally infective, although the hyacinth strain does infect onions. Hyacinths have yellow flecks or blotches on the leaves, which are often twisted, short, and split. In narcissus there are pustules or blisters, called spikkels, in leaves, which can be felt when the leaf is drawn through the fingers. Nematodes in such pustules probably enter leaves as they push through the soil. Bulbs badly diseased at planting produce no foliage, or a few leaves that are premature, twisted, and bent.

When leaves are dry, nematodes are inactive; but when the foliage is moist and decayed, they revive and pass into the soil or the neck of the bulb. They enter bulb scales, move down to the basal plate, and then enter the base of other scales. Infected scales are brown, and since there is little lateral movement of nematodes, the cut surface of a bulb shows one or more brown rings contrasting with healthy tissue. Eggs, larvae, and adults are all present in the brown areas. Male and female adults are wormlike, up to 1.9 mm long. Infective larvae issue in large numbers in whitish tufts in a break between basal plate and scales, and work through the soil to invade adjacent plants. They are also spread in irrigation water, on tools, and by animals. Some winter in weed hosts, some in seed of composites. In moist soil they die in a year or so, but they have been recovered from plants after 5 or 6 years.

The strain on phlox attacks campanula, sweet william, evening primrose, goldenrod, schizanthus, anemone, foxglove, and orchids. The leaves are very narrow, crinkled, and waved, often brittle, with a tendency to lengthen petioles. Stems may be swollen near the top or bent sidewise; plants are stunted, often fail to bloom, and may die prematurely. The nematodes enter through stomata of young shoots and work upward as the stems develop. They infest seed of phlox and other composites, and may be so disseminated.

In onions the inner bulb scales are enlarged, causing a split onion that seldom flowers and sometimes rots at the base. Seedlings are twisted, stunted, covered with yellow spots. On plants grown from sets, a slight stunting and flaccid condition of outer leaves is followed by leaf-tip necrosis and continued stunting. The larvae may live long in infested soil and may be carried in set onions.

Control. Commercial growers routinely treat narcissus bulbs in hot water, 4 hours at 110°F to 112°F. All infected plants, parts, and debris should be removed from fields and destroyed; a 2- to 4-year rotation may be tried. Take up and burn infested phlox or similar plants. Put new plants in a new location or in fumigated soil.

Ditylenchus gallicus. On elm.

Ditylenchus iridis. Probably a strain of *D. dipsaci*, on bulbous iris. Mildly infected plants dry up prematurely and have poor root systems. Heavily infected plants are stunted, having few if any roots, and the bulbs decay before harvest. Treat bulbs with hot water as for narcissus, but soak only 3 hours and as soon after curing as possible.

Dolichodorus

Criconematidae. Awl nematodes similar to sting nematodes with long stylet with well-developed knobs; coarsely annulated cuticle; both sexes wormlike; male tail has a bursa (lateral extension); female has two ovaries; ectoparasites.

Dolichodorus heterocephalus. Awl nematode; causing decline of celery, bean, tomato, corn, pepper, and water chestnut in the Southeast, also recorded on pecan. It feeds largely on root tips and sometimes along the side of roots, causing necrotic lesions. It also feeds on germinating seeds and hypocotyls, sometimes penetrating the seedcoat to reach the embryo. Poor seedling emergence may be due to this nematode.

Dolichodorus obtusus. On arctostaphylus, in California.

Dorylaimus

Dorylaimidae. Spear nematodes, with an odontostylet (hollow tooth), bottle-shaped esophagus; cuticle with longitudinal ridges; both sexes wormlike, tails rounded to cone-shaped; not proven plant parasites.

Dorylaimus spp. Found in soil near soybean, sweet potato, and other plants but doubtful as pathogens.

Helicotylenchus

Hoplolaimidae. Spiral nematodes, ectoparasites or semiendoparasites; long strong stylet with basal knobs; cuticle annulated. The head is inserted in a root, but the body remains outside in a ventrically curved spiral with one or more turns.

Helicotylenchus dinysteria. On gardenia, corn, and bluegrass.

Helicotylenchus erythrinae. Zimmerman's spiral nematode; rather common in Florida around roots of grasses. Present in other states on blueberry, boxwood, cauliflower, cedar, clovers, corn, cranberry, turf grasses, oak, oat, pachysandra, pepper, pieris, pine, rhubarb, soybean, strawberry, wheat, and yew.

Helicotylenchus multicinctus. Cobb's spiral nematode; associated with roots of many plants, including azalea, cherry, cranberry, marsh grass, hibiscus, peach, pine, spruce, and yew.

Helicotylenchus nannus. Steiner's spiral nematode; a small species common in the Southeast. Found damaging roots of apple, azalea, boxwood, asparagus fern, calathea, camellia, centipede grass, civet bean, gardenia, peperomia, philodendron, rubber plant, royal palm, laurel oak, soybean, peanut, and tomato. There is a gradual decline, stunting, and failure to form flower buds.

Helicotylenchus pseudorobustus. On corn, grape, tomato, and soybean.

Hemicriconemoides

Criconematidae. Ectoparasites; female with cuticular sheath, stylet with anteriorly concave knobs; males without sheath or stylet. Commonly associated with turf and woody plants in warm climates, but pathogenicity not yet demonstrated.

Hemicriconemoides biformis. Oak sheathoid nematode; on roots of oak, Florida.

Hemicriconemoides chitwoodi. Associated with stunting of camellias.

Hemicriconemoides floridensis. Pine sheathoid nematode; on pine.

Hemicriconemoides gaddi. On camellias.

Hemicriconemoides wessoni. On myrica, in Florida.

Hemicycliophora

Criconematidae. Sheath nematodes; ectoparasites with sedentary habits; female retains last molt as an extra cuticle; knobs of stylet with anteriorly directed processes; males rare, without stylet.

Hemicycliophora arenaria. Causing root galls on rough lemon, also reproducing in tomato, pepper, celery, squash, and bean. Celery has large, multibranched galls.

Hemicycliophora brevis. On California laurel.

Hemicycliophora obtusa. On beet, in Utah.

Hemicycliophora parvana. Tarjan's sheath nematode; damaging celery in Florida, also recorded on corn, beans, and dracaena.

Hemicycliophora similis. Grass sheath nematode; also causes small galls on roots of blueberry and cranberry.

Heterodera

Heteroderidae. Cyst nematodes, highly specific, attacking members of but few genera in a given plant family, partially endoparasitic, quite sedentary, attached to root by neck only. The female is lemon-shaped to globoid, white, yellow, or brown, 0.5 to 0.75 mm. Eggs are deposited or retained in body of mother, whose leathery wall forms a true cyst. Eggs remain alive for years in cysts, which are spread by wind or in soil around nonhost plants. Males are slender worms, up to 1.75 mm. Root-knot nematodes, formerly all classed as *Heterodera marioni*, have been reclassified as various species of *Meloidogyne*. The stylet *Heterodera* is twice as long as that in *Meloidogyne*, and the latter does not form true cysts.

Heterodera avenae. Oat cyst nematode; on pea.

Heterodera cacti. Cactus cyst nematode; obtained from various localities in Mexico, where it is probably indigenous, and likely to occur on cacti wherever grown. The cyst is lemon-shaped.

Heterodera carotae. Carrot cyst nematode.

Heterodera cruciferae. Cabbage cyst nematode; closely related to the sugarbeet nematode. On crucifers in California. Hosts include broccoli, Brussels sprouts, cabbage, cauliflower, kale, kohlrabi, mustard, radish, rutabaga, seakale, lobularia, sweet alyssum, wallflower, and garden cress.

Heterodera fici. Fig cyst nematode; on fig in Florida and California.

Heterodera glycines. Soybean cyst nematode; causing yellow dwarf disease. An immigrant from Japan and Korea, first noted in North Carolina in 1954; then it spread to Arkansas, Florida, Illinois, Kentucky, Louisiana, Mississippi, Missouri, Tennessee, and Virginia. Plants are yellow, stunted; roots are small and dark with few or no bacterial nodules but with lemon-shaped brown cysts clearly visible. This nematode reproduces only in roots of lespedeza, vetch, tomato, and bean, besides soybean, but the cysts occur as contaminants of narcissus bulbs and gladiolus corms grown in infested soil and may be so disseminated.

Infested areas are under federal and state quarantines. Soil fumigation temporarily reduces nematode populations and increases plant growth and yield.

Heterodera humuli. Hop cyst nematode; on bean, pea, and cucumber.

Heterodera mothi. Cyst nematode; on nutsedge.

Heterodera punctata (*Punctodera punctata*). Grass cyst nematode; found on wheat and small grains, also associated with bent grasses in North Dakota, Michigan, and Minnesota, and turf grass in New Jersey.

Heterodera rostochiensis (*Globodera rostochiensis*). Golden nematode; on white potatoes, also eggplant, tomato, and other members of the Solanaceae, but not on tobacco. It was first discovered in the United States on Long Island in 1941, and it was kept there, by a rigorous quarantine, until 1967, when it was found at a single location in upstate New York. In 1968, it was found on a potato farm in Delaware. Known as "potato sickness," the disease has been serious in the British Isles for many years. Crops do not show much damage until heavy populations have built up in the soil; then there is midday wilting, stunting, poor root development, early death, with up to 85% reduction in potato yield. The eggs live in the soil inside cysts barely visible to the naked eye. Each may contain up to 500 eggs, and some hatch one year, some another. Cysts have remained viable 17 years. In spring, when soil temperature is around 60°F, a chemical given off by potato or tomato roots stimulates hatching, and the larvae (which have had a first molt inside the egg) leave the cysts and migrate to host plants, entering the roots. The females become stationary, swell to pear shape, and break through the roots, remaining attached by a thin neck. The cylindrical males work out of the roots and cluster around to mate with the females. Eggs are formed, and the dead female becomes the cyst, first white, then gold, orange, finally brown. Cysts detached from roots remain in the soil or may be spread in potato bags, crates, machinery, even in trouser cuffs of farm workers. Lily-of-the-valley pips, cacti, and other plants intercepted at quarantine have had golden nematode cysts in fragments of soil around the roots.

Control. A quarantine restricts movement of potatoes, nursery stock, root crops and top soil from infested land. Healthy potatoes are sold in paper bags to prevent reinfestation from secondhand burlap bags. The Peconic strain of potato is said to be resistant; Rosa, Elba, and NY71 are also resistant.

Heterodera schactii. Sugar-beet nematode; occurring in sugar-beet areas from California to Michigan, also infesting table beets and crucifers—cabbage, broccoli, rape, turnip, rutabaga, and radish. The females, numerous white specks clinging to roots, contain 100 to 600 eggs. Slender larvae puncture root cells with their strong stylets and pass through three molts inside the roots. The wormlike males then leave the roots to search for the flask-shaped females, which are attached to the roots only by their heads. Eggs are deposited in a gelatinous mass. These soon hatch to start other generations, but the females die with more eggs inside their bodies, which turn brown and become cysts. Eggs inside cysts may remain viable 5 or 6 years. Control depends on a very long crop rotation or soil fumigation.

Heterodera tabacum. Tobacco cyst nematode; reported from Connecticut on tobacco, tomato, and other solanaceous plants, but not potato; also reported on Jerusalem cherry, eggplant, and pepper in Virginia.

Heterodera trifolii. Clover cyst nematode; on clover and other legumes except peas. Spinach, beet, soybean, and carnation are minor hosts. Cysts are brown, lemon-shaped.

Heterodera zeae. Corn cyst nematode; on sweet corn, field corn, and barley.

Hoplolaimus

Hoplolamidae. Lance nematodes, somewhat migratory, some species tropical or subtropical, of moderate length; strong stylet with basal knobs; often in a spiral or C-shape position.

Hoplolaimus galeatus (*H. coronatus*). Crown-headed lance nematode; widespread. On turf grasses, zoysia, nursery crops, corn, sugarcane, citrus, tomato, sweet potato, pine seedlings, and carnation. This species may feed from the outside, burying the head only, or it may enter the root completely, destroying the cortex, which is sloughed off, and feeding on the phloem.

Hoplolaimus uniformis. On various ornamentals, reported from Rhode Island.

Hypsoperine

Heteroderidae. Similar to *Meloidogyne*, the root-knot nematode, but female body oval rather than pear-shaped.

Hypsoperine graminiae. Described in 1964 from roots of grass and forming inconspicuous galls, primarily on members of the Gramineae. St. Augustine grass may become chlorotic and die. Bermuda grass may decline. Also present on zoysia.

Longidorus

Dorylaimidae. Needle nematodes; relatively large ectoparasites with long, slender stylet; similar to *Xiphinema* but not causing galls.

Longidorus elongatus. On grape, causing necrosis and excessive root branching.

Longidorus maximus. Reported associated with celery, leek, lettuce, and parsley.

Longidorus sylphus. Thorne's needle nematode; fairly common in the Pacific Northwest, causing severe stunting of peppermint.

Meloidodera

Heteroderidae. A new genus, a link between *Heterodera* and *Meloidogyne*; eggs are retained in the female, but there is no distinct cyst stage; second-stage larvae invade roots but no galls are formed.

Meloidodera floridensis. In roots of slash pine in Florida.

Meloidoderita

Meloidoderita sp. On grapes. Males developed in soil and have a degenerate esophagus that lacks a stylet.

Meloidogyne

Heteroderidae. Root-knot nematodes, formerly considered one species, now known to be several, distinguished by slight morphological differences such as striations, perineal pattern of the tail, type of galls formed, host preferences, and somewhat by locality. Females are white, pear-shaped to sphaeroid with elongated necks, slender stylets with well-developed basal knobs; males are slender, wormlike. Females deposit eggs in a gelatinous mass, and the body is not turned into a cyst as in *Heterodera* (see Fig. 41).

Root knot is the best known nematode disease, with over 2,000 plant species susceptible to one or more forms of Meloidogyne. Root knot was first reported in England, in 1865, on cucumbers; in 1876, it was recorded in the United States on violet. Infected plants are stunted; they often wilt, turn yellow, and die. The chief diagnostic symptom is the presence of small or large swellings or galls in the roots (see Fig. 42). They are nearly round or long and irregular, but they are an integral part of the root and cannot be broken off, which differentiates them from beneficial nodules, formed on legume roots by nitrogen-fixing bacteria that can be readily broken off.

Root-knot nematodes occur in practically every state. We used to think they were killed by northern winters, but some species can survive extreme cold. They do have fewer generations in the North and do not build up such large populations as in southern sandy or peat soils. Grasses and grains are about the only plants immune or resistant to root knot.

The long, thin, young larva takes form inside the egg, breaks out, and migrates through the soil to a root. It moves in to the axial cylinder and there becomes sedentary. It injects a secretion of its esophageal glands into the

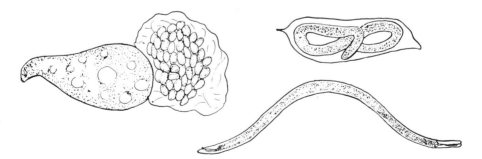

Figure 41. Root-knot nematode; pear-shaped female with egg sac; encysted young larva; and wormlike adult male.

Figure 42. Root-knot nematode galls on potato.

tissue by means of its short buccal stylet, which stimulates the formation of three to five giant cells around the injection point. The nematode absorbs its food from these nectarial cells the rest of its life. As it feeds, the larva swells rapidly into a sausage-shaped body, which, in the female, becomes whitish and pear-shaped, large enough to be just visible to the naked eye. The male changes into a threadlike cylindrical form, folded inside the larval molt, from which it finally escapes.

The female deposits its eggs in an extruded yellow-brown jelly. There may be up to 3,000; the average is nearer 300 to 500. The larvae develop inside the eggs and become free in the soil when the host root cracks or decays. They may attack the same root in a new place or another root. At 80° F a generation takes only 25 days; at 67° F the cycle averages 87 days, and below 55° F activity ceases. Root-knot nematodes may be injurious by their feeding punctures even if typical swellings are not formed. Some have been shown to increase *Fusarium* and bacterial wilts, and they almost surely complicate the crown-gall problem.

Control. Rotation of crops may be practical for species with a narrow host range, and some varieties of vegetables, fruits, and ornamentals have been developed resistant to particular species. In general, soil fumigation before planting is the best control; see page 41 for satisfactory chemicals. These usually kill larvae free in the soil but not all of those inside knots.

Meloidogyne arenaria. Root-knot nematode; causes stunting and root gall on dwarf gardenia, compacta holly, and japanese boxwood.

Meloidogyne arenaria thamesii. Thames' root-knot nematode; occurring naturally in Florida, on Chinese silk plant (Boehmeria); found elsewhere in greenhouses. Also reported on tomato and scindapsus.

Meloidogyne chitwoodi. Columbia root-knot nematode; has been associated with alfalfa, potato, tomato, sugar beet, wheat, and corn, which is significant because wheat and corn are commonly grown in rotation with potato and sugar beets to reduce *M. hapla*. Both monocotyledonous and dicotyledonous plant species are good hosts, indicating a wide host range for this nematode.

Meloidogyne graminicola. Rice root-knot nematode; on purple nutsedge and yellow nutsedge.

Meloidogyne hapla. Northern root-knot nematode; common on many outdoor crops in the North and in florist and nursery stock. Hosts include abelia, anoda, barberry, bean, blueberry, boxwood, California laurel, cantaloupe, carrot, cherry, clematis, clovers, cocklebur, corn, cress, cucumber, dog fennel, eggplant, escarole, forsythia, geranium, germander, gladiolus, grape hyacinth, goldenchain, jimsonweed, kale, lettuce, marigold, mock orange, morning glory, mulberry, myrtle, mustard, parsnip, pachysandra, pansy, peanut, peony, pepper, periwinkle, potato, privet, rose, sainfoin, sequoia, soybean, spirea, spurge, strawberry, sugar beet, tomato, velvetleaf, viburnum, wheat, and weigela.

This species is a particular pest of peanut and is probably the most important nematode on strawberries. It causes galls, reduces growth of main roots, resulting in excessive branch roots; plants are stunted and may die. Injury is more serious in sandy soils. Yields have been increased by using granular Nemagon, mixed with fertilizer, as a side-dressing or by planting in fumigated beds. Rotation with corn and some grains may be practical.

Meloidogyne incognita. Root-knot nematode; on sequoia, society garlic, sweet potato, and jacquemontia.

Meloidogyne incognita acrita. Cotton root-knot nematode; a southern native associated with many plants—forage crops, bean, cabbage, cantaloupe, carrot, celery, chard, corn, cucumber, grape, lettuce, pepper, potato, radish, rhubarb, soybean, New Zealand spinach, squash, tobacco, tomato, turnip, watermelon; also on azalea, boxwood, camellia, calthea, coleus, collinsia, daylily, gardenia, hibiscus, hollyhock, iris, India love grass, nephthytis, roystonea, schefflera, and scindapsus. It was reported on iris in 1955, from New York and Texas, the first instance of rhizomatous iris credited as host to a root-knot nematode. Tips of leaves turn yellow, then brown, with whole leaf gradually dying. There are some resistant soybean varieties, and asparagus, strawberry, and peanut can be used in a rotation.

Meloidogyne incognita incognita. Southern root-knot nematode; native to the South and common there, but overwintering as far north as New Jersey. It is the most important root-knot species on peach; it is also recorded on abelia, banana, bean, carrot, coleus, corn, cucumber, daylily, eggplant, gardenia, geranium, hibiscus, onion, okra, sweet potato, pepper, tomato, watermelon, and willow. It causes stunting and chlorosis of gardenia, but does not occur on peanuts or strawberries, and these may be used in a rotation. Resistant crotalaria and oats can be used as cover crops in peach orchards, and some peach understocks are highly resistant. Nemagon injected into soil around peach trees has given control.

Meloidogyne javanica. Javanese root-knot nematode; common in southern peach orchards and nurseries, widespread in Georgia on peaches such as Yunnan and Shali that are otherwise resistant to root knot. Found in northern greenhouses. May be associated with azalea, bean, beet, cabbage, calendula, carrot, carnation, corn, *Cocos plumosa*, cucurbits, eggplant, impatiens, radish, sequoia, snapdragon, soybean, and tomato. Resistant peanut, strawberry, cotton, and pepper can be used in the rotation.

Naccobus

Tylenchidae. Males wormlike; females swollen in the middle, saclike, with a short, narrow tail; eggs extruded in a gelatinous matrix or held within the body; stylet with small basal knobs; endoparasites.

Naccobus batatiformis. False root-knot nematode; important in western sugar-beet fields, also present on garden beets, cacti, carrot, crucifers, gaillardia, lettuce, and salsify. Root galls are similar to those caused by *Meloidogyne*, and may be fairly large.

Naccobus dorsalis. Reported on heronsbill (erodium), probably on other hosts.

Nacobbodera

Nacobbodera chitwoodi. Reported on fir and spruce.

Nothanguina

Nothanguina phyllobia. Foliar nematode; on nightshade.

Paratylenchus

Criconematidae. Pin nematodes, related to ring nematodes but thinner, primarily ectoparasites; minute; cuticle finely annulated; female with long stylet, body ventrally curved but too short for a spiral.

Paratylenchus anceps. On California-laurel.

Paratylenchus dianthus. Carnation pin nematode; first reported on carnation in 1955 in Maryland, now well distributed through the Northeast.

Paratylenchus (*Cacopaurus*) **epacris.** California sessile nematode; associated with a decline of black-walnut trees in California.

Paratylenchus hamatus. Celery pin nematode; fig pin nematode; on azalea, bean, boxwood, celery, chrysanthemum, clover, corn, fig, geranium, gladiolus, turf grasses, hemlock, holly, horseradish, iris, oak, onion, parsley, peach, pieris, pine, mountain pink, prune, rose, soybean, strawberry, and tomato. This species is responsible for celery losses in New England, plants being stunted and chlorotic, and with decline of fig in California, symptoms being chlorosis and leaf drop and undersized fruit. On mint it has caused one-third reduction in growth. The nematodes can be starved out of celery fields by a 2-year rotation with lettuce and spinach.

Paratylenchus projectus. Reported from Maryland in 1955 on pasture grasses, also found on roots of alfalfa, bean, clover, corn, and soybean. A serious decline of celery and parsley in New Jersey was attributed to this species. Preplanting fumigation has produced a striking growth response.

Pratylenchus

Tylenchidae. Lesion nematodes, sometimes called root-lesion or meadow nematodes, widely distributed migratory endoparasites; males and females wormlike, small, 0.3 to 0.9 mm, with short stylet. Conspicuous necrotic spots are formed on roots, and eggs are deposited in root tissues or in soil. Feeding punctures afford entrance to pathogenic microorganisms.

Pratylenchus brachyurus (*P. leiocephalus*). Godfrey's meadow nematode; smooth-headed meadow nematode; on corn, grasses, cereals, asparagus, avocado, citrus, collinsia, dogwood, peanut, pieris, pine, pineapple, potato, soybean, strawberry, and tomato. Unsightly lesions are formed on peanut shells, and the nematode survives through curing. Preplanting soil fumigation has increased yield.

Pratylenchus coffeae (*P. musicola*). Associated with strawberry black root and decline, in Arkansas.

Pratylenchus crenatus. Associated with many kinds of nursery plants.

Pratylenchus hexincisus. Described from corn roots, in Maryland.

Pratylenchus minyus. On pear and grape, in California.

Pratylenchus penetrans. Cobb's meadow nematode; associated with decline in alfalfa, amaranth, apple, arborvitae, azalea, bean, blackberry, blueberry, boxelder, cabbage, carrot, cedar, celeriac, celery, cherry, chrysanthemum, clover, corn, cucumber, eggplant, fern, garden balsam, gayfeather, gladiolus, grass, hemlock, holly, horseradish, lettuce, lily, maple, mock orange, onion, parsnip, peach, pear, peony, pepper, pieris, pine, mountain pink, phlox, plum, peach, pear, potato, raspberry, rose, safflower, sequoia, soybean, spinach, spirea, strawberry, sweet potato, tobacco, tomato, turnip, zinnia, and yarrow. This species is distributed throughout the United States.

Apples have necrotic black or amber spots on white rootlets; roots may be stunted and distorted; tree vigor is reduced; leaves are small. The disease has been called "little leaf" and "rosette." The nematodes invade cortex only; secondary fungi may play a part in symptoms. Control measures include root dips and soil fumigation, hot-water treatment for strawberry stock plants, and removal of all old roots on lilies before forcing. Marigolds produce a chemical toxic to nematodes and can be used in rotations.

Pratylenchus pratensis. De Man's meadow nematode; important on grasses, strawberry, lily, and narcissus; reported on a great many other hosts, but there may have been some confusion with other species.

Pratylenchus safaenis. On soybean, corn, cotton, millet, rice, and sorghum.

Pratylenchus scribneri. Scribner's meadow nematode; first reported on potatoes in 1889 in Tennessee. Associated with amaryllis, hibiscus, strawberry in Florida, roses in California, and in New Jersey, in clover, corn, dahlia, orchids, parsnip, peach, potato, raspberry, rose, soybean, and tomato.

Pratylenchus subpenetrans. Described from pasture grasses, in Maryland.

Pratylenchus thornei. Thorn's meadow nematode; on wheat, other grains, and grasses.

Pratylenchus vulnus. Walnut meadow nematode; described in 1951 from California as an important parasite of walnut and rose on the West Coast, also present elsewhere. It may affect avocado, boxwood, almond, fig, forsythia, gayfeather, apricot, citrus, peach, plum, raspberry, loganberry, rose, sequoia, strawberry, Japanese boxwood, spiny Greek juniper, blue rug juniper, and yew. Soil fumigation has increased growth of roses by 400%.

Pratylenchus zeae. Corn meadow nematode; associated with corn, also alfalfa, bean, chrysanthemum, cucumber, grasses, pea, phlox, potato, soybean, tobacco, and tomato.

Pratylenchus spp. Lesion nematode; probably as widespread as a group as root-knot nematodes and even more serious, though less readily recognized. The brown or black root condition usually comes from secondary fungi entering and rotting the roots after cells are pierced and torn by the nematodes. In boxwood and other ornamentals there is often a brush or witches' broom of new surface roots to compensate for old roots sloughed off. First symptoms are usually yellow, black, or brown lesions on fine feeder roots. Boxwood becomes sickly, stunted; foliage is dark brown to orange, sometimes drops; some branches may be killed. Tuberous begonias may be heavily infested in roots and tubers, with poor growth. Where possible, fumigate soil before planting. Help plants to recover from root injury by mulching, adequate watering, and feeding.

Radopholus

Hoplolaimidae. Burrowing nematodes; endoparasites with entire life cycle inside plants, including copulation and egg deposition. Male and female wormlike, with short stylet. Female with flat lip region, two ovaries; 0.6 mm long; male with rounded lip region.

Radopholus similis. Burrowing nematode; associated with **spreading decline of citrus.** The most important citrus disease in Florida; a subtropical species, first reported in 1893 from banana roots in the Fiji Islands. Citrus decline was known for many years before the nematode connection was made in 1953. This species is also responsible for **avocado decline** and in 1963 was reported as infesting 237 plants in many families. Possible hosts include acanthus, allamanda, aluminum plant, calathea, Barbados cherry, banana, castor bean, cocculus, hibiscus, Japanese boxwood, Japanese persimmon, ixora, jacobinia, ginger lily, loquat, *Momordica*, pandanus, peperomia, philodendron, periwinkle, pothos, podocarpus, palms, guava, as well as corn, pepper, tomato, and other vegetables, and various trees. Asparagus, marigold, and crotalaria are among the few nonhosts. The burrowing nematode has been found in Louisiana as well as central Florida.

The nematodes enter the cortical parenchyma of young succulent roots just back of the tip, and form burrows, leaving behind avenues of infection for soil fungi and bacteria. Infected trees seldom die outright, but have poor growth and cease to produce a profitable crop. The disease spreads in all directions from an infected specimen, but somewhat unevenly, the distance ranging from 25 to 200 feet in a year, averaging about 50 feet. Long-distance spread is by transplants from nurseries.

Control. Living trees, once infected, cannot be restored to vigor. Diseased trees in quarantine areas are pulled and burned, including two trees beyond those known to be infested in an orchard, and the soil is treated with D-D. Bare-rooted nursery stock can be treated with hot water, 10 minutes at 122° F. After the "pull and treat," nonhosts are grown for 2 years before citrus is replanted. There is some hope of resistant varieties.

Rotylenchulus

Tylenchidae. Reniform nematodes, partially endoparasitic root parasites. Female swollen, kidney-shaped; two ovaries; male wormlike, unable to feed.

Rotylenchulus reniformis. Reniform nematode; first described from pineapple roots in Hawaii, now found in Florida and other warm states on turf, cotton, peanut, sweet potato, tomato, gardenia, jacquemontia, and other ornamentals. The head of the female, with elongated neck, goes in the cortical parenchyma of the rootlet, and her kidney-shaped body projects outside. It is covered with a gelatinous material containing eggs and larvae, so that soil particles adhere.

Rotylenchus

Hoplolaimidae. Spiral nematodes, worldwide in temperate and tropical climates; mostly ectoparasitic but partially endoparasitic, somewhat migratory; body wormlike but held in shape of a spiral; long stylet; female with two ovaries; 0.5 to 1 mm long.

Rotylenchus buxophilus. Boxwood spiral nematode; associated with boxwood decline in Maryland and nearby states; also found with barberry, privet, and peony. The roots have minute brown spots, and the root system is much reduced.

Rotylenchus uniformis. Reported on many ornamental trees and shrubs in New Jersey nurseries.

Scutellonema

Hoplolaimidae. Spiral nematodes, similar to *Rotylenchus*.

Scutellonema blaberum (*Rotylenchus blaberus*). West African spiral nematode; on banana, yam, red spider lily, and African violet.

Scutellonema brachyurum. Caroline spiral nematode; working at crown and roots of African violet, destroying root cells, depositing eggs in cortical tissues. Also on amaryllis.

Scutellonema bradys. Yam nematode.

Scutellonema christiei. Christie's spiral nematode; common on lawn grasses in Florida, also reported on apple and grasses in Maryland and West Virginia.

Sphaeronema

Sphaeronema sp. Decline; of Alaska cedar.

Tetylenchus

Tylenchidae. Male and female wormlike, stylet short.

Tetylenchus joctus. On blueberry.

Trichodorus

Dorylaimidae. Stubby-root nematodes; migratory ectoparasites with wide host ranges; thick-bodied, cylindrical; 0.5 to 1.5 mm long; smooth cuticle; tail short, bluntly rounded; long, slender stylet is a grooved tooth.

Trichodorus allius. Reported reducing onion yield in Oregon.

Trichodorus christiei. Christie's stubby root nematode; widespread in southern states but also present elsewhere feeding on many plants in many different plant families. These include azalea, avocado, blueberry, bean, beet, cabbage, citrus, corn, cranberry, chayote, onion, potato, squash, tomato, and turf grasses—St. Augustine, Bermuda, and zoysia. On tomato there is general stunting and formation of short lateral roots. The stubby effect is apparently caused by a secretion and not just mechanical piercing by the stylet; there is reduced cell multiplication. The host list is too long for crop rotation to be practical, and soil fumigation is not as effective as with some other species. Asparagus and poinsettia are nonhosts, and asparagus has a nematicidal effect. This nematode and some other *Trichodorus* species are vectors of tobacco rattle virus, cause of potato corky ringspot.

Trichodorus obtusus. Cobb's stubby root nematode; on Bermuda grass.

Trichodorus pachydermis. Seinhorst stubby root nematode; on turf and dahlia.

Trichodorus primitivus. On azalea.

Tylenchorhynchus

Hoplolaimidae. Stylet nematodes, sometimes called stunt nematodes, primarily ectoparasites, somewhat migratory, common in roots of nursery stock and cultivated plants. Male and female wormlike, 0.6 to 1.7 mm long; stylet variable in length with well-developed knobs; female has rounded tail, two ovaries; male tail is pointed; cuticle coarsely annulated.

Tylenchorhynchus capitatus. Causes stunting and chlorosis of pepper, bean, tomato, and sweet potato.

Tylenchorhynchus claytoni. Tesselate stylet nematode; common and widespread through southeastern and eastern states. Associated with andromeda, apple, arborvitae, azalea, bean, blueberry, boxwood, broccoli, cherry, cereals, clovers, corn, cranberry, dogwood, forsythia, grape, grasses, hemlock, holly, lettuce, lilac, maple, peach, pine, potato, raspberry, rhododendron, soybean, strawberry, tomato, tulip tree, veronica, willow, and yew. Azaleas may be severely injured, with reduced root system, short twigs, leaf chlorosis, and increased susceptibility to winter injury. Soil treatment with the standard fumigants and also with systemics gives adequate control. Nonhosts include peanut, pepper, cucumber, and crotalaria.

Tylenchorhynchus dubius. Reported on cereals, grasses, clovers, also azalea and carnation.

Tylenchorhynchus martini. Sugar-cane stylet nematode; on sugar cane, rice, soybean, and sweet potato.

Tylenchorhynchus maximus. On turf.

Tylenchulus

Tylenchulidae. Female sedentary, with elongated anterior portion entering the root and swollen, flask-shaped posterior outside the root; well-developed stylet with large basal knobs; male remains small, cylindrical; does not feed.

Tylenchulus semipenetrans. Citrus nematode; first noted in California in 1912, now widespread in citrus regions; important in California and Florida, present also in Arizona and Texas. Hosts other than citrus include olive, persimmon, grape, and lilac. Citrus trees exhibit a slow decline resulting from reduced root activity. Symptoms also include twig dieback, chlorosis and dying of foliage, wilting under moisture stress, and reduced fruit production. Control measures include resistant rootstock, and hot-water treatment of nursery stock, 25 minutes at 113° F or 10 minutes at 116° F.

Tylenchus

Tylenchidae. This genus, described in 1865, originally contained most species with stomato stylets, but many of these have been transferred to other genera. Those left are common in soil around plants but apparently not important parasites.

Xiphinema

Dorylaimidae. Dagger nematodes; very common migratory ectoparasites; very long, males and females both wormlike; long, slender stylet from a bottle-shaped esophagus.

Xiphinema americanum. American dagger nematode; a native, first described in 1913 from specimens taken around roots of corn, grasses, and citrus trees. Found all over the United States associated with many kinds of plants, including ash, azalea, bean, boxwood, clover, camellia, citrus, dogwood, elm, geranium, melon, palm, pea, pecan, peach, pepper, pine, poplar, rose, soybean, strawberry, sweet potato, viburnum, and walnut. In addition to its causing decline and sometimes winterkill by its feeding on roots, this species is believed to transmit tomato ringspot, peach yellow bud mosaic, and grape yellow vein viruses and to increase the incidence of Cytospora canker on spruce. Dagger nematodes may be introduced into greenhouses with virgin soil from the woods and may destroy almost all the feeder roots of plants. There may be very high soil populations.

Xiphinema bakeri. Dagger nematode; on sequoia.

Xiphinema chambersi. Chamber's dagger nematode; causing a decline in strawberries, with stunting and sunken, reddish brown root lesions.

Xiphinema diversicaudatum. European dagger nematode; a proven pathogen of rose, strawberry, peanut, fig, tomato, soybean, garden balsam, and other plants. This species is very common in commercial rose greenhouses, reducing vigor, causing chlorosis. Galls are formed on rose roots; they are similar to root-knot galls but more elongate and nearer the tip of the root, causing it to curl. Cleaning up a greenhouse infestation means disposal of all plants in a bed, careful sterilization of soil, and replanting with clean stock.

Xiphinema index. California dagger nematode; reported on grape, fig, and rose. Feeding in root tips causes a terminal swelling with angling of main roots, death of lateral roots.

Xiphinema radicicola. Pacific dagger nematode; reported on oak, in Florida.

Nematicides

Bedrench, allyl alcohol and ethylene dibromide. Used for a seedbed drench with plenty of water; wait 2 weeks before planting.

Chloropicrin, trichloronitromethane, tear gas. Larvacide; Picfume. Use special applicator to inject 6 to 8 inches deep at 10-inch intervals, in moist soil, temperature above 60°F. Apply a water seal immediately; wait 2 to 4 weeks before planting. Do not treat soil near living plants; wear gas mask while working.

Cynem, O,O-diethyl O-2-pyrazinyl phosphorothioate. Zinophos (discontinued by American Cyanamid); Nemaphos. For preplanting treatment and sometimes around living plants; also used as a bare-root dip. Available as liquid, dust, or granules. Highly toxic; follow all precautions.

DBCP, dibromochloropropane (discontinued by Anevac Chemical). Nemagon, Fumazone, for pre- and postplanting. Used as a drench for turf, shrubs, and some perennials; available as liquid or granules and relatively safe for home garden use. Apply in furrow or holes 6 inches deep for preplanting application; no cover is necessary.

D-D Soil Fumigant, mixture of dichloropropanes and dichloropropenes (a similar compound sold as Vidden-D [discontinued by Dow Chemical Co.]). Preplanting soil treatment for home gardens or commercial growers. Make furrows 6 to 8 inches deep and 12 inches apart, and dribble in 1 pint D-D to 150 to 175 feet. Use a fruit jar with two holes punched in metal cap as an applicator; stop, rake over, and tamp soil after each 100 feet of treatment. No other cover is necessary; wait 2 to 4 weeks before planting; do not apply within 30 inches of living plants.

Di-Syston, O-O-diethyl S-2(ethylthio) ethyl phosphorothioate, systemic insecticide sometimes used as a soil drench or bare-root dip for nematodes. Highly toxic, not recommended for amateurs.

DMTT, 3,5-dimethyltetrahydro-1,3,5,3H-thiadiazine-2-thione. Mylone. Preplanting drench, powder, or granules. Disc, rake, or rototill into soil; use a water seal or plastic cover. Wait 3 to 4 weeks before planting.

Dorione, mixture of dichloropropene (Telone) and ethylene dibromide. Use as D-D but not where onions will be grown.

EDB, ethylene dibromide. Dowfume W-85; Bromofume. Preplanting treatment, 6 to 8 inches deep at 10- to 12-inch intervals, preferably in fall.

Methyl Bromide. Dowfume MC-2, Bromo-O-Gas, Brozone; Weedfume, etc. Extremely hazardous, recommended for commercial growers only; usually sold with chloropicrin added as a warning agent. Apply with special applicator under tarpaulin or other gas-proof cover. Dosage 1 to 2 pounds per 100 square feet; wait 7 to 10 days before planting. Do not use on soil planned for carnation, delphinium, salvia, or snapdragon.

MIT, methyl isothiocyanate, Vorlex. Preplanting treatment; apply as chloropicrin with injections 8 inches apart and water seal or plastic cover. Cultivate soil after 4 days; wait 2 to 3 weeks before planting.

Parathion, O,O-diethyl-O-*p*-nitrophenyl thiophosphate. Highly toxic systemic insecticide. Recommended for commercial growers only as a spray to control foliar nematodes on chrysanthemum.

Phorate, O,O-diethyl S-(ethylthio) methyl phosphorodithioate. Thimet. Very poisonous systemic insecticide sometimes used by commercial growers as a bare-root dip or a drench around some living plants, or applied at planting time.

Sarolex. Special formulation of diazinon, used for turf.

SMDC, sodium methyl dithiocarbamate. Vapam. Preplanting treatment that can be applied with a sprinkling can, 1½ to 2 quarts per 100 square feet, and covered with a water seal. Cultivate to prevent soil crusting; wait 3 weeks before planting.

TCTP, tetrachloro thiophene. Penphene (discontinued by Pennwalt Corp.). Available as emulsifiable concentrate or granules for preplant treatment.

Telone, dichloropropene. Similar to D-D Soil Fumigant.

VC-13 Nemacide, O-2,4-dichlorophenyl O,O-diethyl phosphorothioate. Safe to use around living plants, fairly safe for the operator but avoid skin contact and inhaling. For *fallow soil,* use 15 to 25 gallons per acre; work into top 6 inches; wait 2 weeks before planting. Drench *turf* with 1 gallon of 75% emulsifiable concentrate diluted with 25 to 50 gallons of water for 1,600 square feet; water well. For *shrubs,* add 2 to 3 teaspoons to 1 gallon of water and apply to 4 square feet, in holes 9 inches deep, 12 inches apart. For *potting soil,* use 1 teaspoon to 1 quart of water for 1 cubic foot of soil.

Before using any nematicide read the label very carefully, noting not only proper dosage but plants that may be injured by treatment and all safety precautions for the operator.

NONPARASITIC DISEASES

Plants in poor health from one or more environmental conditions far outnumber those afflicted with diseases caused by parasites—bacteria, fungi, and nematodes. When foliage turns yellow from lack of nitrogen, or from unavailability of iron in an alkaline soil, or from lack of oxygen in a waterlogged soil, we call it a physiological or physiogenic or nonparasitic disease. The adverse condition may be continuing, as it is with a nutrient deficiency, or it may be transitory, an ice storm, perhaps, lasting but a day but with resultant dieback continuing for the next two years. It may be chemical injury from injudicious spraying or fertilizing or from toxic substances in the atmosphere. It may be due to a toxin injected by an insect.

Trees and crops can be insured against hail, hurricanes, lightning, and other acts of nature, but not the misguided zeal of gardeners. Years of working in gardens in my own state and visiting gardens in other states from coast to coast have convinced me that plants often suffer more from their owners than from pests and diseases. Azaleas die from an overdose of aluminum sulfate applied to correct acidity, when the original cause of ill health was a too-wet soil. Rhododendron die when a deep, soggy mass of maple or other "soft" leaves is kept around the trunks. Roses die when the beds are edged with a spade and soil is mounded in the center, burying some plants too deeply and exposing roots of others. Seedlings die from an overdose of fertilizer in hot weather. Trees die from grading operations.

Spray injury is exceedingly common, with the gardener thinking the red or brown spots are fungus leaf spots and increasing the chemical dosage until all foliage is lost. Weed killers take their unexpected toll of nearby ornamentals.

Either a deficiency or an excess of plant nutrients can cause a physiological disease. Greenhouse operators and commercial growers in the field must watch nutrition very carefully. The backyard farmer gets along pretty well by using a "complete" fertilizer containing nitrogen, phosphorus, and potassium in large amounts and minor elements in trace amounts. There are kits available for amateur diagnosticians who wish to check soil deficiencies and acidity, but you may prefer to send a soil sample to your state experiment station for a correct interpretation of nutrients and soil acidity. Take a slice through the soil to spade or trowel depth from several places in the garden, mix those samples together, and send a small sample of the mixture.

Acidity, Excess. Soil acidity or alkalinity is measured on a pH scale that runs from 0 to 14. When the number of acid or hydrogen ions balances the number of alkaline or hydroxyl ions, we have pH 7.0 or neutral. Above pH 7.0 the soil is alkaline and may contain free lime; below it, the soil is acid. Few crop plants will grow below pH 3.5 or above pH 9.0. If the soil becomes very acidic, roots are poorly developed and may decay, growth is slow, and foliage is mottled or chlorotic. This result is due either to actual excess of hydrogen ions or to physical structure of the soil and solubility of nutrients.

Most flowering plants, fruits, and vegetables do well in a soil just slightly acidic, in a pH range of 6 to 7 or 6 to 8. Plants flourishing in a very acidic soil, pH 4 to 5, are few: alpines, azalea, arbutus, andromeda, bunchberry, wild calla, camellia, *Chamaecyparis* (white cedar), a few ferns, wild orchids, pitcher plants, galax, and mountain ash. In the pH 5 to 6 list are arbutus tree, azalea, bleeding heart, birch, blueberry, bent grasses, bracken, camellia, Carolina jessamine, *Clarkia*, cranberry, cypress, *Daphne odora* (but not *D. mezerium*, which is in the 6 to 8 group), hemlock, juniper, mountain laurel, some ferns, some orchids, some oaks, pine, rhododendron, sour gum, spruce, silver-bell tree, *Styrax*, strawberry, sweet potato, and yew.

The small kits for home testing of soils include a booklet giving the pH preferences of a long list of plants and the amount of lime required to correct the acidity. This amount varies with the type of soil and the original pH. To bring a sandy soil from pH 4 to above 6 takes only ½ pound of hydrated lime; it takes 2 pounds of lime to effect the same change in a clay soil.

Air Pollution. Polluted air is not confined to cities. Even in the country crops suffer when sunlight plus automobile exhaust produce ozone and other gases. Air pollutants come from smelters, pulp mills, factories, power plants, incinerators, and other sources. Ozone injury is common in pine, resulting in chlorotic and needle mottling, tipburn, blight, needle flecking, and stunting; in tobacco, causing "weather fleck"; in spinach, with oily areas followed by white necrotic spots on upper leaf surface; in grape, with a dark stippling. Other sensitive plants include bean, celery, corn, tomato, carnation, orchid, radish, marigold, and petunias. Some varieties are more susceptible than others. Smog occurs from a chemical reaction of unburned hydrocarbons, as from automobiles, ozone, sunlight, and, usually, thermal inversion. Tremendous losses in California orchid houses come when smog appears when plants are in the budding stage.

Chrysanthemums may be prevented from flowering by ethylene in the atmosphere; tomatoes are also very sensitive. Injury from sulfur dioxide, a product of fuel combustion, is at a high level in the colder months. Foliage has white spots, tips, or margins. Soot particles entering houses from smoke-stacks cause necrotic spots.

Control. For orchids and other high-priced greenhouse crops, air can be passed through a filter of activated charcoal. Taller smokestacks reduce injury from gases and soot. Increasing the vitamin C content of plants by treating them with a substance such as potassium ascorbate may reduce injury from ozone. Installation of purification devices in automobiles and industrial plants may provide some future relief.

Alkali Injury. Some semiarid soils are nearly barren from excess of chemi-cals with a basic reaction. Composition varies, but three common salts are sodium chloride, sulfate of soda, and carbonate of soda; these salts become concentrated at the soil surface with a whitish incrustation. Other soils are black alkali, where the organic matter has been dissolved. Applications of gypsum or sulfur, cultivation, and mulching are correctives.

Alkalinity. Either aluminum sulfate or sulfur, or both mixed together, can be used to reduce the pH for plants doing best in a somewhat acidic soil.

Aluminum Toxicity. This problem is occasional, if aluminum is used in excess. Browning, dieback, sometimes death of azaleas and other plants may occur.

Arsenical Injury. Leaves of peaches, apricots, and other stone fruits are readily spotted or burned with lead arsenate unless lime or zinc sulfate is added as a corrective. There may be similar leaf spotting and defoliation when these tender fruits are grown in old apple land that has accumulated a residue of lead arsenate over a period of years. Even apple trees can be severely injured by arsenical sprays under some conditions.

Baldhead. In beans this problem causes loss of the growing point, due to mechanical injury in threshing seed.

Bitter Pit. On apples this problem is called stippen or Baldwin spot and is characterized by small, circular, slightly sunken spots on fruit, increasing in storage, especially at warm temperatures, most frequent on varieties Jonathan, Baldwin, Spy, Rhode Island Greening. It seems to be related to fluctuation of the moisture supply in soil and increased by abundant rainfall shortly before harvest. On pear, bitter pit is sometimes associated with moisture deficiency; in olives, with overnutrition.

Black End. In pear, the whole blossom end of the fruit may turn black and dry; the disease appears when oriental pear rootstocks are used in poor soil. In walnut, black end of nuts is probably drought injury.

Black Heart. In beets, this problem is generally caused by boron defi-ciency (see below); occasionally it is caused by potassium or phosphorus deficiency. In apple wood it may be freezing injury; in potatoes, lack of oxygen; in celery, fluctuating soil moisture.

Black Root. Defective soil drainage and accumulation of toxins are associ-ated with black roots, but so too are soil fungi and root nematodes.

Blasting. This problem causes influorescence and failure to produce seeds. These symptoms seem associated with extremes of soil moisture, too wet or too dry, at blossom time. Onion blast, prevalent in the Connecticut Valley, appears within a few hours after bright sunshine follows cloudy, wet weather. Leaf tips are first white, then brown.

Blindness. This problem occurs in tulips and other bulbs. Failure to flower may be due to botrytis blight or other disease, but it may come from root failure in dry soil or from heating of bulbs in storage or transit. Too early forcing may result in blindness.

Blossom-End Rot. This problem is very common on tomatoes, also on pepper, squash, watermelon. The tissues at the blossom end of the fruit shrink, causing a dark, flattened or sunken, leathery spot, which may include nearly half the fruit (see Fig. 43). The disease is most common on plants that have had an excess of rainfall in the early part of the season, followed by a period of drought. There are, however, various contributing factors, the most important being a deficiency of calcium, which is needed for synthesis of rigid cell walls of the tomato. Adding calcium oxide to the soil or spraying with 1% calcium chloride has reduced the disease. For home gardens, deep soil preparation, use of a complete balanced fertilizer, and mulching to conserve moisture should help.

Bordeaux Injury. Both the copper and the lime in bordeaux mixture can be injurious to some crops. Cucurbits are stunted, and blossoming and fruit setting are delayed in tomatoes. Red spotting of foliage of roses and apples is followed by yellowing and defoliation. See Copper Spray Injury; Lime-Induced Chlorosis.

Boron Deficiency. A small quantity of boron is required for normal growth of most plants. For some there is not much leeway between necessary and toxic amounts; other plants require or tolerate large amounts. Deficiency symptoms vary with the crop.

Fruit trees. Internal and external cork of apples, dieback, rosette; dieback,

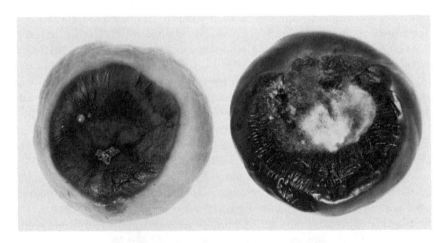

Figure 43. Blossom-end rot on tomato.

blossom blight of pear; stunting, excessive branches, internal necrosis of peaches. Apple leaves on terminal shoots turn yellow, are convex with red veins; twigs die back from tip; dwarfed, thickened, brittle leaves are in tufts at nodes; internodes are abnormally shortened. Fruit has dry corky lesions throughout the flesh or diffuse brown lesions and bitter taste. McIntosh, Baldwin, Rome, Northwestern Greening, and Jonathan exhibit external cork with severe russeting of surface. Control by applying borax, 1 ounce per each inch of diameter of tree trunk, in a 1-foot band outside the drip of the branches. Apply only once in 3 years, and reduce the amount by half for peaches and other stone fruits and for very sandy soils.

Beets, turnips, other root crops. Black heart, brown heart. Roots have dark spots; plants are gradually stunted and dwarfed; leaves are small, variegated, twisted. The interior of the beet or turnip has a dark brown to nearly black water-soaked area, sometimes with a hollow center. The amount of borax that can be added without injury depends on type of soil and moisture content.

Celery. Cracked stem. Leaves have a brownish mottling; stems are brittle, cracked with brown stripes.

Lettuce. There is malformation of young leaves, death of growing point.

Ornamentals. Terminal flower bud dies; top leaves are thick and brittle. Application of boron in fritted form has prevented splitting in carnations, and has increased flower production in greenhouse roses.

Boron Toxicity. Retardation or prevention of germination, death or stunting of plants, bleaching or yellowing of tops, disappearance of color along midrib and veins—all are indications of excess boron. Beans are extremely sensitive to boron, with injury from as little as 4 pounds borax broadcast per acre. If borax has been used for root crops, boron-tolerant cabbage should follow before beans in the rotation.

Brown Bark Spot. This problem occurs in fruit trees. Perhaps it is the result of arsenical injury from residue in the soil.

Brown Heart. This problem occurs in turnip, cabbage, and cauliflower. See Boron Deficiency.

Bud Drop. In sweet pea very young flower buds turn yellow and drop off when there is a deficiency of phosphorus and potassium during periods of low light intensity. Water sparingly at such periods; avoid excess of nitrogen. Gardenias often drop their buds when taken from greenhouses to dry homes, but there is also bud drop in greenhouses with high soil moisture, high temperature, and lack of sunlight in winter.

Calcium Chloride Injury. Trees may be damaged when this dust-laying chemical is washed off country roads or driveways down to roots.

Calcium Deficiency. All plants require calcium, which is built into walls of cells, neutralizes harmful by-products, and maintains a balance with magnesium and potassium. Calcium is leached out of the soil as calcium carbonate and should be replaced by adding ground limestone, or dolomite (calcium magnesium carbonate), or gypsum (calcium sulfate), which does not increase the pH of the soil.

In fruits, calcium deficiency shows first in the roots, which are short and stubby with a profuse growth behind the tips that have died back. Basal immature peach leaves sometimes have reddish discolorations, and twigs may die back. Corn and legumes require large amounts of calcium, which may become unavailable under conditions of high soil acidity.

Catface. This disease results in fruit deformity, due to insects or growth disturbances.

Chlorine Injury. A tank of chlorine gas for the swimming pool carelessly opened too close to trees and shrubs causes foliage browning and sometimes death. Leaf margins are sometimes killed by chlorine gas from manufacturing processes.

Chlorosis. Yellowing or loss of normal green color may be due to deficiency of nitrogen, magnesium, or manganese. Occasionally boron deficiency or toxicity, insufficient oxygen to the roots in a waterlogged soil, or alkali injury may cause chlorosis, but in the majority of cases, and particularly with broad-leaved evergreens, it occurs because iron is unavailable in an alkaline soil. See Iron Deficiency.

Copper Deficiency. This problem results in exanthema or dieback of fruits—apple, apricot, citrus, olive, pear, prune; failure of vegetables on muck soils. Copper deficiency in fruits is widespread in Florida and occurs frequently in California. Leaves are unusually large and dark green, or very small and quickly shed, on twigs that die back, with a reddish brown gummy discharge. Citrus fruits are bumpy and drop, or have insipid flavor and dry pulp. Application of copper sulfate to the soil corrects the deficiency, but often spraying trees once or twice in the spring with bordeaux mixture provides sufficient copper indirectly. Spraying almonds with a copper chelate has prevented shriveling of kernels. Muck or peat soils in New York, formerly unproductive, now grow normal crops of onions and lettuce with the addition of copper sulfate. On copper-deficient Florida soils, many truck crops fail to grow or are stunted, bleached, and chlorotic.

Copper Spray Injury. Some fixed copper sprays are less injurious than bordeaux mixture, but all coppers may be harmful to some plants under some conditions. Foliage spots are small, numerous, reddish, sometimes brown. In peach leaves the centers of the spots may fall out, leaving shot holes. Rosaceous plants follow spotting with yellowing and dropping of leaves. Even mild coppers may be injurious if the temperature is below 55° F, or if the weather continues to be rainy or cloudy. Treated leaves are often harsher than normal and more subject to frost injury. Dwarfing and stunting are important symptoms on many crops, especially cucurbits. Tomato flowering is injured or delayed; apple and tomato fruits are russeted. Tree roots are injured by overflow from pools treated with copper for algae.

Cork. This problem is the result of boron deficiency, in apple.

Cracked Stem. This problem is the result of boron deficiency, in rhubarb and celery.

DDT Injury. This problem occurs in the foliage of some plants—cucurbits particularly, roses occasionally—turning foliage yellow or orange, often with stunting. Certain camellia varieties have been injured when shrubs are under trees sprayed with DDT. Continued spraying with DDT builds up a residue in the soil that may eventually have a toxic effect on the root system, the effect varying with the type of soil and plant.

Dieback. This problem is due to deficiency or excess of moisture, nutrients; winter injury; also cankers, nematodes, borers.

Drought. The effects of a prolonged dry period may be evident in trees and shrubs for two or three years thereafter.

End Spot. This disease occurs in avocado. Unequal maturity in both ends of the fruit seems to be a factor in withering, spotting, and cracking at lower end. Pick promptly, instead of leaving on trees.

Exanthema. This problem is the result of copper deficiency, in fruits.

Frost Injury. This injury is caused by low temperature after plants have started growth in spring or before they are dormant in fall (see Winter Injury for freezing during the dormant period). Yellow color of some leaves in early spring is due to temperatures unfavorable for chlorophyll formation. Some leaves, including those of rose, are reddened or crinkled with frost (see Fig. 44). Blossom buds of fruit trees are critically injured by frost late in spring. In the South, where plants come out of dormancy early, orchard heaters, smudge fires, power fans, and airplanes flying low to stir up the air are all used to help save the crop. Many ornamentals are injured when a long, warm autumn ends in a sudden very cold snap, or warm weather in February or March is followed by heavy frosts. Cracks in tree trunks come from such temperature fluctuations.

Gas Toxicity. Illuminating gas escaping from aging gas mains causes slow decline or sudden death, depending on the plant. Tomatoes are extremely sensitive and indicate the slightest trace of gas by leaves and stems bending sharply downward. Plane trees develop "rosy canker"—long, narrow cankers

Figure 44. Frost injury on holly.

near the trunk base with inner bark watermelon-pink and swollen. With large amounts of gas escaping, foliage wilts and browns suddenly, followed ʋy death of twigs and branches; with slow leaks, the symptoms appear gradually over a year or two. After the leak is repaired, it is sometimes possible to save trees by digging a trench to aerate the roots, applying large quantities of water, burning out severely injured roots, then replacing soil and feeding to stimulate new growth. Natural gas is, apparently, not as injurious.

Girdling Roots. Unfavorable conditions sometimes deflect roots from their normal course, and one or two may grow so closely appressed to a tree as to almost strangle it. If one side of a tree shows lighter green leaves with tendency to early defoliation, dig down on that side to see if a root is choking the trunk under the soil surface. The root should be severed and removed, then all cut surfaces painted.

Grading Injuries. Many shrubs die when they are planted much deeper than the level at which they were grown in the nursery. Similarly, many trees die when they are covered with fill from house excavations. Roots require oxygen for survival, and a sudden excess of soil cuts off most of the supply. A tree expert should be on hand to give advice before any digging starts— afterward is too late. And if grading means filling in soil around trees, a little well around the trunk is not enough. There must be radial and circular trenches laid with tile, and then crushed stone and gravel, before the top soil goes in place. Consult *Tree Maintenance* by P. P. Pirone (1959) for clear descriptions and diagrams for protecting trees from contractors.

Graft Incompatibility. Lilacs are sometimes blighted from incompatibility of the lilac scion on privet stock. Walnut girdle is due to incompatibility of scions on black walnut roots.

Gummosis. Formation of gum on bark of fruit trees is commonly formed in cases of bacterial canker, brown rot, crown rot, and root rots from soil fungi and in connection with the peach tree borer, but other cases of gummosis seem connected with adverse sites and soil moisture conditions irrespective of parasitic organisms.

Heart Rot. This disease is caused by boron deficiency, in root crops.

Heat Injury. There are many ways in which excessive high temperatures can injure plants, ranging from death to retarded growth or failure of mature flowers and fruit. Sunstroke, outright killing of plants, is a limiting factor in flower and vegetable production in summer in the South. Seedlings, especially tree seedlings and beans, may have heat cankers with stem tissues killed at the soil line. See also Leaf Scorch; Sunscald; Tipburn.

Hollow Heart. This disease is sometimes due to excessive soil moisture.

Hopperburn. Marginal chlorosis, burning and curling of leaves of potatoes and dahlias is due to leafhoppers.

Internal Browning or Cork. This disease is caused by boron deficiency, in apple.

Iron Deficiency. Iron is seldom, or never, actually deficient in the soil, but it is often in such an insoluble form in neutral or alkaline soils that plants

Figure 45. Iron deficiency in chrysanthemum.

cannot absorb it, or it may be precipitated as insoluble iron phosphate where excessive amounts of phosphates are added to the soil. Chlorosis is an indication of the lack of iron, for it is necessary for the formation of chlorophyll, the green pigment (see Fig. 45). In acid soils iron is usually available; in alkaline soils leaves turn yellowish green, often remaining green along the veins but yellowing in interveinal areas. Terminal growth of twigs is small, and the shrub or tree is generally stunted.

To obtain a quick response it is possible to spray leaves with a solution of ferrous sulfate. More lasting is a soil treatment of a 50–50 mixture of ferrous sulfate and sulfur. Rather recent is the use of chelated iron, sold as Sequestrene and under other trade names. In this form the iron cannot be combined with soil elements and remains available to the plant even under alkaline conditions. The solution, prepared according to directions on the package, is poured on the soil around the unthrifty bush, and often the green color returns in a matter of days. Iron chelates are now extensively used for citrus and for ornamentals.

Leaf Scorch. This disease occurs in maple, horse chestnut, beech, walnut, and other trees. Scattered areas in the leaf, between the veins or along the margins, turn light or dark brown, with all the leaves on a branch affected more or less uniformly. The canopy of the tree looks dry and scorched; leaves may dry and fall, with new leaves formed in summer. Lack of fruiting bodies distinguishes scorch from a fungus leaf blotch. It appears during periods of high temperature and drying winds and often after a rainy period has produced succulent growth.

Leaf scorch of Easter lilies has been a problem for years but can be prevented by keeping the pH of soil near 7.0 with lime, adequate nitrogen, but low phosphorus. It may have some connection with root rots.

Leaf scorch of iris has puzzled amateur growers in the past few years; it is more serious in the Southwest but has appeared in gardens elsewhere. Leaves turn bright reddish brown at the tips in spring before flowering, and in a few days the whole fan is scorched and withered, and the roots have rotted with a reddish discoloration (see Fig. 46). Many theories, including nutrition and nematodes, have been advanced, but there is no general agreement as to cause.

Lightning Injury. Trees may be completely shattered or a narrow strip of bark and a shallow layer of wood torn down the trunk. Tall trees or those growing in the open are most likely to be struck. Valuable trees can be protected with lightning conductors, installed by a competent tree expert.

Lime-Induced Chlorosis. Plants are sickly, with yellow foliage, in calcareous soils or near cement foundations. See Iron Deficiency.

Little Leaf. On almond, apricot, avocado, and other fruits. See Zinc Deficiency.

Magnesium Deficiency. Large areas in the Atlantic and Gulf coasts truck crop regions are low in magnesium because of natural lack of magnesium rock, extensive leaching from heavy rainfall, removal of large quantities in crops, and use of fertilizers lacking this element. In tomatoes, veins remain dark green while rest of leaf is yellow or chlorotic. Cabbages have lower leaves puckered, chlorotic, mottled, turning white at the margin and in center. In strawberries, leaves are thin, bright green, then with necrotic blotches. On fruit trees, fawn-colored patches are formed on mature, large leaves, with affected leaves dropping progressively toward the tip. In flowering plants there are a greatly reduced rate of growth, yellowing between veins of lower leaves, sometimes dead areas between veins, sometimes puckering.

Control by using dolomitic limestone, or with fertilizers containing magnesium, or with Epsom salts (magnesium sulfate) around azaleas and other shrubs in home gardens.

Manganese Deficiency. Top leaves become yellow between veins, but even smallest veins retain green color, giving a netted appearance. Lower the pH below 7 and add manganese sulfate to the soil.

Marginal Browning. This problem is the result of potassium deficiency or hopperburn.

Mercury Toxicity. Roses are extremely sensitive to mercury vapor and have been gravely injured when paints containing mercury were used to paint sash bars in greenhouses. Covering the paint with a paste of dry lime sulfur mixed with lime, flour, and water reduces the amount of toxic vapor.

Molybdenum Toxicity. This problem is the cause of whiptail in broccoli and cabbage, and the cause of chlorosis of citrus in Florida and of grapes in Michigan. Citrus leaves have large interveinal yellow spots with gum on undersurface and may fall. Injecting the trunk with sodium molybdenate has corrected the condition quickly. On grapes chlorosis of terminal leaves was

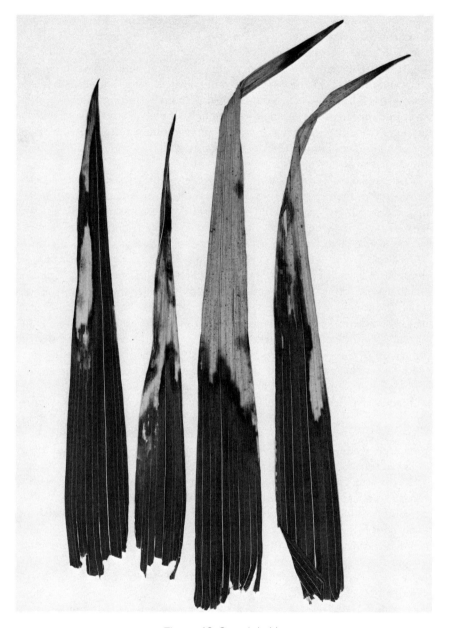

Figure 46. Scorch in iris.

attributed to molybdenum deficiency correlated with nitrogen toxicity and was corrected by adding 0.01 ppm molybdic acid to nutrient solutions.

Mottle Leaf. This problem is the result of zinc deficiency.

Nitrogen Deficiency. Symptoms are paleness or uniform yellowing of leaves and stems, firing or burning of lower leaves, sometimes red pigments along veins, stunted growth, reduced yield with small fruit. Immediate results can be obtained by side-dressing with a quickly available nitrogenous fertilizer, but long-range planning includes use of legumes in the rotation, green manure crops, and balanced fertilizers. Urea is recommended for turf, one application providing a slow release through the season.

Nitrogen Excess. Too much nitrogen leads to overdevelopment of vegetative growth at the expense of flowers and fruit; to bud drop of roses, sweet peas, and tomatoes; and, in high concentrations, to stunting, chlorosis, and death. Excessive nitrogen decreases resistance to winter injury and to such diseases as fire blight, powdery mildew, and apple scab.

Oedema. Small, wartlike, sometimes corky, excrescences are formed on underside of leaves of many plants—cabbage, tomatoes, geraniums, begonia, camellias, and so on. When roots take up more water than is given off by leaves, the pressure built up may cause enlarged mesophyll cells to push outward through the epidermis. This condition is rare outdoors but is found in greenhouses and sometimes on house plants where they have been overwatered. Copper sprays sometimes produce similar intumescences. Camellias frequently have corky swellings on bottom surface of leaves, often due to water relations, sometimes to a spot anthracnose fungus.

Oxygen Deficiency, Asphyxiation. Overwatered house plants and crops in poorly drained low situations often show the same symptoms as those caused by lack of water, for the roots cannot respire properly and cannot take up enough water. Improve drainage; lighten soil with compost and sand; avoid too much artificial watering.

Phosphorus Deficiency. Young leaves are dark green; mature leaves are bronzed; old leaves are mottled light and dark green. In some plants there is yellowing around leaf margins. Stems and leaf stalks develop reddish or purplish pigments; plants are stunted, with short internodes; growth is slow, with delayed maturity. Most complete commercial fertilizers have adequate phosphorus, but it can be added separately in the form of superphosphate. In preparing rose beds, apply a liberal amount at the second spade depth as well as in the upper soil.

Potassium Deficiency. Marginal browning, bronzing, or scorching appears first on lower leaves and advances up the plant, which is stunted. Leaves are often crinkled, curl inward, develop necrotic areas; the whole plant may look rusty. The lack of potassium can be made up with a complete fertilizer containing 5% to 10% potash. Wood ashes also help to supply potassium.

Ring Spot. Yellow rings on African-violet foliage come from breaking down of the chloroplasts when the leaf temperature is suddenly lowered, as in watering with water considerably colder than room temperature.

Rosette. This problem is caused by zinc deficiency in pecan and walnut, and boron deficiency in apple.

"Rust." This term is used by amateur gardeners for any rust discolorations — for a leaf blight of phlox of unknown origin (probably a water relation), a spot necrosis of gladiolus, red-spider injury, and many other troubles that have nothing to do with true fungus rusts.

Salt Injury. Trees and shrubs along the seacoast are injured by ocean spray, and after hurricanes and high winds traces of injury can be found 35 to 40 miles inland. Conifers are usually affected most; they appear damaged by fire, with needles bright yellow, or orange-red. Eastern white pine is very susceptible; Austrian and Japanese black pines, blue spruce, and live oak are highly resistant. Roses have often survived submersion in salt water during hurricanes. Roadside trees, and especially maples, may be injured by salt used on highways during the winter. Either sodium chloride or calcium chloride may be harmful.

Scald. This problem occurs in apples. It is the result of asphyxiation injury to fruit in storage from accumulation of harmful gases; most important when immature fruit is stored without adequate ventilation at too high temperature and humidity. Wrapping fruit in oiled paper or packing with shredded oiled paper, and storage near 32°F, with a high concentration of carbon dioxide at the start, control scald.

Scorch. See Leaf Scorch.

Shot Berry. This problem occurs in grapes and is the result of defective pollination.

Smog Injury. Unsaturated hydrocarbons and ozone in the atmosphere are the cause, with many kinds of plants injured in the Los Angeles area. Tan lesions appear on fern leaves in 24 hours with necrosis in 24 additional hours (see Fig. 47). Many ornamentals and vegetables are injured, with an annual loss $3 million. Spraying carnations in greenhouses with vitamin C prevents sleepiness from smog. Some greenhouses have installed activated-carbon filters for polluted air.

Smoke Injury. The most important agent in smoke injury is sulfur dioxide, a colorless gas with a suffocating odor released from smelters and many industrial processes. Acute smoke injury shows in rapid discoloration of foliage, defoliation, sometimes death. Conifer needles turn wine red, in whole or part, then brown. Leaves of deciduous trees have yellow to dark brown dead areas between veins, with tissue next to larger veins remaining green. Chronic injury results in unhealthy, stunted trees, but less apparent discoloration and defoliation. Roses, grapes, and legumes are seriously injured. Gladiolus leaves appear burned from the tips down.

Control of injurious smoke must be at the source — by filtering, using tall smokestacks, neutralizing the acid gases, or using them in the manufacture of sulfur and sulfuric acid.

Soot Injury. City trees and shrubs acquire an accumulation of soot, the solid residue of smoke, which screens out the sunlight. Evergreens can be

Figure 47. Ozone injury on tobacco.

sprayed with a soapy solution of Calgon (sodium hexametaphosphate), followed by syringing with clear water.

Stigmonose. This disease is the result of dimpling of fruit by insect punctures.

Sulfur Injury. Sulfur sprays and dusts are likely to burn foliage in hot weather, when the temperature is much higher than 85°F. There is often a browning of the tip or margin of leaves. Lime sulfur is injurious to some plants in any weather, russeting peach foliage, causing apple drop, and so forth. When roses or other plants are continuously dusted with sulfur over a period of years, the soil may become too acidic and require lime as a corrective.

Sunscald. Trees with smooth bark are subject to sunscald when trunks or branches are suddenly exposed to the sun, as when the next tree is removed. Young trees are subject to sunscald the first year or two after planting and should have trunks wrapped in burlap or sprayed with a protective wax to prevent the cambium under the thin bark from drying out.

Boxwood foliage is subject to sunscald in spring after winter covering is removed, particularly if this is done on a sunny day with drying winds. Sunscald is common on green tomatoes when fruits are exposed to sun in hot dry weather (see Fig. 48). It happens when foliage is lost through disease or excessive irrigation, or when too much is removed in training tomatoes to a single stem. A yellow or white patch appears on the side of the tomato nearest the sun, often developing into a blister, then into a large, flattened spot with a papery white surface darkened by the growth of secondary fungi and internal decay.

Sunstroke. It is the cause of outright killing in excessive heat.

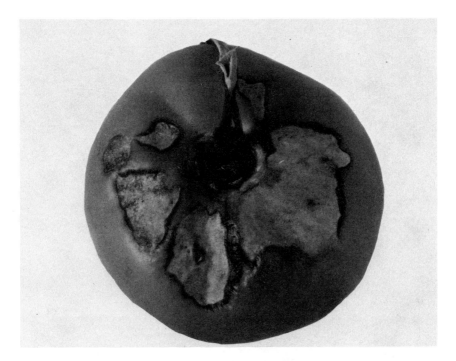

Figure 48. Tomato sunscald.

Tipburn. Potassium deficiency may produce a tipburn, but more often tipburn is a reaction to heat, common in potatoes and particularly in lettuce, which shows marginal browning of leaves and small brown or black spots in tissues near larger veins. A regular supply of moisture and avoidance of excessive fertilization in warm weather reduce tipburn, but more reliance should be placed on growing varieties resistant to summer heat.

Topple. This problem occurs in gladiolus. Toppling over is apparently due to calcium deficiency. It can be reduced by using a spray of 2% calcium nitrate.

Variegation. Chlorophyll deficiency, genetic factors, and virus diseases can produce variegated plants.

Water Deficiency. Practically all of the injury laid to excessive heat or cold is basically due to lack of water. Winter winds and summer sun evaporate it from cells faster than it can be replaced from roots, so that the cells collapse and die.

Weed-Killer Injury. There has always been some unintentional injury to neighboring plants due to the use of weed killers of the kill-all variety on driveways; but since we have had 2,4-D as a selective weed killer for lawns, the damage to innocent bystanders has been enormous, not only from spray drift and volatile material in the atmosphere but from use of equipment for other spraying purposes. It is impossible to adequately clean out such a sprayer;

Figure 49. Weed-killer injury; tomato and oak.

it should be marked with red paint and kept for weed prevention only. Symptoms of injury are curling, twisting, and other distortions; there is often a fern-leaf effect instead of normal-size foliage (see Fig. 49). I have seen roses seriously malformed when a factory several hundred feet away mixed 2,4-D. I have seen tall oaks with all leaves unrecognizable after powdered 2,4-D was applied to the lawn. I have seen chrysanthemum in a greenhouse utterly deformed when 2,4-D was used on an outside lawn. Fortunately, unless the dose is extremely heavy, the plants gradually grow back to normal.

Winter Injury. Most winter browning of evergreens is due to rapid evaporation of water in sudden warm or windy spells. Copious watering late in the fall, a mulch, and windbreaks are helpful for broad-leaf evergreens, as is spraying them with a waxy material, Wilt-Pruf, which prevents evaporation.

Sudden ice storms cause obvious breaking in trees; in boxwood and similar shrubs they result in bark sloughing off and gradual dieback for months, even years afterward. I have seen symptoms on azaleas long after the ice was gone.

Yellows. This term is used for some deficiency disease but also for various virus diseases and Fusarium wilts.

Zinc Deficiency. This problem causes little leaf of almond, apricot, apples, grape, peach, plum. Foliage is small, narrow, more or less crinkled, chlorotic at tips of new growth, with short internodes producing rosettes of leaves. Defoliation progresses from base to top of twigs. The method of supplying zinc depends somewhat on the fruit. Spray apples, peaches, plums, and pears during dormant period with zinc sulfate. Swab grape vines immediately after winter pruning.

Mottle Leaf. This problem occurs in citrus. Leaves are small, pointed, with a sharply contrasting pattern of green along midrib and main laterals and light green or yellow between veins.

Rosette. This problem occurs in pecans and walnuts. Narrow, crinkled leaflets with dead or perforated areas have a rosette appearance; trees often bear no nuts. Pecan growers in southeastern states broadcast zinc sulfate on soil under each tree in winter. Variety Money-maker is resistant to zinc deficiency.

Vegetable crops. Corn, beans, tomato, and soybean have been protected by amending the soil with 23 pounds zinc sulfate per acre.

POWDERY MILDEWS

Mildew is a disease in which the pathogen is seen as a growth on the surface of plants. The same word is used for the fungus causing the disease. Mildews are Ascomycetes, members of the order Erysiphales. Black mildews are parasites in the family Meliolaceae with a dark mycelium to give a sooty effect. They are common in the South or on tropical plants in greenhouses (see under Black Mildew). Powdery mildews are plant parasites in the family Erysiphaceae.

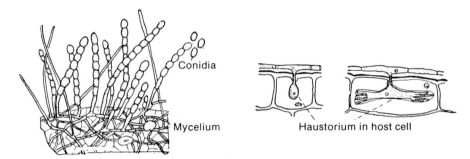

Figure 50. Powdery mildew; mycelium and conidia formed on surface of a leaf and two types of the haustoria in host cells.

They have white mycelium, in a delicate weft or thick felt, made up of a criss-cross tangle of hyphae. Colorless spores borne in chains on upright conidophores give the white powdery effect (see Fig. 50). False or downy mildews are Phycomycetes, and the conspicuous growth is not vegetative mycelium but fruiting structures and conidia protruding through stomata or epidermis to give a white frosty appearance in moist weather (see Downy Mildews).

True powdery mildews—and in speaking of them we usually eliminate the word powdery—are widely distributed but sometimes more abundant in semiarid regions than in areas of high rainfall, where other diseases flourish. Unlike those of most other fungi, powdery mildew spores do not require free water for germination. Some species require high humidity, but it is usually provided at the leaf surface when cold nights change to warm days or when plants are grown in crowded, low, or shady locations without sufficient air circulation. Spores of other species can germinate with very low humidity.

When a mildew spore lands on a leaf and puts out its germ tube, it does not make its nearest way inside the leaf but produces a tangle of septate threads—hyphae—on the surface. Special sucking organs—haustoria—penetrate the epidermal cells, occasionally the subepidermal cells, in search of food (see Fig. 50). The penetrating tube is slender, but once inside the cells, the haustorium becomes a round or pear-shaped enlargement or a branched affair, with greatly increased absorbing surface.

Conidiophores, growing at right angles from the mycelium, produce one-celled conidia in rows or chains of somewhat barrel-shaped hyaline cells, which become oval as they are dislodged from the top of the chain and disseminated by wind. Mildews known only in this imperfect state are called by the form genus name *Oidium*. This state requires the sexual fruiting bodies—perithecia—to place mildews in their proper genera.

Perithecia are round with a dark membranous wall, technically cleistothecia because they have no beak or ostiole, and rupture irregularly to free the asci.

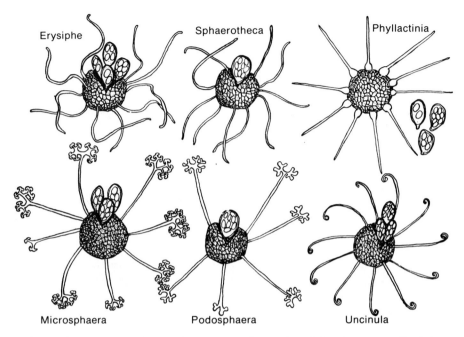

Figure 51. Powdery mildews. Perithecia (cleistothecia) of the six genera: *Erysiphe*, simple appendages and several asci; *Sphaerotheca*, same with one ascus; *Mycrosphaera*, dichotomously branched appendages and several asci; *Podosphaera*, same with one ascus; *Phyllactinia*, appendages bulbous at base; *Uncinula*, appendages coiled at tip.

They are held in place in the mycelium by appendages. The form of these appendages and the number of asci in the perithecium are the chief characters differentiating the six genera important in this country (see Fig. 51). *Sphaerotheca* and *Erysiphe* both have simple appendages; but the former has only one ascus, the latter several. *Podosphaera* has appendage tips dichotomously branched and one ascus; *Microsphaera* has the same type of appendage but several asci. *Phyllactinia* has lancelike appendages swollen at the base; those of *Uncinula* are coiled at the tip. Both have more than one ascus.

Powdery mildews are obligate parasites, having no saprophytic growth periods in dead plant parts, although the perithecia carry the fungus through the winter on either living or dead tissue. Mycelium sometimes winters in buds. Symptoms of mildew are dwarfing and stunting, often with a slight reddening and curling of leaves before the white mycelium is noticeable. There may be deformation of flower buds. Such symptoms are due to the withdrawal of plant foods by the fungus and to excessive respiration.

Sulfur dust and lime sulfur sprays have long been considered specific remedies for powdery mildews; some copper sprays are effective. Many of the newer organics are ineffective.

Erysiphe

Cleistothecia globose, or globose-depressed, sometimes concave; asci several, two- to eight-spored; appendages floccose (cottony), simple or irregularly branched; sometimes obsolete, usually similar to mycelium and interwoven with it; mycelium brown in rare cases.

Erysiphe aggregata. Alder powdery mildew; perithecia large, asci with eight spores, rarely six or seven.

Erysiphe cichoracearum. Powdery mildew of cucurbits; and many ornamentals, mostly composites, perhaps best known to gardeners as the **Phlox mildew.** Asci are two-spored, perithecia rather small, haustoria not lobed. There are nearly 300 hosts including cucumber, squash, pumpkin, gourds, cantaloupe, watermelon, lettuce, endive, Jerusalem artichoke, pepper, potato, salsify, *Achillea, Anchusa, Artemisia,* aster, begonia, *Boltonia,* calendula, campanula, chrysanthemum, clematis, coreopsis, cosmos, dahlia, delphinium, *Eupatorium,* gaillardia, golden glow, goldenrod, *Helenium,* hollyhock, *Inula,* mallow, *Mertensia,* phlox, rudbeckia, *Salpiglossis,* salvia, sunflower, stokesia, and zinnia.

There are various strains of the fungus, the form on cucurbits not affecting ornamentals, the strain of phlox (see Fig. 52) reportedly limited to that host, the strain on zinnia with a wide range of host plants. The lettuce strain, perhaps a mutation of the form on wild lettuce, was not reported on cultivated lettuce before 1951 and is important only in California and Arizona.

Powdery mildew was reported on cucurbits in North America in 1890, but did not gain much prominence until 1926, when it suddenly reduced the melon crop in the Imperial Valley of California by 5,000 carloads. By 1939, mildew-resistant Cantaloupe 45 had been developed to meet the situation, but in another decade the fungus had produced a different strain to which Cantaloupe 45 was susceptible. Plant breeders can never rest on their laurels because fungi that are obligate parasites seldom stay long outwitted. Other varieties, Cantaloupes 5, 6, and 7, were bred resistant to both strains of the fungus.

Powdery mildew is the principal disease of cucumbers in greenhouse culture, with tiny white superficial spots on leaves and stems enlarging and becoming powdery. Young watermelon fruits in greenhouses have small pimples or warts under the area covered by mildew mycelium.

Phlox mildew is only too familiar to gardeners. The white coating often appears on variety Miss Lingard in June, but on other varieties (in New Jersey) more prominently in July and August. The mycelium is present on both leaf surfaces and forms a thick felt on stems. In late summer black perithecia are formed in great abundance. Powdery mildew on zinnias and chrysanthemums usually starts so late in the season that it is more conspicuous than harmful.

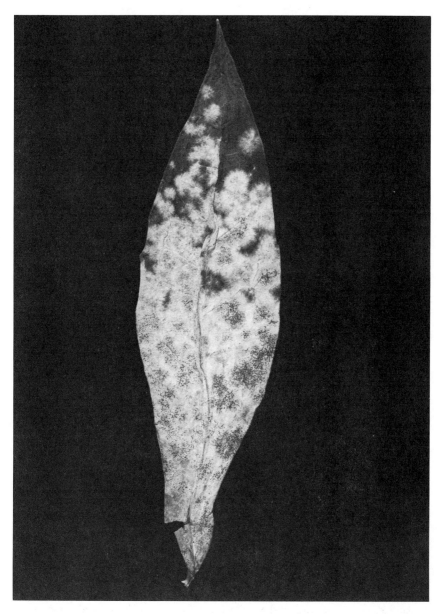

Figure 52. *Erysiphe cichoracearum* on phlox.

Control. Sulfur dust gives excellent control but is phytotoxic to some melons and other cucurbits. Sulfur-tolerant cantaloupes have been produced. The variety Homegarden is supposedly resistant to mildew and other pests. Keep phlox and other ornamentals well spaced, and dust with sulfur at the first sign of white growth. Because the perithecia winter on old stems and leaves, plants should be cut at ground level in autumn and burned.

Erysiphe graminis. Powdery mildew of cereals and grasses; economically important on wheat, oats, barley, and rye; aesthetically important on lawn grasses, wheatgrass, fescue, and bluegrass. The foliage is yellow or chlorotic with a white weft or mealy coating on upperside of leaf, which later turns yellow and is often studded with tiny brown perithecia. Seed from infected plants is small and shriveled. Apply sulfur dust or wettable sulfur sprays.

Erysiphe heraclei. Powdery mildew; on carrot.

Erysiphe lagerstroemiae. Crape myrtle powdery mildew; on crape myrtle only, from Maryland to Florida and Texas, the most serious disease of this shrub. The perithecia have been found only in Florida, but presumably it is the same species throughout the host range. The disease appears on young shoots in early spring, later infecting leaves and different parts of the inflorescence. Affected parts are covered heavily with a white mealy to dusty growth; young leaves are stunted, often less than one-third normal size but abnormally thickened. Internodes are short, flower stems stunted; buds often fail to develop flowers. Infected portions often have a reddish discoloration under the white coating. Diseased leaves and buds drop in a week or two, but stems may sprout again and sometimes produce normal growth in hot weather.

The fungus winters as mycelium in dormant buds and in spring covers such buds with a dense white coating of conidia, the source of primary infection, which starts as small, circular white patches on young leaves. Spores produced in abundance on these patches account for rapid spread of the disease until midsummer heat.

Control. Spray with lime sulfur when buds burst in spring and repeat 2 weeks later. If the initial infection is not checked, spray wettable sulfur or dust with sulfur.

Erysiphe polygoni. Powdery mildew of legumes; and many other vegetables and ornamentals, about 200 species in 90 plant genera. Appendages are long or short, interwoven with the mycelium, but the perithecia are not immersed in it. Asci have three to six spores. Peas exhibit a white powdery coating over leaves and pods, with the latter often discolored. Leaves are sometimes yellowish and deformed. The disease may be severe on peas in arid sections of western states, particularly on late homegarden varieties.

On beans, the mildew is grayish. It is prevalent in California in cloudy weather or in autumn when humidity is increased, but it is more important in the Southeast. Other vegetables infected by this species include lima bean, soybean, cabbage, turnip, radish, horseradish, and carrot. Spores germinate at quite low humidity.

The legume mildew is widespread on lupine, occasional on sweet pea. Other ornamental hosts include acacia, anemone, arrowleaf clover, candytuft, calendula, California poppy, iceland poppy, China aster, clematis, columbine, dahlia, delphinium, *Erigeron*, gardenia, geranium, hydrangea, honeysuckle, locust, matrimony vine, peony, and tulip tree.

Control. Choose resistant vegetable varieties or dust with sulfur. Spray or dust ornamentals with Karathane or sulfur.

Erysiphe taurica. On leaves of mesquite.

Erysiphe trina. Oak powdery mildew; on tanbark oak and coast live oak, in California, causing witches' brooms (but see *Sphaerotheca lanestris* for the common live-oak mildew). Perithecia are small, yellow-brown, with appendages lacking or rudimentary; asci have two, rarely three, spores.

Leveillula

Cleistothecia globose, or globose-depressed, sometimes concave; asci two- to eight-spored; appendages floccose (cottony), simple or irregularly branched; usually similar to mycelium and interwoven with it.

Leveillula taurica. Commonly causes powdery mildew on tomatoes in Eastern Mediterranean region and reported on fresh market tomato in California and Utah; also on onion, guar, wild tobacco, cotton, and desert bird of paradise.

Microsphaera

Cleistothecia globose to globose-depressed; appendages branched dichotomously at apex, often ornate; asci several, with two to eight spores.

Microsphaera alni. Named for the alder (*Alnus*), on which it is widespread, but best known to gardeners as the Lilac mildew. It also infects many other trees, shrubs, and vines, including azalea, beech, bittersweet, birch, catalpa, dogwood, elder, elm, euonymus, forestiera, hazelnut, magnolia, mountain holly, plane, New Jersey tea, privet, trumpetvine, and viburnum. According to some taxonomists the proper name of this species is *Microsphaera penicillata*, but *M. alni* is more familiar and still widely used.

Mildew is prevalent on lilac in late summer and fall, sometimes in dry seasons, almost completely covering foliage with a thick white coating; but, because it comes so late in the season, it is not very injurious. It is also common on deciduous azaleas in late summer, forming a very thin grayish white coating with numerous prominent dark perithecia. This species is more prevalent than the legume mildew on sweet peas, but it is chiefly a greenhouse problem in spring, when temperatures and humidity are less uniform. The foliage may be malformed, dropping prematurely or drying out and shriveling.

Among tree hosts oaks are probably most susceptible, but it would seldom pay to attempt control measures except in nursery rows. On pecans the white coating starts forming on leaves and nuts in July with occasional defoliation, shuck splitting, and shriveled kernels. Most commercial pecan varieties are mildew-resistant.

Blueberry mildew is caused by a special strain of lilac mildew, reported as *Microsphaera alni* var. *vaccinii*, *M. vaccinii*, and *M. penicillata* var. *vaccinii*. Varieties Pioneer, Cabot, and Wareham are said to be particularly susceptible; Concord, Jersey, and Rubel are intermediate; and Stanley, Rancocas, Harding, and Katherine are highly resistant. On some blueberries the mycelium is conspicuous on upper leaf surfaces, on others it is barely visible on the underside. Midsummer defoliation weakens the bushes. Cranberries, farkleberries, trailing arbutus, and lyonia are possible hosts to this strain.

Control. Bordeaux mixture is recommended for pecans—two applications, June and July. Dust blueberries with sulfur. Use sulfur on lilacs and other ornamentals.

Microsphaera diffusa. General on snowberry, widespread on wolfberry, coralberry, occasional on black locust, lima bean, kidney bean, and soybean. Appendages are two to four times the diameter of the perithecia, with ultimate branches long, forming a narrow fork.

Microsphaera euphorbiae. On lima bean, euphorbia, roselle.

Microsphaera grossulariae. European powdery mildew; occasional on currant, gooseberry. There is a light weft of mycelium mainly on upper surface of leaves. For the more important American mildew see *Sphaerotheca mors-uvae*.

Microsphaera penicillata. Powdery mildew; on *Leucothoë axillaris*.

Oidium

This term is used for mildews known solely from the conidial stage. In some cases the type of conidial fructitication may suggest correct genera, but until perithecia are found, *Oidium* is preferred.

Oidium begoniae. Begonia mildew; especially important on tuberous begonias on the West Coast, though it may also occur on fibrous-rooted begonias.

Oidium euonymus japonici. Euonymus mildew; general throughout the South and on the Pacific Coast on *Euonymus japonicus*. The mycelium forms a thick felt on the leaf surface, causing some yellowing and defoliation (see Fig. 53). I have seen this disease rampant in foggy coast towns like Beaufort, South Carolina, or Mobile, Alabama, and equally severe in semiarid El Paso, Texas. The washing effect of a water spray applied with pressure, either by adjusting the hose nozzle or putting the thumb over a portion of the orifice, is a deterrent to this mildew. Sulfur dust can be used, probably Karathane.

Oidium obductum. On oriental plane.

Oidium pyrinum. On crabapple.

Oidium tingitaninum. Citrus mildew; common in Java, Ceylon, India, but

Figure 53. Powdery mildew on euonymus; prevalent in the South.

in this country causing only limited injury to tangerine trees in California. White patches are formed on upper surface of leaves, the tissue underneath first a darker, watery green, then losing color, turning yellowish.

Oidium sp. On greenhouse snapdragons, a white powdery growth on both leaf surfaces, sometimes on young stems. Control with Karathane or sulfur.

Odium sp. On avocado, occasionally in Florida, in nurseries or on young trees in shaded locations. Tips of shoots are killed back; dark green spots appear on upper leaf surfaces with white mycelium on the underside. The mildew can be controlled with lime sulfur.

Phyllactinia

Perithecia are large; appendages are lancelike with a bulbous base. Mycelium does not send haustoria into epidermal cells of host but forms special branches that pass through stomata into intercellular spaces; each of these intercellular branches or hyphae sends a single haustorium into the adjacent cell.

Phyllactinia angulata. Powdery mildew; on elm.

Phyllactinia corylea (*P. guttata*). **Powdery mildew of trees;** named for the hazelnut or filbert but prevalent on many other trees and shrubs, such as amelanchier, ash, barberry, beech, birch, boxwood, catalpa, chinaberry, crabapple, currant, blackberry, raspberry, gooseberry, crape myrtle, dogwood, buttonbush, chestnut, elm, elder, fringe tree, hawthorn, hickory, hornbeam, holly, linden, oak, plane trees, quince, rose, sassafras, tulip tree, walnut, and willow. Mildew is seldom serious enough on shade trees to warrant control measures, but in the nursery dusting with sulfur may be advisable. It is common on filberts in Oregon, but comes so late in the season that it does not affect yield.

Podosphaera

Perithecia globose; one ascus, with eight spores; appendages dark brown or colorless, dichotomously branched at tip; rarely an extra set of basal appendages present.

Podosphaera leucotricha. Powdery mildew of apple; also crabapple, pear, quince, photinia. First noted in Iowa on seedling apples in 1871, this mildew became more important in orchards when organic fungicides, ineffective for mildew, were substituted for sulfur and copper in the apple-scab schedule. Twigs, foliage, blossoms, and fruits may be disfigured, stunted, deformed, or killed. Gray to white felty patches are formed on leaves, usually on underside. Leaves are crinkled, curled, sometimes folded longitudinally and covered with masses of powdery spores. They soon turn brittle and die, resulting in decreased yield. The same powdery growth starts on 1-year twigs, but in midsummer it is transformed into a brown, felty covering, in which minute, dark perithecia are embedded in dense aggregations. Infected twigs are stunted or killed. The fungus winters as dormant mycelium on twigs or in buds. Such buds produce shriveled blossoms and no fruit. Fruit produced on infected twigs is stunted or russeted. Jonathan variety is especially susceptible.

Podosphaera oxyacanthae (*P. clandestina* var. *tridactyla*). **Powdery mildew of cherry;** occasional on plum, peach, apricot, apple, pear, quince, hawthorn, serviceberry, spirea. Budded sour cherry is most severely attacked, but the disease is seldom serious except on nursery stock. Young leaves and twigs are covered with a white mycelium and powdery spores. Leaves are curled upward; terminal leaves are smaller; twig growth is stunted. Sulfur sprays or dusts will control.

Podosphaera tridactyla. Recently reported on almond in California and the most common mildew on apricot, causing large nonnecrotic lesions on leaves.

Sphaerotheca

Appendages simple, flexuous, resembling hyphae; only one ascus in a perithecium.

Sphaerotheca castagnei. On buffaloberry, spirea.

Sphaerotheca fuliginea. Powdery mildew; of summer squash and cucurbits.

Sphaerotheca lanestris. Powdery mildew of coast live oak; on *Quercus agrifolia* in California, reported also on white, southern red, bur, and post oaks. The disease is most destructive in the narrow coastal plain. The most conspicuous symptom is a powdery white, stunted growth developing from certain terminal or lateral buds. The shoots are swollen, fleshy, with much shortened internodes. Foliage on such shoots is often reduced to pale yellow, bractlike leaves, which turn brown, dry, and shrivel; these shoots resemble witches' brooms. On leaves developing from normal buds and shoots, the fungus forms a dense layer on both surfaces, more abundant on the lower side. This species is sometimes called the brown mildew because the grayish-white mycelium changes to tan and then brown with age. Perithecia are formed in the brown felt, abundantly in some years, rarely in others. In southern California the fungus may winter in the conidial state, with widespread leaf and shoot infections coming from wind-borne spores.

Control is not easy. Spraying with lime sulfur in March and October is fairly effective but may be phytotoxic at high temperatures and low humidity. Wettable sulfur has not been consistently effective. Removal of witches' brooms by pruning back to normal lateral branches is effective only if the tree is slightly susceptible and conditions for reinfection are unfavorable. Heavy pruning stimulates new growth and increases the amount of mildew. The Holm or holly oak is apparently resistant to mildew and well adapted to the coastal region.

Sphaerotheca macularis (*S. humuli*). Hop mildew; also on fruits, blackberry, dewberry, gooseberry, raspberry, strawberry, rose (probably rarely in this country), and other ornamentals, including *Agastache*, betony, buffaloberry, delphinium, *Epilobium*, *Erigeron*, gaillardia, geranium, geum, gilia, hawksbeard, hawkweed, *Hydrophyllum*, kalanchoë, matricaria, meadowsweet, ninebark, *Polemonium*, phlox, sumac, spirea, tamarisk, and *Vernonia*.

This mildew may be important on Latham variety of raspberry, appearing on new canes when they are 2 to 3 feet high. The tip leaves are dwarfed, mottled, and distorted, almost as if they had mosaic. The undersurface of leaves is water-soaked or has the familiar white coating. There is no specific control except to space plants for free air circulation.

The powdery mildew sometimes serious on strawberries in northeastern and Pacific Coast states is probably a special strain. The edges of affected leaves curl upward, exposing the lower surface, where the powdery frosty growth is evident. Fruit, stems, and berries may be affected, with fruit often failing to color. Resistant varieties include Sparkle, Puget Beauty, Siletz, and India.

Sphaerotheca mors-uvae. American gooseberry mildew; also on currant; sometimes the limiting factor in gooseberry production. Fruits dry up with a brown, felty covering. Leaves and canes are stunted with the usual white coating. Perithecia are formed on canes, and ascospores are discharged in early May as fruit is set. Conidia for secondary infection are produced within 10 days. Spray with lime sulfur immediately after bloom.

Sphaerotheca pannosa var. **persicae.** Peach mildew; general on peach, also on almond, apricot, nectarine, matrimony vine, and *Photinia*. The mycelium is pannose (ragged) or in dense patches, persistent, usually satiny, shining white, or sometimes grayish or brown. Immature fruits are highly susceptible. They have brown blotches and are scabby and malformed. The fungus winters in shoots. Nonglandular varieties Peak and Paloro are more affected than glandular Walton, Johnson, Halford, and Stuart, at least in California. Sulfur in the spray schedule for brown rot should control mildew without additional treatments. Karathane is effective but very slightly phytotoxic. Lime sulfur is recommended.

Sphaerotheca pannosa var. **rosae.** Rose mildew; general on rose; distinct from peach mildew but apparently not confined to rose, since apricots growing near roses have been infected. More than one strain may be involved. Rose mildew is found wherever roses grow. Always a problem with greenhouse roses, it was enhanced when aerosol treatments for red spiders and other pests were substituted for old-fashioned syringing. Mildew increased in garden roses when ferbam and other new organics replaced the old sulfur and copper in the blackspot sprays. Rose mildew is omnipresent along the Pacific Coast and is serious in the semiarid Southwest. In the East, it appears on small-flowered ramblers such as Dorothy Perkins and Crimson Rambler in May, and may be quite serious on hybrid teas and some floribundas in late summer, with the advent of cool nights.

The first symptom may be a slight curling of leaves, with the mycelial growth such a light and evanescent weft as to be almost unnoticed. Later the white coating is conspicuous from the chains of conidia produced lavishly over the surface. The coating may cover buds, resulting in no bloom or distorted flowers. Leaves often have a reddish or purplish cast under the white mycelium and sometimes turn black. They may be slightly blistered. On canes, the growth is heavier and more felty, especially near thorns. Toward the end of the season perithecia may be found on canes, but they are not common, and I have not seen them on leaves except on a Rugosa rose at Ithaca, New York. Mildew is prevalent on soft succulent shoots, fostered by an excess of nitrogen.

Control. Sulfur dusts have been standard treatment for garden roses for many years; to be effective dusting must be started at the first sign of mildew, before the mycelium gets too thick. Sulfur may be injurious to roses in very hot weather. Choice of variety of rose is important. Shiny-leaved climbers like Dr. Van Fleet seldom have mildew, and the shrub polyantha, The Fairy, is very resistant. Many red roses, hybrid teas, and fluoribundas are especially susceptible, but the orange-red floribunda Spartan remains free from mildew (in my own experience). Garden planning avoids a lot of mildew trouble. Keep the plants well spaced, in beds away from buildings, and not surrounded by tall hedges or walls.

Sphaerotheca phytophila. Associated with gall mites causing witches' brooms on hackberry. The mycelium is evanescent; perithecia are formed inside loose scales of enlarged buds.

Sphaerotheca sp. On *Tolmiea;* pick-a-back plant, in greenhouse.

Uncinula

Perithecia globose; appendages uncinate, slightly coiled at tips; several asci, with two to eight spores.

Uncinula circinata. On maples, Virginia creeper, western soapberry.

Uncinula clintonii. General on American linden.

Uncinula flexuosa. Horse chestnut powdery mildew; on *Aesculus* spp., including red, yellow, and Ohio buckeye, widespread in central and eastern states. This mildew gives a very thin coating on the leaf surface, supposedly mostly on the underside although I have seen it on the upper. Perithecia are numerous, small, barely discernible with the naked eye. Control is usually unnecessary except in nurseries. A copper spray used for blotch will also control mildew.

Uncinula macrospora. General on American and winged elms.

Uncinula necator. Grape powdery mildew; general on grapes, also on *Ampelopsis;* common in late summer on eastern grapes but not serious; a major problem in California. Leaves, canes, and young fruits are covered with white patches; growth is often distorted. Late in the season the white mycelium disappears and the spots appear brown or black; berries are russeted or scurfy, failing to mature.

Control. Keep California grapes covered with a light coating of sulfur dust. Apply when new shoots are 6 to 8 inches long; when they are 12 to 16 inches, 14 days later; when shoots are 2 to 3 feet; when fruit is half-grown; when fruit begins to ripen. If some of the applications are omitted, and mildew gets a head start, wettable sulfur is used as an eradicant spray. Karathane is also effective. Copper sprays are often used in the East, if any are necessary.

Uncinula parvula; U. polychaeta. Widespread on hackberry and southern hackberry.

Uncinula prosopodis. On mesquite.

Uncinula salicis. Willow powdery mildew; also on pussy willow and poplar, sometimes causing defoliation but not often serious. The growth is in diffused or circumscribed patches on both leaf surfaces.

ROTS

A rot is a decay, a decomposition or disintegration of plant tissue. It may be a hard dry decay or a soft and squashy one. It may affect root or rhizome, stem, tree trunk, blossom, or fruit. Some rots also affect leaves, but diseases that are primarily of foliage are more often designated leaf spots or blights. Rots caused by bacteria are discussed under Bacterial Diseases.

There are a great many wood rots of trees, recognized by the sporophores or conks of the various species of *Fomes, Polyporus,* and other shelving or bracket fungi. By the time these signs appear, it is usually too late to do anything about the disease. The tree-rot fungi enter through unprotected wounds—either pruning cuts or breaks due to wind and ice storms. For proper pruning methods and treatment of wounds, see U.S. Department of Agriculture *Farmers' Bulletin* (1966), *Care of Damaged Shade Trees, Tree Maintenance* by P. P. Pirone (1959), or *Tree Experts Manual* by Richard R. Fenska (1954). The fact that tree wound dressings are now available in convenient aerosol bombs should make it easier for home gardeners to protect pruning cuts from wood-rotting fungi.

Acanthorhynchus

Ascomycetes, Sphaeriales, Sphaeriaceae

Perithecia separate, innate, beaked; spores one-celled, dark.

Acanthorhynchus vaccinii. Cranberry blotch rot; a common fruit rot thriving in warmer sections, more important in New Jersey than in Massachusetts. The rot starts as a small, light-colored spot on the berry, spreading to destroy the whole fruit, with dark blotches on the skin. The fungus may invade leaves, but it seldom fruits on them until they have fallen. Cranberry bogs in New Jersey may need three or four sprays of bordeaux mixtures starting at midbloom, but in Massachusetts two are sufficient.

Alternaria

See under Blights.

Alternaria alternata. Fruit rot; on tomato and black pit disease on potato tubers (stored).

Alternaria citri. Alternaria rot of citrus fruits; navel-end rot, black rot, widespread, prevalent in warm and dry sections, but not too serious. In oranges the rot is most common in the Washington Navel variety—a firm, dry, black rot at the navel end, often in only one segment, with fruit coloring prematurely, appearing sound on the outside. In lemons the disease is a soft, dark, internal rot of old or weak fruit in storage. Firm dark brown spots are formed on the rind. Grapefruit sometimes has a dark internal storage rot, not readily discernible externally.

Control. Chemical treatment after picking is not very satisfactory. Produce sound fruit in the orchard; avoid holding too long on the tree; avoid holding weak or old fruit too long in storage; store at low temperatures.

Alternaria mali. Fruit rot; widespread storage rot of apple, sometimes quince. Also a weak parasite enlarging injured spots on foliage. Try captan at 6- to 14-day intervals.

Alternaria radicina (*Stemphylium radicinum*). Black rot of carrots; a soft storage rot of roots held over winter. Rot may start at the crown or from some wound on the side of the root. Initial infection may be in field or in storage house; a black mycelial weft with large brown muriform spores develops over the rotted tissue. There is no control except to choose firm, healthy roots for storage and to store at low temperatures.

Alternaria solani. Collar rot of tomato; also fruit rot and early blight, general on tomato with the collar rot stage most frequent in the South. See under Blights.

Alternaria zinniae. Stem rot; on *Ageratum.*

Alternaria sp. Flower rot; of Vanda orchids, causing infection in transit along with *Botrytis.*

Alternaria sp. On Schefflera, in Florida.

Alternaria sp. Calyx-end rot; on apple.

Aphanomyces

Phycomycetes, Saprolegniales, Saprolegniaceae

Thallus composed of cylindrical branching hyphae without definite constrictions; sporangium cylindrical, threadlike, swarmspores arranged in a single row and encysting at the mouth; saprophytic or parasitic, living in the soil and causing root rots or damping-off.

Aphanomyces cladogamus. Causing rootlet necrosis of tomato, pepper, spinach, and a severe root rot of pansy.

Aphanomyces cochlioides. A seedling disease of sugar and table beets, part of the complex called black root; causing tip rot, a wilting of tops. Crop rotation and proper fertilization are helpful.

Aphanomyces euteiches. Pea root rot; also on bean, sweet pea, and perennial pea. The fungus is also a weak parasite in roots of many nonlegumes. First described in 1925, the fungus probably existed earlier in various root disease complexes and was responsible for giving up land formerly devoted to canning peas. Considered the most important of the pea root rots, found in every district, it is particularly destructive in eastern and central states.

The fungus is parasitic on subterranean parts, causing root and stem rot in peas of all ages, symptoms and crop yield varying with the time of infection. If the root system is invaded when only three or four nodes are formed, the plant may wilt and die suddenly; later invasion results in dwarfing and drying out of foliage from the ground upward. When seedlings are pulled out of the ground, the roots do not break off but come out as a fibrous string or vascular cylinder freed from the cortex. The fungus invades only the cortex or roots and base of stem, causing softening and rapid decay of tissue. Large numbers of thick-walled oospores are formed in the cortex; these may remain viable in the soil more than one season.

Control. A well-drained soil with low moisture content decreases rot. When soil moisture is at 45% of saturation, there is no disease; at 75% there may be more than 70% infection. Nitrogenous fertilizers are helpful.

Aphanomyces raphani. Radish black root; and damping-off, widespread; more important on long-rooted icicle varieties. Also on Abyssinian mustard, cabbage, Chinese cabbage, Chinese kale, honesty, mustard green, rape, rocuet salad, sea kale, Spanish mustard, wild radish, and *Brassica robertiana*. Small, steel-gray to black areas appear around point of emergence of secondary roots. Enlarging roots are constricted and turn black. Rotation is essential for control. Choose globe rather than long varieties.

Armillaria

Basidiomycetes, Agaricales, Agaricaceae

One of the mushrooms, cap-shaped on a stalk with an annulus or ring but no volva (cup) at the base; gills attached to the stem; spores white.

Armillaria mellea. Mushroom root rot; of trees and shrubs, also known as Armillaria root rot or toadstool disease, first described in America in 1887, known in Europe a hundred years earlier. The fungus is called honey mushroom, honey agaric, oak fungus, and shoestring fungus. Although the honey-colored toadstools are often seen in the East around rotting tree stumps, and may occasionally cause death to weak ornamental trees, the chief damage is west of the Rocky Mountains, especially in California, where most fruit and nut crops and ornamental trees and shrubs are menaced.

The decay is of the roots and root crown. Sheets of tough, fan-shaped mycelium are found between bark and wood, the latter changing to light tan, becoming soft and watery in texture. Clumps of toadstools are often found at the base of dead or dying trees, especially in autumn, but do not always appear in dry seasons (see Fig. 54). They are honey-colored or light tan, with a stalk 4 to 6 inches or more high and a cap 2 to 4 inches across, often dotted with brown scales.

Basidiospores formed along the gills are wind-borne. They can establish themselves in old stumps and dead trees but cannot infect healthy trees. The

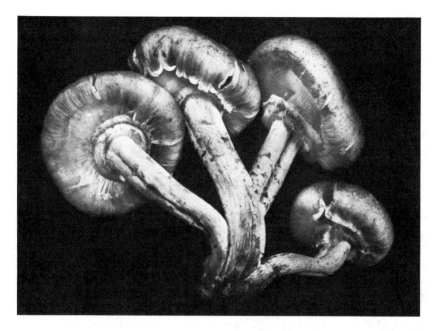

Figure 54. Mushroom root-rot fungus, *Armillaria mellea.*

latter are infected in the ground by means of black or brown cordlike rhizomorphs, the "shoestrings," which grow from infected roots a short distance through the soil. On meeting and penetrating a healthy root, the fungus progresses along the cambium layer, working up to and girdling the root crown. Leaves are dwarfed, turn yellow, or fall prematurely; on small trees all foliage may die simultaneously. On conifers, particularly pines, there is an abnormal flow of resin from the root collar.

Trees subnormal in vigor and suffering from drought are most injured. Orchards of citrus and other fruits on lands recently cleared of oaks are liable to enormous damage unless resistant rootstocks are used. The rot is found less often on dry hillsides than in valleys near streambeds, where flood waters deposit soil and infected debris around root crowns, or in places kept too wet by artificial watering. Ornamental trees and shrubs are often injured when extra soil is added in grading and terracing, and are then kept too wet by watering the lawn frequently.

The list of susceptible plants is far too long to be given in entirety. A representative selection includes almond, apple, apricot, avocado, cherry, citrus, currant, grape, incense cedar, peach, pear, plum and raspberry, hickory, filbert and walnut, California peppertree; oaks, pines, spruce and sycamore; azalea, rhododendron, boxwood and rose (root, crown, and stem rot on); African daisy and (stem rot on) sunflower; and sometimes other herbaceous plants such as begonia, carnation, dahlia, narcissus, peony, rhubarb, and strawberry.

Control. Use resistant plants where possible. Of fruits, only French pear, Northern California black walnut, fig, and persimmon are sufficiently resistant to grow safely on infested soil. Some plants can be grafted onto resistant rootstock such as Myrobalan 29. The University of California has prepared a list of resistant or moderately resistant ornamental shrubs. Some on the list are *Acacia decurrens* var. *mollis*, *A. verticillata*, *Buxus semipervirens*, *Ilex aquifolium*, *Lonicera nitida*, *Prunus ilicifolia* (hollyleaf cherry), *P. lyoni* (Catalina cherry), *Pyracantha coccinea* var. *lalandii* but not *P. angustifolia*, which is susceptible.

Moderately resistant shrubs include *Abelia grandiflora*, Darwin, Japanese, and Mrs. Wilson barberry, Mexican orange (*Choisya*), *Elaeagnus argentea*, *Euonymus japonica*, Japanese privet, *Myrtus communis*, *Pittosporum tobira*, and *Spiraea prunifolia*.

Mechanical measures are often helpful. Excavate and expose the root crown; remove diseased portions of bark and affected small roots. Paint wounds with a pruning wound compound. Leave the treated roots exposed until cool weather in autumn. Trenching or digging a ditch around a plot will restrict the disease temporarily, but roots will grow through the ditch in time.

Carbon disulfide is still recommended as a soil disinfectant, applied in staggered rows, in holes 18 inches apart each way, which should be immediately closed by tamping. Hand applicators are available for injecting the disulfide 6 to 7 inches deep. This treatment is for land where valuable trees have been removed; one cannot go closer to a healthy tree than the edge of the branch spread. After treatment, the land should remain fallow for at least 60 days, and then be ploughed before planting.

Ascochyta

See under Blights.

Ascochyta pinodella. Foot rot of peas; of the three species that make up the Ascochyta blight complex, this one produces most definitely a foot rot, with infection at the root crown or base of stem.

Aspergillus

Fungi Imperfecti, Moniliales, Moniliaceae

Conidiophores have a round head at the top, with radially arranged bottle-shaped sterigmata that bear conidia in chains; spores are one-celled, globose to ellipsoid, hyaline. Bread molds are in this genus. When, rarely, a sexual fruiting body (cleistothecium) is formed, the species is placed in the order Eurotiales.

Aspergillus alliaceus. Cladode rot; stem and branch rot; on *Cereus* and *Opuntia* cacti. This is a high temperature species. Spores are yellow in mass.

Aspergillus niger. Calyx end rot of dates; fig smut; bunch mold of grapes; pomegranate rot; black mold of peach; crown rot of peanut; also market and storage rot of shallot, onion, apple, and potatoes. The fungus is a weakly parasitic black mold invading ripe tissue wounds. In dates, the interior of the fruit is filled with a black dusty mass of spores, spread to a large extent by the dried-fruit beetle. Practice orchard sanitation; keep decaying fruits cleaned up so insects cannot carry spores.

Aspergillus niger var. **floridanus.** Wound parasite on *Dracaena.* Lower stem black, rotted, with dark brown spore masses.

Aspergillus spp. Green and yellow molds causing secondary rots of many fruits and some vegetables in storage.

Botryosphaeria

See under Blights.

Botryosphaeria dothidea. Fruit rot; of peach.
Botryosphaeria obtusa. Fruit rot; of peach.
Botryosphaeria rhodina. Fruit rot; of peach.
Botryosphaeria ribis (*Dothiorella gregaria*). Dothiorella rot of avocado and citrus; black fruit rot of apple and pear; nut rot of tung oil. On avocado it is a soft, rapidly spreading surface rot, starting from small spots when fruit begins to soften. The fruit may be covered with decay spots by the time it is usable. The fungus winters in dead twigs, in tip-burned leaves, and enters the avocado while it is still on the tree. Two sprays, mid-September and early October, using bordeaux mixture, have given fairly good control. Remove dead wood from trees, to reduce source of inoculum, and pick fruit early.

On lemons and other citrus fruits the rot starts as a discoloration around the button, becoming a brown, leathery but pliable decay. When fruit is entirely involved it becomes olivaceous black. On tung, brown lesions appear on green fruit, which drops prematurely. See further under Cankers and Diebacks.

Botryotinia

See under Blights.

Botryotinia convoluta. Botrytis crown rot of iris; gray mold rot; on rhizomatous iris, first recorded in Canada in 1928 and apothecia later produced in culture. The chief diagnostic character is the presence of many shining black sclerotia, much convoluted and agglomerated into large clusters on rotting rhizomes. These are often found in spring on plants that started into the winter apparently healthy, for the fungus is active in cool, wet weather.

Conidiophores are brown, formed in fascicles, and bear dense clusters of light brown ovate or slightly pyriform conidia. They appear in spring growing from or near sclerotia. Affected plants do not start spring growth.

Botrytis

See under Blights.

Botrytis allii. Gray mold neck rot of onion; also shallot and garlic; widespread. It is usually found on bulbs after harvest, infection taking place through neck tissue and scales appearing sunken and "cooked." Sclerotia are first white, then dark, 2 to 4 mm across. Conidiophores and conidia forming the gray mold are produced directly from mycelium in tissue or from sclerotia. Artificially cure bulbs after harvest to cause rapid dessication of neck tissue; store at low temperature. Colored varieties keep better than white.

Botrytis byssoidea. Mycelial neck rot of onion; the fungus is much like *B. allii* but produces more mycelium and less profuse gray mold.

Botrytis cinerea. Gray mold fruit rot; cosmopolitan on peach, cherry, plum, pomegranate, quince, pear, grape, strawberry, pepper, tomato, and eggplant. Also causing a leaf rot of hothouse rhubarb and a rot of carrot, lettuce, celery, and onion. See further under Blights.

Botrytis gladiolorum. Botrytis neck rot; corm rot; blight; of gladiolus. See under Blights.

Botrytis (*Botryotinia*) **squamosa.** Small sclerotial neck rot of onion; elliptical leaf lesions with withering of tips.

Brachysporium

Fungi Imperfecti, Moniliales, Dematiaceae

Conidiophores brown, erect, usually solitary, septate; conidia dark, unequally two- or more-septate; attached to apical cell of conidiophore by a short narrow cell; saprophytic.

Brachysporium tomato. Fruit rot of tomato.

Calonectria

See *Cylindrocladium* under Blights.

Calonectria crotalariae. Basal stem rot; of oleander.
Calonectria sp. (*Cylindrocladium*). Crown and collar rot; on papaya.

Catenularia

Fungi Imperfecti, Moniliales, Dematiaceae

Hyphae dark; conidophores simple or sparingly branched, with terminal chains of conidia; spores dark, one-celled.

Catenularia fuliginea. Fruit rot of date.

Centrospora

Fungi Imperfecti, Moniliales, Moniliaceae

Spores hyaline, filiform, with long, whiplike tapering beaks, several cross walls and a swordlike appendage from basal cell; mycelium dark.

Centrospora acerina. Black crown rot of celery; storage rot of carrot; on celery the disease appears 7 or 8 weeks after stock has been placed in cold storage with pale, ochraceous lesions at the crown end, gradually turning black, sometimes reddish. The fungus lives in the soil; use infested fields for early celery to be marketed without storage. The same species also causes leaf spot of pansy.

Cephalosporium

See under Leaf Spots.

Cephalosporium carpogenum. Fruit rot; on apple in storage, reported from Washington and Pennsylvania.

Cephalosporium gregatum. Brown stem rot of soybean; a vascular disease of major importance in the Midwest, also present in Florida, North Carolina, and Virginia. It has been controlled with a long rotation—5 years corn, 1 year soybeans.

Ceratocystis

See under Cankers.

Ceratocystis fimbriata (*Endoconidiophora fimbriata*). Sweet potato black rot; found wherever sweet-potatoes are grown, most destructive in storage but present also in seedbed and field. Round, blackish spots extend into vascular ring or deeper; sprouts are sickly with black cankers belowground, or they are killed. The fungus winters in storage houses, on wild morning glory and other weeds near the field and in soil, where it remains viable for several years. Spores are spread by the sweet-potato weevil and in wash water if potatoes are washed before storing. This fungus also infects *Jacquemontia*.

Control. Standard treatment has been disinfection of planting stock in a solution of borax. Using pulled sprouts provides plants free from black rot. Plan a 4-year rotation; sort carefully before storage; cure quickly at high temperature and humidity. Yellow Jersey is highly susceptible; some varieties are quite resistant.

A strain of this fungus is reported causing **Black cane rot**; in propagating bed of *Syngonium auritum* (*Philodendron trifoliatum*) in a California nursery. Brown to black water-soaked girdling cankers, often on parts in contact with the soil, cause yellowing and death of foliage. The fungus can be eradicated by treating canes with hot water, 120° F for 30 minutes.

Ceratocystis wageneri. **Root rot**; of lodgepole pine and ponderosa pine.

Ceuthospora

Fungi Imperfecti, Sphaeropsidales, Sphaerioidaceae

Pycnidia in a valsoid stroma; conidia oblong to bacillar, extruded in tendrils; conidiophores obsolete or none.

Ceuthospora lunata. **Black rot of cranberry**; developing in berries after picking. The fruit turns dark and soft. The disease is more important in Washington and Oregon. Spraying for other cranberry diseases largely controls this rot. Pick berries when dry; avoid bruises; keep them cool.

Chalaropsis

See under Molds.

Chalaropsis thielaviopsis. **Root rot**; on poinsettia.

Clitocybe

Basidiomycetes, Agaricales, Agaricaceae

One of the mushrooms, with gills typically decurrent (running down the stem), cap homogenous and confluent with fleshy stripe, which has neither ring nor cup; spores white or very lightly colored.

Clitocybe monadelpha. On privet, apple.

Clitocybe tabescens. Mushroom root rot; clitocybe root rot; of citrus, pecan, and other fruits and many ornamentals. This root rot is as devastating in Florida as Armillaria rot is in California and very similar (some say the pathogen is identical). It is important in the decline of citrus groves, on orange, grapefruit, lemon, tangerine, and lime on rough lemon stock; is very destructive to Australian pine (*Casuarina*); and has been reported on more than 200 species in 59 plant families, including *Acalypha*, avocado, arborvitae, apricot, camellia, castor bean, cherry laurel, crape myrtle, cotoneaster, cypress, dogwood, *Eugenia*, eucalyptus, grape, guava, glorybush, *Hamelia*, holly, *Ligustrum*, juniper, jasmine, loquat, oleander, poinciana, pomegranate, pear, *Parkinsonia*, rose, viburnum, and wax myrtle. In recent years Clitocybe root rot has become economically important on Georgia peach trees and has killed many lychee trees in Florida. It is said to account for 75% of rose mortality in some sections.

Symptoms of decline do not ordinarily develop until the pathogen has been working a number of years and has killed a large part of the root system. Often mushrooms are present at the base of trees before the tops show more than a slight yellowing or lack of vigor; but if soil is removed from the root crown, many lateral roots are found dead, and often the taproot is also gone. Infection starts at some point on the lateral roots, spreads to the base of the tree, and then to other roots. Sometimes there is gumming at the crown extending upward on the trunk. Mycelial fans or sheets are present between bark and wood; the clusters of mushrooms developing at the base are similar to those of *Armillaria*, but the black shoestring rhizomorphs are lacking. Instead, there are sometimes black, hard stromatic outgrowths from fissures in bark of infected roots. The fruiting clusters develop in fall, from mid-September to December. The caps are light tan to honey-colored, 2 to 3½ inches in diameter. The rot is most prevalent on land cleared of oaks and other hardwoods, also on sandy, well-drained land subject to drought.

Control. Citrus trees on sour orange stock are quite resistant. Surgical treatment for fruits and ornamental trees is often quite successful. Remove the soil at least 2 or 3 feet from the trunk, working carefully to avoid injuring healthy roots. Cut off all dead roots, flush with the root crown, and remove any infected oak or foreign roots in the vicinity. Cut out dead and infected bark at the root crown or the base of the trunk, being sure to collect all chips (on heavy paper placed under exposed roots) for burning. Paint all exposed surfaces with a pruning wound compound and fill in partially, disinfesting the soil with bordeaux mixture. The root crown can be left exposed to aeration and drying, or if too large a proportion of the root system has been lost, new roots can be stimulated by mounding the soil around the base to a height of several inches above the partial girdle. The new roots will come from callus formed at the margin of living bark.

Trenches 2 or 3 feet deep will aid in preventing spread to healthy trees. Fallow soil can be treated with carbon disulfide; see *Armillaria mellea*.

Colletotrichum

See under Anthracnose.

Colletotrichum capsici. Ripe rot of pepper; boll rot of cotton.

Colletotrichum circinans. Onion smudge; surface rot, also on shallot, garlic, and leek. Bulb or neck has a dark green or black smudge, often covered with stiff bristles of the acervuli of the fungus. Smudge is more prominent in white onions; it is confined to the neck of colored bulbs. The fungus winters on mature onions, on sets or in soil. It develops in the field at a fairly high temperature and soil moisture, with most of the damage just before harvest. Cure rapidly after harvest; rotate crops; clean up debris; change to colored onions if the rot is too serious on white.

Colletotrichum coccodes. Root rot and wilt; of greenhouse tomato.

Colletotrichum lilii. Black scale rot of Easter lily; brown scale. First noticed in Louisiana in 1937, the rot immediately threatened the lily industry in that section. Bulbs are brown to nearly black when dug, with outer scales most affected. Young lesions start as irregular light brown areas, then become black and sunken owing to collapse of epidermal cells and subepidermal layers. Oldest lesions are nearly black, with tissue dry and shriveled. Stems and roots are not affected. The acervuli are small, gregarious, with many dark brown setae and continuous hyaline conidia.

Colletotrichum nigrum. Fruit rot of pepper; probably general on pepper in the South and East. The fungus is a wound parasite on pepper pods. The spots are irregular, indefinite, depressed, blackish. Numerous acervuli with stout setae are scattered over spots.

Collybia

Basidiomycetes, Agaricales, Agaricaceae

Margin of young cap turned in; gills not decurrent; stipe central; no annulus or volva; spores white or light; causing wood rots.

Collybia velutipes. Heart rot; white sapwood rot; of hardwoods. The fungus is a small toadstool with central stem, base covered with dark brown velvety hairs, cap yellowish or brownish. The disease is a soft spongy white rot of sapwood of living hardwoods, particularly basswood, horse chestnut, American elm, and catalpa. The toadstools are formed in clusters at wounds.

Coniophora

Basidiomycetes, Agaricales, Thelephoraceae

Pileus resupinate, effuse; hymenium with one layer, cystidia lacking; spores dark; wood-destroying.

Coniophora cerebella. Brown cubical rot; of conifers and sometimes hardwoods—on slash, building timbers, and sometimes living trees. The crustlike, fleshy, fruiting bodies are a little over 2 inches in diameter, olive to brown with whitish margins and smooth to slighty waxy surface.

Coniophora corrugis. Sapwood rot of alpine fir.

Coniothyrium

See under Cankers.

Coniothyrium diplodiella. White rot of grapes; appearing spasmodically on grapes but not one of the more important diseases. Small pycnidia appear on outside of fruit cuticle as shiny, rosy points, also on leaves. Infection is usually through wounds. Spots on ripe grapes are grayish, with brown borders.

Coprinus

Basidiomycetes, Agaricales, Agaricaceae

Inky cap mushrooms; hymenium lining gills; gills deliquesce into a black, inky liquid.

Coprinus urticicola. Fruit rot; of pear.

Corticium

Basidiomycetes, Agaricales, Thelephoraceae

Pileus resupinate, effuse; hymenium with one layer, cystidia lacking; spores hyaline. *Corticium vagum* and other species with a thin film of mycelium with short, broad cells on substratum have been transferred to *Pellicularia*. Species with cystidia have been placed in *Peniophora*. See also *Corticium* under Blights.

Corticium centrifugum. Fisheye fruit rot of apple; generally distributed. A dry, spongy rot often following scab.

Corticium fuciforme. Pink patch of turf; red thread. Grass is first water-soaked, then dead, in isolated patches, 2 to 15 inches in diameter, with pinkish red gelatinous strands of the fungus matting the blades together and growing into coral red horns, ⅛ inch to 2 inches long. These turn brittle, breaking into pieces to spread the pathogen. Velvet bent grasses are more susceptible than colonial and creeping bents. Cadmium compounds will control if applied as protectants before the disease appears.

Corticium galactinum. White root rot of apple; also recorded on blackberry, dewberry, wineberry, peach, and many ornamentals—baptisia, dogwood, holly, flowering almond, flowering plum, iris, winter jasmine, kalmia, pearl bush, peony, spirea, sumac, viburnum, and white campion. The fungus also causes a root rot of white pine and a decay of firs, affecting also western white cedar and spruce. The disease starts at the collar or on larger roots and advances rapidly outward on smaller roots. The collar may be girdled and killed while distal portions are still alive. A dense weft of white mycelium covers roots and penetrates to wood, causing the white rot. The disease is prevalent on lands recently cleared of oaks.

Corticium radiosum. White butt rot; on subalpine fir in Colorado.

Cryptochaete

Basidiomycetes, Agaricales, Thelephoraceae

Basidocarp cartilaginous or coriaceous, erumpent, at first tuberculiform; gloecystidia yellowish or hyaline; cystidia present or lacking; spores hyaline, curved-cylindrical to allantoid, smooth.

Cryptochaete (*Corticium*) **polygonia.** White rot; on aspen in Colorado.

Cylindrocarpon

Fungi Imperfecti, Moniliales, Tuberculariaceae

Conidia on sporodochia; spores with several cells, like *Fusarium* but more nearly cylindrical with rounded ends; cosmopolitan in soil, occasionally pathogenic.

Cylindrocarpon liriodendri. Root rot; of tulip poplar.

Cylindrocarpon radicicola. Sometimes listed as cause of scale-tip rot of Easter lily in Pacific Northwest, but probably secondary. True cause of rot unknown.

Cylindrocladium

See under Blights.

Cylindrocladium crotalaria (perfect stage, *Calonectria crotalaria*). Cylindrocladium black rot; on peanut. Root rot; on tulip tree and kiwi.

Cylindrocladium floridanum. Root rot; on peach and tulip tree.
Cylindrocladium heptaseptatum. Postharvest decay; on leatherleaf fern.
Cylindrocladium pteridis. Postharvest decay; on leatherleaf fern.
Cylindrocladium scorparium. Root rot; on sweet gum and tulip tree.

Daedalea

Basidiomycetes, Agaricales, Polyporaceae

Pileus dimidiate to caplike and stipitate; pores waved, mazelike, or somewhat resembling gills; without cystidia; hymenium labyrinthine.

Daedalea confragosa. White mottled wound rot; of hardwoods, also on fir. It is a white soft rot, a slash destroyer in eastern hardwood forests but sometimes on living trees, especially willows, near wounds. Annual leathery to rigid conks (sporophores) are shelf-shaped, up to 6 inches wide, and may occasionally encircle a small, dead stem. The upper surface is gray to brown, smooth, concentrically zoned. Mouths of tubes on undersurface are elongated, wavy in outline.

Daedalea quercina. Brown cubical rot; of dead timber. Heart rot; of living trees in immediate vicinity of butt wounds, usually on oak, chestnut, sometimes on maple, birch, and hickory. In advanced stages the wood is reduced to a yellow-brown friable mass, with a tendency to break into small cubes. Conks are corky and shelf-shaped, up to 7 inches wide, grayish to almost black with smooth upper surface and cream to brownish undersurface. Mouths are large, elongated, irregular. The conks are more or less perennial.

Daedalea unicolor. Heart rot; canker; of maples and other living hardwoods, including alder, ailanthus, amelanchier, birch, chestnut, and hackberry. Decayed wood is yellow at first, later white and soft. Conks are small, corky, often occuring in clusters, varying from brown to gray.

Daldinia

Ascomycetes, Sphaeriales, Xylariaceae

Perithecia in a globoid to pulvinate, concentrically zoned stroma, carbonaceous to leathery, 3 to 5 cm across; spores one-celled; dark.

Daldinia concentrica. Wood rot; of ash, beech, various hardwoods, and occasionally citrus. There is a superficial white rot on dead parts of living trees. On English ash the decay is called calico wood and is strikingly marked with irregular brown to black bands. Stroma containing perithecia are hemispherical, black, carbonaceous.

Diaporthe

See under Blights.

Diaporthe batatatis. Sweet potato dry rot; if diseased potatoes are planted, the sprouts are affected, but the disease shows little in the field. The roots, infected at the stem end, continue to rot in storage. They are shrunken, often mummified, covered with papillae, which are pycnidia under the skin massed in a coal-black stroma. Optimum temperature for the fungus is 75° F to 90° F. Use cool storage.

Diaporthe citri. Phomopsis stem end rot; melanose; general on citrus; stem rot of mango. The rot on fruits is a leathery, pliable, buff to brown area at the button end. The melanose is a superficial marking of fruits with yellow or brown, scabby, waxy dots or crusts, on leaves, twigs, and fruit, often in streaks. On lemon trees, especially variety Eureka, there is a condition known as decorticosis or shell bark. The outer bark dies, loosens, peels off in longitudinal strips. New bark forms below this layer, and the tree may recover only to develop the disease again in 4 or 5 years. Some leaves and twigs die; the fungus winters in dead wood.

Control. A single copper spray, bordeaux or a neutral copper, applied within 1 to 3 weeks after fruit is set, controls melanose. Copper applied in summer induces excessive cork formation in the melanose lesions, a condition known as star melanose. Applied early, it is noninjurious.

Diaporthe phaseolorum. Fruit rot of pepper and tomato; also pod blight of lima bean. See under Blights.

Dichlotomophthora

See under Cankers.

Dichlotomophthora portulacae. Black stem rot; on common purslane.

Diplodia

See under Blights.

Diplodia natalensis. Diplodia collar and root rot; fruit rot; gummosis; general on citrus, sometimes peach, mango, and avocado. On fruit, the rot resembles phomopsis rot in being a leathery pliable decay of the stem end. It can be prevented by spraying with bordeaux mixture, adding 1% oil to check the increase in scale insects after the copper kills entomogenous fungi keeping them in check. The collar rot may girdle young trees and produce some gumming. Trees affected with root rot seldom recover and should be removed.

Diplodia opuntia. Cladode rot of cactus.

Diplodia phoenicum. Leaf and stalk rot; of date palms; fruit rot. The disease is sometimes fatal to transplanted offshoots. Leaves decay and die prematurely; spores are produced in great abundance. Infection is through wounds. Remove diseased tissue as far as possible and apply copper-lime dust.

Diplodia pinea. Collar rot of pine.

Diplodia theobromae. Sometimes considered a synonym of *D. natalensis* but differentiated by pycnidia developed in a stroma instead of on a subiculum and by darker spores. Causing rots of tropical fruits, stem-end rot of avocado, and collar rot of peanuts. The peanut rot appears in Georgia, Florida, and Alabama. Runners and central stem are invaded; they are brown at first, then black with pycnidia.

Diplodia tubericola. Java black rot; general on sweet potatoes, especially in the South. So named because the first diseased specimens came from Java; it is strictly a storage rot. The inner part of the tuber is black and brittle; innumerable pycnidia are produced under the skin, giving it a pimply appearance. The potato is finally mummified. Use care in handling so skins are not broken or bruised; cure properly after harvest; have suitable temperature in the storage house.

Diplodia zeae. Diplodia corn ear rot; root and stalk rot; seedling blight. This fungus is one of several commonly causing ear rot in corn. The rot is dry, varying from a slight discoloration of kernels to complete rotting of the ear. Seedlings and inner stalks have a dry, brown decay. Another species (*D. macrospora*) is similar but less common, found in more humid, warmer regions. The rot is greater in smutted plants. Treat seed before planting with Spergon.

Diplodina

See under Leaf Spots.

Diplodina persicae. Fruit rot of peach; found in Louisiana in 1952, affecting stem and leaves as well as fruit. All varieties are susceptible.

Echinodontium

Basidiomycetes, Agaricales, Hydnaceae

Hymenium in the form of teeth with spiny serrate margins; pileus caplike to crustose.

Echinodontium tinctorium. The Indian paint fungus causes Brown stringy rot; heartwood rot, of living conifers—balsam fir, hemlock, Engelmann spruce, and Douglas fir—chiefly in the West, often with large losses in forest stands. Light brown to tan spots are produced in heartwood accompanied by small radial burrows resembling insect galleries. Rusty streaks follow the grain. In older trees rot can extend entire length of heartwood and into roots. External signs of decay are hard, woody, hoof-shaped perennial conks, the upper surface dull black, cracked, the undersurface gray, covered with coarse teeth, the interior rust are brick red with a pigment used by the Indians for paint. Even one fruiting body is indicative of extensive decay.

Favolus

Basidiomycetes, Agaricales, Polyporaceae

Pileus usually stipitate; lamellae forking irregularly to form elongate, rhomboidal pores.

Favolus alveolaris. Heart rot of hickory.

Fomes

Basidiomycetes, Agaricales, Polyporaceae

Pileus woody, perennial, with tubes in layers; common cause of wood decay. Spores hyaline to brown to nearly black.

Fomes annosus. Heart rot; root and butt rot; spongy sap rot of conifers, sometimes hardwoods; also root rot on juniper and rhododendron. Infection is through wounds. Tissue thin, mycelial felts are formed between bark and wood, which is pinkish to violet in incipient states. In advanced stages white pockets are formed in wood. Perennial conks are bracket-shaped to flat layers, upper surface zonate, light to dark grayish brown, undersurface biege with small pores. Infection is sometimes through dead roots from mycelium growing through soil, sometimes by spores washed by rain or carried by rodents.

Fomes applanatus. See *Ganoderma applanatum*.

Fomes connatus. White spongy rot; of heartwood of living hardwoods, most prevalent on maples, especially red and sugar maples. Entrance is through wounds or branch stubs, but fruiting is usually on basal stems or scars. Conks appear annually but are perennial, small, less than 6 inches wide, hoof-shaped, corky to woody, white to yellowish, the upper surface covered with moss or algal growth. There is usually a limited area of decay.

Fomes everhartii. Yellow flaky heart rot; of living hardwoods, including birch and beech and especially oaks. Infection is usually limited to the lower trunk, and the flaky character is because the decay is more rapid between rays. There are narrow, dark brown zone lines. Gnarled swellings on the trunk indicate sapwood invasion. The conks are perennial, hard, woody, shelf-shaped, up to a foot wide, with the yellow-brown upper surface becoming black, charred, rough, concentrically grooved with age. The undersurface is reddish brown.

Fomes fomentarius. White mottled rot; of birch, beech, poplar, maple, and other hardwoods. This fungus mostly decays dead timber; sometimes it attacks living trees. The wood is brownish, firm in early stages of decay, but in advanced stages is yellowish white, soft, spongy, with narrow dark zone lines and small radial cracks filled with yellow mycelium, giving a mottled effect. Decay starts in upper part of the bole and progresses downward. Conks are profuse on dead trees. They are hard, perennial, hoof-shaped, up to 8 inches wide, with a smooth concentrically zoned upper surface, gray to brown undersurface. The interior is brown, punky, with tubes encrusted with white.

Fomes fraxinophilus. White mottled rot of ash; a heartwood rot most common on white ash, also on green ash and willow. Conks are up to a foot wide, with dark, rough upper surface, brownish underneath, appearing first when wood has decayed only a short distance. Infection is usually through branch stubs.

Fomes igniarius (*Phellinus igniarius*). **White spongy rot**; white trunk rot; heart rot; on a wide variety of hardwoods but not on conifers. Aspen and birch are particularly susceptible. Decay is mostly confined to heartwood, but in yellow birch living sapwood is killed, causing cankers on the trunk. In an advanced stage the decay is soft, whitish, with fine black lines running through it. The conks are perennial, hard, woody, thick, usually hoof-shaped, up to 8 inches wide, the upper surface gray to black, becoming rough and cracked with age; undersurface is brown and the interior rusty brown with many layers of tubes, the oldest stuffed with white. Infection is through branch stubs and open wounds. A single conk may indicate 15 linear feet of rot in heartwood.

Fomes officinalis (*Fomitopsis officinalis*). **Brown trunk rot**; of conifers infecting heartwood of living larch and other trees. Intensely white spore surface; very bitter, known as the quinine fungus.

Fomes pini (*Trametes pini*). **Red ring rot**; white pocket rot; of conifers, especially Douglas fir, larch, pine, and spruce, causing heavy forest losses. Decay starts as a purplish or red discoloration of the heartwood, but in an advanced stage there are many soft, white fibrous pockets separated by sound wood. Sporophores vary from shelf- to bracket- to hoof-shaped, averaging 4 to 8 inches across, rough gray to brownish black with light brown margin on upper surface and gray to brown underneath. Tube mouths are circular to irregular. On living trees conks are formed at knots or branch stubs.

Fomes pinicola. Brown crumbly rot; of many conifers and some hardwoods—maple, birch, beech, hickory, peach—usually on dead trees, occasionally in heartwood of living trees. Sporophores are shelf- to hoof-shaped, 2 to 10 inches across, sometimes up to 2 feet, upper surface gray to black, often with a red margin, underside white to yellow when fresh.

Fomes rimosus. Heart rot; on locust.

Fomes robustus. Heart rot; of cacti and other desert plants; of oak, fir, juniper, in different strains. Context of sporophores bright yellow-brown; spores hyaline.

Fomes roseus. Brown pocket rot; cubical rot of heartwood of living conifers, particularly Douglas fir. Decay originates in upper part of bole. Wood is yellow to reddish brown, soft, breaking into irregular cubes. Woody bracket conks, up to 6 inches wide, have black tops and rose undersurface. Infection is through dead branch stubs and broken tree tops.

Fusarium

Fungi Imperfecti, Moniliales, Tuberculariaceae

Mycelium and spores generally bright in color. Macroconidia fusoid-curved, septate, on branched conidia in slimy masses, sporodochia; smaller microconidia with one or two cells; resting spores, chlamydospores, common (see Fig. 55). Perfect stage when known usually in Hypocreales, *Nectria*, or *Gibberella*. Cause of many important rots, wilts, and yellows diseases. Classification difficult, with different systems and synonyms, many forms and races.

Fusarium avenaceum. Associated with cereal diseases, fruit and storage rots, but now included in *F. roseum* by many pathologists.

Fusarium culmorum. Also on cereals, included in *F. roseum* by many.

Fusarium moniliforme (*Gibberella fujikuroi*). Ripe rot of figs; carried by the pollinating fig wasp. Root, stalk, pink kernel rot of corn; the rotted kernels are pink to reddish brown; the stalks have brown lesions, may break over or ripen prematurely.

Fusarium orthoceras. Reported as causing a new disease of soybean in Missouri and Iowa. Root rot, with rapid wilting and drying of leaves; most severe on seedlings.

Fusarium oxysporum. Root rot; on apple. Tomato hypocotyl rot; on sugar pine, red and white firs. Stem rot; on zygocactus. Rot of stone plant. This pathogen may also be seedborne and pathogenic on Douglas fir.

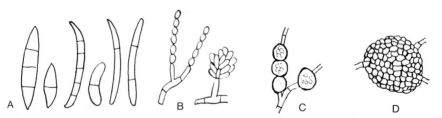

Figure 55. Forms of *Fusarium*. *A*, septate macroconidia; *B*, microconidia in chains or a head; *C*, chlamydospores; *D*, sclerotium.

Fusarium oxysporum. Iris basal rot; on bulbous varieties of iris. Plants fail to emerge, or turn yellow, wilt, and die. Roots are few or none. The bulb is infected at the base, which shrinks; the husk adheres firmly, sometimes with a white or reddish mass of mycelium. The rot is more serious in warm climates and on yellow rather then blue varieties. De Wit is very susceptible; Wedgewood is quite resistant.

Control. Avoid bruising bulbs in digging; sort and discard diseased bulbs right after digging; dry bulbs rapidly.

Fusarium oxysporum. Tulip basal rot; leaves growing from diseased bulbs turn red, wilt, and die; roots are few or none. Bases of bulbs have a rather firm rot with white or pink felty masses of spores. The diseased area usually turns chalky. It is primarily a storage disease in bulb sheds and warehouses.

Fusarium oxysporum f. batatas. Sweet potato stem rot; a widespread field disease, destroying more than 50% of plants in some fields. This fungus also infects *Jacquemontia*. The stem rot is conspicuous about 2 weeks after sprouts are set. Sprouts are yellow or dead, and the vines brown or black, often split near the ground. Some plants develop new roots above the decayed section and so survive. Sweet potatoes from infected plants are small, decayed at the stem end, with vascular tissues brown. The fungi winter in stored roots and can live indefinitely in soil. Varieties Big Stem Jersey, Little Stem Jersey, Maryland Golden, and Nancy Hall are very susceptible; Porto Rico is intermediate; Southern Queen, Triumph, and Yellow Strassburg are quite resistant.

Fusarium oxysporum f. cepae. Bulb rot; basal rot of onion; shallot and garlic. In the field there is progressive yellowing and dying back from tips, the roots commonly turning pink and gradually decaying. The rot is often associated with wounds of maggots and other insects. In storage the rot is most active at room temperature or above.

Fusarium oxysporum f. **chrysanthemi.** Fusarium wilt of chrysanthemum.

Fusarium oxysporum f. **gladioli** (*F. orthoceras* var. *gladioli*). Fusarium brown rot; yellows of gladiolus; a major disease in some sections. Most infection takes place in the field, but subsequent decay appears in storage. Corm lesions are first small, reddish brown, more often on lower half of corm. They enlarge in storage to irregular to circular, sometimes zonate brown areas, which do not infrequently advance until the whole corm is a hard, dry, brownish black mummy. Infection comes from old corms, the fungus penetrating through the basal plate and the center of the new corm. The latter may be entirely decayed in storage, with the fungus advancing from the center to the outside, causing brown to black surface lesions.

Symptoms of yellows, a vascular disease, include bending of young leaf stalks, cupping of leaf stalks in older plants, crooked flower stems, often greener than normal and a curving of growth away from the side of the corm showing rot. There is gradual yellowing and dying of foliage, starting with the oldest leaves. Picardy and Spotlight varieties are particularly susceptible. Nitrogenous fertilizers and manures, especially where phosphorus is low, increase corm rot.

Control. Cure immediately after digging at 95° F to develop wound periderm and cuticle resistant to the fungus; use resistant varieties where possible or a 3- to 4-year rotation.

Fusarium oxysporum f. lilii. Basal rot of lily; on bulbs, roots, stems of garden and native lilies. Corm rot of crocus, also on freesia and cactus (*Cereus*). A chocolate rot at base of scales next to the basal plate progresses until the scales fall away. The disease is more destructive to Madonna and some other garden lilies; it is seldom a problem with Easter lilies grown in the Northwest. Keep bulbs cool in storage, and plant in cool soil. Infection comes from contaminated soil as well as diseased bulbs.

Fusarium oxysporum f. narcissi (*F. bulbigenum*). **Narcissus basal rot;** general on hardy varieties, rare on polyanthus varieties. Rot begins at the root plate at base of bulbs and spreads through central portions first, extension of the rot being more rapid in affected scales than across to adjacent healthy scales. Rotted tissue is chocolate or purplish brown, the mycelium a delicate weft of white or pink threads. The rot is dry, spongy, with little external evidence; it is primarily a storage or transit disease, but it may occur in the field late in the season. When lightly infected bulbs are planted, there is no root development, and plants are stunted. Basal rot is spread in hot-water treatment for nematodes. It is more prevalent where soil temperatures are above 65° F and on large trumpet varieties. Golden Harvest is much more susceptible than King Alfred.

Control. Discard all bulbs showing rot, or ones that are soft when pressed; if disease has occurred previously, plant in a new location.

Fusarium oxysporum f. radicis-lycopersici. Root and crown rot; of tomato.

Fusarium poae. Carnation bud rot; silver spike disease of bluegrass. The interior of carnation buds is brown or pink, decayed, moldy, and often infested with grass mites, which have introduced the spores. The disease is favored by excessive dampness. Pick and destroy diseased buds; control mites.

On bent grasses, fescues, and especially Kentucky bluegrass, seed heads wither before they are fully expanded, appearing silvery. Seeds are aborted, and in moist weather copious mycelium grows from decayed areas in culms. The pathogen is disseminated and grass inoculated by the grass mite (*Siteroptes graminum*). Burning over dead grass is a practical means of control.

Fusarium roseum. Peppermint root and rhizome rot; reported from Oregon as part of a complex with *Rhizonctonia solani* and *Pythium* sp. Necrotic lesions girdle rhizomes; new shoots damp-off. Fall-plowed mint gave stronger stands. This pathogen also causes seedling stem rot on Douglas fir.

Fusarium roseum f. cerealis. Stem rot of carnation and cereals; roots and stems of cuttings and young plants rot; in older stock the diseased tissue turns brownish red or crimson. Infection is only through injured, weak, or old tissue.

Fusarium solani. Tuber rot; on caladium. Stem rot; on chrysanthemum, Fraser fir, Douglas fir, dieffenbachia (cutting rot), and sweet potato (root rot); shefflera is susceptible with no symptoms. This pathogen also causes root rot of apple.

Fusarium solani (Perfect Stage, *Nectria haematococca*). Stem rot and wilt; of *Exacum*.

Fusarium solani f. **cucurbitae** (*Hypomyces solani*). Fusarium root rot of cucurbits; primarily pumpkin and squash, occasionally muskmelon, watermelon, and cucumber. The fungus usually girdles the plant at ground level with a soft dark decay, resulting in a striking wilt of the entire vine. Fruits on the ground may be rotted and the fungus carried on seed to infest clean soil. Do not plant cucurbits in land known to be contaminated.

Fusarium solani f. **phaseoli.** Dry root rot of bean; and lima bean, common but most important in New York, Idaho, and other areas intensively cropped for many years. Indefinite reddish lesions or streaks on taproot and subterranean stem turn dark brown to black. Lateral roots are reduced and plants stunted. It is a late-season disease favored by warm soil. The fungus winters in crop refuse and soil and may be carried in dust on seed. The best control is a long rotation between crops.

Fusarium solani f. **pisi.** Root rot; on chick pea, spruce, pine, fir, and hemlock.

Fusarium sp. Root and seed rot of bird-of-paradise (*Strelitizia*); part of a fungus complex. Controlled by treating seed in hot water, 135°F for 30 minutes, and immediately cooling in cold water and treating planting medium with methyl bromide or steam.

Fusarium sp. Dill root rot; wilt; discovered in Ohio in 1949. Symptoms include browning of roots, necrosis of vascular system, yellowing, wilting, and death. Young plants are most susceptible. Seed treatment did not give satisfactory control.

Fusarium sp. Root rot of sweet peas; reported as prevalent in Montana. Plants turn yellow when in bloom, with necrosis of vascular system, which leads to drying up of plant.

Ganoderma

Basidiomycetes, Agaricales, Polyporaceae

Differing from *Fomes* in having spores truncated at one end and two-layered, the spines of the brown endospore projecting into hyaline exospore. Sporophore has a hard crust, formed by a layer of thick-walled, elongated cells.

Ganoderma applanatum (*Fomes applanatus*). White mottle rot; widely distributed on hardwoods, maple, beech, alder, acacia, birch, horse chestnut, hawthorn, and hickory, and sometimes on conifers. The rot is ordinarily on dead timber, but the fungus can attack living trees through wounds and destroy heartwood for a few feet. In early stages the wood is somewhat bleached, surrounded by a dark brown band. This shelf fungus is called artists' conk because the white undersurface immediately turns brown when bruised and can be used for writing or etching pictures. The upper surface is smooth, zoned, gray or gray-black; up to 2 feet wide.

Ganoderma cudisii. Perennial, with several layers of pores.

Ganoderma lucidum (*Polyporus lucidus*). The varnish or lacquer fungus causes **Heart rot** of eastern hardwoods and conifers, especially hemlock, reported also on boxwood, hackberry, sassafras, maples, and citrus. This fungus may be an important facultative parasite on city shade trees. The rot is white, spongy, with black spots scattered throughout. The conks are annual, with a reddish, shiny, lacquered upper surface and a short, thick, lateral stalk; common on logs, stumps, standing or fallen trees.

Ganoderma sulcatum. **Butt rot of queen palms;** in Florida.

Ganoderma zonatum. On mesquite, in Texas.

Gibberella

See under Blights.

Gibberella zeae (*Fusarium graminearum*). **Corn root rot; stalk rot; ear rot;** also Fusarium head blight or scab of cereals and grasses. Corn is attacked at all ages, with both roots and kernels rotted. Conidia are pinkish in mass; black perithecia are numerous on overwintered corn stalks and residues. Hybrid corn with loose husks exposing the ear tip or varieties, with upright ears retaining water, are more apt to be infected. Rotation and clean plowing aid in control.

Gloeosporium

See under Anthracnose.

Gloeosporium foliicolum. (*Glomerella cingulata*). **Fruit rot;** on citrus fruits.

Glomerella

See under Anthracnose.

Glomerella cingulata. **Bitter rot of apple and pear; fruit rot of peach;** also stem rot, canker, dieback of many fruits and ornamentals; ripe rot of grapes. Bitter rot is a late-season disease of apple, often destructive in central and southern states. The fruits have light brown circular spots, which gradually enlarge; they cover rotting flesh, which has a bitter taste. Lesions become concave and have concentric rings of pink to dark spore pustules in sticky masses. Spores are splashed by rain or carried by flies and other insects. Eventually apples turn into dry, shriveled mummies, in which the fungus overwinters and where the ascospore stage is produced. Large limbs have oval, roughened, sunken cankers. The disease is favored by hot muggy weather.

Apple varieties vary greatly in resistance, and some, like Yellow Newtown, are resistant to the canker but susceptible to fruit rotting. Varieties somewhat resistant include Delicious, Rome Beauty, Stayman Winesap, Winesap, and York Imperial. Ripe rot starts on grapes as they mature and gives a bitter taste to the pulp. To control disease remove mummies from trees and prune out dead twigs and cankers.

Glomerella cingulata var. **vaccinii.** Cranberry bitter rot; a field and storage rot. A soft brownish yellow discoloration develops on fruit late in the season, most serious in a hot July and August.

Godronia

Ascomycetes, Helotiales, Dematiaceae

Apothecia coriacious, pitcher-shaped; spores filiform, hyaline.

Godronia cassandrae (*Fusicoccum putrefaciens*). Cranberry end rot; general on cranberry, with the ascospore stage also found on dead branches of leatherleaf (*Cassandra*). The rot appears late, often after picking and packing, and is enhanced by injuries during harvesting and screening. It starts at either blossom or stem end of the berry; the fruit becomes soft and light-colored.

Godronia cassandrae f. **vaccinii.** On blueberry.

Guignardia

See under Blotch.

Guignaria bidwellii. Black rot of grapes; widespread, principal cause of failure of European grapes in eastern United States, causing more loss than all other grape diseases combined. All parts of the vine are attacked. On leaves, reddish brown dead spots are sprinkled with black pycnidia. Rot starts on half-grown fruit as a pale spot, soon turning brown and involving the entire berry, which shrivels into a black wrinkled mummy, dropping or remaining in the cluster (see Fig. 56). Some berries shatter if attacked early. Ovoid conidia and sometimes microconidia (spermatia) are formed on leaves, berries, and

Figure 56. Black rot on grapes.

canes. Ascospores are produced in overwintered mummied berries. Primary infection in spring comes from either spore form.

Control. If mildew is also a problem, use a fixed copper. Cultivate in early spring so as to cover old mummies with soil and so eliminate that source of inoculum.

Guignardia vaccinii. Cranberrry early rot; scald, blast, general on cranberry and sometimes on huckleberry. All aerial plant parts are attacked, but the disease is more destructive to the fruit. Young fruit may blast and shrivel, but more often rot starts as a light-colored soft spot when fruit is half grown. The berry mummifies, turns black, and is covered with small pycnidia. Leaves have reddish brown spots, sometimes drop prematurely.

Helicobasidium

Basidiomycetes, Tremellales, Auriculariaceae

An exposed cottony hymenium or fruiting layer; basidia transversely septate; spores coiled like a watch spring.

Helicobasidium corticioides. Brown pocket rot; on subalpine fir, in Colorado.
Helicobasidium purpureum (*Rhizoctonia crocorum*). Violet root rot; of

potato, sweet potato, asparagus, beet, carrot and some ornamentals—ash, catalpa, chinaberry, crocus, elm, mulberry, parthenocissus, and western soapberry. The fungus invades roots from the soil, turning them reddish or violet. The disease is confined to underground parts unless continuously wet weather allows the reddish-purple mycelium to grow up the stem. Small, darker sclerotia are embedded in this purplish mat, which turns brown with age.

Helminthosporium

See under Blights.

Helminthosporium cactivorum. Stem rot of cacti; basal or top rot of seedling cacti, which turn into a shrunken brown mummy covered with spores. Initial symptoms are yellow lesions; rotting may be complete in 2 to 4 days.

Helminthosporium sesami. Stem rot; on sesame, in Texas.

Helminthosporium turcicum. Crown rot of sweet corn; leaf blight. See under Blights.

Hericium

Basidiomycetes, Agaricales, Hydnaceae

Fleshy, branched or unbranched, with subulate spines long and pendant; spores spherical or subspherical, staining blue with iodine. Like *Hydnum* but sporophore formed on wood, not on the ground.

Hericium erinaceus (*Hydnum erinaceus*), hedgehog fungus. White heart rot; occasional on living oak, maple, and other trees. The soft, white, spongy rot may entirely decompose the tissue, leaving large hollows lined with yellowish mycelium. Sporophores are annual; soft, white, browning with age, globular with a hairy top and long slender teeth on the lower surface.

Hypholoma

Basidiomycetes, Agaricales, Agaricaceae

Margin of cap with a curtainlike veil; stipe with incomplete or vanishing ring; spores purple.

Hypholoma perplexum. Root rot of currant.

Idriella

Fungi Imperfecti, Moniliales, Moniliaceae

Mycelium hyaline to brown; conidiophores brown, simple, nonseptate, narrowed above, with prominent spore scars; conidia (sympodulospores) lunate to falcate, with pointed ends, produced in clusters near apex of the conidiophore; aleuriospores brown, several-celled.

Idriella lunata. Root rot; on strawberry.

Irpex

Basidiomycetes, Agaricales, Polyporaceae

Resupinate, effused-reflexed, or shelflike; younger parts of hymenophore are poroid; with increasing age produce flattened teeth.

Irpex tulipiferae. Wood rot and decline; of apple.

Isaria

Fungi Imperfecti, Moniliales, Stilbaceae

Conidiophores equally distributed on a synnema, erect fascicle of hyphae; conidia hyaline, one-celled, ovoid; some species in insects.

Isaria clonostachoides. Isaria rot of tomato; fruits are partly covered with cottony mycelium, white turning pink or orange and becoming granular, but rot remaining firm. Reported from around Washington, D.C.

Kluyveromyces

Ascomycetes, Endomycetales, Saccharomycetaceae

Kluyveromyces marxianus var. **marxianus.** Soft rot; of onion caused by a true yeast on bulbs.

Lentinus

Basidiomycetes, Agaricales, Agaricaceae

Gills are notched or serrate at edge, decurrent, stipe often lateral or lacking cap, tough-fleshy to leathery; spores white.

Lentinus lepideus. Scaly cap; causing a brown cubial rot of coniferous wood and sometimes decaying heartwood of living pines.

Lentinus tigrinus. Sapwood rot; white mottled butt rot of living hardwoods, commonly associated with fire scars and one of the most important decay fungi in the Mississippi Delta. Fruiting body is white with cap depressed in center, more or less covered with blackish brown hairy scales, rarely developing on living trees.

Lenzites

Basidiomycetes, Agaricales, Polyporaceae

Pores elongated radially to resemble gills; pileus shelflike; woolly and zonate above.

Lenzites betulina. Heart rot; of birch and cypress.

Lenzites saepiaria (*Gloeophyllum saepiaria* to those who separate forms with a rust-brown trama). **Timber rot; brown pocket rot**; usually of dead sapwood, occasionally a heart rot, rarely on living trees. It is the common destroyer of coniferous slash; it is found on telephone poles and other timber. Fruiting bodies are long narrow shelves coming from cracks, the upper surface a yellow red to dark reddish brown.

Leptosphaeria

See under Blights.

Leptosphaeria korrae. Root and crown rot; of turf grasses (necrotic ring spot).

Macrophomina

Fungi Imperfecti, Sphaeropsidales, Sphaerioidaceae

Spores hyaline, one-celled, in pycnidia.

Macrophomina phaseoli (sterile stage *Sclerotium* or *Rhizoctonia bataticola*). **Charcoal rot; ashy stem blight**; on many plants in warm climates and sometimes in temperate zones. The name for the sterile stage comes from sweet potato, and the term *charcoal rot* is used because the interior of the potato becomes jet black.

The fungus lives in the soil, is particularly prevalent in warm soils, and attacks roots and stems of a varied list of hosts, including bean, lima bean, soybean, beet, corn, cowpea, cabbage, eggplant, garlic, gourds, pepper, strawberry, and watermelon; also chrysanthemum, dahlia, garden mallow, mountain laurel, marigold, and zinnia. In most cases the pycnidial stage is not formed. The mycelium spreads through the soil, and very small black sclerotia are formed in great abundance on or in lower stems and roots. On beans, black sunken cankers appear just below the cotylendonary node, and the lesion may extend up the stem, ashy gray in the center. Stems may break over, or the growing point may be killed. In sweet potatoes the disease is a storage rot, the tissue becoming a dark red-brown with the outer zone black from the formation of myriads of sclerotia. The decay is spongy, then hard, mummified. The fungus is spread in irrigation water, crop debris, imported soil, and on seed.

Control. Use bean seed grown in western disease-free regions. Keep plants growing vigorously with proper food and water; practice general sanitation.

Macrophomina phaseolina. Charcoal rot; on soybean, sunflower, *Amaranthus, Euphorbia* spp., *Ipomea, Sonchus,* and *Tidestrominia.* Root rot; on caper spurge.

Melanconium

See under Leaf Spots.

Melanconium fuligineum. Bitter rot of grapes; widespread but especially serious on Muscadine grapes in Georgia. Decayed berry pulp has a bitter taste; up to 30% of fruit is reduced to dry, hollow shells.

Control. Spray with bordeaux mixture three times at 14-day intervals beginning after fruit is set. The later sprays for black rot should control bitter rot.

Monilinia

See under Blights.

Monilinia fructicola (*Sclerotinia fructicola*). Brown rot of stone fruits; blossom blight, general on peach, plum, and cherry; also on apricot, almond, beach plum, Japanese quince, and rarely, apple and pear. The fungus is distinct from the species in Europe (*Sclerotinia fructigena*), causing brown rot of stone fruits and also a serious apple rot. In the United States, brown rot is the most destructive stone-fruit disease, causing an annual peach loss of over $5 million. *Monilina fructicola* is the usual causative agent east of the Rocky Mountains; *M. laxa* causes a similar rot and blossom blight on the Pacific Coast. See also under Blights.

Flowers turn brown prematurely, rot in moist weather; the calyx cup is blackened, and the discoloration may extend down into the pedicels. Infrequently there is a leaf and twig blight; cankers are formed on the larger limbs, with exudation of gum. The fruit rot is the familiar stage seen in any backyard with a fruit tree and usually in baskets of peaches, plums, or cherries purchased for preserving and held over to the next day. The rot starts as a small, circular, brown spot but spreads rapidly to take in the entire fruit, with the rotted surface covered with gray to light brown spore tufts or cusions (sporodochia), sometimes in concentric rings (see Fig. 57). Conidia are formed in chains on the sporodochia. The fruit finally shrinks and mummifies and either falls to the ground or remains clinging to the tree.

Figure 57. Brown rot on plums.

The fungus and decayed tissue together form a stroma that acts as a sclerotium; in spring, if the mummy has been kept moist and partially or wholly covered with soil, cup-shaped brown apothecia are produced. Primary infection is from ascospores, forcibly ejected and carried up to blossoms by air currents or from a new crop of conidia formed on mummies hanging on trees. Secondary infection is from conidia wind-borne from blossom to blossom and later from fruit to fruit. Entrance is often through wounds made by the plum curculio, oriental fruit moth, and other insects. Rotting and conidial production continue after picking.

The rot is favored by wet weather, conidia germinating only in a film of water. Acidic soil is said to increase apothecial production from mummies on the ground. In a normal season reduction from blossom blight is not important because some thinning is advantageous; but if blossom blight is not prevented, inoculum is provided for the fruit rot that causes such enormous losses.

Control. Sanitary measures are important. In the small garden rake up and burn or bury deeply the fallen mummies; pick mummies from trees; cut out twigs showing gum; in summer remove infected fruit before conidia form. Standard control has been wettable sulfur sprays or sulfur dust, applied every 3 or 4 days during bloom to control blossom blight; when shucks are falling; 2 or 3 weeks after shuck fall; and 2 to 4 weeks before fruit ripens. In some instances the newer organic fungicides are preferred to sulfur, and sometimes they are used with it. Control of the plum curculio is very important. For one or two trees in a home garden one of the all-purpose fruit sprays or dusts now available under various trade names may be satisfactory. Consult your county agent for the schedule right for your locality.

Monilinia laxa (*Sclerotinia laxa*) **Brown rot**; green and ripe fruit rot; blossom blight; on almond, apple, apricot, cherry, peach, plum, pear, nectarine, quince, and Japanese quince in Washington, Oregon, and California; also reported from Wisconsin and Michigan. Although this disease is similar to the one caused by *M. fructicola*, the blossom and twig blight phase is more important than the brown rot. Sulfur, which can be used in later sprays for most stone fruits, may injure apricots.

Monilinia oxycocci (*Sclerotinia oxycocci*). **Cranberry hard rot**; tip blight; in Pacific Northwest and Wisconsin. Young growing tips wilt and dry just before blossoming; grayish spore tufts are formed on tips. Fruit is attacked through blossoms or wounds. The berries are yellowish white, firm, leathery, cottony inside, turning dark and mummifying late in the season. The disease is too erratic to justify cost of regular spraying; clean harvest will prevent overwintering.

Monilinia urnula (*Sclerotinia vaccinii-corymbosi*). **Blueberry brown rot**; mummy berry; twig blight; of high bush blueberry, similar to hard rot of cranberry. Varieties differ in susceptibility, with June and Rancocas often showing severe primary infection.

Monilochaetes

Fungi Imperfecti, Moniliales, Monileaceae

Conidiophores dark, erect, slender, usually simple; septate; conidia hyaline or becoming pigmented in age, borne singly at apex or produced in chains under conditions of high humidity.

Monilochaetes infuscans. Root rot; of weed species of genus *Ipomoea*.

Mucor

Phycomycetes, Mucorales, Mucoraceae

Mycelium profusely developed. Sporangiophores erect, simple or branched, all branches terminated by sporangia that are globose to pyriform with a columella and thin wall; gametangia essentially alike, suspensors without definite outgrowths; hyaline chlamydospores sometimes formed.

Mucor racemosus. Storage rot of sweet potato; occasional after chilling; fruit rot of citrus. Control with low temperatures and dry atmosphere in the storage house.

Mucor mucedo. Postharvest rot; of tomato.

Mucor piriformis. Postharvest rot; of tomato.

Myrothecium

See under Leaf Spots.

Myrothecium roridum. Ring rot of tomato; crown rot of snapdragon; crowns of greenhouse snapdragons appear water-soaked, then covered with a thin white mycelium and numerous black sporodochia. Irregular brown spots on tomato fruits are surrounded by slight depressions. Also causes root rot of red clover and alfalfa.

Myrothecium sp. On Bells of Ireland, causing crown necrosis. Stems are girdled at ground level; tops wilt; basal branches die.

Nematospora

Ascomycetes, Endomycetales, Saccharomycetaceae

This is a yeast or budding fungus, following after insect injury; asci, with 8 to 16 spores, derived directly from vegetative mycelial cells; spores elongate, fusiform to needle-shaped, flagellate.

Nematospora coryli. Yeast spot; of soybean. **Dry rot;** of pomegranate, citrus. **Pod spot;** of pepper, bean, and soybean. **Cloudy spot;** of tomato. **Kernel spot;** of pecan. The yeast is almost always associated with plant bug injury. The western leaf-footed plant bug carries the fungus from pomegranate to citrus. On pomegranates depressed light spots in flesh around seeds are followed by general browning and collapse. In citrus, the juice sacs just inside the rind dry out with a brownish to reddish stain. Cloudy spot on tomato fruit is associated with pumpkin bugs and leaf-footed plant bugs. Brown areas are formed on pecan kernels.

Nematospora phaseoli. Yeast spot of lima bean; a seed disease, destructive from Maryland southward. Infection follows puncture of pods by the southern green stinkbug and possibly other insects. The seed lesions are dark brown, sunken, wrinkled.

Neurospora

Ascomycetes, Sphaeriales, Sphaeriaceae (Fimetariaceae)

Perithecia flask-shaped, membranous; ascospores dark, one-celled with gelatinous coating; conidial stage monilioid.

Neurospora sitophila. Ripe rot of pear; the fungus is the same one causing pink bakery mold on bread. There is a luxuriant pink growth over fruit; conidia are formed in chains.

Nigrospora

Fungi Imperfecti, Moniliales, Dematiaceae

All hyphae more or less creeping, hyaline; conidiophores short, dark, cells somewhat inflated; conidia black, one-celled, situated on a flattened, hyaline vesicle at top of the conidiophore.

Nigrospora oryzae. Ripe fruit rot of tomato; nigrospora cob rot of corn; corn cobs are shredded, with the pith completely disintegrated; kernels are filled with masses of black spores. Corn on poor soil is more susceptible; stalks break at any point. Rapid drying checks infection of seed corn.

Olpidium

Phycomycetes, Chytridiales, Olpidiaceae

Endobiotic, living in host cells or tissues, living or dead.

Olpidium brassiacae. Sometimes found in outer cells of rootlets of cabbage and other crucifers, tomato, lettuce, and other plants, producing zoosporangia and resting spores in the cells. The effect on the host is usually merely a slight unthriftiness. Olpidium has been found associated with big vein, a disease of lettuce, now thought due to a virus.

Omphalia (Omphalina)

Basidomycetes, Agaricales, Agaricaceae

Gills decurrent, cap sunken in center, somewhat funnel-shaped; central cartilaginous stem; spores white.

Omphalia pigmentata; O. tralucida. Decline disease of date palms; growth is retarded; roots decay; leaves die prematurely; fruit is worthless. Deglet Noor variety is most susceptible. Select thrifty offshoots from healthy plants for new date gardens. Soil can be treated with carbon disulfide, as for Armillaria rot.

Oospora

Fungi Imperfecti, Moniliales, Moniliaceae

Slender branched or unbranched mycelium breaking into ellipsoidal or spherical hyaline or light-colored conidia called "oidia."

Oospora citri-aurantii. Sour rot of citrus; a soft, putrid slimy rot of fruit, mostly of stored lemons, where it is spread by contact. The mycelium forms a thin, compact, somewhat wrinkled layer over the surface. Fruitflies help to spread the spores. Fruit should be stored as short a time as possible and frequent inspections made during storage.

Oospora lactis. Sour rot; watery fruit rot of tomato; common in transit and market, especially on fruit from the South. There is a velvety or granular coating over the surface or a fluffy growth along the margin of cracks, and a disagreeable odor and flavor. The rot is common on ripe fruit touching the ground, occasional on green fruit. The fungus is a weak parasite, entering through wounds.

Paecilomyces

Fungi Imperfecti, Moniliales, Moniliaceae

Conidiophores and branches more divergent than in *Penicillium*; conidia (phialospores) in dry basipetal chains, one-celled, ovoid to fusoid, hyaline.

Paecilomyces buxi. Root rot and decline; on boxwood.

Pellicularia

See under Blights.

Pellicularia filamentosa (considered by some *Botryobasidium*). **Rhizoctoniose; black scurf of potatoes**; stem canker and soil rot of beans (see under Blights for web blight of beans and other plants); rhizoctonia dry rot canker of beets, crown and crater rot of carrots; rhizoctonia disease of celery, crucifers, cucurbits; bottom rot of lettuce; damping-off of pepper and eggplant; root rot of onion; root and basal stem rot of pea; crown rot of rhubarb. The sterile stage of this fungus, *Rhizoctonia solani*, was first named in 1858 in a German textbook and is still the most familiar term for a fungus with many pathogenic strains causing many types of diseases.

Any cook has seen signs of the pathogen on potato tubers—small brown to black hard flecks, sclerotia, on the skin. They look like particles of dirt but do not scrub off when potatoes are washed. There may be only one or two sclerotia, or they may nearly cover the whole surface of the tuber. When such potatoes are planted, the growing point may be killed. Some sprouts renew growth after being girdled, which may be repeated until they die. Larger plants have stems decayed just below the soil line, interrupting the downward transfer of food and resulting in a cluster of green or reddish aerial tubers. Roots may be killed back extensively. Most of the tubers are small, often with a brown jelly rot at the stem end.

Under moist conditions a white cobwebby weft of mycelium is formed at the base of potato stems, and the basidial stage is produced as a powdery crust on this weft. The fungus winters as mycelium or sclerotia in soil or tubers. The mycelium can grow saprophytically long distances in the soil independent of any plant. Infection is favored by cool temperatures; the disease is most serious in wet seasons on heavy soils. The average yearly loss for the country is about 10 million bushels, 2% to 3%, but individual losses may be from 5% to 50%. For control use healthy tubers for seed.

Pellicularia filamentosa (*Rhizoctonia solani*). **Brown patch of turf**; root and leaf rot of lawn grasses, wheat grass, bent grass, fescues, ryegrass, Kentucky bluegrass (infrequently on Canada bluegrass), St. Augustine grass, and Bermuda grass. Brown or blackish patches on the turf resemble sunscald or chinch bug injury. The areas are roughly circular, from an inch to 3 feet across, sometimes up to 20 feet. The fungus works outward with a "smoke ring" of grayish black mycelium at the advancing margin. The leaves are first water-soaked, black, then collapsed, dry, and light brown, but the roots are seldom killed. The disease develops most rapidly during warm humid periods and with an excess of nitrogen.

Pellicularia filamentosa (*Rhizoctonia solani*). **Root and stem rot; damping-off**; of ornamentals. In wet weather cobwebby mycelium develops on lower portions of stems; the lower leaves rot and upper portions of plants wilt and die. Seedlings and older plants so rotted include *Aconitum*, abelia, *Achillea*,

Ageratum, aster, artichoke, begonia, calendula, campanula, carnation, endive, dahlia, delphinium, geranium, iris, lettuce, lupine, orchids, platycodon, poinsettia, salsify, sunflower, and tulip. For control avoid excessive use of manure.

Penicillium

See under Cankers.

Various species cause blue, green, occasionally pink molds, including the common blue-green mold on jellies. Some produce antibiotics, *Penicillium notatum* being the one used for production of penicillin.

Penicillium digitatum. Green mold of citrus fruit; clove rot of garlic; on lemons and other citrus, olive-green powdery spore masses, forming a dust cloud when disturbed, cover fruit except for a band of white mycelium outside the green area. Garlic plants are yellow and stunted. Avoid injury in harvesting and packing. Commercial growers use chemicals in the wash water to prevent decay.

Penicillium expansum. Blue mold rot; of many fruits. Soft rot; of apple, pear, avocado, pomegranate, Japanese persimmon, quince, and feijoa. The decay on avocados is slow, and often the affected portions can be trimmed off. This fungus causes 80% to 90% of the decay of storage apples. The rotted portions are light-colored, soft, watery, with a disagreeable moldy taste and odor. A few rotted apples spoil all the others in a container. Use great care in harvesting and grading to avoid wounds; keep temperature as low as possible.

Penicillium gladioli. Blue mold rot; penicillium dry rot of gladiolus; also found in imported bulbs—scilla, tritonia (montbretia). It is a storage rot. Light to dark brown sunken lesions appear on any part of corms with border of the decayed area water-soaked and greenish. Small grayish sclerotia are formed, and under moist conditions masses of blue mold. Dry rapidly after harvest, 80° F for 10 to 14 days, then store at low temperature; avoid wounds and bruises; sort before planting.

Penicillium italicum. Blue contact mold of citrus; fruit rot. The mold is blue in the older portion but powdery white at margins. It spreads readily from fruit to fruit by contact, through uninjured skin.

Penicillium martensii. Crown rot of asparagus; a seedling disease recently prevalent in Washington, following freezing injury. Bright blue spore masses appear on diseased crowns. Protect seedlings for winter by slight hilling in fall; avoid mechanical injury in harvesting; prevent drying out of crowns between digging and replanting.

Penicillium roseum. Fruit rot; of citrus and of dates. A pink mold, found on lemons but not oranges.

Penicillium vermoeseni. Bud rot of palms; the terminal bud is killed and base of leaf stalks rotted. Affected trees of very susceptible *Washingtonia filifera* should be replaced with resistant *Washingtonia robusta*, Mexican fan palm. See also under Cankers and Diebacks.

Peniophora

Basidiomycetes, Polyporales, Thelephoraceae

Like *Corticium* but with cystidia.

Peniophora luna. Brown rot; in lodgepole pine, Rocky Mountain area.

Pestalotia

See under Blights.

Pestalotia longisetula. Root, stolon, and petiole rot; on strawberry.

Phialophora

Fungi Imperfecti, Moniliales, Dematiaceae

Conidiophores dark, short, single or clustered; phialides broader near middle, tapering toward ends, producing conidia endogenously, spores subhyaline to dark, one-celled.

Phialophora malorum. Storage rot of apples.

Phlebia

Basidiomycetes, Aphyllophorales, Corticiaceae

Basidiocarp effuse, typically monomitic; spores even in general outline, hyaline or pale in color, typically nonamyloid.

Phlebia chrysocrea. Heart rot; on oak.

Pholiota

Basidomycetes, Agaricales, Agaricaceae

Spores ochre yellow to rusty brown; gills attached to stipe, which has an annulus but no cup at the base.

Pholiota adiposa. Brown mottled heart rot; of maple and other living hardwoods—basswood, birches, poplars, and more rarely conifers. The wood has brown mottled streaks. The sporophores are formed in clusters on trunks and stumps—mushrooms with yellow central stems and caps, sticky yellow slightly scaly upper surface, yellow to brown gills.

Phoma

See under Blackleg.

Phoma apiicola. Phoma root rot of celery; occasionally serious, especially in Golden Self Blanching, also on carrot, parsnip, parsley, and caraway. The disease appears first in the seedbed, a black rot of the crown or base of leafstalks. Plants are stunted, outer leaves or entire plant killed, falling over as roots rot off. Spores are produced in tendrils from black pycnidia and spread in rains and irrigation water. Use clean seed, grown in California, where the disease is rare; sterilize seedbed soil or use a fresh location.

Phoma betae (*Pleospora betae*). Phoma rot of beets; causing black root of seedlings, necrotic streaks on seedstalks, brown spots on old leaves and rot of fleshy roots. The fungus is seed-borne and winters in roots carried over for seed production and in debris. Crop rotation is essential.

Phoma destructiva. Phoma rot of tomato and pepper; nearly general, especially in the South, but not in North Central states. Small, irregular dark spots appear on leaves in great numbers; zonate markings are similar to those of early blight. Severely infected leaves turn yellow, wither. Fruit spots in field are small, ⅛ inch, slightly depressed, with numerous tiny black pycnidia. After harvest, spots enlarge to ½ inch to 1½ inches and become black, leathery, with minute pustules. The fungus winters in decaying refuse in soil; seedbed infection is common, and the disease reaches the field via infected seedlings. Masses of spores produced on leaves are washed to fruits by rain or spread by workers and are distributed during harvesting and packing.

Control. Locate seedbeds away from land that has previously grown tomatoes; spray as for early blight; do not harvest when wet.

Phoma macdonaldii (Perfect Stage, *Leptosphaeria lindquistii*). Stem rot; on sunflower.

Phomopsis

See under Blights.

Phomopsis mali. Fruit and core rot (postharvest); on apple.
Phomopsis vaccinii. Fruit rot; on blueberry.
Phomopsis sp. Fruit rot; on peach.

Phymatotrichum

Fungi Imperfecti, Moniliales, Moniliaceae

Conidiophores stout with inflated tips bearing loose heads of conidia; spores hyaline; one-celled, produced on surface of soil.

Phymatotrichum omnivorum. Texas root rot; Phymatotrichum root rot; cotton root rot. It is the most destructive plant disease in Texas, a limiting factor in gardening and crop production. It occurs in the Red River counties of Oklahoma, the southwestern half of Arizona, the southeastern edge of Nevada and California, the southeastern corner of Arkansas and Utah, the northwestern corner of Louisiana, and in most of Texas except the Panhandle.

The list of susceptible plants—flowers, vegetables, fruits, field crops, and trees—is much longer than that of plants resistant to this omnivorous fungus, so aptly named. At least 1,700 plant species are attacked, more than by any other known pathogen. Because of the wide host range and destructiveness, the economic losses are enormous, $100 million a year in Texas alone, with perhaps $50 million in adjacent states.

Crops that either are resistant or escape the disease are the cereals and grasses, annuals grown in winter only, and sweet alyssum, amaranth, sweet basil, beauty berry, bee balm, collinsia, diosma, calceolaria, calla lily, California poppy, candytuft, canna, chicory, cranberry, cucumber, currant, cyclamen, daffodil, dahoon, deutzia, dill, fennel, fern, staghorn, foxglove, freesia, goldentuft, mustang grape, gypsophila, hackberry, hoarhound, hyacinth, iris, lily, nigella, marsh marigold, mignonette, mints, mimulus, muskmelon, mustard, nasturtium, oak, osage orange, oxalis, Indian paintbrush, palms, pansy, petunia, phlox, Chinese pink, pitcher plant, pomegranate, poppy, portulaca, primrose, pumpkin, red cedar, sage, scarlet brush, snapdragon, snowdrop, stock, strawberry, strawflower, tuberose, valerian, verbena, violet, wallflower, wandering jew, watercress, watermelon, yaupon, yucca, and zinnia.

Phymatotrichum root rot occurs from July until frost. It kills plants in more or less circular spots, ranging from a few yards to an acre or more. Death may come within a few days of first wilt symptoms, and just preceding the wilt, plants actually run a fever. If plants next to the wilted ones are pulled out, these apparently healthy plants will often be found to be covered with yellow to buff mats of mycelium, and under moist conditions spore mats appear on the surface of the soil around diseased plants. Such mats are 2 to 12 inches in diameter, first snow white and cottony, later tan and powdery from spores produced in quantities. The fungus spreads through the soil by means of rhizomorphs, smooth, dark brown strands. The rate of spread may be 2 to 8 feet a month in an alfalfa field, 5 to 30 feet a season in a cotton field, or around fruit trees.

Sclerotia are formed along the mycelial strands. They are small, roundish, light at first, then dark and warty. The fungus winters either as sclerotia in soil, persisting several years in the absence of live hosts, or as dormant mycelium in living roots. The disease is most common and severe on heavy, alkaline soils. Abundant organic material reduces rot by favoring antagonistic soil saprophytes.

Control. In ornamental plantings replace diseased plants with some of those given in the resistant list. Monocotyledons are generally resistant. In locating new orchards, make sure that root rot has not been present previously by growing an indicator crop of cotton for a year. Grow immune crops in rotation with susceptible crops, and grow susceptible annuals in winter rather than summer. Try heavy manuring.

Ammonium sulfate can sometimes save a valuable ornamental tree or shrubs already infected with root rot. Prune back the top, make a circular ridge about the plant at the edge of the branch spread, and work ammonium sulfate into the soil within the ridge; then fill the basin with water to a depth of 4 inches. The chemical treatment and watering is repeated in 5 to 10 days, then no more chemical the same season. Follow through with frequent watering.

Physalospora

See under Cankers.

Physalospora mutila. Black rot of apple; in the West, similar to disease by *P. obtusa* in the East.

Physalospora obtusa. Black rot of apple; New York apple tree canker; frog-eye leaf spot; general on apple and crabapple, from Atlantic Coast to the Great Plains; also widespread on pear, mountain ash, peach, quince, currant, and various woody species. The fungus, in its imperfect stage (*Sphaeropsis malorum*), was first reported as causing apple rot in 1879.

The lesions start as small brown spots, frequently at a worm hole, but they darken and turn black as they expand. There is usually one lesion to an apple, often at the calyx end, with concentric zones of black and brown, and minute black pycnidia. The rot eventually takes in the whole fruit, which is shriveled and wrinkled and finally mummifies. The pycnidia are black, carbonaceous, and may contain three types of spores—large one-celled brown spores, large hyaline spores, and two-celled colored spores. Perithecia, sometimes formed in cankers or on twigs, apparently play little part in the life history, the fungus wintering as dormant mycelium or in the pycnidial state. Conidia, entering through wounds, start primary infection in spring on leaves with the small "frog-eye" leaf spots.

Control. Use the same spray schedule as for apple scab, starting with an application at the time of petal fall. Clean up mummified apples; avoid bruising; cut out cankers.

Physalospora rhodina. Diplodia rot of citrus; also on fig, rubber tree, and pear, possibly apple. The conidial stage is a *Diplodia*, probably *D. natalensis*, with dark, two-celled spores.

Phytophthora

See under Blights.

Phytophthora cactorum. Stem rot; foot rot; of lily, *Photinia*, tulip, *Hydrastis*, blue laceflower, baby's breath, *Centaurea*, peony, clarkia, rhubarb, and tomato;

leather rot of strawberries; collar rot of dogwood, walnut, apple, and pear. With foot rot, lilies suddenly fall over, wilt, and die; the lower part of the stem is shrunken. Plant only healthy bulbs and where the disease has not occurred previously.

Strawberry leather rot occurs when berries come in contact with soil, starting with a brown rotted area on green fruit and a discoloration of vascular bundles. Ripe fruit has a bitter taste. Crown rot of rhubarb starts with slightly sunken lesions at base of petiole, which enlarge until the entire leaf collapses. Spraying crowns with bordeaux mixture is helpful. Start new beds with healthy plants. Collar rot on English walnut is a bark disease starting below the ground with irregular dark brown or black cankers and soft, spongy areas at the crown, a black fluid in cambial cavities. Trees are stunted, with sparse yellow-green top growth. There may be an unusually heavy crop of nuts, but the tree dies the next season. Grow walnuts grafted on Persian or Paradox rootstocks. See under Cankers and Diebacks for symptoms on apple and dogwood.

Stem rot and wilt of snapdragon starts with water-soaked lesions on the stem; these turn yellow, brown, enlarge to girdle the stem; plant wilts. Sterilize soil before planting.

Phytophthora capsici. One of the species causing buckeye rot of tomato. See under Blights for pepper rot and blight.

Phytophthora cinnamomi. Avocado root rot; pine little leaf; collar rot of hardwoods and conifers, seedling root rot, on more than 100 hosts, including firs, cedars, cypress, juniper, Japanese umbrella tree, larch, pine, spruce, arborvitae, heaths, heather, azalea, highbush blueberry, rhododendron, camellia, birch, western swordfern, walnut, oak, locust, yew, and gold-dust plant. In conifers root rot is dry with resin flow; needles gradually lose color. Infected tissue of hardwoods turns reddish brown except in black walnut, where it is black; seedlings die. The disease is aggravated in pine by poor aeration and low fertility.

Root rot is the most serious avocado disease in California, present also in Florida and Texas. It occurs on soils with poor drainage, excess moisture being necessary. As the roots rot, leaves become light-colored and wilt even if soil is moist; trees decline over a period of years. The fungus can be spread with seed if fruit is allowed to lie on the ground. Treat suspected seed with hot water, 120° F to 125° F for 30 minutes; use nursery stock grown in fumigated soil; prevent movement of soil water from infested areas; plant on well-drained soil; water trees individually to avoid excess moisture.

Phytophthora citricola. Root rot; of Fraser fir seedlings in Christmas tree plantings; also fruit rot of avocado.

Phytophthora citrophthora. Brown rot; gummosis; foot rot of citrus; masses of amber gum break out from the trunk near crown; the bark is killed above and below ground; foliage turns yellow; trees may die. The disease is prevalent where excess water stands around the tree after irrigation or where there is poor drainage. Brown rot of fruit is a decay with no visible surface mold, except in moist air, but a slightly rancid, penetrating odor. Lemons and oranges may be affected on the tree, on branches near the ground, and there is much loss in storage. The fungus lives in the soil; spores are splashed up in rainy weather and are spread in the washing tank. Lemons are most susceptible to gummosis, then lime, pumelo, grapefruit, sweet orange, and finally sour and trifoliate oranges. The latter two are used as fairly resistant understocks.

Control. Plant susceptible trees high, with lateral roots barely covered; expose the root crown of infected trees with a basin 6 inches deep and 4 feet across. Once a year cover crown and lower trunk with bordeaux paste. To control fruit rot, spray ground and lower branches, up to 3 feet, with bordeaux mixture just before rains begin. If fumigation is to be practiced, substitute a copper-zinc-lime spray for the bordeaux.

Phytophthora cryptogea. Collar rot; of rhododendron, China aster, marigold, gloxinia, and zinnia; root, crown, and stem rot on watercress, juniper, African daisy, chicory, and ice plant; stem rot on sunflower, pink rot of potato. Stems and roots appear water-soaked, then black from a soft rot. Sterilize soil.

Phytophthora cryptogea var. **richardiae. Root rot of calla**; the feeder roots rot from tips back to rhizomes, leaving the epidermis a hollow tuber. New roots sent out from the rhizome rot in turn. Leaves turn yellow and drop, starting with outer leaves; plants do not flower, or the tips of blossoms turn brown. Rot in the rhizome is dry and spongy, not wet and slimy. Clean old rhizomes thoroughly; cut out rotted spots. Grow in sterilized pots rather than benches.

Phytophthora drechsleri. Root rot; on fir. Sometimes associated with tomato buckeye rot, basal decay of sugar beets, tuber rot of potato, root rot of safflower.

Phytophthora erythroseptica. Pink watery rot of potato; rot of calla lily and golden calla, crown and root rot of wild rice. The rot starts at stem end of potatoes; affected tissues exude water under pressure. When tubers are cut, flesh turns pink or red, then black. The fungus can exist in soil for 4 years.

Phytophthora fragariae. Strawberry red stele disease; brown core rot; a very serious strawberry disease, first noticed in Illinois in 1930, now widespread in northern strawberry sections and in California. A strain of this pathogen causes root rot of loganberry. The fungus attacks roots only, destroying fine feeding roots first, then invading the central cylinder, stele, which turns dark red. New spring leaves on badly affected plants are small, bluish, have short petioles; large leaves from the previous season dry up; little or no fruit is produced; plants die in the first dry period or are stunted.

The fungus is most active in cold, wet soil, in rainy periods in late fall, winter, and early spring except when ground is frozen. Zoospores produced on roots are spread by water; resting spores formed in the red stele carry the pathogen in a dormant state through the heat of summer. There are at least three physiological races, and once the fungus infests a field it is worthless for strawberries for 10 years.

Control. Buy clean, certified plants. Aberdeen and Stelemaster varieties are resistant; Temple, Sparkle, Fairland, Redcrop, and Pathfinder, fairly so.

Phytophthora lateralis. Cypress root rot; on Lawson cypress (*Chamaecyparis lawsoniana*) often called Port Orford cedar, and Hinoki cypress (*C. obtusa*), killing thousands of trees in Oregon nurseries and landscape plantings. It is also reported from Washington and apparently native to the Northwest. The fungus enters through the roots and spreads to lower part of main trunk, killing the tissues. Blue cypress changes to purple, green, finally tan, and dies. The color changes take several months in cool, damp weather, only 2 or 3 weeks in hot, dry weather.

There is no practical chemical control, and Lawson cypress seems to be incompatible with resistant rootstocks. Grow disease-free propagating stock in new soil. Avoid large plantings of Lawson cypress such as windbreaks or hedges. Remove and destroy infected plants, getting the entire root system.

Phytophthora megasperma. Root rot; occasional on cabbage, cauliflower, brussel sprouts, carrot, artichoke, stock, citrus, soybean, and wallflower. Diseased plants wilt suddenly; leaves turn red to purple; underground stems and roots rot. The disease is more prevalent in winter plantings in California and in low, poorly drained areas. Level ground properly before planting to avoid waterlogged spots.

Phytophthora megasperma f. sp. **glycinea.** Root and stem rot; on soybean.

Phytophthora nicotianae var. **nicotianae.** Crown rot, root rot, and stem canker; on flannel bush.

Phytophthora nicotianae var. **parasitica.** Crown rot; on petunia and poinsettia (stem rot).

Phytophthora palmivora. Palm bud rot; leaf drop; wilt; of coconut, *Washingtonia*, and queen palm; also root rot on English ivy. The fungus is an omnivorous tropical species, presumably the one causing stem rot of dieffenbachia and peperomia. It has been prevented in nurseries by using cuttings from healthy plants in pasteurized soil.

Phytophthora parasitica (*P. terrestris*). Brown rot of citrus; in Florida. Buckeye rot of tomato; also on lily roselle, sempervivum, potato (tuber rot), zebra plant (stem rot), and Christmas cactus (root rot). The disease appears on the lowest tomato fruits, where water stands after rains. The lesions have concentric, narrow, dark brown bands alternating with wide, light brown bands. The decay is rapid and the internal tissue is semiwatery, though the exterior is firm. Control by staking tomatoes; avoid poorly drained soil or plant on ridges. This species is often present with *P. citrophthora* in cases of citrus foot or collar rot.

Phytophthora parasitica var. **nicotianae.** Root rot; on pine.

Phytophthora sojae. Root and stem rot of soybean; a relatively new disease reported from Illinois, Indiana, Missouri, North and South Carolina, and Ohio. Serious in cool rainy weather, causing pre- and postemergence damping-off.

Plectospira

Phycomycetes, Saprolegniales, Saprolegniaceae

Sporangium with much inflated branching; swarm spores are formed in basal portion and cut into a single row in an elongate filamentous apical portion, which acts as an exit tube. Swarm spores encyst at the mouth as in *Aphanomyces.* Oogonium terminal or intercalary, accompanied by up to 65 antheridia.

Plectospira myriandra. Rootlet necrosis on tomato; the fungus is weakly parasitic on roots.

Plenodomus

Fungi Imperfecti, Sphaeropsidales, Sphaerioidaceae

Pycnidia coriaceous or carbonaceous, more or less sclerotoid. Conidiophores obsolete or none; conidia one-celled, hyaline.

Plenodomus destruens. Foot rot of sweet potato; one of the more important field diseases and sometimes a storage rot. The base of the stem turns brown from just under the soil surface to 4 or 5 inches above; leaves turn yellow and drop off; vines wilt unless adventitious roots are put out; pycnidia are numerous. The root has a firm brown rot, not affecting the whole potato but enough to make it worthless for food. The fungus winters in old plant refuse but not in soil. Use clean seed potatoes; rotate crops. This fungus also infects *Jacquemontia.*

Pleospora

See under Leaf Spots.

Pleospora lycopersici. Fruit rot of tomato; a firm dark rot develops in fruit after picking, starting from infections through cracks near stem end of fruit. Progress is most rapid at 65° F to 70° F and is checked by storage at 45° F.

Pleurotus

Basidiomycetes, Agaricales, Agaricaceae

Stipe off center or lacking; cap sometimes inverted; gills more or less fleshy and separable into two layers, edges acute; spores white.

Pleurotus ostreatus, oyster cap. White flaky sapwood rot; of maple and other hardwoods, sometimes on living trees. A light-colored decay is surrounded by a narrow brown zone. Fleshy annual conks are shelving, sessile, or with a short, stout excentric stalk. The upper surface is smooth, white, or grayish, gills extending onto the stalk, an edible fungus. Infection is through open wounds.

Pleurotus ulmarius. Brown heart rot; sapwood wound rot; of elm, maple, and other living hardwoods. Rot starting in heartwood may extend into sapwood; infected wood separates along annual rings. Annual sporophores have a long excentric stalk, and white to yellow to brown smooth upper surface. They issue from crotches and pruning wounds.

Polyporus

Basidiomycetes, Agaricales, Polyporaceae

Pileus tough, thick, with a stipe, or as a shelf; pores rounded, small, tubes crowded.

Polyporus abietinus (*Hirshioporus abietinus*) Pitted sap rot; hollow pocket; white pocket rot on fir. May attack dead sapwood in wounds of living trees.

Polyporus anceps. Red ray rot; on western conifers, causing heart rot of living trees but beneficial as a cause of rapid decay of slash in forests. Fruiting bodies rarely develop on living trees.

Polyporus balsameus. Balsam butt rot; of living balsam fir, eastern hemlock, northern white cedar, western red cedar, also prevalent on dead trees. Advanced decay is brown, brittle, breaking into large cubes, easily crushed to a clay-colored powder. In living trees the rot column is usually only 3 or 4 feet from ground. Sporophores are shelving, up to 2 inches wide, with pale brown upper surface with concentric zones, white underneath.

Polyporus betulinus (*Piptoporus betulinus*). Brown cubical rot; of dead or dying gray and paper birches. Conks have smooth grayish upper surface with incurved margin.

Polyporus dryadeus. White root rot; occasional in oaks and conifers in the West. Roots are killed; tree dies. Decayed wood is white to cream; bark is loosened and shredded.

Polyporus gilvus. White sapwood rot; prevalent on dead trees, occasional on living trees. Small, annual, yellow to red, brown leathery to corky sporophores, developed in profusion.

Polyporus hispidus. White spongy heart rot; of living trees of black ash, oak, maple, and birch; does not decay dead trees. Heartwood in upper portion of trunk is reduced to soft spongy yellow or white mass. Shelf sporophores, up to 10 inches wide, have dark brown, coarse, velvety to hairy upper surface and golden brown undersurface, turning dark brown with age. They are formed at branch stubs, frost cracks, or trunk cankers.

Polyporus lucidus. Root rot; on redbud.

Polyporus pargamenus. White pocket rot; of dead sapwood in eastern United States but sometimes on living maple and other hardwoods.

Polyporus schweinitzii. Root rot; on pine.

Polyporus squamosus. White mottled heart rot; on maple, buckeye, birch, and occasional on living trees near wounds. Conks are annual, fleshy, white to dingy yellow with a short, thick, lateral stalk, upper surface with broad appressed scales, up to 18 inches wide.

Polyporus sulphureus (*Laetiporus sulphureus*), sulphur fungus. Red brown heart rot; brown cubical rot in heartwood of maple and other living hardwoods and conifers, widespread on oak, balsam, Douglas fir, and spruce. The annual, shelflike, fruiting bodies are most conspicuous—soft, fleshy, moist when fresh, with bright orange-red upper surface and brilliant yellow underneath, formed in overlapping clusters. When old they are hard, brittle, dirty white (see Fig. 58). Infection is through dead branch stubs and wounds.

Polyporus tomentosus var. **circinatus** (=*Inonotus circinatus*). Root rot; of sand pine.

Polyporus versicolor (*Coriolus versicolor*), rainbow conk. Sapwood rot; it is the most common fungus on hardwood slash in woods and sometimes on conifers. The rot is soft, white, spongy. Heartwood of living catalpa may

Figure 58. Shelf fungus *(Laetiporus sulphureus)* on oak.

be decayed, the fungus entering through wounds and dead branches. The conks are thin, tough, leathery, annual, up to 2 inches wide with a hairy or velvety surface multicolored white, yellow, brown, gray, and black. The undersurface is yellow or white. This pathogen also causes wood decay and decline of apple and has been reported as *Trametes versicolor.*

Poria

Basidiomycetes, Agaricales, Polyporaceae

Pileus resupinate, thin, membranous; tubes wartlike, separate.

Poria cocos. Root rot; on roots of various trees, especially pine, in southeastern United States. Huge sclerotia, weighing up to 2 pounds, are formed; this stage is known as *Pachyma cocos.*
Poria laevigata. Red mottle rot; on *Prunus* spp.
Poria luteoalba. Brown rot; of coniferous wood.
Poria prunicola. White rot; of cherry and other *Prunus* spp.
Poria subacida. Feather rot; spongy root rot, string butt rot of living conifers and dead hardwood. Decay rarely extends more than 6 to 10 feet in the trunk. Irregular pockets run together forming masses of white fibers; annual rings separate readily. Sporophores are white to straw-colored to cinnamon-buff crusts forming sheets several feet long on underside of fallen trunks or on underside of root crotches or exposed roots of living trees.
Poria weirii. Douglas fir root rot; the disease is most destructive to trees 70 to 150 years old, which are killed in groups. The fungus can persist in dead roots for a century. Less susceptible conifers should be planted with judicious cutting of infected stands.

Pyrenochaeta

See under Blights.

Pyrenochaeta lycopersici. Root rot; on tomato.
Pyrenochaeta terrestris. Pink root of onions; widespread on onions, garlic, and shallot; also on grasses. Roots of affected plants shrivel and turn pink. New roots replacing the old are infected in turn; plants are stunted, bulbs small. The fungus persists indefinitely in the soil and is distributed on onion sets and transplants. Yellow Bermuda is the most resistant of commercial onion varieties. The green Beltsville Bunching onion, Nebuka strain of Welsh onion, Evergreen variety of shallot, and leeks and chives are resistant. In Arizona, Granex gives a better yield than other onions despite pink root.

Pythium

Phycomycetes, Peronosporales, Pythiaceae

Wall of sporangium smooth; discharging swarmspores in imperfectly formed state into thin-walled vesicle, which later ruptures to allow spores to escape. Sporangia terminal or intercalary. Species live in moist soil causing damping-off and root rots.

Pythium acanthicum; P. myriotylum; P. periplocum. Causing rot of watermelon fruit.

Pythium aphanidermatum. Leak; root rot; damping-off of muskmelon, cucumber, squash, also papaya, bean, radish, spinach, sugar beet, guayule, caper spurge, and ice plant. There is a watery decay with a yellow-brown liquid leaking out when fruit is pressed. Lesions are brown and wrinkled. The fungus lives in the soil; primary infection is in the field, secondary from contact in transit or storage. Sort carefully before packing. Refrigerate at 45° F to 50° F in transit.

Phythium aristosporum. Root rot; of bean.

Pythium arrhenomanes. Root rot; on tomato.

Pythium carolinianum. Root and stem rot; of parrotfeather (*Myriophyllum*).

Pythium catenulatum. Root rot; of bean.

Pythium debaryanum. Damping-off of seedlings; watery leak of potatoes; leak starts as a brown discoloration around a wound and soon spreads to include the whole potato, which is soft, easily crushed, and drips a brown liquid with the slightest pressure. Entrance to the tuber is usually through harvest wounds. *Pythium* hyphae grow through the soil in great profusion and can enter seedlings through either stomata or unbroken epidermis. See Damping-off, for rot of seedlings.

Pythium dissotocum. Root rot; of bean and spinach.

Pythium irregularae. Associated with melon root rot and fruit rots of other cucurbits in cool weather and seed decay of corn.

Pythium myriotylum. Root rot; on tomato.

Pythium splendens. On Chinese evergreen, peperomia, and philodendron.

Pythium ultimum. Fruit rot of muskmelon; often with luxuriant white fungus growth. Damping-off. Root rot; of many seedlings in greenhouse and field.

Pythium spp. Most soils contain several species of *Pythium* ready to perform at optimum moisture and temperature. Exact determination is not always practical. Nematode wounds often dispose plants to rot. Diseases include African violet rot, aloe root rot, black rot of orchids, begonia root rot, coleus black leg, geranium cutting rot (see Fig. 59), bean and parsley root rot, rhubarb crown rot, mottle necrosis of sweet potato, and other rots. Some plants can be treated with hot water, 115° F for 30 minutes. Sterilize soil before use; avoid excessive watering. See Damping-off, for seedling rots.

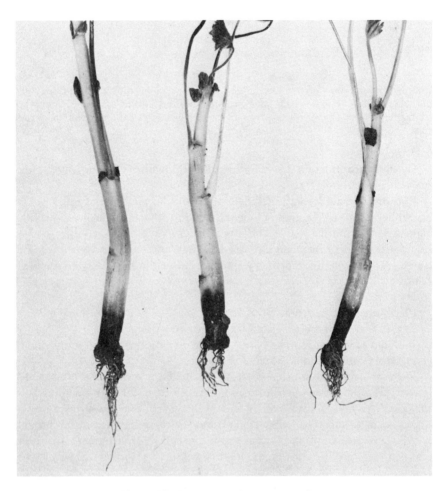

Figure 59. *Pythium* blackleg on geranium.

Rhizina

Ascomycetes, Pezizales, Pezizaceae

Cup-shaped apothecia with rhizoids underneath; asci operculate, opening with a lid, eight-spored; spores fusoid, spindle-shaped, paraphyses present.

Rhizina inflata. Seedling root rot; damping-off; coniferous seedlings in the Pacific Northwest are sometimes killed in isolated circular patches 2 to 4 feet in diameter, particularly in burned areas. Infected roots are matted together with white mycelium. More or less resinous annual fructifications are formed on the ground. They are irregular, an undulating brown upper surface with narrow white margin, 2 to 3 inches across. There is no control, but the disease is minor.

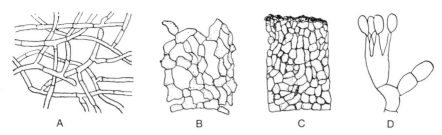

Figure 60. Forms of *Rhizoctonia solani. A*, young mycelium, constricted at branches; *B*, loosely formed angular to barrel-shaped colored cells; *C*, section through sclerotium formed from aggregation of cells in *B*; *D*, basidium and spores of *Pellicularia*, perfect stage of *R. solani.*

Rhizoctonia

Fungi Imperfecti, Mycelia Sterilia

Sclerotial form of *Pellicularia, Corticium, Macrophomina,* and *Helicobasidium.* Young mycelium colorless; branches constricted at points of origin from main axis; older mycelium colored, wefts of brownish yellow to brown strands, organized into dense groups of hyphae, sclerotia, made up of short, irregular, angular or somewhat barrel-shaped cells (see Fig. 60).

Rhizoctonia sp. Postharvest decay; on leatherleaf fern.
Rhizoctonia bataticola. Charcoal rot; see *Macrophomina phaseoli.*
Rhizoctonia crocorum. Violet root rot; see *Helicobasidium purpureum.*
Rhizoctonia solani. Black scurf of potatoes; brown patch of turf; see *Pellicularia filamentosa.*
Rhizoctonia solani. Root and stem rot of poinsettia; and other ornamental plants, including begonia, camellia, calla, carnation, chrysanthemum, coleus, cornflower, geranium, gloxinia, lily, pansy, pothos, peperomia, primrose, caper spurge, and sainfoin (crown rot). Although *Pythium* flourishes best in the low oxygen content of poorly drained soils, *Rhizoctonia,* causing similar root rots, is serious in well-drained soils. On poinsettia, dark brown lesions at or above soil level are often covered with brown mycelium; the leaves turn yellow and drop, the roots rot, and the plant dies.
Rhizoctonia tuliparum. Gray bulb rot of tulips; in northeastern and Pacific states. The most conspicuous sign of this disease is a bare patch in spring where tulip shoots should be showing. Occasionally an infected bulb will produce some aboveground growth, but the plants are slow and often wither and die before flowering. Bulbs rot from the top down; mycelium forms felty masses between scales; on bulbs and in surrounding soil are masses of brown to black, flattened sclerotia, composed of the yellow-brown, thin-walled irregular cells typical of *Rhizoctonia.* These can survive in soil for years, germinating to infect bulbs after planting or in very early spring. Occasion-

ally sclerotia are transported on bulbs, but the bulbs are usually so noticeably diseased that they are not sold.

Control. Remove soil and plants from affected area and for at least 6 inches beyond. Destroy all infected bulbs at harvest. Use a 4- to 5-year rotation.

Rhizopus

Phycomycetes, Mucorales, Mucoraceae

Sporangium large, globose, multispored, with a columella and a thin wall; sporangiola and conidia lacking. Sporangiophores arise in fascicles from aerial arching stolons, which develop rhizoids at point of contact with substratum (Fig. 5).

Rhizopus arrhizus. Soft rot; on gladiolus corms, light brown.

Rhizopus nigricans (*R. stolonifer*), the common black bread mold. **Soft rot of sweet potato**; and other vegetables. **Rhizopus rot**; "Whiskers," Leak of peach, strawberry, and other fruits. It is one of the more serious storage rots of sweet potato, soft, watery, progressing rapidly, with rotting complete inside 5 days after visible infection. The tuber is brownish within, covered with a coarse, whiskery, mycelial growth; there is a mild odor. Cucurbits, crucifers, carrots (see Fig. 61), beans, lima beans, onions, peanuts, potatoes, Jerusalem artichoke, and guava are susceptible to this black mold. Nancy Hall and Southern Queen are among the more resistant varieties of sweet potato. To prevent rot, cure at 80° F to 85° F for 10 to 14 days, at high humidity, to permit rapid corking over of wounds; then store at 55° F.

The fungus is a weak parasite on ripe fruit—peach, fig, strawberry, citrus, persimmon, pear, avocado, and melons. A coarse cottony mold appearing in wounds and over the surface is covered with sporangia, white when young, black at maturity. A watery fluid with an offensive odor leaks from the soft fruit. Avoid wounding in harvesting; do not pack overripe fruit; keep at low temperature in transit and market. Amaryllis, lily, and tulip bulbs may be infected.

Rhizopus oryzae. Head rot; on sunflower.

Rhizopus stolonifer. Soft rot; of *Euphorbia trigona*.

Roesleria

Ascomycetes, Helotiales

Mycelium inconspicuous; apothecia cup-shaped, opening more or less completely; asci disappearing early, leaving a persistent mass of spores and paraphyses. Spores hyaline, one-celled, globose.

Roesleria hypogaea. Root rot; of grape.

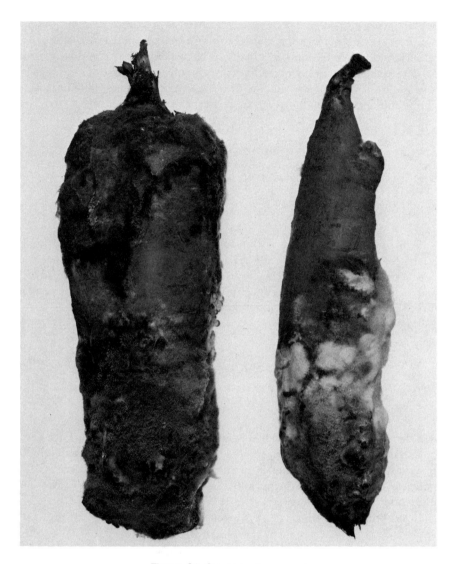

Figure 61. Storage rot on carrot.

Rosellinia

Ascomycetes, Sphaeriales, Sphaeriaceae

Perithecia smooth, ostiole simple or with a low papilla; with a subicle under the fruiting layer; paraphyses present; spores olive to brown, one-celled.

Rosellinia necatrix (*Dematophora*). **White root rot**; of fig, grape, avocado, apricot, cherry, apple, pear, peach, walnut, holly, osmanthus, privet, and poplar. It is like Armillaria root rot in that all trees in a certain area are killed, but there are no rhizomorphs or toadstools formed. A white mycelial growth on surface of affected roots turns black and cobwebby. During wet weather a delicate mold forms on surface of bark and on soil around base of tree. Foliage is sparse and wilting, growth slow or none. Crabapples are quite resistant. Plums and apricots can be grown on resistant understock.

Schizophyllum

Basidiomycetes, Agaricales, Agaricaceae

Pileus leathery; stipe lateral or none; edge of gills split; spores white.

Schizophyllum commune. Wound rot; common on dead parts of living trees—maple, boxelder, almond, acacia, ailanthus, birch, catalpa, hickory, peach, pecan, citrus, fig. Fruiting bodies are small, thin, sometimes lobed, up to 2 inches wide, fan-shaped with gray-white downy upper surface, brownish forked gills on underside, common on fruit trees. This pathogen also causes wood decay and decline of apple.

Sclerotinia

See under Blights.

Sclerotinia homeocarpa. Dollar spot; small brown patch of turf; on bent grasses, fescues and bluegrass. Spots are brown at first, then bleached and straw-colored, about 2 inches in diameter but coalescing to large irregular patches. While leaves are being killed, a fine, white, cobwebby growth of mycelium can be seen in early morning when dew is present.

Sclerotinia intermedia. Stem rot; market disease of celery, carrot, and salsify.

Sclerotinia minor. Stem rot; of lettuce, celery, carrot, and cocklebur. Also Root and pod rot; of peanut. Resembles rot due to *S. sclerotiorum*, but sclerotia are much smaller.

Sclerotinia narcissicola. Narcissus smoulder; perhaps the fungus should be transferred to *Botryotinia*, since there is a conidial stage. The disease is a decay of stored narcissus bulbs, also known on snowdrop, and a rot of foliage and flowers in the open, especially during cold wet seasons. Leaves are distorted, stuck together as they emerge from soil. Sclerotia are small, black, flattened bodies, up to ½ inch long when several grow together, just below outer papery bulb scales. In prolonged storage there is a yellow-brown rot.

Control. Remove and destroy diseased plants as soon as noticed; destroy weeds to provide air circulation; spray with bordeaux mixture; discard rotting bulbs at harvest; change location every year.

Sclerotinia sclerotiorum (*S. libertiana*). **Lettuce drop**; watery soft rot of endive. **Pink rot of celery; cottony rot of bean**; carrot, parsnip, cabbage, and other crucifers and cucurbits. In lettuce, older leaves wilt and fall flat on the ground, leaving center leaves erect, but these are soon invaded by mycelium and reduced to a slimy wet mass. In continued moisture a thick, white cottony mold is formed, bearing large black sclerotia up to the size of peas (see Fig. 62). They winter in the soil, send up groups of apothecia in spring. These are brown, cup- to saucer-shaped, up to an inch across, on a stalk. Ascospores are ejected in a veritable cloud; there is no known conidial stage.

Control. In commercial celery fields deep plowing or flooding is used to inhibit apothecial production. Sterilize seedbed soil before planting.

Sclerotinia sclerotiorum. Stem rot; of pepper, cocklebur, tomato, and many ornamentals—aconite, calendula, chrysanthemum, cynoglossum, dahlia, daisy, delphinium, gazania, hollyhock, peony, snapdragon, sunflower, zinnia, and others. The same sort of cottony mold is formed on flower stems as on vegetables, but here the sclerotia are usually inside the pith and so are rather long and thin. You can feel them by running thumb and finger along the stem; sometimes the cottony mycelium, or cracks in the stem, or one or two external sclerotia indicate their presence. They are common in peony stems. When sclerotia are formed in flowers, the shape corresponds to floral parts. Sunflowers have large compound sclerotia.

Control. Cut out and destroy affected parts, trying to keep sclerotia from falling onto soil. Dusting with sulfur sometimes checks rapid spread of mycelium.

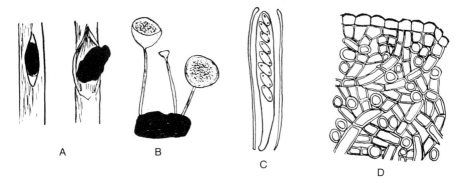

Figure 62. *Sclerotinia sclerotiorum.* A, sclerotium formed in pith of stem and one falling out from broken stem; B, apothecia produced from sclerotium on ground; C, ascus with paraphyses; D, section through sclerotium, with colorless medulla and dark wall on rind cells.

Sclerotinia sclerotiorum. Black rot; of bulbous iris, hyacinth, narcissus, and tulip. Iris fails to start growth, or plants turn yellow, wilt, and die, often in clumps. Bulbs are covered with thin gray masses of mold with black irregular sclerotia between scales. Tulip leaves develop reddish color early in spring, wilt, and die; stems and bulbs are rotted into a crumbly mass of fragments and black sclerotia. It is a cool-temperature fungus that stops action about the time the heat-loving *Sclerotium rolfsii* starts in. Remove diseased plants and surrounding soil as soon as noticed. Discard all small bulbs at harvest; plant healthy bulbs in clean soil. The pathogen supposedly dies out after 2 years in soil without suitable host.

Sclerotinia sclerotiorum. Calyx end rot; on apple. **Root and pod rot;** of peanut.

Sclerotinia sclerotiorum. Green fruit rot; of almond, peach, apricot, fig, and strawberry. **Rhizome rot;** of ginseng. In almond, young shoots and fruits are killed and wither soon after petals fall. Infection takes place through jackets from apothecia produced under trees where weeds or crop plants have been previously infected with cottony rot. Losses are serious only when there is continuous wet weather during and after blooming. Spraying for brown rot helps to control green rot. Shaking or jarring trees after bloom to remove jackets from young fruits is suggested.

Sclerotinia trifoliorum. Root rot; stem rot; of cocklebur.

Sclerotium

See under Blights.

Sclerotium cepivorum. White rot of onion; of shallot and garlic. Affected plants die from a rotting at the neck, at which point there is a surface crust of small black sclerotia and a thin weft of white mycelium. The sclerotia are smaller and rounder than those of *Botrytis*. Roots are often rotted off, and sometimes spots in a field covering several square yards are infested.

Sclerotium rolfsii (including *Sclerotium delphinii*). **Crown rot;** of delphinium, iris, ajuga, aconite, quinoa, sainfoin, kiwi, and many other ornamentals and vegetables; wet scale rot of narcissus; southern blight. For a full discussion see *Pellicularia rolfsii* under Blights.

Seaverinia

Ascomycetes, Helotiales, Sclerotiaceae

Apothecia shallow, cup- to disc-shaped; a stroma formed but no definite sclerotia; conidia botryose.

Seaverinia geranii (*Sclerotinia geranii*). **Rhizome rot;** on geranium.

Steccherinum

Basidiomycetes, Agaricales, Hydnaceae

This genus has been separated from *Hydnum*. The pileus is sessile or substipitate and laterally attached, on a woody substratum; spines are terete or flattened; cystidia present; spores white, smooth.

Steccherinum abietis (*Hydnum abietis*). **Brown pocket rot**; of heartwood of living firs and western hemlock in Pacific Northwest. Elongated pockets, empty or with white fibers, are separated by firm reddish brown wood. Sporophores are like coral, white to cream, up to 10 or 12 inches high and wide, usually on dead trees, sometimes in wounds of living trees.

Steccherinum septentrionale (*Hydnum septentrionale*). **White spongy rot**; of heartwood of living maples, beech, hickory, and other hardwoods. A zone of brown discolored wood is around the white rot area, and there are fine black zone lines. The fruiting bodies are large, soft, soggy, creamy white, in very large, bracket-shaped clusters on trunks.

Stereum

Basidiomycetes, Agaricales, Thelephoraceae

Effused-reflexed to stipitate; spore-bearing surface smooth, pale brown, upper surface with a velvety coating of hairs, formed in several distinct layers; gloeocystidia and cystidia present or lacking; spores smooth, colorless.

Stereum fasciatum (*S. ostrea*). **Brown crumbly rot**; mostly on slash, sometimes on maple and birch. Thin, leathery, grayish sporophores with undersurface light brown, smooth.

Stereum hirsutum. Wood rot; sapwood wound rot; occasionally near wounds of living trees—birch, maple, hickory, mountain mahogany, eucalyptus, peach, and others. Thin, leathery, crustlike sporophores have hairy, buff to gray upper surface, smooth gray undersurface.

Stereum purpureum. Silver leaf; sapwood rot; common on plums and other fruit trees, sometimes important on apples, occasional on shade and ornamental trees, widespread but more serious in the Northwest. The fungus enters through wounds; grows first in heartwood, and then kills sapwood and bark; infected branches develop foliage with dull leaden or metallic luster. If the disease is not checked, the entire tree may be lost. The sporophores appear after death, resupinate to somewhat shelf-shaped, with purple undersurface.

Control. Remove branches and burn at first sign of silvering. Protect trees from wounds; paint pruned surface with bordeaux paste or other disinfectant; keep brush removed from orchard.

Stereum sanguinolentum. Red heart rot; of slash and living conifers—firs and eastern white pine. Fruiting bodies are small, not over 2 inches wide; upper surface is a silky pale olive buff; lower surface "bleeds" readily when wounded, dries to grayish brown. Sporophores are produced in profusion on dead wood, occasionally on dead branches of living trees.

Streptomyces

Schizomycetes, Actinomycetales, Streptomycetaceae

Intermediate form between bacteria and fungi. Much-branched mycelium that does not fragment in bacillary or coccoid forms; conidia in chains on sporophores; primarily soil forms, some parasitic.

Streptomyces ipomoea (*Actinomyces ipomoea*). **Soil rot or pox of sweet potatoes**; general New Jersey to Florida and in the Southwest. This pathogen also infects *Jacquemontia*. Leaves are small, pale green to yellow; plants are dwarfed, make little or no vine growth, and may die before end of the season; the root system is poorly developed with most roots rotted off, or breaking off if plant is pulled from the soil. Small dark lesions are formed on stems below the soil line. Pits with jagged or roughened margins, often coalescing, are formed on mature roots. The rot is found in soils at pH 5.2 or above, and is worse in dry soils and seasons. Variety Porto Rico is very susceptible.

Control. Apply sulfur to acidify soil to pH 5.0.

Stromatinia

Ascomycetes, Helotiales, Sclerotiniaceae

Apothecia arising from a thin, black, subcuticular, effuse sclerotium or stroma; small black sclerotia are borne free on mycelium, not giving rise to apothecia. There is no conidial stage; apothecia resemble those of *Sclerotinia*.

Stromatinia gladioli (*Sclerotinia gladioli*). **Dry rot of gladiolus**; also found on crocus, freesia, and tritonia. Lesions on corms start as reddish specks, with slightly elevated darker border; spots enlarge, and centers become sunken, dark brown to black with lighter raised edges; they grow together into irregular areas. On husks the lesions are tobacco brown. Very small black sclerotia are formed on husks, in corm lesions, and on dead stems. Plants in the field turn yellow and die prematurely owing to decay of leaf sheath. Corms may appear normal when dug, the rot developing in storage. The disease is more prevalent in heavy soils, and the fungus can survive several years in soil. Apothecia have been produced artificially by fertilizing receptive bodies on sclerotia with spermatia (microconidia). They are densely crowded, 3 to 7 mm broad, on stipes 6 to 10 mm high.

Control. Use soil with good drainage and a 4-year rotation. Removing husks before planting helps to reduce gladiolus rot diseases. Cure corms rapidly after harvest.

Stromatinia narcissi. Large-scale speck fungus on narcissus and zephyranthes. Black, thin, round, flat sclerotia ½ to 1 mm, adhere firmly to outer scales. The fungus is mostly on bicolor varieties and seems to be saprophytic without causing a definite disease.

Thielaviopsis

Fungi Imperfecti, Moniliales, Dematiaceae

Hyphae dark; two kinds of conidia: small, cylindrical, hyaline endogenous spores and large, ovate, dark brown exogenous spores, both formed in chains.

Thielaviopsis basicola. Black root rot; seedling root rot of tobacco and many vegetables—bean, carrot, corn, chickpea, lentil, okra, onion, pea, tomato, and watermelon; and ornamentals—begonia, cyclamen, gerbera, elm, oxalis, lupine, pelargonium, peony, poinsettia, pansy, scindapsus, and others. There is blackening and decay of roots; young plants damp-off and die; older plants are stunted, with the decay proceeding until all roots are destroyed. Stem discoloration extends 2 to 3 inches above the soil line. The fungus lives in soil as a saprophyte, entering through nematode wounds. Hyaline conidia produced inside conidioles are forced out through hyphal tips. Chlamydospores are larger, dark, club-shaped, with several cells; they break up so that each pillbox acts as a spore. This disease is especially serious on poinsettia, dwarfing plants, causing misshapen leaves and flower bracts.

The rot is most destructive in heavy, cold, slightly acidic to alkaline soils well supplied with humus. Long wet periods after transplanting increase rot. Soils with pH lower than 5.6 or sandy soils low in organic matter are less conducive to disease.

Control. Sterilize soil for seedbeds; use clean pots for poinsettias and other greenhouse plants; reduce pH with sulfur or by using half peat moss and half soil.

Trametes

Basidiomycetes, Agaricales, Polyporaceae

Pileus without stipe, sessile to effuse-reflexed, firm; hymenium white or pallid, punky to corky, not friable when dry; tubes unequally sunken.

Trametes suaveolens. White wood rot; of willow and poplar, after wounding. A dry, corky decay with an anise odor begins in lower trunk and progresses upward. Leathery to corky sporophores 6 inches wide are white when young, gray to yellow with age.

Trichoderma

Fungi Imperfecti, Moniliales, Moniliaceae

Conidia in heads on conidiophores divided into two or three tips, a single head on each tip; spores hyaline, one-celled.

Trichoderma viride. Green mold rot; cosmopolitan on narcissus, also on shallot, garlic, occasional on citrus, but saprophytic. This fungus has an antibiotic or antagonistic effect on *Rhizoctonia*, *Pythium*, and other damping-off fungi and is quite helpful in reducing Armillaria root rot and crown rot due to *Sclerotium rolfsii*.

Trichoderma harzianum. Fruit rot; of apples in storage.

Trichothecium

Fungi Imperfecti, Moniliales, Moniliaceae

Conidiophores long, unbranched; conidia two-celled, hyaline or bright, single, at apex of conidiophore; upper cell usually larger than basal cell; mostly saprophytic.

Trichothecium roseum. Fruit, storage rot; on tomato, fig, celery, carrot, occasional on quince and pear; a pink mold.

Ustulina

Ascomycetes, Sphaeriales, Xylariaceae

Stroma globoid, cupulate to pulvinate; carbonaceous, black, somewhat hollow; spores dark, one-celled.

Ustulina vulgaris. White heart rot; a brittle white rot with prominent black zones in butts of living hardwoods; prevalent on sugar maple sprouts. Black crusts appear on stumps, logs, and on flat cankered areas of American beech.

Valsa

See under Cankers and Diebacks.

Valsa leucosomoides. Causing decay around holes of tapped sugar maples.

Verticicladiella

Fungi Imperfecti, Moniliales, Moniliaceae

Conidiophores upright, tall, brown, branched only near apex, penicillate; conidia (sympodulospores) hyaline, one-celled, ovoid to clavate, often curved, apical on sympodially formed new growing points, in slime droplets.

Verticicladiella abietina. Root rot; on white pine.
Verticicladiella penicillata. Root rot; on white pine.
Verticicladiella procera. Root rot and decline; of eastern white pine, sand pine, and red pine.
Verticicladiella wagenerii. Root rot; on fir and pine.

Xylaria

Ascomycetes, Sphaeriales, Xylariaceae

Stroma is upright, simple or branched; perithecia, immersed laterally, are produced after conidia; spores dark, one-celled.

Xylaria hypoxylon. Root rot; of hawthorn and gooseberry.
Xylaria mali. Black root rot of apple; also honey locust. Wood is soft, spongy, dirty white, with narrow conspicuous black zones forming fantastic patterns. Roots are covered with thin, compact, white mycelium, which changes to black incrustations. Fruiting bodies are dark brown to black, club-shaped, 1 to several inches high, united at the base, extending upward like a fan. The disease is not common, and where it does occur, only a few trees are killed.
Xylaria polymorpha. On decaying wood, identified by cylindrical thumblike fruiting bodies.

RUSTS

Rust fungi belong to the Uredinales, a highly specialized order of the Basidiomycetes. In common with mushrooms they have spores of the sexual stage borne in fours on a club-shaped hypha known as a basidium, but they differ very decidedly from woody and fleshy Basidiomycetes. The term *rust* is applied both to the pathogen and to the disease it inflicts. There are more than 4,000 species of rusts, all obligate parasites on ferns or seed plants. Many are heteroecious, completing their life cycle on two different kinds of plants; but some are autoecious (monoecious), having all spore forms on a single host species. There are only two families, Melampsoraceae and Pucciniaceae.

Many rusts show physiological specialization, the existence within a species of numerous strains or races that look alike but attack different varieties of crop plants, thus greatly complicating the problem of breeding for rust resistance. Rusts with a complete life cycle have five different spore forms, numbered 0 to IV.

0. *Pycniospores* (spermatia) formed in *pycnia* (spermagonia). The pycnia resemble pycnidia of Ascomycetes, are usually on upperside of leaves. They discharge one-celled pycniospores with drops of nectar, and these, usually distributed by insects attracted to the sweet secretion, function in fertilization.
 I. *Aeciospores* (aecidiospores), one-celled, orange or yellow, formed often in chains, in a cuplike sorus or *aecium*, which has a peridium (wall) opening at or beyond the surface of the host.
 II. *Urediospores* (uredospores, summer spores, red rust spores), one-celled, walls spiny or warty, reddish brown, on stalks or in chains in a *uredium* (uredinium or uredosorus), over which the epidermis of the host is broken to free the spores. Resting II spores, formed by some rusts, have thicker and darker walls.
 III. *Teliospores* (teleutospores, winter spores, black rust spores), one or more cells, in *telia* (teleuto sori), either on stalks, as in the family Pucciniaceae, or sessile, in crusts or cushions as in the Melampsoraceae.
 IV. *Basidiospores* (sporidia) on a basidium or promycelium formed by the germinating teliospore. Basidium is usually divided transversely into four cells, with one sporidium formed from each cell at the tip of a sterigma.

In heteroecious rusts spore stages 0 and I are formed on one host and II and III on another, and are so indicated in the information given with each species. Stage IV always follows III on germination. Although most autoecious rusts have all spore forms, on one host, there are a few short-cycle (microcylic) rusts with some spore stages dropped out. For a detailed life history of a heteroecious rust, see *Puccinia graminis*.

Gardeners frequently mistake a reddish discoloration of a leaf, perhaps due to spray injury or weather or a leaf-spot fungus, for rust. True rust is identified by the presence of rust-colored spores in powdery pustules or perhaps gelatinous horns. With rusts, the discoloration of tissue is yellowish, not red, and it is due to increased evaporation from the broken epidermis. Plants are often stunted.

Losses in food crops due to rust have been enormous since the beginning of history. The Romans had a festival to propitiate the rust gods. Now we try to do it by removing the alternate host, barberry to save wheat, black currants to save white pine; or by developing more and more resistant varieties for the ever-increasing rust strains; or by the use of fungicides, classically sulfur, latterly some of the carbamates, and, in a few cases, antibiotics.

Achrotelium

Melampsoraceae. Telia on underside of leaves; spores one-celled at first, four-celled on germination, stalked.

Achrotelium lucumae. II, III on lucuma and egg fruit, in Florida.

Aecidium

This is a form genus, a name applied to the aecial stage where the full cycle is unknown and 0 and I are the only spores. Aecia have a peridium and catenulate spores. There are many species.

Aecidium avocense. On poppy-mallow, probably aecial stage of *Puccinia avocensis.*

Aecidium conspersum. On houstonia and galium, in Wisconsin.

Aecidium rubromaculans. On viburnum, in Florida.

Angiospora

See *Physopella.*

Aplopsora

Melampsoraceae. Teliospores sessile, hyaline, one-celled, in a single layer; aecia unknown.

Aplopsora nyssae. On tupelo, II, III.

Baeodromus

Pucciniaceae. Spores one-celled; telia pulvinate, erumpent; short chains of spores.

Baeodromus californicus. On senecio, III.

Baeodromus eupatorii. On eupatorium.

Bubakia (*Phakopsora*)

Melampsoraceae. Telia indehiscent, lenticular, spores formed in irregular succession, one-celled. Uredia without peridium or paraphyses.

Bubakia erythroxylonis. On erythroxylon.

Caeoma

Form genus. Aecia with catenulate spores but no peridium.

Caeoma faulliana. Needle rust; on alpine fir. Aecia orange-yellow, on needles of current year.

Caeoma torreyae. On torreya, in California.

Cerotelium

Pucciniaceae. Spores one-celled; teliospores in a many-layered mass; hyaline, not exserted through stomata; aecia with peridium; uredia with paraphyses; spores borne singly.

Cerotelium dicentrae. 0, I on bleeding heart; II, III, on *Urticastrum*.

Cerotelium fici (*Physopella fici*). **Fig rust**; II, III, on common fig, in Florida strangler fig and osage orange, in Alabama, Florida, Louisiana, Minnesota, South Carolina, Texas.

Chrysomyxa

Melampsoraceae. Teliospores in cylindrical or branching chains; promycelium exserted; urediospores typically in short chains; uredia without peridium.

Chrysomyxa arctostaphyli. On bearberry, III.

Chrysomyxa chiogenis. II, III, on creeping snowberry; 0, I, on spruce.

Chrysomyxa empetri. II, III, on crowberry; 0, I, on red and white spruce. Aecia on upper and lower surfaces of needles.

Chrysomyxa ilicina. II, III, on American holly.

Chrysomyxa ledi. 0, I, on black, red, and Norway spruce; II, III, on underside of leaves of *Ledum* spp.

Chrysomyxa ledi var. **cassandrae. Spruce needle rust**; 0, I, on black, red, blue, and Engelmann spruce; II, III, on bog rosemary (*Chamaedaphne*). May become epidemic on spruce, causing considerable defoliation.

Chrysomyxa ledi var. **groenlandici.** On Labrador-tea, in Michigan, New Hampshire.

Chrysomyxa ledi var. **rhododendri.** II, III, on rhododendron, in Washington. A European rust first noted on Pacific Coast in 1954, apparently entering despite quarantine on nursery stock. Yellow uredia on leaves.

Chrysomyxa ledicola. 0, I, on white, black, red, blue, Engelmann, and Sitka spruce; II, III, on upper side of leaves of *Ledum* spp. Spruce needles may be so discolored that trees appear yellow.

Chrysomyxa moneses. On Sitka spruce and moneses.

Chrysomyxa piperiana. 0, I, on Sitka spruce; II, III, on underside of leaves of *Rhododendron californicum*, in California, Oregon, Washington.

Chrysomyxa pirolata (*C. pyrolae*). 0, I, on cones of black, blue, Engelmann, Norway, red, and white spruce; II, III, on pyrola. Aecia are on upper side of cone scales; infected cones turn yellow, produce no seed.

Chrysomyxa weirii. Spruce needle rust; III, on Engelmann and red spruce. Waxy orange to orange-brown elongate or elliptical telia occur on 1-year needles. It is the only spore stage known; teliospores can reinfect spruce.

Coleosporium

Melampsoraceae. Pycnia and aecia are on pines; uredia and telia on dicotyledons. Pycnia subepidermal or subcortical, flattish, linear, dehiscent by a slit; aecia on needles, erumpent, with prominent peridium, spores ellipsoid or globular; uredia erumpent, powdery without peridia; urediospores globose or oblong, catenulate, with verrucose (warty) walls; telia indehiscent, waxy, gelatinous on germination; spores sessile or obscurely catenulate, one-celled, smooth but with thick and gelatinous walls (see Fig. 63).

Coleosporium apocyanaceum. 0, I, on loblolly, longleaf, and slash pines; II, III, on *Amsonia* spp. in the Southeast.

Coleosporium asterum (*C. solidaginis*). Needle blister rust of pine; 0, I, on all two- and three-needle pines in eastern United States; II, III, on aster and goldenrod, on China aster (except far South), on golden aster (*Chrysopsis*), erigeron, grindelia, seriocarpus, and other composites. This blister rust on pine needles has pustules higher than they are long, in clusters or short rows. The rust is fairly common on ornamental pines in gardens, wintering on aster and related composites. Older needles of young pines may be severely infected, with white aecia conspicuous in spring and early summer. Aster leaves have bright orange-yellow spore pustules on undersurface. Destroy goldenrod near pines.

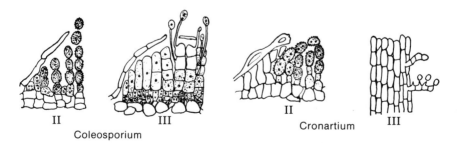

II III
Coleosporium

II
Cronartium III

Figure 63. Pine rusts. *Coleosporium asterum*, uredial (II) and telial (III) stages on aster, teliospores germinating *in situ. Cronartium ribicola*, II and III stages on currant.

Coleosporium campanulae. 0, I, on pitch, red, and Virginia pines; II, III, on campanula, lysimachia, and specularia. Underside of bluebell leaves are covered with orange to reddish brown pustules. Leaves dry; plants are stunted.

Coleosporium crowellii. III, only stage known; on needles of pinon and limber pines, Arizona, Colorado, New Mexico, Utah, Nevada, and California.

Coleosporium delicatulum. Pine needle rust; 0, I, on two- and three-needle pines; II, III, on goldenrod and euthamia.

Coleosporium elephantopodis. 0, I, on two- and three-needle pines in South; II, III, on elephantopus.

Coleosporium helianthi. Sunflower rust; 0, I, on pitch and short-needle pines; II, III, on wild and cultivated sunflower, Jerusalem artichoke, and heliopsis. Sunflower leaves, with brown rust pustules, dry up and drop. Control is not easy.

Coleosporium ipomoeae. 0, I, on southern and Chihuahua pines; II, III, on moonflower, morning glory, sweet potato, jacquemontia, and quamoclit; most abundant in warmer regions. The uredia are orange-yellow, telia deep reddish orange on sweet potato.

Coleosporium jonesii. 0, I, on pinon pine; II, III, on flowering currant and gooseberry.

Coleosporium lacinariae. 0, I, on loblolly, longleaf, and pitch pines; II, III, on liatris.

Coleosporium madiae. 0, I, on Monterey, Coulter, and Jeffrey pines; II, III, on marigold, sunflower, tarweed, and other composites.

Coleosporium mentzeliae. On mentzelia.

Coleosporium minutum. 0, I, on loblolly and spruce pines; II, III, on forestiera.

Coleosporium occidentale. 0, I, unknown; II, III, on senecio.

Coleosporium pinicola. III, on Virginia or scrub pine.

Coleosporium sonchi. 0, I, on Scotch pine; II, III, on sow thistle.

Coleosporium terebinthinaceae. 0, I, on two- and three-needle pines, especially in the Southeast; II, III, on silphium and parthenium.

Coleosporium vernoniae. 0, I, on various two- and three-needle pines; II, III, on ironweed.

Coleosporium viburni. 0, I, unknown; II, III on *Viburnum* spp.

Cronartium

Causing Blister Rusts.

Melampsoraceae. Heteroecious; pycnia and aecia on trunk and branches of pine; uredia, telia on herbaceous or woody dicotyledons.

Pycnia on stems, caeomoid, forming blisters beneath host cortical layer; dehiscent by longitudinal slits in bark; aecia on trunks, erumpent, with peridium sometimes dehiscent at apex, more often splitting irregularly or circularly at side; aeciospores ellipsoid with coarsely warted walls, sometimes with smooth spot on one side. Uredia on underside of leaves or on stems of herbaceous hosts; delicate peridium, dehiscent at first by a central pore; urediospores borne singly on pedicels, ellipsoidal with spiny walls; telia erumpent, often coming from uredia; catenulate, one-celled teliospores often form an extended cylindrical or filiform column, horny when dry (see Fig. 63).

Blister rusts are characterized by swellings that are globose, subglobose, or fusiform, depending on species. A rust on a pine stem is invariably a *Cronartium*, although this stage has often gone under the name of *Peridermium*.

Cronartium appalachianum (*Peridermium appalachianum*). I, on Virginia pine, in North Carolina, Tennessee, Virginia, West Virginia. Girdling bark lesions with a columnar aecia.

Cronartium coleosporioides (*C. filamentosum*). Western gall rust; paintbrush blister rust; 0, I, on lodgepole, ponderosa, and Jeffrey pines, in West; II, III, on Indian paintbrush, birds beak, owls clover, and wood betony. Slight swellings are formed on twigs, trunks, and branches; many lodgepole pine seedlings are killed.

Cronartium comandrae. Comandria blister rust; 0, I, on ponderosa, Arizona, and lodgepole pines in the West, and pitch, mountain jack, loblolly, Austrian, Scotch, and maritime pines in the East; II, III, on bastard toadflax (*Comandra* spp.). Destructive effect is limited to distribution of toadflax, which is widespread but locally restricted to small areas. Ponderosa pine suffers most severely, with many seedlings and saplings destroyed; occasionally a large tree is attacked.

Cronartium comptoniae. Sweet fern blister rust; 0, I, on two- and three-needle pines; II, III, on sweet fern and sweet gale in northern pine regions and south to North Carolina, and on Pacific wax myrtle on Pacific Coast. Young pines may be girdled and killed, but are fairly safe after attaining a trunk diameter of 3 inches. Losses in nurseries and plantations are high, especially among lodgepole and ponderosa pines. Affected stems swell slightly near the base with long fusiform swellings or depressed streaks on eastern hard pine; pitch oozes out from insect wounds in these areas. Killing of main stem often results in multiple-stemmed shrublike trees. Orange aecia appear on 3-year seedlings, preceded by pycnia the year before; spores are wind-borne many miles to herbaceous hosts.

Control. Remove *Myrica* species for several hundred yards around nurseries or pine plantations, and allow no large groups within a mile.

Cronartium conigenum. Pine cone rust; 0, I, on cones of Chihuahua pine; II, III, on oaks in the Southwest. Cones develop in large galls, producing aecia with distinct, erumpent peridium 2 or 3 years after infection.

Cronartium filamentosum. Ponderosa pine rust; widespread in Rocky Mountains; II, III, on Indian paintbrush.

Cronartium fusiforme. Southern fusiform rust; 0, I, on hard pines in southern states, especially loblolly, slash, and pitch pine; II, III, on evergreen oaks on underside of leaves. Pine stems have pronounced spindle-shaped swellings, sometimes with witches' brooms. Branch infections that do not reach the main trunk are not serious, but those that go on to the trunk may kill the tree. Longleaf pines are rather resistant, and shortleaf *P. echinata* almost immune. Pines well spaced in good locations grow more rapidly and may have more rust than those in poor sites. It has also been reported on oaks.

Control. Prune branches yearly before swellings reach main stem.

Cronartium harknessii. Western gall rust; 0, I, on Jeffrey, ponderosa, lodgepole, and digger pines; II, III, on Indian paintbrush, lousewort, owls clover, or omitted, with direct infection from pine to pine. Galls are globose, with large, confluent aecia; bark sloughs off in large scales; witches' brooms are formed. A variety of this species, alternate stage unknown, occurs on Monterey and knobcone pines in California.

Control. Remove trees with galls for a distance of 300 yards around nurseries. Do not ship infected trees from nurseries.

Cronartium occidentale. Pinon blister rust; 0, I, in pinon and Mexican pinon; II, III, on currant, gooseberry and flowering currant. This rust cannot be told from whitepine blister rust on *Ribes* hosts, but is differentiated by the type of pine attacked. Aecia on Mexican or singleleaf pinon are distinct sori; on pinon they form broad layers under bark.

Cronartium quercuum (*C. cerebrum*). **Eastern gall rust;** 0, I, on pines, especially scrub and shortleaf in the South; II, III, on chestnut, tanbark, and oak. Globose to subglobose galls are formed on pine stems; in spring aecia break through the bark in more or less cerebroid (brainlike) arrangement.

Cronartium ribicola. White pine blister rust; 0, I, on eastern white pine from Maine to Virginia and Minnesota, on western white pine in the Pacific Northwest, on sugar pine in California; II, III, on currant, flowering currant and gooseberry. Occurs also on limber pine in North Central and Southeastern Wyoming.

This dread disease is supposed to have originated in Asia, whence it spread to Europe, where the eastern white pine introduced from America was very susceptible. White pine blister rust was found in Russia in 1854, and by 1900, had spread over most of Europe. It was recorded on *Ribes* at Geneva, New York, in 1906, but probably was there some years previously. In 1909, it was found on pine, at which time it was learned that infected pines from a German nursery had been widely planted throughout the Northeast. The next year the disease reached Vancouver, British Columbia, in a shipment from a French nursery, whence it spread to Washington, Oregon, northern California, Idaho, and western Montana. Thus, from cheap stock brought in for forest planting has come one of our greatest forest hazards. Our present quarantine laws are designed to prevent such introductions.

The western white sugar and whitebark pines are even more susceptible to blister rust than eastern white pine; but in either case robust, dominant trees are more severely attacked, with frail individuals lightly infected. This situation is partly explained by more vigorous trees having more needles to receive spores. Of the *Ribes* species, black currant is most susceptible and dangerous. Cultivated red currants are somewhat resistant, causing a minimum of pine infection; Viking and Red Dutch varieties are practically immune. Wild gooseberries and skunk currant are highly susceptible in the Northeast, as are western black currant, stink currant, and red flowering currant. The greater the susceptibility of the *Ribes* species, the more spores are produced to inoculate pines, with proportionate damage.

When a spore arrives on a pine needle from a currant, the first sign of infection is a small golden yellow to reddish brown spot. The next season, or possibly in two years, the bark looks yellowish, often with an orange tinge to the margin of the discolored area, and there may be a spindle-shaped swelling. If such symptoms appear early in the season, pycnia are formed in bark by July or August; but if discoloration is delayed until midsummer, they appear the next year. The male fruiting bodies are small, honey yellow to brown patches, swelling to shallow blisters and rupturing to discharge drops of a yellowish, sweet liquid. After the liquid is eaten by insects or washed away by rain, the lesions turn dark. The next spring or summer aecia push through the bark in the same region. These are white blisters, rupturing to free orange-yellow aeciospores, which are carried away by wind. The bark then dries out and cracks, with death of cambium and underlying wood. The disease has taken 3 to 6 years to reach this stage.

Production of aecia continues yearly until stem is killed beyond the lesion. Dead foliage assumes a conspicuous red-brown color. This "flag" of brown on a green background is the most conspicuous symptom of blister rust before death of the pine. Infection progresses downward from small to larger branches and into the trunk. Swellings are not apparent on stems much over 2 inches in diameter on eastern white pine, but in the West they sometimes show up in stems 5 inches through. Larger limbs and trunks sometimes show constriction in the girdled area.

The aeciospores, large, ellipsoidal, with thick, warty walls, are carried by wind great distances to *Ribes* species (they cannot reinfect pine). They send their germ tubes into a currant or gooseberry leaf through stomata, and within 1 to 3 weeks pinhead-size blisters appear in clusters on yellowed leaf tissue. These uredia rupture to release large, ellipsoidal, yellow urediospores with thick, colorless walls and short, sharp but sparse spines. The spores are somewhat moist and sticky, and are wind-borne short distances to other *Ribes* bushes nearby. There may be up to seven generations in a summer, or the spores may remain viable over winter in uredia; this stage can infect only currant.

In late summer telia follow uredia in the same or new leaf lesions, appearing as short brown bristles on underside of leaves or looking like a coarse felt. Each felty bristle is composed of vertical rows of broad, spindle-shaped spores, which germinate in situ to a five-celled promycelium with each of the four upper cells bearing at the point of a sterigma a small, thin-walled, round basidiospore. This basidiospore cannot reinfect currant and soon dies from exposure to the sun unless the wind blows it immediately to a pine needle. The effective range is around 300 feet except for spores from black currants, which can be carried a mile. The spores from pine to currant can be carried up to 300 miles. Blister rust is more important at elevations of 1,000 feet or more, where it is increased by lower temperatures and more rainfall.

Control. Eradication of the *Ribes* host is definitely effective in controlling white pine blister rust. Complete removal is necessary of black currants and local removal of cultivated red and wild currants and gooseberries within 300 or 900 feet of pines, according to state regulations, taking care to get all the root system capable of resprouting.

Blister rust is seldom found on ornamental pines in cities; the smoke and fumes are unfavorable to the fungus. Elsewhere valuable ornamentals can be saved by cutting off infected branches and cleaning out trunk infection, stripping off diseased bark and a 2-inch side margin, 4-inch margin at top and bottom, of healthy bark. If the cankers are nearer to the trunk than 6 inches, the bark should be excised around the branch stub. The red currant Viking is immune to blister rust, and a couple of black currant hybrids are resistant. Some white pines are exhibiting resistance.

Cronartium stalactiforme. 0, I, on lodgepole pines in Rocky Mountain regions; II, III, on Indian paintbrush. The rust enters pine trunks through small twigs, producing diamond-shaped lesions that elongate an average of 7 inches a year, but grow laterally less than ½ inch. Removal of diseased trees is the only known control.

Cronartium strobilinum. Pine cone rust; 0, I, on cones of longleaf and slash pines; II, III, on evergreen oak. Cones are swollen, reddish; 25% to 90% drop.

Cumminsiella

Pucciniaceae. Autoecious; teliospores two-celled; pycnia and other sori subepidermal; aecia cupulate.

Cumminsiella mirabilissima. 0, I, II, III, on barberry and mahonia in the West, Arizona, California, Colorado, Idaho, Montana, Nebraska, New Mexico.

Cumminsiella texana. On barberry, in Texas.

Desmella

Pucciniaceae. Uredia and telia subepidermal, protruding in tufts; uredia without peridium or paraphyses. Spores globoid, on pedicels, two-celled.

Desmella superficialis. On Boston fern, in Florida.

Endocronartium

Badisiomycete, Uredinales, Pucciniaceae

Endocronartium harknessii. Western gall rust or pine-pine gall rust; on pine.

Endophyllum

Pucciniaceae. Teliospores in form of aeciospores; telia with cupulate peridium.

Endophyllum sempervivi. III, on houseleek and hen-and-chickens. Succulent leaves may be covered with reddish pustules. It is not common, but may be serious. Clean out infected parts.

Endophyllum tuberculatum. III, on hollyhock and checkermallow.

Frommea

Pucciniaceae. Teliospores two- to many-septate; aecia and uredia erumpent.

Frommea obtusa duchesneae (= *Frommella duchesnea*). II, III, on mock strawberry, false strawberry, or Aztec Indian berry.

Gymnoconia

Pucciniaceae. Uredia lacking; aecia present but without peridium; teliospores two-celled, one pore in each cell.

Gymnoconia peckiana (*G. interstitialis*). Orange rust of blackberry; 0, I, III, on blackberry, dewberry, and black raspberry, first described from eastern United States in 1822, present from Canada to Florida and from Alaska to southern California. Very bright orange spores cover underside of leaves in spring. The mycelium is perennial in the bush, living throughout the year between cells of the stem, crown, and roots, each season invading new tissue as new growth begins. Shoots may be bunched, often with a witches' broom effect; plants are dwarfed. Spraying is useless; infected plants never recover. Plant only healthy stock, obtained from a nursery where the disease is unknown. Remove infected plants showing upright habit of growth, yellow color, and glistening yellowish dots of pycnia before the orange spore stage appears. Blackberry varieties Eldorado, Orange Evergreen, Russell, Snyder, Ebony King, dewberry *Leucretia*, and boysenberries are quite resistant.

Gymnosporangium

Pucciniaceae. All but one species heteroecious. Picnia and aecia usually on trees and shrubs of the apple family; telia confined to cedars and junipers except for one species on cypress; uredia wanting. Teliospores thick- or thin-walled, various in form but mostly flat, tongue-shaped, expanding greatly when moistened, usually with two cells; walls smooth, one to several pores in each cell; pedicel colorless, usually with outer portion swelling and becoming jellylike when moistened. Aecia are highly

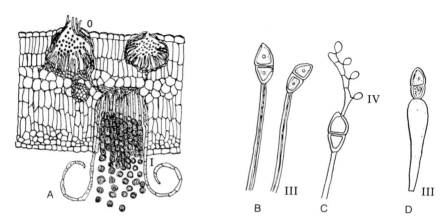

Figure 64. Cedar-apple rust. *Gymnosporangium juniperi-virginianae. A*, section through crabapple leaf with pycnia (0) on upper surface and aecium (I) with prominent peridium and aeciospores in chains on undersurface; *B*, two-celled teliospores on gelatinous stalks, which help form the jellylike telial horns on cedar galls; *C*, teliospores germinating with a promycelium and basidiospores; *D*, teliospore of *G. clavipes*, the quince rust.

differentiated and conspicuous, with catenulate aeciospores, deeply colored with verrucose walls (see Fig. 64).

The life cycle is similar in all juniper leaf rusts. In early summer, small, slightly swollen spots appear on leaves of the pomaceous host, then small raised specks in this area on the upper surface, openings of flask-shaped pycnia embedded in leaf tissue. After exuding an orange liquid containing pycniospores, the specks are black. Later, aecia push out on the underside of the same spots as dingy white columns, rostelia, with the outer coating rupturing to release a powdery mass of yellow to brown aeciospores. The ruptured segments sometimes make the open aecium look star-shaped, but in the common cedar apple, rust aecia are cup-shaped. Aecia are also formed on fruit and tender green stems. Aeciospores released during summer are wind-borne to junipers. Mycelium winters in the juniper needle or stem, and in spring galls are started that take a year or more to produce teliospores in cushions or horns.

Gymnosporangium bermudianum. 0, I, III, on stems of eastern and southern red cedar in the Gulf states. No alternate host; aecia precede telia on small galls.

Gymnosporangium bethelii. III, on Rocky Mountain juniper; 0, I, on fruits of hawthorn.

Gymnosporangium biseptatum. III, on stems of *Chamaecyparis thyoides*; 0, I, on amelanchier. Spindle-shaped swelling in stem; trees may die.

Gymnosporangium clavariiforme. III, on common and mountain juniper; 0, I, on chokeberry, amelanchier, pear, and quince. Slender telia 5 to 10 mm high produced on long fusiform swellings on branches.

Gymnosporangium clavipes. Quince rust; III, on eastern red cedar, dwarf, mountain, and prostrate junipers; 0, I, on fruits and young stems of amelanchier, apple, chokeberry, crabapple, hawthorn, mountain ash, quince, Japanese quince, and pear. Short slight swellings, somewhat spindle-shaped, occur in cedar twigs and branches, many of which die. On the main trunk, infected areas are black rough patches or rings around the bark. Mycelium is perennial, confined to the outer layer of living bark; it can sometimes be scraped out by scraping the bark. On pomaceous hosts, the disease is most frequent on fruits, often causing distortion. Rust sometimes affects twig and buds but seldom leaves. Aecia are particularly prominent on hips of English hawthorn, with long whitish perithecium around orange spores.

Control. Some apple varieties susceptible to apple rust are rather resistant to quince rust, including Jonathan, Rome, Ben, Davis, and Wealthy. Red Delicious is quite susceptible. Destroy cedars in neighborhood of orchards; spray as for apple rust.

Gymnosporangium cofusum. III, on Savin Juniper; 0, I, on hawthorn.

Gymnosporangium corniculans. III, on juniper and red cedar; 0, I, on leaves of amelanchier.

Gymnosporangium cornutum (*G. auriantiacum*). **Juniper gall rust;** III, on leaves and stems of common juniper; 0, I, on mountain ash.

Gymnosporangium cupressi. III, on Arizona cypress; 0, I, on amelanchier.

Gymnosporangium davisii. III, on mountain and common juniper; 0, I, on leaves of red and black chokecherry. Telia are usually on upper surface of needles, sometimes at base of stems.

Gymnosporangium effusum. III, on eastern red cedar; 0, I, on chokeberry. Fusiform swellings on cedar trunk and branches.

Gymnosporangium ellisii. Witches' broom rust; III, on southern white cedar (*Chamaecyparis*); 0, I, on sweet fern, gale, bayberry, wax-myrtle leaves, fruits, and young stems. Aecia are cluster cups; telia are cylindrical, filiform, 3 to 6 mm high, appearing on leaf blade or axil the first season after infection, thereafter only on stems, invading inner bark and wood. Witches' brooms are abundant; even large trees die if heavily broomed.

Gymnosporangium exigum. III, on leaves of alligator and Mexican junipers, eastern red cedar; 0, I, on leaves, fruits of hawthorn.

Gymnosporangium exterum. III, on stems of eastern red cedar; 0, I, on gillenia. Flattened telia anastomose over short fusiform swellings with roughened bark on cedars. Also galls on stems of juniper.

Gymnosporangium floriforme. III, on red cedar; 0, I, on leaves of hawthorn. Cedar galls are small.

Gymnosporangium fratemum (*G. transformans*). III, gall on *Chamaecyparis thyoides*; 0, I, on chokeberry.

Gymnosporangium globosum. Hawthorn rust; III, generally on eastern red cedar, also on dwarf, prostrate, and Rocky Mountain junipers; 0, I, mostly on hawthorn, also on apple, crabapple, pear, and mountain ash. Leaf galls on cedar are very similar to those of common cedar-apple rust, but are smaller, seldom over ½ inch, nearer mahogany red in color, and not perennial, producing telial horns for one season only. Apple and pear foliage may be slightly affected but not the fruit; aecia are common on hawthorn pips.

Gymnosporangium gracile. III, witches' brooms on juniper; 0, I, on hawthorn, quince, and shadbush.

Gymnosporangium haraeanum. III, on leaves of Chinese juniper; 0, I, on Chinese flowering quince and pear.

Gymnosporangium harknessianum. III, on western juniper; 0, I, on amelanchier, chiefly on fruits, sometimes stems. Papery margins of aecia are usually long.

Gymnosporangium hyalinum. III, on southern white cedar; 0, I, on hawthorn and pear leaves. Slight swellings are formed on small twigs and branches of white cedar.

Gymnosporangium inconspicuum. III, on Utah juniper; 0, I, on fruits, mostly of amelanchier and squaw-apple. Juniper leaves turn yellow; rarely telia appear on branches.

Gymnosporangium japonicum (*G. photiniae*). III, gall on stems of Chinese juniper; 0, I, on photinia.

Gymnosporangium juniperi-virginianae. Cedar-apple rust; III, generally on red cedar, eastern and southern, on prostrate and Rocky Mountain junipers; 0, I, generally on apple and crabapple east of Great Plains. The fungus is a native of North America and does not occur elsewhere. It is more important commercially in the apple-growing regions of the Virginias and Carolinas and certain states in the Mississippi Valley. It is important in many areas on ornamental crabapples in home plantings.

The cedar "apples" or galls vary from $\frac{1}{16}$ inch to over 2 inches across. Leaves are infected during the summer, and by the next June a small, greenish brown swelling appears on upper or inner leaf surface. This swelling enlarges until by autumn the leaf has turned into a chocolate brown, somewhat kidney-shaped gall covered with small circular depressions. The next spring in moist weather orange telial horns are put forth from the pocketlike depressions. The teliospores are enveloped in a gelatinous material that swells vastly, a gall covered with horns sometimes reaching the size of a small orange. They germinate in place to produce the basidiospores, which are carried by wind to infect apple or other deciduous host.

By midsummer, apple leaves show yellow areas with amber pustules on upper surface; but after pycnia have exuded drops of sticky liquid, they appear as black dots in a rather reddish circle. On the undersurface of these spots small cups are formed, with recurved fimbriate margins. These aecia may also appear near stem end of apples and are common on swollen twigs of crabapple. Spores from these cups are blown back to the cedar in late summer, the entire cycle thus taking 2 years: 18 to 20 months on the cedar, 4 to 6 on the apple host.

Chief injury is to the apple host, the rust causing premature defoliation, dwarfing, and poor-quality fruit. On very susceptible crabapples, such as Bechtel's crab, repeated infection may cause death of the branches or of the entire tree. All our native crabapples are susceptible; most Asiatic varieties are resistant.

Control. Care in planning is most important. Don't let your landscape architect or gardener put cedars and native crabapples or hawthorns close together. Keep them separated as far as possible with a windbreak of some tall nonsusceptible host in between. Some states have laws prohibiting red cedars within a mile of commercial apple orchards, but for practical garden purposes a few hundred yards is sufficient, the danger markedly decreasing with distance, especially with a house or hedge as a windbreak.

If junipers are already planted, it is possible in late winter to go over small specimens and remove galls before spore horns are formed. Spraying in spring inhibits telial development and germination of teliospores. Spray red cedars in August to prevent infection from crabapples.

Fairly resistant apple varieties are Baldwin, Delicious, Rhode Island and Northwestern Greening, Franklin, Melrose, Red Astrachan, Stayman, and Transparent. Avoid susceptible Jonathan, Rome, Wealthy, and York Imperial. Most junipers susceptible to apple rusts are cultivars of *Juniperus virginiana* and *J. scopulorum*. Many cultivars of *J. chinensis* and *J. horizontalis* are resistant, and there are even some resistant forms of *J. virginiana*.

Gymnosporangium kernianum. III, on alligator, Utah, and western junipers; 0, I, on amelanchier and pear. Telia arise between leaves on green twigs, but mycelium is perennial in stems, causing dense witches' brooms 6 to 18 inches in diameter.

Gymnosporangium libocedri. III, on incense cedar; 0, I, on leaves, fruits, of amelanchier and hawthorn, also apple, crabapple, pear, quince, Japanese quince, and mountain ash. Aecium is a cluster cup on foliage; telia are always on leaves; witches' brooms and swellings are produced on branches, rarely on trunks. The fungus is said to persist in the mycelial stage up to 200 years.

Gymnosporangium multiporum. III, on stems of western, one-seed, and Utah juniper between leaves; 0, I, unknown.

Gymnosporangium nelsoni. III, on one-seed, prostrate, Rocky Mountain, Utah, and western junipers; 0, I, on hawthorn, quince, Oregon crab, pear, squaw apple, and Pacific mountain ash. Galls are firm, woody, round, up to 2 inches in diameter.

Gymnosporangium nidus-avis. Witches' broom rust; III, on eastern and southern red cedars, on prostrate and Rocky Mountain junipers; 0, I, on fruit, young stems, leaves of apple, hawthorn, mountain ash, quince, Japanese quince, amelanchier or serviceberry. Trunks and branches of large trees have witches' brooms and long spindle-shaped swellings. Aecia are on both leaf surfaces.

Gymnosporangium nootkatense. Gall rust; II, III, on Alaska cedar; 0, I, on mountain ash and Oregon crabapple. It is the only *Gymnosporangium* species with uredial stage. Uredia are bright orange fading to pale yellow; teliospores appear later in the same pustules. Aecia are cluster cups.

Gymnosporangium speciosum. III, on alligator, one-seed, and Utah junipers; 0, I, on leaves of syringa (*Philadelphus*) and fendlera. Telia are in longitudinal rows on long fusiform swellings on juniper branches, which are girdled and die. In severe infections the whole tree dies.

Gymnosporangium trachysorum. III, on stem of eastern red cedar; 0, I, on hawthorn leaves. Swellings on cedar are abruptly fusiform to globoid with prominent telia 6 to 10 millimeters high.

Gymnosporangium tremelloides (*G. juniperinum*). III, stem gall on mountain juniper; 0, I, on Pacific mountain ash. On smaller branches swellings are subglobose galls up to ¾ inch in diameter; hemispherical swellings on larger branches are covered with flattened telia.

Gymnosporangium tubulatum. III, on stems of prostrate and Rocky Mountain junipers; 0, I, on leaves, fruit of hawthorn. Telia are 3 to 4 mm high on irregular galls on cedar twigs and branches.

Gymnosporangium vauqueliniae. Witches' broom rust; III, on one-seed juniper; 0, I, on *Vauquelinia californica*. This rust is the only *Gymnosporangium* causing witches' brooms on the aecial host.

Hyalopsora

Melampsoraceae. Telia on ferns, teliospores several-celled, in epidemis; urediospores of two kinds, with pores.

Hyalopsora aspidiotus. Fir fern rust; 0, I, on balsam fir; II, III, on oak fern (*Phegopteris dryopteris*). Pycnia are slightly raised orange-yellow spots on needles; aecia are yellow to white, columnar, on 2-year needles.

Hyalopsora cheilanthus. Fir fern rust; 0, I, on balsam fir; II, III, on rock brake, parsley fern, and cliff brake.

Hyalopsora polypodii. Fir fern rust; general in northern and western states on polypody fern and woodsia.

Kuehneola

Puciniaceae. Teliospores two- to many-celled; wall faintly colored or colorless.

Kuehneola malvicola. II, III, on hibiscus and malvaviscus.

Kuehneola uredinia. Yellow rust, cane rust; 0, I, II, III, on blackberry, dewberry, and raspberry. The disease appears to be increasingly prevalent, especially on leaves, but there is a great difference in varietal susceptibility. Eldorado, Foster, Jumbo, Lawton blackberries are highly susceptible; Nantichoke, Austin Thornless, Boysen Brainerd, Burbank Thornless, Jersey Black are resistant. European varieties are generally resistant.

Kunkelia

Pucciniaceae. Pycnia subcuticular; telia subepidemal, caeomoid; teliospores catenulate, one-celled.

Kunkelia nitens. Short-cycle orange rust of blackberry; I, general on blackberry but more common in the South and West, also on dewberry and black, but not on red raspberry. It is a perennial rust, a systemic disease with only the aecial stage present. Underside of leaves may be covered with quantities of orange-yellow spores. Remove infected bushes.

Melampsora

Melampsoraceae. Telia more or less indefinite; teliospores sessile, subcuticular or subepidermal, forming crusts of a single layer; aecia when present with rudimentary peridium; uredia erumpent, pulverulent; spores globoid or ellipsoid, single on pedicels. Species heteroecious when telia are on woody plants; autoecious if telia are on herbaceous plants (see Fig. 65).

Malampsora abieti-capraearum. Fir willow rust; 0, I, on balsam, white, and alpine firs; II, III, on willows, widespread. Yellow spots on willow leaves in early summer are followed by dark pustules when the telial stage is produced. There may be some defoliation.

Melampsora abietis-canadensis. Hemlock-poplar rust; 0, I, on eastern hemlock; II, III, on various poplars. Cones have golden powdery masses of spores over the surface; later shrivel, turn black, and hang as mummies; no viable seed produced. Uredia are golden powdery pustules on undersurface of poplar leaves; in late summer telia are formed in orange-yellow crusts that change to black; in spring basidiospores reinfect hemlock.

Melampsora albertensis. Douglas fir needle rust; 0, I, on Douglas fir, big-cone spruce; II, III, on native poplars. Pycnia are on upper surface of current-year needles; aecia, of the caeoma type, are orange-yellow on the undersurface. The rust is often epidemic on young trees but with little permanent ill effect.

Melampsora arctica. 0, I, on saxifrage; II, III, on willow.

Melampsora farlowii. Needle and cone rust of hemlock; 0, I, unknown; III, on hemlock. Reddish slightly raised telia are on undersurface of needles,

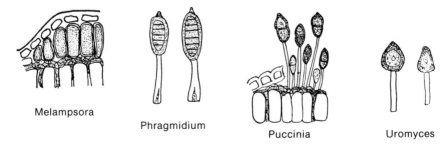

Melampsora

Phragmidium

Puccinia

Uromyces

Figure 65. Teliospores. *Melampsora,* sessile in crust under host epidermis; *Phragmidium,* stalked, with several cells; *Puccinia,* stalked, two-celled; *Uromyces,* stalked, one-celled.

shoots of the current year, and on cones. Young shoots may be twisted and killed. Injury may occur in nurseries and in ornamental hedges.

Melampsora hypericorum (*Mesopsora hypericorum*). On St. Johnswort, Montana.

Melampsora medusae. Larch needle rust; 0, I, on larch in northeastern states; II, III, on native and introduced poplars except in far South.

Melampsora occidentalis. Poplar rust; 0, I, unknown; II, III, on native poplars in the West.

Melampsora paradoxa (*M. bigelowii*). Larch willow rust; 0, I, on larch; II, III, on many species of willow. The damage to larch is insignificant. The fungus winters on willow as mycelium in catkins, terminal buds, and young stems and can maintain itself on willow in the uredial stage without larches.

Melampsora ribesii-purpureae. 0, I, on currant, flowering currant, and gooseberry; II, III, on willow species.

Melampsorella

Melampsoraceae. Heteroecious on fir, spruce, and dicotyledons; pycnia subcuticular, aecia and uredia subepidermal, telia in epidemal cells. Only one species in United States.

Melampsorella caryophyllacearum (*M. cerastii*). Yellow witches' broom rust; 0, I, on many firs; II, III, on chickweed. Infected evergreen branches develop numerous upright lateral shoots from one point, forming a compact witches' broom; twigs are dwarfed, and needles turn yellow and drop, leaving brooms bare. The fungus is perennial in stems, and shoots develop with yellow leaves. Pycnia appear in raised orange spots on both surfaces of dwarfed leaves in spring; aecia form in summer on underside, in two rows of orange blisters. The disease is seldom serious enough for control measures. In forest practice remove trees with main stem infections early in life of the stand.

Melampsoridium

Melampsoraceae. Heteroecious, on larch and dicotyledonous shrubs and trees; pycnia subcuticular; other sori subepidermal; teliospores sessile, one-celled.

Melampsoridium betulinum. Birch leaf rust; 0, I, on larch; II, III, on birches. Uredia on underside of birch leaves are small reddish yellow powdery pustules, followed later in summer by telia, first waxy yellow, then dark brown to nearly black.

Milesia

Melampsoraceae. Heteroecious on firs and ferns. All spores are colorless; urediospores obovate or laceolate; teliospores in epidemal cells.

Milesia fructuosa. 0, I, on balsam fir; II, III, on *Dryopteris* spp. Aecia are white on current needles, maturing by midsummer.

Milesia laeviuscula. Needle rust; 0, I, on grand fir; II, III, on licorice fern, in the West.

Milesia marginalis. 0, I, on balsam fir; II, III, on *Dryopteris marginalis*. Pycnia are on both sides of needles, aecia of needles of current year, maturing by midsummer.

Milesia pycnograndis (*M. polypodophila*). 0, I, on balsam fir; II, III, on *Polypodium virginianum*. Hyphae are perennial in needles and small stems of balsam fir; aecia on needles 3 to 9 years old.

Nyssopsora

Puccinaceae. Autoecious; teliospore with three cells.

Nyssopsora clavellosa. III, on *Aralia hispida*.

Peridermium

A form genus with 0, I, on Gymnosperms. Aecia have peridia and are cylindrical, tonguelike or bullate.

Peridermium bethelii. On dwarf mistletoe.

Peridermium ornamentale. 0, I, on white, alpine, and noble firs.

Peridermium rugosum. 0, I, on Pacific silver and lowland white firs.

Phakopsora

Melampsoraceae. Telia indehiscent, lenticular; spores formed in irregular succession, not in chains.

Phakopsora cherimoliae. On cherimoya.

Phakopsora jatrophicola. On cassava.

Phakopsora pachyrhizi. On soybean.

Phakopsora zizyphi-vulgaris. On *Zizyphus jujuba*, in Florida.

Phragmidium

Pucciniaceae. Autoecious. Pycnia subcuticular, other sori subepidermal; aecia caeomoid; teliospores large, conspicuous, of one to ten or more cells, each with two or three lateral pores; walls somewhat layered, inner layer colored, outer layer nearly colorless, smooth or verrucose; pedicel colorless except near spore; often swelling in lower portion (see Fig. 65). Aecia with catenulate globoid or ellipsoid verrucose spores; uredia when present circled with paraphyses; urediospores single on pedicels, walls verrucose or echinulate with indistinct scattered pores.

Phragmidium americanum. 0, I, II, III, on leaves of native and cultivated roses. Teliospores with eight to eleven cells.

Phragmidium fusiforme (*P. rosae-acicularis*). 0, I, II, III, on several hosts species. Teliospores with five to eleven cells, walls chocolate brown, verrucose.

Phragmidium montivagum. 0, I, II, III, on many species of roses. Teliospores with six to nine cells.

Phragmidium mucronatum (*P. disciflorum*). **Leaf rust of rose;** 0, I, on leaves and stems; II, III, on leaves of cultivated roses, eastern states to the Rocky Mountains and on the Pacific Coast. This disease is the common rust of hybrid teas and other roses with large, firm leaflets. It is not much of a problem in the East, although sometimes found in New York and New England gardens, but it is a serious menace along the Pacific Coast. Aecia appear on leaves as small, roughly circular spots, $1/25$ inch across, bright orange on the underside of leaf, from the spore masses, light yellow on the upper surface, sometimes bordered with a narrow green zone. Leaf lesions may be slightly cup-shaped viewed from the upper surface. Stem lesions are long and narrow.

The summer uredial stage has reddish orange spores in very small spots, $1/50$ inch, over underside of leaves. This stage may repeat every 10 to 14 days in favorable weather, with wilting and defoliation. In mild climates the uredial stage continues; in cooler areas the telial stage is formed toward autumn — black pustules of stalked dark spores, rough, with a point, five to nine cells.

The leaf surface must be continuously wet for 4 hours for rust spores to germinate and enter the leaf, which means liquid water conditions and not high humidity as with mildews. High summer temperatures adversely affect infection, summer spores retaining viability for only a week at 80° F. In southern California temperatures are uniformly favorable for rose rust, and from October to April there is sufficient rainfall. In drier months fog may provide requisite moisture.

Control. Removing infected leaves during the season and all old leaves left at the time of winter or early spring pruning may be somewhat helpful.

Phragmidium rosae-arkansanae. 0, I, II, III, on *Rosa arkansana* and *R. suffulta.* Teliospores with five to eight cells.

Phragmidium rosae-californicae. 0, I, II, III, on many rose species. Teliospores with eight to eleven cells.

Phragmidium rosicola. III, on *Rosa engelmanii* and *R. suffulta*. Teliospores one-celled, nearly round.

Phragmidium rubi-idaei. Leaf and cane rust of raspberry; western yellow rust, general but important only in the Pacific Northwest. 0, I, II, III, on red raspberries, sometimes black but not on blackberries. Small, light yellow spore pustules appear in young leaves, with black teliospores following in the same spots later in the season. Deep, cankerous lesions are formed on canes in the fruiting year, Cuthbert variety being particularly susceptible. Spring infection probably comes from sporidia formed in telia on fallen leaves. A dormant spray may be helpful, along with cleaning out infected canes at winter pruning.

Phragmidium speciosum. 0, I, on stems and leaves; III, on stems of cultivated and native roses, throughout United States except far South.

Phragmidium subcorticium. Obsolete name. Some specimens formerly recorded as this species belong to *P. mucronatum*, others to *P. rosae-pimpinellifoliae*.

Phragmidium tuberculatum. On *Rosa* sp., in Connecticut and Alaska.

Phragmopyxis

Pucciniaceae. Teliospores colored, two- to many-septate; wall three-layered, the middle layer swelling in water; aecia, uredia, and telia with a border of paraphyses.

Phragmopyxis acuminata. 0, III on *Coursetia*.

Physopella (*Angiopsora*)

Pucciniaceae. Only uredia and telia known. Telia indehiscent, lenticular; teliospores in chains.

Physopella ampelopsidis (*Phakopsora vitis*). On ampelopsis and grape, in Florida.

Physopella compressa. On paspalum, southern ornamental grass.

Pileolaria

Pucciniaceae. Autoecious, on members of family Anacardiaceae. Teliospores stipitate, dark, with pores, one-celled; pycnia subcuticular; uredia present.

Pileolaria cotini-coggyriae. On smoke tree.
Pileolaria patzcuarensis. 0, I, II, III, on sumac.

Prospodium

Pucciniaceae. Autoecious on Bignoniaceae and Verbenaceae in warm climates.

Prospodium appendiculatum. On tecoma, in Florida and Texas.
Prospodium lippiae. On lippiae, in Arizona.
Prospodium plagiopus. On tabebuia, in Florida.
Prospodium transformans. On tecoma, in Florida.

Puccinia

Pucciniaceae. A very large genus, comprising nearly half of all known rusts; autoecious and heteroecious. Teliospores smooth, two-celled with apical pores, firm pedicels, colored; aecia cluster cups with peridium (see Fig. 65). The species listed here are a small selection of those on garden plants; others are listed in host section.

Puccinia allii (*P. porri*). Autoecious on onion, garlic, and shallot, but 0, I stages rare. Occasional on cultivated onion, more common on garlic and wild onion. Uredia are yellowish, telia black.
Puccinia amphigena (*Aecidium yuccae*). On yucca.
Puccinia andropogonis, with various strains. 0, I, on lupine, Indian paintbrush, and turtlehead; II, III, on andropogon.
Puccinia antirrhini. Snapdragon rust; II, III, general on snapdragon, also on linaria, corydylanthus; 0, I, unknown. Pustules of spores on underside of leaves are chocolate brown, often in concentric circles (see Fig. 66). The area over the pustule is pale or yellow on upper surface. Spores also appear on stems; there is a drying and stunting of whole plant. The rust is spread by wind-blown spores and on cuttings. For infection, plants need to be wet with rain or dew 6 to 8 hours with day temperatures around 70° F to 75° F. Spores are killed above 94° F. There are at least two races.
Control. Purchase only rust-resistant varieties. Bordeaux mixture controls secondary fungi following rust but not the rust itself. Sulfur dust is still useful, or a spray made by adding 1 ounce rosin soap to a gallon of water and then adding 1 ounce dry lime sulfur.
Puccinia arachidis. Peanut rust; occasional in Alabama, Florida, and Texas.
Puccinia aristidae and varieties. II, III, on wild grasses, *Aristides*, and *Distichlis*; 0, I, on eriogonum, greasewood, beet, spinach, western wallflower, garden cress, radish, California bluebell, heliotrope, cleome, primrose, sand verbena, and others.
Puccinia asparagi. Asparagus rust; II, III, general on susceptible varieties; 0, I, not reported in natural infections. Also on onion. Asparagus rust reached America in 1896 from Europe and spread with devastating suddenness from Boston and New Jersey to California, reaching there by 1912, one of the fastest cases of disease spread in our history. If tops are attacked several years

Figure 66. Rust on snapdragon.

in succession, the root system is so weakened that shoots fail to appear in spring or are culls.

The first symptom is a browning or reddening of smaller twigs and needles, with the discolored area spreading rapidly until the whole planting looks as if it had ripened prematurely. The reddish color is due to numerous small pustules of urediospores that give off a dusty cloud when touched. These appear in successive generations until autumn, or a spell of drought, when they are replaced by black teliospores, either in the same or a new fruiting body. They remain on old stems until spring, germinating then to infect new shoots as they emerge from the ground.

Control. For a long time resistant varieties Mary Washington and Martha Washington were the answer to the rust problem, but the fungus has developed resistant strains. Waltham Washington, Seneca Washington, and California 500 have some resistance. Clean up volunteer or wild asparagus around beds. A parasitic fungus, *Darluca filum*, helps keep rust in check.

Puccinia canaliculata. Rust; on purple nutsedge and yellow nutsedge.

Puccinia carduorum. Rust; on *Carduss tenniflorus*.

Puccinia caricina (*P. caricis* var. *grossulariata*, *P. pringsheimia*). 0, I, on currant, flowering currant, gooseberry; II, III, on *Carex* spp. Common only on wild species or in neglected gardens. Leaves are thickened, sometimes curled in reddish cluster cup areas; there are enlargements on stems and petioles, red spots on berries. Control by eliminating the sedge host.

Puccinia carthemi. Widely distributed on safflower in Great Plains and California. Spores carried on seed or persisting in soil infect seedlings, which often die.

Puccinia claytoniicola. On claytonia, in Wyoming.

Puccinia conoclinii. On ageratum, in Ohio.

Puccinia coronata. Crown rust of oats; orange leaf rust of oats; 0, I, on buckthorn and rattan vine; II, III, on oats and grasses. There are several varieties and many physiological races of this rust, which is as destructive to oats as leaf rust is to wheat. Redtop, meadow fescue, ryegrass, and bluegrass are among the lawn grasses that may show orange or black pustules on leaves.

Puccinia crandalli. 0, I, on snowberry, wolfberry, coralberry; II, III, on grasses, fescues, bluegrass.

Puccinia cynodontis. On Bermuda grass, in New Mexico.

Puccinia cypripedii. On orchids.

Puccinia dioicae (*P. extensicola*) in many varieties. 0, I, on aster, goldenrod, erigeron, senecio, lettuce, oenothera, rudbeckia, and helenium; II, III, on *Carex* spp.

Puccinia dracunculi. On artemisia, in Wisconsin to the Pacific Coast.

Puccinia flaveriae. On calendula.

Puccinia graminis. Stem rust of grains and grasses; 0, I, on barberry and mahonia, especially in north central and northeastern states; II, III, on wheat and other cereals and wild and cultivated grasses.

This disease is the classic example of rust, the one used in school textbooks and known through the ages as the major limiting factor of wheat production. Proof of the connection between barberry and wheat in the life cycle was not made until 1864, but long before that farmers had noticed that wheat suffered when barberry plants were near. France in 1660, Connecticut in 1726, and Massachusetts in 1755, enacted laws requiring the destruction of barberry near grain fields.

There are six commonly recognized varieties of stem rust:

> *Puccinia graminis avenae* — on oats, sweet vernal grass, brome grasses, some fescues.
>
> *P. graminis agrostidis* — on redtop and other *Agrostis* spp.
>
> *P. graminis phlei-pratensis* — on timothy and some related grasses.
>
> *P. graminis poae* — on Kentucky and other bluegrasses.
>
> *P. graminis secalis* — on rye, some wheat, and barley grasses.
>
> *P. graminis tritici*, wheat rust — on wheat, barley, rye, and many grasses.

Stem rust occurs wherever wheat is grown, but is most serious in northern states. It is dependent on weather conditions, with epidemics and disastrous losses in certain seasons. The amount depends on the maturity of the crop when rust strikes, but losses may run 25% of expected yield for the nation and much higher for individual states. There are a great many physiological races.

On grains and grasses the first rust appears as long, narrow streaks on stems, leaf sheaths, leaf bases, and distal portions of blades. These streaks are uredial sori, the epidermis being torn back to form a white collar around a dark red powdery mass of one-celled urediospores. Later the same sori turn black as dark, two-celled teliospores replace summer urediospores. Stems may be broken at this stage.

The summer spores appear about 10 days after infection. This stage can be repeated, the spores reinfecting wheat, and, since they are carried by wind from one plant to another, one state to another, even to hundreds of miles, they account for large outbreaks of disease. In Mexico and southern Texas, this II stage continues through the winter and causes spring infection without the intervention of barberry. Waves of urediospores coming from the South may start northern infection.

Normally in the North, spring infection starts on barberry from sporidia (basidiospores) produced on a promycelium put forth by a teliospore wintered on a wheat stem. Two sexes occur in this rust, designated + and − rather than male and female. A young teliospore contains two nuclei, one + and the other −; as the spore matures, these fuse to a single nucleus, which divides twice in the production of the four-celled basidium (promycelium). Each cell produces a sporidium; two of these are + and two are −. A sporidium falling on a barberry leaf germinates, sends in an infection thread, and develops a mononucleate (haploid)-feeding mycelium and finally a flask-shaped pycnium containing pycniospores, which correspond to the sex of the sporidium starting infection. The pycnia are in reddish lesions on the upper leaf surface. Hyphal threads, receptive hyphae, extend through the mouth of the pycnium. Aided by insects, which are attracted by a sweet nectar, pycniospores (spermatia) of one sex are brought into contact with receptive hyphae of the opposite sex, and sexual union takes place, without which there is no further development of the rust.

The dicaryotic or binucleate mycelium formed from the fertilized hypha grows through the cells of the barberry leaf and masses together on the underside to produce aecia filled with a yellowish waxy layer of aeciospores in cluster-cup formation. These spores, unable to reinfect barberry or mahonia, are wind-borne to the cereal or grass host, the subsequent mycelium continuing binucleate until the fusion in the teliospore. New crops of urediospores can be produced every 10 to 14 days.

Control. Resistant varieties are of primary importance, but they are difficult to maintain because the sexual process in rusts allows the continuous development of new strains. More than 200 strains are known, but only a dozen or so are important in any one year. Race 15B is prevalent most years and can attack all varieties of wheat grown in this country. Eradication of the barberry eliminates the alternate host and also the breeding place of new rust varieties. Most barberry and mahonia species are under quarantine, but some have been designated rust-resistant by the U.S. Department of Agriculture and may be shipped interstate under permit.

Puccinia helianthi. Sunflower rust; 0, I, II, III, generally on sunflower, Jerusalem artichoke, and heliopsis. Numerous brownish pustules in which repeating spores are formed develop on underside of leaves, which may dry and drop.

Puccinia heterospora. III on abutilon, hollyhock, mallow, and malvaviscus.

Puccinia heucherae. III, on coral bells, woodland star, saxifrage, bishops cap, and foam flower.

Puccinia hieracii. 0, I, II, III, widespread on endive and hawksbeard. Endive leaves are spotted and blighted with dusty spore pustules. The crop is occasionally lost, but no control has seemed practical.

Puccinia horiana. White rust; III, IV on chrysanthemum; no alternate host known. First reported in England in 1964; became widespread there in 1976. Found in amateur chrysanthemum plantings in New Jersey and Pennsylvania in 1977.

Puccinia iridis. Iris rust; 0, I, II, III, on bulbous iris, serious in the Southeast, uncommon in the Northwest. Small, oblong to oval, red or dark brown powdery spots, often surrounded by a yellow margin, are present on leaves and stems, which may die prematurely. In inoculation tests with Dutch iris, varieties Early Blue, Gold and Silver, Golden West, Imperator, Lemon Queen, and Texas Gold were resistant.

Puccinia malvacearum. Hollyhock rust; III, generally on hollyhock, also on mallow, and lavatera. This rust is so common and destructive that it limits the use of hollyhocks as ornamentals. Stems, leaves, and bracts may be attacked. There are yellow areas on the upper surface of leaves, orange-red spore pustules on the underside, and elongated lesions on stems. Spore pustules are sometimes grayish from formation of sporidia, but the alternate host is unknown. In severe infections leaves dry and hang along the stem. The fungus winters in pustules in basal leaves and in old stems.

Control. Cleaning up all infected plant parts in fall and again very early in spring is most important; infection starts early in the season, and once it is under way, it is very difficult to curb with a fungicide.

Puccinia menthae. Spearmint rust; 0, I, II, III, on spearmint, peppermint, also horse mint, mountain mint, dittany, bee balm, yerba buena, and germander; especially serious for mint farmers in Middle West and Northwest. In spring and early summer the disease appears as light yellow to brown raised spots on deformed stems and leafstalks, sometimes on main veins; golden to chocolate brown spots appear in late summer and fall. Affected leaves dry, and the yield of oil is reduced. The pathogen has at least 10 races. Dusting with sulfur and early cutting are recommended.

Puccinia pelargonii-zonalis. Pelargonium rust; the uredinial stage of a rust, presumably this species, was found on geranium in New York and California in 1967. It has now been reported in Pennsylvania and Florida.

Brown spore pustules appear on leaves, petioles, and stems; leaves turn yellow and drop. Destroy infected plants.

Puccinia phragmitis. 0, I, on rhubarb; II, III, on reed grass, sometimes present in California but not serious. Aecia are white, on underside of rhubarb leaves, surrounded by pycnia.

Puccinia poae-nemoralis (*P. poae-sudeticae*). Bluegrass leaf rust; yellow leaf rust; II, III, on turf grasses, mostly Canada and Kentucky bluegrass; 0, I, unknown; general east of the Rocky Mountains. The uredia are orange-yellow with numerous peripheral paraphyses. Telia are covered rather permanently with epidermis; spores are dark brown with short pedicels. The wheat stem rust is more important on Merion bluegrass.

Puccinia polysora. Southern corn rust; 0, I, unknown; II, III, on corn and grasses. Present in the South, requiring higher temperatures than common corn rust; not very important. Urediospores are yellow to golden, teliospores chestnut brown, angular; often parasitized by *Darluca filum*.

Puccinia psdii. Rust; on allspice (*Pimenta dioica*) and *Syzygium jambos*.

Puccinia pygmaea. Rust; on grasses.

Puccinia recondita (*P. rubigo-vera*). Leaf rust of cereals and grasses; with several varieties:

> *P. recondita tritici* (*P. triticina*). II, III, on wheat (but not grasses); 0, I, on meadow rue. This rust is worldwide and more serious than stem rust in the southern half of the American wheat belt, sometimes epiphytotic with losses up to 30%. The leaf tissue is progressively destroyed through the season, resulting in a reduced number of kernels, shriveled grain, low weight, and low protein content. Rust pustules breaking through the epidermis greatly increase transpiration losses. Orange uredial pustules are followed later by gray telial sori, but urediospores are the effective spore form and can survive southern winters. There are many physiological races.
>
> *P. recondita agropyri*. II, III, on wheatgrasses and wild ryegrasses; 0, I, on clematis, buttercup, columbine, larkspur, and other Ranuculaceae. Common in Rocky Mountain area.
>
> *P. recondita agropyrina*. Similar to the above but occurring outside mountainous areas.
>
> *P. recondita apocrypta*. II, III, on wheat and wild grasses; 0, I, on waterleaf and mertensia.
>
> *P. recondita impatientis*. II, III, on redtop and related grasses; 0, I, on touch-me-not.
>
> *P. recondita secalis*. II, III, on rye; 0, I, on bugloss (*Lycopsis*).

Puccinia ripulae. On baccharis, in Texas.

Puccinia solheimi. On dodocatheon, in Wyoming.

Puccinia sorghi. Corn rust; 0, I on oxalis; II, III, on corn, sweetcorn, general in northeastern and north central states. Cinnamon brown spore pustules cover both leaf surfaces with black pustules toward autumn. The disease is not often serious enough for control measures.

Puccinia sparganoides (*P. peridermiospora*). **Ash rust**; 0, I, generally on ash east of the Great Plains; II, III, on marsh and cord grasses (*Spartina* spp.). Ash twigs and petioles are swollen and leaves distorted. Cluster cups filled with yellow powdery aeciospores are formed in the swellings. In New England, where rust is often severe, the most important infection period on ash is from May 15 to June 20, with 6 to 8 hours of damp air necessary. Marsh grasses are infected and reinfected from July 20 to August 20.

Puccinia stenotaphri. On St. Augustine grass, in Florida.

Puccinia striiformis (*P. glumarum*). **Stripe rust of wheat**; II, III, on wheat, barley, rye, redtop, orchardgrass, and many other grasses. Uredial stage is yellow, and pustules are formed in streaklike clusters on leaves; telia are in black streaks.

Puccinia substriata. Rust; on eggplant.

Puccinia taneceti. Chrysanthemum rust; II, general; 0, I, II, III on *Tanacetum* and *Artemisia*. Small blisters of pinhead size appear on underside of leaves and occasionally on upper surface. The spore mass is dark reddish brown and powdery. The rust is more common in greenhouses than outdoors. Optimum germination is at 60° F to 70° F; spores are killed at high temperatures.

Puccinia thaliae (*P. cannae*). II, III, on edible canna, garden canna, and maranta.

Pucciniastrum

Melampsoraceae. Heteroecious with perennial mycelium, pycnia and aecia on conifers: firs and spruces; pycnia subcuticular, other sori subepidermal; telia may be intraepidermal; aecia and urediospores yellow.

Pucciniastrum americanum. Late leaf rust of raspberry; 0, I, on white spruce; II, III, on red raspberry, not black. This rust appears late in the season on Cuthbert and other susceptible varieties, in northern half of the country, most common east of the Mississippi. Fine light yellow powdery masses of spores appear on basal leaves, leaf petioles, shoots, and even fruit.

Pucciniastrum epilobii. Fuchsia rust; the alternate hosts are species of Abies.

Pucciniastrum goeppertianum. Fir-huckleberry rust; blueberry witches' broom; 0, I, on firs; III, on low and high bush blueberries. The fungus is systemic and perennial in blueberries, producing short swollen twigs in a witches' broom effect, and telia forming a polished red layer around the shoots. Destroy diseased bushes; keep blueberry plantations some distance from firs.

Pucciniastrum hydrangeae. 0, I, on eastern and Carolina hemlock; II, III, on hydrangea.

Pucciniastrum vaccinii (*P. myrtilli*). **Hemlock rust; leaf rust of blueberry**; widespread; 0, I, on eastern hemlock; II, III, on azalea, blueberry, cranberry, lyonia, menziesia, and rhododendron. It is the most common hemlock rust, but often only a single leaf or twig is infected. Aecia are formed on current-year needles. Blueberries have yellow pustules, on leaves only, with defoliation in mid- or late summer.

Ravenelia

Pucciniaceae. Autoecious, tropical with only a few species in United States. Teliospores more or less muriform, with compound stalks.

Ravenelia dysocarpae. On mimosa, in Arizona.
Ravenelia humphreyana. On poinciana, in Florida and Texas.
Ravenelia indigoferae. On indigofera, in Arizona.

Scopella

Pucciniaceae. Tropical. Uredia and telia subepidermal. Teliospores one-celled, on pedicel.

Scopella sapotae (*Uredo sapotae*). On sapodilla in Florida, infecting leaves in winter and early spring.

Sphenospora

Pucciniaceae. Tropical. Telia and peridia subepidermal, then erumpent; teliospores waxy, two-celled, on pedicel.

Sphenospora mera. On bletilla, in Florida.

Sphaerophragmium

Pucciniaceae. Teliospores stalked, four- to several-celled, with transverse and horizontal septa; on legumes.

Sphaerophragmium acaciae. On lebbek, in Florida.

Tranzschelia

Pucciniaceae. Teliospores two-celled, stalked; uredia with pseudoparaphyses; on Ranunculaceae and *Prunus*.

Tranzschelia discolor (*T. pruni-spinosae* var. *discolor*). Rust of stone fruits; peach rust; 0, I, on *Anemone coronaria*; II, III, on apricot, peach, plum, prune, almond, and cherry, in late summer. Yellow angular spots appear on leaves with powdery spore pustules on underside, reddish on peach, dark brown on almonds; sometimes with late season defoliation. Peach fruit may have round sunken green spots; twigs may have oval blisters in early spring. Urediospores wintering on sucker shoots can start spring infection without the alternate host. The Drake variety of almond is most susceptible.

Tranzschelia pruni-spinosae var. **typica.** 0, I, on anemone, hepatica, thalictrum, and buttercup; II, III, on wild species of *Prunus*.

Triphragmium

Pucciniaceae. Teliospores stalked, with three cells forming a triangle, each with a single pore.

Triphragmium ulmariae. 0, I, II, III, on meadowsweet.

Uredinopsis

Melampsoraceae. Telia on ferns; teliospores scattered irregularly in mesophyll, rarely in subepidermal crust, typically several-celled; aecia white.

Uredinopsis osmundae. Fir fern rust; 0, I, on balsam fir, widespread; II, III, on *Osmunda* spp.

Uredinopsis phegopteridis. Fir fern rust; 0, I, on balsam fir; II, III, on *Phegopteris dryopteris*.

Uredinopsis pteridis (*U. macrosperma*). Fir fern rust; 0, I, on various firs; II, III, on *Pteridium aquilinum*. Aecia are on 1- to 5-year needles of Pacific silver, white, lowland white, alpine, and noble firs.

Uredinopsis struthiopteridis. Fir fern rust; 0, I, on balsam, lowland white, alpine, and noble firs; II, III, on ostrich fern.

Uredo

Form genus; uredia with or without peridia.

Uredo coccolobae. On sea-grape, in Florida.
Uredo ericae (*Pucciniastrum ericae*). On erica, in California.
Uredo phorodendri. On mistletoe.

Uromyces

Pucciniaceae. Like *Puccinia* but teliospores with one cell, yellow to dark; aecia when present with a persistent peridium (see Fig. 65).

Uromyces betae. Beet rust; II, III, on beets and swiss chard, in California, Oregon, occasionally Arizona and New Mexico. Reddish brown pustules may be numerous on foliage in late summer or in wet seasons. Control is seldom attempted for table beets; some sugarbeet varieties are resistant. The seed-borne fungus also persists in volunteer plants and debris.

Uromyces ciceris-arietini. Rust; on chickpea.

Uromyces costaricensis. Rust; on wild bamboo.

Uromyces dianthi (*U. caryophyllinus*). Carnation rust; 0, I, on euphorbia (but not in United States); II, III, generally on carnation and sweet william, a serious disease under glass. Chocolate brown pustules, varying from $\frac{1}{16}$ to $\frac{1}{4}$ inch, break out on both sides of leaves and on buds and stems. Leaves curl up, often die; infected plants are stunted.

Control. Use surface watering where possible, avoiding syringing; keep greenhouses properly ventilated; use rust-free cuttings.

Uromyces fabae. Pea rust; 0, I, II, III, on pea, peavine, occasionally on broad bean; not very serious.

Uromyces galii-californici. On galium, in California.

Uromyces phaseoli var. **typica.** Bean rust; 0, I, rare on bean; II, III, generally on dry beans, widespread but infrequent on lima bean, scarlet runner bean. It is the true bean rust, an old disease reported as far back as 1798 and quite distinct from anthracnose which is sometimes called rust. It is particularly serious and prevalent on Kentucky Wonder pole beans.

Small rust pustules are formed on leaves most frequently, sometimes on stems and pods. The reddish brown sori are most numerous on underside of leaves, with the upper surface yellowing in the same areas. There may be nearly complete defoliation. In late summer in the North, dark telia replace summer spores, but in the South, urediospores survive the winter and start early spring infection. Rust spores are spread by wind and on tools and clothing. Some even cling to supporting poles and can start a fresh outbreak of rust if poles are not disinfested before reuse.

Control. No bean variety is resistant to all of the more than 30 races so far identified. Most snapbeans are highly tolerant of rust; and pole beans White Kentucky Wonder, U.S. 4 Kentucky Wonder, Potomac, and Rialto are fairly tolerant.

Uromyces trifolii, in several varieties. 0, I, II, III, on clovers. Pale brown pustules surrounded by torn epidermis appear on underside of leaves and on petioles and stems.

Uropyxis

Pucciniaceae. Autoecious. Teliospores two-celled, on pedicels; uredia with paraphyses.

Uropyxis eysenhardtiae. On dalea and eysenhardtia in Arizona.

SCAB

Diseases characterized by an overgrowth of tissue in a limited area are commonly called scab. The hyperplastic scablike lesions correspond to the necrotic or dead areas of leaf spots and cankers. Diseases called scab caused by *Elsinoë*, or its imperfect state, *Sphaceloma*, are treated under Spot Anthracnose.

Cladosporium

See under Blotch Diseases.

Cladosporium bruneo-atrum. Possible cause of russeting of citrus fruit hitherto attributed solely to citrus mite.

Cladosporium carpophilum. Peach scab; general on peach, widespread on apricot, nectarine, cherry, and plum. The form on cherry and European plum has been attributed to *Venturia cerasae* (*Cladosporium cerasi*). Small, round, olive black spots appear on infected fruits about 6 weeks after petals have fallen. These are usually on upper side of fruit, and cracking may follow. Twigs show nearly circular yellow-brown blotches with gray or bluish borders; cambium may be killed and twig may die. Leaf spots are brown, scattered, with tissue drying and falling out, leaving circular holes.

Control. The brown-rot spray schedule should also control scab, a sulfur spray 4 to 6 weeks after petal fall being especially important. A fungicide can be combined with an insecticide spray for curculio.

Cladosporium carpophilum (*Fusicladium carpophilum*). Apparently a different strain from peach scab fungus. **Almond scab**; water-soaked symptoms on young shoots turn brown; leaves turn black, drop prematurely; circular, olivaceous spots coalesce on fruit.

Cladosporium coreopsidis. Reported on coreopsis in Wisconsin, causing stunting and suppression of flowering.

Cladosporium cucumerinum. Cucumber scab; general on cucumber in greenhouses, an important transit and storage decay of muskmelon, sometimes serious on late-planted squash. The disease was first noted in New York in 1887. Leaves with water-soaked spots may wilt, stems have slight cankers, but most injury is to the fruit. First symptoms, while cucumbers are still small, are gray, slightly sunken spots, sometimes exuding a gummy substance. They darken with age, and the collapsed tissue forms a pronounced cavity, lined with a dark green velvety layer of greenish mycelium, short conidiophores, and dark, one- to two-celled spores. On leaves, these fruiting fascicles are extruded through stomata. The disease becomes epidemic after midsummer, when night temperatures are cold, or with heavy dews and fog.

Control. Resistant cucumber varieties include Maine No. 2, Wisconsin SR 10, SR 6, and Highmoor. A long rotation is advised.

Cladosporium effusum. Pecan scab; leaf spot; generally on pecan and hickory. Scab is perhaps the most important limiting factor in pecan production in the Southeast. All varieties are somewhat susceptible, even those like Stuart that have been quite resistant in the past. Crop losses may reach 75% to 95%.

The fungus attacks rapidly growing tissue in leaves, shoots, and nuts; mature growth seems to be immune. On Schley and other highly susceptible varieties, primary infection shows in elongated olive brown lesions on veins and underside of leaves. With secondary infection leaves appear almost black, as a result of coalescing of spots; defoliation follows. On more resistant varieties, such as Moore and Stuart, infection is often delayed until the leaves are nearly mature, and so scab spots are confined to nuts. Nut lesions are small, black, circular, slightly raised at first, then sunken. Spots may be so close together that the entire surface turns black; the nuts drop prematurely or remain attached to shoots indefinitely. Infection is correlated with spring and early summer rainfall, continuous moisture for 6 to 8 hours being required for the spores to germinate and enter the host. First lesions appear in 1 or 2 weeks.

Control. Knock off old shucks and leaf stems before trees leaf out in spring. When they are wet after a rain, a slight jarring of branches will make such diseased material drop. Prune off low limbs for better air circulation. Four protectant sprays are required in Georgia, five in Florida.

Cladosporium pisicolum. Pea scab; black spot of pea. Dark spots covered with velvety mold are formed in moist weather on leaves, stems, where black streaks may develop into cankers and pods may be distorted. The fungus is seed-borne, and lives in soil in plant refuse.

Fusarium

See under Rots.

Fusarium heterosporium. Head scab; of tall fescue.

Fusicladium

See under Leaf Spots.

Fusicladium dendriticum. Conidial stage of the apple-scab fungus; see *Venturia inaequalis.*

Fusicladium eriobotryae. Loquat scab; widespread on leaves, stems, fruit of loquat. It is similar to pear and apple scab. Dark velvety spots cause more or less deformation of fruit, but the disease is seldom important.

Fusicladium photinicola. Christmasberry scab; on *Photinia arbutifolia.* Brown velvety spots appear on leaves, flower stalks, and green berries; the berries being disfigured when mature. Prune in winter to remove dead wood and foliage. Spray before blossoming with bordeaux mixture.

Fusicladium pyracanthe. Pyracantha scab; widespread on leaves and fruit. The unsightly black scabs spoil the appearance of bright berries. The fungus winters in the mycelial state in attached leaves. Frequent spraying with bordeaux mixture controls scab but causes some defoliation.

Fusicladium saliciperdum (*Venturia chlorospora*). **Willow scab;** blight; first noticed on willow in Connecticut in 1927, apparently introduced from

Europe. Repeated defoliation has killed thousands of trees in the Northeast. Young leaves are attacked and often killed in spring, almost within a few hours, and from the leaf blades the fungus enters twigs, kills back young shoots, and causes cankers. Olive green felty spore masses are formed on the long veins on underside of leaves. Overwintering is as dormant mycelium in twigs infected the previous spring. Another fungus, *Physalospora miyabeana*, is found with the scab fungus, and the two together form the disease complex known as willow blight. *Physalospora* usually attacks later in the season than *Fusicladium* and causes cankers on larger stems.

Control. Prune heavily to remove diseased parts. Spray with bordeaux with excess lime.

Spongospora

Phycomycetes, Plasmodiophorales, Plasmodiophoraceae

Spores in a hollow sphere with several openings; zoosporangia formed; zoospores anteriorly biflagellate; sexual fusion of myxamoebae.

Spongospora subterranea. Powdery scab of potatoes; canker; spongy scab. Indigenous to South America and introduced into Europe more than a century ago, potato scab was not noticed in North America before 1913, in Maine. Ordinarily not important, it causes economic loss in some seasons.

Slightly raised pimples appear on tubers when they are less than an inch in diameter; they are varying shades of brown on the surface, faintly purple underneath. The epidermis, not growing as fast as the pimple, breaks and curls over the pustule, which by this time is a brown powdery mass of spore balls and decomposed plant tissue. The lesions are often "corked off," but under favorable conditions large depressed cankers may form. The fungus winters on stored tubers or in soil, remaining viable for many years. In the presence of a potato tuber and enough moisture each spore in the ball germinates by swarmspores, which stay grouped together in a plasmodium, dissolving cuticle and killing cells. When the food supply diminishes, the plasmodium again breaks into spore balls.

Control. Avoid low soggy ground; if such soil must be used, acidify it with sulfur as for common scab.

Streptomyces

See under Rots.

Streptomyces scabies (*Actinomyces scabies*). **Common scab of potatoes**; beet scab; corky scab; actinomycosis; generally on potatoes, widespread on beets, also reported on carrot, parsnip, radish, rutabaga, and turnip. This disease may have been in America as long as potatoes have been grown, but the causal organism was not described until 1890. Scabby potatoes, by lowering the market grade, mean an annual loss of several million dollars. Chief symptoms are the tuber lesions, starting as minute brown specks and progressing to scabs that are warty or with corky ridges, or are pitted and depressed with the skin cracking open. Such potatoes can be eaten, but have poor customer appeal and are wasteful because of the deep peeling required. On beets, the scabs are similar but more bulging.

The pathogen can be found even in virgin soil. It invades young tubers and may sometimes be seen as a grayish coating on freshly dug potatoes. It is most destructive in soils with pH 5.7 and over, with its activity sharply limited in soils slightly more acidic. Although its optimum temperature is 72° F to 86° F, the fungus can withstand great extremes of temperature and moisture and can pass through the digestive tract of animals, returning to the field in manure.

Control. Seed tubers have been treated with formalin, but the organism is so prevalent in potato soils that such treatment may have little result. Soils already slightly acidic may be further acidified with sulfur. Enough sulfur to acidify highly alkaline soil would be too expensive and too injurious to potatoes. Alkaline materials—lime, wood ashes, and manure—should not be applied to potato soil. Somewhat resistant varieties include Menonimee, Ontario, Cayuga, and Seneca.

Venturia

Ascomycetes, Sphaeriales, Mycosphaerellaceae

Perithecia setose, often only near apex, papillate; paraphyses absent; spores unequally two-celled, olive.

Venturia cerasi. On cherry and European plum, perhaps a strain of the peach scab fungus but not infecting peach.

Venturia inaequalis (*Fusicladium dendriticum*). **Apple scab**; scurf; black spot; generally on apple except in far South, widespread on crabapple; reported also on mountain ash and hawthorn, but probably other species of *Fusicladium* infect these hosts. Scab is the world's top-ranking apple disease and is probably coextensive with the host. In this country it takes a fourth or more of the crop in a favorable scab year, the average national loss running around 8%, or over 10 million bushels. Scab is somewhat less important in the South and in irrigated sections of Washington, D.C., but it is important in the humid coastal areas. The pathogen was first described and named by Fries in Sweden in 1819, and was recognized in New York and New Jersey in 1834, apparently having arrived with some European imports.

The first symptom of scab on leaves is a dull smoky area that changes to an

Figure 67. Apple scab on leaf and fruit.

olive-drab moldy spot, ¼ inch or more in diameter, without a sharp outline. Sometimes the leaf is raised or domed in the vicinity of the spot; sometimes it turns brown and drops prematurely. Similar spots may be formed on blossom pedicel, calyx, and petals, followed by dropping of young fruit. Scabby lesions sometimes appear on twigs, but are less common.

On fruits, spots are small, more or less raised, rounded, dark olive areas (see Fig. 67). As they increase in size, the cuticle ruptures to form a white rim around a dark, velvety center, and still later the center may be raised, corky, and tan in color, after dark mycelium and spores have disappeared. Lesions are usually most abundant near calyx end of fruit; if they are too numerous, the fruit splits.

The fungus winters in dead fallen leaves, producing small, dark, flask-shaped perithecia and, toward spring, asci with eight brown ascospores, unequally two-celled, with the upper cell wider than the lower (see Fig. 68). The ascospores mature about the time blossoms show pink, and are forcibly expelled during warm spring rains. Each ascus elongates, protrudes its tip through the mouth of the perithecium, and explodes its spore content. When a spore, carried by wind, arrives on a young leaf or bud, it penetrates the cuticle with a germ tube and develops a layer of branching mycelium just under it. The scab spot is evident in about 10 days, when brown conidiophores bearing olive brown, one-celled, somewhat pointed spores appear on the surface.

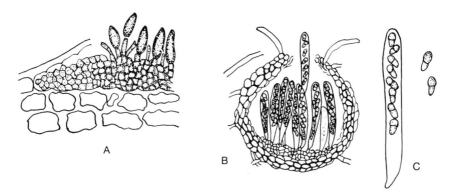

Figure 68. *Venturia inaequalis*, the apple-scab fungus. *A*, one-celled dark conidia of *Fusicladium* stage; *B*, perithecium with two-celled ascospores.

Secondary infection occurs when these conidia are carried to new infection courts.

The expulsion of ascospores proceeds in a series of discharges over a rather long period, up to 3 months, starting in February on the West Coast, but a shorter period, beginning in April in New York. Germination and infection take place from 41° F to 79° F. Length of wetting period necessary for primary infection decreases as the temperature rises—13 to 18 hours of continuous wetting at 43° F and only 4 to 6 hours at 70° F. Secondary infection from conidia continues all season in rainy periods and even in storage scab may show up on apples infected just before picking.

Control. No varieties are immune to scab. Resistance varies with the season and the part of the country. McIntosh apples are very susceptible; Baldwins are fairly resistant but may scab badly during some years. There is more than one strain of the fungus. Nitrogenous fertilizers increase yield of the fruit but also susceptibility to scab.

Protective spraying, having a chemical film on blossom, fruit, or foliage at all times when weather makes infection probable, is the only real answer to scab. Thus, more than a dozen applications may be necessary in a wet year and a minimum of five applications in any season, a program more suited to the commercial grower than to the amateur. Timing is all-important, and most states have a spray warning service that tells of imminent discharge of ascospores. Any spray schedule must be tailored for the locality, the season, and apple varieties grown. The apple grower gets this specific help from county agents.

Venturia pyrina (*Fusicladium pyrinum*). **Pear scab**; generally on pear, also on quince, similar to apple scab. The pear species of *Venturia* overwinters in fallen leaves and also in affected twigs; the perithecia mature somewhat later than those of apple scab. Conidia are formed on pear twigs and washed to leaves and fruit.

Pear scab is not serious except on such varieties as Flemish Beauty, Winter Nelis, Seckel, Anjou, Bosc, and Duchess. Bartlett pears are rather resistant.

SCURF

Two diseases, one of sweet potatoes and one of potatoes, are commonly called scurf.

Monilochaetes

Fungi Imperfecti, Moniliales, Dematiaceae

Hyphae and conidiophores dark, spores hyaline, one-celled, oblong-cylindric, in chains.

Monilochaetes infuscans. Sweet potato scurf; small, circular, brown or black spots are formed on all underground parts, often forming a uniform patch over the whole potato or a black patch on red-skinned varieties. The skin cracks, and potatoes shrink in storage. The black conidiophores stick up from the surface of the lesions like bristles. The fungus winters on the roots and on decaying vines.

Control. Scurf, formerly present in 50% of New Jersey sweet potatoes, is now rare because of proper care. Set only healthy sprouts, grown from potatoes bedded in sand that has not grown sweet potatoes before.

Spondylocladium

Fungi Imperfecti, Moniliales, Dematiaceae

Conidiophores dark, straight, septate, the upper cells bearing whorls of conidia; conidia dark with three or more cells.

Spondylocladium atrovirens. Silver scurf of potatoes; scab; dry rot; present in almost all potato districts but not too important. Light brown lesions become somewhat blistered, giving the skin a marked silvery appearance. The disease is only skin deep, and control measures are seldom used.

SLIME MOLDS

Slime molds belong to the Myxomycetes, a group intermediate between bacteria and fungi. Their assimilative phase is a plasmodium, which is transformed into distinct fructifications on a substratum. They are not parasitic and are often found in rotting logs. Sometimes they are a nuisance in lawns, since the plasmodium after ingesting decayed organic matter or microorganisms for food moves up a grass blade for fruiting. Their spores are produced on or in aerial sporangia and are spread by wind. On absorbing water the spore cracks open and the contents escape as a swarmspore, sometimes two, with two flagella. The swarmspore ingests food like an amoeba, divides by fission into a myxamoeba, unites with another to form a zygote, which enlarges, with mitotic division, into a multinucleate plasmodium. There are many species. Two only are listed here, as being common on turf.

Mucilago spongiosa. Cream to yellow plasmodium forms large grayish white structures, 2 to 6 cm long by 1 to 6 cm wide, that are lobed and branched sporangia filled with a dark mass of purple, spiny spores.

Physarum cinereum. Plasmodium colorless, watery-white or yellow. Fruiting bodies small, gray, sessile, crowded on grass blades, and scattered in groups or rings over an area of several feet. Spores are purple brown in mass. The sporangia develop during humid weather in summer and autumn. Use a stream of water to wash the spore masses off the grass.

SMUTS

Smuts, of the fungus order Ustilaginales, are named for their sooty black spore masses. Like the rusts, they belong to the Basidiomycetes and are all plant parasites, of most economic importance on cereals and grasses, but they differ from rusts in having a less complicated life history and in being able to live part of their lives saprophytically in rich organic matter or in culture media. There are two spore forms. The teliospore, usually called a chlamydospore, is formed by the rounding up of a hyphal cell. In addition to a thin inner endospore wall, it has a thick outer exospore wall, usually dark, smooth or ornamental. Teliospores are formed singly or united into balls. They can be distributed long distances by wind, and spores of some species remain viable for years. Some have to ripen several months before they can germinate.

Occasionally the teliospore puts out a germ tube that penetrates host tissue directly. More often it produces a promycelium that gives rise to sporidia, which can bud to more sporidia. Classically true smuts have been divided into two families on the type of sporidial formation: Ustilaginaceae, with sporidia produced on the sides of a four-celled promycelium, and Tilletiaceae, with sporidia produced at the end of a one- or two-celled promycelium. Fischer, however, points out that there are so many variations that it is preferable to include all species in a single family, Ustilaginaceae, and to differentiate the species on the basis of morphological characters and the host family. This differentiation is logical, but we include here the families as they are given in most textbooks and also the false smuts, Graphiolaceae, which have an uncertain taxonomic position.

There are three types of infection with smuts, with control measures modified according to type. The mycelium always penetrates the young host tissue directly; it does not enter through stomata.

1. Infection of seedlings as the seed germinates, from spores adhering to the outside of the seed or present in soil; controlled by dusting seed and planting in noninfested soil.
2. Seedling infection by mycelium within the seed as a result of ovary infection from spores germinating on the stigma; controlled by treating seed with hot water.
3. Infection of any actively growing meristematic tissue (roots, shoot, tassels, or young ears) by spores transported by wind from decaying plant material; partially controlled by spraying or dusting susceptible plants.

Burrillia

Tilletiaceae. Sori in various host parts, often in leaves, rather permanently embedded. Spore balls with a central sterile mass surrounded by fertile teliospores but no sterile cortex (surface layer). Teliospore hyaline to yellowish, rather firmly united. On water plants.

Burrillia decipiens. Leaf smut; of floating heart (*Nymphoides*).

Cintractia

Ustilaginaceae. Sori usually in ovaries, black, more or less agglutinated spore masses with a peridium. Teliospores single, olive to reddish brown, formed from a fertile stroma surrounding a central columella of host tissues. On Cyperaceae and Juncaceae.

Doassansia

Tilletiaceae. Sori usually in leaves; spore balls large and conspicuous, with a sterile layer around fertile cells. Teliospores pale yellowish brown to hyaline, thin-walled. Germination often in situ. On water plants.

Doassansia epilobii. Leaf smut; on epilobium.

Entyloma

Tilletiaceae. Sori generally in leaves forming light spots, giving the name white smut, or slightly raised darker blisters. Teliospores produced singly but often adhering in irregular groups—hyaline to pale green, yellow, or brown. Sporidia formed on the surface give the white powdery appearance.

Entyloma calendulae. Calendula white smut; spots are pale yellow, turning brown to black, ¼ inch in diameter. The smut is common but not very serious in commercial calendula plantings around San Francisco. Plant debris should be cleaned up, perhaps the location changed.

Entyloma compositarum, White smut; of composites, boltonia, calendula, erigeron, eupatorium, gnaphalium, golden glow, helenium, and prairie coneflower.

Entyloma dactylidis (*E. crastophilum* and *E. irregulare*). **Bluegrass blister smut**; on *Poa* spp.; in Oregon, Washington, Minnesota, and North Dakota. Gray-black, blister areas in leaves from subepidermal masses of chlamydospores. A series of fine dotlike masses of sporidia (conidia) appear scattered in rows along surface of the blisters.

Entyloma dahliae. Dahlia leaf smut; a European disease occasionally reported here. It showed up in one location in California where overhead watering was used, but disappeared when the practice was discontinued. Leaves are marked by more or less circular spots, first yellow-green, then brownish and dry. Primary spores germinate in leaves and send projections to the outside, where secondary spores are formed to spread the disease. Late planting seems to increase disease incidence.

Entyloma ellisii. Spinach smut; an occasional disease with infected leaves pale and worthless. Spores are produced in irregular, marginal necrotic lesions.

Entyloma lineatum. Smut; of wild rice.

Entyloma polysporum. Leaf smut; of gaillardia, golden glow, senecio, sylphium, and sunflower.

Graphiola

Graphiolaceae. This family and genus are sometimes included in the smuts, sometimes not. The sori are erumpent, enclosed in a compact black peridium on leaves of palms. The spores are formed in parallel chains, and bud laterally to form two or more sporidia, which become somewhat colored with thickened walls.

Graphiola phoenicis. False smut of palms; leaf spot; on queen, canary date, royal and Washington palms, and palmetto. Leaves are yellow-spotted with small black scabs or warts having a dark, horny outer surface and long, flexuous sterile hyphae protruding from an inner membrane containing powdery yellow or light brown spore masses. Seriously infected leaves may die. The disease occurs on date palm where humidity is continuously high, but is checked in desert areas best suited to date culture. Kustawy variety is less susceptible than some others.

The disease also appears on small ornamental palms in greenhouses and conservatories. Cut out and burn infected leaves or parts.

Mycosyrinx

Ustilaginaceae. Spores united in pairs; sori with a double peridium in swollen pedicels and peduncles. Mostly tropical.

Mycosyrinx osmundae. Inflorescence smut; on osmunda fern.

Neovossia

Tilletiaceae. Sori in ovaries, semiagglutinated to powdery. Teliospores borne singly, each with a long pedicel appendage, and producing many sporidia.

Neovossia iowensis. On grains, affecting kernels in the dough stage.

Schizonella

Ustilaginaceae. Sori in leaves; short to long striae; black, agglutinated teliospores in pairs, germinating with three- to four-celled promycelium with lateral sporidia. Two species on Cyperaceae.

Sorosporium

Ustilaginaceae. Spores loosely united into balls, readily separable by pressure, in various hosts, more often in reproductive parts. Germination by promycelium and sporidia or germ tube. Mostly on grains.

Sorosporium saponariae. Flower smut; of silene.

Sphacelotheca

Ustilaginaceae. Sori in various host parts but mostly in inflorescence; granular to powdery, covered at first by a peridium. Teliospores single, formed around a central columella. Germination usually with sporidia. Most species on grains and grasses, sometimes causing severe stunting.

Sphacelotheca cruenta. Loose kernel smut; on sorghum, causing smutting and excessive branching. Controlled by seed treatment and resistant varieties.
Sphacelotheca reiliana. Head smut of corn; in Pacific states and scattered locations in South. Galls on tassels and ears breaking into loose dark brown spore masses. Do not plant in a smutted field for 2 years; use certified seed, resistant hybrids.
Sphacelotheca sorghi. Covered kernel smut; kernels replaced by smut galls.

Thecaphora

Ustilaginaceae. Sori in various host parts, mostly inflorescence; powdery or granular. Spores firmly united into balls, with no sterile cells. Chiefly on Leguminosae and Convolvulaceae.

Thecaphora deformans. Seed smut; of lupine.

PLATE 1

A, B, lethal yellowing on coconut palm caused by a mycoplasma pathogen. C, D, tulip break on tulip caused by lily latent mosaic virus. E, F, ringspot on Vanda orchid caused by Vanda ringspot virus.

PLATE 2
A, B, rust on rose caused by **Phragmidium mucronatus**. C, cedar-apple rust on apple caused by **Gymnosporangium juniperi-virginianae**. D, cedar-apple rust on cedar caused by **Gymnosporangium juniperi-virginianae**.

PLATE 3

A, stunt on chrysanthemum caused by chrysanthemum stunt viroid. Var. dark pink orchid queen. **B,** green flowers on chrysanthemum caused by aster yellows mycoplasma. **C,** phyllody on hydrangea caused by a mycoplasma pathogen. **D,** mosaic on rose caused by prunus necrotic ringspot virus. **E,** foliar symptoms on chrysanthemum (variety Bonnie Jean) caused by (clockwise from upper left) chrysanthemum chlorotic mottle viroid, healthy leaf, potato spindle tuber viroid, chrysanthemum stunt viroid, and potato spindle tuber viroid (mild strain).

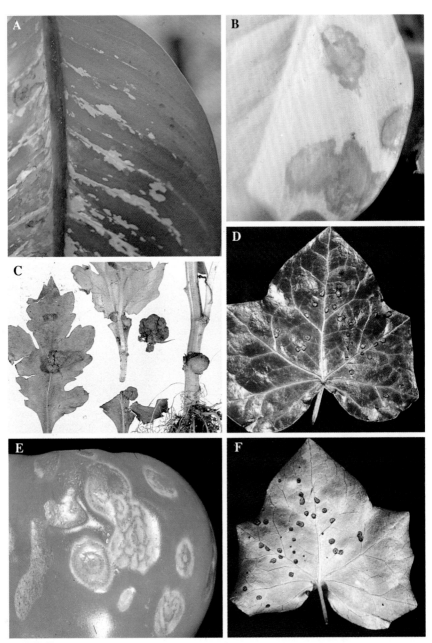

PLATE 4

A, bacterial leaf rot on Dieffenbachia caused by **Erwinia chrysanthemi**. B, bacterial leaf rot on philodendron caused by **Erwinia chrysanthemi**. C, crown gall on chrysanthemum caused by **Agrobacterium tumefaciens**. D, F, common leaf spot on boston ivy caused by **Guignardia bidwellii**. E, ringspot on tomato fruit caused by cucumber mosaic virus.

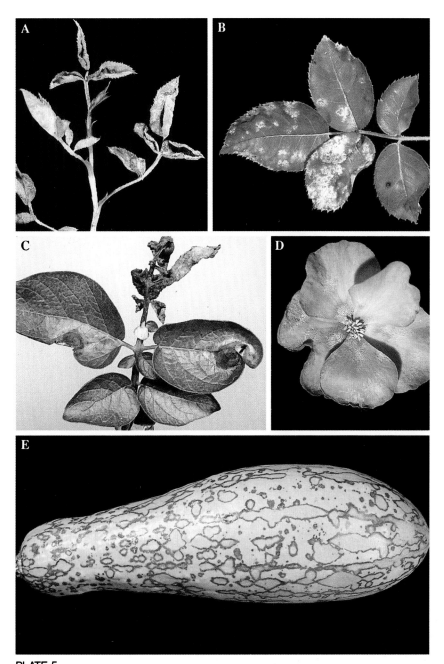

PLATE 5
A, B, powdery mildew on rose caused by **Sphaerotheca pannosa**. C, late blight on potato caused by **Phytophthora infestans**. D, powdery mildew on begonia caused by **Erysiphe cichoracearum**. E, mosaic on squash caused by cucumber mosaic virus.

PLATE 6

A, B, snowmold blight on turf caused by **Fusarium nivale**. C, copper injury on rose caused by sprays containing copper. D, blackspot on rose caused by **Diplocarpon rosae**. E, foliar nematode on chrysanthemum caused by **Aphelenchoides ritzema-bosi**. F, apple scab on apple caused by **Venturia inaequalis**.

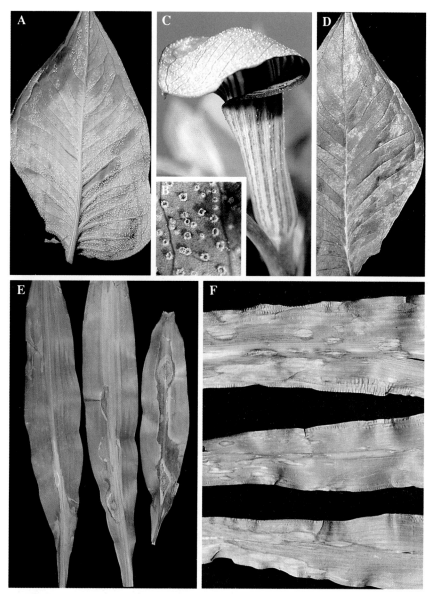

PLATE 7
A, B, C, D, rust on jack-in-the-pulpit caused by **Uromyces caladii**. E, leaf spot on dracaena caused by **Physalospora dracaenae**. F, yellow leaf blight on corn caused by **Phyllosticta maydis**.

PLATE 8

A, blossom blight on rose caused by **Botrytis cinerea**. B, flower blight on peony caused by **Botrytis cinerea**. C, flower and leaf blight on geranium caused by **Botrytis cinerea**. D, scab on citrus caused by **Elsinoë fawcettii**. E, brown rot on cherry caused by **Monilinia fructicola**.

Tilletia

Tilletiaceae. Sori mostly in ovaries, occasionally in vegetative parts of host forming a powdery or semiagglutinated spore mass, often foetid. On grains and grasses, called bunt; interior of seed a solid mass of spore balls.

Tilletia buchloëana. Bunt; of buffalograss.

Tilletia caries. Dwarf bunt of wheat; plants a fourth or half size of healthy plants.

Tilletia foetida. Stinking smut; common bunt of wheat; on wheat and wheat grasses wherever grown, occasionally on rye. A major agricultural disease, especially in Pacific Northwest, it is of historical importance as the first disease controlled by seed disinfection. In 1670, a ship was wrecked off the coast of England, but the cargo of wheat was salvaged, free from bunt because of its saltwater bath. Dark smut balls replace kernels, and there is a fishy odor. Plants are stunted, but not as much as with dwarf bunt. Spore balls are broken in threshing and seed is contaminated. Many materials are offered for seed treatment. Seed dealers treat seed for farmers in special machinery at low cost.

Tilletia pallida. Bunt; on velvet and creeping bent grass, in Oregon and Rhode Island. Seeds are filled with black spores; plants are stunted. The disease is serious where grass is grown for seed, with up to 80% nonviable seed.

Tuburcinia (*Urocystis*)

Tilletiaceae. Sori mostly in leaves and stems, blackish; embedded in host tissues. Spore balls permanent, without sterile cortex but sometimes with a layer of hyaline, hyphal fragments. On Liliaceae, Primulaceae.

Tuburcinia trientalus. Leaf and stem smut; of starflower.

Urocystis

Tilletiaceae. Sori usually in leaves, stem sheaths, occasionally in flowers; dark brown to black, powdery to granular. Spore balls with distinct sterile spores on the surface, only a few fertile spores. Sori without peridium.

Urocystis agropyri. Flag smut of wheat; also on wheatgrass, red top, and bluegrasses. Symptoms are similar to those of stripe smut.

Urocystis anemones (including *U. hepaticae-trilobae*). **Leaf and stem smut;** of anemone, hepatica, and trautveteria.

Urocystis carcinodes. Smut; of aconite, baneberry, clematis, and cimicifuga.

Urocystis cepulae (*U. colchici*). **Onion smut;** generally on onion, also on leek, shallot, garlic, and chives. It is the most destructive onion disease, found in the Connecticut Valley as early as 1861, and then spreading to all northern onion-growing sections, but more important where onions are grown from seed rather than sets as in most home gardens.

Black elongated blisters or pustules of spores break out on scales or leaves of young plants. Many plants die; others survive and have black or brown smut pustules on the cured bulbs. Plants are stunted but not rotten, although smut may be followed by secondary rot organisms.

The spores can live in soil for years, but infection is possible only in young plants from the second day after seed germination until the seedling is in first leaf, a period of 10 to 15 days. The spore is able to penetrate the onion through root and cotyledon but cannot enter a true leaf. After entrance it spreads through the seedling until it reaches the leaves to form fruiting pustules just below the epidermis. When the pustules rupture, spores are dropped, to be disseminated by running water and tools, on feet of persons and animals, and on roots of transplanted vegetables. Onion smut is confined to states with cool summers, optimum soil temperature for infection being 72° F.

Urocystis colchici (Fischer includes *U. cepulae* in this species). **Leaf smut**; of autumn crocus, camassia, Solomons seal and false Solomons seal.

Urocystis gladiolicola. **Gladiolus smut**; this disease had been intercepted several times at quarantine and appeared once in California fields, in 1950, but apparently is eradicated there. Growers should be on the lookout for corms with low blister swellings, with ridges paralleling veins, bluish black, breaking open to expose dense black spore balls. Seedlings exhibit blistering, shredding, and necrosis of stem and leaf tissues; they die if the disease is severe.

Urocystis kmetiana. **Floral smut**; of field pansy (*Viola bicoler*).

Urocystis tritici. **Flag smut of wheat**; plants are dwarfed with twisted leaf blades; sheaths are marked with grayish-black stripes; diseased tissues dry up and are shredded. Infected plants rarely produce heads.

Ustilago

Ustilaginaceae. Sori in various host parts; spore masses powdery to agglutinated; usually dark brown to black, in some species yellow to purple without a peridium. Spores single, not united in balls (see Fig. 69).

Ustilago avenae (including *U. perennans*). **Black loose smut**; on oats and some grasses. Individual flowers in panicle are largely replaced by a spore mass. The young seedling is diseased from the seed, and the fungus grows systemically in the plant.

Ustilago bullata. **Head smut**; on many grasses.

Ustilago heufleri. **Erythronium smut**; large dusty pustules lead to cracking and dying of leaves of dogtooth violet.

Ustilago hordei. **Covered smut of barley**; heads are converted into hard, black, smutted masses, enclosed within thin membranes.

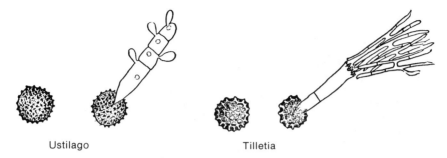

Ustilago   Tilletia

Figure 69. Smut spores. *Ustilago* (left), spiny chlamydospore germinating with promycelium and sporidia formed at sides; *Tilletia* (right), reticulate chlamydospore with long H-shaped sporidia formed at end of promycelium and sometimes forming small secondary sporidia.

Ustilago kolleri. **Covered smut of oats**; spore balls remain intact within glumes until threshing, when spores are distributed over surface of seed, ready to infect young seedlings.

Ustilago maydis (*U. zeae*). **Corn smut**; boil smut, general on corn but most destructive to sweet corn. The average annual loss is 3% to 5%, but it can be 100% in any one field. The fungus was described in Europe in 1754, and not reported in the United States until 1822, but it may be native along with its host. There are many physiological races, and smut resistance is likely to be correlated with lack of vigor, so that it has been hard to breed desirable resistant varieties.

Any plant parts aboveground may be attacked—stalks, prop roots, leaves, tassels, husks, and ears (see Fig. 70). Large boils are formed, at first covered with a greenish white membrane, said to be good eating when boiled or fried. Later the membrane breaks and releases myriads of dark chlamydospores. The plant is often distorted. Infections are local; each boil is formed where a spore lands, and there is no systemic growth through the plant. The fungus is not seed-borne, and germinating seedlings are not affected. Chlamydospores winter in soil, corn debris, and manure. They produce sporidia, which may bud to form secondary sporidia, and these are carried by wind and other agencies to corn plants, which are 1 foot to 3 feet high. Mycelium from spores of two sexes is needed for active development. Spores formed in the first boils provide inoculum for secondary infection of ears.

Corn smut thrives in warm weather, optimum temperature for spore germination being 80° F to 92° F. Heaviest infection occurs when scant rainfall in early stages of growth is followed by moderate rainfall as corn approaches maturity. Vigorous plants are most susceptible, but may escape the most serious effects because of their rapid growth. Spores retain viability for 5 to 7 years. They remain viable in passage through an animal into manure, but are killed by the acids in silage.

Figure 70. Corn smut.

Control. Seed treatment is not effective. Some hybrid varieties are rather resistant. Most reliance in home gardens should be placed on cleanliness, cutting off and burning all smutted parts before the boils break open to release spores.

Ustilago mulfordiana. Fescue smut; on fescue grasses.

Ustilago nigra. Nigra loose smut; generally on barley.

Ustilago nuda. Nuda loose smut; normal heads replaced by black powdery masses.

Ustilago striiformis. Stripe smut; generally on grasses—wheatgrasses, redtop, bent grasses, fescues, ryegrass, and bluegrass; does not occur on cereals. Long, dark, narrow striations develop in leaves; as the sori mature, spores are freed, and the blade splits into ribbons. Plants are systemically infected, make poor growth, and inflorescences are stunted or absent. Perennial mycelium may overwinter in the plant.

Ustilago violacea. Anther smut of carnation; dianthus, lychnis, and silene. Infected plants grow slowly, produce many weak axillary shoots; stem internodes are shortened; flower buds are short and squatty; calyxes tend to split; flowers are sprinkled with black sooty dust from the anthers, whose pollen grains are replaced by smut spores. The fungus enters through flowers or injured surfaces and grows systemically. Spores are spread on cuttings. Control by roguing diseased plants before flowering. Do not take cuttings from plants with grassy or bushy habit.

SNOWMOLD

Northern lawns and turf of golf greens often show round light patches as the snow melts in early spring. Such a disease is called snowmold and may be due to one of several fungi, sometimes to two appearing together.

Fusarium

See under Rots.

Fusarium nivale (*Calonectria nivale*). Pink snowmold; Fusarium patch, most important on bent grass on golf courses but infecting other turf grasses and winter wheat and winter rye. Irregularly circular patches, from 1 inch to 2 inches to a foot or more, appear as snow is melting. They are whitish gray, often with a pinkish tinge, and several patches may run together to cover large areas. Individual plants have a bleached appearance, feel slimy when wet. Spores are formed in salmon-pink sporodochia over stomata in leaves. They are sickle-shaped, one- to three-septate. Perithecia are produced on the luxuriant white mycelial mat.

Abundant moisture in the fall, snow falling on unfrozen ground, deep snow, and a prolonged, cold wet spring are predisposing factors, but the presence of snow is not a requisite for the disease. Severity is increased by applying fertilizer in late autumn and an excess of organic matter in the soil. Reports differ as to susceptibility, but Colonial, Washington, and Metropolitan bent grasses appear to be more resistant than Seaside bent.

Sclerotium

See under Blights.

Sclerotium rhizodes. Frost scorch; string of pearls; in northern states. Not exactly a snowmold but appearing in early spring with bleached, withered leaves covered with rows of tiny sclerotia. Collect clippings when mowing diseased areas to remove sclerotia on leaf tips.

Typhula

Basidiomycetes, Agaricales, Clavariaceae

Fruit body erect, simple, like a little club, on a long stipe from a sclerotium; basidia with four sterigmata and simple, hyaline spores.

Typhula itoana. Snowmold of turf and lawn grasses; typhula blight; common in eastern United States. As the snow disappears in spring, a felty white mycelial mat is seen over grass and adjacent soil. Plants wither and turn light brown or tan in roughly circular patches, very conspicuous against the green of the rest of the lawn. The chief diagnostic character is the presence of very small, tawny to hazel brown spherical sclerotia in large numbers over affected parts. These can be made to fruit in the laboratory into rose-colored sporophores up to 1 inch tall.

Control. The disease gradually disappears as moisture decreases and temperature and sunlight increase, so control seldom seems necessary. Six weeks after striking cases of snowmold, lawns are often uniformly green and show little sign of having been affected. Phosphate fertilizers are said to decrease injury from *Typhula*.

Typhula idahoensis. Snowmold; on wheat and grasses in Idaho and Montana. Sclerotia are chestnut brown, sporophores fawn to wood brown, less than ½ inch high.

SOOTY MOLD

Sooty mold is a black coating on surface of leaves or fruit composed of a weft of dark mycelial threads. As used here, the term applies to saprophytic fungi that live on insect honeydew and harm plants only indirectly. See Black Mildew for the true parasites with dark mycelium and spores giving a sooty appearance to foliage.

Capnodium

Ascomycetes, Dothideales, Capnodiaceae

Mycelium superficial, dark; spores muriform, in perithecium-like conceptacles at tips of branches of a carbonaceous stroma; associated with insect secretion on living plants.

Capnodium citri. Sooty mold; on citrus; on honeydew secreted by scale insects, aphids, whiteflies, especially abundant following whiteflies in Florida, black scale in California. A black, velvety, membranous coating is formed over leaves, twigs, and fruit. If honeydew is slight, the coating appears in spots; but if the insect secretion is abundant, the entire surface may be covered by a dense continuous membrane resembling black tissue paper. With age, under dry conditions it may be blown off in fragments. The black membrane is made up of hyphae that are individually olive green to deep brown, with wide short cells. Branches may crisscross or be cemented together. There are several spore forms: simple conidia that are cut off from upright hyphae; others formed in small, black pycnidia; stylospores in very long flask-shaped conceptacles; and muriform brown ascospores in perithecia.

Although sooty molds do not obtain food from the plant, the black membrane interferes greatly with photosynthesis and food manufacture. Affected fruit is smaller, with coloring retarded; it is more likely to decay than normal fruit.

Control. Treatment is directed against the insects, either by spraying with insecticides or by using entomogenous fungi and insect parasites. Oil sprays kill the insects and help to clean the trees of the sooty covering.

Capnodium elongatum. Sooty mold; of tulip tree, oleander, holly osmanthus, and others. Foliage of tulip trees very frequently has a black coating, often on honeydew secreted by the tulip tree aphid, sometimes following attacks of tulip-tree scale. A dormant oil spray controls the latter.

Capnodium spp. **Sooty mold**; on gardenia, fig, crape myrtle, azaleas, and many other plants. Gardenias are especially subject to sooty mold following whiteflies, crape myrtle after aphids, azaleas after mealybugs, and magnolias after scales. A summer oil spray helps to control the insects and loosens the black coating so that it is more readily washed off.

Very often rhododendrons and other broad-leaved evergreens growing inside the branch spread of tulip and other trees afflicted with aphids and scales are covered with sooty mold growing in the honeydew dropped on foliage from the tree overhead.

Fumago

Fungi Imperfecti, Moniliales, Dematiaceae

Mycelium dark, creeping over surface of leaves; conidiophores dark, variable, bearing conidia terminally or laterally; conidia variable, dark, muriform, frequently in chains; saprophytic on honeydew; probably conidial stage of *Capnodium*.

Fumago vagans. A heavy black moldlike growth on leaves of linden and many other ornamental trees and shrubs, also on house plants in honeydew from aphids, mealybugs, and scale insects.

Scorias

Ascomycetes, Dothideales, Capnodiaceae

Mycelium with parallel walls, forming a thick spongy mass; perithecium long-stalked, round; spores four-celled.

Scorias spongiosa. Sooty mold; often on trees such as alder, beech, and pine.

SPOT ANTHRACNOSE

Diseases caused by species of *Elsinoë* imperfect stage *Sphaceloma* are characterized by some overgrowth of tissue. When this hyperplasia is pronounced, the disease is usually called scab; when the overgrowth is scarcely noticeable (merely a slightly raised border around a necrotic center), the disease has been commonly known as anthracnose. Recently, the term *spot anthracnose* was introduced to differentiate a *Sphaceloma* malady from anthracnoses caused by fungi with slime spores (*Colletotrichum*, *Glomerella*) and from the *Venturia* type of scab diseases. Consequently, all spot anthracnose diseases are included in this section, but with the common designation, scab or anthracnose, also listed.

Elsinoë

Ascomycetes, Myriangiales, Elsinoaceae

Asci are borne singly, at different levels, in an effused stroma, having a gelatinous interior and crustose rind, under or within the epidermis, which ruptures to expose the asci. The imperfect stage is a *Sphaceloma*.

Elsinoë ampelina. Grape anthracnose; bird's eye rot; widespread on grape. Small sunken spots with dark margins and light centers occur on fruit, young shoots, tendrils, petioles, and leaf veins. Leaves may be distorted and ragged from diseased portions dropping out. Outbreaks are sporadic in eastern states. Varieties Catawba, Campbell Early, Diamond, Norton, and Salem are quite susceptible; Concord, Delaware Moore Early, and Niagara are resistant. The fungus winters on canes.

Control. Apply a dormant lime-sulfur spray and four or five sprays of bordeaux mixture as follows: when new shoots are 7 to 8 inches long; just after bloom; 7 to 10 days later; and when berries are half grown.

Elsinoë cinnamomi. Camphor tree scab; inconspicuous brown leaf spots, sometimes dropping out; elongated raised lesions on veins, petioles, and stems. Reported from South Carolina.

Elsinoë corni. Dogwood spot anthracnose; a serious threat to flowering dogwood from Delaware to Florida, also reported from Louisiana. Infected buds do not open, or they produce stunted, malformed "flowers," marked with numerous small, circular to elongated spots with light tan centers, purple to brown borders, up to 50 on a bract. Leaf spots are 1 to 2 mm, slightly raised at the margin, purple paling to yellow-gray at centers, which may be broken in a shot-hole effect. There may be 100 spots on a leaf, scattered or concentrated at tip, margin, or midrib. Spots on petioles, fruit clusters, and stems are similar to leaf spots.

Elsinoë diospyri. Spot anthracnose; on leaves of native persimmon, reported from Florida, 1955.

Elsinoë euonymi-japonici. Spot anthracnose; on evergreen euonymus, in Florida. Small, roundish spots, mostly on upper surface of leaves, brown with raised, orange-cinnamon margin; stem cankers circular to elliptical, wrinkled or fissured, grayish white with raised orange margins.

Elsinoë fawcetti. Sour orange scab; citrus scab; verrucosis; on citrus fruits, except rare on sweet orange. Lemons, sour orange, King orange, bitter orange, and calamondin are very susceptible; Mandarin and Satsuma oranges, tangerines, and all grapefruit except Royal and Triumph are moderately susceptible. Climatic conditions play a part. Grapefruit and lemons in the Rio Grande Valley are less susceptible than in Florida, but Satsumas in Alabama are more susceptible than those in Florida. Known in the Orient since ancient times, scab is believed to have come to Florida on Satsumas from Japan. It was first recorded there in 1885; the fungus was identified as a *Sphaceloma* in 1925.

Tender growth is most readily infected, and the disease is most important on young trees. On leaves, minute, semitranslucent spots change to raised excrescences with corky crests, pale yellow to pinkish, then dull olive drab with a conical depression opposite the crust. Foliage may be wrinkled or stunted. Fruits have slightly raised scabs or are warty with corky crests, which may grow together into large irregular patches. Scabs on grapefruit may flake off as the fruit matures, with the area remaining green.

Spores are spread by wind, rain, dew-drip, possibly by insects. The young fruit of grapefruit is very susceptible right after petal fall, but becomes progressively resistant and is practically immune when it reaches ¾ inch in diameter. Temperature range for severe infection is from 59° F to 73° F. Excessive nitrogen increases scab. The pathogen winters on infected leaves, sometimes fruits.

Control. Apply a neutral copper spray or bordeaux mixture just before growth starts in spring. A second copper spray, just after flowers shed, controls melanose as well as scab.

Elsinoë ilicis. Chinese holly spot anthracnose; numerous black spots, 1 to 2 millimeters, coalesce to large black patches on upperside of leaf, mostly the apical half, with distortion. Shoots and berries have brown to gray lesions with slightly raised margins.

Elsinoë jasminae. Jasmine scab; reported from Florida. Spots numerous, round or irregular, up to 2 millimeters.

Elsinoë ledi. Ledum spot anthracnose; widespread on ledum, Labrador tea, and salal in Northwest, leucothoë in Florida. Leaf spots are grayish white with red-brown borders and purple margins. The disease is not serious.

Elsinoë lepagei. Scab; on sapodilla and canistel in Florida (found on young nursery stock in cans). Small, raised spots, gray at center.

Elsinoë leucospila. Camellia scab; also recorded on ternstroemia in Florida. Some corky excrescences on camellia foliage are due to this pathogen, others to moisture relations.

Elsinoë magnoliae. Magnolia scab; on *Magnolia grandiflora* from Georgia to Louisiana. Spots are circular to angular, with black papillae in centers, on upper leaf surface along midrib, margin, or tip. Infected leaves may drop.

Elsinoë mangiferae. Mango scab; spots usually originate on underside of young mango leaves but become visible above. They are circular to angular, dark brown to black with olive buff centers. Spots on mature leaves are larger, slightly raised with narrow brown margins and dirty white centers. Stems have irregular grayish blotches; fruit, gray to brown spots with dark margins.

Elsinoë mattirolianum. Spot anthracnose; on madrona and strawberry tree (*Arbutus* spp.) in California.

Elsinoë parthenocissi. Virginia creeper soft anthracnose; leaf spots are few to numerous, circular, scattered or along midribs and veins; they have buff centers with narrow brown margins; fruit spots are grayish white; lesions on petioles are somewhat raised. Also reported from pepper vine.

Elsinoë phaseoli. Lima bean scab; first U.S. report from North Carolina probably from imported seed. Lesions on pods, stems, and leaves.

Elsinoë piri. Pome fruit spot anthracnose; on pear, apple, and quince in moist sections of western Washington and Oregon, more prevalent in home gardens than commercial orchards. Fruit spots are small, up to 2 millimeters, red or reddish purple with pale centers, upward of 100 on an apple.

Elsinoë quercicola. Spot anthracnose; on water oak, Florida.

Elsinoë quercus-falcatae. Spot anthracnose; on southern red oak, Georgia, South Carolina. Blackish brown leaf spots are few to abundant, scattered over upper surface.

Elsinoë randii. Pecan anthracnose; nursery blight; on pecans in the Southeast, an important nursery disease, limiting factor in production of budded pecans in wet seasons. Small reddish lesions develop on both leaf surfaces, those on the upper surface later turning ash gray. Diseased tissues become brittle and fall out, leaving ragged margins and perforations. Spray trees with bordeaux mixture when first leaves are half grown; follow with three sprays of bordeaux at 3- to 4-week intervals.

Elsinoë rosarum. Rose anthracnose; widespread on rose, collected on wild roses as early as 1898, in most areas more important on climbing roses than on hybrid teas. Leaf spots are scattered or grouped, sometimes running together, usually circular, up to ¼ inch. Young spots are red, varying brown or dark purple on upper leaf surface, showing up 2 to 6 days after inoculation but not visible on undersurface for 2 to 4 weeks, then dull reddish brown to pale purple. On aging, the center of the spot turns ashen white, with a dark red margin. Leaves may turn yellow or reddish in area of spots, may have slits or perforations as the centers fall out.

Cane spots are circular to elongated, raised, brown or purple, with depressed light centers and acervuli in barely visible dark masses. The fungus winters in cane spots; spores are produced and spread only in rainy periods. A single leaf lesion may produce 10,000 spores within an hour after wetting and will continue production as long as the rain lasts.

Control. Where possible, prune out infected canes in spring. Keep foliage protected as for blackspot. Sulfur and copper compounds are effective.

Elsinoë solidaginis. Goldenrod scab; in Florida, South Carolina, and Georgia. New growth is affected as it develops. Lesions formed on midrib, veins, petioles, and leaf blades are raised on one surface, sunken on the other, with white to gray centers and brown borders.

Elsinoë tiliae. Linden spot anthracnose; reported from Nova Scotia and Virginia. Gray spots with black margins are numerous on leaf blades and petioles.

Elsinoë veneta. Bramble anthracnose; general on blackberry, dewberry, raspberry, being most common on black raspberry. Circular, reddish brown sunken spots with purple margins and light gray centers, up to ⅜ inch in diameter, appear on young shoots. On older canes these grow together into large cankers. Similar spots, not always with purple margins, are formed on fruit, leaf, and flower stalks. Leaf spots are first yellowish, then with a red margin around a light center, which may drop out. Leaves may drop prematurely; fruit may dry up as a result of loss of water from infected canes. Primary spring infection comes from ascospores produced in old lesions on canes; secondary spread is by conidia.

Control. Cut old canes or "handles" from black raspberries after setting; remove and burn old fruiting canes after harvest. In some cases the single late dormant spray has controlled anthracnose without later sprays; in others three foliage sprays have been effective without a dormant spray. Black raspberry Quillen is quite resistant.

Sphaceloma

Fungi Imperfecti, Melanconiales, Melanconiaceae

Acervuli disc- or cushion-shaped, waxy; conidiophores simple, closely grouped or compacted, arising from a stromalike base; spores one-celled, hyaline, ovoid or oblong. Perfect stage where known is *Elsinoë*.

Sphaceloma araliae. Aralia scab; on Hercules club (*Aralia spinosa*), in Maryland and Missouri.

Sphaceloma cercocarpi. Spot anthracnose; of birch-leaf mahogany, in California. Leaf spots are nearly circular, up to 3 millimeter across, with pale centers and slightly elevated purple margins.

Sphaceloma hederae. English ivy scab; leaf spots are raised with red-brown margins, pale depressed centers, often numerous.

Sphaceloma lippiae. Lippia spot anthracnose; on fog fruit. Closely resembling mint anthracnose and found in same fields in Indiana, also reported from Florida. Numerous spots on leaves and stems are scattered or grouped and nearly confluent; centers are depressed, buff-colored, with purple margins.

Sphaceloma menthae. Mint anthracnose; leopard spot disease; circular, oval or irregular spots on leaves, stems, and rootstocks are black with white centers, up to 5 millimeters. Formerly serious, this disease is now controlled in commercial mint fields by fall plowing, covering plants deeply.

Sphaceloma morindae. Morinda scab; buff-colored spots on leaves, stems, and petioles, in Florida.

Sphaceloma murrayi. Gray scab of willow; leaf spots are round, irregular, somewhat raised, grayish white with narrow, dark brown margins, often confluent, sometimes fragmenting; long narrow patches along veins.

Sphaceloma oleanderi. Oleander scab; leaf spots spherical to irregular, densely grouped over entire surface, whitish with brownish black margin, slightly elevated, 1 to 4 mm.

Sphaceloma perseae. Avocado scab; one of the most important avocado diseases in Florida, some years with nearly 100% infection; also occurring in Texas. Leaf lesions are mostly on upper surface, very small red spots with a dark olive conidial growth. Fruit lesions are corky, raised, brownish, oval, but often coalescing, giving a russeted appearance; sometimes cracking to allow entrance of fruit-rotting fungi. Avoid highly susceptible varieties like Lulu. Spray with bordeaux mixture as for blotch.

Sphaceloma poinsettiae. Poinsettia scab; light raised lesions on stems, veins, and midribs, pale buff at center with purple to nearly black margins.

Sphaceloma psidii. Guava scab; reported in feijoa in Florida.

Sphaceloma punicae. Pomegranate spot anthracnose; very small purple spots with paler centers on both leaf surfaces.

Sphaceloma ribis. Gooseberry scab; in Washington. Leaf spots numerous, very small, raised, and grayish.

Sphaceloma spondiadis. Mombin scab; on purple mombin (*Spondias*) in Florida.

Sphaceloma symphoricarpi. Snowberry anthracnose; widespread on snowberry, impairing beauty of ornamental plants, first described from New York in 1910; also on coralberry. Leaf spots appear in early spring, minute, dark purple to black, aging with dirty gray centers, coalescing into large areas subject to cracking. Leaves may be misshapen from early marginal infections. Spotting is inconspicuous in flowers but pronounced on berries, with purple areas becoming sunken and pinkish. Secondary infection by an *Alternaria* shrivels fruit into brown mummies. A dormant lime-sulfur spray followed by foliage sprays may help.

Sphaceloma viburni. Snowball spot anthracnose.

Sphaceloma violae. Violet scab; pansy scab; widespread on violet and pansy from Connecticut to Louisiana and Texas, a limiting factor in maintaining violet collections. Reddish spots with white centers change to irregular to elongated raised scabs on leaves and stems, often with much distortion. Remove and burn old leaves.

Sphaceloma spp. Undetermined species have been reported on *Bignonia*, catalpa, camellia, and sambucus, in Louisiana, on buttonwood in Florida, on rhododendron in Washington.

VIRUS DISEASES

The classification of plant viruses is still in a state of chaos, many systems having been proposed and none universally accepted. Holmes in 1939, and McKinney in 1944, proposed Linnaen-style latinized binomials tied to classifications based on the natural vectors of the viruses, and on the type of symptoms caused. This scheme has not been widely adopted. In 1965, the International Provisional Virus Nomenclature Committee met in Paris and decided that the kingdom VIRA would no longer be classified by hosts and symptoms but on the basis of nucleic acid, particle structure, serology, and other technical details. Most virologists still prefer vernacular names of the kind *tobacco mosaic virus* and this system will be used throughout the book (see Chapter 2). Following are virus diseases in alphabetical order by common names.

Abutilon Infectious Variegation; Abutilon Mosaic. A single variegated seedling found among green plants imported into England from the West Indies in 1868 was propagated vegetatively as an ornamental variety. The bright yellow mottling on green leaves tends to disappear in subdued light. Transmission is by grafting, occasionally by seed, and, in native Brazil, by whitefly *Bemisia tabaci*. Plants may recover if variegated leaves are persistently removed but may be reinfected.

Alfalfa Mosaic. Potato, celery calico; bean yellow dot. Various strains of the alfalfa virus are transmitted by cotton, pea, and other aphids to bean, clovers, pea, cucumber, potato, tomato, zinnia, tumbleweed, poison hemlock, wild carrot, Japanese pachysandra, and other hosts. Calico is a minor potato disease in California and Idaho. Leaf spots are irregular, brilliant yellow to gray; yield may be reduced. Celery has a conspicuous yellow-green mosaic; bean has small, necrotic lesions.

Almond Bud Failure. Virus on Drake almond, in California, is transmissible by grafting. Limbs have many branches, some dead at the end; leaves are darker green, more upright, retained longer than normal; fruits few, often misshapen.

Almond Calico. On almonds, in California, graft transmissible. Chlorotic blotches in leaves.

Apple, Dapple. Fruit with circular islands or patches remaining green; on trees with Virginia crab or Robusta V bodystock; first noted in New Hampshire.

Apple Green Mottle. On Duchess variety, in New York. Fruit with discolored rings, of little value.

Apple Mosaic. Occurring naturally only on apples; no insect vector known; transmitted by budding. Small irregular cream to yellow leaf spots coalesce to large chlorotic areas, with or without vein-banding. Three strains cause severe mosaic, vein-banding mosaic, and mild mosaic. There is no marked reduction in yield. Hosts other than members of Rosaceae family include filbert, hop, birch, and horse chestnut.

Apple Mosaic, Tulare. Reported from California; has a wider host range than apple mosaic.

Apple Stem-pitting. Wood-pitting, in Virginia Crab bodystock, sometimes followed by bark cracks, dwarfing, early fruit production.

Apricot Gummosis. First noted in Washington in 1947; transmitted by budding. Profuse gumming on branches and trunks, necrosis and dieback of new shoots; dead, punky areas in fruit.

Apricot Ring Pox. In California and Colorado. Irregular ring spots with marked vein-clearing in some varieties, chlorotic mottling in others; dead tissue may fall out leaving shot holes. Fruit bumpy or with reddish brown necrotic spots extending into flesh.

Artichoke (Globe) Curly Dwarf. In California on artichoke, cardoon, and zinnia, milk thistle. Leaves curled, plants dwarfed, killed.

Ash Ring Spot. On white ash, in New York. Chlorotic green and reddish spots, rings, line patterns; stunting; dieback.

Ash Witches' Broom. Reported from Louisiana on Arizona ash. Yellowish leaves, a fourth to a third normal size; multiple, spindly shoots.

Asparagus Virus 1. On asparagus.

Aster Ring Spot. In Florida on China aster, pepper, pansy, and other plants. Yellow ring, line, and oakleaf patterns. A strain of tobacco rattle virus.

Avocado Sun Blotch. There are long, narrow, shallow, longitudinal grooves, buff-colored on stems, whitish on green fruit, reddish purple on purple fruit.

Shoots may be twisted and abnormal. Transmitted through seeds. This disease is caused by a viroid.

Barley Yellow Dwarf. Occurs on tall fescue, and various *Poa* and *Festuca* spp.

Bayberry Yellows. Wavy margins and tips on young apical leaves, distorted margins on older leaves, which are pale, yellow, small. Plant is stunted, with shortened internodes, few or no fruits.

Bean Mosaic. Found wherever beans are grown, transmitted by many species of aphids—bean, cotton, cowpea, cabbage, peach, spirea, turnip—and in seed. First leaves are crinkled, stiff, chlorotic; older leaves have chlorotic mottling, often with leaf margins rolled down. Mosaic-resistant varieties include Robust, Great Northern, U.S. No. 5 Refugee, Idaho Refugee, and Wisconsin Refugee.

A strain known as bean greasy pod virus causes a greasy appearance of the pods in some western states. The asparagus-bean mosaic is a light and dark green mosaic with leaf rolling transmitted by seed and by the pea aphid. The virus may be a strain of bean mosaic virus or a different virus.

Bean Pod Mottle. Systemic mottling in some varieties; circular, light brown local lesions on pods of other varieties. May also be seed transmitted in soybean.

Bean Southern Mosaic. Chlorotic mottling or localized necrosis of foliage; pods with dark green blotches or shiny areas, slightly malformed, short, curled at end. Virus present in new seed but not in seed stored 7 months.

Bean Yellow Mosaic. Mild mosaic of gladiolus. On beans, peas, sweet peas, clover, Tahitian bridal veil, gladiolus, and freesia. In beans there is a coarse yellow mottling and distortion of leaves, which are pointed downward; proliferation of stems; shortening of nodes and general stunting; reduced pod production; delayed maturity. In pea and sweet pea there is veinal chlorosis, with slight ruffling. Gladiolus flowers are striped or flecked, young leaves have an angular green mottling, but symptoms are mild compared with the disease on beans and freesia, which should not be planted near gladiolus and clovers. The virus is transmitted by bean and pea aphids but not through seed. Rogue infected plants as soon as noticed.

Bean Yellow Stipple. Mild mottle and chlorotic spots on bean leaves, sometimes coalescing.

Beet Curly Top. Confined to North America, curly top is especially important in the commercial sugar-beet industry west of the Continental Divide, but it is common on many plants. Vegetables include bean, beet, carrot, celery, cabbage and other crucifers, cucumber, melon, squash, pumpkin, eggplant, spinach, chard, New Zealand spinach, horseradish, and tomato. Ornamentals include alyssum, blue flax, campanula, carnation, columbine, coreopsis, cosmos, delphinium, foxglove, geranium, larkspur, nasturtium, pansy, petunia, poppy, portulaca, pyrethrum, scabiosa, Shasta daisy, stock, strawflower, veronica, and zinnia.

In beets there are clearing of veins, leaf curling, with sharp protuberances from veins on underside of leaf, increase in number of rootlets. In tomato, where the disease is called western yellow blight or tomato yellow, seedlings turn yellow and die. Older plants show twisting and upward rolling of leaflets, stiff and leathery foliage, with leaf petioles curling downward; branches and stems are abnormally erect; the whole plant is dull yellow, often with purple veins; roots are killed, few fruits formed.

In cucurbits, tips of runners bend upward; old leaves are yellow, tip leaves and stems abnormally deep green. In beans, there is a thickening and downward curling of first true leaf, which becomes brittle. The plant stops growing and may die. Older plants survive until the end of the season, with puckering and downward curling at the top of the plant, reduction in size of new leaves, shortened internodes.

Ornamentals grown near diseased beets are usually infected. Zinnias have shortened internodes, chlorotic secondary shoots arising from leaf axils. Cosmos leaflets are twisted and curled, petioles bent down. Geranium leaves are chlorotic between veins with protuberances on lower surface.

The virus is confined to phloem in plants and is transmitted by the beet leafhopper (*Circulifer tenellus*). The insects winter on weed hosts, laying eggs and maturing the first generation there before migrating in swarms, often hundreds of miles, to sugar-beet fields. When the beets are plowed out, the hoppers migrate to neighboring gardens.

Control. Destruction of weed hosts helps somewhat, as does early planting. There are resistant varieties of sugar beets, none of table beets. Tomatoes are sometimes protected with temporary muslin tents. Infected plants must be destroyed.

Beet Latent Virus. A symptomless virus in beets.

Beet Mosaic. On sugar and table beets, spinach. Discrete yellowish lesions, then chlorotic mottling, darkening of vascular tissue; leaves bend back near the tips, which sometimes die. Transmission is by pea, peach, bean, and other aphids.

Beet Pseudo-yellows. Yellowing of sugar beet, carrot, spinach, cucumber, lettuce, and ornamentals; transmission by greenhouse whitefly.

Beet Ring Mottle. On sugar beet and spinach; stunting, distortion, mottling; transmission by aphids.

Beet Savoy. Leaves are dwarfed, curled down, with small veins thickened; roots have phloem necrosis. A plant bug (*Piesma cinerea*) is the vector.

Beet Yellow Net. On beets and chard. Leaves have a yellow network of veins against a green background. Transmission by the peach aphid.

Beet Yellows. On beets and spinach. Outer and middle leaves are yellowed, thickened, brittle, with chlorotic areas waxy. Vectors are peach and bean aphids.

Bidens Mottle. On rudbeckia, zinnia, and ageratum.

Blackberry Dwarf. See Loganberry Dwarf.

Blackberry Dwarfing. On brambles in California.

Blackberry Mosaic. Mottling, crinkling, vein-clearing, and distortion.

Blackberry Variegation. On raspberry and blackberry. Infected leaves are nearly white at maturity.

Blackeye Cowpea Mosaic. On urd bean.

Blueberry Necrotic Ring Spot. A strain of tobacco ring spot virus, causing stunting and distortion; transmitted by dagger nematodes.

Blueberry Red Ring Spot. Red spots and rings, oak leaf patterns.

Blueberry Ring Spot. A minor disease chiefly on Cabot with red rings and dots in leaves.

Blueberry Shoestring. On highbush bluberry.

Blueberry Stunt. Bushes are dwarfed with small leaves, yellowing in summer, brilliant red in fall; berries are small, poor. Transmission by a leafhopper (*Scaphytopius magdalensis*). Variety Rancocas is quite resistant; Harding is tolerant. Eliminate wild blueberries near cultivated ones.

Broad Bean Wilt. Causes leaf mottle, ring spots, and poor growth of fibrous-rooted begonia; also found on clockvine, bean, lettuce, spinach, lambsquarter, ajuga, and dogwood.

Brome Grass Mosaic. Recently reported on Kentucky bluegrass, in Kansas.

Brome Mosaic. On cowpea.

Cabbage Black Ring Spot. See Turnip Mosaic.

Cabbage Ring Necrosis. On cabbage and other crucifers, also infecting petunia, zinnia, calendula, cucumber, beet, chard, and spinach. Concentric necrotic rings on leaves and irregular, dark, slightly sunken lesions on stems.

Cactus Virus X. On night-blooming cactus (*Hylocereus*).

Camellia Yellow Mottle Leaf. Infectious variegation; color-breaking. A graft-transmissible disease with at least four strains: CV1—large white spots on petals; CV2—small white flecks on petals, leaf variegation; CV3—feathery mottling of flowers; CV4—flower variegation and leaf mottle.

Canna Mosaic. Leaves with irregular pale yellow stripes from midrib to margin, parallel with veins, may be wrinkled and curled with chlorotic areas often dusty brown. Stems, sepals, and petals have yellow bands. Transmitted by peach and other aphids.

Carnation Etched Ring. Causes an etched-ring pattern on carnation leaves.

Carnation Latent Disease. This virus is serologically related to potato viruses S and M, chrysanthemum virus B, and other carlaviruses.

Carnation Mosaic. Widespread on carnation and sweet william. Light and dark green mottling in young leaves is followed by yellow or necrotic spots or streaks. Flowers may be spotted or striped. The vector is the peach, not the carnation, aphid. Greenhouse fumigation helps in control.

Carnation Mottle. Common in commercial carnations but producing only faint leaf mottling and flower streaks; transmitted by root contact or cutting knife.

Carnation Necrotic Fleck. Severe necrotic flecking and streaking to mild yellow flecks on carnation.

Carnation Ring Spot. Concentric rings on sweet william with vein clearing, then general mosaic; necrotic rings on carnation, often combined with reddening and curling of older leaves. Sap-transmissible on cutting knife.

Carnation Streak. Considered by some due to a strain of aster yellows virus. Yellow or reddish spots and streaks parallel to veins; lower leaves turn yellow and die. Graft-transmissible.

Carnation Yellows. Foliage mottling, flower streaking due to combination of streak and mosaic viruses.

Carrot Motley Dwarf. An Australian disease now present in California. Leaflets are small, chlorotic, with twisted petioles; plants are stunted, roots unmarketable. Vectors are aphids.

Cauliflower Mosaic. Widespread on crucifers, broccoli, brussels sprouts, cauliflower, chinese cabbage, collard, kale, annual stock, and honesty. Clearing of veins in cauliflower is followed by mild chlorotic mottling, with veins usually banded by dark green necrotic flecks. Midribs are curled, leaves distorted, plants stunted, terminal heads dwarfed. Stock is rosetted, with shortened internodes. Transmission is by cabbage, false cabbage, peach, and other aphids.

Celery Calico. On cucumber, crookneck squash, tomato, celery, and larkspur. In celery there are vein clearing, puckering, and downward cupping of younger leaves, green islands in yellows areas of outer leaves, yellow and green zigzag bands on leaflets. Basal and middle leaves of delphinium have orange-amber or lemon-yellow areas, chlorotic ring and line patterns; younger leaves are normal green. Transmission is by many species of aphids— celery, celery leaf, rusty banded, cotton, erigeron root, foxglove, greenpeach, and honeysuckle.

Celery Mosaic. On celery, celeriac, and carrot in California. Young leaflets are mottled green and yellow, in advanced stages narrow, twisted, cupped; plants are stunted, with central leafstalks shortened, outer in a horizontal position with rusty necrotic spots. Transmission is by many species of aphids. A crinkle leaf strain of this virus causes a yellow mottle with raised blister areas and crinkling on celery leaves.

Celery Yellow Spot. On celery and parsnip in California. Yellow spots and stripes, mostly along veins; circular white spots on petioles; transmission by honeysuckle aphid.

Chickpea Filiform. On chickpea.

Cherry Albino. On sweet cherry in Oregon. Branches die back in spring; leaves golden brown with up-rolled margins; late summer, new growth small and rosetted; fruit small and white; trees soon killed. Transmission by tissue union.

Cherry Bark Splitting. On sour cherry and apricot.

Cherry Black Canker. On sweet cherry in Oregon. Black, rough cankers on twigs and branches.

Cherry Buckskin. See Peach Western X-Disease (under Bacterial Diseases, Mycoplasma-like Organisms).

Cherry Bud Abortion. On sweet and sour cherry.

Cherry Chlorosis. On Malaheb and chokecherry.

Cherry Freckle Fruit Disease. On sweet cherry.

Cherry Green Ring Mottle. On sour cherry.

Cherry Gummosis. On sour cherry. Dieback of terminal shoots, excessive gumming of branches.

Cherry Little Cherry. On sweet and sour cherry; fruits are half-size. Flowering cherry acts as a reservoir for the virus.

Cherry Midleaf Necrosis. Dark brown necrosis starting in midvein; heavy leaf fall; trees small and less vigorous but fruit normal. On sour cherry, in Oregon.

Cherry Mora. Abnormal delay of a month or more in ripening fruit; leaves yellowish, small, rosetted, twisted on fruit spurs. On sweet cherry, in Oregon.

Cherry Mottle Leaf. On sweet cherry, in the Northwest. Leaves show chlorotic mottling, are puckered, wrinkled, distorted but not perforated. Fruit is small, hard, insipid, uneven or delayed in ripening; crop is reduced. Branches are shortened, trees stunted. Transmission by grafting or budding with no insect vector known, but the disease spreads from wild bitter cherry to sweet cherry. Remove diseased trees. Propagate with scions from virus-free trees.

Cherry Necrotic Rusty Mottle. On sweet cherry. Foliage and blossoms delayed in spring; brown necrotic or rusty chlorotic spots, often with shot holes and defoliation.

Cherry (Sour) Pink Fruit. Fruit bitter, pink; tree stunted; foliage pale.

Cherry Pinto Leaf. On sweet cherry. Chlorotic patterns of varying size with diseased tissue changing to bright yellow or white. Transmission by budding.

Cherry Rasp Leaf. On sweet cherry, in the Northwest. Enations, elongated protuberances from underside of leaves with depressed lighter areas on upper surface.

Cherry Ring Spot. On sour cherry, sweet cherry (tatter leaf), peach, plum, prune, widespread in Northeast. Chlorotic or necrotic rings and spots on leaves, with lacerations to give the tatter-leaf effect. Transmission by grafting or budding and to a small extent by seed. Control by testing budwood sources on a differential host, such as Shirfugen variety of *Prunus serrulata*, sensitive to all strains.

Cherry (Flowering) Rough Bark. On Kwanzan flowering cherry. Internodes shortened; leaves in clusters and arched downward from necrosis and cracking of midribs; bark deep brown, rough with longitudinal splitting; trees dwarfed.

Cherry Rugose Mosaic. On sweet cherry. General chlorosis of leaf between midvein and margin, with distortion; fruit yield reduced; fruits flattened, angular. Transmission by grafting; incubation 9 months.

Cherry Rusty Mottle. On sweet cherry. Many leaves turn bright yellow to red with islands of green, and drop before harvest; remaining leaves have yellow-brown spots, rusty appearance; fruit is small, late, insipid. Remove diseased trees. Select grafting material from virus-free trees.

Cherry Twisted Leaf Virus. On Bing cherry, severe stunting, leaves small, distorted, distal portion bent abruptly downward; sometimes defoliation.

Cherry Vein Clearing. Sweet cherry crinkle. A viruslike disease but not transmissible; probably genetic. Clearing of veins throughout leaves or in localized areas; margins irregular; some leaves with elongated, slotlike perforations; small blisters on lower side of veins, upper silvery. Leaves may fold along midrib, wilt and drop in midsummer. Rosetting of some branches. Blossoms abnormally abundant, but fruit reduced, pointed, flattened on one side with swollen ridge.

Cherry Yellows. Widespread on sour cherry. Yellow areas enlarge to cover whole leaves; defoliation. Diseased leaves and fruit larger than on healthy trees, but yield reduced by half. Transmission by budding and through seed.

Chrysanthemum Aspermy. See Tomato Aspermy.

Chrysanthemum Chlorotic Mottle. Widespread in greenhouses and gardens. Bonnie Jean, Ridge, and Delaware varieties are used as indicators. Caused by a viroid-type pathogen.

Chrysanthemum Flower Distortion. Apparently not widespread in United States. Virus is carried without symptoms in leaves of White Wonder, but if it is grafted to Friendly Rival, flowers are extremely dwarfed and distorted, with ray florets short, narrow, incurved, or irregularly curved.

Chrysanthemum Mosaic. Noordam's B, Keller's Q, and other virus strains are widespread in chrysanthemum with mild to severe leaf mottling and sometimes a brown streaking of flowers. Transmission is by grafting and by aphids. Control by indexing tips from heat-treated plants on reliable test varieties to make sure they are virus-free.

Chrysanthemum Ring Spot. Reported from Alabama in plants also afflicted with stunt. Large yellowish chlorotic ring patterns, severe leaf dwarfing and distortion.

Chrysanthemum Rosette. A strain from symptomless Ivory Seagull produces vein-banding, crinkle, distortion, rosetting on Blazing Gold.

Chrysanthemum Stunt. Widespread in greenhouses and gardens. Symptoms vary with variety, but plants are dwarfed, with small flowers and leaves, and bloom earlier than normal or later in some varieties. Blazing Gold, Blanche, Mistletoe, Dauntless, and Bonnie Jean often used as indicator varieties. Leaves of Blanche are crinkled, and Mistletoe has a "measles" pattern. Transmitted by dodder, sap inoculation, grafting with incubation period 6 weeks or longer; no insect vector is known. Many plant species have been infected experimentally. Commercial growers go to great lengths to select and reselect a virus-free stock, and great care is taken to prevent recontamination. Garden varieties are now indexed and available. Stunt is now known to be a viroid-type infectious agent.

Cineraria Mosaic. Mottling, dwarfing, and distortion of leaves, transmitted by seed, mechanically, and by *Aphis marutae*.

Citrus Exocortis. Probably same as rangpur lime disease; in Florida and Texas on red grapefruit and sweet orange trees, on Rangpur lime and trifoliate orange rootstocks. Trees are stunted with bark shelling, caused by a viroid-type pathogen.

Citrus Psorosis. Found wherever citrus is grown. Leaf symptoms are small, elongated, white or yellow areas near veins. Bark symptoms are scales or small pustules with irregular growth and gum deposits. With B strain of the virus, leaves have dots, rings, or large translucent areas and small corky pustules; fruit has surface rings. In the concave gum strain, cavities develop on trunks and larger limbs. The blind-pocket strain usually produces troughlike depressions in bark, sometimes bark scaling. The crinkly leaf strain, usually on lemon, causes warping and pocketing of mature leaves, and rough, bumpy fruit. Transmission is by budding or through natural root grafts.

Remove trees with advanced infection; use budwood from trees known to be free from psorosis. Sometimes bark can be scraped, going several inches beyond the margin of affected areas and painting the scraped areas with bordeaux paste.

Citrus Stubborn Disease. Oranges have multiple buds, abnormal branching, acorn-shaped fruit, which is sour and bitter at the navel end.

Citrus Tatter Leaf. Reported from California and Texas on meyer lemon and lime. Blotchy spotting of younger leaves and ragged margins.

Citrus Tristeza. Quick decline. In California, Florida, and Texas; usually in trees on sour orange rootstock. First symptoms are partial or complete suppression of new flushes of growth. Older leaves are dull or bronzed, later yellow. Defoliation continues progressively from base of twigs to tip. Rootlets and then roots die. Limbs die back, and weak shoots are produced from main limbs and trunk. Transmission is by melon and other aphids. Make new plantings with stock-scion combinations known to be resistant. Best rootstocks are sweet orange, rough lemon, rangpur lime, and sweet lime.

Citrus Vein Enation. In California on sour orange, Mexican lime, and other citrus. Veins swell and enations develop on lower surface. Transmission by grafts and aphids.

Citrus Xyloporosis. Cachexia. In Florida, chiefly on sweet lime rootstock, but also on mandarins and some of the tangelo oranges. Symptoms include stunting of 2- or 3-year trees, small yellow leaves, partial leaf drop, early blooming and fruiting; horizontal growth of branches in middle section of trees; dieback, followed by decay of entire trunk and roots. Fruits are more rounded, with a thicker rind. Transmission is by budding, possibly through seed. Use resistant rootstocks, as sour oranges of Israel and Bagdad, Valencia orange.

Citrus Yellow Vein. In California on limequat. Petioles and veins are bright yellow.

Clerodendron Zonate Ring Spot. In Florida on bleeding-heart vine. Chief symptoms are cleared veins.

Clover Club Leaf. On crimson clover. Young leaves are light-colored, have club leaf appearance due to delayed opening. Yellow margins of leaves turn red or purple during summer. Transmission by a leafhopper.

Clover (Alsike) Mosaic. On pea, causing chlorotic spotting and dark green banding of veins, leaves slightly cupped or distorted. Leaf puckering and plant stunting on bean. Pea aphid is vector.

Clover (Red) Vein Mosaic. On garden pea, causing pea stunt, broad bean, sweet pea, and red clover. Vein-clearing and chlorosis are chief symptoms on peas, curling of leaves and rosetting of younger shoots, wilting, and collapse. Vein-clearing is the only symptom on sweet pea. Broad beans may be stunted and killed. Transmission is by the pea aphid without incubation period or long retention. The Wisconsin pea stunt virus may be a strain of the red clover virus.

Clover Wound Tumor. Big vein disease, causing enlargement of veins, sometimes with enations, woody tumors on roots, sometimes stems. The virus was discovered accidentally in leafhopper nymphs and has been transmitted experimentally to many plants beside clovers.

Clover Yellow Vein Mosaic. On winged bean, wild carrot, poison hemlock, and red bean.

Coleus Mosaic. Reported from Illinois on coleus, symptoms varying with variety. Leaves may be puckered, crinkled, asymmetrical, with oak-leaf markings or ring spots or small necrotic spots.

Corn Leaf Fleck. On field and sweet corn in California. Small, circular, pale spots on leaves with tip and marginal burning, leaves dying 7 to 10 days after initial symptoms. Transmission by corn, peach, and apple grain aphid, which retain the virus for their entire lives.

Corn (Sweet) Mosaic. Leaves have broken or continuous interveinal chlorosis.

Cowpea Chlorotic Mottle. In peanut.

Cowpea Mosaic. Clearing of veins is followed by chlorotic mottling, slight convex cupping of leaflets, shortened internodes, abortion of flowers, twisting of petioles, and delayed maturity. Yield is reduced. Vectors are potato, pea, and cotton aphids. Another cowpea mosaic, known in Trinidad and probably the same as one in the United States, is transmitted by bean leaf beetles. May infect soybeans, hoary-tick clover (*Desmodium canescens*).

Cranberry False Blossom. The most serious cranberry disease in Massachusetts, New Jersey, and Wisconsin; known also on the Pacific Coast. American and European cranberries are the only natural hosts, but the virus has been dodder-transmitted to other plants. Cranberry flowers are erect, instead of pendent, with calyx lobes enlarged, petals short, streaked with red or green, stamens and pistils abnormal. Flowers may be replaced by leaves or short branches. Axillary buds produce numerous erect shoots forming witches' brooms; diseased fruits are small and irregular. Transmission is by the blunt-nosed leafhopper. Select strains resistant to the vector or flood the bogs after leafhoppers have hatched. Spray with pyrethrum.

Cucumber Mosaic. General with many strains in cucumber, squash, melon, winged bean, periwinkle, wild violets, and a wide range of other plants, including spinach, where the disease is called blight; tomato, causing shorestring disease with filiform leaflets; pepper, petunia, and tobacco. Wintering is on ground cherry, milkweed, pokeweed, catnip, Texas bluebell and *Peristroph* sp. and other weed hosts. Transmission is by peach, cotton, potato, and lily aphids and, in some cases, through seed.

In cucurbits there is a yellow-green systemic mottling, with leaves small, distorted, curled, plants dwarfed with shortened internodes, few fruits set and those mottled and misshaped, a condition called "white pickle." The lily mosaic strain produces a masked infection or chlorotic mottling and necrosis when mixed with lily symptomless virus. The lima bean, southern celery mosaic, and cowpea strains cause chlorotic mottling.

Geraniums are stunted and mottled; gladiolus flowers are color-broken; dahlia foliage has oakleaf patterns; periwinkle (myrtle) has a streaky mottle, down-curved leaves, small flowers with a white streak in the blue color. Petunias have distorted leaf blades, few or no blossoms. In delphinium, which is very susceptible, the disease is called ring spot, stunt, witches' broom.

Control. Resistant varieties of spinach, cucumber, and squash are available. Diseased lilies and other flowers should be rogued immediately. Control aphids by systemic ground treatments or sprays; repel aphid vectors by an aluminum foil mulch. Lilies may possibly be freed of the virus by scale propagation.

Currant (Red) Mosaic. Irregular, light green circular spots along midrib and larger veins enlarge to bands. Canes are stunted; plants decline.

Dahlia Mosaic; Stunt. General. Bands along midrib and veins remain yellow-green. In some varieties leaves are distorted and blistered; in others, leaves are yellowed with margins up-rolled; in others, plants are very short and bushy with short flower stems. Transmission is by peach and other aphids.

Dahlia Oakleaf. May be a separate entity or a strain of tomato spotted wilt. A pale chlorotic line across the leaf suggests the outline of an oak leaf.

Dahlia Ring Spot. Caused by a strain of tomato spotted wilt virus. Leaves have concentric rings or irregular zigzag markings. In Utah a yellow strain causes bright yellow rings and zigzags.

Delphinium Ring Spot. Faint chlorotic rings around green and yellow centers appear on young leaves, irregular necrotic spots or rings with yellow bands on mature leaves.

Dodder Latent Mosaic. Three species of dodder transmit mosaic to cantaloupe, potato, tomato, and celery.

Elderberry Disease. A new virus reported from golden elderberry can infect various stone fruits and is considered a potential threat to the fruit industry.

Elm Mosaic. On American elm in Ohio and eastern states. Some leaves are larger than normal, others small, distorted, with yellow and green mottling. There may be some branch brooming, gradual decline in vigor. Transmission is by grafting; no insect vector is known.

Elm Zonate Canker. On American elm, New Jersey, Ohio, Missouri. Zonate cankers appear in bark as rings of dead and living tissue in cortex or phloem. Some leaves develop brown necrotic spots. Transmission is by bark patch grafts; no insect vector is known.

Euonymus Mosaic. Infectious variegation. Persistent yellowing along veins; transmission by grafting and budding.

Fig Mosaic. Systemic chlorotic mottling is accompanied by severe leaf distortion; fruits have light circular areas or rusty spots, may be deformed and drop prematurely. Transmission is by grafting and the fig midge (*Aceria ficus*).

Filaree Red Leaf. On erodium in California. Early symptoms are mild vein-clearing, outward curvature of petioles, inward cupping of leaflets. Later leaflets cup outward, with reddish discoloration, are brittle, with petioles stiffly upright; flowers are dwarfed or suppressed. Aphids are vectors.

Geranium Chlorotic Spot. General on geranium. It is caused by tomato and tobacco ring spot viruses.

Geranium Crinkle. Pelargonium leaf curl. General on geranium. Hyaline spots are small, circular to irregular, sometimes star- or tree-shaped, with brown centers. Young leaves are crinkled, small, sometimes puckered and split; severely infected leaves turn yellow and drop. Petioles and stems have corky, raised necrotic streaks; tops may die. The disease is most severe in spring, inconspicuous in summer. Transmission is by grafting (not by knife-prepared cuttings) and probably by whiteflies.

Geranium (Pelargonium) Mosaic. Leaf breaking. Leaves smaller, with purple spotting along veins, and suppression of horseshoe pattern in foliage of some geranium varieties.

Gladiolus Mosaic. See Bean Yellow Mosaic.

Grape Fanleaf. Infectious degeneration in California; Court-noué and Roncet in Europe. New growth is severely stunted; leaves are dwarfed and puckered or with deep indentations and folded like a half-closed fan; fruit set is poor. The virus is present in soil and can be transmitted, apparently, by nematodes.

Grape Leaf Roll. White Emperor disease. In California, restricted to Emperor variety. Fruit is greenish yellow or pink rather than normal red; leaves are darker than normal, turning bronze or reddish along veins, yellow between veins.

Grape Yellow Mosaic (Panachure?). Yellowing of leaves of young shoots in some varieties; various types of leaf mottling; blossom shedding. Transmission by grafting and in soil.

Grape Yellow Vein. Can be transmitted by dagger nematodes.

Henbane Mosaic. Clearing or yellowing of veins of youngest leaves, then a yellow mosaic and dark green vein-banding.

Hollyhock Mosaic. Pronounced yellow and green mottle on hollyhock and malva.

Holodiscus Witches' Broom. On ocean spray. Diseased branches form clusters of thin, wiry shoots with abnormally short internodes, crowded small leaves; foliage turns bronze red early. Transmission by the spirea aphid and by grafting.

Hydrangea Phyllody. Witches' broom, "green" flowers. Thought to be caused by mycoplasma-like pathogen.

Hydrangea Ring Spot. Chlorotic blotches and rings, brown rings and oak-leaf patterns are common in florist's hydrangea. A probable cause of hydrangea "running out." Transmission is by cutting knife. Virus can infect snapdragon, sweet william, and globe amaranth.

Iris Mosaic. Widespread on bulbous iris, especially serious on Pacific Coast. Plants are stunted with yellowish streaks on leaves and dark, teardrop markings on white, blue, or lavender flowers, clear feathery markings on yellow flowers. Transmission is by peach and potato aphids. Establish disease-free foundation stock; rogue all diseased plants; spray for aphids.

Ixia Mosaic. Perhaps iris mosaic.

Laburnum Mosaic. Infectious variegation. Bright mottling of foliage, often with veins picked out in yellow.

Lettuce Big Vein. Now considered a virus disease although associated with *Olpidium brassicae* in roots. This fungus is now known to transmit the virus. Vein-clearing followed by enlarging and bleaching.

Lettuce Mosaic. Widespread on lettuce. Leaves mottled, deformed, yellowed, browned; plants stunted or dead. Transmission is by peach and root aphids and in seed. Control vectors, use virus-free seed; rogue seedbeds.

Lilac Ring Spot. Pale green to yellow spots, lines, broad diffuse rings, and bands on lilac leaves, often with distortion and holes in tissue.

Lilac Witches' Broom. On lilac, privet in Maryland. Brooming symptoms; lateral buds produce two to six slender shoots, which branch freely, with very small leaves on Japanese lilac. In common lilac and Regal privet there is yellow vein-clearing with less prominent brooming. Transmission by grafting, and by dodder; no insect vector known.

Lily Color Adding.

Lily Color Removing.

Lily Fleck. Caused by lily symptomless virus and cucumber mosaic virus. Yellow flecks on Easter lily leaves change to gray or brown, elongating parallel to veins; surface is depressed but unbroken. Plants are dwarfed with curled leaves, flowers small with brown streaks.

Lily Latent Mosaic. In Easter lily and tulip, symptoms masked or systemic chlorotic mottling.

Lily Ring Spot. Possibly a form of cucumber mosaic. There is only a faint mottling on some species, but on *Lilium tigrinum* and *L. regale* dark ring markings develop into necrotic areas. The growing point is killed; no flowers are formed; whole plant is twisted, stunted. Peach aphid is the vector.

Lily Rosette. Yellow flat. On Easter lily. Leaves curl downward; plants are dwarfed, yellowed, mature early; bulbs are small. Transmission by the melon aphid, not by seed. Rogue diseased plants; spray for aphids.

Lily Symptomless Virus. Present in Easter lilies wherever grown commercially but producing no symptoms alone; in combination with cucumber mosaic virus causing necrotic fleck. Transmission by melon aphid.

Locust Witches' Broom. Brooming disease on locust from Pennsylvania to Georgia, Ohio, and Tennessee. Vein-clearing is followed by reduction in size of new leaves, growth of spindly shoots to witches' brooms. Roots are more brittle, shorter, and darker than normal; rootlets branch excessively to root brooms. Transmission is by budding and grafting; no insect vector is known.

Loganberry Dwarf. Blackberry dwarf. On loganberry and phenomenal blackberry in Northwest. Leaves are small, obovate, rigid, with new canes short and spindly. Young plants have crinkled leaves with some chlorosis or necrosis along veins. Flowers are small, drupelets ripen unevenly and tend to fall apart when fruit is picked. Transmission is by aphids.

Lonicera Infectious Variegation. Vein yellowing and variegation on honeysuckle; graft-transmitted.

Maize Dwarf Mosaic. First noted in Ohio in 1962 and since devastating to corn in many states. Red to purple streaks in upper leaves, ears usually incomplete; plants dwarfed with great reduction in yield. Transmission by corn leaf and peach aphids. Occurs also on *Sorghum* sp. and *Triticum* sp.

Maize Stunt. See Corn Stunt.

Maize White Line Mosaic. On field corn and weed hosts including *Panicum*, *Setaria*, and *Digitaria*.

Muskmelon Mosaic. Widespread on melon. First leaves have dark green bands parallel with main leaf veins; later leaves are mottled, sometimes deformed. Transmitted by seed and sap; insect vectors unknown.

Mustard Mosaic. On black mustard, California. Small, brown, local lesions are followed by a general mottling.

Narcissus Chocolate Spot. Often present with white streak in a decline complex.

Narcissus Flower Streak. Strong breaking of flowers but normal foliage in Oregon bulb crops.

Narcissus Mosaic. Widespread on narcissus, but with mild symptoms, seldom apparent before plants bloom; has been confused with yellow stripe.

Narcissus White Streak. Silver leaf. Paper tips and white streaks in leaves are primary symptoms, with wilting and falling over of foliage long before harvest so bulbs are small. Causes decline combined with chocolate spot. Transmission by aphids. Replant only the largest bulbs.

Narcissus Yellow Stripe. Strong yellow streaking and mottling of foliage, often roughened near veins and with a peculiar twist. Flowers are streaked. Transmission by several species of aphids. Select the best plants for a mother block, with final selection during bloom; rogue plantings early before symptoms are masked by hot weather.

Nasturtium Mosaic. Vein-clearing, ruffling and cupping of young leaves, dark green vein-banding in older leaves, sometimes chlorotic spots or white rings between veins. Flower color may be broken, petals crinkled. Transmitted by several aphids.

Nothoscordum Mosaic. False garlic (wild amaryllis) mosaic transmitted through bulbs but not seed. Typical mosaic mottling of foliage.

Onion Yellow Dwarf. Yellow streaks develop at base of leaves, with yellowing, crinkling, and flattening of new leaves. Leaves may be prostrate, flower stalks bent, twisted, and stunted; yield is reduced. Some species are relatively tolerant; tree onions are symptomless. Bean, apple grain, corn leaf, and other aphids are vectors. Control is by indexing, growing sample lots of sets and mother bulbs in greenhouse beds or production of virus-free stocks in areas where disease is absent, and roguing of infected volunteer onions. Some varieties are resistant to the onion strain of the virus but not to the strain from shallot or garlic.

Orchid (Cattleya) Blossom Brown Necrotic Streak. Brown spots, streaks of whole flower; leaves may have yellow streaks; transmission by knife. In removing flower spikes use "hot knife," with attached propane torch.

Orchid (Cattleya) Mosaic. Flower-break. On *Cattleya* and other orchids. There are apparently two diseases: mild color break, with variegation in the flower but no distortion, and severe color break, with flowers distorted or twisted as well as variegated. Leaves are mottled and sometimes twisted. The virus may be present in apparently healthy plants but can be detected with antisera, and infected plants removed. Transmission is by the green peach aphid.

Orchid (Cymbidium) Mosaic. Black streak; Cattleya leaf necrosis. The most common virus disease on many kinds of orchids. On *Cymbidium* there is initially a mosaic mottle, then necrotic spots, streaks, and rings on leaves but no effect on flowers. In *Cattleya* there are sunken brown to black leaf patterns, sometimes rings, more often elongated streaks on older leaves. If leaves are killed prematurely, flowers are fewer and smaller but normal in form and color. No insect vector is known.

Orchid (Odontoglossum) Ring Spot. On *Odontoglossum* only. Small, necrotic spots and rings on older leaves, light green to yellow areas on young leaves. Leaves may turn yellow and drop in 2 or 3 months or persist longer. There are no flower symptoms; no insect vector is known.

Orchid (Oncidium) Ring Spot. On mature leaves of *Oncidium*; round to irregular, slightly sunken yellow areas on both leaf surfaces; becoming necrotic with age.

Orchid (Vanda) Ring Spot.

Ornithogalum Mosaic. On ornithogalum, galtonia, hyacinth, lachenalia, agapanthus, hebe, fine light and dark green leaf mottling becomes gray or yellow as leaves mature. Flower stalks are marked with light and dark green blotches; there are thin longitudinal streaks on perianth segments. Transmission is by melon, peach, potato, and lily aphids.

Panicum Mosaic. On St. Augustine grass.

Pea Enation Mosaic. On pea, sweet pea, broad bean, soybean, and sweet clover. Symptoms are yellowish spots on leaves, which are later white, with crinkling and savoying. Very susceptible varieties like Alderman have necrotic spots and proliferations or enations from underside of leaves. Pods may be markedly distorted and twisted with seeds small and yellow. Transmission is by pea, potato, and peach aphids.

Pea Mosaic. On pea, sweet pea, red clover, and broad beans. Sweet pea has leaf mottling, chlorosis, breaking of flower color. Garden pea has vein-clearing followed by mottling or general chlorosis and stunting. Transmission is by pea, peach, and bean aphids. Perfection and Horal varieties are resistant to this virus but not to pea enation mosaic.

Pea Mottle. Fairly widespread on garden pea, snapbean, white clover, and broad bean. On pea a severe systemic mosaic may be fatal. Some plants have chlorotic mottling of leaves and stipules, but stems, pods, and seeds are normal. Bean and pea aphids are probably vectors.

Pea Streak. Light brown to purple, oblong, necrotic lesions are scattered along stems and petioles with stems often girdled. Leaves and pods are roughened with light brown necrotic areas.

Pea Wilt. Causing severe streak in pea if pea-mottle virus is also present.

Peach Asteroid Spot. Discrete, chlorotic lesions spread along veins forming starlike spots; some chlorophyll is retained in lesions as leaves turn yellow.

Peach Calico. Leaves are first mottled, then yellowed, then papery white. Creamy white streaks develop on twigs. Fruit is shorter, rounder, with creamy white to red patches. Transmission is by budding.

Peach Dwarf. Only on Muir peach. Profusion of large, flat, dark green leaves, closely appressed on short twigs; witches' broom showing in dormant period; fruit larger than normal, misshapen.

Peach Golden Net. Probably identical with line pattern.

Peach Little Peach. Related to peach yellows, and in same host range, eastern United States.

Peach Mosaic. In Southwest on peach, apricot, nectarine, plum, and capable of infecting almond. Spring growth of peach has short internodes, with sometimes flower breaking, chlorotic mottling, and foliage distortion early in the season, with masking of symptoms or dropping out of affected areas in midsummer. Fruit is small, irregular in shape, unsalable. Apricot stones have white rings and blotches. Transmission is by budding, grafting, a mite (*Eriophyes insidiosus*), and perhaps the plum aphid. Removal of infected trees, nursery inspection, and quarantine reduce the incidence of mosaic.

Peach Mottle. Known only in Idaho.

Peach Necrotic Leaf Spot. On peach but with sweet cherry as a symptomless carrier. Light brown, dead, membranous areas in unfolding leaves fall out, leaving holes. The disease is recurrent on peach.

Peach Phony Disease. The most important peach disease in the Southeast. Trees are dwarfed; foliage is abnormally green, fruit small; there are flecks in wood, especially in roots. Phony trees have short terminal and lateral twigs; profuse lateral branching. Growth starts in spring earlier than on normal trees. Production gradually decreases, with trees worthless in a few years. Transmission is by root grafting and sharpshooter leafhoppers. Control has been by eradication and by quarantine to restrict movement of nursery stock.

Peach Red Suture. Probably a form of yellows. On peach and Japanese plum. Fruit ripens prematurely with softening, swelling, and red blotching on the suture, flesh coarse and watery while rest of fruit is hard and green. Eradicate diseased trees; propagate from healthy budwood.

Peach Ring Spot. See Cherry Ring Spot.

Peach Rosette. On peach and plum. Trees suddenly wilt and die, or there are abnormally short stems bearing dwarfed leaves, with veins cleared and thickened; death follows in a few months. The virus can be inactivated by heating at 122° F for 10 minutes.

Peach Rosette Mosaic. Of minor importance on peach, highbush blueberry, and plum. Delayed foliation, chlorotic mottling, rosetting of shoots, dark green color; transmission by grafting and through soil. Eradicate trees; do not replant in same soil without fumigation.

Peach Stubby Twig. A new disease of peach and nectarine in California. Chlorotic leaves, stubby twig growth, decreased fruit production; transmitted with infected budwood.

Peach Wart. Foliage is normal but fruits are blistered, welted, and have conspicuous raised warty outgrowths. Tissues are light tan to red, rough, cracked, and russeted or smooth, with severe gumming. Transmission by budding or inarching.

Peach Yellow Bud. Winter's peach mosaic. On peach, apricot, and almond in California. Pale yellow, feather-edged blotches along the midvein with leaf distortion, and defoliation near base of shoots. Transmission is by grafting. In field spread is only to adjacent trees, perhaps through soil.

Peanut Mottle. On wild peanut (*Arachis chacoense*).

Peanut Stunt. First noted on peanuts in Virginia in 1964 and also occurs in bean, red and white clover. Severe dwarfing and malformation of foliar parts and suppression of fruit development. Transmission by grafting and green peach aphid.

Pear Decline. A relatively new and devastating disease in California, Oregon, and Washington, trees showing a slow decline or rapid collapse. First thought due to a toxin of the pear psylla (*Psylla pyricola*), now considered a virus disease transmitted by the psylla.

Pear Stony Pit. On Bosc and other pears in Pacific Northwest. Dark green areas appear just beneath epidermis of fruit, 10 to 20 days after petal fall, resulting in deeply pitted or deformed fruit at maturity, with corky or necrotic hard tissue at base of pits. The fruit is gnarled, hard to cut. Transmission is by grafting; no insect vector is known. Bosc and Anjou pears can be top-worked with Bartlett to reduce losses from stony pit.

Peony Leaf Curl. Plants half normal height, with crooked flower stalks, curled leaves. Transmission is by grafting but not contact; no insect vector is known.

Peony Ring Spot. Marked yellow mosaic, irregular or in rings, sometimes small necrotic spots.

Peperomia Ring Spot. Concentric brown, necrotic rings on leaves, which may be cupped, curled, or twisted and may fall. Severely affected plants are stunted. Grower losses in Florida may be 25%. Take cuttings from healthy, vigorous plants.

Pepper Mottle. On pepper.

Pepper Strain of Alfalfa Mosaic.

Pepper Vein Banding Mosaic. Probably caused by potato virus Y, a new disease in Florida. Plants are stunted with up to 50% loss of marketable fruit. Vein-clearing and banding on leaves, fruit roughened with chlorotic spots or stripes. Transmission is by green peach and melon aphids. Eradicate deadly nightshade as a weed host for 150 feet from peppers, or use sunflower as a barrier.

Phlox Streak. Streaks evident in leaves and stems. Clearing of veins is followed by necrosis in leaf veins and petioles. Graft-transmissible.

Plum Line Pattern. On plum, oriental cherry, widespread. Some plum varieties have yellow vein-banding, brilliant green and yellow patterns of the oakleaf type, formed by single or multiple irregular lines or bands; in early summer the yellow fades to creamy white. In other varieties patterns are faint or absent. On flowering cherries discolored areas are bounded by a chlorotic band. Transmission is by budding or grafting.

Plum White Spot. Small pale yellow to white spots, mostly aggregated near leaf tips on Santa Rosa plum.

Potato Acropetal Necrosis. Caused by potato viruses Y and X.

Potato Aucuba Mosaic. Bright yellow mottle in most varieties, sometimes necrosis of tubers.

Potato Bouquet Disease. Caused by tobacco ring spot virus.

Potato Calico. Caused by strain of alfalfa mosaic virus.

Potato Crinkle. Mild mosaic. Due to potato virus X plus A. Leaf mottling and crinkling are often inconspicuous, but plants die prematurely. Plant healthy tubers; isolate seed plots. Varieties Katahdin, Chippewa, Houma, and Sebago are resistant.

Potato Green Dwarf. Caused by a strain of beet curly top virus. Terminal growth is dwarfed and deformed; leaflets cupped upward.

Potato Leaf Roll. Important wherever potatoes are grown. Symptoms show about a month after plants appear aboveground. Leaves are thick, leathery, rolled, with excessive starch; sometimes with a reddish or purple discoloration on the underside. Plants are dwarfed; tubers are few, crisp, with net necrosis—brown strands of dead tissue—in some varieties; sprouts are spindling; yield may be reduced by one-half. Transmission is by peach and other aphids.

Use certified seed potatoes. These come from a foundation stock obtained by indexing. Seed pieces or tubers are planted consecutively in a row, and if any show virus symptoms, the whole unit is destroyed.

Potato Leaf Rolling Mosaic. Leaves are mottled, flaccid, with some upward rolling but without distinct rolling, rigidity of leaf roll. Transmission by peach, potato, and geranium aphids.

Potato Mottle. Caused by potato virus X.

Potato Rugose Mosaic. Caused by potato virus Y, often with X. Leaves are crinkled, mottled; lower leaves with black veins; plants are stunted, die prematurely. Control by careful roguing.

Potato Spindle Tuber. General on all tested varieties of potatoes. Plants are more erect than normal but spindly, lacking vigor. Stems are stiff, leaves small, dark green; tubers are elongated, pointed at the end, the eyes "staring." Symptoms are accentuated by high soil moistures. Transmission is by contaminated knives in cutting, by contact between freshly cut seed pieces, and by insects — aphids, grasshoppers, flea beetles, and tarnished plant bugs (this is not well documented). It is now known to be caused by a viroid-type infectious agent. Control by using certified seed.

Potato Vein Banding (Potato Virus Y). On potato and many other hosts, transmitted mechanically and by many aphids. On some varieties there is leaf drop and necrotic streak or chlorotic mottling; on others there is no sign of disease.

Potato Virus A. Present in nearly symptomless form in some varieties, causing crinkle with virus X.

Potato Virus X. Almost universally present in commercial potato stocks. Cause of latent mosaic.

Potato Witches' Broom. Apical leaves are slightly rolled, upright, light green with reddish or yellowing margins. There is proliferation of axial buds with tendency to bloom and set fruit; there are aerial tubers and numerous small subterranean tubers. Such tubers put out spindle shoots without a rest period and produce dwarfed, very bushy plants with small, round, or heart-shaped leaves. Use certified seed potatoes.

Potato Yellow Dwarf. Formerly causing heavy losses in Northeast but now mostly controlled by seed certification. Potato leaves are rolled and yellowed; the plant is dwarfed with split stems showing rusty flecks. Transmission is by clover leafhoppers; overwintering hosts are groundcherry, oxeye daisy, vinca, and other plants.

Potato Yellow Spot. Reported from Maine, mostly on Katahdin variety. Spots are small, circular, bright yellow, chiefly on lower leaves.

Primrose Mosaic. Plants are chlorotic, stunted, rugose, with upward, sometimes downward, cupping of leaves. Petioles and peduncles are shortened; flowers are white-streaked; leaves are coarsely mottled yellow-green, with green islands; tips of leaves are narrowed. No insect vector is known.

Privet Ring Spot. Reported on privet in Texas. Leaves are smaller, lighter green, drop early.

Prune (Standard) Constricting Mosaic. Spots are concentrated in a band across tip of the leaf; this area is killed, and all tissue except the midvein drops out.

Prune Diamond Canker. Symptoms expressed only on French prune — diamond or oval excrescences on secondary branches, often excess sprouts from body of tree.

Prune Dwarf. On prune, plum, cherry, and peach. Leaves are small, narrow, rugose, distorted, glazed. Internodes are short, but some branches escape and appear normal. Blossoms are numerous, but mature fruits few; pistils are aborted, petals narrowed. Most injurious to Italian prune, symptomless in Bradshaw and damson plums. Transmission by grafting and budding.

Radish Mosaic. Chlorotic spotting and mottling of foliage; plants not stunted.

Raspberry Alpha Leaf Curl. Common on red raspberry. Veins are retarded in growth, causing downward curling and crinkling of leaves. Foliage is dark green, but bronzed in late summer with glistening surface. Berries are small, poor; diseased canes are readily winter-killed. Transmission is by small raspberry aphid (*Aphis rubiphila*). Cuthbert variety is most susceptible. Rogue diseased plants.

Raspberry Beta Leaf Curl. Infecting blackcaps, especially Cumberland, but also causing severe curling on Cuthbert and hybrid purple Columbian.

Raspberry Decline. On red raspberry.

Raspberry (Red) Mosaic. Green mottle; mild mosaic; yellows. Widespread on red and black raspberries, dewberry, and blackberry. Symptoms vary greatly, but usually mottled areas are darker green than rest of leaf tissue; there may be blistering and curling downward. On blackcaps, tips are stunted, fruiting laterals shortened, fruit seedy or with poor flavor. Foliation of diseased plants is delayed. Transmission by aphids.

Raspberry (Black) Necrosis. On red and black raspberries and perhaps related to red raspberry mosaic. Leaves are curled down, have necrotic spots.

Raspberry Streak. Eastern blue streak, rosette. On black raspberry. Plants are stunted, smaller in successive seasons, leaves usually curled, close together on canes, dark green, often twisted upside down. New canes have bluish dots or streaks near the base and sometimes on branches of fruiting spurs. Fruit is small, poor; plants are short-lived. Symptoms are less severe in the mild streak strain. Roguing aids in control.

Raspberry Yellow Mosaic. Black raspberries are severely dwarfed; fruiting laterals are more upright than normal; foliage is yellow; leaflets are long and narrow. Symptoms show at high temperatures, while those of red raspberry mosaic are masked. Plants are weakened, die in 2 or 3 years. Transmitted by the raspberry aphid.

Rhubarb Chlorotic Ring Spot. Chlorotic spots and rings, necrotic stippling and rings on leaves of rhubarb, reported from Oregon.

Rhubarb Ring Spot. Caused by turnip mosaic virus.

Robinia Brooming. See Locust Witches' Broom.

Rose Mosaic. Infectious chlorosis. Common on garden roses on the Pacific Coast, sometimes on greenhouse roses in the East and on garden roses originating in the West. Chlorotic areas feather away from midribs of leaflets, often with local distortion, sometimes with ring, oakleaf, and watermark patterns. Plants are dwarfed, with buds often bleached, imperfect, on short stems. The virus is carried in understock and infects tops after budding or grafting; no insect vector is known. More than one virus is probably involved. Much rose mosaic seems to be due to the Prunus necrotic ring spot and apple mosaic viruses. Rootstocks can be heat-treated to provide a virus-free source.

Rose Rosette. On species roses, Wyoming, California. Leaflets and flower parts are misshapen, stems dwarfed, with precocious growth of lateral buds, indefinite chlorotic pattern in leaves, increase in thorniness of stems. The general effect resembles 2, 4-D injury. Graft and mite transmissible, but the disease develops slowly.

Rose Streak. On rose in eastern United States. Leaves have brownish or reddish ring and vein-banding patterns; stems have ring patterns and sometimes necrotic areas near inserted buds, causing girdling, wilting of foliage. Transmission is by grafting.

Rose Yellow Mosaic. Chlorotic areas are brighter and lighter yellow than in typical rose mosaic; there is less puckering of leaves.

Sonchus Yellow Net. On lettuce.

Soybean Mosaic. Widespread on soybean. Leaves are distorted, narrow, with margins turning down, some with ruffling along main veins; plants are often stunted, pods misshapen with fewer seeds. Transmission by peach, pea, and other aphids and in seed. Control by roguing.

Soybean Yellow Mosaic. Caused by the bean yellow mosaic virus. Younger leaves show chlorotic mottling, followed by necrotic spots. Soybean bud blight, due to tobacco ring spot virus, is serious in the Midwest, causing losses up to 100%. Tip buds turn brown, dry brittle; plant is dwarfed, produces no seed.

Sparaxis Mosaic. Strong leaf mottling and crinkling.

Spinach Blight. Caused by cucumber mosaic virus.

Spinach Yellow Dwarf. In California, confined to spinach, with vein-clearing, curvature of midrib; young leaves with mottling, puckering, curling, blisters; old leaves with yellow blotches becoming necrotic. All varieties are equally susceptible. Mechanical transmission and by aphids.

Squash Mosaic. On squash and muskmelon, mostly in California. Foliage is severely mottled and malformed with dark green blisters. Transmission by banded, western striped and 12-spotted cucumber beetles.

Squash (Southern) Mosaic. On squash in Florida, infecting also cucumber and watermelon. The cucumber mosaic virus also infects squash.

Stock Mosaic. On stock, reported from California. Definite mottling with dark green islands conspicuous against light green areas. Plants are stunted; seed pods are small; flowers are broken with petals undersized. Cut-flower fields are often a total loss, but there are resistant varieties.

Strawberry Crinkle. Chlorotic and necrotic spotting with crinkled leaves and vein-clearing. Transmission by strawberry aphid.

Strawberry Latent Ringspot. Infects strawberry and rose; latent and seedborne in parsley.

Strawberry Latent Virus. Causing no distinct symptoms but intensifying those caused by other viruses.

Strawberry Leaf Curl. Caused by strawberry veinbanding virus plus strawberry latent virus.

Strawberry Leaf Roll. Leaflets are rolled down, pale green, small, on spindly petioles.

Strawberry Mild Crinkle. Caused by strawberry vein chlorosis virus with or without strawberry mottle virus.

Strawberry Mild Yellow Edge Chlorosis. Slight chlorosis of leaf margin.

Strawberry Mottle. Chlorotic spotting, leaf distortion.

Strawberry Multiplier Disease. Resembling witches' broom and stunt; transmitted by leaf grafting.

Strawberry Necrotic Shock. Blackish spots on leaves and petioles; whole crown may be killed, but plants recover, and virus becomes latent.

Strawberry Severe Crinkle. Due to strawberry mottle virus plus strawberry crinkle virus.

Strawberry Stunt. In the Pacific Northwest. Plants are erect but short; leaves at first folded, later open, dull with a papery rattle; leaflets cupped or with margins turned down; midveins tortuous; petioles short; fruits small, hard, seedy. Transmission by the strawberry aphid.

Strawberry Veinbanding. Diffuse banding along veins; leaflets with epinasty, mild crinkling, wavy margins. Transmission by several aphids, grafting, dodder.

Strawberry Witches' Broom. Leaves are numerous, light in color with spindly petioles; margins of leaflets are bent down; runners are shortened, plants dwarfed; flower stalks spindly and unfruitful. Transmission by the strawberry aphid.

Strawberry Yellow Edge. Central leaves dwarfed, with yellow edges.

Strawberry Yellows. A complex caused by mild yellow edge, crinkle, and mottle viruses. June yellows is a genetic leaf variegation, not due to a virus. For control of strawberry viruses buy certified plants. Nurseries on the Maryland eastern shore provide 37 varieties virus-free from a foundation stock of indexed plants.

Streptanthera Mosaic. Mottled foliage. Caused by bean yellow mosaic virus.

Sweetpotato Feathery Mottle. First symptom is a yellowing along veins or small diffuse yellow spots. Some leaves are abnormally dark green with feathery yellow areas along veins. Leaves may be slightly rugose and dwarfed. Transmitted by aphids, whiteflies, and sprouts.

Sweetpotato Internal Cork. First recognized in South Carolina in 1944, now in most sweet potato areas, most prevalent in Georgia and the Carolinas. Dark brown to blackish corky spots in flesh of roots, which appear normal outside. Some are present when sweet potatoes are dug, but cork spots increase in number and size during storage, especially at temperatures higher than the recommended 55°F to 60°F. Foliage symptoms are vein feathering and mottling followed by reddish to purple blotching sometimes in ring form. Quality of Porto Rico variety is severely affected, but not yield. Transmission is by peach and potato aphids and by grafting, with morning glories used as index plants. Control insects to reduce disease; there is little spread to new plantings 100 yards or more from diseased fields. Cure immediately after digging at 85°F with 90% humidity; then store at 55°F except seed stocks, which should be kept at 70°F so that lots with internal cork can be selected and discarded.

Sweetpotato Mosaic. Transmitted by fleshy-core and sprout grafts and by sweetpotato whitefly.

Sweetpotato Russet Crack. Dark lesions and fine cracks in skin of fleshy roots.

Teasel Mosaic. On Fuller's teasel and scabiosa, with vein-clearing, asymmetry, strong mosaic pattern, malformation, death of plant. Transmission by peach and rose aphids.

Tigridia Mosaic. Pale to yellow-green irregular streaks and blotches in leaves and flower bracts, occasionally dark streaks in flowers. Transmission by lily and melon aphids.

Tobacco Broad Ring Spot. In tobacco, Wisconsin, experimentally to other plants. Chlorotic or necrotic rings, sometimes concentric; young leaves puckered at first.

Tobacco Etch. Mild and severe strains widespread on tobacco, tomato, pepper, petunia, potato, and other plants. Symptoms are vein-clearing with fine necrotic etching, usually toward base of leaves. Plants are stunted with smaller, mottled leaves. Transmission is by peach, lily, bean, and other aphids.

Tobacco Mosaic. Tomato mosaic; pepper mosaic. General in gardens, fields, greenhouses on tobacco, tomato, pepper, eggplant, petunia, Moraine ash, *Achimenes, Aeschynanthus, Chirita, Codononthe, Episcia,* gloxinia, *Kohleria, Nematanthus, Streptocarpus, Smithantha, Rhoeo,* and nearly all solanaceous plants. Tomato foliage has a light and dark green mottling, accompanied by some curling and malformation of leaflets, often with a fernleaf effect. A yellow strain of the virus causes striking yellow mottling of leaves, sometimes stems and fruits. Yield is greatly reduced. In pepper, yellowish chlorotic lesions are followed by systemic chlorosis. Spinach has some mottling, stunting, necrosis. Eggplant is often killed.

Transmission is by mechanical means—by handling, on tools, through soil, by grafting, possibly but not probably by seed. The virus can be transmitted by feeding of grasshoppers, but apparently there is little spread by the usual aphid vectors. Tobacco mosaic is the most resistant and highly infectious of all viruses. It withstands heat, even alcohol and various germicides, and retains infectivity in a dried state for many years. The most common source of inoculum is smoking tobacco. Gardeners contaminate their hands by smoking and then infect plants as they transplant, tie, prune, and so forth, the virus entering through scratches or broken hairs. The first symptoms appear in 8 to 10 days. In greenhouses, even doorknobs, faucets, and flats can be contaminated after handling virus-infected plants and remain a source of infection.

There are many strains of the virus, causing cowpea mosaic, tomato aucuba mosaic, tomato enation mosaic, tomato streak, orchid aucuba, and so on.

Control. Remove and burn any suspicious plants in the seedbed along with neighboring plants. Destroy weeds, especially groundcherry and other solanaceous species. Never smoke while working with plants, and before handling young seedlings or healthy plants always wash hands thoroughly with soap. Also wash after handling tobacco in any form or after touching diseased plants.

Tobacco Necrosis. On tobacco, tomato, aster, geranium, and bean, confined to roots, or systemic without symptoms, or systemic with symptoms. In Holland, the virus causes a severe crippling of tulips called Augusta disease, often preventing flowering, resulting in death.

Tobacco Rattle. On Romaine lettuce and transmitted by *Paratrichodorus christiei.*

Tobacco Ring Spot. General on tobacco, petunia, potato, cucumber, celery, Moraine ash, and geranium, causing pimple disease of watermelon, bud blight of soy bean, in gladiolus, iris, and Easter lily without symptoms. Causing large chlorotic areas on spinach leaves; faint zigzag lines on beet; pinpoint necrotic spots with yellow haloes on cucurbits and fruits first pitted, then with elevated pimples; eggplant yellows, "bouquet disease" of potatoes, with stems curved, shortened, sometimes with black lesions on underside of veins. Petunia seedlings are stunted, first leaves are mottled, and seed pods have few seeds. Mint is stunted. Transmission is through seed of petunia (but not of tobacco), by nematodes and in some crops by grasshoppers.

Tobacco Streak. On tobacco, soybean, sweet clover, tomato, common yellow mustard, wild radish, milk thistle, and experimentally a wide range of hosts. Irregular spots, lines, and rings.

Tomato Aspermy. Chrysanthemum aspermy. On tomato the growing point of the main stem is inhibited, axillary shoots giving the plants a bushy appearance; fruit production is curtailed; there may be failure to set seed. The disease was introduced into North America on European and Asiatic varieties of chrysanthemums, which have mottled leaves. Transmission is by foxglove, green peach, and green and black chrysanthemum aphids. Perennial chrysanthemums near tomato fields are a source of infection.

Tomato Big Bud. Leaves curl and hang down; stems are shortened and calyxes greatly enlarged.

Tomato Enation Mosaic. Caused by a strain of tobacco mosaic virus.

Tomato Fernleaf. Shoestring. Caused by cucumber mosaic virus sometimes with tobacco mosaic.

Tomato Mosaic. Circular, water-soaked necrotic spots on leaves; black streaks on veins; concentric sunken rings on fruit.

Tomato Ring Spot (Tobacco Ring Spot Virus 2). Curling and extensive necrosis of shoot terminals; brown rings and streaks on leaflets, stems, fruits, more pronounced at high temperature. Causes leaf streaks in iris, crumbly fruit of red raspberry and chlorotic spot of geranium. May infect chicory, healall, black medic, moth mullein, impatiens, apple, common cinquefoil, and dandelion. Transmission is by dagger nematodes. Plants that may serve as reservoir plants for tomato ringspot virus are chickweed, henbit, dandelion, woodsorrel, plantain, strawberry, sorrel, and red clover.

Tomato Spotted Wilt. More serious on the Pacific Coast, but also occurring in Texas and some central and eastern states in greenhouses and sometimes outdoors on plants such as ragwort, purslane, nightshade, and puncture vine. Spotted wilt is common on tomato, potato, tobacco, lettuce, pea, pepper, celery, and other vegetables. Ornamentals include amaryllis, aster, begonia, calendula, calla, chrysanthemum, dahlia, delphinium, fuchsia, gaillardia, gloxinia, nasturtium, geranium, primrose, petunia, Rieger Begonia, hydrangea, stephanotis, salvia, stock, verbena, and zinnia.

In tomato there are bronze, ringlike secondary lesions; plants are stunted with some necrosis; there may be a yellowish mosaic with leaf distortion. Fruits are often marked with concentric rings of pale red, yellow, or white. Potatoes have zonate necrotic spots on upper leaves, streaks on stems, which collapse at the top; plants are stunted, with small yield. Lettuce is yellowed, with retarded growth, brown blemishes on central leaves; affected spots are like parchment but with brown margins. Peas have purplish necrotic spots on stems and leaves following mottling, and circular spots and wavy lines on pods. Spots on outer stalks of celery are first yellow, then necrotic with pockets of dead tissue inside petioles; plants are stunted and worthless.

China asters have dead tan areas in leaves, brown surface blotches on stems. Calla lillies have whitish, then brown spots and streaks. On sweet pea, reddish brown to purple streaks may run full length of the stem. Circular to oval leaf spots with diffuse margins are followed by yellowing and death of leaves and stems. Blossoms sometimes develop a circular pattern in the pigment. Delphiniums may have numerous distinct double rings. The viruses causing oak leaf and ring spot in dahlia are probably strains of the spotted wilt virus.

Transmission is by onion and flower thrips. Only the larvae can become viruliferous by feeding on infected plants, but then there is an incubation period of 5 to 9 days, during which the insect becomes adult, before the virus can be transmitted to healthy plants.

Tomato Streak. Double streak. Caused by tobacco mosaic virus plus potato virus X. Leaves are mottled green with numerous small, grayish brown papery spots, may wither and dry. Later growth is mottled green and yellow with small chocolate brown spots and dark brown streaks on stems; fruit has brown greasy spots. The disease is more important in greenhouses; workers should refrain from handling tobacco or potatoes while working with tomatoes.

Tomato Western Yellow Blight. See Beet Curly Top.

Tomato Yellow Net. Pronounced yellow necrosis of veins and veinlets. Transmission by the peach aphid.

Tomato Yellow Top. Leaflets small, curled, yellow; purplish in cool weather.

Tritonia Mosaic. Mottling at base of young leaves.

Tulip Breaking. Due to lily latent mosaic virus often present with tulip color-adding virus. Broken tulips appear wherever hybrids are grown. There is little or no obvious effect on foliage and little interference with growth, but there are marked color patterns on the flowers; differences in named broken varieties possibly due to the proportion of color-breaking and color-adding viruses present. Most pure white flowers do not change; some turn pink or red. Pink and bright red flowers have strong color changes; very dark tulips turn even darker. There may be dark stripes due to pigment intensification.

Transmission is by aphids, and roguing should take place early, before insects are active. Broken varieties should not be grown near those with solid colors.

Turnip Mosaic. Cabbage black ring spot; watercress mottle; nasturtium mosaic. On turnip, rutabaga, rape, mustard, cabbage, collard, horseradish, watercress, garden balsam, nasturtium, stock, lady's slipper, impatien, safflower, and sweet rocket. Turnip shows a systemic chlorotic mottling with crinkling, leaf distortion, stunting of plants. Cabbage has numerous small, black necrotic rings or spots; cauliflower and broccoli have a diffuse systematic mottling. Horseradish has blotchy mottling, necrotic rings, flecks, and streaks on petioles and leaf veins. Variegated flowers appear on nasturtium stock, wallflower, and sweet rocket; the last may be severely crippled or killed. Transmission is by peach and cabbage aphids. Protect seedlings by spraying or screening seedbeds.

Walnut Brooming Disease. Bunch disease. Presumably virus.

Watermelon Mosaic. In Florida and probably other states. Symptoms include mild interveinal chlorosis, stunt, distortion, mottle, consisting of green bands along veins or raised green blisters. Leaf apices often form long, narrow, sometimes twisted projections, "shoestrings." A yellow strain of the virus causes more yellow spotting with less shoestring effect.

Wheat Streak Mosaic. On wild rice (*Zizania*); foliar streak symptoms with chlorotic areas becoming necrotic with eventual leaf death.

White Line Mosaic. A mosaic disease of sweet and dent corn with short chlorotic lines along leaf veinal tissue.

Wisteria Mosaic. Diffuse yellowish blotches with scattered green islands; mature leaflets twisted.

Wisteria Vein Mosaic. On wisteria.

Zucchini Yellow Mosaic. On cucurbits, melon, squash, pumpkin, and watermelon.

WHITE RUSTS

White rusts are all members of a single genus, *Albugo*, in the Phycomycetes, and are apparently obligate parasites like the true rusts. They form a white blister just underneath the epidermis.

Albugo (*Cystopus*)

Phycomycetes, Peronosporales, Albuginaceae

Sporangia are borne in chains at apex of a short, clavate, usually unbranched sporangiophore, forming a limited sorus beneath the host epidermis and exposed by its rupture. The mycelium is intercellular except for small, knoblike haustoria. The sporangia dry to a white powder and are disseminated by wind, germinating by swarmspores. Fertilization of a globose oogonium and a clavate antheridium produces a single oospore, also germinating by swarmspores.

Albugo bliti. White rust; or white blister, on beet, amaranth, globe amaranth, and smooth pigweed. Blisterlike white pustules formed in leaves change to reddish brown when mature. Flowers and stems are dwarfed, distorted. The fungus winters in seed coats. Destroy infected plants and debris at end of season. Change location of plantings.

Albugo candida. White rust of crucifers; on cabbage, Chinese cabbage, radish, horseradish, turnip, watercress, garden cress, peppergrass, salsify, mustard, arabis, sweet alyssum, boerhavia, draba, hesperis, candytuft, stock, wallflower, and western wallflower.

Blisters appear on any part of the plant except root. They vary in size and shape and are often confluent in extended patches. There seem to be two types of infection: general or systemic, resulting in stunting of entire plant and formation of pustules on all parts; or local, with direct invasion of single leaves, stems, or flowers. Upper surface of leaves often has yellow areas with white pustules on the underside. The latter are powdery when mature, and the epidermis is ruptured to free chains of sporangia that are carried by wind to moist surfaces. They germinate by 6 to 18 zoospores, swarmspores, which settle down, produce germ tubes, and enter plants through stomata.

Stems have localized or extended swellings, sometimes sharp bends, proliferation from lateral buds giving a bushy growth. Various flower parts are deformed with pronounced distortion of flower pedicels. When these thickened parts die, oospores are formed to survive the winter in crop refuse. The disease flourishes in cool, wet weather; the spores germinate better when slightly chilled.

Control. Remove infected parts of ornamentals as noticed. Clean up all vegetable refuse at end of season and all cruciferous weeds nearby. Spraying is impractical.

Albugo ipomoeae-panduratae. White rust; general on sweet potato, also on morning glory, moonflower, *Jacquemontia*, and quamoclit. The disease is usually late on sweet potato, after vines have made their growth, but it is very conspicuous with irregular yellow areas on upper surface and white cheesy pustules on lower surface. Oospores wintering in host tissue are liberated by decay in spring. There are no control measures.

Albugo occidentalis. White rust of spinach; after a report from Virginia in 1910, the disease went unrecorded until 1937, when it appeared in epidemic form in Texas; it has since been serious in Oklahoma and Arkansas and has attacked all commercial varieties tested at the University of Wisconsin. The white blisters are small, usually on underside of leaves, sometimes on upper. Infected leaves become chlorotic, then brown; the entire crop may be lost.

Albugo platensis. White rust; on trailing four o'clock, common four o'clock, and boerhavia.

Albugo portulacae. White rust of portulaca; swollen and deformed branches bear white pustules. Shoots tend to become more erect and spindling.

Albugo tragopogonis. White rust of salsify; also on antennaria, artemisia, centaurea, feverfew, matricaria, senecio, and sunflower. Light yellow areas appear on leaves. The epidermis, forced into domelike swellings, bursts to show chalky sori of spores. Foliage may die; plants are dwarfed. There is no control.

WILT DISEASES

To wilt means to lose freshness or to become flaccid. Wilting in plants may be temporary, due to too rapid transpiration; or it may be permanent, due to continued loss of water beyond the recovery point. Disease organisms, by reducing or inhibiting water conduction, may cause permanent wilting.

Because wilt diseases are systemic, and tied up with the entire vascular system of a plant, they are usually more important and harder to control than localized spots or cankers. In many cases the fungus enters the plant from the soil through wounds or root hairs and cannot be controlled by protective spraying. Often, although the fungus is present only near the base of a plant, the first symptom is a flagging or wilting or yellowing of a branch near the top. Many species of *Fusarium* are responsible for important wilts and "yellows." *Verticillium* is a common cause of wilt in maples, other trees, and shrubs, but most important among the wilt pathogens are two species of *Ceratocystis*, one causing oak wilt, the other Dutch elm disease.

Cephalosporium

See under Leaf Spots.

Cephalosporium diospyri. Persimmon wilt; a lethal disease of common persimmon. Wilt appears in scattered localities from North Carolina to Florida and west to Oklahoma and Texas, but most infection is in north central Florida and central Tennessee. Spread is rapid and death is quick. First notice of the disease was in Tennessee in 1933. By 1938, only 5% of the persimmons in the infected stand were alive. Topmost branches wilt suddenly, then the rest of the tree, with defoliation and death. The fungus fruits in salmon-colored spore masses in cracks in dead bark of dying trees or under bark of dead rings. Fine, blackish streaks are present in five or six outer rings of trunk, branches, and roots. No control is known.

Cephalosporium sp. Sunflower wilt.

Ceratocystis

Ascomycetes, Sphaeriales, Ceratostomataceae

Perithecia enlarged at base, with thin walls, and long slender neck, ascus wall evanescent, ascospores hyaline. Conidial stage may be *Chalara* with endogeneous spores or *Graphium* with external conidia or conidiophores united into a dark stalk (synnema).

Ceratocystis fagacearum (*Chalara quercinum, Endoconidiophora faga-cearum*). Oak wilt; our most serious disease of oaks, now known in 20 states from Texas and Oklahoma east to Pennsylvania and South Carolina. It has also been reported in Florida. Although apparently present in the Upper Mississippi Valley for many years, the disease did not cause concern, and the fungus was not described until 1943; since then it has become a major threat to our forest economy and to trees in residential areas. All native oak species are susceptible, also chinquapin, chestnut, lithocarpus (and apples in experimental inoculation); but red oaks succumb most rapidly. Scouting for the disease has been done largely by airplane, the discolored foliage being visible up to a half mile.

First symptoms are a slight crinkling and paling of leaves, followed by progressive wilting, bronzing, and browning of leaf blades from margins toward midribs and defoliation progressively downward and inward throughout the tree. Red oaks almost never recover and may be killed within 4 to 8 weeks after symptoms appear. White and burr oaks may persist for some years, with affected branches dying in a staghead effect.

The first internal symptoms are the formation of gums and tyloses in the xylem. After wilting, mycelial mats are formed between the bark and wood, and the bark cracks from the pressure exerted. Perithecia are formed in these mats, which have a sour odor and attract insects. Nitulid beetles, fruit flies, brentids, springtails, bark beetles, and possibly other insects get conidia and ascospores on or in their bodies as they feed, and can inoculate other trees through wounds. We know that ascospores remain viable several months on insects and can be distributed through fecal pellets, but we do not yet know how great a role they play in the spread of oak wilt. Birds have been suspected as carriers but are not yet indicted. Local spread is largely by root grafts, one tree infecting others within 50 feet and with grafts possible between red and white oaks, not limited to the same species.

Control. In residential areas infected trees should be removed. In forests, felling may wound other trees and spread the disease more than letting the dead tree remain but treating it so that it is not infective. Different states handle the problem in different ways. In Pennsylvania, each infected tree is cut, with all other oaks within 50 feet, and ammate crystals are placed on each stump. In North Carolina stumps and felled trees are thoroughly sprayed. In West Virginia the trees are left standing, but a deep girdle into the heartwood dries out the tree so that mycelial mats and spores do not form.

Ceratocystis (*Ceratostomella*) **ulmi** (*Graphium ulmi*). Dutch elm disease; on American, Siberian, Slippery, and European elms in 31 states, Maine to North Carolina and west to Oklahoma, and on cedar. This fatal disease is not really of Dutch origin but is so named because it was first investigated in Holland. It was noticed in Europe about 1918, first in France, then in Belgium and Holland. It spread throughout central and southern Europe, then into England and Wales. In many places it virtually exterminated the elms, including those on the famous avenues at Versailles. It is suspected that the fungus came to Europe from Asia during World War I.

Dutch elm disease was discovered in Ohio in 1930 and in New Jersey in 1933. It has spread north through New England and has become very serious in the Midwest. In 1948, the disease was found in Denver, Colorado, and in 1976, in California. It is now fairly widespread in reports of its occurrence in the United States. The spread of the fungus is linked with the presence of the large and small European bark beetles, *Scolytus scolytus* and *S. multistriatus.* Only the latter is established in this country, having arrived in Boston about 1919. Patient detective work established the fact that the fungus came here in elm burl logs imported for furniture veneer. After one such infected elm burl was found in Baltimore in 1934, months of scouting went on in the vicinity of ports of entry, railroad distributing yards, and veneer plants. Such backtracking showed the infected material had come in at four ports of entry and had been carried by 16 railroads over 13,000 miles in 21 states. From this source the disease got its start in at least 13 areas in 7 states.

Elm nursery stock is, of course, quarantined, and elm burls are embargoed; but who would have believed that dishes could have anything to do with killing our elms? Dishes have to be crated, however, and several times since 1933 English dishes crated with elm wood carrying bark beetles and *Ceratocystis* have been intercepted. All American and European elms are susceptible. Asiatic elms, *Ulmus parvifolia* and *U. pumila*, are resistant. A seedling elm, named Christine Buisman for its Netherlands' discoverer, is highly resistant, though not immune, and is now available. Other promising seedlings have been tested by the U.S. Department of Agriculture.

Symptoms are apparent from the latter part of May until late fall. The acute form of the disease is characterized by sudden and severe wilting. First the young leaves, then all leaves wilt and wither, sometimes so rapidly that they dry, curl, and fall while still green, before they can turn the usual brown of dead leaves. Sometimes terminal twigs are curled into a shepherd's crook. Chronic disease symptoms are gradual, often taking all summer for complete defoliation. In many cases individual branches or "flags" appear, the yellowed leaves conspicuous against the rest of the tree; but sometimes all leaves gradually turn yellow. In another type of chronic disease, trees leaf out late in spring, with sparse chlorotic foliage and a staghead appearance.

When an affected twig is cut across, the vessels or water-conducting tubes show dark brown or black, being clogged with bladderlike tyloses and brown gummy substances (see Fig. 71). The production of these substances is thought to be stimulated by a toxin secreted by the fungus and carried in the sap stream. Symptoms are not dependent on the physical presence of fungal hyphae in all parts of the tree. The fungus lives in the sapwood, fruiting in cracks between wood and loosened bark and in bark beetle galleries under the bark. This fruiting is of the imperfect stage, spores being produced in structures called coremia. These are black stalks about 1 mm high with enlarged heads bearing vast numbers of minute, pear-shaped spores embedded in a translucent drop of sticky liquid. Spores in the vessels increase in a yeastlike manner. The perithecial stage, not found in nature, has been produced in culture by crossing plus and minus strains of the fungus.

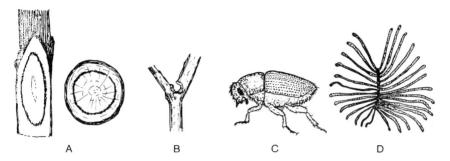

Figure 71. Dutch elm disease. *A,* branch cut to show discoloration of wood; *B,* wound in twig crotch due to beetle feeding; *C,* bark-beetle carrier of the fungus; *D,* egg and larval galleries of the beetle engraved on sapwood.

Although the smaller European elm bark beetle is chiefly responsible for spread of the pathogen, at times the native elm bark beetle, *Hylurgopinus rufipes,* is the agent. When the adult beetles emerge from under the bark of dead or dying trees, they bring along sticky spores on their bodies and deposit them as they feed in the crotches of young twigs or leaf axils of nearby healthy trees. Although the beetles feed on healthy wood, usually within 200 feet of their original tree, they breed only on weakened or dying wood and may fly some distance for it. The European female tunnels out a brood gallery 1 or 2 inches long in the wood, and when the larvae hatch, they tunnel at right angles across the wood (see Fig. 71). There is a second brood in August and September, but the overwintering one, emerging in May, is to be feared most. Because the disease often follows traffic routes, automobiles probably account for a good deal of long-distance spread. So far as we know, the only other natural means of infection is by root grafts, made when trees are planted so close together that their roots touch. This is another argument for diversified planting, rather than streets closely lined with but one type of tree.

Control. In the first few frantic years an enormous amount of money (more than $26 million) was spent on trying to eradicate the disease by removing and burning diseased trees; and while this was undoubtedly helpful, it did not stop the spread of wilt. The federal government has now left the control of Dutch elm disease up to the communities and is restricting its efforts to research. Many towns have taken a laissez-faire attitude, thinking that our elms are doomed anyway, so why waste money? Other more enlightened communities have proved that a sustained control program keeps the disease down to a negligible 1% or 2%, or less, and that the cost is far, far less than that of continuous removal of dead trees.

The midwestern chapter of the National Shade Tree Conference, in its *Guide for Community-Wide Control of Dutch Elm Disease,* suggests:

1. Survey of the total elm tree population to be protected.
2. Symptom scouting for detection of diseased trees and sanitation scouting for badly weakened elms and wood piles containing elm wood.

3. Destruction of known sources of elm wood actually or potentially hazardous for spread of disease. Elm wood piles should be destroyed completely, or each log stripped of bark and the bark destroyed. Diseased trees should be burned, on site if possible, or thoroughly sprayed. Wood chips from diseased elms may still carry the fungus; material should be burned, not used for mulches.
4. Spraying of healthy trees to prevent infection.
5. Maintenance of elms in healthy condition to prevent invasions of bark beetles, including proper watering and fertilizing, spraying to control summer foliage pests if necessary. A single annual dormant spray is now considered sufficient to protect healthy elms from bark beetles if enough material is used and complete coverage is obtained. This spray was originally a very heavy dosage of DDT, which caused some bird mortality and other environmental problems. Some communities, for example, Greenwich, Connecticut figure that it costs less to spray for control than to remove a dead tree. Where dormant spraying and sanitation have been combined consistently, the annual loss from Dutch elm disease has been kept to 1% or less.

Chemotherapy — injection of chemicals that will inactivate the fungus — has been a promising line of research for many years. A parasitic European wasp is now being bred at several laboratories for release against the bark beetles. To have elms in our future we must keep on planting them. Some forms, such as the Christine Buisman and Groeneveld elms, are quite resistant although not immune. Chinese and Siberian elms are resistant.

Dothiorella

See under Cankers.

Dothiorella ulmi. Dothiorella or Cephalosporium wilt of elms; dieback, rather common on American elms, occasional on slippery and Siberian elms in central and eastern states. The names are confusing. In culture the fungus develops spores as in *Cephalosporium*, but in nature *Dothiorella*-type pycnidia are developed on bark of killed twigs. The fungus has also been classified as *Deuterophoma*. Spores are extruded in a sticky mass and are disseminated by wind, rain, possibly insects. Infection is through insect or other wounds on foliage. The mycelium proceeds from leaf petioles into wood, where it is confined to the vessels. The foliage wilts and yellows; there are gradual dying back of the crown and a brownish discoloration in outer rings of the wood. Without laboratory diagnosis the disease cannot be positively separated from Dutch elm disease, but the elliptical cankers on the stems, with small black specks of pycnidia, provide one diagnostic symptom. Older trees die 3 to several years after first symptoms; nursery trees, in 1 or 2 years. Some trees recover, and some remain infected for many years without showing much effect.

Control. Prune out infected branches a foot or more below the lowest point of discoloration. Promote vigor by feeding, watering, aerating soil. The inclusion of a fungicide in sprays for elm-leaf beetles or cankerworms might be helpful.

Fusarium

See under Rots.

Fusarium annuum (*F. solani*). **Fusarium wilt of chili pepper;** underground stems are dry, brown, but the roots soft and water-soaked; plants wilt and die rapidly. Spores are spread in irrigation water and with wind-blown particles of soil. Avoid heavy, poorly drained soils.

Fusarium oxysporum. Wilt; on pyracantha.

Fusarium oxysporum f. sp. **apii.** Celery wilt; yellows; general in northern celery districts. There are three strains of the fungus, all causing stunting, vascular discoloration, crown and root rot, but one form causes the entire plant to turn yellow at high temperatures, producing brittle stalks with a bitter taste. Another strain causes downward curling of young heart leaves, and the third produces no aboveground symptoms except stunting. The fungus persists indefinitely in soil. Golden, self-blanching varieties are more susceptible. Grow green petiole celery or somewhat resistant Michigan Golden, Cornell 19, Tall Golden Plume, Golden Pascal, or Emerson Pascal.

Fusarium oxysporum f. sp. **asparagi.** Fusarium wilt of asparagus; a major factor in asparagus decline in California, found in most plantings. The fungus lives in soil and may be distributed on seed.

Fusarium oxysporum f. sp. **barbati.** Fusarium wilt of sweet william; new growth is yellowed; plants are stunted; leaves point downward and are tinged with tan as they die. Roots and lower stem are discolored brown. Plant in new or sterilized soil.

Fusarium oxysporum f. sp. **batatas.** See under Rots.

Fusarium oxysporum f. sp. **callistephi.** Aster wilt; one of the most serious diseases of China aster, unless resistant seed is used. Plants wilt, wither, and die at any age from seedlings to full bloom. Older plants are often stunted, with a one-sided development and a brown discoloration of the vascular system. Sometimes all lower leaves are wilted, with blackening at base of stem, often with a pink spore mass at ground level. Plants in full bloom may suddenly droop their heads. Such symptoms are in contrast to the mycoplasma-like disease, aster yellows, where the plant remains upright, although stunted and yellow. The fungus is seed-borne and persists in the soil many years.

Control. Sterilize soil for seedbeds. Some seedsmen provide seed of wilt-resistant varieties, but maintaining resistance means continuous selection from asters grown on heavily infested soil under conditions highly favorable for infection, and this process is expensive.

Fusarium oxysporum f. sp. **cattleyae.** Wilt of cattleya orchids; the fungus was isolated from a private collection in Ohio. Leaves wilted, roots abscised and decayed; flowers fewer, smaller, short-lived.

Fusarium oxysporum f. sp. **chrysanthemi.** Fusarium wilt; on chrysanthemums.

Fusarium oxysporum f. sp. **conglutinans.** Cabbage yellows; fusarium wilt; general on cabbage and other crucifers, probably the most destructive disease of such hosts in the Midwest, perhaps other sections. It is serious on cabbage, kohlrabi, and collards. Brussels sprouts, cauliflower, and broccoli are moderately susceptible in hot and dry seasons. The fungus, which can live many years in the soil, enters through the roots, usually right after transplanting or at the first hot weather, with potassium deficiency as well as heat thought to favor infection. The fungus progresses upward in the xylem, not invading other elements until the plant dies.

The most striking symptom is the dull yellow to greenish color of the foliage, together with a warping or curling of basal leaves. Leaves are killed and shed from the base up; the woody tissue in the stem is brown, with a water-soaked appearance. The fungus is spread by soil clinging to farm implements, drainage, water, wind, animals, and infected seedlings. Once the disease is established, general sanitation and crop rotation are of little help against a fungus that can survive so long without a susceptible host.

Control. Once soil is infested resistant varieties offer the only hope. Many have been developed, including Jersey Queen, Marion Market, Wisconsin Golden Acre, Resistant Detroit, resistant strains of Early Jersey Wakefield, Charleston Wakefield, Globe, Wisconsin All Season, and Wisconsin Hollander.

Fusarium oxysporum f. sp. **cucumerinum.** Cucumber wilt; a newly recognized form of *Fusarium* highly pathogenic to cucumber and muskmelon in Florida, only slightly pathogenic to watermelon.

Fusarium oxysporum f. sp. **cyclaminis.** Fusarium wilt; on cyclamen.

Fusarium oxysporum f. sp. **cubense.** Wilt; of banana.

Fusarium oxysporum f. sp. **dianthi.** Carnation fusarium wilt; yellows; branch rot; general. The first symptom is a slow withering of shoots, often accompanied by change of color from normal deep green to lighter green to pale straw yellow. Plants appear wilted, especially during the warmer part of the day. Only one side of the plant may be affected, resulting in distortion and tendency to curl. If the stem is split, a brownish streak is seen in the vascular system. There may be a dry, shreddy rot of affected wood and cortex. Plants may be infected at any age, but succumb faster if attacked when young. This species of *Fusarium* does not rot roots; see *F. roseum* under Rots for the form causing stem and root rot on carnation.

Control. Sterilize greenhouse soil and benches; take cuttings from healthy mother block; avoid overwatering. Drenching newly flatted or benched plants has reduced the number of wilted plants but does not replace steaming or otherwise sterilizing soil.

Fusarium oxysporum f. sp. **gladioli.** See under Rots.

Fusarium oxysporum f. sp. **hebae.** Fusarium wilt; of *Hebe buxifolia*, and veronica. Reported as killing nursery plants in California.

Fusarium oxysporum f. sp. **lycopersici.** Fusarium wilt of tomato; general, in many sections the most damaging tomato disease in field and greenhouse. Chief losses are in states where air temperatures are rather high during most of the season, susceptible varieties dying or producing little fruit. Losses go up to 30,000 tons of canning tomatoes, or 10% to 35% of the crop in many states.

In seedlings there is downward curvature of the oldest leaves followed by wilting and death. In older plants the disease is most evident as fruit begins to mature, lower leaves turning yellow, first on one side of the stem or leaflets on one side of the petiole. One shoot may be killed before the rest of the plant shows symptoms.

The fungus enters through roots and grows into the stem, where it produces the toxic substances causing wilting and eventual death. The vascular system in the stem shows a dark brown discoloration. In severe infections the fungus grows into fruit and seeds, but such fruits usually drop, and seed is not used. Almost all original infection comes from the soil, the *Fusarium* operating best in light sandy soils and at temperatures between 80° F and 90° F, but the disease is spread widely in transplants. It is encouraged by low potassium and high nitrogen nutrients.

Control. Start seedlings in clean soil; do not grow in the same land more than once in 4 years. The use of resistant varieties is the chief means of control. Marglobe, Pritchard, and Rutgers are moderately resistant, but infestation by nematodes may predispose even these to wilt. Pan America, Southland, Homestead, and Jefferson are more highly resistant. Treating soil with nematicides may reduce incidence of wilt even though the wilt pathogen is not killed.

Fusarium oxysporum f. sp. **melonis.** Muskmelon fusarium wilt; similar to that of watermelon, important in Minnesota, New York, New Jersey, and Maryland. Seeds rot in soil; seedlings damp-off; vines wilt. Fungus persists in soil and is carried internally in seed. Varieties Golden Gopher and Iroquois are quite resistant.

Fusarium oxysporum f. sp. **niveum.** Watermelon wilt; general on watermelon, also on citron. The fungus is transported in and on seed and persists in soil 15 to 18 years. It rots seeds or seedlings, causes wilting of plant, sometimes with cottony mycelium on surface of dying vines. Resistant varieties include Improved Kleckley Sweet and Klondike.

Fusarium oxysporum f. sp. **peniciosum.** Mimosa wilt; on mimosa from New Jersey and Maryland to Florida. This extremely pernicious wilt started about 1930 at Tryon, North Carolina, and mimosas have wilted and died at a rapid rate ever since. The wilt appeared in one city block at Morgantown, North Carolina, in 1943, and by 1947, trees were dead and dying on 232 blocks.

The first external symptom is a wilting and yellowing of leaves on some of the branches, causing foliage to hang down, then die and drop. Death of the tree follows from a month to a year after first infection. The trunk has a brown ring of discolored sapwood, usually in the current annual ring, and the color may extend into the branches. The xylem is plugged with brown gummy substances. Small branches may have a one-sided wilting with the bark flattened over collapsing tissue. The disease has been spreading in Maryland since 1947, in Florida since 1952.

As with other *Fusaria*, this soil fungus enters through the roots, and eradication of diseased trees has no effect on spread of the wilt. Nematodes, by their wounds, may increase the incidence of wilt. Out of a great many seedlings grown from seed collected from Maryland to Louisiana, inoculated several times with the fungus and planted in infested soil, some have remained mostly disease-free. These have been propagated by the U.S. Department of Agriculture. Released for commercial sale are Charlotte and Tryon.

Fusarium oxysporum f. sp. **pisi.** Pea wilt; caused by race 1 of this pathogen and near wilt; caused by race 2. Race 1, confined to pea, produces stunted plants, pale yellow green, with leaves curled downward, stem thickened and brittle near the ground. Plants wilt and die prematurely. The disease may cause more or less circular bare spots in the field, enlarging each year if peas are planted continuously, encouraged by high soil temperature. Some commercial pea varieties are resistant to race 1 but not to race 2. Delwiche Commando was the first variety introduced resistant to both races.

Fusarium oxysporum f. sp. **raphani.** Radish wilt; young plants turn yellow and die; others are stunted, with discoloration of roots.

Fusarium oxysporum f. sp. **spinaciae.** Fusarium wilt of spinach; plants are pale; leaves roll inward, gradually die. The wilt is serious in Texas and Virginia. One form of the mosaic-resistant Savoy spinach is also resistant to wilt.

Hendersonula

Fungi Imperfecti, Sphaeropsidales, Sphaerioidaceae

Pycnidia dark, separate; spores dark with several cells.

Hendersonula toruloidea. Branch wilt of walnut; canker, destructive to Persian walnuts but associated with sunburn of affected branches. The fungus is a wound parasite.

Phomopsis

See under Blights.

Phomopsis sp. Wilt; on ice plant.

Phytophthora

See under Blights.

Phytophthora cactorum. Wilt; of blue laceflower and baby's breath.

Phytophthora cinnamomi. Rhododendron wilt; a wilt of young stock, grafted plants 2 to 3 years old, seldom on older shrubs, most severe on *Rhododendron ponticum*. The foliage is first dull yellow, then permanently wilted, roots are decayed; stems are brown at soil level and below. Remove infected stock from frames immediately; avoid excessive irrigation; keep soil acidity at pH 4.0 to 4.5; provide shade and mulch for young plants. This pathogen also causes wilt of Japanese umbrella tree. See under Rots for this fungus at work on many other plants.

Pythium

See under Rots.

Pythium myriotylum. Wilt; on peanut.

Pythium tracheiphilum. Wilt; on lettuce and also leaf blight.

Verticillium

Fungi Imperfecti, Moniliales, Moniliaceae

Conidia one-celled, hyaline, globose to ellipsoid, formed at tips of whorled branches and separating readily from tips.

Verticillium albo-atrum. Verticillium wilt; maple wilt; of many ornamental trees, shrubs, fruits, flowers, and vegetables. The fungus was first isolated from potatoes in Germany in 1870, but apparently was present in California as early as 1850. It attacks nearly 300 cultivated plants of widely diverse types and may persist as a saprophyte in the soil for 15 years or more.

Of the ornamental tree hosts silver maples are most susceptible, then sugar and red maples, elms, with occasional reports on ailanthus, alfalfa, aspen, ash, boxelder, beech, black locust, camphor tree, carob, catalpa, Chinaberry, cucumber, deerbrush, dogwood, goldenrain, horse chestnut, India hawthorn, redbud, linden, magnolia, oak, osage orange, olive, pistachio, persimmon, periwinkle, Russian olive, sassafras, strawberry, smoke tree, tulip tree, walnut, mango, sunflower, and hickory. Maples may wilt suddenly in midsummer, often a large branch or one side of the tree drying and dying while the other side stays fresh. The sapwood of the infected side has greenish streaks, and sometimes slime flux develops on the bark. The disease can be chronic, progressing slowly for several seasons, or acute, affecting the entire tree in a few weeks. In elms the leaves may be smaller than normal, with a drooping flaccidity in hot hours of the day. Later there is a slight yellowing, deepening until the foliage is a striking lemon yellow. Defoliation starts at time of first yellowing, and quite often branchlets drop as well as leaves. Sapwood discoloration is brown, and the disease cannot be told positively from Dutch elm disease without laboratory cultures. Tyloses and gums are formed in the wood as with other toxin-producing fungi. The fungus always progresses upward through the xylem vessels so there is little danger of downward infection of the main trunk from pruning operations. Progress is slowed by adequate moisture and by high nitrogen fertilizers, ammonium sulfate being particularly helpful.

Verticillium wilt is also a problem on rose understock. Ragged Robin, Odorata, and Multiflora are very susceptible, Dr. Huey less susceptible, and Manetti resistant. In fruit trees the wilt is often known as black heart or verticillosis. It is common in apricots, less so in almonds and peaches; branches may drop their leaves and die. Also susceptible are sweetcherry, sourcherry, avocado, plum, and prune.

On bush fruits—raspberry, blackberry, dewberry, and youngberry—the disease is commonly known as blue stem. The symptoms appear late in the season, leaves turning pale, cane tips bending downward, canes taking on a bluish color, lower leaves wilting and drying. Death is often delayed until the season after first infection. Black raspberries are more susceptible than red. The disease is sometime serious on strawberries, especially in California, but cannot always be separated from root rots. Plants may collapse in large areas at the beginning of hot weather.

Verticillium wilt is very destructive to mint in Michigan and Indiana, also reported though not as serious in Oregon and Washington. Infected plants are stunted, defoliated, and killed; yield of oil is greatly reduced. The fungus attacks all species of mint, but peppermint is most susceptible. There are some

Figure 72. Verticillium wilt on tomato.

resistant hybrids. Deep plowing, inverting the soil, has reduced the amount of wilt.

Verticillium is especially damaging to tomatoes in Utah and California. First symptoms are yellowing of older leaves and wilting of tips during the day; later, margins of all leaves curl upward, then leaves drop (see Fig. 72). Plants are stunted; fruit is small. Moderately resistant varieties Riverside and Essar have been developed for California. Symptoms on potatoes are rather indefinite, but often there is yellowing of lower leaves, shortening of internodes, and rosetting of the top (see Fig. 73). Resistant varieties may be symptomless

Figure 73. Verticillium wilt on potato.

hosts. Verticillium wilt is common on eggplant and okra, rather rare on pepper. It occurs on Chinese yard-long bean, rhubarb, and New Zealand spinach.

On herbaceous perennials in eastern gardens I find Verticillium wilt common on aconite and chrysanthemum, with leaves turning dark brown and hanging along the stem. When the stem is cut across near the base, a circle of

Figure 74. Verticillium wilt on snapdragon.

black dots indicates the fungus in the vessels. Such plants seldom die immediately but flower poorly and gradually peter out. Wilt was serious on greenhouse chrysanthemums until a wholesale commercial concern started to provide healthy propagating stock from cultured cuttings. Other ornamental hosts include abutilon, aralia, barberry, begonia, China aster, carnation, dahlia, fremontia, geranium, marguerite, peony, poppy, snapdragon (see Fig. 74), stock, and viburnum.

Control. Sometimes it is possible to prune out an infected maple and still save the tree, but often the dying tree must be taken out. Neither maple nor elm should be replanted in the same spot. Do not transfer plants from areas where wilt has appeared. Do not set raspberries following potatoes or tomatoes; do not use tomatoes after eggplant or potatoes without a long rotation. Proper fertilization and adequate watering may help trees to recover from wilt.

Verticillum dahliae. Considered by some as a synonym of *V. albo-atrum* and by others as a distinct species; reported as causing wilt of dahlias, mint, marigold, ice plant, barley, wheat, oat, potato, and other plants. This form has microsclerotia and grows on agar at slightly higher temperatures.

Verticillium fungicola. Dry bubble; of oxyster mushroom. Infection of sporophores at pin or button stage cause development of typical dry bubbles; mature sporophores show cracking and curling of tissues, and depressed, brown, necrotic areas.

WITCHWEED

A parasitic weed, *Striga asiatica*, new to the Western Hemisphere, was reported from North Carolina in 1956 and later from South Carolina, although apparently it was first seen in the latter state in 1951 following construction of a power line across a farm. The plant is an obligate root parasite of corn and crabgrass, perhaps other plants. It is 2 to 15 inches high, foliage varying from dark to light green, with linear leaves curving downward, tubular flowers with two-lipped corolla, cardinal red on the upper surface with a yellow eye, straw yellow on the lower surface. The numerous brown seeds are very minute.

Witchweed is reported from other countries on 63 plant species, 56 of them members of the Gramineae (grains and grasses). Tests in the United States with 77 nongramineous hosts found none parasitized by witchweed, but 45 species of our grasses and grains are susceptible to this new pest. To help in eradication, report suspicious weeds immediately to your county agent or extension pathologist.

4

Host Plants and
Their Diseases

The information telescoped in this section is taken in large part from the records of the Plant Disease Survey as given in the *Plant Disease Reporter*, *Plant Diseases*, and the *Index of Plant Diseases in the United States*, Agriculture Handbook 165, U.S. Department of Agriculture. I have added recent records as I have run across them in the literature and a few personal observations. Inevitably I have missed some, and there will be many more appearing while this text is in press, so the lists cannot be regarded as complete.

The hosts selected for inclusion here are those trees, shrubs, vines, flowers, as well as vegetables likely to be grown in home gardens. Native plants sometimes grown in wild gardens are here, as well as some forest trees if they are sometimes used as ornamentals. Some plants more often grown for profit are included if they have a place around the home. Cereals, cotton, and other strictly field crops are omitted. Hosts are listed alphabetically by common names except where the scientific names mean less confusion. Often there are several common names, and the Latin name is more likely to be generally recognized.

The diseases are those reported from Maine to Florida, from New York to California, and some from Alaska. Tropical diseases are included only as they affect plants in southern Florida. Diseases peculiar to Puerto Rico, the Canal Zone, and Hawaii are omitted for lack of any personal experience with them, as well as lack of space.

The geographical distribution of diseases can be taken only as a general guide. It is likely that a disease present in New York is also present in neighboring states but has not been officially reported, or that I have missed seeing the report, or that a long list of states would take too much room. Diseases listed as "general" are prevalent throughout the host range; "widespread" means found over a wide area but not prevalent; "occasional" means of infrequent occurrence.

Fungi that are possibly parasitic, and have been recorded as present on leaves or woody plant parts but not as causing a specific disease, have been omitted. Brief comments, following some of the listings, sort out a few of the more important problems, but specific descriptions and control measures are to be found in Chapter 3. In that chapter the diseases are grouped according to the names by which they are commonly known—rot, wilt, blight, blackspot, and so on—and then by the name of the pathogen, the agent causing the disease. In this section on host plants the key word, for example, rot or Blight, is given first, followed by the name of the pathogen in **boldface**. In the disease section (Chapter 3) the pathogens are likewise listed in boldface, but in alphabetical order under each heading, such as rot or blight, and then the common name of the disease is given.

For instance, your acacia seems to be dying, and you think it may have a root rot; perhaps you can see objects like toadstools at the base. You look up ACACIA and check the possibilities until you come to the line: Rot, mushroom root. **Armillaria mellea,** occasional; **Clitocybe tabescens,** FL. "Occasional" means that this rot might be found wherever acacias grow. You live in California so you turn to the section head ROTS (starting on page 362) and thumb down through the A's until you come to **Armillaria.** Under the name is the classification of the genus, but you can leave that to the pathologist and go on to:

"**Armillaria mellea;** mushroom root rot; of trees and shrubs, also known as Armillaria root rot or toadstool disease." You learn that this disease is especially common in California, and that the honey-colored mushrooms or toadstools are not always present for diagnosis but that black shoestrings are also telltale characters. You conclude that this is your fungus, and you read on to see what can be done to the soil to prevent a recurrence of the problem. But before you do anything too drastic, you should discuss the whole situation with someone at the University of California, for you could be mistaken.

It cannot be expected that a gardener can make accurate diagnosis of disease from reading this book any more than reading a medical book can turn a layperson into a doctor. It takes years of experience to recognize diseases on sight, from macroscopic symptoms, and it takes some technical training to recognize diseases by studying the fungus under the microscope and perhaps growing it in culture. For airtight identification of a bacterium or fungus with a new disease the organism must be repeatedly isolated in culture; the disease must be produced in healthy plants by inoculating them with a pure culture of the organism; and then the fungus, or bacterium, must be reisolated from the artificially infected plant.

In some cases the small number of known diseases for a plant together with their distinctive type and geographical distribution makes layperson identification relatively reliable. In other cases, specific identification, other than to know that it is a leaf spot, is unnecessary. And in still other cases, specimens should be sent to your state experiment station for diagnosis. It is my hope that the overburdened extension pathologist, receiving some unusual specimen, will find this list of host plants and their diseases of value in speeding up identifications.

Following is a list of the headings under which the diseases are described in Chapter 3. The numbers given with the rusts refer to spore stages (see under Rusts).

Anthracnose—dead spots with definite margins, often with pinkish slimy spore masses, on leaves, stems, or fruit.

Bacterial diseases—all types of diseases, galls, blights, rots, leaf spots, caused by bacteria.

Black knot—black, knotty enlargement of woody tissue.

Blackleg—darkening at the base of a plant.

Black mildew—superficial dark growth caused by parasitic fungi.

Blackspot—a dark leaf spot on rose.

Blights—general killing of leaves, flowers, stems.

Blotch diseases—irregular necrotic areas on leaves or fruit.

Broomrapes—leafless herbs parasitic on roots.

Cankers and dieback—localized lesions on stems or trunks, sometimes accompanied by dying back from the top.

Club root—distorted swollen roots.

Damping-off—sudden wilting of seedlings or rotting of seeds in soil.

Dodder—parasitic seed plant with orange tendrils.

Downy mildews—with internal mycelium but fruiting structures protruding to form white, gray, or violet patches.

Fairy rings—mushrooms growing in circles.

Fruit spots—blemishes on fruit.

Galls—noticeable enlargements of leaves, stems, or roots.

Leaf blister, leaf curl diseases—leaf deformities.

Leaf scorch—discoloration as if by intense heat.

Leaf spots—delimited dead areas in leaves.

Lichens—occasional on trees or shrubs.

Mistletoe—semiparasitic seed plant, forming leafy tufts in trees.

Molds—conspicuous fungus growth on leaves, seeds, or grafts.

Needle casts—conspicuous shedding of evergreen foliage.

Nematodes—causing decline diseases.

Nonparasitic diseases—due to environmental conditions rather than specific organisms.

Powdery mildews—superficial white felty or powdery growth on leaves and flowers.

Rots—soft or hard decay or disintegration of plant tissues.

Rusts—with reddish or rust-colored spore masses.

Scab—raised or crustlike lesions on leaves or fruit.
Scurf—flaky or scaly lesions.
Slime molds—found in lawns.
Smuts—with sooty black spore masses.
Snowmold—light patches in turf, especially early spring.
Sooty mold—superficial black mycelium growing in insect exudate.
Spot anthracnose—light spots with raised darker borders or scabby lesion caused by *Elsinoë* species.
Virus diseases—mosaics, ring spots, yellows, wilt caused by viruses.
White rusts—white blisters in leaves.
Wilts—systemic diseases, with wilting, death of leaves, and branches.
Witchweed—weed parasitic on roots.

ABELIA

Leaf spot. **Cercospora abeliae,** IL.
Nematode, root knot. **Meloidogyne arenaria; M. hapla; M. incognita.**
Nonparasitic. **Chlorosis,** due to soil alkalinity, TX.
Powdery mildew. **Oidium** sp., TX.
Rot, root. **Phymatotrichum omnivorum,** TX; **Rhizoctonia solani,** TX; **Pythium** sp.

ABUTILON
(Flowering Maple, Indian Mallow, Velvet Leaf)

Blight, foliage. **Colletotrichum coccodes,** VT.
Leaf spot. **Alternaria** sp., TX, IL, IN, NJ, NY; **Cercospora avicennae,** KS, MO, VA; **Cladosporium herbarum,** NY, KS; **Colletotrichum malvarum,** IA; **Phyllosticta althaeina,** TX.
Nematode, root knot. **Meloidogyne** sp., AL, FL; **M. incognita,** IN; **M. hapla.**
Rot, root. **Phymatotrichum omnivorum,** TX; **Armillaria mellea,** CA.
Rot, stem. **Macrophomina phaseoli,** IL.
Rust. **Puccinia heterospora** (III), FL to AZ.
Virus. **Abutilon Infectious Variegation; Mosaic,** universal.
Wilt. **Verticillium albo-atrum,** NJ; **V. dahlia,** WI.

The leaf spots and rot occur on Indian mallow and velvet leaf. Variegation in flowering maple is a true mosaic disease, although fostered as a desirable ornamental quality.

ACACIA

Canker, twig and branch. **Nectria ditissima,** CA, SC.
Leaf spot. **Physalospora fusca,** FL; **Cercospora** sp., GA; **Phyllachora texana,** TX.

Leaf spot, algal. **Cephaleuros virescens,** FL.
Mistletoe. **Phoradendron californicum,** CA, TX; **P. serotinum (flavescens),** TX.
Nematode, root knot. **Meloidogyne** sp.
Nonparasitic. **Chlorosis,** due to excess lime, CA. **Gummosis,** due to deficient or irregular moisture, CA.
Powdery mildew. **Erysiphe polygoni,** CA.
Rot, heartwood. **Ganoderma applanatum,** CA.
Rot, mushroom root. **Armillaria mellea,** occasional; **Clitocybe tabescens,** FL.
Rot, root. **Phymatotrichum omnivorum,** TX.
Rot, sapwood. **Schizophyllum commune,** CA.
Rust. **Ravenelia australis,** TX; **R. gooddingii,** AZ; **R. hieronymi,** witches' broom, TX; **R. igualica,** TX; **R. roemerianae,** TX; **R. siliquae,** NM, TX; **R. subtortuosae,** witches' broom, TX; **R. thornberiana,** witches' broom, AZ, TX; **R. versatilis,** AZ, CA, NM, TX.

ACALYPHA
(Copper Leaf)

Downy mildew. **Plasmopara acalyphae,** WI.
Gall, leaf. **Synchytrium aureum,** WI.
Leaf spot. **Cercospora acalyphae,** NY to AL, OK, TX, WI; **Phyllosticta** sp., NJ; **Ramularia acalyphae,** TX.
Nematode, root knot. **Meloidogyne** sp., MD, GA; **M. incognita,** IN.
Nematode, lesion. **Pratylenchus** sp., FL.
Powdery mildew. **Erysiphe cichoracearum,** WI.
Rot, mushroom root. **Clitocybe tabescens,** FL.
Rot, root. **Phymatotrichum omnivorum,** TX; **Rhizoctonia solani,** IL.

ACANTHOPANAX
(Five-Leaf Aralia)

Leaf spot. **Alternaria** sp., MO.
Rot, root. **Phymatotrichum omnivorum,** TX.

ACHIMENES

Virus. **Tobacco Mosaic,** CA, CT, FL, OH, WA, DC.

ACTINOMERIS
(Yellow Ironweed)

Leaf spot. **Cercospora anomala,** IA; **Gloeosporium** sp., WV.
Powdery mildew. **Erysiphe cichoracearum,** PA to VA and KS.

Rust. **Puccinia verbesinae** (0, I, II, III), MD, ME, TN.
Virus. **Tobacco Ring Spot,** VA.

ADOXA
(Musk Root)

Gall, leaf. **Synchytrium anomalum,** IA.
Leaf spot. **Phyllosticta adoxae,** CO.
Rust. **Puccinia adoxae** (III), CO, UT, WY; **P. argentata** (0, I), IA, MN, WI. II,
III on impatiens.

AESEHYNANTHUS

Virus. **Tobacco Mosaic,** CA, CT, FL, OH, WA, DC.

AFRICAN DAISY
(*Arctotis*)

Leaf spot, blotch. **Cercospora** sp., FL.
Nematode, root knot. **Meloidogyne** sp., FL.
Rot, crown and stem. **Whetzelinia sclerotiorum,** CA.
Rot, root. **Phymatotrichum omnivorum,** TX.
Rot, root, crown and stem. **Phytophthora cryptogea,** CA.

AFRICAN VIOLET
(*Saintpaulia*)

Bacterial leaf blight. **Pseudomonas** sp., OR.
Blight, gray mold. **Botrytis cinerea,** cosmopolitan.
Nematode, leaf. **Aphelenchoides ritzema-bosi;** lesion, **Pratylenchus** sp., MD,
NJ, OH.
Nematode, root knot. **Meloidogyne arenaria,** general.
Nematode, spiral. **Rotylenchus brachyurus,** MD.
Nonparasitic. **Ring Spot,** due to wetting foliage with cold water, general.
Petiole Rot, from touching rim of salt-encrusted pot.
Powdery mildew. **Oidium** sp., occasional in greenhouses.
Rot, root. **Rhizoctonia solani; Fusarium solani,** MD, NY; **Cylindrocarpon
radicicola,** MD, NY.
Rot, root and crown. **Pythium ultimum,** CA, probably general; **Phytophthora**
sp., MD, NC.

Bright yellow ring patterns appear if there is too steep a temperature
gradient between leaf and water. Yellowing may be due to excessive bright

sunlight. Nonflowering may be due to improper light. Root-knot nematodes as well as mites cause thickened, blistered leaves.

AGAPANTHUS
(African Lily)

Virus. **Ornithogalum Mosaic.**

AGASTACHE
(Giant Hyssop)

Downy mildew. **Peronospora lophanthi,** IL, IA, WI.
Leaf spot. **Ramularia lophanthi,** CA, MT; **Septoria lophanthi,** IL, MO, OH, WI.
Leaf spot; stem spot. **Ascochyta lophanthi,** IL, IA, WI.
Powdery mildew. **Sphaerotheca macularis** (*S. humuli*), MO, WI, UT, WA.
Rust. **Puccinia hyssopi** (III), NY to IA, MO, WI.
Virus. **Mosaic.** Unidentified, IN.

AGERATUM

Blight, southern. **Sclerotium rolfsii,** NJ, NC, probably widespread.
Powdery mildew. **Erysiphe cichoracearum,** MD.
Rot, root. **Pythium mamillatum,** CA.
Rot, root and stem. **Rhizoctonia solani,** IL, NJ, NM.
Rust. **Puccinia conoclinii** (II, III), GA, MS, NC; 0, I unknown.
Stem rot. **Alternaria zinniae,** IL.
Virus. **Biddens Mottle,** FL.

AGRIMONY
(*Agrimonia*)

Blight, stem. **Phoma herbarum,** TX.
Downy mildew. **Peronospora potentillae,** NY, WI, IA.
Leaf spot. **Cercospora** sp., IL; **Phyllosticta decidua,** NC, WI; **Septoria agrimoniae,** IL, IN, IA, MO, NC, NY, TX, WI; **S. agrimoniae-eupatoriae,** IA.
Powdery mildew. **Sphaerotheca macularis,** MA, NC, NE, WI.
Rot, root. **Phymatotrichum omnivorum,** TX.
Rust. **Puccinia agrimoniae** (II, III), MO; 0, I unknown.
Rust. **Pucciniastrum agrimoniae** (II, III), general ME to FL, NM, ND; 0, I unknown.
Virus. **Mosaic.** Unidentified, NY.

AILANTHUS
(Tree of Heaven)

Black mildew. **Dimerosporium robiniae,** DC.

Blight, twig. **Gibberella baccata** (*Fusarium lateritium*), VA; **Diplodia ailanthi,** TX; **D. natalensis,** TX; **Phoma ailanthi,** TX.

Canker; dieback. **Nectria cinnabarina,** KS, NJ, SC; **N. coccinea,** occasional; **Physalospora obtusa,** KS, MI, NY.

Leaf spot. **Cercospora glandulosa,** widespread; **Gloeosporium ailanthi,** WV, LA, TX; **Phyllosticta ailanthi,** VA; **Cristulariella pyramidalis,** FL.

Rot, butt. **Daedalea unicolor,** occasional.

Rot, root. **Armillaria mellea,** Northeast; **Phymatotrichum omnivorum,** TX.

Rot, wound. **Polyporus lacteus; P. versicolor,** occasional in living trees; **Schizophyllum commune.**

Wilt. **Verticillium albo-atrum,** NY, PA, VA.

Ailanthus is well adapted to city smoke and not often troubled by disease except for wilt, which can be serious and has killed many trees in the Philadelphia area.

ALDER
(*Alnus*)

Blight, gray mold. **Botrytis cinerea,** WA.

Canker, brown felt. **Septobasidium filiforme,** NC; **S. peckii,** NY.

Canker; dieback. **Nectria coccinea,** NY, PA; **N. galligena,** NH; **Hymenochaete agglutinans,** MI, NY, PA; **Melanconis alni,** ME to NJ and WI; **Phomopsis alnea,** KY.; **Physalospora obtusa,** SC, VA; **Solenia ochracea,** widespread on bark; **Didymosphaeria oregonensis,** ID, OR, WA.

Dodder. **Cuscuta compacta,** MD, VA.

Gall, root. Cause unknown.

Leaf blister, catkin hypertrophy. **Taphrina robinsoniana,** widespread; **T. japonica,** CA, OR, AK; **T. occidentalis,** CA, OR, ID, MT, WA.

Leaf spot. **Cercospora alni,** WI; **Cercosporella alni,** OR, WA, AK; **Dothidella alni,** black spot; **Gloeosporium tubiformis; Cylindrosporium vermiforme,** WI; **?Ophiodothis alnea,** tar spot, NH, PA; **Phyllosticta alnea,** AK; **Septogloeum variegatum,** CA; **Septoria alni,** NY, WI, CA, OR, WA, AK; **Hypospila californica,** CA.

Mistletoe. **Phoradendron serotinum (flavescens),** AZ, NM; **P. villosum,** CA; **Viscum album,** CA; on *Alnus rubra* (Red Alder).

Mistletoe, European. **Viscum album,** CA.

Nematode, root knot. **Meloidogyne** sp.

Powdery mildew. **Erysiphe aggregata,** ME to NJ, MN, CA, WA: **Phyllactinia corylea,** ME to WI, AL, SC; **Microsphaera alni,** widespread.

Rot, heart. **Daedalea unicolor; Fomes igniarius; F. pinicola; Ganoderma applanatum; Polyporus sulphureus.**

Rot, root. **Armillaria mellea,** CA; **Phymatotrichum omnivorum,** TX.
Rot, wood. **Daedalea confragosa; Daldinia occidentalis; Fomes scutellatus; Lenzites sepiaria; L. trabea; Pholiota adiposa; Pleurotus serotinus; Polyporus adustus; P. hirsutus; P. versicolor; Schizophyllum commune; Steccherinum ochraceum; Stereum** spp., mostly on dead wood.
Rust. **Melamsoridium alni** (II, III), CA, TX.
Sooty mold. **Scorias spongiosa,** NY to NC and WV.

ALLAMANDA

Nematode, burrowing. **Radopholus similis,** FL.
Nonparasitic. **Chlorosis,** magnesium deficiency in overlimed or acid soil, FL.

ALLIONIA
(Trailing Four-O'Clock)

Downy mildew. **Peronospora oxybaphi,** KS, SD.
Leaf spot. **Ascochyta oxybaphi,** IA, WI; **Cercospora oxybaphi,** IA, KS, WI.
Rot, root. **Phymatotrichum omnivorum,** TX.
Rust. **Puccinia aristidae,** AZ, NM, TX.
White rust. **Albugo platensis,** AZ, NM, TN, TX.

ALLSPICE
(*Pimenta*)

Rust. **Puccinia psidii,** FL.

ALMOND
(*Prunus amygdalus*)

Anthracnose; kernel rot. **Gloeosporium amygdalinum,** CA.
Bacterial crown gall. **Agrobacterium tumefaciens,** AL, AZ, CA, NC.
Bacterial leaf spot. **Xanthomonas pruni,** NE.
Bacterial shoot blight; blast. **Pseudomonas syringae,** CA.
Blight, blossom; brown rot. **Monilinia laxa,** CA, OR; **Botrytis cinerea.**
Blight, leaf; shot-hole disease. **Coryneum carpophilum,** CA, ID, OR.
Canker; dieback; crown rot. **Phytophthora cactorum,** CA, **P. citrophthora,** CA; **Botryosphaeria dothidea** (Dothiorella canker) CA; **Ceratocystis fimbriata,** CA.
Canker, pruning wound. **Phytophthora syringae,** CA.
Decline, almond. Unknown etiology, CA. "Golden death."
Leaf spot. **Cercospora circumscissa,** CA, OR; **Hendersonia rubi,** CA.

Nematode, lesion. **Pratylenchus vulnus,** CA.
Nematode, root knot. **Meloidogyne** sp., AZ.
Nonparasitic. **Bud Failure,** seed transmitted, increased by pruning. **Chlorosis,** iron deficiency. **Little Leaf,** zinc deficiency.
Powdery mildew. **Podosphaera tridactyla,** CA; **Sphaerotheca pannosa,** CA.
Rot, green fruit. **Sclerotinia sclerotiorum,** CA.
Rot, heart. **Polyporus versicolor,** OR.
Rot, hull. **Rhizopus** spp.
Rot, root. **Armillaria mellea,** CA, NC; **Phymatotrichum omnivorum,** TX.
Rot, wound. **Schizophyllum commune,** CA.
Rust. **Tranzschelia discolor** (II, III); 0, I on anemone, CA.
Scab. **Cladosporium carpophilum,** CA, OR, CT.
Virus. **Almond Bud Failure; Almond Calico,** CA: **Peach Mosaic; Peach Ring Spot; Peach Rosette; Peach Yellow Bud Mosaic; Peach Yellows.**
Wilt. **Verticillium albo-atrum,** CA, IL.

Crown rot and Armillaria rot are often limiting factors in almond production. At least five sprays are necessary to control brown rot, leaf blights, and scab, and often a zinc sulfate spray for little leaf. The California Agricultural Experiment Station Extension Service provides each year a revised "Spray, Dust and Fumigation Program for Almonds."

ALMOND, FLOWERING
(*Prunus triloba*)

Bacterial fire blight. **Erwinia amylovora,** IN.
Bacterial spot. **Xanthomonas pruni,** NJ.
Blight, blossom and twig. **Botrytis cinerea,** NY; **Monilinia fructicola,** CT, KS.
Powdery mildew. **Podosphaera oxyacanthae,** IA.
Rot, mushroom root. **Armillaria mellea,** MS; White Root, **Corticium galactinum,** MD.

ALOE

Rot, root. **Pythium ultimum,** CA.

Immerse nursery plants of *Aloe variegata* in hot water at 115° F for 30 minutes; place in cold water; dry; replant.

ALTERNANTHERA

Leaf spot. **Phyllosticta amaranthi,** NJ.
Nematode, root knot. **Meloidogyne** sp.

Rot, root. **Rhizoctonia solani,** NJ.
Rot, root; wilt. **Fusarium oxysporum,** NJ.

ALYSSUM
(Goldentuft, Yellowtuft)

Club root. **Plasmodiophora brassicae,** NJ.
Damping-off. **Rhizoctonia solani,** NJ.

AMARANTHUS
(Love-Lies-Bleeding, Princes-Feather, Joseph-Coat)

Bacterial, MLO. **Aster Yellows,** NY, MD.
Damping-off. **Pythium debaryanum,** CT.
Leaf spot. **Cercospora canescens,** MD.
Nematode, root knot. **Meloidogyne** sp., FL, MD.
Rot, root. **Phymatotrichum omnivorum,** TX.
Rot, stem. **Macrophomina phaseolina,** AZ.
Virus. **Beet Curly Top,** CA, TX.; **Alfalfa Mosaic,** WA.
White rust. **Albugo bliti,** MA to FL; TX, NE.

AMARYLLIS
(Includes *Hippeastrum*)

Blight, gray mold; bulb rot. **Botrytis cinerea,** occasional, chiefly in outdoor
  plantings after chilling.
Blight, southern; bulb rot. **Sclerotium rolfsii,** FL, TX.
Leaf scorch; red blotch. **Stagonospora curtisii,** general.
Leaf spot. **Cercospora amaryllidis,** AL, LA: **Colletotrichum crassipes; Fusarium
  bulbigenum; Melanospora fallax; Epicoccum purpurascens,** secondary.
Nematode, lesion. **Pratylenchus scribneri,** FL. **Scutellonema brachyurum,**
  FL.
Rot, bulb. **Rhizopus stolonifer.**
Rot, seedling root. **Pythium debaryanum,** FL, TX; Root, **Armillaria mellea,**
  CA.
Virus. **Cucumber Mosaic,** FL; unidentified **Mosaic,** CA, FL, OK, WI; **Tomato
  Spotted Wilt,** CA, TX.

Although amaryllis is subject to red spotting from various physiological
causes, mite, and insect injuries, the fungus leaf scorch or red blotch is fairly
common, with red spots on leaves, flower stalks, and bulb scales.

AMARYLLIS, WILD
(*Nothoscordum bivalve; N. inodorum*)

Virus. **Nothoscordum Mosaic,** LA.

AMELANCHIER
(Serviceberry, Juneberry)

Bacterial fire blight. **Erwinia amylovora,** widespread.
Black mildew; witches' broom. **Apiosporina collinsii,** widespread.
Blight, leaf. **Fabraea maculata,** widespread.
Blight, fruit and leaf. **Monilinia gregaria,** WA, CO, IA.
Canker. **Nectria cinnabarina,** occasional; **Pezicula pruinosa,** widespread; **Gloeosporium perennans,** OR.
Canker, blister. **Nummularia discreta,** IA, KS, MA, OH.
Leaf blister; witches' broom. **Taphrina amelanchieri,** CA.
Leaf spot. **Coccomyces tumidus,** MO, MT; **Cristulariella pyramidalis,** NC; **Phyllosticta innumerabilis,** IA, NE, ND, MN; **P. paupercula,** KS, MT; **P. virginiana,** NY, PA, WI.
Powdery mildew. **Erysiphe polygoni,** ID, MT; **Phyllactinia corylea,** OR, WA; **Podosphaera oxyacanthae,** occasional.
Rot, brown; fruit. **Monilinia amelanchieris,** NY; **M. fructicola,** MI, OH.
Rot, root. **Phymatotrichum omnivorum,** TX.
Rot, wood butt. **Daedalea unicolor,** cosmopolitan.
Rust. **Gymnosporangium biseptatum** (I); III on *Chamaecyparis,* northeastern and Middle Atlantic states. Horned galls, underside leaves. **G. clavariae-forme** (I) on leaves, fruits, stems; III on juniper. **G. clavipes** (I) on fruits, stems; III on juniper. **G. corniculans** (I); III on juniper. **G. cupressi** (0, I); III on cypress. **G. gracile** (0, I); III on juniper. **G. harknessianum** (I) on fruits, CA, OR; III on juniper. **G. inconspicuum** (I) on fruits; III on juniper. **G. juvenescens** (I); III on juniper, in West (may be **G. nidusavis**). **G. kernianum** (I) on leaves, AZ, CO, OR; III on juniper. **G. libocedri** (I) on leaves, CA, OR; III on juniper. **G. nelsoni** (I) on leaves, fruits; III on juniper. **G. nidusavis** (I) on leaves, fruits, stems; III on juniper, central and eastern states.

AMORPHA
(Leadplant, Indigobush)

Canker, twig. **Cytospora amorphae.**
Leaf spot. **Cylindrosporium passaloroides,** widespread; **Diplodia amorphae,** TX.
Powdery mildew. **Erysiphe polygoni,** IA, MN, WY.
Rust. **Uropyxis amorphae,** general.

AMPELOPSIS
(*A. cordata* and Other Species)

Canker; dieback. **Nectria cinnabarina,** occasional.
Downy mildew. **Plasmopara viticola,** NY, WI.
Leaf spot. **Cercospora truncata,** LA; **C. vitis,** LA, AL; **Guignardia bidwellii,** widespread; **Linospora psederae,** WV; **Phleospora ampelopsidis,** WI.

Nematode, dagger. **Xiphinema index.**
Powdery mildew. **Uncinula necator,** widespread.
Rot, root. **Helicobasidium purpureum,** TX.

AMSONIA

Leaf spot. **Mycosphaerella** sp., GA.
Rust. **Coleosporium apocyanaceum** (II, III) on leaves, FL, GA, SC; 0, I on
  pine. **Puccinia seymouriana** (0, I), MO; II, III on marsh grass.

ANAPHALIS
(Pearl Everlasting)

Leaf spot. **Septoria margaritaceae,** NY, OR, WI.
Rust. **Uromyces amoenus** (III), CA, ID, MI, MT, OR, WA, WY.

ANCHUSA
(Buglos, Alkanet)

Damping-off. **Rhizoctonia solani,** CT.
Leaf spot. **Stemphyllium** sp., NY.
Rust. **Puccinia recondita** var. **secalis** (0, I), IN, MI: II, III on rye.
Virus. **Mosaic,** unidentified, IN.

ANDROMEDA
(Bog Rosemary)

See Lyonia and Pieris for shrubs commonly called Andromeda.

Black mildew. **Asterina clavuligera,** GA, NJ.
Gall; red leaf spot. **Exobasidium vaccinii,** ME, NY, WI, AK.
Leaf spot. **Venturia arctostaphyli,** NY.
Leaf spot, tar. **Rhytisma andromedae,** ME to NC, MI, MN, AK.
Powdery mildew. **Microsphaera alni** var. **vaccinii,** MA, NJ.
Rust. **Chrysomyxa cassandrae** (II, III); 0, I on pine.

ANEMONE
(Cultivated Forms, Japanese)

Bacterial, MLO. **Aster Yellows,** CA.
Blight, collar rot. **Botrytis cinerea,** NJ, PA.
Blight, southern. **Sclerotium rolfsii,** CA, MD, MA.
Leaf spot. **Gloeosporium** sp., NY; **Phyllosticta anemones,** MD.

Nematode, leaf. **Aphelenchoides fragariae; A. ritzema-bosi,** CA.
Nematode, root. **Aphelenchus agricola,** CA.
Rust. **Tranzschelia cohaesa** (0, I, II, III), TX; **T. discolor** (0, I), CA, OR; II, III on *Prunus* spp.
Virus. **Mosaic.** Unidentified, CA.

ANEMONE
(Native Species)

Downy mildew. **Plasmopara pygmaea,** widespread.
Gall. **Synchytrium anemones,** IA, MA, MI, MN, VT, WI.
Leaf spot. **Cercospora pulsatillae,** CO; **Cercosporella filiformis,** WI; **Didymaria didyma,** IA, MI, WI; **Phleospora anemones,** IA; **Phyllosticta anemonicola,** IL, MI, ME, NE, VA, WI; **Ramularia ranunculi,** NY, WI; **Septoria anemones,** IL, IA, MS, MO, TX, VT, WI; **S. cylindrica,** MT, VA; **S. punicea,** MA, MI, MS, NE, KS.
Powdery mildew. **Erysiphe polygoni,** IL, IA, IN, MI, MN, ND, NJ, WI.
Rust. **Puccinia anemones-virginianae** (III), ME to MS to NE, ND; **P. gigantispora** (0, I, III), CO, IL, ID, MT, ND, WI, WY; **P. magnusiana** (0, I), NY to KS, ND; II, III on reed grasses; **P. pulsatillae** (III), CO, IA, ND, SD; **P. recondita** (0, I), NY to TX; II, III on grasses; **P. retecta,** CO, AK; **P. vesiculosa,** AK.
Rust. **Tranzschelia cohaesa** (0, I, II, III), TX; **T. pruni-spinosae** (0, I), KS, IA, NE to TX, AL; II, III on *Prunus* spp.; **T. fusca** (0, III), MA to VA, CA, and Northwest; **T. suffusca** (0, III), CO, IA, MT, ND, SD; **T. tucsonensis** (0, I, II, III), AZ.
Smut, leaf and stem. **Urocystis anemones,** ME to DE, CO, KS, IA, MN, ND, WI, AK; **U. sorosporioides,** AK.
Smut, white. **Entyloma ranunculi,** WI.

ANGELICA

Leaf spot. **Cercospora apii** var. **angelicae,** AK; **C. thaspii,** AL, TX; **Fuscladium angelicae,** general; **Gloeosporium angelicae,** SC; **Phyllosticta angelicae,** CA, WY; **Piggotia depressa,** MT; **Ramularia angelicae,** CO; **Septoria dearnessii,** WI, NC.
Rot, root. **Phymatotrichum omnivorum,** TX.
Rust. **Puccinia angelicae** (0, I, II, III), NY, OR, PA, WA; **P. bistortae,** AK; **P. coelopleuri** (0, I, II, III), AK; **P. ellisii,** CA, ID, NV, OR, WA; **P. ligustici,** CO, WA; **P. poromera,** UT.

ANISE
(Pimpinella anisum)

Leaf spot. **Cercospora malkoffi,** VA.
Rot, root. **Phymatotrichum omnivorum,** TX.

Rot, stem. **Sclerotinia sclerotiorum**, CA, TX.
Rust. **Puccinia pimpinellae** (0, I, II, III), CA.

ANISE-TREE
(*Illicium floridanum*)

Black mildew. **Lembosia illiciicola**, AL, MS.
Leaf spot, algal; green scurf. **Cephaleuros virescens**, LA, SC.
Sooty mold. **Capnodium footii**, MS.

ANODA

Nematode, cyst. **Meloidogyne hapla.**
Powdery mildew. **Oidium erysipheoides**, NM.
Rust. **Puccinia** sp., TX.

ANTIDESMA
(Chinese Laurel)

Leaf spot, algal; green scurf. **Cephaleuros virescens**, FL.

ANTHURIUM

Bacterial blight. **Xanthomonas campestris** pv. **dieffenbachiae**, CA.
Nematode, lesion. **Pratylenchus pratensis**, LA.

APHELANDRA
(*Zebra plant*)

Leaf spot. **Myrothecium roridum**, FL.
Nonparasitic. **Physiological,** leaf crinkle, shortened internodes, axillary bud proliferation, TX.
Rot, stem. **Phytophthora parasitica**, FL.

APPLE
(*Malus sylvestris*)

Anthracnose, northwestern; canker; fruit rot. **Neofabraea malicorticis**, prevalent OR, WA, occasional CA, IL, MA, ME, MI, NE, OK.
Bacterial blast, of flowers, shoots. **Pseudomonas syringae**, AR, CA; blister spot of bark and fruit. **P. syringae** pv. **papulans**; rot, **P. melophthora.**

Bacterial crown gall. **Agrobacterium tumefaciens,** general; hairy root, **A. rhizogenes,** central states, ID, NY.

Bacterial fire blight. **Erwinia amylovora,** general.

Blight, limb. **Corticium laetum,** LA, NC; **C. salmonicolor,** FL, LA.

Blight, southern. **Sclerotium rolfsii,** on seedlings.

Blight, thread. **Pellicularia koleroga,** WV and IN to Gulf states.

Blotch, fruit; Leaf spot; twig canker. **Phyllosticta solitaria,** general except New England and the far South.

Blotch, sooty, of fruit. **Gloeodes pomigena,** eastern and central states.

Canker, bark; fruit rot. **Myxosporium corticola,** New England to MD, IL, MI, OK, OR, SD.

Canker, blister. **Nummularia discreta,** east of the Rocky Mountains.

Canker, crater. Cause unknown, sunken bark, crown rot, decline. WA.

Canker; dieback. **Botryosphaeria ribis,** East and South; **Cytospora** spp., widespread; **Fusarium** spp., also fruit rot, Pacific Northwest, secondary to drought, winter injury, insect punctures; **Coryneum foliicola,** widespread; **Glutinium macrosporium,** OR; **Leptosphaeria coniothyrium,** East, central states, WI; **Plenodomus fuscomaculans,** CA, MI.

Canker, European. **Nectria galligena,** East and central states to NC and MS; Pacific Coast.

Canker; Leaf spot. **Monochaetia mali,** IL, NJ to SC, MO, WV.

Canker, perennial; bull's-eye fruit rot. **Neofabraea perennans,** ID, MT, OR, WA.

Canker, sapwood rot. **Hymenochaete agglutinans,** CT; **Hypoxylon** spp.

Canker, scurf bark. **Clasterosporium** sp.

Canker, twig. **Coryneum foliicolum,** widespread; **Gibberella baccata; Nectria cinnabarina,** northern U.S.

Decline. **Irpex tulipiferae,** MN; **Coriolus versicolor,** also dieback, MN, WA; **Schizophyllum commune,** MN.

Fruit spot, black pox; canker; leaf spot. **Helminthosporium papulosum,** IN, MA, MS, NJ, OH, PA, WV, GA.

Fruit spot, brooks. **Mycosphaerella pomi,** New England to NC, AR, MO, IA.

Fruit spot, fly speck. **Microthyriella rubi,** general.

Leaf spot. **Cercospora mali,** Gulf states, VT; **Coniothyrium pyrinum; Diaporthe pernisiosa,** AR; **Fabraea maculata,** IA; **Illosporium malifoliorum,** PA to NC, IN; **Mycosphaerella sentina,** IL, NJ, PA; **Pestalotia** spp., MD to NC, IN, ID; ghost spot, **Alternaria tenuis,** GA; **Cristulariella pyramidalis,** FL, on *Malus pumila.*

Mistletoe. **Phoradendron serotinum (flavescens),** NC, TX; **Viscum album,** CA.

Nematode, lesion. **Pratylenchus penetrans; P. pratensis.**

Nematode, root knot. **Meloidogyne** spp., MS, TX, UT.

Nonparasitic

    Bitter Pit, Stippen, Baldwin Spot, general in storage.

    Black End. Probably drought injury, AL, OR, WA.

    Black Heart, of wood. Perhaps freezing injury; widespread.

    Brown Core, in MacIntosh variety. Excessive nitrogen, low storage temperature.

Callus Knot. Wound overgrowth from defective union of stock and scion.

Chlorosis. Usually iron deficiency in alkaline western soils.

Collar Rot. Winter injury.

Cork; Rosette. Boron deficiency, northeastern states to IN, KY, Pacific Coast.

Internal Bark Necrosis. Cause unknown, possibly same as measles; general.

Jonathan Spot. Associated with dry weather, delayed storage.

Leaf Scorch. Magnesium deficiency.

Measles, reddish pimples in bark. Widespread; cause unknown, possibly boron deficiency in part. False measles due to manganese toxicity.

Ozone Injury. Pitted area in flesh.

Rosette, little leaf. Usually zinc deficiency.

Scald. Discoloration of fruit skin by volatile respiratory products, general in storage. Controlled by oiled paper wraps.

Soft Scald and **Soggy Breakdown.** Associated with delayed storage and low temperature.

Stigmonose. Fruit dimpling, distortion from insect punctures; widespread.

Sunburn, of fruit. Heat or light injury.

Sunscald, of bark. Freezing injury of trunk and larger branches on side exposed to sun.

Water Core. Deficient or irregular water supply.

Powdery mildew. **Podosphaera leucotricha** and **P. oxyacanthae,** general.

Rot, bitter, of fruit; canker. **Glomerella cingulata,** general, especially in South, rare in West; **G. rubicola,** IL.

Rot, black; frog-eye leaf spot; canker. **Physalospora obtusa,** general to the Great Plains; **P. mutila,** CA, MT, OR, WA; **P. rhodina,** KY.

Rot, black root. **Xylaria mali,** eastern and central states.

Rot, brown. **Monilinia fructicola,** general except for far South; **M. laxa,** OR, WA.

Rot, calyx-end. **Alternaria** sp., **Sclerotinia sclerotiorum,** NH, WA.

Rot, crown. **Phytophthora cactorum,** MA.

Rot, fruit. **Alternaria mali; Aspergillus niger; Cephalosporium carpogenum; Corticium centrifugum,** Pacific Northwest, occasional in East; **Chaetomella** sp., WA; **Cladosporium** spp.; **Endomyces mali; Epicoccum granulatum,** WA; **Gloeosporium** spp.; **Gliocladium viride,** IL; **Hormodendron cladosporiodes,** WA; **Mucor piriformis,** WA; **Oospora** sp.; **Phialophora malorum,** IN, VA, WA; **Pleospora fructicola,** WA; **P. herbarum,** Pacific states; **Penicillium** spp., cosmopolitan; **Phoma** spp.; **Stemphylium congestum,** WA; **Tricothecium roseum,** general; **Botrytis** spp.; **Colletotrichum fructus; Rhizopus nigricans,** cosmopolitan.

Rot, fruit (postharvest). **Phomopsis mali,** core rot, CA, NY; **Trichoderma harzianum,** MD.

Rot, heartwood. **Fomes fomentarius; F. igniarius; F. pinicola; Ganoderma applanatum.**

Rot, mushroom root. **Armillaria mellea,** prevalent on Pacific Coast; **Clitocybe tabescens,** AR, FL, TX, VA.

Rot, root. **Phymatotrichum omnivorum,** AZ, AK, NM, TX; **Fusarium oxysporum,** ID; **F. solani,** ID.

Rot, silverleaf. **Stereum purpureum,** KS, ME, MN, NY, Pacific Northwest.

Rot, white root. **Corticium galactinum,** DE to VA; AR, IL; **Rosellina necatrix,** CA.

Rot, wood. **Daedalea confragosa; Poria** spp.; **Pleurotus ulmarius; Pholiota adiposa; Polyporus** spp.; **Stereum** sp.; **Schizophyllum commune; Trametes** spp.

Rust. **Gymnosporangium libocedri.** (0, I), on leaves, fruit; III on incense cedar, CA, OR; **Gymnosporangium nidus-avis** (0, I), on leaves, fruit, stems, IN, MD, MS, NJ; III on red cedar.

Rust, cedar apple. **Gymnosporangium juniperi-virginianae** (0, I) on leaves, fruit; general east of Great Plains; III on red cedar and Rocky Mountain juniper.

Rust, hawthorn. **Gymnosporangium globosum** (0, I) on leaves; III on red cedar, ME to AK, KS, NE.

Rust, quince. **Gymnosporangium clavipes** (0, I), on fruit; III on common juniper and red cedar.

Scab. **Venturia inaequalis,** general.

Sooty mold. **Fumago vagans,** occasional.

Spot anthracnose. **Elsinoë piri,** OR, WA.

Virus, **Apple Mosaic; Tulare Apple Mosaic,** CA; **Tobacco Mosaic; Dapple Apple,** NH; **Green Mottle,** NY; **Stem-pitting; Tomato Ringspot, Decline** and **Graft Union Necrosis,** WA

Wilt. **Verticillium albo-atrum.**

If this appalling list of diseases should make you think twice before planting apples in the backyard with the expectation of getting cheap and abundant fruit, that is all to the good. There is no easy or cheap road to perfect fruit. The commercial grower may, in a wet season, apply nearly 20 sprays to keep scab under control. The homeowner thinks one or two are enough. State experiment stations offer abbreviated schedules for home gardeners, tailored for the area, and if these are followed carefully, a fairly good crop can be expected. There are also all-purpose mixtures available for fruit trees, which may work reasonably well. Scab is the most important apple disease, and proper timing of early season sprays is most essential. Bitter rot, black rot, sometimes rust need attention. Fire blight control is primarily a question of proper pruning to remove infected wood, with an antibiotic spray during bloom.

APPLE-OF-PERU
(*Nicandra*)

Leaf spot. **Cercospora physaloides,** IN.

Rot, Root. **Phymatotrichum omnivorum,** TX.

Virus. **Mosaic.** Unidentified, ID, IA, KY, WA, WI. Experimentally infected with several viruses.

APRICOT
(*Prunus armeniaca*)

Bacterial canker; gummosis. **Pseudomonas syringae,** CA, OR.

Bacterial crown gall. **Agrobacterium tumefaciens,** widespread. Japanese apricot (*Prunus mume*) is resistant.

Bacterial fire blight. **Erwinia amylovora,** CO, FL, NE, PA, TX.

Bacterial leaf spot. **Xanthomonas pruni,** IL to TX, NE.

Black knot. **Dibotryon morbosum,** CO, IA, NY.

Blight, blossom, twig; brown rot. **Monilinia laxa,** Pacific Coast states.

Blight, shoot; shot hole; fruit spot. **Coryneum carpophilum,** widespread.

Canker, felt fungus. **Septobasidium pseudopedicellatum,** MS.

Canker, trunk. **Phytophthora cactorum,** CA; **P. citrophthora,** CA.

Canker, trunk and limb gall. **Monochaetia rosenwaldia,** CA.

Canker, twig; dieback. **Cytospora** spp., AZ, NY; **Valsa leucostoma,** MO, TX; Cytosporina dieback, **Eutypa armeniaceae,** CA; **Ceratocytis fimbriata,** CA.

Canker; dieback; coral spot. **Nectria cinnabarina,** IN, WA.

Fruit spot. **Venturia ?cerasi** (See under Scab).

Leaf curl. **Taphrina ?deformans,** SC.

Leaf spot. **Cercospora circumscissa,** TX; **Coccomyces** sp., CA, MA, TX, VT; **Phyllosticta circumscissa,** CA.

Nematode, dagger. **Xiphinema** sp., WV, NY.

Nematode, lesion. **Pratylenchus vulnus,** CA; **P. penetrans,** NY.

Nematode, root knot. **Meloidogyne** sp., AZ, TX.

Nonparasitic

    Arsenical Injury, from accumulation in soil, CA.

    Chlorosis, alkali injury. Mineral deficiency, AZ, CA.

    Exanthema. Copper deficiency, CA.

    Gummosis, sour sap. Adverse soil and moisture conditions, AZ, CA, NJ, WA.

    Internal Browning, fruit cracking. Boron deficiency, WA.

    Little Leaf. Zinc deficiency, CA, OR, WA.

    Mottle Leaf. Manganese deficiency, CA.

Powdery mildew. **Podosphaera oxyacanthae,** CA, IA; **Sphaerotheca pannosa,** NY, CA.

Rot, blossom end: fruit spot. **Alternaria** sp., and **A. citri,** CA.

Rot, brown; blossom blight. **Monilinia fructicola,** widespread on ripe fruit.

Rot, green fruit; gummosis. **Botrytis cinerea,** CA.

Rot, green fruit; twig blight. **Sclerotinia sclerotiorum,** CA; **Lambertella pruni,** CA.

Rot, heart. **Schizophyllum commune,** after freezing, TX, WA; **Trametes hispida,** CO.

Rot, mushroom root. **Armillaria mellea,** CA, TX; **Clitocybe tabescens,** FL.

Rot, root. **Phymatotrichum omnivorum,** AZ, OK, TX.

Rot, silver leaf. **Stereum purpureum,** CA.

Rust, **Tranzchelia discolor** and **T. pruni** (II, III), CA, MS, NM, TX.

Scab; freckle; twig canker. **Cladosporium carpophilum,** widespread.
Virus. **Apricot Gummosis,** WA; **Apricot Pucker Leaf,** UT; **Apricot Ring Pox,**
CA, CO, WA; **Peach Mosaic,** Southwest; **Peach Phony,** GA; **Peach Ring
Spot,** western U.S.; **?Peach Rosette; Peach Yellow Bud Mosaic,** CA; **Peach
Yellows,** occasional in East; **Stem Pitting.**
Wilt. **Verticillium albo-atrum,** CA, UT, WA.

Apricots are very susceptible to Armillaria root rot and should be on
resistant Myrobalan rootstock. Bacterial canker is an epidemic disease in many
seasons, with activity starting in late autumn, ceasing in early summer. Sprays
are usually needed in California for zinc deficiency, brown rot, green or jacket
rot, shot hole. Write to the California Agricultural Experiment Station for a
special spray schedule.

AQUATIC PLANTS
(*Caboriaba* sp., *Limnophila* sp., *Replis diandra*, and *Potamogeton* sp.)

Nematode, foliar. **Aphelenchoides fragariae,** FL.

ARABIS
(Rock-Cress)

Blight, gray mold. **Botrytis cinerea,** WA.
Club root. **Plasmodiophora brassicae.**
Damping-off. **Rhizoctonia solani,** NJ.
Downy mildew. **Peronospora parasitica,** AL, CO, IN, MI, TX, WI, AK.
Leaf spot. **Septoria arabidis,** CO, TX.
Rot, root. **Phymatotrichum omnivorum,** TX.
Rust. **Puccinia thlaspeos** (*P. holboelli*) (0, III), on numerous native but not
cultivated species in Rocky Mountains and Pacific states; **P. monoica** (0, I)
on native species, WI to CO, NM, CA, WA; II, III on *Koehleria* and *Trisetum*.
White rust; white blister. **Albugo candida,** NY to VA, TX, CO, WA.

ARALIA, HERCULES CLUB
(*Aralia spinosa*)

Canker, dieback. **Botryosphaeria ribis,** GA, VA, WV; **Nectria cinnabarina,**
VA, WV.
Leaf spot. **Cercospora atromaculans,** LA, TX; **Phyllosticta araliae,** TX;
P. everhartii, TX, WV, **Stagonospora** sp., WV.
Rot, root. **Phymatotrichum omnivorum,** TX; **Rhizoctonia solani,** TX.
Rot, sapwood. **Polyporus tulipiferus,** MD.
Spot anthracnose, scab. **Sphaceloma araliae,** MD, MO.

ARALIA, MING
(*Polyyscias*)

Blight. **Alternaria panax,** FL.

ARALIA, SARSAPARILLA, AMERICAN SPIKENARD
(*Aralia hispida, A. nudicaulis, A. racemosa*)

Leaf spot. **Alternaria** sp., NY; **Ascochyta marginata,** WI; **Cercospora leptosperma,** IA, MI, NY, WI; **Phyllosticta decidua,** WI; **Ramularia repens,** WI; **Sclerotium deciduum,** WI.
Powdery mildew. **Phyllactinia corylea,** MI, NE.
Rust. **Nyssospora clavellosa** (III), MN, OR, TX, CA, ME to PA.
Wilt. **Verticillium albo-atrum,** NY.

ARALIA, UDO
(*Aralia cordata*)

Blight. **Alternaria** sp., DE, NJ.
Rot, stem. **Sclerotinia sclerotiorum,** MD.
Wilt. **Verticillium albo-atrum,** MD, PA.

ARAUCARIA
(Monkey-Puzzle, Norfolk-Island-Pine)

Bacterial crown gall. **Agrobacterium tumefaciens,** CA.
Blight. **Cryptospora longispora.**
Leaf spot. **Pestalotia funerea,** CA, and **P. micheneri,** PA, probably secondary; **Stictis araucariae,** CA.

ARBORVITAE
(*Thuja*)

Blight, fire. **Cercospora thujina,** on oriental arborvitae, AR, LA, MS, TX, GA.
Blight, gray mold. **Botrytis cinerea,** NJ.
Blight, leaf. **Didymascella thujina,** TX, VT to WI.
Blight, nursery. **Phomopsis juniperovora,** IN, KY, OH, PA, VA; **P. occulta,** secondary.
Blight, snow. **Phacidium infestans,** Northeast.
Blight, twig. **Coryneum berckmanii** on oriental arborvitae, OR, WA.
Blight, twig. **Alternaria** sp.; **Mycosphaerella** sp; **Pestalotia funerea; Phytophthora** sp.; **Heterosporium** sp., OR.
Canker, bark patch. **Aleurodiscus amorphus; A. nivosus,** other species.
Canker, cypress. **Coryneum cardinale,** CA.

Canker; dieback. **Diplodia** sp., AL, FL.

Damping-off. **Rhizoctonia solani,** NM, NY, TX, VA.

Needle cast. **Lophodermium thujae,** ME, NH, NY, WI.

Rot, mushroom root. **Armillaria mellea,** MI, MS, NY, TX; **Clitocybe tabescens,** FL, LA.

Rot, root. **Phymatotrichum omnivorum,** TX; **Fusarium solani,** TX.

Rot, wood. **Fomes annosus,** MI; **F. roseus,** ME; **Poria vaporaria, P. weirii,** Great Lakes states; **Lenzites saepiaria,** MN; **Polyporus** spp.; **Schizophyllum commune,** ME; **Trametes** spp.

Giant arborvitae used for timber has many more wood rots than those listed. Oriental arborvitae in the South often looks as if blighted by fire, with nursery losses higher than in gardens. A copper spray, monthly from June to September, controls this blight. In the Northwest **Coryneum berckmanii** causes discoloration and shedding of branches, while **C. cardinale** is sporadically injurious.

ARCTOTIS

Leaf blotch. **Cercospora** sp., FL.

Nematode, root knot. **Meloidogyne** sp., FL.

Rot, root. **Phymatotrichum omnivorum,** TX.

ARDISIA

Leaf spot, algal. **Cephaleuros virescens,** FL.

ARGYREIA

Nematode, root knot. **Meloidogyne** sp., SC.

Rot, root. **Phymatotrichum omnivorum,** TX.

ARMERIA
(Sea-Pink, Thrift)

Rust. **Uromyces armeriae** (0, I, II, III), CA, OR.

ARNICA

Leaf spot. **Ovularia hughesiana,** MT; **Phyllosticta arnicae,** CO, MT, UT, WY.

Powdery mildew. **Erysiphe cichoracearum,** CO; **Sphaerotheca macularis** (*S. humuli* var. *fuliginea*), CA, WA, WY, AK.

Rust. **Puccinia arnicalis** (II, III) AK, to MT, CO, CA; 0, I unknown; **Uromyces junci** (0, I), CA, CO, MT, OR, SD, WY; II, III on *Juncus* spp.
Smut, white. **Entyloma arnicale,** CO, ID, MT, UT, WA, WY, AK.

ARROW-ARUM
(*Peltandra*)

Leaf spot. **Cercospora callae,** AL, DE, FL, NY; **Colletotrichum** sp., AL; **Gloeosporium paludosum,** DE, IN, MA, NY; **Pestalotia aquatica,** secondary; **Ramularia** sp., MI.
Rust. **Uromyces ari-triphylli** (*U. caladii*) (0, I, II, III), MA to FL, IL, IN, IA.

ARROWHEAD
(*Sagittaria*)

Leaf spot. **Cercospora alismatis,** VT to AL, TX, WI; **Didymaria alismatis, Gloeosporium confluens,** IA, MA, TX, WI; **Marssonina** sp., LA.
Smut, leaf. **Burrillia pustulata,** IL, NE, WI; **Doassansia deformans,** MA to NJ, MO, SD, FL, TX; **D. furva,** WI; **D. intermedia,** IA, MN, NH, ND, WI; **D. obscura,** CT, MA; **D. opaca,** MA to DE, IL, WI; **D. sagittariae,** CT to KS, TX, AR, MT, WY.

ARROWROOT
(*Maranta*)

Leaf spot. **Glomerella cingulata,** MD, NJ.
Rust. **Puccinia thaliae** (*P. cannae*) (II, III), FL.

ARTEMISIA
(Wormwood)

Blight, leaf. **Cercospora olivaceae,** ND, NJ, NY; **Systrema artemisiae,** black pustule, PA.
Downy mildew. **Peronospora leptosperma,** IA, MN, ND, SD, WI.
Gall, leaf. **Synchytrium aureum,** WI.
Leaf spot. **Cercospora ferruginae,** NY, WI; **Gloeosporium heterophyllum,** CA.
Nematode, root knot. **Meloidogyne** sp., FL.
Powdery mildew. **Erysiphe cichoracearum,** CA, SD, IA, TX.
Rot, root. **Phymatotrichum omnivorum,** TX.
Rust. **Puccinia tanaceti** (*P. absinthii*) (0, I, II, III), CA; **P. atrofusca** (0, I) ND; II, III on *Carex* spp.; **P. millefolii** (III), CA.
White rust. **Albugo tragopogonis,** IA, MT, ND, SD, TX, WI.

ARTICHOKE, GLOBE
(*Cynara scolymus*)

Blight, gray mold. **Botrytis cinerea,** CA, NY.
Blight, southern. **Sclerotium rolfsii,** GA.
Leaf spot. **Cercospora obscura,** CA, TX: **Cladosporium** sp., CA, SC; **Ramularia cynarae,** CA, prevalent.
Nematode, root knot. **Meloidogyne** sp., CA.
Powdery mildew. **Erysiphe cichoracearum,** CA, NJ.
Rot, root. **Phytophthora megasperma,** CA; **Phymatotrichum omnivorum,** TX.
Rot, root, stem. **Rhizoctonia solani,** MS, TX. **Sclerotinia sclerotiorum,** OR.
Virus. **Artichoke Curly Dwarf.**

ARTICHOKE, JERUSALEM
(*Helianthus tuberosus*)

Bacterial leaf spot. **Pseudomonas syringae pv. tagetis,** MN
Bacterial spot. **?Pseudomonas helianthi,** IL.
Blight, southern. **Sclerotium rolfsii,** FL, LA, MS, SC, TX.
Downy mildew. **Plasmopara halstedii,** VT and NJ to KS, SD.
Leaf spot. **Cercospora helianthi,** KS; **Septoria helianthi,** IL, IA, WI.
Nematode, root knot. **Meloidogyne** spp., CA, FL, MD, NY, SC, TN.
Powdery mildew. **Erysiphe cichoracearum,** general.
Rot, root. **Phymatotrichum omnivorum,** TX, AZ; **Rhizopus stolonifer,** MS.
Rot, stem, wilt. **Sclerotinia sclerotiorum,** MA, MN, WA.
Rust. **Coleosporium helanthi** (II, III), AL, IL, NY, NC, SC, OK, PA, TN, VA; **Puccinia helianthi** (0, I, II, III), general; **Uromyces junci,** (0, I), NE, ND.

ARTILLERY PLANT, ALUMINUM PLANT
(*Pilea*)

Bacterial leaf spot. **Xanthomonas campestris,** FL.
Blight, leaf, stem, bud. **Phytophthora parasitica,** FL.
Leaf spot. **Myrothecium roridum,** FL; **Septoria pileae,** IL, IN, IA, MI, MO, NY, WI.
Nematode, root knot. **Meloidogyne** sp., FL.
Rot. **Rhizoctonia** sp., FL.

ARUNDO
(Giant Reed)

Leaf spot; cane "Anthracnose"; dieback. **Papularia sphaerosperma,** general. **P. odorae,** KS; **Phyllosticta tuberosa,** OK.

Leaf spot; stem speckle.. **Selenophoma donacis,** CA.

Rot, root. **Armillaria mellae,** MD

Rust. **Puccinia chloridis** (*P. bartholomaei*) (0, I), OK, TX, KS; II, III, on *Bouteloua* spp.; **Uromyces asclepiadis** (II, III), KS, NM, TX.

Rust, crown. **Puccinia coronata,** CA.

ASH
(*Fraxinus*)

Anthracnose, leaf scorch. **Gloeosporium aridum,** eastern and central states.

Bacterial decline. **MLO,** IA, IN, IL, NY.

Bacterial hairy root. **Agrobacterium rhizogenes,** IA, NE to OK.

Black mildew. **Dimerosporium pulchrum** (*Sarcinella heterospora*).

Blight, seedling. **Rhizoctonia solani,** OK.

Canker, bark patch. **Aleurodiscus** spp., eastern states; **Felt Fungus, Septobasidium** spp., NC to FL.

Canker, branch, trunk. **Dothiorella fraxinicola,** IA, KS, NE; **Sphaeropsis** sp., widespread; **Nectria cinnabarina; N. coccinea,** Northeast.

Canker; dieback. **Cytospora annularis,** north central states; **Diplodia infuscans,** Northeast; **Cytophoma pruinosa** and **Fusicoccum** sp., may be secondary, Northeast.

Dodder. **Cuscuta** sp.

Leaf spot. **Cercospora fraxinites,** AL, FL, LA, TX; **Cylindrosporium fraxini** (including reports of *Marssonina fraxini, Piggotia fraxini*), widespread; **Mycosphaerella effigurata,** widespread; **M. fraxinicola; Actinopelte dryina; Cytospora** sp., VA.

Mistletoe. **Phoradendron serotinum (flavescens),** south central to Pacific states.

Mistletoe. **Viscum album,** CA, on *Fraxinus velutina* (Arizona ash).

Mistletoe. European. **Viscum album** on Arizona ash, CA.

Nematode. **Meloidogyne** spp., AZ, MD, OK.

Nematode, dagger. **Xiphinema americanum,** SD on green ash.

Powdery mildew. **Phyllactinia corylea,** northeastern, central, Pacific Coast states. ?**Uncinula circinata,** IA; **Microsphaera alni,** IL.

Rot, collar. **Helicobasidium purpureum,** TX.

Rot, root. **Phymatotrichum omnivorum,** AZ, TX.

Rot, sapwood. **Lentinus tigrinus,** MS.

Rot, white mottled heart, stem. **Fomes fraxinophilus,** eastern and central states to Great Plains; many other species of *Fomes*.

Rot, wood. **Daldinia concentrica,** cosmopolitan; **Daedalea** spp.; **Ganoderma lucidum,** MS, LA; **Polyporus** spp.; **Poria** spp.; **Schizophyllum commune,** cosmopolitan; **Trametes** spp.

Rust. **Puccinia sparganoides** (0, I), general east of Great Plains; II, III on marsh grass (*Spartina* spp.).

Virus. **Ash Ring Spot,** NY; **Tobacco Mosaic Virus,** MA, NY; **Tobacco Ring Spot Virus,** NY.
Wilt. **Verticillium albo-atrum,** CO.

Ash rust is epidemic in New England in many seasons, causing defoliation and sometimes death of trees. Anthracnose and leaf spots may be important in a wet season.

ASH, MORAINE
(*Fraxinus holotriocha*)

Virus. **Tobacco Mosaic,** NY; **Tobacco Ringspot,** NY.

ASPARAGUS
(*Asparagus officinalis*)

Anthracnose; canker. **Colletotrichum** sp., AL, CT, IL.
Bacterial soft rot. **Erwinia carotovora,** general.
Blight, ashy stem. **Macrophomina phaseoli,** TX.
Blight, branchlet; dieback. **Alternaria** sp., IL, MA, NY, OK, SC; **Stemphylium botryosum,** secondary; **Ascochyta** sp., DE, TX.
Blight, gray mold, shoot. **Botrytis cinerea,** CA, IL, MA, NY, WV.
Canker, stem spot. **Phoma asparagi.**
Damping-off, stem canker. **Rhizoctonia solani,** occasional.
Leaf spot. **Cercospora asparagi,** general; **Pleospora herbarum; P. allii** anamorph, **Stemphylium vesicarium,** OK, MI, WA.
Nematode, root knot. **Meloidogyne** sp., SC. Usually resistant.
Rot, crown. **Penicillium martensii,** WA; **Penicillium** sp., blue mold rot.
Rot, mushroom root. **Armillaria mellea,** OR.
Rot, root; stem wilt; decline. **Fusarium oxysporum** f. sp. **asparagi,** SC, CA, WA; **Fusarium** sp.
Rot, stem. **Diplodia asparagi; Phytophthora** sp., CA.
Rot, watery soft. **Sclerotinia sclerotiorum,** occasional in South.
Rust. **Puccinia asparagi** (0, I, II, III), general on susceptible varieties.
Rust is the most important asparagus disease, and resistant varieties are sometimes disappointing.
Virus. Asparagus I, NJ.

ASPARAGUS FERN
(*Asparagus plumosus*)

Canker, stem; blight. **Ascochyta asparagina,** FL, TX; **Phoma** sp., FL.
Leaf mold. **Cladosporium** sp., FL, MS, TX.

Nematode, root knot. **Meloidogyne** spp., FL.
Nematode, spiral. **Helicotylenchus nannus.**
Nonparasitic "rust." Cause undetermined, FL.
Rot, root; wilt. **Fusarium** sp., FL, WA.

ASPARAGUS, FLORISTS' SMILAX
(*Asparagus asparagoides*)

Blight, gray mold. **Botrytis cinerea**, AK.
Leaf spot. **Stagonospora smilacis**, WI.
Rot, root; wilt. **Fusarium** sp., NJ.

ASPARAGUS, SPRENGER
(*Asparagus sprengeri*)

Bacterial crown gall; fasciation. **Agrobacterium tumefaciens** (possibly con-
fused with *Clavibacter fascians*).
Nematode, root knot. **Meloidogyne** spp., CA.
Rot, root. **Rhizoctonia solani**, NY.

ASPIDISTRA

Blight, leaf. **Labrella aspidistrae**, IL, LA.
Leaf spot. **Ascochyta aspidistrae**, MN, NJ; **Colletotrichum omnivorum**, CA,
MO, NJ, PA, WV.

ASTER, CHINA
(*Callistephus*)

Bacterial, MLO. **Aster Yellows**, general.
Anthracnose. **Colletotrichum** sp., FL.
Blight, bud, stem; gray mold. **Botrytis cinerea**, CA, CT, IL, NJ, NY, PA, WI, AK.
Blight, southern. **Sclerotium rolfsii**, MS, NC.
Canker, stem. **Phomopsis callistephi**, IL, WI.
Dodder, **Cuscuta** spp., widespread.
Downy mildew. **Basidiophora entospora**, FL, TX.
Leaf spot. **Ascochyta asteris**, CA, NY, ND, OH; **Septoria callistephi**, AL, DE,
MI, MO, NJ, NY, OH, PA.
Mold, seed. **Pleospora herbarum**, cosmopolitan; **Alternaria** sp.
Nematode, root knot. **Meloidogyne** sp., CT, FL, TX, WA.
Powdery mildew. **Erysiphe cichoracearum**, DE, MN, NE, NC, VT, WA;
E. polygoni, NJ.
Rot, foot. **Phytophthora cryptogea; Pythium ultimum.**

Rot, root, stem; leaf blight. **Rhizoctonia solani,** widespread, chiefly in north-eastern and central states; **Phymatotrichum omnivorum,** TX.

Rust. **Coleosporium asterum** (*C. solidaginis*) (II, III), general except far South; II, III on two- and three-needle pines.

Virus. **Beet Curly Top,** OR.

Virus. **Tobacco Rattle** (aster strain of); **Tomato Spotted Wilt,** CA; **Biddens Mottle,** FL.

Wilt. **Verticillium** sp., CA, CT, IL, MA, NY.

Wilt, stem rot. **Fusarium oxysporum** f. sp. **callistephi,** general.

Fusarium wilt and aster yellows are the two principal diseases. Choose wilt-resistant varieties or plant in a new location. Rogue plants with yellows immediately; use systemic insecticides to control insect vectors.

ASTER, PERENNIAL
(*Aster* spp.)

Bacterial crown gall. **Agrobacterium tumefaciens,** CT.

Black knot. **Gibberidea heliopsidis,** NY, NE, ND, WI.

Blight, gray mold. **Botrytis cinerea,** CT, NJ, AK.

Dodder. *Cuscuta* spp., eastern and central states.

Downy mildew. **Basidiophora entospora,** IL, IN, MD, MO, NE, WI.

Gall, leaf. **Synchytrium nigrescens,** WI.

Leaf spot. **Alternaria** sp., MI, TX, VT; **Ascochyta compositarum,** WI; **Cercospora asterata,** AL, TX; **Cercosporella cana,** OR, WI; **Leptothyrium doelligeriae,** NY; **Ovularia asteris,** WY; **O. virgaureae,** CO, MS, WI; **Phyllachora sterigena,** KS, NE; **Ramularia asteris,** IA, MI, NE, TX, WI, WY; **Septoria angularis** and other species; **Phyllosticta astericola,** TX.

Leaf spot, black, tar. **Placosphaeria haydeni** (?*Discosphaerina pseudimantia*), NY, IA, NJ, ND.

Nematode, leaf. ?**Aphelenchoides ritzema-bosi,** CT.

Nematode, root knot. **Meloidogyne** sp.

Parasitic lichen. **Strigula elegans, S. camplanata,** LA, Southern U.S.

Powdery mildew. **Erysiphe cichoracearum,** general.

Rot, root. **Phymatotrichum omnivorum,** TX.

Rot, stem. **Sclerotinia sclerotiorum,** CT.

Rust. **Coleosporium asterum** (II, III); **Puccinia asteris** (III), general; **P. dioicae** var. **asteris** (0, I); II, III, on sedge; **P. grindeliae** (III), CO, KS, NV, WY; **P. stipae** (0, I), CO, IA, KS, NE, ND, SD; **Uromyces compactus,** AZ, NM, TX; **U. junci** (0, I), NH.

Smut, white. **Entyloma aster-seriaceanum,** WI; **E. compositarum,** ME, MA, SD, WI.

Virus. **Mosaic.** Unidentified, CA.

Virus. **Tomato Spotted Wilt,** CA.

Wilt. **Verticillium albo-atrum,** CT.

ASTILBE

Powdery mildew. **Erysiphe polygoni,** MA.
Wilt. **Fusarium** sp., WA.

AUCUBA
(Gold-Dust Tree)

Anthracnose; leaf spot. **Gloeosporium** sp. (?*Glomerella cingulata*), NJ, PA,
 SC, WA. Probably includes reports of *Colletotrichum pollaci*, MS, NJ.
Leaf spot. **Pestalotia aucubae,** secondary; **Phyllosticta aucubae,** CA, MS, SC.
Rot, root. **Phytophthora cinnamomi, P. citricola,** NC.
Virus. **Tobacco Mosaic,** Aucuba strain.
Wilt. **Verticillium albo-atrum,** NJ.

AUTUMN CROCUS
(*Colchicum*)

Blight; leaf spot. **Botrytis eiliptica,** WA.
Smut, leaf. **Urocytis colchici,** DE, NY, OH, PA, WA.
Virus. **Tobacco Ring Spot Virus,** MD.

AVOCADO
(*Persea americana*)

Anthracnose; leaf and fruit spot; black rot. **Glomerella cingulata,** general.
Bacterial crown gall. **Agrobacterium tumefaciens,** CA.
Bacterial fruit blast. **Pseudomonas syringae,** CA.
Black mildew. **Irene perseae,** FL.
Blight, seedling. **Sclerotium rolfsii,** FL; **Phytophthora palmivora,** FL.
Blotch, fruit spot. **Cercospora perseae,** FL; **C. purpurea.**
Canker, branch; dothiorella fruit rot. **Botryosphaeria ribis** var. **chromogena,**
 CA.
Leaf spot, **Phyllosticta micropuncta,** AL, FL.
Leaf spot, algal; green scurf. **Cephaleuros virescens,** FL.
Leaf spot; fruit spot; seedling blight. **Pestalotia** spp., general.
Leaf spot, smudgy spot on twigs. **Helminthosporium** sp., CA.
Nematode, burrowing. **Radopholus similis,** FL.
Nematode, dagger. **Xiphinema americanum.**
Nematode, meadow. **Pratylenchus brachyurus; P. vulnus.**
Nematode, root knot. **Meloidogyne** sp., resistant to.

Nematode, stubby root. **Trichodorus christiei.**
Nonparasitic
  Carapace Spot. Abrasion of young fruits.
  Dieback. Copper deficiency.
  End Spot. Dessication of young fruits; overmaturity.
  Little Leaf; Rosette. Zinc deficiency.
  Melanorhiza. Defective drainage and aeration.
  Mottle Leaf. Nutritional deficiency.
  Tipburn. Sometimes salt accumulation in poorly drained soil.
Rot, blue mold. **Penicillium expansum, CA.**
Rot, collar; trunk canker. **Phytophthora cactorum** CA; **P. parasitica** FL;
  Sclerotinia sclerotiorum, CA.
Rot, fruit. **Diplodia theobromae, FL; Phomopsis** sp., FL, TX; **Phytophthora
  citrophthora, CA; P. citricola, CA; Rhizopus nigricans; Alternaria** sp.;
  Fusarium sp.
Rot, root. **Armillaria mellea, CA; Clitocybe tabescens, FL; Phymatotrichum
  omnivorum,** TX; Dematophora Rot, **Rosellinia necatrix.**
Rot, root; decline. **Phytophthora cinnamomi, CA; Pythium** spp., CA.
Rot, seed and root. **Rhizoctonia solani, CA.**
Spot anthracnose; scab. **Sphaceloma perseae, FL, TX.**
Virus. **Avocado Sun-Blotch, CA.**
Wilt. **Verticillium albo-atrum, CA, FL.**

Avocado scab may cause heavy damage in susceptible varieties of West
Indian stock. Cercospora blotch attacks both leaves and fruit. Decline, root
rot, is most serious in wet soils, killing trees if they are waterlogged 6 to 8 days.

AZALEA
(*Rhododendron*)

Anthracnose. **Colletotrichum** sp., LA.
Blight, bud and twig. **Briosia azaleae** (*Pycnostysanus*), MA, NH, NJ, NC.
Blight, cutting. **Cylindrocladium scoparium,** AL, FL, NY, OH; **C. floridanum.**
Blight, flower; seedling. **Botrytis cinerea,** often after frost, cosmopolitan;
  Alternaria tenuis; Cladosporium herbarum; Epicoccum purpurascens.
Blight, petal; flower spot. **Ovulinia azaleae,** MD, VA, NC, SC, GA, FL, AL,
  MS, LA, TX, CA, PA; also NY, NJ in greenhouses.
Blight, shoot. **Monilinia azaleae,** GA, MA, NY.
Blight, shoot and stem. **Phytophthora citrophthora, P. citricola, P. nicotianae**
  var. **parasitica,** OH, FL.
Blight, thread. **Pellicularia koleroga,** LA.
Canker, stem. **Gloeosporium** sp., OR. **Phomopsis** sp., SC.
Damping-off, leaf blight. **Rhizoctonia solani,** cosmopolitan.
Dodder. **Cuscuta** sp., FL, SC.

Gall, leaf; shoot hypertrophy. **Exobasidium vaccinii,** general; **E. burtii, Yellow Leaf Spot,** ID, OR, WA, NJ; **Synchytrium vaccinii,** NJ.

Leaf scorch; angular leaf spot. **Septoria azaleae,** widespread; severe in CA.

Leaf spot. **Cercospora rhododendri,** MD; **Pestalotia** spp., general but secondary; **Phyllosticta** sp.; **Ramularia angustata,** MS, NY; **Colletotrichum azaleae,** FL; **Septoria solitaria,** CA, OR, TX; Tar Spot, **Melasmia menziesii,** WA; **Corynespora cassicola,** GA, FL.

Nonparasitic. **Chlorosis,** usually iron deficiency, general in alkaline soils, sometimes defective drainage.

Nematode, leaf. **Aphelenchoides fragariae,** FL.

Nematode, root knot. **Meloidogyne incognita acrita.**

Nematode, spiral. **Helicotylenchus nannus; Rotylenchus robustus.**

Nematode, stubby root. **Trichodorus christiei; T. primitivus.**

Nematode, stunt. **Tylenchorhynchus claytoni.**

Powdery mildew. **Erysiphe polygoni,** CA, NJ; **Microsphaera alni,** GA, NJ, NY, PA, RI, VA.

Rot, root. **Armillaria mellea,** CA, NJ, WA; **Clitocybe tabescens,** FL; **Phymatotrichum omnivorum,** TX; **Pythium** spp.

Rot, root, stem; wilt. **Phytophthora cinnamomi,** AL, MD, MO.

Rust. **Pucciniastrum myrtilli** (II, III), ME to FL, TX; 0, I, on hemlock.

Wilt. **Cylindrocarpon radicicola,** MA.

Azalea flower spot or petal blight devastates azalea blooms in the South in rainy or foggy weather starting when buds show color. Bud and twig blight has killed some shrubs in Massachusetts. Leaf galls are unsightly but not too serious. Leaf scorch may be prevalent in a rainy season. Powdery mildew is common on deciduous azaleas in late summer.

AZARA

Rot, stem. **Sclerotium rolfsii,** CA.

BABIANA

Virus. **Iris Mosaic,** CA.

BALD CYPRESS
(*Taxodium*)

Blight, twig. **Pestalotia funerea,** TX.

Canker, felt fungus. **Septobasidium** spp., LA.

Rot, brown pocket heart. **Fomes geotropus,** cause of "pecky cypress," FL to LA; **F. extensus; Ganoderma applanatum.**

Rot, root. **Phymatotrichum omnivorum,** TX.
Rot, wood. **Lenzites** spp., **Polyporus** spp., **Poria** spp.

BALM
(*Melissa*)

Blight, gray mold. **Botrytis cinerea,** AK.
Leaf spot. **Phyllosticta decidua,** NY.

BALSAM-APPLE, BALSAM-PEAR
(*Momordica*)

Anthracnose. **Colletotrichum lagenarium,** IN.
Downy mildew. **Pseudoperonospora cubensis,** IA.
Leaf spot; blight. **Ramularia momordicae,** TX.
Nematode, burrowing. **Radopholus similis,** FL.
Nematode, root knot. **Meloidogyne** sp., FL.
Powdery mildew. **Erysiphe cichoracearum,** WI.

BALSAM-ROOT
(*Balsamorhiza*)

Leaf spot. Septoria sp., WA.
Nematode, leaf gall. **Anguina balsamophila,** UT.
Powdery mildew. **Erysiphe cichoracearum,** WY.
Rust. **Puccinia balsamorhizae** (0, I, II, III), general.

BAMBOO
(*Bambusa, Phyllostachys*)

Black mildew. **Meliola tenuis.**
Blight, tip. **Diplodia bambusae,** LA, TX.
Leaf mold. **Cladosporium gramineum,** OR, SC.
Leaf spot. **Helminthosporium** sp., FL; **Mycosphaerella** sp., CA; culm spot,
   Selenophoma donacis, CA.
Nematode, burrowing. **Radopholus similis,** FL.
Rust. **Puccinia melanocephala** (II, III), FL, GA, MS, TX; **Uredo ignava** (II),
   GA, FL.
Rust. **Uromyces costaricensis,** FL. on wild bamboo (*Lasiacis divaricata*).
Smut. **Ustilago shiraiana,** CA, FL, LA, TX.
Virus. **Sugarcane Mosaic Virus,** LA, on *Arundinaria gigantea*.

BANANA, DWARF
(*Musa nana*)

Anthracnose; fruit rot. **Gloeosporium musarum**, FL, TX.
Bacterial leaf blight. **Pseudomonas solanacearum**, FL.
Nematode, burrowing. **Radopholus similis**, FL, LA.
Nematode, lesion. **Pratylenchus musicola**, FL.
Nematode, root knot. **Meloidogyne** spp., FL, TX.
Rot, root. **Clitocybe tabescens**, FL.
Virus. **Cucumber Mosaic**, FL.
Wilt. **Fusarium oxysporum** f. sp. **cubense**, FL.

BANEBERRY, COHOSH
(*Actaea*)

Leaf spot. **Ascochyta actaeae**, WI; **Ramularia actaeae**, CO, IA, NM, VT, WI.
Rust. **Puccinia recondita** (0, I), NY to VA, IL, WA; II, III, on grasses.
Smut, leaf and stem. **Urocystis carcinodes**, ID, PA, WV, UT.

BAPTISIA
(False Indigo)

Leaf spot. **Cercospora velutina**, KS, IL, WI; **Marssonina baptisiae**, IA;
   Septoria baptisiae, SC, TX; **Stagonospora baptisiae**, SC.
Powdery mildew. **Erysiphe polygoni**, prevalent; **Microsphaera alni**, WI.
Rot, root. **Armillaria mellea**, CA; **Phymatotrichum omnivorum**, TX.
Rust. **Puccinia andropogonis**, KS, NE, NC, OK.

BARBERRY
(*Berberis*)

Bacterial leaf spot. **Pseudomonas berberidis**, general.
Blight, gray mold. **Botryris cinerea**, CA, MO.
Canker; dieback. **Botryosphaeria ribis**, FL.
Damping-off. **Pythium debaryanum**, CA.
Leaf spot. **?Gloeosporium berberidis**, CT, MA, MN, OH, WI; **Phyllosticta berberidis**, KY.
Nematode, root knot. **Meloidogyne hapla**.
Nematode, spiral. **Rotylenchus buxophilus**.
Powdery mildew. **Erysiphe polygoni**; **Phyllactinia corylea**, MA, VT.
Rot, heart. **Poria punctata**, MD.
Rot, root, **Phymatotrichum omnivorum**, TX.
Rust. **Cumminsiella sanguinea** (0, I, II, III), OR; **Puccinia koehleriae** (0, I);
   II, III on Koehleria; **P. montanensis** (0, I); II, III, on grasses.

Rust, wheat. **Puccinia graminis** (0, I), general; II, III, on cereals and grasses.
Wilt. **Verticillium albo-atrum,** CT to VA, IL, and MI.

All interstate movement of barberry is under quarantine because of the wheat rust, but resistant cultivars may be shipped under permit. Common barberry is eradicated near wheat fields; Japanese barberry is resistant to rust.

BARREN-STRAWBERRY
(*Waldsteinia*)

Leaf spot. **Ramularia waldsteiniae,** WI; **Septoria waldsteiniae,** MI, NY, VT.
Rust. **Puccinia waldsteiniae,** ID, MI, NY, VT, WI.
Smut. **Urocystis** (*Whetzella*) **waldsteiniae,** NY, WI.

BASIL
(*Ocimum*)

Nematode, root knot. **Meloidogyne** sp., FL; **M. incognita acrita,** CA.
Virus. **Bromegrass Mosaic Virus,** general, IA, SD.

BAUHINIA
(Orchid-Tree, Mountain Ebony)

Leaf spot. **Colletotrichum** sp., TX; **Phyllosticta** sp., FL.
Powdery mildew. **Microsphaera diffusa,** MD.

BAYBERRY
(*Myrica carolinensis*)

Leaf spot. **Mycosphaerella myricae,** GA, MS; **Phyllosticta myricae,** NY.
Rust. **Gymnosporangium ellissii** (0, I), MA to NY, VA; III, on *Chamaecyparis.*
Virus. **Bayberry Yellows,** NJ.

BEAN, ADZUKI
(*Phaseolus angularis*)

Bacterial stem rot. **Pseudomonas adzukicola,** MN.

BEAN, ASPARAGUS, YARDLONG
(*Vigna sesquipedalis*)

Bacterial spot. **Pseudomonas syringae,** IN, NY.
Leaf spot. **Mycosphaerella cruenta,** VA.

Leaf spot; pod spot. **Cladosporium vignae,** IN.

Powdery mildew. **Erysiphe polygoni,** CA.

Virus. **Potato Virus X,** work at the University of Wisconsin on *V. unguiculata* spp. *cylindrica* "Catjang," *V. unguiculata* spp. *sesquipedalis, V. unguiculata* spp. *unguiculata.*

BEAN, KIDNEY, LIMA
(*Phaseolus vulgaris, P. limensis*)

Anthracnose. **Colletotrichum lindemuthianum,** general in East; **C. truncatum,** PA to AL, TX, IA.

Bacterial blight. **Xanthomonas phaseoli,** general in East, rare Pacific Coast.

Bacterial halo blight; grease spot. **Pseudomonas syringae** pv. **phaseolicola,** general in East, rare on Pacific Coast.

Bacterial northern wilt. **Clavibacter flaccumfaciens,** Northeast.

Bacterial soft rot. **Erwinia carotovora,** cosmopolitan in market.

Bacterial southern wilt; brown rot. **Pseudomonas solanacearum,** AL, FL, GA, OK.

Bacterial spot, leaf and pod. **Pseudomonas syringae,** widespread.

Bacterial, stem rot. **Pseudomonas adzukicola,** MN.

Bacterial "stickiness." **Pseudomonas coadunata,** CA; **P. ovata,** VA.

Bacterial "wildfire." **Pseudomonas syringae** pv. **tabaci,** MA, NC.

Blight, ashy stem; charcoal rot. **Macrophomina phaseoli,** MD to GA, TX, CO, CA.

Blight, gray mold. **Botrytis cinerea,** occasional.

Blight, pod. **Diaporthe phaseolorum,** CT to FL, LA, OK, OH on lima bean.

Blight, southern. **Sclerotium rolfsii,** VA to FL, TX, AR, CA.

Damping-off. **Pythium ultimum; P. debaryanum; Rhizoctonia solani.**

Downy mildew. **Phytophthora phaseoli,** on lima bean, East, central states.

Leaf and pod spot. **Ascochyta boltshauseri,** OR, NC, PA; **A. phaseolorum,** WA.

Leaf spot. **Alternaria** spp.; **Aristastoma oeconomicum,** VA, GA; **Cercospora canescens,** southeastern states to NY; **Epicoccum** sp., secondary on lima bean; **Isariopsis griseola,** Angular Leaf Spot, ME to FL, TX, OK; **I. laxa,** IN, NJ; **Mycosphaerella cruenta,** NJ to FL, AR, TX, WI; **Phyllosticta phaseolina,** NY to FL, IN, TX; **Stemphylium botryosum; Stagonospora phaseoli,** TN.

Leaf spot, zonate. **Cristulariella pyramidalis,** WV.

Nematode, awl. **Dolichodorus heterocephalus.**

Nematode, hop cyst. **Heterodera humudi,** OR; **H. glycines,** soybean cyst, IL.

Nematode, root knot. **Meloidogyne arenaria; M. hapla; M. incognita; M. javanica;** root lesion, **Pratylenchus pratensis.**

Nematode, stubby root. **Trichodorus christiei.**

Nematode, stunt. **Tylenchorrynchus claytoni.**

Nonparasitic

Baldhead. Mechanical injury to seed growing point.

Blossom Drop. High temperature, low humidity.

Bronze Leaf. Excessive salt concentration, CO, MT, WY.

Chlorosis. Deficiency of copper, FL; magnesium, FL, MA, SC, VA; manganese, FL; zinc, FL, CA. Soil alkalinity, West.

Ozone Injury. Air pollution.

Seed Pitting. Plant bug injury.

Variegation. Genetic leaf abnormality.

Wind Whip.

Pod spot, seed mold. **Cladosporium herbarum,** on lima bean, CA, FL.

Pod spot, yeast spot. **Nematospora phaseoli,** on lima bean; **Pullularia pullulans,** NY.

Powdery mildew. **Erysiphe polygoni,** East, South, CA: **Microsphaera diffusa,** GA, MD, IL; **M. euphorbiae,** IL.

Rot, black root. **Thielaviopsis basicola,** AL, CA, NH, NJ, NY.

Rot, brown stem. **Cephalosporium gregatum,** IL.

Rot, root. **Fusarium solani** f. **phaseoli,** general; **Phoma terrestris,** secondary, CA; **Phymatotrichum omnivorum,** AZ, OK; **Pythium aristorum,** WI; **P. catenviatum,** WI; **P. dissotocum,** WI; **Pythium** spp., widespread; **Aphanomyces eutiches,** NY.

Rot, root and stem; web blight. **Pellicularia filamentosa,** general.

Rot, soft. **Rhizopus stolonifer** and **Pythium aphanidermatum,** leak, of market beans.

Rot, stem; wilt. **Sclerotinia sclerotiorum,** general, especially in South and West; **S. minor; S. ricini; S. intermedia,** in market beans. **Pythium myriotylum,** FL, GA, SC.

Rust. **Uromyces phaseoli** var. **typica,** II, III, general; 0, I, rare.

Spot anthracnose; lima bean scab. **Sphaceloma phaseoli,** NC.

Virus. **Alfalfa Mosaic,** WA.

Virus. **Bean Common Mosaic,** WA.

Virus. **Bean Mosaic,** general; **Southern Bean Mosaic,** CA, CO, LA, MD, MI; **Bean Pod Mottle; Bean Phyllody; Bean Red Node;** CO, ID, WY; **Bean Yellow Mosaic,** ND; **Bean Yellow Stipple,** IL; **Bean Yellow Dot** = **Alfalfa Mosaic,** WA; **Beet Curly Top,** West; **Cucumber Mosaic; Tobacco Necrosis; Tobacco Ring Spot.**

Virus. **Bromegrass Mosaic,** general, IA, SD, on broad bean (*Vicia fava* var. *minor*).

Virus. **Lima Bean Mild Mottle,** GA.

Virus. **Lettuce Mosaic,** NY.

Virus. **Potato Virus X,** work at the University of Wisconsin on *P. lathyroides, P. lunatus, P. aboriginens, P. vulgaris* "Pinto"; **Peanut Stunt,** AR.

Wilt; yellows. **Fusarium oxysporum** f. sp. **phaseoli,** CA, CO, ID, MT; **F. oxysporum** f. sp. **vasinfectum,** FL, AL.

Anthracnose and bacterial blight, often erroneously called "rust," are common and destructive bean diseases best avoided by purchasing healthy seed grown in disease-free arid sections of California and the Northwest. True rust is prevalent in the Southwest and sometimes in the East on susceptible Kentucky Wonder pole beans. Use resistant varieties; clean and disinfest poles each season.

Downy mildew is common on lima beans in moist summers. Resistant varieties are the best solution to virus problems. Avoid picking or cultivating beans when foliage is wet.

BEAN, MUNG
(*Phaseolus aureus*)

Rot, sprout. **Cylindrocephalum** sp., WA.

BEAN, SCARLET RUNNER
(*Phaseolus coccineus*)

Anthracnose. **Colletotrichum lindemuthianum**, NY.
Bacterial blight. **Xanthomonas phaseoli**, IN, NJ, TX.
Bacterial halo spot. **Pseudomonas phaseolicola**, NY.
Leaf spot. **Ascochyta boltshauseri**, OR; **Cercospora cruenta**, AL.
Powdery mildew. **Erysiphe polygoni**, CA.
Rot, root. **Fusarium solani** f. **phaseoli**, NY.
Rust. **Uromyces phaseoli** (II, III), AL, MA, NH.

BEAN, TEPARY
(*Phaseolus acutifolius*)

Blight, Southern. **Sclerotium rolfsii**, AL.
Powdery mildew. **Erysiphe polygoni**, CA.
Rot, root. **Fusarium solani**, CA; **Phymatotrichum omnivorum**, TX.
Rust. **Uromyces phaseoli** (II, III), CA, TX.
Virus. beet curly top, CA.

BEAN, URD
(*Vigna*)

Virus. **Cowpea Mosaic**, NY

BEARBERRY
(*Arctostaphylos uva-ursi*)

Black mildew. **Asterina gaultheriae**, WI; **A. conglobata**, ME.
Gall, leaf; red leaf spot. **Exobasidium vaccinii**, CO, ID, MT, MA, NJ, NY, VT, WI.
Leaf spot. **Cercospora gaultheriae**, WI.

Rust. **Chrysomyxa arctostaphyli** (III), CO, MT, UT, WI, AK; **Puccinia sparsum** (II, III), WI.

BEAUTY-BUSH
(Kolkwitzia)

Leaf spot. **Cercospora kolkwitziae,** AL, OK.

BEECH
(Fagus)

Broomrape. **Conophilis americana; Epifagus virginiana,** beechdrops.
Canker, beech bark disease. **Nectria coccinea** var. **faginata,** associated with woolly beech scale, destructive northern New England and New York; **N. galligena,** New England, NY, WV; **Cryptococcus fagesuiga,** VA, OH.
Canker, bleeding. **Phytophthora cactorum.** MA, NY, RI.
Canker, dieback. **Cytospora** spp.; **Strumella coryneoides,** New England; **Phomopsis** sp., MI, NY, PA.
Canker, felt fungus. **Septobasidium** spp.; **S. cookeri, S. curtisii,** FL, NC.
Canker, perennial. **Endothia gyrosa,** NY.
Leaf spot. **Cercospora** sp., NY; **Gloeosporium fagi,** CT to WI; **Microstroma** sp., IL, NJ; **Phyllosticta faginea,** MA to WV.
Mistletoe. **Phoradendron serotinum (flavescens),** occasional VT to IN and southward.
Nonparasitic. **Leaf Scorch,** common in Northeast but cause unknown, associated with high temperature and water deficiency. **Leaf Mottle,** cause unknown, virus suspected, CT, NJ, NY, PA.
Powdery mildew. **Microsphaera alni,** MA to AL, WI; **Phyllactinia corylea,** New England to AL, IL and WI.
Rot, heart, butt, wound. **Fomes** spp.; **Ganoderma applanatum; Polyporus** spp.; **Schizophyllum commune; Stereum** spp.
Rot, mushroom root. **Armillaria mellea,** CT, NJ, NY, OH.
Rot, sapwood. **Hericium** spp.; **Ustulina deusta.**
Rot, wood. **Daedalea** spp.; **Daldinia** spp.; **Lenzites betulina,** widespread.
Sooty mold. **Scorias spongospora,** MA to AL and MO.

BEET
(Beta vulgaris)

Bacterial, symtomless. **Clavibacter sepedonicum,** ND.
Bacterial black streak. **Pseudomonas syringae** pv. **aptata,** CA, OR, UT, WA.
Bacterial crown gall. **Agrobacterium tumefaciens,** occasional.
Blight, southern. **Sclerotium rolfsii,** NC to FL, TX, AZ, CA.
Damping-off. **Pythium** spp.; **Rhizoctonia solani,** general.
Dodder. **Cuscuta** spp., occasional when beets follow legumes.

Downy mildew. **Peronospora schachtii,** CA, MN, NJ, NY, OR, WA.

Leaf spot. **Alternaria** sp., general, secondary; **Gloeosporium betae,** MS, MT, **Ramularia beticola,** CA, OR, WA; **Septoria betae,** DE, IN, MA, OH.

Leaf spot; blight. **Cercospora beticola,** general.

Nematode, leaf and stem. **Ditylenchus dipsaci,** KS.

Nematode, root knot. **Meloidogyne arenaria; M. javanica; M. chitwoodi,** Pacific NW; **M. hapla,** Pacific NW; Sugar beet, **Heterodera schachtii;** clover cyst, **H. trifolii;** gall, **Naccobus batatiformis.**

Nematode, sheath. **Hemicycliophora obtusa.** Sting, **Belonolaimus gracilis;** stubby root, **Trichodorus christiei;** lesion, **Pratylenchus penetrans.**

Nonparasitic

Black Heart. Boron deficiency, general; phosphorus deficiency, occasional.

Bronzing. Potassium deficiency occasional.

Chlorosis. Alkalinity, iron, manganese deficiency.

Girdle. Strangling constriction of tap root.

Tipburn. Black tips, from high nitrogen with low light intensity.

Powdery mildew. **Erysiphe polygoni,** CA.

Rot, black heart. **Phoma betae,** general, occasional as leaf spot.

Rot, black root. **Aphanomyces cochlioides,** general.

Rot, charcoal. **Macrophomina phaseoli,** CA.

Rot, crown. **Sclerotinia sclerotiorum,** CT, IL.

Rot, root. **Phytophthora drechsleri,** CA, CO, ID; **Physalospora rhodina,** AL; **Phymatotrichum omnivorum,** TX; **Pythium deliense,** TX, AZ.

Rot, violet root. **Helicobasidium purpureum,** occasional.

Rot, wound, storage. **Rhizopus** spp., cosmopolitan; **Cylindrocarpon radicola,** NY; **Penicillium** spp.; **Fusarium** spp.

Rust. **Puccinia aristidae** (0, I), CO, KS, NM, UT; II, III, on grasses; **Uromyces betae** (II, III), AZ, CA, NM, OR.

Scab. **Streptomyces scabies,** widespread.

Virus. **Beet Curly Top; Beet Latent; Beet Mosaic; Beet Pseudo-Yellows; Beet Ring Mottle; Beet Savoy; Beet Yellow Net; Beet Yellows; Beet Yellow Vein; Cucumber Mosaic; Beet Necrotic Yellow Vein,** CA, NE, TX, WA; **Beet Necrotic Yellow Vein** (Vector, *Polymyxa beta*), CA, TX, WA.

White rust. **?Albugo bliti,** IA, OH.

Wilt. **Verticillium albo-atrum,** CO.

Wilt; yellows. **Fusarium conglutinans (F. orthoceras)** var. **betae.**

Most of the troubles are more important for commercially grown sugar beets. Cercospora leaf spot or blight is common on garden beets although spraying is not often practical. Boron deficiency can be prevented by treating soil with borax.

BEGONIA

Bacterial crown gall. **Agrobacterium tumefaciens,** CT, MS, TX.

Bacterial spot. **Xanthomonas begoniae,** general.

Blight, gray mold. **Botrytis cinerea,** cosmopolitan.

Leaf spot, anthracnose. **Gloeosporium** sp., FL, LA, MA, MS, TX, NJ; ?**Cercospora** sp., FL, GA, MS, NJ, TX; **Penicillium bacillosporium,** ?secondary; **Phyllosticta** sp., NJ, PA.

Nematode, leaf. **Aphelenchoides fragariae,** cosmopolitan in greenhouses; lesion, **Pratylenchus** sp.

Nematode, root knot. **Meloidogyne** spp., cosmopolitan in gardens in the South, greenhouses in the North.

Nonparasitic. **Oedema,** a water-soaked spotting, frequent in house plants.

Powdery mildew. **Erysiphe cichoracearum,** CA; **Oidium** sp., FL, NC, CA; *Oidium begoniae,* OH.

Rot, root. **Armillaria mellea,** CA; **Thielaviopsis basicola,** MA, OH.

Rot, root and stem. **Pythium** spp., CA, MO; **Rhizoctonia solani,** cosmopolitan; **Sclerotinia sclerotiorum,** CA: **Sclerotium rolfsii,** IL.

Virus. **Tomato Spotted Wilt,** CA, MO, TX; **Broad Bean Wilt,** MN.

Wilt. **Verticillium albo-atrum,** CT, NY.

The dry air of the average living room keeps leaf diseases at a minimum. Overwatering may foster root rots and physiological oedema. In greenhouses botrytis blight, bacterial spot, leaf nematode, and pythium rots may become problems. Powdery mildew is serious on tuberous begonias.

BELLS-OF-IRELAND
(*Molucella*)

Rot, crown. **Myrothecium** sp., TX.

BIDENS
(*Bur-Marigold*)

Bacterial, MLO. **Aster Yellows,** CA.

Powdery mildew. **Sphaerotheca macularis,** general.

Rust. **Uromyces bidenticola** (0, I, II, III), CA, FL, NM.

Virus. **Bidens Mottle,** FL.

BIGNONIA
(*Crossvine*)

Black mildew. **Asterina bignoniae,** LA; **Meliola bidentata,** Gulf states; **M. furcata,** FL.

Leaf spot. **Cercospora capreolata,** AL, MS; **Leptostromella bignoniae,** TN.

Nematode, root knot. **Meloidogyne** sp.

Sooty mold. **Capnodium elongatum,** AL, MS.

Spot anthracnose. **Sphaceloma** sp., LA.

BIRCH
(*Betula*)

Anthracnose. Gloeosporium sp.; **Glomerella cingulata,** VA, on European white birch.

Canker, bark patch. **Aleurodiscus oakesii,** VT; **Solenia ochracea,** cosmopolitan.

Canker, bleeding, **Phytophthora cactorum,** NJ.

Canker, trunk. **Hymenochaete agglutinans,** MI, PA; **Nectria coccinea; N. galligena;** twig canker, **N. cinnabarina.**

Dieback. **Melanconis stilbostoma,** MA to IN on white birch; other species; **Cylindrosporium orthosporum.**

Leaf blister. **Taphrina americana,** witches' broom, NH, VT, WI; **T. carnea,** NH, ME; **T. flava,** ME, MA, NH, WI.

Leaf spot. Gloeosporium betularum; **G. betulae-luteae,** NY, PA; **Cylindrosporium betulae,** DE, NY, WI; **Phyllosticta betulina; Septoria betulicola; S. betulae.**

Mistletoe. **Phoradendron serotinum (flavescens),** FL, IN, TX; **Viscum album,** CA, on *Betula verrucosa* (weeping white birch).

Powdery mildew. **Microsphaera alni,** MA, NH, PA; **Phyllactinia corylea,** widespread.

Rot, heart. Fomes spp.; **Stereum purpureum; Polyporus hispidus.**

Rot, root. **Armillaria mellea; Phytophthora cinnamomi,** of seedlings; **Phymatotrichum omnivorum,** TX.

Rot, sapwood. **Polyporus betulinus,** general on gray and paper birches; **P. gilvus; Steccherinum** spp.

Rot, wood. **Polyporus** spp., **Stereum** spp.; **Daedalea unicolor** and other spp.

Rot, wound. **Schizophyllum commune; Pleurotus serotinus.**

Rust. **Melampsoridium betulina** (II, III); 0, I, on larch.

?Virus. Dieback, vein-clearing, New England.

BIRD-OF-PARADISE
(*Strelitzia*)

Bacterial leaf spot, blight. **Xanthomonas campestris,** FL.

BISCHOFIA

Leaf spot, algal. **Cephaleuros virescens,** FL.

BISHOPS-CAP
(*Mitella*)

Leaf spot. Cercospora mitellae, MI; **Phyllosticta mitellae,** NY, WI; **Ramularia mitellae,** IN, IL, MI, NY: **Septoria mitellae,** MI, WI.

Powdery mildew. **Sphaerotheca macularis,** WI.
Rot, leaf. **Sclerotium decidium,** WI.
Rust. **Puccinia heucherae** (III), widespread.

BITTERSWEET
(*Celastrus*)

Bacterial crown gall. **Agrobacterium tumefaciens,** CT.
Canker; dieback. **Glomerella cingulata,** NH, NC.
Leaf spot. **Ascochyta** sp., CT; **Asteria celastri,** KS, ME, MI; **Cercospora melanochaeta,** IA, KS, NE, ND; **Marssonina** sp., NY; **Phyllosticta celastri,** IL, KS, MA, NY, WV; **Ramularia celastri,** VT to MS, TX, ND.
Powdery mildew. **Microsphaera alni,** WI; **Phyllactinia corylea,** ME to VA, TX, SD.
Rot, root. **Phymatotrichum omnivorum,** TX.

BIXA
(*Annato-Tree*)

Leaf spot. **Cercospora bixae; Phyllosticta bixae.**
Leaf spot, algal. **Cephaleuros virescens,** FL.

BLACK BEARBERRY
(*Arctous*)

Rust. **Pucciniastrum sparsum** (II, III), AK.

BLACKBERRY
(*Rubus*)

Bacterial crown gall. **Agrobacterium tumefaciens,** general; cane gall, **A. rubi,** NY, OR, PA, WA, WI; hairy root, **A. rhizogenes,** OR.
Blight, cane. **Gnomonia rubi,** MD, ME, NY, PA, VT; **Leptosphaeria coniothyrium,** NY to NC, WI, TX, Pacific Northwest.
Blight, spur. **Didymella applanata,** VA, WI.
Blight, stamen; dry berry. **Hapalosphaeria deformans,** OR, WA.
Blight, thread. **Pellicularia koleroga,** LA.
Blotch, sooty. **Gloeodes pomigena,** MD to NC, IN, TX.
Canker. **Glomerella cingulata,** MD, VA; **Phomopsis** sp., WA; **Botryosphaeria obtusa,** OH; **B. dothidea,** MD; **Leptosphaeria coniothyrium,** OH; **Gnomonia rubi,** OH;
Canker, cane spot. **Ascospora ruborum,** WA.
Downy mildew. **Peronospora rubi,** MD, WI.

Fruit spot; fly speck. **Leptothyrium pomi,** PA to NC, IL.

Leaf spot. **Mycosphaerella confusa,** VA to FL; TX, IL, IN; **M. rubi,** NC, WI; **Pezizella oenotherae,** also fruit rot, OH, MD, VA; **Cylindrosporium rubi,** NC, TX; **Phyllosticta spp.,** FL, IL, NH; **Septoria darrowi; S. rubi,** general.

Leaf spot, algal. **Cephaleuros virescens,** FL.

Powdery mildew. **Sphaerotheca macularis,** CT to MD, IL, MN, Pacific Northwest; **Phyllactinia corylea,** MI.

Rosette, double blossom. **Cercosporella rubi,** NY to FL, TX, CA.

Rot, gray mold. **Botrytis cinerea,** general on fruit, bud, shoot.

Rot, root. **Armillaria mellea,** TX, WA; **Phymatotrichum omnivorum,** TX; **Rhizoctonia solani.**

Rot, white root. **Corticium galactinum,** near apple trees, AR, MD, VA.

Rust, orange. **Gymnoconia peckiana** (0, I, III), ME to GA and west to Pacific; **Kunkelia nitens** (I), general, more common in South.

Rust, yellow cane. **Kuehneola uredinia** (0, I, II, III), ME to FL, LA, WI.

Spot anthracnose. **Elsinoë veneta,** general.

Virus. **Blackberry Dwarfing; Blackberry Mosaic; Blackberry Variegation; Loganberry Dwarf,** OR; **Raspberry Beta-Leaf Curl; Red Raspberry Mosaic,** MA to VA, IA, WI, Pacific Northwest; **Raspberry Streak,** OH, PA, WA.

Wilt. **Verticillium albo-atrum,** CA, MN, NY, WA.

Sanitation is the best approach to home garden blackberry diseases. Plants with crown gall, orange rust, or virus diseases should be removed and burned, replanting with clean stock.

BLACKBERRY-LILY
(*Belamcanda*)

Leaf spot. **Alternaria** sp., KS, VA; **Didymellina macrospora,** CA, KS, IA, NY OK, VA, VT.

Rust. **Puccinia iridis,** FL.

BLADDER-SENNA
(*Colutea*)

Blight, twig. **Diplodia coluteae,** PA.

Powdery mildew. **Erysiphe polygoni.**

Rot, root. **Ganoderma** sp., OK; **Phytophthora cactorum,** MO.

Rust. **Uromyces colutea** (II, III), KS.

BLEEDING-HEART
(*Dicentra spectabilis*)

Rot, stem. **Sclerotium rolfsii,** NY; **Sclerotinia sclerotiorum,** MN.

Wilt. **Fusarium** sp., NJ.

BLEEDING-HEART VINE
(*Clerodendrin*)

Virus. Zonate ring spot, FL.

BLEPHILIA

Leaf spot. **Cercoseptoria blephiliae**, WI; **Septoria menthicola**, WI.
Rust. **Puccinia menthae** (0, I, II, III), IL, IN, IA, MD, MI, MO, TN, WI.

BLOODROOT
(*Sanguinaria*)

Blight, gray mold. **Botryris cinerea**, NY.
Leaf spot. **Cercospora sanguinariae**, MD, MO, NY, PA, TX, WI; **Cylindro-sporium circinans**, MD, MO, WI; **Gloeosporium sanguinariae**, OH, TX.
Phyllosticta sanguinariae, MO, TX, WV.

BLUEBERRY
(*Vaccinium*)

Anthracnose. **Gloeosporium** sp., FL, NJ.
Bacterial canker. **Pseudomonas** sp., OR.
Bacterial crown gall. **Agrobacterium tumefaciens**, MA, MI, WA.
Black mildew, **Meliola nidulans**, AL, GA.
Blight, blossom, fruit, twig. **Botrytis cinerea**, ME, MA, NJ, CA, OR, WA; **Diaporthe vaccinii**, MA, ME, NJ.
Blight, stem. **Botryosphaeria dothidea** (*B. ribis*), NC; **Phomopsis vaccinii**, NC.
Canker, blight of stem, MI. **Glomerella cingulata, Alternaria** sp., **Sordaria** sp., **Epicoccum** sp., **Tympanis** sp., **Papulospora** sp., **Nectria cinnabarina, Conio-thyrium** sp., **Pestalotia** sp., **Verticillium** sp., **Godronia cassandrae** f. **vaccinii, Phoma** sp., **Dendrophoma** sp., **Diaporthe (Phomopsis) vaccinii, Melano-spora** sp., **Pullularia** spp., **Coryneum microstictum, Sphaeronema** sp., **Fusarium** spp., **Bispora** sp., **Botrytis cinerea, Cephalosporium** sp., **Cylindro-carpon** sp., **Pyrenochaeta** sp.
Canker, cane. **Physalospora corticis**, AL, FL, MS, NC; **Godronia cassandrae**, ME, MI; **Botryosphaeria corticis**, NC, AL, FL, GA, MS; **Coryneum microstictum**, MA, MI.
Canker, twig. **Dothidella vacciniicola**, NC; **Fusicoccum putrefaciens**, MA, ME, MI, **Phomopsis vaccinii**, MI.
Dodder. **Cuscuta** sp., FL.
Gall, red leaf. **Exobasidium vaccinii**, general; **Synchytrium vaccinii**, ME, MS.
Leaf spot. **Gloecercospora inconspicua**, GA, NC, MD; **Phyllosticta** sp., GA, AL; **Phyllostictina vaccinii**, GA, MD, MI, MC; **Piggotia vaccinii**, WI;

Discohainesia oenotherae, NC, NJ; **Ramularia vaccinii**, FL, MD, MI, NJ, NY; **Septoria albopunctata; Alternaria tenuissima**, NC; **Cristulariella pyramudates**, MD.

Leaf spot, double spot. **Dothichiza caroliniana**, NC.

Leaf spot, tar. **Thytisma vaccinii**, widespread.

Nematode. **Helicotylenchus** sp.; **Hemicycliophora similis; Tetylenchus joctus; Xiphinema americanum.**

Powdery mildew. **Microsphaeria alni** var. **vaccinii**, widespread.

Rot, berry; dieback. **Alternaria** sp., MA, NJ, NC; **Glomerella cingulata**, NJ, NY; **Phomopsis vaccinii**, NC.

Rot, brown; mummy berry. **Monilinia vaccinii-corymbosi**, IN, ME, MA, NJ, MI, NY, NC, PA.

Rot, root. **Phytophthora cinnamomi**, NJ, AR.

Rust, witches' broom. **Pucciniastrum geoppertianum** (III) ME to MN; 0, I, on fir; Leaf, **P. myrtilli** (II,III) ME to PA, WI; 0, I, on spruce.

Virus. **Blueberry Necrotic Ring Spot; Blueberry Red Ring Spot; Blueberry Stunt**, MI, MA, NJ, NY, NC; **Blueberry Shoestring**, WA.

Virus. **Tobacco Ring Spot**, IL; **Peach Rosette Mosaic**, MI.

BLUE COHOSH
(*Caulophyllum*)

Blight, leaf. **Botrytis ?streptothrix**, NY, NJ.

Leaf spot. **Cercospora caulophylli**, VT to VA, MO, WI.

BLUE CURLS
(*Trichostema*)

Leaf spot. **Septoria trichostematis**, NY.

BLUE-EYED GRASS
(*Sisyrhinchium*)

Blight, leaf. **Kellermania sisyrhinchii**, CA, ND, NM.

Nematode, lesion. **Pratylenchus pratensis.**

Rust. **Uromyces houstoniatus** (II, III), ME, WV; 0, I, on *Houstonia* spp.; **U. Probus** (I, II, III), ID, OR, UT, WA, TX; **Aecidium residuum** (0, I), OK, TX.

BLUE LACE-FLOWER
(*Trachymene*)

Bacterial, MLO. **Western Aster Yellows**, CA.

Nematode, root knot. **Meloidogyne** sp., FL.

Rot, root; stem. **Fusarium** sp., CT, NJ; **Rhizoctonia solani,** NJ.
Wilt. **Phytophthora cactorum,** MA.

BLUESTEM
(*Andropogon*)

Leaf spot. **Ascochyta brachypodii,** PA, NY.

BOEHMERIA

Nematode. **Paratylenchus elachistus,** FL.

BOISDUVALIA

Rust. **Puccinia glabella** (II, III), NV, OR, UT; 0, I unknown; **P. oenotherae**
(0, I, II, III), CA, ID, NV, OR, WA; **P. vagans** var. **epilobi-tetragoni** (0, I, II, III),
CA, NV, ID, OR, UT.

BOLTONIA

Leaf spot. **Septoria erigerontis** var. **boltoniae,** IA, WI.
Powdery mildew. **Erysiphe cichoracearum,** SD.
Rust. **Puccinia dioicae,** var. **asteris** (II, III), IA, NE, ND, SD; 0, I, on sedge;
 Uromyces compactus (0, I, II, III), TX.
Smut, white. **Entyloma compositarum,** WI.

BORAGE
(*Borago*)

Leaf spot. **Ramularia** sp., CA; **Stemphylium** sp.

BOTTLE-BRUSH
(*Callistemon*)

Rot, root. **Armillaria mellea,** CA.

BOUGAINVILLEA
(*Buginvillaea*)

Blight, foliage. **Phytophthora parasitica,** FL.
Leaf spot. **Cladosporium arthrinioides,** TX: **Cercospora bougainvilleae,** FL.
Virus. **Mosaic.** Undetermined, FL.

BOUVARDIA

Nematode, leaf. **Aphelenchoides fragariae.**
Nematode, root knot. **Meloidogyne** sp., NY.
Rust. **Puccinia bouvardiae** (0, I, III), AZ.

BOXELDER
(*Acer negundo*)

Anthracnose. **Gloeosporium apocryptum,** widespread.
Bacterial leaf spot. **Pseudomonas aceris,** CA.
Blight, leaf. **Coryneum negundinis,** MO; on twigs, ME.
Blight, twig. **Coniothryium negundinis,** IL, OK; **Coryneum negundinis,** ME.
 Nectria cinnabarina.
Canker, felt fungus. **Septobasidium** spp., NC.
Leaf spot. **Alternaria** sp.; **Ascochyta negundinis,** IL, NC: **Cercospora negun-
 dinis,** NE, WI, KS; **Piggotia negundinis,** WI; **Phyllosticta minima,** general;
 P. negundinis, ME to AL, TX, WI; **Septoria aceris,** general; **S. negundinis;
 Cylindrosporium negundinis; Cristulariella pyramidalis.**
Leaf spot, tar. **Rhytisma punctatum,** NY, CA; **R. acerinum,** OR, WA.
Powdery mildew. **Phyllactinia corylea,** SD: **Microsphaera alni, Uncinula
 circinata.**
Rot, root. **Phymatotrichum omnivorum,** CA, TX; **Helicobasidium pur-
 pureum,** TX.
Wilt. **Verticillium albo-atrum,** occasional.

BOXWOOD
(*Buxus*)

Blight, leaf cast. **Hyponectria buxi,** general; **Verticillium buxi,** cosmopolitan
 on dead leaves, often associated with *Hyponectria.*
Blight, leaf tip. **Phoma conidiogena,** MD, NJ, NY, OK.
Canker, "Nectria"; leaf blight. **Volutella buxi,** general; considered imperfect
 stage of *Pseudonectria rouselliana* but unconfirmed connection.
Canker; dieback. **Fusarium buxicola** (*Nectria desmazierii*), AL, MD, PA;
 F. lateritium (see *Gibberella baccata*), secondary, MD, SC, VA.
Leaf spot. **Phyllosticta auerswaldii,** MD, MA, NJ, NY, VA, WA; **Macrophoma
 candollei,** prominent, general on dead leaves following winter injury or
 disease.
Nematode, dagger. **Xiphinema americanum.**
Nematode, lesion. **Pratylenchus pratensis; P. vulnus,** NC.
Nematode, root knot. **Meloidogyne hapla; M. incognita; M. incognita acrita;
 M. javanica; M. arenaria,** NC.

Nematode, spiral. **Helicotylenchus nannus; Rotylenchus buxophilus.**
Nematode, stubby root. **Trichodorus** sp.
Nonparasitic. **Freezing,** ice standing on stems causes bark to slough off and
branches to die back for months thereafter. **Sunscald,** Injury in late winter
or early spring when covering is removed in bright sun or high wind.
Rot, root. **Armillaria mellea,** NJ; **Fusarium oxysporum** and **F. solani,** MD,
perhaps secondary; **Phymatotrichum omnivorum,** TX; **Phytophthora parasitica,** MD, NC; **Pythium** sp., MA; **Rhizoctonia solani,** cosmopolitan after
nematode injury.
Rot, heart; trunk. **Ganoderma lucidum,** VA; **Fomes ignianius,** VA; **Poria
punctata,** VA.
Rot, root; decline. **Paecilomyces buxi** and some **Fusaria,** VA.
?Virus. **Variegation,** cause unknown, MD, NY, VA.

Salmon-pink pustules appearing on backs of leaves, along twigs, and on
main stems are indications of *Volutella buxi,* controlled by thorough cleaning,
then spraying. Much of the dieback, bronzing, general unhealth of boxwood
is due to nematodes, especially spiral and root lesion or meadow.

BOYSENBERRY

Subject to most blackberry diseases: anthracnose, crown gall, cane gall, cane
canker, dieback, leaf spots, mosaic.

BRACHYCOMBE
(Swan River Daisy)

Bacterial, MLO. **Western Aster Yellows,** CA.

BRACHYPODIUM
(Slender False-Brome)

Anthracnose. **Colletotrichum graminicola,** MD.
Bacterial spot. **Pseudomonas coronafaciens** var. **atropurpurea,** ND.
Rot, root. **Curvularia geniculata** and **Fusarium scirpi,** secondary, ND.

BRICKELLIA
(Brickle-Bush)

Dodder, **Cuscuta exaltata,** TX.
Leaf spot. **Cercospora coleosanthi,** CA, CO.
Rot, root. **Phymatotrichum omnivorum,** TX.

Rust. **Aecidium arcularium** (0, I), AZ, CO, NM; **Coleosporium aridum** (II), CA; **Puccinia kuhniae,** AZ, FL; **P. subdecora** (0; I, II, III), AZ, CO, NM, UT; **Uredo arida** (II), CA.

BRISTLEGRASS
(*Setaria*)

Blight. **Beniowskia sphaeroidea,** TX, GA.

BROCCOLI

See Cabbage.

BRODIAEA

Rust. **Puccinia carnegiana** (0, I, III), AZ; **P. dichelostemmae** (0, I, III), WA, OR, CA; **P. moreniana** (III), CA; **P. nodosa** (0, I, II, III), CA; **P. pattersoniana,** ID, UT, WA; II, III, on grasses; **P. subangulata** (0, I, III), WA; **Uromyces brodiaeae** (0, I, III), CA, OR, WA.

BROMEGRASS, SMOOTH
(*Bromus*)

Bacterial. Halo Blight. **Pseudomonas syringae pv. coronafaciens,** AK.
Rust. **Puccinia recondita,** PA.

BROMELIA

Leaf spot. **Gloeosporium** sp., FL, MD; **Helminthosporium rostratum,** FL.

BROOM
(*Cytisus*)

Canker. **Diaporthe** spp., NJ, NY; **Dothidea tetraspora,** CA; **Gloeosporium** sp., NJ; **Nectria coccina,** OR; **Pestalotia polychaeta,** CA; **Phomopsis** sp., MA, NJ; **Physalospora obtusa,** AL. Many are secondary.
Nematode, lesion. **Pratylenchus pratensis.**
Rot, root. **?Thielaviopsis basicola,** WI.

BROOM, SPANISH
(*Spartium*)

Canker. **Diplodia sarothamni,** CA; **Pestalotia polychaeta,** CA; **Phomopsis sarothamni,** CA. All possibly secondary infections.

BROUSSONETIA
(Paper-Mulberry)

Canker; dieback. **Nectria cinnabarina,** AL, NY; **Fusarium solani,** OH.
Mistletoe. **Phoradendron serotinum (flavescens),** TX.
Nematode, root knot. **Meloidogyne** sp., MD.
Rot, root. **Phymatotrichum omnivorum,** TX; **Armillaria mellea,** CA.

BROWALLIA
(*Streptosolen jamesonii*)

Bacterial canker, vascular. **Clavibacter michiganense,** WY.
Nematode, root knot. **Meloidogyne** sp.
Virus. **Tomato spotted wilt,** CA.
Wilt. **Fusarium** sp., DE.

BRUNFELSIA

Bacterial canker, vascular. **Clavibacter michiganense,** WY.

BRUSSELS SPROUTS

See Cabbage.

BRYONOPSIS

Bacterial spot. **Pseudomonas lachrymans,** WI.
Downy mildew. **Pseudoperonospora cubensis,** MA, OH.

BUCKEYE
See Horse-Chestnut.

BUCKLEYA

Rust. **Cronartium comandrae** (II, III), TN; 0, I, on pine.

BUCKTHORN
(*Rhamnus*)

Leaf spot. **Cercospora aeruginosa,** MO, NE, SC; **C. bacilligera; C. rhamni,** LA, NJ, NE, TX, WI; **Cylindrosporium rhamni,** ID; **Marssonina rhamni,** WA; **Phaeosphaerella rhamni,** CA; **Phyllosticta rhamnigena; Septoria blasdalei,** CA, OR, ID, TX.
Powdery mildew. **Microsphaera alni,** WI.
Rot, heart. **Daedalea unicolor,** WA; **Fomes igniarius,** ID.
Rot, root. **Phymatotrichum omnivorum,** TX.
Rust. **Puccinia coronata** (0, I), widespread; II, III, on grasses, cereals, widespread; **P. mesneriana** (III), CA.
Sooty mold. **Capnodium** sp., CA.

BUCKWHEAT-TREE
(*Cliftonia*)

Black mildew. **Morenoella cliftoniae,** MS.
Leaf spot. **Coccomyces** sp.; **Pestalotia cliftoniae,** MS.

BUDDLEIA
(Butterfly-Bush)

Canker, stem. **Phomopsis buddleiae,** AZ.
Nematode, root knot. **Meloidogyne** spp., AL, MS, TX.
Rot, root. **Phymatotrichum omnivorum,** TX.
Scab. **Cladosporium heugelinianum,** VA, DE.

BUFFALOBERRY
(Shepherdia)

Damping-off. **Pythium ultimum,** NE; **Rhizoctonia solani,** NE.
Leaf spot. **Septoria shepherdiae,** UT, WI, ID, MT, AK; **Cylindrosporium** sp., Great Plains.
Powdery mildew. **Phyllactinia corylea,** UT; **Sphaerotheca macularis** (*S. humilis*); CO, MT, TX, WY, AK.
Rot, root. **Phymatotrichum omnivorum,** TX.
Rot, white heart. **Fomes fraxinophilus,** CO, MT, NM, SD, WY.

Rust. **Puccinia caricis-shepherdiae** (0, I), Rocky Mts., MI, NY, OR, WA, AK; II, III on *Carex*; **P. coronata** (0, I), SD to NM, WA, AK.

BUFFALOGRASS
(*Buchloe*)

Smut. **Tilletia buchloëana,** KS, NE, OK, TX.

BUGLEWEED
(*Ajuga*)

Blight, southern; crown rot. **Sclerotium rolfsii,** CA, CT, KS, NJ, NY. Serious in warm, muggy weather.
Nematode, southern root knot. **Meloidogyne incognita,** GA.

BUMELIA

Leaf spot. **Cercospora lanuginosa,** TX; **Phyllosticta bumeliifolia,** AL, TX; **P. curtisii,** FL, MO; **Septoria bumeliae,** MS.
Rot, root. **Helicobasidium purpureum,** TX; **Phymatotrichum omnivorum,** TX.

BUNCHFLOWER
(*Melanthium*)

Leaf spot. **Septoria allardii,** VA.
Rust. **Puccinia atropuncta** (II, III), NC, TN, VA.

BUNDLEFLOWER
(*Desmanthus*)

Leaf spot. **Cercospora desmanthi,** KS, LA, MO, NM, SD, TN.
Powdery mildew. **Erysiphe polygoni,** IL, MS.
Rust. **Ravenelia texensis** (II, III), TX.

BURNET
(*Sanguisorba*)

Leaf spot. **Graphium sessile,** NY; **Ovularia bulbigera,** IL, AK.
Powdery mildew. **Podosphaera oxyacanthae,** IA; **Sphaerotheca macularis,** MA, NY, PA, AK.
Rust. **Xenodochus carbonarius** (I, III), AK; **X. minor,** AK.

BUTTERFLY-FLOWER
(*Schizanthus*)

Bacterial, MLO. **Aster Yellows**, NJ.
Anthracnose. **Colletotrichum schizanthi**, NY.
Bacterial canker, vascular. **Clavibacter michiganense**, WY.
Damping-off; root rot. **Pythium ultimum**, MO; **Rhizoctonia solani**, NY.
Nematode, root knot. **Meloidogyne** sp.
Rot, stem. **Sclerotinia sclerotiorum**, MS, MO.
Virus. **Tomato Spotted Wilt**, TX.

BUTTERFLY-PEA
(*Centrosema*)

Leaf spot. **Cercospora clitoriae**, AL; **Colletotrichum** sp., on pods.

BUTTERFLY PEA
(*Clitoria*)

Leaf spot. **Cercospora clitoriae**, AL, FL; **C. cruenta**, FL.
Rot, root. **Phymatotrichum omnivorum**, TX.

BUTTERFLY WEED
(*Asclepias*)

Blight, stem. **Phoma asclepiadea**, UT.
Leaf spot. **Cercospora asclepiadorae**, AL, TX; **C. clavata**, general; **C. venturioides**, WA; **Phyllosticta tuberosa**, IL, NJ.
Rot, root. **Phymatotrichum omnivorum**, TX.
Rust. **Puccinia bartholomaei** (0, I), ND, SD; II, III, on *Spartina*; **Uromyces asclepiadis** (II III), widespread.

BUTTONBUSH
(*Cephalanthus*)

Blight, leaf. **Cercospora perniciosa**, TX.
Blight, thread. **Pellicularia koleroga**, FL.
Dodder. **Cuscuta compacta**, FL; **C. gronovii**, NY.
Leaf spot. **Ascochyta cephalanthi**, LA: **Coniothyrium cephalanthi**, LA; **Phyllosticta cephalanthi**, TX; **Ramularia cephalanthi**, AL, KS, LA, NY, WI; **Septoria cephalanthi**, KS, WI.
Powdery mildew. **Microsphaera alni**, widespread. **Phyllactinia corylea**, IN.

Rust. **Puccinia seymouriana** (0, I), New England to FL and central states; II, III, on *Spartina*, **Uredo cephalanthi** (II), FL.

CABBAGE
(*Brassica oleracea*, including Broccoli, Brussels Sprouts, Cauliflower, Kale, Kohl-rabi)

Bacterial black rot. **Xanthomonas campestris**, general.

Bacterial leaf spot. **Pseudomonas maculicola**, widespread; **P. cichorii**, FL and in market.

Bacterial, MLO. Western **Aster Yellows**, CA.

Bacterial soft rot; stump rot; **Erwinia carotovora**, cosmopolitan.

Blackleg; leaf spot. **Phoma lingam**, general east of Rocky Mts., also OR, WA.

Blight, gray mold. **Botrytis cinerea**, FL; Pacific states, frequent.

Blight, southern. **Sclerotium rolfsii**, NC to FL, TX, occasional.

Club root. **Plasmodiophora brassicae**, general.

Damping-off; bottom rot. **Pythium** spp.; **Rhizoctonia solani**, general.

Downy mildew. **Peronospora parasitica**, general.

Leaf spot, gray. **Alternaria brassicae**, general; black spot, head browning, **A. brassicicola**, general; **A. oleraceae**, occasional; **Cercospora brassicola**, CA, DE, FL, IL, OK, MS, NC; **Phyllosticta brassicicola**, CA; **Leptosphaeria maculans**, WI.

Leaf spot; white spot. **Cercosporella brassicae**, IN, PA, OR; ring spot, **Mycosphaerella brassicicola**, AL, CA, IL, NY, OR, WA, TX; **Pseudocercosporella capsellae**, CA.

Mold, leaf. **Heterosporium variabile**, MT, NY; seed, **Curvularia geniculata; Stemphylium botryosum.**

Nematode. **Naccobus batatiformis.**

Nematode, cabbage cyst. **Heterodera cruciferae.** Sugar beet, **H. schachtii.**

Nematode, lesion. **Pratylenchus pratensis.** Sting, **Belonolaimus gracilis; B. longicaudatus;** stubby root, **Trichodorus christiei.**

Nematode, root knot. **Meloidogyne arenaria; M. javanica.**

Nonparasitic

    Brown Heart. Probably boron deficiency.

    Chlorosis. Magnesium or manganese deficiency.

    Oedema. Excessive water tension, or copper sprays.

    Pink Head. Probably genetic.

    Tipburn. Potassium deficiency.

    Whip-tail. Spindly growth, failure to head, due to acid soil, mineral deficiencies.

Powdery mildew. **Erysiphe polygoni**, AZ, CA, CT, FL, MD, MA, OR.

Rot, cottony; drop. **Sclerotinia sclerotiorum**, AZ, MA, NY, TX.

Rot, firm head. **Sclerotinia sclerotiorum**, CA.

Rot, root. **Macrophomina phaseoli**, CA; on *Brassica campestris*.

Rot, root. **Phymatotrichum,** TX; **Phytophthora megasperma,** CA, OR.
Rot, root; damping-off. **Aphanomyces raphani,** WI. Also **Brasscia alboglabra**
— Chinese kale; **B. carinata** — Abyssinian mustard; **B. napus** — rape; **B. per-
viridis** — Spanish mustard; **B. robertiana; Eruca sativa** — racuet salad;
Raphanus raphanistruno — wild radish.
Rot, soft; black mold. **Rhizopus stolonifer,** occasional.
Rot, sprout. **Alternaria tenuis, Rhizoctonia solani,** and **Fusarium** sp., on
brussels sprouts.
Virus. **Cabbage Ring Necrosis; Cauliflower Mosaic,** GA; **Turnip Mosaic,**
GA; **Beet Curly Top; Tomato Spotted Wilt.**
White rust; blister. **Albugo candida,** CA, ID, KY, NE, NC, OH, TX.
Wilt, yellows. **Fusarium oxysporum** f. sp. **conglutinans,** general.

A general control program starts with choosing varieties resistant to
Fusarium yellows, purchasing disease-free seed, or having it hot-water treated
against blackleg and black rot, growing in soil free from club root and careful
cleaning up of all vegetable refuse at the end of the season.

CACTUS
(*Cereus*)

Anthracnose. **Mycosphaerella opuntiae,** TX.
Bacterial soft rot. **Erwinia carotovora,** TX.
Blight, gray mold; rot. **Botrytis cinerea,** occasional when too damp indoors.
Leaf scorch. **Hendersonia opuntiae,** TX.
Leaf spot. Black, **Diplotheca** (*Stevensea*) **wrightii,** TX; stem, **Septoria cacti-
cola,** TX.
Nematode, cyst. **Heterodera cacti; Naccobus batatiformis.**
Nematode, root knot. **Meloidogyne** sp.
Rot, root; stem. **Fusarium oxysporum,** AZ, CA; **Helminthosporium cactivorum,**
TX, CA. **Phymatotrichum omnivorum,** TX.
Rot, stem and branch. **Aspergillus alliaceus,** TX; dry rot, **Poria** sp., CA.

CACTUS
(*Leuchtenbergia principis and Schlumbergera
gaertneri* "Makoyana")

Nematode, cyst. **Heterodera cacti.**

CACTUS, FISHHOOK, PINCUSHION
(*Mammillaria*)

Anthracnose; zonate spot. **Gloeosporium cactorum; Mycosphaerella opuntiae;**
spine spot, **Phoma mammillariae,** MT.

Nematode, root knot. **Meloidogyne incognita,** Bermuda Botanical Garden; **Meloidogyne** sp., TX; cyst, **Heterodera cacti.**
Rot, root. **Phymatotrichum omnivorum,** TX.

CACTUS, GIANT, SAGUARO
(*Carnegiea*)

Bacterial blight. **Erwinia carnegiaeana,** AZ.
Bacterial crown gall. **Agrobacterium tumefaciens,** AZ.
Rot, dry. **Poria carnegieae;** Heart, **Fomes robustus.**
Rot, seedling. **Fusarium solani; F. oxysporum.**

The bacterial blight is spread over the whole giant cactus area in Arizona, with mortality heaviest in magnificent specimens 150 to 200 years old.

CACTUS, PRICKLY PEAR
(*Opuntia*)

Anthracnose; zonate spot. **Gloeosporium cactorum,** FL, MS; black rot, **Mycosphaerella opuntiae,** AL, FL, LA, NY, SC, TX.
Bacterial rot. **Erwinia aroideae,** FL, MS, OK, TX.
Black mildew. **Lembosia cactorum,** FL.
Leaf scorch. **Hendersonia opuntiae,** AL, NJ, KS, MT, TX.
Leaf spot, black. **Diplotheca wrightii,** FL, TX.
Nematode, root knot. **Meloidogyne** sp.
Nonparasitic. **Oedema,** causing glassiness or scab, from overwatering.
Rot, charcoal. **Macrophomina phaseoli,** TX.
Rot, cladode. **Aspergillus alliaceus,** TX; **Diplodia opuntiae,** MD, PA, KS; **Physalospora obtusa,** NY; **P. rhodina,** FL; **Phyllosticta cacti,** NM; **P. concava,** MO, NJ, OK, TX; **Septoria fici-indicae,** TX.
Rot, root. **Armillaria mellea,** CA.
Rot, stem. **Phytophthora parasitica,** NY; **Pythium debaryanum,** CA.
Virus. **?Mosaic,** MD; **Chlorotic Ring Spot,** AZ, CA, ID, MT, NM, UT, WA, WY.

CACTUS, STAR, SEA-URCHIN, BARREL
(*Echinocactus*)

Anthracnose. **Mycosphaerella opuntiae,** TX.
Leaf scorch; scald. **Hendersonia opuntiae,** TX.
Leaf spot, black. **Diplotheca** sp., TX.
Rot, root. **Phymatotrichum omnivorum,** TX; Stem, **Aspergillus alliaceus,** TX.

CAESALPINIA

Anthracnose. **Gloeosporium** sp., FL.
Bacterial crown gall. **Agrobacterium tumefaciens,** FL.
Canker; dieback. **Botryosphaeria ribis** var. **chromogena,** FL, TX.
Rot, root. **Clitocybe tabescens,** FL; **Phymatotrichum omnivorum,** TX.
Rust. **Ravenellia hymphreyana** (II, III), FL.

CALADIUM

Bacterial soft rot. **Erwinia carotovora,** FL.
Blight, southern. **Sclerotium rolfsii,** FL.
Blight; tuber rot. **Botrytis ricini,** FL.
Leaf spot. **Gloeosporium ?thuemeni,** FL.
Nematode, citrus. **Tylenchulus semipenetrans,** FL.
Nematode, lesion. **Pratylenchus** sp., FL.
Nematode, root knot. **Meloidogyne** sp., FL, MS; **M. javanica.**
Rot, tuber. **Fusarium solani,** FL.

CALATHEA

Leaf spot. **Glomerella cincta,** NJ; **Phyllosticta** sp.; **Alternaria alternata,** FL;
   Drechslera setariae, FL.
Nematode, burrowing. **Radopholus similis,** FL. Spiral, **Helicotylenchus nannus.**

CALCEOLARIA
(Slipperwort)

Blight, gray mold. **Botrytis cinerea,** AK.
Nematode, leaf. **Aphelenchoides fragariae.**
Nonparasitic. **Boron Deficiency,** leaf necrosis, CA.
Rot, root. **Pythium mastophorum** and **P. ultimum,** CA.
Rot, stem. **Sclerotinia sclerotiorum,** NY, WA.
Virus. **Tomato Spotted Wilt,** CA.
Wilt. **Verticillium albo-atrum,** NY, WA.

CALENDULA
(Pot Marigold)

Bacterial, MLO. **Aster Yellows,** CT, DE, ME, NJ, PA, VA; **California Aster
   Yellows,** CA;
Blight, gray mold. **Botrytis cinerea,** MO, NJ, NY, AK.

Blight, southern. **Sclerotium rolfsii**, TX.
Leaf spot. **Alternaria** sp., NY; **Cercospora calendulae**, PA, TX, VA; **Colletotrichum gloeosporioides**, VA.
Nematode, root knot. **Meloidogyne** spp., WV, TX, TN; **M. javanica.**
Powdery mildew. **Erysiphe cichoracearum**, CA, NY; **E. polygoni**, PA.
Rot, root. **Rhizoctonia solani**, IN, NJ, NC, TX; **Pythium ultimum**, CA; **Phymatotrichum omnivorum**, TX.
Rot, stem; wilt. **Sclerotinia sclerotiorum**, CA, FL, LA, MO, OH, TX.
Rust. **Puccinia flaveriae** (III), IL, IN, IA, KS, MO, NE, TX.
Smut, white, **Entyloma calendulae**, CA, NH, OR; **E. compositarum**, WA.
Virus. ?Cucumber Mosaic, Tomato Spotted Wilt, CA, MI, TX; Biddens Mottle, FL.

CALIFORNIA BLUEBELL
(*Phacelia*)

Leaf spot. **Cylindrosporium phaceliae**, MT, TX.
Powdery mildew. **Erysiphe cichoracearum**, CA, MT, NM, TX.
Rust. **Puccinia aristidae** (0, I); II, III, on wild grasses; **P. Phaceliae** (III); **P. recondita** (0, I), MT to CO, OR, CA, NM; II, III, on bromegrass; **Uredo contraria** (II), CA.
Virus. **Beet Curly Top**, CA.

CALIFORNIA LAUREL
(*Umbellularia*)

Anthracnose. **Kabatiella phorodendri** f. **umbellulariae**, CA; leaf blight, **Colletotrichum gloeosporioides.**
Bacterial leaf spot. **Pseudomonas lauraceum**, CA.
Black mildew. **Asterina anomala**, CA.
Canker; dieback. **Nectria cinnabarina; N. coccinea**, CA.
Nematode, pin. **Paratylenchus anceps.** Ring, **Criconemoides xenoplax**, CA; Sheath, **Hemicycliophora brevis**, CA.
Nematode, root knot. **Meloidogyne hapla**, CA.
Rot, wood. **Fomes** spp., **Lenzites betulina; Polyporus versicolor; Porio ambigua; P. ferruginosa; Stereum albobadium; Schizophyllum commune.**

CALIFORNIA PEPPER-TREE
(*Schinus*)

Dodder. **Cuscuta subinclusa**, CA.
Nematode, root knot. **Meloidogyne** sp., TX.
Rot, heart. **Fomes applanatus**, CA; **Polyporus dryophilus**, CA; **P. farlowii**, AZ, CA; **P. sulphureus; P. versicolor**, CA.

Rot, root. **Armillaria mellea,** CA; **Phymatotrichum omnivorum,** TX.
Rot, wood. **Ganoderma polychromum,** CA; **Stereum hirsutum; Trametes hispida,** CA; **Schizophyllum commune.**
Wilt, **Verticillium albo-atrum,** CA.

CALIFORNIA PITCHER-PLANT
(*Darlingtonia*)

Leaf spot. **Mycosphaerella sarraceniae,** CA; **Septoria darlingtoniae,** OR.

CALIFORNIA POPPY
(*Eschscholtzia*)

Bacterial blight. **Xanthomonas papavericola,** TX.
Bacterial, MLO. **Aster Yellows,** NJ, NY; and **California Aster Yellows,** CA.
Blight, gray mold. **Botrytis cinerea,** AK.
Leaf spot, mold. **Heterosporium eschscholtziae,** CA.
Nematode, root knot. **Meloidogyne** sp., FL, TX.
Powdery mildew. **Erysiphe polygoni,** CA.
Rot, collar. **Alternaria** sp., TX.
Smut, leaf. **Entyloma eschscholtziae,** CA.
Wilt. **Verticillium albo-atrum,** CA.

CALIFORNIA-ROSE
(*Convolvulus japonicus*)

Leaf spot. **Phyllosticta batatas,** SC; **Septogloeum convolvuli,** CA, WI; **Septoria calystegiae,** CA; **S. convolvuli,** WI; **S. flagellaris,** NY.
Rust. **Coleosporium ipomoeae** (II, III), LA, MA, 0, I on pine; **Puccinia convolvuli** (0, I, II, III), CA, OR.

CALLA, COMMON, GOLDEN, PINK
(*Zantedeschia*)

Bacterial soft rot. **Erwinia aroidea,** general; **E. carotovora.**
Blight, gray mold. **Botryris cinerea,** NJ, AK.
Blight, southern. **Sclerotium rolfsii,** CA, FL, OR.
Leaf spot. **Alternaria** sp., secondary; **Cercospora richardiaecola,** AL, MS; **Coniothecium richardiae,** CA, FL, MA, NJ, NY, OR; **Gloeosporium callae,** WA.
Nematode, root knot. **Meloidogyne** sp., CA, FL.
Nonparasitic. **Chalk rot,** cause uncertain but may be immaturity of rhizomes, CA, OR

Rot, rhizome, **Phytophthora cryptogaea** var. **richardiae,** CA, FL, IL, MA, NJ, NY, OH, OR, PA, WA; **P. erythroseptica,** CA; **Phoma** sp., CA, OR.
Rot, root. **Armillaria mellea,** CA; **Rhizoctonia solani,** CA.
Virus. **Tomato Spotted Wilt,** CA, IL, IN, MD, NY, OR, TX, WA.

Root and rhizome rots are controlled by treating rhizomes before planting and growing in pots rather than benches. Specimens showing spotted wilt must be destroyed before thrips spread the virus.

CALLA, WILD
(*Calla palustris*)

Leaf spot. **Cercospora callae,** MA, NY, WI; **Marssonina callae,** NY.

CALLIANDRA
(False Mesquite)

Rot, root. **Clitocybe tabescens,** FL.
Rust. **Ravenelia reticulatae** (II, III), AZ.

CALLICARPA
(Beauty-Berry; French-Mulberry)

Black mildew. **Meliola cookeana,** FL, LA.
Leaf spot. **Cercospora callicarpae,** SC to TX.
Nematode, burrowing. **Radopholus similis,** FL.

CALYCANTHUS
(Sweetshrub, Carolina Allspice)

Bacterial crown gall. **Agrobacterium tumefaciens,** MS, NY.
Canker. **Botryosphaeria calycanthi,** NC, VA.
Powdery mildew. **Phyllactinia corylea,** CA.

CAMASS
(*Camassia*)

Blight, gray mold. **Botrytis cinerea,** OR.
Leaf spot. **Septoria chlorogali,** OR.
Rot, root. **Phymatotrichum omnivorum,** TX.
Smut, leaf. **Urocystis colchici,** OR, IN.

CAMELLIA

Bacterial crown gall. **Agrobacterium tumefaciens,** WA.

Black mildew. **Meliola camelliae.**

Blight, flower. **Sclerotinia camelliae,** CA, GA, LA, MS, NC, SC, VA, TX; **S. sclerotiorum,** NC; **Pestalotia** sp., **Penicillium** sp., AL.

Blight, gray mold; bud and flower. **Botrytis cinerea,** general after frost.

Blight, petal. **Pestalotia** sp., AL.

Canker; dieback. **Glomerella cingulata,** widespread; **Botryosphaeria ribis,** MS; **Phomopsis** sp., FL.

Canker, felt fungus. **Septobasidium castaneum; S. conidiophorum; S. pseudopedicellatum,** SC.

Dodder. **Cuscuta** sp.

Gall, leaf. **Exobasidium camelliae,** and var. **gracilis** on *Camellia sasanqua*, FL, LA, MS, TX; **E. monosporum,** AL.

Leaf spot. **Cercospora** sp., GA; **C. theae,** LA; **Hendersonia subalbicans,** GA; **Pestalotia quepini,** also twig blight, widespread; **Phyllosticta camelliae** and **P. camelliaecola,** Southeast; **Sporonema camelliae.**

Leaf spot, algal. **Cephaleuros virescens,** Gulf states.

Nematode, burrowing. **Radopholus similis,** FL.

Nematode, lesion. **Pratylenchus coffeae.**

Nematode, ring. **Criconemoides komabaensis,** FL; **Hemicriconemoides gaddi,** LA, GA.

Nematode, root knot. **Meloidogyne incognita-acrita.**

Nematode, spiral. **Helicotylenchus nannus.**

Nonparasitic. **Bud Drop, Dieback,** malnutrition, freezing, desiccation, widespread. **Oedema,** corky excrescences, from disturbed water relations. **Sunscald,** light circular spots, often with secondary fungi.

Rot, root. **Phytophthora cinnamomi,** major problem, AL, CA; **Clitocybe tabescens,** FL, LA.

Spot anthracnose, scab. **Elsinoë leucospila,** FL, GA, LA.

Virus. **Camellia Yellow Mottle Leaf** (color-breaking strains).

Flower blight is a devastating disease that has spread with plants in cans or pots. Order camellias bare-rooted, with all flower buds showing color picked off. Dieback is the subject of much controversy among camellia fans. Drastic surgery and a copper spray early in the season to prevent infection through bud scars seems to be helpful. Chelated iron may mask color-breaking virus symptoms.

CAMOMILE
(*Anthemis*)

Bacterial, MLO. **California Aster Yellows,** CA.

Damping-off. **Rhizoctonia solani,** WA.

Nematode, root knot. **Meloidogyne** sp., MD, FL.
Rot, root. **Phymatotrichum omnivorum,** TX.

CAMPANULA
(Bellflower, Canterbury Bells)

Bacterial, MLO. **Aster Yellows,** PA.
Blight, gray mold. **Botrytis cinerea,** AK.
Blight, southern. **Sclerotium rolfsii,** IL, NJ.
Leaf spot. **Ascochyta bohemica,** WI; **Cercoseptoria minuta,** WI; **Phyllosticta alliariifoliae,** NJ, NY; **Ramularia macrospora,** AK; **Septoria campanulae,** IL, IA, KS, MO, WI.
Nematode, leaf. **Aphelenchoides** sp.
Nematode, root knot. **Meloidogyne** sp.
Powdery mildew. **Erysiphe cichoracearum,** PA.
Rot, root. **Fusarium** sp., NY, NJ; **Rhizoctonia solani,** IL, CT, TX.
Rot, stem. **Sclerotinia sclerotiorum,** MD, WA.
Rust. **Coleosporium campanulae** (II, III), widespread; 0, I, on pine; **Aecidium campanulastri** (0, I), IA, MN; **Puccinia campanulae** (III), CA, MT, NY, OR, WA.
Virus. **Tomato Spotted Wilt,** CA.
Wilt. **Verticillium albo-atrum.**

CAMPHOR-TREE
(*Cinnamomum*)

Anthracnose. **Glomerella cingulata,** Gulf states.
Black mildew. **Lembosia camphorae,** FL.
Canker; dieback. **Diplodia camphorae, D. natalensis, D. tubericola,** widespread; **Gloeosporium camphorae** and **G. ochraceum,** also leaf spot, Gulf states.
Leaf spot, algal. **Cephaleuros virescens,** FL to LA.
Mistletoe. **Phoradendron serotinum (flavescens),** FL.
Nematode, lesion. **Pratylenchus pratensis,** FL.
Nonparasitic. **Chlorosis,** manganese deficiency, FL.
Powdery mildew. **Microsphaera alni** var. **cinnamomi,** LA.
Rot, root. **Armillaria mellea,** FL; **Phymatotrichum omnivorum,** TX; **Clitocybe tabescens,** FL.
Spot anthracnose. **Elsinoë cinnamomi,** MS, SC.

CANDLESTICK SHRUB
(*Cassia*)

Leaf spot. **Tubakia dryina,** LA.

CANDYTUFT
(*Iberis*)

Blight, gray mold. **Botrytis cinerea,** AK.
Club root. **Plasmodiophora brassicae,** MS, NJ.
Dodder. **Cuscuta indecora,** TX.
Downy mildew. **Peronospora parasitica,** CA.
Nematode, root knot. **Meloidogyne** sp., AL.
Powdery mildew. **Erysiphe polygoni,** CA.
Rot, root. **Phoma lingam,** CA; **Pythium oligandrum,** ME.
Virus. **Lettuce Mosaic,** NY
White Rust. **Albugo candida,** CA.

CANNA

Blight, southern. **Sclerotium rolfsii,** TX.
Leaf spot. **Alternaria** sp., MI, SC, TX.
Nematode, burrowing. **Radopholus similis.**
Rot, rhizome. **Fusarium** sp., MN, MO, PA.
Rust. **Puccinia cannae** (II, III), FL, OH, TX.
Virus. **Cucumber Mosaic,** RI; **Mosaic;** unidentified, DE, MD, MI, NY.

CANTALOUPE

See Melon.

CAPE-COWSLIP
(*Lachenalia*)

Virus. **Ornithogalum Mosaic,** AL.

CAPE-HONEYSUCKLE
(*Tecomaria*)

Anthracnose. **Colletotrichum gloeosporioides,** TX.
Rot, root. **Armillaria mellea,** CA; **Clitocybe tabescens,** FL.

CAPE-MARIGOLD
(*Dimortheca*)

Bacterial, MLO. **Aster Yellows,** NJ, NY.
Blight, gray mold. **Botrytis cinerea,** CT, AK.

Downy mildew. **Plasmopara halstedii**, IA.
Nematode, root knot. **Meloidogyne** sp., FL.
Rot, charcoal. **Macrophomina phaseoli**, CA.
Rot, root. **Pythium ultimum**, CA; **Rhizoctonia solani**, IL.
Rust. **Puccinia flaveriae** (III), IL, IN, NJ, NE.
Virus. **Curly Top Virus (Beet)** on *Dimorphotheca pluvialis* var. *fingens* and
   D. sinuata.
Wilt. **Fusarium** sp., FL.

CAPEWEED, CAPE DANDELION
(*Arctotheca calendula*)

Blight; Rot, root, crown and stolon. **Sclerotium rolfsii**, CA.

CAPER
(*Capparis*)

Black mildew. **Asterina lepidigenoides**, FL; **A. radians**, FL.

CARAWAY
(*Carum*)

Bacterial, MLO. **California Aster Yellows**, WA.
Dodder. **Cuscuta** sp., WA.
Nematode, root knot. **Meloidogyne** sp.
Rot, stem. **Sclerotinia sclerotiorum**, WA.

CARDOON
(*Cynara cardunculus*)

Leaf spot. **Cercospora obscura**, CA, TX; **Ramularia cynarae**, CA.
Powdery mildew. **Erysiphe cichoracearum**, CA.

CARISSA

Canker; dieback. **Physalospora obtusa** and **P. rhodina**, FL.
Leaf spot. **Colletotrichum gloeosporioides**, FL; **Macrophoma** sp., CA; **Septoria**
   sp., CA.
Nematode, root knot. **Meloidogyne** sp., FL, CA.
Rot, root. **Phymatotrichum omnivorum**, TX.

CARISSA
(Natal-Plum)

Cankers; gall. **Sphaeropsis tumefaciens,** FL.

CARNATION
(*Dianthus caryophyllus*)

Anthracnose. **Colletotrichum** sp., NJ, NY, TX.

Bacterial crown gall. **Agrobacterium tumefaciens,** MD.

Bacterial fasciation; witches' broom. **Clavibacter fascians,** CA, OH.

Bacterial, MLO. **Aster Yellows.**

Bacterial pimple. **Xanthomonas oryzae.**

Bacterial spot. **Pseudomonas woodsii,** MS to GA, IN, MI, ND, OK, OR, WA.

Bacterial wilt. **Pseudomonas caryophylli,** IL, IN, IA, MA, MO, WA.

Blight, branch and collar rot. **Alternaria dianthi,** general.

Blight, gray mold. **Botryris cinerea,** cosmopolitan in high humidity.

Blight, petal. **Stemphylium floridanum,** FL.

Blight, southern. **Sclerotium rolfsii,** FL, MS, TX.

Blight, web. **Pellicularia koleroga,** NC.

Blotch, greasy. **Zygophiala jamaicensis,** CA, PA.

Downy mildew. **Peronospora dianthicola,** CA.

Leaf spot, mold; fairy ring. **Heterosporium echinulatum** occasional in greenhouses, CA, TX; **Cladosporium herbarum,** secondary, general.

Leaf spot. **Septoria dianthae,** VT to SC, TX, MI, CA.

Nematode, cyst. **Heterodera trifolii.**

Nematode, lance. **Hoplolaimus coronatus.**

Nematode, pin. **Paratylenchus dianthus.**

Nematode, ring. **Criconemoides curvatum.**

Nematode, root knot. **Meloidogyne arenaria; M. hapla; M. incognita-acrita; M. javanica.**

Nonparasitic. **Yellow Spotting,** potassium deficiency, NJ, NY.

Petal spot. **Bipolaris** (*Helminthosporium*) **setariae,** FL.

Powdery mildew. **Oidium** sp., FL, NC.

Rot, bud. **Fusarium tricinctum** f. **poae,** MA to VA, KS, NE, WA.

Rot, root. **Phymatotrichum omnivorum,** TX; **Pythium** sp., IL; **Rhizoctonia solani,** general; **Armillaria mellea,** CA.

Rot, stem. **Armillaria mellea,** CA; **Fusarium roseum** f. **ceralis,** general.

Rust. **Uromyces dianthi** (II, III), general.

Smut. Anther. **Ustilago violaceae,** MA.

Virus. **Carnation Latent; Carnation Mosaic; Carnation Mottle; Carnation Ring Spot; Carnation Streak; Carnation Yellows; Carnation Etch-Ring; Beet Curly Top,** CA, TX; **Carnation Necrotic Fleck,** CA.

Wilt. **Fusarium oxysporum** f. sp. **dianthi; Verticillium albo-atrum** (*dahliae*), NJ; **Phialophora** (*Verticillium*) **cinerescens**, NY.

A rigid sanitation program is necessary for healthy carnations, taking cuttings high on the plant, breaking instead of cutting, planting in sterile medium, controlling aphids and other insect vectors of virus diseases and controlling mites that spread bud rot. The mother block system controls wilts; heat cures viruses.

CAROB, ST. JOHNS BREAD
(*Ceratonia*)

Canker. **Botryosphaeria ribis**, CA.
Nematode, root knot. **Meloidogyne** spp.
Rot, root. **Phymatotrichum omnivorum**, TX; **Phytophthora cactorum**, CA.
Wilt. **Verticillium albo-atrum**, CA.

CAROLINA JESSAMINE
(*Gelsemium*)

Black mildew; black spot. **Asterina somatophora**, FL.
Leaf spot. **Phyllosticta gelsemii**, NJ.
Rot, root. **Phymatotrichum omnivorum**, TX.
Sooty mold. **Capnodium grandisporum**, MS, TX.

CAROLINA MOONSEED
(*Cocculus*)

Leaf spot. **Cercospora cocculicola**, OK; **C. menispermi**, MS, TX.
Nematode, burrowing. **Radopholus similis**, FL.
Rot, root. **Phymatotrichum omnivorum**, TX.

CARROT
(*Daucus carota* var. *sativa*)

Bacterial blight. **Xanthomonas carotae**, CA, ID, IA, WI.
Bacterial core rot. **Erwinia chrysanthemi** and/or **E. carotovora** var. **carotovora**, IL.
Bacterial, MLO. **Aster Yellows**, widespread and it is the California strain, CA, OR, WA, ID.
Bacterial soft rot. **Erwinia carotovora**, general.

Blight, early; leaf spot. **Cercospora carotae,** general.

Blight, late. **Alternaria dauci,** general.

Blight, southern. **Sclerotium rolfsii,** GA to FL, TX, CA.

Canker, root; storage rot. **Rhizoctonia** spp., MI, NY, OR, WA.

Damping-off. **Pythium** sp., ID; **Rhizoctonia solani,** general.

Dodder. **Cuscuta** sp., ID, MS, NM, NY, TX, WV.

Leaf spot. **Ramularia** sp., KS; **Alternaria tenuis,** secondary, also seed mold.

Nematode. **Naccobus batatiformis.**

Nematode, cyst. **Heterodera carotae.**

Nematode, root knot. **Meloidogyne arenaria; M. incognita; M. javanica.**

Nonparasitic. **Black Heart,** cause unknown, WI. **Chlorosis,** magnesium deficiency in acid soil. **Root Girdle,** cause unknown.

Powdery mildew. **Erysiphe polygoni,** CA; **E. heraclei,** NC.

Rot, black. **Alternaria radicina,** ID, MA, NY, PA, WA.

Rot, black mold. **Rhizopus** spp.; blue mold, **Penicillium** sp.; gray mold, **Botrytis cinerea;** pink mold, **Trichothecium roseum,** IN.

Rot, dry. **Fusarium** spp., ID, NY, associated with scab.

Rot, root. **Helicobasidium purpureum,** OR, WA; **Phymatotrichum omnivorum,** AZ, LA, TX; **Phytophthora megasperma,** CA.

Rot, storage. **Centrospora acerina,** NY; **Aspergillus niger; Pellicularia filamentosa; Typhula** sp.

Rot, watery soft. **Sclerotinia sclerotiorum,** general; **S. intermedia, S. minor.**

Rust. **Uromyces scirpi** (0, I), OR; II, III, on *Scirpus.*

Scab. **Streptomyces scabies,** CA, MI, PA, WA.

Virus. **Alfalfa Mosaic,** CA, WA; **Clover Yellow Vein,** WA.

Virus. **Beet Curly Top,** OR, UT; **Carrot Motley Dwarf; Celery Mosaic.**

Carrots in home gardens require a deeply dug friable soil even more than treatment for diseases. Root-knot nematodes are rather common.

CARROT, WILD
(*Daucus carota*)

Virus. **Alfalfa Mosaic,** WA; **Clover Yellow Vein,** WA.

CASHEW
(*Anacardium*)

Blight, seedling. **Sclerotium rolfsii,** FL.

CASSABANA, CURUBA
(*Sicana*)

Anthracnose. **Colletotrichum lagenarium,** FL.

CASSIA
(Senna)

Blight, seedling. **Alternaria cassiae,** MS.
Dieback. **Diplodia natalensis,** TX.
Leaf spot. **Cercospora nigricans,** MS; **Septoria cassiicola; Tubakia dryina,** LA.
Nematode, root knot. **Meloidogyne** sp., CA, TX.
Powdery mildew. **Microsphaera alni,** MD; **Erysiphe polygoni.**
Rot, root. **Clitocybe tabescens,** FL; **Phymatotrichum omnivorum,** TX.
Wilt. **Fusarium oxysporum** f. sp. **cassiae,** NC, SC, FL, GA.

CASSIOPE

Leaf gall. **Exobasidium vaccinii,** WA.

CASTOR-BEAN
(Ricinus communis)

Bacterial crown gall. **Agrobacterium tumefaciens.**
Bacterial leaf spot. **Xanthomonas ricinicola,** MD, OK, TX.
Bacterial wilt. **Pseudomonas solanacearum,** AL, GA, FL, MI.
Blight, influorescence; gray mold. **Botryotinia ricini,** GA to FL, TX.
Blight, southern. **Sclerotium rolfsii.**
Damping-off. **Rhizoctonia solani,** FL, KS.
Gall. **Synchytrium** sp., TX.
Leaf spot. **Alternaria** sp., FL, LA, TX; **Cercospora canescens,** AL, KS, MT;
    White Spot, **C. ricinella; Corynespora cassicola.**
Nematode, burrowing. **Radopholus similis,** FL.
Nematode, root knot. **Meloidogyne** sp., OK.
Rot, root. **Clitocybe tabescens,** FL; **Phymatotrichum omnivorum,** TX.
Rot, stem. **Sclerotinia sclerotiorum,** FL; **Phytophthora parasitica,** FL.

CASUARINA
(Australian-Pine)

Nematode, root knot. **Meloidogyne** sp., FL.
Rot. **Armillaria mellea,** CA; **Clitocybe tabescens,** FL.

   Casuarina is particularly susceptible to Clitocybe root rot but *C. cunninghamiana* is more resistant than other species.

CATALINA CHERRY
(Prunus lyoni)

Leaf spot; shot hole. **Coryneum beierincki.**

CATALPA

Anthracnose. **Gloeosporium catalpae,** MD, MA, NJ, NY, PA.
Blight, southern. **Sclerotium rolfsii,** on seedlings.
Damping-off. **Rhizoctonia solani,** NE; **Pythium ultimum** (somewhat resistant).
Leaf spot. **Alternaria catalpae,** widespread; **Cercospora catalpae,** MA to FL, TX and IA; **Phyllosticta catalpae,** general.
Nematode, root knot. **Meloidogyne** sp., southern states to OH.
Nonparasitic. **Chlorosis,** soil alkalinity. **Leaf Scorch,** heat, drought.
Powdery mildew. **Microsphaera alni** var. **vaccinii,** MA to AL, TX, IL, NE; **Phyllactinia corylea,** IL, IN, KY, OH, MA, SC, VA.
Rot, root. **Armillaria mellea,** WA; **Helicobasidium purpureum,** OH; **Phymatotrichum omnivorum,** TX; **Thielaviopsis basicola,** of seedlings.
Rot, wood. **Collybia velutipes,** IN; **Polyporus** spp.; **Schizophyllum commune; Stereum** spp.; **Trametes sepium.**
Sooty mold. **Capnodium axiliatum,** LA, SC.
Spot anthracnose. **Sphaceloma** sp.
Wilt. **Verticillium albo-atrum,** IL, IN, KS, NJ, NY, MA, OH, VA.

Leaf spots may cause some defoliation in a wet season, but many years the expense of spraying may be unjustified on a limited budget. Verticillium wilt kills street trees.

CATHA
(Arabian-Tea)

Blight, leaf tip. **Colletotrichum gloeosporioides,** FL.

CATNIP
(*Nepeta*)

Bacterial leaf spot. **Pseudomonas tabaci,** WI.
Blight, gray mold. **Botrytis cinerea,** AK.
Blight, southern. **Sclerotium rolfsii,** TX.
Leaf spot. **Ascochyta nepetae,** WI; **Cercospora nepetae,** IL, TX; **Phyllosticta decidua,** IL, NY, OH, WI; **Septoria alabamensis,** AL; **S. nepetae,** WI.
Rot, root. **Rhizoctonia solani,** TX.
Virus. **Cucumber Mosaic,** IN, IA, KS, MI, WI.
Wilt. **Fusarium** sp., GA.

CATS-CLAW
(*Doxantha*)

Rot, root. **Phymatotrichum omnivorum,** TX.

CAT TAIL
(*Typha*)

Leaf spot. **Stagonospora typhoidearum,** TX, WI; **Phyllosticta typhina,** NE, NY, OR, TX, WI; **Scolecotrichum typhae,** CO.
Mold, leaf. **Cladosporium** spp., general.
Rot, culm. **Ophiobolus** sp., AR.
Rot, leaf. **Phythiogeton autossytum,** OH; **Pythium helicoides,** OH.

CEANOTHUS

Bacterial crown gall. **Agrobacterium tumefaciens,** WA.
Leaf spot. **Cercospora ceanothi,** KS, WI; **Septoria ceanothi,** ID; **Cylindrosporium ceanothi,** Pacific Coast; **Phyllosticta ceanothi,** MS.
Powdery mildew. **Microsphaera alni,** widespread.
Rot, root. **Armillaria mellea,** CA.
Rot, sapwood. **Schizophyllum commune,** CA.
Rust. **Puccinia tripsaci** (0, I), KS, NE, WI; II, III, on grasses.
Wilt. **Verticillium albo-atrum,** CA.

CEDAR
(*Cedrus* spp; Atlas Cedar, Deodar, Cedar of Lebanon)

Canker; dieback. **Diplodia pinea,** AL.
Rot, heart. **Fomes pini,** occasional.
Rot, root, **Armillaria mellea,** MS; **Clitocybe tabescens,** FL; **Phymatotrichum omnivorum** TX; **Phytophthora cinnamomi,** CA.

CELANDINE
(*Chelidonium*)

Leaf spot. **Septoria chelidonii,** TX.
Rot, root. **Phymatotrichum omnivorum,** TX.

CELERY, CELERIAC
(*Apium graveolens*)

Bacterial leaf spot. **Pseudomonas jaggeri** pv. **apii,** CT, DE, FL, IN, MI, MN, ND, NY, OH; leaf blight, **P. cichorii,** FL.
Bacterial, MLO. **California Aster Yellows,** CA, ID, UT, WA.
Bacterial soft rot. **Erwinia carotovora,** cosmopolitan.
Blight, early. **Cercospora apii,** general.

Blight, late. **Septoria apiicola** (includes *S. apii* large leaf spot, and *S. apiigraveolentis*), general.

Damping-off. **Pythium** spp.; **Rhizoctonia solani; Aphanomyces euteiches,** cosmopolitan.

Leaf spot. **Phyllosticta apii,** DE, NJ; **Stemphylium** sp. and **Alternaria** sp., secondary.

Nematode, awl. **Dolichodorus heterocephalus,** red root, FL.

Nematode, pin. **Paratylenchus hamatus,** New England, CA; **P. projectus,** NJ.

Nematode, root knot. **Meloidogyne incognita-acrita.**

Nematode, sheath. **Hemicycliophora parvana,** FL.

Nematode, stem. **Ditylenchus dipsaci,** CA.

Nematode, sting. **Belonolaimus gracilis,** FL; **B. longicaudatus.**

Nematode, stubby root. **Trichodorus christiei.**

Nonparasitic

  Black Heart. Wide fluctuations of soil moisture, general.

  Brown stem. Overaged plants.

  Cracked stem. Boron deficiency.

  Hollow stem. Pithiness. Sometimes chilling; sometimes genetic.

Rot, gray mold. **Botrytis cinerea,** cosmopolitan.

Rot, pink; watery soft. **Sclerotinia sclerotiorum,** general; **S. intermedia; S. minor.**

Rot, root; scab on celeriac. **Phoma apiicola,** CA, MI, NY, OH, WI.

Rot, stem; brown spot. **Cephalosporium apii,** CA.

Rot, storage. **Centrospora acerina,** NY; **Typhula variabilis,** NY; **Tricothecium roseum,** pink mold, NY.

Slime mold. **Physarum** spp., CA.

Virus. **Celery Calico,** CA, FL, NY, OH; **Celery Mosaic,** CA, WA, AL, FL; **Cucumber Mosaic, Beet Curly Top,** OR; **Celery Yellow Spot,** CA: **Tomato Spotted Wilt; Lettuce Mosaic,** NY; **Broad Bean Wilt,** NY.

Wilt. **Verticillium albo-atrum,** CA.

Wilt, yellows. **Fusarium oxysporum** f. sp. **apii,** TX CA, MI, NY, TX, general.

In many areas soil treatment for nematodes increases yield. Yellow-resistant varieties are on the market. Seed should be treated for leaf blights unless more than 2 years old. In Florida, development of apothecia of *Sclerotinia* is inhibited by flooding fields. Western growers fight mosaic by celery-free periods and by controlling insect vectors.

CELTUCE
(*Lactuca serriota* var. *sativa*)

Downy mildew. **Bremia lactucae,** ND, PA, WA.

Leaf spot. **Septoria lactucae.**

Rot, watery soft; drop. **Sclerotinia sclerotiorum,** MA.

Celtuce is a kind of lettuce and subject to some of the same diseases.

CENTAUREA
(Bachelors-Button, Basketflower; Cornflower; Dusty-miller)

Bacterial, MLO. **Aster Yellows,** widespread; **California Aster Yellows,** CA.
Blight, southern. **Sclerotium rolfsii,** CT, MD, NJ, TX.
Dodder. **Cuscuta** sp.
Downy mildew. **Bremia lactucae,** CA; **Plasmopara halstedii,** IA, TX.
Nematode, root knot. **Meloidogyne** sp., CA, FL, OH.
Powdery mildew. **Erysiphe cichoracearum,** CA, CT.
Rot, root. **Phymatotrichum omnivorum,** TX; **Pythium** sp.; **Rhizoctonia solani,** IN, IL, NJ, NY, TX.
Rot, stem. **Sclerotinia sclerotiorum,** CA, IN, MS, MO, TX; **Phytophthora cactorum,** NY.
Rust. **Puccinia cyani** (0, I, III), MA to NC, IN, CA, OR, WA; **P. irrequiseta** (II, III), TX; 0, I, unknown.
White rust. **Albugo tragopogonis,** TX.
Wilt. **Verticillium albo-atrum,** NY.
Wilt, stem rot. **Fusarium oxysporum** f. sp. **callistephi,** MI.

CENTIPEDE GRASS
(*Eremochloa*)

Anthracnose. **Colletotrichum graminicola,** FL.
Leaf mold. **Curvularia** sp., LA; **Stachybotrys** sp., MD.

CENTURY PLANT
(*Agave*)

Anthracnose. **Glomerella cingulata** (*Colletotrichum agaves*) occasional.
Blight, gray mold. **Botrytis cinerea,** after chilling.
Leaf scorch; blight. **Stagonospora gigantea,** TX, NM.
Leaf spot. **Coniothryium concentricum, C. agaves,** common.
Leaf spot, black patch. **Dothidella parryi,** CA.

CEPHALOTAXUS
(Japanese Plum-Yew)

Blight, nursery. **Phomopsis juniperovora,** NY.

CESTRUM

Bacterial canker, vascular. **Clavibacter michiganense,** WY.

CHAEROPHYLLUM

Broomrape. Orobanche ramosa, TX.

CHAMAECYPARIS
(Atlantic White-Cedar; Port Orford White-Cedar; Alaska Yellow-Cedar; Hinoki Cypress; Sawara Cypress-Retinospora)

Bacterial crown gall. **Agrobacterium tumefaciens**, CA.
Blight, nursery. **Phomopsis juniperovora**, widespread.
Blight, tip. **Pestalotia funerea**, MI, NJ, TX; **Didymascella chamaecyparissi**, NY.
Canker, bark patch. **Aleurodiscus nivosus**, NJ.
Nematode. **Sphaeronema**, AK; **Crossonema**, AK.
Nonparasitic. **Scorch**, sun, freezing, drought, mites.
Rot, collar and root. **Phytophthora lateralis**, OR, WA, serious; **P. cinnamomi.**
Rot, heart. **Fomes pini**, occasional; **F. pinicola**, AK; **F. subroseus**, NJ, NC.
Rot, root. **Armillaria mellea**, VA; **Clitocybe tabescens**, FL; **Phymatotrichum ominivorum**, TX; **Pythium ultimum**, CA; **Phytophthora lateralis**, CA.
Rot, wood. **Lenzites saepiaria**, VA; **Poria** spp.; **Polyporus** spp.; **Steccherinum balloui**, NJ; **Trametes isabellina**, CA; **Fomes annosus.**
Rust, gall. **Gymnosporangium biseptatum** (III), ME, NH to NJ, AL; 0, I, on serviceberry; **G. nootkatense** (III), OR, WA, AK; 0, I, on crabapple, pear, mountain-ash; **G. fraternum** (III), ME, MA, NJ; 0, I, on chokeberry.
Rust, witches' broom. **Gymnosporangium ellisii** (III), ME to FL and AL; 0, I, on sweet fern, bayberry, wax-myrtle.

CHAMAEDAPHNE
(Cassandra, Leatherleaf)

Gall, leaf. **Exobasidium vaccinii**, widespread; **Synchytrium vaccinii**, NJ.
Leaf spot. **Ascochyta cassandrae**, NY to WI; **Venturia arctostaphyli**, NH, NY.
Rust. **Chrysomyxa ledi** var. **cassandrae** (II, III), ME to PA and MN; AK; 0, I, on pine.

CHAYOTE
(*Sechium*)

Anthracnose. **Colletotrichum lagenarium**, FL, TX.
Blight, southern. **Sclerotium rolfsii**, TX.
Leaf spot. **Cercospora sechii**, FL, TX.
Nematode, root knot. **Meloidogyne** sp.; **M. incognita-acrita.**

Nematode, stubby root, Trichodorus sp.
Rot, fruit. **Glomerella cingulata,** LA.

CHECKER MALLOW
(*Sidalcea*)

Leaf spot. **Ramularia sidalceae,** CO, CA, WY.
Rust. **Endophyllum tuberculatum** (III), CO, WY; **Puccinia interveniens** (0, I),
CA, CO, MT, ID, OR, WA, WY; II, III, on *Stipa*; **P. schedonnardi** (0, I), CO;
P. sherardiana (0, III), CA, CO, NV, AZ, OR, WA.

CHENOPODIUM

Damping-off. **Sclerotium rolfsii,** CA.
Virus. Commonly used as test plant (bioassay) for viruses.

CHERIMOYA, CUSTARD-APPLE
(Annona)

Anthracnose, fruit rot. **Glomerella cingulata,** FL.
Blight, stem. **Diplodia natalensis,** TX.
Rot, root. **Armillaria mellea,** CA; **Clitocybe tabescens,** FL; **Phymatotrichum
omnivorum,** TX.
Rust. **Phakopsora cherimoliae** (II, III), FL, TX.

CHERRY
(*Prunus* spp.)

Bacterial black spot; canker; gummosis. **Xanthomonas pruni,** NY to MI,
GA and TX.
Bacterial canker. **Pseudomonas syringae,** MA to MI; Pacific states and
P. *morsprunorum.*
Bacterial crown gall. **Agrobacterium tumefaciens,** CA, TX, WA.
Bacterial fire blight. **Erwinia amylovora,** OR, WA.
Bacterial, MLO. **Peach X-Disease; Peach Yellow Leaf Roll (Peach Western
X-Disease),** Northwest.
Black knot. **Dibotryon morbosum,** eastern states.
Blight, blossom, brown rot. **Monilinia laxa,** general, Pacific states, NY, MI,
WI.
Blight, seedling, twig. **Monilinia rhododendri,** VT to GA, AR, IA; **M. fructicola,**
shoot and leaf, GA.
Blight, shoot; shot hole. **Coryneum carpophilum** (C. *beijerinckii*), CA, ID,
OR, WA.

Canker, felt fungus. **Septobasidium retiforme,** GA.

Canker, trunk and collar. **Phytophthora cactorum, P. citrophthora,** CA.

Canker, twig. **Phomopsis padina; Cytospora leucostoma,** NY, PA; **Valsa leucostoma,** widespread; **Nectria** sp., NY.

Leaf blister. **Taphrina farlowii,** VT to FL, TX; **Taphrina cerasi,** ME to NJ, MN.

Leaf spot. **Cercospora circumscissa** (*Mycosphaerella cerasella*), NJ, PA, VA to FL, TX; **Alternaria citri** var. **cerasi,** CA; **Phyllosticta pruni-avium,** OR; **Cercospora graphioides** and **Phyllosticta serotina** on wild blackcherry only; **Coccomyces lutescens,** VA.

Leaf spot; blight; shot hole. **Coccomyces hiemalis,** general.

Mistletoe. **Phoradendron serotinum (flavescens).**

Nematode, lesion. **Pratylenchus** sp.; **P. vulnus,** CA.

Nematode, root knot. **Meloidogyne** sp; **M. incognita-acrita; M. javanica.**

Nonparasitic. **Brown Bark Spot,** possibly caused by arsenical poisoning of soil, ID, MT, WA. **Chlorosis,** alkali injury, CA, TX. **Little Leaf,** zinc deficiency, CA, OR, WA.

Powdery mildew. **Podosphaera oxyacanthae,** general.

Rot, brown. **Monilinia fructicola,** general.

Rot, fruit. **Alternaria** sp.; **Botrytis cinerea; Cladosporium herbarum,** CA to WA and ID; **Lambertella** sp., OR; **Microstroma tonellianum,** MA; **Penicillium expansum; Rhizopus stolonifer,** occasional in market; **Pullularia** sp., Northwest.

Rot, heart; wood. **Fomes fomentarius,** Northeast; **Polyporus** spp.; **Poria ambigua,** CA.

Rot, root. **Armillaria mellea,** NM, OK, OR; **Phymatotrichum omnivorum,** TX; **Xylaria** spp., VA; **Poria ambigua,** CA.

Rot, silver leaf. **Stereum purpureum,** NY, MT.

Rust. **Tranzschelia discolor** (II, III), GA, MA, NC, NE, NY, OK, TX; 0, I, on anemone.

Scab. **Fusicladium cerasi,** NY to IA and WI.

Virus. **Cherry Rough Fruit,** UT on sweet cherry (*Prunus avium*).

Virus. **Cherry Albino; Cherry Bark-Splitting; Cherry Black Canker; Cherry Buckskin; Cherry Bud Abortion; Cherry Chlorosis; Cherry Freckle Fruit Disease; Cherry Green Ring Mottle; Cherry Gummosis; Cherry Little Cherry; Cherry Midleaf Necrosis; Cherry Mora; Cherry Mottle Leaf; Cherry Necrotic Rusty Mottle; Cherry Pink Fruit; Cherry Pinto Leaf; Cherry Rasp Leaf; Cherry Ring Spot; Cherry Rugose Mosaic; Cherry Rusty Mottle; Cherry Twisted Leaf; Cherry Vein Clearing; Cherry Yellows; Prune Dwarf; Tomato Ring Spot; Stem Pitting.**

CHERRY, FLOWERING, ORIENTAL
(*Prunus serrulata*)

Bacterial fire blight. **Erwinia amylovora,** GA, OH.

Bacterial leaf spot. **Xanthomonas pruni,** NJ.

Canker; dieback. **Botryosphaeria ribis,** GA.

Decline. Cause, unknown; complex of insect, poor soil, viral infection and air pollutants suspected, VA.

Leaf blister; witches' broom. **Taphrina cerasi,** MD, NJ.

Leaf spot. **Coccomyces hiemalis,** MA, NJ.

Scab. **Cladosporium carpophilum,** MS.

Virus. **Cherry Vein Clearing; Cherry (flowering) Rough Bark; Cherry Little Cherry.**

CHERRY, JAPANESE FLOWERING
(Prunus subhirtella)

Leaf spot. **Cristulariella pyramidalis,** FL.

CHERRY-LAUREL
(Prunus laurocerasus)

Bacterial spot. **Xanthomonas pruni,** GA, MS, NJ, SC.

Blight, blossom; brown rot. **Monilinia fructicola,** CA; **M. laxa,** CA; **Alternaria** sp., TX.

Leaf spot. **Cercospora circumscissa,** CA; **C. cladosporioides,** LA, TX; **Coccomyces lutescens,** MS; **Phyllachora beaumontii,** AL; **Phyllosticta laurocerasi,** CA, FL, NJ; **Septoria ravenelii,** SC.

Mistletoe. **Phoradendron serotinum (flavescens),** FL.

Rot, root. **Armillaria mellea,** CA; **Clitocybe tabescens,** FL; **Phymatotrichum omnivorum,** TX.

Wilt. **Verticillium albo-atrum,** CA.

CHESTNUT
(Castanea)

Blight, canker. **Endothia parasitica,** general, with American chestnut practically exterminated by it.

Blight, twig. **Cytospora** sp.; **Phomopsis** sp.; **Diplodia longispora.**

Canker, bark patch. **Aleurodiscus aceris.**

Canker; dieback. **Cryptodiaporthe castanea; Strumella coryneoidea.**

Leaf spot. **Actinopelta dryina; Cylindrosporium castaneae; Marssonina ochroleuca,** general; **Monochaetia desmazierii; M. kansensis; Phyllosticta castanea; Exosporium fawcettii; Scolecosporium fagi.**

Mistletoe. **Phoradendron villosum,** CA.

Powdery mildew. **Microsphaera alni; Phyllactinia corylea.**

Rot, root. **Armillaria mellea; Phytophthora cinnamomi; Phymatotrichum omnivorum.**

Rot, wood. **Fomes** spp.; **Polyporus** spp.; **Poria** spp.; **Stereum** spp.

Japanese and Chinese chestnuts are resistant to chestnut blight. Plant breeders are trying to develop hybrids between Asiatic and native species that will be resistant to *Endothia*.

CHICKORY

See Endive.

CHICK-PEA, GARBANZO
(*Cicer*)

Bacterial blight. **Pseudomonas andropogonis**, ME.
Blight. **Ascochyta radiei**, ID, WA.
Nematode, root knot. **Meloidogyne** spp.
Rot, root. **Fusarium solani** f. **pisi; Thielaviopsis basicola**, ID, WA.
Rot, root; damping-off. **Pythium ultimum**, CA; **Rhizoctonia solani**, CA.
Rust. **Uromyces ciceris-arietini**, GA.
Virus. **Mosaic,** unidentified, CA; **Chickpea filiform**, WA.
Wilt. **Fusarium lateritium** f. **cicerii; Verticillium albo-atrum.**

CHICKWEED
(*Stellaria*)

Virus. **Tomato Ringspot**, PA.

CHINABERRY
(*Melia*)

Black mildew. **Meliola** sp.
Blackleg. **Phoma lingam**, WA.
Blight, limb; twig. **Pellicularia koleroga**, FL; **Eutypella stellulata**, OK, TX; **Fusarium lateritium**, TX.
Canker. **Nectria coccinea**, MS, SC.
Downy mildew. **Peronospora parasitica**, general.
Leaf spot. **Cercospora leucosticta**, Gulf states; **C. meliae; C. subsessilis; Phyllosticta azedarachis**, AL; **P. meliae**, LA, TX.
Mistletoe. **Phoradendron serotinum (flavescens)**, TX.
Nematode, root knot. **Meloidogyne** sp.
Powdery mildew. **Phyllactinia corylea**, MS.

Rot, root. **Helicobasidium purpureum,** TX; **Phymatotrichum omnivorum,** AZ, TX.

Rot, wood. **Fomes meliae,** AL; **Polyporus versicolor,** GA.

CHINESE CABBAGE
(*Brassica pekinensis, B. chinensis*)

Anthracnose. **Colletotrichum higginsianum,** FL.

Bacterial black rot. **Xanthomonas campestris,** IN, MD, TX.

Bacterial leaf spot. **Pseudomonas maculicola,** VA.

Bacterial, MLO. **Aster Yellows.**

Blight, southern. **Sclerotium rolfsii,** TX.

Club root. **Plasmodiophora brassicae,** CT, MA, NJ, OH, PA.

Leaf spot, black, **Alternaria oleracea,** CA, CT, FL, MA, NH, TX; gray, **A. brassicae,** CA, CT, FL, IN, MD; **Cercospora brassicola,** FL, GA, NH, NJ; **Cercosporella brassicae,** white spot, widespread.

Leaf spot; white spot. **Pseudocercosporella capsellae,** CA.

Nematode, root knot. **Meloidogyne** sp.

Powdery mildew. **Erysiphe polygoni,** AZ, MA; **Peronospora parasitica.**

Rot, root; damping-off. **Aphanomyces raphani,** WI.

Rot, watery soft. **Sclerotinia sclerotiorum,** AZ, MA, NY, TX.

Virus. **Turnip Mosaic; Cauliflower Mosaic.**

White rust. **Albugo candida,** AL.

CHINESE EVERGREEN
(*Aglaonema*)

Anthracnose. **Colletotrichum** sp., **Gloeosporium** sp., WA.

Bacterial leaf spot. **Erwinia aroideae; Xanthomonas dieffenbachiae,** FL.

Glebal masses. **Sphaerobolus stellatus (Gasteromycetous fungus),** TX, FL.

Leaf spot. **Myrothecium roridum,** FL.

Nematode, lesion. **Pratylenchus coffeae,** FL.

Nematode, root knot. **Meloidogyne javanica.**

Rot, root; leaf. **Pythium splendens,** FL.

The root rot is serious, with all roots rotted, plants stunted, dying. Destroy infected plants. Take tip cuttings well above ground.

CHINESE LANTERN
(*Physalis alkekengi*)

Leaf spot. **Alternaria solani; Phyllosticta** sp., OK; **P. physaleos,** CT.

Smut, white. **Entyloma australe,** CT, NY.

Virus. Cucumber Mosaic; Tobacco Mosaic; Potato Mottle.
Wilt. Verticillium albo-atrum, NY.

CHINESE TALLOWTREE
(*Sapium*)

Leaf spot. Cercospora stillingiae, LA; Phyllosticta stillingiae, LA.
Rot, root. Clitocybe tabescens, FL; Phymatotrichum omnivorum, TX.

CHINESE WAXGOURD
(*Beincasa*)

Anthracnose. Colletotrichum lagenarium, IN.
Downy mildew. Pseudoperonospora cubensis, MA, OH.
Nematode, root knot. Meloidogyne sp.

CHINQUAPIN
(*Castanopsis*)

Blight, chestnut; canker. Endothia parasitica, general.
Canker, brown felt. Septobasidium pseudopedicellatum.
Leaf blister. Taphrina castanopsidis, CA.
Leaf spot. Dothidella castanopsidis, CA, OR; ?Mycosphaerella sp., CA, OR.
Powdery mildew. Microsphaera alni, NC.
Rot, heart; sapwood; wood. Fomes igniarius, OR; Ganoderma oregonensis;
    Peniophora sanguinea; Polyporus hirsutus; P. versicolor.
Rot, root. Armillaria mellea, OR.
Wilt, oak. Ceratocystis fagacearum.

CHIOGENES
(Creeping Snowberry)

Rust. Chrysomyxa chiogenis (II, III), MI, NH, NY, WI; 0, I unknown.

CHIONODOXA
(Glory-of-the-Snow)

Nematode, bulb. Ditylenchus dipsaci.

CHIRITA

Virus. Tobacco Mosaic, CA, CT, DC, FL, OH, WA.

CHIVES
(*Allium schoenoprasum*)

Downy mildew. **Peronospora destructor,** CA.
Rust. **Puccinia porri** (II, III), CT, NY, WA.
Smut. **Urocystis cepulae,** MA.

CHLOROGALUM
(Soap-Plant)

Leaf spot. **Heterosporium gracile,** CA.
Rust. **Uromyces aureus** (0, I, III), CA.

CHOKEBERRY
(*Aronia*)

Bacterial fire blight. **Erwinia amylovora,** MI, TX, WV.
Dodder. **Cuscuta compacta,** FL.
Leaf spot. **Ascochyta pirina,** WI; **Cercospora mali,** AL, TX; **C. piri,** MI, WI, NH; **Mycosphaerella arbutifolia,** NY; **Phyllosticta arbutifolia,** NJ.
Rot, brown. **Monilinia fructicola,** WI.
Rot, root. **Phymatotrichum omnivorum,** TX.
Rust. **Gymnosporangium clavariaeforme** (0, I), MA; III on juniper; **G. clavipes** (0, I), CT, ME, MA, TX; **G. davisii** (0, I), ME, MI, NH, WI; **G. fraternum** (0, I), DE, MA, ME, NJ, PA; III on chamaecyparis.

CHOKECHERRY
(*Prunus virginiana*)

Bacterial leaf spot. **Xanthomonas pruni,** IL, NY, MT, WY.
Bacterial, MLO. **Peach X-Disease,** NH to VA, IL, WI.
Black knot. **Dibotryon morbosum,** general.
Blight, fruit and shoot. **Monilinia demissa,** ID, WA; ?**M. padi** (*Sclerotinia angustior*), VT to KS, ND.
Canker; dieback. **Cytospora chrysosperma,** MT.
Leaf blister; fruit, shoot hypertrophy. **Taphrina confusa,** widespread.
Leaf spot. **Cylindrosporium nuttalli,** OR; **Gloeosporium prunicola,** NY; **Lophodermina prunicola,** tar spot, CO; **Mycosphaerella cerasella,** KS; **Phyllosticta circumscissa; P. virginiana,** leaf blotch, NY to KS and MT; **Septoria pruni,** MI.
Leaf spot; shot hole. **Cercospora circumscissa,** New England, IA, WI; ND to KS, and MT; **Cercospora lutescens,** general.
Nematode, root knot. **Meloidogyne** sp.

Powdery mildew. **Podosphaera oxyacanthae,** widespread; **Phyllactinia corylea,** WA.
Rot, brown heart. **Fomes fulvus,** ND, SD.
Rust. **Tranzschelia pruni-spinosae** (II, III), CT to IL, WI.
Virus. **Western X-Disease,** OR, ID, UT, WA; **Tatter Leaf (?Peach Ring Spot).**

Chokecherries should be eliminated near peach orchards to control the peach X-disease.

CHRISTMAS CACTUS
(*Zygocactus truncatus*)

Rot, basal stem. **Fusarium oxysporum,** MA.
Rot, root. **Phytophthora parasitica,** FL.

CHRISTMAS ROSE
(*Helleborus niger*)

Blight, flower spot. **Botrytis cinerea,** NJ; **Gloeosporium** sp., NJ.
Leaf spot, black. **Coniothyrium hellebori,** MD, NY, OR.
Rot, stem. **Sclerotium delphinii** (*S. rolfsii*), NY.

CHRYSANTHEMUM
(*Dendranthenia grandiflora*)

Bacterial, MLO. **Phloem Necrosis.** MLO, widespread; **Aster Yellows.**
Bacterial blight; wilt. **Erwinia chrysanthemi,** serious in greenhouses; **E. carotovora,** OK; **Pseudomonas cichorii,** FL.
Bacterial crown gall. **Agrobacterium tumefaciens,** CT, NJ, TX, FL.
Bacterial fasciation. **Clavibacter fascians,** NJ, MI, NY, OH.
Blight, gray mold. **Botrytis cinerea,** cosmopolitan.
Blight, petal. **Itersonilia perplexans,** MN, FL.
Blight, ray. **Ascochyta chrysanthemi** (*Mycosphaerella ligulicola*), MD to FL, MS, OH, CA; Ray Speck, **Stemphylium floridanum,** FL; **Alternaria** spp.
Blight, southern. **Sclerotium rolfsii,** FL, VA.
Blight, leaf; blotch. **Septoria leucanthemi,** widespread.
Dodder. **Cuscuta** sp., MI, NJ, NY, WA, WV; **C. arvensis** and **C. indecora,** TX.
Leaf spot. **Alternaria** sp., OK, TX; **Cercospora chrysanthemi,** AL, LA, MD, PA, TX; **Cylindrosporium chrysanthemi,** MA to AL and KS; **Phyllosticta chrysanthemi,** FL, MA, MS, VA.
Leaf spot. **Septoria chrysanthemella** (*S. chrysanthemi*) and **S. obesa,** general.
Nematode, leaf. **Aphelenchoides ritzema-bosi,** widespread.
Nematode, lesion. **Pratylenchus pratensis,** NJ, TX.
Nematode, root knot. **Meloidogyne** spp.

Nonparasitic. **Crackneck,** in greenhouses, probably overwatering and insufficient ventilation. **Air Pollution Injury. Yellow Strapleaf,** amino acid imbalance.

Petal spot. **Bipolaris** (*Helminthosporium*) **setaniae,** FL.

Powdery mildew. **Erysiphe cichoracearum,** general.

Rot, charcoal stem. **Macrophomina phaseoli,** OK.

Rot, flower. **Fusarium tricinctum** f. **poae,** NJ, LA.

Rot, root. **Fusarium** sp.; **Rhizoctonia solani,** general; **Pythium** sp.

Rot, root. **Phymatotrichum omnivorum,** AZ, TX.

Rot, stem. **Fusarium solani.**

Rot, stem; drop. **Sclerotinia sclerotiorum,** AZ, LA, MI, VA.

Rust. **Puccinia taneceti** (II), general.

Rust, white. **Puccinia horiana** (III, IV), NJ, PA.

Viroid. **Chrysanthemum Stunt,** serious; **Chrysanthemum Chlororic Mottle,** NY, FL, widespread.

Virus. **Chrysanthemum Flower Distortion; Chrysanthemum Mosaic; Chrysanthemum Ring Spot; Chrysanthemum Rosette; Tomato Aspermy; Tomato Spotted Wilt.**

Wilt. **Fusarium oxysporum** f. sp. **callistephi; F. oxysporum** f. sp. **tracheiphilum,** SC.

Wilt. **Fusarium oxysporum** f. sp. **chrysanthemi,** SC, FL, widespread.

Wilt. **Verticillium albo-atrum.**

When garden chrysanthemums have foliage browning and dying progressively up the stem, the cause can be leaf nematodes, Septoria leaf spots, or Verticillium wilt. To reduce spread of nematodes home gardeners should take tip cuttings rather than crown divisions. Commercial growers should procure stock free from stunt and other virus diseases.

CHRYSOPSIS
(Golden Aster)

Leaf spot. **Cercospora macroguttata,** AL, MS; **Ramularia chrysopsidis,** NY.

Powdery mildew. **Erysiphe cichoracearum,** MT, WY.

Rust. **Coleosporium asterum** (II, III), CO, FL, NE; 0, I, on pine; **Puccinia grindeliae** (III), CA, CO, OK, UT, WY; **P. stipae** (0, I), AZ, CO, FL, MT, NE, WY; II, III, on grasses.

CIMICIFUGA
(Bugbane, Black Cohosh)

Leaf spot. **Ascochyta actaeae,** CT, NY; **Ectostroma afflatum,** VA.

Nematode, root knot. **Meloidogyne** spp., NJ.

CINCHONA

Leaf spot. **Cercospora cinchonae,** LA.
Nematode, root knot. **Meloidogyne** sp.

CINERARIA
(*Senecio*)

Bacterial, MLO. **Aster Yellows,** NY.
Blight. **Botrytis cinerea,** IN, MO, NJ, PA, AK.
Damping-off. **Rhizoctonia solani,** IL, NY.
Downy mildew. **Plasmopara halstedii,** NY.
Nematode, root knot. **Meloidogyne** sp.
Powdery mildew. **Erysiphe cichoracearum,** MA.
Rot, root. **Pythium** sp., MD; **P. ultimum; Thielaviopsis basicola,** MA.
Rot, stem. **Fusarium** sp., PA; **Phytophthora** sp., NJ; **Sclerotinia sclerotiorum,** WA.
Virus. **Cineraria Mosaic,** WA; **Tomato Spotted Wilt; Chrysanthemum Stunt** (viroid).
Wilt. **Verticillium albo-atrum,** NJ, NY, WA.

CINNAMON-TREE
(*Cinnamomum zeylandicum*)

Anthracnose. **Glomerella cingulata,** FL.
Leaf spot, algal. **Cephaleuros virescens,** FL.

CIRSIUM
(Plumed Thistle)

Leaf spot. **Alternaria chrysanthemi,** MT; **Cercospora** spp.; **Phyllosticta cirsii; Septoria cirsii,** WI.
Powdery mildew. **Erysiphe cichoracearum,** general; **Sphaerotheca macularis.**
Rot, root. **Rhizoctonia solani,** IL; **Phymatotrichum omnivorum,** TX.
Rust. **Puccinia cirsii** (0, I, II, III), PA to NC, TX, CA, OR; **Uromyces junci** (0, I), MT, NE, ND.
White rust. **Albugo tragopogonis,** NY to IA, TX, WY.
Smut, Inflorescence. CO, UT.

CISSUS

Leaf spot. **Cercospora viticola,** LA; **C. arboreae,** TX; **Phyllosticta cissicola,** TX.

Rot, root. **Phymatotrichum omnivorum,** TX.

Rust. **Aecidium mexicanum** (0, I), OK.

Smut. **Mycosyrinx cissi,** FL.

CITRUS FRUITS
(Grapefruit, Lemon, Lime, Orange)

Anthracnose, lime; withertip. **Gloeosporium limetticolum,** on lime only, CA.

Anthracnose; withertip, fruit spot. **Glomerella cingulata,** general.

Bacterial blast; blight; black pit. **Pseudomonas syringae,** CA; after cold rains.

Bacterial canker. **Xanthomonas campestris** pv. **citri,** FL.

Bacterial crown gall. **Agrobacterium tumefaciens,** AZ, CA.

Bacterial, MLO. **Citrus Stubborn Disease.**

Blight, brown spot; fruit and young shoots. **Alternaria citri,** FL.

Blight, leaf and stem. **Phytophthora syringae,** CA.

Blight, seedling; fruit rot. **Sclerotium rolfsii.**

Blight, thread. **Pellicularia koleroga,** Gulf states.

Blight, twig. **Gibberella baccata** (*Fusarium lateritium*), CA; **Tryblidiella rufula,** TX.

Blotch, sooty. **Gloeodes pomigena,** Gulf states.

Canker, branch knot. **Sphaeropsis tumefaciens,** FL.

Canker, branch wilt. **Exosporina fawcetti,** CA.

Canker; dieback. **Aspergillus foetidus,** CA.

Canker, felt fungus. **Septobasidium** spp., Gulf states.

Canker, trunk. **Botryodiplodia theobromea,** TX.

Canker, wound. **Hendersonula toruloidea,** CA.

Damping-off. **Pythium** spp.; **Rhizoctonia solani,** cosmopolitan.

Dodder. **Cuscuta americana,** CA, FL; **Cassytha filiformis,** a dodderlike plant, FL.

Leaf spot. **Cercospora aurantia,** TX, FL; greasy spot, **C. citri-grisea,** FL; tar spot, **C. gigantea,** FL; **Cladosporium oxysporum,** FL; **Mycosphaerella lageniformis,** CA, **Pleospora** sp., CA; **Alternaria citri,** on Rangpur lime; fruit spot, **Septoria citri,** CA, TX; **S. limonium,** CA.

Leaf spot, algal; red rust. **Cephaleuros virescens,** Gulf states.

Leaf spot; greasy spot. **Mycosphaerella citri,** FL.

Mistletoe. **Phoradendron** sp., Gulf states.

Nematode, burrowing. **Radopholus similis,** FL.

Nematode, citrus. **Tylenchus semipenetrans,** FL, CA.

Nematode, lance. **Hoplolaimus coronatus.**

Nematode, lesion. **Pratylenchus brachyurus; P. vulnus.**

Nematode, ring. **Criconema civellae; Criconemoides citri,** FL.

Nematode, stubby root. **Trichodorus christiei.**

Nonparasitic

    Blight. Boron deficiency, CA, FL.

    Bronzing. Magnesium deficiency, FL.

    Cancroid Spot. Genetic abnormality.

    Chlorosis. Iron deficiency, AZ, CA, FL; manganese deficiency, FL.

Exanthema. Copper deficiency, augmented by excessive nitrogen fertilization and bad drainage, CA, FL.

Greasy Spot. Black melanose, cause unknown, CA, TX; chiefly grapefruit.

Gummosis. Gum spot; gummosis. Environmental injuries in part.

Leprosis. Florida scaly bark; nailhead rust. Cause unknown, FL.

Mottle Leaf. Zinc deficiency, AZ, CA, FL.

Oleocellosis. Rind-oil spot. Chemical injury from release of oil in rind.

Rumple. Rind network on lemon; cause unknown.

Silver Scurf. Thrips injury on fruit, widespread; silvering. Rust mite injury; stigmonose. Insect punctures.

Wilt; Blight; leaf curl. Irregular water supply, FL.

Powdery mildew. **Oidium** sp., CA, FL; **O. tingitaninum.**

Rot, black, of fruit. **Alternaria citri,** widespread.

Rot, brown; gummosis. **Phytophthora citrophthora,** AZ, CA, FL; **P. parasitica,** "mal di gomma," CA, FL.

Rot, charcoal. **Macrophomina phaseoli,** AZ, CA.

Rot, cottony fruit; twig blight. **Sclerotinia sclerotiorum,** CA, FL, TX.

Rot, Dothiorella; bark canker. **Botryosphaeria ribis,** widespread.

Rot, Diplodia; collar; twig blight. **Diplodia natalensis,** general.

Rot, fruit. **Aspergillus** spp.; **Fusarium** spp.; **Mucor** spp.; **Oospora citriaurantii,** sour rot, **Nematospora coryli,** dry rot; **Candida krusei.**

Rot, fruit. **Gloeosporium foliicolum** (*Glomerella cingulata*), FL.

Rot, gray mold; twig blight. **Botrytis cinerea,** CA.

Rot, green mold. **Penicillium digitatum;** blue mold, **P. italicum;** pink mold, **P. roseum,** cosmopolitan; black mold, **Rhizopus stolonifer; Trichothecium viride,** occasional.

Rot, melanose; phomopsis; decorticosis; shell bark. **Diaporthe citri,** general.

Rot, mushroom root. **Armillaria mellea,** CA; **Clitocybe tabescens,** FL.

Rot, root. **Fusarium solani; Phymatotrichum omnivorum,** TX; **Poria vaporaria,** CA; **Thielaviopsis basicola.**

Rot, wood. **Daldinia concentrica,** occasional; **Ganoderma lucidum,** FL; **Polyporus** spp.; **Schizophyllum commune; Trametes hydnoides; T. hispida.**

Sooty mold, **Capnodium citri; C. citricola.** Gulf states.

Spot anthracnose; citrus scab. **Elsinoë fawcetti,** AL, FL, LA, MS, CA.

Virus. **Algerian Navel Orange,** FL; **Citrus Ringspot,** CA, FL.

Virus. **Citrus Exocortis** (viroid); **Citrus Psorosis,** scaly bark; **Citrus Tatter Leaf; Citrus Tristeza,** quick decline; **Citrus Vein Enation; Citrus Xyloporosis,** cachexia; **Citrus Yellow Vein.**

In the home garden oil sprays for scale help to get rid of sooty mold. A neutral copper spray just before growth starts is recommended in Florida for citrus scab and one just after flowering for melanose. Citrus trees in the burrowing nematode area are being pulled and the soil is treated. Consult the California and Florida Agricultural Experiment Stations for the latest information on specific citrus problems.

CLARKIA

Anthracnose. **Colletotrichum** sp., PA.
Bacterial, MLO. **Aster Yellows**, NY, and **California Aster Yellows**, CA.
Blight, gray mold; canker. **Botrytis cinerea**, CA, NY.
Damping-off. **Pythium debaryanum**, CA; **Rhizoctonia solani**, CT.
Downy mildew. **Peronospora arthuri**, CA.
Gall, leaf. **Synchytrium fulgens**, CA.
Leaf spot. **Alternaria tenuis**, secondary.
Rot, stem. **Fusarium** sp., CA; **Phytophthora cactorum**, NY.
Rust. **Puccinia oenotherae** (0, I, II, III), CA, ID, NV, OR, WA; **Pucciniastrum pustulatum** (II, III), NY, AK.
Wilt. **Verticillium albo-atrum**, CA.

CLAUSENA
(Wampi)

Leaf spot. **Glomerella cingulata**, MD.
Nematode, citrus. **Tylenchulus semipenetrans**, FL.
Spot anthracnose; citrus scab. **Elsinoë fawcetti**, FL.

CLAYTONIA
(Spring Beauty)

Downy mildew. **Peronospora claytoniae**, CA, IA, MD, WA, TX.
Gall, leaf. **Physoderma claytoniana**; MI, WI.
Leaf spot. **Ramularia claytoniae**, CA.
Rot, root. **Phymatotrichum omnivorum**, TX.
Rust. **Puccinia marie-wilsoniae** (0, I, III), NH to VA, MO and WI, CO; UT, WA, **P. agnita** (0, III); **Uromyces claytoniae** (0, I, III), NY.

CLEMATIS
(Including Virgins-Bower)

Bacterial crown gall. **Agrobacterium tumefaciens**, MN, TX.
Blight, leaf. **Phleospora adusta**, TX.
Leaf spot. **Ascochyta clematidina**, widespread; **Cercospora rubigo**, CA, WA; **C. squalidula**, widespread; **Cylindrosporium clematidis**, East and South; **Glomerella cingulata**, FL; **Phyllosticta clematidis**, MT, VA; **Ramularia clematidis**, MT; **Septoria clematidis**, WA, WI; **Sphaerella applanata**, MT, TX.
Nematode, root knot. **Meloidogyne hapla**, ND.

Powdery mildew. **Erysiphe polygoni,** widespread.

Rust. **Puccinia recondita** (0, I), Rocky Mountains and Pacific Coast; II, III, on grasses; **P. pulsatillae** (III), CA; **P. stromatica** (III), AL; **Tranzschelia viornae** (II, III), TX.

Smut. **Urocystis carcinodes,** UT.

CLEOME
(Spider-Flower)

Downy mildew. **?Peronospora parasitica,** LA.

Leaf spot. **Cercospora cleomis,** MI, NJ; **C. conspicua; Heterosporium hybridum,** IA, MT.

Nematode, root knot. **Meloidogyne** sp.

Rust. **Puccinia aristidae** (0, I), AZ, CO, IN, MT, NE, NM; II, III, on grasses.

CLERODENDRON
(Glorybower)

Canker, stem. **Kutilakesa pironii,** FL.

Leaf spot. **Septoria phylctaenioides,** SC; **Cercospora apii** f. sp. **clerodendri,** FL.

Nematode, root knot. **Meloidogyne** sp., MD.

Virus. **Zonate Ring Spot,** FL.

CLETHRA
(Sweet Pepperbush, White-Alder)

Gall, red leaf. **Synchytrium vaccinii,** NJ.

Leaf spot. **Phyllosticta clethricola,** MD, NJ, TX.

Rot, root. **Corticium galactinum.**

CLINOPODIUM
(Basil-Weed)

Leaf spot. **Phyllosticta decidua,** NY.

Rust. **Puccinia menthae** (0, I, II, III), MA to VA, CO and WI.

CLINTONIA

Gall, leaf; false rust. **Synchytrium aureum,** WI.

Rot, leaf. **Ceratobasidium anceps,** ME, NY, WI.

Rust. **Puccinia mesomajalis** (III), MI, MN, NH, NY, TN, VA, WI, CA, ID, MT, OR, WA.

CLIVIA

Nematode, spiral. **Rotylenchus brachyurus.**

COCCOLOBA
(Sea-Grape, Dove-Plum)

Black mildew. **Lembosia coccolobae; L. philodendri; L. portoricense; L. tenella,** FL.
Leaf spot. **Pestalotia coccolobae,** FL.
Rust. **Uredo coccolobae; U. uviferae** (II), FL.
Rust. **Puccinia canaliculata,** GA.

COCKLEBUR
(*Xanthium*)

Nematode, root knot. **Meloidogyne hapla.**
Rot, stem. **Sclerotinia trifoliorum,** MD; **S. sclerotiorum,** MD; **S. minor,** MD.

COCKSCOMB
(*Celosia argentia*)

Damping-off. **Rhizoctonia solani,** CT.
Leaf spot. **Cercospora celosiae,** AL, OK; **Phyllosticta** sp., NJ; **Alternaria** sp., NJ.
Nematode, root knot. **Meloidogyne** sp., KS, OH, TX.
Rot, charcoal. **Macrophomina phaseoli,** TX.
Virus. **Beet Curly Top,** CA, TX.
Virus. **Cucumber Mosaic,** NY.

COCOA-PLUM
(*Chrysobalanus*)

Leaf spot, **Cercospora chrysobalani,** FL.
Leaf spot, algal. **Cephaleuros virescens,** FL.

COCOYAM
(*Xanthosoma*)

Bacterial leaf spot. **Xanthomonas campestris** pv. **dieffenbachiae,** FL.

CODONANTHE

Virus. **Tobacco Mosaic,** CA, CT, DC, FL, OH, WA.

COFFEE-BERRY
(*Rhamnus californicus*)

Leaf spot. **Phaeosphaerella rhamni,** CA.
Rust. **Puccinia mesneriana** (III), CA.
Sooty mold. **Capnodium** sp., CA.

COLEUS

Damping-off; cutting rot. **Pythium** spp., CA, MD; **Rhizoctonia solani,** FL.,
   IL, NY, TX.
Broomrape. **Orobanche ramosa,** NY.
Leaf spot. **Alternaria** sp., NJ; **Phyllosticta** sp., NJ.
Nematode, leaf. **Aphelenchoides fragariae,** NJ.
Nematode, root knot. **Meloidogyne** spp., AL, CA, CT, MD, MO, NJ, NY, OK.
Nonparasitic. **Crinkle,** genetic leaf deformity.
Rot, gray mold; leaf blight. **Botrytis cinerea,** MO, AK.
Slime mold. **Badhamia panicea,** KS.
Virus. **Coleus Mosaic.**
Wilt. **Verticillium** sp., CT.

COLLARDS

See Cabbage. Subject to downy mildew and Sclerotinia rot.

COLLINSIA
(Blue-Lips, Blue-Eyed Mary)

Leaf spot. **Septoria collinsiae,** IL.
Rot, root. **Pythium mammillatum,** CA.

Rust. **Aecidium insulum** (0, I), UT; **Puccinia collinsiae** (0, I, II, III), CA, OR, UT, WA.
Smut, white. **Entyloma collinsiae**, CA, OR.

COLLINSONIA
(Horse-Balm)

Black spot, on stem. **Phyllachora** sp., PA.

COLLOMIA

Powdery mildew. **Sphaerotheca macularis**, CA, CO, ID, MT, ND, WA, WY.
Nematode, bulb. **Ditylenchus dipsaci**, Pacific Northwest.
Rust. **Puccinia giliae** (II, III), CA, WA; 0, I, unknown; **P. plumbaria** (0, I, III), NV; **Uromyces acuminatus** var. **polemonii** (0, I), CO, NE, ND; II, III, on marsh grass.

COLTSFOOT
(*Tussilago*)

Leaf spot. **Mycosphaerella tussilaginis**, NY; **Septoria farfaricola**, TN.

COLUMBINE
(*Aquilegia*)

Blight, gray mold. **Botryris cinerea; ?B. streptothrix.**
Damping-off; root rot. **Rhizoctonia solani**, IL.
Leaf spot, **Ascochyta aquilegiae**, CT, IL, IA, NJ, NY, PA, TX, WI; **Cercospora aquilegiae**, KS, OR; **Septoria aquilegiae**, CT, IN, MI, NY, OH, VA, VT, WI; **Haplobasidium pavoninum**, AK.
Nematode, root knot. **Meloidogyne** sp.
Powdery mildew. **Erysiphe polygoni**, IL, IN, IA, NJ, NY, OH, PA, TX, UT, WY, WI, AK.
Rot, root. **Phymatotrichum omnivorum**, TX; **Pythium mamillatum**, CA.
Rot, stem. **Phoma** sp., PA; **Sclerotinia sclerotiorum**, DE, OH, PA, TX: **?Phyllosticta aquilegicola**, WA.
Rust. **Puccinia recondita** (0, I), CO, CA, ID, NM, OR, WA; II, III, on grasses.
Smut, leaf and stem. **Urocystis sorosporioides**, UT.
Virus. **Mosaic,** unidentified, IA, KS.

COLUMBO
(*Frasera*)

Leaf spot. **Asteroma fraserae,** black mildew, CO, ID; **Cercospora fraserae,** CO, UT; **Marssonina fraserae,** ID, WA; **Phyllosticta fraserae,** CO.
Rust. **Uromyces speciosus** (II, III), CO, NM; 0, I, unknown.

CONFEDERATE-JASMINE
(*Trachelospermum*)

Black mildew; sooty mold. **Dimerosporium pulchrum,** LA.
Leaf spot. **Cercospora repens,** LA.
Rot, root. **Clitocybe tabescens,** FL.

CORAL-BELLS

See Heuchera.

CORALBERRY
(*Symphoricarpos orbiculatus*)

Canker; stem gall. **Phomopsis** sp., MD.
Leaf spot. **Cercospora symphoricarpi,** KS, NE, TX.
Powdery mildew. **Microsphaera diffusa,** general.
Rot, berry. **Alternaria** sp., CT.
Rot, root. **Helicobasidium purpureum,** TX; **Phymatotrichum omnivorum,** TX.
Rust. **Puccinia crandalli** (0, I), KS, MO, OK.
Spot anthracnose. **Sphaceloma symporicarpi.**

CORDIA

Powdery mildew. **Erysiphe cichoracearum,** TX.
Rot, root. **Clitocybe tabescens,** FL; **Phymatotrichum omnivorum,** TX.

COREOPSIS
(Tickseed)

Bacterial, MLO. **Aster Yellows,** NJ, NY, and **California Aster Yellows,** CA.
Blight, gray mold. **Botrytis cinerea,** AK.
Blight, southern. **Sclerotium rolfsii,** FL, TX.
Broomrape. **Orobanche ramosa,** TX.
Dodder. **Cuscuta** sp., NJ.

Leaf spot. **Cercospora coreopsidis,** OK; **Phyllosticta coreopsidis,** WI; **Septoria coreopsidis,** IA, WI, TX.

Nematode, root knot. **Meloidogyne** sp., FL.

Powdery mildew. **Erysiphe cichoracearum,** NY, MD, MN, WI.

Rot, root. **Rhizoctonia solani,** NY, MN, TX; **Phymatotrichum omnivorum,** TX.

Rot, stem. **Sclerotinia sclerotiorum,** OH, WA.

Rust. **Coleosporium inconspicuum** (II, III), GA, MD, NC, SC, OH, TN, VA, WV; 0, I, on pine.

Scab. **Cladosporium coreopsidis.**

Virus. **Beet Curly Top,** CA.

Wilt. **Verticillium albo-atrum,** NY.

CORIANDER
(*Coriandrum*)

Anthracnose. **Gloeosporium** sp., MD.

Nematode, root knot. **Meloidogyne** sp., FL.

CORN, SWEET
(*Zea mays* var. *saccharata*)

Bacterial leaf stripe. **Pseudomonas andropogonis,** FL.

Bacterial spot. **Pseudomonas syringae,** MA.

Bacterial stalk rot. **Erwinia dissolvens,** FL, WV.

Bacterial wilt. **Erwinia stewartii,** general.

Blight, leaf. **Helminthosporium turcicum;** Middle Atlantic and southern states, occasional in central states; **H. maydis** (*Cochliobolus heterostrophus*), southern leaf blight.

Blight, seedling; stalk rot; ear rot. **Penicilllium oxalicum.**

Leaf spot. **Cercospora zeae-maydis,** VA; **Physoderma maydis,** brown spot, in South; eyespot, **Kabatiella zeae.**

Nematode, lesion. **Pratylenchus penetrans.**

Nematode, root knot. **Meloidogyne chitwoodi,** Pacific Northwest; **M. hapla,** Pacific Northwest.

Rot, black bundle disease. **Cephalosporium acremonium,** MT.

Rot, dry ear. **Nigrospora sorghi,** IA.

Rot, ear, root and stalk; seedling blight. **Diplodia zeae,** ME to VA, TX, SD; **Fusarium** spp.; **F. moniliforme,** cosmopolitan.

Rot, root. **Gibberella zeae,** chiefly east of Mississippi River; **Pythium** spp., IL, IA, OH, TX; **Helminthosporium pedicellatrum,** CA.

Rust. **Puccinia sorghi** (II, III), general; 0, I, on oxalis; **P. polyspora,** southern.

Smut. **Ustilago maydis,** general; **Sphacelotheca reiliana,** head smut.

Virus. **Corn Leaf Fleck; Corn Mosaic, Corn Stunt,** AL, AZ, CA, GA, IN, KY, MO, SC; **Wheat Streak Mosaic,** IN, MI, OH; **Cucumber Mosaic,** celery strain.

Virus. **Maize Dwarf Mosaic,** AL, AR, AZ, CA, GA, IA, ID, MO, NM, NY, OH, NJ, PA, WA.

Witchweed. **Striga asiatica,** on roots, NC.

Smut is the most conspicuous corn disease in home gardens. Bacterial wilt is dependent on survival of insect vectors and may be serious after a warm winter unless resistant hybrids are used.

CORNCOCKLE
(*Agrostemma*)

Leaf spot. **Gloeosporium** sp., IN; **Marssonina delastrei,** IL, IN, MI, MO, MS; **Septoria lychnidis** var. **pusilla,** ND.

Rot, root. **Phymatrotrichum omnivorum,** TX.

Rot, stem. **Fusarium** sp., IN.

CORN-MARIGOLD
(*Chrysanthemum segetum*)

Bacterial, MLO. **California Aster Yellows,** CA.

Leaf spot. **Septoria chrysanthemi,** NY.

CORYDALIS

Downy mildew. **Peronospora corydalis,** IN, MA, MD, OK, TX.

Leaf spot. **Septoria corydalis,** TX, WI.

Nematode. **Meloidogyne** sp., FL.

Rust. **Puccinia aristidae** (0, I), CO, KS, NE; II, III, on grasses; **P. brandegei** (III) CO, WA.

COSMOS

Bacterial, MLO. **Aster Yellows,** DE, NJ, NY, and **California Aster Yellows,** CA.

Bacterial wilt. **Pseudomonas solanacearum,** NC.

Blight, southern. **Sclerotium rolfsii,** MS.

Canker; stem blight. **Diaporthe stewartii,** CT to NJ; KS, SD, CA, TX.

Dodder. **Cuscuta** sp., NY.

Leaf spot. **Cercospora** sp., TX; **Septoria** sp., CT.

Nematode, root knot. **Meloidogyne** sp.
Powdery mildew. **Erysiphe cichoracerum,** CA, MD, NC, NE, TX.
Rot, root. **Phymatotrichum omnivorum,** AZ, TX; **Macrophomina phaseoli,** TX; **Pythium** sp., MD; **Rhizoctonia solani,** CT, MD, TX.
Virus. **Beet Curly Top; Mosaic,** unidentified, FL, TX; **Tomato Spotted Wilt,** TX.
Wilt. **Fusarium** sp., NJ.

COTONEASTER

Bacterial fire blight. **Erwinia amylovora,** widespread.
Bacterial hairy root. **Agrobacterium rhizogenes,** central states.
Canker; twig blight. **Physalospora obtusa,** NY to OH, TX; **Diplodia** sp., TX; **Gibberella baccata,** CA.
Leaf spot. **Fabraea maculata,** CA, IA; **Phyllosticta cotoneastri,** MD; **P. cydoniae,** MS.
Powdery mildew. **Podosphaera** sp., CA.
Rot, root. **Armillaria mellea,** CA; **Clitocybe tabescens,** FL; **Phymatotrichum omnivorum,** AZ, TX; **Phytophthora** sp.
Scab. **Venturia** sp., WA.

COURSETIA

Rust. **Phragmopyxis acuminata** (0, III), AZ, CA.

COWANIA

Rust. **Phragmidium andersonii** (I, II, III), ID.

COWPEA
(*Vigna sinensis*)

Virus. **Brome Mosaic,** CA.

Cowpeas are of most interest to gardeners as a green manure crop, and there is little need of repeating here the long list of possible diseases, most of which are given under Bean. Ashy stem blight, charcoal rot, is fairly serious. The root-knot nematode is general, but varieties Iron and Bragham are almost immune. Fusarium wilt, general in the South, is largely controlled by using resistant varieties. Leaf spots are numerous, and some may cause defoliation. Rust is widespread on Blackeye and related varieties, but many varieties are resistant.

CRABAPPLE, FLOWERING
(*Malus*)

Anthracnose, northwestern. **Neofabraea malicorticis,** OR.
Bacterial fire blight. **Erwinia amylovora,** widespread.
Blight, thread. **Pellicularia koleroga,** LA.
Blotch, sooty. **Gloeodes pomigena,** IN.
Canker, blister. **Nummularia discreta,** WV.
Canker, coral spot; twig blight. **Nectria cinnabarina,** AK.
Canker; dieback. **Valsa leucostoma,** WV; **Physalospora obtusa,** MI.
Leaf spot. **Cercosporella pirina,** IL, WI; **Coniothyrium pirinum,** WI; **Fabraea maculata,** MD, AK; **Illosporium malifoliorum,** WV; **Marssonina coronaria,** IN, IA, WI, MO; **Phyllosticta** sp., IA; **P. solitaria,** KS, OH, TX, IN, IA, WV; **P. zonata,** IA; **Septoria pyri.**
Mistletoe, European. **Viscum album,** CA.
Powdery mildew. **Oidium pyrinum,** WI; **Phyllactinia corylea,** WA; **Podosphaera leucotricha,** IL, IA; **P. oxyacanthae,** WA, WI.
Rot, black, fruit, leaf spot. **Physalospora obtusa,** eastern and central states.
Rot, heart. **Ganoderma applanatum.**
Rot, root. **Armillaria mellea,** WA; **Phymatotrichum omnivorum,** TX.
Rust. **Gymnosporangium clavipes,** NJ, CT; **G. globosum,** AL, IN, KS, NJ, SC, VA; **G. juniperi-virginianae,** cedar-apple rust, general; **G. libocedri,** OR, AK; **G. nelsoni,** AK, WA; **G. nootkatense,** AK.
Scab. **Venturia inaequalis,** general.
Scab, twig infections. **Venturia inaequalis,** MA, OH, PA, RI.

Cedar-apple rusts are common and injurious on most native crabapples; Asiatic varieties are usually resistant. Remove red-cedars in the vicinity.

CRANBERRY
(*Vaccinium,* subgenus *Oxycoccus*)

Dieback. **Diaporthe vaccinii,** WI.
Gall, red leaf. **Synchytrium vaccinii,** NJ.
Gall; shoot hypertrophy; rose-bloom. **Exobasidium vaccinii,** general.
Leaf spot. **Discohainesia oenotherae,** also storage rot, MA, NJ, OR, WA, WV; **Mycosphaerella nigromaculans,** black spot, OR, WA; **Phyllosticta putrefaciens,** also berry rot, MA, NJ; **Ramularia multiplex,** MI, NY, WI; **Venturia compacta,** leaf smudge, general; **Cladosporium oxycocci,** leaf mold, NJ, WA.
Nematode. **Atylenchus decalineatus.**
Nematode, sheath. **Hemicycliophora similis.**
Nematode, stubby root. **Trichodorus christiei.**
Powdery mildew. **Microsphaera alni** var. **vaccinii,** AL, NJ, OH.
Rot, berry; blotch. **Acanthorynchus vaccinii,** ME to NC and WI, OR, WA;

Ceuthospora lunata, black rot, MA, NJ, OR, WA, WI; **Curvularia inaequalis,** IL, NJ, WI; **Sphaeronema pomorum,** NJ; **Sporonema oxycocci,** ME, MA, NJ, WI, OR, WA, AK.

Rot, berry speckle. Several fungi associated.

Rot, bitter; leaf spot. **Glomerella cingulata** var. **vaccinii,** general.

Rot, early; scald, blast. **Guignardia vaccinii,** general.

Rot, end. **Godronia cassandrae,** general.

Rot, fairy ring; root. **Psilocybe agrariella** var. **vaccinii,** MA, NJ.

Rot, hard; twig blight. **Monilinia oxycocci,** MA, ME, OR, WA, WI.

Rot, storage. **Penicillium** spp., cosmopolitan; **Diaporthe vaccinii,** general; **Botryris cinerea,** gray mold, also blossom blight; **Alternaria** sp.; **Dematium** spp.; **Melanospora destruens,** in market; **Gloeosporium minus,** MD, NJ; **Pestalotia vaccinii; Stemphylium ilicis.**

Rot, witches' broom. **Naevia oxycocci,** ME, MI, NH, NY.

Rust. **Pucciniastrum vaccinii** (II, III), OR, UT, WA; 0, I, on hemlock.

Virus. **Cranberry False-Blossom,** ME to NJ and WI, OR, WA.

?Virus. **Cranberry Ring Spot,** WI.

CRAPE-MYRTLE
(*Lagerstroemia*)

Blight, thread. **Rhizoctonia ramicola,** FL.

Leaf spot. **Cercospora lythracearum,** TX; **Cercospora** sp., blotch, FL, TX; **Phyllosticta lagerstroemiae,** tip blight, LA, TX.

Nonparasitic. Chlorosis. Manganese deficiency.

Powdery mildew. **Erysiphe lagerstroemiae,** general; **Phyllactinia corylea,** AL.

Rot, root. **Clitocybe tabescens,** FL; **Phymatotrichum omnivorum,** TX.

Sooty mold. **Capnodium** spp., in aphid honeydew.

Mildew is serious on crape-myrtle, calling for sprays as the buds break.

CRASSULA

Anthracnose. **Gloeosporium** sp., NJ.

Leaf spot. **Phomopsis** sp., CT.

Nematode, leaf. **Aphelenchoides** sp.

Rot, root. **Armillaria mellea,** CA; **Pythium** sp., NJ, NY.

CREOSOTE BUSH
(*Larrea*)

?Blight. **Omphalia** sp.

Mistletoe. **Phoradendron californicum,** TX to CA.

CRINUM

Leaf scorch; red blotch. **Stagonospora curtisii,** CA, NY.
Leaf spot. **Cercospora pancratii,** AL, FL, MS.
Virus. **Mosaic.** Unidentified.

CROCUS

Bacterial scab. **Pseudomonas marginata,** occasional on imported stocks.
Rot. Corm. **Fusarium oxysporum,** NY, PA; **Stromatinia gladioli,** dry rot,
   widespread; **Penicillium** sp., blue mold.
Virus. **?Iris Mosaic,** CA, MD.

CROTALARIA
(*C. retusa*)

Blight, seedling. **Alternaria cassiae,** MS.
Powdery mildew. **Oidium erysiphoides** var. **crotalariae,** LA, MS.
Virus. **Potato Virus X,** on *C. juncea.*
Wilt; root rot. **Fusarium** sp., GA, TX.

This species is sometimes grown as an ornamental. Crotalaria as a cover
crop has more diseases, but *C. spectabilis* is immune to root-knot nematode
and thus is particularly useful between susceptible crops.

CROTON
(*Codiaeum*)

Anthracnose. **Gloeosporium** spp., FL, NJ.
Rot, root. **Phymatotrichum omnivorum,** TX.

This is the croton of florists.

CROTON
(*Croton*)

Dodder. **Cuscuta indecora,** TX.
Gall, leaf. **Kutilakesa pironii,** FL.
Leaf spot. **Cercospora capitati,** TX; **C. crotonicola,** TX, SC; **C. crotonis,** AL,
   FL, SC; **C. crotonophila,** WI; **C. maritima,** MS; **C. crotonifolia,** SC.
Nematode, lesion. **Pratylenchus** sp., FL.
Rot, root. **Phymatotrichum omnivorum,** TX.
Rust. **Bubakia crotonis** (II, III), KY to AL; AZ, CA, FL, NE, TX.

CROWBERRY
(Empetrum)

Rust. Chrysomyxa empetri (II, III), ME, NH, NY.

CROWN VETCH
(Coronilla varia)

Nematode. Meloidogyne sp., VA.
Rot, stem. Sclerotinia trifoliorum; Fusarium roseum.

CRYPTANTHA

Gall, leaf. Synchytrium myosotidis, AZ, CA.
Powdery mildew. Erysiphe cichoracearum, CA, NV.
Rust. Puccinia aristidae (0, I); II, III, on native grasses; P. cryptanthes, CA, WA.

CRYPTANTHUS

Anthracnose. Gloeosporium sp., WA.

CRYPTOMERIA

Blight, leaf and twig. ?Diaporthe eres; Phyllosticta cryptomeriae, VA and general U.S.
Leaf spot. Pestalotia cryptomeriae, SC; P. funerea, NJ.

CUCUMBER
(Cucumis sativus)

Anthracnose. Colletotrichum lagenarium, general.
Bacterial angular leaf spot. Pseudomonas syringae pv. lachrymans, general.
Bacterial soft rot. Erwinia aroideae; E. carotovorus.
Bacterial spot. Xanthomonas cucurbitae, MA, MI.
Bacterial wilt. Erwinia tracheiphila, general.
Blight, blossom; fruit rot. Choanephora cucurbitarum, FL, GA, NJ, OK, RI, TX.
Blight, gummy stem; black fruit rot. Mycosphaerella citrullina, widespread; M. melonis, AZ.
Blight, leaf. Alternaria cucumerina, general.

Blight, southern. **Sclerotium rolfsii,** VA to FL, TX.

Damping-off. **Rhizoctonia solani,** also stem rot; **Pythium** spp.

Dodder. **Cuscuta gronovii,** NY.

Downy mildew. **Pseudoperonspora cubensis,** general.

Leaf spot. **Cercospora cucurbitae,** AL, DE, IA, NJ, TX, WI: **Ascochyta** sp., OR; **Gloeosporium** sp. IL; **Phyllosticta cucurbitacearum,** DE, OH, TX; **Septoria cucurbitacearum,** DE, MA, NH, PA; **Stemphylium cucurbitacearum,** OH; ?**Alternaria consortiale,** WA.

Mold. Seed. **Alternaria tenuis,** cosmopolitan; **Curvularia trifoli,** NJ.

Nematode, hop cyst. **Heterodera humuli,** OR.

Nematode, root knot. **Meloidogyne incognita; M. incognita-acrita; M. javanica.**

Nonparasitic. **Chlorosis,** nitrogen, manganese, or potassium deficiency.

Powdery mildew. **Erysiphe cichoracearum,** general; **Sphaerotheca fuligenea,** CA, NC.

Rot, charcoal. **Macrophomina phaseoli,** IL.

Rot, fruit. **Fusarium** spp., TX; **Rhizopus stolonifer,** occasional. **Diplodia natalensis; Botrytis cinerea.**

Rot, root. **Fusarium solani** f. **radicicola,** CT, OR, WA; **Phymatotrichum omnivorum,** TX.

Rot, stem; fruit. **Sclerotinia sclerotiorum,** occasional.

Scab. **Cladosporium cucumerinum,** general.

Virus. **Bromegrass Mosaic Virus,** general.

Virus. **Cucumber Mosaic,** general; **Beet Curly Top,** CA, ID, OR, TX, UT, WA; **Tobacco Ring Spot,** MD, PA, VA; **Zucchini Yellow Mosaic,** CA, FL, NY.

Wilt. **Fusarium oxysporum** f. sp. **cucumerinum,** FL; **F. oxysporum** f. sp. **cucurbitae; Verticillium albo-atrum,** CA.

Starting cucumbers under Hotkaps and then treating to control insect vectors helps to reduce bacterial wilt and virus diseases. Choose varieties resistant to mosaic and scab.

CULVERS-ROOT
(*Veronicastrum*)

Leaf spot. **Cercospora leptandrae,** WI; **Phyllosticta decidua,** TX, WI; **Ramularia veronicae,** TX; **Septoria veronicae,** WI.

Powdery mildew. **Erysiphe cichoracearum,** IL; **Sphaerotheca macularis,** CT, IL, IA, MD, MI, MO, WI.

Rot, root. **Phymatotrichum omnivorum,** TX; **Rhizoctonia solani,** TX.

Rust. **Puccinia veronicarum** (III), IA, TX, WI.

CUNNINGHAMIA

Leaf spot. Necrotic. **Phomopsis** sp., OR.

CUPHEA

Blight, gray mold. **Botrytis cinerea,** occasional in greenhouses.
Leaf spot. **Septoria maculifera,** NY, PA, VA, WV.
Nematode, root knot. **Meloidogyne** sp., MD.
Powdery mildew. **Erysiphe polygoni,** MD, VA.
Rot, root. **Rhizoctonia solani,** IL.

CURRANT
(*Ribes* spp.)

Anthracnose; fruit spot. **Pseudopeziza ribis,** general.
Blight, cane. **Bostryosphaeria ribis** var. **chromogena,** general.
Blight, gray mold; fruit spot. **Botryris cinerea,** Northeast, OR, WA.
Blight, thread. **Pellicularia koleroga,** FL.
Canker, cane knot. **Thyronectria berolinensis,** CT to IN, KS, UT.
Canker; dieback. **Nectria cinnabarina,** ME to CO, WA; **N. ditissima,** MN,
  NY; Black pustule, **Phragmodothella ribesia,** Northeast, Pacific Northwest.
Downy mildew. **Plasmopara viticola,** WI.
Leaf spot. **Alternaria** sp., MI; **Cylindrosporium ribis,** WI; **Mycosphaerella
  ribis,** ME to MD, AR, OR, WA.
Leaf spot, angular. **Cercospora angulata,** NY to VA, KS, MN; **C. ribis,** AL,
  IN, IA.
Powdery mildew. **Microsphaera grossulariae,** MT, NE, NH; **Sphaerotheca
  morsuvae,** CA, CT, MT, NE, OR, WA, AK; **Phyllactinia corylea,** MI.
Rot, berry. **Glomerella cingulata,** CT, PA.
Rot, collar. **Fomes ribis,** NY to IN, MN, UT.
Rot, root. **Armillaria mellea,** CA, OR, WA; **Phymatotrichum omnivorum,**
  TX; **Hypholoma perplexum,** NY.
Rust, white pine blister. **Cronartium ribicola** (II, III), ME to VA, IL and MN,
  OR, WA. **Puccinia caricis** (0, I), CT, IN, MD, NY, II, III on Carex.
Virus. **Currant Mosaic,** NY, MD.
Wilt. **Verticillum** sp., NY.

Black currant is the most important host of white pine blister rust and
should not be grown; red currants should be kept at least 300 feet from pines.
Cut out canes with cane blight.

CURRANT, FLOWERING
(*Ribes*)

Anthracnose. **Pseudopeziza ribis,** IA, MN, MT, IL.
Bacterial leaf spot. **Pseudomonas ribicola.**
Blight; dieback. **Botrytis cinerea,** IN, AK.

Canker; dieback. **Botryosphaeria ribis,** KS; **Phragmodothella ribesia,** NY; **Thyronectria berolinensis,** KS; **Nectria cinnabarina,** AK, KS.

Leaf spot. **Cercospora angulata,** MN; **C. ribicola,** OR, WA; **Marssonina ribicola,** CO; **Mycosphaerella aurea,** NY to KS and SD, WA; **M. ribis,** NY to KS, MN and UT; **Phyllosticta grossulariae,** IN; **Septoria sanguinea,** WA.

Rust. **Coleosporium jonesii,** MN; **Cronartium occidentale,** MT to NM, CA, WA; **C. ribicola,** ME to MD, CO, MN, Pacific states; **Melampsora ribesii-purpureae;** II, III on willow; **Puccinia caricina,** NY to IA, CA, OR, AK; **P. micrantha; P. parkerae.**

CYCAD, SAGO-PALM
(*Cycas*)

Coralloid roots. **Anabaena cycadearum** and **Nostoc commune,** algae, and bacterium *Azotobacter* are associated with roots but are mostly innocuous.

Leaf spot. **Ascochyta cycadina,** MO, TX; **Pestalotia cycadis** (?secondary), CT, FL.

A destructive blight of unknown cause (but with *Gloeosporium* and *Phoma bresadolae* often associated with it) causes pale green areas on pinnae of young leaves, which are curled out of the flat plane and dieback. The disease is apparently systemic and increases annually until death. Eradication of blighted plants is the only control suggested.

CYCLAMEN

Bacterial tuber rot. **Erwinia carotovora,** OH, NJ, NY.

Blight; bud and leaf rot; petal spot. **Botrytis cinerea,** cosmopolitan.

Blight, leaf and bud. **Glomerella cingulata,** IN, MA, MO, NJ, OH, PA, TX, VA.

Leaf spot. **Phyllosticta cyclaminicola,** IL, OH, TX; **P. cyclaminis,** VA; **Systoria cyclaminis,** CA.

Leaf spot; stunt; white mold. **Ramularia cyclaminicola** (*Cladosporium cyclaminis*), CA, IL, MN, NJ, NY, OH, PA.

Nematode, leaf. **Aphelenchoides fragariae.**

Nematode, lesion. **Pratylenchus pratensis,** VA.

Nematode, root knot. **Meloidogyne** sp., cosmopolitan.

Rot, root. **Thielaviopsis basicola,** CT.

Wilt. **Fusarium oxysporum** f. sp. **cyclaminis,** CA, NJ; **Fusarium** sp., NJ.

Discard plants with stunt; sterilize soil and pots, and benches where diseased plants have grown. Avoid splashing to reduce botrytis blight and leaf nematodes. Spray for blight and leaf spots with zineb or ferbam.

CYNOGLOSSUM
(Hounds-Tongue)

Blight, southern. **Sclerotium rolfsii**, FL.
Downy mildew. **Peronospora cynoglossi**, MD, IL, TX.
Leaf spot. **Cercospora cynoglossi**, IN; **Phyllosticta decidua**, WI; **Ramularia lappulae**, WI, TX.
Nematode, root knot. **Meloidogyne** sp.
Powdery mildew. **Erysiphe cichoracearum**, UT, WY, VA.
Rot, root. **Phymatotrichum omnivorum**, TX.
Rot, stem. **Sclerotinia sclerotiorum**, WA.

CYPRESS
(*Cupressus*)

Bacterial crown gall. **Agrobacterium tumefaciens**, AZ, CA, FL.
Blight, nursery. **Phomopsis juniperovora**, MD, NC, VA.
Blight, seedling. **Fusarium solani**, TX.
Blight, twig. **Coryneum asperulum**, AL; **C. berckmanii**, OR; **Cercospora thujina**, AL, GA, LA; **Pestalotia funerea**, CA, TX; **Botryosphaeria** sp., GA, on Arizona cypress.
Canker; dieback. **Coryneum cardinale,** on Monterey, sometimes Italian cypress; **Cytospora cenisia**, CA, on Italian cypress; **Macrophoma cupressi**, AL, CA, FL, TX; **Monochaetia unicornis**, GA, SC.
Mistletoe. **Phoradendron** sp., AZ, CA, OR.
Needle cast. **Lophodermium** sp., NJ.
Nematode, lesion. **Pratylenchus thornei.**
Rot, heart. **Fomes pini**, CA.
Rot, root. **Clitocybe tabescens**, FL; **Phymatotrichum omnivorum**, TX; **Phytophthora cinnamomi**, LA.
Rot, wood. **Stereum taxodii,** cause of "pecky" cypress; **Coniophora puteana,** CA; **Lenzites saepiaria,** cosmopolitan; **Polyporus** spp.; **Poria subacida;** Steccherhinum ochraceum.
Rust. **Gymnosporangium cupressi** (III), on Arizona cypress.

   Coryneum canker has killed thousands of Monterey cypress trees in California, and Cytospora canker has been fatal in a narrow belt along the coast.

CYPRESS-VINE
(*Quamoclit*)

Nematode, root knot. **Meloidogyne** sp., AL.
Rot, root. **Phymatotrichum omnivorum**, TX.

Rust. **Coleosporium ipomoeae** (II, III), IL, SC, TN.
White rust. **Albugo ipoemeae-panduratae,** NM, MS.

CYRILLA
(Leatherwood)

Canker, brown felt. **Septobasidium sinuosum,** FL.
Leaf spot. **Phyllosticta cyrillae,** FL.
Rust. **Aecidium cyrillae,** FL, LA, MS.

DAHLIA

Bacterial crown gall. **Agrobacterium tumefaciens,** CT, IL.
Bacterial, MLO. **Aster Yellows,** MD.
Bacterial rot. **Erwinia carotovora,** MS, WA; **E. cytolitica,** NY.
Bacterial wilt. **Pseudomonas solanacearum,** DE, MI, MS, NJ, NC, OK, TX.
Blight, blossom. **Choanephora americana,** FL; **Stemphylium floridanum.**
Blight, gray mold. **Botrytis cinerea,** cosmopolitan on flower buds.
Blight, southern. **Sclerotium rolfsii,** FL, KS, MS, NJ, NC, TX.
Leaf spot. **Alternaria** sp., VA to AL, MO, and MI, ?secondary; **Cercospora**
sp., FL, MS.
Nematode, bulb. **Ditylenchus destructor,** OR.
Nematode, leaf. **Aphelenchoides ritzema-bosi,** CA.
Nematode, root. **Trichodorus pachydermus,** MI.
Nematode, root knot. **Meloidogyne** sp., NC to AL, TX, MO, AZ, CA.
Powdery mildew. **Erysiphe cichoracearum,** general; **E. Polygoni,** CA, DE,
GA, IA, MO, NJ, PA, VA; **Uncinula** sp., NC.
Rot, charcoal. **Macrophomina phaseoli,** SC, OK.
Rot, root. **Armillaria mellea,** CA; **Phymatotrichum omnivorum;** TX.
Rot, stem. **Sclerotinia sclerotiorum,** CA, ME, NY.
Rot, stem, root, cutting. **Pythium** spp.; **Rhizoctonia solani; Fusarium ?roseum.**
Scab. **Streptomyces scabies,** NC.
Smut, leaf. **Entyloma dahliae,** CA, NJ, OR.
Virus. **Dahlia Mosaic,** general; **Tomato Spotted Wilt,** CA, MI, NY, NJ, TX,
WI; **Cucumber Mosaic; Tobacco Ring Spot.**
Wilt. **Fusarium oxysporum; Verticillium albo-atrum** (*V. dahliae*) IL, MI, MO,
NJ, OH, TX.

Leafhopper injury, hopperburn, looks like a true disease. Margins of leaves
turn brown, and there may be general stunting and yellowing. The spotted
wilt virus causes yellow ring spots in dahlia foliage. Heavy, wet soil contrib-
utes to bacterial and fungus rots and wilts. Mildew is often prevalent in late
summer.

DAISY, OXEYE
(*Chrysanthemum leucanthemum*)

Bacterial, MLO. **Aster Yellows,** NY, KS, NJ.
Blight, southern. **Sclerotium rolfsii,** TX.
Leaf spot. **Cylindrosporium chrysanthemi,** OK; **Septoria chrysanthemella;**
 S. leucanthemi, blotch, CT, NY.
Nematode, root knot. **Meloidogyne** sp.
Nematode, stem. **Ditylenchus dipsaci,** NY.
Rot, stem. **Fusarium roseum** and **F. solani,** TX; **Sclerotinia sclerotiorum,** TX.
Rot, root. **Phymatotrichum omnivorum,** TX.
Virus. **Potato Yellow Dwarf,** NY.

DALIBARDA
(Dewdrop)

Leaf spot. **Septoria dalibardae,** ME, MI, NH, NY, VT.

DANDELION
(*Taraxacum*)

Virus. **Tomato Ringspot,** PA.

DAPHNE

Blight, twig. **Botrytis** sp., Northeast, Northwest.
Dieback; wilt. **Fusarium** sp., NJ.
Leaf spot; leaf drop. **Gloeosporium mezerei,** WA; **Marssonina daphnes.**
Rot, collar, stem. **Phytophthora cactorum,** NY, CA; **Rhizoctonia solani,** NY.
Rot, stem; wilt. **Sclerotium rolfsii,** FL.

DATURA

Bacterial canker, vascular. **Clavibacter michiganense,** WY. On *Datura* sp.,
 D. innoxia, D. metal, D. meteloides, D. mollis, D. stramonium.
Blight, southern. **Sclerotium rolfsii,** FL.
Leaf spot; pod blight. **Alternaria crassa,** FL, WI.
Nematode, root knot. **Meloidogyne hapla.**
Rot, root. **Thielaviopsis basicola,** WI.
Virus. **Bromegrass Mosaic,** general.
Virus. **Tomato Spotted Wilt,** CA, TX.

DAYLILY
(*Hemerocallis*)

Blight, gray mold. **Botrytis** sp., MD.

Blight, leaf. **Kabatiella** sp., ?secondary, MD.

Leaf spot. **Cercospora hemerocallis,** IL; **Heterosporium gracilis,** NJ, NY, TX; Leaf streak, **Gloecephalus hemerocalli,** MS, and **Collecephalus hemerocalli,** MS, PA, LA.

Nematode, root knot. **Meloidogyne incognita; M. incognita-acrita.**

Rot, root. **Sclerotium** sp., IN; **Phymatotrichum omnivorum,** TX.

DECUMARIA

Leaf spot. **Cercospora decumariae,** MS.

DELPHINIUM
(Larkspur)

Bacterial black spot, leaf spot. **Pseudomonas syringae** pv. **delphinii,** ME to VA, TX, MN, rare in Pacific Coast states.

Bacterial collar rot. **Erwinia carotovora,** CA, NY.

Bacterial crown gall. **Agrobacterium tumefaciens,** WA.

Bacterial foot rot; blackleg. **Erwinia carotovora** pv. **atroseptica,** general.

Bacterial, MLO. **California Aster Yellows** Stunt, "Greens," Pacific Coast and Rocky Mountains.

Blight, gray mold, bud rot. **Botrytis cinerea,** CT, MA, MN, MS, NY, WV, WI.

Blight, southern; crown rot. **Sclerotium rolfsii** (*S. delphinii*), general.

Canker, stem. **Diaporthe arctii,** MD, NC, NY, OH, PA; **Fusarium** spp., widespread; **F. oxysporum** f. sp. **delphinii,** wilt, NY; **Volutella** sp., MD; **Phoma** sp., CT, NJ, NY.

Damping-off; root rot. **Pythium** spp. and **Rhizoctonia solani,** cosmopolitan.

Gall, leaf. **Synchytrium aureum,** IA.

Leaf spot. **Ascochyta aquilegiae,** CT; **Cercospora delphinii,** CO, MD; **Ovularia delphinii,** WY; **Phyllosticta** sp., NY; **P. delphinii,** CO; **Ramularia delphinii,** CA, CO, UT; **Septoria delphinella,** IL, KS, WI.

Nematode, leaf and stem. **Ditylenchus dipsaci,** OR, WA.

Nematode, lesion. **Pratylenchus pratensis.**

Nematode, root knot. **Meloidogyne** spp., AZ, NJ, NY, VA, WA.

Nonparasitic. **Chlorosis,** low temperature and wet soil. **Variegation,** non-infectious, seed-transmitted leaf-color anomalies.

Powdery mildew. **Erysiphe cichoracearum,** MA, MN, NY, WA; **Erysiphe polygoni,** general but some varieties resistant; **Sphaerotheca macularis,** CA.

Rot, collar; leaf spot. **Diplodina delphinii,** CA, NY.

Rot, crown. **Sclerotium delphinii.** See Blight, southern.

Rot, root. **Phymatotrichum omnivorum,** TX; **Pythium aphanidermatum; P. ultimum; P. vexans.**

Rot, stem. **Phytophthora** sp., MN; **Sclerotinia sclerotiorum,** widespread.

Rust. **Puccinia delphinii** (III), CA; **P. recondita** (0, I), NE to NM, CA.

Smut, leaf and stem. **Urocystis sorosporioides,** CA, KY, VA.

Smut, white. **Entyloma winteri,** CA; **E. wyomingense,** WY.

Virus. **Delphinium Ring Spot,** CA; **Celery Calico,** CA, ID, WA; **Beet Curly Top,** CA; **Cucumber Mosaic; Tomato Spotted Wilt,** CA.

Wilt. **Verticillium albo-atrum,** NY, WA.

One of the chief delphinium problems is a condition known as "blacks," which looks like a disease and is often confused with bacterial black spot but is caused by cyclamen mites. Plants are stunted and deformed; buds turn black. The bacterial disease causes black tar spots on leaves but no deformity, no stunting of the whole plant. Crown rot or southern blight is often fatal to delphiniums. When yellowing and wilting appear, check the soil around the crown for reddish sclerotia and white mycelium and take immediate sanitary measures. Many foot, collar, and root rots and stem cankers afflict delphinium. Because of these, many gardeners grow hybrid delphinium as biennials, rotating locations, choosing well-drained sites. Virus diseases are more important along the Pacific Coast. Use virus-free planting stock and rogue out infected individuals.

DEMORPHOTHECA

Virus. **Biddens Mottle,** FL.

DESERT BIRD OF PARADISE
(*Caesalpinia*)

Powdery mildew. **Leveillula taurica,** AZ.

DESERT-CANDLE
(*Eremurus*)

Leaf spot. **Myrothecium roridum,** OH.

DESERT-PLUME
(*Stanleya*)

Leaf spot. **Cercospora nasturtii,** KS.

Rust. **Puccinia aristidae,** CO, NV.

DESERT-WILLOW
(*Chilopsis*)

Damping-off. **Pythium ultimum**, NE; **Rhizoctonia solani**, NE, TX.
Leaf spot. **Phyllosticta erysiphoides.**
Rot, root. **Phymatotrichum omnivorum**, TX.

DESMODIUM
(Arrowleaf)

Virus. **Peanut Mottle**, GA.

DEUTZIA

Leaf spot. **Cercospora deutziae**, DE, IA, TX; **Phyllosticta deutziae**, AL, IA, NJ, TX.
Nematode, root knot. **Meloidogyne** spp.
Rot, root. **Armillaria mellea**, CA.

DEVILS-CLUB
(*Oplopanax*)

Blight, gray mold. **Botrytis cinerea**, AK.
Leaf spot. **Cercospora daemonicola**, OR.

DEVILWOOD
(*Osmanthus americanus*)

Black mildew. **Asterina asterophora**, FL, GA; **A. discoidea; A. purpurea**, FL; **Lembosia oleae**, MS; **Meliola amphitricha**, FL to MS.
Leaf spot. **Phyllosticta oleae**, FL, NC.
Mistletoe. **Phoradendron serotinum (flavescens)**, FL.

DEWBERRY
(*Rubus*)

Bacterial cane gall. **Agrobacterium rubi**, NY, OR.
Bacterial crown gall. **Agrobacterium tumefaciens**, general.
Bacterial hairy root. **Agrobacterium rhizogenes**, OR.
Black mildew. **Irenina sanguinea**, AL, LA.
Blight, cane; dieback. **Leptosphaeria coniothyrium**, general; **L. thomasiana**, OR, WA.

Blight, spur. **Didymella applanata,** OR, WA.
Blight, stamen; dry berry. **Hapalosphaeria deformans,** OR, WA.
Blotch, sooty. **Gloeodes pomigena,** NC, PA.
Canker; cane spot. **Ascospora ruborum,** AL, CA.
Canker; fruit rot. **Glomerella cingulata,** GA, IL, MD, MS.
Downy mildew. **Peronospora rubi,** FL, MD, WI, WA; **P. potentillae,** CT, IL, LA.
Fruit spot; flyspeck. **Leptothyrium pomi,** NC.
Gall, yellow leaf. **Synchytrium aureum,** WI.
Leaf spot. **Mycosphaerella confusa,** NJ to FL, TX, IL; **Pezizella oenotherae,** MD to NC; **Phyllosticta ruborum,** NY; **P. dispergens,** IL; **Septoria rubi (Mycosphaerella rubi),** general; **S. darrowi,** NY.
Powdery mildew. **Sphaerotheca macularis,** IL, MN, IN, OH, PA, Pacific Northwest.
Rosette; double blossom. **Cercosporella rubi,** NC to AL, LA, MS, TX, IL.
Rot, collar. **Rhizoctonia solani,** TX, WA.
Rot, fruit. **Botrytis cinerea; Phyllosticta carpogena,** MD, NJ, NY, NC.
Rot, root. **Armillaria mellea,** OR; **Collybia dryophila,** NC; **Corticium galactinum,** MD, VA, TX; **Helicobasidium purpureum,** NC, TX; **Phymatotrichum omnivorum,** TX.
Rust. **Gymnoconia peckiana,** orange (0, I, III), ME to VA, MO, MN, CA; **Kuehneola uredinis,** yellow (0, I, II, III), ME to FL, TX, KS, CA, WA; **Kunkelia nitens,** orange (I), CA, OR, CT to FL, TX, IA; **Mainsia rubi** (II, III), TX.
Spot anthracnose. **Elsinoë veneta,** general.
Virus. **Raspberry Beta Leaf Curl,** MI, OH, TX; **Raspberry Mosaic,** CT, MI, NJ, NY, Pacific Coast; **Loganberry Dwarf,** especially on loganberry, CA, OR, WA.
Wilt. **Verticillium albo-atrum,** CA, OR, WA.

Use virus-free, bacterial-free, planting stock; spray for anthracnose.

DIANTHUS
(Garden Pinks)

Blight; stem rot. **Alternaria dianthi,** widespread.
Blight, gray mold. **Botrytis cinerea,** AK.
Blight, southern. **Sclerotium rolfsii,** IL, TX.
Leaf spot. **Ascochyta dianthi,** NY, MS; **Heterosporium echinulatum,** CA, NY, OR; **Septoria dianthi,** AL, MI, MS, NC, NJ, NY.
Nematode, root knot. **Meloidogyne** sp., AL, MS, TX.
Rot, bud. **Fusarium poae,** NY.
Rot, root. **Phymatotrichum omnivorum,** TX; **Pythium ultimum,** CA.
Rot, stem. **Rhizoctonia solani,** widespread.
Rust. **Puccinia arenariae,** NY; **Uromyces dianthi** (II, III), NE, MS, NY.
Virus. **Beet Curly Top,** CA; **Carnation latent,** NY.

DICHONDRA
(Lawn-Leaf)

Blight, southern. **Sclerotium rolfsii**, CA.
Gall, leaf. **Synchytrium edgertonii**, LA.
Nematode, root knot. **Meloidogyne** sp., CA.
Rust. **Puccinia dichondrae** (II, III), LA, MS, NC, TX.
Virus. **Cucumber Mosaic**, CA.

DIEFFENBACHIA

Anthracnose. **Glomerella cincta**, NJ; **Gloeosporium** sp., WA; **Colletotrichum** sp., WA.
Bacterial stem and leaf rot. **Erwinia dieffenbachiae**, FL.
Leaf spot. **Cephalosporium dieffenbachiae**, FL, NY; **Leptosphaeria** sp., FL; **Myrothecium roridum**, FL.
Rot, root. **Pythium splendens**, FL; **Rhizoctonia** sp., FL.
Rot, stem. **Phytophthora palmivora**, CA, FL.
Rot, stem, leaf, cutting. **Fusarium solani**, FL.

   Propagate from disease-free canes.

DIERVILLA
(Bush Honeysuckle)

Leaf spot. **Cercospora weigeliae** (?*C. diervillae*) ME, TX; **Phyllosticta diervillae**, WI; **Ramularia diervillae**, ME, NH, NY, WI; **Septoria diervillae**, IA, MA, MN, WI.
Nematode, root knot. **Meloidogyne** sp.
Powdery mildew. **Microsphaera alni**, NJ, WA.
Rot, root. **Phymatotrichum omnivorum**, TX.
Virus. **Tobacco Ringspot**, MD.

DILL
(*Anethum*)

Bacterial, MLO. **Aster Yellows**, NY, TX.
Damping-off. **Rhizoctonia solani**, GA.
Dodder. **Cuscuta** sp., GA.
Leaf spot; stem spot. **Cercospora anethi**, ND, OR, TX; **Phoma anethi**, CT, IN, IA.
Rot, root, wilt. **Fusarium** sp., OH; **Phymatotrichum omnivorum**, TX.
Rot, stem. **Sclerotinia sclerotiorum**, TX.

DITTANY, STONEMINT
(*Cunila*)

Leaf spot. **Septoria cunilae,** IL.
Rust. **Puccinia menthae** (0, I, II, III), NY to VA, AR and IL.

DIZYGOTHECA

Leaf spot. **Alternaria panax,** CA.

DODECATHEON
(Shooting-Star)

Leaf spot. **Heterosporium** sp., AK; **Phyllosticta dodecathei,** TX, WI.
Rust. **Puccinia melanconioides** (0, I, III), CA, OR; **P. ortonii** (0, I, II, III), CA,
OR, SD, UT, WA, AK; **P. solheimi** (III) WY; **Uromyces acuminatus** var.
steironematis (0, I), NE, ND; II, III, on marsh grasses.

DOGBANE
(*Apocynum*)

Leaf spot. **Cercospora apocyni,** north central states, TX, VA; **Phyllosticta
apocyni,** IA, MS, NJ, NY, OR, PA, WI; **Septoria littorea,** MI, KS, NE, ND,
OH; **Stagonospora apocyni,** IL, IN, IA, NY, WI, VA.
Rot, root. **Phymatotrichum omnivorum,** TX.
Rust. **Puccinia seymouriana** (0, I), IN, IL, KS, NY, NJ, NE, OK, SD; **P. smilacis**
(0, I), IL, KS, MD, NC, TN, VA, WI; II, III on smilax.

DOGWOOD, DWARF, BUNCHBERRY
(*Cornus canadensis*)

Leaf spot. **Ceratobasidium anceps,** rot, NH; **Discohainesia oenotherae,** also
stem spot, ME, NH; **Phyllosticta** sp., NY; **Ramularia** sp., NY; **Septoria
canadensis,** ME, WA, AK; **Glomerularia corni,** ME to WI, OR.
Powdery mildew. **Phyllactinia corylea,** WA.
Rust. **Puccinia porphyrogenita** (III), ME to WA and AK.

DOGWOOD, FLOWERING
(*Cornus florida*)

Anthracnose. **Colletotrichum gloeosporioides,** TX; **Discula** sp., ID, OR,
Pacific NW.
Bacterial crown gall. **Agrobacterium tumefaciens,** MO.
Black mildew. **Dimerosporium pulchrum** and **Meliola nidulans,** Southeast.

Blight, flower and leaf. **Botrytis cinerea,** NJ, NY, MA, MD, probably general in wet springs.

Blight, foliar. **Phytophthora parasitica,** FL.

Blight, thread. **Pellicularia koleroga,** LA.

Canker, crown; bleeding; collar rot. **Phytophthora cactorum,** MA, MD, NJ, NY, WA.

Canker; dieback. **Botryosphaeria ribis,** PA; **Cytospora** sp., NJ; **Cryptostictis,** NY; **Sphaeropsis** sp.

Canker, felt fungus. **Septobasidium** spp., VA to FL, LA.

Leaf spot. **Cristulariella pyramidalis,** FL; **Phyllosticta cornicola,** PA to VA, TN and KS; **Ascochyta cornicola,** NC, PA; **Cercospora cornicola,** NC to FL; **Septoria cornicola,** NY to GA and IA.

Mistletoe. **Phoradendron serotinum (flavescens),** FL.

Nematode, dagger. **Xiphinema americanum.**

Nematode, lance. **Hoplolaimus uniformis,** RI.

Nematode, sting. **Belonolaimus longicaudatus,** FL.

Nonparasitic. **Scorch,** water deficiency, frequent in Southeast.

Powdery mildew. **Microsphaera alni,** MA to NC, IL, WI; **Phyllactinia corylea,** general.

Rot, root. **Armillaria mellea,** NY; **Clitocybe tabescens,** FL, GA; **Phymatotrichum omnivorum,** TX; **Corticium galactinum,** VA; **Pythium** sp.

Rot, wood. **Daedalia confragosa,** MD, NC, PA, TN, VA, WV; **Daldinia vernicosa,** MD; **Lenzites betulina,** NC; **Polyporus** spp.; **Poria** spp., MD, PA.

Spot anthracnose. **Elsinoë corni,** DE, MD, GA, NC, SC, FL, LA, VA.

Virus. **Tobacco Ring Spot,** MD. **Witches' Broom Disease,** NJ.

Wilt. **Verticillium** sp., MA.

The most serious dogwood disease in the East is crown or bleeding canker, which attacks trees after transplanting or injury. In a wet season spot anthracnose badly disfigures leaves, twigs, berries. In wet weather botrytis blight is conspicuous as flowers fade and petals rot onto leaves.

DOGWOOD, PACIFIC
(*Cornus nuttalli*)

Canker, bleeding; collar rot. **Phytophthora cactorum,** WA.

Canker, trunk. **Nectria galligena,** OR, WA.

Powdery mildew. **Phyllactinia corylea,** general.

Rot, root. **Armillaria mellea,** WA; heart, **Fomes igniarius,** OR.

DOGWOOD, PAGODA, GRAY, RED OSIER, WESTERN OSIER
(*Cornus* spp.)

Most of the diseases listed for flowering dogwood occur on these shrub dogwoods.

DOLICHOS
(Twinflower, Hyacinth Bean)

Anthracnose. **Colletotrichum dematium** f. sp. **truncata**, GA.
Black mildew. **Parodiella perisporoides**, NC.
Leaf spot. **Cercospora canescens**, FL.
Nematode, root knot. **Meloidogyne** spp., FL, SC.
Powdery mildew. **Microsphaera euphorbiae**, IN.
Rot, root. **Phymatotrichum omnivorum**, TX.
Virus. **Mosaic,** unidentified, MI.
Virus. **Potato Virus X,** Work at Wisconsin on *Dolichos biflorus, D. lablab.*

DORONICUM
(Leopards bane)

Nematode, leaf. **Aphelenchoides** sp.
Nematode, root knot. **Meloidogyne** sp., CA, MD.
Powdery mildew. **Erysiphe cichoracearum**, CA.

DOUGLAS-FIR
(*Pseudotsuga*)

Bacterial gall. **Agrobacterium pseudotsugae,** CA.
Blight. **Phoma eupyrena,** CA.
Blight, brown felt. **Herpotrichia nigra,** Rocky Mountains. and Pacific Northwest.
Blight, gray mold; snow mold. **Botrytis cinerea,** cosmopolitan.
Blight, needle. **Rosellinia herpotrichioides,** CA.
Blight, seedling. **Rhizina undulata,** Pacific Northwest; **Fusarium oxysporum,** OR.
Blight, seedling smother. **Thelephora terrestris,** Pacific Northwest.
Blight, snow. **Phacidium infestans,** ID.
Canker, bark. **Brunchorstia** (*Cryptosporium*) **boycei,** WA; **Chondropodium pseudotsugae,** OR; **Aleurodiscus** spp., weakly parasitic.
Canker, branch. **Dermia pseudotsugae,** CA.
Canker, branch, trunk. **Dasyscypha pseudotsugae; D. ellisiana,** twig, MA, NC, RI; **Phomopsis lokoyae,** Pacific Coast; **Cytospora** sp., twig, CO, NJ, OR; **Phaciopycnis** (*Phomopsis*) **pseudotsugae; Phomopsis lokoyae,** CA.
Canker; dieback. Collar rot; seedling blight. **Diplodia pinea,** CA, KS, NJ, NY.
Damping-off. **Rhizoctonia solani,** cosmopolitan.
Mistletoe; witches' broom. **Arceuthobium douglasii,** MT to CO, OR, WA.
  Phoradendron serotinum (flavescens), TX.
Needle cast. **Adelopus gaumanni,** Pacific states, Northeast; **Rhabdocline pseudotsugae,** general; **R. pseudotsugae** subsp. **pseudotsugae,** PA; **R. weirii,** PA; **Rhabdogloeum hydrophyllum,** AZ, NM.
Needle cast, swiss. **Phaeocryptopus gaumanni,** general, MN (also Lake states).

Needle, flyspeck. **Leptothyrium pseudotsugae,** CO.

Nematode, ring. **Criconemoides crotaloides; Meloidodera** sp., OR.

Rot, heart, wood. **Poria weirii,** destructive, OR, WA; **Echinodontium tinctorium; Fomes** spp.; **Polyporus** spp.; **Lenzites saepiaria,** widespread; **Stereum** spp.; **Trametes** spp.

Rot, root. **Armillaria mellea,** cosmospolitan; **Sparassis radicata,** Pacific Northwest; **Phytophthora cinnamomi,** OR, WA, Southeast; **Phymatotrichum omnivorum,** TX; **Verticicladiella wagenerii,** CA, MT.

Rot, root. **Armillaria mellea,** PA; **Fusarium solani,** PA; **F. oxysporum,** PA; **F. avenaceum,** PA.

Rot, stem, seedling. **Fusarium roseum** f. **avenaceum,** OR; **F. roseum** f. **sambucinum,** OR.

Rot, white pocket. **Ganoderma oregonense,** MT, OR, WA; **Hydnum coralloides,** Pacific Northwest.

Rust. **Melampsora albertensis** (0, I), MT to CO, UT, WA; II, III, on poplar.

DRABA
(Whitlow-Grass)

Downy mildew. **Peronospora parasitica,** CO, IL, IA, KS, NE, SD, TX, WI.

Rust. **Puccinia aristidae** (0, I), AZ; **P. drabae** (III), AK, UT, WY; **P. holboellii** (0, III), CA, CO, UT; **P. monoica** (0, I), CA, NM.

White rust. **Albugo candida,** KS, WY.

DRACAENA

Anthracnose. **Colletotrichum** sp., **Gloeosporium** sp., WA.

Bacterial, leaf spot. **Erwinia herbicola,** FL; **E. carotovora** pv. **carotovora,** FL.

Blight, tip. **Physalospora dracaenae,** WV; **P. rhodina,** MD.

Leaf spot; tip blight. **Glomerella cincta** (*Colletotrichum dracaenae*), general; **Phyllosticta dracaenae,** NJ, FL; **P. draconis; P. maculicola; Gloeosporium polymorphum** and **G. thuemeni,** widespread; **Lophodermium dracaenae,** black spot, CA.

Nematode, lance. **Hoplolaimus bradys,** FL.

Nematode, root knot. **Meloidogyne** sp., FL.

Nematode, sheath. **Hemicycliophora parvana.**

Rot, stem. **Aspergillus niger** var. **floridanus,** FL.

DRAGONHEAD
(*Dracocephalum*)

Downy mildew. **Peronospora** sp., WI, WY.

Blight, southern. **Sclerotium rolfsii,** IL, TX.

Leaf spot. **Phyllosticta dracocephali,** TX; **Septoria dracocephali,** TX, WI.

DURANTA

Blight, seedling. **Sclerotium rolfsii,** FL.
Leaf spot, black. **Phyllachora fusicarpa,** FL.

DUTCHMANS-BREECHES, SQUIRREL-CORN
(*Dicentra*)

Downy mildew. **Peronspora dicentrae,** IN, MD, MI, MO, NY, VA, WI.
Rust. **Cerotelium dicentrae** (0, I), NY to MD, KS and SD; II, III, on wood
nettle.

DUTCHMANS-PIPE
(*Aristolochia*)

Blight, gray mold. **Botrytis cinerea,** CT, MD.
Leaf spot. **Cercospora guttulata,** WV, IL; **Gloeosporium** sp., MA; **Ovularia
aristolochiae,** WV; **Phyllosticta aristolochiae,** NJ.
Rot, root. **Diplodia radicicola,** VA.

DYSCHORISTE

Rust. **Aecidium tracyanum** (0, I), FL.

EASTER CACTUS
(*Rhipsalidopsis*)

Rot, stem. **Drechslera cactivora** (*Helminthosporium cactivorum*), FL.

ECHEVERIA

Leaf spot. **Stemphylium bolicki,** FL.
Nematode, root knot. **Meloidogyne** sp., CA.
Rust. **Puccinia echeveriae** (III), CA.

ECHINACEA
(Purple Coneflower)

Leaf spot. **Cercospora rudbeckii,** IA; **Septoria lepachydis,** WI.
Rot, root. **Phymatotrichum omnivorum,** TX.
Virus. **Mosaic,** unidentified, NY.

EGGPLANT
(*Solanum melongena*)

Anthracnose. **Colletotrichum** spp.; **Gloeosporium melongenae,** NJ to FL, TX, IA; **Glomerella cingulata,** IA.

Bacterial canker. **Clavibacter michiganense,** WY.

Bacterial soft rot. **Erwinia carotovora,** NJ.

Bacterial wildfire. **Pseudomonas syringae pv. tabaci.**

Bacterial wilt. **Pseudomonas solanacearum,** general.

Blight, early. **Alternaria solani,** occasional, NY to FL, LA, WI.

Blight, late. **Phytophthora infestans,** FL, NY.

Blight, phomopsis; fruit rot. **Phomopsis vexans,** general.

Blight, southern. **Sclerotium rolfsii,** VA to FL, LA, WI.

Damping-off. **Pythium debaryanum,** CT, LA, NY; **Rhizoctonia solani,** also stem and fruit rot, general.

Dodder. **Cuscuta** spp., KS, NJ, PA, VA.

Downy mildew. **Peronospora tabacina,** SC.

Fruit spot. **Diplodia natalensis,** FL.

Leaf spot. **Ascochyta lycopersici,** DE, IL, IN, NY; **Cercospora melongenae,** CA; **Phyllosticta solani,** LA; **P. hortorum,** NJ, LA; **Septoria lycopersici,** IN, MD, NC, VA; **Stemphylium solani,** FL.

Nematode, golden. **Heterodera rostochiensis** (*Globodera rostochiensis*).

Nematode, lesion. **Pratylenchus pratensis,** TX.

Nematode, root knot. **Meloidogyne hapla; M. incognita; M. javanica.**

Nematode, tobacco cyst. **Heterodera tabacum,** VA.

Powdery mildew. **Erysiphe cichoracearum,** NJ, VA.

Rot, charcoal. **Macrophomina phaseoli,** NJ.

Rot, cottony leak. **Pythium aphanidermatum,** CA, FL, TN.

Rot, fruit. **Colletotrichum truncatum,** MS; **Phytophthora parasitica,** FL, IN; **Rhizopus stolonifer,** CA, IN, TX.

Rot, gray mold. **Botrytis cinerea,** CA, CT, MA, NJ, VA, WA.

Rot, root. **Phymatotrichum omnivorum,** TX.

Rust. **Puccinia substriata,** FL, IA, TX, GA, AL.

Virus. **Eggplant Mosaic; Cucumber Mosaic; Beet Curly Top,** OR, TX, WA, **Tomato Spotted Wilt.**

Wilt. **Fusarium** sp.; **Verticillium albo-atrum,** general.

Phomopsis blight and Verticillium wilt are the two most important eggplant diseases. Choose varieties resistant to blight and for wilt plant a long rotation that does not include tomatoes, potatoes, or raspberries.

ELAEAGNUS
(Russian-Olive, Silverberry)

Bacterial crown gall. **Agrobacterium tumefaciens,** GA; hairy root, **A. rhizogenes,** IA.

Blight, southern. **Sclerotium rolfsii,** TX.

Blight, stem and branch, bark and cambium necrosis. **Botryodiplodia theobromae,** ND, SK, Southern Plains, NE, Great Plain States.

Blight, thread. **Rhizoctonia ramicula,** FL.

Blight, tip. **Gloeosporium fructigenum.**

Canker. **Nectria cinnabarina,** CA; **Fusarium** sp., WY; **Phytophthora cactorum,** AZ, IL; **Fusicoccum elaeagni,** IL; **Tubercularia ulmea,** ND, SD.

Canker, phomopsis. **Phomopsis elaeagni,** (*Fusicoccum elaeagni*) OH, DE; **P. arnoldia** (syn. *P. elaeagni*), ND, MI, SD.

Leaf spot. **Cercospora carii,** TX; **C. elaeagni,** FL, MS, OK, TX; **Phyllosticta argyrea,** MD, NC; **Septoria argyrea,** IA, ND, NE, WI; **S. elaeagni,** KS.

Mistletoe. **Phoradendron serotinum (flavescens),** FL.

Powdery mildew. **Phyllactinia corylea,** OR; **P. elaeagni,** WY.

Rot, root. **Clitocybe tabescens,** FL.

Rust. **Puccinia caricis-shepherdiae** (0, I), Northern Plains; II, III, on *Carex;* **P. coronata** (0, I), MT, ND; II, III, on *Calamagrostis.*

Wilt. **Verticillium** sp., WA.

ELDER
(*Sambucus*)

Blight, thread. **Pellicularia koleroga,** LA; Web, **P. filamentosa,** FL.

Canker, branch. **Cytospora sambucicola,** IL; **C. chrysosperma.**

Canker; dieback. **Botryosphaeria ribis,** FL, GA: **Diplodia** spp.; **Nectria cinnabarina,** widespread; **N. coccinea,** MD, MI, WA; **Sphaeropsis sambucina.**

Gall, leaf. **Synchytrium sambuci,** LA.

Leaf spot. **Ascochyta sambucina,** AK; **A. wisconsina,** WI, NY; **Cercospora catenospora,** AL, KS, MS; **C. depazeoides,** general; **Cercosporella prolificans,** CA, NM, OR; **Gloeosporium tineum,** MS, TX; **Phyllosticta sambuci,** MO, NY, WI; **Ramularia sambucina,** MO, NY, WI, WA; **Mycosphaerella** sp., NM; **Septoria sambucina,** VT to FL, TX, CA, OR, WA.

Powdery mildew. **Microsphaera alni** and **M. grossulariae,** general; **Phyllactinia corylea,** MI; **Sphaerotheca macularis,** MA.

Rot, heart, wood. **Fomes igniarius,** ID; **Hymenochaete agglutinans,** WY; **Polyporus** spp.

Rot, root. **Helicobasidium purpureum,** TX; **Phymatotrichum omnivorum,** TX; **Xylaria multiplex,** TX.

Rust. **Puccinia bolleyana** (0, I), ME to FL, TX, and MN; II, III on *Carex.*

Spot anthracnose. **Sphaceloma** sp., LA.

Virus. **Tobacco Mosaic.**

Wilt. **Verticillium albo-atrum,** MD.

ELDERBERRY

Virus. **Tobacco Ring Spot.**

ELEPHANTS-EAR
(*Colocasia*)

Bacterial soft rot. **Erwinia carotovora**, FL; **E. aroideae**, FL, TX.
Blight, southern. **Sclerotium rolfsii**, FL, NY.
Nematode, root knot. **Meloidogyne** sp., FL, TX.
Rot, black, of tuber. **Diplodia** sp., FL, SC, TX; Gray, **Fusarium solani**, FL, TX.
Rot, root. **Pythium debaryanum**, CA.

ELM
(*Ulmus*)

Bacterial, leaf spot. **Pseudomonas syringae**, PA.
Bacterial, MLO. **Phloem Necrosis**, AL, AR, GA, IN, IL, IA, KS, KY, MO, MS, NE, NJ, NY, OH, OK, PA, TN, WV.
Bacterial wetwood. **Erwinia nimipressuralis**, VA, on Siberian elm (*Ulmus pumila*).
Bacterial wetwood; slime flux; **Erwinia nimipressuralis.**
Blight, twig. **Phomopsis oblonga**, VA, on Chinese elm (*Ulmus parvifolia*).
Blight, twig. **Septogloeum parasiticum**, MI; **Phomopsis oblonga**, MA; **Fusarium** spp.
Canker, bleeding. **Phytophthora cactorum**, RI; Pit, **P. inflata**, CT, MA, NY, PA.
Canker, felt fungus. **Septobasidium pseudopedicellatum**, NC.
Canker, stem. **Botryosphaeria ribis.**
Canker, twig. **Sphaeropsis ulmicola**, VA, on Chinese elm (*Ulmus parvifolia*).
Canker, twig dieback. **Apioporthe apiospora**, IA; **Coniothyrium** spp., IL, MA, MI; **Cytospora ludibunda**, CT, PA; **Nectria coccinea**, NJ, NY; **N. cinnabarina**, coral spot, widespread; **Phoma** sp., **Phomopsis** sp., Northeast to SC, IL, MN; **Sphaeropsis** sp., CT to MS; **Cytosporina ludibunda**, IL, KS.
Damping-off. **Rhizoctonia solani**, cosmopolitan; **Pythium** sp.
Dieback. **Cephalosporium** sp., VA, on English elm (*Ulmus procera*).
Leaf blister. **Taphrina ulmi**, CT to MS, MO, WI.
Leaf spot. **Cercospora sphaeriaeformis**, LA, TX; **Cylindrosporium tenuisporium**, TX; **Coryneum tumoricola**, NY; **Gloeosporium ulmicola**; **Monochaetia desmazierii**, GA; **Phyllosticta confertissima**, PA; **Mycosphaerella ulmi**, MA to AL; **Septogloeum profusum**, AL; **Coniothryium ulmi**, WV; **Ceratophorum ulmicola**, KS, NE.
Leaf spot; anthracnose. **Gloeosporium inconspicuum**, MA to VA; OK, MN.
Leaf spot; black spot. **Gnomonia ulmi**, general.
Mistletoe. **Phoradendron serotinum (flavescens)**, IN, TX.; **P. tomentosum**, TX.
Mistletoe, European. **Viscum album**, CA.
Nematode, dagger. **Xiphinema americanum.**

Nematode, root knot. **Meloidogyne** spp.

Nematode, stem. **Ditylenchus gallicus;** leaf, **Aphelenchoides fragariae.**

Powdery mildew. **Microsphaera alni,** IL, IA, MS, OH; **Phyllactinia corylea,** NC to TX, IA; **P. ungulata,** GA; **Uncinula macrospora,** general.

Rot, heart. **Collybia velutipes,** widespread; **Daedalea confragosa,** widespread; **Fomes** spp.; **Ganoderma curtisii,** NY.

Rot, root. **Helicobasidium purpureum,** TX; **Phymatotrichum omnivorum,** TX; **Xylaria** spp.; **Armillaria mellea,** MO; **Clitocybe tabescens,** FL.

Rot, wood. **Daldinia concentrica,** widespread; **Lenzites betulinum,** IN, MA, MD; **Pleurotus ostreatus,** widespread; **Polyporus** spp.; **Schizophyllum commune,** cosmopolitan; **Ustulina vulgaris,** MD.

Wilt. **Dothiorella** (*Cephalosporium*) **ulmi,** general; **Verticillium albo-atrum,** ME to VA, OR, WI, MS.

Wilt; Dutch elm disease. **Ceratocystis ulmi,** general.

The Dutch elm disease and phloem necrosis have taken a heavy toll of elms in many states. A dormant spray for the bark beetles that spread Dutch elm disease is the present recommendation, combined with general sanitation. Chemotherapy is still promising but not yet practical. Some seedling elms are highly resistant but not immune.

EMILIA
(Tasselflower, Floras-Paintbrush)

Rust. **Puccinia emiliae,** FL.

Virus. **Cucumber Mosaic,** in part, FL; **Tomato Spotted Wilt,** CA.

ENCELIA

Nematode, root knot. **Meloidogyne** sp., CA.

Rust. **Puccinia enceliae** (0, I, III), CA.

ENDIVE, ESCAROLE, WITLOOF CHICORY
(*Cichorium*)

Bacterial, apical rot. **Pseudomonas cichorii,** CA.

Bacterial center rot. **Pseudomonas cichorii** and **P. intybus,** AZ, CA, FL, MT, TX, WA.

Bacterial, MLO. **Aster Yellows** and **California Aster Yellows.**

Bacterial soft rot. **Erwinia carotovora,** MA, NY.

Blight, southern. **Sclerotium rolfsii,** FL, TX.

Damping-off; bottom rot. **Rhizoctonia solani,** CT, FL, NY, TX.

Downy mildew. **Bremia lactucae,** FL, PA.

Leaf spot. **Alternaria** sp., CT, FL, NY; **Cercospora cichorii,** TX; **Marssonina panattoniana,** TX; **Ramularia cichorii,** NY.

Nematode, root knot. **Meloidogyne** sp., TN, MA, NJ.

Nematode, sting. **Belonolaimus gracilis.**

Nonparasitic. **Brown Heart,** boron deficiency, in part, NJ, NY. **Tipburn,** high temperature and excessive transpiration.

Powdery mildew. **Erysiphe cichoracearum,** ID, NJ.

Rot, gray mold. **Botrytis cinerea,** CA, FL, NY, PA.

Rot, root. **Pythium debaryanum,** CT, FL, NY, PA; **Phymatotrichum omnivorum,** AZ.

Rot, watery soft. **Sclerotinia sclerotiorum,** AR, AZ, CA, FL, LA, MT, PA, TX.

Rust. **Puccinia hieracii** (0, I, II, III), CA, CT, MA, NY.

Slime mold. **Fuligo septica,** NJ.

Virus. **Mosaic,** unidentified, FL; **Tomato Spotted Wilt; Tomato Ringspot,** VT.

ENGELMANNIA
(Engelmann Daisy)

Broomrape. **Orobanche ramosa,** TX.

Gall, leaf. **Synchytrium taraxaci,** TX.

ENGLISH DAISY
(*Bellis perennis*)

Bacterial, MLO. **Aster Yellows,** NJ, NY.

Blight, gray mold. **Botrytis cinerea,** AK.

Leaf spot. **Cercospora** sp., MN.

Nematode, root knot. **Meloidogyne** sp., FL.

Rot, crown. **Sclerotinia sclerotiorum,** NJ.

Rot, root. **Phymatotrichum omnivorum,** TX; **Pythium mastophorum,** MD.

EPIGAEA
(Mayflower, Trailing Arbutus)

Leaf spot. **Cercospora epigaeae,** NY, NC, WI; **Phyllosticta epigaeae,** MA, NY.

Powdery mildew. **Microsphaera alni** var. **vaccinii,** CT to VA and WI.

EPILOBIUM
(Willow-Herb; Fireweed)

Bacterial, MLO. **California Aster Yellows,** CA.

Blight, gray mold. **Botrytis cinerea,** AK.

Blight, southern, **Sclerotium rolfsii,** TX.

Downy mildew. **Plasmopara epilobii,** AK, IL, NY.

Leaf spot. **Cercospora montana,** widespread; **Discosia bubaki,** NY, WI; **Phyllosticta chamaeneri,** OR; **P. wyomingensis,** WY; **Ramularia cercosporoides,** MT, TX, WA, WY, AK; **Septoria epilobii,** CA, DE, IL, VT, WI.

Nematode, root knot. **Meloidogyne** sp.

Powdery mildew. **Sphaerotheca macularis,** widespread; **Erysiphe polygoni,** WA; **Microsphaera** sp., IL.

Rot, root. **Phymatotrichum omnivorum,** TX.

Rust. **Puccinia epilobii** (III), MI, WY; **P. dioicae** (0, I), CO; **P. gigantea** (III), ID, MT, TX, WA, WY; **P. scandica** (III), UT, WA, WY; **P. oenotherae** (0, I, II, III), CA; **P. pulverulenta** (0, I, II, III), ND to NM, CA; **P. veratri** (0, I), NH, MT to WA; **Pucciniastrum pustulatum** (II, III) widespread; 0, I, on fir.

Smut, leaf. **Doassansia epilobii,** CO, NH.

EPISCIA

Leaf spot. **Myrothecium roridum,** FL.

Virus. **Tobacco Mosaic,** CA, CT, DC, FL, OH, WA.

ERANTHEMUM

Leaf spot. **Phyllosticta** sp., NJ.

ERIGERON
(Fleabane)

Bacterial, MLO. **Aster Yellows,** KS, MD, MS, NJ, NY, OK.

Blight, gray mold. **Botrytis cinerea,** AK.

Broomrape. **Orobanche ramosa,** TX.

Downy mildew. **Basidiophora entospora,** AL, IL; **Plasmopara halstedii,** MD, IA.

Gall, leaf. **Synchytrium erigerontis,** LA.

Leaf spot. **Cercospora cana,** LA; **Cercosporella colubrina,** WA; **Septoria erigerontis,** ME to MD, NE, MI.

Powdery mildew. **Erysiphe cichoracearum,** CO, MI, MT, NM, PA, SD, WY; **Phyllactinia corylea,** WA; **Sphaerotheca macularis,** IN, KY.

Rot, stem. **Sclerotium rolfsii,** IL.

Rust. **Puccinia cyperi** (0, I), MO; II, III, on sedge; **P. dioicae** (0, I), East, South; **P. grindeliae** (III), CO, UT, NV, WY; **P. stipae** (0, I), CO, WY; **Coleosporium asterum** (II, III), CA, AK; 0, I, on pine.

Smut, white. **Entyloma compositarum,** MI, ND, UT, WA, WI, WY.

Virus. **Mosaic,** unidentified, IN.

Wilt. **Verticillium albo-atrum,** MA.

ERIOPHYLLUM

Nematode. Meloidogyne sp., CA.
Rust. **Puccinia eriophylii** (II, III), WY, OR to CA; **Uromyces junci** (0, I), CA; II, III, on *Juncus*.

ERYNGIUM

Leaf spot. **Cylindrosporium eryngii**, IA, KS, TX, WI; **Septoria eryngicola**, WI.
Rot, root. **Phymatotrichum omnivorum**, TX.
Rot, stem. **Macrophomina phaseoli**, OK.
Smut, white. **Entyloma eryngii**, IA.

ERYTHRINA

Blight, thread. **Pellicularia koleroga**, FL; **Rhizoctonia ramicola**, FL.
Nematode, root knot. **Meloidogyne** sp.
Rot, root. **Clitocybe tabescens**, FL; **Phymatotrichum omnivorum**, TX.
Wilt. **Verticillium albo-atrum**, CA.

ERYTHRONIUM
(Dogs-Tooth Violet, Adders-Tongue, Trout-Lily)

Blight. **Botrytis** sp., IL, NY, VT, WA; **B. elliptica**, WA; **Ciborinia gracilis**, IL, NE; **C. erythronii**, NY.
Leaf spot; black spot. **Asteroma tenerrimum** var. **erythronii**, ID, MT, WA.
Rust. **Uromyces heterodermus** (0, III), CA, CO, ID, MT, OR, TX, UT, WA, WY.
Smut. **Ustilago heufleri**, DE, MD, MI, MO, NJ, NY, PA; **Urocystis erythronii**, CT, NY.

EUCALYPTUS
(Gum-Tree)

Bacterial crown gall. **Agrobacterium tumefaciens**, CA.
Blight, seedling. **Fusarium oxysporum** f. sp. **aurantiacum**, CA.
Canker. **Diaporthe cubensis**, FL, HI, PR.
Canker; dieback, seed capsule abortion. **Botryosphaeria ribis**, FL.
Canker, felt fungus. **Septobasidium curtisii**, NC.
Leaf spot. **Actinopelte dryina**, LA; **Mycosphaerella molleriana**, CA; **Phyllosticta extensa**, CA.

Nonparasitic. **Chlorosis,** iron deficiency, CA.
Rot, heart; wood. **Ganoderma applanatum; Polyporus** spp.; **Stereum hirsutum.**
Rot, root. **Clitocybe tabescens,** FL; **Armillaria mellea,** CA; **Phymatotrichum omnivorum,** TX.

Many other fungi may be found on leaves, twigs, and branches but are not reported as causing specific diseases.

EUCHARIS
(Amazon-Lily)

Blight, gray mold. **Botrytis cinerea,** FL.
Leaf scorch; red blotch. **Stagonospora curtisii,** CA.

EUGENIA

Black mildew. **Asterinella puiggarii,** FL.
Leaf spot. **Pezizella oenotherae,** NY.
Rot, root. **Clitocybe tabescens,** FL.

EUONYMUS
(Burning-Bush, Spindle-Tree)

Anthracnose. **Colletotrichum griseum,** NJ, NY, AL and GA to TX and AR; **Gloeosporium frigidum,** AR, MS.
Bacterial crown gall. **Agrobacterium tumefaciens,** CT, MS, NJ, MI, SC, TX.
Blight, thread. **Pellicularia koleroga,** LA.
Dieback, canker, basal. **Whetzelinia sclerotiorum** (*Sclerotinia*), RI.
Leaf spot. **Cercospora destructiva,** VA to TX; **C. euonymi,** PA to WI; **Exosporium concentricum,** AL, MS, SC, TX, VA; **Phyllosticta euonymi,** NY to MS and TX; **P. pallens,** AL; **Septoria euonymi,** MS, VA; **S. atropurpurea,** IL; **Ramularia euonymi,** CA, IA, KS, MO; **Marssonina thomasiana,** OH to WI, MO; **Myocentrospora** sp., OH.
Nematode, root knot. **Meloidogyne** sp., MD, TX.
Powdery mildew. **Microsphaera alni,** NJ to SD and southward; **Oidium euonymi-japonici,** AL, CA, IA, LA, MS, NJ, SC, TX, WA.
Rot, root. **Phymatotrichum omnivorum,** TX; **Fusarium scirpi,** NJ.
Virus. **Euonymus Mosaic;** Infectious Variegation, MA.

The Oidium mildew is prevalent throughout the South and in California. Crown gall is common, with conspicuous knobs along the vines, but seldom fatal.

EUPATORIUM
(Boneset, Blue Mist-flower, White Snakeroot, Joe-Pye Weed)

Blight, gray mold; canker. **Botrytis cinerea,** NJ.

Downy mildew. **Plasmopara halstedii,** NY to MD, KS, MO, TX, WI, WV.

Leaf spot. **Ascochyta compositarum,** WV, WI; **Cercospora ageratoides,** AL, MS, NJ, TX, WV; **P. eupatorina,** IL, NJ: **Septoria eupatorii,** IL, MS, MD, NJ, TX; **S. eupatoriicola,** IL.

Nematode, root knot. **Meloidogyne** sp., AL, FL; **M. hapla.**

Powdery mildew. **Erysiphe cichoracearum,** general in East to TX and MN.

Rot, root. **Phymatotrichum omnivorum,** TX.

Rot, stem. **Sclerotium rolfsii,** CT, IL, MD, NJ, TX; **Rhizoctonia solani,** NJ.

Rust. **Puccinia conoclinii** (II, III) MD to AL, TX and IL; **P. eleocharidis** (0, I), widespread in eastern and central states; II, III, on *Eleocharis;* **P. tenuis** (0, I, III), MA to NC, NE and MN; **P. tolimensis** (III), NY.

Smut, white. **Entyloma compositarum,** IL, IA, MS, WV, WI.

Virus. **Yellows.** Apparently distinct from Aster Yellows, central states.

Wilt. **Fusarium** sp., NJ.

EUPHORBIA TRIGONA

Rot, soft. **Rhizopus stolonifera,** CA.

EUSTOMA
(Prairie Gentian, Texas Bluebell)

Blight, stem. **Alternaria** sp., TX; **Sclerophoma eustomonis,** TX.

Leaf spot. **Cercospora eustomae,** CO, NE, TX; **C. nepheloides,** CA, TX; **Phyllosticta** sp., TX.

Virus. **Cucumber Mosaic,** NY.

EVOLVULUS

Rot, root. **Phymatotrichum omnivorum,** TX.

Rust. **Puccinia lithospermi** (0, I, II, III), CO, KS, NE, TX.

Virus. **Tobacco Streak,** WI.

EVERLASTING
(*Antennaria*)

Leaf spot. **Phoma antennariae,** CO; **Phyllosticta antennariae,** DE, WI; **Septoria lanaria,** NY.

Rot, stem. **Nectria haematococca,** CA.
White rust. **Albugo tragopogonis,** IL, NE.
Wilt. **Nectria haematococca,** CA.

EXACUM

Blight, stem canker. **Botrytis cinerea,** KS.

FARKLEBERRY, TREE-HUCKLEBERRY
(*Vaccinium arboreum*)

Canker, felt fungus. **Septobasidium sinuatum,** FL.
Gall, leaf. **Exobasidium vaccinii,** AL, FL.
Leaf spot. **Cylindrosporium** sp., TX; **Ophiodothella vaccinii,** MD to GA,
 TX; **Pestalotia vaccinicola,** secondary, FL; **Phyllosticta vaccinii,** AL, FL,
 MS, TX; **Septoria albopunctata,** FL, SC, TX; Tar Spot, **Rhytisma vaccinii,**
 FL, OK, TX.
Rot, root. **Phymatotrichum omnivorum,** TX.

FEIJOA

Blight, thread. **Rhizoctonia ramicola,** FL.
Rot, fruit. **Botrytis cinerea,** CA; **Colletotrichum gloeosporioides,** CA; **Penicil-
lium expansum,** CA.
Rot, root. **Phymatotrichum omnivorum,** TX.
Spot anthracnose. **Sphaceloma psidii,** FL.

FENDLERA

Rust. **Gymnosporangium speciosum** (0, I), CO and UT to NM and AZ; II, III,
 on juniper.

FENNEL
(*Foeniculum*)

Bacterial soft rot. **Erwinia carotovora,** IL.
Damping-off; stem pitting. **Rhizoctonia solani,** GA, NJ.
Nematode, root knot. **Meloidogyne** sp., FL.
Rot, stem. **Sclerotinia sclerotiorum,** IL, NJ, TX.

FERN, ADDERS-TONGUE
(*Ophioglossum*)

Blight, leaf. **Curvularia crepini,** OH.

FERN, BIRDS-NEST
(*Asplenium*)

Bacterial leaf spot. **Pseudomonas asplenii,** CA.
Bacterial, leaf spot, blights. **Pseudomonas gladioli,** FL.
Leaf spot. **Cercospora** sp., CA.
Nematode, leaf. **Aphelenchoides fragariae,** CT, NJ, NY, PA, FL.

FERN, BLADDER
(*Cystopteris*)

Leaf blister. **Taphrina cystopteridis,** KS, IN, WI.
Rust. **Hyalospora polypodii** (II, III), general in North and West; **Uredinopsis ceratophora** (II, III), IN, NY, WI; **U. glabra** (II, III), NM.

FERN, BOSTON
(*Nephrolepis*)

Anthracnose; tip blight. **Glomerella nephrolepidis,** NY, OH.
Blight, gray mold. **Botrytis cinerea,** AK.
Damping-off. **Rhizoctonia solani,** FL.
Leaf spot. **Cercospora** sp., IN; **Cylindrocladium pteridis,** FL; **Phyllosticta** sp., NY.
Rust. **Desmella superficialis,** FL.

FERN, BRACKEN
(*Pteridium*)

Leaf spot; tar spot; black mildew. **Phyllachora** (*Catacauma*) **flabellum,** GA, MD, NJ, PA, SC, TN, WV, WI; **Cryptomycina pteridis,** also leaf roll, widespread, reported under various conidial names; **Phyllosticta pteridis,** ME, NJ.
Rot; canker. **Rhizoctonia** sp., including **Sclerotium decidium,** OR, WI.
Rust. **Uredinopsis aspera** (II, III), CA; **U. macrosperma** (II, III), AL, CA, FL, GA, ID, MS, MT, NM, OR, WA, WI; **U. virginiana** (II, III), NY to NC and TN, GA to TX.

FERN, BRAKE
(*Pteris*)

Blight, tip. **Phyllosticta pteridis, MS, NJ.**
Damping-off. **Completoria complens, NY; Pythium intermedium, NY; Trichothecium roseum, IN.**
Nematode, leaf. **Aphelenchoides fragariae, CT, NJ, NY.**

FERN, CHRISTMAS
(*Polystichum*)

Leaf blister. **Taphrina faullinana, OR; T. polystichi, ME to NC and TN.**
Leaf spot. **Cylindrocladium pteridis, FL;** Tar Spot, **Trabutiella filicina, AK.**
Nematode, leaf. **Ditylenchus dipsaci, OR; Aphelenchoides fragariae, OR;** Lesion, **Pratylenchus penetrans, FL.**
Rot, root. **Phytophthora cinnamomi, CA**
Rust. **Milesia polystichi** (II, III), ID, MT, OR, WA; **M. vogesiaca** (II, III), OR.

FERN, CLIFF-BRAKE
(*Pellaea*)

Rust. **Hyalopsora cheilanthis** (II, III), CA, TX.

FERN, HOLLY
(*Cyrtomium*)

Blight, gray mold. **Botrytis cinerea, AK.**
Damping-off. **Completoria complens, NY.**

FERN, LADY, SILVERY SPLEENWORT
(*Athyrium*)

Blight, gray mold. **Botrytis cinerea, AK.**
Leaf spot. **Septoria asplenii, MI.**
Rust. **Uredinopsis copelandii** (II, III), CA; **U. longimucronata** (II, III), ME to PA and WI; 0, I, on balsam fir; **U. longimucronata** f. **cyclosora,** CA, ID, MT, OR, WA, AK; 0, I, on alpine fir; **U. longimucronata** f. **acrostichoides** (II, III), NH, NY, WI.

FERN, LEATHERLEAF
(*Rumohra*)

Rot, postharvest. **Rhizoctonia** sp., FL; **Cylindrocladium pteridis,** FL; **C. heptaseptatum,** FL.

FERN, MAIDENHAIR
(*Adiantum*)

Leaf spot. **Mycosphaerella** sp., FL.
?Rot. **Sclerotium decidium** (?*Rhizoctonia* sp.), WI.

FERN, OSMUNDA
(*Osmunda;* Cinnamon, Interrupted, and Royal Ferns)

Leaf blister. **Taphrinia higginsii,** GA.
Leaf spot. **Gloeosporium osmundae,** MI.
Rust. **Uredinopsis osmundae** (II, III), northeastern and Great Lakes states, and to FL, and AL; 0, I, on balsam fir.
Smut, inflorescence. **Mykosyrinx osmundae,** NY, MI, WI.

FERN, OSTRICH
(*Pteretis*)

Leaf blister. **Taphrinia hiratsukae,** WI.
Rot; stem necrosis. **Ceratobasidium anceps** (*Rhizoctonia?*), WI.
Rust. **Uredinopsis struthiopteridis** (II, III), NY, VT, WI; 0, I, on balsam fir.

FERN, POLYPODY
(*Polypodium*)

Leaf spot. **Alternaria polypodii,** ?secondary, NY; **Cercospora phyllitidis,** FL; **Phyllosticta** sp., VA.
Nematode, leaf. **Aphelenchoides fragariae.**
Rust. **Milesia laeviuscula** (II, III), CA, and **M. glycyrrhiza,** OR, WA, AK; **M. polypodophila** (II, III), CT, ME, MA, NH, NY, PA, TN, VT.

FERN, ROCK
(*Woodsia*)

Rust. **Hyalopsora polypodii** (II, III), MI, ID; 0, I unknown.

FERN, ROCK-BRAKE
(*Cryptogramma*)

Rust. **Hyalopsora cheilanthis** (II, III), IA, MI, MT, WI; **Milesia darkeri** (II, III), CA, OR.

FERN, SENSITIVE
(*Onoclea*)

Dodder. **Cuscuta gronovii,** NY.
Leaf blister. **Taphrinia filicina,** NY; **T. hiratsukae,** NY, PA.
Rust. **Uredinopsis mirabilis** (II, III), ME to VA, NE and MN; 0, I, on balsam fir.

FERN, TREE
(*Cibotium*)

Leaf spot. **Pestalotia cibotii,** NJ.

FERN, WALKING
(*Camptosorus*)

Leaf spot. **Cercospora camptosori,** WI.

FERN-WOOD-SHIELD
(*Dryopteris*)

Leaf blister, gall. **Taphrina californica,** CA, OR; **T. filicina,** NY, PA; **T. fusca,** NJ, VT, WV; **T. gracilis,** NY; **T. lutescens,** ME, MN, NY, WI.
Leaf spot. **Cylindrocladium pteridis,** FL; tar spot, **Cryptomycina pteridis,** FL; **Herpobasidium filicinum,** white mold.
Nematode, leaf. **Aphelenchoides fragariae.**
Rust. **Hyalopsora aspidiotus** (II, III), ME to NC, WI and WA; 0, I, on balsam fir; **Milesia dilatata** (II, III), OR; **M. fructuosa** (II, III), ME, MA, NH, NY, VT; **M. marginalis** (II, III), MA, NH, VT; **Uredinopsis atkinsonii** (II, III), ME to MS, NE, ND; **U. phegopteris** (II, III), ME, NH, WI; 0, I, on balsam fir.

FERN, WOODWARDIA, CHAIN
(*Woodwardia*)

Rust. **Uredinopsis arthurii** (II, III), VT to AL, IN and MI, and var. **maculata,** ME to AL; 0, I unknown.

FEVERFEW
(*Chrysanthemum parthenium*)

Powdery mildew. **Erysiphe cichoracearum,** NY.
Rot, root and stem. **Rhizoctonia solani,** WA.

FIG
(*Ficus carica*)

Anthracnose; fruit rot. **Colletotrichum gloeosporioides** (*Glomerella cingulata*), NC to TX.

Bacterial crown gall. **Agrobacterium tumefaciens,** TX, CA, FL.

Blight, limb. **?Corticium salmonicolor,** Gulf states.

Blight, southern. **Sclerotium rolfsii,** FL.

Blight, thread. **Pellicularia koleroga,** FL, LA, MS; Web, **P. microsclerotia,** FL to LA, TX.

Blight, twig. **Gibberella baccata,** CA.

Blotch, leaf. **Cercospora fici,** NC to FL and TX.

Canker; dieback. **Botryosphaeria ribis; Diplodia sycina,** NC, OR; **Macrophoma fici,** also fruit dry rot, LA, NC, TX; **Nectria cinnabarina,** TX; **Megalonectria pseudotrichia,** LA, TX; **Physalospora rhodina,** AL, FL, TX; **Sclerotinia sclerotiorum,** CA, TX.

Canker, felt fungus. **Septobasidium** spp., MS.

Canker, phomopsis. **Phomopsis cinerescens,** CA.

Leaf spot. **Colletotrichum elastica,** FL; **Corynespora cassiicola,** FL; **Ascochyta caricae,** CA; **Alternaria** sp., SC; **Cephalosporium ?acremonium,** LA; **Eutypa** sp., TX; **Mycosphaerella bolleana,** NC to TX; **Ormathodium fici,** LA.

Nematode, dagger. **Xiphinema index,** CA.

Nematode, fig cyst. **Heterodera fici,** CA.

Nematode, lesion. **Pratylenchus vulnus, P. musicola** and **P. pratensis,** CA.

Nematode, pin. **Paratylenchus hamatus,** CA.

Nematode, root knot. **Meloidogyne** spp., southern states to CA.

Nonparasitic. **Chlorosis,** manganese deficiency, FL. **Little Leaf,** zinc deficiency, CA. **Sunburn,** low temperature injury to trunk and branches, CA.

Rot, fruit. **Alternaria tenuis,** in market; **Aspergillus niger,** black mold; **Botrytis cinerea; Choanephora cucurbitarum,** TX; **Cladosporium herbarum,** market; **Diplodia natalensis,** TX; **Fusarium moniliforme,** CA; **Oospora** sp., sour rot, TX; **Rhizopus nigricans,** CA, Gulf states; **Trichothecium roseum,** Gulf states.

Rot, root. **Armillaria mellea,** CA; **Dematophora necatrix,** NC; **Phymatotrichum omnivorum,** TX; **Rhizoctonia** sp., TX.

Rust. **Cerotelium** (*Physopella*) **fici** (II, III), NC to FL and TX, AR.

Sooty mold. **Capnodium** sp., TX; **Fumago vagans,** Gulf states.

Virus. **Fig Mosaic,** CA, GA, TX, VA?.

Wilt, branch. **Hendersonula toruloides,** CA.

FIG, FLORIDA STRANGLER
(*Ficus aurea*)

Anthracnose. **Colletotrichum gloeosporioides,** FL.

Bacterial crown gall. **Agrobacterium tumefaciens,** FL.

Leaf spot. Ophiodothella fici, FL; Phlyctaena ficuum, FL; Phyllosticta physopellae; P. roberti, FL.

Nematode, root knot. Meloidogyne spp.

Rot, fruit. Fusarium moniliforme, CA.

Rust. Cerotelium fici (II, III), FL.

FIGWORT
(*Scrophularia*)

Downy mildew. Peronospora sordida, CA, IL, NY, VA to KS.

Leaf spot. Cylindrosporium scrophulariae, IL, OK, PA; Mycosphaerella sp., KS; Septoria scrophulariae, CA, OR, WA, NY to MS and CO.

FILAREE, RED-STEM
(*Erodium cicutarium*)

Nematode, lesion. Pratylenchus minyus, CA.

Virus. Filaree Red Leaf, CA.

FILBERT

See Hazelnut.

FIR
(*Abies*)

Black mildew. Dimerosporium abietis, ID, OR, WA; Adelopus nudus, ME, NY, NC, WI.

Blight. Phoma eupyrena, CA.

Blight, brown felt. Herpotrichia nigra, on high western firs.

Blight, needle. Acanthostigma parasiticum; Rehmiellopsis balsameae, Northeast; Cenangium ferruginosum, MI, PA; Rhabdogloeum abietinum, NC; Macrophoma parca, CO, OR, ID, MT, WA; Phoma eupyrena, CA.

Blight, needle. Furcaspora pinicola, CA.

Blight, seedling smother. Thelephora terrestris.

Blight, snow. Phacidium infestans, Northeast; P. balsameae; P. abietinellum.

Canker. Aleurodiscus amorphus, general; Cytospora pinastri, ME, WI; C. cylindroides and C. abietis, OR, WA; Cephalosporium sp., MN, WI; Cryptosporium macrospermum, New England; Dasyscypha resinaria, MN; Phomopsis boycei, ID, MT; P. fokoyae, CA.; P. montanensis; Ophionectria scolecospora, widespread; Scleroderris abieticola, OR; Sphaeropsis abietis, MI; Valsa (*Cytospora*) kunzei; Thyronectria balsamea, CO.

Dieback, twig. **Sydowia polyspora,** CA.

Mistletoe. **Arceuthobium campylopodium,** widespread in West; **A. douglasii,** NM, OR, WA; **Phoradendron pauciflorum,** AZ, CA.

Needle cast; blight. **Bifusella abietis,** ID, MT, WA; **B. faullii,** ME, MI, NH; **Hypoderma robustum,** CA, ID, OR, WA; **Hypodermella abietis-concoloris,** widespread on western firs; **H. mirabilis,** MI; **H. nervata,** ME, NH, VT; **H. punctata,** OR, ID; **Lophodermium piceae,** tar spot, widespread, weakly parasitic; **L. autumnale,** MI, CA, OR, ID, UT, WY; **L. consciatum; L. decorum; L. uncinatum; L. lacerum,** NH, NY, PA, VT.

Needle cast; blight. **Tirula nervisequa conspicuous,** CA.

Nematode. **Nacobbodera chitwoodi,** OR.

Rot, heart. **Stereum sanguinolentum,** widespread; **Polyporus sulphureus,** general.

Rot, hypocotyl. **Fusarium oxysporum,** CA.

Rot, root. **Armillaria mellea,** New England, NY; **Fusarium solani,** PA; **F. oxysporum,** PA; **F. avenaceum,** PA; **Phytophthora cinnamomi,** NC; **P. citricola,** NC.

Rot, root. **Phytophthora drechsleri,** NC.

Rot, root. **Poria weirii,** OR.

Rot, root. **Verticicladiella wagenerii,** Pacific Coast.

Rot, wood. **Coniophora puteana,** New England, NY; **C. corrugis,** AZ, CO, WY; **Fomes annosus,** CO; **Echinodontium tinctorium; Hydnum abietis; H. balsameum; Polyporus** spp.; **Poria** spp.

Rust, fir-broom. **Melampsorella caryophyllacearum.**

Rust, fir-fern. **Milesia fructuosa** (0, I), on new needles, ME, NH, NY; II, III, on *Dryopteris;* **M. marginalis,** MA, NH, NY; **M. polypodophila** (0, I), ME, NY, NH, on old needles; II, III, on *Polypodium,* **Hyalopsora aspidiotus** (0, I), on 2-year needles; II, III, on *Dryopteris;* **Uredinopsis mirabilis** (0, I), general; II, III, on sensitive fern; **U. osmundae** (0, I), widespread; II, III, on *Osmunda;* **U. struthiopteridis,** MI, ID, OR, WA; II, III, on ostrich-fern; **U. phegopteris** (0, I); II, III, on *Dryopteris.*

Rust, fir-fireweed. **Pucciniastrum pustulatum,** widespread; II, III, on Epilobium.

Rust, fir-huckleberry. **Pucciniastrum goeppertianum,** widespread on western firs, and ME, PA, WI; II, III, on *Vaccinium.*

Rust, fir-willow. **Melampsora abieti-capraearum,** widespread; II, III, on willow.

Rust, needle. **Caeoma faulliana,** OR; **Peridermium ornamentale,** ID, MT, OR, WA; **P. rugosum,** CA, OR, WA.

Rust, witches' broom. **Melampsorella cerastii,** general; II, III, on chickweed.

Despite this long list of possibilities the gardener should not have much trouble with ornamental firs. Rehmiellopsis tip blight yellows needles of new growth and causes twig dieback of native balsam firs in the Northeast but can be controlled with copper sprays, which will also aid in preventing needle-cast diseases. Avoid bark and branch injuries that induce cankers. Rust is taken care of, if necessary, by eliminating the proper alternate host, but only a specialist can identify the many different rust species.

FIRETHORN

See Pyracantha.

FITTONIA

Blight, leaf. **Rhizoctonia** sp., OH.
Leaf spot. **Myrotherium roridum,** FL.

FLAX, FLOWERING
(*Linum*)

Damping-off. **Rhizoctonia solani,** IL.
Nematode, root knot. **Meloidogyne** sp., CA.
Rot, stem. **Sclerotinia sclerotiorum,** CA.

FOAM-FLOWER
(*Tiarella*)

Powdery mildew. **Sphaerotheca macularis,** AK.
Rust. **Puccinia heucherae** (III), CT to NC and TN, CO, MI, WI, CA, ID, MT, OR, WA, AK.

FORESTIERA
(*Swamp-Privet*)

Mistletoe. **Phoradendron serotinum (flavescens),** TX.
Powdery mildew. **Microsphaera alni,** IL, TX.
Rot, root. **Phymatotrichum omnivorum,** TX.
Rust. **Coleosporium minutum** (II, III), FL, TX; 0, I, on pine; **Puccinia peridermiospora** (0, I), FL, TX; II, III, on marsh grass.

FORGET-ME-NOT
(*Myosotis*)

Bacterial, MLO. **California Aster Yellows,** CA.
Blight, gray mold. **Botrytis cinerea,** cosmopolitan.
Downy mildew. **Peronospora myosotidis,** IL, MI, MS, WI.

Leaf spot. **Stemphylium** sp., NY.
Rot, crown. **Sclerotinia sclerotiorum,** IL, WA.
Rust. **Puccinia eatoniae** var. **myosotidis** (0, I), IL, IN, MS, MO, NC, WI;
   II, III, on *Sphenopholis;* **P. mertensiae** (III), CO.
?Virus. **Chlorosis, GA.** Perhaps eastern strain of Aster Yellows.

FORSYTHIA
(Goldenbells)

Anthracnose. **Gloeosporium** sp., OK.
Bacterial grown gall. **Agrobacterium tumefaciens,** MS, NJ, TX.
Blight, cane. **Botryosphaeria ribis.**
Blight, southern. **Sclerotium rolfsii,** GA.
Blight, twig. **Sclerotinia sclerotiorum,** NC.
Canker, stem gall. **Phomopsis** sp., KY.
Leaf spot. **Alternaria** sp., IA; **Phyllosticta discincola,** MD.
Nematode, lesion. **Pratylenchus vulnus.**
Nematode, root knot. **Meloidogyne** sp., TX; **M. hapla; M. incognita-acrita.**
Rot, root. **Phymatotrichum omnivorum,** TX.
Virus. **Tobacco Ring Spot,** MD.

FOUQUIERIA
(Ocotillo, Candlewood)

Rot, root. **Phymatotrichum omnivorum,** TX.
Rust. **Aecidium cannonii** (0, I), AZ.

FOUR-O'CLOCK
(*Mirabilis*)

Leaf spot. **Cercospora mirabilis,** TX.
Nematode, root knot. **Meloidogyne** sp., FL.
Rot, root. **Phymatotrichum omnivorum,** TX.
Rust. **Aecidium mirabilis** (0, I), AZ, NM; **Puccinia aristidae** (0, I), AZ.
Virus. **Beet Curly Top,** CA.
White rust. **Albugo platensis,** TX.

FOXGLOVE
(*Digitalis*)

Anthracnose. **Colletotrichum fuscum,** CT, MA, OR, PA, WI.
Blight, leaf; inflorescence. **Alternaria** sp., MD.
Leaf spot. **Cladosporium** sp., NJ; **Phyllosticta digitalis,** NY, TX, AK; **Ramularia variabilis,** OR.

Nematode, leaf and stem. **Ditylenchus dipsaci,** CT.
Nematode, root knot. **Meloidogyne** spp., CA, MD.
Rot, root and stem. **Rhizoctonia solani,** NJ; **Fusarium** sp., CA, NH.
Rot, stem wilt. **Sclerotinia sclerotiorum,** NY; **Sclerotium rolfsii,** IN, NJ, TX.
Virus. **Tobacco Mosaic,** WI.
Wilt. **Fusarium** sp., CA, NH; **Verticillium albo-atrum,** NY.

FREESIA

Bacterial scab. **Pseudomonas marginata,** occasional in imported stock.
Leaf spot. **Heterosporium iridis,** CT.
Nematode, root knot. **Meloidogyne** sp., CA.
Rot, corm wilt. **Fusarium** spp., including **F. oxysporum; F. solani,** CA, FL, TX.
Rot, dry. **Stromatinia gladioli,** NJ, NY.
Rot, blue mold. **Penicillium ?gladioli.**
Rust. Cause unknown, not true rust.
Virus. **?Iris Mosaic,** CA; **Bean Yellow Mosaic.**

Fusarium corm rot causes wilting and death. Corms should be inspected before planting, and all those showing pinkish lesions should be discarded.

FREMONTIA
(Flannel Bush)

Leaf spot. **Ascochyta fremontiae,** CA.
Rot; stem girdle. **Phytophthora cactorum,** CA.
Rot, crown, root. **Phytophthora nicotianae** var. **nicotianae,** CA.
Wilt. **Verticillium albo-atrum,** CA.

FRINGE-TREE
(*Chionanthus*)

Leaf spot. **Cercospora chionanthi,** NJ to NC and WV; **Phyllosticta chionanthi,** NJ, WV; **Septoria chionanthi** and **S. eleospora,** SC and TX.
Powdery mildew. **Phyllactinia corylea,** MD.
Rot, root. **Phymatotrichum omnivorum,** TX.
Rot, wood. **Daedalea confragosa,** MD.

FRITILLARIA

Leaf spot. **Phyllosticta fritillariae,** CA.
Rust. **Uromyces miurae** (III), WA, AK.
Virus. **Mosaic.** Undetermined, NY.

FROELICHIA

Leaf spot. **Cercospora crassoides,** OK, TX, WI.
Nematode, root knot. **Meloidogyne** sp., FL.
Rot, root. **Phymatotrichum omnivorum,** TX.
White rust. **Albugo froelichiae,** NE, TX.

FROSTWORT
(*Crocanthemum*)

Leaf spot. **Cylindrosporium eminens,** WI.

FUCHSIA

Blight, gray mold. **Botrytis cinerea,** WV, AK.
Dieback. **Phomopsis** sp., VA.
Leaf spot. **Septoria** sp., OK.
Nematode, root knot. **Meloidogyne** sp.
Rot, mushroom root. **Armillaria mellea,** CA.
Rot, root. **Pythium rostratum; P. ultimum,** CA; **Phytophthora parasitica,**
    OR; **Thielaviopsis sp.,** OR.
Rust. **Pucciniastrum epilobii,** NC, OR; **Uredo fuchsiae** (II), OH.
Virus. **Tomato Spotted Wilt,** CA.
Wilt. **Verticillium albo-atrum,** CA.

Verticillium wilt is common in garden plantings of fuchsia in California.

GAILLARDIA

Bacterial, MLO. **Aster Yellows,** NJ, NY, PA, and **California Aster Yellows,** CA.
Leaf spot. **Septoria gaillardiae,** IA, OK, TX.
Nematode. **Naccobus batiformis,** NE.
Powdery mildew. **Erysiphe cichoracearum,** MT, OK, TX; **Sphaerotheca mac-**
    ularis, MT, WA, WY.
Rot, root. **Phymatotrichum omnivorum,** TX; **Pythium ultimum,** CA.
Rust. **Coleosporium asterum** (II, III), CA; 0, I, on pine; **Puccinia gaillardiae**
    (0,I), CA, II, III unknown.
Smut, white. **Entyloma polysporum,** KS, MN, NE.
Virus. **Tomato Spotted Wilt,** CA; **Biddens Mottle,** FL.

GALAX

Leaf spot. **Clypeolella leemingii,** black spot, MD to GA and MS; **Discohainesia**
    oenotherae, NC; **Phyllosticta galactis,** NC, VA, WV.

GALIUM
(Bedstraw)

Dodder. **Cuscuta cuspidata,** TX.

Downy mildew. **Peronospora calotheca,** IA, ND, WI.

Leaf spot. **Cercospora galii,** AK, AL, IA, NH, NY, OR; **Pseudopeziza repanda,** CA, CT, GA, IL, IA, ND, NY, OR, WI; **Melasmia galii,** IA; **Septoria cruciatae,** IN, MI, NJ, NY, WI, WV.

Powdery mildew. **Erysiphe cichoracearum,** CA, MT, OR, PA, WA; **E. polygoni,** KS.

Rot, root. **Phymatotrichum omnivorum,** TX.

Rust. **Puccinia difformis** (0, I, III), OH to KS, MT, Pacific Coast; **P. punctata** (0, I, II, III); CT, NC, MS, ND to CA, WA; **P. punctata** var. **troglodytes** (0, I), CT to MO and SD, WA; **P. rubefaciens** (III), IA and WI to CA, WA, AK; **Pucciniastrum galii,** NY and PA to CO, CA and OR; **Uromyces galii-californici** (II, III), CA.

GALTONIA
(Summer-Hyacinth)

Virus. **Ornithogalum Mosaic,** OR.

GARDENIA
(Cape-Jasmine)

Bacterial leaf spot. **Xanthomonas maculifolium-gardeniae,** CA.

Blight; bud rot. **Botrytis cinerea,** in greenhouses; outdoors in CA.

Canker, stem gall. **Phomopsis** (*Diaporthe*) **gardeniae,** CA, widespread in greenhouses, outdoors in CA, FL.

Leaf spot. **Myrothecium roridum,** PA; **Pestalotia langloisii,** FL, AL; **Phyllosticta** sp., MS, NJ, TX; **Rhizoctonia** sp., NJ; **Mycosphaerella gardeniae,** GA.

Leaf spot, algal. **Cephaleuros virescens,** Gulf states.

Nematode, burrowing. **Radopholus similis,** FL.

Nematode, reniform. **Rotylenchulus reniformis.**

Nematode, root knot. **Meloidogyne hapla; M. incognita; M. incognita-acrita; M. javanica;** general; **M. arenaria,** NC.

Nematode, spiral. **Helicotylenchus nannus.**

Nonparasitic. **Bud Drop,** excessive soil moisture; temperature fluctuation. **Chlorosis,** soil too alkaline or soil temperature too low. **Dieback,** yellow vein-banding, defoliation from root smothering by overwatering.

Powdery mildew. **Erysiphe polygoni,** TX.

Rot, root. **Phymatotrichum omnivorum,** TX

Sooty mold. **Capnodium** spp., Gulf states.

In the Deep South gardenia foliage is disfigured with sooty mold growing in whitefly honeydew. Gardenias are difficult house plants, dropping buds with uneven humidity and temperature. To control phomopsis canker take cuttings near top of plants. Substitute the more resistant Veitchii variety for susceptible Belmont and Hadley. Avoid syringing, which spreads bacterial leaf spot.

GARLIC
(*Allium sativum*)

Subject to diseases of Onion, which see. White rot is serious in Louisiana; pink root is a common problem. Clove rot is caused by *Penicillium digitatum*.

GARRYA
(Tassel-Tree, Silk-Tassel Bush)

Leaf spot. **Cercospora garryae,** CA, TX; **Phyllosticta garryae,** CA, TX; **Dothichiza garryae,** CA.
Rot, root. **Phymatotrichum omnivorum,** TX.
Sooty mold. **Lembosia lucens,** CA.

GAULTHERIA
(Checkerberry, Teaberry; Source of Oil of Wintergreen)

Blotch, sooty. **?Gloeodes pomigena,** WI.
Fruit spot; black speck. **?Leptothryium pomi,** WI; **Schizotherium gaultheriae,** ME to VA and WI.
Leaf spot. **Cercospora gaultheriae,** NJ, WI; **Discohainesia oenotherae,** VA; **Mycosphaerella gaultheriae** (*Phyllosticta gaultheriae*), general; **Venturia arctostaphyli,** MD, MA, NJ, NY, VA.
Powdery mildew. **Microsphaera alni,** MD.

GAURA

Bacterial, MLO. **California Aster Yellows,** CA.
Broomrape. **Orobanche ramosa,** TX.
Downy mildew. **Peronospora arthuri,** KS, NE.
Gall, leaf. **Synchytrium fulgens,** TX.
Leaf spot. **Cercospora gaurae,** NY, OK, TX; **Septoria gaurina,** IL, KS, NE, ND, OK, TX.
Powdery mildew. **Erysiphe polygoni,** CO, TX.
Rot, root. **Phymatotrichum omnivorum,** TX.
Rust. **Puccinia extensicola** var. **oenotherae** (0, I), CO, NE, TX; II,˙III, on Carex spp.; **Uromyces plumbarius** (0, I, II, III), NY to VA, TX and WI, ND and MT to MS and NM.

GAZANIA

Rot, crown. **Rhizoctonia solani,** CA.

GENISTA
(Wood-waxen)

Dieback. **Diplodia** sp., CA, NJ.
Powdery mildew. **Erysiphe polygoni,** MO.
Rust. **Uromyces genistae-tinctoriae** (II, III), CA.

GENTIAN
(*Gentiana*)

Blight; stem canker. **Botrytis cinerea,** NY.
Leaf spot. **Asteromella andrewsii** (?*Mycosphaerella andrewsii*), DE, IL, IA, ND, NE, NJ, PA, WI, WV; **Cercospora gentianae,** ND, NY, VT; **C. gentianicola,** DE, WI.
Rot, root. **Fusarium solani,** MD.
Rust. **Puccinia gentianae** (0, I, II, III), NY to IN, NE, MN; western states to CA, WA, AK; **P. haleniae,** WY; **Uromyces gentianae** (II, III), CO, IA, NV, NM, WA, WY, NC, VT; **Pucciniastrum alaskanum,** AK.

GERANIUM
(*Pelargonium*)

Bacterial crown gall. **Agrobacterium tumefaciens,** MA, MD, OH.
Bacterial fasciation. **Clavibacter fascians,** MA, OH, IN, OR.
Bacterial leaf spot. **Xanthomonas pelargonii,** MA to VA, MS, OH, CA, WA; **Pseudomonas cichorii,** FL.
Bacterial wilt, southern. **Pseudomonas solanacearum,** NC.
Blight, blossom; gray mold; cutting rot. **Botrytis cinerea,** cosmopolitan.
Leaf spot. **Alternaria** sp., secondary; **Ascochyta** sp., CT, NJ; **Cercospora brunkii,** FL, TX, MD, OH, NH; **Pleosphaerulina** sp., PA.
Mold, leaf. **Botryosporium pulchrum,** occasional in greenhouses.
Nematode, dagger. **Xiphinema americanum.**
Nematode, leaf. **Aphelenchoides** sp., NY.
Nematode, root knot. **Meloidogyne** spp., CT, FL, NJ, OH, OR.
Nonparasitic. **Oedema,** dropsy; intumescence from excessive soil moisture and retarded transpiration. **Crook-Neck,** chimeral mutation.
Petal and leaf spot. **Bipolaris** (*Helminthosporium*) **setariae,** FL.
Rot, blackleg, stem and cutting rot. **Pythium debaryanum, P. mammilatum, P. splendens, P. ultimum, P. vexans,** cosmopolitan; **Rhizoctonia solani,** cosmopolitan; **Aspergillus fischeri,** CA; **Fusarium** sp., IN, NY, WA.

Rot, root. **Thielaviopsis basicola,** CT; **Armillaria mellea,** CA.

Rust, pelargonium. **Puccinia pelargonii-zonalis** (II), CA, NY, FL, PA.

Virus. **Pelargonium Leaf Curl** (Geranium Crinkle), general; **Geranium Mosaic; Cucumber Mosaic; Beet Curly Top; Tomato Spotted Wilt,** CA, TX; **Tomato Ring Spot; Tobacco Ring Spot; Pelargonium Flower Break.**

Wilt. **Verticillium albo-atrum,** CA, NY, OR.

The dry air of the average home makes foliage diseases due to pathogenic organisms unlikely but sometimes a water-logged soil and cloudy weather, with less evaporation, leads to oedema, small swellings in leaves, corky ridges on petioles. Botrytis blight and bacterial leaf spots may be expected in greenhouses unless plants are spaced widely, have proper air circulation, little overhead watering, and all infected plants or parts are speedily removed. Start with clean cuttings from a culture-indexed mother block.

GERANIUM
(Cranesbill, Herb-Robert)

Bacterial leaf spot. **Pseudomonas erodii,** FL, IL, IN, OR, TX; **Xanthomonas geranii,** NY.

Blight, leaf spot; stem rot. **Botrytis cinerea,** KS, MO.

Broomrape. **Orobanche ramosa,** TX.

Downy mildew. **Plasmopara geranii,** NJ to FL, KS, TX, MA to IA, MT, UT, WI.

Gall, leaf. **Synchytrium geranii,** LA, OK, TX.

Leaf spot. **Cercospora geranii,** CO, KS, IA, MO, MT, NY, TX, UT, WI; **Cylindrosporium geranii,** LA; **Dilophospora geranii,** WI; **Pestolozziella subsessilis,** MS, MO, NJ, WI; **Phyllosticta geranii,** LA, TX; **Ramularia geranii,** CA, LA, WA, WY; **Septoria expansa,** KS, TX; **Stigmatea geranii,** black leaf speck, MD; **Venturia circinans,** mold, AK.

Powdery mildew. **Erysiphe polygoni,** CO, IL, IN, OH, PA, WV, WI, WY; **Sphaerotheca macularis,** CA, ID, MN, MT, NE, PA, WA, WI, WY.

Rot, rhizome. **Seaverinia geranii,** NY, OH, WI.

Rot, root. **Phymatotrichum omnivorum,** TX.

Rust. **Puccinia leveillei** (III), CO, MT, UT, WA, WY; **P. polygoni-amphibii** (0, I), CT to WI, MO, MN, MT, KS, TX; II, III, on Polygonum; **Uromyces geranii** (0, I, II, III), ME, WY, AK.

Virus. **Cucumber Mosaic,** FL.

GERBERA
(Transvaal Daisy)

Blight, flower; ray speck. **Alternaria dauci** f. sp. **solani,** FL.

Blight, gray mold. **Botrytis cinerea,** FL, NY.

Leaf spot; stem rot. **Gloeosporium** sp., NY.

Nematode, root knot. **Meloidogyne** sp., AL, CA, MD, NC, NY.
Powdery mildew. **Erysiphe cichoracearum,** CA, OK.
Rot, crown. **Sclerotinia sclerotiorum,** MD, NY.
Rot, root and stem. **Phytophthora cryptogea** and **P. drechsleri,** CA, NJ, NY.
Virus. ?Tobacco Rattle, FL.

GERMANDER
(*Teucrium*)

Downy mildew. **Peronospora** sp., OK.
Leaf spot. **Cercospora teucrii,** NY to MS, TX, WI; **Phyllosticta decidua,** TX, WI.
Nematode, root knot. **Meloidogyne hapla.**
Powdery mildew. **Erysiphe cichoracearum,** IL, PA, WI.
Rust. **Puccinia menthae** (0, I, II, III), PA.

GEUM
(*Avens*)

Bacterial, MLO. **California Aster Yellows,** CA.
Downy mildew. **Peronospora potentillae,** CA, IL, IN, IA, KS, NE, WI.
Gall, leaf. **Synchytrium aureum,** WI.
Leaf spot. **Cercospora gei,** WI; **Cylindrosporium gei,** NH, WI; **Marssonina adunca,** MT, WA; **Phyllosticta** sp., WI, WV; **Ramularia gei,** WI, MO; **Septoria gei,** DE, IL, MI, NY, NE, OH, AK.
Nematode, root knot. **Meloidogyne** sp., CA.
Powdery mildew. **Erysiphe polygoni,** AK; **Sphaerotheca macularis,** IN, MD, NE, NY, ND, OH, PA, WA, WI.
Rot, root. **Phymatotrichum omnivorum,** TX.
Rust. **Puccinia sieversii,** CO.
Smut, leaf. **Urocystis** (*Whetzelia*) **waldsteinae,** MT, WA.

GILIA
(Skyrocket)

Bacterial, MLO. **California Aster Yellows,** CA.
Downy mildew. **Peronospora giliae,** TX.
Leaf spot. **Ramularia giliae,** OR.
Nematode, lesion. **Pratylenchus pratensis,** TX.
Nematode, root knot. **Meloidogyne** sp., CA.
Nematode, stem. **Ditylenchus dipsaci.**
Powdery mildew. **Sphaerotheca macularis,** CA, TX, WA.
Rot, root. **Phymatotrichum omnivorum,** TX.

Rust. **Puccinia aristidae** (0, I), AZ, CO; **P. giliae** (II, III), AZ, CA, CO, NE, OR, WA; 0, I unknown; **P. plumbaria** (0, I, III), CA, CO, NE, UT, WY; **P. yosemitana** (0, III), CA, CO.

GILLENIA
(American Ipecac, Indian Physic)

Rust. **Gymnosporangium exterum** (0, I), IN, KY, MO, NC, TN, VA, III, on red-cedar.

GINGER, WILD
(*Asarum*)

Gall, leaf. **Synchytrium asari**, MN, CA, WI.
Leaf spot. **Ascochyta versicolor**, ID; **Plaglostoma** (*Laestadia*) **asarifolia**, SC.
Rot, rhizome. **Sclerotinia sclerotiorum**, NY.
Rust. **Puccinia asarina**, CA, ID, OR, WA.

GINKGO
(Maidenhair-Tree)

Anthracnose. **Glomerella cingulata**, MD, TX.
Leaf spot. **Phyllosticta ginkgo**, PA; **Epicoccum purpurascens**, IL.
Nematode, root knot. **Meloidogyne** sp., MS.
Rot, root. **Phymatotrichum omnivorum**, TX.
Rot, sapwood. **Polyporus hirsutus, P. lacteus; P. tulipiferous; P. versicolor; Fomes connatus**, MD.
Rot, seed. **Xylaria longeana**, PA.

GINSENG
(*Panax*)

Blight. **Alternaria panax**, general; **Botryis cinerea**, NY to NC and MI, WA.
Damping-off. **Pythium debaryanum**, NY; **Rhizoctonia solani**, AR, IN, MI, NJ, NY, WA.
Leaf spot. **Septoria** sp.; **S. araliae**, WI; **Colletotrichum dematium**, secondary, NY to NC, MO, MN.
Nematode, root knot. **Meloidogyne** sp., CT, MI, NY, OH, PA, WI.
Nonparasitic. **Papery Leaf**, moisture deficiency, sunscald, MI, MO, NJ, NY, PA.
Rot, rhizome. **Sclerotinia sclerotiorum**, white rot, MI, NY, OH, PA, WA, WI; **S. smilacina**, black rot, MI, MN, NY, WI.

Rot, root. **Armillaria mellea,** WA; **Fusarium scirpi,** NY to AL, MO and WI, WA; **Thielaviopsis basicola,** black root, IL, MI, NJ, NY, OH; **Ramularia** spp., MI, NY, OR, WA, WI.

Rot, stem, root; downy mildew. **Phytophthora cactorum,** CT to NC, IA and MI, WA.

Rust. **Puccinia araliae** (III), MA, PA.

Wilt. **Verticillium albo-atrum,** IN, KY, MI, NJ, NY, OH, PA, TN, WI.

GLADIOLUS

Bacterial leaf blight. **Xanthomonas gummisudans,** NY to MO, and ND, WA.

Bacterial, MLO. **Aster Yellows,** MD.

Bacterial scab; neck rot; leaf spot. **Pseudomonas marginata,** general.

Bacterial soft rot. **Erwinia carotovora,** MI.

Blight, leaf; flower spot; corm rot. **Botrytis gladiolorum,** AK, CA, FL, MA, MD, NJ, NY, OR, WA, WI.

Blight, seedling. **Botrytis elliptica,** WA.

Blight, brown spot of leaves, flowers. **Curvularia lunata** (*Cochliobolus lunata*), FL, MD, MI, NC, NY, VA, WI.

Blight, southern. **Sclerotium rolfsii,** FL.

Leaf spot. **Alternaria** spp., cosmopolitan; **Cladosporium herbarum,** cosmopolitan but secondary; **Heterosporium** sp., MD; **Stemphylium** sp., leaf and stem spot, FL, MI, NJ, NY.

Nematode, root knot. **Meloidogyne** spp., NC to FL and TX, CA; **M. hapla.**

Nematode, sheath. **Hemicycliophora oracilis,** OR.

Nonparasitic

Brown Tip. From waterlogged soil.

Ink Spot. On husks and corms. Cause unknown.

Leaf Scorch. Atmospheric fluorides, WA.

"Rust." Spot necrosis on leaves, from sun on water drops.

Topple. Partially controlled by spraying flowers with calcium nitrate.

Rot, corm, basal; yellows; wilt. **Fusarium oxysporum** f. sp. **gladioli,** general.

Rot, dry, of corms; leaf and stalk rot. **Stromatinia gladioli,** general.

Rot, hard; leaf spot. **Septoria gladioli,** general.

Rot, root, collar, leaf base. **Rhizoctonia solani,** NJ, ND, TX.

Rot, storage. **Penicillium gladioli,** general in the North, occasional in the South; **P. funiculosum,** core rot; **Rhizopus arrhizus,** soft, occasional.

Smut. **Urocystis gladiolicola,** CA.

Virus. **Cucumber Mosaic; Yellow Bean Mosaic** (mild mosaic); **White Break,** NY to IL, CA and WA; **Tomato Ring Spot.**

The backyard gardener with a few rows of gladiolus probably sprays for thrips and forgets about diseases, but the serious grower has much to combat. Control starts with choosing varieties resistant to Fusarium yellows, treating corms after digging and before planting to control rots and scab, with removal of husks usually giving a healthier crop, and field spraying for Botrytis, Curvularia, and Stemphylium flower blights and leaf spots. Mild mosaic is spread from beans to gladiolus by aphids; so they should be kept widely separated. Rogue plants with white-break mosaic. An aluminum mulch repels aphid vectors of virus diseases.

GLOBE-AMARANTH
(*Gomphrena*)

Leaf spot. **Cercospora gomphrenae**, GA, OK, TX.
Nematode, root knot. **Meloidogyne** sp., FL.
Virus. **Beet Curly Top**, TX.
White rust. **Albugo bliti**, NM.

GLOBE-MALLOW
(*Sphaeralcea*)

Powdery mildew. **Erysiphe polygoni**, ID.
Rust. **Puccinia interveniens**, ID, WA; **P. schedonnardi**, AZ, NM; **P. sherardiana**, AZ, CA, CO, ID, NE, NM, TX, UT; **P. sphaeralceae** (I, III), CA.

GLOBE-THISTLE
(*Echinops*)

Rot, crown. **Sclerotium rolfsii**, CT.

GLORY-BUSH
(*Tibouchina*)

Rot, mushroom root. **Clitocybe tabescens**, FL.

GLOXINIA
(*Sinningia*)

Bacterial leaf blight. **Pseudomonas** sp., OR.
Bacterial leaf spot. **Pseudomonas alcaligenes**, FL.
Bacterial, MLO. **Aster Yellows**, CA.
Blight, bud rot. **Botrytis cinerea**, CA, MO.
Nonparasitic. **Dieback,** wilt; boron deficiency, CA.
Rot, crown; flower, leaf blight. **Sclerotinia sclerotiorum**, CA.
Rot, leaf. **Cladosporium herbarum**, NJ.

Rot, root and crown. **Phytophthora cryptogaea,** CA; **Pythium ultimum,** CA; **Myrothecium roridum,** FL.
Virus. **Tomato Spotted Wilt,** CA, MO, TX; **Tobacco Mosaic,** CA, CT, DC, FL, OH, WA.

GNAPHALIUM
(Cudweed)

Bacterial, MLO. **California Aster Yellows;**
Canker, stem. **Phoma erysiphoides,** TX, WI.
Downy mildew. **Plasmopara halstedii,** AL, MS.
Leaf spot. **Cercospora gnaphaliacea,** MS, KS; **Cylindrosporium gnaphalicola,** AL, TX; **Septoria cercosperma,** TX.
Nematode, root knot. **Meloidogyne** sp.
Rot, root. **Phymatotrichum omnivorum,** TX.
Rust. **Puccinia gnaphaliicola** (II, III), AL, LA, NC, SC; 0, I unknown; **P. investita** (0, I, III), AZ, CA, CT, MA, NY, OH, PA, TN, VT, WV, WI.
Smut, white. **Entyloma compositarum,** AL, MD.
Virus. **Beet Curly Top,** CA.

GOATS-BEARD
(*Aruncus*)

Leaf spot. **Cercospora** sp., OR; **Ramularia ulmariae,** AK.

GODETIA

Bacterial, MLO. **California Aster Yellows,** CA.
Blight, gray mold. **Botrytis cinerea,** AK.
Damping-off. **Rhizoctonia solani,** IL.
Downy mildew. **Peronospora arthuri,** CA.
Nematode, root knot. **Meloidogyne** sp.
Rot, root. **Pythium ultimum,** MO; **Phytophthora cryptogaea.**
Rust. **Puccinia oenotherae** (0, I, II, III); **P. pulverulenta** (0, I, II, III), CA; **Pucciniastrum epilobii** (II, III), AK; 0, I, on fir.
Virus. **Tomato Spotted Wilt,** CA.

GOLDEN-CHAIN
(*Laburnum*)

Blight, twig. **Gibberella baccata,** NJ; **Fusarium** sp., MD, OH.
Leaf spot. **Cercospora laburni,** OK.

GOLDEN-CLUB
(*Orontium*)

Blight, leaf. **Botrytis streptothrix, NJ.**
Leaf spot. **Mycosphaerella sp.**, NJ; **Phyllosticta orontii,** NJ, TX; **Ramularia orontii,** NJ; **Volutella diaphana,** NJ.

GOLDEN-EYE
(*Viguiera*)

Powdery mildew. **Erysiphe cichoracearum, UT.**
Rot, root. **Phymatotrichum omnivorum,** TX; **Helicobasidium purpureum,** TX.
Rust. **Puccinia abrupta** (II, III), AZ, TX; **P. turgidipes** (II, III), AZ, CA.
Smut, white. **Entyloma compositarum, UT.**

GOLDEN-GLOW
(*Rudbeckia lacinata*)

Bacterial, MLO. **Aster Yellows; California Aster Yellows.**
Blight, southern. **Sclerotium rolfsii, FL.**
Downy mildew. **Plasmopara halstedii,** IA, NE, NC, ND, WI.
Gall, leaf. **Synchytrium aureum,** IL, WI.
Leaf spot. **Cercospora rudbeckiae,** NY; **Phyllosticta rudbeckiae,** IA, NY, WI; **Ramularia rudbeckiae,** VT to MS, CO and ID; **Septoria rudbeckiae,** KS, NE, WI.
Powdery mildew. **Erysiphe cichoracearum,** general.
Rot, stem. **Sclerotinia sclerotiorum,** CT.
Rust. **Puccinia dioicae** (0, I), MO, SD; **Uromyces perigynius** (0, I), MD to MO and MT; II, III, on *Carex*; **U. rudbeckiae** (III), MD to MS, NM and MT.
Smut, white. **Entyloma compositarum,** IA, MO, OH, WI.
Virus. **Potato Yellow Dwarf; Mosaic,** unidentified.

Powdery mildew is commonly present.

GOLDEN-LARCH
(*Pseudolarix*)

Canker. **Dasyscypha willkomii,** MA.

GOLDENRAIN-TREE
(*Koelreuteria*)

Canker. **Nectria cinnabarina,** CA, CT.
Leaf spot. **Cercospora sp.,** FL.

Rot, root. **Phymatotrichum omnivorum,** TX.
Wilt. **Verticillium** sp., NJ.

GOLDENROD
(*Solidago*)

Black knot. **Gibberidia heliopsidis,** CT, MO, ND, NY.
Blight, thread. **Pellicularia koleroga,** LA.
Canker, stem. **Botryosphaeria ribis,** IA.
Dodder. **Cuscuta** spp., occasional.
Downy mildew. **Basidiophora entospora,** IL, WI; **Plasmopara halstedii,** IL, WI.
Gall, leaf. **Rhodochytrium spilanthidis,** LA.
Leaf spot. **Ascochyta compositarum,** WI; **Asteroma solidaginis,** black scurf; **Cercospora parvimaculans,** WI; **Colletotrichum solitarium,** KS, NE, WI; **Macrophoma sphaeropsispora; Phyllosticta solidagnicola,** IL, WI; **Placosphaeria havdeni,** black spot; **Ramularia serotina,** CO, IL, TX, WI, WY; **Septoria** spp.
Mold, leaf. **Cladosporium astericola,** WI.
Nematode, lesion. **Pratylenchus pratensis,** FL.
Powdery mildew. **Erysiphe cichoracearum,** general; **Phyllactinia corylea,** WA; **Sphaerotheca macularis,** IN; **?Uncinula** sp., NY.
Rot, root. **Phymatotrichum omnivorum,** TX
Rust. **Coleosporium delicatulum** (II, III), ME to VA, KS; 0, I, on pine; **C. asterum** (II, III), general; 0, I, on 2- and 3-needle pines; **Puccinia dioicae** (0, I), general; II, III, on *Carex*; **Puccinia virgae-aureae** (III), IL, NH, MA, MI, NY; **P. grindeliae** (III), IL, WI to CA, WA; **P. stipae** (0, I), CO, MT, NE, NM, ND; II, III, on *Stipa*; **Uromyces perigynius** (0, I), ME; II, III, on *Carex*; **U. solidaginis** (III), CO, ID.
Smut, inflorescence. **Thecaphora cuneata,** KS.
Spot anthracnose. **Elsinoë solidaginis,** FL, GA, SC.
Virus. **Mosaic.** Unidentified, NY.

GOLDENSEAL
(*Hydrastis*)

Blight, leaf. **Alternaria** sp., MI, NY, OH; **Botrytis** sp., CT to NC, IN, WA, WI.
Nematode, root knot. **Meloidogyne** spp., MI, OH, WA.
Rot, root. **Phymatotrichum omnivorum,** TX; **Rhizoctonia solani,** NC.
Rot, stem. **Phytophthora cactorum,** NC.
Virus. **Mosaic,** unidentified, CT.
Wilt. **Fusarium** sp., IL, NY, OH, WA.

GOLDENTOP
(*Lamarckia*)

Rust. Puccinia coronata (II, III), CA; P. graminis, CA.
Virus. ?Mosaic, IA.

GOLDTHREAD
(*Coptis*)

Leaf spot. Mycosphaerella coptis, ME, NY, VT; Septoria coptidis, ID, MI, NY, VT, WI, WA; Phyllosticta helleboricola var. coptidis, AK; Vermicularia coptina, NY.

GOPHER PLANT or CAPER SPURGE
(*Euphorbia*)

Rot, root. Macrophomina phaseolina, AZ; Pythium aphanidermatum, AZ; Rhizoctonia solani, AZ.

GOOSEBERRY
(*Ribes*)

Anthracnose. Pseudopeziza ribis, general.
Blight, cane. Botryosphaeria ribis, NJ, VA; Leptosphaeria coniothyrium, IN, MO.
Blight, thread. Pellicularia koleroga, FL.
Canker; dieback. Nectria cinnabarina, coral spot; Botrytis cinerea, CT, OR, WA; Phragmodothella ribesia, twig knot; Physalospora obtusa, NY to VA, KS.
Dodder, Cuscuta sp., MN, NY.
Downy mildew. Plasmopara ribicola, MT, OR, WI, WV.
Leaf spot. Marssonina grossulariae, OH, WI; Cercospora angulata, NY to MI and MO; Mycosphaerella ribis, MA to VA, KS, MN, OR, AL; Phyllosticta grossulariae, WI, CT, NJ, WA; Ramularia sp., MI.
Nematode, leaf, bud. Aphelenchoides ribes, CA.
Nonparasitic. Leaf Blotch, magnesium deficiency, OR. Leaf Scorch, potassium deficiency.
Powdery mildew. Sphaerotheca macularis, MN; S. mors-uvae, general; Microsphaera grossulariae, CA; Phyllactinia corylea, NY.
Rot, root. Armillaria mellea, OR, WA; Dematophora sp.; Xylaria hypoxylon, OR; Phymatotrichum omnivorum, TX.
Rust. Puccinia caricina var. grossulariata (0, I) leaves, fruit, ME to MD, MS, KS, ID, AK, II, III, on *Carex*; P. Caricina var. uniporula, IA, MD, NY, WI; Coleosporium jonesii (II, III), CO, MO, MN, NM, WI, WY; Cronartium

occidentale, AZ, CO, UT; **C. ribicola,** white pine blister rust (II, III), European varieties are resistant.
Spot anthracnose, scab. **Sphaceloma ribis,** WA.
Virus. **Mosaic.** Unidentified, IL, NY.

Powdery mildew caused by *Sphaerotheca mors-uvae* is probably the most important gooseberry disease but is readily controlled by a lime-sulfur spray immediately after bloom, followed by bordeaux mixture spray for leaf spots.

GORDONIA
(Franklinia and Loblolly-Bay)

Black mildew, **Meliola cryptocarpa,** FL, LA.
Leaf spot. **Phyllosticta gordoniae,** FL.
Rot, root. **Phymatotrichum omnivorum,** TX.

GOUANIA

Rust. **Puccinia invaginata** (II, III), FL.

GOURD
(*Lagenaria, Luffa, Trichosanthes*)

Anthracnose. **Colletotrichum lagenarium,** CT, IL, IN, IA, MN, MD, NE, PA.
Bacterial, angular leaf spot. **Pseudomonas lachrymans,** WI.
Blight, thread. **Pellicularia koleroga,** FL.
Downy mildew. **Pseudoperonospora cubensis,** CT, FL, MA, NC, OH.
Fruit spot. **Macrophoma trichosanthis,** AL; **Phoma subvelata,** TX; **Stemphylium** sp., NY.
Leaf spot. **Cercospora cucurbitae,** AL, IN.
Nematode, root knot. **Meloidogyne** sp., AL, FL, OH.
Powdery mildew. **Erysiphe cichoracearum,** CT, WI.
Rot, fruit. **Mycosphaerella citrullina** (*Laestadia cucurbitacearum*), PA.
Rot, root. **Phymatotrichum omnivorum,** TX.
Virus. **Cucumber Mosaic,** IN, NY.
Virus. **Tobacco Ring Spot,** TX, on *Luffa acutangula.*

GRAPE
(*Vitis*)

Bacterial crown gall. **Agrobacterium tumefaciens,** widespread.
Bacterial leaf spot. **Xanthomonas** sp., NY.
Bacterial, xylem-limited. **Pierce's Disease,** NC.

Blight, shoot. **Sclerotinia sclerotiorum,** CA.

Blotch, leaf. **Briosia amphelophaga,** TX.

Canker; deadarm; branch necrosis. **Cryptosporella** (*Phomopsis*) **viticola,** widespread.

Dieback. **Eutypa armeniacae,** NY, CA.

Downy mildew. **Plasmopara viticola,** general, serious in East.

Fruit spot; fly speck. **Leptothyrium pomi,** PA, WV.

Leaf spot. **Mycosphaerella personata** (*Isariopsis clavispora*) widespread; **Septoria ampelina,** NY, TX, VA; **Septosporium heterosporum,** CA.

Leaf spot, zonate. **Cristulariella pyramidalis,** WV.

Nematode. **Meloidoderita** sp., NY.

Nematode, citrus. **Tylenchulus semipenetrans,** CA.

Nematode, cyst. **Heterodera punctata,** MI.

Nematode, dagger. **Xiphinema index.**

Nematode, lesion. **Pratylenchus pratensis,** CA; **P. minyus,** CA; **P. coffeae,** CA; **P. vulnus,** CA.

Nematode, pin. **Paratylenchus hamatus.**

Nematode, ring. **Criconemoides xenoplax,** CA.

Nematode, root knot. **Meloidogyne** spp.

Nematode, spiral. **Helicotylenchus pseudorobustus,** CA.

Nematode, stubby root. **Trichodorus christiei.**

Nonparasitic. **Little Leaf,** zinc deficiency, CA. **Shot Berry,** defective pollination. **Skin Blanching,** sulfur dioxide injury.

Powdery mildew. **Uncinula necator,** general.

Rot, bitter. **Melanconium fuligineum,** widespread, often secondary after black rot.

Rot, black. **Guignardia bidwellii,** general.

Rot, charcoal. **Macrophomina phaseoli,** TX.

Rot, fruit. **Alternaria** sp., CA; **Aspergillus niger,** black mold, CA, OR; **Botrytis cinerea,** gray mold, CA; **Cladosporium** sp., green mold; **Penicillium** spp., blue mold, cosmopolitan; **Glomerella cingulata,** ripe rot; **Pestalotia** sp., **Phoma** spp.

Rot, root. **Armillaria mellea,** AR, CA, MO, TX, WA; **Phymatotrichum omnivorum,** TX; **Clitocybe tabescens,** SC to TX and OK; **Roesleria hypogaea,** NY to VA, MO, IA; **Rosellina necatrix,** AL, IN, MI, NY, OH.

Rot, summer bunch. **Diplodia viticola** (?*D. natalensis*).

Rot, white; dieback. **Coniothyrium diplodiella,** MA to FL and TX.

Rot, wood. **Schizophyllum commune,** VA; **Stereum** spp.; **Poria** spp.; **Polyporus** spp.

Rust. **Physopella ampelopsidis** (*P. vitis*) (II), FL, SC.

Scorch, leaf. **Pseudopezicula tetraspora,** NY, PA.

Spot anthracnose; bird's eye rot. **Elsinoë ampelina,** widespread.

Virus. **Alfalfa Dwarf; Grape Fanleaf; Grape Leaf Roll** (White Emperor Disease); **Grape Yellow Mosaic.**

Virus. **Tobacco Ring Spot,** NY; **Tomato Ring Spot,** NY.

Black rot is the most destructive grape disease in most sections of the country, often causing total loss of fruit in home gardens. Pierce's disease has destroyed many vineyards in California; the pathogen is transmitted by grafting and by leafhoppers.

GRAPEFRUIT

See Citrus Diseases.

GRAPE-HYACINTH
(*Muscari*)

Rot, dry. **Sclerotium** sp., MO, WA.
Smut, flower. **Ustilago vaillantii,** MA, WA.

GRASSES, LAWN, TURF

Includes *Agrostis alba*, redtop; *A. canina, A. palustris, A. stolonifera, A. tenuis*, bentgrasses; *Cynodon dactylon*, Bermuda grass; Buffalograss; Bromegrass; *Festuca* spp., fescues; *Poa compressa* and *P. pratensis*, Canada and Kentucky bluegrasses; *Stenotaphrum*, St. Augustine grass; *Dactylis glomerata*, orchardgrass.

Anthracnose. **Colletotricum graminicolum,** general.
Bacterial blight. **Pseudomonas syringae pv. coronafaciens,** AK.
Bacterial stunting. **Clavibacter xyli** subsp. **cyodontis,** FL.
Bacterial, xylem-limited. **Rickettsialike organism,** MI.
Blight. **Fusarium roseum** f. sp. **cerealis,** MD, NJ, NY, OH, PA; **F. tricinctum** f. sp. **poae; Leptosphaeria korrae,** CA, NY, NJ, and **Phialophora graminicola,** CA, NY, NJ. (New identification for Fusarium blight).
Blight, cottony. **Pythium aphanidermatum,** general.
Blight, leaf. **Piricularia grisea,** on bent and St. Augustine; **Pellicularia fila-mentosa** f. sp. **sasakii; Drechslera catenaria,** OH; **Rhizoctonia solani,** FL.
Blight, melting out. **Curvularia geniculata; C. inaequalis; C. lunata.**
Blight, southern. **Sclerotium rolfsii,** FL, NJ.
Blight (zonate leaf spot). **Drechslera gigantea** (*Helminthosporium gigantea*), RI.
Blotch, purple leaf. **Septoria macropoda; S. agropyrina,** brown; **S. elymi,** northern U.S.
Damping-off. **Cladochytrium graminis,** on bentgrass.
Dodder. **Cuscuta** sp., MO.
Downy mildew. **Sclerophthora macrospora,** MS, TX, LA, on St. Augustine grass.
Downy mildew. **Sclerospora farlowi,** OK, on Bermuda grass.

Ergot. **Claviceps purpurea,** NY to KY, TX, ND, MT, OK, OR, WI; **C. micro-cephala,** MD, MI, OH.

Fairy ring. **Marasmius oreades; Psalliota compestris; Lepiota morgani.**

Leaf spot. **Ascochyta agropyrina,** Northwest; **A. desmazieri;** CA, OR, WA; **A. elymi; A. graminae** on Bermuda grass; **A. hordei; A. sorghi; A. graminicola; A. utahensis; Cercospora seminalis,** TX; **C. fuscomaculans; C. poagena; Cylindrosporium glyceriae,** NY; **Leptosphaeria korrae,** ringspot, WI; **Macro-phoma** sp., OR; **Ovularia pulchella,** OR, UT; **Phaeoseptoria** sp., OR; **Placo-sphaeria graminis; Septogloeum oxysporum,** char spot, widespread; **Sta-gonospora intermixta.**

Leaf spot, black, tar. **Phyllachora graminis,** widespread; **P. sylvatica,** Northwest.

Leaf spot, brown stripe. **Scoletotrichum graminis,** widespread, except on ryegrass.

Leaf spot, copper. **Ramulispora** (*Gloeocercospora*) **sorghi,** CA, LA, PA, RI, on bentgrass.

Leaf spot, frog-eye. **Selenophoma donacis,** northern states; **S. everhartii; S. obtusa.**

Leaf spot, gray leaf speckle. **Septoria triseti,** WA, OR on redtop, bent; **S. calamagrostidis,** OR, WY, AK; **S. loligena,** CA; **S. nodorum; S. oude-mansii,** northern U.S.; **S. secalis,** white leaf spot; **S. tenella,** northern plains; **S. tritici** var. **lolicola.**

Leaf spot; melting-out. **Helminthosporium vagans,** on bluegrass; **H. sativum; H. giganteum,** zonate eyespot, general; **H. dictyoides,** fescue netblotch; **H. siccans,** brown blight; **H. stenacrum,** leaf mold; **H. triseptatum,** gray leaf mold; **H. cynodontis,** Bermuda grass leaf blotch; **H. erythrospilum; H. rostratum; H. stenophilum.**

Leaf spot, red eye-spot. **Mastigosporium rubricosum,** ME, OR, WA, WY.

Leaf spot, yellow ring. **Trechispora alnicola,** IL.

Mold. **Cladosporium herbarum; Fusarium heterosporium.**

Nematode, cyst. **Punctodera punctata,** NJ.

Nematode, dagger. **Xiphinema americanum.**

Nematode, gall. **Anguina agrostis.**

Nematode, lance. **Hoplolaimus coronatus.**

Nematode, lesion. **Pratylenchus brachyurus; P. pratensis; P. subpenetrans; P. thornei.**

Nematode, pin. **Paratylenchus projectus.**

Nematode, pseudo root knot. **Hypsoperine graminis,** AL, MD, TN.

Nematode, ring. **Criconemoides lobatum; C. rusticum, C. xenoplax, C. orna-tum, C. parvum, C. curvatum,** OH; **Criconemella xenoplax,** SC.

Nematode, root knot. **Meloidogyne** sp.; **M. incognita-acrita.**

Nematode, spiral. **Helicotylenchus nannus; H. dihystera; Rotylenchus christiei,** FL.

Nematode, sting. **Belonolaimus gracilis; B. longicaudatus.**

Nematode, stubby root. **Trichodorus christiei; T. obtusa.**

Nematode, stylet. **Tylenchorhynchus brevideus; T. claytoni; T. dubius.**
Nonparasitic. **Burn,** mowing too close in hot weather. **Spring Dead Spot,** too much thatch.
Powdery mildew. **Erysiphe graminis,** general.
Rot, crown. **Drechslera catenaria,** OH.
Rot, culm. **Fusarium culmorum; Gibberella zeae.**
Rot; dollar spot; small brown patch. **Sclerotinia homeocarpa,** general.
Rot, foot. **Ophiobolus graminis,** NY, WA.
Rot; large brown patch. **Pellicularia filamentosa** (*Rhizoctonia solani*); widespread, more important on bentgrass than on bluegrass.
Rot; pink patch; red thread. **Corticium fuciforme,** general.
Rot, root. **Pythium debaryanum; P. ultimum; Fusarium** spp.; **Olpidium brassicae; Pyrenochaeta terristris,** IL.
Rust. **Puccinia pygmaea,** NY; **Uromyces mysticus** (II, III), OR; **U. dactylidis,** IN, MI, SD, VI.
Rust, bluegrass. **Puccinia poae-nemoralis,** widespread; **P. recondita,** on bromegrass, PA; **P. striiformis,** stripe rust, OR; **P. cynodontis,** on Bermuda grass; **P. stenotaphri,** on St. Augustine grass; **P. piperi.**
Rust, crown. **Puccinia coronata** (II, III), cosmopolitan; **P. glumarum.**
Rust, stem; wheat. **Puccinia graminis,** cosmopolitan.
Scab. **Fusarium heterosporum,** MO.
Slime mold. **Mucilago spongiosa; Physarum cinereum; Fuligo septica.**
Smut, bluegrass blister. **Entyloma irregulare** and other spp.
Smut, flag. **Urocystis agropyri,** WI, OH.
Smut, head; bunt. **Tilletia pallida,** OR, NJ, OH; **T. fusca; T. buchloëana,** OK, KS, NE, TX.
Smut, inflorescence. **Sorosporium syntherisme,** CA.
Smut, stripe. **Ustilago striiformis,** widespread; **U. cynodontis** on Bermuda grass in South.
Snowmold, blight. **Fusarium nivale,** AK, MI, MN, OR, WA; **Calonectria graminicola; Sclerotinia borealis,** AK; **Sclerotium rhizodes,** white tip, CT, MA, NJ, PA, WI; **Typhula itoana,** MA, MN, NJ, NY, PA; **T. idahoensis,** ID.
Stem eyespot. **Phleospora idahoensis,** OH.
Virus. **Bromegrass Mosaic.** General, IA, IL; **Clover Yellow Vein,** GA; **Maize Dwarf Mosaic; Peanut Mottle,** GA; **St. Augustine Decline Virus (Panicum Mosaic),** AR, LA, TX; **Sugarcane Mosaic,** FL; **Barley Yellow Dwarf,** MO, IN.

The possibility of any large proportion of these diseases appearing in the average suburban lawn is remote. Snowmold occurs occasionally after a winter when the snow cover has been rather continuous, but the light tan areas disappear by late spring. Large brown patch is fairly general in humid summer weather along with dollar spot, and Helminthosporium leaf spots or melting-out. Merion bluegrass is resistant to the latter but not to bluegrass rust.

GRASS-OF-PARNASSUS
(*Parnassia*)

Powdery mildew. **Erysiphe polygoni**, NY.
Rust. **Puccinia parnassiae** (III), UT; **P. caricina** var. **uliginosa**, (I), AK; II, III, on *Carex*.

GREVILLEA
(Silk-Oak)

Dieback; gum disease. **Diplodia** sp. (?*Physalospora rhodina*), FL.
Leaf spot, algal. **Cephaleuros virescens**, FL.
Nematode, root knot. **Meloidogyne** sp., CA.
Rot, root. **Phymatotrichum omnivorum**, AZ.

GRINDELIA
(Gumweed)

Leaf spot. **Cercospora grindeliae**, CA, TX, WI; **Septoria grindeliae**, CO, KS, TX; **S. grindeliicola**, WI.
Powdery mildew. **Erysiphe cichoracearum**, MN to NM, CA, MT.
Rot, root. **Phymatotrichum omnivorum**, TX.
Rust. **Coleosporium asterum** (II, III), CA, CO, WA, WI; 0, I, on pine; **Puccinia dioicae** (0, I), KS, NE, TX; II, III, on *Carex*; **P. grindeliae** (III), NE to TX, CA, MT; **P. stipae** (0, I), CO, KS, NE, ND, SD; II, III, on *Stipa* and other grasses; **Uromyces junci** (0, I); II, III, on *Juncus*.
Smut, inflorescence. **Thecaphora californica**, UT, CA; **T. Cuneata**, CO, KS, NE, NM.

GROUND-CHERRY, HUSK-TOMATO
(*Physalis*)

Bacterial angular leaf spot. **Pseudomonas angulata**, KY.
Bacterial canker. Vascular. **Clavibacter michiganense**, WY. Also **Pseudomonas longifolia**.
Bacterial, wildfire. **Pseudomonas syringae** pv. **tabaci**, PA.
Blight, seedling. **Gloeosporium fructigenum**, NY.
Blight, southern. **Sclerotium rolfsii**, FL.
Leaf spot. **Alternaria solani**; **Cercospora diffusa**, IL, KS, WI; **C. physalicola**, CT to GA, TX; **Leptosphaeria physalidis**, KY; **Phyllosticta** sp., OK; **Septoria** sp., NE; **Stemphylium solani**, FL.
Nematode, leaf and stem. **Ditylenchus dipsaci**, CA.
Nematode, root knot. **Meloidogyne** spp., FL, AL.

Rot, root. **Phymatotrichum omnivorum,** TX.
Rust. **Aecidium physalidis,** WI to TX, NM, CO; **Puccinia physalidis,** CO, IA, MN, NE, WI.
Smut, white. **Entyloma australe,** MA to MS, NM and ND.
Virus. **Beet Curly Top; Mosaic** (part cucumber, part tobacco mosaic); **Ring Spot** (?Tobacco).

GROUND-CHERRY, PURPLE-FLOWERED
(*Quincula*)

Leaf spot. **Cercospora physalidis,** KS.
Rust. **Puccinia aristidae** (0, I), CO.

GROUND-SMOKE
(*Gayophytum*)

Leaf spot. **Cercospora gayophyti,** CA.
Rust. **Puccinia vagans** (0, I, II, III), ND to NM, CA, WA.
Smut, seed. **Ustilago gayophyti,** CA, NV, OR, UT.

GUAR
(*Cyamopsis*)

Bacterial blight. **Xanthomonas campestris** pv. **cyamopsidis,** AZ.
Powdery mildew. **Leveillula taurica,** AZ.

GUAVA
(*Psidium*)

Anthracnose; leaf and fruit spot; ripe rot. **Glomerella cingulata,** FL.
Blight, thread. **Pellicularia koleroga,** FL.
Leaf spot. **Cercospora psidii,** FL.
Leaf spot, algal; green scurf. **Cephaleuros virescens,** FL.
Nematode, burrowing. **Radopholus similis,** FL.
Nematode, root knot. **Meloidogyne** sp., FL.
Rot, fruit. **Alternaria citri,** CA; **Rhizopus stolonifera,** HI.
Rot, root. **Clitocybe tabescens,** FL; **Phymatotrichum omnivorum,** TX.
Rot, wound. **Polyporus versicolor,** FL.

GUAYULE
(*Parthenium*)

Rot, root. **Pythium aphanidermatum,** AZ.

GYPSOPHILA
(Babys-Breath)

Bacterial fasciation. **Clavibacter fascians,** OH.
Bacterial, MLO. **Aster Yellows,** NY, and **California Aster Yellows,** CA.
Bacterial root and stem gall. **Erwinia herbicola** (*Agrobacterium gypsophilae*), NJ.
Blight, gray mold. **Botrytis cinerea,** NJ.
Damping-off. **Pythium debaryanum,** CT; **Rhizoctonia solani,** CT.
Nematode, root knot. **Meloidogyne** sp., FL.
Wilt. **Phytophthora cactorum,** MA.

HACKBERRY, SUGARBERRY
(*Celtis*)

Blight, leaf. **Cylindrosporium defoliatum,** TX.
Blight, thread. **Pellicularia koleroga,** FL.
Canker, felt fungus. **Septobasidium burtii,** TX; **S. sydowii,** TX.
Downy mildew. **Pseudoperonospora celtidis,** MD, GA.
Leaf spot. **Cercosporella celtidis,** AL, TX, central states; **Cylindrosporium celtidis,** AL; **Phleospora celtidis,** MA to TX; **Phyllosticta celtidis,** general; **Septogloeum celtidis,** NY.
Mistletoe. **Phoradendron tomentosum** and **P. serotinum (flavescens),** southeastern and Gulf states.
Powdery mildew. **Sphaerotheca phytoptophila,** associated with gall mites causing witches' brooms, widespread, especially in central states; **Uncinula macrospora,** FL, GA; **U. parvula,** widespread; **U. polychaeta,** in South.
Rot, root. **Armillaria mellea,** occasional; **Helicobasidium purpureum,** seedling blight, TX; **Phymatotrichum omnivorum,** TX; **Poria ambigua,** TX.
Rot, wood. **Daedalea** spp.; **Fomes** spp.; **Ganoderma lucidum; Polyporus** spp.; **Stereum bicolor,** cosmopolitan.
Virus. **Island Chlorosis,** IL; **Hackberry Leaf Mosaic.**

Hackberry witches' brooms, so prominent in a winter landscape, are caused by gall mites in association with powdery mildew. There is no real control, except to cut off unsightly branches. Chinese hackberry and southern hackberry are less susceptible.

HALESIA
(Silver-Bell, Snowdrop-Tree)

Leaf spot. **Cercospora halesiae,** TN.
Rot, wood. **Polyporus halesiae,** GA.

Other fungi are reported on twigs and branches but not as causing specific diseases.

HAMELIA
(Scarlet-Bush)

Rot, root. **Clitocybe tabescens,** FL.
Rust. **Uredo hameliae,** FL.

HARBINGER-OF-SPRING
(*Erigenia*)

Rust. **Puccinia erigeniae** (0, I, III), OH.

HARDENBERGIA

Nematode, root knot. **Meloidogyne** sp.

HARDY ORANGE
(*Poncirus*)

Anthracnose; dieback. **Glomerella cingulata,** Gulf states.
Canker; dieback. **Diaporthe citri,** melanose, Gulf states; **Diplodia natalensis,**
   gummosis, general.
Rot, root. **Phymatotrichum omnivorum,** TX; **Xylaria polymorpha,** LA.
Spot anthracnose; citrus scab. **Elsinoë fawcettii,** MS.

HARES-TAIL
(Lagurus)

Rust. **Puccinia coronata** (II, III), crown; **P. graminis,** stem, WA.

HAWKBIT, FALL DANDELION
(*Leontodon*)

Bacterial, MLO. **Aster Yellows,** ME, NY
Rust. **Puccinia hieracii** (0, I, II, III), ME, NH

HAWKSBEARD
(*Crepis*)

Leaf spot. ?**Cercospora stromatis,** CO; **Phyllosticta eximia,** CO; **Ramularia
   crepidis,** NM.
Powdery mildew. **Erysiphe cichoracearum,** NE, PA; **Sphaerotheca macularis,**
   WY.

Rust. **Puccinia crepidis-montanae** (0, I, II, III), CO, ID, MT, OR, UT, WA, WY; **P. hieracii,** ND to CO, WA; II, III, on *Carex*.

HAWKWEED
(*Hieracium*)

Blight, stem. **Phoma hieracii,** TX.
Downy mildew. **Bremia lactucae,** WI.
Leaf spot. **Cercospora hieracii,** AL, NC; **Phyllosticta decidua,** WI; **Septoria cercosperma,** TX; **S. hieracicola,** NY.
Powdery mildew. **Erysiphe cichoracearum,** PA, TX; **Sphaerotheca macularis,** IL.
Rot, root. **Phymatotrichum omnivorum,** TX.
Rust. **Puccinia dioicae** (0, I), PA to IL, WI, CA, OR, MT; II, III, on *Carex*; **P. fraseri** (III), MT, NE, NY, NH, PA, TN, VA, WV; **P. hieracii** (0, I, II, III), TX, ME to VA, IL, WA, FL, CA, CO; **P. columbiensis,** OR, TX, WA; **Aecidium columbiense** (0, I), CA, ID, OR, WA.

HAWTHORN
(*Crataegus*)

Bacterial fire blight. **Erwinia amylovora,** widespread, especially on English hawthorn.
Blight, leaf. **Fabraea thuemenii** (*F. maculata, Entomosporium thuemenii*), East and central states, southward.
Blight, leaf; fruit spot. **Monilinia johnsonii,** NY to MN.
Blight, seedling. **Sclerotium rolfsii,** FL.
Canker, felt fungus. **Septobasidium** spp., Southeast and Gulf states.
Leaf spot. **Cercospora apiifoliae,** TX; **Cercosporella mirabilis,** CO, NY, WI; **Coniothyrium pyrinum,** IL; **Cylindrosporium brevispina,** CA to MT, WA; **C. crataegi,** WA; **Gloeosporium crataegi,** WA; **Hendersonia crataegicola,** AL, TX; **Monochaetia crataegi,** FL; **Phyllosticta** spp., widespread; **Septoria crataegi,** MI to ND; **Mycosphaerella** sp., GA.
Mistletoe. **Phoradendron serotinum (flavescens),** TX.
Mistletoe. **Viscum album,** CA.
Powdery mildew. **Podosphaera oxyacanthae,** East and central states; **Phyllactinia corylea,** general.
Rot, gray mold, on fruit. **Botrytis cinerea,** AK.
Rot, root. **Armillaria mellea,** OK; **Phymatotrichum omnivorum,** TX. **Xylaria digitata; X. hypoxylon,** IN, OH.
Rot, wood. **Daedalea confragosa,** cosmopolitan; **Polyporus versicolor,** cosmopolitan; **Fomes** spp.

Rust. **Gymnosporangium bethelii** (0, I), on fruits; ND to WA, CO, NM; III, on juniper; **G. clavipes,** quince rust (0, I), on leaves, fruits, general, east of Rocky Mountains; **G. exiguum,** on leaves, fruits, TX; II, III, on juniper; **G. floriforme,** on leaves, SC to FL, OK, TX; III, on red-cedar; **G. globosum,** hawthorn rust (0, I), on leaves, ME to ND, FL, TX; II, III, on juniper; **G. hyalinum** (0, I), leaves, Atlantic Coast, NC to FL; III, unknown; **G. libocedri** (0, I), leaves and fruit, Pacific Coast; III on *Libocedrus*; **G. nelsoni** (0, I), on leaves, fruits, WY; III, on juniper; **G. trachysorum** (0, I), on leaves, Atlantic and Gulf coasts; III, on red-cedar; **G. tubulatum** (0, I), on leaves, Rocky Mountains to OR and WA.

Scab. **Venturia inaequalis,** widespread; **V. crataegi,** FL.

Rusts are common on hawthorn and particularly so in the central states, where junipers should be kept at a distance. Leaf blight, beginning in a moist spring, causes spotting and defoliation in August. When branches die back from fire blight, cut them off several inches below the visible blighted portion.

HAZELNUT, FILBERT
(*Corylus*)

Bacterial blight. **Xanthomonas corylina,** OR, WA.

Bacterial crown gall. **Agrobacterium tumefaciens,** WA.

Bacterial spot. **Pseudomonas colurnae,** IL.

Blight, eastern filbert. **Anisogramma anomala,** NY, WA, OR.

Canker; twig blight. **Apioporthe anomala,** eastern and central states; **Hymenochaete agglutinans,** ME; **Gleosporium** sp.

Filbert stunt. Undetermined causal agent, OR.

Kernel spot; yeast spot. **?Nematospora coryli,** OR.

Leaf blister. **Taphrina coryli,** CT, MA, WI.

Leaf spot. **Cercospora corylina,** OK; **Cylindrosporium vermiformis,** WI; **Gloeosporium coryli,** ME to NJ, PA, OR, WA; **Phyllosticta coryli,** widespread; **Septogloeum profusum,** IN, MA, MS; **Septoria corylina,** MA to NJ and WI, OR, WA; **Sphaerognomonia carpinia,** GA; **Gnomoniella coryli,** general.

Nonparasitic. **Bitter Pit** and **Brown Stain** of nuts, cause unknown, OR, WA. **Shrivel,** "Blanks," defective pollination, OR, WA.

Powdery mildew. **Phyllactinia corylea,** northeastern and north central states, OR, WA; **Microsphaera alni,** northeastern and north central states.

Rot, root. **Armillaria mellea,** CA, OR, WA; **Phymatotrichum omnivorum,** TX.

Rot, wood. **Poria ferrea,** CA.

Virus. **Apple Mosaic,** OR.

Bacterial blight often kills young trees in Washington and Oregon. Powdery mildew is important in the Northwest.

HEATH
(*Erica*)

Powdery mildew. **Erysiphe polygoni,** CA.
Rot, collar. **Phytophthora cinnamomi,** CA, NY.
Rot, stem. **Ascochyta** sp., NJ.
Rust. **Pucciniastrum ericae** (II), CA.

HEATHER
(*Calluna*)

Nonparasitic. **Chlorosis,** iron deficiency, corrected by spraying with ferrous
   sulfate.
Rot, root, collar. **Phytophthora cinnamomi,** CA, OR.

HEBE

Bacterial fasciation. **Clavibacter fasciens,** CA.
Leaf spot. **Septoria exotica,** CA.
Rot, root. **Armillaria mellea,** CA.
Wilt. **Fusarium oxysporum** f. sp. **hebe,** CA.

HEDGE PARSLEY
(*Torilis*)

Gall, leaf. **Protomyces macrosporus,** AR.

HELENIUM
(Sneezeweed)

Bacterial, MLO. **Aster Yellows,** NY, and **California Aster Yellows,** CA.
Leaf spot. **Cercospora helenii,** AL, TX; **Septoria helenii,** IA, PA, TX.
Powdery mildew. **Erysiphe cichoracearum,** CO, OR, TX, UT, WA, WI, WY.
Rot, root. **Phymatotrichum omnivorum,** TX.
Rust. **Puccinia conspicua** (0, I), AZ, CO, NM; II, III, on *Koeleria,* **P. dioicae**
   (0, I), NE, CO, TX; II, III, on *Carex.*
Smut. **Entyloma compositarum,** TX, WI; **E. polysporum,** MT.

HELIOPSIS

Black knot. **Gibberidea heliopsidis.**
Leaf spot. **Phyllosticta pitcheriana,** NY; **Septoria helianthi,** IN; **S. heliopsidis,**
   WI.

Powdery mildew. **Erysiphe cichoracearum,** IA, MN, NE, ND, NM, WI; **Leveil-lula taurica,** IA.

Rot, root. **Phymatotrichum omnivorum,** TX.

Rust. **Coleosporium helianthi** (II, III), NC; 0, I, on pine; **Puccinia batesiana** (0, I, III), DE, PA, IA, KS, MD, MN, NE, WI; **P. helianthi** (0, I, II, III), IN, MN, NC, PA, VA.

Virus. **Mosaic.** Unidentified, IA.

HELIOTROPE
(Heliotropium)

Blight, shoot; leaf spot. **Botrytis cinerea,** cosmopolitan; **Stemphylium** sp., NY.

Leaf spot. **Cercospora heliotropii,** NM, OR.

Nematode, root knot. **Meloidogyne** sp., MA, WA.

Rust. **Puccinia aristidae** (0, I), AZ, CA, NV, NM, TX, UT; II, III, on grasses.

Wilt. **Verticillium albo-atrum,** MD.

HEMLOCK
(Tsuga)

Blight, needle. **Didymascella tsugae,** MA, NH, WI; **Rosellinia herpotri-chioides,** NC.

Blight, snow. **Herpotrichia nigra,** MT to OR and AK.

Blight, twig. **Botrytis cinerea,** ID.

Canker, bleeding. **Phacidiopycnis pseudotsugae** (*Phacidiella coniferarum*) on eastern hemlock.

Canker, branch. **Discocainia treleasi** on western hemlock, WA, AK.

Canker, stem girdle, of saplings. **Hymenochaete agglutinans,** PA.

Canker, twig. **Cytospora** sp., MD, VA; **Dermatea balsamea,** GA, NY, TN, VA.

Damping-off. **Rhizoctonia solani,** cosmopolitan.

Mistletoe, dwarf. **Arceuthobium campylopodium,** ID, AK; **A. cyanocarpum,** OR; **A. tsugense,** AK.

Nonparasitic. **Canker,** bark split, heavy soil and poor drainage.

Rot, root. **Armillaria mellea,** MA to PA and MI, OR, WA; **Phymatotrichum omnivorum,** TX; **Fusarium solani,** PA; **F. oxysporum,** PA; **F. avenaceum,** PA.

Rot, seedling root. **Cylindrocladium scoparium,** NY; **Rhizinia undulata,** NY.

Rot, wood. **Fomes** spp.; **Ganoderma lucidum,** Northeast; **Polyporus** spp., general.

Rust, needle, cone. **Caeoma dubium** (0, I), ID, MT, OR, WA; **Melampsora abietis-canadensis** (0, I), New England to PA; II, III on poplar; **M. farlowi** (III), New England to NC, WI; **Pucciniastrum hydrangeae** (0, I), IN, MD,

PA, TN; II, III, on hydrangea; **P. myrtilli** (0, I), ME to AL, IN, WI; II, III, on Ericaceae; **Uraecium holwayi** (0, I), AK, MT, OR, WA.

HEMP
(*Cannabis*)

Blight, gray mold. **Botrytis cinerea,** OR, VA.
Blight, southern. **Sclerotium rolfsii,** SC, TX.
Broomrape. **Orobanche ramosa,** CA, IL, KY, WI.
Canker; stem rot. **Gibberella quinqueseptata; G. saubinetti,** IN, VA.
Canker; stem wilt. **Botryosphaeria marconii,** MD, VA.
Leaf curl, wilt. **Cercospora cannabina,** MS.
Leaf spot. **Cylindrosporium** spp., MD; **Septoria cannabis,** MD to KY, IA, MN, FL, TX; **Cercospora cannabis,** MS.
Nematode, root knot. **Meloidogyne** sp., TN.
Rot, charcoal. **Macrophomina phaseoli,** IL.
Rot, root. **Phymatotrichum omnivorum,** TX; **Hypomyces cancri,** MD.
Rot, stem; wilt. **Sclerotinia sclerotiorum,** MT.

HENBIT
(*Laminum*)

Virus. **Tomato Ringspot,** PA.

HEPATICA

Downy mildew. **Plasmopara pygmaea,** IA, IL, NY, PA, WI.
Leaf spot. **Discosia artocreas,** secondary, IA; **Septoria hepaticae,** MI, NC.
Rust. **Transzchelia pruni-spinosae** (0, I), MA to MD, TN and MN; II, III on wild plum.
Smut, leaf and stem. **Urocystis anemones,** NY to IN, MO, MN, WI.

HERACLEUM
(Cow-Parsnip)

Leaf spot. **Cylindrosporium heraclei,** CA, CO, ID, MT, ND, TX, UT, WA, WY; **Fusicladium angelicae,** WI; **Phyllosticta heraclei,** OR, TX, AK; **Ramularia heraclei,** general.
Rot, root. **Phymatotrichum omnivorum,** TX.

HERCULES-CLUB
(*Aralia spinosa*)

Canker; dieback. Botryosphaeria ribis, GA, VA, WV; Nectria cinnabarina, VA, WV.

Leaf spot. Cercospora atromaculans, LA, TX; Phyllosticta araliae, TX; P. everhartii, TX, WV; Stagonospora sp., WV.

Rot, root. Phymatotrichum omnivorum, TX; Rhizoctonia solani, TX.

Rot, sapwood. Polyporus tulipiferus, MD.

Spot anthracnose. Sphaceloma araliae, MD, MO.

HERCULES-CLUB
(*Zanthoxylum clava-herculis*)

Leaf spot. Cercospora zanthoxyli, FL, GA, TX; Septoria pachyspora, TX.

Mistletoe. Phoradendron serotinum (flavescens), TX.

Rust. Puccinia andropogonis var. zanthoxyli (0, I) FL, TX: II, III, on *Andropogon.*

HERONSBILL
(*Erodium*)

Bacterial, MLO. California Aster Yellows.

Blight, southern. Sclerotium rolfsii, TX.

Downy mildew. Pseudoperonospora erodii, MT, TX.

Gall, leaf. Synchytrium papillatum, AZ, CA, OR, TX.

Rot, root. Rhizoctonia solani, TX.

Virus. Beet Curly Top, CA.

HESPERIS
(Dames-Rocket)

Club root. Plasmodiophora brassicae, NJ.

Downy mildew. Peronospora parasitica, NY, PA.

Virus. Cucumber Mosaic, OR.

White rust; white blister. Albugo candida, NY.

HEUCHERA
(Alum-Root, Coral-Bells)

Leaf spot. Cercospora heucherae, IN, OH, PA, VA, WV, IL, IN, IA, WI; Phyllosticta excavata, ID, AK; Ramularia mitellae var. heucherae, WA; Septoria heucherae, IN.

Leaf nematode. **Aphelenchoides fragariae,** CA.
Powdery mildew. **Erysiphe cichoracearum,** CA; **Sphaerotheca macularis,** MT, NM; **Phyllactinia corylea,** MT.
Rot, root. **Pythium hypogynum** and **P. ultimum.**
Rust. **Puccinia heucherae** (III), PA to NC and SD; western states, AK.
Smut, leaf and stem. **Urocystis lithophragmae,** UT.

HIBBERTIA
(Guinea-Gold-Vine)

Virus. **Cucumber Mosaic,** CA.

HIBISCUS
(Arborescent Forms; Rose-of-Sharon, Confederate-Rose, Chinese Hibiscus)

Bacterial crown gall. **Agrobacterium tumefaciens,** MS.
Bacterial leaf spot. **Pseudomonas syringae,** FL, CA; **P. syringae** pv. **hibisei,** FL; **Xanthomonas campestris** pv. **malvacearum,** FL.
Bacterial wilt. **Xanthomonas solanacearum,** FL.
Blight, blossom. **Choanephora infundibulifera,** FL.
Blight, leaf stem. **Phytophthora palmivora,** LA; **P. parasitica,** FL.
Blight, thread. **Pellicularia koleroga,** FL; web, **P. filamentosa,** FL.
Canker; dieback. **Colletotrichum hibisci,** FL, TX.
Leaf spot. **Alternaria tenuis,** IN, NJ, PA; **Cercospora hibisci,** FL, OK, TX; **C. amalayensis,** GA; **Phyllosticta hibiscina,** LA; **Cristulariella pyramidalis,** FL, MD, MS, and GA.
Nematode, leaf. **Aphelenchoides fragariae,** FL.
Nematode, root knot. **Meloidogyne** spp., MS, TX.
Nonparasitic. **Strapleaf,** molybdenum deficiency, FL.
Rot, root. **Clitocybe tabescens,** FL; **Phymatotrichum omnivorum,** TX.
Rot, stem. **Fusarium** sp., FL; **Phytophthora parasitica,** TX; **P. cactorum,** LA; **P. palmivora,** LA.
Rust. **Kuehneola malvicola** (II, III), Gulf states.
Virus. **Mosaic,** unidentified, FL.

HICKORY
(*Carya*)

Anthracnose; leaf spot. **Gnomonia caryae,** general (see under Leaf Spots).
Bacterial crown gall. **Agrobacterium tumefaciens,** MD, KS, TX.
Blotch, leaf. **Mycosphaerella dendroides,** widespread East and South.
Canker. **Nectria galligena,** eastern states; **Strumella coryneoidea,** PA, MD.

Canker, bark patch. **Aleurodiscus candidus,** MO, OH, PA; **Solenia ochracea,** MA.

Canker, felt fungus. **Septobasidium** spp., Southeast.

Canker, heart rot. **Poria spiculosa,** destructive to pignut, shagbark, PA, and South.

Canker, twig. **Rosellinia caryae,** MI; **Botryosphaeria berengeriana.**

Gall, trunk, branch. **Phomopsis** sp., eastern states.

Leaf spot. **Fusarium carpineum,** WI; **Hendersonia davisii,** WI; **Marssonina juglandis** (*Gnomonia leptostyla*), NJ, NC to IA; **Microstroma juglandis,** witches' broom; **Monochaetia desmazierii,** MD to NC, TN; **Mycosphaerella carigena,** WI; **Pestalotia sphaerelloides,** LA; **Phyllosticta caryae,** widespread; **Septoria caryae,** DE, MI; **S. hicoriae,** TX; **Cercospora fusca.**

Mistletoe. **Phoradendron serotinum (flavescens),** IN, TX; **Viscum album,** CA.

Nematode, root knot. **Meloidogyne** sp.

Nonparasitic. **Rosette,** zinc deficiency, southern states.

Powdery mildew. **Phyllactinia corylea,** MI, WI; **Microsphaera alni,** cosmopolitan.

Rot, heart. **Favolus alveolaris,** VA, VT, NY; **Fomes** spp., cosmopolitan.

Rot, root. **Armillaria mellea,** cosmopolitan; **Clitocybe parasitica,** OK; **Phymatotrichum omnivorum,** AZ, TX; **Helicobasidium purpureum.**

Rot, wood. **Daedalea** spp.; **Ganoderma curtissii; Polyporus** spp.; **Steccherinum septentrionale; Schizophyllum commune; Stereum hirsutum; Trametes rigida.**

Scab; leaf spot. **Cladosporium effusum,** general.

Spot anthracnose. **Elsinoë randii.**

Wilt. **Verticillium** sp., VA.

Of the various leaf spots on hickory, anthracnose caused by *Gnomonia caryae* is the most destructive. There are no recommended control measures for witches' brooms appearing on shagbark hickory except cutting them out.

HELICHRYSUM

Virus. Biddens Mottle, FL.

HIPPEASTRUM

See Amaryllis.

HOARHOUND
(*Marrubium*)

Gall, leaf. **Synchytrium marrubii,** TX.

Leaf spot. **Cercospora marrubii,** OK, TX.

Nematode, root knot. **Meloidogyne** sp., AL.

HOARY-TICK CLOVER
(*Desmodium*)

Virus. Cowpea Mosaic, IL.

HOLLY
(*Ilex; Ilex opaca*, American; *I. equifolium*, English; *I. cornuta*, Chinese; *I. crenata*, Japanese).

Bacterial blight. **Clavibacter ilicis,** MA.

Black mildew. **Asterina** (*Englerulaster*) **orbicularis,** DE to FL and TX; **A. ilicis; A. pelliculosa; Morenoella** (*Lembosia*) **ilicis,** Gulf states.

Blight, flower. **Botrytis cinerea,** NJ.

Blight, leaf and twig. **Phytophthora ilicis,** OR, WA.

Blight, thread. **Pellicularia koleroga,** LA; silky thread, **Rhizoctonia ramicola;** Web, **Cylindrocladium scoparium,** AL.

Canker; dieback. **Boydia insculpta,** secondary, OR, WA; **Diaporthe eres,** ?secondary; **Diplodia** sp., MD; **Nectria coccinea; Phomopsis** sp.; **Physalospora ilicis** on twigs, also leaf spot, NC, NJ, NY, TX, SC, WV, CA, WA; **Gloeosporium** sp., AL.

Canker, felt fungus. **Septobasidium** spp., FL, LA, NC.

Dodder. **Cuscuta compacta,** FL; **C. exaltata,** TX.

Leaf spot. **Cercospora ilicis,** NJ and Gulf states; **Gloeosporium aquifolli,** NJ, TX, WA; **Macrophoma phacediella,** NJ; **Microthyriella cuticulosa,** black spot; **Phyllosticta concomitans,** LA; **P. terminalis,** MS; **Septoria ilicifolia,** NJ; **?Cuticularia ilicis,** OR; **Sclerophoma** sp., OR.

Leaf spot. **Sclerophoma** sp., Kentucky on English Holly (*Ilex aquifolium*).

Leaf spot, tar. **Phacidium curtisii,** MA to FL and TX; **Rhytisma ilicinicolum,** VA to GA and TN; **R. velatum,** GA, MS, SC.

Leaf spot; twig dieback. **Cylindrocladium avesiculatum,** GA.

Nematode, lance. **Hoplolaimus uniformis,** RI.

Nematode, root knot. **Meloidogyne incognita; M. incognita-acrita; M. arenaria,** NC; **M. hapla; M. javanica.**

Nonparasitic.
  Boron Deficiency.
  Chlorosis. Mineral deficiency from excess of lime.
  Copper Spray Injury.
  Leaf Blotch. Purplish blotches associated with soil deficiencies and drought or winter injury.
  Spine Injury.

Powdery mildew. **Microsphaera alni,** AL, FL, IL, NC; **Phyllactinia corylea,** TN.

Rot, leaf. **Rhizoctonia solani,** of cuttings.

Rot, root. **Clitocybe tabescens,** FL; **Phymatotrichum omnivorum,** TX; **Corticium galactinum,** white root, MD.

Rot, wood. **Daldinia vernicosa,** MD; **Daedalea confragosa,** MD; **Polyporus** spp.; **Stereum** spp., MD; **Ustulina deusta,** MD; **Poria** spp.; **Pleurotus ostreatus,** MD.

Rust. **Chrysomyxa ilicina** (II, III), TN, NC, WV, 0, I unknown.

Sooty mold. **Capnodium elongatum,** Gulf states; **Fumago vagans,** OR, WA.

Spot anthracnose. **Elsinoë ilicis,** PA, on Chinese holly.

None of these diseases is as important to the average gardener as the holly leaf miner. Spraying is necessary for that insect but seldom obligatory for the various leaf spots. The most common leaf discoloration is a purplish blotch due to environment rather than a fungus. Proper growing conditions and correct soil acidity are important. Pick off occasional spotted leaves; prune out blighted branches.

HOLLYHOCK
(*Althaea*)

Anthracnose. **Colletotrichum malvarum,** NY to MS, IA, TX.

Bacterial hairy root. **Agrobacterium rhizogenes,** WI.

Bacterial wilt. **Pseudomonas solanacearum,** NY, WV.

Blight, southern. **Sclerotium rolfsii,** AR.

Blight, web. **Pellicularia filamentosa,** TX.

Canker. **Nectria cinnabarina,** OK.

Leaf spot. **Alternaria** spp., secondary; **Ascochyta althaeina,** IN, MD, NJ, NY, PA, WV; **Cercospora althaeina,** eastern and central states to AL, TX, SD; **C. kellermanii,** IN, MN, MO, MD, NJ, OH; **Myrothecium roridum,** MD; **Septoria malvicola,** MI, MN, NY, OH, OK, VT, WI.

Nematode, lesion. **Pratylenchus pratensis,** TX.

Nematode, root knot. **Meloidogyne** spp., FL, KS, MS, OK, TX.

Powdery mildew. **Erysiphe cichoracearum,** CA, MS; **E. polygoni,** IA.

Rot, crown. **Phytophthora megasperma,** MD, VA; **Sclerotinia sclerotiorum,** MT, NJ.

Rot, root. **Phymatotrichum omnivorum,** TX.

Rust. **Puccinia malvacearum** (III), general; **P. heterospora** (III), CA, KS, MS, TX; **P. lobata,** AZ, NM, TX; **P. schedonnardi** (0, I), KS, MS, NE, ND; **P. sherardiana** (0, III), CA; **Endophyllum tuberculatum** (III), CO, KS, NE, OK.

Virus. **Hollyhock Mosaic.**

Rust is by far the most destructive disease of hollyhocks, disfiguring foliage in most seasons and some years causing nearly all leaves to shrivel and die. Remove infected leaves and stalks as noticed.

HOLLY OSMANTHUS
(*Osmanthus ilicifolius*)

Leaf spot. **Phyllosticta oleae**, TX.
Nematode, root knot. **Meloidogyne** sp.
Rot, root. **Phymatotrichum omnivorum**, TX; **Rosellinia necatrix**, CA.
Sooty mold. **Capnodium elongatum**, TX; **Fumago salicina**, TX.
Wilt. **Verticillium albo-atrum**, VA.

HOLODISCUS
(Ocean Spray, Rock-Spirea)

Canker, twig; coral spot. **Nectria cinnabarina**, WA.
Leaf spot. **Cylindrosporium ariaefolium**, OR; **Rhopalidium cercosporel-loidis**, ID, **Septogloeum schizonoti**, WA.
Powdery mildew. **Podosphaera oxyacanthae**, ID, **Phyllactinia corylea**, WA.
Virus. **Holodiscus Witches' Broom**, OR, WA.

HOMALOMENA

Leaf spot. **Glomerella cincta**, NJ.

HONESTY
(*Lunaria*)

Club root. **Plasmodiophora brassicae**, NJ.
Leaf spot. **Alternaria oleracea**, black spot of leaves, pods; **Helminthosporium lunariae**, MA.
Rot, root; Damping-off. **Aphanomyces raphani**, WI.

HONEY LOCUST
(*Gleditsia*)

Bacterial hairy root. **Agrobacterium rhizogenes**, central states.
Canker. **Dothiorella** sp., MS; **Cytospora gleditschiae**, IL; **Physalospora obtusa**.
Canker, felt fungus. **Septobasidium curtisii**, AR, LA.
Canker, stem and branch. **Kaskaskia gleditsiae**, IL; **Tubercularia ulmea**, ND; **Nectria cinnabarina**, MN; **Thyronectria austro-americana**, KS.
Canker; wilt. **Thyronectria austro-americana**, CA, MA; **T. denigrata**, MA to SC, LA and NE, CA.
Leaf spot. **Cercospora condensata**, IL to KS, NE, WI.
Leaf spot, tar. **Linospora gleditsiae**, SC, NC to NE, TX.

Mistletoe. **Phoradendron serotinum (flavescens),** IN, TX.
Powdery mildew. **Microsphaera alni,** widespread.
Rot, crown. **Phytophthora citrophthora,** CA.
Rot, root. **Phymatotrichum omnivorum,** OK, TX; **Xylaria mali,** VA.
Rot, wood. **Fomes** spp., widespread; **Daedalea ambigua; D. elegans; Gano-
derma curtisii** and **G. lucidum; Polyporus** spp.; **Schizophyllum commune.**
Rust. **Ravenelia opaca** (III), IL.
Virus. **Robinia Brooming,** OH, KY.

HONEY PLANT, BITTERBUSH
(*Picramnia*)

Spot, tar. **Phyllachora domingensis,** FL.

HONEYSUCKLE
(*Lonicera*)

Bacterial crown gall. **Agrobacterium tumefaciens,** CT.
Bacterial hairy root. **Agrobacterium rhizogenes,** central states.
Blight, leaf. **Herpobasidium deformans,** CT, IA, NY, WI.
Blight, gray mold. **Botrytis cinerea,** CT, AK.
Blight, thread. **Pellicularia koleroga,** LA.
Blight, twig. **Phoma mariae,** MA, CT, NY.
Leaf spot. **Cercospora antipus,** MI to MT, TX; **C. varia,** TX; **Guignardia
lonicerae,** CA; **Kabatia lonicerae,** CA; **Lasiobotrys affinis,** CA; **L. lonicerae,**
Black Spot, OR; **Leptothyrium periclymeni,** widespread; **Marssonina loni-
cerae,** CA; **Mycosphaerella clymenia,** VA; **Septoria sambucina,** MT, WI;
S. xylostei, WI.
Nematode, root knot. **Meloidogyne** sp.
Powdery mildew. **Microsphaera alni,** general; **Erysiphe polygoni,** CA, WY.
Rot, collar, wood. **Fomes** spp.
Rot, root. **Armillaria mellea,** NJ; **Phymatotrichum omnivorum,** TX.
Rust. **Puccinia festucae** (I), IL, IA.
Virus. **Mosaic,** MD.

HOP
(*Humulus*)

Anthracnose; leaf spot. **Colletotrichum** sp.; **Glomerella cingulata,** IN, KS,
MD, NY, OR, WA, WI.
Bacterial crown gall. **Agrobacterium tumefaciens,** CA, OK, OR, WA.
Downy mildew. **Pseudoperonospora humuli,** CA, NY, OR, WA, WI.

Leaf spot. Cercospora sp., NE; **Cylindrosporium humuli,** NY to NC, IA, WI; **Phyllosticta decidua,** IA, WI; **P. humuli,** IA, MA, MI; **Mycosphaerella erysiphina,** CA; **Septoria humuli,** NH; **S. lupulina,** KS.

Nematode, root knot. **Meloidogyne** sp., CA.

Powdery mildew. **Erysiphe cichoracearum; Sphaerotheca macularis,** general.

Rot, root. **Armillaria mellea,** OR.

Rust. **Aecidium** sp., WA.

Sooty mold. **Fumago vagans,** CA, OR, WA.

Virus. **?Hop Mosaic,** NY, OR, WA.

Wilt. **Verticillium albo-atrum,** ME, OH, OR, WI.

HOP-HORNBEAM, IRONWOOD
(*Ostrya*)

Canker. **Nectria** sp., NY; **Strumella coryneoidea,** MD, PA, WV.

Canker, bark patch. **Aleurodiscus** spp.; Felt Fungus, **Septobasidium** spp.

Leaf blister. **Taphrina virginica,** NH to FL, TX and WI.

Leaf spot. **Cylindrosporium dearnessi,** VA; **Gloeosporium robergei,** PA, WI; **Septoria ostryae,** IA, NY, WI.

Powdery mildew. **Microsphaera alni,** widespread; **Phyllactinia corylea,** NY to FL, TX, WI; **Uncinula macrospora,** MI, WI.

Rot, root. **Armillaria mellea,** cosmopolitan; **Clitocybe tabescens,** FL; **Phymatotrichum omnivorum,** TX.

Rot, wood. **Daedalea confragosa,** cosmopolitan; **Fomes** spp.; **Pleurotus similis,** NY; **Poria** spp.; **Polyporus** spp.; **Stereum** spp.; **Trametes mollis,** NY, VT.

Rust. **Melampsoridium carpini** (II, III), NY; 0, I unknown.

HOP-TREE
(*Ptelea*)

Leaf spot. Cercospora afflata, IN, MO, TX; C. **pteleae,** TX; **Phleospora pteleae,** TX; **Phyllosticta pteleicola,** IL; **Septoria pteleae,** WI.

Rot, root. **Phymatotrichum omnivorum,** TX.

Rust. **Puccinia windsoriae** (0, I), NY to AL, KS; II, III on grasses.

HORNBEAM
(*Carpinus*)

Blight, twig, dieback. **Gibberella baccata,** AL.

Canker, bark patch. **Aleurodiscus oakesii,** NY.

Canker, branch, trunk. **Nectria galligena,** CT, NY; **Pezicula carpinea,** MA to GA and OK.

Canker, felt fungus. **Septobasidium** spp., VA to FL and LA.

Leaf blister. **Taphrina australis,** AL, CT, KY.

Leaf spot. Clasterosporium cornigerum, MD, NY, WI; **Cylindrosporium dearnessi,** MI; Phyllosticta sp., OK; Sphaerognomonia carpinaea, GA, NY, PA, WV, WI;, **Cercoseptoria caryigena,** WI.

Powdery mildew. **Microsphaera alni,** MA, IN, IA, MI, TX; **Phyllactinia corylea,** AL, IN, OH, TX, WI.

Rot, root. Armillaria mellea, FL; **Phymatotrichum omnivorum,** TX.

Rot, wood. Daedalea confragosa, NC; **Polyporus** spp., general; **Poria** spp., general; **Steccherinum** sp., CT; **Stereum** spp., general; **Daldinia concentrica,** MD; **Schizophyllum commune,** cosmopolitan.

HORSE-CHESTNUT, BUCKEYE
(*Aesculus*)

Anthracnose; leaf blight. **Glomerella cingulata,** CT, MD, MO, NJ, NY, TX.

Blight, twig. **Botryosphaeria ribis** var. **chromogena,** MD, NY, GA.

Blotch, leaf. **Guignardia aesculi** (*Phyllosticta paviae*), general.

Canker; dieback. **Nectria cinnabarina,** cosmopolitan; **Phytophthora cactorum,** bleeding canker, RI; **Gibberella acuminatum,** CA.

Leaf blister, yellow. **Taphrina aesculi,** CA, TX.

Leaf spot. **Cercospora aesculina** FL, VA; **Macrosporium baccatum,** KS; **Monochaetia desmazieri,** NC; **Micosphaerella maculiformis** var. **hippocastani,** CA; **Septoria glabra,** IN; **S. hippocastani,** PA, VT.

Mistletoe. **Phoradendron serotinum (flavescens),** South, central states; **P. villosum,** CA, OR, TX.

Mistletoe. **Viscum album,** CA, on *Aesculus californica* (California Buckeye).

Nonparasitic. **Scorch,** leaf scald, response to drought and heat.

Powdery mildew. **Phyllactinia corylea,** CA, TX; **Uncinula flexuosa,** widespread.

Rot, root. **Armillaria mellea,** cosmopolitan; **Phymatotrichum omnivorum,** TX.

Rot, wood. **Collybia velutipes,** RI; **Ganoderma applanatum,** cosmopolitan; **Polyporus** spp.

Rust. **Aecidium aesculi,** IN, KS, NE, MO.

Wilt. **Verticillium albo-atrum,** PA.

In a wet season leaves are blotched, turn brown, and drop from the *Guignardia* fungus; in a dry season leaves look scorched or blotched, turn brown, and drop from drought and heat. The minute black fruiting bodies of the fungus distinguish the parasitic form from the physiogenic disease.

HORSE-GENTIAN
(*Triosteum*)

Leaf spot. **Cladosporium triostei,** IL, IA, MO, NE, WV, WI; **Cylindrosporium triostei,** IL, OK, KS, WI; **Cercospora triostei,** WI.

Powdery mildew. **Phyllactinia corylea,** MI.
Rust. **Aecidium triostei** (0, I), MO.

HORSE PURSLANE
(*Trianthema*)

Leaf spot. **Gibbago trianthemae,** AR.

HORSE-RADISH
(*Armoracia*)

Bacterial crown gall. **Agrobacterium tumefaciens,** NJ, NY.
Bacterial leaf spot. **Xanthomonas campestris** pv. **armoraciae,** CT, IL, IA,
    MD, MO, SD, VA; **X. phaseoli,** IL.
Bacterial, MLO. Root, brittle, disease, virescence. **Spiroplasma citri,** CA,
    IL, MD.
Bacterial soft rot. **Erwinia carotovora,** in stored roots, CT, NY.
Club root. **Plasmodiophora brassicae,** IL.
Downy mildew. **Peronospora parasitica,** AL, IL, NJ, SD, WI.
Leaf spot. **Alternaria brassicae,** CT to DE, IL, IA; **A. oleracea,** CT to DE,
    MO, TX, NE, TX; **Cercospora armoraciae,** general; **Phyllosticta decidua,**
    TX, WA, WI; **P. orbicula,** NY; **Ramularia armoraciae,** general.
Nematode root knot. **Meloidogyne** sp., IL, MS, OK.
Powdery mildew. **Erysiphe polygoni,** CA.
Rot, blue mold. **Penicillium hirsutum,** IL, NJ.
Rot, collar, root. **Pellicularia filamentosa,** IL, MI, NJ, NY, TX, WA.
Rot, root. **Phymatotrichum omnivorum,** TX; **Thielaviopsis basicola,** KS, NJ.
Virus. **Beet Curly Top,** CA, OR, WA, IL, KS, UT; **Turnip Mosaic,** widespread.
White rust. **Albugo candida,** probably general.
Wilt. **Verticillium albo-atrum,** MI, WA.

HOSTA
(*Plantain-Lily*)

Leaf spot. **Alternaria** sp., NY; **Phyllosticta** sp., NJ.
Rot, crown. **Botrytis cinerea,** AK, NJ; **Sclerotium rolfsii** (*S. delphinii*), CT,
    MD, MN, NJ, NY.
Rot, root. **Rhizoctonia solani,** NJ.

HOUSTONIA
(*Bluets*)

Downy mildew. **Peronospora calotheca,** IL; **P. seymourii,** AL, AR, IL, IA,
    MS, TX.
Leaf spot. **Cercospora houstoniae,** DE; **Septoria** sp., TX.

Rot, root. **Phymatotrichum omnivorum,** TX.
Rust. **Puccinia lateritia,** TX; **Uromyces houstoniatus** (0, I), MA to MS, MO, IL, WI; II, III, on blue-eyed grass; **U. peckianus** (0, I), AL, MS, TX; II, III, on grasses.

HOYA

Leaf spot. **Myrothecium roridum,** FL.

HUCKLEBERRY
(Gaylussacia)

Black mildew. **Dimerosporium ellissi,** NJ, MS.
Dodder. **Cuscuta** sp., PA.
Gall, leaf. **Exobasidium vaccinii,** ME to VA, WI, AL, FL; **Synchytrium vaccinii,** red lead, NJ.
Leaf spot. **Ophiodothella vaccinii,** TX; **Pestalotia vaccinii,** WV; **Phyllosticta** sp., NJ, PA; **Ramularia effusa,** WI; **Rhytisma vaccinii,** tar spot, MA, SC, OK.
Nematode, root knot. **Meloidogyne** sp., OK.
Powdery mildew. **Microsphaera alni** var. **vaccinii,** widespread.
Rust. **Pucciniastrum myrtilli** (II, III), ME to VA and WI; 0, I, on hemlock.

HUISACHE, SWEET ACACIA
(Acacia farnesiana)

Rot, root. **Clitocybe tabescens,** FL; **Phymatotrichum omnivorum,** TX.
Rust. **Ravenelia australis,** TX; **R. hieronymi,** witches' broom, TX; **R. siliquae,** NM, TX.

HYACINTH
(Hyacinthus)

Bacterial soft rot. **Erwinia carotovora,** cosmopolitan.
Bacterial yellows. **Xanthomonas hyacinthi,** occasional in imported bulbs.
Blight, gray mold. **Botrytis hyacinthi,** WA; **B.?cinerea,** NC, after frost.
Nematode, bulb; ring disease. **Ditylenchus dipsaci,** NJ, Pacific states.
Nonparasitic. **Loose Bud, Stem Break,** excessive water intake during early shoot growth.
Rot, black. **Sclerotinia ?sclerotiorum,** Pacific Northwest; **S. bulborum,** black slime.
Rot, bulb. **Fusarium** sp., CO, MO, NJ, RI, TX, WA; **Penicillium** spp., blue mold, cosmopolitan.
Rot, root. **Phytophthora** sp., NJ.
Virus. **Ornithogalum Mosaic,** CA, MD, NY, OR, TX, WA.

HYDRANGEA

Bacterial, MLO. **Hydrangea virescence.**
Bacterial wilt. **Pseudomonas solanacearum,** NY.
Blight, flower; gray mold. **Botrytis cinerea,** cosmopolitan, especially after frost.
Blight, southern. **Sclerotium rolfsii,** FL.
Leaf spot. **Ascochyta hydrangeae,** AK, NJ; **Cercospora arborescentis,** IL, OK; **C. hydrangeae,** MD, VA to FL, AL, TX; **Colletotrichum** sp., MD, NJ; **Phyllosticta hydrangeae,** IN; **Septoria hydrangeae,** CT, MS, OH; **Corynespora cassiicola,** FL, GA.
Nematode, leaf. **Aphelenchoides** sp.
Nematode, lesion. **Pratylenchus pratensis,** TX.
Nematode, root knot. **Meloidogyne** spp., MS, OK, TX.
Nematode, stem, bulb. **Ditylenchus dipsaci.**
Nonparasitic. **Chlorosis,** iron deficiency, excess lime.
Powdery mildew. **Erysiphe polygoni,** general.
Rot, root. **Phymatotrichum omnivorum,** TX; **Armillaria mellea,** CA.
Rot, stem. **Rhizoctonia solani,** MD; wound, **Polyporus versicolor,** CT, MD, IA.
Rust. **Pucciniastrum hydrangeae** (II, III), PA to AR, IL; 0, I, on hemlock.
Virus. **Hydrangea Ring Spot; Tomato Spotted Wilt.**

HYDROPHYLLUM
(*Waterleaf*)

Downy mildew. **Peronospora hydrophylli,** MD, IL, IA, NY, WA, WI.
Leaf spot. **Ascochyta hydrophylli,** WA; **Gloeosporium hydrophylli,** NH; **Ramularia hydrophylli,** WA; **Septocylindrium hydrophylli,** NY; **Septoria hydrophylli,** NY.
Powdery mildew. **Erysiphe cichoracearum,** occasional; **E. polygoni,** ID, WA; **Sphaerotheca macularis,** OH, ND, SD.
Rot, leaf and stem. **Ceratobasidium anceps,** NY, WI.
Rust. **Puccinia hydrophylli** (III), CA, CO, NY to NE, ND, UT; **P. recondita** (0, I), MT, widespread; II, III, on grasses.

HYMENOPAPPUS

Downy mildew. **Plasmopara halstedii,** OK.
Rot, root. **Phymatotrichum omnivorum,** TX.
Rust. **Puccinia grindeliae** (0, III), OK.

HYSSOP
(*Hyssopus*)

Nematode, root knot. **Meloidogyne** sp., MI.

ICE PLANT
(*Carpobrotus*)

Rot, root. **Pythium aphanidermatum,** CA; **Phytophthora cryptogea,** CA.
Wilt. **Verticillium dahliae,** CA; **Phomopsis** sp., CA.

IMPATIENS
(Garden Balsam, Sultan)

Bacterial fasciation. **Clavibacter fasciens,** CA.
Bacterial wilt. **?Pseudomonas solanacearum,** WI.
Damping-off. **Pythium** sp., WA; **Rhizoctonia solani,** FL.
Leaf spot. **Cercospora fukushiana,** KS, FL; **Septoria noli-tangeris,** OH;
  Phyllosticta sp., NJ.
Gall, leaf. **Synchytrium impatientis,** LA.
Nematode, lesion. **Pratylenchus penetrans.**
Nematode, root knot. **Meloidogyne** sp.
Rot, mushroom root. **Armillaria mellea,** CA.
Rot, stem. **Sclerotium rolfsii; Myrothecium roridum.**
Virus. **Cucumber Mosaic; Turnip Mosaic,** NY; **Tobacco Ringspot,** IA,
  MN.
Wilt. **Verticillium albo-atrum,** NY.

INCENSE-CEDAR
(*Libocedrus* [= *Calocedrus decurrens*])

Bacterial crown gall. **Agrobacterium tumefaciens,** AZ, CA.
Blight, brown felt. **Herpotrichia nigra,** CA.
Canker, branch. **Coryneum cardinale,** CA.
Mistletoe. **Phoradendron juniperinum,** CA, NV, OR.
Needle cast. **Lophodermium juniperinum,** CA, OR.
Rot, root. **Phymatotrichum omnivorum,** TX; **Armillaria mellea,** OR.
Rot, wood. **Fomes pini** and **F. pinicola,** CA, OR; **Lenzites saepiaria,** general;
  Polyporus spp.; **Trametes isabellina,** CA.
Rust. **Gymnosporangium libocedri,** gall, witches' broom (II, III), CA, NV,
  OR; 0, I, on apple and other Malaceae.

INDIA-HAWTHORN
(*Raphiolepis*)

Blight, southern. **Sclerotium rolfsii,** FL.
Leaf spot. **Entomosporium maculatum** (*Fabraea maculata*), CA.

INDIAN CUCUMBER-ROOT
(*Medeola*)

Gall, stem hypertrophy. **Medeolaria farlowii,** ME, MA, NJ.
Leaf spot. **Phyllosticta medeolae,** NY.

INDIAN GRASS
(*Sorghastrum*)

Leaf spot. **Ascochyta brachypodii,** NY, PA.

INDIGO
(*Indigofera*)

Rot, root. **Phymatotrichum omnivorum,** TX.
Rust. **Ravenelia laevis,** TX; **Uromyces indigoferae** (II, III), FL, TX.

INDIGO-BUSH, LEAD-PLANT
(*Amorpha*)

Leaf spot. **Cylindrosporium passaloroides,** widespread; **Diplodia amorphae,**
TX.
Powdery mildew. **Erysiphe polygoni,** IA, MN, WY.
Rot, root. **Phymatotrichum omnivorum,** TX.
Rust. **Urophyxis amorphae** (0, I, II, III), general.

INKBERRY
(Ilex glabra)

Black mildew. **Asterina ilicis,** NJ; **Morenoella ilicis,** FL, GA, MS.
Blight, thread. **Pellicularia koleroga,** LA.
Canker, felt fungus. **Septobasidium** spp., FL.
Dodder. **Cuscuta compacta,** FL.
Leaf spot. **Cercospora ilicis,** AL, FL, MS, NJ; **Phacidium curtisii** and **P.
sphaerodieum,** tar spot, MA to FL, and TX.
Rot, wood. **Poria versipora,** AR, LA.

INULA
(*Elecampane*)

Leaf spot. **Ramularia** sp., MI.
Powdery mildew. **Erysiphe cichoracearum,** NY, WI, NJ, probably general.
Rust. **Puccinia hieracii,** WI.

Inula is very susceptible to powdery mildew, which coats the foliage white in late summer.

IRESINE
(Blood-Leaf)

Leaf spot. **Septoria iresines,** OK.
Nematode, root knot. **Meloidogyne** sp., MD, FL.
Rot, root. **Helicobasidium purpureum,** TX; **Rhizoctonia solani,** IL.
Smut, inflorescence. **Thecaphora iresine,** IN.

IRIS
(Bulbous; English, Spanish, Dutch)

Blight, blossom. **Botrytis cinerea; Glomerella cingulata,** MD.
Blight, leaf; ink disease of bulbs. **Mystrosporium adustum,** NC, OR, VA.
Blight, southern; white bulb rot. **Sclerotium rolfsii,** CA, CT, FL, MD, NC, SC, OR, TX.
Leaf spot, blight. **Didymellina macrospora,** CA, NC, TX, WA.
Nematode, bulb. **Ditylenchus dipsaci** (*D. iridis*), FL, NY, NC, SC, OR, VA, WA.
Nematode, root knot. **Meloidogyne** sp., NC.
Nematode, root plate. **Aphelenchoides parietinus,** MI, NY, NC, TX, WA.
Nonparasitic. **Blindness,** complex physiological causes.
Rot, basal. **Fusarium oxysporum,** AZ, CA, NC, NY, OR, TX, VA, WA.
Rot, black. **Sclerotinia bulborum,** bulb rot, Northwest; **Rhizoctonia tuliparum,** occasional; Blue Mold, **Penicillium** spp., general.
Rot, root, neck. **Rhizoctonia solani,** NJ, Pacific Northwest.
Rust. **Puccinia iridis** (II, III), IN, CA, LA, NC.
Virus. **Iris Mosaic,** general.

IRIS
(Rhizomatous; German, Siberian, Native Species)

Bacterial soft rot. **Erwinia carotovora,** general.
Blight, blossom. **Botrytis cinerea,** MA.
Blight, southern; crown rot. **Sclerotium rolfsii** (*S. delphinii*), general.
Leaf spot. **Alternaria iridicola,** IL, WI, MT, TX, WA; **Didymellina macrospora** (*Heterosporium iridis*); **Ascochyta iridis** MD, NY; **Cladosporium herbarum,** secondary; **Kabatiella microsticta,** secondary; **Phyllosticta iridis,** FL, MI, NY, OH, TX, WI; **Stictopatella iridis,** IL.
Nematode, root knot. **Meloidogyne incognita-acrita.**
Nonparasitic. **Scorch,** cause unknown, sometimes associated with nematodes.
Rot, gray mold; crown. **Botryotinia convoluta** MN, NJ, NY, WI.
Rot, root; damping-off. **Rhizoctonia solani,** occasional.

Rust. **Puccinia iridis** (II, III), widespread on native species; most garden varieties are resistant. **P. sessilis** (0, I), ME to NE and MN; II, III on *Phalarus*.

The most common disease of bearded iris in home gardens is bacterial soft rot following iris borer. Crown rot or southern blight gets started when iris is too crowded. Didymellina leaf spot or blight may be conspicuous in a wet summer. The physiological (?) disease, scorch, is causing concern, particularly in central states.

IRONWEED
(*Vernonia*)

Black mildew. **Stigmella vernoniae**, MO.
Downy mildew. **Plasmopara halstedii**, KS, MO.
Leaf spot. **Cercospora noveboracensis**, MO; **Ascochyta treleasei**, WI; **Mycosphaerella** sp.; **Septoria** sp., MO; **C. oculata**, WV to AL, TX, NE, WI.
Powdery mildew. **Erysiphe cichoracearum**, general ; **Sphaerotheca macularis**, MO.
Rot, root. **Phymatotrichum omnivorum**, TX.
Rust. **Coleosporium vernoniae** (I, III), general; 0, I, on pine; **Puccinia vernoniae** (0, I, II, III), widespread.

ITCHGRASS
(*Rottobellia*)

Blight. **Curvularia cymbopogonis**, LA.

IVESIA

Rust. **Phragmidium horkeliae** (III), UT; **P. ivesia**, CA; **P. jonesii**, NV, OR, UT.

IVY, BOSTON
(*Parthenocissus tricuspidata*)

Canker; dieback. **Cladosporium herbarum**, IN, NJ, OH, PA.
Downy mildew. **Plasmopara viticola**, NJ, TX.
Leaf spot. **Cercospora amelopsidis**, MS; **C. pustulata**, TX; **Guignardia bidwellii** f. parthenocissi widespread; **Phleospora ampelopsidis**, IA; **Sphaeropsis hedericola**, NJ.
Nematode, dagger. **Xiphinema index**, CA.
Powdery mildew. **Uncinula necator**, TX.

Rot, root. **Phymatotrichum omnivorum,** TX.
Rot, stem. **Pellicularia filamentosa,** CT.

The Guignardia leaf spot is common and disfiguring.

IVY, ENGLISH
(*Hedera helix*)

Anthracnose. **Amerosporium** (*Colletotrichum*) **trichellum,** MA to SC, TX and OK, OR, WA.
Bacterial spot. **Xanthomonas hederae,** GA, IL, MD, NJ, NY, VA, WA.
Dodder. **Cuscuta** sp., AZ, NJ.
Leaf spot. **Glomerella cingulata,** CT, MD, NY, TX; **Phyllosticta concentrica,** MA to AL, TX and NE, CA, WA; **Ramularia hedericola,** TX; **Sphaeropsis hedericola,** NY, WV; **Macrophoma** sp., TX.
Mold, leaf. **Cladosporium brunneolum,** CA.
Nonparasitic. **Winter Injury,** sunburn, freezing.
Powdery mildew. **Erysiphe cichoracearum,** OK.
Rot, root. **Phytophthora palmivora,** CA.
Rot, root. **Rhizoctonia solani,** CT; **Phymatotrichum omnivorum,** TX.
Sooty mold. Common under trees, on insect exudate.
Spot anthracnose; ivy scab. **Sphaceloma hederae,** CA, NC, VA.

Leaf spots are not often important in the garden but browning of foliage from winter injury is conspicuous in early spring. English ivy, used as a ground cover, is often black, with sooty mold growing in honeydew dropped from aphids in trees overhead. Dodder sometimes gets a strangle hold and is difficult to eradicate.

IXIA

Rot, corm. **Fusarium oxysporum** f.sp. **gladioli; Sclerotium** sp.
Virus. **?Iris Mosaic,** OR, WA.

IXORA

Nematode, burrowing. **Radopholus similis,** FL.
Nematode, root knot. **Meloidogyne** sp., FL.
Rot, root. **Clitocybe tabescens,** FL.

JACARANDA

Rot, root. **Armillaria mellea,** CA; **Clitocybe tabescens,** FL; **Phymatotrichum omnivorum,** TX.

JACK-BEAN, SWORD BEAN
(*Canavalia*)

Leaf spot. **Cercospora ternateae**, AL.
Nematode, root knot. **Meloidogyne** spp., FL. Usually resistant to.
Pod spot. **Vermicularia capsici**, FL; **V. polytricha**, AL.
Virus. **Potato Virus X**, WI.

JACK-IN-THE-PULPIT
(*Arisaema*)

Blight, leaf and stalk. **Streptotinia arisaemae** (*Botrytis streptothrix*), IL, IA, MD, NY, PA, WI.
Leaf spot. **Cladosporium** sp., mold, VA; **Volutella** sp., IN.
Rust. **Uromyces caladii** (0, I, II, III), NY to FL and TX; Pacific Northwest.

JACOBINIA

Nematode, burrowing. **Radopholus similis**, FL.

JACQUEMONTIA

Leaf spot. **Cercospora alabamensis**, MS.
Nematode, reniform. **Rotylenchulus reniformis**, GA.
Nematode, root knot. **Meloidogyne** sp., AL; **Rotylenchulus reniformis**, LA; **Meloidogyne incognita**, LA.
Rot, root. **Streptomyces ipomoea**, LA; **Ceratocystis fimbriata**, LA; **Fusarium oxysporum** f. sp. **batatas**, LA; **Plenodomus destruens**, LA; **Monilochaetes infuscans**, LA.
Rust. **Coleosporium ipomoeae** (II, III), LA.
White rust. **Albugo ipomoeae-panduratae**, AL.

JACQUINIA

Black spot. **Asterella paupercula**, FL.

JAMESIA
(Cliffbrush)

Blight. **Ovularia edwiniae**, CO.

JASMINE
(*Jasminum*)

Bacterial crown gall. **Agrobacterium tumefaciens,** MD.
Blight, blossom. **Choanephora infundibulifera,** FL.
Blight, southern. **Sclerotium rolfsii,** FL.
Canker, stem gall. **Phomopsis** sp., FL; **Phoma** sp., TX.
Leaf spot. **Colletotrichum gloeosporioides,** FL, TX.
Leaf spot, algal; green scurf. **Cephaleuros virescens,** FL.
Nematode, root knot. **Meloidogyne** sp., FL.
Rot, root. **Clitocybe tabescens,** FL; **Corticium galactinum,** MD.
Spot anthracnose. **Elsinoë jasminae,** FL.
Virus. **Tobacco Ring Spot,** MD.
Virus. Variegation; infectious chlorosis.

JATROPHA

Black mildew. **Meliola jatrophae.**
Leaf spot. **Colletrotrichum gloeosporioides,** FL.
Rot, root. **Clitocybe tabescens,** FL; **Phymatotrichum omnivorum,** TX.
Rust. **Phakopsora jatrophicola** (II, III), TX.

JERUSALEM-CHERRY
(*Solanum pseudocapsici;* also *Solanum capsicastrum,* false Jerusalem- cherry; *S. dulcamara,* Bittersweet; *S. integrifolium,* Scarlet Eggplant)

Bacterial canker. **Clavibacter michiganense,** WY. Also on *S. dulcamara* (Bittersweet).
Bacterial crown gall. **Agrobacterium tumefaciens,** CT.
Blight, late. **Phytophthora infestans,** MD, NY.
Dodder. **Cuscuta gronovii,** NY.
Leaf spot. **Alternaria solani; Mycosphaerella solani,** OH; **Ascochyta lycopersici,** NY, OH; **Cercospora dulcamerae,** MI, NY, WI; **Phyllosticta pseudocapsici,** LA; **Stemphylium solani,** FL.
Nematode, root knot. **Meloidogyne** spp., CA, OR, MD.
Nematode, tobacco cyst. **Heterodera tabacum,** VA.
Parasitic lichen. **Strigula elegans** and **S. Camplanata,** southern U.S., LA.
Virus. **Beet Curly Top** on false Jerusalem-cherry (*S. capsicastrum*)
Virus. ?**Tobacco Mosaic,** IA, VA; **Tomato Spotted Wilt,** tip blight, OR.
Wilt. **Verticillium albo-atrum,** NY.

JETBEAD
(*Rhodotypos*)

Anthracnose. Gloeosporium sp., IL.
Blight, twig; coral spot. **Nectria cinnabarina**, MA.
Leaf spot. **Ascochyta rhodotypi**, IL.

JOBS-TEARS
(*Coix lachryma-jobi*)

Smut, head. **Ustilago coicis**. Intercepted on imported seed.
Virus. **Chlorotic Streak.**

JUJUBE
(*Zizyphus*)

Leaf spot. **Cercospora jujubae**, FL.
Rust. **Phakopsora zizyphi-vulgaris**, FL.

JUNIPER, RED-CEDAR
(*Juniperus*)

Bacterial crown gall. **Agrobacterium tumefaciens**, MS, FL.
Black mildew. **Apiosporium pinophilum**, OR; **Dimerium juniperi**, CA; **Asterina cupressina**.
Blight, brown felt. **Herpotrichia nigra**, northern Rockies to Pacific Northwest.
Blight, leaf. **Chloroscypha juniperina**, IA; **Pestalotia funerea**, secondary; **Cercospora sequoiae** var. **juniperi**, CT, IA, KY, MO, NE, OK, WI; **Exosporium glomerulosum**, NC, SC; **Stigmina juniperina**.
Blight, nursery, **Phomopsis juniperovora**, MA to FL, KS, MN.
Blight, tip dieback. **Sclerophoma pythiophila**, WI.
Blight, twig. **Phomopsis juniperovora**, VA.
Canker. **Coryneum cardinale**, CA; **Caliciopsis nigra**, gall, NY.
Canker, bark patch. **Aleurodiscus nivosus**, AL, OR, TX.
Mistletoe. **Phoradendron densum** and P. **juniperinum**, widespread.
Needle cast; leaf spot. **Lophodermium juniperinum**, widespread, secondary; **Cylindrocarpon** sp., OR; **Kriegeria** sp., OR.
Nematode, lesion. **Pratylenchus vulnus**, NC.
Parasite, "false foxglove." **Aureolaria grandiflora** var. **serrata**, TX.
Rot, heart. **Fomes juniperinus**, PA to KY, TN; F. **earlei**, Southwest; **R. subroseus**, general; F. **texanus**, AZ, NM, TX.

Rot, root. **Armillariella mellea,** CA.

Rot, root. **Clitocybe tabescens,** FL; **Phymatotrichum omnivorum,** OK, TX; **Phytophthora cactorum; P. citrophthora,** CA.

Rot, root. **Fomes annosus,** VA.

Rot, root and crown. **Phytophthora cinnamomi,** CA; **P. cryptogea,** CA.

Rot, wood. **Lenzites saepiaria,** occasional; **Coniophora corrugis,** Pacific Northwest; **Daedalea juniperina,** SC to AR.

Rust. **Gymnosporangium bermudianum** (III), gall on stems, Gulf states; **G. betheli** (III), on stems, ND to OK, NM, WA; 0, I, on hawthorn; **G. clavariiforme** (III), gall on stems, ME to AL west to MT; 0, I on *Amelanchier,* chokeberry, quince, pear; **G. clavipes** (III), quince rust, on stems, ME to IL, MT; 0, I, on *Amelanchier,* hawthorn, quince, apple; **G. corniculans** (III), gall on stems, ME, MI, NY, WI; 0, I, on *Amelanchier;* **G. cornutum** (III), gall on stems, leaves, CO, ME, MI; 0, I, on mountain-ash; **G. davissii** (III), leaf gall, ME, WI; 0, I, on chokeberry; **G. effusum** (III), gall on stems, NY to SC; 0, I, on chokeberry; **G. exiguum** (III), on leaves, OK, TX; 0, I, on hawthorn; **G. exterum** (III), gall on stems, KY; 0, I, *Gillenia;* **G. floriforme** (III) gall, on leaves, stems, SC to FL, OK, TX; 0, I, on hawthorn; **G. globosum,** hawthorn rust (III); 0, I, on hawthorn, apple, pear, mountain-ash; **G. harknessianum** (III), on western juniper; 0, I, on *Amelanchier;* **G. inconspicuum** (III), CA, CO, UT; 0, I, on *Amelanchier;* **G. gracile,** witches' broom (III), 0, I, on sand pear.

Rust. **Gymnosporangium clavipes,** galls on stems, VA; **G. globusum,** gall on stems, VA; **G. exterum,** gall on stems, VA.

Rust. **Gymnosporangium juvenescens** (III), gall on stems, witches' brooms, MN, NE, WI; 0, I, on *Amelanchier;* **G. japonicum** (III), CT, MA, NJ, WA; 0, I, on *Photinia;* **G. kernianum,** witches' broom, ID, OR to AZ, NM; 0, I, on *Amelanchier,* pear; **G. multiporum** (III), CO to NM, CA; 0, I unknown; **G. nelsoni** (III), gall on stems, MT, SD; 0, I, on hawthorn, crabapple, mountain-ash, quince, *Amelanchier;* **G. nidus-avis** (III), gall on stems; witches' brooms, East and South; 0, I, on *Amelanchier,* apple, hawthorn, mountain-ash, quince; **G. speciosum** (III), on stems, AZ, CO, NV, NM; 0, I, on *Fendlera, Philadelphus;* **G. trachysorum** (III), on stems, FL, LA, MS, SC; **G. tremelloides** (III) gall on stems, CO to Pacific Northwest; 0, I, on mountain-ash; **G. tubulatum** (III), gall on stems, SD to OR, WA; 0, I, on hawthorn; **G. vauqueliniae** (III), on one-seed juniper; 0, I, on *Vauquelinia,* causing witches' brooms.

Rust, cedar-apple. **Gymnosporangium juniperi-virginianae** (III), gall on leaves of red-cedar, prostrate and Rocky Mountain junipers, general; 0, I, on apple, crabapple.

Of this lengthy list of juniper rusts the three common apple rusts are most important in the garden, not so much for the damage to this host as for the harm to the deciduous fruit or ornamental. These three are the cedar-apple rust caused by *Gymnosporangium juniperi-virginianae*, the hawthorn rust caused by *G. globosum*, and the quince rust caused by *G. clavipes*. The latter is perennial in juniper and may produce spores each spring for as long as 20 years. Varieties of *Juniperus chinensis* and *J. communis* and some other forms are resistant to cedar-apple and cedar-hawthorn rusts. If you have only a few red-cedars, it is quite feasible to cut out galls in late winter, before spores are produced.

KAGENECKIA

Scab. **Spilocaea botryae**, CA.

KALANCHOË

Bacterial crown gall. **Agrobacterium tumefaciens.**
Bacterial wilt and soft rot. **Erwinia carotovora pv. carotovora**, FL.
Blight, flower. **Stemphylium bolicki**, FL; **S. floridanum** f. sp. **kalanchoe**, also leaf spot, FL.
Leaf spot. **Cercospora** sp., MS.
Rot, crown, stem; wilt. **Phytophthora ?cactorum**, NJ, NY; **Diplodia natalensis**, AL.
Virus. **Mosaic**, FL.

KENTUCKY COFFEE-TREE
(*Gymnocladus*)

Leaf spot. **Cercospora gymnocladi**, north central states; **Marssonina** sp., Northeast; **Phyllosticta gymnocladi**, IL.
Rot, root. **Phymatotrichum omnivorum**, OK, TX.
Rot, wood. **Polyporus pulchellus**, IN, MI.

KERRIA

Blight, twig. **Phomopsis japonica**, OH, NJ, TX.
Canker, coral spot. **Nectria cinnabarina**, OR, WA, NJ.
Leaf spot; leaf and twig blight. **Coccomyces kerriae**, widespread, eastern states to IA and TX; **Septoria** sp., MD, NJ.
Rot, root. **Phymatotrichum omnivorum**, TX.

KIDNEY VETCH, LADYS-FINGERS
(*Anthyllis*)

Blight, leaf and stem. **Fusarium** sp., NC.
Leaf spot. **Phyllosticta** sp., NC.

KIWI
(*Actinidia*)

Bacterial canker. **Pseudomonas syringae**, CA.
Rot, root and wilt. **Cylindrocladium crotalaria**, SC.

KNIPHOFIA
(Tritoma, Torch-Lily, Poker-Plant)

Leaf spot. **Alternaria** sp., AL.
Nematode, root knot. **Meloidogyne** sp., CA.

KNOTROOT BRISTLEGRASS
(*Setaria geniculata*)

Blight. **Beniowskia sphaeroidea**, GA.

KOCHIA
(Summer-Cypress)

Damping-off; root rot. **Pythium debaryanum**; SD; **Rhizoctonia solani**, TX.
Rot, root. **Phymatotrichum omnivorum**, TX.
Rust. **Puccinia aristidae** (0, I), CO, NE, TX.
Virus. **Beet Curly Top**, CA.

KOHLERIA

Virus. **Tobacco Mosaic**, CA, CT, DC, FL, OH, WA.

KRIGIA
(Dwarf Dandelion)

Downy mildew. **Bremia lactucae**, MS, MO, OK, WI.
Leaf spot. **Mycosphaerella krigiae**, IL, WI; **Septoria krigiae**, KY, NY, WI.

Rot, root. **Phymatotrichum omnivorum,** TX.
Rust. **Puccinia dioicae** (0, I), IA, WI, IL; II, III, on *Carex*; **P. hieracii** (0, I, II, III), NC; **P. maculosa** (III), IL, MI, MS, MO, PA, TN.

KUDZU
(*Pueraria*)

Bacterial blight. **Pseudomonas phaseolicola,** CT to FL, LA, IN; **P. syringae,** NY.
Blight, web. **Pellicularia filamentosa,** GA, MS.
Damping-off. **Rhizoctonia solani.**
Leaf spot. **Alternaria** sp., secondary; **Mycosphaerella pueraricola,** MS, AL, GA.
Nematode, root knot. **Meloidogyne** spp., general.
Rot, charcoal. **Macrophomina phaseoli,** GA; Stem, **Fusarium** sp., LA.
Rot, root. **Phymatotrichum omnivorum,** TX.

KUHNIA
(False-Boneset)

Leaf spot. **Pleospora compositarum,** NM, TX.
Rot, root. **Phymatotrichum omnivorum,** TX.
Rust. **Puccinia kuhniae** (0, I, II, III), IN to AL, NE, ND, FL, TX, MT.

KUMQUAT
(*Fortunella*)

Dodder. **Cuscuta campestris,** FL.
Leaf spot. **Cephaleuros virescens,** algal spot, Gulf states; **Phyllosticta citricola,** MS.
Nematode, citrus. **Tylenchulus semi-penetrans,** FL.
Rot, black. **Alternaria citri;** stem-end, **Diaporthe citri,** CA.

LANTANA

Black mildew. **Meliola cookeana,** FL.
Leaf spot. **Alternaria** sp., TX.
Nematode, leaf. **Aphelenchoides fragariae,** NJ.
Nematode, root knot. **Meloidogyne** spp., widespread.
Rot, root. **Phymatotrichum omnivorum,** TX.
Rust. **Puccinia lantanae** (III), FL.
Wilt. **Fusarium** sp., NJ.

LAPPULA
(Hackelia)

Downy mildew. Peronospora echinospermi, IA, TX; P. myosotidis, MT.
Leaf spot. Cercospora cynoglossi, WI; Ovularia asperifolii var. lappulae, WI; Phyllosticta decidua, WI, TX.
Powdery mildew. Erysiphe cichoracearum, IA, MI, MN, MO, OH, WI, TX; Microsphaera sp., IL.
Rot, root. Phymatotrichum omnivorum, TX.
Rust. Puccinia mertensiae (III), TX, UT.
Smut, leaf. Entyloma serotinum, UT.

LARCH
(*Larix*)

Blight, needle. Hypodermella laricis, Great Lakes, Pacific Northwest. Meria laricis, ID.
Blight, seedling. Thelephora caryophyllea, ID, MT, girdle, smother; Rhizina undulata, ID; Meria laricis, WA.
Canker. Aleurodiscus amorphus, bark patch, Northwest; A. spinulosus; A. werii; Lachuellula willkommii, European larch canker, MA, eradicated, ME; Valsa kunzei var. kunzei.
Canker. Phomopsis pseudotsugae (*Phacidiella coniferarum*) on western larch, Northwest.
Damping-off. Rhizoctonia solani, cosmopolitan.
Mistletoe, dwarf. Arceuthobium campylopodum, MT to OR, WA; A. pusillum, Northeast, MN.
Needle cast. Lophodermium sp., WA; L. laricinum, MT to OR; L. laricis, ID; Cladosporium sp., ME; Mycosphaerella laricina, IA, VT, WI.
Rot, heart. Fomes spp., widespread.
Rot, root. Armillaria mellea, cosmopolitan; Phymatotrichum omnivorum, TX.
Rot, sapwood. Lenzites saepiaria.
Rot, seedling. Botrytis ?douglasii, gray mold, Northwest; Cylindrocladium scoparium, NJ; Phytophthora cinnamomi, MD; Sparassis radicata, MT to OR, WA.
Rot, wood. Polyporus spp.; Poria spp.; Stereum spp., widespread.
Rust. Melampsora paradoxa (0, I), northern U.S. including AK; II, III, on willow; M. medusae (0, I), New England to MI and IN; II, III, on poplar; Melampsoridium betulinum (0, I), CT, WI; II, III, on birch.

LARKSPUR

See Delphinium. Crown rot, southern blight, is prevalent on annual larkspur.

LAUREL

See California-Laurel, Cherry-Laurel, Mountain-Laurel.

LAUREL, SWEET BAY
(*Laurus*)

Blight, thread. **Pellicularia koleroga,** SC.

LAURESTINUS
(*Viburnum tinus*)

Downy mildew. **Plasmopara viburni,** GA.
Leaf spot. **Hendersonia tini,** LA; **Leptosphaeria tini,** LA.
Leaf spot, algal. **Cephaleuros virescens,** FL, LA.
Nematode, root knot. **Meloidogyne** sp., CA.
Wilt. **Verticillium albo-atrum,** OR.

LAVATERA
(Treemallow)

Anthracnose, leaf spot. **Colletotrichum malvarum,** CA, TX.
Damping-off. **Rhizoctonia solani,** IL.
Rot, root. **Phymatotrichum omnivorum,** TX.
Rust. **Puccinia malvacearum** (III), CA.
Virus. **Abutilon Infectious Variegation.**

LAVENDER
(*Lavandula*)

Leaf spot. **Septoria lavandulae,** OH, OK.
Nematode, root knot. **Meloidogyne** sp.
Rot, root. **Armillaria mellea,** TX.

LAWNS

See Grasses.

LAYIA
(Tidy-Tips)

Powdery mildew. **Erysiphe cichoracearum,** CA.
Virus. **Tomato Spotted Wilt,** CA.

LEADTREE
(*Leucaena*)

Rot, root. **Ganoderma sulcatum,** TX.
Rust. **Ravenelia leucaenae** (II, III), TX.

LEATHERWOOD
(*Dirca*)

Rust. **Puccinia dioicae** (I), ME to MN, MO, AL; II, III on *Carex*.

LEBBEK
(*Albizzia lebbek*)

Leaf spot, algal. **Cephaleuros virescens,** FL.
Rust. **Sphaerophragmium acaciae,** FL.

LEDUM
(Labrador-Tea)

Gall, leaf. **Exobasidium vaccinii,** OR, WA, AK; **Synchytrium vaccinii,** red
spot, ME.
Leaf spot. **Ascochyta ledi,** WI; **Cryptostictis arbuti,** CA, OR; **Rhytisma
?andromedae,** tar spot, ID.
Powdery mildew. **Microsphaera alni,** WA.
Rust. **Chrysomyxa ledi** (II, III), CA, CT, ID, MT, NV, MI, NH, NY, WI, WY;
C. ledicola (II, III), ME, NH, NY, WA, WI, AK; 0, I, on spruce.
Spot anthracnose. **Elsinoë ledi,** CA, OR, WA, ME, MI, MN, NY, PA, WI.

LEEK

See Shallot.

LEMON

See Citrus Fruits.

LEMON GRASS, CITRONELLA GRASS
(*Cymbopogon*)

Leaf spot, eye-spot. **Helminthosporium sacchari,** FL.
Tangle-top. **Myriogenospora paspali,** FL.

LENTIL
(*Lens*)

Rot, root. **Thielaviopsis basicola,** ID, WA.

LEPTOSPERMUM

Rot, root. **Armillaria mellea,** CA.

LETTUCE
(*Lactuca*)

Anthracnose. **Marssonina panattoniana,** NY to FL, TX, MI, CA, OR, WA.

Bacterial marginal leaf blight. **Pseudomonas fluorescens** pv. **marginalis,** KS, MO, NJ, NY; **P. cichorii,** NY.

Bacterial, MLO. **Aster Yellows** (white heart, Rio Grande disease) and **California Aster Yellows,** widespread.

Bacterial rot. **Pseudomonas virdilivida,** LA, VA, WA, DE, NH, NY, MI.

Bacterial soft rot. **Erwinia carotovora,** cosmopolitan in market.

Bacterial wilt. **Xanthomonas vitians,** NJ, NM, NY, PA, SC, VA.

Blight, southern. **Sclerotium rolfsii,** CA, FL, NC, SC, TX, VA.

Damping-off; stump wilt; stunt. **Pythium** spp., cosmopolitan.

Downy mildew. **Bremia lactucae,** general.

Leaf spot. **Alternaria** sp., secondary; **Cercospora longissima,** FL, IL, IN, TX, VA, WI; **Septoria lactucae,** occasional in East and central states, to FL, CO, MN.

Mold, leaf; seed. **Pleospora herbarum,** FL, KY, NY.

Nematode. **Naccobus batatiformis.**

Nematode, lesion. **Pratylenchus** sp.

Nematode, root knot. **Meloidogyne hapla.**

Nematode, sting. **Belonolaimus gracilis.**

Nonparasitic. **Brown Blight,** cause unknown, AZ, CA. **Tipburn,** high temperature and excessive transpiration.

Powdery mildew. **Erysiphe cichoracearum,** CA, MI.

Rot, bottom; damping-off. **Rhizoctonia solani,** general.

Rot, gray mold. **Botrytis cinerea,** chiefly in greenhouses, sometimes outdoors.

Rot, root. **Fusarium** sp., KY, OH; **Phymatotrichum omnivorum,** AZ, TX.

Rot, watery soft; drop. **Sclerotinia sclerotiorum,** widespread; **S. minor,** in transit and market.

Rust. **Puccinia dioicae** (0, I), IN, MN, ND, WI; II, III on *Carex*; **P. hieracii** (II), CA.

Slime mold. **Physarum cinereum,** occasional under glass.

Virus. **Cucumber Mosaic,** NY; **Sonchus Yellow Net,** FL; **Tobacco Rattle,** CA; **Lettuce Mosaic,** general; **Radish Yellows; Lettuce Big Vein,** associated

with *Olpidium brassicae* in roots, FL; **Tomato Spotted Wilt,** CA, TX; **Tobacco Necrosis.**
White rust. **Albugo** sp., TX.
Wilt, leaf blight. **Pythium tracheiphilum,** WI.

Nonparasitic tipburn is the most general of lettuce diseases, prevalent at high temperatures when soil is deficient in moisture. Some hot-weather varieties are rather resistant. Spacing plants well apart in a well-drained soil will reduce bottom rot and drop. Eliminate wild weed hosts and spray for leafhopper vectors to reduce aster yellows.

LEUCOJUM
(Snowflake)

Leaf scorch; red blotch. **Stagonospora curtisii,** CA.
Nematode, meadow. **Pratylenchus** sp.
Rot; scale speck, of bulbs. **Botrytis** sp., CA.

LEUCOTHOË

Black mildew. **Asterina diplodioides,** AL.
Blight, leaf spot. **Cylindrocladium avesiculatum,** GA.
Canker, felt fungus. **Septobasidium pseudopedicellatum,** FL.
Gall, leaf. **Exobasidium vaccinii,** MA, MS, NC.
Leaf spot. **Cercospora leucothoës,** NJ, NY; **Cryptostictis** sp.; **Guignardia leucothoës,** MD, NC, SC, RI, TN, VA; **Mycosphaerella leucothoës,** NJ; **Phyllosticta terminalis,** NY to FL; **Ramularia andromedae,** NJ; **Rhytisma decolorans,** tar spot, CT, FL, TN, VA. **Pestalotia leucothoës,** NJ.
Powdery mildew. **Microsphaera penicillata,** GA.
Spot anthracnose. **Elsinoë ledi,** FL.

LEWISIA
(Bitterroot)

Rust. **Uromyces unitus** (I, III), CA, MT, WA.

LIATRIS
(Gayfeather)

Dodder. **Cuscuta glomerata,** OK.
Leaf spot. **Phyllosticta liatridis,** WI; **Septoria liatridis,** MN, ND, TX, WI.
Nematode, root knot. **Meloidogyne** sp.

Powdery mildew. **Erysiphe cichoracearum,** OK.
Rot, root. **Phymatotrichum omnivorum,** TX.
Rot, stem. **Sclerotinia sclerotiorum,** NY.
Rust. **Coleosporium laciniariae** (II, III), NJ to FL, AR, TX; 0, I, on pines;
　　Puccinia liatridis (0, I), KS, IN, ND, NE, MT, WI to CO; II, III, on grasses.
Wilt. **Verticillium albo-atrum,** NJ.

LIGUSTRUM

See Privet.

LILAC
(*Syringa*)

Anthracnose, shoot blight. **Gloeosporium syringae,** CT, MA.
Bacterial crown gall. **Agrobacterium tumefaciens,** CT.
Bacterial blight; twig canker. **Pseudomonas syringae,** northeastern states
　　to AL, IL, Pacific Coast.
Bacterial; MLO. **witches' broom,** IL.
Blight, blossom, shoot. **Phytophthora cactorum,** IA, MD, MA, MN, NJ; **P.
　　syringae,** MD, NY; **Sclerotinia sclerotiorum,** WA.
Blight, cutting. **Cylindrocladium scoparium** and **C. floridanum,** FL, OH.
Blight, gray mold. **Botrytis cinerea.** Pacific Northwest, Northeast.
Blight, thread. **Pellicularia koleroga,** FL, MS, NC.
Blotch, leaf. **Heterosporium syringae,** NJ; **Cladosporium herbarum,** ?second-
　　ary, cosmopolitan.
Canker; dieback. **Physalospora obtusa,** MA to VA, OH.
Canker; stem girdle. **Hymenochaete agglutinans,** CT.
Leaf spot. **Cercospora lilacis,** widespread; **Macrophoma halstedii,** CT, NJ,
　　NY; **Phyllosticta** sp., MA; **Pleospora herbarum,** secondary, MD.
Nematode, citrus. **Tylenchulus semipenetrans.**
Nonparasitic. **Blight,** graft incompatibility of lilac scion on privet stock.
Powdery mildew. **Microsphaera alni,** general.
Rot, root. **Phymatotrichum omnivorum,** TX; **Armillaria mellea,** CA, MS;
　　Thielaviopis basicola, CT.
Rot, wood. **Polyporus gilvus,** MD; **Stereum purpureum,** OK.
Virus. **Lilac Ring Spot,** MI, MN; **Lilac Witches' Broom,** MD.

Powdery mildew is the most general and conspicuous disease of lilacs,
but it comes too late in the season to damage the bushes materially. It can
be controlled with repeated application of sulfur or Karathane, where the
time and expense are justified. In wet seasons bacterial and Phytophthora

blights may be important, but dieback is more often due to borers than to diseases.

LILY
(*Lilium*)

Bacterial soft rot. **Erwinia carotovora,** GA, MA, NJ, WV.

Blight, botrytis; leaf spot. **Botrytis elliptica,** general; **B. cinerea,** general; **B. liliorum,** CA.

Blight, bud. **Sporotrichum** sp., ?secondary, VA.

Blight, southern; bulb rot. **Sclerotium rolfsii,** cosmopolitan.

Canker, stem. **Rhizoctonia solani,** CA.

Damping-off. Pythium debaryanum; Rhizoctonia solani, cosmopolitan.

Leaf spot. Cercospora sp., FL; **Cercosporella lilii,** CT, NY; **Ramularia** sp., WA; **Heterosporium** sp., MD.

Mold, leaf and bulb. **Cladosporium** sp., cosmopolitan.

Nematode, leaf; bunchy top; dieback. **Aphelenchoides fragariae,** CA, OR, WA and in greenhouses.

Nematode, lesion. **Pratylenchus pratensis, P. penetrans,** Northwest; **P. vulnus,** OR.

Nematode, root knot. **Meloidogyne** spp., FL.

Nonparasitic

Bud Blast. Insufficient light; also causes drop of lower leaves.

Chlorosis. Iron deficiency. Sometimes result of systemic insecticide.

Limber Neck. Physiological.

Scorch, of Easter lilies. Acid soil, low fertility, ?excessive phosphorus.

Rot, basal. **Fusarium oxysporum** f. sp. **lilii,** general.

Rot, black scale. **Colletotrichum lilii,** LA, MS; brown scale, **Colletotrichum** sp., OR, WA; scale tip, **Cylindrocarpon radicicola,** secondary.

Rot, bulb. **Penicillium** spp., blue mold; **Rhizopus** sp., soft rot, cosmopolitan.

Rot, charcoal. **Macrophomina phaseoli,** CA.

Rot, root. **Pythium** spp.; **Rhizoctonia solani,** widespread.

Rot, stem, foot. **Phytophthora cactorum,** MD, MN, NJ, NC, NY, OH, WA, WI; **P. parasitica,** Top Rot, IN, MD, NJ, NY; **Rhizoctonia tuliparum,** WA; **Sclerotinia sclerotiorum,** CT, FL, TX, WA.

Rust. Puccinia sporoboli (0, I), NE, ND; II, III, on *Sporobolus*; **Uromyces holwayi** (0, I, II, III), ME to NJ, NE, MI, MN, CA, ID, OR, WA.

Virus. Curl-Stripe Disease, OR.

Virus. Lily Fleck; Lily Latent Mosaic; Lily Ring Spot; Lily Rosette; Lily Symptomless; Cucumber Mosaic; Tulip Breaking Mosaic; Streak, on Easter lily, virus from wild cucumber.

Garden lilies are particularly subject to botrytis blight and mosaic. Madonna lilies are most susceptible, with leaves often completely blackened in wet weather. Copper sprays are perhaps most effective. The only sure way to be free from mosaic and other virus diseases is to grow lilies from seed in an isolated portion of the garden.

LILY-OF-THE-VALLEY
(*Convallaria*)

Blight, gray mold; rhizome rot. **Botrytis paeoniae,** IL, ME, PA.
Blight, southern. **Sclerotium rolfsii,** ME.
Leaf blotch. **Ascochyta majalis** (*Mycosphaerella convallaris*), PA.
Leaf spot. **Gloeosporium convallariae,** NY; **Kabatiella microsticta,** ?secondary, MD; **Phyllosticta** sp., NJ, NY.
Nematode, meadow. **Pratylenchus pratensis,** associated with forcing failures.
Nematode, root knot. **Meloidogyne** sp., occasional.

LINARIA
(Blue Toadflax; Butter and Eggs)

Anthracnose. **Colletotrichum vermicularioides,** MA, NJ, NY, TX, WI.
Bacterial, MLO. **Aster Yellows,** CA.
Blight, gray mold. **Botrytis cinerea,** AK.
Blight, southern. **Sclerotium rolfsii,** TX.
Downy mildew. **Peronospora linariae,** FL, MA, OK, WI.
Leaf spot. **Alternaria** sp., MI; **Septoria linariae,** WI.
Nematode, root knot. **Meloidogyne** spp.
Nematode, stem and leaf. **Ditylenchus dipsaci,** NY.
Powdery mildew. **Erysiphe cichoracearum,** CA.
Rot, root. **Phymatotrichum omnivorum,** TX; **Rhizoctonia solani,** IL; **Thielaviopsis basicola,** CT.
Rot, stem. **Sclerotinia sclerotiorum,** AZ.
Rust. **Puccinia antirrhini** (II, III), CA; **Aecidium** sp., WI.
Smut, white. **Entyloma linariae,** CT, NJ, PA.

LINDEN, BASSWOOD
(*Tilia*)

Anthracnose; leaf spot. **Gnomonia tiliae** (*Gloeosporium tiliae*), CT to VA, IA, MN.
Blight, leaf. **Cercospora microsora,** general.
Canker, bark. **Aleurodiscus acerinus; A. griseo-canus,** PA, IA, MO.
Canker; dieback. **Botryosphaeria** sp., MD; **Nectria** spp., NY, PA, VA; **Strumella** sp., NJ.

Canker, felt fungus. **Septobasidium fumigatum,** FL.
Leaf spot. **Phlyctaena tiliae,** NY; **Phyllosticta praetervisa,** WI; **Sphaeropsis** sp., OK.
Mistletoe. **Phoradendron serotinum (flavescens),** South.
Powdery mildew. **Microsphaera alni,** MN; **Phyllactinia corylea,** MN; **Uncinula clintonii,** general.
Rot, heart. **Daedalea confragosa,** VT; **Fomes** spp.; **Steccherinum septentrionale,** AL, MI; **Pholiota adiposa,** MA, PA, TN.
Rot, root. **Phymatotrichum omnivorum,** TX; **Ustulina vulgaris,** NY.
Rot, sapwood. **Collybia velutipes,** occasional; **Pleurotus ostreatus,** cosmopolitan.
Rot, wood. **Daldinia concentrica,** MN, NY; **Lenzites betulina,** NY, VT; **Schizophyllum commune,** MN; **Polyporus** spp.; **Stereum** spp.; **Trametes mollis,** VT.
Sooty mold. **Fumago vagans.**
Spot anthracnose. **Elsinoë tiliae,** VA.
Wilt. **Verticillium albo-atrum,** IL.

Anthracnose and Cercospora leaf blight are common diseases.

LINNAEA
(Twin-Flower)

Black mildew. **Halbaniella linnaeae,** NY.
Leaf spot. **Phyllachora wittrockii,** tar spot, MI, MT, NM, NY; **Venturia dickei,** ID, MI, MT, NM, NY, OR, WA, WI. **Septoria breviuscula,** NY.

LIONS-EAR
(*Leonotis*)

Leaf spot. **Cercospora leonotidis,** LA; **Septoria breviuscula,** NY.
Rust. **Puccinia leonotidis** (0, I, II, III), FL.

LIPPIA
(Fog-Fruit, Lemon-Verbena)

Bacterial crown gall. **Agrobacterium tumefaciens,** AZ.
Black mildew. **Meliola lippiae,** FL, AZ.
Blight, southern. **Sclerotium rolfsii,** AZ, CA.
Leaf spot. **Cercospora lippiae,** widespread; **Cylindrosporium lippiae,** TX.
Nematode, root knot. **Meloidogyne** sp., AZ.
Spot anthracnose. **Sphaceloma lippiae,** IN, FL.

LIPSTICK VINE
(*Aeschynanthus*)

Leaf spot. **Corynespora casiicola**, FL; **Myrothecium roridum**, FL.

LITHOCARPUS
(Tanbark Oak)

Blight, leaf. **Pestalotia castagnei**, CA.
Leaf spot. **Ceuthocarpum conflictum**, CA.
Powdery mildew. **Erysiphe trina**, CA.
Rot, wood. **Poria spp.; Stereum hirsutum.**
Rust. **Cronartium quercuum** (II, III), CA; 0, I, on pine.

LITHOPHRAGMA
(Woodland-Star)

Rust. **Puccinia heucherae** (III), CA, UT, WA.
Smut, leaf and stem. **Urocystis lithophragmae**, UT.

LITHOSPERMUM
(Gromwell, Puccoon)

Leaf spot. **Septoria lithospermi**, WI.
Powdery mildew. **Erysiphe cichoracearum**, PA, TX.
Rust. **Aecidium hesleri**, TN; **Puccinia recondita** (0, I), NE, ND, SD, TX.

LITSEA
(Pond-Spice)

Leaf spot. **Cercosphora olivacea**, GA.

LOBELIA
(Cardinal-Flower, Blue Lobelia)

Blight, gray mold. **Botrytis cinerea.**
Damping-off. Pythium debaryanum, MA, NY.
Leaf spot. **Cercospora lobeliae**, AL, KS, IL, IN, MD, TX; **C. effusa**, IA, TX;
    Phyllosticta bridgesii, IN; **Septoria lobeliae**, ME to VA, TX, WI.
Nematode, root knot. **Meloidogyne** spp., NE, NY, FL, MD.
Rot, root. **Phymatotrichum omnivorum**, TX; **Rhizoctonia solani**, NY, OH.

Rot, stem. **Sclerotium rolfsii**, NJ.
Rust. **Puccinia lobeliae** (III), AR, MI, WI, NY to NC, TX, WA.
Smut, leaf. **Entyloma lobeliae**, ME to PA, MO, WI.
Virus. **Beet Curly Top**, TX; **Tomato Spotted Wilt**, TX.

LOCUST
(*Robinia*)

Bacterial, MLO. **Witches-Broom**, VA.
Blight, seedling; leaf. **Alternaria** sp., NC to AL, MO; **Fusicladium robiniae**, MD to AL, MO, WI.
Canker. **Nectria cinnabarina**, VA.
Canker; twig blight. **Aglaospora anomala**, ME to GA; **Fusarium sarcochroum**, IA; **Diaporthe oncostoma**, NY to GA, IL.
Damping-off. **Rhizoctonia solani**, ME to AL, NE, TX; **Pythium spp.**, NE, TX.
Dodder. **Cuscuta** sp.; **C. arvensis**, widespread.
Leaf spot. **Cladosporium epiphyllum**, TN, VA, WV; **Cylindrosporium solitarium**, TX; **Gloeosporium revolutum**, NJ; **Phleospora robiniae**, NY to OH; **Phyllosticta robiniae**, LA.
Mistletoe. **Phoradendron serotinum (flavescens)**, NC, AZ, NM, TX; **Viscum album**, CA.
Nematode, lesion. **Pratylenchus** sp., OR.
Nematode, root knot. **Meloidogyne** spp., OK, TX.
Nonparasitic. **Chlorosis**, iron deficiency, NE, TX. **Little Leaf**, zinc deficiency, CA.
Powdery mildew. **Erysiphe polygoni**, CA; **Microsphaera diffusa**, IL, NC; **Phyllactinia corylea**, NM.
Rot, heart. **Fomes** spp.; **Polyporus** spp.
Rot, heart. **Fomes rimosus**, VA.
Rot, root. **Armillaria mellea; Phymatotrichum omnivorum**, OK, NM; **Pythium myriotylum**, NC; **Fusarium** sp., AL, GA.
Rot, seedling stem. **Phytophthora cinnamomi**, MD; **Rhizoctonia bataticola**, AL, NC; **Sclerotium bataticola**, TX.
Rot, wood. **Daedalea unicolor**, WI; **Poria** spp.
Virus. **Locust Witches' Broom**, Robinia Brooming Disease, PA to GA and AR, OH.
Wilt, seedling. **Phytophthora parasitica**, AL, NC, VA; **Verticillium alboatrum**, IL.

LOGANBERRY

See Blackberry.

LOMATIUM
(Biscuit-Root)

Downy mildew. **Plasmopara nivea,** MT.
Leaf spot. **?Phyllachora** sp., WA.
Rust. **Puccinia asperior** (0, I, III), CA, OR, WA; **P. jonesii** (0, I, III), KS, NE, UT, WA; **P. ligustici,** ID, WA.

LOOSESTRIFE, FRINGED
(*Steironema*)

Leaf spot. **Cylindrosporium steironematis,** NY; **Mycosphaerella** sp., NY; **Phyllosticta lysimachiae,** NY; **Ramularia lysimachiae,** NC, WI; **Septoria conspicua,** ME to MS, CO and WI.
Rot, root. **Phymatotrichum omnivorum,** TX.
Rust. **Puccinia dayi,** NY to WV, IL, MI, MT, WI; **P. distichlidis** (0, I), CO, ND; II, III, on marsh grass; **Uromyces acuminatus** (0, I), CT to CO, SD, ND.

LOOSESTRIFE, MONEYWORT
(*Lysimachia*)

Blight, leaf and stem necrosis. **Ceratobasidium anceps,** WI.
Gall, leaf. **Synchytrium aureum,** WI.
Leaf spot. **Cercospora lysimachiae,** NJ; **Cladosporium lysimachiae,** MA; **Ramularia lysimachiae,** WI; **Septoria conspicua,** IA, NY, VT.
Nematode, root knot. **Meloidogyne** sp., TX.
Nematode, stem. **Ditylenchus dipsaci.**
Rot, crown stem. **Sclerotium rolfsii,** KS; **Phymatotrichum omnivorum,** TX.
Rust. **Coleosporium campanulae** (II, III), TN; 0, I, on red pine; **Puccinia limosae** (0, I), MA to NC and MI, WI, NE.

LOQUAT
(*Eriobotrya*)

Anthracnose; flower blight; withertip. **Colletotrichum gloeosporioides,** CA, FL, TX.
Bacterial fire blight. **Erwinia amylovora,** Gulf states, AZ, CA.
Blight, leaf; blotch. **Fabraea maculata,** FL.
Leaf spot. **Phyllosticta eriobotryae,** FL; **Pestalotia** sp., secondary; **Septoria eriobotryae,** FL.
Leaf spot, algal. **Cephaleuros virescens,** FL.
Nematode, burrowing. **Radopholus similis,** FL.

Rot, collar, crown. Phytophthora cactorum, CA.
Rot, root. **Armillaria mellea**, CA; **Clitocybe tabescens**, FL; **Phymatotrichum omnivorum**, TX.
Scab. **Fusicladium eriobotryae**, on leaves, stems, fruit, Gulf states; **Spilocaea eriobotryae** (*Fusicladium photinicola*).

LOTUS
(*Nelumbo*)

Blight. **Dothiorella nelumbonis**, on flower parts, DE, MD.
Leaf spot. **Alternaria nelumbii**, MD, NJ, NY, OK, PA, TX; **Cercospora nelumbonis**, IN,TX.

LUCUMA
(Canistel, Egg-Fruit)

Anthracnose; fruit spot. **Colletotrichum gloeosporioides**, FL.
Canker. **Physalospora obtusa**, SC.
Rust. **Achrotelium lucumae** (II, III), FL.

LUDWIGIA
(False Loosestrife)

Leaf spot. **Cercospora ludwigiae**, AL; **Pezizella oenotherae**, VA; **Phyllosticta ludwigiae**, NY, WI; **Septoria ludwigiae**, MD, MS, PA.
Rust. **Puccinia jussiaeceae** (0, I, III), DE to FL, MS, LA, TX, CA, OH, WI.

LUPINE
(*Lupinus*)

Anthracnose. **Colletotrichum trifolii**, NC; **C. fragariae**, NC.
Blight, gray mold. **Botrytis cinerea**, MA, MT, NY.
Blight, leaf. **Hadrotrichum globiferum**, CA, OR, CO, WA, WY.
Blight, seedling. **Pleiochaeta setosa, Alternaria** sp., **Aspergillus flavus, Aspergillus niger, Curvularia** sp., **Rhizopus stolonifer.**
Blight, southern. **Sclerotium rolfsii**, probably general.
Blight, stem necrosis. **Ascochyta** sp., WI, CT.
Damping-off. **Rhizoctonia solani**, CT, TX.
Downy mildew. **Peronospora trifoliorum**, WI.
Leaf spot. **Alternaria** sp., MA; **Cercospora longispora**, FL, MO, NY, WI; **C. lupini**, FL, OR, SC; **C. lupinicola**, TX; **Corynespora cassicola**, TX;

Cylindrosporium lupini, CA; **Mycosphaerella pinodes,** WI; **Ovularia lupinicola,** WA; **Phoma lupini,** also stem spot, CO to NM, CA; **Phyllosticta ferax,** CA to WA, WY, SD to CO; **P. lupini,** CA; **Ramularia lupini,** TX; **Septogloeum lupini,** MI, CA; **Septoria lupinicola,** WI; **Stictochlorella lupini,** CA, WA.

Nematode, leaf. **Aphelenchoides ritzema-bosi.**

Nematode, lesion. **Pratylenchus pratensis,** CA.

Nematode, root knot. **Meloidogyne** sp., FL.

Nematode, sting. **Belonolaimus gracilis,** GA; ring, **Criconema** sp., CA.

Powdery mildew. **Microsphaera** sp.; **Erysiphe polygoni,** widespread.

Rot, root. **Armillaria mellea,** CA; **Fusarium** sp., NJ, VA; **Phymatotrichum omnivorum,** TX; **Pythium ultimum,** CA; **Thielaviopsis basicola,** IA, WI.

Rot, stem. **Macrophomina phaseoli,** charcoal rot; **Sclerotinia sclerotiorum,** LA, NY, TX; **Pythium debaryanum,** TX, CA.

Rust. **Puccinia andropogonis** var. **onobrychidis** (0, I), MI, MN, NY, WI; II, III, on *Andropogon;* **Uromyces lupini** (0, I, II, III), CA, OR, WA, NE, MT; **U. occidentalis** (II, III), on native lupine, MT to NM, CA, WA.

Smut. **Thecaphora deformans,** on seed, CO, WY.

Virus. **Bean Yellow Mosaic,** FL; **Bidens Mottle,** FL; **Ring Spot.** Unidentified; **Tomato Spotted Wilt,** TX; **Peanut Mottle,** GA.

These diseases are of lupines grown as ornamentals. Lupines as ground covers and soil preservers have their own troubles. In gardens powdery mildew is prevalent, and leaf spots are not often important.

LYCHEE

Rot, mushroom rot. **Clitocybe tabescens,** sometimes fatal in FL.

LYCHNIS
(Campion)

Blight, shoot and flower. **Botrytis cinerea,** AK.

Blight, southern. **Sclerotium rolfsii,** TX.

Leaf spot. **Alternaria dianthi,** AK; **Leptothyrium lychnidis,** AL; **Phyllosticta lychnidis,** IA, TX; **Septoria ?lychnidis,** MA.

Mold, leaf. **Heterosporium** sp., AK.

Rot, root. **Phymatotrichum omnivorum,** TX; **Rhizoctonia solani,** IL; **Corticium galactinum,** MD.

Rot, stem. **Phytophthora cactorum,** IN.

Rust. **Puccinia arenariae** (III), PA; **Uromyces suksdorfii** (I, II, III), UT; **U. verruculosus,** IN, MI, NY, TX.

Smut, anther. **Ustilago violacea,** MN, WI, WY.

LYCIUM
(Desert-Thorn, Christmasberry)

Leaf spot. **Cercospora lycii**, OK.
Rot, root. **Phymatotrichum omnivorum**, TX.
Rust. **Aecidium lycii** (0, I), AZ; **Puccinia globosipes** (II, III), UT to NM,
CA; 0, I unknown; **P. tumidipes** (II, III), TX to AZ, UT.

LYCORIS

Leaf scorch; red spot. **Stagonospora curtisii**, CA.
Nematode, bulb scale rot. **Ditylenchus dipsaci**, NC, VA; **Aphelenchoides
fragariae**, FL, GA, ND, SD.
Nematode, root. **Hoplolaimus** sp., NC.

LYONIA
(Maleberry, Fetterbush, Staggerbush)

Black mildew. **Asterina lepidigena**, FL; **Lembosia andromedae**, AL; **Moreno-
ella dothideoides**, FL.
Blight, leaf. **Ramularia cylindriopsis**, ME.
Dodder. **Cuscuta compacta**, FL.
Gall, leaf; shoot hypertrophy. **Exobasidium vaccinii**, MA to FL, AL.
Leaf spot. **Ceuthocarpon ferrugineum**, FL; **Pestalotia vaccinii**, ?secondary;
Septoria pulchella, GA.
Leaf spot, tar. **Rhytisma andromedae**, widespread; **R. decolorans**, MS.
Powdery mildew. **Microsphaera alni** var. **vaccinii**, AL.
Rot, wood. **Poria versispora**, ME.
Rust. **Pucciniastrum myrtilli** (II, III), DE to AL, AR; 0, I, on hemlock.

LYSILOMA

Rot, white pocket heart. **Fomes extensus**, FL.
Rust. **Ravenelia annulata** (II, III), FL; **R. lysilomae**, FL.

LYTHRUM
(Winged, Purple Loosestrife)

Gall, leaf. **Synchytrium lythrii**, LA.
Leaf spot. **Cercospora lythri**, WI; **C. lythracearum**, MS; **Pezizella oenotherae**,
NY, MI; **Septoria lythrina**, KS, NY, WI.
Rot, root. **Rhizoctonia solani**, IL.

MAACKIA

Rot, root. **Phymatotrichum omnivorum**, TX.

MACADAMIA

Canker. **Phytophthora cinnamomi**, CA.

MADRONE
(*Arbutus menziesii*)

Canker. **Hendersonula toruloidea**, WA.
Canker, trunk. **Phytophthora cactorum**, CA, WA.
Gall, leaf; red leaf spot. **Exobasidium vaccinii**, CA, OR, WA, TX.
Leaf spot. **Ascochyta hanseni**, CA; **Cryptostictis arbuti**, CA, OR; **Didymosporium arbuticola**, OR; **Mycosphaerella arbuticola**, CA, OR, WA; **Phyllosticta fimbriata**, OR.
Leaf spot, tar. **Rhytisma arbuti**, CO, OR, TX, WA.
Rot, heart. **Fomes subroseus**, OR.
Rot, root. **Phymatotrichum omnivorum**, TX.
Rot, wood. **Lenzites saepiaria**, OR, TX; **Polyporus** spp., OR, TX; **Trametes sepium**, CA.
Rust. **Puccinastrum sparsum** (II, III), CA, OR, WA.
Spot anthracnose. **Elsinoë mattirolanum**, CA.

MAGNOLIA

Bacterial leaf spot. **Pseudomonas syringae**, IL; **P. cichorii**, AL.
Black mildew. **Dimerosporium magnoliae**, TX; **Irene araliae**, MS; **Meliola amphitricha**, Gulf states; **M. magnoliae; Trichodothis comata**, South.
Blight, seedling. **Rhizoctonia solani**, NJ.
Blight, thread. **Pellicularia solani**, NJ.
Canker, felt fungus. **Septobasidium langoisii, S. tenui, S. leprieurii**, Gulf states.
Canker; twig blight. **Nectria** sp., WV; **N. magnoliae**, NJ, TN, WV; **Tubercularia** sp., MI.
Leaf spot. **Alternaria tenuis**, TX; **Cladosporium fasciculatum**, GA to TX; **Cercospora magnoliae** (*Mycosphaerella milleri*), NC, SC, FL, NJ, WV; **Cristulariella pyramidalis; Colletotrichum** sp., FL, GA, SC; **Coniothyrium** sp., TX; **C. olivaceum**, TX; **Epicoccum nigrum**, TX; **Exophoma magnoliae**, FL, TX; **Glomerella cingulata**, AL, FL, GA, LA, MS, SC, TX; **Hendersonia magnoliae**, VA; **Heterosporium magnoliae**, FL, NC, TX; **Micropeltis alabamensis**, AL; **Phyllosticta cookei**, NY to Gulf states, CA; **P. magnoliae**, large leaf spot, NY to Gulf states, CA; **Septoria magnoliae**, SC to FL and TX.

Leaf spot; algal; green scurf; "red rust." **Cephaleuros virescens,** general.
Nematode, burrowing. **Radopholus similis,** FL.
Nematode, citrus. **Tylenchulus semipenetrans,** FL.
Nematode, lance. **Hoplolaimus uniformis,** RI.
Nematode, lesion. **Pratylenchus pratensis.**
Nematode, root knot. **Meloidogyne** sp.
Nematode, sting. **Belonolaimus longicaudatus,** FL.
Parasitic lichen. **Strigula elegans** and **S. camplanata,** Southern U.S., LA.
Rot, flower. **Cibornia** (*Sclerotinia*) **gracilipes,** MD.
Rot, heart. **Fomes fasciatus; F. geotropus.**
Rot, root. **Phymatotrichum omnivorum,** TX; **Clitocybe tabescens,** GA.
Rot, wood. **Daldinia concentrica,** cosmopolitan; **Polyporus** spp.; **Poria** spp.,
Stereum spp.
Spot anthracnose. **Elsinoë magnoliae,** FL, GA, LA, MS.
Wilt. **Verticillium albo-atrum,** IN, CA.

The many leaf spots need not be alarming; they seldom cause premature defoliation. In the Gulf states lichens appear as small round gray spots on leaves, and the parasitic alga *Cephaleuros* often forms a velvety coating with hairlike outgrowths.

MAHOGANY
(*Swietenia*)

Blight, twig, stem. **Phomopsis** sp., OR.
Leaf spot. **Pestalotia swieteniae,** FL; **Phyllachora swieteniae,** FL.

MAHONIA
(Oregon-Grape)

Canker; leaf blotch. **Leptosphaeria berberidis,** ID, MD.
Leaf spot. **Cercospora** sp., LA; **?Gloeosporium berberidis,** WA; **Phomopsis**
sp., ?secondary; **Phyllosticta** spp., CT, AL, WA; **Cylindrocladium** sp., FL, GA.
Nematode, root knot. **Meloidogyne** sp.
Nonparasitic. **Scald,** in eastern states winter injury to foliage is severe.
Parasitic lichen. **Strigula elegans** and **S. camplanata,** Southern U.S., LA.
Rot, root. **Phymatotrichum omnivorum,** TX.
Rust. **Cumminsiella mirabilissima** (0, I, II, III), common Western Great Plains
to Pacific Coast; **C. texana,** TX; **C. wootoniana** (II, III), AZ, NM; **Puccinia
graminis** (0, I), CA, MI, ND; II, III, on cereals and grasses; **P. koeleriae**
(0, I), CO, ID, MT, OR; II, III on *Koeleria*; **P. oxalidis** (0, I), NM; II, III on
Oxalis.

Rust due to *Cumminsiella* is usually inconspicuous, but in a wet season there is a general blighted effect. *Mahonia*, like barberry, is under wheat-rust quarantine. *Mahonia repens* is immune.

MAIANTHEMUM

Blight. **Botrytis** sp., NY.
Leaf spot. **Ramularia rubicunda,** MI, NY, OH, PA, WI, CA to AK; **Sphaeropsis cruenta,** AK, WA.
Rust. **Puccinia sessilis** (0, I), MA, MI, MN, NY, PA, WI; II, III, on *Phalaris*; **Uromyces acuminatus** var. **magnatus** (0, I), MI, NY, WI; II, III, on marsh grass.

MALACHRA

Leaf spot. **Cercospora malachrae,** TX.

MALACOTHRIX

Gall, leaf. **Synchytrium innominatum,** CA.
Rust. **Puccinia harknessii** (III), CA; **P. hieracii** (0, I, II, III), CA.

MALLOTUS

Leaf spot. **Cercospora malloti,** MS.

MALLOW, GARDEN
(*Malva*)

Bacterial, MLO. **California Aster Yellows,** CA; **Beet Curly Top,** CA.
Leaf spot. **Alternaria** sp.
Rot, root. **Macrophomina phaseoli,** CA on *Malva parviflora*.
Rot, root. **Phymatotrichum omnivorum,** TX.
Rust. **Puccinia heterospora** (III), KS, TX; **P. malvacearum,** CA, CO, OR, WV.

MALVASTRUM
(False-Mallow, Bush-Mallow)

Rot, root. **Phymatotrichum omnivorum,** TX.
Rust. **Puccinia interveniens** (0, I), CA; III on *Stipa*; ?**P. malvacearum; P. schedonnardi** (0, I), MT, ND to NM; II, III, on wild grasses; **P. sherardiana** (0, III), MT, ND to TX, NM; **P. heterospora** (III), TX.

MALVAVISCUS

Blight, twig. **Sclerotinia sclerotiorum,** TX.
Leaf spot. **Phyllosticta malvavisci,** TX.
Rot, root. **Clitocybe tabescens,** FL; **Helicobasidium purpureum,** TX; **Phymatotrichum, omnivorum,** TX.
Rust. **Kuehneola malvicola** (II, III), TX; **Puccinia heterospora** (III), TX.

MANFREDA
(Spice-Lily, Wild Tuberose)

Leaf spot. **Cercospora amaryllidis,** TX; **Phyllosticta hymenocallidis,** TX.
Rust. **Aecidium modestum,** TX.

MANGO
(*Mangifera*)

Anthracnose; flower and twig blight; fruit rot. **Glomerella cingulata,** cosmopolitan.
Blight, twig. **Phomopsis** sp., FL; **Physalospora** (*Diplodia*) spp.
Canker, brown felt. **Septobasidium pilosum** and **S. pseudopedicellatum,** FL.
Leaf spot. **Phyllosticta mortoni,** FL, TX; **Septoria** sp., FL.
Leaf spot, algal, green scurf. **Cephaleuros virescens,** general.
Nonparasitic. **Little Leaf,** zinc deficiency, FL. **Soft Nose,** on Indian varieties.
Powdery mildew. **Oidium mangiferae,** CA, FL.
Rot, root. **Phymatotrichum omnivorum,** TX.
Rot, stem-end fruit. **Diplodia** sp.
Sooty mold. **Capnodium** spp., general.
Spot anthracnose; mango scab. **Elsinoë mangiferae,** FL.
Wilt. **Verticillium albo-atrum,** FL.

MANGROVE
(*Rhizophora*)

Leaf spot. **Cercospora rhizophorae,** FL.

MANIHOT
(Cassava, Manioc)

Anthracnose; withertip. **Gloeosporium ?manihotis,** FL, LA, TX.
Dieback. **Physalospora abdita** and **P. rhodina,** FL.
Leaf spot. **Cercospora henningsii,** FL, TX.

Nematode, root knot. **Meloidogyne** sp., AL, FL.
Rot, root. **Phymatotrichum omnivorum**, TX; **Rhizoctonia solani**, FL.

MANZANITA
(*Arctostaphylos*)

Black mildew. **Meliola** sp., CA.
Gall, red leaf spot. **Exobasidium vaccinii**, widespread; shoot gall, **E. vaccinii-uliginosi**, CA, OR.
Leaf spot. **Cryptostictis arbuti**, OR; **Phyllosticta amicta**, CA, OR.
Mistletoe. **Phoradendron villosum**, CA, OR.
Rot, root. **Phymatotrichum omnivorum**, TX.
Rot, wood. **Fomes annosus**, CA; **F. arctostaphyli**, and **F. igniarius**, general; **Poria ferruginosa.**
Rust. **Pucciniastrum sparsum** (II, III), CA, OR.

MAPLE
(*Acer*)

Anthracnose; leaf blight. **Gloeosporium apocryptum,** large blotches, general; **G. acerinum**, small spots; **G. saccharinum; G. aceris.**
Bacterial crown gall. **Agrobacterium tumefaciens**, MI, OK, TX, VA.
Bacterial leaf spot. **Pseudomonas aceris**, on Japanese maple, CA; **P. acernea.**
Blight, inflorescence. **Ciborinia acerina**, MA, NY.
Blight, leaf. **Didymosporina aceris**, CT, PA.
Blight, seedling; smother. **Thelephora albido-brunnea**, NC, VA.
Blight, twig. **Macrophoma** sp., AR, NY.
Canker, bark. **Aleurodiscus acerinus**, MA, VT; **Hymenochaete agglutinans,** CT, MT, MI.
Canker, basal, crown. **Phytophthora cinnamomi**, NJ; **P. cambivora**, NJ.
Canker, bleeding. **Phytophthora cactorum**, CT, MA, NJ, NY, RI,VA.
Canker; dieback. **Coniothyrium negundinis**, IL, OK; **Cytospora** spp., MI; **Eutypella parasitica**, MI, MN, NH, NY, VT, WI; **Hypoxylon morsei**, MI, MN; **Fusarium** spp.; **Nectria cinnabarina**, cosmopolitan, often secondary; **N. coccinea; N. galligena**, widespread trunk canker; **Phomopsis** sp., OH; **P. acerina**, NY; **Physalospora** spp., ?secondary; **Sphaeropsis albescens**, IL, KS, IA, NY, ND, SD, WI; **Strumella coryneoidea**, MI, PA; **Schizoxylon microsporum**, Lake states; **Steganosporium acerinum**, NJ.
Canker, felt fungus. **Septobasidium** spp., AL, FL, KY, LA, NC, SC, TN, VA.
Damping-off. **Rhizoctonia solani**, cosmopolitan.
Dodder. **Cuscuta gronovii**, on seedlings, NY.
Leaf blister. **Taphrina carveri**, AL, KY, MI, MO; **T. bartholomaei**, UT; **T. darkeri**, OR; **T. dearnessii**, black, GA, MI, MN, NC, NY, OK, PA, VA;

T. sacchari, on sugar maple, AR, GA, IN, KS, ME, MI, MO, NH, NY, OH, PA, TN, WV, WI.

Leaf spot. **Actinopelte dryina,** IL; **Alternaria** sp., ?secondary; **Cercospora negundinis,** KS, NE, WI; **C. saccharini,** MA; **Cercosporella aceris,** WA; **Cristulariella depraedens,** CT, NY; **C. pyramidalis,** FL; **Monochaetia desmazierii,** GA, NC, OK, TN; **Illosporium maculicola,** WA; **Laestadia brunnea,** NC, SC; **Leptothryium acerinum,** OK; **Marssonina truncatula,** OR; **Pezizella oenotherae,** NC, NY, VA; **Phyllosticta minima,** eyespot, general; **P. negundinis,** ME to AL, TX, WI; **Piggotia negundinis,** leaf blotch, WI, WY; **Stilbella acerina,** PA; **Septoria aceris,** general; **Venturia acerina,** NY, PA, VA, WV, WI.

Leaf spot, tar. **Rhytisma acerinum,** general; **R. punctatum,** speckled, general.

Mistletoe. **Phoradendron serotinum (flavescens),** common from NJ to FL, TX and MO.

Mistletoe. **Viscum album** on big leaf maple (*A. macrophyllum*) and silver maple (*A. saccharinum*), CA.

Nematode. **Criconemoides** sp.; **Hemicycliophora** sp.; **Pratylenchus thornei; Tylenchorhynchus** sp.; **Xiphinema** sp.

Nematode, root knot. **Meloidogyne** sp., OR; **M. ovalis,** WI.

Nonparasitic. **Decline,** frequently roadside salt injury. **Frost Crack,** gas injury. **Leaf Scorch.** Common on street and lawn trees, often associated with high temperature after a moist spring.

Parasitic lichen. **Strigula elegans** and **S. camplanata,** southern U.S., LA.

Powdery mildew. **Phyllactinia corylea,** IA, NC, CA, OR, VT, SD; **Uncinula circinata,** ME to AL, TX, MO, MI; **Microsphaera alni.**

Rot, heart, sapwood, wound. **Collybia velutipes,** CT, MA; **Daedalea** spp., northeastern and north central states, southward; **Daldinia concentrica,** cosmopolitan; **Fomes applanatus; F. connatus; F. igniarius; Ganoderma lucidum,** fatal to some street trees, NY and NJ; **Hericeum erinaceous,** VT to MD, MI, MN; **Lenzites** spp.; **Pholiota adiposa; Pleurotus** spp.; **Polyporus** spp.; **Poria** spp.; **Steccherhinum septentrionale,** general; **Stereum** spp.; **Schizophyllum commune,** cosmopolitan; **Ustuline vulgaris,** northeastern and north central states; **Valsa leucosomoides,** on tapped sugar maples.

Rot, root. **Armillaria mellea,** general; **Clitocybe tabescens,** GA, MO; **Helicobasidium purpureum,** TX; **Phymatotrichum omnivorum,** TX.

Rot, sapstreak. **Endoconidiophora virescens; Ceratocystis coerulescens.**

Rot, seedling, charcoal. **Macrophomina phaseoli,** IL.

Virus. **Peach Rosette; Tobacco Necrotic Ring Spot.**

Wilt. **Verticillium albo-atrum,** widespread in cultivated trees, especially Norway maple.

Verticillium wilt is the most destructive maple disease and is particularly prevalent in street trees. The wilting may be confined to a single branch, which can be cut out, or may kill the whole tree. In removing dead trees, get all of the root system and replace with a different kind of tree. Leaf scorch is common on sugar maple in hot, windy weather; anthracnose may be conspicuous in wet weather.

MARANTA
(Calathea)

Leaf spot. **Glomerella cincta,** NJ; **Phyllosticta** sp., NJ; **Drechslera setariae,** FL.
Rust. **Puccinia cannae** (II, III), FL.

MARGUERITE
(Chrysanthemum frutescens)

Bacterial crown gall. **Agrobacterium tumefaciens,** IA, MD, NJ, NY, VA.
Bacterial, MLO. **Aster Yellows,** KS, NJ, NY, and **California Aster Yellows,** CA.
Nematode, root knot. **Meloidogyne** sp.
Powdery mildew. **Erysiphe cichoracearum,** NJ.
Virus. **Beet Curly Top,** CA.
Wilt. **Verticillium albo-atrum,** NJ.

MARIGOLD
(Tagetes)

Bacterial leaf spot. **Pseudomonas syringae** pv. **tagetis,** NC.
Bacterial, MLO. **Aster Yellows,** CT, NJ, NY, PA, WI; **California Aster Yellows,** CA.
Bacterial wilt. **Pseudomonas solanacearum,** FL.
Blight. **Alternaria tagetica,** SC.
Blight, head. **Botrytis cinerea,** CT, NJ, PA, AK; **Helminthosporium** sp., TX.
Blight, southern. **Sclerotium rolfsii,** FL, NJ, VA.
Leaf spot. **Alternaria tagetica,** SC, NJ; **Cercospora** sp., CT, **C. tageticola,** FL; **Septoria tageticola,** FL.
Nematode. **Aphelenchoides tagetae,** MD; **Paratylenchus micoletzkyi,** MD.
Nematode, root knot. **Meloidogyne hapla,** VA.
Nonparasitic. Air pollution, NO_2, SO_3, O_3, NC.
Rot, charcoal. **Macrophomina phaseoli,** OK.
Rot, root. **Pythium ultimum,** CA; **Rhizoctonia solani,** TX.
Rot, stem; wilt. **Sclerotinia sclerotiorum,** NY; **Phytophthora cryptogaea,** NY; **Fusarium** sp., CA, NJ, NY.
Rust. **Coleosporium madiae** (II, III), CA; 0, I, on pine; **Puccinia tageticola** (II, III), TX.
Virus. **Cucumber Mosaic,** FL.
Wilt. **Fusarium oxysporum** f. sp. **callistephi,** CA; **Verticillium albo-atrum,** NY; **V. dahliae,** AZ.

Marigolds are easy to grow without paying too much attention to disease. Cut fading flower heads off into a paper bag before the gray mold of botrytis blight gets started.

MARIPOSA-LILY, GLOBE-TULIP
(*Calochortus*)

Rust. **Puccinia calochorti** (0, I, III), CA, OR, WA to NE and NM.

MARSH-MARIGOLD
(*Caltha*)

Gall, leaf. **Synchytrium aureum,** WI.
Leaf spot. **Cercospora calthae,** WI; **Cylindrosporium** sp., NY; **Fabraea rousseauana,** CA, WI; **Ramularia calthae,** NY, WI.
Powdery mildew. **Erysiphe polygoni,** MI, OH, WI.
Rust. **Puccinia areolata** (0, I, II, III), CA, CO, WA, AK; **P. calthae** (0, I, II, III), NY, NJ to IA and ND; **P. calthicola** (0, I, II, III), NY to IA and MN; **P. gemella** (III), CA, ID, MT, OR, WA, AK; **P. treleasiana** (III), CO, NV, UT, WA, WY.

MATELEA

Parasitic lichen. **Strigula elegans; S. complanata,** Southern U.S., LA.

MATRICARIA
(False Chamomile)

Bacterial, MLO. **California Aster Yellows,** CA.
Nematode, root knot. **Meloidogyne** sp., CA.
Powdery mildew. **Erysiphe cichoracearum,** WA; **Sphaerotheca macularis,** WA.
White rust. **Albugo tragopogonis,** CA, ND, OR.

MATRIMONY-VINE
(*Lycium halimifolium*)

Leaf spot. **Alternaria** sp., IA; **Cercospora lycii,** IA; **Phyllosticta lycii,** OH, NY.
Powdery mildew. **Erysiphe polygoni,** CT, DE, MD, NJ, PA; **Microsphaera diffusa,** OH, PA, UT; **Sphaerotheca pannosa,** ID, WA.
Rust. **Puccinia tumidipes** (II, III), NY to AL, TX, SD; **P. globosipes** (II, III), PA.

MAURANDYA

Leaf spot. Septoria antirrhinorum, TX.

MAURITIUS-HEMP
(*Furcraea*)

Nematode, root knot. **Meloidogyne** sp., MD.

MAY-APPLE
(*Podophyllum*)

Blight, gray mold. **Botrytis cinerea,** NJ.
Blight, leaf. **Septotinia podophyllina,** DE, MD, MO, NJ, NY, VA, WV.
Leaf spot. **Cercospora podophylli,** IL; **Glomerella cingulata,** DE; **Pezizella oenotherae,** VA; **Phyllosticta podophylli,** NY to AL, AR and WI; **Vermicularia podophylli,** TX, VA.
Rot, stem. **Rhizoctonia** sp., MO.
Rust. **Puccinia podophylli,** general.

MEADOW-BEAUTY
(*Rhexia*)

Leaf spot. **Cercospora erythrogena,** AL, DE, MS, TN; **Colletotrichum rhexiae,** DE; **Phyllosticta rhexiae,** FL.

MEADOW-RUE
(*Thalictrum*)

Downy mildew. **Phytophthora thalictri,** WI, CT, NY.
Leaf spot. **Ascochyta clematidina** f. **thalictri,** WI; **Cercospora fingens,** WI, IL; **Cercosporella filiformis,** WI; **Cylindrosporium thalictri,** IN, KS, WI; **Gloeosporium thalictri,** WI; **Mycosphaerella thalictri,** NJ, NY, VT, WI; **Septoria thalictri,** KS.
Powdery mildew. **Erysiphe polygoni,** MA to PA, IL and ND.
Rust. **Puccinia recondita** (0, I), CO, MI, northeastern and north central states; II, III on grasses; **P. septentrionalis** (0, I), CO, IN, IA, CA; II, III, on *Polygonum,* **Tranzschelia pruni-spinosae** (0, I), CO, IN, IA, KS, NE, ND, SD, OH, PA; II, III on *Prunus,* **T. thalictri** (0, III), eastern and central states to MS, CA, NM, ID.
Smut, leaf and stem. **Urocystis sorosporioides,** AZ, MA, NY, UT.
Smut, white. **Entyloma thalictri,** CT, IL, IN, NY, WI.

MEADOWSWEET
(Filipendula)

Leaf spot. **Cylindrosporium** sp.; **Septoria ulmariae,** CT.
Powdery mildew. **Sphaerotheca macularis,** IN, NY, VT.
Rust. **Triphragmium ulmariae** (0, I, II, III), IN.

MEDICAGO
(Black Medic)

Virus. Tomato Ringspot, VT.

MEDLAR
(Mespilus)

Bacterial fire blight. **Erwinia amylovora,** NY.
Leaf spot. **Fabraea maculata,** CA.
Rust. **Gymnosporangium clavipes,** NY.

MELILOTUS

Broomrape. **Orobanche ramosa,** TX.

MELON, MUSKMELON,
CANTALOUPE, CASSABA
(Cucumis melo)

Anthracnose. **Colletotrichum lagenarium,** general in East and South to CO,
AZ, ND; **Marssonina melonis,** NY.
Bacterial angular leaf spot. **Pseudomonas lachrymans,** CA, DE, IA, MD,
MI, NJ, PA; **P. pseudoalcaligenes** subsp. **citrulli,** GA.
Bacterial rind necrosis. **Erwinia** sp., TX.
Bacterial soft rot. **Erwinia aroideae** and **E. carotovora.**
Bacterial wilt. **Erwinia tracheiphila,** general east of Rocky Mountains; AZ to
ID, WA.
Blight, gummy stem; black rot. **Mycosphaerella citrullina** (*M. melonis*), DE,
FL, MA, NJ, NY, TX.
Blight, leaf; black mold. **Alternaria cucumerina,** general.
Blight, southern. **Sclerotium rolfsii,** VA and OH to FL and TX.
Blight, stem gumming and stem-end rot. **Diplodia natalensis,** TX.
Damping-off. **Rhizoctonia solani,** CA, GA, NJ; fruit rot, FL, TX; **Pythium
debaryanum,** CA, CT, IA, NJ, NY; **Fusarium equiseti,** CA.

Dodder. **Cuscuta arvensis,** MD.

Downy mildew. **Pseudoperonospora cubensis,** general.

Leaf spot. **Cercospora** sp., CO, GA, TX; **Phyllosticta** sp., GA, OH, TX; **Septoria cucurbitacearum,** DE, MA, MI, NH, NY, PA, VT, WI.

Nematode, root knot. **Meloidogyne arenaria, M. hapla,** NJ to FL, CA.

Nonparasitic. **Leaf Spot,** magnesium deficiency, NY.

Powdery mildew. **Erysiphe cichoracearum,** general.

Rot, charcoal. **Macrophomina phaseoli,** CA, OR, TX.

Rot, fruit. **Fusarium** spp., general in market, also **Alternaria** spp.; **Monilia sitophila,** IN, NY; **Mucor** sp., NY; **Penicillium** spp., blue mold; **Phytophthora** spp.; **Rhizopus stolonifer,** cosmopolitan; **Trichoderma viride,** green mold; **Trichothecium roseum,** pink mold, occasional in market.

Rot, root. **Phymatotrichum omnivorum,** TX; **Pythium periplocum,** CA.

Rot, root; cottony leak. **Pythium aphanidermatum,** AZ, CA, TX.

Rot, stem. **Sclerotinia sclerotiorum,** AR, MA, TX; **Cephalosporium** sp.

Scab. **Cladosporium cucumerinum,** occasional, East and central states.

Virus. **Beet Curly Top,** AZ, CA, ID, OR, TX, WA; **Cucumber Mosaic,** general; **Muskmelon Mosaic,** general; **Tobacco Ring Spot,** KS, MD, NC, PA, WI; **Squash Mosaic; Watermelon Mosaic; Zucchini Yellow Mosaic,** CA, FL, NY.

Wilt. **Fusarium oxysporum** f. sp. **melonis,** general; **Verticillium albo-atrum,** CA, OR.

Melons belong to the cucurbit family and in general have the same diseases as cucumbers. Downy mildew is a problem on the moist East Coast, and powdery mildew may be a limiting factor in the arid Southwest. Although there are varieties resistant to powdery mildew, different physiological races of the fungus keep things complicated. Sulfur dust should not be used except on sulfur-resistant melon varieties. Mosaic is transmitted both by seed and aphids; purchase virus-free seed, eliminate weeds, and keep down insects.

MELOTHRIA

Downy mildew. **Pseudoperonospora cubensis,** GA, OH, TX.

Nematode, root knot. **Meloidogyne** sp., FL.

Powdery mildew. **Erysiphe cichoracearum,** WI.

MENTZELIA
(Blazing Star)

Leaf spot. **Phyllosticta mentzeliae,** KS, TX; **Septoria mentzeliae,** KS, WA, TX.

Rot, root. **Phymatotrichum omnivorum,** TX; **Rhizoctonia solani,** NJ.

Rust. **Puccinia aristidae** (0, I), AZ, CO; II, III, on grasses; **Uredo floridana,** FL.

MENZIESIA

Gall, leaf. **Exobasidium vaccinii,** Pacific Northwest, AL, NC, VA, WV.
Leaf spot, tar. **Melasmia menziesii,** MT, and WY to OR and AK; **Rhytisma** sp.
Powdery mildew. **Microsphaerea alni** var. **vaccinii,** AK, VA.
Rust. **Pucciniastrum myrtilli** (II, III), WV; 0, I, on hemlock.

MERTENSIA
(Bluebells, Virginia Cowslip)

Downy mildew. **Peronospora** sp., MT.
Leaf spot. **Septoria poseyi,** OR.
Powdery mildew. **Erysiphe cichoracearum,** AL, CO, MT, PA, NM, NV, UT, WY.
Rot, stem. **Sclerotinia sclerotiorum,** CO.
Rust. **Puccinia mertensiae** (III), CO, NV, UT, WY; **P. recondita** (0, I), ID, MT,
   OR; II, III, on grasses.
Smut, leaf. **Entyloma serotinum,** IN, IA, MD, VA, WY.
Virus. **Cucumber Mosaic,** IL, NJ.

MESEMBRYANTHEMUM
(Fig-Marigold)

Nematode, root knot. **Meloidogyne** sp., AL, TX.
Sooty mold. **Torula herbarum,** CA.

MESQUITE
(*Prosopis*)

Bacterial crown gall. **Agrobacterium tumefaciens,** TX.
Blight, leaf. **Cercospora prosopidis; Scleropycnium aureum,** AZ, TX.
Leaf spot. **Napicladium prosopodium,** TX; **Phyllosticta juliflora,** also pod
   spot, OK, TX; **Gloeosporium leguminum,** pod spot, TX.
Mistletoe. **Phoradendron californicum; P. serotinum (flavescens),** TX to CA,
   and **P. tomentosum,** TX.
Powdery mildew. **Leveilulla taurica; Uncinula prosopodis,** TX.
Rot, heart. **Polyporus texanus,** CA, TX; **Fomes everhartii,** AZ, TX; **Schizo-
   phyllum commune,** TX.
Rot, root. **Phymatotrichum omnivorum,** TX.
Rust. **Ravenelia arizonica** (II, III), TX to CA; **R. holwayi** (0, I, II, III), TX to CA.

MIGNONETTE
(*Reseda*)

Damping-off, root rot. **Rhizoctonia solani,** CT.
Leaf spot. **Cercospora resedae,** MA to MS, MO, IA.

Nematode, root knot. **Meloidogyne** sp., FL.
Wilt. **Verticillium albo-atrum, NY.**

MILK THISTLE
(*Silybum*)

Virus. **Tobacco Streak, CA.**

MIKANIA
(Climbing Hempweed)

Leaf spot. **Cercospora mikaniae,** MS; **Septoria mikanii,** CT, TX.
Rust. **Puccinia spegazzinii,** AL, FL, MS, NC.

MILKWORT
(*Polygala*)

Anthracnose. **Gloeosporium ramosum,** IN, NJ, WI.
Leaf spot. **Cercospora grisea,** MS, NJ, VA; **Septoria consocia,** IN, MI; **S. polygalae,** NY.
Rust. **Aecidium renatum,** NM; **Puccinia andropogonis** var. **polygalina** (0, I), IA, MI, WI; II, III, on *Andropogon*; **P. pyrolae** (III), CT, ME, MI, NH, NY, WI.

MIMOSA, SILK-TREE
(*Albizzia julibrissin*)

Canker; dieback. **Nectria cinnabarina,** DC, NC, VA.
Nematode, root knot. **Meloidogyne arenaria** and var. **thamesii; M. incognita; M. javanica; M. hapla.**
Nematode, stubby root. **Trichodorus primitivus.**
Rot, heart. **Ganoderma lucidum;** Root, **Armillaria mellea.**
Wilt. **Fusarium oxysporum** f. sp. **perniciosum,** AL, GA, FL, NC, SC, NJ, VA, AR, MS.

The mimosa wilt is one of the most devastating tree disease on record. The fungus is in the soil with no possibility of control by aerial spraying. The incidence of wilting is probably increased by nematodes. Resistant varieties Tryon and Charlotte have been released, but occasional specimens succumb to wilt. The fungus may also be seed-transmitted.

MIMULUS
(Monkey-Flower)

Bacterial, MLO. **California Aster Yellows, CA.**
Blight, gray mold. **Botrytis cinerea, AK.**

Leaf spot. **Cercospora mimuli,** MO; **Ramularia mimuli,** CA, NY, OH, WY; **Septoria mimuli,** PA to MS, MO, NE.

Nematode, leaf. **Aphelenchoides ritzema-bosi.**

Powdery mildew. **Erysiphe cichoracearum,** CA, PA, UT.

Rust. **Puccinia andropogonis** (0, I), MO, WI; II, III on *Andropogon*; **Uredo** sp. (II), WI.

MINT
(Mentha)

Canker, stem. **Fusarium** sp., MI, WA; **Alternaria** sp., MI; **Phoma menthae,** OR.

Leaf spot. **Cercospora menthicola,** IL, TX; **Phyllosticta decidua,** occasional, ME to OH, IA, WI; **Ramularia menthicola,** CA, ME, MT, OR; **Septoria menthae** (*S. menthicola*), IN, WI.

Nematode. **Aphelenchoides parietinus; Longidorus sylphus,** OR; **Paratylenchus macrophallus,** OR.

Nematode, root knot. **Meloidogyne hapla,** OR.

Powdery mildew. **Erysiphe cichoracearum,** CO, IA, UT, WA; **E. galeopsidis,** IA; **E. polygoni,** TX; **Sphaerotheca macularis,** WA.

Rust. **Puccinia menthae** (0, I, II, III), East and central states to TX and Pacific Coast; **P. angustata** (0, I), CA, SD, WI; II, III, on grasses.

Spot anthracnose. **Sphaceloma menthae,** IN, MI, MD.

Virus. **Tobacco Ring Spot,** IN.

Wilt. **Verticillium albo-atrum** f. **menthae** (*V. dahliae*), IN, MI, OR.

Rust is serious in mints grown commercially, in greenhouses or in the field. Overwintering spores can be killed by treating rhizomes with hot water. Spot anthracnose is largely controlled by thorough coverage when mint is plowed under in the fall. Mint in the backyard is too prolific for worry about disease.

MISTLETOE
(Phoradendron)

Black mildew. **Asterina phoradendricola,** FL.

Blight, brown felt. **Herpotrichia juniperi,** CA.

Blight, leaf. **Sphaeropsis visci,** SC, TX; **Phyllosticta phorodendri,** CA.

Blight, twig. **Nectria cinnabarina,** TX.

Canker. **Cystospora pinicola,** OR, WA; **Hymenochaete agglutinans,** AL.

Canker, felt fungus. **Septobasidium pseudopedicellatum,** FL.

Dodder. **Cuscuta exaltata,** TX.

Leaf spot. **Cercospora struthanthi,** FL; **Exosporium phoradendri,** TX.

Rust. **Peridermium bethelii,** CO; **Uredo phoradendri,** CA, OR.

If you treasure mistletoe for Christmas greens, you will be sorry it has diseases; but if you consider mistletoe a pest, you probably wish the above list were longer.

MISTLETOE, DWARF
(*Arceuthobium* spp.)

Blight, brown felt. **Herpotrichia juniperi,** CA.
Rust. **Peridermium bethelii,** CO.

MOCK-CUCUMBER
(*Echinocystis*)

Anthracnose. **Colletotrichum lagenarium,** FL.
Downy mildew. **Plasmopara australis,** IA, KS, MN, OH, WI; **Pseudoperono-spora cubensis,** OH.
Leaf spot. **Cercospora echinocystis,** NY to FL, NE, WI; also fruit spot; **Septoria** spp., widespread; **Alternaria** sp., FL.
Powdery mildew. **Erysiphe cichoracearum,** WI.
Virus. **Beet Curly Top,** CA; **Cucumber Mosaic,** CA, CO, IN, IL, MI, NY, WI.
Wilt. **Fusarium** sp., FL.

MOCK-ORANGE
(*Philadelphus*)

Bacterial blight. **Pseudomonas syringae,** MN.
Blight, flower and shoot. **Botrytis cinerea,** cosmopolitan.
Blotch, sooty, **Sarcinella heterospora,** FL.
Leaf spot. **Ascochyta philadelphi,** NY; **Cercospora angulata,** MO, TX; **Ramularia philadelphi,** TX, WA; **Septoria philadelphi,** ID, MT, IA.
Nematode, root knot. **Meloidogyne** sp.
Powdery mildew. **Phyllactinia corylea,** MT, WA, GA.
Rot, root. **Phymatotrichum omnivorum,** TX.
Rust. **Gymnosporangium speciosum** (0, I), CO, NM, UT; III, on juniper.

MOCK-STRAWBERRY
(*Duchesnea*)

Downy mildew. **Peronospora potentilae,** NC.
Gall, leaf. **Synchytrium globosum,** SC.
Leaf spot. **Pezizella oenotherae,** VA.
Rust. **Frommea obtusa duchesneae** (0, I, II, III), NH to FL and KY; **Frommella duchesneae,** IN.

MONARDA
(Horse-Mint, Bee-Balm)

Blight, southern. **Sclerotium rolfsii,** TX.
Gall, leaf. **Synchytrium holwayi,** IA, WI.

Leaf spot. **Cercospora** sp., OK; **Phyllosticta decidua,** KS, NE, OK, TX, WI; **P. monardae,** KS; **Ramularia brevipes,** AL, TX.
Rust. **Puccinia angustata** (0, I), NE, WI; **P. menthae** (0, I, II, III), general from ME to MS; TX, ID.
Virus. **Mosaic.** Unidentified, IN.

MONARDELLA

Leaf spot. **Phyllosticta monardellae,** CA.
Rust. **Puccinia menthae** (0, I, II, III), CA, NV, OR, NM, UT.

MONESES
(Wood-Nymph)

Rust. **Chrysomyxa pirolata** (II, III). CO, ME, MI, MT, NM, WA, WY, AK; 0, I, on spruce.

MONKSHOOD, ACONITE
(*Aconitum*)

Bacterial leaf spot. **Pseudomonas delphinii,** ME, NJ.
Downy mildew. **Plasmopora pygmaea,** AK.
Nematode, root knot. **Meloidogyne** sp., NY, VT.
Powdery mildew. **Erysiphe polygoni,** NY, TX, WV.
Rot, root. **Phymatotrichum omnivorum,** TX; **Rhizoctonia solani,** CT, NJ.
Rot, stem. **Sclerotinia sclerotiorum,** CO; **Sclerotium rolfsii,** CT, DE, MN, NJ, NY.
Rust. **Puccinia recondita** (0, I), CO, AK; II, III, on grasses; **Uromyces lycoctoni** (0, I, II, III), CA, CO, TX, UT, WY.
Smut, leaf and stem. **Urocystis carcinodes,** UT; **U. sorosporioides,** UT.
Virus. **Mosaic,** Unidentified, NY.
Wilt, **Verticillium albo-atrum,** MA, NJ, NY, OH.

Verticillium wilt is more widespread in monkshood than the official reports indicate. The leaves dry along the stem, flowers are poor, and when the stem is cut across, blackened bundles are readily seen. The clumps do not die immediately but decline over a period of years.

MONKSHOOD VINE
(*Ampelopsis aconitifolia*)

Dieback. **Tubercularia nigricans,** NH, TX.
Nematode, dagger. **Xiphinema index.**

MONSTERA

Anthracnose. **Gloeosporium** sp., WA; **Colletotrichum** sp.
Leaf spot. **Macrophoma philodendri,** FL, MI.

MONTIA
(Indian Lettuce)

Smut, seed. **Ustilago claytoniae,** WA.

MOONFLOWER
(*Calonyction*)

Leaf spot. **Phyllosticta** sp., NJ.
Nematode, leaf. **Aphelenchoides fragariae,** NJ.
Nematode, root knot. **Meloidogyne** sp., NJ, SC.
Rot, root. **Phymatotrichum omnivorum,** TX.
Rust. **Coleosporium ipomoeae** (II, III), AL, NC, SC, TX; 0, I, on pine.
White rust. **Albugo ipomoeae-panduratae,** FL.

MOONSEED
(*Menispermum*)

Leaf spot. **Cercospora menispermi,** NY to VA, KS, WI; **Colletotrichum
sordidum,** WI; **Phyllosticta menispermicola,** IL; **Septoria abortiva,** IL, KS.
Powdery mildew. **Microsphaera alni,** widespread.
Smut, leaf. **Entyloma menispermi,** PA to VA, KS, ND.

MOREA

Rust. **Puccinia iridis,** FL.

MORINDA
(Royoc, Indian-Mulberry)

Leaf spot. **Cercospora morindicola,** FL.
Spot anthracnose. **Sphaceloma morindae.**

MORNING-GLORY
(*Ipomoea*)

Blight, blossom, blight. **Choanephora compacta,** GA.
Blight, southern. **Sclerotium rolfsii,** TX.

Blight, thread. **Pellicularia koleroga,** FL.
Canker, stem. **Vermicularia ipomoearum,** NY, PA.
Leaf spot. **Alternaria** sp., VA; **Cercospora alabamensis,** AL, FL, NJ; **Phyllosticta ipomoeae,** FL, KS, MS ; **Septoria convolvuli,** KS, FL, PA, TX, WI.
Nematode. **Rotylenchulus reniformis,** LA.
Nematode, root knot. **Meloidogyne** sp.; **M. hapla,** AL, OK; **M. incognita,** LA.
Rot, root. **Phymatotrichum omnivorum,** TX; **Macrophomina phaseolina,** AZ; **Streptomyces ipomoea,** LA; **Ceratocystis fimbriata,** LA; **Fusarium oxysporum** f. sp. **batatas,** LA; **Plenodomus destruens,** LA; **Monilochaetes infuscans,** LA.
Rust. **Puccinia crassipes** (I, III), FL, GA, LA, SC; **Coleosporium ipomoeae** (II, III), NJ to FL, TX, KS; 0, I on pine; **Uredo laeticolor** (II), FL.
Virus. **?Cucumber Mosaic,** FL.
White rust. **Albugo ipomoeae-panduratae,** NJ to AZ, NE.
Wilt. **Fusarium** sp., TX.

MOTHERWORT
(*Leonurus*)

Black mildew. **Dimerosporium hispidulum,** TX.
Leaf spot. **Ascochyta leonuri,** LA; **Phyllosticta decidua,** OH, TX, WI; **Septoria lamii,** PA.
Virus. **Mosaic.** Unidentified, IN.

MOUNTAIN-ASH
(*Sorbus*)

Bacterial fire blight. **Erwinia amylovora,** widespread, VA.
Bacterial crown gall. **Agrobacterium tumefaciens,** CT, NJ.
Blight, leaf. **Fabraea maculata,** AK, WI, MN, WV.
Blight, twig. **Nectria cinnabarina,** corla spot, NC, AK; **Phomopsis** sp., MA; **Valsa leucostoma,** OH, WV.
Canker. **Cytospora** sp., VA.
Canker, blister. **Nummularia discreta,** IA, MA.
Canker, branch; fruit rot. **Glomerella cingulata,** IN.
Canker; dieback. **Cytospora chrysosperma,** MN, MT, NE, NJ, WA; **C. leucostoma,** MT; **C. massariana,** ID; **C. microspora,** MT; **Fusicoccum** sp., IL.
Canker, trunk; black rot. **Cytospora rubescens,** IA; **Physalospora obtusa,** IN, MI, OH, CT to VA.
Leaf spot. **Alternaria** sp., IA; **Graphium sorbi,** NY, WI; **Phyllosticta globigera,** ID, WA; **P. sorbi,** IL, ME, IA, MO, OK, TX; **Septoria sorbi,** IA; **S. sitchensis,** ID.
Mistletoe, European. **Viscum album,** CA.
Powdery mildew. **Podosphaera oxyacanthae** var. **tridactyla,** WA.

Rot, heart. **Polyporus hirsutus,** MI, WA; **P. versicolor,** WI.
Rot, root. **Armillaria mellea,** NJ; **Phymatotrichum omnivorum,** TX.
Rust. **Gymnosporangium cornutum** (0, I), ME to NJ and WI, MT to CO, WA, AK; III, on juniper; **G. globosum** (0, I), OR; III, on *Libocedrus*; **G. nelsoni** (0, I), MT, WA, WY; III, on juniper; **G. nootkatense** (0, I), OR, WA, AK; III, on *Chamaecyparis*; **G. tremelloides** (0, I), MT to CO, WA, AK.
Scab. **Venturia inaequalis,** IL, MN, NY, WA.

Mountain-ash is quite susceptible to fire blight, but affected branches can usually be pruned out. Rust may appear on foliage in midsummer. If the mountain-ash is more desirable than the junipers nearby, the latter can be eradicated. Fertilizing will help the tree recover from a bout with Cytospora canker but may increase susceptibility to fire blight.

MOUNTAIN-HEATHER
(*Phyllodoce*)

Blight, brown felt. **Herpotrichia nigra,** ID.
Gall, leaf. **Exobasidium vaccinii-uliginosi,** OR.

MOUNTAIN-HOLLY
(*Nemopanthus*)

Leaf spot. **Ramularia nemopanthus,** NY.
Leaf spot, tar. **Rhytisma ilicis-canadensis,** ME to WV and MI.
Powdery mildew. **Microsphaera alni,** NY, WI.
Rot, wood. **Poria inermis,** PA.

MOUNTAIN-LAUREL
(*Kalmia*)

Blight, flower. **Ovulinia azaleae,** SC, AL.
Blight, leaf. **Phomopsis kalmiae,** NY to NC.
Canker; felt fungus. **Septobasidium** sp., TX, AL.
Gall, leaf. **Exobasidium vaccinii,** AK; **Synchytrium vaccinii,** red spot, NJ.
Heart, rot, wood rot. **Fomes annosus,** NC.
Leaf spot. **Cercospora kalmiae,** CT to AL and TN; **Mycosphaerella colorata** (*Phyllosticta kalmicola*), CT to AL, IN, MI, TX; **Pestalotia kalmicola,** ?secondary, DE, NJ, PA, TX, WA; **Septoria angustifolia,** MA to AL, OH; **Rhytisma andromedae,** tar spot, VT.
Nonparasitic. **Chlorosis,** usually iron deficiency.
Powdery mildew. **Microsphaera alni** var. **vaccinii,** WI.
Rot, root. **Armillaria mellea,** MD; **Corticium galactinum,** MD; **Phymatotrichum omnivorum,** TX.
Rot, wood. **Polyporus versicolor,** VA; **Stereum rameale,** VA.

Phomopsis leaf blight or blotch and Mycosphaerella (Phyllosticta) leaf spot are common and rather disfiguring on bushes in shade or under tree drip. In light cases, removal of spotted leaves is sufficient.

MOUNTAIN-MAHOGANY
(Cercocarpus)

Leaf spot. **Septogloeum cercocarpi,** CA.
Rot, wood. **Stereum hirsutum,** OR.
Spot anthracnose. **Sphaceloma cercocarpi,** CA.

MOUNTAIN-MINT
(Pycnanthemum)

Gall, leaf. **Synchytrium cellulare,** WI.
Leaf spot. **Cerceseptoria blephiliae,** WI; **Cercosporella pycnanthemi,** AL.
Rust. **Puccinia angustata** (0, I), IN; **P. menthae** (0, I, II, III), MA to VA, TX and IA, AL, OK, CA.

MOUNTAIN-SORREL
(Oxyria)

Rust. **Puccinia oxyriae** (II, III), CA, CO, ID, OR, UT, AK; 0, I unknown.
Smut, floral. **Ustilago vinosa,** CA, CO, WA, WY, AK.

MULBERRY
(Morus)

Bacterial hairy root. **Agrobacterium rhizogenes,** NE.
Bacterial leaf spot. **Pseudomonas syringae** pv. **mori,** general.
Bacterial scorch. **Xylem limiting bacteria,** Mid-Atlantic and Southern U.S.
Blight, berry; popcorn disease. **Ciboria carunculoides,** NC to FL, TX, VA.
Blight, twig. **Myxosporium diedickei,** TX, WA.
Canker; twig blight. **Cytospora** sp., NJ, TX; **Dothiorella** sp.; **D. mori,** NJ, TX; **Gibberella baccata** var. **mori,** widespread; **Nectria** sp., widespread; **N. cinnabarina,** widespread; **Sclerotinia** sp., TX.
Leaf spot. **Cercospora moricola,** PA to FL, NE, TX; **Cercosporella mori,** NE, OK, TX; **Exosporium** sp., FL; **Mycosphaerella arachnoidea,** false mildew, GA, NC; **M. mori,** widespread; **Phyllosticta moricola,** KS; **Cytospora** sp., VA.
Nematode, root knot. **Meloidogyne** spp., NJ, NC, OK.
Powdery mildew. **Phyllactinia corylea,** OH; **Uncinula geniculata,** NY to KS, AL.
Rot, heart. **Polyporus farlowii,** AZ, NM; **P. hispidus,** CT; **Ganoderma applanatum; Hymenochaete agglutinas,** MD.

Rot, root. **Armillaria mellea,** DE, NC, OK; **Helicobasidium purpureum,** TX; **Phymatotrichum omnivorum,** TX.

Rot, wood. **Schizophyllum commune,** CA; **Stereum cinerescens,** MA.

Rust. **Cerotelium fici** (II), LA.

Bacterial leaf spot damages nursery trees having overhead irrigation; Mycosphaerella leaf spot sometimes defoliates older trees.

MULLEIN
(*Verbascum*)

Downy mildew. **Peronospora sordida,** NJ.

Leaf spot. **Cercospora verbasicola,** TX; **Phyllosticta verbasicola,** IN, KS, TX; **Ramularia variabilis,** NY to MS, TX, WA; **Septoria verbasicola,** TX, NY to AL, MO, TX.

Nematode, root knot. **Meloidogyne** sp.

Powdery mildew. **Erysiphe cichoracearum,** PA; **Oidium** sp., NJ.

Rot, root, **Phymatotrichum omnivorum,** TX.

Virus. **Tomato Ringspot,** VT.

MUSHROOM, OYSTER
(*Pleurotus ostreatus*)

Dry bubble. **Verticillium fungicola,** PA

MUSK-ROOT
(*Adoxa*)

Gall, leaf. **Synchytrium anomalum,** IA.

Leaf spot. **Phyllosticta adoxae,** CO.

Rust. **Puccinia adoxae** (III), CO, UT, WY; **P. argentata** (0, I), IA, MN, WI; II, III, on *Impatiens*.

MUSTARD GREENS
(*Brassica juncea*)

Bacterial black rot. **Xanthomonas compestris,** FL, OH.

Bacterial, MLO. **Aster Yellows.**

Bacterial, yellows. **Spiroplasma citri,** IL.

Club root. **Plasmodiophora brassicae,** CA, CT, OH, TX, WA.

Damping-off. **Rhizoctonia solani,** cosmopolitan.

Downy mildew. **Peronospora parasitica,** CT, FL, IA, TX.

Leaf spot. **Cercospora brassicola,** IN, LA, NJ; **Cercosporella brassicae,** CA, VA.

Nematode, root knot. **Meloidogyne** sp., FL, MO, TX; **Heterodera schactii,** UT.
Powdery mildew. **Erysiphe polygoni,** AZ, CA, FL, TX.
Rot, crown; drop. **Sclerotinia sclerotiorum,** TX.
Rot, root; damping-off. **Aphanomyces raphani,** WI.
Rust. **Puccinia aristidae** (0, I), CO; II, III, on grasses.
Virus. **Cauliflower Mosaic; Turnip Mosaic; Tobacco Stread,** CA.
Wilt. **Fusarium** sp.

MYRTLE
(*Myrtus*)

Leaf spot. **Pestalotia decolorata,** LA.
Rot, stem. **Sclerotium rolfsii,** FL.

NANDINA

Anthracnose. **Glomerella cingulata,** TX.
Leaf spot. **Cercospora nandinae,** AL, SC.
Nematode, root knot. **Meloidogyne** sp., NC, TX.
Nonparasitic. **Chlorosis,** alkaline soil, TX.
Rot, root. **Phymatotrichum omnivorum,** TX.
Virus. **Cucumber Mosaic,** MD, GA.

NARCISSUS
(Daffodil, Jonquil)

Bacterial streak; stem rot. Unidentified, WA.
Blight, leaf; fire. **Sclerotinia** (*Botrytis*) **polyblastis,** OR, WA; **Botrytis cinerea;
Botryotinia polyblastis,** WA.
Leaf scorch. **Stagonospora curtisii,** general, especially in East and South.
Leaf spot, blight; white mold. **Ramularia vallisumbrosae,** OR, WA.
Nematode, bulb. **Aphelenchoides fragariae,** FL, GA, NC, SC; **A. subtenuis,**
Pacific Coast; **Aphelenchus avenae,** ?secondary.
Nematode, bulb; brown-ring disease; leaf "spikkel." **Ditylenchus dipsaci,** in
all commercial narcissus areas.
Nematode, lesion. **Pratylenchus pratensis,** OH, WA.
Rot, basal. **Fusarium oxysporum** f. sp. **narcissi,** general on hardy varieties.
Rot, black bulb. **?Sclerotinia sclerotiorum;** crown; wet scale, **Sclerotium
rolfsii,** CA, FL, NY, VA.
Rot, large scale speck. **Stromatinia narcissi,** general in northern bulb areas.
Rot, leaf and stem. **Gloeosporium** sp., LA, NC.
Rot, neck; smoulder. **Sclerotinia narcissicola,** NJ, NY, OR, VA, WA; probably
general except on polyanthus varieties.

Rot, root and bulb. **Armillaria mellea,** CA, OR, WA; **Aspergillus** spp., black mold; **Penicillium** spp., blue mold, in wounds; **Trichoderma viride,** green mold in scales, cosmopolitan after sunscald; **Rhizopus stolonifer,** soft rot, cosmopolitan after sunscald; **Cylindrocarpon radicicola,** secondary root rot.

Rot, small scale speck; neck rot. **Sclerotium** sp., general, especially in southern bulb districts.

Virus. **Narcissus Mosaic,** mild; **Narcissus Flower Streak; Narcissus Chocolate Spot; Narcissus Yellow Stripe,** gray disease, often called mosaic, general; **White Streak,** general.

Control of narcissus diseases rests with the grower, who should, and usually does, supply the gardener with sound, healthy bulbs. Inspect all bulbs carefully before planting, making sure there are no dark sclerotia on the scales or the chocolate brown of Fusarium rot at the base. The bulb and stem nematode is controlled by treating in hot water.

NASTURTIUM
(*Tropaeolum*)

Bacterial fasciation. **Clavibacter fascians,** CA.

Bacterial leaf spot. **Pseudomonas syringae** pv. **aptata,** ME, MN, MS, NJ, PA, VA.

Bacterial, MLO. **California Aster Yellows,** CA.

Bacterial wilt. **Pseudomonas solanacearum,** FL, MD, NJ, NC, VA.

Blight, gray mold. **Botrytis cinerea,** AK.

Dodder. **Cuscuta** sp., MO, NH.

Leaf spot. **Cercospora tropaeoli,** AL; **Heterosporium tropaeoli,** CA, NY; **Pleospora** sp., MS, NJ, OH.

Nematode, root knot. **Meloidogyne** spp., NJ, TX; **Heterodera schactii,** root gall.

Rust. **Puccinia aristidae** (0, I), UT; II, III, on grasses.

Virus. **Beet Curly Top,** CA, TX; **Tomato Spotted Wilt,** CA, MD, TX.

Compared with the almost inevitable affliction of black aphids, nasturtium diseases are insignificant.

NECTARINE
(*Prunus persica* var. *nectarina*)

Bacterial canker. **Pseudomonas syringae,** CA.

Bacterial crown gall. **Agrobacterium tumefaciens,** MO.

Bacterial leaf spot; canker. **Xanthomonas pruni,** OK.

Bacterial, MLO. **Peach X-Disease.**

Canker. **Valsa leucostoma,** DC.

Leaf curl. **Taphrina deformans,** CA, OR, TX, WA.

Leaf spot. **Cristulariella pyramidalis,** FL.

Leaf spot; shot hole. **Coryneum carpophilum,** CA, OR, WA.
Nematode, lesion. **Pratylenchus thornei,** CA.
Nematode, root knot. **Meloidogyne** sp., CA; usually resistant.
Powdery mildew. **Podosphaera oxyacanthae,** ID, WA; **Sphaerotheca pannosa,**
  NY, ID, WA.
Rot, brown; twig blight. **Monilinia laxa,** CA, WA; **M. fructicola,** CT, NY, TX.
Scab. **Cladosporium carpophilum,** CT, DE, IL, NY, PA, TX.
Virus. **Peach Mosaic; Peach Yellows.**

NEMATANTHUS

Leaf spot. **Myrothecium roridum,** FL.
Virus. **Tobacco Mosaic,** CA, CT, DC, FL, OH, WA.

NEMOPHILA
(Baby Blue-Eyes)

Powdery mildew. **Erysiphe cichoracearum,** CA, NV, TX, WA.

NEPHTHYTIS

Leaf spot. **Cephalosporium cinnamomeum,** MD, FL.
Nematode, root knot. **Meloidogyne incognita-acrita.**
Rot, root. **Pythium splendens,** FL.

  Spraying with maneb should control the leaf spot.

NERINE
(Guernsey-Lily)

Leaf scorch; red blotch. **Stagospora curtisii,** CA.
Nematode, lance. **Hoplolaimus coronatus,** NC.

NEW ZEALAND FLAX
(*Phormium tenax*)

Rot, root. **Armillaria mellea,** CA.

NEW ZEALAND SPINACH
(*Tetragonia*)

Bacterial, MLO. **Aster Yellows,** NJ, NY.
Leaf spot. **Cercospora ?tetragoniae,** IN, MA; **Helminthosporium** sp., TX.

Nematode, root knot. **Meloidogyne** sp., TX.
Virus. **Beet Curly Top,** CA; **Mosaic,** DE, and **Rosette,** IN, unidentified.

NICOTIANA
(Flowering Tobacco)

Bacterial blackfire. **Pseudomonas tabaci,** WI; **P. angulata,** WI.
Black shank. **Phytophthora parasitica,** var. **nicotianae,** CT.
Downy mildew. **Peronospora tabacina,** CA, TX.
Leaf spot. **Alternaria longipes,** TX; **Rhizoctonia solani,** NC.
Mold, blue. **Peronospora tabacina,** GA.
Nematode, root knot. **Meloidogyne** sp., FL.
Powdery mildew. **Oidium** sp., KY.
Rot, root. **Phymatotrichum omnivorum,** TX.
Stunt. **Glomus macrocarpum,** KY.
Virus. **Beet Curly Top,** TX; **Tomato Spotted Wilt,** TN; **Tobacco Mosaic; Tobacco Ring Spot; Tobacco Vein-Mottling,** NC.

NIGHT-BLOOMING CEREUS
(*Hylocereus*)

Virus. **Cactus Virus X,** CA.

NIGHTSHADE, Silverleaf
(*Solanum elaeagnifolium*)

Weed found in dry, open woods, prairie, waste places, and disturbed soil in Southwestern, U.S.

Nematode, foliar. **Nothanguina phyllobia,** TX.
Virus. **Tomato Spotted Wilt,** CA.

NINEBARK
(*Physocarpus*)

Leaf spot. **Cercospora spiraea,** IN; **Marssonina neilliae,** CA, TX, WI; **M. lonicerae,** OR; **Phyllosticta opulasteris,** ID; **Ramularia spiraeae,** MI, NY, WI.
Powdery mildew. **Sphaerotheca macularis,** MA to WI.
Rot, root. **Phymatotrichum omnivorum,** TX.
Rot, wood. **Fomes conchatus,** NY.

NOTHOSCORDUM
(False Garlic)

Anthracnose. **Colletotrichum circinans,** OK.
Rust. **Uromyces hordeinus** (0, I), KS, OK, TX; II, III, on grasses; **U. prima-verilis** (0, I, III), IL, MO, TX.
Virus. **Nothoscordum Mosaic, LA.**

NUTSEDGE
(*Cyperus rotunders*)

Blight, flower. **Balansia cyperi,** LA.
Nematode, cyst. **Heterodera mothi,** GA.
Nematode, root-knot. **Meloidogyne graminicola,** GA.
Rust. **Puccinia canaliculata,** GA.

NYMPHOIDES
(Floating-Heart)

Rust. **Puccinia scirpi** (0, I); II, III, on *Scirpus,* FL.
Smut, leaf. **Burrillia decipiens,** NJ.

OAK
(*Quercus*)

Anthracnose; leaf and twig blight. **Gnomonia quercina (Gloeosporium quercinum),** general.
Bacterial canker. Unidentified, IL.
Bacterial crown gall. **Agrobacterium tumefaciens,** MD, MI.
Bacterial drippy nut. **Erwinia quercina,** CA, on live oak.
Bacterial leaf scorch. **Xylella fastidiosa,** GA.
Bacterial wetwood. **Erwinia nimipressuralis,** Va.
Black mildew. **Morenoella quercina,** SC to FL, GA, TX; **Irenina manca,** MS.
Blight, twig. **Cryptocline cinerescens,** CA; **Diplodia longispora,** NY to NC, IL and WI; **D. quercina,** CA; **Discula quercina,** CA.
Canker, bark. **Aleurodiscus oakesii,** NY to IL, IA, CA; **A. candidus; A. acerinus; A. griseo-canus; Dichaena quercina,** NJ.
Canker, bleeding. **Phytophthora cactorum,** MA, NY, FL.
Canker, branch. **Endothia gyrosa,** VA.
Canker, branch; decline; dieback. **Botryodiplodia gallae,** MI; **Cephalosporium** sp.; **Hyalodendron** sp.
Canker, chestnut. **Endothia parasitica,** SC, VA, MS.
Canker, felt fungus. **Septobasidium** spp., NC to FL to LA.

Canker, trunk. **Nectria galligena; Strumella coryneoides** (*Urnula craterium*).

Canker, twig. **Cytospora chrysosperma,** NJ, RI; **Endothia parasitica,** occasional; **Dothiorella quercina,** MD, VA; **Fusarium solani; Physalospora glandicola,** MD; **P. obtusa,** MN, VA; **P. rhodina,** VA; **Pseudovalsa longipes; Pyrenochaete venuta.**

Decline. **Hypoxylon atropunctatum, H. punctulatum, H. mediterraneum,** VA.

Dodder. **Cuscuta** spp., occasional in forest nurseries.

Leaf blister. **Taphrina caerulescens,** Northeast to north central and Gulf states; also CO, UT, WY, CA, GA.

Leaf spot. **Actinopelte dryina,** occasional; **Acantharia echinata,** black, on live oak; **Cercospora macrochaeta,** CA; **Cylindrosporium microspilum, C. kelloggii,** CA; **Cibornia** (*Sclerotinia*) **candolleana; C. hirtella;** Dothiorella phomiformis, widespread; **Gloeosporium septorioides; Marssonina martini,** general; **Leptothryium californicum,** on live oak, CA; **Monochaetia desmazierii,** widespread; **Mycosphaerella** sp; **Phyllosticta** spp.; **Septogloeum defolians,** CA; **S. querceum,** WI; **Septoria** spp.; **Venturia orbicula,** NY to VA and OH.

Mistletoe. **Phoradendron serotinum (flavescens),** NC to FL and TX; **P. villosum,** CA.; **P. tomentosum,** TX.

Mold, leaf. **Cladosporium brevipes; C. herbarum.**

Nematode, dagger. **Xiphinema americanum,** Southeast.

Nematode, ring. **Criconemoides annulatum; C. teres; Hemicriconemoides biformis.**

Nematode, root. **Hoplolaimus coronatus; Pratylenchus** sp.; **Meloidogyne** sp.

Nonparasitic. **Chlorosis,** iron deficiency, especially in pin oaks.

Parasitic lichen. **Strigula elegans** and **S. camplanata,** southern U.S., LA.

Powdery mildew. **Erysiphe trina,** witches' broom, on live oak, CA; **Microsphaera alni,** widespread; **Phyllactinia corylea,** widespread; **Sphaerotheca lanestris,** "brown mildew," serious on coast live oak, AL, CA, MS, NC; **Saccardia quercina,** AZ, GA.

Rot, heart. **Daedalea quercina,** widespread, other species; **Fistulina hepatica; Fomes** spp.; **Hericium erinaceous; Polyporus** spp.; **Stereum** spp.

Rot, heart. **Phlebia chrysocrea,** eastern U.S. (PA to FL) and WI to MS.

Rot, root. **Armillaria mellea,** widespread; **Clitocybe tabescens,** FL, MO, OK; **C. olearia,** CA; **Corticium galactinum** and other spp.; **Phytophthora cinnamomi,** of seedlings.

Rot, wood. **Daldinia concentrica; D. vernicosa; Hypoxylon atropunctatum; Lentinus tigrinus; Linzites betulina** and other species; **Schizophyllum commune; Steccherhinum ochraceum.**

Rust. **Cronartium quercuum** (*C. cerebrum*) II, III, widespread; 0, I, on pines; **C. fusiforme** (II, III), southern states; **C. strobilinum** (II), AR, FL, IL, IA, KS, MO, MS; 0, I, on long-leaf pines; **C. conigenum** (II, III), AZ; 0, I, on pine cones.

Rust, fusiform. **Cronartium fusiforme,** GA.

Spot anthracnose. **Elsinoë quercus-falcatae,** on southern red oak, GA, NC; **E. quercicola,** FL.

Wilt, dieback. **Fusarium oxysporum** or **F. solani,** VA.

Wilt, oak. **Ceratocystis fagacearum** (*Chalara quercina*), AR, IL, IN, IA, KS, KY, MD, MI, MN, MO, NE, NC, PA, TN, VA, WV, WI, OH, OK, SC, MO, FL.

Witches' broom. **Articularia quercina** var. **minor.** AZ, NM, UT.

Oak wilt is our most serious disease with red and black oaks often dying during the first season that symptoms appear. Anthracnose is general, most severe on white oak, defoliating in wet seasons. Leaf blister, important in the South, can be prevented by a single dormant spray. Powdery mildew, due to *Sphaerotheca lanestris,* is important in California, where it produces witches' brooms on live oaks. The honey mushroom, *Armillaria mellea,* sometimes called the oak fungus, causes "shoestring" root rot. Strumella canker is frequent in forest trees, sometimes found in ornamentals.

OCOTILLO, COACH-WHIP, CANDLEWOOD
(*Fouquieria*)

Rot, root. **Phymatotrichum omnivorum,** TX.

Rust. **Aecidium cannonii** (0, I), AZ.

OENOTHERA
(Evening-Primrose)

Blight, gray mold. **Botrytis cinerea,** AK.

Dodder. **Cuscuta arvensis,** OK.

Downy mildew. **Peronospora arthuri,** MA to MS, KS, MT, NE, OK, SD.

Gall, leaf. **Synchytrium fulgens,** AL, KS, LA, MS, NC, OK, TX to CA.

Leaf spot. **Alternaria tenuis,** ?secondary, NJ; **Cercospora oenotherae,** AL, KS, TX, WV; **C. oenotherae-sinutae,** AL, NC; **Pezizella oenotherae,** GA, MD, NC, SD, VA; **Pestalotia oenotherae,** OH, OK; **Septoria oenotherae,** ME to FL, OK and SD, NM, CA, UT.

Powdery mildew. **Erysiphe polygoni,** general.

Rot, root. **Phymatotrichum omnivorum,** TX; **Rhizoctonia solani,** TX.

Rust. **Aecidium anograe** (0, I), NE; **Puccinia aristidae** (0, I), AZ, NV; II, III, on grasses; **P. dioicae** (0, I), ME to AL, CO, ND, TX, CA; II, III, on *Carex*; **P. oenotherae** (0, I, II, III), CO, CA to MT, WA; **Uromyces plumbarius** (0, I, II, III), general.

Virus. **Mosaic.** Unidentified, PA.

OKRA
(*Hibiscus esculentus*)

Anthracnose. **Colletotrichum gloeosporioides,** pod spot, FL, PA; **C. hibisci,** dieback, TX.

Blight, blossoms. **Choanephora cucurbitarum,** FL, GA, TX.

Leaf spot. **Ascochyta abelmoschi,** GA, MD, NJ, NY, pod spot; **Alternaria** sp., ?secondary, FL, OH, PA, SC, UT; **Cercospora abelmoschi** (*C. hibisci*), NC to FL, TX; **C. malayensis,** VA to FL, TX, OK; **Corynespora cassicola; Phyllosticta hibiscina,** AL, IL, NJ, NC, OK.

Nematode, root knot. **Meloidogyne incognita** and **M. incognita-acrita,** general.

Powdery mildew. **Erysiphe cichoracearum,** CT, NJ, NC, PA.

Rot, charcoal. **Macrophomina phaseoli,** TX.

Rot, pod. **Botrytis** sp., NY.

Rot, root. **Phymatotrichum omnivorum,** TX; **Rhizoctonia solani,** damping-off, AL, FL; **Thielaviopsis basicola,** NJ.

Rot, stem. **Sclerotinia sclerotiorum,** MA.

Virus. **Tobacco Ring Spot,** GA, VA.

Wilt. **Fusarium oxysporum** f. sp. **vasinfectum,** CT to FL, TX, AZ; **Verticillium albo-atrum,** widespread.

OLEANDER
(*Nerium*)

Anthracnose; leaf spot. **Gloeosporium** sp., MA, MS, NJ, TX.

Bacterial knot. **Pseudomonas syringae** pv. **tonelliana,** AZ, CA, CT.

Blight, stem and leaf. **Pseudomonas syringae,** CA.

Canker; witches' broom. **Sphaeropsis** sp., FL.

Dodder. **Cuscuta indecora,** FL.

Leaf spot. **Alternaria** sp. (*Macrosporium nerii*), CA, AL, FL, GA, MS; **Cercospora neriella,** also pod spot, AL, LA, FL, TX; **Phyllosticta nerii,** FL, LA, MI, TX; **Septoria oleandrina,** CA, FL, LA.

Rot, root. **Clitocybe tabescens,** FL; **Phymatotrichum omnivorum,** TX.

Rot, stem. **Calonectria crotalariae,** CA.

Sooty mold. **Capnodium elongatum,** AL, FL.

Spot anthracnose, scab. **Sphaceloma oleandri,** LA.

OLIVE
(*Olea*)

Anthracnose. **Gloeosporium olivarum,** CA.

Bacterial knot. **Pseudomonas syringae** pv. **savastanoi,** CA.

Black mildew; leaf spot. **Asternia oleina,** FL, GA.

Dodder. **Cuscuta indecora,** CA.

Leaf spot. **Cycloclonium oleaginum,** peacock spot, CA; **Cercospora ?caldosporioides,** also fruit spot, CA.

Nematode, citrus. **Tylenchulus semipenetrans.**

Nematode, lesion. **Pratylenchus musicola** (*P. coffeae*), CA.

Nematode, root knot. **Meloidogyne** sp., CA.

Nonparisitic

    Bitter Pit, dry rot, of fruit. Overnutrition?

    Exanthema. Dieback. Deficiency of organic matter.

    Fruit Pit. Boron deficiency.

    Soft Nose. Blue nose. ?Moisture supply. On variety Sevillano only.

Rot, root. **Armillaria mellea,** CA, TX; **Phymatotrichum omnivorum,** TX; **Polyporus olaea,** CA.

Virus. **Sickle Leaf.**

Wilt. **Verticillium albo-atrum,** CA.

ONCOBA

Rot, root. **Phymatotrichum omnivorum,** TX.

ONION
(*Allium cepa*)

Bacterial, MLO. **Aster Yellows** and **California Aster Yellows,** widespread.

Bacterial rot; slippery skin. **Pseudomonas alliicola,** MA, NY, WA; **P. cepacia,** sour skin, scale rot, NY.

Bacterial soft rot. **Erwinia carotovora,** widespread in field, transit, storage.

Bacterial, yellows. **Spiroplasma citri,** CA.

Blight, southern. **Sclerotium rolfsii,** AL, CA, GA, OK, TX.

Blotch, purple, brown. **Alternaria porri,** ME to MS, TX, CO, UT; **A. tennis,** CO; **A. tenuissima,** CO.

Damping-off. **Pythium** spp., ID, MA, NY; **Rhizoctonia solani,** occasional.

Dodder. **Cuscuta** spp., widespread.

Downy mildew. **Peronospora destructor,** general.

Leaf spot. **Heterosporium allii,** CA, CO, WA; **Phyllosticta allii,** IL, NM; **Stemphylium vesicarium,** TX.

Nematode, bulb; onion bloat. **Ditylenchus dipsaci,** NY, TX.

Nematode, root knot. **Meloidogyne incognita** and **M. incognita-acrita.**

Nematode, sting. **Belonolaimus gracilis.**

Nematode, stubby root. **Trichodorus christiei.**

Nonparasitic

    Blast, of inflorescence. Appears in Connecticut Valley when bright sun follows cloudy weather. Air pollution in NJ.

    Chlorosis. Copper deficiency, NY and FL; manganese deficiency, NY and RI.

Scald. High temperature, general in summer.

Stain, alkali spot. Contact with alkaline materials or ammonia fumes in transit.

Powdery mildew. **Oidiopsis taurica,** CA.

Rot, basal. **Fusarium oxysporum** f. sp. **cepae,** other species, widespread.

Rot, black stalk, tip blight; seed mold. **Stemphylium botryosum,** AR, CA, LA, NH, TX, WA.

Rot, charcoal. **Macrophomina phaseoli,** TX, CA, OK.

Rot, dry. **Diplodia natalensis,** TX; **Helminthosporium allii,** also canker.

Rot, neck; gray mold. **Botrytis allii,** LA, NJ, CA, TX; **B. cinerea;** small sclerotial; leaf blight, **B.** (*Botryotinia, Sclerotinia*) **squamosa,** occasional.

Rot, pink root. **Pyrenochaeta tenestris,** widespread.

Rot, root. **Thielaviopsis basicola,** TX.

Rot, smudge. **Colletotrichum circinans,** general.

Rot, soft. **Kluyveromyces marxianus,** WA.

Rot, various, of bulbs. **Aspergillus niger,** black mold, general in market; **Penicillium** spp., blue mold; **Rhizopus stolonifer,** soft, after sunscald or freezing; **Sclerotinia sclerotiorum,** watery soft.

Rot, white. **Sclerotium cepivorum,** CA, KY, LA, NJ, NY, OH, OR, PA, TX, VA.

Rust. **Puccinia asparagi** (0, I, II, III), CA, CT, IA, KS, MN; **P. allii** (*P. porri*) (II, III), CA, NE, CT.

Smut. **Urocystis cepulae,** general; **U. colchici.**

Virus. **Onion Yellow Dwarf.**

Smut is the most general onion disease, but it seldom afflicts onions grown from sets, the usual method for a small garden. Growing colored instead of white onions avoids smudge and neck rots to some extent. Sweet Spanish onions are resistant to pink root and yellow dwarf.

ONOSMODIUM
(Marbleseed)

Rot, root. **Phymatotrichum omnivorum,** TX.

Rust. **Puccinia recondita** (0, I), CO, KS, NE, ND.

OPLISMENUS
(Basket-grass)

Leaf spot, tar. **Phyllachora punctum,** FL, LA.

ORANGE

See Citrus Fruits.

ORCHIDS
(Imported Species)

Anthracnose; leaf and stem spot. **Colletotrichum cinctum** and **C. gloeo-sporioides** (*Glomerella cincta* and *G. cingulata*), general; **Gloeosporium affine,** FL; **G. cattleyae,** VA.

Bacterial brown rot. **Erwinia cypripedii.**

Bacterial brown spot. **Pseudomonas cattleyae,** common on *Cattleya,* severe on *Phalaenopsis.*

Bacterial leaf scorch and pseudobulb rot. Unidentified.

Bacterial soft rot. **Erwinia carotovora,** serious on *Cattleya* and other general.

Blight, petal. **Botrytis cinerea** (*Sclerotinia fuckeliana*), spotting common on older flowers; **Glomerella** sp., black spot of *Vanda* orchids.

Blotch, sooty. **Gloeodes pomigena.**

Dodder. **Cuscuta gronovii;** laurel dodder. **Cassytha filiformis.**

Leaf spot. **Cercospora epipactidis; C. cypripedii; C. peristeriae; C. dendrobii; C. odontoglossi; Chaetodiplodia** sp.; **Phyllostictina pyriformis; Physalospora** spp., also stem decay; **Diplodia laeliocattleyae; Phyllosticta** spp.; **Selenophoma** spp.; **Septoria selenophomoides; Volutella albido-pila.**

Flyspeck. **Microthyriella rubi.**

Nematode, lesion. **Pratylenchus pratensis; P. scribneri.**

Nonparasitic. **Cattleya Dry Sepal,** industrial fumes.

Rot, black; leaf and heart. **Phytophthora cactorum; Pythium ultimum,** serious on *Cattleyas;* **P. splendens.**

Rot, dry. **Nectria bulbicola,** ?secondary.

Rot, root. **Rhizoctonia solani;** soft, **Fusarium moniliforme,** CA.

Rot, stem; southern blight. **Sclerotium rolfsii.**

Rust. **Uredo behnickiana** (II), NJ, NY, FL; **U. epidendri** (II) FL; **U. guacae** (II), FL; **U. nigropuncta** (II), FL; **Sphenospora kevorkianii,** FL; **S. mera,** FL; **S. saphena,** FL.

Snow mold. **Ptychogaster** sp. Potting fiber mold.

Sooty mold. **Capnodium citri.**

Virus. **Cymbidium Mosaic; Mild Cattleya Color-Break; Severe Cattleya Color-Break; Cattleya Blossom Brown Necrotic Streak; Odontoglossum Ring Spot; Oncidium Ring Spot; Vanda Ring Spot.**

Wilt. **Fusarium oxysporum** f. sp. **cattleyae,** CA, FL, OH; **F. moniliforme.**

When dividing plants use a "hot knife" or disinfest between cuts; keep new orchids isolated; sterilize or fumigate the potting medium and disinfest the bench. Destroy plants seriously affected by rusts.

ORCHIDS
(Native Species)

Leaf spot. **Cercospora cypripedii,** NY, WI; **Fusicladium aplectri,** DE; **Mycosphaerella cypripedii,** NY; **Phyllosticta aplectri,** DE; **Septoria calypsonis,** MI.

Rust. **Puccinia praegracilis** (*Aecidium graebnerianum*) (I), CA, MT, OR, WA, AK; **P. cypripedii** (II, III), IN, NJ, IA, MI, VA, WI; **Uredo goodyerae** (II, III), CA, CO, NM, OR, WA.

OSAGE-ORANGE
(*Maclura*)

Blight, leaf. **Sporodesmium maclurae**, MO, SC, TX; **Botrytis cinerea**, OR, gray mold on stems.
Damping-off. **Pythium ultimum**, NE; **Rhizoctonia solani**, NE.
Leaf spot. **Cercospora maclurae**, AL; **Ovularia maclurae**, AL, LA, TX; **Phyllosticta maclurae**, MO, NJ.
Mistletoe. **Phorodendron tomentosum**, TX; **P. serotinum (flavescens)**, TX.
Rot, root. **Phymatotrichum omnivorum**, TX.
Rust. **Cerotelium fici** (II), SC to FL and TX.
Wilt. **Verticillium** sp., CT.

OSIER, BASKET-WILLOW
(*Salix*)

Canker. **Cryptomyces maximus,** blister; **Valsa salicina,** twig, branch, CA, IA.
Rust, leaf. **Melampsora abieti-capraearum**, NY, PA.

OSOBERRY
(*Osmaronia*)

Leaf spot. **Cylindrosporium nuttallii,** CA to OR, WA; **Gloeosporium osmaroniae**, WA.
Powdery mildew. **Phyllactinia corylea**, OR.

OWLS CLOVER
(*Orthocarpus*)

Leaf spot. **Ascochyta garrettiana**, OR, UT.
Rust. **Cronartium coleosporioides** (II, III), CO, ID, UT; 0, I, on pine.

OXALIS
(Wood-Sorrel)

Blight, gray mold. **Botrytis cinerea**, AK.
Leaf spot. **Cercospora oxalidiphila**, WI; **Phyllosticta guttulatae**, VT to NJ, IN, and WI, **Ramularia oxalidis**, NH, NE, NM, OR, PA, TN, VT; **Septoria acetosella**, NY; tar spot, **Phyllachora oxalina**, DE, ME, VT.

Nematode, root knot. **Meloidogyne** spp.
Nematode, stem. **Ditylenchus dipsaci,** NY.
Powdery mildew. **Microsphaera russellii,** ME to WV, KS and MN, WA.
Rot, root. **Thielaviopsis basicola,** CT.
Rust. **Puccinia andropogonis** (0, I), OK, TX, II, III, on *Andropogon*; **P. oxalidis** (II, III), FL, GA, LA, MS, NM, SC, TN, TX, WI; 0, I, on mahonia; **P. sorghi** (0, I), KS, IN, IA, MI, MS, ND, SD, OK, TX; II, III, on corn and *Euchlaena.*
Smut, seed. **Ustilago oxalidis,** CT to MS, TX, OH, WI.
Virus. **Beet Curly Top,** TX; **Tomato Ringspot,** PA.

OXYDENDRON
(Sourwood, Sorrel-Tree)

Blight, twig. **Sphaerulina polyspora,** NC.
Leaf spot. **Cercospora oxydendri,** AL, MS, TX, WV; **Mycosphaerella caroliniana,** GA, NC, TX, WV.
Rot, root. **Phymatotrichum omnivorum,** TX.
Rot, wood. **Poria punctata,** WV.

OYSTER MUSHROOM
(*Pleurotus ostreatus*)

Dry bubble. **Verticillium fungicola,** PA.

PACHISTIMA

Leaf spot. **Mycosphaerella pachystimae,** ID.

PACHYSANDRA
(Japanese-Spurge)

Blight, leaf; stem canker. **Volutella pachysandrae** (*Pseudonectria pachysandricola*), CT, DE, NJ, NY, PA, VA; **Sphaeropsis** sp., tip blight, secondary.
Leaf spot. **Gloeosporium** sp., VA; **Phyllosticta pachysandrae,** NC, NY, PA, VA; **Septoria pachysandrae,** FL, TN.
Nematode, root knot. **Meloidogyne** sp.
Virus. **Alfalfa Mosaic,** NJ.

Volutella leaf blight is fairly common after injury or when plants are too crowded; pinkish spore pustules appear on stems, large brown areas on leaves.

PACIFIC WAX MYRTLE
(*Myrica californica*)

Leaf spot. **Phyllosticta myricae,** CA.
Rust. **Cronartium comptoniae** (II, III), OR.

PAINTED CUP, INDIAN PAINTBRUSH
(*Castilleja*)

Leaf spot. **Cercospora** sp., AK.
Powdery mildew. **Erysiphe polygoni,** CO, WA; **Sphaerotheca macularis,** CO, WA, WI.
Rust. **Cronartium coleosporioides** (II, III), CA, CO, SD to NM, UT, WY, WA; 0, I, on pine; **Puccinia andropogonis** (0, I), CA, IA MT, NM, WI; II, III, on *Andropogon*; **P. castillejae** (II, III), CA, UT.

PALM
(*Chamaedorea*)

Leaf spot. **Bipolaris setariae,** FL; **Exserohilum** (*Helminthosporium*) **rostratum,** FL; **Phaetrichoconis crotalariae,** FL.
Rot, stem. **Gliocladium vermoeseni,** FL.

PALM ARECA
(*Chrysalidocarpus*)

Leaf spot. **Bipolaris setariae,** FL; **Exserohilum** (*Helminthosporium*) **rostratum,** FL; **Phaetrichoconis crotalariae,** FL.

PALM, COCONUT
(*Cocos*)

Bacterial, MLO. **Lethal Yellowing,** FL.
Blight, thread. **Pellicularia koleroga,** FL.
Leaf scorch; leaf-bitten disease; stem bleeding. **Ceratocystis** (*Endoconidiophora, Thielaviopsis*) **paradoxa,** FL.
Leaf spot, gray; leaf break. **Pestalotia palmarum,** secondary, FL.
Rot, bud; leaf drop; wilt. **Phytophthora palmivora,** FL; **Pythium** sp., wilt, FL.

PALM, DATE
(*Phoenix*)

Blight, inflorescence; fruit rot. **Fusarium** spp., AZ.
Canker. **Penicillium vermoeseni,** CA.

Fruit spot, brown. **Helminthosporium molle**, AZ; **Alternaria citri**, AZ, CA, TX.

Leaf scorch; black heart; bud rot. **Ceratocystis paradoxa**, AZ, CA.

Leaf spot. **Exosporium palmivorum**, Gulf states; **Pestalotia** sp., CA, FL; **P. palmarum**, FL; **Alternaria** sp., FL, AZ, CA, TX; **Annellophora phoenicis**, TX.

Nematode, root knot. **Meloidogyne** spp., AZ, CA.

Nonparasitic. **Black Nose**, fruit checking from rain and high humidity. **Chlorosis**, manganese deficiency, FL.

Rot; decline disease. **Omphalia pigmentata** and **O. tralucida**, CA.

Rot, fruit. **Aspergillus niger**, calyx-end, CA, WA; **Catenularia fuliginea**, AZ, CA; **Penicillium roseum**, CA, AZ; **Pleospora herbarum**, CA; **Phomopsis phoenicola**, CA; **Alternaria stemphylioides**, CA.

Rot, leaf-stalk; shoot blight; fruit rot. **Diplodia phoenicum**, AZ, CA.

Rot, root. **Armillaria mellea**, CA; **Clitocybe tabescens**, FL; **Ceratostomella radicola**, CA.

Rot, wood. **Poria** spp., AZ, CA.

Smut, false. **Graphiola phoenicis**, widespread.

PALM, FISHTAIL
(*Caryota*)

Blight, leaf and stem. **Glomerella cingulata**.

Leaf spot. **Bipolaris setariae**, FL; **Exserohilum** (*Helminthosporium*) **rostratum**, FL; **Phaetrichoconis crotalariae**, FL.

PALM, QUEEN, PLUMY COCONUT
(*Arecastrum*)

Canker; gummosis. **Dothiorella gregaria**, CA; **Penicillium vermoeseni**, CA.

Leaf spot. **Exosporium palmivorum**, FL; **Glomerella cincta**, NJ; **Septoria cocoina**, MO.

Nonparasitic. **Chlorosis**, manganese deficiency, FL.

Rot, bud; wilt. **Phytophthora palmivora**, FL; **Pythium** sp., wilt.

Rot, butt. **Ganoderma sulcatum**, FL.

Smut, false. **Graphiola phoenicis**, FL.

PALM, Rhapis
(*Rhapis*)

Leaf spot. **Bipolaris setariae**, FL; **Exserohilum** (*Helminthosporium*) **rostratum**, FL; **Phaetrichoconis crotalariae**, FL.

PALM, ROYAL
(*Roystonea*)

Anthracnose; petiole spot. **Colletotrichum gloeosporioides,** FL, TX.
Leaf spot. **Alternaria** sp., FL; **Diplodia** sp., FL; **Epicoccum neglectum,** FL; **Helminthosporium** sp., leaf stripe, FL.
Little leaf. Cause unknown, FL.
Nematode, root knot. **Meloidogyne** sp., FL.
Nematode, spiral. **Helicotylenchus nannus.**
Smut, false. **Graphiola phoenicis,** FL.
Wilt. **Phytophthora palmivora,** FL.

PALM, SUGAR
(*Arenga*)

Smut, false. **Graphiola phoenicis,** occasional.

PALM, WASHINGTON
(*Washingtonia*)

Bacterial leaf spot. **Pseudomonas washingtoniae,** AZ.
Leaf spot. **Auerswaldia** sp., CA; **Cercospora** sp., FL; **Colletotrichum** sp., FL; **Cylindrocladium macrosporium,** FL; **Pestalotia palmarum,** FL; **Phoma palmicola,** ?secondary, TX.
Nematode, root knot. **Meloidogyne** spp., AZ, FL.
Rot, bud. **Penicillium vermoeseni,** CA; **Phytophthora palmivora,** AZ, FL.
Rot, root. **Clitocybe tabescens,** FL; **Phymatotrichum omnivorum,** FL.
Rot, trunk. **Phytophthora** sp., CA.
Smut, false. **Graphiola phoenicis,** FL, TX.
Wilt. **Pythium** sp., FL.

The Penicillium bud rot causes serious losses to *Washingtonia filifera* in California; *W. robusta* is resistant.

PALMETTO, CABBAGE PALM
(*Sabal*)

Black mildew. **Asterina sabalicola,** FL, GA; **Meliola** spp., Gulf states.
Canker, felt fungus. **Septobasidium sabalis,** LA; **S. sabal-minor,** FL.
Leaf spot. **Helminthosporium apiculiferum,** LA, MS; **Mycosphaerella serrulata,** FL, SC; **Phyllachora** (*Catacauma*) **sabal,** Black Spot, FL, GA, TX; **Phyllosticta palmetto,** LA, MS; **Myrianginella sabaleos,** Black Speck, FL, GA.
Smut, false. **Graphiola congesta,** AL, FL, SC; **G. phoenicis,** FL, MS; **G. thaxteri,** FL.

PALOVERDE
(*Cercidium*)

Mistletoe. **Phoradendron californicum,** TX to CA.
Rot, root. **Phymatotrichum omnivorum,** TX.

PAMPAS GRASS
(*Cordaderia*)

Leaf spot. **Helminthosporium** sp., GA; **Hendersonia culmiseda,** leaf mold; **Phyllosticta** sp., KY.

PANDANUS
(Screw Pine)

Leaf spot. **Heterosporium iridis,** IA; **Macrophoma pandani,** CA, FL; **Melanconium pandani,** FL, MD; **Phomopsis** sp., NJ; **Pestalotia palmarum,** ?secondary, FL; **Volutella mellea,** NY.
Nematode, burrowing. **Radopholus similis,** FL.

PANSY
(*Viola tricolor*)

Anthracnose. **Colletotrichum violae-tricoloris,** ME to FL, IN and MI, WA, PA.
Blight, gray mold. **Botrytis cinerea,** AK, LA, NJ, probably general.
Blight, southern. **Sclerotium rolfsii,** FL, VA.
Damping-off; root rot. **Pythium** spp., CA, CT, MO, NJ; **Rhizoctonia solani,** DE, IL, MN, NY.
Downy mildew. **Peronospora violae,** AL, IL, MS, NE.
Leaf spot. **Alternaria violae,** NJ, NY, PA; **Cercospora violae,** CT, IN, MI, NY, TX, WI; **Phyllosticta rafinesquii,** AL, IL; **Ramularia agrestis,** OR; **R. lactea,** WA.
Nematode, root knot. **Meloidogyne** sp., NY, TX.
Powdery mildew. **Sphaerotheca macularis,** IA, KS, WA.
Rot; leaf spot. **Centrospora acerina,** CA, GA.
Rot, root; wilt. **Aphanomyces cladogamus,** MD; **Fusarium oxysporum,** CT, MI, NE, NJ, NY, OH, TX; **Thielaviopsis basicola,** CT.
Rust. **Puccinia ellisiana** (0, I), KS, NE; II, III, on *Andropogon*; **P. violae** (0, I, II, III), CT, FL, KS, NJ, ND, SC; **Uromyces andropogonis** (0, I), CT; II, III, on *Andropogon*.
Smut, seed. **Urocystis kmetiana,** AR, MO, TN.
Spot anthracnose; pansy scab. **Sphaceloma violae,** KY, MD, NJ.
Virus. **Beet Curly Top,** CA, OR, TX; **Mosaic,** unidentified, MD; **Western Cucumber Mosaic** and **Cherry Calico** cause flower break and leaf mottle in CA.

PAPAYA
(*Carica*)

Anthracnose; fruit, stem, and leaf spot. **Glomerella cingulata,** general.
Black mildew. **Asterina caricarum,** FL.
Blight, flower; leaf spot. **Choanephora americana,** FL.
Blight, southern. **Sclerotium rolfsii,** TX.
Damping-off. **Pythium aphanidermatum,** CA, FL; **P. debaryanum,** TX.
Leaf spot. **Asperisporium caricae,** FL; **Mycosphaerella caricae,** target spot,
    FL; **Phyllosticta caricae-papayae,** target spot, FL.
Powdery mildew. **Erysiphe cichoracearum,** CA; **Oidium caricae,** FL, TX.
Nematode. **Meloidogyne** sp., FL, TX.
Rot, crown, collar. **Calonectria** sp. (*Cylindrocladium*), HI.
Rot, fruit. **Diplodia** sp., TX.
Rot, root. **Phymatotrichum omnivorum,** TX.
Rot, stem. **Fusarium** sp., CA, TX.
Virus. **?Tobacco Ring Spot.**

PARKINSONIA

Leaf spot. **Cylindrosporium parkinsoniae,** TX; **Phyllosticta parkinsoniae,** TX.
Mistletoe. **Phoradendron californicum,** CA, TX.
Rot, root. **Clitocybe tabescens,** FL; **Phymatotrichum omnivorum,** AZ, TX.
Sooty mold. **Capnodium** sp., TX.

PARROTFEATHER
(*Myriophyllum*)

Rot, root and stem. **Pythium carolinianum,** CA.

PARSLEY
(*Petroselinum*)

Bacterial, MLO. **Aster Yellows,** NY, CO; **California Aster Yellows,** CA.
Bacterial soft rot. **Erwinia aroideae,** CA; **E. carotovora,** Fl.
Blight, gray mold. **Botrytis cinerea,** AK.
Blight, leaf. **Alternaria dauci,** CT; **Septoria petroselini,** CA, CT, NJ, NY.
Damping-off; root rot. **Pythium** sp., NJ; **Rhizoctonia solani,** NJ, NY.
Dodder. **Cuscuta** sp., TX.
Leaf spot. **Cercospora petroselini,** NJ; **Stemphylium** sp., NJ.
Nematode, bulb. **Ditylenchus dipsaci,** CA.
Nematode, pin. **Paratylenchus projectus,** MD.
Nematode, root knot. **Meloidogyne** spp., FL, GA, KS, TX, VA.

Rot, root. **Phymatotrichum omnivorum**, TX.
Rot, stem. **Sclerotinia sclerotiorum**, CT, GA, LA, NY, PA, TX, VA.
Virus. **Alfalfa Mosaic, CA; Strawberry Latent Ringspot**, CA.
Virus. **Beet Curly Top**, CA.

PARSNIP
(*Pastinaca*)

Bacterial crown gall. **Agrobacterium tumefaciens**, VA.
Bacterial, MLO. **Aster Yellows**, IL, ME, NY, PA, SD, TX, WI, and **California Aster Yellows**, CA, WA.
Bacterial rot. **Pseudomonas pastinacae**, NY; **Erwinia carotovora**, occasional in storage.
Blight, leaf; canker. **Itersonilia perplexans**, GA, MA, NY.
Downy mildew. **Plasmopara nivea.**
Leaf spot. **Alternaria** sp., NJ; **Cercospora pastinacae**, MD, NE, NY, TX, WV; **C. pastinacina**, CA, IN, MI, NY; **Cylindrosporium pastinacae**, UT, WI; **Ramularia pastinacae**, IN, MA, NY, ND, OH; **Septoria pastinacae**, TX; **Cercosporella pastinacae**, MA; **Phomopsis dichenii.**
Nematode, root knot. **Meloidogyne** spp., NJ to KS and southward.
Nonparasitic. **Heart Rot**, boron deficiency, NY.
Powdery mildew. **Erysiphe unbelliferarum.**
Rot, black, crown. **Centrospora acerina.**
Rot, black mold. **Rhizopus** spp., occasional in storage.
Rot, gray mold. **Botrytris cinerea**, occasional.
Rot, root. **Phymatotrichum omnivorum**, AZ, TX; **Rhizoctonia solani**, black scurf, NC, TX; **Phytophthora parasitica**, dry rot, PA; **Phoma** sp.
Rot, watery soft. **Sclerotinia sclerotiorum**, CA, IN, LA, MA.
Scab. **Streptomyces scabies**, WA.
Virus. **Mosaic**, unidentified, OR, UT.

PARTRIDGE-BERRY
(*Mitchella*)

Black mildew. **Meliola mitchellae**, AL, FL, MS, PA.
Rot, stem. **Sclerotium rolfsii**, MD.

PASSION FLOWER
(*Passiflora*)

Blight, southern. **Sclerotium rolfsii**, FL.
Leaf spot. **Colletotrichum gloeosporioides**, FL, also stem spot; **Cercospora biformis**, AR, NC; **C. fuscovirens**, ME to MO, TX; **C. regalis**, TX; **C. truncatella**, SC, TX; **Gloeosporium fructigenum**, LA; **Phyllosticta** sp., NJ.

Nematode, root knot. **Meloidogyne** sp.
Rot, collar. **Sclerotinia** sp., CA.
Rot, root. **Phymatotrichum omnivorum,** TX.
Virus. **Cucumber Mosaic,** CA.

PAULOWNIA
(Princess-Tree)

Leaf spot. **Ascochyta paulowniae,** MD; **Phyllosticta paulowniae,** AL, MD, NY, OK.
Powdery mildew. **Phyllactinia corylea; Uncinula clintonii.**
Rot, root. **Phymatotrichum omnivorum,** TX; **Armillaria mellea.**
Rot, wood. **Polyporus spraguei,** AL; **P. versicolor,** MD; **P. robiniophilus.**

PAWPAW
(*Asimina*)

Blotch, leaf. **Phleospora asiminae,** IL, KS, MO, OH, WV.
Canker; dieback. **Nectria cinnabarina; Polyporus amplectrens,** FL, GA; **Sphaeropsis asiminae,** MD, WV; **Valsa ambiens,** VA, WV.
Leaf spot. **Cercospora asiminae,** widespread; **Phyllosticta asiminae,** widespread; **Mycosphaerella** sp., IN, OH, WV; **Septoria asiminae,** TX.
Parasitic lichen. **Strigula elegans** and **S. camplanata,** Southern U.S., LA.
Rot, root. **Phymatotrichum omnivorum,** TX.
Rot, sapwood. **Poria isabellina,** VA, WV.

PEA
(*Pisum*)

Anthracnose, leaf and pod spot. **Colletotrichum pisi,** CT, GA, IA, ME, MS, TX.
Bacterial blight. **Pseudomonas syringae** pv. **pisi,** general, especially East and South.
Bacterial leaf spot. **Pseudomonas syringae,** ID.
Blight, ashy stem, charcoal rot. **Macrophomina phaseoli,** TX.
Blight, foot rot. **Mycosphaerella pinodes,** widespread, usually with **Ascochyta pinodella.**
Blight, leaf and stem. **Choanephora conjuncta,** GA.
Blight, southern. **Sclerotium rolfsii,** FL.
Damping-off; root rot, pod rot. **Pythium** spp. and **Rhizoctonia solani,** general.
Dodder. **Cuscuta** sp., MO.
Downy mildew. **Peronospora viciae** (*P. pisi*), general, especially northern central and Pacific Coast states.

Leaf spot. **Alternaria** sp., seedling blight, DE, NH; **A. alternata,** blight, WI; **Ascochyta pisi,** general but rare in Northwest; **Cercospora pisi-sativae,** GA; **Fusicladium pisicola,** black leaf, ID, UT; **Pleospora hyalospora,** MS; **Septoria flagellifera,** MN, ND, SD, WI; **S. pisi,** widespread; **Stemphylium polymorphum,** ME.

Nematode, hop cyst. **Heterodera humuli,** OR.

Nematode, oat cyst. **Heterodera avenae,** WA.

Nematode, pea cyst. **Heterodera gottingiana.**

Nematode, root knot. **Meloidogyne** spp., AZ, CA, FL, NC, SC, TX, UT, WI.

Nematode, sting. **Belonolaimus gracilis.**

Nonparasitic. **Chlorosis,** manganese or zinc deficiency, FL, TX, WA. **Intumescence,** pod swellings, CA, NJ, WA. **Seed Spotting,** cause unknown.

Powdery mildew. **Erysiphe polygoni,** general.

Rot, black mold. **Rhizopus stolonifer,** cosmopolitan; gray mold, **Botrytis cinerea,** occasional.

Rot, foot. **Aphanomyces euteiches,** general except in far north.

Rot, root. **Aphanomyces euteiches,** ID; **Fusarium solani** f. **pisi,** widespread; **Pellicularia filamentosa,** stem canker, general; **Phoma** sp., NJ, WI; **Pyrenochaete terrestris,** IA; **Phymatotrichum omnivorum,** TX; **Phytophthora** sp., CA, CT; **Thielaviopsis basicola,** AR, CA.

Rot, stem; wilt. **Sclerotinia sclerotiorum,** CA, DE, FL, ID, MT, NJ, PA, TX, VA, WA.

Rust. **Uromyces favae** (0, I, II, III), CA, ID, ME, MA, MN, NE, ND, WA, WY.

Scab; black spot; leaf mold. **Cladosporium pisicola,** CA, ME, OR, UT, WA, TX.

Virus. **Pea Enation Mosaic; Pea Mosaic; Pea Mottle; Pea Streak; Wisconsin Pea Streak; Pea Wilt; Clover** (red) **Vein Mosaic; Tomato Spotted Wilt.**

Virus, seed-borne; **Pea Stunt,** WI, **Turnip Mosaic Virus,** MN.

Wilt. **Fusarium oxysporum** f. sp. **pisi,** NH to SC, IL, MN, NE, CA, CO, ID, MT, OR, WA; and race 2, Near Wilt; **Fusarium oxysporum** f. sp. **medicaginis,** MS.

Resistant varieties are the answer to Fusarium wilt and some virus diseases. Using clean seed, preferably grown in the West, is the best way to avoid bacterial blight. A well-drained, fertile soil, 3- to 5-year rotation, cleaning up or plowing under pea refuse immediately after harvest, all help to produce healthy peas. Some organic soil amendments reduce pea wilt.

PEACH
(Prunus persica)

Bacterial crown gall. **Agrobacterium tumefaciens,** general.

Bacterial hairy root. **Agrobacterium rhizogenes,** general.

Bacterial leaf spot. **Xanthomonas pruni,** eastern, central, southern States.

Bacterial, MLO. **Peach X-Disease,** CT, MA, MI, NY, OH, PA.

Bacterial shoot blight; canker; gummosis. **Pseudomonas syringae,** CA, OR.

Blight, blossom; green fruit rot. **Sclerotinia sclerotiorum,** CA.

Blight, leaf. **Fabraea maculata,** CA, NJ.

Blight; shot-hole disease; pustular spot. **Coryneum carpophilum,** general.

Blight, twig. **Coniothyrium** sp., TX; **Cyphella marginata,** OR; **Nectria cinnabarina,** AL.

Canker, crown. **Phytophthora citrophthora,** CA; **P. cactorum,** stem, AR, CA; **P. syringae,** OR.

Canker; dieback; gummosis. **Botryosphaeria dothidea,** AL, FL, GA, LA, TX, TN.

Canker; dieback; gummosis. **Botryosphaeria ribis** var. **chromogena,** FL, GA; **Valsa** (*Cytospora*) **cincta** and **V. leucostoma,** widespread East and central states; **Coniothyrium fuckelii,** WV; **Ceratocystis fimbriata,** CA.

Canker; peach; constriction disease. **Fusicoccum amygdali,** East and South. Sometimes reported as **Phoma persicae.**

Damping-off. **Rhizoctonia solani,** AR, CT.

Leaf curl. **Taphrina deformans,** general.

Leaf spot. **Phyllosticta circumscissa,** widespread; **P. persicae,** MD, NE, OH; **Mycosphaerella persicae,** frosty mildew, widespread.

Leaf spot; shot-hole. **Cercospora circumscissa,** general; **C. consobrina,** IL, LA.

Nematode, dagger. **Xiphinema americanum.**

Nematode, lesion. **Pratylenchus penetrans; P. vulnus.**

Nematode, ring. **Criconemoides simile, C. xenoplax,** NJ.

Nematode, root knot. **Meloidogyne incognita, M. incognita-acrita; M. javanica,** on "nematode-resistant" varieties; **M. arenaria.**

Nematode, sting. **Belonolaimus gracilis.**

Nonparasitic.

> **Catface.** Insect blemish on fruit, often from plant bugs.
>
> **Chlorosis.** Iron and magnesium deficiency in alkaline soils, Southwest.
>
> **Gummosis.** Winter injury, bad drainage.
>
> **Internal Bark Necrosis.** ?Manganese toxicity, IL.
>
> **Little Leaf.** Zinc deficiency, CA.
>
> **Stem Pitting.** Cause unknown, WV.
>
> **Suture Spot.** Cause unknown. Decline, PA.

Powdery mildew. **Podosphaera oxyacanthae,** general; **Sphaerotheca pannosa** var. **persicae,** general.

Rot, brown. **Monilinia fructicola,** general.

Rot, brown; blossom and twig blight. **Monilinia laxa,** Pacific states.

Rot, bud and twig blight. **Fusarium lateritium,** other spp., CA, GA, KS.

Rot, fruit. **Botryosphaeria obtusa,** AL, GA; **B. dothidea,** AL, GA; **Aspergillus niger,** black mold, cosmopolitan; **Botrytis cinerea,** gray mold, cosmopolitan; **Cephalothecium roseum,** pink mold, widespread; **Choanephora persicaria,** in market; **Diplodia natalensis,** gumming disease; **D. persicae,** GA; **Fusarium** spp., CA, IL, NY, TX; **Gibberella persicaria,** CA; **Glomerella** cingulata, occasional, AL, GA; **Rhizopus nigricans,** cosmopolitan; **Pho-mopsis** sp., NJ.

Rot, root. **Armillaria mellea,** cosmopolitan; **Clitocybe monadelpha,** AR, FL, MO, OK; **C. parasitica,** OK; **C. tabescens,** FL, GA; **Ganoderma curtisii,** NC, VA; **Phymatotrichum omnivorum,** AZ, TX; **Cylindrocladium floridanum,** GA.

Rot, seedling stem. **Sclerotium rolfsii,** South; wilt, **Cylindrocladium floridanum; Penicillium** sp.

Rot, silver leaf. **Stereum purpureum,** occasional.

Rot, wood. **Fomes** spp.; **Lenzites saepiaria; Polyporus hirsutus, P. lacteus** and **P. versicolor,** cosmopolitan; **Stereum hirsutum; Shizophyllum commune,** wound rot.

Rust. **Tranzschelia discolor** (II, III), general; 0, I, on anemone, hepatica.

Scab. **Cladosporium carpophilum,** general.

Sooty mold; fruit stain. **Fumago vagans,** cosmopolitan.

Virus. **Peach Asteroid Spot,** OK and TX to southern CA, OR, UT, WA; **Peach Calico,** ID, WA; **Peach Dwarf; Peach Golden Net,** CO; **Peach Little Peach,** eastern US, CT to NC to MI, MO; **Peach Mosaic,** OK and TX to CA, CO, UT; **Peach Mottle,** ID; **Peach Necrotic Leaf Spot,** MI; **Peach Phony Disease,** Southeast to MO and TX; **Peach Red Suture,** IN, MD, MI; **Peach Ring Spot,** CA, WA, **Peach Rosette,** SC to FL, TN, MS, MI to AR, OK; **Peach Rosette Mosaic,** MI, NY; **Peach Stubby Twig; Peach Wart,** AZ, CA, ID, OR, WA; **Peach Yellow Bud Mosaic,** CA, **Peach Yellow Leaf Roll** (western X-disease, cherry buckskin), CA, CO, ID, OR, UT, WA; **Peach Yellows; Mule's Ear Disease** (Drake almond bud failure); **Prunus Ring Spot; Cherry Yellows; Green Ring Mottle.**

Virus. **Stem Pitting.**

Wilt. **Verticillium albo-atrum,** occasional.

Brown rot is the major peach enemy, and spray schedules are built around it, although they start with a dormant spray for leaf curl. Get the latest advice and spray schedule from your county agricultural agent. Peach foliage is sensitive to arsenicals; do not allow a spray prepared for shade trees to touch peaches. The commercial propagator must take many precautions to provide stock free from the numerous viruses.

PEANUT
(*Arachis*)

Bacterial wilt. **Pseudomonas solanacearum,** AL, FL, NC, VA.

Blight, seedling. **Rhizoctonia** spp., NC to FL, CA, OK, TX.

Blight, southern; stem and nut rot. **Sclerotium rolfsii,** general.

Blight, stem. **Diaporthe sojae,** VA, WV.

Blotch, web. **Phoma arachidicola,** VA.

Damping-off. Synergistic interaction of **Pythium myriotylum, Fusarium solani, Meloidogyne arenaria,** FL.

Leaf spot. Alternaria sp., FL, MO, NJ, NM, SC; **Ascochyta** sp., AR; **Phoma** sp., MO, VA; **Phyllosticta** sp., AL, AR, MS; **Pleospora** sp., AR, OK; **Stemphylium** sp., ND; **Leptosphaerulin arachidicola,** GA, TX.

Leaf spot, brown, halo. **Mycosphaerella arachidicola,** general; **M. berkeleyi** (*Cercospora personata*), general.

Nematode. Panagrolaimus subelongatus, associated with shoot elongation.

Nematode, leaf. **Aphelenchoides** sp.

Nematode, lesion. **Pratylenchus brachyurus,** GA, AL, VA, FL, SC.

Nematode, reniform. **Rotylenchus reniformis.**

Nematode, ring. **Criconemoides cylindricum.**

Nematode, root knot. **Meloidogyne arenaria; M. hapla; M. javanica,** GA.

Nematode, spiral. **Heliocotylenchus nannus.**

Nematode, sting. **Belonolaimus gracilis; B. longicaudatus.**

Nematode, stubby root. **Trichodorus christiei.**

Nematode, stunt. **Tylenchorhynchus claytoni.**

Nonparasitic.

    Blue Stain. Seedcoat discoloration.

    Chlorosis. Excess lime; magnesium, manganese, or iron deficiency.

    Necrotic Spot. May be caused by nutrient deficiency.

    Pops. Empty pods. May be caused by nutritional deficiency.

    Pouts. Stunting and chlorotic spotting from thrips.

Rot, Black Mold, of Pods. **Cladosporium herbarum,** cosmopolitan.

Rot, blue mold, of pods and nuts. **Penicillium** sp., cosmopolitan.

Rot, charcoal. **Macrophomina phaseoli,** CO, NC, SC, OK, TX.

Rot, collar. **Diplodia gossypina.**

Rot, crown. **Aspergillus niger,** GA, NM, TX.

Rot, Cylindrocladium black rot. **Cylindrocladium crotalariae,** VA, NC.

Rot, gray mold, leaf. **Botrytis cinerea,** CT, MD, MS, TN, VA.

Rot, peg, pod, root. **Calonectria** (*Cylindrocladium*) **crotalariae,** GA.

Rot, root. **Curvalaria inaequalis,** SC; **Helminthosporium** sp., OK; **Phymatotrichum omnivorum,** AZ, TX; **Pythium** sp., CA, GA, NC; **Thielaviopsis basicola,** NC.

Rot, root and pod. **Sclerotinia minor, S. sclerotiorum,** NC, VA.

Rot, seed. **Rhizopus** spp.; **Trichoderma viride,** SC.

Rot, stem, pod. **Fusarium** spp., also root rot, wilt; **Physalospora rhodina,** FL, GA; **Rhizoctonia solani.**

Rust. Puccinia arachidis (II), occasional, FL, GA, TX, NC, VA.

Virus. Peanut Stunt, VA, NC, AL, GA; **Peanut Mottle,** OK; **Cowpea Chlorotic Mottle,** SD.

Virus. Tomato Spotted Wilt, TX.

Wilt. Pythium myriotylum, VA.

Wilt. Verticillium sp., NM.

PEAR
(Pyrus)

Anthracnose, northwestern. **Neofabraea malicorticis,** OR, WA.

Bacterial blossom, twig blight; canker. **Pseudomonas syringae,** AR, CA, CT.

Bacterial crown gall. **Agrobacterium tumefaciens,** general.

Bacterial fire blight. **Erwinia amylovora,** general.

Bacterial fruit rot. **Erwinia carotovora,** MA.

Blight, leaf; black fruit spot. **Fabraea maculata,** general.

Blight, thread. **Pellicularia koleroga,** NC to FL, TX, WV.

Blight, twig. **Corticium salmonicola,** FL, LA; **Fusarium** spp., occasional; **Phomopsis ambigua,** widespread; **Valsa leucostoma,** WA.

Blotch, sooty. **Gloeodes pomigena,** eastern states to OK, TX.

Canker. **Botryosphaeria dothidea,** AL; **Cytospora** spp., OR, VA, WA; **Nectria cinnabarina,** coral spot; dieback; **N. galligena,** trunk canker; **Nummularia discreta,** DE, IA.

Canker, bark. **Glutinium microsporum,** OR; **Helminthosporium papulosum,** blister canker, black pox; **Myxosporium corticolum,** NY to MI, MS, OR.

Canker, felt fungus. **Septobasidium** spp., NC to FL and TX.

Canker, perennial. **Neofabraea perennans,** OR, WA.

Fruit spot; flyspeck. **Leptothyrium pomi,** eastern states.

Leaf spot. **Cercospora minima,** FL to TX; **C. pyri,** MI; **Coniothyrium pyrinum,** MA to AL, TX, IA; **Coryneum foliicolum,** IN; **Hendersonia cydoniae,** NY; **Phyllosticta pyrorum,** IL, MS, SC; **Mycosphaerella sentina,** ashy leaf spot, fruit spot, widespread, especially in East.

Mistletoe. **Phoradendron serotinum (flavescens),** AZ, TX.

Mistletoe, European. **Viscum album,** CA.

Nematode, lesion. **Pratylenchus minyus; P. pratensis,** CA.

Nematode, root knot. **Meloidogyne** sp., CA.

Nonparasitic

  Bitter Pit. Moisture irregularity, Pacific Coast, NY.

  Black End; Hard end. Oriental pear rootstocks on shallow, poor soil.

  Black Leaf; Brown Bark Spot; Brown Blotch, of Kieffer fruit. Undetermined.

  Chlorosis. Mineral deficiency, soil alkalinity, Pacific Coast.

  Cork; drought spot; fruit pitting. Boron deficiency, Pacific Coast, TX.

  Exanthema. Copper deficiency, CA, FL.

  Little Leaf; Rosette. Zinc deficiency, sometimes boron, CA.

  Marginal Leaf Blight; Scorch. Potassium or calcium deficiency, ID, WA.

  Red Leaf. In Oriental pear, undetermined cause.

  Scald. Immaturity; deficient ventilation.

  Stigmonose. Insect punctures during growth, widespread.

  Target Canker, Measles. Undetermined, GA, NY, VA, CA, WA.

Powdery mildew. **Podosphaera leucotricha,** CO, OR, WA; **P. oxycanthae,** NJ.

Rot, bitter; twig, branch canker. **Glomerella cingulata,** widespread but not destructive.

Rot, black; canker; leaf spot. **Physalospora obtusa,** widespread.

Rot; blossom and twig blight. **Botrytis cinerea,** widespread.

Rot, brown. **Monilinia fructicola,** eastern states, TX, WA.

Rot, brown; blossom blight. **Monilinia laxa.** Pacific states.

Rot, collar, root. **Phytophthora cactorum,** widespread; **P. citrophthora,** CA.

Rot, fruit. **Alternaria** sp., black mold; **Aspergillus** sp., cosmopolitan; **Botryosphaeria ribis,** black rot, VA; **Cephalosporium carpogenum,** storage; **Trichothecium roseum,** occasional pink mold; **Cladosporium** sp., occasional; **Coprinus urticicola,** OR; **Gloeosporium** sp., widespread **Neurospora sitophila,** ripe rot, NC; **Penicillium** spp., blue mold, widespread; **Phialophora malorum,** storage; **Phoma exigua; P. mali; Rhizopus nigricans,** black mold, cosmopolitan; **Pleospora fructicola; Sclerotinia sclerotiorum,** CA, WA; **Sporotrichum malorum,** storage rot, OR, WA.

Rot, heart. **Fomes igniarius; F. pinicola; Polyporus** spp.

Rot, root. **Armillaria mellea,** widespread; **Phymatotrichum omnivorum,** TX, OK, AZ; **Xylaria** sp., ID, IN; **Xylaria mali,** VA; **Phytophthora cinnamomi.**

Rot, silver leaf. **Stereum purpureum,** NY, OR; **S. hirsutum,** trunk rot.

Rot, trunk. **Schizophyllum commune,** cosmopolitan.

Rust, **Gymnosporangium clavariiforme** (0, I), leaves and fruit, SC: **G. clavipes** (0, I), chiefly on fruit; **G. fuseum,** CA; **G. globosum** (0, I), on leaves and fruit, eastern states to IA, MN; **G. hyalinum** (0, I), FL; **G. kernianum** (0, I), on leaves and fruit, AZ, CO; **G. nootkatense** (0, I), on Asiatic pear, AK.

Scab. **Venturia pyrina,** general.

Spot anthracnose, scab. **Elsinoë pini,** OR, WA.

Virus. **Apple Chlorotic Leafspot, Spy Epinasty Decline, Virginia Crab Stem Pitting and Brownline, Apple Mosaic, Flat Limb, Pear Vein Yellows/Red Mottle, Pear Ring-Pattern Mosaic, Pear Latent** on Bradford Pear (*Pyrus calleryana*), MD.

Virus. **Pear Stony Pit,** Pacific Coast states, NY; **Pear Decline,** CA, OR, WA; **Pear Leaf Curl,** CA.

Virus? **Pear Bark Measles,** CO.

Fire blight is the limiting factor in pear production, many orchards having been abandoned because of this devastating bacterial disease. Kieffer pears and some Asiatic varieties are resistant and are being used in breeding; some resistant forms are now available. Sprays during blossoming, usually an antibiotic, sometimes combined with weak copper, are used in conjunction with sanitary measures for fire-blight control. Pear decline, a virus disease transmitted by the pear psylla, is serious on the Pacific Coast.

PEA-TREE
(*Caragana*)

Bacterial hairy root. **Agrobacterium rhizogenes,** KY to NE, OK.

Blight, leaf. **Ascochyta** sp., OH; **Septoria** sp., MN.

Blight, pod. **Botrytis cinerea, MA.**
Damping-off. **Rhizoctonia solani, ND.**
Leaf spot. **Phyllosticta gallarum, AK, WI.**
Rot, root. **Pellicularia filamentosa, ND; Phymatotrichum omnivorum, TX; Phytophthora cactorum,** wilt, of seedlings.

PECAN
(*Carya illinoensis*)

Bacterial crown gall. **Agrobacterium tumefaciens,** widespread.
Blight, thread. **Pellicularia koleroga, FL, NC.**
Blotch, leaf. **Mycosphaerella dendroides,** South.
Canker. **Cytospora** sp., AZ.
Canker; black bark spot. **Myriangium tuberculans, GA, MS.**
Canker; dieback. **Botryosphaeria berengeriana, AZ, SC to FL.**
Canker, stem. **Microcera** (*Fusarium*) **coccophila, LA, TX.**
Leaf spot. **Microstroma juglandis, GA to TX; Phyllosticta convexula, OK; Septoria caryae, TX; Pestalotia uvicola, FL, TX.**
Leaf spot. algal. **Cephaleuros virescens, FL.**
Leaf spot. brown. **Cercospora fusca,** prevalent through pecan belt.
Leaf spot; downy spot. **Mycosphaerella caryigena, GA and FL to TX.**
Leaf spot; liver spot. **Gnomonia caryae** var. **pecanae, AR, GA, LA, MS, TX, AL.**
Leaf spot; vein spot. **Gnomonia nerviseda, AR, IL, LA, MS, TX.**
Leaf spot; zonate. **Cristulariella pyramidalis, AL.**
Mistletoe. **Phoradendron serotinum (flavescens),** widespread Gulf states.
Nematode. **Paratylenchus** sp.; **Trichodorus** sp.
Nematode, awl. **Dolichodorus obtusus.**
Nematode, dagger. **Xiphinema americanum,** Southeast.
Nematode, root knot. **Meloidogyne** spp.; **M. incognita.**
Nonparasitic
  Black Pit; kernel spot. Feeding punctures of plant bugs.
  Leaf Scorch. Low fertility and soil moisture capacity.
  Rosette. Zinc deficiency.
  Sand Burn of Seedlings. High temperature.
Powdery mildew. **Microsphaera alni,** occasional.
Rot, heart. **Schizophyllum commune,** after drought injury, OK.
Rot, on nuts. **Aspergillus chevalieri,** storage mold; **Trichothecium roseum,** pink mold.
Rot, root. **Armillaria mellea, CA; Helicobasidium purpureum, TX; Phymatotrichum omnivorum, TX; Pestalotia uvicola, FL, TX; Clitocybe tabescens, GA.**
Scab. **Cladosporium effusum,** general.
Spot anthracnose; nursery blight. **Elsinoë randii,** southeastern and Gulf states.
Virus. **Bunch Disease.** Undetermined.

Pecan scab is the limiting factor in nut production in southeastern states, with strains of the fungus sometimes attacking varieties long considered immune; four or five sprays are required for control. In Arizona severe infection from Phymatotrichum root rot has come where trees are intercropped with lucerne, which provides a rapid transit medium for the fungus. Getting rid of the lucerne and treating with ammonium sulfate saves some trees. Zinc sulfate, added to the soil or to foliar sprays, controls rosette.

PENNISETUM

Leaf spot. **Helminthosporium giganteum**, MD.
Smut, seed. **Ustilago penniseti**, VA.
Virus. **Maize Chlorotic Dwarf Virus**, OH.

PENSTEMON
(Beard-Tongue)

Black mildew. **Dimerium alpinum**, CA.
Leaf spot. **Cercospora penstemonis**, AL, KS, IN, MT, ND, SD, NE, OK, TX, WI; **Cercosporella nivosa**, CO, ID, OH, WA; **Phyllosticta antirrhini**, IL; **Ramularia penstemonis**, CA; **Septoria penstemonis**, CA, IL, ME, MI, MS, MO, NY, OK, TX, WA, WI.
Leaf spot; stem spot. **Ascochyta penstemonis**, CA.
Nematode, root knot. **Meloidogyne** sp.
Nematode, stem. **Ditylenchus dipsaci**, VA.
Powdery mildew. **Erysiphe cichoracearum**, WA.
Rot, crown, stem. **Sclerotium rolfsii**, CT, IL, NJ, MA, TX.
Rot, root. **Phymatotrichum omnivorum**, TX.
Rust. **Puccinia andropogonis** (0, I), widespread; II, III, on *Andropogon*; **P. confraga** (III), AZ; **P. palmeri** (0, I, III), AZ, CA, ID, WA, MT, to NM; **P. penstemonis** (III), AZ, CA, OR, UT.

PEONY
(*Paeonia*)

Anthracnose. **Gloeosporium** sp., KS, IL, MA, MD, NC, NJ, PA, VA.
Bacterial crown gall. **Agrobacterium tumefaciens**, MI, MD.
Blight; early bud rot. **Botryris paeoniae**, general.
Blight, late; gray mold. **Botrytis cinerea**, also leaf rot, general.
Blight, southern. **Sclerotium rolfsii**, MS, TX.
Blight, tip; crown rot. **Phytophthora cactorum**, CT, IL, IN, KS, NJ, NY, OH.
Blotch, leaf, stem; measles. **Cladosporium paeoniae**, general.
Canker, stem wilt. **Coniothryium** sp., CA on tree peonies.

Leaf spot. **Alternaria** sp., occasional in northeastern and central states; **Cercospora paeoniae,** IL; **Pezizella oenotherae,** MD, PA; **Phyllosticta** sp., NJ, PA, VA, **Septoria paeoniae,** also stem canker, ME, MI, MN, NJ, NY, OR, RI, WA, WI; **Cryptostictis paeoniae,** IL.

Nematode, leaf and stem. **Ditylenchus dipsaci,** NJ, WA.

Nematode, root knot. **Meloidogyne** spp., widespread; **M. hapla; M. incognita.**

Nematode, spiral. **Rotylenchus buxophilus.**

Nonparasitic. **Bud Blast,** various causes, sometimes potassium deficiency. **Le Moine Disease,** club root, cause unknown, possibly virus.

Powdery mildew. **Erysiphe polygoni,** TX.

Rot, root. **Armillaria mellea,** CA, IA, MI, OR; **Fusarium** sp., CO, IN, MO, NE, NJ, OK; **Phymatotrichum omnivorum,** AZ, TX; **Rhizoctonia solani,** CT, IL, MN, NY, PA, VA; **Thielaviopsis basicola,** CT.

Rot, stem; wilt. **Sclerotinia sclerotiorum,** IL, MD, MN, NJ, NY, OK, OH.

Virus. **Peony Ring Spot,** MA to VA, KS and MI, CA, WA.

Virus. **Witches' Broom, Crown Elongation.** Cause unknown, MD, NY, VA.

Wilt. **Verticillium albo-atrum,** IL, KS, MD, NY, OH.

Botrytis blight is the best-known peony disease. Young shoots are rotted at base; buds turn black, flowers are blasted. Cutting down all tops at ground level in fall and spraying in spring when reddish shoots first show will reduce blight. Anthracnose, blotch, sometimes leaf spot, may be serious occasionally. In humid summers *Sclerotinia sclerotiorum* frequently kills stalks, filling the pith with very large sclerotia. Sickly plants may have root-knot nematodes. Lack of bloom may be due to too-deep planting, growing in shade, botrytis blight, or nematodes.

PEPEROMIA

Anthracnose. **Colletotrichum** sp., WA; **Gloeosporium** sp., WA.

Leaf spot. **Myrothecium roridum,** FL.

Nematode. **Aphelenchoides ritzema-bosi; Pratylenchus** sp.; **Radopholus similis,** FL.

Rot, cutting; root. **Rhizoctonia** sp.; **Pythium splendens,** CA; **Phytophthora ?palmivora,** CA; **P. nicotianae** var. **parasitica,** OH.

Virus. **Ring Spot; Oedema,** pimples in leaves, graft-transmissible, probably due to a virus.

The virus ring spot distorts leaves, which have chlorotic or brown necrotic rings; plants are stunted.

PEPPER
(*Capsicum*)

Anthracnose; fruit. Leaf and stem spot. **Gloeosporium piperatum,** MA to FL, TX, IL; **Glomerella cingulata,** fruit rot, CT to FL, TX, KS, OH.

Bacterial canker, vascular. **Clavibacter michiganense,** WY, on **C. annuum** var. **cerasiforme,** on *C. frutescens* (Bell Pepper), CA.

Bacterial fruit and stem spot; seedling blight. **Xanthomonas vesicatoria,** general in South and East.

Bacterial leaf spot. **Xanthomonas vesicatoria,** NC; **Pseudomonas syringae;** **P. viridiflara,** FL.

Bacterial soft rot. **Erwinia aroideae,** and **E. carotovora, pv. carotovora,** CA, FL, NJ; **Pseudomonas marginalis,** CA, FL, NJ, occasional in market.

Bacterial wilt. **Pseudomonas solanacearum,** PA to FL.

Blight; blossom rot. **Choanephora cucurbitarum,** FL, NC.

Blight; fruit rot. **Phytophthora capsici,** CA, CO, FL, LA, MO, NM, NY, OH, TX, VA; **P. parasitica,** IL, IN.

Blight, southern. **Sclerotium rolfsii,** NC to FL.

Damping-off; stem and root rot. **Pythium** spp., cosmopolitan; **Rhizoctonia solani,** cosmopolitan.

Dodder. **Cuscuta** sp., GA, NJ, VA.

Downy mildew. **Peronospora tabacina,** GA, NC, SC, TX.

Fruit spot, black. Nonparasitic, TX.

Leaf spot. **Ascochyta capsici,** WA, NY to FL; **Cercospora capsici,** frog-eye; stem-end rot; **C. unamunoi,** CA and Gulf states; **Stemphylium solani,** FL; **S. botryosum** f. sp. **capsicum,** NY.

Mold, leaf. **Cladosporium herbarum,** CA, GA, TX; Seed Mold, **Stemphylium botryosum,** CT, FL.

Nematode, awl. **Dolichodorus heterocephalus.**

Nematode, burrowing. **Radopholus similis.**

Nematode, lesion. **Pratylenchus penetrans.**

Nematode, root knot. **Meloidogyne arenaria; M. incognita; M. incognita-acrita; M. hapla.**

Nematode, spiral. **Helicotylenchus erythrinae.**

Nematode, stylet. **Tylenchorhynchus claytoni; T. capitatus.**

Nematode, tobacco cyst. **Heterodera tabacum,** VA.

Nonparasitic. **Blossom-End Rot,** deficient water supply, also lack of calcium. **Sunscald,** high temperature, often after defoliation.

Powdery mildew. **Leveillula taurica,** UT.

Rot, black; internal mold, early blight. **Alternaria** spp., general after sunscald and blossom-end rot.

Rot, charcoal. **Macrophomina phaseoli,** CA, GA, KS, NJ, TX.

Rot, fruit. **Colletotrichum capsici** and **C. nigrum,** general; **Diaporthe phaseolorum,** MS, MO; **Phoma destructiva,** AL, DE, FL, GA, MS, NY; **Penicillium** sp., GA; **Rhizopus stolonifer,** FL, TX, WA; **Sclerotinia sclerotiorum,** also stem rot, CA, CT, FL, MA; **Fusarium** spp.

Rot, gray mold. **Botrytis cinerea,** occasional in market or field.

Rot, pod. **Curvularia lunata,** FL; **Nematospora coryli,** yeast spot after plant bug injury.

Rot, root. **Aphanomyces** sp., NJ, of seedlings; **Phymatotrichum omnivorum,** AZ, TX.

Virus. Pepper Mottle, CA; **Pepper Strain of Alfalfa Mosaic; Pepper Vein-Banding Mosaic** (?Potato Y virus); **Potato Virus X; Cucumber Mosaic; Tobacco Mosaic; Tobacco Etch; Aster Ring Spot; Beet Curly Top; Tomato Spotted Wilt**, TN.
Virus. **Potato Virus Y** and **Tobacco Ring Spot**, TX.
Wilt. **Fusarium annuum**, AZ, CO, LA, ME, MS, NJ, NM, OK, and TX; **Verticillium albo-atrum**, CA, CO, CT, NY, TX, LA.

To control bacterial spot and anthracnose, rotate crops, avoiding land growing potatoes, tomatoes, or eggplant the previous year, and do not grow next to other solanaceous crops. To avoid sunscald, keep fruits shaded; spray to control leaf spots to prevent defoliation. For virus diseases, obtain healthy seed or use resistant varieties. Do not start plants in greenhouses with petunias or Jerusalem cherries; do not smoke or handle tobacco near plants; control aphid vectors; control weed hosts before the crop is planted.

PEPPER-GRASS, GARDEN CRESS
(*Lepidium*)

Damping-off. **Pythium debaryanum**, TX; **Rhizoctonia solani**, TX.
Downy mildew. **Peronospora lepidii**, IA, SD, TX.
Rot, crown. **Sclerotinia sclerotiorum**, MA.
Rot, root. **Pyrenochaeta terrestris**, pink root, ND and ST; **Phymatotrichum omnivorum**, TX.
Rust. **Puccinia aristidae** (0, I), TX; II, III, on grasses.
Virus. **Bidens Mottle**, FL.
White rust; White Blister. **Albugo candida**, general.

PEPPER VINE, CISSUS
(*Ampelopsis arborea*)

Blight, thread. **Pellicularia koleroga**, LA.
Dodder. **Cuscuta compacta**, FL.
Leaf spot. **Cercospora arboreae**, TX; **C. vitis**, LA; **Guignardia bidwellii** f. **parthenocissi**, MS, NJ.

PERSIMMON
(*Diospyros*)

Anthracnose; fruit, twig, blight. **Gloeosporium diospyri**, East and South to KS.
Bacterial crown gall. **Agrobacterium tumefaciens**, CA.
Bacterial, witches' broom. **MLO**, TX.
Blight, thread. **Pellicularia koleroga**, FL, on Japanese persimmon.
Blight, twig. **Phoma diospyri**, FL, SC; **Physalospora obtusa**, Gulf states.
Blotch, leaf. **Mycosphaerella diospyri**, Gulf states.

Canker; dieback. **Diplodia natalensis,** AL, TX; **Botryosphaeria ribis,** AL, LA, MD.

Fruit spot; fly speck. **Leptothyrium pomi,** FL; **Macrophoma diospyri,** AL, NJ, TX.

Leaf spot, algal. **Cephaleuros virescens,** FL.

Leaf spot; brown. **Cercospora diospyri,** ME, IL, MS, SC, TX, VA; Black, **C. fuliginosa,** AL, FL, GA, IL, MS, TX; **Pestalotia** sp., ?secondary; **Phyllosticta** sp., FL, IN; **Ramularia** sp., FL; **Fusicladium levieri,** CT, FL, MS.

Mistletoe. **Phoradendron serotinum (flavescens),** FL, TX.

Mistletoe. **Viscum album,** CA, on *Diospyros kakai* (Japanese persimmon).

Nematode, citrus. **Tylenchulus semipenetrans,** CA.

Nematode, root knot. **Meloidogyne** sp., FL, SC, TX.

Powdery mildew. **Podosphaera oxyacanthae,** TX.

Rot, fruit. **Alternaria** sp., occasional on Japanese persimmon; **Botryris cinerea,** gray mold; **Penicillium expansum,** blue mold, cosmopolitan; **Physalospora** spp., AZ, AL, GA, TX, NY; **Rhizopus stolonifer,** TX.

Rot, root. **Armillaria mellea,** CA; **Phymatotrichum omnivorum,** TX.

Rot, wood. **Daedalea ambigua,** LA, MS; **Daldinia concentrica,** LA, VA; **Fomes** spp.; **Hericium erinaceus,** LA; **Lentinus tigrinus,** LA, MS; **Pleurotus ostreatus,** LA, MS; **Polyporus** spp.; **Schizophyllum commune,** KY, TN.

Wilt. **Cephalosporium diospyri,** AL, FL, AR, GA, NC, SC, MS, OK, TN, TX; **Verticillium albo-atrum,** TX.

Persimmon wilt started in Tennessee in 1933, and at the end of 5 years, only 5% of the persimmons in that native stand were alive. Oriental persimmons are resistant.

PETALOSTEMON
(Prairie-Clover)

Rot, root. **Phymatotrichum omnivorum,** TX.

Rust. **Puccinia andropogonis** (0, I), ND to KS, CO and WY, TX; **Uropyxis petalostemonis** (0, I, II, III), WI to ND, CO and NM.

PETASITES
(Butter-Bur)

Gall, leaf. **Synchytrium aureum,** WI.

Leaf spot. **Ramularia variegata,** WI; **Stagonospora petasitidis,** WI.

Rust. **Puccinia conglomerata,** MI, MN, NY, WI; **P. poarum,** AK.

PETUNIA

Bacterial fasciation. **Clavibacter fascians,** CA, OH, PA.

Bacterial, MLO. **Aster Yellows.**

Bacterial wilt. **Pseudomonas solanacearum,** FL.
Blight, of old flowers. **Choanephora conjuncta,** GA.
Blotch, leaf. **Cercospora petuniae,** FL, OK.
Damping-off. **Rhizoctonia solani,** FL, NJ, NY, NC, OK, PA, TX.
Dodder. **Cuscuta** spp., MD, NJ, NY, OK, TX, WV.
Leaf spot. **Ascochyta petuniae,** TX.
Nematode, root knot. **Meloidogyne** spp., general in South, occasional in greenhouses.
Nonparasitic. **Air Pollution Injury.**
Powdery mildew. **Oidium** sp., MN, NY, VA.
Rot, black stem. **Stemphylium botryosum,** secondary, TX.
Rot, crown. **Phytophthora nicotianae** var. **parasitica,** FL.
Rot, crown. **Phytophthora parasitica,** CO.
Rust. **Puccinia aristidae** (0, I), AZ.
Virus. **Beet Curly Top,** CA, OR; **Tobacco Etch; Tobacco Mosaic; Cucumber Mosaic,** general; **Tobacco Ring Spot; Tomato Spotted Wilt,** CA; **Potato Rugose Mosaic; Biddens mottle,** FL; **Bromegrass Mosaic,** general.
Wilt. **Fusarium** sp., WA; **Sclerotinia sclerotiorium,** WA; **Verticillium albo-atrum,** CA.

Dodder is common on petunias, reported in window boxes as well as in garden beds. Plants started in greenhouses may get infected with tobacco mosaic. Smoking around petunias, or a tobacco-stem mulch, may also foster mosaic.

PHILIBERTIA

Powdery mildew. **Phyllactinia corylea,** FL.
Rust. **Puccinia bartholomaei** (0, I), AZ; **P. obliqua** (III), FL, AZ, CA, NM, TX.

PHILODENDRON

Anthracnose. **Gloeosporium** sp., WA; **Colletotrichum** sp., WA.
Bacterial leaf rot. **Erwinia chrysanthemi,** FL.
Blight, southern. **Sclerotium rolfsii,** CA.
Leaf spot. **Colletotrichum philodendri,** NJ.
Nematode, burrowing. **Radopholus similis,** FL; Root Knot. **Meloidogyne** sp.
Nematode, spiral. **Helicotylenchus nannus:** Lesion **Pratylenchus,** sp., FL.
Nonparasitic. **Exudation** of sugars, CA.
Rot, root. **Pythium splendens,** FL.
Rot, stem. **Phytophthora** spp., FL.
Sooty mold. **Capnodium** sp.

PHLOX

Bacterial crown gall. **Agrobacterium tumefaciens,** NJ.

Bacterial fasciation. **Clavibacter fascians,** CA.

Bacterial, MLO. **Aster Yellows,** NJ, NY, PA, and **California Aster Yellows,** CA.

Blight, gray mold. **Botrytis cinerea,** AK.

Blight, southern; crown rot. **Sclerotium rolfsii,** CT, FL, IL, MD, NJ, NY, OH, TX, VA.

Blight, stem. **Pyrenochaeta phlogis,** NY.

Canker, stem. **Colletotrichum** sp., FL.

Downy mildew. **Peronospora phlogina,** IA, WI.

Leaf spot. **Ascochyta phlogis** var. **phlogina,** NY, MA, TX; **Cercospora omphakodes,** NY to IA, WI; **Macrophoma cylindrospora,** CA; **Phyllosticta** sp., WA; **Septoria** spp.; **Volutella phlogina,** LA; **Ramularia** sp., WA; **Stemphylium botryosum** (*Pleospora herbarum*), NJ.

Nematode, leaf. **Aphelenchoides fragariae,** MD.

Nematode, leaf and stem. **Ditylenchus dipsaci,** CA, CT, MD, NJ, NY, OH, TX, WA.

Nematode, root knot. **Meloidogyne** spp., KS, MD, MA, NJ, OH, TX, WA.

Nonparasitic. **Leaf Drop,** "Rust," blight, cause uncertain but includes inability of old stems of some varieties to take up enough water.

Powdery mildew. **Erysiphe cichoracearum,** general; **Sphaerotheca macularis,** KS, NH, NY, OH, WA.

Rot, charcoal. **Macrophomina phaseoli,** IL.

Rot, root. **Thielaviopsis basicola; Phymatotrichum omnivorum,** TX.

Rust. **Puccinia douglasii** (0, I, III), NJ, PA, CO, MT, NE, NM, OR, UT, WA, WY; **P. plumaria** (0, I, III), IL, IA, MO, TX, WY to NM, CA and WA; **Uromyces acuminatus** var. **polemonii** (0, I), IL, IA, MN, MS, SD, WI.

Virus. **Mosaic,** MD, NC, NY.

Wilt. **Verticillium albo-atrum,** NY, MN.

When gardeners talk about "rust" on phlox, they usually mean the physiological blight, for true rusts are uncommon on perennial summer phlox. Powdery mildew is general and especially disfiguring when plants are shaded or crowded with little air circulation. Sulfur dust is still good.

PHOENIX-TREE
(*Firmiana simplex*)

Blight, web. **Rhizoctonia microsclerotia,** FL.

Canker; coral spot. **Nectria cinnabarina,** OK.

Rot, root. **Phymatotrichum omnivorum,** TX.

PHOTINIA
(Christmasberry, Toyon and Oriental Species)

Anthracnose. Gloeosporium sp., MS.
Bacterial fire blight. Erwinia amylovora, CA, NJ.
Blight, leaf. Fabraea maculata, CA, LA.
Dieback. Cytospora sp.
Leaf spot. Cercospora sp., GA; C. heteromeles, CA, TX; C. photiniae-serrulata; C. eriobotryae; Lophodermium heteromeles, CA; Phyllosticta heteromeles, CA, TX; Pestalotia sp., NJ; Septoria photiniae, CA, PA.
Powdery mildew. Podosphaera leucotricha, CA; Sphaerotheca pannosa, OR.
Rot, crown. Phytophthora cactorum, CA.
Rot, root. Clitocybe tabescens, FL; Phymatotrichum omnivorum, TX.
Rust. Gymnosporangium japonicum (0, I), III on juniper; G. clavipes, MA.
Scab. Photinia photinicola (?Spilocaea eriobotryae), on leaves, berries, CA.

PHYSOSTEGIA
(False Dragonhead)

Blight; southern crown rot. Sclerotium rolfsii, VA to OK, KS, CT, NJ, NY, WI.
Downy mildew. Plasmopara cephalophora, WI.
Leaf spot. Mycosphaerella phystostegiae, GA, VA; Septoria physostegiae, IL, WI.
Rot, stem. Sclerotinia sclerotiorum, ME.
Rust. Puccinia physostegiae (III), IN, NH, MT, NY.

PICK-A-BACK
(Tolmiea)

Powdery mildew. Sphaerotheca sp., OR.

PIERIS
(Mountain, Japanese Andromeda)

Blight, twig. Cytospora sp., OR.
Dieback. Phytophthora sp., PA.
Leaf spot. Pestalotia sp., CT, NJ; Phyllosticta andromedae, NJ; P. maxima, CT and NJ; Alternaria tenuis, RI.
Leaf spot, tar. Rhytisma andromedae, FL, GA.
Rot, root. Armillaria mellea, NJ.

PIGEON PEA
(*Cajanus*)

Bacterial, MLO. Witches' broom, FL.

PINE
(*Pinus*)

Black mildew. **Lembosia acicola,** CA.

Blight, brown felt. **Herpotrichia nigra,** Northwest, on snow-buried foliage at high altitudes; **Neopeckia coulteri,** MT, CA.

Blight, brown spot needle. **Systremma** (*Scirrhia*) **acicola,** on hard pines in South, on Red pine, WI.

Blight, needle. **Cytospora pinastri,** ME, NJ, PA; **Hendersonula pinicola,** NC, TN, WY; **Pullaria pullulans,** after insect injury; **Dothistroma pini,** NE, OH, IL, IA, OK, MD; **Septoria spadicea,** NH, NY, VT, VA, MN; **Lecanosticta** sp.

Blight, needle. **Dothistroma pini,** MN, PA.

Blight, needle. **Lophodermella cerina,** LA, MS, AL, FL.

Blight, seedling. **Botrytis cinerea,** cosmopolitan; **Cylindrocladium scoparium,** NJ; **Fusarium** spp.; **Rhizina undulata,** MN, ME, CA, northern Rocky Mountains; **Thelephora terrestris,** ME, NJ, OH, PA, KS, northern Rocky Mountains; **Scleroderris lagerbergii,** MI, WI, NY, also canker.

Blight, shoot. **Fusarium moniliforme** var. **subglutinans,** FL, GA; **Diplodia gossypina,** GA; **Gremmeniella abienta,** NH, NY.

Blight, snow. **Phacidium infestans,** occasional, New England ; **P. convexum,** NC.

Blight, tip; twig; collar rot. **Diplodia pinea** (*Sphaeropsis ellisii* and *S. Sapinea*), New England to VA, CA, IA, KS, ND, SD, and WI. **Cenangium ferruginosum** (*C. abietis*), "pruning twig blight," widespread, sometimes secondary; **Monochaetia pinicola.**

Canker. **Atropellis pinicola,** Northwest; **A. piniphila,** Northwest, NM, AL, TN; **A. tingens,** GA, MA, NH, NC, PA, VA; **A. arizonica,** AZ; **Caliciopsis pinea,** New England to SC, TN; **Dasyscypha ellisiana,** on introduced pines in eastern states; **D. pini,** on five-needle pines, Pacific Northwest, MI; **Nectria** spp., New England to NC, IA; **Rhabdospora mirabilissima,** of seedlings, NY; **Tympanis confusa,** VA.

Canker. **Scleroderris lagerbergii** on red pine MI, MN, NY, VT, WI.

Canker, bark. **Aleurodiscus amorphus,** widespread; Felt, **Septobasidium** spp.

Canker, bleeding. **Sphaeropsis ellisii,** PA.

Canker, pitch. **Fusarium lateritium** f. **pini,** FL, VA; **F. moniliforme** var. **subglutinans,** FL, NC.

Damping-off. **Rhizoctonia solani** and **Pythium** spp., cosmopolitan.

Dieback, tip. **Sirococcus strobilinus,** CA, MI, MN, WI.

Mistletoe, dwarf. **Arceuthobium americanum,** Rocky Mountain states; Pacific Northwest; **A. campylopodum,** western dwarf, Rocky Mountain states to

Pacific Coast, TX; **A. vaginatum,** southern Rocky Mountain states; **A. vaginatum** subsp. **cryptopodum,** CO; **A. cyanocarpum,** CO, OR, WY; **A. occidentale,** CA.

Needle cast. **Bifusella linearis,** tar spot, widespread; **B. striiformis,** CA; **Elytroderma deformans,** also witches' broom, SD to AZ, CA and WA, GA; **Hypoderma desmazierii;** ME to NC, GA, and WI; **H. hedgecockii,** Southeast; **H. lethale,** gray blight, on hard pines, New England to FL, LA; **H. pedatum,** CA; **H. pini,** CA, NV; **H. saccatum,** tar spot, CO, NM; **Hypodermella arcuata,** CA, OR; **H. ampla,** tar spot, Great Lakes states; **H. cerina,** CA; **H. concolor,** CO to ID, MT, OR; **H. lacrimformis,** CA, OR; **H. limitata,** CA; **H. medusa,** CA, CO; **H. montana,** CA to ID, OR; **H. montivaga,** CA to MT, OR; **Lophodermium nitens,** New England to GA and MI and Pacific Northwest, on five-needle pines; **L. pinastri,** widespread; **L. ponderosae,** CA.

Needle cast. **Bifusella saccata,** CA, CO; **Hypodermella** sp., OR and CA; **Lophodermium durilabrum,** OR; **L. seditiosum,** MI.

Needle droop. Abiotic inability of poor roots to acquire sufficient water, MN, WI, MI.

Nematode, dagger. **Xiphinema americanum.**

Nematode, lance. **Hoplolaimus coronatus.**

Nematode, leaf. **Aphelenchoides fragariae,** FL.

Nematode, pinewood. **Bursaphelenchus xylophilus,** vectored by pine sawyer beetle, IL, MO; also **B. lignicolus,** MO.

Nematode, ring. **Criconemoides lobatum;** FL; **C. simile,** FL, NC.

Nematode, root knot. **Meloidogyne** sp., NM; **M.javanica,** FL; **Meloidodera floridensis,** on slash pine, FL.

Nematode, sting. **Belonolaimus gracilis.**

Nonparasitic. **Ozone Injury, Sulfur Dioxide Injury,** NJ, IN, WI. **Needle Curl,** high temperature. **Chlorotic Dwarf.**

Rot, charcoal. **Macrophomina phaseoli.**

Rot, heart; wood. **Fomes** spp., general; **Fomes annosus; Lentinus lepideus,** widespread; **Polyporus** spp., widespread; **Poria** spp., **Lenzites saepiaria; Stereum** spp.; **Trametes** spp., widespread.

Rot, hypocotyl. **Fusarium oxysporum,** CA.

Rot, root. **Armillaria mellea,** widespread; **Ceratocystis wageneri,** OR, Pacific NW; **Clitocybe tabescens,** FL; **Cylindrocladium scoparium,** seedling blight, NJ, PA, WA; **Fusarium solani,** PA; **F. oxysporum,** PA; **F. avenaceum,** PA.; **Phytophthora cactorum,** seedling blight, Northeast; **Phymatotrichum omnivorum,** TX; **Sparassis radicata,** MT to OR and WA; **Verticicladiella wagenerii,** CA, MT, western states; **V. procera,** FL, MN; **V. penicillata,** ID; **V. abietina,** ID.

Rot, root. **Fomes annosus,** VA; **Polyporus schweinitzii,** VA; **Phytophthora parasitica** var. **nicotianae,** CA; **Iononutus circinatus,** FL.

Rot, root; little leaf. **Phytophthora cinnamomi,** NC, TN, VA.

Rust. **Coleosporium crowellii,** UT, NV, CA.

Rust, comandra blister. **Cronartium comandrae** (0, I), swellings in twigs, trunks of hard pines; II, III, on bastard toadflax.

Rust, cone. **Cronartium conigenum,** hypertrophy of cones, especially in South; II, III, on oak; **C. strobilinum;** II, III, on evergreen oaks.

Rust, eastern gall. **Cronartium quercuum,** galls on trunk, branches, witches' brooms on two- and three-needle pines East to Rocky Mountains especially Southeast; II, III, on oak, rarely chestnut.

Rust, fusiform. **Cronartium fusiforme,** swellings in trunk, branches, in South; II, III, on evergreen oaks.

Rust, lodgepole pine blister; western fusiform. **Cronartium coleosporioides** (*Peridermium harknessi*) (0, I), swellings on twigs, branches; trunk cankers; widespread in West; II, III, on painted cup, birdbeak, owl-clover, wood-betony; **C. stalactiforme** (0, I); II, III, on Indian paintbrush.

Rust, needle. **Coleosporium apocyanaceum** (0, I), Southeast; II, III, on *Amsonia*; **C. asterum** (*C. solidaginis*) (0, I), on all two- and three-needle pines in eastern U.S., western from CO to MT, WA; II, III on aster, goldenrod, and other composites; **C. campanulae** (0, I), Northeast to IN, NC; (I, III, on bellflower, loosestrife, Venus-looking-glass; **C. crowellii** (III), on pinon and limber pines, AZ, CO, NM; **C. delicatulum** (0, I), New England to FL and west to Great Plains; II, III, on goldenrod; **C. elephantopodis** (0, I), NJ to FL, TX; II, III, on *Elephantopus*; **C. helianthi** (0, I), NY to GA, OH; II, III, on sunflower; **C. inconspicuum** (0, I), MD to GA, TN, OH; II, III, on coreopsis; **C. ipomoeae** (0, I), NJ to FL, IL, AZ; II, III, on morning-glory; **C. jonesii** (0, I), AZ, CO, NM; II, III, on currant and gooseberry; **C. laciniariae,** NJ to FL; II, III, on *Liatris*; **C. madiae** (0, I), CA, OR; II, III, on composites; **C. minutum** (0, I), on loblolly and spruce pines, FL; II, III, on *Forestiera*; **C. pinicola,** DE to NC, TN; **C. senecionis,** CO, RI; II, III, on *Senecio*; **C. sonchiarvensis** (0, I), on Scotch pine, CT to NE; II, III, on sowthistle; **C. terebinthinaceae** (0, I), Southeast; II, III, on *Parthenium* and *Silphium*; **C. vernoniae** (0, I), on two- and three-needle pines East and South; II, III, on ironweed.

Rust, pine-pine gall. **Endocronartium harknessii,** MN, PA.

Rust, piñon blister. **Cronartium occidentale.**

Rust, stem. **Cronartium appalachianum** (0, I), NC, TN, VA; II, III, on *Buckleya*; **C. filamentosum; Peridermium weirii.**

Rust, sweet-fern blister. **Cronartium comptoniae,** swellings on trunk and branches of two- and three-needle pines from northeastern to central and Great Lakes states; II, III, on sweet-fern and sweet gale.

Rust, western gall. **Endocronartium harknessii,** VA, also Twig necrosis and witches' broom, MA, ND.

Rust, white pine blister. **Cronartium ribicola** (0, I), swellings on trunk and branches of eastern white pine from New England to VA and MN; on western white pine in Pacific Northwest; on sugar pine, CA; on lumber pine, WY; II, III, on gooseberry, currant, CA.

Sooty mold. **Fumago vagans; Capnodium pini,** widespread; **Scorias spongiosa.**

Witches' broom. **Elytroderma deformans,** CA.

White pine blister rust is, of course, our foremost disease of pines, and full details are given under Rusts. Black currants are banned entirely in infected areas, red currants within 300 feet of pines. Of the other possible rusts on various pines, the only one seen in northern gardens is the aster rust (*Coleosporium asterum*), which is slightly disfiguring to the needles but not very damaging to general health. Fusiform rust is serious in the South. Brown needles may be due to one of the needle blights or needle cast fungi and also to winter drying. New shoots of Austrian pine turn brown from Diplodia tip blight, which should not be confused with the common discoloration caused by the pine shoot moth.

PINEAPPLE
(*Ananas*)

Nematode, lesion. **Pratylenchus brachyurus.**
Nematode, root knot. **Meloidogyne** sp., FL.
Nonparasitic. **Spike,** Long leaf, FL.
Rot, leaf base; white leaf spot. **Ceratocystis paradoxa,** FL.
Wilt. Toxic effect of mealybug feeding.

PIPSISSEWA
(*Chimaphila*)

Leaf spot. **Mycosphaerella chimaphilina,** NY, OR, PA, WA; **Septoria chimaphilae,** DE.
Rust. **Pucciniastrum pyrolae** (II, III), NY to NC, TN and WI, CA, MT, OR, WA.

PISTACHIO
(*Pistacia*)

Blight, shoot. **Botrytis cinerea,** CA.
Blight, thread. **Pellicularia koleroga,** FL, TX.
Leaf spot. **Phyllosticta lentisci,** TX; **Septoria pistaciarum,** TX.
Nematode. **Meloidogyne** sp., CA; **Xiphinema index,** CA.
Rot, root. **Phymatotrichum omnivorum,** AZ, CA, TX.
Rot, sapwood. **Pleurotus ostreatus,** CA; **Schizophyllum commune,** CA.
Wilt. **Verticillium albo-atrum,** CA.

PITCHER-PLANT
(*Sarracenia*)

Blight, southern. **Sclerotium rolfsii,** TX.
Leaf spot. **Colletotrichum gloeosporioides,** NJ, TX; **Helminthosporium sar-**

racenia, secondary, MN; **Mycosphaerella sarraceniae,** ME, MI, MN, NY, PA, GA, MS, SC; **Pestalotia aquatica,** secondary, MN, MD.

Rot, root. **Pythium graminicola,** NC; **Rhizoctonia solani,** TX.

PITHECELLOBIUM
(Blackbead, Catsclaw)

Blight, twig. **Phomopsis** sp., FL.

Leaf spot. **Colletotrichum erythrinae,** TX; **Pestalotia funerea,** TX; **Phyllosticta pithecolobii,** TX.

Nematode. **Meloidogyne** sp.

Rust. **Ravenelia gracilis** (0, I, II, III), TX; **R. pithecolobii** (II, III); **R. siderocarpi** (II, III), TX.

PITTOSPORUM

Blight, southern; wilt. **Sclerotium rolfsii,** FL, TX.

Blight, thread. **Pellicularia koleroga,** LA; **Rhizoctonia ramicola,** FL.

Leaf spot, angular. **Cercospora pittospori,** FL, LA, SC to TX; **Phyllosticta** sp., AL; **Alternaria tenuissima,** FL.

Nematode, root knot. **Meloidogyne** sp.

Rot, foot. **Diplodia** sp., FL.

Rot, root. **Phymatotrichum omnivorum,** TX.

Virus. **Mosaic,** undetermined, CA; **Rough Bark,** undetermined, CA; **Variegation.** Variegated forms may be due to a virus.

Wilt. **Verticillium albo-atrum,** CA.

PLANE-TREE, SYCAMORE
(*Platanus;* American Sycamore, *P. occidentalis;* California Sycamore, *P. racemosa;* London Plane, *P. acerifolia;* and Oriental Plane, *P. orientalis*)

Anthracnose; leaf and twig blight. **Gnomonia platani** (*G. veneta, Gloeosporium platani, G. nervisequum*), general.

Bacterial scorch. **Xylem-limited Rickettsialike** bacteria, DC, LA, TX.

Blight, leaf. **Phleospora multimaculans,** IN, TX.

Blight, twig. Canker. **Massaria platani,** NJ, GA, IN, IA, KS, CA.

Canker. **Dothiorella** sp., NY; **Botryodiplodia theobromae,** MS; **Hypoxylon tinctor,** NC, GA, LA.

Canker stain; London plane blight. **Ceratocystis fimbriata** f. **platani,** DE, MD, NJ, NY, PA, MS, MO, NC, OH, TN, VA, WV.

Leaf spot. **Cercospora platanicola** (*Mycosphaerella platanifolia*) NC to GA, TX, IA; spermatial stage is **Phyllosticta platani; Mycosphaerella stigmina-**

platani, NC; Septoria platanifolia, GA, IA, MD, SC, TX, WV; **Stigmella platani-racemosae**, CA; **Cristutariella pyramidalis**, FL; **Tubakia dryina**, LA.

Mistletoe. **Phoradendron serotinum (flavescens)**, OK, TN, TX, AZ, NM.

Nonparasitic. **Rosy canker,** illuminating gas in soil, NJ, MD, NY. **Decline,** moisture deficiency, MS.

Powdery mildew. **Microsphaera alni,** widespread; **Oidium obductum,** PA, VA, WV; **Phyllactinia corylea,** IN.

Rot, heart, trunk. **Fomes** spp.; **Steccherinum** (*Hydnum*) **erinaceus,** sometimes on living trees.

Rot, root. **Armillaria mellea,** MD, TX, WV; **Phymatotrichum omnivorum,** TX, AZ;, **Phytophthora cinnamomi,** MD; **Clitocybe tabescens,** CA; **C. olearia.**

Rot, wood. **Daedalea** spp., widespread; **Sterum** spp.; **Polyporus** spp.

The canker stain of London plane and American sycamore flared up in epidemic form around Philadelphia in 1935, killing thousands of street and ornamental trees from Newark to Baltimore before it was learned that the fungus was spread in pruning and in tree paint as well as by certain beetles. Pruning is now restricted to winter months, and a disinfectant is added to the wound dressing. Sycamore anthracnose is serious in wet seasons, particularly when the mean daily temperature for 2 weeks after bud-break is below 55° F.

PLANTAIN, Common
(*Plantago*)

Virus. **Tomato Ringspot,** PA.

PLATYCODON
(Balloon-Flower)

Blight. **Phytophthora ?cactorum,** MN.
Rot, root. **Rhizoctonia solani,** CT, PA; **Phymatotrichum omnivorum,** TX.

PLUM (GARDEN), PRUNE
(*Prunus domestica*)

Bacterial crown gall. **Agrobacterium tumefaciens,** widespread.
Bacterial fire blight. **Erwinia amylovora,** occasional, OR, WA.
Bacterial leaf spot; black spot. **Xanthomonas pruni,** eastern and southern states to WI, TX.
Bacterial shoot blight; gummosis. **Pseudomonas syringae,** CA.
Black knot. **Dibotryon morbosum,** widespread except far West.
Blight, blossom, twig. **Monilinia laxa,** CA, OR, WA; **Botrytis cinearea.**
Blight, thread. **Pellicularia koleroga,** LA.

Blight, twig. **Diplodia** spp., secondary.

Blotch, leaf. **Phyllosticta congesta,** GA, TX.

Canker. **Phytophthora cactorum,** CA, IN; **Ceratocystis fimbriata,** CA; **Valsa** (*Cytospora*) **leucostoma,** dieback, widespread.

Leaf curl; witches' broom. **Taphrina** spp., occasional; **T. communis,** plum pockets; **T. pruni,** pockets, bladder plums; **T. insititiae.**

Leaf spot; shot hole. **Cercospora circumscissa,** CA, FL, MA, TX, WA; **Coccomyces prunophorae,** widespread; **Coryneum carpophilum,** CA to ID and WA; **Phyllosticta circumscissa,** IA, WA; **Septoria pruni,** TX.

Mistletoe. **Phoradendron serotinum (flavescens),** TX.

Mistletoe. **Viscum album,** CA, on *Prunus salicina* ("Santa Rosa" Plum).

Nematode, lesion. **Pratylenchus pratensis; P. vulnus.**

Nematode, root knot. **Meloidogyne** spp., FL, TX.

Nonparasitic

    Brown Bark Spot. ?Arsenical poisoning, MT.

    Chlorosis. Alkaline soil, mineral deficiency, CA, FL.

    Exanthema. Copper deficiency, CA, FL.

    Gum Spot, drought spot. Irregular water supply, NY, Pacific Northwest.

    Little Leaf. Zinc deficiency, CA, OR, WA.

    Marginal Scorch. Fluorine injury, WA.

    Myrobalan Asteroid Spot. Cause unknown, occasional.

    Myrobalan Mottle. Genetic abnormality, occasional in seed stocks.

Powdery mildew. **Podosphaera oxyacanthae** var. **tridactyla,** occasional.

Rot, brown; blossom blight. **Monilinia fructicola,** general.

Rot, fruit. **Alternaria** sp., OR; **Botrytis cinerea,** CA, WA; **Cladosporium** sp., ID, OR; **Lambertella pruni,** CA.

Rot, heart. **Fomes applanatus,** OR; **F. fulvus,** widespread; **Lenzites saepiaria,** Pacific Northwest; **Polyporus hirsutus; P. versicolor,** widespread.

Rot, root. **Armillaria mellea,** widespread; **Phymatotrichum omnivorum,** TX.

Rot, silver leaf. **Stereum purpureum,** WA.

Rust. **Tranzschelia discolor** (II, III), widespread, especially in South and Pacific Coast states; 0, I, on anemone; **T. pruni-spinosae** (II, III); 0, I, on anemone, hepatica, thalictrum, buttercup.

Scab. **Cladosporium carpophilum,** widespread.

Virus. **Plum Line Pattern; Plum White Spot; Prune Diamond Canker; Prune Constricting Mosaic; Prune Dwarf; Cherry Vein Clearing; Little Peach; Peach Mosaic; ?Peach Rosette; Peach Yellows; Ring Spot; Plum Rusty Blotch; Apricot Ring Pox; Stem Pitting.**

Wilt; seedling black heart. **Verticillium albo-atrum.**

Plums are even more subject to brown rot than peaches and take the same spray schedule, including insect control to avoid infection through feeding injuries. Black knot is sometimes conspicuous on twigs, but diseased portions can be pruned out and a dormant lime-sulfur spray applied.

PLUMEGRASS
(*Erianthus*)

Anthracnose. Colletotrichum falcatum, LA.
Ergot. Claviceps purpurea, AL, OK.
Leaf spot; mold. Cladosporium erianthi, SC; Curvularia sp., FL; Helminthosporium sp., FL; Phyllachora erianthi, tar spot, AL, FL, GA, SC.
Rust. Puccinia virgata (II, III), GA; P. polysora (II, III), 0, I unknown.

PLUMERIA
(*Frangipani*)

Leaf spot. Cercospora plumeriae, FL.
Mistletoe. Phoradendron serotinum (flavescens), FL.
Rot, root. Phymatotrichum omnivorum, TX.
Rust. Coleosporium domingense (II, III), FL.

PODOCARPUS

Nematode, burrowing. Rodopholus similis, FL.
Rot, root. Clitocybe tabescens, FL.

POINCIANA
(*Caesalpinia*)

Anthracnose. Gloeosporium sp., FL.
Bacterial crown gall. Agrobacterium tumefaciens, FL.
Rot, root. Clitocybe tabescens, FL; Phymatotrichum omnivorum, TX.
Rust. Ravenelia humphreyana (II, III), FL.

POINSETTIA
(*Euphorbia pulcherrima*)

Bacterial, bract spot. Clavibacter cassiicola, FL.
Bacterial canker; leaf spot. Clavibacter poinsettiae, MD, NJ, NY, PA.
Bacterial, greasy canker. Pseudomonas viridiflava, CA.
Bacterial leaf spot. Xanthomonas poinsetriaecola, FL.
Bacterial stem rot. Erwinia carotovora pv. chrysanthemi, OH, CT, MI, PA, WV.
Blight, tip; stem canker. Botrytis cinerea, MO, TX, WA.
Canker, stem. Fusarium solani, WI.

Leaf spot. **Cercospora pulcherrima**, TX.

Nematode, root knot. **Meloidogyne** sp., NY; **M. incognita-acrita.**

Nonparasitic. **Chlorosis,** possibly due to cloudy weather. **Stunt,** water-logged soil.

Rot, crown and stem. **Phytophthora nicotianae** var. **parasitica**, FL.

Rot, root. **Chalaropsis thielaviopsis**, IL; **Clitocybe tabescens**, FL; **Phymato-trichum omnivorum**, TX, AZ; **Rhizoctonia solani**, FL, IL, NJ, TX; **Thie-laviopsis basicola,** serious, common.

Rot, stem; wilt. **Fusarium** sp., FL, NJ; **Pythium debaryanum,** OK; **P. perni-ciosum,** root rot, CA; **P. ultimum; Phytophthora** sp., NJ; **Sclerotinia sclerotiorum**, WA.

Rust. **Uromyces euphorbiae** (0, I, II, III), OK, TX.

Spot anthracnose; poinsettia scab. **Sphaceloma poinsettiae**, FL.

Keep soil acid, pH 4.8 to 5, to reduce trouble with *Thielaviopsis. Rhizoc-tonia* is prevalent at high temperatures; *Pythium* at low. Steaming soil is safer than using chemicals.

POISON HEMLOCK
(*Conium*)

Virus. **Alfalfa Mosaic,** WA; **Clover Yellow Vein,** WA.

POLEMONIUM
(Jacobs-Ladder, Greek-Valerian)

Leaf spot. **Cercospora omphakodes**, PA; **Septoria polemonii**, MO, WI; **S. polemoniicola**, CT, IN, MO.

Powdery mildew. **Erysiphe cichoracearum**, UT; **Sphaerotheca macularis**, WA.

Rust. **Puccinia gulosa** (III), CA; **P. polemonii** (III), CA, ID, IN; **Uromyces acuminatus** var. **polemonii** (0, I), IL, IN, IA, WI.

Wilt. **Fusarium** sp., NJ; **Verticillium albo-atrum**, NJ.

POMEGRANATE
(*Punica granatum*)

Anthracnose; fruit spot. **Colletotrichum** sp., FL.

Blight, thread. **Pellicularia koleroga**, FL.

Blotch, leaf; fruit spot. **Mycosphaerella lythracearum**, FL to MS, TX.

Nematode. **Meloidogyne** sp., FL.

Rot, fruit. **Alternaria** sp., CA; **Aspergillus niger,** AZ, CA, TX; **Botrytis cinerea,** gray mold, cosmopolitan; **Nematospora coryli,** dry rot, CA; **Penicillium expansum,** blue mold, cosmopolitan; **Coniella granati**, NC.

Rot, root. **Clitocybe tabescens**, FL; **Phymatotrichum omnivorum**, TX.

Spot anthracnose. **Sphaceloma punicae**, FL, LA, TX.

POND-SPICE
(Litsea)

Leaf spot. Cercospora olivacea, GA.

POPLAR, ASPEN, COTTONWOOD
(Populus)

Bacterial limb gall. **Agrobacterium tumefaciens,** CT, IA, MN, NE, TX.
Bacterial wetwood. **Clavibacter humiferum.**
Blight, shoot. **Venturia populina** (*Didymosphaeria populina*), Northeast to WI; **V. tremulae; Colletotrichum gloeosporioides,** PA.
Canker, branch and trunk. **Dothichiza populea,** most serious on Italian varieties, widespread; **Hypoxylon pruinatum,** Southwest, Northeast, and Great Lakes states, AZ, CA; **Nectria cinnabarina; N. galligena; Dothiora polyspora,** CO, NM, UT.
Canker; dieback. **Botryosphaeria ribis** var. **chromogena,** widespread, especially in the South; **Cytospora chrysosperma** (*Valsa sordida*); prevalent on ornamental poplars; **Valsa nivea; Ceratocystis tremullo-aurea,** Rocky Mountains; **C. fimbriata,** MN, PA; **Fusarium** (*Hypomyces*) **solani,** IA; **Phomopsis macrospora,** MS.
Canker; dieback. **Ceratocystis** sp., Rocky Mountains, MN, PA, on aspen.
Canker, sooty bark. **Cenangium singulare,** Rocky Mountains.
Leaf blister, yellow. **Taphrina aurea,** SC; **T. populina,** New England to Great Lakes states; **T. johansonii,** catkin deformity, widespread; **T. populisalicis,** CA, OR.
Leaf blotch. **Septotinia populiperda,** MD, ME, MA, NY, VT.
Leaf spot. **Marssonina** spp., widespread; **Mycosphaerella** (*Septoria*) **populicola; M. populorum** (*Septoria musiva*), also twig canker. **Phyllosticta** spp.; **Cercospora populina,** LA, AL, MO; **C. populicola,** TX; **Stigmina populi,** DE; **Marsonina brunnea,** central U.S.
Leaf spot; ink spot. **Cibornia bifrons** (*Sclerotinia whetzelii*) in Northeast; **C. confundens** (*Sclerotinia bifrons*), CO to WY, OR, WA.
Mistletoe. **Phoradendron serotinum (flavescens),** AZ, NM, TX.
Mistletoe, European. **Viscum album,** CA, on *Populus Fremontii* (Fremont Cottonwood) and *P. tremuloides* (Quaking Aspen).
Nematode, dagger. **Xiphinema americanum,** SD, on cottonwood.
Nematode, ring. **Criconemoides crotaloides.**
Nonparasitic. **Chlorosis,** iron deficiency, WY.
Powdery mildew. **Uncinula salicis,** widespread ; **Erysiphe cichoracearum,** UT.
Rot, heart. **Daedalea** spp., sometimes on living trees; **Fomes igniarius,** widespread; **F. pini; Pholiota adiposa** and **P. destruens,** New England states.
Rot, root. **Armillaria mellea,** occasional; **Phymatotrichum omnivorum,** OK, TX; **Cylindrocladium scoparium,** of seedlings, GA.

Rot, wood. **Collybia velutipes,** sometimes living trees, Rocky Mountain states; **Fomes** spp.; **Polyporus** spp.; **Lenzites saepiaria,** widespread.

Rot, wound. **Schizophyllum commune,** cosmopolitan.

Rust, leaf. **Melampsora bietis-canadensis** (II, III), New England to Great Plains; 0, I, on hemlock; **M. albertensis** (II, III), CA, MT, NM; 0, I, on Douglas-fir; **M. medusae** (II, III), through U.S. except far South; 0, I, on larch; **M. populnea** (III), CO, RI, Pacific Coast; **M. occidentalis** (II, III), MT to CA, WA; 0, I unknown.

Spot anthracnose. **Sphaceloma populi,** OK.

Cytospora canker is rather common on poplars lacking in vigor, but the Rio Grande cottonwood in the West is resistant. Avoid wounding, prune out twigs that have died back, and promote better growing conditions. Dothichiza canker may kill Lombardy poplars. Seriously diseased trees should be destroyed.

POPPY
(*Papaver*)

Anthracnose. **Gloeosporium** sp. (*Glomerella cingulata*), NJ.

Bacterial blight. **Xanthomonas papavericola,** CT, ME, MO, NJ, NY, OH, OR, VA.

Blight, gray mold. **Botrytis cinerea,** MD, WI, AK.

Leaf spot; pod spot. **Cercospora papaveri,** AL, FL, TX; **Septoria** sp., IA.

Nematode, leaf. **Aphelenchoides fragariae,** NJ.

Nematode, root knot. **Meloidogyne** sp.

Powdery mildew. **Erysiphe polygoni,** OR.

Rot, root, stem; damping-off. **Rhizoctonia solani,** KS, ID, IN, ME, NJ, NY.

Smut, leaf. **Entyloma fuscum,** IA, ME, TX.

Virus. **Beet Curly Top,** TX; **Tomato Spotted Wilt,** CA.

Wilt. **Verticillium albo-atrum,** NY.

POPPY-MALLOW
(*Callirhoë*)

Broomrape. **Orobanche ramosa,** TX.

Leaf spot. **Cercospora althaeina,** KS, IL, NE, WI ; **Vermicularia sparsipila,** TX.

Rot, root. **Phymatotrichum omnivorum,** TX.

Rust. **Endophyllum tuberculatum** (III), KS, IN, NE, OK, TX; **Puccinia interveiniens** (0, I), NE TX, NY; II, III, on Stipa; **P. schedonnardi,** CO, KS, NE, OK, TX, UT.

Wilt. **Verticillium albo-atrum,** NY.

PORTULACA

Blight, stem. **Dichotomophthora indica,** VA.
Damping-off. **Rhizoctonia solani,** IL; **Helminthosporium (Bipolaris) portulacae,** NC.
Nematode, root knot. **Meloidogyne** sp., AL.
Rot, black stem. **Dichotomophthora portulacae,** CA, TX.
Virus. **Beet Curly Top,** CA; **Tomato Spotted Wilt,** NY
White rust. **Albugo portulacae,** MT to GA, TX and MT.

POTATO
(Solanum tuberosum)

Anthracnose; black dot disease. **Colletotrichum atramentarium** (*C. coccodes*), probably general after wilt; **Gloeosporium** sp., IN, OH.
Bacterial blackleg. **Erwinia carotovora** pv. **atroseptica,** general.
Bacterial canker. Vascular. **Clavibacter michiganense,** WY.
  Also on *Solanum ciliatum, S. gilo, S. guineense, S. khasianum, S. nigrum, S. atropurpureum, S. avioculare, S. carolinense, S. indicum, S. nodiflorum, S. quitoense.*
Bacterial, MLO. **Aster Yellows,** purple top.
Bacterial ring rot. **Clavibacter sepedonicum,** general.
Bacterial soft rot. **Erwinia carotovora** and **E. aroideae,** cosmopolitan in transit and market.
Bacterial wilt. **Pseudomonas solanacearum,** chiefly in the South.
Blight, early. **Alternaria solani,** general.
Blight, late. **Phytophthora infestans,** general, common in Northeast, Middle Atlantic and north central states, occasional elsewhere, FL.
Blight, southern. **Sclerotium rolfsii,** NC to FL, AZ, OH.
Blotch, leaf. **Cercospora concors,** GA, IN, MI, NY.
Dodder. **Cuscuta** sp., DE, NE, NJ, WA.
Leaf spot. **Ascochyta lycopersici,** OR.
Nematode, golden. **Heterodera rostochiensis (Globodera rostochiensis),** DE, NY.
Nematode, lesion. **Pratylenchus brachyurus; P. scribneri.**
Nematode, ring. **Criconemoides mutabile.**
Nematode, root knot. **Meloidogyne** spp., general; **M. arenaria; M. chitwoodi,** CO, Pacific Northwest; **M. hapla,** Pacific Northwest.
Nematode, sting. **Belonolaimus gracilis; B. longicaudatus.**
Nematode, tuber. **Ditylenchus destructor,** ID.
Nonparasitic
  Black Heart. Oxygen deficiency.
  Blackening after Cooking. Drought, heat, deficient light during tuber growth; potassium deficiency, chilling.

Blackening before Cooking. Mechanical injury.

Checking. Skin rough, scruf. Partly fertilizer injury, alkalinity.

Chlorosis. Tip blight, boron deficiency; Leaf drop, magnesium deficiency.

Dimple. Depression at bud end and **Dimple skin,** pits. Cause unknown.

Elephant Hide. May be caused by fertilizer burn, on Russet Burbank.

Fasciation (tubers and aerial parts). Probably genetic.

Giant Hill. Oversized, late-maturing plants. Genetic factors.

Glassy End. Starch deficiency; high water content.

Growth Cracks. Fluctuating moisture.

Hollow Heart. Excessive soil moisture and fertility.

Hopperburn. Marginal necrosis from leafhoppers. General.

Internal Brown Spot. Various physiological factors.

Knobbiness, "second growth." Extreme fluctuations of soil moisture.

Lenticel Enlargement. Wet soil or oxygen deficiency.

Mahogany Browning. Low temperature, ME.

Marginal Browning, bronzing. Potassium deficiency.

Ozone injury. Purple-black specks from air pollution.

Pitting, spot necrosis of tubers. Oxygen deficiency in storage.

Pointed Ends. Irregular growth conditions.

Psyllid Yellows. Insect injury, western states.

Ring Spot. Zonate, depressed lesions. Cause unknown.

Scald of tubers. Overheating, sunburn, frequent in South.

Spraing. Internal concentric necrosis. ?Virus.

Sprout Tubers; "little potato." Overheating in storage; planting in dry, cold soil.

Stem-end Browning. Vascular necrosis. ?Virus.

Stem Necrosis; Defoliation. Manganese toxicity in acid soil.

Tipburn. Abrupt transition from cool, moist to hot, dry weather, general.

Powdery mildew. **Erysiphe cichoracearum,** KY, NJ; **Oidium** sp.; MD, NJ, PA, UT, OH, western U.S.

Rot, charcoal. **Macrophomina phaseoli,** CA, GA, IL, OK, TX.

Rot, gray mold; shoot blight. **Botrytis cinerea,** CT, ID, ME, OH, AK.

Rot, pink, watery. **Phytophthora erythroseptica,** ID, LA, MA, ME, OK, DE, PA; **P. parasitica,** TX; **P. cryptogea,** TX.

Rot, root. **Armillaria mellea,** CA, FL, MI, OR, WA, WI, WY; **Phymatotrichum omnivorum,** TX; **Helicobasidium purpureum,** MA, MT, NE, NY, ND, OR, OK, TX, WA.

Rot, silver scurf. **Spondylocladium atrovirens,** general but less frequent in the South.

Rot, stem. **Sclerotinia** sp., FL, ME, SC; **S. minor,** CA; **S. sclerotiorum,** FL, MA, MT, NY, OR, TX, WA.

Rot, stem. **Sclerotinia** sp., FL, ME, SC; **S. minor,** CA; **S. sclerotiorum,** FL, MA, MT, general.

Rot, tuber. **Aspergillus niger; Fusarium** spp.; **F. solani,** jelly end-rot; **Gliocladium** sp., secondary; **Oospora pustulans,** skin spot; **Penicillium** spp., blue mold; **Phoma tuberosa; Phomopsis tuberivora; Phytophthora drech-**

sleri; Rhizopus spp.; **Trichothecium roseum,** pink mold; **Xylaria apiculata.**
Rot, watery leak. Chiefly **Pythium debaryanum,** common in West; **Pythium** spp.

Scab. **Streptomyces scabies,** general; powdery scab, **Spongospora subter-ranea,** occasional; russet scab, **Streptomyces** sp.

Virus. Apical **Leaf Roll,** unidentified; **Potato Aucuba Mosaic; Potato Calico** strain of **Alfalfa Mosaic; Potato Crinkle; Mild Mosaic** (virus X plus A); **Green Dwarf; Beet Curly Top; Potato Leaf Roll,** general; **Potato Leaf Rolling Mosaic; Mottle** (potato virus X); **Potato Rugose Mosaic** (potato virus Y), general; **Potato Spindle Tuber** (viroid-type), general; **Potato Vein Banding; Potato Witches' Broom; Potato Yellow Dwarf; Potato Yellow Spot; Tobacco Ring Spot; Potato Corky Ring Spot** (Tobacco Rattle).

Wart, potato. **Synchytrium endobioticum,** MD, PA, VA; see under Galls.

Wilt. **Verticillium albo-atrum,** widespread.

Wilt; stem-end rot. **Fusarium** spp.; **F. oxysporum,** general but more frequent East and South.

Potato growing is a highly specialized business not well adapted to most small gardens. Use certified seed, resistant varieties where possible. Control insects spreading virus diseases.

POTENTILLA
(*Cinquefoil*)

Downy mildew. **Peronospora potentiallae,** IA, NJ, TX.

Leaf spot. **Fabraea dehnii,** IA, NY, VT; **Marssonina potentiallae,** CA, CO, MA, MI, NM, NY, WI; **Phyllosticta anserinae,** IL; **Ramularia arvensis,** WI.

Powdery mildew. **Erysiphe polygoni,** CO; **Sphaerotheca macularis,** CO, WI.

Rust. **Phragmidium andersonii** (I, II, III), general.

Virus. **Tomato Ringspot,** NY.

POTHOS, IVY-ARUM
(*Scindapsus*)

Bacterial blight. **Pseudomonas cichorii,** FL.

Nematode, burrowing. **Radopholus similis,** FL.

Nematode, lesion. **Pratylenchus** sp.

Nematode, root knot. **Meloidogyne arenaria thamesii; M. incognita-acrita.**

Rot, root. **Pythium splendens,** FL; **Rhizoctonia** sp., foot rot, MO.

PRENANTHES
(Rattlesnake-Root)

Downy mildew. **Bremia lactucae,** IA, MA, MN.

Gall, leaf. **Synchytrium aureum,** WI.

Leaf spot. **Cercospora brunnea,** AR, NC, WI; **C. prenanthis,** AL, KS, IN; **C. tabacina,** WI; **Laestadia prenanthis,** AL; **Septoria nabali,** NY to IA, WI, ME to VA.

Powdery mildew. **Erysiphe cichoracearum,** IL, MD, NC, PA; **Sphaerotheca macularis,** NY to AL, IL, MN, OH, PA.

Rust. **Puccinia atropuncta** (0, I), PA, VA; II, III, on *Amianthium;* **P. dioicae** (0, I), IL, MN, NY, VA, WI; II, III, on *Carex;* **P. insperata** (I, II, III), OR; **P. orbicula** (0, I, II, III), ME to TN and ND.

PRICKLY-ASH
(*Xanthoxylum americanum*)

Canker, stem. **Diplodia natalensis,** TX.

Leaf spot. **Cercospora xanthoxyli,** IN, TX; **Septoria pachyspora,** IA, NE.

Powdery mildew. **Phyllactinia corylea,** widespread.

Rot, white heart. **Fomes igniarius.**

Rust. **Puccinia andropogonis** var. **xanthoxyli** (0, I), KS, IA, MO, NE, TX, WI; II, III, on *Andropogon.*

PRICKLY-POPPY
(*Argemone*)

Downy mildew. **Peronospora arborescens,** TX.

Leaf spot. **Alternaria lancipes,** KS, TX; **Gloeosporium argemonis,** KS, TX; **Septoria argemones,** NE, OK, TX.

Rot, root. **Phymatotrichum omnivorum,** TX.

Rust. **Aecidium plenum** (0, I), TX.

PRIMROSE
(*Primula*)

Bacterial leaf spot. **Pseudomonas primulae,** CA.

Bacterial, MLO. **California Aster Yellows.**

Blight, gray mold. **Botrytis cinerea,** frequent in greenhouses, occasional in gardens.

Leaf spot. **Ascochyta primulae,** WI; **Asteroma garretrianum,** black spot, CO, UT; **Cercosporella primulae,** WA; **Colletotrichum primulae,** FL; **Mycosphaerella** sp., AZ; **Ramularia primulae,** CA, CT, DE, NY.

Nematode, leaf and stem. **Ditylenchus dipsaci,** MD, PA, CA.

Nematode, root knot. **Meloidogyne** spp., occasional in greenhouses.

Nonparasitic. **Chlorosis,** excessive soil acidity or iron or magnesium deficiency.

Powdery mildew. **Erysiphe polygoni,** CT, NJ, VA.

Rot, root. **Pythium irregulare,** CA; **Rhizoctonia solani,** FL, IL, TX; **Phymatotrichum omnivorum,** TX.

Rot, stem. **Alternaria** sp., CT; **Sclerotinia sclerotiorum,** MD.

Rust. **Puccinia aristidae** (0, I), ME; **Uromyces apiosporus,** CA, NV.

Virus. **Primrose Mosaic; Cucumber Mosaic; Tobacco Necrosis; Tomato Spotted Wilt.**

PRIVET
(*Ligustrum*)

Anthracnose; canker; dieback. **Glomerella cingulata,** general, especially on common privet.

Bacterial crown gall. **Agrobacterium tumefaciens,** occasional

Blight, leaf. **Ramularia** sp., NJ, WA.

Blight, thread. **Pellicularia koleroga,** FL.

Dodder. **Cuscuta** sp., FL.

Gall. **Phomopsis** sp., MD, TX; **P. ligustri-vulgaris,** blight, PA; **Phoma** sp., TX.

Leaf spot. **Cercospora adusta,** DE to AL, TX; **C. ligustri,** Gulf states; **Exosporium concentricum,** TX; **Phyllosticta ovalifolii,** MD, MS, TX; **Corynespora cassiicola,** FL.

Leaf spot, algal. **Cephaleuros virescens,** Gulf states.

Nematode, leaf. **Aphelenchoides fragariae.**

Nematode, root knot. **Meloidogyne** spp., southern states.

Nonparasitic. **Chlorosis,** manganese deficiency, FL.

Parasitic lichen. **Strigula elegans** and **S. camplanata,** Southern U.S., GA.

Powdery mildew. **Microsphaeria alni,** IN, LA, NJ, OH, on *Ligustrum japonicum* (Wax-leaf Ligustrum).

Rot. **Rosellinia necatrix,** CA; **Ganoderma applanatus,** collar rot.

Rot, root. **Armillaria mellea,** AR, CA, MS, TX; **Clitocybe monadelpha,** AR; **C. tabescens,** FL; **Phymatotrichum omnivorum,** AZ, OK, TX.

Rot, wood. **Stereum hirsutum,** cosmopolitan; **Polyporus versicolor,** cosmopolitan.

Sooty mold. Common after whiteflies in the South.

Virus. **Variegation.** Graft transmitted. **Chlorotic Spot,** LA.

PRUNELLA
(Self-Heal, Heal-All)

Blight, southern. **Sclerotium rolfsii,** TX.

Leaf spot. **Gibberidea abundans,** Tar Spot, ME, WA; **Linospora brunellae,** AK, ID, WA; **Phyllosticta brunellae,** TX; **Ramularia brunellae,** IL, IN, NY, OH, TX, VA, WI; **Septoria brunellae,** general.

Powdery mildew. **Erysiphe cihoracearum,** PA; **Sphaerotheca macularis,** IL, IN, MD, MS, WA, WI.

Rot, root. **Pythium palingenes; P. polytylum,** VA.

Virus. **Tomato Ringspot,** VT.

PUMPKIN

See Squash.

PUNCTURE VINE
(*Tribulus*)

Virus. **Tomato Spotted Wilt,** General.

PYRACANTHA
(Firethorn)

Bacterial fire blight. **Erwinia amylovora,** widespread.
Blight, leaf. **Fabraea maculata,** LA.
Blight, silky thread. **Rhizoctonia ramicola,** FL.
Blight, twig. **Diplodia crataegi,** PA.
Canker, felt fungus. **Septobasidium cokeri** and **S. marianai,** on bark scales.
Canker; dieback. **Botryosphaeria ribis,** LA.
Mistletoe, European. **Viscum-album,** CA.
Rot, root. **Armillaria mellea,** CA; **Phymatotrichum omnivorum,** TX.
Scab. **Fusicladium pyracanthae,** widespread on leaves and fruit.
Wilt. **Fusarium oxysporum,** FL.

Fire blight is the most common disease. It infects all species, but some are relatively resistant. Scab, often disfiguring on berries, can be prevented by spraying at bud-break and 10 and 20 days later.

PYRETHRUM
(*Chrysanthemum cinerariifolium, C. coccineum*)

Bacterial fasciation. **Clavibacter fascians,** CT, MD.
Bacterial, MLO. **Aster Yellows,** NY, KS, NJ.
Blight, gray mold. **Botryris cinerea,** PA.
Damping-off. **Gloeosporium** sp., PA; **Pythium** sp., root rot, CO.
Nematode, root knot. **Meloidogyne** sp.
Rot, root. **Phymatotrichum omnivorum,** TX; **Rhizoctonia solani,** NJ.
Rot, stem. **Sclerotinia sclerotiorum,** VA.

PYROLA
(Shinleaf)

Blight, gray mold. **Botrytis cinerea,** MD.
Leaf spot. **Mycosphaerella chimaphilae,** MI; **Ovularia pyrolae,** WI; **Ramularia pyrolae,** WI; **Phyllosticta pyrolae,** DE, MT, WI; **Septoria pyrolae,** MI.

Rust. **Chrysomxya pirolata** (II, III), general from ME to MN and from MT to CA and AK; 0, I, on spruce; **Pucciniastrum pyrolae** (II, III), general in West.

QUINCE
(*Cydonia*)

Anthracnose, Northwestern. **Neofabraea malicorticis,** OR, WA.

Bacterial crown gall. **Agrobacterium tumefaciens,** general.

Bacterial fire blight. **Erwinia amylovora,** general.

Bacterial hairy root. **Agrobacterium rhizogenes,** ME to NC; Pacific Coast.

Blight, Dothiorella twig. **Botryosphaeria ribis,** TX.

Blight, leaf; black spot. **Fabraea maculata,** general.

Blight, thread. **Pellicularia koleroga,** NC, Gulf states.

Canker, perennial. **Neofabraea perennans,** OR.

Canker, trunk. **Nectria galligena,** OR; twig, **Valsa leucostoma.**

Fruit spot. **Leptothryium pomi,** MO; **Mycosphaerella pomi,** New England to OH.

Leaf spot. **Phyllosticta** sp., DE.

Nematode, root knot. **Meloidogyne** sp., TX.

Powdery mildew. **Phyllosticta corylea,** VA; **Podosphaera leucotricha,** CA., WA; **P. oxyacanthae,** NY, WV to IN.

Rot, bitter; canker. **Glomerella cingulata,** eastern and central states southward.

Rot, black; canker; leaf spot. **Physalospora obtusa,** eastern states to AL, TX.

Rot, brown. **Monilinia fructicola,** eastern states, MS, OR, TX; **M. laxa,** also blossom and twig blight, Pacific Coast.

Rot, fruit. **Alternaria mali,** IN; **Botrytis cinerea,** occasional; **Cephalothecium roseum,** pink mold, occasional; **Penicillium expansum,** cosmopolitan; **Phoma cydoniae,** pale rot, IL, MI; **P. mali,** IN.

Rust. **Gymnosporangium clavariiforme** (0, I), leaves, fruit, stems, CT, ME, NH; III, on juniper; **G. clavipes** (0, I), orange rust, quince rust, on fruit, stems; III, on juniper; **G. gracile** (0, I); III, on juniper; **G. libocedri** (0, I), fruit stems, OR; III, on incense-cedar; **G. nelsoni** (0, I), leaves, stems, AZ, CO; III, on juniper; **G. nidus-avis** (0, I), leaves, fruit, stems, CT, NY; III, on juniper.

Scab. **Venturia pirina,** CT.

Spot anthracnose. **Elsinoë piri,** WA.

Quinces are subject to fire blight; infected branches should be cut out with the usual precautions. For leaf blight, spray when blossoms show pink, again when last of the petals are falling, and perhaps twice more at 2-week intervals. Brown rot is not very important on quince. To prevent rust, remove nearby susceptible junipers or spray them in spring as spore horns are developing on galls. See under Rusts.

QUINCE, FLOWERING, JAPANESE, CHINESE
(*Chaenomeles*)

Bacterial crown gall. **Agrobacterium tumefaciens,** occasional.
Bacterial fire blight. **Erwinia amylovora,** occasional.
Blight, leaf. **Fabraea macula,** AL, CT, NY, NJ.
Canker, felt fungus. **Septobasidium burtii,** MS; **S. mariani,** NC.
Canker; twig blight. **Botryosphaeria ribis,** AL, TX; **Phoma** sp., MD, TX; **Physalospora obtusa,** TX, eastern states.
Fruit spot. **Mycosphaerella pomi,** MD, IL, also leaf blotch.
Leaf spot. **Cercospora cydoniae,** AL, GA.
Nematode, root knot. **Meloidogyne** sp.
Rot, brown. **Monilinia fructicola,** leaf blight, MI; **M. laxa,** also blossom and twig blight, CA.
Rust, quince. **Gymnosporangium clavipes** (0, I), on stems; CT, NJ; III, on juniper; **G. libocedri** (0, I), on leaves; III, on incense-cedar.

RABBITBRUSH
(*Chrysothamnus*)

Leaf spot. **Phleospora bigeloviae,** CA.
Powdery mildew. **Erysiphe cichoracearum,** CA, MT, UT, WY; **E. polygoni** var. **sepulta,** CO, MT, UT, WY.
Rust. **Puccinia dioicae** (0, I), NM and CA; **P. grindeliae** (III), MT to NM, CA; **P. stipae** (0, I), MT to NM, CA; II, III, on grasses.
Smut, inflorescence. **Thecaphora pilulaeformis,** AZ.

RADISH
(*Raphanus*)

Bacterial black rot. **Xanthomonas campestris,** IN, IA, MI, NJ, NY, OH, PA, TX.
Bacterial, MLO. **California Aster Yellows,** CA;
Bacterial soft rot. **Erwinia carotovora,** cosmopolitan.
Bacterial spot. **Xanthomonas vesicatoria** pv. **raphani,** IN.
Blotch; black pod. **Alternaria raphani,** CA, MI, MN, NJ, OH, PA.
Club root. **Plasmodiophora brassicae,** occasional in North, MA to NJ, MN, WA.
Damping-off. **Pythium debaryanum,** MA, MN, NJ, WY; **Rhizoctonia solani,** cosmopolitan.
Downy mildew. **Peronospora parasitica,** northeastern and central states to MS and TX, CA.
Leaf spot. **Alternaria brassicae,** gray leaf spot; **A. oleracea,** black leaf spot, CT, NJ; **Cercospora cruciferarum,** AL, IL, MO, TX; **C. atrogrisea,** NJ.

Nematode, leaf and stem. **Ditylenchus dipsaci,** NY.
Nematode, root knot. **Meloidogyne arenaria; M. javanica,** AL, MS, OR, PA, TX.
Nonparasitic. Air pollution, **NO$_2$, O$_3$, SO$_2$,** NC.
Powdery mildew. **Erysiphe polygoni,** CA, MO, TX.
Rot, black root. **Aphanomyces raphani,** ME to FL, OK, OR, IA, CA; **Pythium aphanidermatum,** IN, KS, MA, MI, NY, OH, OK, PA, WI, SC.
Rot, crown, watery soft. **Sclerotinia sclerotiorum,** CA, IN, MN, NJ, TX.
Rot, pod. **Phoma lingam,** FL, CA.
Rot, root. **Ascochyta** sp.; **Phymatotrichum omnivorum,** TX.
Rust. **Puccinia aristidae** (0, I), AZ, CO.
Scab. **Streptomyces scabies,** IN, MI, NJ, OH, TX, WI.
Virus. **Radish Mosaic; Beet Curly Top; Tobacco Streak,** CA.
White rust. **Albugo candida,** general.
Wilt; yellows. **Fusarium oxysporum** f. sp. **raphani; F. oxysporum** f. sp. **conglutinans,** CA.

Radishes are so easily grown in home gardens that not many gardeners worry about disease control. Seed should, however, be treated for damping-off and root rots.

RAGWEED
(*Ambrosia*)

Bacterial blight. **Pseudomonas syringae** pv. **tagetis,** WI.

RAIN-LILY
(*Cooperia*)

Leaf spot. **Cercospora amaryllidis,** TX.
Rust. **Puccinia cooperiae** (0, I, II, III), TX.

RANUNCULUS
(Buttercup, Crowfoot)

Blight, gray mold. **Botrytis cinerea,** CA, NY, WI.
Bacterial, MLO. **California Aster Yellows.**
Downy mildew. **Peronospora ficariae,** occasional MA to MD, IA, MN; also CA.
Gall, leaf. **Synchytrium anomalum,** IA; **S. aureum,** IL, CA, WI; **S. cinnamomeum,** WI.
Leaf spot. **Ascochyta infuscans,** WI; **Cercospora ranunculi,** IA, WI; **Cylindrosporium ficariae,** WA; **Didymaria didyma,** IL, IN, IA, MA, MI, MS, NY, WI; **Fabraea ranunculi,** CA, NE, NY, WI; **Ovularia decipiens; Ramularia**

aequivoca, IL, IA, OR, WI; **Septocylindrium ranunculi**, IL, NY, WA, WI; **Septoria** sp.

Nematode, leaf and stem. **Ditylenchus dipsaci**, OR.

Powdery mildew. **Erysiphe polygoni**, frequent in eastern and central states; **Sphaerotheca macularis**, CO.

Rot, leaf. **Ceratobasidium anceps**, WI.

Rot, root. **Phymatotrichum omnivorum**, TX.

Rot, stem. **Sclerotinia sclerotiorum**, AZ, CA; **Pythium** sp., CA; **Sclerotium rolfsii**, CA.

Rust. **Puccinia andina** (III), IL, IN; **P. eatoniae** var. **ranunculi** (0, I), CT to SC, MS, ND, CA, CO; II, III, on *Sphenopholis*; **P. ranunculi** (III), AZ, CO, UT, WA, WY, **P. recondita** (0, I), WI to TX, CA, WA; III, on *Hordeum*: **Uromyces dactylidis** (0, I), CO, TX, MA; **U. jonesii** (0, I, III), CA, CO, MT, WY.

Smut, leaf. **Doassansia ranunculina**, IN, WI; **Urocystis anemones**, IL, UT, WY.

Smut, white. **Entyloma microsporum**, IL, IN, IA, KY, VA, WI; **E. ranunculi**.

Virus. **Beet Curly Top; Ranunculus Mottle; Ranunculus Mosaic.**

RASPBERRY
(*Rubus*)

Anthracnose; dieback; gray bark. **Gloeosporium allantosporum**, OR, WA.

Bacterial blossom blight. **Pseudomonas** sp., OR.

Bacterial crown gall. **Agrobacterium tumefaciens,** general; **A. rubi,** cane gall, NY, PA to IL, and WI, OR.

Bacterial fire blight; flower and twig blight. **Erwinia amylovora**, ME, NH, PA, WA, WI.

Bacterial hairy root. **Agrobacterium rhizogenes,** OR.

Bacterial, witches' broom. **MLO,** OR.

Blight, cane. **Leptosphaeria coniothyrium,** general; **Physalospora obtusa,** IA, MD, MI, MO, ND; **Sclerotinia** sp., NY.

Blight, spur. **Didymella applanata**, general.

Canker; cane spot. **Ascospora ruborum**, MA, OR, WI.

Canker; dieback. **Glomerella cingulata,** AR, KY, MD, MI, MO, NJ, OH, RI, WV; **G. rubicola,** white bud, IL, NJ; **Botryosphaeria ribis,** FL; **Macrophoma rubi,** IL.

Dodder. **Cuscuta gronovii,** CT, IL, WI.

Downy mildew. **Peronospora rubi**, WA.

Fruit spot, fly speck. **Leptothyrium pomi**, MA, IN, KY.

Leaf spot. **Cylindrosporium rubi** (*Sphaerulina rubi*), also cane spot, common east of Rocky Mountains; **Mycosphaerella confusa** (*Cercospora rubi*), NJ to FL, TX, IL; **M. rubi,** general; **Pezizella oenotherae,** MD, MO, VA; **Septoria darrowi** (perhaps same as *Cylindrosporium rubi*), **Stigmatea rubicola,** black leaf and cane spot, MT, NM, NY, VT, WI.

Nematode, lesion. **Pratylenchus vulnus.**

Nematode, sheath. **Hemicycliophora** sp.

Nonparasitic. **Chlorosis,** iron deficiency in West.

Powdery mildew. **Phyllactinia corylea,** MI; **Sphaerotheca macularis,** Northeast, Northwest, IL, MN, CA.

Rosette, double blossom. **Cercosporella rubi,** KY, IL, MD, NY, PA.

Rot, fruit. **Botrytis cinerea,** gray mold, cosmopolitan; **Phyllostictina carpogena,** MD; **Monilinia fructicola,** brown rot, IL; **Alternaria** sp., MA; **Rhizopus nigricans,** black mold, cosmopolitan.

Rot, root. **Armillaria mellea,** OR, WA; **Rhizoctonia solani,** CO, ID, WA; **Xylaria** sp., WA; **Phytophthora erythroseptica,** Pacific Northwest.

Rust, late. **Pucciniastrum americanum** (II, III), Northeast to NC, OH, IL, ID; 0, I, on spruce.

Rust, leaf. **Phragmidium rubi-idaei** (0, I, II, III), northeastern and central states to CO; Pacific Northwest.

Rust, orange. **Gymnoconia peckiana** (0, I, III), on black raspberry, Northeast to MN, Pacific Northwest; **Kunkelia nitens** (I), IL, IN, MI, OH.

Rust, yellow. **Kuehneola uredinis** (0, I, II, III), IL, PA, DE to WI.

Spot anthracnose. **Elsinoë veneta,** general but less common on red than black raspberry.

Virus. **Raspberry Leaf Curl; Red Raspberry Mosaic; Black Raspberry Necrosis; Raspberry Streak; Raspberry Yellow Mosaic,** general; **Tomato Ring Spot,** cause of crumbly fruit, NY; **Tobacco Ring Spot,** NC.

Wilt. **Verticillium albo-atrum,** MA to NJ, OH, OR, WA.

Virus diseases are important on raspberries and cannot be controlled by spraying. Purchase healthy plants and set, if possible, 500 feet away from old patches. Inspect at least three times the first year, roguing all diseased plants, after first searing them with a blow torch or flame thrower so aphids will not carry the virus to nearby healthy bushes. Plants seldom recover from Verticillium wilt and never from orange rust, which is systemic. Crown gall is important on red raspberries; if infected plants are found, raspberries should not be replanted in the same soil for several years.

RATIBIDA
(Prairie Coneflower)

Downy mildew. **Plasmopara halstedii,** IA.

Leaf spot. **Cercospora ratibida,** KS, WI; **Physalospora lepachydis,** MT; **Ramularia rudbeckiae,** ID; **Septoria infuscata,** MI, MO, WI.

Powdery mildew. **Erysiphe cichoracearum,** ND, TX.

Rot, violet root. **Helicobasidium purpureum,** TX; **Phymatotrichum omnivorum,** TX.

Rust. **Uromyces perigynius** (0, I), ND, TX; II, III, on *Carex*.

Smut, white. **Entyloma compositarum,** NE, MI to IN, KS, MN.

RATTAN VINE
(*Berchemia*)

Rust. **Puccinia coronata** (0, I), VA to LA; II, III, on oats and wild grasses.

REDBUD, JUDAS-TREE
(*Cercis*)

Canker; dieback. **Botryosphaeria ribis** var. **chromogena**, DE, MD, NC, NJ, TX, VA.

Dodder. **Cuscuta exaltata**, TX.

Downy mildew. **Plasmopara cercidis**, TN.

Leaf spot. **Cercospora cercidis**, OK; **Cercosporella chionea**, IL, IN, KS, NC; **Mycosphaerella cercidicola**, general; **Phyllosticta cercidicola**, IN, WV; **Pestalotia guepini**, IL.

Rot, root. **Phymatotrichum omnivorum**, TX; **Polyporus lucidus**, VA.

Rot, wood. **Polyporus adustus; P. versicolor; Ganoderma lucidum** (*Polyporus lucidus*).

Wilt. **Verticillium** sp., OH; of seedlings, **Cylindrocladium** sp.

Wilt. **Verticillium albo-atrum**, VA.

RED-BAY, SWAMP-BAY
(*Persea borbonia*)

Black mildew. **Asterina delitescens**, VA to FL, TX; **Irenopsis martiniana**, AL, MS, TX; **Lembosia rugispora**, MS, NC; **Meliola amphitricha**, FL, MS; **Englerula carnea**, FL.

Leaf spot. **Cercospora purpurea**, GA to FL and MS; **Phyllosticta micropuncta**, MD to FL, TX; **Pestalotia** spp., general.

Leaf spot, algal. **Cephaleuros virescens**, Gulf states.

Rot, wood. **Polyporus hirsutus; P. mutabilis.**

RHEXIA
(Deergrass, Meadow-Beauty)

Leaf spot. **Cercospora erythrogena**, AL, DE, MS, TN; **Colletotrichum rhexiae**, DE; **Phyllosticta rhexiae**, FL.

RHODESGRASS
(*Chloris gayana*)

Leaf stripe, culm stripe. **Helminthosporium hawaiiense**, FL.

RHODODENDRON

Bacterial crown gall. **Agrobacterium tumefaciens,** OH.

Blight, bud, twig. **Briosia azalea,** GA, NC, NJ, NY, PA, TN, VA.

Blight, cutting. **Cylindrocladium scoparium** and **C. floridanum,** OH, FL.

Blight; flower spot. **Ovulinia azaleae,** SC, CA.

Blight, gray. **Pestalotia macrotricha,** general after winter injury.

Blight, silky thread. **Rhizoctonia ramicola,** FL.

Blight; twig canker. **Phomopsis** sp., OR, CT, NJ, NY.

Canker; dieback. **Botryosphaeria ribis,** MD, MA, NJ, NY; **Gloeosporium** sp., MD, NY; **Phytophthora cactorum,** MD, MA, NJ, NY, OH, PA, RI; **Glomerella cingulata,** MD.

Damping-off. **Alternaria** sp., CT; **Rhizoctonia solani,** CT, NJ, NY.

Dodder. **Cuscuta gronovii,** NJ.

Gall. **Exobasidium vaccinii,** leaf and shot, MA to FL, and MS; **E. burtii,** yellow leaf spot, NJ; **E. vaccinii-uliginosae,** witches' broom, NJ.

Leaf spot. **Cercospora handelii,** FL, NJ, NC, VA; **Coryneum rhododendri,** NC, PA, TN, VA; **Cryptostictis mariae,** KY, NY, TN, VA; **Discosia artocreas,** secondary, MD, NY; **Guignardia rhodorae,** CT, MD, MA, NJ, NY, PA, VA; **Gloeosporium ferrugineum,** NC; **Hendersonia concentrica,** NC, TX, WV; **Lophodermium schweinitzii,** NY to NC and TX; **Pestalopeziza rhododendri,** TN, WV; **Mycosphaerella clintoniana,** NJ, NY, NC, OR, WA; **Phyllosticta rhododendri,** NJ; **Physalospora rhododendri,** PA, TN, VA; **Phomopsis rhododendri,** NJ; **Septoria rhododendri,** ME, NC; **S. solitaria,** NJ; ?**Venturia rhododendri,** MD, VA.

Leaf spot, tar. **Melasmia rhododendri,** AK.

Nonparasitic. **Chlorosis,** mineral deficiency, usually iron, widespread. **Sun-scald,** windburn, severe winter injury in exposed locations. **Walnut Toxicity,** poisoning by root emanation from *Juglans nigra.*

Powdery mildew. **Microsphaera alni,** MD, NJ, NY.

Rot, heart; wood rot. **Fomes annosus,** NC.

Rot, root. **Armillaria mellea,** CA, NJ, NY; **Phymatotrichum omnivorum,** TX.

Rust. **Chrysomyxa ledi** var. **rhododendri,** CA, WA; **C. roanenis** (II, III), NC, TN; 0, I, on spruce; **C. piperiana** (II, III), CA, OR, WA; **Pucciniastrum vaccinii** (II), CT, NJ, RI.

Spot anthracnose. **Sphaceloma** sp., WA.

Wilt. **Phytophthora cinnamomi,** NJ, NY, PA, MD, OH; **P. citricola,** OH.

Most rhododendron leaf spots are not worth worrying about. Some come after winter injury; some are definitely parasitic but not serious. Winter and early spring sun will turn some of the foliage brown. Do not prune out supposedly dead twigs and branches too soon; wait for new growth to start. An accumulation of matted wet leaves around the trunk fosters root and collar rot. Be cautious in the use of aluminum sulfate to acidify soil; sulfur is somewhat safer.

RHOEO

Virus, Tobacco Mosaic, MD.

RHUBARB
(*Rheum*)

Anthracnose. **Colletotrichum** sp., IL, MO, PA, WV, WI.

Bacterial crown gall. **Agrobacterium tumefaciens**, IA, MA, NY.

Bacterial soft rot. **Erwinia carotovora**, occasional in market; **E. rhapontici**, crown rot, OK.

Blight, southern. **Sclerotium rolfsii**, FL, MS, TX, VA.

Damping-off; crown rot. **Pythium** spp., CA, MD.

Downy mildew. **Peronospora rumicis**, CA.

Leaf spot. **Ascochyta rhei**, eastern and central states to MS, KS; **Alternaria** sp., CA, MN, NE, NJ, PA; **Cercospora** sp., DE, MD, NE; **C. rhapontici**, IL; **Cladosporium** sp., CA, WA; **Macrophoma straminella**, general; **Ramularia rhei**, CA; **Septoria rhapontici**, IA.

Nematode, root knot. **Meloidogyne** spp., CA, MD, NY, OK.

Nonparasitic. **Crack Stem**, boron deficiency, WA.

Rot, gray mold. **Botrytis cinerea**, serious in greenhouses.

Rot, root. **Armillaria mellea**, CA, TX; **Phymatotrichum omnivorum**, AZ, TX; **Phytophthora cactorum**, crown rot, CA, MO, OK, PA; **P. parasitica**, IL, KS, LA, MD, MO, NY, TX, VA; **Rhizoctonia solani**, CA, CT, IL, MN, MO, NY, OK, TX, WA.

Rust. **Puccinia phragmites** (0, I), CA, MN, NE; II, III, on *Phragmites*.

Virus. **Rhubarb Chlorotic Ring; Turnip Mosaic; Beet Curly Top.**

Macrophoma leaf spot is common but seldom calls for control measures beyond removal of old stalks in late fall. Plants with crown rot should be dug and burned.

RIBBON-BUSH
(*Homalocladium*)

Powdery mildew. **Erysiphe polygoni**, NY, PA, WI.

RIBBON-GRASS
(*Phalaris*)

Ergot. **Claviceps purpurea**, NY.

Leaf blotch (spot). **Stagonospora foliicola**, PA.

Leaf spot. **Helminthosporium catenarium**, PA on Reed Canary grass (*P. arundinacea*)

Leaf spot, zonate eye-spot. **Helminthosporium giganteum**, MD.

RICE-PAPER PLANT
(*Tetrapanax*)

Nematode, root knot. **Meloidogyne** sp., FL.

ROCK-JASMINE
(*Androsace*)

Downy mildew. **Peronospora candida,** KS.
Leaf spot. **Mycosphaerella primulae,** NM.
Rust. **Puccinia volkartiana** (III), AK.

ROCK-ROSE
(*Cistus*)

Rot, root. **Armillaria mellea.**

ROLLINIA

Dieback, fruit rot. **Glomerella cingulata,** FL.

ROMANZOFFIA

Rust. **Puccinia romanzoffiae** (III), OR.

ROSE
(*Rosa*)

Bacterial blast. **Pseudomonas syringae.**
Bacterial crown gall. **Agrobacterium tumefaciens,** general.
Bacterial hairy root. **Agrobacterium rhizogenes,** PA, TX, MD, VA, New England.
Blackspot. **Diplocarpon rosae,** general.
Blight, blossom. **Botrytis cinerea,** cosmopolitan; gray mold on canes in storage; **Dothiorella** sp., LA, VA.
Blight, cane. **Physalospora obtusa,** CT to AL, KS, TX; **Gloeosporium** spp., widespread.
Blight, southern. **Sclerotium rolfsii,** FL, KS, TX.
Blight, thread. **Pellicularia koleroga,** FL, LA.
Canker, brand. **Coniothyrium wernsdorffiae,** MN, NY, PA (also reported from CO, IN, MS, TX, but probably mistaken for *C. fuckelii*).
Canker, brown. **Cryptosporella umbrina,** MA to FL, TX, NE, and MI, CA, ID.

Canker, common, graft. **Leptosphaeria coniothyrium** (*Coniothyrium fuckelii*), general; Graft, **C. rosarum**, CA, IA, MA, MN, NJ, PA, TX.

Canker, crown. **Cylindrocladium scoparium,** in greenhouses, MA to GA, TX and IL.

Canker; dieback. **Botryosphaeria ribis** var. **chromogena,** AL, MD, TX, VA; **Cryptosporium minimum,** OR, PA; **Diplodia** spp., probably secondary; **Griphosphaeria corticola** (*Coryneum microstictum*), NH to AL, ND; Pacific Northwest; **Nectria cinnabarina,** coral spot, MA to VA, WA, AK; **Cytospora** sp., KY, PA, VA, WA; **Didymella sepincoliformis,** MD; **Glomerella cingulata,** MD, NJ, VA; **Macrophoma** sp., TX, VA; **Botryodiplodia theobromae,** MN; **Trichothecium roseum,** MN.

Dodder. **Cuscuta indecora** and **C. paradoxa,** FL, TX.

Downy mildew. **Peronospora sparsa,** mostly under glass, ME to FL, IA, CA; reported outdoors in DE.

Leaf spot. **Alternaria** sp., VA to AL, TX; **A. brassicae** var. **microspora,** TX; **Cercospora puderi,** FL, GA; **Mycosphaerella rosicola** (*Cercospora rosicola*), general; **M. rosigena,** doubtfully distinct from *M. rosicola,* reported from South; **Monochaetia compta,** AK, IA, KS, MD; **Pezizella oenotherae,** also cane spot, NJ to FL, TX, MI; **Phyllosticta rosae,** NY to FL, IN; **P. rosae-setigerae,** IN; **Septoria rosae,** MS, NJ, SC.

Mold, black, of grafts. **Chalaropsis thielavioides,** CA, IL, NY, PA on understock from OR and WA.

Mold, leaf and bud. **Cladosporium** sp. and **C. fuscum,** AK, CA, MD, MN, MS, OK, TX.

Nematode, dagger. **Xiphinema diversicaudatum,** greenhouses in Northeast; **X. americanum; X. krugi.**

Nematode, leaf. **Aphelenchoides** spp.; lance, **Hoplolaimus** sp.

Nematode, lesion. **Pratylenchus pratensis; P. scribneri; P. vulnus.**

Nematode, ring. **Criconemoides** sp.; Pin, **Paratylenchus** spp.

Nematode, root knot. **Meloidogyne** sp.; **M. hapla.**

Nematode, spiral. **Helicotylenchus** spp.; Sheath, **Hemicycliophora** spp.

Nematode, sting. **Belonolaimus gracilis;** stubby root, **Trichodorus** spp.

Nematode, stylet. **Tylenchorhynchus** spp.

Nonparasitic

Boron Deficiency. Leaves distorted, greenhouse.

Chlorosis. Iron deficiency, upper leaves yellow, with green veins; nitrogen deficiency, lower leaves pale; potassium deficiency, leaves grayish, may drop, stems weak.

Leaf Scorch. Marginal, potash deficiency. In greenhouses scorch may be boron and calcium deficiency.

Mercury Toxicity. In greenhouses when paint containing mercury used on sash.

Weed-killer Injury. Leaves fernlike, twisted when 2,4-D used in vicinity.

Pedicel Necrosis. Collapse of flower stem, cause unknown.

Petal spot. **Bipolaris** (*Helminthosporium*) **setariae**, FL.

Powdery mildew. **Sphaerotheca pannosa** var. **rosae**, general; **S. macularis**, not readily distinguished from *S. pannosa*; **Phyllactinia corylea**, WA.

Rot, root. **Armillaria mellea**, CA, MS, OR, TX, WA; **Clitocybe tabescens**, FL;, **Fusarium** spp., occasional, especially in the South; **Phymatotrichum omnivorum**, AZ, TX; **Ramularia macrospora**, MD.

Rust. **Phragmidium americanum** (0, I, II, III), on leaves of cultivated and native roses, ME to NC, TX and ND; **P. montivagum** (0, I, II, III), on native species, SD to NM, AZ, WA; **P. mucronatum** (*P. disciflorum*), the common rust of cultivated roses, possibly on native species (0, I), on leaves and stems; II, III, on leaves, eastern states to Rocky Mountains; Pacific Coast; **P. fusiforme** (*P. rosae-acicularis*) (0, I, II, III), on native species, MI to CO, WY, CA; **P. rosae-californicae** (0, I, II, III), on natives, AZ, CA, MT, OR; **P. rosae-pimpinellifoliae** (*P. subcorticium*) (0, I), on stems; II, III, on leaves of brier and sweetbrier groups, northern U.S.; **P. rosicola** (III), CO, MT, NE, on native spp.; **P. speciosum** (0, I), on leaves and stems, III, on stems of cultivated and native roses, general except far South.

Spot anthracnose. **Elsinoë rosarum**, ME to FL, MI, MO, KS, TX; Pacific Coast.

Unknown. **Speckle.** Chlorotic flecks in leaves, not transmitted by grafting, MD, NJ, NY, PA, TX, VA.

Virus. **Rose Mosaic** (in part Prunus Necrotic Ring Spot and apple mosaic), Pacific Coast and eastern states in greenhouses and in gardens on plants shipped from the West Coast; **Rose Rosette** (?witches' broom), CA, KS, MO; **Rose Streak**, MD, NJ, NY, TX, VA; **Rose Yellow Mosaic**, IL, MD, NY, PA, VA, CA; **Crinkle.** On Manetti understock, sometimes garden roses, Pacific Coast, TX, MD, NY, PA, VA; **Rose Leaf Curl**, CA; **Rose Spring Dwarf**, CA.

Wilt. **Verticillium albo-atrum**, AR, CA, IL, NJ, NY; probably widespread.

Blackspot, brown canker, powdery mildew, and rust are the big four diseases of garden roses. Blackspot is almost inevitable except in some dry western states, and shows up even there when overhead watering is substituted for the usual irrigation. It can be controlled by regular weekly spraying or dusting.

Powdery mildew, a problem on the Pacific Coast, is increasing in eastern gardens. Brown canker and other cane diseases are best controlled at spring pruning, by cutting out infected canes and cutting other canes just above a bud, not leaving any stub to die back. Cankers are increased by excessive winter protection. Where temperatures permit, as in the Central Atlantic region, eliminate soil mounding and other special winter treatment.

Roses are sensitive to many chemicals; it is important to distinguish spray injury from blackspot and not increase the dosage because you think you are not getting control. Combination sprays or dusts should take care of most diseases, as well as insects, in one operation. The bacterial crown gall is occasionally present on plants purchased from a nursery. Ask for a replacement; do not contaminate your soil by planting such a bush.

ROSE-ACACIA
(*Robinia hispida*)

Leaf spot. **Alternaria fasciculata,** ND.
Rot, root. **Phymatotrichum omnivorum,** TX.

ROSELLE
(*Hibiscus sabdariffa*)

Anthracnose, pod spot. **Colletotrichum gloeosporioides,** FL.
Blight, gray mold. **Botrytis cinerea,** MD.
Blight, southern. **Sclerotium rolfsii,** TX.
Leaf spot. **Cercospora hibisci,** TX.
Nematode, root knot. **Meloidogyne** sp., TX.
Powdery mildew. **Microsphaera euphorbiae,** AL, FL.
Rot, fruit, stem. **Fusarium** sp., FL; **Phytophthora parasitica,** TX.
Rot, root. **Phymatotrichum omnivorum,** TX; **Rhizoctonia solani,** TX.

ROSE-GENTIAN
(*Sabatia*)

Anthracnose. **Gloeosporium** sp., OK.
Leaf spot. **Cercospora sabbatiae,** DE, MS, NC, OK, TX.

ROSE-MALLOW
(*Hibiscus palustris*)

Bacterial crown gall. **Agrobacterium tumefaciens,** MS.
Dieback. **Colletotrichum hibisci,** NJ, NY, TX.
Leaf spot. **Ascochyta abelmoschi,** NY; **?Cercospora kellermanii,** IN; **Phyllosticta hibiscina,** CT, FL, LA, MD, NJ, NY; **Septoria** sp., NJ.
Rot, root. **Phymatotrichum omnivorum,** TX.
Rust. **Puccinia schedonnardi** (0, I), CT to AL, NE, TX; II, III, on grasses.

ROSEMARY
(*Rosmarinus*)

Rot, root. **Phymatotrichum omnivorum,** TX.

ROSE-OF-SHARON, SHRUB-ALTHAEA
(*Hibiscus syriacus*)

Leaf spot. **Cercospora malayensis,** GA; **Phyllosticta hibiscina,** OK; **P. syriaca,** NY.

Nematode, root knot. **Meloidogyne** sp., MS, TX.

Rot, root. **Phymatotrichum omnivorum,** TX.

Rust. **Kuehneola malvicola** (II, III), Gulf states.

ROUGE-PLANT
(*Rivina*)

Leaf spot. **Cercospora flagellaris,** FL, TX; **Septoria rivinae,** TX.

Rot, root. **Helicobasidium purpureum,** TX; **Phymatotrichum omnivorum,** TX.

Rust. **Puccinia raunkaerii** (0, I, II, III), FL, TX.

RUBBER-PLANT
(*Ficus elastica*)

Anthracnose. **Glomerella cingulata,** general; **Gloeosporium** sp., WA.

Bacterial crown gall. **Agrobacterium tumefaciens,** CA, TX.

Canker; dieback. **Physalospora rhodina,** GA.

Leaf spot. **Alternaria** sp., IN, OH, TX; **Leptostromella elastica,** NY, northeastern states; **Mycosphaerella bolleana,** GA; **Phyllosticta** sp., MD, NY; **Phyllosticta roberti,** Gulf states, MD, NY; **Stemphylium elasticae,** ?secondary; **Trabutia** (*Phyllachora*) **ficuum,** black spot, FL.

Nematode, leaf. **Aphelenchoides besseyi,** FL.

Nematode, root knot. **Meloidogyne** spp.; **M. incognita-acrita.**

Nematode, spiral. **Helicotylenchus nannus.**

RUDBECKIA
(Golden-Glow, Coneflower, Black-Eyed Susan)

Bacterial, MLO. **Aster Yellows,** NY, and **California Aster Yellows,** CA.

Blight, southern. **Sclerotium rolfsii,** FL, NJ.

Downy mildew. **Plasmopara halstedii,** IA, NC, ND, NE, NY, WI.

Gall, leaf. **Synchytrium aureum,** IL, WI.

Leaf spot. **Cercospora rudbeckiae,** NY; **C. tabacina,** WI, IL, NY; **Phyllosticta rudbeckiae,** IA, NY, WI; **Ramularia rudbeckiae,** VT to MS, CO, ID, MT, VA, VT, WV; **Septoria rudbeckiae,** DE, KS, NE, WA, WI.

Powdery mildew. **Erysiphe cichoracearum,** general.

Rot, root. **Phymatotrichum omnivorum,** TX.

Rot, stem. **Sclerotinia sclerotiorum,** CT.

Rust. **Aecidium batesii** (0, I), NE; **Puccinia dioicae** (0, I), MD, SD; **P. rud-beckiae** (III) TX; **Uromyces perigynius** (0, I), MD to MT; II, III, on *Carex;* **U. rudbeckiae** (III), MT, MD to MS, NM, TX.

Smut, white. **Entyloma compositarum,** IA, MO, OH, WI.

Virus. **Potato Yellow Dwarf,** NY; **Mosaic,** unidentified, IL, IN; **Biddens Mottle,** FL.

Wilt. **Verticillium albo-atrum,** NY.

RUE ANEMONE
(*Anemonella*)

Leaf spot. **Cercospora caulophylli,** MO.

Powdery mildew. **Erysiphe polygoni,** IA.

Rust. **Puccinia recondita** (0, I), IN, IA, MO.

Smut, leaf and stem. **Urocystis anemones,** NY.

RUELLIA

Leaf spot. **Cercospora consociata,** AL, IL, MS, MO, IA.

Rot, root. **Phymatotrichum omnivorum,** TX.

Rust. **Puccinia lateripes** (0, I, II, III), MD to FL, KS, MO, TX; **Uromyces ruelliae** (0, I, II, III), TX.

RUMEX
(Garden Sorrel)

Bacterial, MLO. **California Aster Yellows,** CA.

Gall, leaf. **Synchytrium anomalum,** IA.

Leaf spot. **Cercospora acetosellae,** LA, TX; **Phyllosticta** sp., NY; **Gloeo-sporium rumicis,** NY, TX; **Septoria pleosporioides,** TX.

Nematode, ring. **Criconemella xenoplax,** SC.

Nematode, root knot. **Meloidogyne** sp., FL.

Rot, root. **Rhizoctonia solani,** TX.

Rust. **Puccinia acetosae** (II, III), ME to FL.

Virus. **Tomato Ringspot,** PA.

RUSSIAN-OLIVE, SILVERBERRY
(*Elaeagnus*)

Canker. **Fusicoccum elaeagni,** IL, MO.

RUTABAGA

See Turnip.

SAFFLOWER
(Carthamus)

Anthracnose; blight. **Gloeosporium carthami,** IN, TX, VA.
Bacterial blight. **Pseudomonas syringae,** CA.
Blight. **Botrytis cinerea; Rhizoctonia** sp., NM.
Leaf spot. **Alternaria** spp., IN, AZ, CA, MT, NE, ND; **Septoria carthami,**
  IN, TX; **Stemphylium** sp., FL.
Powdery mildew. **Erysiphe ?cichoracearum,** CA.
Rot, root. **Phytophthora drechsleri,** CA.
Rot, stem; wilt. **Sclerotinia sclerotiorum,** IN, ND, VA.
Rust. **Puccinia carthami** (II, III), CO, MA, MT, NE, ND.
Virus. **Cucumber Mosaic,** AZ, CA; **Turnip Mosaic,** CA.
Wilt. **Fusarium oxysporum** f. sp. **carthami; Verticillium albo-atrum.**

SAGE
(Salvia; includes Blue, Scarlet, Black Ornamental Forms)

Bacterial, MLO. **California Aster Yellows,** CA.
Blight, southern. **Sclerotium rolfsii,** IL.
Damping-off. **Pythium debaryanum,** OH; **Rhizoctonia solani,** CT, IL, NJ, OH.
Downy mildew. **Peronospora lamii,** IA; **P. swinglei,** KS.
Leaf spot. **Cercospora salviicola,** OK, TX; **Ramularia salviicola,** OK.
Nematode, leaf. **Aphelenchoides fragariae,** DE, NJ.
Nematode, root knot. **Meloidogyne** spp., AZ, NJ; **M. incognita-acrita; M. javanica.**
Powdery mildew. **Erysiphe cichoracearum,** CA.
Rot, charcoal. **Macrophomina phaseoli,** SC.
Rot, root. **Phymatotrichum omnivorum,** TX; Stem, **Sphaeropsis salviae,** MS.
Rust. **Aecidium subsimulans,** AZ; **Puccinia ballotaeflorae** (II, III), TX; 0, I
  unknown; **P. caulicola** (0, I, II, III), IA to TX, NM; **P. farinacea** (0, I, II, III),
  AL, AZ, KS, MS, MO, NE, OK, TX; **P. salviicola** (0, I, II, III), FL, TX;
  P. vertisepta (0, I, III), AZ, NM; **P. melliflora** (I, III), CA.
Virus. **Tomato Spotted Wilt.**

SAGE-BRUSH
(Artemisia)

Blight, gray mold. **Botrytis cinerea,** AK.
Blight, stem. **Sclerotium** sp., OR.

Canker, stem gall, black knot. **Syncarpella tumefaciens,** CA, MT, NV.
Dodder. **Cuscuta** sp., TX.
Downy mildew. **Peronospora leptosperma,** CA, IA, KS, ND, WI.
Leaf spot. **Cercospora ferruginea,** WI; **C. olivacea,** NY; **Cylindrosporium artemisiae,** WA, WI; **Heterosporium** sp., AK; **Phyllosticta raui,** MT; **Ramularia artemisiae,** NY, WI; **Septoria artemisiae,** WA.
Nematode, root knot. **Meloidogyne** sp., AL.
Powdery mildew. **Erysiphe cichoracearum,** WI to NM, CA, WA.
Rot, root. **Phymatotrichum omnivorum,** TX.
Rust. **Puccinia atrofusca,** IA to TX, CA, OR; II, III, on *Carex;* **P. millefolii** (III), ND to TX, CA, WA, AK; **P. tanaceti** (0, I, II, III), WI to TX, CA, WA; 0, I, II, III on *Tanacetum* and II on chrysanthemum; **Uromyces oblongisporus** (III), WY.

ST. ANDREWS CROSS, ST. PETERSWORT
(*Ascyrum*)

Leaf spot. **Cladosporium gloeosporioides,** AL.
Rust. **Uromyces triquetrus** (0, I, II, III), MS, NJ, TX.

ST. AUGUSTINEGRASS
(*Stenotaphrum*)

Virus. **Panicum Mosaic,** AR.

ST. JOHNSWORT
(*Hypericum*)

Black knot. **Gibberidea heliopsidis,** MD.
Blight, leaf. **Rosellina** (*Dematophora*) **necatrix.**
Leaf spot. **Cercospora hyperici,** IL; **Cladosporium gloeosporioides,** AL, NJ, NY, WI.
Nematode, root knot. **Meloidogyne** spp.
Powdery mildew. **Erysiphe cichoracearum.**
Rust. **Melampsora hypericorum** (II), MT; **Uromyces triquetrus** (0, I, II, III), ME to AL, and IA, TX, WI.

SALAL
(*Gaultheria shallon*)

Black mildew. **Meliola** sp.
Leaf spot. **Dasyschypha gaultheriae,** CA, OR, WA; **Mycosphaerella gaultheriae,** Pacific Coast, AK; **Pestalopezia brunneo-pruinosa,** CA, OR, WA; **Phyllosticta gaultheria,** general.

Powdery mildew. **Microsphaera alni,** OR.
Spot anthracnose. **Elsinoë ledi,** OR, WA.

SALPIGLOSSIS
(Painted-Tongue)

Bacterial canker, vascular. **Clavibacter michiganense,** WY.
Bacterial, MLO. **California Aster Yellows,** CA.
Nematode, lesion. **Pratylenchus pratensis,** NY.
Nematode, root knot. **Meloidogyne** sp., NY.
Wilt. **Fusarium** sp., WA; **Verticillium albo-atrum,** NY.

SALSIFY
(*Tragopogon*)

Bacterial, MLO. **California Aster Yellows,** CA; **Aster Yellows,** MD, NY, PA, WI.
Bacterial soft rot. **Erwinia carotovora,** CT, TX.
Blight, leaf. **Sporodesmium scorzonerae,** AL, MD, NY, PA, VA, WV.
Blight, southern. **Sclerotium rolfsii,** TX.
Leaf spot. **Cercospora tragopogonis,** MT, OK; **Stemphylium botryosum,** NY;
  Alternaria tenuis.
Nematode, leaf and stem. **Ditylenchus dipsaci,** CA.
Nematode, root knot. **Meloidogyne** sp., NY to AL, TX, WA.
Powdery mildew. **Erysiphe cichoracearum,** general.
Rot, root. **Phymatotrichum omnivorum,** TX, AZ; **Rhizoctonia solani,** TX, WA.
Rot, stem crown. **Sclerotinia intermedia,** IL; **S. sclerotiorum,** IL.
Virus. **Beet Curly Top; Lettuce Mosaic,** NY.
White rust. **Albugo tragopogonis,** CA.
Wilt. **Verticillium albo-atrum,** NY.

SALSIFY, BLACK
(*Scorzonera*)

Bacterial, MLO. **California Aster Yellows,** CA.
Nematode, root knot. **Meloidogyne** sp., FL.
White rust. **Albugo tragopogonis,** CA.

SALT BUSH
(*Atriplex*)

Downy mildew. **Peronospora farinosa,** MT, TX.
Gall, leaf, stem. **Urophlyctis pulposa,** ND, TX.
Leaf spot. **Cercospora dubia,** widespread; **Stagonospora atriplicis,** NJ, NY,
  PA, KS, NE.

Nematode, root knot. **Meloidogyne** sp., CA; Root Gall, **Heterodera schachtii**, UT.
Rust. **Puccinia aristidae** (0, I); II, III, on grasses ; **Uromyces shearianus** (0, I, III).
Virus. **Beet Curly Top.**

SAINFOIN
(*Onobrychis*)

Damping-off. **Rhizoctonia solani**, TX; **Sclerotium rolfsii**, TX.
Nematode, root knot. **Meloidogyne hapla**, WY.
Rot, stem. **Rhizoctonia solani**, TX; **Sclerotium rolfsii**, TX.

SANCHEZIA

Rot, mushroom root. **Clitocybe tabescens**, FL.

SAND-MYRTLE
(*Leiophyllum*)

Gall, leaf. **Exobasidium vaccinii**, NC, NJ.

SAND-VERBENA
(*Abronia*)

Downy mildew. **Peronospora oxybaphi**, TX.
Leaf spot. **Heterosporium abroniae**, CA, TX.
Rust. **Puccinia aristidae**, AZ, CA, CO, NM; II, III on grasses.

SANDVINE
(*Ampelanus*)

Black mildew. **Meliola bidentata**, NC.
Downy mildew. **Plasmopara gonolobi**, SC.
Leaf spot. **Cercospora gonolobi**, OK; **Septoria** sp., LA.
Rust. **Puccinia obliqua** (III), OK, TX.

SANDWORT
(*Arenaria*)

Leaf spot. **Hendersonia tenella**, TX.
Powdery mildew. **Erysiphe polygoni**, CA.

Rot, root. **Phymatotrichum omnivorum,** TX.
Rust. **Puccinia arenariae** (III), CA, FL, MT, NY, TX, WI; **P. tardissima,** CO,
NM, UT, WY; **Uromyces inaequialtus** (0, I, II, III), CO, UT.
Smut, anther. **Ustilago violacea,** ME, NH, NY, VT.

SANGUISORBA
(Burnet)

Leaf spot. **Graphium sessile,** NY; **Ovularia bulbigera,** AK, IL.
Powdery mildew. **Podosphaera oxyacanthae,** IA; **Sphaerotheca macularis,**
AK, KS, NY, PA.
Rust. **Xenodochus carbonarius** (I, III), AK; **X. minor,** AK.

SANSEVIERIA
(Bowstring-Hemp, Snake Plant)

Bacterial soft rot. **Erwinia aroideae** and **E. carotovora,** AZ, FL, NJ, MD.
Leaf spot. **Fusarium moniliforme,** FL, MO, WA; **Gloeosporium sansevieriae,**
FL, WA.
Nematode, lesion. **Pratylenchus** sp., CA.
Nematode, root knot. **Meloidogyne** sp.; **M. javanica.**
Nonparasitic. **Wilt,** overfertilization or toxic salts.
Rot. **Aspergillus niger; Fusarium** sp.

SAPODILLA
(*Achras*)

Gall, limb. **Pestalotia scirrofaciens,** FL, TX.
Leaf spot. **Phyllosticta** sp., FL; **Septoria** sp., FL.
Rot, root. **Phymatotrichum omnivorum,** TX.
Rust. **Scopella** (*Uredo*) **sapotae** (II), FL.
Spot anthracnose. **Elsinoë lepagei,** FL.

SASSAFRAS

Canker, branch, trunk. **Nectria** sp., CT to WV, MS.
Canker; dieback. **Physalospora obtusa,** NY to GA.
Leaf spot. **Septoria** sp., NY; **Actinothyrium gloeosporioides** (*Actinopella
dryina*); **Cristulariella pyramidalis; Phyllosticta illinoiensis,** IL, MA; **P.
sassafras,** NY to GA, TX, IL.
Mistletoe. **Phoradendron serotinum (flavescens),** TX.
Powdery mildew. **Phyllactinia corylea,** MI.

Rot, heart, trunk. **Daedalea confragosa,** IN, NY; **Fomes igniarius,** OH, VA; **F. ribis,** MO.

Rot, root. **Armillaria mellea,** PA; **Phymatotrichum omnivorum,** TX.

Rot, wood. **Daldinia vernicosa; Hymenochaete agglutinans; Hypoxylon** spp.; **Polyporus** spp., sometimes on living trees; **Poria ferruginosa; Schizophyllum commune,** NY; **Trametes sepium,** IN.

Virus. **Mosaic,** NY; **Yellows,** TX, unidentified.

The undetermined yellows disease causes fasciation of tops, leafroll, and dwarfing of leaves.

SAURURUS
(Swamp-Lily, Water Dragon)

Gall, leaf. **Physoderma** sp., VA.

Leaf spot. **Cercospora saururi,** AL, FL, IL, IN, LA, NY, TX; **Ramularia sauturi,** OK.

Nematode, root knot. **Meloidogyne** sp.

SAXIFRAGE
(*Saxifraga*)

Blight, gray mold. **Botrytis cinerea,** AK.

Leaf spot. **Cercosporella saxifragae,** WI; **Phyllosticta saxifragarum,** WY; **Septoria albicans,** WI; **Ramularia** sp., AK.

Powdery mildew. **Sphaerotheca macularis,** NY, AK, CO, PA, WY.

Rust. **Melampsora artica** (0, I), CO, AK; II, III, on willow; **Puccinia heucherae** (III), AK, MT to NM, ID, WA, WY, NY to IL, MT; **P. pazschkei** (III), ID, MT, WA; and var. **tricuspidatae,** CO, UT.

SCABIOSA

Bacterial, MLO. **Aster Yellows,** CT, NJ, NY, VA; and **California Aster Yellows,** CA.

Blight, southern. **Sclerotium rolfsii.**

Powdery mildew. **Erysiphe polygoni.**

Rot, root. **Phymatotrichum omnivorum,** TX.

Rot, stem. **Sclerotinia sclerotiorum,** NY.

Virus. **Beet Curly Top,** CA.

SCARBOROUGH-LILY
(*Vallota*)

Leaf scorch; red spot. **Stagonospora curtisii,** LA.

SCHEFFLERA
(*Brassaia actinophylla* = *Schefflera actinophylla*)

Bacterial blight. **Pseudomonas cichorii,** FL.
Leaf spot. **Alternaria panax,** CA, FL.
Nematode, root knot. **Meloidogyne incognita-acrita.**
Rot, stem, leaf, cutting. **Fusarium solani,** FL.
Virus. ?**Ghost Ring.**

SCHEFFLERA, DWARF
(*Schefflera arboricola*)

Bacterial leaf blight. **Pseudomonas cichorii,** FL.
Leaf spot. **Alternaria** sp. FL.

SCHRANKIA

Rust. **Ravenelia morongiae,** TX.
Stem spot. **Cercospora morongiae,** MS.

SCILLA
(Squill)

Nematode, bulb. **Ditylenchus dipsaci,** VA.
Rot, blue mold. **Penicillium gladioli,** on imported bulbs.
Rot, bulb. **Sclerotium rolfsii,** WA.
Smut, flower. **Ustilago vaillantii,** MA, WA.
Virus. ?**Ornithogallum Mosaic,** NY.

SCINDAPSUS

See Pothos.

SEA-GRAPE, DOVE-PLUM
(*Coccoloba*)

Spot, tar. **Phyllachora simplex,** FL.

SEA-KALE
(*Crambe*)

Leaf spot, black. **Alternaria oleracea,** VA.
Rot, root; damping-off. **Aphanomyces raphani,** WI.

Virus. Beet Western Yellows Virus, CA.
Wilt; yellows. **Fusarium oxysporum** f. sp. **conglutinans,** IN.

SEDUM
(Stonecrop)

Blight, southern. **Sclerotium rolfsii,** KS, NJ, VA.
Leaf spot. **Septoria sedi,** IL, IA, ME, NY; **Pleospora** sp., NY; **Stemphylium bolicki,** FL.
Nematode, root knot. **Meloidogyne** sp.
Rot, stem. **Colletotrichum** sp. (*Vermicularia benficiens*), VA, NY; **Phytophthora** sp., NY; **Rhizoctonia solani,** IL, NJ.
Rust. **Puccinia rydbergii** (III), UT; **P. umbilici** (III), CO, WY.
Wilt. **Fusarium oxysporum** f. sp. **sedi,** CA.

SEMPERVIVUM
(Houseleek)

Rot, leaf and stem. **Phytophthora parasitica,** NY.
Rot, root. **Pythium** sp., IA.
Rust. **Endophyllum sempervivi** (III), MA, NJ, NY.

SENECIO
(Groundsel)

Bacterial, MLO. **California Aster Yellows.**
Gall, leaf. **Synchytrium aureum,** WI.
Leaf spot. **Cercospora senecionicola,** WI; **C. senecionis,** TX; **Gloeosporium senecionis,** CA; **Phyllosticta garrettii,** OR, UT, WY; **Ramularia filaris,** CO, MT; **R. pruinosa,** CO, WY; **R. senecionis,** CA, CO; **Septoria cacaliae,** AL, IN, TX; **S. senecionis,** CA.
Nematode, leaf. **Aphelenchoides ritzema-bosi.**
Powdery mildew. **Erysiphe cichoracearum,** ID, MN, NE, VA, WA; **Sphaerotheca macularis,** CO, MT, UT, WY.
Rot, root. **Phymatotrichum omnivorum,** TX; **Rhizoctonia solani,** IL, NJ.
Rot, stem. **Phytophthora** sp., NJ; **Sclerotinia sclerotiorum,** LA.
Rust. **Coleosporium occidentale** (II, III), CA, CO, ID, MT, OR, WA, WY; 0, I unknown; **C. senecionis** (II, III), CO, RI; **Puccinia angustata** var. **eriophori** (0, I), CT, IA, MN, NH, UT, VT; II, III, on *Eriophorum* and *Scirpus*; **P. expansa,** CA, UT, WA, WY; **P. recedens,** CO, CT to NC, TN, IA, ND to OR, WA, WY; **P. dioicae** (0, I), NE, NM, TX; II, III, on *Carex*; **P. stipae** (0, I), CO, WY, NE; II, III, on *Stipa*; **P. subcircinata** (0, I, III), NE, NV, ND, ID, UT, WA, NM; **Baeodromus californicus** (III), CA.

Smut, white. **Entyloma compositarum,** KS, MD, NE, PA, TX, WI.
Virus. **Tomato Spotted Wilt.**
White rust. **Albugo tragopogonis,** CA, CO, IN, MO, MT, NE, UT, WA.
Wilt. **Fusarium** sp., NJ; **Verticillium albo-atrum,** WA.

SEQUOIA
(Redwood and Giant Sequoia)

Blight, needle. **Chloroscypha chloromela; Cercospora sequoiae,** MD, PA; **Mycosphaerella sequoiae; Pestalotia funerea,** TX.
Blight, seedling. **Botrytis douglasii,** CA, OH, PA.
Blight, twig. **Phomopsis juniperovora** (?); **Botrytis cinerea.**
Burls, galls, on trunk. Cause unknown.
Canker. **Botryosphaeria dothidea,** VA. [Redwood, Dawn] [U.S. National Arboretum, Washington, DC]; **B. dothidea** (= **B. ribis**), CA.
Canker, bark. **Dermatea livida.**
Nematode. **Meloidogyne hapla, M. incognita, M. javanica, Pratylenchus penetrans, P. vulnus, Xiphinema bakeri,** CA.
Rot, charcoal. **Macrophomina phaseoli.**
Rot, root. **Armillaria mellea; Phymatotrichum omnivorum,** TX.
Rot, trunk; heart. **Fomes annosus; Ganoderma sequoiae; Poria sequoiae; P. albipellucida.**
Rot, wood. **Hymenochaete tabacina; Lenzites saepiaria; Merulius hexagonoides; Polyporus** spp.; **Schizophyllum commune; Stereum** spp.; **Trametes** spp.

SERIOCARPUS
(White-Topped Aster)

Rust. **Coleosporium asterum** (II, III), CT; **Puccinia dioicae** (0, I), NC, IN, TN.

SESAME
(*Sesamum*)

Bacterial leaf spot. **Pseudomonas sesami,** KS, TX.
Bacterial wilt. **Pseudomonas solanacearum,** AZ.
Blight. **Corynespora cassiicola,** MS, Southeast.
Leaf spot. **Alternaria sesami,** Southeast; **Cercospora sesami,** FL, GA, SC; **Cylindrocladium sesami,** FL, SC; **Helminthosporium sesami,** TX.
Rot, charcoal. **Macrophomina phaseoli,** CA, TX.
Wilt. **Verticillium albo-atrum,** NM.

SESUVIUM

Nematode, root knot. **Meloidogyne** sp., AL, FL.
Rot, root. **Phymatotrichum omnivorum**, TX.
Rust. **Puccinia aristidae** (0, I), TX; II, III, on grasses.
White rust. **Albugo trianthemi**, TX.

SHALLOT
(*Allium ascalonicum*)

Bacterial, MLO. **Aster Yellows**, LA.
Blight, southern. **Sclerotium rolfsii**, LA.
Blotch, purple. **Alternaria porri**, LA, TX.
Downy mildew. **Peronospora destructor**, LA.
Nematode, root knot. **Meloidogyne** sp., TX.
Rot, neck. Gray mold, **Botryris allii**, LA; smudge, **Colletotrichum circinans**, IL, LA, WI.
Rot, pink root. **Pyrenochaete terrestris**, CO, LA, TX; **Fusarium solani**, TX.
Rot, white. **Sclerotium cepivorum**, LA, VA.
Smut. **Urocystis cepulae**, MA.
Virus. **Mosaic**, unidentified.

Pink root and white rot are prevalent in Louisiana. Losses from white rot are heavy if plants are set late; September setting may give a good crop.

SHASTA DAISY
(*Chrysanthemum maximum*)

Bacterial fasciation. **Clavibacter fascians**, CA.
Leaf blotch. **Septoria leucanthemi**, CA, OR.
Leaf spot. **Cercospora chrysanthemi**, OK.
Nematode, root knot. **Meloidogyne** spp., AL, MS, FL.
Rot, root. **Pythium** sp., NJ.
Rot, stem. **Rhizoctonia solani**, MD; **Sclerotinia sclerotiorum**, MT, WA; **Fusarium roseum** and F. **solani**, TX.

SHEPHERD'S PURSE
(*Capsella*)

Bacterial, yellows. **Spiroplasma citri**, IL.

SHORTIA
(Oconee-bells)

Leaf spot. **Pezizella oenotherae**, NC, SC.

SIDA

Blight, southern. **Sclerotium rolfsii**, FL.
Leaf spot. **Cercospora sidicola; Colletotrichum malvarum**, KS, TX, UT; **Phyllosticta spinosa**, KS, TX; **Ramularia sidarum**, FL.
Nematode, root knot. **Meloidogyne** spp., AL, MS, FL.
Rot, root. **Phymatotrichum omnivorum**, TX.
Rust. **Puccinia heterospora**, FL to TX, MO, IN; **P. lobata**, AZ, CA, NM, TX, UT; **P. schedonnardi**, NM.
Virus. **Abutilon Mosaic**, FL.

SILENE
(Catchfly, Cushion-Pink, Campion)

Broomrape. **Orobanche ramosa**, TX.
Damping-off. **Rhizoctonia solani**, IL.
Downy mildew. **Peronospora silenes**, KS, IL, NE, TX, WI.
Leaf spot. **Ascochyta silenes**, MT, OK, WI; **Marssonina delastrei**, WI; **Phyllosticta nebulosa**, MT, NY, WI; **Septoria dimera**, NE, WI; **S. silenes**.
Rust. **Uromyces silenes** (0, I, II, III), PA, CA, IA, KS, MT, WA; **U. suksdorfii**, CA, ID, NM, UT, WA; **Puccinia aristidae** (0, I), AZ, TX.
Smut, flower. **Sorosporium saponariae**, CO, NV, UT; **Ustilago violacea**, anther smut, CA, MT, NH, TX, VA, WA, WY.

SILK-TASSEL BUSH
(*Garrya*)

Black mildew. **Lembosia lucens**, CA.
Leaf spot. **Cercospora garryae**, CA, TX; **Dothichiza garryae**, CA; **Phyllosticta garryae**, CA, TX.
Rot, root. **Phymatotrichum omnivorum**, TX.

SILPHIUM
(Compass Plant, Indian-Cup)

Downy mildew. **Plasmopara halstedii**, IL, IA, WI to AR, KS, MN.
Leaf spot. **Ascochyta compositarum**, WI; **Cercospora silphii**, AL, KS, IL, TX, WV, WI; **Colletotrichum silphii**, WI; **Septoria alba**, IL, KS.
Powdery mildew. **Erysiphe cichoracearum**, CT, MD.
Rot, root. **Phymatotrichum omnivorum**, TX; **Rhizoctonia solani**, ME.
Rust. **Puccinia silphii** (III), NC to AL, TX, ND; **Coleosporium terebinthinaceae** (II, III), PA to FL, TX, KS; 0, I, on pines; **Uromyces silphii** (0, I), OH to MO, KS and WI; II, III, on *Juncus*.
Smut, white. **Entyloma compositarum**, TX, WI.

SKIMMIA

Virus. Tobacco Ring Spot, NY.

SKULLCAP
(*Scutellaria*)

Leaf spot. **Cercospora scutellariae,** IL, MS, MO, TX; **Phyllosticta decidua,** TX, WI; **Septoria scutellariae,** ME to IA, CA, CO, MS, OK.
Powdery mildew. **Erysiphe ?galeopsidis,** MI, NY, IL, IN, IA, KS, OH, WI; **Microsphaera** sp., IL.
Rot, root. **Phymatotrichum omnivorum,** TX; **Rhizoctonia solani,** TX.
Rot, stem. **Botrytis cinerea,** WA.

SKUNK-CABBAGE
(*Symplocarpus*)

Blight, leaf. **Botryris ?streptothrix,** CT, IL, NJ, NY.
Leaf spot. **Cercospora symplocarpi,** MA to VA, IN, WI; **Septoria spiculosa,** MD, NY, PA, WI.

SLENDERFLOWER THISTLE
(*Carduus*)

Rust. **Puccinia carduorum,** CA.

SMALL FLOWER GALINSOGA
(*Galinsoga parviflora*)

Rot, stem. **Whetzelinia sclerotiorum,** MD.

SMELOWSKIA

Rust. **Puccinia aberrans** (0, III), CO, MT, NE, UT, WA; **P. holboelli,** NV; **P. monoica,** CO, WY.

SMILACINA
(False Solomons Seal)

Leaf spot. **Cercosporella idahoensis,** ID; **Cylindrosporium smilacinae,** CO, OR, UT, CA; **?Heterosporium asperatum,** WY; **Phleospora vagnerae,** MT; **Ramularia smilacinae,** MT, WY, WA; **Septoria smilacinae,** general; **Sphaeropsis cruenta,** CA, NM.

Rot, rhizome. **Stromatinia smilacinae,** NY.
Rust. **Puccinia sessilis,** CA, KS, IA, ID, ND, SD, MT, NE, NY, OK, PA, WA; **Uromyces acuminatus** var. **magnatus,** CO, MN, MT, NE, WI, IL, ND, SD; II, III on *Spartina.*
Smut, leaf. **Urocystis colchici,** MT.

SMILAX
(Greenbrier Cat-Brier; for Florists' Smilax, see Asparagus)

Canker, felt fungus. **Septobasidium pseudopedicellatum,** FL.
Gall. **Synchytrium smilacis.**
Leaf spot. **Ascochyta confusa,** NY, WI; **Cercospora smilacina,** CT to FL, TX; **C. smilacis,** MA to FL, TX, and MN; **Colletotrichum smilacis,** IL; **Cylindrosporium smilacis,** AL; **Dothiorella smilacina,** MA to LA, TX; **Mycosphaerella smilacicola,** GA, SC; **Pestalotia clavata,** NY to AL, LA; **Phyllosticta subeffusa,** KS, IL, WV; **Ramularia subrufa,** IA, MS, NE, WI; **Septogloeum subnudum,** IL, WI; **Septoria smilacis,** WV; **Sphaeropsis cruenta,** IA; **Stagonospora smilacis,** CT to MD, ND, TX.
Nematode, burrowing. **Radopholus similis,** FL.
Powdery mildew. **Phyllactinia corylea,** MI.
Rot, root. **Helicobasidium purpureum,** TX.
Rust. **Puccinia amphigena** (0, I), KS, MI, NE, ND; II, III, on *Calamovilfa;* **P. arundinariae** (0, I), NC, SC; **P. macrospora** (0, I), DE, NJ, NY; II, III, on *Carex;* **P. smilacis** (II, III), MA to FL, TX, and NE; 0, I, on *Apocynum.*

SMITHANTHA

Virus. Tobacco Mosaic, CA, CT, DC, FL, OH, WA.

SMOKE TREE
(*Cotinus*)

Leaf spot. **Cercospora rhoina,** AL; **Pezizella oenotherae,** MD; **Septoria rhoina,** CT, MA, NY, VA; **Gloeosporium** sp., IL.
Rot, root. **Phymatotrichum omnivorum,** TX.
Rust. **Pileolaria cotini-coggyriae,** GA, RI.
Wilt. **Verticillium albo-atrum,** CT, IL, NE, NJ, NY.

SMOKE-TREE
(*Dalea*)

Leaf spot. **Cercospora daleae,** KS.
Mistletoe. **Phoradendron californicum,** CA to TX.

Rot, root. **Phymatotrichum omnivorum,** TX.

Rust. **Pileolaria cotini-coggyriae,** AR.

Rust. **Puccinia andropogonis** (0, I), SD to KS; II, III, on *Andropogon*; **P. paroselae** (II, III), CA.

SNAPDRAGON
(*Antirrhinum*)

Anthracnose. **Colletotrichum antirrhini,** general in eastern and southern states to CO and TX.

Bacterial crown gall. **Agrobacterium tumefaciens,** NY.

Blight, gray mold. **Botrytis cinerea,** cosmopolitan in greenhouses.

Blight, southern; stem rot. **Sclerotium rolfsii,** CA, FL, MS, NJ, NY, TX.

Canker, stem and crown. **Myrothecium roridum,** IL.

Damping-off; root rot. **Pythium** spp., cosmopolitan; **Rhizoctonia solani,** also collar rot.

Dodder. **Cuscuta** sp., WA.

Downy mildew. **Peronospora antirrhini,** CA, NY, OK, OR, PA.

Leaf spot. **Cercospora antirrhini,** FL, IL.

Leaf spot; stem rot; canker. **Phyllosticta antirrhini,** general in eastern and north central states, also TX and WA.

Nematode, lesion. **Pratylenchus pratensis.**

Nematode, root knot. **Meloidogyne** spp., general in South and in northern greenhouses; **M. javanica,** MD.

Nematode, pin. **Paratylenchus penetrans,** CA.

Nonparasitic. **Fasciation,** probably genetic. **Tip Blight,** cause unknown, MD, OK, VA, CA; injury from peach aphid.

Petal spot. **Bipolaris** (*Helminthosporium*) **setariae,** FL.

Powdery mildew. **Oidium** sp., MA, NY, PA.

Rot, charcoal. **Macrophomina phaseoli,** OK.

Rot, root. **Phymatotrichum omnivorum,** TX; **Thielaviopsis basicola,** CT, NJ.

Rot, stem; wilt. **Fusarium** sp., perhaps secondary; CT, FL, GA, OK, TN, WA; **Phytophthora cactorum,** CA, IL, MN, NJ, NY; **P. cryptogea,** CA, OK; **P. parasitica,** GA; **Sclerotinia sclerotiorum,** CA, IN, MI, PA, TX; **S. minor,** CT.

Rust. **Puccinia antirrhini** (II, III), general; 0, I unknown.

Virus. **Cucumber Mosaic, Mosaic,** unidentified, KS, NY, OH, PA; **Ring Spot,** unidentified, OK.

Wilt. **Verticillium albo-atrum,** CA, CT, ME, MA, MN, NY, NJ, PA.

Rust is the most generally important disease and can be prevented, to some extent, by purchasing rust-resistant seed, but not all such seed is resistant to all strains of rust.

SNOWBERRY
(*Symphoricarpos*)

Anthracnose; black berry rot; twig canker. **Glomerella cingulata,** widespread, MA to VA, IL and WI.

Bacterial crown gall. **Agrobacterium tumefaciens,** MD.

Bacterial hairy root. **Agrobacterium rhizogenes,** IA.

Leaf spot. **Ascochyta symphoricarpophila,** NY; **Cercospora symphoricarpi,** MT, SD; **Phyllosticta symphoricarpi,** NY, NM, WA; **Lasiobotrys symphoricarpi,** black spot, CO, UT, WY; **Septoria oedospora,** CO; **S. signalensis,** WY; **S. symphoricarpi,** ND to CO, CA and WA.

Powdery mildew. **Microsphaera diffusa,** general; **Podosphaera oxyacanthae,** WA.

Rot, berry. **Alternaria** sp., CO, CT, MA, NY; **Botrytis cinerea,** CT, MA, NY.

Rot, collar. **Fomes ribis,** KS, MT.

Rot, root. **Helicobasidium purpureum,** TX.

Rust. **Puccinia crandalli** (0, I), ND, CO, CA, WA, ID, MT, UT, WY; II, III, on grasses; **P. symphoricarpi** (III), MT to CA and AK.

Spot anthracnose; scab. **Sphaceloma symphoricarpi,** ME to VA, AR and WI, CA, CO, OR, WA.

Anthracnose and spot anthracnose, scab, commonly disfigure berries.

SNOWDROP
(*Galanthus*)

Blight, botrytis. **Botrytis galanthina** on imported bulbs.

SNOW-ON-THE-MOUNTAIN
(*Euphorbia marginata*)

Blight, gray mold. **Botrytis cinerea,** NJ.

Leaf spot. **Alternaria** sp., KS, TX; **Cercospora euphorbiicola,** NE; **C. pulcherrimae,** OK; **Phyllosticta** sp., NJ.

Powdery mildew. **Microsphaera euphorbiae,** IN, IA, KS, MO.

Rot, root. **Phymatotrichum omnivorum,** TX.

Rust. **Puccinia panici** (0, I), MS to TX, CO, and SD; **Uromyces euphorbiae,** IA and SD to TX and CO.

SOAPBERRY, SOUTHERN
(*Sapindus saponaria*)

Blight, thread. **Pellicularia koleroga,** FL.

Leaf spot; dieback. **Glomerella cingulata,** FL; **Phyllosticta sapindii,** FL.

SOAPBERRY, WESTERN
(*Sapindus drummondii*)

Blight, leaf. **Cylindrosporium griseum,** OK, TX.
Leaf spot; dieback. **Glomerella cingulata,** TX; **Mycosphaerella sapindii,** MO.
Mistletoe. **Phoradendron serotinum (flavescens),** AZ, NM, TX.
Powdery mildew. **Uncinula circinata,** TX.
Rot, root. **Helicobasidium purpureum,** TX.
Virus. **Mosaic,** TX.

This is one of the few plants reported resistant to Texas root rot.

SOAPWORT
(*Saponaria*)

Leaf spot. **Alternaria saponariae,** also stem spot, CT to MD, IN, MN; **Cylindrosporium officinale,** IN; **Phyllosticta tenerrima,** NJ, TX; **Septoria noctiflorae,** IL.
Rot, root. **Phymatotrichum omnivorum,** TX.
Rust. **Puccinia aristidae** (0, I), CO.

SOCIETY GARLIC
(*Tulbaghia*)

Nematode, root knot. **Meloidogyne incognita,** FL.

SOLOMONS-SEAL
(*Polygonatum*)

Leaf spot. **Colletotrichum liliacearum,** ?secondary, cosmopolitan; **Sphaeropsis cruenta,** CT, IN, IA, NY, OH, VA, WI.
Rot, rhizome. **Stromatinia smilacinae,** NY.
Rust. **Puccinia sessilis** (0, I), AL, CT, ID, IA, MN, NY, OH, PA, WI, WY; II, III, on *Phalaris*; **Uromyces acuminatus** var. **marginatus** (0, I), IL, IA, MN, NE, ND, SD; II, III, on *Spartina*.
Smut, leaf. **Urocystis colchici,** IA.
Virus. **Mosaic.** Unidentified, ME.

SONCHUS

Rot. **Macrophomina phaseolina,** AZ.

SOPHORA
(Pagoda Tree, Silky Sophora, Mescalbean)

Blight, twig. **Nectria cinnabarina,** CT, NY.
Canker. **Fusarium lateritium,** NJ.
Damping-off. **Rhizoctonia solani,** CT.
Dieback. **Diplodia sophorae,** OH.
Leaf spot. **Phyllosticta sophorae,** OK, TX.
Mistletoe. **Phoradendron serotinum (flavescens),** TX.
Nematode, root knot. **Meloidogyne** sp., MD.
Powdery mildew. **Microsphaera alni,** CT.
Rot, root. **Phymatotrichum omnivorum,** TX.
Rust. **Uromyces hyalinus** (0, I, II, III), SD to TX, AZ, WY.
Virus. **Brooming Disease.** Unidentified.

SOYBEAN
(Glycine Max)

Anthracnose. **Glomerella glycines,** IA, MI, NC to FL, TX and NE; **Colletotrichum truncatum.**
Bacterial blight. **Pseudomonas syringae** pv. **glycinea,** East and South to TX, MN.
Bacterial, MLO. Bud proliferation, LA.
Bacterial leaf crinkle. Unidentified, Midwest.
Bacterial pustule; pustular spot. **Xanthomonas glycines (phaseoli** var. **sojense),** general.
Bacterial, tan spot. **Clavibacter flaccumfaciens,** IA.
Bacterial wildfire. **Pseudomonas tabaci,** MD to AL, LA, NE.
Bacterial wilt. **Pseudomonas solanacearum,** NC.
Blight, aerial. **Rhizoctonia solani,** LA.
Blight, ashy stem; charcoal rot. **Macrophomina phaseoli,** NJ to SC, TX and NE, OH, WA, WI.
Blight, gray mold; leaf spot. **Botrytis cinerea,** CT, OH.
Blight, pod and stem. **Diaporthe sojae,** NY, MI to GA, LA, KS.
Blight. **Sclerotinia minor, S. sclerotiorum,** VA.
Blight, southern. **Sclerotium rolfsii,** VA to FL, TX, IA.
Canker, stem. **Diaporthe phaseolorum** vars. **batatatis** and **caulivora,** AR, FL, MD, OH, SC, TX; **Phyllosticta sojaecola,** AR, MD, stem and pod canker.
Damping-off; root rot. **Pythium** spp., IL, IA, MN, MO, NC, ND; **P. aphanidermatum.**
Downy mildew. **Peronospora manshurica,** East and South to LA, IA.
Leaf spot. **Alternaria** sp., widespread, secondary; **Cercospora canescens,** also on pods, stems, AL, IL, MD, MS, NC, TX, WV; **C. kikuchii,** also purple

stain of seed, IN, MD, NC, VA; **C. sojina,** frog-eye spot, NY and MI to FL, OK and IA; **Coryneospora casiicola** (*Helminthosporium vignicola*), target spot, AL, AR, FL, GA, LA, MS, NC, SC; **Helminthosporium vignae,** zonate leaf spot, NC; **Mycosphaerella cruenta,** GA, MS; **Myrothecium roridum,** secondary, LA; **Phyllosticta glycinea,** IL, MD, MO, NC, VA; **Pleosphaerulina sojaecola,** ME, NJ, NY, WI.

Leaf spot, brown spot. **Septoria glycines,** AR, DE, IN, IA, MD, NC, WI.

Nematode, cyst. **Heterodera glycines,** soybean cyst, AR, IA, KS, KY, LA, MI, MN, MO, NC, NE, OH, TN, TX; **H. gottingiana,** pea cyst; **H. trifolii,** clover cyst; **Cactodera estonica,** WI.

Nematode, dagger. **Longidorus sp.**

Nematode, lesion. **Pratylenchus safaensis,** GA.

Nematode, root knot. **Meloidogyne hapla; M. javanica.**

Nematode, spiral. **Helicotylenchus erythrinae; H. nannus.**

Nematode, stem. **Ditylenchus dipsaci,** NY.

Nematode, sting. **Belonolaimus gracilus; B. longicaudatus.**

Nonparasitic. **Baldhead,** mechanical injury. **Chlorosis,** interveinal, manganese deficiency, AL, NC, TN. **Yellowing,** potassium deficiency, iron deficiency.

Powdery mildew. **Erysiphe polygoni,** DE, IA, NC, SC; **Microsphaera diffusa,** DE, MN, NC.

Rot, root. **Phymatotrichum omnivorum,** TX; **Pythium aphanidermatum,** AZ; **P. dissotocum,** AZ.

Rot, root and stem. **Phytophthora megasperme** f. sp. **glycinea,** DE, Northeast, MN, SD.

Rot, seed. **Fusarium scirpi,** NE; **F. graminearum,** IL; **Nematospora coryli,** NC, SC, OK, VA; **Aspergillus spp; Phomopsis sojae,** NY.

Rot, stem. **Cephalosporium gregatum,** brown stem, IL, IN, IA, KY, MO, MN, NC, VA, OH; **Pellicularia filamentosa,** also root rot, canker, general; **Sclerotinia sclerotiorum,** AZ, MD, IA, NY, VA; **Phytophthora megasperma** var. **sojae; Neocosmospora vasinfecta,** AL.

Rust. **Phakospora pachyrhizi,** MD.

Virus. **Soybean Mosaic; Tobacco Ring Spot,** causing bud blight, midwestern U.S.; **Beet Curly Top; Bean Pod Mottle,** NE; **Soybean Yellow Mosaic,** Midwest; **Cowpea Mosaic,** IL; **Tobacco Streak,** OK.

Wilt. **Fusarium oxysporum** f. sp. **tracheiphilum,** PA to FL, TX, and NE; also CA.

Edible soybeans are well suited to home garden culture and usually produce a good crop without control measures beyond dusting seed with a protectant before planting. Commercial growers find a number of diseases of economic importance: bacterial pustule, the various blights, leaf spots, downy mildew, wildfire, virus diseases. Some areas are now under quarantine for the soybean cyst nematode.

SPANISH MOSS
(*Tillandsia*)

Blight, stem and leaf. **Fusarium solani.** Southeast, U.S.

SPARAXIS
(*Wandflower*)

Virus. ?**Iris Mosaic,** CA, OR.

SPATHIPHYLLUM

Leaf spot. **Myrothecium roridum,** FL.

SPECULARIA
(Venus Looking-Glass)

Dodder. **Cuscuta** sp., TX.
Gall, leaf. **Synchytrium** sp., TX.
Leaf spot. **Cercospora speculariae,** LA; **Septoria speculariae,** KS, PA to AL,
    TX, and WI.
Rot, root. **Phymatotrichum omnivorum,** TX.
Rust. **Coleosporium campanulae** (II, III), NC, PA.
Smut, seed. **Ustilago speculariae,** OK.

SPHACELE
(Pitcher-Sage)

Rust. **Uredo sphacelicola** (II), CA.

SPICE-BUSH
(*Lindera*)

Canker. **Botryosphaeria ribis** var. **chromogena,** MD.
Leaf spot. **Phyllosticta linderae,** DE, IN, WV; **P. lindericola,** WV.
Mistletoe. **Phoradendron serotinum (flavescens),** eastern states.
Rot, root. **Phymatotrichum omnivorum,** TX.

SPIDER-LILY
(*Hymenocallis*)

Leaf blotch; red spot. **Stagonospora curtissi,** CA, TX.

Leaf spot. **Cercospora amaryllidis,** TX; **C. pancratii,** FL, LA, TX; **Gloeosporium hemerocallidis,** TX.

Nematode, spiral. **Rotylenchus blaberus.**

Virus. **Mosaic.** Unidentified, CA.

SPIDERLING, WINE-FLOWER
(*Boerhaavia*)

Bacterial leaf spot. **Xanthomonas campestris,** TX.

Leaf spot. **Ascochyta boerhaaviae,** TX; **Cercospora boerhaaviae,** TX.

Nematode, root knot. **Meloidogyne** spp.

Rot, root. **Phymatotrichum omnivorum,** TX.

White rust. **Albugo platensis,** AZ, FL, NM, TX.

SPINACH
(*Spinacea*)

Anthracnose. **Colletotrichum spinaciae,** CT, LA, MS, NJ, NY, TX, VA; **C. spinacicola** (*Gloeosporium spinaciae*).

Bacterial, MLO. **Aster Yellows.**

Bacterial soft rot. **Erwinia carotovora,** general in transit, market.

Damping-off; root rot. **Rhizoctonia solani,** general; **Pythium** spp., pre-emergence seed decay.

Downy mildew. **Peronospora effusa,** general.

Leaf spot. **Cercospora beticola,** CA, GA, IL, IA, MA, NY, TX; **Phyllosticta chenopodii,** DE, NJ, NY, VA; **Stagonospora spinaciae,** SD; **Alternaria spinaciae,** MA.

Mold, leaf. **Cladosporium macrocarpum,** secondary, DE, OK, PA, TX; **Heterosporium variabile,** pinhead "rust," general.

Mold, seed. **Pleospora herbarum** (*Stemphylium botryosum*); **Alternaria** sp., and secondary leaf spot; **Curvularia inaequalis.**

Nematode, clover cyst. **Heterodera trifolii.**

Nematode, root knot. **Meloidogyne** spp.

Nonparasitic. **Yellows,** nutrient deficiency.

Rot, root. **Aphanomyces cladogamus,** NJ, VA; **Phymatotrichum omnivorum,** TX; **Pyrenochaete terrestris,** IA; **Olpidium brassicae.**

Rot, root, crown; wilt. **Fusarium solani; F. oxysporum** f. sp. **spinaciae,** general; **Phytophthora** sp., AZ, IN, NJ, NY; **P. megasperma,** CA, NC; **Sclerotinia sclerotiorum,** GA, NY.

Rust. **Puccinia aristidae** (0, I), AZ, CA, CO, OR, WA; II, III, on grasses.

Smut, leaf. **Entyloma ellisii,** NJ, WA.

Virus. **Cucumber Mosaic** (spinach blight); **Spinach Yellow Dwarf; Beet Curly Top; Beet Ring Mottle; Beet Pseudo-Yellows; Tomato Spotted Wilt; Lettuce Mosaic,** NY.

White rust. **Albugo occidentalis,** AR, OK, TX.

Wilt. **Verticillium** sp., NY.

Downy mildew is the outstanding spinach disease, with all varieties susceptible in some degree because of physiologic races, although Dixie Market, Savoy Supreme, and others are considered resistant. To reduce blight, use virus-tolerant varieties.

SPIRAEA
(Native Hardhack, Meadowsweet)

Blight, seedling; stem girdle. **Thelephora terrestris,** ID.

Canker. **Cryptodiaporthe macounii,** NY.

Leaf spot. **Cercospora rubigo,** CA, KS, OR, WI; **Cylindrosporium** spp., NY, ID; **Phleospora salicifoliae,** NY to KS, TX, WA.

Powdery mildew. **Podosphaera oxyacanthae,** widespread; **Sphaerotheca macularis,** CT, MI, NY, PA.

SPIREA, ORIENTAL FLOWERING
(*Spiraea*)

Bacterial fire blight. **Erwinia amylovora,** MD, NJ, NC, VA.

Bacterial hairy root. **Agrobacterium rhizogenes,** IA.

Leaf spot. **Cylindrosporium filipendulae,** IA, OR, WA, WI.

Nematode, root knot. **Meloidogyne** spp., FL, MS; **M. hapla.**

Powdery mildew. **Microsphaera alni,** CT; **Podosphaera oxyacanthae,** widespread; **Sphaerotheca macularis,** TX.

Rot, root. **Phymatotrichum omnivorum,** AZ, TX.

SPONDIAS
(Mombin)

Nematode, root knot. **Meloidogyne** sp., FL.

Spot anthracnose; mombin scab. **Sphaceloma spondiadis,** FL.

SPRUCE
(Picea)

Blight, brown felt. **Herpotrichia nigra,** MT, Northern Rocky Mountains, Pacific Northwest; **Neopeckia coulteri.**

Blight, needle. **Sirula macrospora,** ND.

Blight, seedling smother. **Thelephora terrestris,** ID, MN, OH; seedling, **Rosellinia herpotrichioides,** WA.

Blight, shoot. **Sirococcus strobilinus,** KS, NC, WI; **Phomopsis occulta,** WI.

Blight, snow. **Phacidium infestans,** New England states; **Botrytis cinerea,** NJ; **Lophophacidium hyperboream,** MN.

Blight, twig. **Ascochyta piniperda,** NC, ME.

Brooming. **Rhizosphaera kalkhoffii,** Central Rocky Mountains.

Canker, bark. **Aleurodiscus amorphus,** widespread.

Canker; twig blight. **Cytospora kunzei** (*Valsa kunzei* var. *piceae*), New England to NJ, IL, and MN.

Damping-off. **Cylindrocladium scoparium,** NJ, MI; **Phytophthora cinnamomi,** VA, MD, NY; **P. cactorum,** NY, VA; **Pythium ultimum,** widespread; **Rhizoctonia solani,** cosmopolitan; **Aphanomyces euteiches; Caloscypha fulgens,** OR, WA.

Mistletoe; witches' broom. **Arceuthobium campylopodium,** Rocky Mountains, TX, NM; **A. pusillum,** New England, especially ME, NH, to Great Lakes states, MN, WI.

Needle cast. **Rhizosphaera kalkhoffii,** northeastern U.S., AZ, IN, MI, MN, WI, PA.

Needle cast; tar spot. **Lophodermium filiforme,** NY, CO, AK; **L. piceae,** New England to Great Lakes states, Pacific Northwest; **Lophodermina septata,** OR; **Rhizophaera kalkhoffii,** CT, NY, VA; **Bifusella crepidiformis,** MT; **Cladosporium** sp., ME.

Nematode. **Nacobbodera chitwoodi,** OR.

Nematode. **Paratylenchus projectus; Pratylenchus penetrans; Tylenchus marginatus.**

Nematode, dagger. **Xiphinema americanum,** WI.

Rot, collar. **Diplodia pinea,** NJ, NY; **Sphaeropsis ellisii,** NJ.

Rot, heart. **Fomes annosus,** CA to WA; **F. pini,** widespread; **F. pinicola,** widespread; **F. roseus; F. subroseus,** widespread; **Polyporus** spp.

Rot, root. **Armillaria mellea,** AZ, CO, NM, PA, OR, UT, WA; **Phymatotrichum omnivorum,** TX; **Sparassis radicata,** ID, MT, OR, WA; **Phytophthora cinnamomi,** NC; **Fusarium solani,** PA; **F. oxysporum,** PA; **F. avenaceum,** PA.

Rust, cone. **Chrysomyxa pirolata** (0, I), AK, ME, MA, MI, NY, NH, PA, VT, CO, MT, OR; II, III, on Moneses and Pyrola.

Rust, needle. **Chrysomyxa empetri** (0, I), ME; **C. ledi** (0, I), New England to Great Lakes; II, III, on Ledum; **C. ledi** var. **cassandrae** (0, I), New England to Great Lakes; II, III, on Chamaedaphne; **C. ledicola** (0, I), northern U.S. from ME to WA, and AK, CO; II, III, on Ledum; wild rosemary;

C. piperiana (0, I), CA, OR; II, III, on *Rhododendron californicum*,
C. roanensis (0, I), NC, TN; II, III, on *Rhododendron catawbiense*,
C. weirii, TN, VT, WV; **C. arctostaphyli** (0, I); III, on bearberry.
Rust; witches' broom. **Melampsorella cerastii**, widespread.

Cytospora canker frequently kills lower branches of ornamental spruces.
There is little control except to remove affected portions.

SPURGE CAPER
(*Euphorbia lathyris*)

Rot, root. **Macrophomina phaseolina, AZ, Rhizoctonia solani, AZ; Phythium aphanidermatum, AZ.**

SPURGE, CYPRESS
(*Euphorbia cyparissias*)

Anthracnose. **Sphaceloma poinsettiae, FL.**
Leaf spot. **Cercospora euphorbiae, TX.**
Rot, root. **Phymatotrichum omnivorum**, TX; **Rhizoctonia** sp.
Rust. **Melampsora euphorbiae** (0, I, II, III), ME to PA, IN, WI.

SPURGE, FLOWERING
(*Euphorbia corollata*)

Leaf spot. **Cercospora euphorbiae**, KS, TX; **C. heterospora**, WI; **Phyllosticta** sp.
Mold, leaf. **Cercosporidium fasciculatum, IA.**
Powdery mildew. **Microsphaera euphorbiae**, MD to GA, IN, WI.
Rot, root. **Phymatotrichum omnivorum, TX.**
Rust. **Puccinia panici** (0, I), OH to AL, TX, MN; II, III, on Panicum.

SPURGE, PAINTED
(*Euphorbia heterophylla*)

Rust. **Uromyces euphorbiae** (0, I, II, III), IN to FL, TX, KS.
Smut, stem. **Tilletia euphorbiae, LA.**

SPURGE, SPOTTED
(*Euphorbia maculata*)

Nematode, root knot. **Meloidogyne hapla.**

SQUASH and PUMPKIN
(*Cucurbita*)

Anthracnose. **Colletotrichum lagenarium,** CT to NJ, TX, KS.

Bacterial, MLO. **Aster Yellows.**

Bacterial soft rot. **Erwinia carotovora** pv. **carotovora,** cosmopolitan; **E. aroideae.**

Bacterial spot. **Xanthomonas cucurbitae,** CT, GA, IL, IN, MA, MD, MI, WA, WI; **Pseudomonas syringae** pv. **lachrymans.**

Blight, blossom. **Choanephora cucurbitarum,** brown rot of fruit, ME to FL, MI, OK, TX.

Blight, gummy stem; black rot of fruit. **Mycosphaerella citrullina** (*M. melonis*), CT, FL, GA, MA, MI, NJ, NY.

Blight, leaf. **Alternaria cucumerina,** DE, MN, NC, NJ, NY, UT; **Phytophthora capsici,** also stem rot, NC, CA.

Blight, southern. **Sclerotium rolfsii,** AL, FL, GA.

Damping-off. **Pythium debaryanum,** CT, WI.

Downy mildew. **Pseudoperonospora cubensis,** ME to AL, LA, and TX, CA, IA.

Leaf spot. **Ascochyta** sp.; **Cercospora cucurbitae,** DE, NJ, WI, AL; **C. citrullina,** AL; **Gloeosporium** sp., IL; **Phyllosticta cucurbitarum,** IN, NY; **P. orbicularis,** DE, PA, NY; **Septoria cucurbitacearum,** MA, NY, WI; **Stemphylium cucurbitacearum,** IN, OH.

Mold, seed. **Alternaria tenuis,** cosmopolitan; **A. radicina,** occasional; **Curvularia trifolii,** CT.

Nematode, root knot. **Meloidogyne** spp., MD to FL and TX.

Nematode, stubby root. **Trichodorus christiei.**

Nonparasitic. **Blossom-end Rot,** on summer squash when hot, dry weather follows a cool, rainy spell. **Bronzing,** marginal necrosis, fruit deformity, potassium deficiency. **Chlorosis,** nutrient deficiency: manganese, interveinal chlorosis, nitrogen, leaf yellowing, and chlorosis of bud end of fruit.

Powdery mildew. **Erysiphe cichoracearum,** general; **Sphaerotheca fuligenea,** CA, NC.

Rot, blossom-end; root. **Pythium aphanidermatum,** AZ, CA, MD.

Rot, charcoal. **Macophomina phaseoli,** OR, TX.

Rot, fruit. **Alternaria** sp., MA, NH, NJ, NY, OR, VT, WA; **Diplodia natalensis,** TX; **Coniosporium fairmani,** black mold; **Botrytis cinerea,** gray mold; **Fusarium** spp.; **F. solani** f. **cucurbitae,** also root and stem rot, CA, CT, NY, OR, WA; **Phytophthora cactorum,** AZ; **Pythium ultimum,** CA; **Rhizopus stolonifer,** general in storage after injury; **Sclerotinia sclerotiorum,** ID, MA, ME, NH, NY, WA; **Trichothecium roseum,** pink mold, MA; **Phoma subvelata,** also leaf spot, stem rot, TX.

Rot, root. **Phymatotrichum omnivorum,** TX; **Rhizoctonia solani,** TX.

Scab; leaf spot; storage rot. **Cladosporium cucumerinum,** CT, MD, MA, NJ, NY, OR, WA.

Virus. **Bromegrass Mosaic.** General, IA, SD, on squash (*Cucurbita pepa*); **Lettuce Mosaic,** NY; **Squash Mosaic; Southern Squash Mosaic; Cucumber Mosaic; Beet Curly Top; Tobacco Ring Spot; Prunus Ring Spot; Watermelon Mosaic; Zucchini Yellow Mosaic,** AR, CA, FL, NY, OH.

Wilt. **Fusarium oxysporum** f. sp. **niveum,** CA, IL, MI, WI; **Verticillium alboatrum,** OR.

Wilt, anasa. Feeding injury from the squash bug.

In general, diseases and control measures are the same as for cucumbers. Some squash varieties are injured by sulfur dusts. Acorn and butternut squashes are resistant to bacterial wilt.

SQUASH-BUSH
(*Condalia*)

Mistletoe. **Phoradendron californicum,** CA.

SQUAW-APPLE
(*Peraphyllum*)

Leaf spot. **Septoria peraphylli,** UT.

Rust. **Gymnosporangium inconspicuum** (0, I), CO; III, on juniper; **G. nelsoni** (0, I), on leaves, fruit; CO, UT.

STACHYS
(Betony, Hedgenettle, Woundwort)

Gall, leaf. **Synchytrium stachydis,** LA.

Leaf spot. **Cercospora stachydis,** IA, ME; **Cylindrosporium stachydis,** IL; **Ovularia bullata,** CA; **Phyllosticta decidua,** MA, WI; **P. palustris,** OH, IL; **Ramularia stachydis,** OR; **Septoria stachydis,** CA, IL, MS, NY, WI.

Nematode, root knot. **Meloidogyne** sp.

Powdery mildew. **Erysiphe galeopsidis** (*E. cichoracearum*), OH to CO, MT, NY to IN, WI; **Sphaerotheca macularis,** CA.

Rust. **Puccinia pallidissima,** TX.

STAPHYLEA
(Bladdernut)

Blight, twig. **Hypomyces ipomoeae,** MA; **Coryneum microstictum,** MA.

Leaf spot. **Mycosphaerella staphylina,** GA, KS. **Ovularia isarioides,** NY to MO, IA; **Septoria cirrhosa,** MO.

STARFLOWER
(*Trientalis*)

Gall, leaf. **Synchytrium aureum**, PA.
Leaf spot. **Cylindrosporium magnusianum**, CA, MA, MI, NY, WI; **Septoria increscens**, CA, ME, MI, NY, VT, WI.
Rot, leaf. **Ceratobasidium anceps**, WI.
Rust. **Puccinia caricina** var. **limosae** (0, I), NY, WI, AK; II, III, on *Carex*.
Smut, leaf and stem. **Tuburcinia trientalis**, OR, WA.

STARGRASS
(*Aletris*)

Leaf spot. **Gloeosporium aletridis**, MS.
Rust. **Puccinia aletridis** (II, III), DE, FL, IL, IN, MA, MS, NC, NJ, RI, SC, TN, WI; 0, I unknown.

STARGRASS, GOLDEN
(*Hypoxis*)

Leaf spot. **Cylindrosporium guttatum**, WI; **Septoria hypoxis**, PA.
Rust. **Uromyces affinis** (I, III), MO, CT, MS; **U. necopinus**, NY
Smut, flower. **Urocystis hypoxis**, CT, MA.

STAR-OF-BETHLEHEM
(*Ornithogalum*)

Blight, southern. **Sclerotium rolfsii**, CA.
Leaf spot. **Didymellina ornithogali**, IL, PA, WA ; **Septoria ornithogali**, CT, MA.
Virus. **Ornithogalum Mosaic**, AL, OK, OR.

STATICE, SEA-LAVENDER
(*Limonium*)

Anthracnose; rot, crown. **Colletotrichum gloeosporioides**, FL.
Bacterial, crown and leaf rot. **Pseudomonas caryophylli**, FL.
Bacterial, MLO. **California Aster Yellows**, CA;
Bacterial, MLO. **Yellows and phyllody**, MI.
Blight, flower. **Botrytis cinerea**, FL.
Leaf spot. **Alternaria** sp., CT; **Cercospora** sp., TX; **C. insulana**, FL; **Ascochyta plumbaginicola**, IA; **Fusicladium staticis**, TX; **Phyllosticta** sp., CT; **P. staticis**, NY.

Nematode, root knot. **Meloidogyne** sp.
Rot, crown. **Sclerotium rolfsii**, FL; **Rhizoctonia solani,** FL.
Rust. **Uromyces limonii** (0, I, II, III), ME to MS and TX, NM, CA.
Virus. **Turnip Mosaic, CA.**

STENANTHIUM

Rust. **Puccinia atropuncta** (II, III), GA; **P. grumosa,** OR.

STENOLOBIUM
(Florida Yellow-Trumpet)

Rot, root. **Clitocybe tabescens,** FL; **Phymatotrichum omnivorum,** TX.
Rust. **Prosopodium appendiculatum** (0, I, II, III), FL, TX.

STEPHANOMERIA
(Wire-Lettuce)

Leaf spot. **Cercospora clavicarpa,** CA.
Rust. **Puccinia harknessi** (0, I, III), MT to NM, CA, WA.

STEPHANOTIS

Anthracnose. **Gloeosporium** sp., WA.
Blight, flower. **Botrytis elliptica,** CA.
Virus. **Tomato Spotted Wilt,** OR.

STERNBERGIA
(Fall-Daffodil)

Leaf scorch; red spot. **Stagonospora curtisii,** CA.

STEVIA
(*Piqueria*)

Bacterial fasciation. **Clavibacter fascians,** MI.
Bacterial, MLO. **Aster Yellows,** MI, NJ.
Damping-off. **Rhizoctonia solani,** IL.
Nematode, lesion. **Pratylenchus pratensis,** NJ.
Powdery mildew. **Erysiphe cichoracearum,** IL.
Rot, stem. **Sclerotium rolfsii,** NJ.

STILLINGIA
(Queens Delight)

Dodder. **Cuscuta** sp., OK.
Leaf spot. **Cercospora stillingiae,** TX.
Rot, root. **Phymatotrichum omnivorum,** TX.
Rust. **Uromyces graminicola** (0, I), OK; II, III, on *Panicum.*

STOCK
(*Matthiola*)

Anthracnose; leaf and stem spot. **Colletotrichum gloeosporioides,** TX.
Bacterial black rot. **Xanthomonas incanae,** CA, TN.
Blight, gray mold. **Botrytis cinerea,** TX, AK.
Club root. **Plasmodiophora brassicae,** NJ.
Damping-off. **Pythium** spp.; **Rhizoctonia solani,** cosmopolitan.
Downy mildew. **Peronospora parasitica,** IL.
Leaf spot. **Alternaria raphani,** CA; **Myrothecium roridum,** TX.
Nematode. **Naccobus batatiformis.**
Nematode, root knot. **Meloidogyne** sp., TX.
Rot, crown; wilt. **Sclerotinia sclerotiorum,** CA, MI, PA.
Rot, root. **Fusarium avenaceum,** DE, NJ; **Phymatotrichum omnivorum,** TX:
   Phytophthora megasperma, CA; **P. cryptogea,** stem rot, CA.
Virus. **Turnip Mosaic; ?Cauliflower Mosaic; Beet Curly Top,** CA; **Tomato
   Spotted Wilt,** TX.
Wilt. **Fusarium oxysporum** f. sp. **mathioli,** AZ, CA; **Verticillium albo-atrum,**
   NY, CA.

Mosaic causes flower-breaking as well as mottling of foliage. Verticillium
wilt is prevalent in cut-flower-producing areas of California.

STOKESIA
(Stokes-Aster)

Blight, head. **Botrytis cinerea,** NY.
Leaf spot. **Ascochyta** sp., IA, PA.
Powdery mildew. **Erysiphe cichoracearum,** MD.
Virus. **Mosaic.** Unidentified, IA; **Biddens Mottle,** FL.

STONE PLANT
(*Lithops*)

Rot. **Fusarium oxysporum,** ID.

STRANVAESIA

Bacterial fire blight. **Erwinia amylovora,** NJ.
Rot, root. **Clitocybe tabescens,** FL.

STRAWBERRY
(*Fragaria*)

Anthracnose; crown rot. **Colletotrichum fragariae,** FL, LA; **Gloeosporium** sp., MD.

Anthracnose, fruit. **Colletotrichum gloeosporioides,** FL, OH.

Bacterial angular leaf spot. **Xanthomonas fragariae,** KY, MN, WI, FL.

Bacterial, MLO. **Aster Yellows,** strawberry green petal; **Lethal Disease Decline** or **Peach Western X Disease.** WA, OR.

Bacterial soft rot. **Erwinia carotovora,** MA.

Blight, gray mold. **Botrytis cinerea,** general on fruit.

Blight, leaf. **Dendrophoma obscurans,** angular leaf spot, MA, to FL, and TX, MI, MN, NE, OR.

Blight, southern. **Sclerotium rolfsii,** AL, FL, NC, TX.

Broomrape. **Orobanche** sp., WA.

Downy mildew. **Peronospora fragariae,** IA.

Gall, leaf. **Synchytrium aureum,** on native spp., WI; **S. fragariae,** root gall, CA, WA.

Leaf blotch. **Gnomonia fructicola,** also fruit rot.

Leaf scorch. **Diplocarpon earliana,** general.

Leaf spot. **Cercospora fragariae,** LA; **C. vexans,** NY, CA, WI; **Gloeosporium** sp., IL, MA, PA, MI, NC, UT, associated with black root; **Mycosphaerella fragariae,** common leaf spot, black-seed disease, general; **M. louisianae,** purple leaf spot, LA, MS, NC.

Nematode, dagger. **Xiphinema americanum; X. diversicaudatum; X. chambersi.**

Nematode, leaf and stem. **Ditylenchus dipsaci,** CA, ID, NC, OR, TX, WA.

Nematode, lesion. **Pratylenchus penetrans; P. coffeae; P. pratensis; P. scribneri.**

Nematode, root knot. **Meloidogyne hapla.**

Nematode, spring dwarf. **Aphelenchoides fragariae,** MA to FL and TX, also AR, CA, MI, TN, WA; Summer Dwarf, Crimp. **A. besseyi,** southeastern and Gulf states, also AR, CA, DE, IL, TX.

Nematode, sting. **Belonolaimus gracilis.**

Nematode, stubby root. **Trichodorus christiei.**

Nonparasitic. **Black Root, Brown Root,** winter injury, defective drainage, soil toxins, widespread. **Chlorosis,** iron deficiency. **Variegation,** June yellows. chlorophyll deficiency, especially in Blakemore, Progressive, and related varieties, general.

Powdery mildew. **Sphaerotheca macularis,** general but rare in South, reported in FL.

Rot, crown. **Sclerotinia sclerotiorum,** MD to FL, and TX, IL, CA, IA, MN; **Sclerotium rolfsii,** AL, FL, NC, TX.

Rot, fruit. **Penicillium** spp., secondary; **Pezizella lythri,** also leaf spot, root rot, ME to FL, and OK, IL, IA, MI, OH, OR, WI, AK; **Phytophthora cactorum,** leather rot, AL, AR, IL, KY, LA, MD, MI, MO, OH, TN, VA; **Rhizopus nigricans,** black rot, cosmopolitan; **Rhizoctonia solani,** hard brown rot, widespread; **Sphaeropsis** sp., CA, CO, IL, IA, MN.

Rot, red stele; brown core root rot. **Phytophthora fragariae,** ME to VA, OK and IA; also CO, MI, WI, Pacific states.

Rot, root. **Armillaria mellea,** CA, OR, WA; **Cylindrocladium scoparium,** OR, TN; associated with black root; **Leptosphaeria coniothyrium; Fusarium** spp.; **Olpidium brassicae,** WA; **Pythium** spp., widespread; **Ramularia** spp., ID, OR, WA; **Idriella lunata,** CA, MD.

Rot, root, stolon and petiole. **Pestalotia longisetula,** OR, IL.

Slime mold. **Diachea leucopodia,** CA, IL, KS, LA, MS, MO, TN, TX; **Fuligo septata,** IA, KS, MN, NE, NJ, TX, WA; **Mucilago spongiosa,** KS, MN, MO, NE, OK; **Physarum cinereum,** IL, NE, NJ.

Sooty mold. **Scoria spongiosa,** ME.

Virus. **Strawberry Crinkle,** CA, ID, OR, WA, MN, NY; **Strawberry Yellow Edge,** Xanthosis, CA, OR, WA, NC and other states introduced from West Coast; **Strawberry Leaf Roll,** MD, NJ, NY, VT; **Strawberry Mottle; Strawberry Multiplier Disease,** WI; **Strawberry Necrotic Shock,** CA; **Strawberry Vein Necrosis,** MN; **Strawberry Vein Banding; Strawberry Witches' Broom,** ID, MT, OR, WA, MN, NY; **Tomato Ringspot,** PA.

Wilt. **Verticillium albo-atrum,** CA, FL, NY, OR, WA; **Colletotrichum acutatus,** AR, CA, FL, MO, MS.

Red Stele is, except in the Deep South, of first importance on strawberries. The roots rot, and above-ground parts are stunted and wilted. Choose resistant varieties, such as Stelemaster, Sparkle, Surecrop, Midway, and Fairland. Purchase plants certified free from nematodes and virus diseases. Mulching helps to prevent fruit rots.

U.S. Department of Agriculture Farmers Bulletin **1891** gives an excellent discussion of "Diseases of Strawberries." The American Phytopathological Society, St. Paul, MN, has published a *Compendium of Strawberry Diseases.*

STRAWBERRY-TREE
(Arbutus unedo)

Bacterial crown gall. **Agrobacterium tumefaciens,** CA.

Leaf spot. **Septoria unedonis,** OR.

Spot anthracnose. **Elsinoë mattirolianum,** CA.

STRAWFLOWER
(*Helichrysum*)

Bacterial, MLO. **Aster Yellows,** eastern and central states; **California Aster Yellows, CA.**
Downy mildew. **Plasmopara halstedii,** CA.
Nematode, root knot. **Meloidogyne** sp., FL.
Rot, stem. **Fusarium** sp., FL.
Virus. **Beet Curly Top, CA, OR, WA; Biddens Mottle.**
Wilt, **Verticillium albo-atrum,** CA.

STREPTANTHERA

Virus. **Iris Mosaic, CA, OR.**

STREPTOCARPUS

Virus. **Tobacco Mosaic,** CA, CT, DC, FL, OH, WA.

STREPTOPUS
(Twisted-Stalk)

Leaf spot. **Cercospora streptopi,** WA; **Septoria streptopidis,** MT.
Rust. **Puccinia sessilis** (0, I), NY.
Smut, leaf. **Tuburcinia clintoniae,** WI.

STYRAX
(Snowbell)

Leaf spot. **Mycosphaerella punctiformis,** GA.
Nematode, root knot. **Meloidogyne** sp., MD.

STYLOSANTHES

Leaf spot, canker. **Colletotrichum dematium** f. sp. **truncata,** FL.

SWEDISH IVY
(*Plectranthus australis*)

Nonparasitic. **Brown Leaf Spot,** Boron deficiency.

SUMAC
(*Rhus*)

Canker, stem; dieback. **Botryosphaeria ribis,** "umbrella disease," NY to GA, NE, and MN; **Cryptodiaporthe aculeans,** ME to AL, OK, and IA; **Nectria cinnabarina,** NY, OK; **N. galligena,** NY, PA, VA; **Physalospora obtusa,** canker, inflorescence blight, NY to AL, KS, MI.

Dodder. **Cuscuta exaltata,** TX.

Leaf blister. **Taphrina purpurascens,** MA to GA, TX and KS.

Leaf spot. **Cercospora rhoina,** general; **Pezizella oenotherae,** NY to GA, WV; **Coniothryium rhois,** TX; **Harknessia rhoina,** CA; **Phyllosticta rhoiseda,** CA; **Septoria rhoina,** general; **Cylindrosporium** sp., CO, NE.

Mold, leaf. **Cladosporium aromaticum,** IA, NE, NY, WI, WV.

Parasitic lichen. **Strigula elegans** and **S. camplanata,** LA, southern U.S.

Powdery mildew. **Sphaerotheca macularis,** widespread; **Oidium** sp., WI.

Rot, root. **Armillaria mellea,** CA; **Clitocybe tabescens,** FL; **Corticium galactinum,** white root, VA; **Phymatotrichum omnivorum,** TX.

Rot, wood. **Polyporus** spp.; **Poria punctata,** MI; **Schizophyllum commune,** cosmopolitan; **Steccherinum ochraceum,** PA.

Rust. **Pileolaria effusa** (0, III), AZ; **P. patzcuarensis** (0, I, II, III), CO, NM, OK.

Wilt. **Fusarium oxysporum** f. sp. **rhois,** CT, VA; **Verticillium albo-atrum,** IL, IA, MA, MN.

SUNFLOWER
(*Helianthus*)

Bacterial leaf spot. **Pseudomonas syringae** (*helianthi*).

Bacterial, MLO. **California Aster Yellows,** CA.

Bacterial stem rot. **Erwinia carotovora,** ND.

Bacterial wilt, and apical chlorosis. **Pseudomonas solanacearum,** FL; **P. syringae** pv. **tagetis,** KS, MN, ND, WI; **Erwinia carotovora** pv. **carotovora,** IN.

Black knot, black patch. **Gibberidea heliopsidis,** GA, IL, MS, MO, NC, VA.

Blight, gray mold; bud rot. **Botrytis cinerea,** CA, OR.

Blight, seedling. **Alternaria helianthi,** MN; **Diaporthe helianthi,** TX.

Blight, southern, **Sclerotium rolfsii,** LA, TX.

Canker, stem. **Diaporthe helianthi,** OH, TX.

Dodder. **Cuscuta** sp., OK, TX, WA.

Downy mildew. **Plasmopara halstedii,** NY to MD, KS and MT, TX.

Gall, basal. **Plasmopora halstedii,** ND, MN.

Leaf spot. **Ascochyta compositarum,** WI; **Cercospora helianthi,** KS, IL, MO, OH, TX, WI; **C. pachypus,** AL, KS, OK, TX; **Colletotrichum helianthi,** WI; **Septoria helianthi,** general; **Alternaria zinniae; Alternaria helianthi,** MN, OH.

Nematode, leaf gall. **Tylenchus balsamophilus,** WA.

Nematode, root knot. **Meloidogyne** sp., AL, FL, TX, WV.

Powdery mildew. **Erysiphe cichoracearum,** general, CA.

Rot, charcoal. **Macrophomina phaseoli,** MD, CA.

Rot, head. **Rhizopus oryzae,** CA; **Phoma macdonaldii,** premature ripening, ND.

Rot, root. **Armillaria mellea,** OR; **Helicobasidium purpureum,** violet root, TX; **Phymatotrichum omnivorum,** TX; **Pythium debaryanum; Rhizoctonia solani,** IL, MD, NE, NY, WI.

Rot, stem. **Phytophthora cryptogea,** CA; **Septosphaeria lindquistii,** black stem, MN, ND.

Rot, stem. **Sclerotinia sclerotiorum,** widespread.

Rust. **Coleosporium madia,** CA.

Rust. **Puccinia helianthi,** (0, I, II, III), general, CA; **P. massalis,** NM, TX; **P. canaliculata,** GA; **Coleosporium helianthi** (II, III), NY to FL, LA and OK; II, III, on pine; **Uromyces junci** (0, I), CA, KS, NE, ND, SD, OK, WY; **U. silphii** (0, I), IL, NY, MO, TN, WI.

Smut, leaf. **Entyloma polysporum,** MT.

Virus. **Bromegrass Mosaic,** General, IA, SD; **Mosaic.** Unidentified, IN; **Cucumber Mosaic,** FL, MD; **Biddens Mottle,** FL.

White rust. **Albugo tragopogonis,** IL, MO, WI.

Wilt. **Fusarium oxysporum** f. sp. apii, SC; **Verticillium albo-atrum.**

Wilt. **Sclerotinia sclerotiorum** = **Whetzelinia sclerotiorum,** CA.

Wilt. **Verticillium dahliae,** CA.

SUNROSE
(*Helianthemum*)

Leaf spot. **Cylindrosporium eminens,** TX, WI; **Septoria chamaecisti,** WA.

Rot, root. **Phymatotrichum omnivorum,** TX.

SWEET ALYSSUM
(*Lobularia*)

Bacterial, MLO. **Aster Yellows,** NJ.

Club root. **Plasmodiophora brassicae,** NJ.

Damping-off, root rot. **Rhizoctonia solani,** NJ, NY, VA; **Pythium ultimum,** NJ.

Dodder. **Cuscuta** sp., TX.

Downy mildew. **Peronospora parasitica,** CA, NJ.

Nematode, root knot. **Meloidogyne** sp.

Powdery mildew. **Erysiphe polygoni,** CA.

Rot, crown. **Sclerotinia sclerotiorum,** VA.

Rot, root. **Phoma lingam,** CA.

SWEET-FERN
(*Comptonia*)

Rust, blister. **Cronartium comptoniae** (II, III), ME to MN, NC, and OH; 0, I, on hard pine.

Rust, leaf. **Gymnosporangium ellisii** (0, I), NJ; II, III, on *Chamaecyparis*.

SWEET-FLAG
(*Acorus*)

Leaf spot. **Cylindrosporium acori,** CT, KS; **Ramularia aromatica,** CT to MD, IN, and WI; **Septocylindrium** sp., NY.

Rust. **Uromyces sparganii** (II, III), ME to VA, MS, IL, and MN; 0, I, unknown.

SWEET GALE
(*Myrica gale*)

Blight. twig. **Diplodia** sp., NJ.

Leaf spot. **Ramularia monilioides,** NY: **Septoria myricata,** NY.

Rust. **Cronartium comptoniae,** ME to NY, WA; 0, I, on pine; **Gymnosporangium ellisii** (0, I), ME.

SWEET GUM
(*Liquidambar*)

Anthracnose. **Gloeosporium nervisequum.**

Blight; leader dieback. Cause unknown, killing trees in MD, AL, AR, FL, GA, LA, SC, TX.

Blight, thread. **Pellicularia koleroga,** LA.

Canker; dieback. **Botryosphaeria ribis** var. **chromogena,** NY, IL, MD to FL and LA; **B. dothidea,** MS; **Hymenochaete agglutinans,** MD; **Nectria** sp.; **Dothiorella berengeriana,** bleeding necrosis, NJ, NY.

Canker, felt fungus. **Septobasidium alni; S. apiculatum, S. burtii, S. mariana, S. pseudopedicellatum,** and **S. sinuosum,** southern states.

Leaf spot. **Cercospora liquidambaris,** MD to FL and TX; **C. tuberculans,** FL, LA, MS, MO; **Discosia artocreas,** OK; **Exosporium liquidambaris,** TX; **Leptothyriella liquidambaris,** red leaf spot, IL, MD, NC; **Septoria liquidambaris,** MA to FL and TX; **Actinopelte dryina; Cladosporium** sp., OR.

Mistletoe. **Phoradendron serotinum (flavescens),** OH to NC and TX.

Nematode, root knot. **Meloidogyne** sp.

Nonparasitic, **Blight, Decline,** leader dieback; killing trees in MD, AL, AR, FL, GA, LA, SC, TX. Partly moisture shortage.

Rot, heart. **Polyporus adustus** and **P. gilvus,** widespread.

Rot, root. **Clitocybe tabescens,** FL; **Phymatotrichum omnivorum,** TX.
Rot, root. **Cylindrocladium scoparium,** GA.
Rot, wood. **Daedalea confragosa,** cosmopolitan; **Fomes** spp., **Ganoderma**
spp.; **Hericium erinaceus; Steccherinum ochraceum; S.** pulcherrimum, Gulf
states; **Polyporus** spp., **Poria** spp.; **Pleurotus** spp.; **Schizophyllum commune,**
cosmopolitan; **Stereum** spp.

SWEETLEAF
(*Symplocos*)

Gall, bud. **Exobasidium symploci,** Gulf states to NC and IN.
Leaf spot. **Septoria stigma,** AL; **S. symploci,** FL, MS, NC, OK, TX.
Rot, root. **Phymatotrichum omnivorum,** TX.

SWEET-OLIVE
(*Osmanthus fragrans*)

Black mildew; black spot. **Asterina** sp., MS.
Leaf spot. **Gloeosporium oleae,** MD.
Rot, root. **Armillaria mellea,** CA; **Clitocybe tabescens,** LA.

SWEET PEA and PERENNIAL PEA
(*Lathyrus*)

Anthracnose; blossom and shoot blight. **Glomerella cingulata,** general except
on Pacific Coast.
Bacterial crown gall. **Agrobacterium tumefaciens,** NJ, VA.
Bacterial fasciation. **Clavibacter fascians,** widespread.
Bacterial leaf spot. **Pseudomonas pisi,** IN, WI.
Blight, gray mold. **Botrytis cinerea,** general.
Blight, southern. **Sclerotium rolfsii,** FL, TX.
Damping-off; root and stem rot. **Pythium** spp., CA, CT, MD, MA, NJ,
NC, NY; **Rhizoctonia solani,** general.
Dodder. **Cuscuta indecora,** TX.
Downy mildew. **Peronospora trifoliorum,** FL.
Leaf spot. **Alternaria** sp., secondary, MD, MA, NJ, NY, PA, TX; **Colletotri-
chum pisi,** AL, FL, GA, SC; **Ascochyta** sp., also stem spot, CT, LA, MN,
MO, WV, WI; **A. lathyri,** NJ; **Cercospora lathyrina,** GA, OK; **Isariopsis
griseola,** CT; **Mycosphaerella pinodes,** also stem spot, MN, WI; **Ovularia**
sp., CA; **Phyllosticta orobella,** TX; **Ramularia** sp., NY, MA, NJ, PA, TX,
WA; **R. deusta** f. **odorati,** CA; **Curvularia** sp., CA.
Mold, white. **Erostrotheca multiformis** (*Cladosporium album*) CA, MA, NJ,
NY, PA, TX, WA, in greenhouses.

Nematode, lesion. **Pratylenchus pratensis,** NJ, NY.

Nematode, root knot. **Meloidogyne** spp., FL, MA, NJ, NY, TX.

Nonparasitic, **Bud Drop,** unbalanced nutrition, low light intensity in greenhouses.

Powdery mildew. **Microsphaera alni,** general, especially in greenhouses; **Erysiphe polygoni,** widely reported, sometimes confused with *M. alni.*

Rot, root. **Aphanomyces euteiches,** IN, MI, WI; **Fusarium solani** f. **pisi,** occasional to prevalent under glass and outdoors; **Phymatotrichum omnivorum,** TX, **Phytophthora cactorum,** CT; **Thielaviopsis basicola,** black rot, CT to OH, IL, MS, FL, CO, Pacific Coast.

Rot; stem wilt. **Sclerotinia ?sclerotiorum,** MD, PA.

Virus. **Pea Mosaic; Tomato Spotted Wilt; Sweet Pea Streak.** A complex, components not identified, perhaps in part bacterial.

Wilt. **Verticillium** sp., NY, NJ; **Fusarium oxysporum** f. sp **vasinfectum,** FL, NY.

Control aphids and thrips spreading mosaic and spotted wilt.

SWEET POTATO
(*Ipomoea batatas*)

Bacterial soft rot. **Erwinia carotovora,** CT, SC.

Blight, leaf. **Phyllosticta batatas,** occasional NJ to FL, TX and KS, usually in the South; **Choanephora cucurbitarum,** FL, leaf mold.

Blight, southern; cottony rot. **Sclerotium rolfsii,** general in South.

Canker, stem. **Fusarium solani,** NC.

Leaf spot. **Alternaria** sp., secondary, occasional to general; **Cercospora** sp., FL, OK; **Septoria bataticola,** occasional NJ to AL, TX and IA.

Leaf spot, blight. **Acremonium crotocinigenum** (= *Cephalosporium cinnamomeum*), CA, FL.

Nematode, lance. **Hoplolaimus coronatus;** Lesion, **Pratylenchus** sp.

Nematode, root knot. **Meloidogyne incognita** and **M. incognita-acrita.**

Nematode, stem, bulb. **Ditylenchus dipsaci,** causing brown ring, NJ.

Nematode, sting. **Belonolaimus gracilis.**

Nematode, stunt. **Tylenchorhynchus claytoni.**

Nonparasitic. **Internal Breakdown,** perhaps from chilling. **Internal Brown Spot,** boron deficiency. **Intumescence,** high water intake and retarded transpiration.

Rot, black. **Ceratocystis fimbriata,** general on roots, stems; **Diplodia theobromae** (*Physalospora rhodina*), Java black rot, general in South.

Rot, charcoal. **Macrophomina phaseoli,** NJ to FL, TX, and KS, CA.

Rot, gray mold. **Botrytis cinerea,** on sprouts, cosmopolitan.

Rot, root. **Diaporthe batatis,** dry rot, stem rot, NJ to FL, TX, MO; **Helicobasidium purpureum,** KS, TX; **Pyrenochaeta terrestris,** pink root, CA; **Phymatotrichum omnivorum,** AZ, NM, OK, TX; **Phytophthora** sp., NJ,

PA; **Plenodomus destruens,** foot and storage rot, NJ to FL, LA and IA, CA; **Pythium** spp., mottle necrosis, leak, ring rot; **Fusarium solani,** NC.

Rot, soil; pox. **Streptomyces ipomoea,** general, NJ to FL, TX and IA, also AZ, CA.

Rot, sprout; stem canker. **Rhizoctonia solani,** occasional, NJ to FL and TX, AZ, CA, OH, WA.

Rot, storage. **Fusarium** spp., general; **Mucor racemosus,** occasional after chilling; **Penicillium** sp., blue mold, cosmopolitan; **Rhizopus** spp., soft rot, general; **Sclerotinia** sp., NC; **Trichoderma** spp., DE, NJ; **Aspergillus** spp., secondary.

Rust. **Coleosporium ipomoeae** (II, III), AL, MS; 0, I, on pine.

Scurf. **Monilochaetes infuscans,** general.

Slime mold. **Fuligo violacea,** AL, DE, NJ, TX; **Physarum cinereum; P. plumbeum,** AL, AR, KY, MS, NC, VA.

Virus. **Sweetpotato Feathery Mottle** (yellow dwarf); **Sweetpotato Internal Cork,** GA, LA, MD, MS, NC, SC, TN, VA; **Sweetpotato Mosaic** (?tobacco mosaic); **Tobacco Ring Spot; Beet Curly Top; Sweetpotato Russet Crack.**

White rust. **Albugo ipomoeae-panduratae,** general.

Wilt. **Fusarium oxysporium** f. sp. **batatas,** stem rot, general; **Verticillium albo-atrum,** CA.

Select varieties resistant to Fusarium stem rot, in general the Spanish group. Heat treatment eliminates some viruses. Hot water (10 minutes at 120° F) controls scurf. Variety Nemagold is resistant to root-knot nematodes.

SWEET-ROOT
(*Osmorhiza*)

Leaf gall. **Urophlyctis** (*Physoderma*) **pluriannulata,** MT.

Leaf spot. **Cercospora osmorhizae,** IL, OH, WI; **Fusicladium angelicae,** WA; **Phleospora osmorhizae,** NY, IA, WI; **Ramularia reticulata,** WI, VA; **Septoria aegopodii,** NY, ND, OH, WI.

Rust. **Puccinia pimpinelli** (0, I, II, III), general.

SWEET VETCH
(*Hedysarum*)

Black mildew. **Parodiella perisporioides,** NY.

Leaf spot. **Septogloeum hedysari,** WY.

Nematode, root knot. **Meloidogyne** sp.

Rust. **Uromyces hedysari-obscuri** (0, I, II, III), AK, CO, ID, MT, NM, SD, UT, WY.

SWEET WILLIAM
(*Dianthus barbatus*)

Anthracnose. ?**Volutella dianthi,** DE, IN, NY.
Bacterial, MLO. **California Aster Yellows,** CA.
Blight, cutting. **Cylindrocladium scoparium** and **C. floridanum,** FL, OH.
Blight, southern. **Sclerotium rolfsii,** CT, FL, NC.
Leaf spot. **Phyllosticta** sp. (*Ascochyta dianthi?*), WA; **Septoria dianthi,** AL,
    IA, MI, NY.
Nematode, leaf and stem. **Ditylenchus dipsaci,** OR.
Nematode, root knot. **Meloidogyne** sp., FL, TX.
Rot, root. **Pythium ultimum,** CA; **Phymatotrichum omnivorum,** TX.
Rot, stem. **Rhizoctonia solani,** CT, IL, KS, MA, MS, NJ, NY, PA, TX.
Rust. **Uromyces dianthi** (II, III), IA, TX, NE; **Puccinia arenariae** (III), AL,
    CT, MA, NY, PA.
Virus. ?**Carnation Mosaic; Beet Curly Top,** CA, TX.
Wilt. **Fusarium oxysporum** f. sp. **barbati,** CA, KS; **Fusarium** sp., MA, NJ, SC,
    VA.

Fusarium wilt is one of the more serious diseases with the new growth
yellowing, the leaves pointing downward, and plants stunted. Place new plants
in a new location or sterilized soil.

SWISS CHARD
(*Beta vulgaris* var. *cicla*)

Blight, southern. **Sclerotium rolfsii,** LA, SC.
Damping-off, root rot. **Rhizoctonia solani,** NY; **Pythium aphanidermatum,**
    CA.
Downy mildew. **Peronospora schachtii,** CA.
Leaf spot. **Cercospora beticola,** general; **Ramularia beticola,** WA.
Mold, seed. **Alternaria tenuis,** CA; **Stemphylium botryosum,** CA, WA.
Nematode, root knot. **Meloidogyne** sp., LA.
Nonparasitic. **Heart Rot,** cracked stem, boron deficiency, NY, WA.
Rot, crown. **Sclerotinia** sp., MS.
Rot, root. **Phymatotrichum omnivorum,** TX.
Rust. **Uromyces betae** (II), CA, OR.
Virus. **Beet Curly Top,** AZ, CA, OR; **Beet Mosaic,** AZ, CA, WA; **Yellow Net,**
    unidentified, CA.

SWITCHGRASS
(*Panicum*)

Blotch, spot. **Helminthosporium sativum** (*Bipolaris sorokinianum*), PA.

SYCAMORE

See Plane.

SYNGONIUM

Anthracnose. **Gloeosporium** sp., WA; **Colletotrichum** sp., WA.
Bacterial blight. **Xanthomonas campestris,** MD.
Bacterial leaf spot. **Xanthomonas dieffenbachiae,** FL; **X. vitians,** FL.
Leaf spot. **Cephalosporium cinnamomeum,** NY.
Rot; black cane. **Ceratocystis fimbriata,** CA.

SYNTHYRIS

Leaf spot. **Ramularia** sp., OR.
Rust. **Puccinia acrophila** (III), MT to CO, and UT; **P. welfeniae** (III), CA, ID, OR, WA, WY.

TABEBUIA
(Trumpet-Tree)

Rust. **Prosopodium plagiopus** (II, III), FL.

TABERNAEMONTANA
(Crape-Jasmine)

Leaf mold. **Cladosporium** sp., FL.
Leaf spot. **Gloeosporium tabernaemontanae,** FL.
Leaf spot, algal; green scurf. **Cephaleuros virescens,** FL.
Rot, mushroom root. **Clitocybe tabescens,** FL.

TAENIDIA

Leaf spot. **Fusicladium angelicae,** WI; **Septoria pimpinellae,** MN.
Rust. **Puccinia angelicae** (0, I, II, III), IN, MI, MO, NY, WI.

TAHITIAN BRIDAL VEIL
(*Gibasis*)

Virus. **Bean Yellow Mosaic,** VA.

TAMARIND
(*Tamarindus*)

Nematode, root knot. **Meloidogyne** sp., FL.

TAMARISK, SALT CEDAR
(*Tamarix*)

Bacterial blight. Fatal to some plants in AZ, NM, and TX, where the host has become a noxious weed.
Powdery mildew. **Sphaerotheca macularis**, IN.
Rot, root. **Phymatotrichum omnivorum**, CA, TX.
Rot, wood. **Polyporus sulphureus**, MD.

TANSY
(*Tanacetum*)

Leaf spot. **Ramularia tanaceti**, WI.
Powdery mildew. **Erysiphe cichoracearum**, PA.
Nematode, root knot. **Meloidogyne** sp., FL.
Rust. **Puccinia tanaceti** (0, I, II, III), WY; 0, I, II, III on *Artemisia;* II on chrysanthemum.

TEASEL
(*Dipsacus*)

Blight, southern. **Sclerotium rolfsii**, TX.
Downy mildew. **Peronospora dipsaci**, MO.
Leaf spot. **Cercospora elongata**, MD, NY to MO, WA.
Nematode, leaf and stem. **Ditylenchus dipsaci**, CA, OR.
Powdery mildew. ?**Phyllactinia corylea**, WA.
Rot root. **Phymatotrichum omnivorum**, TX.
Rot stem. **Sclerotinia sclerotiorum**, TX.
Virus. **Teasel Mosaic.**

TEA
(*Thea sinensis*)

Blight, twig. **Pestalotia guepini**, also leaf spot, SC.
Leaf spot, algal; green scurf. **Cephaleuros virescens**, FL, SC.
Spot anthracnose; scab. **Sphaceloma** sp., LA; **Elsinoë leucospila**, FL.

TELLIMA

Powdery mildew. **Sphaerotheca macularis,** AK.
Rust. **Puccinia heucherae** (III), CA, OR, WA, AK.

TERNSTROEMIA

Spot anthracnose. **Elsinoë leucospila, FL.**

TEXASWEED
(*Caperonia*)

Canker, stem. **Amphobotrys recini, TX.**

THALIA

Leaf spot. **Cercospora thaliae,** LA.
Rust. **Puccinia thaliae** (II, III), FL.

THERMOPSIS
(Bush-Pea, Golden-Pea)

Leaf spot. **Cercospora thermopsidis,** CO, MT; **Phoma thermopsidicola,** CA; **Ramularia sphaeroides,** WA; **Stigmina thermopsidis,** CA.
Powdery mildew. **Erysiphe polygoni,** Rocky Mountains and Pacific Northwest.

THISTLE
(*Cirsium*)

Leaf spot. **Cercospora** spp., WA, KS, TX; **Phyllosticta cirsii,** NY, WA, WI; **Stagonospora cirsii,** WI; **Septoria cirsii,** VT to IN, TX, and WI; **Alternaria chrysanthemi,** MT.
Powdery mildew. **Erysiphe cichoracearum,** general; **Phyllactinia corylea,** WA; **Sphaerotheca macularis,** MD.
Rot, root. **Rhizoctonia solani,** IL; **Phymatotrichum omnivorum,** TX.
Rust. **Puccinia cirsii** (0, I, II, III), PA to NC, TX, CA, OR; **P. punctiformis** (0, I, II, III), ME to NJ, OH, CA, WA; **Uromyces junci** (0, I), MO, NE, ND.
Smut, inflorescence. **Thecaphora trailii,** CO, UT.
White rust. **Albugo tragopogonis,** NY to IA, TX, and WY, LA.

THISTLE, BLESSED
(*Cnicus*)

Blight, southern. **Sclerotium rolfsii, GA.**

THOROUGHWAX
(*Bupleurum*)

Nematode, root knot. **Meloidogyne** sp., CA.

THUJOPSIS
(Hiba Arborvitae)

Blight, twig. **Phomopsis occulta, CA.**

THUNBERGIA
(Clockvine)

Bacterial crown gall. **Agrobacterium tumefaciens,** CT, FL.
Bacterial, MLO. **Aster Yellows,** NY.
Nematode, root knot. **Meloidogyne** sp., FL.
Virus. **Broad Bean Wilt,** NY.

THYME
(*Thymus*)

Blight, gray mold. **Botryris cinerea, AK.**
Rot, root. **Rhizoctonia solani, MA.**

TI
(*Cordyline terminalis*)

Leaf spot. **Phytophthora parasitica, HI.**

TIBOUCHINA
(Glory-Bush)

Rot, mushroom root. **Clitocybe tabescens, FL.**

TIDESTROMINIA

Rot. Macrophomina phaseolina, AZ.

TIGRIDIA
(Tiger-Flower)

Bacterial scab. **Pseudomonas marginata,** MD.
Nematode, bulb. **Ditylenchus dipsaci.**
Rot, internal. **Fusarium oxysporum** f. sp. **gladioli;** storage, **Penicillium gladioli.**
Virus. **Tigridia Mosaic,** OR, WA.

TOMATO
(*Lycopersicon*)

Anthracnose. **Glomerella phomoides** (*Colletotrichum coccodes*), chiefly ripe
  rot of fruit, sometimes on leaves, general, especially in Northeast; **Colleto-
  trichum dematium,** var. **truncata,** MS; **C. graminicola,** MS; **C. trichellum,**
  MS; **C. glycines,** MS; **C. gossypii,** MS.
Bacterial blight. **Pseudomonas viridiflava,** FL.
Bacterial canker. **Clavibacter michiganense,** birds-eye spot, general, most
  frequent in North and West.
Bacterial crown gall. **Agrobacterium tumefaciens;** in experiments, **A. rhizo-
  genes,** hairy root.
Bacterial, MLO. **Aster Yellows; Tomato Big Bud,** NY.
Bacterial, pith rot. **Pseudomonas corrugata,** CA, FL, LA; also **Stem necrosis,**
  FL.
Bacterial, seedling blight. **Bacillus polymyxa,** NY.
Bacterial soft rot. **Erwinia aroideae** and **E. carotovora,** cosmopolitan.
Bacterial speck. **Pseudomonas syringae** pv. **tomato,** occasional East and
  central states, also CA, OK, TX.
Bacterial spot. **Xanthomonas vesicatoria,** on fruit, leaves, sometimes stem
  cankers, northeastern, central, and southern states.
Bacterial wildfire. **Pseudomonas syringae** pv. **tabaci,** WI.
Bacterial wilt. **Pseudomonas solanacearum,** general, MA to IL and southward.
Blight, blossom. **Sclerotinia** sp., CA, FL, NY, OH.
Blight, early; collar rot; fruit rot. **Alternaria solani,** general.
Blight, late; fruit rot. **Phytophthora infestans,** general in humid regions and
  seasons, especially East and Southeast.
Blight, southern. **Sclerotium rolfsii,** VA to FL and TX; CA, occasional in
  North.
Broomrape. **Orobanche ludoviciana,** WY; **O. racemosa,** CA, NY, NJ.

Canker, stem. **Helminthosporium** sp., TX; **Myrothecium roridum,** also fruit rot, OH, TX, VA, WI.

Damping-off. **Rhizoctonia solani,** also collar rot, stem canker, cosmopolitan; **Pythium** spp., cosmopolitan.

Dodder. **Cuscuta** spp., CA, ID, MD, NY, TX.

Downy mildew. **Peronospora tabacina,** GA, NC, SC.

Fruit spot. **Pullularia pullulans,** WV.

Leaf spot. **Ascochyta lycopersici,** DE, FL, NJ, NC, OR, VA, WI; **Septoria lycopersici,** general except Northwest; **Phyllosticta hortorum,** NY; **Cercospora** spp.; **Stemphylium solani,** gray leaf spot, FL, GA, IN, LA, NJ, NC, SC, TN, TX, VA; **S. floridanum.**

Leaf spot, target. **Corynespora cassiicola,** FL.

Mold, leaf. **Botryosporium pulchrum,** PA, TX; **Chaetomium bostrychodes,** TX; **Cladosporium fulvum,** general under glass, occasional outdoors.

Nematode, awl. **Dolichodorus heterocephalus.**

Nematode, dagger. **Xiphinema americanum.**

Nematode, golden. **Heterodera rostochiensis** (*Globodera rostochiensis*); tobacco cyst, **H. tabacum;** soybean cyst, **H. glycines,** IA, MN, OH.

Nematode, lance. **Hoplolaimus coronatus.**

Nematode, lesion. **Pratylenchus pratensis,** MD, NJ.

Nematode, reniform. **Rotylenchulus reniformis.**

Nematode, root knot. **Meloidogyne arenaria** var. **thamesii; M. hapla,** Pacific NW; **M. incognita; M. incognita-acrita; M. javanica; M. chitwoodi,** Pacific NW.

Nematode, spiral. **Helicotylenchus nannus.**

Nonparasitic

 Blossom-end Rot. Unbalanced moisture and calcium deficiency, general.

 Blotchy Ripening. Malnutrition, ?potassium deficiency.

 Catface. Fruit abnormalities from growth disturbances.

 Cloudy Spot. Feeding punctures of plant bugs.

 Cuticle Crack. Of green fruit; high soil moisture and air temperature.

 Fasciation. ?Genetic abnormality. ?Unbalanced nutrition.

 Leafroll. Excessive soil moisture with starch congestion in leaves.

 Oedema. Leaf hypertrophy from excessive water absorption and reduced transpiration.

 Pockets; Puffing. Fruit defect from factors adversely affecting pollination and growth.

 Psyllid Yellows. Toxemia from psyllid feeding. Western states.

 Russeting. Mite injury.

 Sunscald. Fruit injury in heat on plants defoliated by disease.

 Walnut Wilt. Toxemia from root excretions of walnut trees.

Powdery mildew. **?Erysiphe polygoni,** on seedlings indoors, NC; **Leveillula taurica,** AZ, CA, UT; **Oidiopsis taurica,** CA.

Rot, brown, root. **Pyrenochaeta lycopersici,** MA.

Rot, buckeye; stem rot. **Phytophthora parasitica,** MA to FL, TX, and IL, AZ, CA; **P. cactorum,** NY, PA, WI; **P. capsici; P. cryptogea; P. drechsleri.**

Rot, charcoal. **Macrophomina phaseoli,** CA, TX.

Rot, fruit. **Alternaria** sp., black mold; **Aspergillus** spp., green and yellow mold; **Alternaria alternata,** NY; **Brachysporium tomato,** KS, TX; **Clado-sporium herbarum,** green mold; **Diaporthe phaseolorum,** MS, TX; **Diplodia theobromae,** AL; **Glomerella cingulata,** FL, LA, ME, MI, NJ, NY; **Isaria clonostachoides,** VA; **Mucor mucedo,** MD; **Nematospora coryli,** cloudy spot, CA, FL, GA; **Nigrospora oryzae,** CA; **Oospora lactis,** sour rot, cosmo-politan; **Phoma destructiva,** black spot, nearly general; **Pleospora lycoper-sici,** CA; **Rhizopus stolonifer,** cosmopolitan in transit; **Sclerotinia minor,** TX; **Sporotrichum** sp., IN, TX; **Trichothecium roseum,** MD, OH, NC; **Gibberella zeae; Phoma** sp.; **Myrothecium** sp.

Rot, gray mold; ghost spot. **Botrytis cinerea,** occasional on foliage, fruit, stems.

Rot, nailhead spot. **Alternaria tomato,** on fruit, stems, CT to FL, TX and ND.

Rot, root. **Aphanomyces cladogamus; Colletotrichum atramentarium,** black dot; **Pyrenochaeta terrestris,** secondary, IA, IL, NJ; **Phymatotrichum omnivorum,** AZ, OK, TX; **Plectospira myriandra,** VA; **Thielaviopsis basicola,** TX.

Rot, root. **Fusarium oxysporum,** FL, NE, NH, OH, (different from wilt); **F. oxysporum** f. sp. **radicis,** crown, FL, NH, OH, PA, TX. Also **F. oxysporum** f. sp. **radicis-lycopersici,** TX.

Rot, root (and stunting of seedlings). **Pythium myriotylum** and **P. arrhen-omones,** FL.

Rot, root and wilt. **Colletotrichum coccodes,** CA.

Rot, stem, fruit. **Sclerotinia sclerotiorum,** occasional in greenhouses and in South; **Pythium myriotylum.**

Scab, powdery. **Spongospora subterranea,** PA.

Virus. **Tobacco Mosaic,** general, strains causing fernleaf and internal browning of fruit; **Cucumber Mosaic,** also causing shoestring; **Beet Curly Top,** western yellow blight; **Tomato Aspermy,** AZ; **Tomato Ring Mosaic; Tomato Ring Spot; Tomato Spotted Wilt,** TN; **Tomato Streak** (tobacco mosaic plus potato virus X); **Tomato Yellow Net; Tomato Yellow Top; Witches' Broom; Tobacco Etch,** may cause severe mosaic; **Rugose Mosaic** (due to potato mottle virus and potato virus Y); **Pseudo Curly Top,** FL; **Tobacco streak,** CA; **Tomato Mosaic,** CA.

Wart. **Synchytrium endobioticum,** PA (see under Galls).

Wilt. **Fusarium oxysporum** f. sp. **lycopersici,** general; **Verticillium albo-atrum,** occasional in all regions.

Choose tomato varieties resistant to Fusarium wilt; purchase seed certified free from virus disease. In setting seedlings discard any showing galls of root-knot nematodes. Do not smoke while handling, and do not use tobacco stems as a mulch, although other mulches are helpful. Spray for late blight and other foliage diseases. State and federal agencies give warning when late blight is imminent and it is time to start spraying. See *USDA Agricultural Handbook* **203** for an excellent discussion of tomato diseases with fine illustrations.

TORENIA

Nematode, root knot. **Meloidogyne** sp., MD.

TORREYA

Leaf spot. **Phomopsis** sp.
Needle spot. **Fusarium lateritium,** FL.

TRADESCANTIA
(Wandering Jew, Spiderwort)

Blight, gray mold. **Botrytis cinerea,** AK.
Leaf spot. **Cladochytrium replicatum,** NY, secondary; **Colletotrichum** sp., TX, NJ; **Cylindrosporium tradescantiae,** IA; **Septoria tradescantiae,** WI to TX and SD.
Nematode, root knot. **Meloidogyne** sp., OR, TX.
Rust. **Uromyces commelinae** (II, III), TX.

TRAUTVETTERIA
(False Bugbane)

Downy mildew. **Peronospora ficariae,** TN.
Leaf spot. **Septoria trautveteriae,** WV.
Rust. **Puccinia pulsatillae,** ID, OR, WA.
Smut, leaf and stem. **Urocystis anemones,** UT.

TREE-POPPY
(*Dendromecon*)

Smut, leaf. **Entyloma eschscholtziae,** CA.

TREE-TOMATO
(*Cyphomandra*)

Bacterial canker. **Clavibacter michiganense,** CA.
Powdery mildew. **Oidium** sp., MD.

TRILLIUM
(Wake-Robin)

Blight, leaf. **Ciborinia trillii.**
Leaf spot. **Colletotrichum peckii,** NY to NC, IL, MN; **Gloeosporium trillii,**
  CA, OR, WA; **Heterosporium trillii,** WA; **Phyllosticta trillii,** NY, PA, WA,
  WI; **Septoria trillii,** New England to SC, OK, WI.
Rot, stem. **Sclerotium rolfsii,** NH, PA.
Rust. **Uromyces halstedii** (0, I), IL, NY; II, III, on *Spartina.*
Smut, leaf. **Urocystis trillii,** ID, OR.

TRITONIA
(*Montbretia*)

Blight, leaf. **Alternaria** sp., ?secondary, NH; **Heterosporium** sp., OR, WA.·
Blight, southern. **Sclerotium rolfsii,** CA.
Rot, corm. **Fusarium oxysporum** f. sp. **gladioli,** yellows; **Stromatinia gladioli,**
  in commercial stocks.
Virus. **?Iris Mosaic,** CA, OR.

TROLLIUS
(Globeflower)

Leaf spot. **Ascochyta** sp., NY; **Cylindrosporium montenegrinum,** WY; **Phyllosticta trollii,** WY.
Smut, leaf and stem. **Urocystis anemones,** MD, NY.

TRUMPETVINE, TRUMPET-CREEPER
(*Campsis*)

Leaf spot. **Cercospora duplicata,** LA; **Mycosphaerella tecomae** (*Cercospora
  sordida*), general; **Phyllosticta tecomae,** MS; **Myrothecium roridum,** TX;
  Septoria tecomae, OK, TX, WV.
Mistletoe. **Phoradendron serotinum (flavescens**), TX.

Powdery mildew. **Erysiphe cichoracearum,** IL; **Microsphaera alni,** MD to AL, TX, and IN.
Rot, root. **Phymatotrichum omnivorum,** TX.

TUBEROSE
(*Polianthes*)

Bacterial soft rot. **Erwinia carotovora,** NC.
Blight; leaf and flower spot. **Botrytis elliptica,** CA.
Leaf spot. **Cercospora** sp., TX; **Helminthosporium** sp., also stem spot, TX.
Nematode, root knot. **Meloidogyne** sp., CA, FL, NC, TX, VA.
Rot, root. **Pythium debaryanum,** TX; **Rhizoctonia solani,** NC.

TULIP
(*Tulipa*)

Anthracnose. **Gloeosporium thumenii** f. **tulipae,** CA.
Bacterial soft rot. **Erwinia carotovora,** WA.
Blight, botrytis; tulip fire. **Botryris tulipae,** general except for South and Southwest; **B. cinerea,** sometimes secondary.
Blight, southern; bulb and stem rot. **Sclerotium rolfsii,** CA, CT, NY, OR.
Nematode, bulb. **Ditylenchus dipsaci,** NY.
Nonparasitic. **Chalking,** stone disease, immaturity of bulbs or injury at digging. **Topple,** sugarstem, collapse of flower stem, often from calcium deficiency.
Rot, basal. **Fusarium** sp., OK, WA.
Rot, black. **?Sclerotinia sclerotiorum,** WA.
Rot, bulb. **Penicillium** spp., blue mold; **Aspergillus** spp., black mold, cosmopolitan; **Rhizopus stolonifer,** mushy rot, cosmopolitan; **Sclerotinia** sp., ME, OH, WA; **S. sativa,** MD, NY; **Pythium ultimum,** secondary.
Rot, gray bulb. **Rhizoctonia tuliparum,** Northeast, Pacific states.
Rot, root, stem, bulb. **Rhizoctonia solani,** MA, NY, WA.
Rot, stem; flower spot. **Phytophthora cactorum,** CA, IA, NJ, PA, SC, WA.
Virus. **Tulip Breaking** (lily latent mosaic and tulip color-adding), general; **Tobacco Necrosis,** WI.

Botrytis blight is extensive with tulips. In a wet spring, leaves are blasted, buds blighted, and open flowers covered with spots, followed by the familiar gray mold. Sanitary measures are all-important, for the sclerotia survive in the soil, ready to blight new, healthy bulbs. Cut off fading flowers into a paper bag. Gradual running out of tulips may be due to virus disease. Breaking of flowers, once considered a desirable ornamental character, is now recognized as a disease that may be harmful in the long run. Unless aphids are controlled, the virus will spread from variegated to solid-color plantings in the garden.

TULIP-TREE, YELLOW POPLAR
(*Liriodendron*)

Bacterial root lesion. **Pseudomonas** sp., OR.

Blight, seedling. **Rhizoctonia solani,** OH, VA.

Canker. **Dothiorella** sp., PA; **Myxosporium** spp., NY; **Nectria** sp., WV to NC and TN; **Fusarium solani,** SC.

Leaf spot. **Cylindrosporium cercosporioides,** MD, WV; **Gloeosporium liriodendri,** CT to NJ and TX.; **Mycosphaerella liriodendri,** GA, MI; **M. tulipiferae,** Middle Atlantic and Gulf states; **Phyllosticta lirodendrica** (conidial stage of *Mycosphaerella*), widespread; **Ramularia liriodendri,** AL, DE.

Leaf spot, tar, black. **Ectostroma liriodendri,** widespread; **Rhytisma liriodendri,** CA, TX, VA.

Nematode, lesion. **Pratylenchus pratensis,** TX.

Powdery mildew. **Phyllactinia corylea,** NY to AL, MO; **Erysiphe polygoni,** widespread.

Rot, heart. **Collybia velutipes,** WV; **Ganoderma applanatum,** occasional.

Rot, root. **Armillaria mellea,** VA; **Phymatotrichum omnivorum,** TX; **Cylindrocladium liriodendri,** CA.

Rot, sapwood; wood. **Daedalea extensa** and **D. unicolor,** sometimes on standing trees; **Daldinia vernicosa,** cosmopolitan; **Polyporus** spp.; **Schizophyllum commune,** wound rot; also many stain and timber rots.

Rot, seedling collar. **Cylindrocladium scoparium,** NJ, NC, TN.

Sooty mold. **Capnodium elongatum,** cosmopolitan.

The most conspicuous fungus on ornamental tulip-trees is the black sooty mold growing in copious honeydew secreted by tulip-tree aphids and scales. In hot, dry weather leaves sometimes turn yellow and drop prematurely. Leaf spots are seldom serious enough for treatment.

TUNG TREE
(*Aleurites*)

Anthracnose. **Glomerella cingulata,** FL.

Blight, southern. **Sclerotium rolfsii,** TX.

Blight, thread. **Pellicularia koleroga,** FL, MS, LA, TX; Web, **P. filamentosa,** LA, MS.

Canker; dieback; nut rot. **Botryosphaeria ribis,** GA, LA; **Physalospora rhodina,** FL, LA, MS, TX.

Canker, felt fungus. **Septobasidium pseudopedicellatum,** LA.

Leaf spot. **Cercospora websteri,** MS; **Gloeosporium aleuriticum,** MS; **Phyllosticta** sp., FL, GA, **Cristulariella pyramidalis,** FL.

Leaf spot, angular. **Mycosphaerella** (*Cercospora*) **aleuritidis,** MS.

Nematode, root knot. **Meloidogyne** sp., AL, FL, MS.

Nonparasitic
  Bronzing. Zinc deficiency, FL.
  Chlorosis. Copper or potassium deficiency, FL.
  Frenching. Manganese deficiency, FL.
  Wetwood. Alcoholic slime flux. Perhaps bacterial in part.
  White Seed. Genetic abnormality.
Rot, root, collar. **Clitocybe tabescens,** serious throughout tung belt; **Cephalosporium** sp., LA, ?secondary; **Phytophthora cinnamomi,** LA.
Rot, root; collar rot. **Cylindrocladium scoparium, C. crotalariae, C. floridanum,** MS.

Spraying is usually necessary for thread blight and nut rot.

TUPELO, SOUR GUM, BLACK GUM
(*Nyssa*)

Blight, thread, **Pellicularia koleroga,** LA.
Canker. **Nectria galligena,** CT; **Strumella coryneoidea,** northern Appalachians; **Botryosphaeria ribis,** IL; **Fusarium solani,** LA.
Canker, felt fungus. **Septobasidium** spp., NJ to Gulf states.
Leaf spot. **Actinopelta dryina,** AL, IL, OK; **Cercospora nyssae,** TX; **Mycosphaerella nyssaecola,** MA to GA, MI; **Phyllosticta nyssae,** southeastern states to TX.
Mistletoe. **Phoradendron serotinum (flavescens**), FL, MD, IN, TX.
Rot, heart. **Ganoderma applanatum; Fomes connatus,** CT; **Hericeum erinaceus,** NC.
Rot, root. **Phymatotrichum omnivorum,** TX.
Rot, wood. **Daedalea confragosa; D. unicolor,** occasional; **Lentinus** spp., cosmopolitan; **Polyporus** spp.; **Stereum** spp.; **Trametes rigida,** Gulf states.
Rust. **Aplopsora nyssae** (II, III), MD to AL, TX.
Wilt, seedling. **Phytophthora cactorum,** MO.

TUPIDANTHUS

Leaf spot. **Alternaria panax,** CA.

TURNIP
(*Brassica rapa*)

Anthracnose. **Colletotrichum higginsianum,** NY to FL and TX.
Bacterial black rot. **Xanthomonas campestris,** ME to FL, TX, and MN.
Bacterial crown gall. **Agrobacterium tumefaciens,** KS.
Bacterial soft rot. **Erwinia carotovora,** general in field, transit, storage.

Bacterial spot. **Pseudomonas maculicola,** CT, GA, TX, MA; **Xanthomonas vesicatoria** pv. **raphani,** IN.

Blackleg. **Phoma lingam,** CT, MA, WA.

Blight, southern. **Sclerotium rolfsii,** FL, TX.

Club root. **Plasmodiophora brassicae,** ME to NC, TX and MN, CA, CO, WA.

Damping-off. **Pythium ultimum,** WI; **Rhizoctonia solani,** also root, stem and storage rot, general.

Downy mildew. **Peronospora parasitica,** MA to FL, TX, and IL.

Leaf spot. **Alternaria brassicae,** gray leaf spot, general; **A. oleracea,** black leaf spot, CT, FL, MD, MA, NJ, NC, TX; **Alternaria raphani,** AZ; **Cercospora brassicicola,** AL, FL, GA, MS; **Mycosphaerella brassicola,** ring spot, OR; **Phyllosticta** sp., TX, WV; **Ramularia** sp., AL, FL, WA; **Septomyxa affine,** AL.

Leaf spot; white spot. **Cercosporella brassicae,** MA to FL, TX and IN, OR; **Pseudocercosporella capsellae,** CA.

Mold, seed. **Alternaria tenuis,** cosmopolitan; **Stemphylium botryosum,** occasional; **Curvularia inaequalis.**

Nematode, root. **Pratylenchus pratensis,** MD.

Nematode, root knot. **Meloidogyne** spp., AL, AZ, FL, OR, TX.

Nonparasitic. **Brown Heart,** boron deficiency, CA, MA, MN, VA, WI.

Powdery mildew. **Erysiphe polygoni,** northeastern states to FL, TX, and CA, WA.

Rot, root. **Phymatotrichum omnivorum,** TX; **Pythium** sp., NY.

Rot, watery soft. **Sclerotinia sclerotiorum,** CT, MD, MS, TX, WA.

Scab. **Streptomyces scabies,** CT, MI, NJ.

Virus. **Turnip Mosaic,** AL; **Beet Curly Top,** CA; **Radish Mosaic.**

White rust; white blister. **Albugo candida,** general.

Wilt; yellows. **Fusarium oxysporum** f. sp. **conglutinans,** IN, MS, TX.

The control of turnip diseases is about the same as for cabbage and other crucifers.

TURPENTINE TREE
(*Syncarpia*)

Nematode, root knot. **Meloidogyne** sp., FL.

TURTLE-HEAD
(*Chelone*)

Leaf spot. **Septoria mariae-wilsonii,** ME to PA, OH, and WI.

Powdery mildew. **Erysiphe cichoracearum,** and **E. polygoni,** widespread.

Rust. **Puccinia andropogonis** var. **penstemonis** (0, I), CT, MA, NJ, NY, PA, TN; II, III, on *Andropogon*; **P. chelonis** (III), OR, WA.

UDO
(*Aralia cordata*)

Blight. **Alternaria** sp., DE, NJ.
Rot, stem. **Sclerotinia sclerotiorum,** MD.
Wilt. **Verticillium albo-atrum,** ME, PA.

UMBRELLA-PINE
(*Sciadopitys*)

Blight, twig. **Diplodia pinea,** NJ.
Leaf spot. **Phyllosticta** sp., RI; **Cytospora** sp., OR.
Rot, root; damping-off. **Rhizoctonia solani,** CT.
Rot, root; wilt. **Phytophthora cinnamomi,** VA.

UMBRELLAWORT
(*Oxybaphus*)

Downy mildew. **Peronospora oxybaphi,** KS, SD.
Leaf spot. **Ascochyta oxybaphi,** IA, WI; **Cercospora oxybaphi,** IL, NE, OH,
IA, KS, TX.
Rot, root. **Phymatotrichum omnivorum,** TX.
White rust. **Albugo platensis,** AL.

UNICORN-PLANT, PROBOSCIS-FLOWER
(*Proboscidea*)

Blight, southern. **Sclerotium rolfsii,** TX.
Leaf spot. **Cercospora beticola,** KS, IA, TX, WI, OK.
Rot, root. **Phymatotrichum omnivorum,** TX.
Rot, stem. **Sclerotinia sclerotiorum,** MA.
Virus. **Bean Yellow Mosaic Virus,** NY. Also susceptible to **Cucumber Mosaic,**
Tobacco Etch, Tobacco Mosaic, Tobacco Necrosis, Tobacco Ring Spot,
Tomato Ring Spot, Alfalfa Mosaic, Broad Bean Wilt, Lettuce Mosaic,
Pea Seedborne Mosaic, Potato Virus Y, Turnip Mosaic, Watermelon Mosaic.

UVULARIA
(Bellwort, Merry-Bells)

Leaf spot. **Sphaeropsis cruenta,** IL, IA, MO, VA, WI, IN, CT, NY.
Rust. **Puccinia sessilis** (0, I), DE, IA, MN, MO, NE, NY, WI; **Uromyces**
acuminatus var. **magnatus** (0, I), MS, WI.

VALERIAN, GARDEN HELIOTROPE
(*Valeriana*)

Leaf spot. **Ramularia centranthi**, CA; **Septoria valerianae**, WI.
Powdery mildew. **Erysiphe cichoracearum**, CO, UT.
Rot, root. **Rhizoctonia solani**, NY.
Rot, stem. **Sclerotinia rolfsii**, CT, NJ.
Rust. **Puccinia commutata** (0, I, III), NY, OR; **P. dioicae** (0, I), CO, NM, UT; II, III, on *Carex*; **P. valerianae** (II, III), AK.

VALERIAN, RED, JUPITERS-BEARD
(*Centranthus*)

Leaf spot. **Ramularia centranthi**, CA.

VALERIANELLA
(Corn-Salad)

Gall, leaf. **Synchytrium aureum**, MS.
Leaf spot. **Septoria valerianellae**, MS, TX.
Rot, root. **Phymatotrichum omnivorum**, TX.

VANCOUVERIA

Leaf spot. **Phragmodothis berberidis**, CA; **Ramularia vancouveriae**, CA, OR.

VANILLA

Black mildew. **Lembosia rolfsii**, FL.
Leaf spot; pod spot. **Botryosphaeria vanillae**, FL; **Volutella vanillae**, FL.

VANILLA-LEAF
(*Achlys*)

Leaf spot. **Ascochyta achlyicola**, WA; **Stagonospora achlydis**, OR.

VELVET BEAN
(*Stizolobium*)

Bacterial spot. **Pseudomonas syringae**, IN; **P. stizolobii**, NC.
Blight, southern. **Sclerotium rolfsii**, AL, FL, GA.

Leaf spot. **Cercospora stizolobii,** AL, FL, GA, NC, SC; **Mycosphaerella cruenta,** GA; **Phyllosticta macunae,** AL.
Nematode, root knot. **Meloidogyne** spp., CA, TX.
Rot, pod spot. **Fusarium** sp., TX.
Rot, root. **Phymatotrichum omnivorum,** AZ, TX; **Phytophthora parasitica,** FL.

VERATRUM
(False-Hellebore)

Leaf spot. **Ascochyta veratrina,** WA; **Cylindrosporium veratrinum,** CA, NY, UT, VA, WA; **Cercosporella terminalis,** NY; **Phyllosticta melanoplaca,** CA, CT, ID, UT.
Leaf spot, tar. **Phyllachora melanoplaca,** NY.
Rust. **Puccinia atropuncta** (II, III), MO, NC, OK, TN, VA, WV; 0, I, on composites; **P. veratri,** widespread; 0, I, on *Epilobium.*

VERBENA, GARDEN
(*Verbena hortensis*)

Blight, flower. **Botrytis cinerea,** MA.
Nematode, root knot. **Meloidogyne** sp., MD.
Powdery mildew. **Erysiphe cichoracearum,** general.
Rot, charcoal. **Macrophomina phaseoli,** OK.
Rot, root. **Phymatotrichum omnivorum,** TX; **Rhizoctonia solani,** NY; **Thielaviopsis basicola,** PA.

VERBENA
(Native Species)

Broomrape. **Orobanche ramosa,** TX.
Dodder. **Cuscuta arvensis,** OK.
Downy mildew. **Plasmopara halstedii,** NM.
Leaf spot. **Ascochna verbenae,** WI; **Cercospora verbenicola,** AL, LA, TX; **Phyllosticta texensis,** TX; **Septoria verbenae,** VT to MS, TX, and SD.
Powdery mildew. **Erysiphe cichoracearum,** general.
Rot, root. **Phymatotrichum omnivorum,** TX.
Rust. **Puccinia aristidae** (0, I), AZ; II, III, on grasses; **P. vilfae** (0, I), IN to OK, and SD; II, III, on *Sporobolus.*
Virus. **Mosaic.** Unidentified; **Biddens Mottle,** FL.

VERBESINA
(Crownbeard)

Downy mildew. **Plasmopara halstedii**, NM, TX.
Leaf spot. **Cercospora fulvella**, TX; **Colletotrichum** sp., also stem spot; **Phyllosticta verbesinae**, TX.
Nematode, root knot. **Meloidogyne** sp., AL.
Powdery mildew. **Erysiphe cichoracearum**, SC, TX, VA.
Rot, root. **Helicobasidium purpureum**, TX; **Phymatotrichum omnivorum**, TX.
Rust. **Coleosporium viguierae** (II, III), AZ, FL, NC, TX; 0, I, unknown; **Puccinia abrupta** (II, III), CA, TX; **P. cognata** (0, I, II, III), AR, LA, TN, TX; **P. verbesinae** (0, I, II, III), MD to AL, LA.

VERONICA
(Speedwell)

Downy mildew. **Peronospora grisea**, CA, GA, KS, IL, IN, MO, NY, TX, WI.
Gall, leaf. **Synchytrium globosum**, LA; **Sorosphaera veronicae**, CO, MS.
Leaf spot. **Cercospora tortipes**, WI; **Gloeosporium veronicae**, NY; **Ramularia veronicae**, OK, TX, WI; **Septoria veronicae**, CA, FL, IA, MI, OH, WI.
Nematode, root knot. **Meloidogyne** spp., FL.
Powdery mildew. **Sphaerotheca macularis**, CT, IA, OR, WI.
Rot, root and stem. **Fusarium** sp., NJ; **Rhizoctonia solani**, IL, MD; **Phymatotrichum omnivorum**, TX.
Rot, stem. **Sclerotium rolfsii**, CT, NJ, OH.
Rust. **Puccinia albulensis** (III), CO, MT, OR, UT, WA, WY; **P. probabilis** (II, III), NM; **P. rhaetica** (III), WA; **P. veronicarum** (III), IA, WI.
Smut, leaf. **Entyloma veronicae**, CO, CT, KS, IL, IA, MS, MO, NY, TX, WI; **Urocystis kmetiana**, GA.
Virus. **Cucumber Mosaic**, AR.

VETCH
(*Vicia*)

Nematode, ring. **Criconemella**, SC.

VIBURNUM

Bacterial crown gall. **Agrobacterium tumefaciens**, PA, WA.
Bacterial leaf spot. **Pseudomonas viburni**, IL, IA, NJ.
Blight, gray mold; shoot. **Botrytis cinerea**, MA, WA.

Blight, thread. **Pellicularia koleroga,** FL, NC.

Canker, stem girdle. **Hymenochaete agglutinans,** PA.

Dodder. **Cuscuta compacta,** FL.

Downy mildew. **Plasmopara viburni,** general.

Gall, stem; dieback. **Phomopsis** sp., MD, NJ, NY.

Leaf spot. **Cercospora opuli,** IA, MS; **C. tinea,** LA; **C. varia,** general; **Helminthosporium beaumontii,** AL, TX; **Hendersonia foliorum** var. **viburni,** FL, TX; **H. tini,** LA; **Leptosphaeria tini,** LA; **Monochaetia desmazieri,** WV; **Phyllosticta lantanoides,** NY; **P. punctata,** FL, IA, WI; **P. tinea,** MD; **Ramularia viburni,** TN, WI.

Leaf spot. **Cristulariella pyramidalis,** MD.

Leaf spot, algal. **Cephaleuros virescens,** FL, LA.

Mold, leaf. **Cladosporium herbarum,** OH, PA.

Nematode, root knot. **Meloidogyne** spp., CA, MD, MS.

Powdery mildew. **Microsphaeri alni,** general.

Rot, root. **Clitocybe tabescens,** FL; **Corticium falactinum,** MD; **Phymatotrichum omnivorum,** TX; **Rosellinia necatrix,** CA.

Rust. **Aecidium rubromaculans** (0, I), FL; **Coleosporium viburni** (II, III), IL, IA, MD, MI, VA, WI; 0, I unknown; **Puccinia linkii** (III), ID, MI, MT, NH, WA, AK.

Spot anthracnose. **Sphaceloma viburni,** CA, WA.

Wilt. **Verticillium albo-atrum,** IL, IN, OR, WA.

Viburnum carlesii is extremely sensitive to sulfur and may be injured even by spray drift from other plants.

VINCA
(Periwinkle, Ground-Myrtle)

Bacterial, MLO. **Aster Yellows,** KS, TX.

Bacterial stunt. **MLO,** AR, OK.

Blight, gray mold. **Botrytis cinerea,** CT, WA.

Canker; dieback. **Phomopsis lirella,** CT, MD, NJ, OH, PA, VA; **Phoma** sp., OR.

Dodder. **Cuscuta indecora,** TX.

Leaf spot. **Alternaria** sp., PA, TX; **Colletotrichum** sp., FL; **Macrophoma vincae,** also dieback, IL, NY; **Phyllosticta** sp., GA, NJ, CT; **P. minor,** MD, NJ, NY, VA; **P. vincae-majoris,** CA; **Septoria vincae,** NJ, NY; **Volutella vincae,** NY, PA.

Mold, leaf. **Cladosporium herbarum,** ME, NY, PA.

Nematode, dagger. **Xiphinema americanum,** WI.

Nematode, root knot. **Meloidogyne** sp., KS, OH, CA.

Rot, root and stem. **Rhizoctonia solani,** IL, MD, NJ, PA. **Phytophthora parasitica,** MD.

Rust. **Puccinia vincae** (0, I, II, III), MA, MI, NY, WA.

Virus. Cucumber Mosaic, NJ; Potato Yellow Dwarf, CA.
Wilt. Verticillium albo-atrum, CA, OR.

VINCETOXICUM
(Milkvine)

Downy mildew. **Plasmopara gonolobii,** MD to FL and TX.
Leaf spot. **Cercospora bellynckii,** TX.
Rot, root. **Phymatotrichum omnivorum,** TX.
Rust. **Puccinia obliqua** (III), FL to KY, OK and AZ; **Uromyces asclepiadis**
(II, III), IN, WV.

VIOLET
(*Viola odorata* and Native Species)

Anthracnose. **Colletotrichum violae-tricoloris,** CT, MA, MS, NJ, NY, OH,
PA.
Blight, southern. **Sclerotium rolfsii,** VA to FL and TX, also CA, CT, NY,
VT.
Downy mildew. **Bremiella megasperma,** FL, NJ, IL, IA, MN.
Gall, leaf. **Synchytrium aureum,** NY, WI.
Leaf spot. **Alternaria violae,** CT to GA, TX, and WI; **Ascochyta violicola,**
AK; **A. violae,** IN, IA, PA, WI; **Centrospora acerina,** CA, AK; **Cercospora
granuliformis,** ME to AL, OK, and SD; **C. violae,** MA to FL, TX and ND;
Cryptostictis violae, IL; **Cylindrosporium violae,** MT; **Heterosporium** sp.,
AK; **Marssonina violae,** MA to SC, IA, and MN; **Phyllosticta violae,**
MA to FL, KS, and MN, CA; **P. nigrescens,** CA; **Ramularia lactea,** CO,
MS, MT, AK; **R. agrestis,** NE, OR; **Septoria violae,** northeastern and north
central states to FL, LA, and KS, AK; **Ciborinia violae.**
Nematode, lesion. **Pratylenchus pratensis.**
Nematode, root knot. **Meloidogyne** spp., FL to CA, occasional in North.
Powdery mildew. **Sphaerotheca macularis,** CA, CO, ND, OR, WI, WY.
Rot, root. **Fusarium oxysporum** f. sp. **aurantiacum,** FL, MS, OH; **Helicoba-
sidium purpureum,** TX; **Phymatotrichum omnivorum,** TX; **Rhizoctonia
solani,** IL, NY, FL, MN, MS; **Thielaviopsis basicola,** CT to MS and OH,
also KS, MA.
Rot, wet; gray mold. **Botrytis cinerea,** MD, OH, AK.
Rust. **Puccinia violae** (0, I, II, III), general; **P. effusa** (III), CA; **P. ellisiana**
(0, I), northeastern and north central states to AL, NM, and WY; II, III,
on *Andropogon;* **P. fergussoni** (III), CO, MT, UT, AK; **Uromyces andropogonis,** (0, I), MS, NC, NJ, PA, TN, WV; II, III, on *Andropogon.*
Sooty mold. **Scorias spongiosa,** ME.
Spot anthracnose; scab. **Sphaceloma violae,** ME to FL and LA, OH, KS, WA.

Smut, leaf and stem. **Urocystis violae,** CA, MN, TX, UT, AK.
Virus. **Beet Curly Top,** TX.

Spot anthracnose is an important violet disease in many gardens with disfiguring scabby lesions on stems and leaves.

VIPERS-BUGLOSS
(*Echium*)

Leaf spot. **Cercospora echii; Stemphylium** sp., NY.
Rot, root. **Rosellinia** sp., CA.

VIRGINIA CREEPER
(*Parthenocissus quinquefolia*)

Blight, thread. **Pellicularia koleroga,** FL.
Canker; dieback. **Coniothryium fuckelii,** WV; **Cladosporium** sp., NJ.
Downy mildew. **Plasmopara viticola,** AL, IA, ME, MN, NJ, NY, TX, WI.
Leaf spot. **Cercospora ampelopsidis,** widespread; **C. psedericola,** IL, VA; **Guignardia bidwellii** f. **parthenocissi,** general; **Phleospora ampelopsidis,** IL, IA, NE, WI.
Parasitic lichen. **Strigula elegans** and **S. camplanata,** LA, Southern U.S.
Powdery mildew. **Uncinula necator,** general.
Rot, root. **Helicobasidium purpureum,** TX; **Phymatotrichum omnivorum,** TX.
Spot anthracnose. **Elsinoë parthenocissi,** FL, MO, NH, PA.

The Guignardia leaf spot is commonly disfiguring in a wet season.

VITEX
(*Chaste-Tree*)

Leaf spot. **Cercospora viticis,** LA, TX.
Rot, root. **Phymatotrichum omnivorum,** TX.

WALLFLOWER
(*Cheiranthus*)

Bacterial, MLO. **California Aster Yellows,** CA.
Blight, gray mold. **Botrytis cinerea,** WA, AK.
Leaf spot. **Heterosporium** sp., OK.
Rot, crown. **Rhizoctonia solani,** NJ.
White blister. **Albugo candida,** MN.

WALLFLOWER, WESTERN
(*Erysimum*)

Club root. **Plasmodiophora brassicae,** NJ.
Downy mildew. **Peronospora parasitica,** CO, ID, IA.
Leaf spot. **Cercospora erysimi,** WI.
Powdery mildew. **Erysiphe polygoni,** CA.
Rot, root. **Rhizoctonia solani,** TX.
Rust. **Puccinia aristidae** (0, I), AZ, CO, ND, NE, UT; **P. consimilis** (I, III), MT; **P. holboellii** (0, III), CO.
Virus. **Tomato Spotted Wilt,** CA.
White rust; white blister. **Albugo candida,** ID, OR, TX.

WALNUT
(*Juglans;* includes Butternut, Black, English, and Japanese Walnuts)

Anthracnose; leaf spot. **Gnomonia leptostyla** (*Marssonina juglandis*), general.
Bacterial blight. **Xanthomonas juglandis,** NY to GA, TX, and Pacific Coast, especially on English (Persian) walnut.
Bacterial canker. **Erwinia nigrifluens,** CA; **E. rubrifaciens,** phloem canker, CA.
Bacterial crown gall. **Agrobacterium tumefaciens,** occasional.
Blight, leaf. **Cylindrosporium juglandis,** AL and TN to TX, CA.
Blight, seedling. **Sclerotium rolfsii,** TX.
Canker; branch wilt. **Hendersonula toruloidea,** connected with sunburn.
Canker; dieback. **Diplodia juglandis,** widespread; **Dothiorella gregaria,** CA; **Exosporina fawcetti,** wilt, CA; **Melanconis juglandis,** widespread, especially in East; **Nectria** spp., widespread; **N. galligena; Cytospora** sp., AZ.
Canker; felt fungus. **Septobasidium curtisii,** NC.
Canker, stem. **Fusarium solani,** KS.
Leaf spot. **Cercospora juglandis,** KS, MA; **Phelospora multimaculans,** TX; **Marssonina californica,** CA; **Ascochyta juglandis,** ring spot; **Grovesinia pyramidalis,** IA, IL, MN, OH, WV; **Mycosphaerella juglandis,** IA, IL, IN, NC.
Leaf spot. **Cristulariella pyramidalis,** IL. Also on *Bidens frondosa* (beggarticks), *Campsis radicans* (trumpet vine), *Chenopodium ambrosioides* (Mexican tea), *Commelina diffusa* (dayflower), *Cuphea petiolota* (blue waxweed), *Desmodium canescens* (tick clover), *Eupatorium coelestinum* (mist-flower), *E. rugosum* (white snakeroot), *Ipomoea hederscea* (morning-glory), *I. lacunosa* (morning-glory), *Lobelia inflata* (Indian tobacco), *L. siphilitica* (blue cardinal-flower), *Perilla frutescens* (beefsteak plant), *Phytolacca arnericana* (poke), *Platanus occidentalis* (sycamore), *Poly-*

gonum pennsylvanicum (smart weed), *P. scandens* (false buckwheat), *Rumex crispus* (yellow dock), *Sida spinosa* (prickly mallow), *Solidago canadensis* (goldenrod), *Vitis palmata* (Catbird grape).

Leaf spot; downy spot; white mold. **Microstroma juglandis,** widespread; **M. brachysporum,** general.

Mistletoe. **Phoradendron serotinum (flavescens),** IN southward, AZ, CA, NM.

Mold, nut. **Alternaria** sp., CA.

Nematode, lesion. **Pratylenchus musicola,** CA; **P. vulnus.**

Nematode, root knot. **Meloidogyne** spp., TX.

Nonparasitic

Black End, of nuts. Probably drought injury, CA, OR.

Black Line, girdle. May be caused by graft incompatibility.

Dieback. Boron deficiency, OR.

Erinose. Leaf galls from blister mites.

Leaf Scorch, Sunscald. NJ, OR, WA.

Little Leaf. Zinc deficiency, CA. Probably also causes Rosette.

Shrivel, Witches' Broom, Yellows. Cause unknown.

Powdery mildew. **Phyllactinia corylea,** IN, OH, OR; **Microsphaera alni,** widespread; **Erysiphe ?polygoni,** CA.

Rot, collar. **Phytophthora cactorum,** CA; **P. cinnamomi,** MD to AL, LA.

Rot, heart. **Fomes igniarius** and **F. everhartii,** widespread; **Polyporus sulphureus,** widespread.

Rot, root. **Armularia mellea,** cosmopolitan; **Phymatotrichum omnivorum,** TX; **Cylindrocladium** sp., TN.

Rot, wood. **Fomes conchatus; Polyporus** spp.; **Poria** spp.; **Schizophyllum commune,** cosmopolitan; **Daedalea confragosa; D. quercina,** widespread.

Scab. **Cladosporium** sp., MD, MN.

Virus. **Brooming Disease,** MD, NY, GA, DC.

Bacterial blight is the most serious disease of walnuts on the Pacific Coast, requiring several sprays. A toxin, juglone, has been considered injurious to many shrubs growing in the vicinity of black walnut roots; the toxin also causes vascular wilt-like symptoms in eggplant and tomato in home gardens.

WATER-CRESS
(*Nasturtium officinale*)

Damping-off. **Rhizoctonia solani,** TX.

Leaf spot. **Cercospora nasturtii,** CA, CT, FL, IN, NH, TX, WI.

Rot, root. **Pythium debaryanum,** TX; **Sporgospora subterranea** f. sp. **nasturtii,** FL, PA; **Phytophthora criptogea,** and stem, FL.

Rust. **Puccinia aristidae,** AZ, CO, TX.

White rust; white blister. **Albugo candida,** MN.

WATER-ELM
(*Planera*)

Rot, wood. **Daedalea ambigua,** SC; **Ganoderma lucidum.**

WATER-HOREHOUND
(*Lycopus*)

Gall, leaf. **Synchytrium cellulare,** WI.
Leaf spot. **Ascochyta lophanthi,** MA, WI; **Cercospora lycopi,** LA; **Phyllosticta decidua,** IA, OK. WI; **Septoria lycopi,** WI.
Rust. **Puccinia angustata** (0, I), ME to MD, KS, ND; II, III, on grasses.

WATER-HYACINTHS
(*Eichhornia*)

Blight. **Aquathanatephorus pendulus** (*Rhizoctonia* stage), LA.
Leaf spot. **Cercospora piaropi,** FL.
Rot, root and crown. **Mycoleptodiscus terrestris,** FL.

WATER-LILY
(*Nymphaea*)

Leaf spot. **Alternaria** sp., TX; **Cercospora exotica,** IL; **C. nymphaeacea,** scattered ME to TX, CA; **Helicoceras nymphaerum,** MD, NJ, NY; **Mycosphaerella pontederiae,** VA, MI; **Ovularia nymphaearum,** MD, NY, WA; **Phyllosticta fatiscens,** VT.
Leaf spot. **Dichotomophthoropsis nymphaerum,** MN; **Sclerotium** sp., MN.
Rot, leaf and stem. **Pythium** spp., MA, NY, WI.
Smut, white. **Entyloma nymphaeae,** MA to VA, OK, and WI.

WATER-LILY, YELLOW PONDLILY
(*Nuphar*)

Leaf spot. **Dichotomophthoropsis nymphaerum,** MN.
Leaf spot. **Mycosphaerella pontederiae,** ME, MI, NY, VA, WI; **Phyllosticta fatiscens,** IA, NJ, NY, WI.
Smut, white. **Entyloma nymphaeae,** CT, IL, MA, WI.

WATERMELON
(*Citrullus*)

Anthracnose. **Colletotrichum lagenarium,** general, Pacific Coast; **Marssonina melonis,** NY.

Bacterial, angular leaf spot. **Pseudomonas pseudoalcaligenes** subsp. **citrulli,** GA.

Bacterial rind necrosis. **Erwinia** sp., TX.

Bacterial soft rot. **Erwinia aroideae,** WV; Wilt, **E. tracheiphila,** rare.

Bacterial spot. **Pseudomonas lachrymans,** MI.

Blight, gummy stem; fruit spot. **Mycosphaerella citrullina,** also stem-end rot, leaf spot, MA to FL, AZ, and MO.

Blight, southern. **Sclerotium rolfsii,** also fruit rot, NC to FL, TX.

Damping-off. **Rhizoctonia solani,** Soil rot, leaf blight; **Pythium** spp., also blossom-end rot, foot rot.

Downy mildew. **Pseudoperonospora cubensis,** occasional from MA to FL, TX, and WI; also CA.

Fruit spot; speck. **Cribropeltis citrullina,** IL.

Leaf spot. **Alternaria cucumerina,** general except Pacific Coast; **Cercospora citrullina,** NJ to FL, TX, and OH; **Corynespora cassicola.**

Nematode, root knot. **Meloidogyne arenaria; M. incognita; M. incognita-acrita; M. javanica.**

Nonparasitic. **Blossom-end Rot,** hot dry weather after cool, moist days. **Internal Browning,** drought and nutritional deficiencies.

Powdery mildew. **Erysiphe cichoracearum,** AZ, CA, FL, GA, NY, NC, TX, VA. May make pimples in young fruit.

Rot. **Diplodia** spp., stem-end, gray; MD to FL, AZ and KS; **Fusarium scirpi,** occasional in market; **Helminthosporium** sp., TX; **Rhizopus** spp., mushy soft rot.

Rot, charcoal. **Macrophomina phaseoli,** TX.

Rot, root. **Phymatotrichum omnivorum,** AZ, TX; **Thielaviopsis basicola,** OR, UT.

Rot, stem and fruit. **Phytophthora cactorum,** AZ; **P. capsici,** CO; **P. citrophthora,** CA; **Sclerotinia sclerotiorum,** NJ, TX.

Scab; leaf mold. **Cladosporium cucumerinum,** MD, NE.

Virus. **Watermelon Mosaic,** NY to FL, TX, and IA, also AZ, CA, MI, WA; **Beet Curly Top; Tobacco Ring Spot; Watermelon Stunt** (?strain of squash mosaic); **Zucchini Yellow Mosaic,** CA.

Wilt. **Fusarium oxysporum** f. sp. **niveum,** general; **Verticillium albo-atrum,** CA, NH, OR.

Fusarium wilt is probably the major disease and resistant varieties are available. For anthracnose and leaf spots treat seed before planting, and start spraying or dusting when vines start to run.

WATER-PRIMROSE
(Jussiaea)

Leaf spot. **Alternaria** sp. OK; **Cercospora jussiaeae,** AL, OK, TX; **Colletotrichum jussiaeae,** AL, TX; **Septoria jussiaeae,** AL, FL, LA, TX.

Rust. **Aecidium betheli**, CA; **Puccinia jussiaeae** (0, I, III), MS; **Uredo guauna-bensis** (II), FL.

WATER SHIELD
(*Brasenia*)

Leaf spot. **Dichotomophthoropsis nymphaerum**, MN.

WATSONIA

Rot, root. **Armillaria mellea**, CA.
Virus. **?Iris mosaic**, CA.

WAX-MYRTLE, CANDLEBERRY
(*Myrica cerifera*)

Black mildew. **Irene calostroma**, Gulf states; **Irenina manca**, MS; **Meliola manca**, FL.
Blight, seedling. **Rhizoctonia solani**, NJ.
Dodder. **Cuscuta compacta**, FL.
Leaf spot. **Cercospora dispersa**, NJ; **Phyllosticta myricae**, NJ to FL, TX; **Septoria myricae**, NJ.
Nematode, ring. **Hemicriconemoides wessoni**, FL.
Parasitic lichen. **Strigula elegans** and **S. camplanata**, LA, southern US.
Rot, root. **Clitocybe tabescens**, FL; **Phymatotrichum omnivorum**, TX.
Rust. **Gymnosporangium ellisii** (0, I), MA to MD; III, on *Chamaecyparis*.
Sooty mold. **Capnodium grandisporum**, FL.

WEIGELA

Bacterial crown gall. **Agrobacterium tumefaciens**, MD, MS.
Blight, twig. **Phoma weigelae**, WA.
Leaf spot. **Cercospora weigelae**, MD, MS, NJ; **Ramularia diervillae**, TN.
Nematode, lesion. **Pratylenchus pratensis**.
Nematode, root knot. **Meloidogyne** sp., CA, MD, MS, TX.
Rot, root. **Phymatotrichum omnivorum**, TX.

WHIPPLEA

Downy mildew. **Peronospora whippleae**, CA.

WHORTLEBERRY, BILBERRY
(*Vaccinium* spp.)

Blight, twig; berry rot. **Monilinia ledi,** NY.

Gall, leaf. **Exobasidium vaccinii,** occasional; **E. parvifolii,** stem gall, OR, WA; **E. vaccinii-uliginosi,** shoot gall, rose-bloom, OR.

Leaf spot, tar. **Rhytisma vaccinii,** AK.

Powdery mildew. **Microsphaera alni** var. **vaccinii,** AK, OR, WA, WY; **Podosphaera oxyacanthae,** AK.

Rust. **Pucciniastrum** sp. (II), OR, WA; **P. goeppertianum** (III), witches' broom, general; 0, I, on fir; **P. vaccinii** (II, III), general; 0, I, on hemlock.

WILD RICE
(*Zizania*)

Anthracnose. **Colletotrichum sublineolum,** MN.

Bacterial leaf spot. **Pseudomonas syringae** pv. **syringae,** MN, Pacific Northwest; **P. syringae** pv. **zizaniae,** Pacific Northwest.

Leaf spot, stem lesion. **Bipolaris sorokiniana,** Pacific Northwest; **Sclerotium oryzae,** Pacific Northwest; **S. hydrophilum,** Pacific Northwest.

Leaf spot, zonate. **Drechslera gigantea,** MN.

Rot, crown, root. **Phytophthora erythroseptica,** CA.

Smut. **Entyloma lineatum,** Pacific Northwest.

Virus. **Wheat Streak Mosaic,** MN.

WILDRYE
(*Elymus*)

Leaf spot. **Pyrenophora trichostoma,** ND; **Septorium spraguei,** ND.

WILLOW
(*Salix;* includes Weeping Willow, Pussy Willow)

Bacterial crown gall. **Agrobacterium tumefaciens,** CT, NJ, TX, VA.

Bacterial wetwood. **Erwinia nimipressuralis.**

Blight, twig. **Diplodia salicina.**

Blight, willow. Complex of scab and black canker.

Canker, black. **Physalospora miyabeana,** New England, NY, WV.

Canker, twig and branch. **Botryosphaeria ribis,** VA to GA and AR, CA; **Cryptodiaporthe salicina,** ME to VA, OK, KS, and SD, CA, WA, AK; **Cryptomyces maximus,** bark blister; NM, UT; **Discella carbonacea,** twig blight, New England; **Dothiora polyspora,** CO; **Dothiorella** sp., AR, ND;

Macrophoma spp., AR, KY, MS, NC, SC, TX; **Phomopsis salicina,** IA, MA, VA; **Physalospora gregaria,** WV; **Valsa** spp., twig canker; **V. sordida** (*Cytospora chrysosperma*), widespread; **V. salicina,** widespread; **V. nivea,** western states.

Dodder. **Cuscuta** spp., IA, NY, WA.

Leaf blister. **Taphrina populi-salicis,** CA.

Leaf spot. **Ascochyta salicis,** CA; **Cercospora salicina,** IL, LA, MD, TX; **Cibornia foliicola; C. wisconsinensis; Cylindrosporium salicinum,** MA to CO, and WI, MS; **Gloeosporium** spp., also twig blight, CT, DE, MA; **G. salicis,** VT to NJ, MS, and WI, OR; **Marssonina** spp., widespread; **M. kriegeriana,** WI; **M. apicalis,** CA; **Myrioconium comitatum,** WI; **Phyllosticta apicalis,** KS, WI; **Ramularia rosea,** CO, MT, WI; **Septogloeum salicinum,** NY, WI, AK; **S. maculans,** CA, MT; **S. salicis-fendlerianae,** ID; **Septoria** spp.; **S. salicicola,** AK, OR; **S. didyma,** WI.

Leaf spot, tar. **Rhytisma salicinum,** MI.

Mistletoe. **Phoradendron serotinum (flavescens),** CA, IN, TX, AZ, NM.

Mistletoe, European. **Viscum album,** CA.

Nematode, lesion. **Pratylenchus vulnus,** CA.

Nematode, root knot. **Meloidogyne** sp.

Powdery mildew. **Uncinula salicis,** general; **Phyllactinia corylea,** WA.

Rot, heart. **Daedalea confragosa,** widespread; **Fomes** spp.; **Polyporus farlowii,** TX to CA; **Trametes suaveolens,** New England to MT and AK.

Rot, root. **Armillaria mellea,** CA, WA; **Helicobasidium purpureum,** TX; **Phymatotrichum omnivorum,** TX.

Rot, sapwood; wood. **Daedalea ambigua; Ganoderma lucidum; Pholiota** spp.; **Pleurotus** spp.; **Polyporus** spp.; **Schizophyllum commune,** cosmopolitan.

Rust. **Melampsora abieri-capraearum** (*M. epitea*) (II, III), general except far North; 0, I, on fir; **M. artica** (II, III), AK, CO, NH; **M. paradoxa** (*M. bigelowii*) (II, III), ME to NC, IA, NM, and AK; 0, I, on larch; **M. ribesii-purpurea** (II, III), MT to CO, CA, and AK; 0, I, on *Ribes*.

Scab, gray; blight. **Fusicladium saliciperdum** (*Venturia chlorospora*), New England to NJ, PA, NC.

Sooty mold. **Capnodium** sp.

Spot anthracnose; scab. **Sphaceloma murrayae,** CA, OR, WA, WI.

Willow scab, followed by black canker, forms a very destructive blight, killing many trees in New England.

WINGED BEAN
(*Psophocarpus*)

Virus. **Cucumber Mosaic,** FL; **Clover Yellow Vein,** MD.

WINTERBERRY
(*Ilex verticillata*)

Leaf spot. **Physalospora ilicis,** NY.
Leaf spot, tar. **Rhytisma concavum,** WI; **R. prini,** ME to MS, IL, and WI.
Powdery mildew. **Microsphaera alni,** WI.

WINTER CRESS
(*Barbarea*)

Bacterial black rot. **Xanthomonas barbareae,** NY.
Bacterial, yellows. **Spiroplasma citri,** IL.
Dodder. **Cuscuta gronovii,** NY.
Downy mildew. **Peronospora parasitica,** TX.
Leaf spot. **Cercospora barbarea,** WI; **Ramularia barbareae,** MA to NJ, OH, and WI, TX; **Alternaria** sp., PA.
Rot, stem. **Sclerotium rolfsii,** TX.
Virus. **Beet Curly Top; Potato Yellow Dwarf.**
White rust. **Albugo candida,** CA, TX.

WINTERGREEN, CHECKERBERRY
(*Gaultheria procumbens*)

Blotch, sooty. **?Gloeodes pomigena,** WI.
Fruit spot, black speck. **?Leptothyrium pomi,** WI.
Leaf spot. **Cercospora gaultheriae,** NY, WI; **Pezizella oenotherae,** VA; **Discosia maculicola,** secondary; **Mycosphaerella gaultheriae,** ME to MD and WV; **Phyllosticta gaultheriae,** general; **Schizothyrium gaultheriae,** ME to VA and WI; **Venturia arctostaphyli,** MD, MA, NJ, NY, VA.
Powdery mildew. **Microsphaera alni,** WI.

WINTERGREEN, WAXFLOWER

See Pipsissewa.

WISTERIA
(*Wistaria*)

Bacterial crown gall. **Agrobacterium tumefaciens,** CT, MD, TX.
Canker, stem. **Nectria cinnabarina,** CT.
Leaf spot. **Phyllosticta wistariae,** MA, MO, NJ, TX; **Septoria wisteriae,** TX.
Nematode, root knot. **Meloidogyne** spp.
Powdery mildew. **Erysiphe ?cichoracearum,** TX.

Rot, heart. **Pleurotus sp., WV.**
Rot, root. **Phymatotrichum omnivorum, TX.**
?Virus. **Wisteria Vein Mosaic, MS, NY, TX. Tobacco Mosaic, RI.**

WITCH-HAZEL
(*Hamamelis*)

Bacterial crown gall. **Agrobacterium tumefaciens, MD.**
Leaf spot. **Discosia artocreas, OK; Gonatobotryum maculicola, NH to WV,** FL and WI; **Graphium hamamelidis, NY to IN and TN; Monochaetia desmazierii, VA to GA and TN; Mycosphaerella sp., WV; Phyllosticta hamamelidis, CT to MS, TN, and WI; Ramularia hamamelidis, NY to WV, OK,** and WI.
Powdery mildew. **Phyllactinia corylea, MI, WI; Podosphaera biuncinata,** New England to IL and southward.
Rot, wood. **Fomes scutellatus,** widespread; **Polyporus spp.**

WOLFBERRY
(*Symphoricarpos occidentalis*)

Black knot. **Dibotryon symphoricarpi,** twig canker, ND.
Leaf spot. **Cercospora symphoricarpi, MT, ND, WA; Septoria symphoricarpi,** IA, MT, ND, WA.
Powdery mildew. **Microsphaera diffusa,** general.
Rot, collar. **Fomes ribis, KS, ND.**
Rust. **Puccinia crandallii** (0, I), CO, MT, ND, WY; II, III on grasses; P. **symphoricarpi, MT.**

WOOD-BETONY, LOUSEWORT
(*Pedicularis*)

Gall, leaf. **Synchytrium aureum, WI.**
Leaf spot. **Ramularia obducens, CA; Septoria cylindrospora, WI.**
Powdery mildew. **Sphaerotheca macularis, CA, CO, MD, MI, MN, WY, WI.**
Rust. **Cronartium coleosporioides** (II, III), CA, ID, MT, NM, WA; **Puccinia clintonii** (III), CO, ID, ME, MI, NM, NY, OR, WA, WI, WY; P. **rufescens** (I, III), CO, CA, NV.

WOODRUSH
(*Luzula*)

Rust. **Puccinia obscura** (II, III), ME to KS, WI, ID, OR, WA.
Smut, inflorescence. **Cintractia luzulae, IN.**

WYETHIA

Leaf spot. **Didymaria conferta,** OR, UT; **Marssonina wyethiae,** CA, WA; **Septoria wyethiae,** CA, UT.
Nematode, leaf gall. **Tylenchus balsamophilus,** UT.
Rust. **Puccinia balsamorrhizae** (0, I, II, III), AZ, CA, CO, UT.

XANTHOSMA

Rot, root. **Armillaria mellea,** CA; **Fusarium solani,** powdery gray rot, FL.

XEROPHYLLUM
(Turkeybeard)

Rust. **Puccinia atropuncta** (II, III), MS.

YAM, CINNAMON-VINE
(*Dioscorea*)

Leaf spot. **Cercospora dioscoreae,** MD, IL, IA, MI, PA, WI; **Colletotrichum dioscoreae,** IL; **Phyllosticta dioscoreae,** SC, VA, WV; **Ramularia dioscoreae,** WI.
Nematode, root knot. **Meloidogyne** sp., NC.
Rot, root. **Phymatotrichum omnivorum,** TX.

YARROW
(*Achillea*)

Bacterial crown gall. **Agrobacterium tumefaciens,** IN.
Dodder. **Cuscuta** sp., NH.
Powdery mildew. **Erysiphe cichoracearum,** MO, MT, PA, SD, VT, WI, AK.
Nematode, root knot. **Meloidogyne** sp., OR.
Rot, root. **Phymatotrichum omnivorum,** TX; **Rhizoctonia solani,** general.
Rust. **Puccinia millefolii** (III), CA, CO, ID, MT, NM, OR, UT, WA, WY.

YAUPON
(*Ilex vomitoria*)

Leaf spot; tar spot. **Rhytisma ilicincola,** VA.
Rot, root. **Phymatotrichum omnivorum,** TX.
Sooty mold. **Capnodium** spp., Gulf states.

YELLOW-ROOT
(*Xanthorhiza*)

Leaf spot. **Phyllosticta xanthorhizae,** WV, NC.

YELLOWWOOD
(*Cladastris*)

Powdery mildew. **Phyllactinia corylea,** OK.
Rot, wood. **Polyporus spraguei,** MD.
Virus. **Bean Yellow Mosaic,** NY.
Wilt. **Verticillium albo-atrum,** IL.

YERBA BUENA
(*Micromeria*)

Rust. **Puccinia menthae** (0, I, II, III), CA, ID, OR, WA.

YERBA SANTA
(*Eriodictyon*)

Blotch, sooty. **Coniothecium eriodictyonis,** CA.

YEW
(*Taxus*)

Blight, needle. **Herpotrichia nigra,** ID; **Sphaerulina taxi,** CA, ID, OR, WA; **S. taxicola,** also twig blight, NY, PA; **Alternaria** sp., CT, NY, RI; **Phyllosticta taxi,** VA; **Macrophoma taxi; Mycosphaerella taxi; Neopeckia coulteri.**
Blight, seedling. **Phytophthora cinnamomi,** IN, MD, VA, Pacific Northwest.
Blight twig. **Phyllostictina hysterella** (*Physalospora gregaria*); **Pestuloria** sp., PA; **P. funerea,** MA; **Sphaeropsis** sp., NJ; **Botryosphaeria ribes.**
Damping-off. **Rhizoctonia solani,** CT.
Rot heart. **Fomes hartigii** OR; **F. roseus,** ID.
Rot, root. **Armillaria mellea,** ID.
Rot, wood. **Polyporus schweinitzii,** ID.

YUCCA
(Adams-Needle, Joshua-Tree, Spanish Bayonet)

Blight, flower. **Cercospora floricola** (?*C. concentrica*), TX.
Blight, leaf. **Kellermannia anomala,** ?secondary, general.
Leaf spot, **Cercospora concentrica,** CT, GA, IA, NJ, OK, TX; **Coniothyrium concentricum,** general; **Cylindrosporium angustifolium,** KS, MS, OK, TX;

Diplodia circinans, TX; **Epicoccum asternium,** TX; **Gloeosporium yucco-genum,** MO, TX; **Neottiospora yuccifolia,** GA, WA; **Pestalozziella yuccae,** secondary, TX; **Phyllosticta** sp., TX; **Stagonospora gigantea,** CA; **Leptosphaeria obtusipora.**
Mold, leaf. **Torula maculans,** AZ, CA, SC, TX; **Alternaria tenuis,** leaf rot.
Nematode. **Meloidogyne** sp., OR.
Rot, stem. **Sclerotium rolfsii,** MD.
Rust. **Puccinia amphigena** (0, I), NE.

Many other fungi are found on dead leaves and stems.

ZAMIA
(Coontie)

Algae. **Anabaena cycadeae,** in coralloid roots, FL.
Nematode. **Meloidogyne** sp., FL.

ZAUSCHNERIA
(Fire-Chalice, California Fuchsia)

Rust. **Puccinia oenotherae** (0, I, II, III), CA, UT.

ZEBRA PLANT
(*Aphelandra*)

Nonparasitic. Leaf crinkle; shortened internodes; axillary bud proliferation, TX.
Rot, stems. **Phytophthora parasitica,** FL.

ZEPHYRANTHES
(Atamasco-Lily; Zephyr-Lily)

Leaf scorch; red spot. **Stagonospora curtisii,** CA.
Leaf spot. **Colletotrichum liliacearum,** NC.
Rot, dry; scale speck. **Sclerotium** sp., OR.
Rust. **Puccinia cooperiae** (0, I, II, III), FL, AL, NC, TX.

ZIGADENUS

Rust. **Puccinia atropuncta** (II, III), IA, MO, ND, TX, WI; 0, I, on composites; **P. grumosa** (0, I, II, III), CO, MT, WY; **Uromyces zygadeni** (0, I, II, III), CA, CO, IA, KS, MT, NV, TX, UT, WA, WY.
Smut, leaf. **Urocystis flowersii,** UT.

ZINNIA

Bacterial; leaf spot, flower spot. **Xanthomonas campestris** pv. **zinniae,** LA.
Bacterial, MLO. **Aster Yellows,** MI, PA, and **California Aster Yellows,** CA.
Bacterial wilt. **Pseudomonas solanacearum,** FL.
Blight. **Alternaria zinniae,** CO, CT, NJ, NY, PA, SC.
Blight, head; stem canker. **Botrytis cinerea,** CA, CT, NJ, OR, PA.
Blight, southern. **Sclerotium rolfsii,** FL, NJ.
Damping-off; root rot. **Rhizoctonia solani,** CA, NJ, TX.
Leaf spot. **Cercospora zinniae,** SC to FL and TX, CO, IN, PA.
Leaf spot, bacterial. **Xanthomonas nigromaculans** f. sp. **zinniae,** OH, NC.
Nematode, leaf. **Aphelenchoides ritzema-bosi,** DE, MA, NJ.
Nematode, lesion. **Pratylenchus nannus,** MD; **P. penetrans,** NJ.
Nematode, root knot. **Meloidogyne** sp., NJ, PA, TX.
Powdery mildew. **Erysiphe cichoracearum,** general.
Rot, blossom. **Choanephora** sp., FL.
Rot, charcoal. **Macrophomina phaseoli,** TX.
Rot, root. **Phymatotrichum omnivorum,** TX.
Rot, stem; wilt. **Fusarium** sp., CO, IA, MO, NY; **Phytophthora cryptogea,**
 NJ; **Sclerotinia sclerotiorum,** CA, CO, MA, MT, OR, PA, WA.
Virus. **Cucumber Mosaic; Beet Curly Top; Tobacco Etch,** FL; **Tomato Spotted
 Wilt; Biddens Mottle,** FL.

Powdery mildew in late summer is the most common zinnia disease.

ZIZIA
(Meadow Parsnip)

Gall, leaf. **Urophlyctis** (*Physoderma*) **pluriannulata,** IA, WI.
Leaf spot. **Ascochyta thaspii,** WI; **Cercospora ziziae,** PA, WI; **Cylindro-
sporium ziziae,** ND, WA, WI; **Septoria ziziae,** ND.
Powdery mildew. **Erysiphe polygoni,** PA, WV.
Rust. **Puccinia ziziae** (III), WA.

ZOYSIA
(Japanese Lawn Grass)

Nematode, pseudo root knot. **Hypsoperine graminis,** MD.
Rot, root and crown. **Rhizoctonia solani,** MD; **Helminthosporium tetramera;
 Curvularia** spp.; **Fusarium** spp.
Rust. **Puccinia zoysiae,** AL, AR, FL, GA, LA, MD, MO, MS, TX.

List of Land-Grant Institutions and Agricultural Experiment Stations in the United States

For help in diagnosing and controlling plant diseases write to the extension plant pathologist at the college of agriculture of your state university or to your state experiment station. Bulletins, circulars, and spray schedules are available free from the bulletin room or mailing clerk.

Alabama: Auburn University, Auburn 36849.

Alaska: University of Alaska, College 99775; Experiment Station, Anchorage 99508.

Arizona: University of Arizona, Tucson 85721.

Arkansas: University of Arkansas, Fayetteville 72701; Cooperative Extension Service, P.O. Box 391, Little Rock 72203.

California: University of California, Berkeley 94720; Riverside 92521; Davis 95616.

Colorado: Colorado State University, Fort Collins 80523.

Connecticut: University of Connecticut, Storrs 06268; Connecticut Agricultural Experiment Station, New Haven 06504.

Delaware: University of Delaware, Newark 19711.

District of Columbia: University of the District of Columbia, Cooperative Extension Service, Washington, D.C. 20002.

Florida: University of Florida, Gainesville 32611.

Georgia: University of Georgia, Athens 30602; Agricultural Experiment Station, Experiment 30212; Coastal Plain Station, Tifton 31793.

Hawaii: University of Hawaii, Honolulu 96822.

Idaho: University of Idaho, Extension Service, Boise 83709; Agricultural Experiment Station, Moscow 83843.

Illinois: University of Illinois, Urbana 61801.

Indiana: Purdue University, West Lafayette 47907.

Iowa: Iowa State University, Ames 50011.

Kansas: Kansas State University, Manhattan 66506.

Kentucky: University of Kentucky, Lexington 40546.

Louisiana: Louisiana State University, University Station, Baton Rouge 70803.

Maine: University of Maine, Orono 04469.

Maryland: University of Maryland, College Park 20742.

Massachusetts: University of Massachusetts, Amherst 01003.

Michigan: Michigan State University, East Lansing 48824.

Minnesota: University of Minnesota, St. Paul 55108.

Mississippi: Mississippi State University, State College 39762.

Missouri: University of Missouri, Columbia 65211.

Montana: Montana State University, Bozeman 59715.

Nebraska: University of Nebraska, Lincoln 68583.

Nevada: University of Nevada, Reno 89557.

New Hampshire: University of New Hampshire, Durham 03824.

New Jersey: Rutgers, The State University, New Brunswick 08903.

New Mexico: New Mexico State University, Las Cruces 88003.

New York: Cornell University, Ithaca 14853; Agricultural Experiment Station, Geneva 14456.

North Carolina: North Carolina State University, Raleigh 27650.

North Dakota: North Dakota State University, Fargo 58105.

Ohio: Ohio State University, Columbus 43210; Ohio Agricultural Research and Development Center, Wooster 44691.

Oklahoma: Oklahoma State University, Stillwater 74078.

Oregon: Oregon State University, Corvallis 97331.

Pennsylvania: The Pennsylvania State University, University Park 16802.

Puerto Rico: University of Puerto Rico, Mayaguez 00708.

Rhode Island: University of Rhode Island, Kingston 02881.

South Carolina: Clemson University, Clemson 29631.

South Dakota: South Dakota State University, Brookings 57007.

Tennessee: University of Tennessee, Knoxville 37901.

Texas: Texas A & M University, College Station, 77843; Agricultural Experiment Station, Lubbock 79401.

Utah: Utah State University, Logan 84322.

Vermont: University of Vermont, Burlington 05405.

Virginia: Virginia Polytechnic Institute, Blacksburg 24061.

Virgin Islands: Virgin Islands Extension Service, Kingshill, St. Croix 00850.

Washington: Washington State University, Pullman 99164; Western Washington Experiment Station, Puyallup 98371.

West Virginia: West Virginia University, Morgantown 26506.

Wisconsin: University of Wisconsin, Madison 53706.

Wyoming: University of Wyoming, Laramie 82071.

Glossary

Acervulus, pl. *Acervuli*. A "little heap," an erumpent, cushionlike mass of hyphae bearing conidiophores and conidia, sometimes with setae; characteristic of the Melanconiales (Fig. 8).

Acicular. Needlelike.

Aeciospore. Rust spore formed in an aecium.

Aecium, pl. *Aecia*. A cluster-cup, or cuplike fruiting sorus in the rusts (Fig. 64).

Aerobic. Living or active only in the presence of oxygen.

Allantoid. Sausage-shaped.

Alternate Host. One or other of the two unlike hosts of a heteroecious rust.

Amoeboid. Not having a cell wall and changing in form like an amoeba.

Annulus. A ring; ringlike partial veil around stipe in the mushrooms.

Antheridium, pl. *Antheridia*. Male sex organ in the fungi.

Anthracnose. A disease with limited necrotic lesions, caused by a fungus producing nonsexual spores in acervuli (Figs. 10 and 11).

Antibiotic. Damaging to life; especially a substance produced by one microorganism to destroy others.

Apothecium, pl. *Apothecia*. The cup- or saucer-like ascus-bearing fruiting body; in the Discomycete section of the Ascomycetes (Figs. 6 and 62).

Appressorium. A swelling on a fungus germ tube for attachment to host in early stage of infection; found especially in anthracnose fungi and rusts.

Ascocarp, or *Ascoma*. Any structure producing asci, as an apothecium, perithecium.

Ascomycetes. One of the three main groups of the fungi, bearing sexual spores in asci.

Ascospore. Produced in ascus by free cell formation.

Ascus, pl. *Asci.* Saclike, usually clavate cell containing ascospores, typically eight (Fig. 6).

Aseptate. Without cross-walls.

Asexual. Vegetative, having no sex organs or sex spores; the imperfect stage of a fungus.

Autoecious. Completing life cycle on one host; term used in rusts.

Bacteria. Microscopic one-celled organisms increasing by fission.

Bactericide. Substance causing death of bacteria.

Basidiomycetes. Class 3 in the Fungi, characterized by septate mycelium, sometimes with clamp-connections, and sexual spores on basidia (Fig. 7).

Basidiospore. Spore produced on a basidium.

Basidium, pl. *Basidia.* Club-shaped structure, which, after fusion of two nuclei, produces four basidiospores (Fig. 7).

Binucleate. Having two nuclei.

Blight. A disease with sudden, severe leaf damage and often with general killing of flowers and stems.

Blotch. A blot or spot, usually superficial.

Breaking, of a virus. Loss of flower color in a variegated pattern, especially in tulips.

Canker. A lesion on a stem; a plant disease with sharply limited necrosis of the cortical tissue (Figs. 25, 26, and 27).

Carrier. Infected plant showing no marked symptoms but source of infection for other plants.

Catenulate. In chains, or in an end-to-end series.

Cerebroid. With brainlike convolutions or folds.

Chemotherapy. Treatment of internal disease by chemical agents that have a toxic effect on the microorganism without injuring the plant.

Chlamydospore. Thick-walled, asexual resting spore formed by the rounding up of any mycelial cell (Fig. 55); also used for smut spores.

Chlorosis. Yellowing of normally green tissue due to partial failure of chlorophyll to develop; often due to unavailability of iron (Fig. 45).

Cilium, pl. *Cilia.* Hairlike swimming organ on bacteria or zoospores.

Cirrhus, pl. *Cirrhi.* A tendril or horn of forced-out spores.

Clamp-connections. Outgrowths of hyphae that form bridges around septa, thus connecting two cells; in Basidiomycetes (Fig. 7).

Clavate. Club-shaped.

Cleistothecium, pl. *Cleistothecia.* A perithecium without a special opening; in powdery mildews (Fig. 51).

Coalesce. Growing together into one body or spot.

Coenocytic. Multinucleate; mycelia having no cell walls.

Columella. Sterile central axis in a mature fruiting body (Fig. 5).

Concentric. One circle within another with a common center.

Conidiophore. Simple or branched hyphae on which conidia are produced.

Conidium, pl. *Conidia.* Any asexual spore except sporangiospore or chlamydospore.

Conk. Term used in forestry for sporophores of Polyporaceae on trees.

Control. Prevention of, or reduction of loss from, plant disease.

Coremium, pl. *Coremia.* Synnema, a cluster of erect hyphae bearing conidia (Fig. 8).

Coriaceous. Like leather in texture.

Culturing. Artificial propagation of organisms on nutrient media or living plants.

Cystidium, pl. *Cystidia.* Sterile, often swollen cell projecting from hymenium in Basidiomycetes.

Damping-off. Seed decay in soil, or seedling blight.

Decumbent. Resting on substratum with ends turned up.

Decurrent. Running down the stipe or stem.

Diagnosis. Identification of nature and cause of a disease.

Dieback. Progressive death of branches or shoots beginning at tips.

Defoliate. To strip or become stripped of leaves.

Dichotomous. Branching, frequently successive, into two more or less equal arms.

Dimidiate. Having one half smaller than the other; of a perithecium, having outer wall covering only top half.

Discomycetes. The cup fungi, a subclass of Ascomycetes; with apothecia.

Disease. A condition in which use or structure of any part of the living organism is not normal.

Disinfection. Freeing a diseased plant, organ, or tissue from infection.

Disinfestation. Killing or inactivating disease organisms before they can cause infection; on surface of seed or plant part, or in soil.

Dissemination. Transport of inoculum from a diseased to a healthy plant.

Disjunctor. Cell or projection connecting spores of a chain.

Duster. Apparatus for applying fungicides in dry form.

Echinulate. Having small, pointed spines; used of spores.

Endoconidium, pl. *Endoconidia.* Conidium formed within a hypha.

Enphytotic. A plant disease causing constant damage from year to year.

Epiphytotic. Sudden and destructive development of a plant disease over an extensive area, an epidemic.

Eradicant fungicide. One that destroys a fungus at its source.

Eradication. Control of disease by eliminating the pathogen after it is already established.

Erumpent. Breaking through surface of substratum.

Excentric. Off center.

Exclusion. Control of disease by preventing its introduction into disease-free areas.

Exudate. Liquid discharge from diseased tissues.

Fasciation. Joining side by side; a plant disease with flattened and sometimes curved shoots.

Fascicle. A small bundle or cluster.

Filiform. Threadlike.

Fimbriate. Fringed, or toothed.

Flag. A branch with dead leaves on an otherwise green tree.

Flagellum, pl. *Flagella.* Whiplike organ on a motile cell; cilium.

Fruiting body. Fungus structure containing or bearing spores; mushroom, pycnidium, perithecium, apothecium, etc.

Fumigant. A volatile disinfectant, destroying organisms by vapor.

Fungicide. Chemical or physical agent that kills or inhibits fungi.

Fungi Imperfecti. Fungi that have not been connected with the perfect or sexual stage; most are imperfect states of Ascomycetes.

Fungistatic. An agent preventing development of fungi without killing them.

Fungus, pl. *Fungi.* An organism with no chlorophyll, reproducing by sexual or asexual spores, usually with mycelium with well-marked nuclei.

Fusiform. Spindle-like, narrowing toward the ends.

Fusoid. Somewhat fusiform.

Gall. Outgrowth or swelling, often more or less spherical, of unorganized plant cells as result of attack by bacteria, fungi, or other organisms.

Gametangium. Gamete mother cell.

Gamete. A sex cell, especially one formed in a gametangium.

Germ Tube. Hypha produced by a germinated fungus spore.

Gill. Lamella or hymenium-covered plate on underside of cap of a mushroom.

Girdle. A canker that surrounds stem, completely cutting off water supply and thus causing death; girdling roots also cause death.

Glabrous. Smooth.

Gleba. Sporulating tissue in an angiocarpous fruit body.

Globose. Almost spherical.

Gram-negative, gram-positive. Not being stained, and being stained, by the gram stain used in classifying bacteria.

Haustorium, pl. *Haustoria.* Special hyphal branch extended into living cell for purpose of absorbing food (Fig. 50).

Heteroecious. Undergoing different parasitic stages on two unlike hosts, as in the rusts.

Heterothallic. Of a fungus, sexes separate in different mycelia.

Holocarpic. Having all the thallus used for a fruiting body.

Homothallic. Both sexes present in same mycelium.

Host. Any plant attacked by a parasite.

Hyaline. Colorless, or nearly transparent.

Hymenium. Spore-bearing layer of a fungus fruiting body.

Hyperplastic. Term applied to a disease producing an abnormally large number of cells.

Hypha, pl. *Hyphae.* Single thread of a fungus mycelium.

Hypoplastic. Term applied to a disease with subnormal cell production.

Hyphopodium, pl. *Hyphopodia.* More or less lobed appendage to a hypha.

Hysterothecium. Oblong or linear perithecium, sometimes considered an apothecium, opening by a cleft.

Immune. Exempt from disease; having qualities that do not permit infection.

Immunization. Process of increasing the resistance of a living organism.

Imperfect Fungus. One lacking any sexual reproductive state.

Imperfect state. State of life cycle in which asexual spores, or none, are produced.

Incubation period. Time between inoculation and development of symptoms that can be seen.

Indehiscent. Of fruit bodies, not opening, or with no special method.

Infection. Process of beginning or producing disease.

Infection court. Place where an infection may take place, as leaf, fruit, petal, etc.

Injury. Result of transient operation of an adverse factor, as an insect bite, or action of a chemical.

Innate. Bedded in, immersed.

Inoculation. Placing of inoculum in infection court.

Inoculum. Pathogen or its part, as spores, fragments of mycelium, and so on, that can infect plants.

Inoperculate. Not opening by a lid.

Intercellular. Between cells.

Intracellular. Within cells.

Intumescence. Knoblike or pustulelike outgrowth of elongated cells on leaves, stems, etc., caused by environmental disturbances.

Lamella. Gill.

Lesion. Localized spot of diseased tissue.

Locule. A cavity, especially one in a stroma.

Macroconidia. Large conidia.

Macroscopic. Large enough to be seen with the naked eye.

Medulla. Loose layer of hyphae inside a thallus; body of a sclerotium.

Microconidia. Very small spores, now considered spermatia of a fungus also having larger conidia.

Micron. 1/1000 millimeter, unit used for measuring spores.

Microscopic. Too small to be seen except with the aid of a microscope; true of most of the fungus structures shown in line drawings in this book.

Mildew. Plant disease in which the pathogen is a growth on the surface.

Molds. Fungi with conspicuous mycelium or spore masses, often saprophytes.

Monoecious. Male and female reproductive organs in same individual; in rusts, all stages of life cycle on single species of plant.

Multinucleate. Several nuclei in same cell.

Mummy. Dried, shriveled fruit, result of disease.

Muriform. Having cross and longitudinal septa.

Mushroom. An agaric fruit body (Fig. 7).

Mycelium, pl. *Mycelia.* Mass of fungus hyphae.

Mycelia Sterilia. Fungi Imperfecti where spores, except for chlamydospores, are not present.

Mycoplasmalike organism (MLO). A wall-less prokaryotic plant pathogen that has a single-unit membrane.

Mycorrhiza, pl. *Mycorrhizae.* Symbiotic, nonpathogenic association of fungi and roots.

Necrosis. Death of plant cells, usually resulting in tissue turning dark.

Necrotic. As an adjective, killing.

Nematicide. Chemical or physical agent killing nematodes.

Nematodes. Nemas, roundworms, eelworms, cause of some plant diseases.

Obligate parasite. A parasite that can develop only in living tissues, with no saprophytic stage.

Obovate. Inversely ovate, narrowest at base.

Obtuse. Rounded or blunted, greater than a right angle.

Oogonium, pl. *Oogonia.* Female sex organ in the Oomycetes (Fig. 4).

Oomycetes. Subclass of the Phycomycetes, gametangia of unequal size.

Oospore. Resting spore formed in a fertilized oogonium.

Operculate. With a cover or lid, as in some asci.

Ostiole. Porelike mouth or openings in papilla or neck of a perithecium or pycnidium.

Papilla, pl. *Papillae.* Small, nipplelike projection.

Paraphysis, pl. *Paraphyses.* A sterile hyphal element in the hymenium, especially in the Ascomycetes, usually clavate or filiform.

Paraphysoids. Threads of hyphal tissue between asci, like delicate paraphyses but without free ends.

Parasite. An organism that lives on or in a second organism, usually causing disease in the latter.

Pathogen. Any organism or factor causing disease.

Pathogenic. Capable of causing disease.

Pedicel. Small stalk.

Perfect state. Stage of life-cycle in which spores are formed after nuclear fission.

Peridium. Wall or limiting membrane of a sporangium or other fruit body, or of a rust sorus.

Perithecium. Subglobose or flasklike ascocarp of the Pyrenomycetes (Fig. 6).

Phialide. A cell that develops one or more open ends from which a basipetal succession of conidia develops without an increase in length of the phialide itself.

Physiogenic disease. Caused by unfavorable environmental factors.

Physiologic races. Pathogens of same variety and species structurally the same but differing in physiological behavior, especially in ability to parasitize a given host.

Phytopathology. Plant pathology, science of plant disease.

Pileus. Hymenium-supporting part of a fruit body of a higher fungus; the cap of a mushroom.

Primary infection. First infection by a pathogen after going through a resting or dormant period.

Prokaryotic. Organisms that lack a true nucleus; includes bacteria and mycoplasmalike organisms.

Promycelium. Basidium of rusts and smuts.

Pulvinate. Cushionlike in form.

Pycnidium, pl. *Pycnidia.* Flasklike fruiting body containing conidia.

Pycnium. Spermagonium in the rusts, the 0 stage, resembling a pycnidium (Fig. 64).

Resistance. Ability of a host plant to suppress or retard activity of a pathogen.

Resting spore. A spore, often thick-walled, that can remain alive in a dormant condition for some time, later germinating and capable of initiating infection.

Resupinate. Flat on the substratum with hymenium on outer side.

Rhizoid. Rootlike structure (Fig. 5).

Rhizomorph. A cordlike strand of fungus hyphae.

Ring spot. Disease symptoms characterized by yellowish or necrotic rings with green tissue inside the ring, as in virus diseases.

Roguing. Removal of undesired individual plants.

Rosette. Disease symptom with stems shortened to produce a bunchy growth habit.

Russet. Brownish roughened areas on skins of fruit, from abnormal production of cork caused by disease, insect, or spray injury.

Rust. A fungus, one of the Uredinales, causing a disease also known as rust.

Saprophyte. An organism that feeds on lifeless organic matter.

Scab. Crustlike disease lesion; or a disease in which scabs are prominent symptoms (Fig. 67).

Sclerotium, pl. *Sclerotia.* Resting mass of fungus tissue, often more or less spherical, normally having no spores in or on it (Figs. 60, 62).

Scorch. Burning of tissue, from infection or weather conditions.

Scutellum. Plate or shieldlike cover, as in Microthyriales.

Septate. Having cross-walls, septa.

Sessile. Having no stem.

Seta, pl. *Setae.* A stiff hair, or bristle, generally dark-colored.

Shothole. A disease symptom in which small round fragments drop out of leaves, making them look as if riddled by shot.

Sign. Any indication of disease other than reaction of the host plant — spores, mycelium, exudate, or fruiting bodies of the pathogen.

Slurry. Thick suspension of chemical; used for seed treatment.

Smut. A fungus of the Ustilaginales, characterized by sooty spore masses; the name also used for the disease caused by the smut.

Sooty mold. Dark fungus growing in insect honeydew.

Sorus, pl. *Sori.* Fungus spore mass, especially of rusts and smuts; occasionally, a group of fruiting bodies.

Species. One sort of plant or animal; abbreviated as "sp." singular, and "spp." plural. A genus name followed by sp. means that the particular species is undetermined. Spp. following a genus name means that several species are grouped together without being named individually.

Spermagonium. Walled structure in which spermatia are produced, a pycnium.

Spermatium, pl. *Spermatia.* A sex cell (+ or −), a pycniospore.

Sporangiole. Small sporangium without a columella and with a small number of spores.

Sporangiophore. Hypha bearing a sporangium.

Sporangium. Organ producing nonsexual spores in a more or less spherical wall (Fig. 4).

Spore. A single- to many-celled reproductive body, in the fungi and lower plants, which can develop a new plant.

Sporidium, pl. *Sporidia.* Basidiospore of rusts and smuts.

Sporodochium, pl. *Sporodochia.* Cluster of conidiophores interwoven on a stroma or mass of hyphae (Fig. 8).

Sporophore. Spore-producing or supporting structure — a fruit body; used especially in the Basidiomycetes (Fig. 58).

Sporulate. To produce spores.

Sprayer. Apparatus for applying chemicals in liquid form.

Sterigma, pl. *Sterigmata.* Projection for supporting a spore.

Stipe. A stalk; *stipitate,* stalked.

Strain. An organism or group of organisms differing in origin or minor aspects from other organisms of same species or variety.

Stroma, pl. *Stromata.* Mass of fungus hyphae often including host tissue containing or bearing spores.

Subiculum, subicle. Netlike woolly or crustlike growth of mycelium under fruit bodies.

Substrate. The substance or object on which a saprophytic organism lives and from which it gets nourishment.

Suscept. A living organism attacked by, or susceptible to, a given disease or pathogen; in many cases a more precise term than host but less familiar.

Susceptible. Unresistant, permitting the attack of a pathogen.

Swarmspore. Zoospore.

Synnema, pl. *Synnemata.* Groups of hyphae sometimes joined together, generally upright and producing spores; coremium.

Systemic. Term applied to disease in which single infection leads to general spread of the pathogen throughout the plant body; or to a chemical that acts through the vascular system.

Teliospore. Winter or resting form of rust spore, from which basidium is produced (Figs. 63, 64, and 65).

Telium. Sorus producing teliospores.

Thallophyte. One of the simpler plants, belonging to the algae, bacteria, fungi, slime molds, or lichens.

Thallus. Vegetative body of a thallophyte.

Tolerant. Capable of sustaining disease without serious injury or crop loss.

Toxin. Poison formed by an organism.

Tylosis, pl. *Tyloses.* Cell outgrowth into cavity of xylem vessel, plugging it.

Urediospore. Summer spore of rusts; one-celled, verrucose (Fig. 63).

Uredium. Sorus producing urediospores.

Valsoid. Having groups of perithecia with beaks pointing inward, or even parallel with surface, as in valsa.

Vector. An agent, insect, human, and so on, transmitting disease.

Vein-banding. Symptom of virus disease in which regions along veins are darker green than the tissue between veins.

Verrucose. With small rounded processes or warts.

Viroid. A small viruslike infectious agent having no protein coat and only a small amount of nucleic acid.

Virulent. Highly pathogenic; with strong capacity for causing disease.

Viruliferous. Virus-carrying; term applied particularly to virus-laden insects.

Virus. An obligate parasite capable of multiplying in certain hosts, ultramicroscopic, recognizable by the effects produced in infected hosts. Has nucleic acid with protein coat.

Wilt. Loss of freshness or drooping of plants due to inadequate water supply or excessive transpiration; a vascular disease interfering with utilization of water.

Witches' broom. Disease symptom with abnormal brushlike development of many weak shoots.

Yellows. Term applied to disease in which yellowing or chlorosis is a principal symptom.

Zoospore. A swimming spore, swarmspore, capable of independent movement (Fig. 4).

Zygomycetes. Subclass of the Phycomycetes, characterized by gametes of equal size.

Zygospore. Resting spore formed from the union of similar gametes (Fig. 5).

Selected Bibliography

In the preparation of *Westcott's Plant Disease Handbook* references have been reviewed that cover nearly sixty years of scientific reporting. An attempt has been made also to keep abreast of current literature. To cite all of the individual articles that have been helpful would fill another book. The bibliography presented here is a selected small sampling of the field surveyed, with emphasis on sources consulted in making nomenclatural decisions.

Periodicals that are regularly reviewed include *Plant Disease, Phytopathology, Review of Applied Mycology, Journal of Economic Entomology, A.I.B.S. Bulletin* (Agricultural Institute of Biological Sciences), *Biological Abstracts, Agricultural Chemicals, NAC News* (National Agricultural Chemicals Association), *Arborist's News, Proceedings of the National Shade Tree Conference, American Fruit Grower, American Vegetable Grower, Farm Journal, The Garden Journal* (New York Botanical Garden), *Plants and Gardens* (Brooklyn Botanic Garden), *The National Gardener* (National Council of State Garden Clubs), publications of many of the state garden clubs, most of the popular garden magazines, and yearbooks and magazines of several single plant societies. In addition, there are numerous bulletins, circulars, and spray schedules from state experiment stations.

The following references provide coverage in depth of the current taxonomy of the major types of plant pathogens.

Bacteria: Volume 1 (1984) and Volume (1986) of *Bergey's Manual of Systematic Bacteriology*, published by Williams and Wilkins, and *Laboratory*

Guide for Identification of Plant Pathogenic Bacterial (1980) by N. W. Schaad, published by APS Press, The American Phytopathological Society.

Viruses: *Classification and Nomenclature of Viruses* (1979) by R. E. F. Matthews, published by Academic Press, and *Descriptions of Plant Viruses*, published by the Commonwealth Mycological Institute and Association of Applied Biologists.

Fungi: *Plant Pathogenic Fungi* (1987) by J. A. von Arx, published by J. Cramer in Berlin, *Ainsworth and Bisby's Dictionary of the Fungi* (1983) by D. L. Hawksworth, B. C. Sutton, and G. C. Ainsworth, 7th edition, published by Commonwealth Mycological Institute, and *Illustrated Genera of Imperfect Fungi* (1972) by H. L. Barnett and H. B. Hunter, 3rd edition, published by Burgess Publishing Company.

Nematodes: *Pictorial Key to Genera of Plant-Parasitic Nematodes* (1975) by W. F. Mai and H. H. Lyon, 4th edition, published by Cornell University Press.

Finally, the APS Press, the publishing group of the American Phytopathological Society, produces a *Compendium of Plant Disease* series that provides information about causes, cycles, and control of plant diseases. There are presently 20 books in this series.

Ainsworth, G. C., and G. R. Bisby. *A Dictionary of the Fungi*, 5th ed. Commonwealth Mycological Institute, Kew, Surrey, England, 1961.

Allen, M. W. A review of the genus *Tylenchorhynchus*. *Univ. Calif. Public. Zoology* **61**(3):129–166, 1955.

Allen, M. W. Taxonomic status of the bud and leaf nematodes related to *Aphelenchoides fragariae* (Ritzema Bos 1891). *Proc. Helm. Soc. Wash.* **19**(2):108–120, 1952.

Allen, M. W., and H. J. Jensen. *Pratylenchus vulnus* (Nematoda, Pratylenchinae) a parasite of trees and vines in California. *Helm. Soc. Sci. Jour.* **18**:47–50, 1951.

Allen, M. W., and H. J. Jensen. A review of the nematode genus *Trichodorus* with descriptions of ten new species. *Nematologica* **2**(1):32–62, 1957.

American Society for Horticultural Science. *The Care and Feeding of Garden Plants*. National Fertilizer Association, Washington, DC, 1954.

Anderson, Harry Warren. *Diseases of Fruit Crops*. McGraw-Hill Book Company, Inc., New York, 1956.

Anonymous. Distribution, symptoms and control of some of the more important plant diseases. *Plant Disease Reporter Suppl.* **221**:106–181, 1953.

Anonymous. Plant pest handbook. *Conn. Agr. Exp. Sta. Bull.* **600,** 1956.

Anonymous. Watch out for witchweed. *U.S. Dept. Agr. Pamphlet* **331,** 1957.

Anonymous. *1968 Pesticides for New Jersey*. N.J. College of Agr. Ext. Serv., 1967.

Anonymous. *1968 Recommended Fungicides and Nematicides*. Ext. Serv., College of Agr., Pennsylvania State University, University Park, PA.

Arthur, J. C. *Manual of the Rusts in United States and Canada*. Purdue Research Foundation, Lafayette, IN, 1934. With supplement by G. B. Cummins. The Hafner Publishing Co., Inc., New York, 1962.

Bailey, L. H., and Ethel Zoe Bailey, compilers. *Hortus Second*. The Macmillan Company, New York, 1941.

Baker, Kenneth F., ed. The U. C. system for producing healthy container-grown plants. *Cal. Agr. Exp. Sta. Ext. Man.* **23,** 1957.

Barnett, H. L., and Barry B. Hunter. *Illustrated Genera of Imperfect Fungi*, 3rd ed. Burgess Publishing Company, Minneapolis, MN, 1972.

Bawden, F. C. *Plant Viruses and Virus Diseases*, 4th ed. The Ronald Press Company, New York, 1964.

Beemster, A. B. R., and Jeanne Dijkstra, eds. *Viruses of Plants*. John Wiley & Sons, Inc., New York, 1966.

Bessey, Ernst Athearn. *Morphology and Taxonomy of Fungi*. The Blakiston Company, Philadelphia, PA, 1950, and Hafner Publishing Co., 1961.

Birchfield, W. The burrowing nema situation in Florida. *Jour. Economic Entomology* **50**:562–566, 1957.

Bisby, G. R. *An Introduction to the Taxonomy and Nomenclature of Fungi*. Commonwealth Mycological Institute, Kew, Surrey, England, 1945.

Boyce, John Shaw. *Forest Pathology*, 3rd ed. McGraw-Hill Book Company, Inc., New York. 1961.

Brandes, Gordon A., Tulio M. Cordero, and Robert L. Skiles, eds. *Compendium of Plant Diseases*. Rohm & Haas Company, Philadelphia, PA, 1959.

Bray, D. F. Gas injury to shade trees. *Scientific Tree Topics* **2**(5):19–22. Bartlett Tree Research Laboratories, Stamford, CT, 1958.

Brierley, Philip. Viruses described primarily on ornamental or miscellaneous plants. *Plant Disease Reporter Suppl.* **150**:410–482, 1944.

Buchanan, R. E., and N. E. Gibbons. *Bergey's Manual of Determinative Bacteriology*, 8th ed. The Williams and Wilkins Company, Baltimore, MD, 1974.

Buhrer, Edna M. Common names of some important plant pathogenic nematodes. *Plant Disease Reporter* **38**:535–541, 1954.

Burkholder, Walter H. In *Bergey's Manual of Determinative Bacteriology*, 7th ed. The Williams and Wilkins Company, Baltimore, MD, 1957, pp. 89–183, 288–292, 349–359, 579–597.

Burnett, Harry C. *Orchid Diseases*. State of Florida Dept. of Agriculture, Div. of Plant Industry, Vol. I, No. 3, Gainesville, FL, 1965.

Cairns, Eldon J., et al. Symposium on new developments and new problems concerning nematodes in the South. *Plant Disease Reporter Suppl.* **227**:75–107, 1954.

Campana, Richard, chairman. *Guide for Community-wide Control of Dutch Elm Disease*. Midwestern Chapter, National Shade Tree Conference, 1957.

Caroselli, Nester E. Verticillium wilt of maples. *RI Agr. Exp. Sta. Bul.* **335**:3–84, 1957.

Carter, J. Cedric. Illinois trees; their diseases. *IL Nat. Hist. Survey Circ.* **46,** 1964.

Carter, J. Cedric. Dutch elm disease in Illinois. *IL Nat. Hist. Survey Circ.* **53,** 1967.

Carter, J. Cedric. The wetwood disease of elm. *IL Nat. Hist. Survey Circ.* **50,** 1964.

Carter, W. *Insects in Relation to Disease*. John Wiley & Sons, New York, 1963.

Chester, K. Starr. *Nature and Prevention of Plant Diseases*, 2nd ed. The Blakiston Company, Philadelphia, PA, 1947.

Carter, J. Cedric. *Nature and Prevention of Cereal Rusts as Exemplified in the Leaf Rust of Wheat*. Chronica Botanica Company, Waltham, MA, 1946.

Chitwood, B. G. Golden Nematode of Potatoes. *U.S. Dept. Agr. Circ.* **885,** 1951.

Chitwood, B. G. Root-knot nematodes. Part I. A revision of the genus *Meloidogyne* Goeldi 1887. *Proc. Helm. Soc. Wash.* **16**:90–104, 1949.

Chitwood, B. G., and W. Birchfield. Nematodes, their kinds and characteristics. *State Plant Board of FL. Vol. II. Bull.* **9**:5–49, 1956.

Chitwood, B. G., C. I. Hannon, and R. P. Esser. A new nematode genus, Meloidodera, linking the genera *Heterodera* and *Meloidogyne*. *Phytopathology* **46**:264–266, 1956.

Christensen, C. M. *Common Fleshy Fungi*. Burgess Publishing Company, Minneapolis, MN, rev. 1965.

Christie, J. R. *Plant Nematodes, Their Bionomics and Control*. University of Florida, Gainesville, FL., 1959.

Christie, J. R., and V. G. Perry. A root disease of plants caused by a nematode of the genus *Trichodorus*. *Science* **113**:491–493, 1951.

Christie, J. R., and A. L. Taylor. Controlling nematodes in the home garden. *U.S. Dept. Agr. Farmers Bull.* **2048**, 1952.

Chupp, Charles. *Manual of Vegetable-garden Diseases*. The Macmillan Company, New York, 1925.

Chupp, Charles. *A Monograph of the Fungus Genus Cercospora*. Published by author, Ithaca, NY, 1953.

Chupp, Charles, and Arden F. Sherf. *Vegetable Diseases and Their Control*. Ronald Press, New York, 1960.

Clements, Frederic E., and Cornelius L. Shear. *The Genera of Fungi*. H. W. Wilson Company, New York, 1931. Reprint Hafner Publishing Company, New York, 1965.

Converse, Richard H. Diseases of raspberries and erect and trailing blackberries. *U.S. Dept. Agr. Agriculture Handbook* **310**, 1966.

Cooke, W. B. Nomenclatorial notes on Erysiphaceae. *Mycologia* **44**:57, 1952.

Corbett, M. K., and H. D. Sisler, eds. *Plant Virology*. University of Florida Press, Gainesville, FL, 1964.

Couch, H. B. *Diseases of Turfgrasses*. Reinhold Publishing Corporation, New York, 1962.

Couch, J. N. *The Genus Septobasidium*. University of North Carolina Press, Chapel Hill, NC, 1938.

Cox, R. E. Some common diseases of flowering dogwood in Delaware. *Del. Agr. Expt. Sta. Circ.* **27**, 1954.

Cummins, George B. *Illustrated Genera of Rust Fungi*. Burgess Publishing Company, Minneapolis, MN, 1959.

Cummins, George B., and John A. Stevenson. A check list of North American rust fungi (Uredinales). *Plant Disease Reporter Suppl.* **240**:109–193, 1956.

Darrow, G. M., J. R. McGrew, and D. H. Scott. Reducing virus and nematode damage to strawberry plants. *U.S. Dept. Agr. Leaflet* **414**, 1957.

Davis, Robert E., Robert F. Whitcomb, and Russell L. Steere. Remission of aster yellows disease by antibiotics. *Science* **161**(3843): 793–795, 1968.

Davis, Spencer H., and C. C. Hamilton. Diseases and insect pests of rhododendron and azalea. *NJ Agr. Exp. Sta. Circ.* **571**, 1955.

Dickson, James G. *Diseases of Field Crops*, 2nd ed. McGraw-Hill Book Company, Inc., New York, 1956.

Dickson, R. C. A working list of the names of aphid vectors. *Plant Disease Reporter* **39**:445–452, 1955.

Dimock, A. W. *The Gardener's ABC of Pest and Disease*. M. Barrows and Company, Inc., New York, 1953.

Dimock, A. W., et al. 1975 *Cornell Recommendations for Commercial Floriculture Crops*. N.Y. State College of Agriculture, Ithaca, NY, 1975.

Dimond, A. E., and E. M. Stoddard. Toxicity to greenhouse roses from paints containing mercury fungicides. *Conn. Agr. Exp. Sta. Bull.* **595**, 1955.

Dittmer, Dorothy S., ed. *Handbook of Toxicology*, Vol. V, *Fungicides*. W. B. Saunders Co., Philadelphia, PA, 1959.

Dochinger, Leon S. Verticillium wilt, its nature and control. *Proc. National Shade Tree Conference* **33**:202–212, 1957.

Doolittle, S. P., A. L. Taylor, and L. L. Danielson. Tomato disease and their control. *U.S. Dept. Agr. Agriculture Handbook* **203**, 1961.

Dowson, W. J. *Plant Diseases Due to Bacteria*, 2nd ed. Cambridge University Press, New York, 1957.

Duddington, C. L. *The Friendly Fungi. A New Approach to the Eelworm Problem*. Faber and Faber, London, 1957.

Elliott, Charlotte. *Manual of Bacterial Plant Pathogens*, rev. ed. The Williams and Wilkins Company, Baltimore, MD, 1952.

Emsweller, S. L., Philip Brierley, and Floyd F. Smith. Roses for the home. *U.S. Dept. Agr. Home and Garden Bull.* **25,** 1963.

Engelhard, Arthur W. Host index of *Verticillium albo-atrum.* Reinke & Berthe. *Plant Disease Reporter Suppl.* **244:**23–49, 1957.

Esau, Katherine. *Plants, Viruses, and Insects.* Harvard University Press, Cambridge, MA, 1961.

Fawcett, Howard S. *Citrus Diseases and Their Control,* 2nd ed. McGraw-Hill Book Company, Inc., New York, 1936.

Felt, Ephraim Porter, and R. Howard Rankin. *Insects and Diseases of Ornamental Trees and Shrubs.* The Macmillan Company, New York, 1932.

Fenska, Richard E. *Tree Experts Manual,* 2nd ed. A. T. De La Mare Company, Inc., New York, 1954.

Filipev, I. N., and J. H. Schuurmans Stekoven. *A Manual of Agricultural Helminthology.* E. J. Brill Co., Leiden, Holland, 1941.

Fischer, G. W. *Manual of the North American Smut Fungi.* The Ronald Press Company, New York, 1953.

Fischer, G. W., and C. S. Holton. *Biology and Control of the Smut Fungi.* The Ronald Press Company, New York, 1957.

Fitzpatrick, Harry Morton. *The Lower Fungi: Phycomycetes.* McGraw-Hill Book Company, Inc., New York, 1930.

Forsberg, Junius L. *Diseases of Ornamental Plants.* Univ. of Illinois, College of Agriculture Special Publ. No. 3, Urbana, IL, 1975.

Fowler, Marvin E., and G. F. Gravatt. Reducing damage to trees from construction work. *U.S. Dept. Agr. Farmers Bull.* **1957,** 1945.

Frear, Donald E. H., ed. *Pesticide Handbook Entoma,* 20th ed. College Science Publishers, State College, PA, 1968. (Revised each year, this edition lists 9,486 pesticides by trade names and formulae.)

Frear, Donald E. H. *Chemistry of the Pesticides,* 3rd ed. D. Van Nostrand Company, Inc., Princeton, NJ, 1955.

Gambrell, F. L., and R. M. Gilmer. Insects and diseases of fruit nursery stocks and their control. *NY (Geneva) Agr. Exp. Sta. Bull.* **776,** 1956.

Garrett, S. D. *Root Disease Fungi.* Chronica Botanica Company, Waltham, MA, 1944.

Gilman, Joseph C. *A Manual of Soil Fungi,* 2nd ed. Iowa State College Press, Ames, IA, 1957.

Golden, A. M. Taxonomy of the spiral nematodes (*Rotylenchus* and *Stelicotylenchus*) and the developmental stages and host parasite relationships of *R. buxophilus* n. sp. attacking boxwood. *Univ. MD Bull.* **A-85,** 1956.

Golden, A. M., and A. L. Taylor. *Rotylenchus christiei,* n. sp., a new spiral nematode associated with roots of turf. *Proc. Helm. Soc. Wash.* **23:**109–112, 1956.

Goodey, T. *Plant Parasitic Nematodes and the Diseases They Cause.* E. P. Dutton and Company, New York, 1933.

Goodey, T. *Soil and Fresh Water Nematodes,* 2nd ed. Methuen & Co., Ltd., London, England, 1963.

Goodman, Robert N., Zoltan Kiraly, and Milton Zaitlin. *The Biochemistry and Physiology of Infectious Plant Disease.* D. Van Nostrand Company, Inc., Princeton, NJ, 1967.

Gould, Charles J., ed. *Handbook on Bulb Growing and Forcing.* Northwest Bulb Growers Association, Mt. Vernon, WA, 1957.

Gould, Charles J., M. R. Harris, and George W. Eade. Disease of ornamentals. *State Coll. Wash. Agr. Ext. Mimeo* **683.**

Gram, Ernest, and Anna Weber. *Plant Diseases in Orchard, Nursery and Garden Crops.* R. W. G. Dennis (ed.). Tr. from the Danish by Evelyn Ramsden. Philosophical Library, New York, 1953.

Granados, Robert R., Karl Maramorosch, and Eishiro Shikata. Mycoplasma: Suspected etiologic agent of corn stunt. *Nat. Acad. Sci. Proc.* **60**(3):841–844, 1968.

Guba, Emil. *Monograph of Monochaetia and Pestalotia*. Harvard University Press, Cambridge, MA, 1961.

Hawksworth, Frank G., and Delbert Wiens. Biology and classification of dwarf mistletoes (*Arceuthobium*). *U.S. Dept. Agr. Agriculture Handbook* **401**, 1972.

Hayley, Denis, ed. Official FDA tolerances. *National Agricultural Chemicals Association News* **25**(3):3–23, 1967.

Heald, Frederick DeForest. *Manual of Plant Diseases*, 2nd ed. McGraw-Hill Book Company, Inc., New York, 1933.

Hedden, O. K., J. D. Wilson, and J. P. Sleesman. Equipment for applying soil pesticides. *U.S. Dept. Agr. Agriculture Handbook* **297**, 1966.

Hepting, George H. *Diseases of Forest and Shade Trees of the United States*. Distributed and processed as *Diagnosis of Disease in American Forest and Shade Trees*. U.S. Dept. Agric. Forest Service, in 5 volumes, dated 1964, 1966, 1968.

Hildebrand, Earl M., and Harold T. Cook. Sweetpotato diseases. *U.S. Dept. Agr. Farmers Bull.* **1059**, rev. 1959.

Holmes, Francis O. Virales. In *Bergey's Manual of Determinative Bacteriology*, 6th ed. The Williams and Wilkins Company, Baltimore, MD, 1948, pp. 1128–1224.

Holmes, Francis W., and Clifford S. Chater. Culture, diseases, injuries, and pests of maples in shade and ornamental plantings. *Univ. Massachusetts Coop. Ext. Serv. Publ.* **443**, 1966.

Holton, C. S., G. W. Fischer, R. W. Fulton, Helen Hart, and S. E. A. McCallan, eds. *Plant Pathology, Problems and Progress*. University of Wisconsin Press, Madison, WI, 1959.

Hopkins, J. C. F., ed. Common names of viruses used in the review of applied mycology. *Rev. Appl. Myc. Suppl.* **35**, 1957.

Hopper, Bruce E., and Eldon J. Cairns. *Taxonomic Keys to Plant, Soil and Aquatic Nematodes*. Alabama Polytechnic Institute, Auburn, AL, 1959.

Horsfall, James G. *Principles of Fungicidal Action*. Chronica Botanica Company, Waltham, MA, 1956.

Horsfall, J. G., and A. E. Dimond, eds. *Plant Pathology, An Advanced Treatise*, 3 volumes. Academic Press, New York, 1959, 1960.

Horsfall, James G., and Raymond J. Lukens. Selectivity of fungicides. *Connecticut Agr. Exp. Sta. Bull.* **676**, 1966.

Horst, R. K. *Viroid*. McGraw-Hill Yearbook Science and Technology. McGraw-Hill Book Company, Inc., New York, 1977.

Horst, R. K., and A. W. Dimock. Diseases of bearded irises. *N.Y. State Agr. Exp. Sta. Bull.* **128**, 1977.

Horst, R. K., and P. E. Nelson. Diseases of chrysanthemums. *N.Y. State Agr. Exp. Sta. Bull.* **85**, 1975.

Hough, Walter S., and A. Freeman Mason. *Spraying, Dusting and Fumigating of Plants*, rev. ed. The Macmillan Company, New York, 1951.

Howard, F. L., J. B. Rowell, and H. L. Keil. Fungus diseases of turf grasses. *RI Agr. Exp. Sta. Bull.* **308**, 1951.

Hubert, F. P. Common names of diseases of woody plants. *Plant Disease Reporter Suppl.* **206**:206–235, 1951.

Hutchinson, Martin T. Crop rustling nematodes cost $15 million yearly. *NJ Agr.* **41**(1):9–12, 1959.

Hutchinson, M. T., et al. Plant parasitic nematodes of New Jersey. *NJ Agr. Exp. Sta. Bull.* **796**, 1961.

Irons, Frank. Hand sprayers and dusters. *U.S. Dept. Agr. Home and Garden Bull.* **63**, 1967.

Jenkins, Anna E. A specific term for diseases caused by *Elsinoë* and *Sphaceloma*. *Plant Disease Reporter* **31**:71, 1947.

Jenkins, W. R., and D. P. Taylor. *Plant Nematology*. Reinhold Publishing Corporation, New York, 1967.

Jenkins, W. R., D. P. Taylor, R. A. Rhode, and B. W. Coursen. Nematodes associated with crop plants in Maryland. *Univ. MD Bull.* **A-89**, 1957.

Johnson, Howard W., Donald W. Chamberlain, and S. G. Lehman. Diseases of soybeans and methods of control. *U.S. Dept. Agr. Circ.* **931**, 1954.

Johnson, W. T., and H. J. Kastl, eds. A guide to safe pest control around the home. *Cornell Misc. Bull.* **74**, 1971.

Klotz, Leo J. *Color Handbook of Citrus Diseases*, 3rd ed. Univ. of California, Div. of Agricultural Sciences, Berkeley, CA, 1961.

Kreitlow, K. W., and F. V. Juska. Lawn diseases. *U.S. Dept. Agr. Home and Garden Bull.* **61**, rev. 1967.

Large, E. C. *The Advance of the Fungi*. Henry Holt and Company, New York, 1940 (Paperback, Dover Publishing, Inc., New York, 1962).

Lawrence, Fred P., and James E. Brogden. Control of insects and diseases of dooryard citrus trees. *Univ. FL Agr. Ext. Serv. Circ.* **139**, 1955.

Leach, J. G. *Insect Transmission of Plant Diseases*. McGraw-Hill Book Company, New York, 1940.

Lentz, Paul L. *Stereum* and allied genera of fungi in the Upper Mississippi Valley. *U.S. Dept. Agr. Mono.* **24**, 1955.

Leukel, R. W., and V. G. Tapke. Cereal smuts and their control. *U.S. Dept. Agr. Farmers Bull.* **2069**, 1954.

Lowe, J. L. Polyporaceae of North America; the genus *Fomes*. *NY State College Forestry Tech. Publ.* **80**, 1957.

McFadden, Lorne A. Palm diseases. *American Horticultural Mag.* **40**:148–150, 1961.

McGrew, J. R. Strawberry diseases. *U.S. Dept. Agr. Farmers Bull.* **2140**, rev., 1966.

Mai, W. F., and M. B. Harrison. The golden nematode, *Cornell Ext. Bull.* **870**, 1959.

Maramorosch, Karl, reporter. A new system of classifying viruses proposed by the International Provisional Virus Nomenclature Committee. *Phytopathology* **55**:1158, 1965.

Maramorosch, K., E. Shikata, and R. R. Granapos. Mycoplasma-like bodies in leafhoppers and diseased plants. *Phytopathology* **58**:886, 1968.

Marshall, Rush P. Care of damaged shade trees. *U.S. Dept. Agr. Farmers Bull.* **1896**. 1942.

Marshall, Rush P., and Alma M. Waterman. Common diseases of important shade trees. *U.S. Dept. Agr. Farmers Bull.* **1987**, 1948.

Martin, G. W. *Outline of the Fungi*. W. C. Brown Co., Dubuque, IA, 1950.

Martin, G. W. Are fungi plants? *Mycologia* **47**:779–792, 1955.

Martin, G. W. Key to the families of fungi. In *A Dictionary of the Fungi*, 5th ed., pp. 497–519, 1961.

May, Curtis. Diseases of shade and ornamental maples. *U.S. Dept. Agr. Handbook* **211**, 1961.

May, Curtis. Rusts of juniper, flowering crabapple and hawthorn. *American Horticultural Mag.* **44**:29–32, 1965.

Meister, R. T., ed. *Farm Chemicals; 1967 Handbook*. Meister Publishing Company, Willoughby, OH, 1967 (revised each year).

Melhus, Irving E., and George C. Kent. *Elements of Plant Pathology*. The Macmillan Company, New York, 1939.

Miller, Howard N. Investigations with antibiotics for control of bacterial diseases of foliage plants. *FL Agr. Exp. Sta. Jour. Ser.* **420**, 1955.

Mills, W. D., and A. A. Laplante. Diseases and insects in the orchard. *Cornell Ext. Bull.* **711**, rev. 1954.

Moore, Agnes Ellis. *Bibliography of Forest Disease Research in the Department of Agriculture Misc. Publ.* **725**, 1957.

Moore, W. C. Diseases of bulbs. *Ministry Agr. Fisheries (Great Britain) Bull.* **117**, 1939.

National Research Council, Subcommittee on Plant Pathogens, Agricultural Board. *Plant Disease Development and Control.* National Academy of Sciences, Washington, DC, 1969.

Nelson, Ray. Diseases of gladiolus. *MI Agr. Exp. Sta. Spec. Bull.* **350**, 1948.

Nienstaedt, Hans, and Arthur H. Graves. Blight resistant chestnuts. *CT Agr. Exp. Sta. Circ.* **192**, 1955.

Noble, Mary, J. de Tempe, and Paul Neergaard. *An Annotated List of Seed-borne Diseases.* Commonwealth Mycological Institute, Kew, Surrey, England, 1958.

Osborn, M. R., A. M. Philips, and William C. Pierce. Insects and diseases of the pecan and their control. *U.S. Dept. Agr. Farmers Bull.* **1829**, 1954.

Parker, K. G., E. G. Fisher, and W. D. Mills. Fire blight on pome fruits and its control. *Cornell Ext. Bull.* **966**, 1956.

Parris, G. K. Diseases of watermelons. *FL Agr. Exp. Sta. Bull.* **491**, 1952.

Pirone, Pascal P. *Tree Maintenance*, 3rd ed. Oxford University Press, New York, 1959.

Pirone, Pascal P. *Diseases and Pests of Ornamental Plants.* 4th ed. The Ronald Press Company, New York, 1970.

Pirone, Pascal P., Bernard O. Dodge, and Harold W. Rickett. *Diseases and Pests of Ornamental Plants.* The Ronald Press Company, New York, 1960. (This is the 3rd ed. of Dodge and Rickett's book of the same name.)

Plakidas, A. G. *Strawberry Diseases.* Louisiana State University Press, Baton Rouge, LA, 1964.

Pyenson, Louis L. *Elements of Plant Protection.* John Wiley & Sons, Inc., New York, 1951.

Pyenson, Louis L. *Keep Your Garden Healthy.* E. P. Dutton & Co., Inc., New York, 1964.

Reed, L. B., and S. P. Doolittle. Insects and diseases of vegetables in the home garden. *U.S. Dept. Agr. Home and Garden Bull.* **46**, 1967.

Roberts, Daniel A., and Carl W. Boothroyd. *Fundamentals of Plant Pathology.* W. H. Freeman and Company, San Francisco, CA, 1972.

Rogers, Donald P. The genus *Pellicularia* (Thelephoraceae). *Farlowia* I(1):95–118, 1943.

Rohde, Richard A., and William R. Jenkins. Basis for resistance of *Asparagus officinalis* var. *altilis* L. to the stubby-root nematode *Trichodorus christiei* Allen. *Univ. MD Bull.* **A-97**, 1958.

Salmon, Ernest S. A monograph of the Erysiphaceae. *Mem. Torrey Bot. Club* 9:1–292, 1900.

Sasser, J. N. Identification and host-parasite relationships of certain root-knot nematodes. *Univ. MD Agr. Exp. Sta. Tech. Bull.* **A-77**, 1954.

Sasser, J. N., and Eldon J. Cairns, eds. *Plant Nematology Notes, from Workshops, 1954 and 1955.* Southern Regional Nematode Project (S-19), 2nd ed., 1958.

Sasser, J. N., and W. R. Jenkins, eds. *Nematology, Fundamentals and Recent Advances with Emphasis on Plant Parasitic and Soil Forms.* Univ. of North Carolina Press, Chapel Hill, NC, 1960.

Scharpf, Robert F., and Frank G. Hawksworth. Mistletoes on hardwoods in the United States. *U.S. Dept. Agr. Forest Pest Leaflet* **147**, 1974.

Schindler, A. F. Parasitism and pathogenicity of *Xiphinema diversicaudatum*, an ectoparasitic nematode. *Nematologica* 2(1):25–31, 1957.

Schroth, Milton N., and D. C. Hildebrand. A chemotherapeutic treatment for selectively eradicating crown gall and olive knot neoplasms. *Phytopathology* **58**:848–854, 1968.

Seaver, Fred J. *The North American Cup-fungi (Operculates).* Published by the author, New York, 1928, 1942 (now available from Hafner Publishing Company).

Seymour, Arthur Bliss. *Host Index of the Fungi of North America.* Harvard University Press, Cambridge, MA, 1929.

Sharvelle, E. G. *The Nature and Uses of Modern Fungicides.* Burgess Publishing Company, Minneapolis, MN, 1961.

Sher, S. A., and M. W. Allen. Revision of the genus *Pratylenchus* (Nematoda, Tylenchidae). *Univ. CA Public. Zool.* **57** (6):441–470, 1953.

Shurtleff, Malcolm C. *How to Control Plant Diseases in Homes and Gardens,* 2nd ed. IA State University Press, Ames, IA, 1966.

Sinclair, J. B., ed. *Fungicide-Nematicide Tests—Results of 1967.* American Phytopathological Society, St. Paul, MN, 1968.

Smith, Kenneth M. *A Textbook of Plant Virus Diseases,* 2nd ed. Little, Brown and Company, Boston, MA, 1957.

Smith, K. M. *Plant Viruses,* 4th ed. Methuen & Co., Ltd., London, 1968.

Smith, Ralph E. Diseases of flowers and other ornamentals. *CA Agr. Ext. Circ.* **118,** 1940.

Smith, Ralph E. Diseases of truck crops. *CA Agr. Ext. Circ.* **119,** 1940.

Smith, Ralph E. Diseases of fruits and nuts. *CA Agr. Ext. Circ.* **120,** 1941.

Snyder, W. C., and H. H. Hansen. The species concept in *Fusarium. Amer. Jour. Bot.* **27:**64–67, 1940 and **32:**657–665, 1945. *Ann. New York Acad. Sci.* **60:**6–23, 1954.

Sprague, Howard B., ed. *Hunger Signs in Crops,* 3rd ed. David McKay Co., Inc., New York, 1964.

Sprague, Roderick. *Diseases of Cereals and Grasses in North America.* The Ronald Press Company, New York, 1950.

Stakman, E. C., and J. George Harrar. *Principles of Plant Pathology.* The Ronald Press Company, New York, 1957.

Steiner, G. Plant nematodes the grower should know. *FL Dept. Agr. Bull.* **131,** 1956.

Spears, Joseph F. The golden nematode handbook. *U.S. Dept. Agric. Agricultural Handbook* **353,** 1969.

Stevens, F. L. *Plant Disease Fungi.* The Macmillan Company, New York, 1925.

Stevens, F. L., and J. G. Hall. *Diseases of Economic Plants.* The Macmillan Company, New York, 1933.

Stevens, N. E., and R. B. Stevens. *Disease in Plants.* Chronica Botanica Company, Waltham, MA, 1952.

Streets, R. B. Phymatotrichum (cotton or Texas) root rot in Arizona. *Univ. Arizona Tech. Bull.* **71,** 1937.

Strong, F. C. Nature and control of anthracnose of shade trees. *Arborists News* **22**(5):33–39, 1957.

Suit, R. F. Parasitic diseases of citrus in Florida. *FL Agr. Exp. Sta. Bull.* **463,** 1949.

Tarjan, Armen C. *Check List of Plant and Soil Nematodes.* University of Florida Press, Gainesville, FL, 1960.

Taylor, A. L., V. H. Dropkin, and G. C. Martin. Perineal patterns of root-knot nematodes. *Phytopathology* **45:**26–34, 1955.

Thorne, Gerald. *Principles of Nematology.* McGraw-Hill Book Company, Inc., New York, 1961.

Thorne, Gerald. On the classification of the Tylenchida, new order (Nematoda, Phasmidia). *Proc. Helm. Soc. WA* **16:**37–73, 1949.

Torgenseon, D. C., Roy A. Young, and J. A. Milbrath. Phytophthora root rot diseases of Lawson cypress and other ornamentals. *OR Exp. Sta. Bull.* **537,** 1954.

Toussoun, T. A., and P. E. Nelson. *A Pictorial Guide to the Identification of Fusarium Species.* The Pennsylvania State Univ. Press, University Park and London, 1968.

U.S. Department of Agriculture. *Trees.* The Yearbook of Agriculture, Washington, DC, 1949.

U.S. Department of Agriculture. *Plant Diseases.* The Yearbook of Agriculture, Washington, DC, 1954.

U.S. Department of Agriculture. *Soil.* The Yearbook of Agriculture, Washington, DC, 1957.

U.S. Department of Agriculture. *Index of Plant Diseases in the United States Agriculture Handbook* **165,** 1960.

U.S. Department of Agriculture. *Index of Plant Virus Diseases Agriculture Handbook* **307,** 1966.

U.S. Department of Agriculture. *Virus Diseases and Other Disorders with Virus-like Symptoms of Stone Fruits in North America Agriculture Handbook* **10,** 1951.

U.S. Department of Agriculture. Power sprayers and dusters. *Farmers Bull.* **2223,** 1966.

Walker, John Charles. *Diseases of Vegetable Crops.* McGraw-Hill Book Company, Inc., New York, 1952.

Walker, John Charles. *Plant Pathology,* 2nd ed. McGraw-Hill Book Company, Inc., New York, 1957.

Walker, J. C., and R. H. Larson. Onion diseases. *U.S. Dept. Agr. Agriculture Handbook* **208,** 1961.

Weiss, Freeman. *Ovulinia,* a new generic segregate from *Sclerotinia. Phytopathology* **30:**236–244, 1940.

Weiss, Freeman. Viruses described primarily on leguminous vegetables and forage crops. *Plant Disease Reporter Suppl.* **155:**82–140, 1945.

Weiss, Freeman. Viruses, virus diseases and similar maladies on potatoes, *Solanum tuberosum. Plant Disease Reporter. Suppl.* **155:**82–140, 1945.

Welch, D. S., and J. G. Matthysse. Control of the Dutch elm disease in New York State. *Cornell Ext. Bull.* **932,** 1955.

Westcott, Cynthia. *Garden Enemies.* D. Van Nostrand Company, Inc., Princeton, NJ, 1953.

Westcott, Cynthia. *Are You Your Garden's Worst Pest?* Doubleday and Company, Inc., New York, 1961.

Westcott, Cynthia. *The Gardener's Bug Book,* 3rd ed. Doubleday and Company, Inc., New York, 1964.

Westcott, Cynthia. *Anyone Can Grow Roses,* 4th ed. D. Van Nostrand Company, Inc., Princeton, NJ. 1965.

Westcott, Cynthia, and Jerry T. Walker, eds. *Handbook on Garden Pests.* Brooklyn Botanic Garden, Brooklyn, NY, 1966.

Whetzel, H. H. A synopsis of the genera and species of the Sclerotiniaceae, a family of stromatic inoperculate Discomycetes. *Mycologia* **37:**648–714, 1945.

Wolf, Frederick A., and Frederick T. Wolf. *The Fungi,* vols. I, II. John Wiley & Sons, Inc., New York, 1947.

Zaumeyer, W. J., and H. Rex Thomas. Bean diseases and their control. *U.S. Dept. Agr. Farmers Bull.* **1692,** 1958.

Zundel, George L. The Ustilaginales of the world. *Dept. Botany, Penn State College Contrib.* **176,** 1953.

Remember that the references cited are a mere sampling of the vast amount of material published on plant diseases. Most plant pathologists will have access to the *Plant Disease Reporter, Plant Disease, Phytopathology, Review of Applied Mycology, Mycologia, Biological Abstracts, Journal of Economic Entomology, Agricultural Chemicals, NAC News, Arborists' News, Proceedings of the International Shade Tree Conference,* and other technical publications. Gardeners will find information on plant diseases in the publications of single plant societies, such as the American Camellia Society and the American Rose Society. State agricultural experiment stations have a wealth of material. Government publications are available from the Superintendent of Documents, U.S. Government Printing Office, Washington, DC 20402.

Index

This is a selective index. It includes common and Latin names of the host plants in Chapter 4, common names of the diseases described in Chapter 3 and Latin names of their pathogens. Entries under plant names are chiefly for providing cross references to disease names and are not to be construed as providing a complete checklist of the diseases of each plant. For that, the host section itself must suffice. Orders and families given in Chapter 2 are indexed, and genera described in Chapter 3. Chemicals listed in Chapter 1 are indexed but not their use in control of specific diseases. Boldface type indicates illustrations.

A7 Vapam, 41
AA tack, 10
Aaterra, 10
Abelia, 519
  leaf spot, 255
Abies, 645–647
  canker, 201, 202
Abronia, 806
Abutilon, 519
  infectious variegation, 475
  leaf spot, 255–283
  mosaic, 475
AC 5223, 10
Acacia, 519–520, 687
  twig canker, 208
Acalypha, 520
  downy mildew, 235
Acanthopanax, 520
Acanthorhynchus vaccinii, 362
Acarelti, 10
Acarelti forte, 10
Acer, 720–722
  negundo, 562
Acetic acid, 10
Achillea, 870
Achimenes, 520
Achlyogetonaceae, 63
Achlys, 855
Achras, 807
Achrotelium lucumae, 423
Acidity, excess, 334
Aconite, 731
Aconitum, 731
Acorus, 836
Acquinites, 11
Acrex, 10
Acrospermaceae, 69
Actaea, 548
Actidione, 10–11

Actinidia, 699
Actinomeris, 520–521
Actinomyces ipomoea, 418
  scabies, 454–455
Actinopelte dryina, 249
Actinothyrium gloeo-
  sporioides, 249
Actispray, 11
Adams-needle, 871–872
Adders-Tongue, 636
Adelopus gaumannii, 301–302
Adiantum, 642
Adoxa, 521, 736
Aecidium, 423
  avocense, 423
  conspersum, 423
  rubromaculans, 423
  yuccae, 442
Aerosol bomb, 53
Aeschynanthus, 710
Aesculus, 685
Aesehynanthus, 521
African daisy, 521
African lily, 522
African violet, 521–522
  Botrytis on, 142
  ring spot, 344
Afugan, 11
Agapanthus, 522
Agaricaceae, 73, 76
Agaricales, 75–76
Agaricus campestris, 238
Agastache, 522
  downy mildew, 233
Agave, 587
Ageratum, 522
Aglaonema, 593
Agrimonia, 522
Agrimony, 522

  downy mildew, 233
Agrimycin, 11, 40
Agristrep 11, 40
Agrisol 5, 11
Agritol, 40
Agrobacterium, 98
  gypsophilae, 107
  rhizogenes, 98
  rubi, 98
  tumefaciens, 98–100
Agrostemma, 608
Agrostis, 665–667
Agrox Strep, 11, 40
Agrox 2-Way, 11
Agrox 3-Way, 11
Ailanthus, 523
  black mildew, 129
  twig blight, 157
  twig canker, 207, 213–214
Airone, 11
Air pollution, 334–335
Ajuga, 567
Akzo, 11
Alaska yellow-cedar, 588
Albizzia julibrissin, 728
  lebbek, 703
Albuginaceae, 65
Albugo, 500
  bliti, 501
  candida, 501
  ipomoeae-panduratae, 501
  occidentalis, 501
  platensis, 501
  portulacae, 501
  tragopogonis, 501
Alder, 523–524
  bark patch, 223
  canker, 217
  catkin hypertrophy, 245

Alder (cont.)
  leaf curl, 245
  powdery mildew, 352
Aletris, 828
Aleurites, 851–852
Aleurodiscus, 192
  acerina, 192
  amorphus, 192
  oakesii, 192
Alfalfa, 230
  dodders, 230
  downy mildew, 234
  dwarf, 476
  mosaic, 476
Algal spot, 253
Alkali injury, 335
Alkalinity, 335
Alkanet, 528
Allamanda, 524
Allionia, 524
Allisan, 11
Allium ascalonicum, 812
  cepa, 745–746
  sativum, 652
  schoenoprasum, 595
Allophylaria, 281
Allspice, 524
Almond, 524–525
  bud failure, 476
  calico, 476
  flowering, 525
  pruning wound canker, 221
  scab, 452
  trunk canker, 219–220
Alnus, 523–524
Aloe, 525
Alternanthera, 525–526
Alternaria, 135, 136
  alternata, 136, 247, 362
  brassicae, 249–250
  brassicicola, 249–250
  cassiae, 136
  catalpae, 250
  citri, 250, 362–363
  cucumerina, 136
  dauci, 136–137
  dianthi, 137
  fasciculata, 250
  helanthi, 137
  longipes, 250
  mali, 363
  oleracea, 250
  panax, 137, 250
  passiflorae, 250
  polypodii, 250
  porri, 187
  radicina, 363
  raphani, 250
  rot of citrus, 362–363

solani, 137–138, 363
sonchi, 250
tagetica, 138, 250
tenuis, 185, 187, 250
tenuissima, 187, 250
tomato, 250–251
violae, 138
zinniae, 138, 363
Alternaria blights, 135–138
Althaea, 681
Aluminum plant, 539
  blight, 179
Aluminum toxicity, 335
Alum-root, 677–678
Alyssum, 526
Amaranthus, 526
Amaryllis, 526
  Botrytis on, 142
  leaf spot, 266
  red blotch (fire), 247
Amazon-lily, 637
Ambrosia, 791
Amelanchier, 527
  canker, 213–214, 217
American dagger nematode,
    331
American holly, 680–681
American ipecac, 656
American spikenard, 535
American sycamore, 776–777
Amerosporae, 77
Amerosporium trichelium, 251
Ammonium sulfate, 11
Amobam, 11
Amorpha, 527, 690
Ampelanus, 806
Ampelopsis aconitifolia, 731
  arborea, 767
  cordata, 527–528
Amphisphaeriaceae, 71
Amphobotrys ricini, 192
Amsonia, 528
Ananas, 775
Anaphalis, 528
Anarcardium, 582
Anchusa, 528
Andromeda, 528
Andropogon, 561
Androsace, 797
Anemone, cultivated, 528–529
  Botrytis on, 142
  downy mildew, 235
  leaf gall, 243
  leaf spot, 263
  native, 529
Anemonella, 802
Anethum, 624
Angelica, 529
Angiospora, 423

Anguina, 309
  agrostis, 309
  balsamophila, 309
  graminis, 309
  tritici, 309
Angular leaf spot, of bean,
    272
  of grapes, 277
Anise, 529–530
Anise-tree, 530
  black mildew, 130
Anisomycin, 11
Annato-tree, 557
Annellophora phoenicis, 251
Annona, 589
Anoda, 530
Antennaria, 638–639
Anthemis, 576–577
Anther smut of carnation,
    467
Anthracnose, 87–98. See also
    Spot anthracnose
  apple, 97
  aspidistra and hosta, 91
  azalea, 92
  bean, 90–91
  butterfly-flower, 91
  cactus, 96–97
  cereal, 89
  citrus, 92
  currant, 97–98
  daylilies, 90
  Dolichos, 89
  eggplant, 92
  fern, 95
  foxglove, 89
  hickory, 95
  hollyhock, 91
  hosta, 91
  lettuce, 96
  lime, 92
  linden, 96
  maple, 92
  melon, 90
  northwestern apple, 97
  oak, 93, 93, 96
  orchid, 89
  pea, 91
  peach, 94
  peony, 94
  pepper, 93
  potato, 88
  privet, 94
  raspberry, 92
  rhubarb, 89
  snapdragon, 88
  spinach, 91
  stem, 91
  strawberry, 89

sweet pea, 94
sycamore, 95
tomato, 95
tulip, 93–94
turnip, 89–90
violet and pansy, 92
walnut, 95–96
Anthurium, 530
Anthyllis, 699
Antibiotic, 7
Antidesma, 530
Antirrhinum, 816
Antracol, 11
Apadodine, 11
Aphanomyces, 363
  cladogamus, 363
  cochlioides, 363
  euteiches, 363–364
  raphani, 364
Aphelandra, 530, 872
Aphelenchoides besseyi,
    309–310
  fragariae, 310–311
  olesistus, 310–311
  oryzae, 310
  parietinus, 311
  ribes, 311
  ritzema-bosi, **311,** 312
  subtenuis, 312
Apioporthe, 193
  anomala, 193
  apiospora, 193
Apiosporina collinsii, 130
Apium graveolens, 585–586
APL-luster, 11
Aplopsora nyssae, 423
Apocynum, 625
Apple, 530–533
  anthracnose, 97
  bitter pit, 335
  bitter rot, 384
  black pox, 238
  black root rot, 421
  black rot, 173, 401
  blister spot, 114
  blotch, 190
  canker, 208, 209–210, 213,
    213–214, 217, 218,
    225
  Cedar-apple rust
    Color Plate 2C, 2D
  cork, 338, 340
  dapple apple, 476
  fisheye rot, 373
  fly speck, 238, 239
  green mottle, 476
  hairy root, 98
  leaf spot, 272
  mosaic, 476

Nectria canker, 214–215
perennial canker, 215–216
powdery mildew, 358
rot, 111
scab, 455–457, **456, 457**
  Color Plate 6F
scald, 345
stem-pitting, 476
storage rot, 398
trunk canker, 219–220
twig canker, 199, 208
white root rot, 374
wood rot and decline,
    388
Apple-of-Peru, 533
Apricot, 534–535
  dieback, 207, 213–214
  gummosis, 476
  ring rot, 476
  trunk canker, 219–220
Aquathanatephorus pendulus,
    180
Aquatic plants, 535
Aquilegia, 605
Arabian-tea, 584
Arabis, 535
Arachis, 759–760
Aralia, 535–536
  scab, 474
Aralia cordata, 536, 854
  5-leaf, 520
  hispida, 536
  nudicaulis, 536
  racemosa, 536
  spinosa, 535, 677
Arasan, 11
Araucaria, 536
  branch blight, 150
Arborvitae, 536–537
  blight (fire), 148
  Botrytis on, 142
  Coryneum blight, 150
  leaf blight, 154–155
Arbotect, 11
Arbutus menziesii, 716
  canker, 209
  leaf spot, 262
  unedo, 832
Arceuthobium, 298
  americanum, 299
  campylopodium, 299
  cyanocarpum, 299
  douglasii, 299
  occidentale, 299
  pusillum, 299
  tsugense, 299
  vaginatum subsp. crypto-
    podum, 299
Arceuthobium subsp., 730

Arctostaphylos, 720
  uva-ursi, 552–553
Arctotheca calendula, 579
Arctotis, 521, 537
Arctous, 557
Ardisia, 537
Arecastrum, 751
Arenaria, 806–807
Arenga, 752
Argemone, 786
Argyreia, 537
Arisaema, 694
Aristastoma oeconomicum,
    251
Aristolochia, 629
Arizona purple top wilt, 124
Armeria, 537
Armillaria mellea, 364–366,
    365
Armoracia, 686
Arnica, 537–538
Aronia, 595
Arrow-arum, 538
  leaf spot, 280
Arrowhead, 538
Arrowleaf, 622
Arrowroot, 538
Arsenical injury, 335
Artemisia, 538, 803–804
  downy mildew, 233
Artichoke, globe, 539
  curly dwarf, 476
  Jerusalem, 539
Artillery plant, 539
  blight, 179
Aruncus, 659
Arundo, 539–540
Asarum, 656
Asclepias, 568
Ascochyta, 138, **252**
  abelmoschi, 251
  althaeina, 252
  armoraciae, 252
  asparagina, 138
  aspidistrae, 252
  asteris, 252
  blight, 253
  boltshauseri, 252
  brachypodii, 252
  cheiranthi, 252
  chrysanthemi, 138
  clematidina, 252–253
  compositarum, 253
  cornicola, 253
  cypripedii, 253
  juglandis, 253
  lycopersici, 253
  phaseolorum, 253
  piniperda, 138

Ascochyta *(cont.)*
  pinodelia, 138–139, 366
  pinodes, 138–139
  pisi, 138–139, 253
  rabiei, 165
Ascocorticiaceae, 72
Ascoideaceae, 68
Ascomycetes, 67–73
  reproduction in, 67
Ascospora ruborum, 193
Ascyrum, 804
Ash, 540–541
  black mildew, 129
  canker, 202, 204, 206,
    213–214
  leaf spot, 269, 274, 278
  ring spot, 476
  rust, 448
  witches' broom, 476
Ash, Moraine, 541
Ashy stem blight, 164, 183,
  389–390
Asimina, 756
Asparagus asparagoides, 542
  officinalis, 541
  plumosus, 541–542
  sprengeri, 542
Asparagus crown rot, 397
  Fusarium wilt, 507
  rust, 442–443
  virus 1, 476
Asparagus fern, 541–542
  stem blight, 138
Aspen, 781–782
  canker, 202, 206, 211
  Nectria canker, 214–215
  sooty bark canker, 195
Aspergillus, 366
  alliaceus, 366
  niger, 367
    var. floridanus, 367
Asphyxiation, 344
Aspidistra, 542
  anthracnose, 91
  leaf blight, 163
  leaf spot, 252
Asplenium, 640
Aspor, 12
Aster, China, 542–543
  perennial, 543
  Botrytis, 142
  downy mildew, 232
  gray mold, 141–143
  leaf spot, 252, 294
  ring spot, 476
  rust, 425
  wilt, 507
  yellows, 121
  Color Plate 3B
Asteridiella, 129

Asterina, 128
  anomala, 128
  delitescens, 128
  diplopoides, 128
  gaultherae, 128
  lepidigena, 128
  orbicularis, 128
Asterinella, 128
  puiggarii, 128
Asteroma, 253
  garretianum, 253
  solidaginis, 253
  tenerrimum, 253
Asteromella lupini, 253
Astilbe, 544
Atamasco-lily, 872
Athyrium, 641
Atichiaceae, 69
Atlantic white cedar, 588
Atlas cedar, 585
Atomizer sprayer, 53
Atriplex, 805–806
Atropellis, 193
  apiculata, 193
  arizonica, 193
  pinicola, 193
  piniphila, 193
  tingens, 193–194
Aucuba, 543
  leaf spot, 268, 281, 283
Aules, 12
Auriculariaceae, 74
Australian-pine, 583
Autumn crocus, 544
Avens, 655
Avicol, 12
Avocado, 544–545
  black mildew, 129
  blotch, 188
  decline, 327–328
  Dothiorella rot, 367
  root rot, 402
  scab, 474
  sun blotch, 476–477
  verticillium wilt, 511–515
Award, 12
Awl nematode, 316
Azalea, 545–546
  anthracnose, 92
  bud and twig blight, 146–147
  flower spot, 165–168
  leaf gall, 240, **241**
  leaf scorch, 247
  petal blight, 166–168
  shoot blight, 164, 174
  tar spot, 276
Azara, 546

Babiana, 546
Baby blue-eyes, 739

Babys-breath, 670
Bachelors-button, 587
Bacillus thuringiensis, 13
  polymyxa, 100
Bacterial blight, of bean, 120
  bird's-nest fern, 109, 110
  carrot, 118
  celery, 110
  chrysanthemum, 107
  garden stocks, 119
  gladiolus, 119
  mulberry, 113–114
  pea, 115
  poppy, 120
  soybean, 113
Bacterial canker
  of ground and Jerusalem
    cherry, 101–102
  of poinsettia, 102
  of stone fruits, 114
Bacterial diseases, 98–121
  canker of tomato, 101
  crown gall, 98–100, **99**
  fire blight, 103–105
  gall or knot of oleander,
    115–116
  knot of olive, 115
  leaf blight of bird's-nest
    fern, 110
  leaf spot, of barberry, 109
    crucifers, 110–111
    dracaena, 107
    English ivy, 119
    geranium, 118, 120
    gloxinia, 109
    poinsettia, 102
    primrose, 111–112
    sesame, 112
    sunflower, 113
    velvet bean, 112
    viburnum, 116
    zinnia, 120
  necrosis of giant cactus,
    105
  pustule of soybean, 118
  ring rot of potato, 102
  rot of chicory, 110
  speck of tomato, 115
  spot, of carnation, 116
    beets, 112–113
    cucurbits, 118
    grasses, 109
    nasturtium, 112–113
    pepper, 121
    stone fruits, 120–121
    tomato, 121
  stripe of corn, 109
  wilt, of bean, 101
    carnation, 109–110
    corn, 108

cucurbits, 108–109
lettuce, 121
Bactericides, 7, 40–41
Bacticin, 40
Baeodromus, 423
californicus, 423
eupatorli, 423
Balansia cyperi, 139
Bald cypress, 546–547
Baldhead of bean, 335
Baldwin spot of apple, 335
Balloon-flower, 777
Balm, 547
Balsam-apple, 547
Balsam fir, butt rot, 406
canker, 192, 203
needle blight, 170, 179–180
needle cast, 302, 304
Balsamorhiza, 547
Balsam-pear, 547
Balsam-root, 547
Bamboo, 547
Bambusa, 547
Banana, dwarf, 548
Baneberry, 548
Banol Turf Fungicide, 12
Banrot, 12
Baptisia, 548
Barbarea, 868
Barberry, 548–549
bacterial leaf spot, 109
Bardac, 12
Bark canker, of balsam firs,
216
of cypress, 198–199
of Douglas-firs, 196–197
Nectria beech, 214
superficial, 217
Bark patch, 192, 223
of oak, 192
Barley yellow dwarf virus,
477
Barquat compounds, 12, 40
Barquat MB-50, 40
Barquat MB-80, 40
Barren-strawberry, 549
Basal canker of maple, 221
Basal rot, of iris, 381
lily, 382
narcissus, 382
onion, 381
tulip, 381
Basamid Granular, 41
BASF-Maneb Spritzpulver, 12
Basfungin, 12
Basidiomycetes, 73–76
reproduction in, 73
Basidiophora, 231, 231
entospora, 232
Basil, 549

Basil-weed, 602
Basketflower, 587
Basket-grass, 746
Basket-willow, 748
Basswood, 708–709
Bauhinia, 549
Bavistin, 12
Bay 25141, 42
Bay 4631, 12
Bay 47531, 12
Bay 49854, 12
Bay 572, 12
Bay 68138, 42
Bay 70143, 42
Bay SRA3886, 42
Bayberry, 549
yellows, 477
Bayclean, 40
Baycor, 12
Bayleton, 12
Bean, adzuki, 549
asparagus (yardlong),
549–550
kidney, 550–552
mung, 552
scarlet runner, 552
tepary, 552
urd, 552
Bean diseases, 550–552
angular leaf spot, 272
anthracnose, 90–91
bacterial blight, 120
bacterial wilt, 101
baldhead, 335
cottony rot, 415
downy mildew of lima,
234–235
dry root rot, 383
halo blight, 114–115
leaf spot, 252, 253
mosaic, 477
pod mottel, 477
phyllody, 121
rust, 451
southern mosaic, 477
yellow dot, 476
yellow mosaic, 477
yellow stipple, 477
zonate leaf spot, 251
Bearberry, 552–553
black mildew, 128
shoot hypertrophy, 240
Beard-tongue, 764
Beauty-berry, 575
Beauty-bush, 553
Bedstraw, 651
Bee-balm, 730–731
Beech, 553
bark canker, 213–214,
214

bleeding canker, 219–220
Nectria canker, 214–215
Beet, 553–554
bacterial spot, 112–113
black heart, 335, 337, 340
curly top, 477–478
dodder, 230
downy mildew, 234
latent virus, 478
leaf spot, 255–256
mosaic, 478
Phoma rot, 399
pseudo-yellows, 478
Rhizoctonia dry rot, 396
ring mottle, 478
rust, 450–451
savoy, 478
yellow net, 478
yellows, 478
Begonia, 554–555
bacteriosis, 117
Botrytis on, 141–143
leaf blight nematode, 312
mildew, 356
Color Plate 5D
Beincasa, 594
Belamcanda, 558
Bellflower, 577
Bellis, 634
Bells-of-Ireland, 555
Bellwort, 854
Belonolaimus, 312
gracilis, 313
longicaudatus, 313
Benincasa, 594
Beniowskia sphaeroidea,
139
Benlate, 12
Benomyl, 12
Bentgrass, 665–667
leaf blight, 179
Bentonite, 12
Benzalkonium chloride, 40
Benzimidazoles, 12
Berberis, 548–549
Berchemia, 794
Bermuda grass, 665–667
downy mildew, 237
leaf blotch, 270
Beta vulgaris, 553–554
var. cicla, 840
Betony, 827
Betula, 556
Bidens, 555
mottle, 478
Bifusella, 302, **302**
abietis, 302
faullii, 302
linearis, 303
saccata, 303

Bignonia, 555
  black mildew, 129, 130
Bilberry, 865
Binapacryl, 12
Bioguard, 13
Bioquin 1, 13
Biotrol-Plus, 13
Biotrol VHZ, 13
Biotrol XK, 13
Birch, 556
  anthracnose, 268
  brown leaf spot, 263
  canker, 209–210, 213–214,
    219–220
  leaf blister, 244, 245
  leaf rust, 438
  Nectria canker, 214–215
Bird-of-Paradise, 556
  root and seed rot, 383
Bird's-nest fern, 640
  bacterial leaf blight, 110
  fungi, 238
Bischofia, 556
Biscuit-root, 712
Bishops-cap, 556–557
Bitterbush, 683
Bitter pit of apple, 335
Bitterroot, 705
Bitter rot, of apple, 384
  cranberry, 385
  grapes, 390
Bittersweet, 557, 695
Bixa, 557
Blackbead, 776
Black bearberry, 557
Blackberry, 557–558
  cane blight, 159
  cane gall, 98
  canker, 194, 207, 208, 211,
    213–214
  downy mildew, 234
  dwarf, 478
  dwarfing, 478
  leaf spot, 292–293
  mosaic, 479
  orange rust, 431
  short-cycle orange rust,
    436–437
  variegation, 479
  yellow rust, 436
Blackberry-lily, 558
Black cane rot, 369–370
Black cohosh, 597
Black crown rot of celery, 369
Black dot disease of potato, 88
Black end of pear, 335
Black-eyed Susan, 801–802
Black gum, 852
Black heart, of beets, 335,
    337, 340
  of fruit trees, 510

Black knot, of plum and cherry,
    124–126, **125**
  of goldenrod and sunflower,
    126
Black leaf speck of geranium,
    295
Black leaf spot, of crucifers,
    249–250
  of peas, 267
  of poplar, 258
Black leg, of crucifers, 127
  of delphinium, 106
  of potato, 106
Black medic, 725
Black mildew, 127–130
  ailanthus, 129
  anise-tree, 130
  ash, 129
  avocado, 129
  bamboo, 130
  bearberry, 128
  bignonia, 129, 130
  blueberry, 130
  boxelder, 129, 130
  cactus, 130
  California-laurel, 128
  callicarpa, 130
  camellia, 130
  chinaberry, 130
  firs, 129
  holly, 130
  juniper, 128
  lantana, 130
  lippia, 130
  lyonia, 128
  magnolia, 129, 130
  oak, 276
  palmetto, 130
  redbay, 129, 130
  sea-grape, 130
  swamp bay, 129
  wax-myrtle, 129, 130
  wintergreen, 130
Black mold, of peach, 367
  of rose, 300
  of spinach, 301
Black patch of goldenrod and
    sunflower, 126
Black pox of apple, 238
Black pustule of currant, 218
Black root, rot, 335
Black rot, of apple, 173, 401
  bulbs, 416
  carrot, 363
  cranberry, 370
  grapes, 385–386, **386**
  wintercress, 116
Black sage, 803
Black salsify, 805
Black scale rot of Easter lily,
    372

Black scorch, 246
Black scurf, of potato, 396, 411
  of goldenrod, 253
Black spot
  cactus, 265–266
  delphinium, 113
  elm, 269
  erythronium, 253
  eugenia, 128
  goldenrod, 253
  holly, 128, 276
  leucothoë, 128
  primrose, 253
  raspberry, 295
  redbay, 128
Blackspot of rose, 130–135,
    132, 133
  Color Plate 6D
  control of, 133
Bladdernut, 827
  twig blight, 151, 161
Bladder-senna, 558
Blasting, 335
Blastocladiales, 63
Blazing star, 726
Bleeding canker, 219–220
Bleeding-heart, 558
Bleeding-heart vine, 559
Blephilia, 559
Blessed thistle, 844
Blight(s), 135–185
  Alternaria, of carrot,
    136–137
    carnation, 137
    cucurbits, 137
    ginseng, 137
    potato, 137–138
    tomato, 137–138
    violet and pansy, 138
    zinnia, 138
  Araucaria branch, 150
  arbor-vitae, 148; leaf, 148,
    150
  Ascochyta, of peas, 138–139
  ashy stem, 164, 183, 389–390
  azalea petal, 166–168
    shoot, 164, 174
  bacterial, of bean, 120
    carrot, 118
    celery, 110
    fern, 109, 110
    filbert, 118
    gladiolus, 119
    lilac, 114
    mulberry, 113–114
    pea, 115
    poppy, 120
    soybean, 113
  blossom, of squash, 149
    of stone fruits, 164, 390,
    392

blueberry twig, 154, 172;
    cane canker, 218
Botrytis, of dogwood, 142
    gladiolus, 144
    hyacinth, 144
    lily, 143–144
    peony, 144–145
    snowdrop, 144
    tulip, 145–146, **146**
boxwood leaf, 162, 187;
        tip, 171
    Volutella, 226
brown, of grass, 271
brown felt, 160, 166
bud and twig, of azalea and
        rhododendron,
        146–147
camellia flower, 170,
        181–182, 183
cane, of brambles, 159
carnation collar, 136
cherry, 259–260
chestnut, 157
chrysanthemum, leaf, 183;
        petal, 162; ray, 138
citrus, 179
Coryneum, of arbor-vitae,
        150
    of stone fruits, 150
cyclamen, 159
Diaporthe, of larkspur, 153
Dogwood, 142; twig, 151
early, of carrot, 148
    celery, 147–148
    potato and tomato,
        137–138
fire, 103–105
gladiolus flower, 151–152
gray, on pines, 161, 303
gray mold, 141–143
gummy stem, of cucurbits,
        165
halo, of bean, 114–115
hemlock needle, 155, 181
inflorescence, of maple, 149
juniper, 172
late, of celery, 183–184
    potato and tomato, **Color
        Plate 5C,** 175–179,
        176, 177, 178
leaf, of aspidistra, 163
    corn, 159–160. **Color
        Plate 7F**
    erythronium, 149
    fig, 168
    hackberry, 153
    hawthorn, 158, 164
    lilac, 160
    linden, 148
    parsley, 187
    parsnip, 162

pear, 157
    strawberry, 153
    sweetpotato, 173
    trumpetvine, 148
    yucca, 162
leaf and stem, of squash,
        174
leaf and twig, of holly, 174
lilac, 114, 174
lima bean pod, 154
limb, 150
marginal, of lettuce, 110
mulberry twig, 166
needle, 155, 156, 161
nursery, 172
pea, 138–139, 165
peach shoot, 150
Phomopsis, of eggplant,
        154, 172
Phytophthora, of pepper,
        174
pine twig, 147
poplar shoot, 156
raspberry cane, 159, 163
    spur, 155
seedling, 186
silky thread, 180
snow, of conifers, 171
southern, 169–170
soybean pod and stem, 154
spruce twig, 138
stem, of asparagus fern,
        138
    phlox, 179
    soybean, 154
    spanish moss, 154
thread, 150, 168–169
tip, of conifers, 170
    crape myrtle, 173
    dracaena, 173
    fern, 173
twig, 151, 156
Volutella, 221
walnut, 119
web, 150, 168
wild rice, 183
Blindness, 335–336
Blister canker, 216
    on osier, 199
Blister spot of apple, 114
Blood-leaf, 691
Bloodroot, 559
    leaf spot, 285
Blossom blight, 149, 164,
        390, 392
Blossom-end rot, 336, **336**
Blotch diseases, 187–191
    apple, 190
    avocado, 188
    greasy, of carnation, 191
    horse-chestnut, 189

iris, 263–265
leaf, of grass, 191, 270
lilac, 188
pecan, 189–190
peony, 188
potato leaf, 188
purple, of onion, 187
sooty, 189
Bluebells, 727
Blueberry, 559–560
    black mildew, 130
    brown rot, 392
    bud-proliferating gall, 242
    cane canker, 218
    canker, 116
    double spot, 266
    gall, 242
    leaf spot (rot), 250, 285
    mildew, 356
    ring spots, 479
    rust, 448–449
    shoestring, 479
    stunt, 124, 479
    twig blight, 154, 172
    witches' broom, 448
Blue cohosh, 560
Blue Curls, 560
Blue-eyed grass, 560
    leaf blight, 162
Blue-eyed Mary, 604–605
Bluegrass, 665–667
    blister smut, 460
    leaf spot, 292
    leaf rust, 447
Blue lace-flower, 560–561
Blue-lips, 604–605
Blue lobelia, 710–711
Blue mist-flower, 638
Blue mold rot, of fruits, 397
    of tobacco, 234
Blue sage, 803
Bluestem, 561
Blue toadflax, 708
Bluets, 686–687
Boehmeria, 561
Boerhaavia, 822
Bog Rosemary, 526
Boisduvalia, 561
Boletaceae, 76
Boltonia, 561
Boneset, 638
Borage, 561
Borago, 561
Borax, 13
Bordeaux injury, 13–14,
        336
Bordeaux mixture, 3–4, 13–14
    paint, 13
    wash, 13
Bordeaux-oil emulsion, 13
Bordocop, 14

Boron deficiency, 336–337
  toxicity, 337
Boston ivy, 693
  leaf spot, 270
Botran, 14
Botrilex, 14
Botryobasidium, 396
Botryodiplodia, 139
  gallae, 194
  theobromae, 194
Botryosphaeria, 139
  dothidea, 194, 367
  obtusa, 194, 367
  ribis, 194, 367
    var. chromogena,
      139–140, 194
Botryosporium pulchrum, 300
Botryotinia, 140
  convoluta, 367–368
  fuckeliana, 140
  polyblastis, 182–183
  ricini, 140
  squamosa, 368
Botrytis, 136, 140
  allii, 368
  byssoidea, 368
  cinerea, 141–143, 194, 368
  douglasii, 143
  elliptica, 143–144
  galanthina, 144
  gladiolorum, 144, 368
  hyacinthi, 144
  narcissicola, 144
  paeoniae, 144–145
  polyblastis, 145
  squamosa, 368
  streptothrix, 145
  tulipae, 145–146, **146**
Botrytis blight
  dogwood, 142
  gladiolus, 144
  hyacinth, 144
  lily, 143–144
  peony, 144–145
    Color Plate 8B
  snowdrop, 144
  tulip, 144–146, **146**
Botrytis crown rot of iris, 367
Bottle-brush, 561
Bougainvillea, 561
  blight, 177
  leaf spot, 256
Bouvardia, 562
Bowstring-hemp, 807
Boxelder, 562
  black mildew, 129, 130
Boxwood, 562–563
  dieback, 212, 214, **215**
  leaf blight, 162, 187
  leaf spot, 274

Nectria canker, 187, 214,
    215, 221
  root rot, 395
  spiral nematode, 328
  tip blight, 171
  twig canker, 208
  Volutella blight, 226
Boysenberry, 563
Brachycombe, 563
Brachypodium, 563
Brachysporium tomato, 368
Bracken tar spot, 262
Bramble spot anthracnose,
    473
Branch and trunk canker of
    pine, 193
Branch gall canker of poplar,
    212
Branched broomrape, 191–192
Brand canker of rose, 197–198
Brasenia, 865
Brassaia actinophylla, 809
Brassica chinensis, 593
  juncea, 736–737
  oleracea, 569–570
  pekinensis, 593
  rapa, 852–853
Brassicol, 14
Bravo, 14
Bremia lactucae, **231,** 232
Brestan, 14
Brestanid, 14
Brickellia, 563–564
Brickle-bush, 563–564
Brifur, 42
Briosia, 146
  azaleae, 146–147
Bristlegrass, 564
Broadbean wilt virus, 479
Broccoli, 564, 569–570
Brodiaea, 564
Bromegrass, 665–667
Brome grass mosaic, 479
Bromegrass, Smooth, 564
Bromelia, 564
  leafspot, 270
Bromofume, 14, 40, 42
Brom-O-Gas, 14, 40, 42
Brom-O-Gaz, 14, 42
Bromoethane, 14, 40, 42
Bromomethane, 42
Brom-O-Sol, 14–15, 40, 42
Bromus, 564
Bronopol, 40
Brooks fruit spot, 239
Broom, 564
  Spanish, 565
Broomrape(s), 191–192
Broussonetia, 565
Browallia, 565

Brown bark spot of fruit
    trees, 337
Brown canker of rose,
    199–201, **200**
Brown crumbly rot, 379, 417
Brown cubical rot, 373, 375
Brown felt blight, 160, 166
Brown heart, 337
Brown leaf blotch of grass,
    191, 291
Brown patch of turf, 396, 411
Brown pocket rot, 380, 386,
    389, 417
Brown rot, bacterial, 112
  of citrus, 404
  of stone fruits, 390–392,
    391
    Color Plate 8E
Brown spot, of celery, 254
  corn, 285–286
  orchids, 110
  passion flower, 250
  tobacco, 250
Brown stringy rot, 377–378
Brown stripe of lawn grass,
    290
Brown trunk rot, 379
Brozone, 42
Brunfelsia, 565
Brussels sprouts, 565, 569–570
Bryonopsis, 565
Bubakia erythroxylonis, 423
Buckloe, 567
Buckeye, 565, 685
  leaf blister, 244
  leaf blotch, 189
Buckleya, 566
Buckthorn, 566
Buckwheat-tree, 566
Bud and leaf nematode, 312
Bud blast of peony, 144–145
Buddleia, 566
Bud drop of sweetpea and
    gardenia, 337
Bud gall, sweetleaf, 240
Budrot of palms, 397, 404
Bud scorch, 246
Buffaloberry, 566–567
Buffalograss, 567, 665–667
Bugbane, 597
  false, 708
Buginvillaea, 561
Bugleweed, 567
Bugloss, 528
Bulb nematode, stem and,
    315–316
Bumelia, 567
Bunchberry, 625
Bunch disease, 500
Bunchflower, 567

Bunch mold of grapes, 367
Bundleflower, 567
Bunt, 463
Bupirimate, 15
Bupleurum, 844
Burcop, 15
Burgundy mixture, 15
Bur-marigold, 555
  downy mildew, 235
Burnet, 567, 807
Burning bush, 637
Burrillia decipiens, 460
Burrowing nematode, 327–328
Bursaphelenchus lignicolus,
  313
Busan 72A, 15
Busan 1020, 42
Bush honeysuckle, 624
Bush-mallow, 718
Bush-pea, 843
Butter-and-eggs, 708
Butter-bur, 768
Buttercup, 791–792
Butterfly-bush, 566
Butterfly-flower, 568
  anthracnose, 91
Butterfly-pea, 568
Butterflyweed, 568
Butternut, 861–862
  dieback, 212–213
Buttonbush, 568–569
Butt rot of queen palms,
  384
Buxus, 562–563

Cabbage, 569–570
  black leaf spot, 249–250
  black leg, 127
  black ring spot, 479
  black rot, 117
  club root, 226–227
  cottony rot, 415
  cyst nematode, 318
  leaf spot, 250, 258
  ring necrosis, 479
  yellows, 508
Cabbage, Chinese, 593
Cabbage palm, 752
Cachexia, citrus, 483
Cacopaurus, 313
  epacris, 325
  pestis, 313
Cactobrosis fernaldialis, 105
Cactus, 570–571
  anthracnose, 96–97
  black mildew, 130
  black spot, 266
  cladode rot, 366
  cyst nematode, 318
  scorch, 247

  stem rot, 387
  virux X, 479
Cactus, barrel, 571
  fishhook, 570–571
  giant, 571
    bacterial necrosis of, 105
  pincushion, 570–571
  prickly-pear, 571
  saguaro, 571
  sea-urchin, 571
  star, 571
Caddy, 15
Cadminate, 15
Cadmium chloride, 15
Cadmium sulfate, 15
Cad-Trete, 15
Caeoma, 424
  faulliana, 424
  torreyae, 424
Caesalpinia, 572, 621, 779
Cajanus, 772
Caladium, 572
  soft rot, 140
  tuber rot, 383
Calathea, 572, 722
  leaf spot, 249
Calceolaria, 572
Calcium chloride injury, 337
Calcium deficiency, 337–338
Calcium sulfide, 15
Calendula, 572–573
  Botrytis on, 142
  leaf spot, 256
  white smut, 460
Caliciopsis pinea, 195
California-bluebell, 573
California dagger nematode,
  329
California fuchsia, 872
California-laurel, 573
  black mildew, 130
  pin nematode, 325
California pepper-tree,
  573–574
California pitcher-plant, 574
California poppy, 574
  capsule spot, 272
California-rose, 574
California sessile nematode,
  325
California sycamore, 776–777
  calixin, 15
  leaf spot, 294
Calla, 574–575
  leaf blight, 174
  leaf spot, 257, 285
  root rot, 403
  soft rot, 105–106
Calla palustris, 575
Calliandra, 575

Callicarpa, 575
Callirhoë, 782
Callistemon, 561
Callistephus, 542–543
Calluna, 674
Calocedrus decurrens, 689
Calochortus, 723
Calonectria crotalariae, 368,
  375
  nivale, 467–468
Calonyction, 732
Caltha, 723
Calycanthus, 575
Calyx-end rot of dates, 367
Camass, 575
Camassia, 575
Camellia, 576
  black mildew, 130
  Botrytis on, 142
  DDT injury, 338–339
  dieback, 208
  flower blight, 170, 181–182,
    183
  leaf gall, 240
  leaf spot, 281, 283, 293
  scab, 472
  yellow mottle leaf, 479
Camomile, 576–577
Campanula, 577
Camphor-tree, 577
  dieback, 205
  scab, 471
Campion, 714, 813
Campsis, 849–850
Camptosorus, 643
Canada bluegrass, 665–667
Canavalia, 694
Candleberry, 865
Candlestick shrub, 577
Candlewood, 648, 743
Candytuft, 578
Cane blight, of brambles, 159,
  163
  rose, 173
Cane gall of brambles, 98–100
Cane-knot canker, 224
Cane spot of raspberry, 193,
  221
Canistel, 713
Canker(s), 192–226
  apple, 213, 221, 225
  ash, 202
  bacterial, of stone fruits,
    114
    of tomato, 101–102
  balsam fir, 192–222
  basal, of maple, 221
  black rot, of tung, 219
  bleeding, 219–220
  blister, 199, 216

Canker(s) *(cont.)*
blueberry, 116, 218
Botryosphaeria, 194
branch gall, of poplar, 212
camellia, 208
citrus, 117–118
Coryneum, of cypress,
198–199
of rose, 209
cowpea, 121
crown, 220
currant cane, 139–140
Cytospora, 202
Dothichiza, of poplar, 206
Dothiorella, of oak, 206
Douglas-fir, 196, 203, 218
Endothia, of chestnut, 155
European Nectria, 214–215
gardenia, 217–218
honey locust, 224
Hymenochaete, 209–210
Hypoxylon, of poplar, 211
larch, 203
London plane, 195–196
madrone, 209
magnolia, 215
Nectria, of boxwood, 226
of beech, 214
perennial, of apple, 215–216
of peach, 225
Phomopsis, 217–218
pine, 195
poplar, 279
rose, brand, 197–198
brown, 199–201
common, 197
Coryneum, 209
crown, 201–202
graft, 197
sooty bark, 195
Sphaeropsis, 218–219, 223
Strumella, of oak and elm,
224
sycamore, 195–196
trunk, of apple, 220
twig, of apple, 199
elm, 193
hazelnut, 193
peach, 208
walnut, 212
willow, 219, 225
Canna, 578
mosaic, 479
Cannabis, 676
Cantaloupe, 578, 725–726
Canterbury bells, 577
Caocobre, 15
Cape-cowslip, 578
Cape-dandelion, 579
Cape-honeysuckle, 578

Cape-jasmine, 651–652
Cape-marigold, 578–579
Caper, 579
Caperonia, 843
Cape spurge, 662, 825
Capeweed, 579
Capnodiaceae, 69
Capnodium, 469
citri, 469
elongatum, 469
Capparis, 579
Capsella, 812
Capsicum, 765–767
Capsule spot of California
poppy, 272
Captafol, 15
Captan, 15
Capthion, 15
Caragana, 762–763
Caraway, 579
Carbam, 15
Carbamate, 16
Carbendazin, 16
Carbon disufide, 16
Carboxin, 16
Cardinal-flower, 710–711
Cardoon, 579
Carduus, 814
Carica, 754
Carissa, 579–580
canker, 223
gall, 242
Carnation, 580–581
anther smut, 467
bacterial spot, 116
bacterial wilt, 109–110
Botrytis on, 142
bud rot, 382
collar blight, 136
downy mildew, 233
etched ring, 479
fairy ring spot, 272
Fusarium wilt, 508
greasy blotch, 191
latent disease, 479
leaf mold, 272
mosaic, 479
mottle, 479
necrotic fleck, 479
pimple, 120
pin nematode, 325
ring spot, 480
rust, 451
stem rot, 382
streak, 480
yellows, 480, 508
Carnegiea gigantea, 571
Carob, 581
Carolina allspice, 575
Carolina jessamine, 581

Carolina moonseed, 581
Carolina spiral nematode, 329
Carpene, 16
Carpinus, 684–685
Carpobrotus, 689
Carrot, 581–582
Alternaria blight, 136
bacterial blight, 118
bacterial soft-rot, 105–106
black rot, 363
crown rot, 396
cyst nematode, 318
downy mildew, 235
early blight, 148
leaf blight, 136–137
mosaic, 477–478, 480
motley dwarf, 480
powdery mildew, 354–355
storage rot, 369, **413**
Carrot, Wild, 582
Carthamus, 803
Carum, 579
Carya, 678–679
illinoensis, 763–764
Caryota, 751
Casein, 16
Cashew, 582
Cassabana, 582
Cassabra, 725–726
Cassandra, 588
Cassava, 719–720
Cassia, 577, 583
Cassiope, 583
Castanea, 591–592
Castanopsis, 594
Castilleja, 750
Castor-bean, 583
gray mold blight, 140
stem canker, 192
stem gall, 243
Casuarina, 583
Catalina cherry, 583
Catalpa, 584
leaf spot, 250, 283
Catchfly, 813
Catenularia fuliginea, 369
Catface fruit deformity, 338
Catha, 584
Catkin hypertrophy of alder,
245
Catnip, 584
Cats-claw, 584, 776
Cat-tail, 585
Cattleya blossom brown
necrotic streak, 489
mosaic, 489
wilt, 508
Cauliflower, 569–570
head browning, 249–250
mosaic, 480

Caulophyllum, 560
Ceanothus, 585
Cedar, 585
  blight, 172, 211
Cedar-apple rust, 434–435
  Color Plate 2C, 2D
Cedar of Lebanon, 585
Cedrus, 585
Cela W524, 16
Celandine, 585
Celastrus, 557
Celeriac, 585–586
Celery, 585–586
  bacterial blight, 110
  bacterial soft rot, 105–106
  black crown rot, 369
  brown spot, 254
  calico, 480
  cracked stem, 338
  early blight, 147–148
  late blight, 183–184
  mosaic, 480
  Phoma root rot, 399
  pink rot, 415
  pin nematode, 325
  Rhizoctonia disease, 396
  wilt, 507
  yellows, 121–122
  yellow spot, 480
Celfume, 16, 42
Celosia, 603
Celtis, 670
Celtuce, 586
Cenangium, 147
  ferruginosum, 147
  piniphilum, 193
  singulare, 195
Centaurea, 587
Centipede grass, 587
Centranthus, 855
Centrosema, 568
Centrospora acerina, 369
Century plant, 587
  Botrytis on, 142
  leaf spot, 260–261
Cephalanthus, 568–569
Cephaleuros virescens, 254
Cephalosporium, 254
  apii, 254
  carpogenum, 369
  cinnamomeum, 255
  dieffenbachiae, 255
  diospyri, 502
  gregatum, 369
Cephalosporium wilt of elms,
  506–507
Cephalotaxus, 587
Ceratocystis, 195, 502
  fagacearum, 195, 503
  fimbriata, 369–370

paradoxa, 246
platani, 195–196
ulmi, 503–506, **505**
wageneri, 370
Ceratonia, 581
Ceratostomataceae, 70
Ceratostomella, 195
  paradoxa, 246
  ulmi, 503–506, **505**
Cercidium, 753
Cercis, 794
Cercobin, 16
Cercobin M, 16
Cercocarpus, 735
Cercospora, **136,** 147
  abeliae, 255
  abelmoschi, 255
  albo-maculans, 255
  althaeina, 255
  angulata, 255
  apii, 147–148
  aquilegiae, 255
  arachidicola, 255, 277
  armoraciae, 255
  beticola, 255–256
  bolleana, 277
  bougainvilleae, 256
  brunkii, 256
  calendulae, 256
  cannabina, 256
  cannabis, 256
  capsici, 256
  carotae, 148
  cercidiocola, 278
  circumscissa, 256
  citrullina, 256
  concors, 188, 256
  cornicola, 256
  cruenta, 278
  fusca, 256–257
  halstedii, 189–190
  lathyrina, 257
  lythracearum, 190, 257
  magnoliae, 257
  melongenae, 257
  microsora, 148
  nandinae, 257
  personata, 257, 277
  piaropi, 257
  pittospiri, 257
  puderi, 257
  punicae, 190
  purpurea, 188
  resedae, 257
  rhododendri, 257
  richardiaecola, 257
  rosicola, 257–258, 279
  smilacis, 258
  sojina, 258
  sordida, 148

symphoricarpi, 258
thujina, 148
Cercosporella, **252,** 258
  brassicae, 255, 258
  pastinacae, 287
Cereal anthracnose, 89
  downy mildew, 237
  powdery mildew, 354
  stem rot, 382
Cereus, 570
Cerotelium, 424
  dicentrae, 424
  fici, 424
Cestrum, 587
  bacterial canker, 101–102
Ceuthospora lunata, 370
CGA 38140, 16
CGA 48988, 16
CGA 71818, 16
Chaenomeles, 790
Chaerophyllum, 588
Chaetomiaceae, 70
Chalara quercinum, 503
Chalaropsis thielavioides, 300,
  370
Chamaecyparis, 588
Chamaedaphne, 588
Chamaedorea, 750
Chambers dagger nematode,
  331
Charcoal rot, 389–390, 411
Chard blue mold, 233
Char spot of grass, 290
Chaste-tree, 860
Chayote, 588–589
Checkerberry, 652, 868
Checker mallow, 589
Cheiranthus, 860
Chelidonium, 585
Chelone, 853
Chem Bam, 16
Chemicals, list of, 10
  mixing of, 56–57
Chem Neb, 16
Chem-O-Bam, 16
Chemsect, 16
Chem zineb, 16
Chenopodium, 589
Cheiranthus, 860
Cherimoya, 589
Cherry, 589–590
  albino, 480
  bacterial canker, 114
  bark splitting, 480
  black canker, 480
  black knot, 124–126, **125**
  Blossom blight, 164
  buckskin, 480
  bud abortion, 481
  canker, 203, 204, 218

Cherry *(cont.)*
  chlorosis, 481
  freckle fruit disease, 481
  green ring mottle, 481
  gummosis, 481
  leaf curl, 244
  leaf spot (blight), 161, 250,
    259–260
  little cherry, 481
  midleaf necrosis, 481
  Mora, 481
  mottle leaf, 481
  necrotic rusty mottle, 481
  pink fruit, 481
  pinto leaf, 481
  powdery mildew, 358
  rasp leaf, 481
  ring spot, 481
  rough bark, 481
  rugose mosaic, 481
  rusty mottle, 481
  trunk canker, 219–220
  twig blight, 164
  twisted leaf, 482
  vein clearing, 482
  witches' broom, 244
  yellows, 482
Cherry, flowering, 590–591;
    oriental, 590–591
Cherry, Japanese flowering,
    591
Cherry-laurel, 591–592
Chestnut, 591–592
  blight, 157
  dieback (canker), 199
  mistletoe, 298
Chick-pea, 592
  blight, 165
  filiforme, 480
Chickweed, 592
Chicory, 592
  bacterial rot, 110
Chili pepper wilt, 507
Chilopsis, 622
Chimaphila, 775
China aster, 542–543. *See
    also* aster
Chinaberry, 592–593
Chinese cabbage, 593
  damping-off, 364
  root rot, 364
Chinese evergreen, 593
Chinese hibiscus, 678
Chinese holly, 680–681
Chinese lantern, 593–594
Chinese-laurel, 530
Chinese quince, 790
Chinese tallowtree, 594
Chinese waxgourd, 594
Chinosol, 16, 41

Chinquapin, 594
  leaf blister, 244
Chiogenes, 594
Chionanthus, 649
Chionodoxa, 594
Chipco 26019, 16
Chipco 26019Flo, 16
Chipco Spot Kleen, 16
Chipco Thiram 75, 16
Chirita, 594
Chives, 595
Chloranil, 16
Chlorine injury, 338
Chloris gayana, 794
Chlorogalum, 595
Chloroneb, 16
Chloro-O-Pic, 16
Chloro-O-Pic 70, 16
Chloropicrin, 17, 42
Chlorosis, 338, 342
  infectious, of rose, 494
Chlorothalonil, 17
Chlorotic mottle, chrysanthe-
    mum, 482
Choanephora, 148
  cucurbitarum, 149
  infundibulifera, 149
Choanephoraceae, 66
Chokeberry, 595
Chokecherry, 595–596
Chondropodium pseudotsugae,
    196–197
Christie's spiral nematode,
    329
  stubby root nematode, 329
Christmasberry, 715, 771
  scab, 453
Christmas cactus, 596
  root rot, 404
Christmas Rose, 596
  leaf spot, 261
Chrysalidocarpus, 750
Chrysanthemum, 596–597
  aspermy, 482, 498
  bacterial blight, 107, 110
  bacterial crown gall, 98–100
    Color Plate 4C
  Botrytis on, 142
  chlorotic mottle, 482
    Color Plate 3E
  flower distortion, 482
  foliar nematode, **311, 312**
  leaf blight, 183
  leaf spot, 263, 291
  mosaic, 482
  petal blight, 162
  ray blight, 138
  ring spot, 482
  rosette, 482
  rust, 444

  stem rot, 383
  stunt, 482
    Color Plate 3A, 3E
  white rust, 446
  wilt, 508
Chrysanthemum cinerarii-
    folium, 788
  coccineum, 788
  frutescens, 722
  leucanthemum, 619
  maximum, 812
  parthenium, 643
  segetum, 608
Chrysobalanus, 603
Chrysomyxa, 424
  arctostaphyli, 355
  chiogenis, 424
  empetri, 424
  ilicina, 424
  ledi, 424
    var. cassandrae, 424
    var. groenlandici, 424
    var. rhododendri, 424
  ledicola, 424
  moneses, 424
  piperiana, 425
  pirolata, 425
  pyrolae, 425
  weirii, 425
Chrysopsis, 597
Chrysothamnus, 790
Chytridiaceae, 63
Chytridiales, 63
Ciboria, 149
  acerina, 149
  carunculoides, 149
Ciborinia, 149
  bifrons, 258
  confundens, 259
  erythronli, 149
  gracilis, 149
Cibotium, 643
Cicer, 592
Cichorium, 633–634
Cierodendrin, 559
Cimicifuga, 597
Cinchona, 598
Cineraria, 598
  mosaic, 482
Cinnamomum, 577
Cinnamomum zeylandicum,
    598
Cinnamon-tree, 598
Cinnamon-vine, 870
Cinquefoil, 785
Cintractia, 460
Cirsium, 598, 843
Cissus, 598–599, 767
Cistus, 797
Citronella grass, 703

Citrullus, 863–864
Citrus, 599–600
  anthracnose, 92
  blast, 114
  blight, 179
  blue contact mold, 397
  brown rot, 402–403, 404
  canker, 118
  dieback, 205
  Diplodia rot, 401
  dodder, 230
  exocortis, 482–483
  foot rot, 402–403
  fruit rot, 384
  green mold, 397
  leaf spot, 278
  melanose, 376
  mildew, 356–357
  mottle leaf, 349
  nematode, 329; ring, 314
  psorosis, 483
  quick decline, 483
  scab, 471–472
    Color Plate 8D
  Septoria spot, 291
  sooty mold, 469
  sour rot, 395
  spreading decline, 327–328
  stubborn disease, 483
  tatter leaf, 483
  tristeza, 483
  twig blight, 157, 186
  vein enation, 483
  xyloporosis, 483
  yellow vein, 483
Cladastris, 871
Cladochytriaceae, 63
Cladode rot of cactus, 366
Cladosporium, 188
  album, 301
  beijerinckii, 150
  bruneo-atrum, 452
  carpophilum, 452
  cerasi, 452
  coreopsidis, 452
  cucumerinum, 452
  effusum, 452–453
  epiphyllum, 259
  fulvum, 300
  herbarum, 188, 301
  macrocarpum, 301
  paeoniae, 188
  pisicolum, 453
Clarkia, 601
Clausena, 601
Clavariaceae, 76
Clavibacter, 101
  agropyri, 101
  fascians, 101
  flaccumfaciens, 101

humiferum, 101
michiganense, 101–102
poinsettiae, 102
sepedonicum, 102
Clavicipitaceae, 70
Claytonia, 601
Clematis, 601–602
  leaf blight, 171
  leaf spot, 263
  leaf and stem spot, 252–253
Cleome, 602
Clerodendrin, 559
Clerodendron, 602
  zonate ring spot, 483
Clethra, 602
Cliffbrush, 694
Cliftonia, 566
Climbing hempweed, 728
Clinopodium, 602
Clintonia, 602–603
Clitocybe, 370
  monadelpha, 371
  root rot, 371–372
  tabescens, 371–372
Clitoria, 568
Clivia, 603
Clockvine, 844
Clorto Caffaro, 17
Clorto Caffaro Flow, 17
Clortocof Ramato, 17
Clortosip, 17
Cloudy spot of tomato, 394
Clover club leaf, 483
  cyst nematode, 319
  dodder, 230
  leaf spot, 287
  mosaic, 484
  vein mosaic, 484
  wound tumor, 484
Clove rot of garlic, 397
Club root, 226–227
Clypeosphaeriaceae, 71
Cnicus, 844
Coach-whip, 743
Cobb's meadow nematode,
  326
  ring nematode, 314
  spiral nematode, 317
  stubby root nematode, 329
Coccoloba, 603, 809
Coccomyces, 259
  hiemalis, 161, 259–260
  kerriae, 161, 260
  lutescens, 260
  prunophorae, 260
Cocculus, 581
Cochliobolus heterostrophus,
  159
Cocklebur, 603
Cockscomb, 603

Cocoa-plum, 603
Coconut palm, 751
Cocos, 750
Cocoyam, 604
Codiaeum, 612
Codonanthe, 604
Coffee-berry, 604
Cohosh, 548
Coix lachryma-jobi, 696
Colchicum, 544
Coleosporium, 425
  apocyanaceum, 425
  asterum, 425, **425**
  campanulae, 426
  croweliii, 426
  delicatulum, 426
  elephantopodis, 426
  helianthi, 426
  ipomoeae, 426
  jonesii, 426
  lacinariae, 426
  madiae, 426
  mentzeliae, 426
  minutum, 426
  occidentale, 426
  pinicola, 426
  solidaginis, 425
  sonchi, 426
  terebinthinaceae, 426
  vernoniae, 426
  viburni, 426
Coleus, 604
  mosaic, 484
Collards, 604
Collar rot of rhododendron,
  403
  of tomato, 363
Colletotrichum, 88
  acutatum, 260
  antirrhini, 88
  atramentarium, 88–89
  bletiae, 89
  capsici, 372
  circinans, 372
  coccodes, 260, 372
  dematium f. sp. truncata,
    89, 260
  destructivum, 230
  elastica, 260
  erumpens, 89
  fragariae, 89
  fuscum, 89
  gloeosporioides, 89, 94,
    260
  graminicolum, 89
  higginsianum, 89–90
  lagenarium, 90
  liliacearum, 90
  lilii, 372
  lindemuthianum, 90–91

Colletotrichum (cont.)
  malvarum, 91
  nigrum, 372
  omnivorum, 91
  phomoides, 95
  pisi, 91
  schizanthi, 91
  spinaciae, 91
  truncatum, 91
  violae-tricoloris, 91
Collinsia, 604–605
Collinsonia, 605
Collomia, 605
Collybia velutipes, 372
Colocasia, 632
Coltsfoot, 605
Columbine, 605
  leaf spot, 255
Columbo, 606
Colutea, 558
Comac, 17
Comandra blister rust, 427
Comes, 76
Compass plant, 813
Compressed air sprayers, 54
Comptonia, 836
Condalia, 827
Coneflower, 801–802
Confederate jasmine, 606
Confederate-rose, 678
Conifers, brown felt blight of,
    160
  leaf spot, 281
  Phomopsis disease, 217
  snow blight, 171
  tip blight, 170
Coniophora, 373
  cerebella, 373
  corrugis, 373
Coniothyrium, **196**, 197
  concentricum, 260–261
  diplodiella, 373
  fuckelii, 163, 197
  hellebori, 261
  pyrina, 261
  rosarum, 197
  wernsdorffiae, 197–198
Conium, 780
Convallaria, 708
Conversion table, 57
Convolvulus, 574
Coontie, 872
Cooperia, 791
Cop-O-Cide, 17
Copper acetate, 17
Copper ammonium carbonate,
    17
Copper carbonate, basic, 17
Copper chloride, basic, 17
Coppercide, 17

Copper compounds, 17–18
Copper-Count-N, 18
Copper-Count-NS, 18
Copper deficiency, 338
Copper-Fixed, 18
Copper hydroxide, 18, 41
Copper injury, 338
    Color Plate 6C
Copper leaf, 520
Copper-lime dust, 18
Copper naphthenate, 18
Copper Nordox, 18
Copper oxides, 18
Copper oxychloride, 18
Copper Power, 18
Copper Pride, 18
Copper-Sandoz, 18
Copper spot of grass, 268,
    288
Copper spray injury, 338
Copper sulfate, 18, 41, 57
Coptis, 662
Coral-bells, 606, 677–678
Coralberry, 606
Coral spot, 213–214
Cordaderia, 753
Cordia, 606
Cordyline terminalis, 844
Coreopsis, 606–607
Coriander, 607
Coriandrum, 607
Coriolis versicolor, 407–408
Cork of apple, 338
Corm rot, 368
Corn, 607–608
  bacterial rot, 107
  bacterial stripe, 109
  bacterial wilt, 108
  brown spot, 285–286
  crown rot, 386
  cyst nematode, 319
  ear rot, 377, 384
  head smut, 462
  leaf blight, northern, 160
    southern, 159–160
    Color Plate 7F
  leaf fleck, 484
  meadow nematode, 327
  mosaic, 484
  Nigrospora cob rot, 394
  pink kernel rot, 380
  root rot, 384
  rust, 448
  smut, 465–467
  southern rust, 447
  stunt, 122, 124
Corncockle, 608
Cornflower, 587
Corn-marigold, 608
Corn-salad, 855

Cornus canadensis, 625
  florida, 625–626
  nuttalli, 626
Coronilla varia, 613
Corozate, 18
Cortaderia, 753
Corticium, 149
  centrifugum, 373
  fuciforme, 374
  galactinum, 374
  koleroga, 150
  microsclerotia, 150, 168
  polygonia, 374
  radiosum, 374
  salmonicolor, 150
  stevensii, 150, 168–169
  vagum, 150
Corydalis, 608
Corylus, 673
Corynebacterium. See
    Clavibacter
Coryneliaceae, 69
Corynespora, 261
  cassicola, 261
Coryneum, 150, **196**
  berckmanii, 150
  cardinale, 198–199
  carpophilum, 150
  foliicola, 199
  microstictum, 150, 209
Coryneum blight, of arbor-
    vitae, 150
  of stone fruits, 150
Coryneum canker, of cypress,
    198–199
  of rose, 209
Cosmos, 608–609
Cotinus, 815
Cotoneaster, 609
Cotton root-knot nematode,
    324
Cotton root rot, 398–399
Cottonwood, 781–782
Coursetia, 609
Covered kernel smut, 462
Covered smut, of barley, 464
  of oats, 465
Cowania, 609
Cow-parsnip, 676
Cowpea, 609
  blackeye mosaic, 479
  canker, 121
  chlorotic mottle, 484
  mosaic, 484
  zonate leaf spot, 251
CP Basic Copper TS-53WP,
    18
Crabapple, flowering, 610
  leaf spot, 272
  mistletoe, 298

scab, 453–457
 Color Plate 6F
Cracked stem, 338
Crag Fungicide 974, 42
Crag Nematicide, 42
Crambe, 809–810
Cranberry, 610–611
 bitter rot, 385
 black rot, 370
 black stem spot, 278
 blotch rot, 362
 early rot, 386
 end rot, 385
 false blossom, 484
 hard rot, 392
 red leaf gall, 243
 rose bloom, 240
Cranesbill, 654
Crape jasmine, 841
Crape-myrtle, 611
 leaf spot, 257
 powdery mildew, 354
 tip blight, 173
Crassula, 611
Crataegus, 672–673
Creeping snowberry, 594
Creosote bush, 611
Crepis, 671–672
Cress, winter, 868
 garden, 767
 downy mildew, 233
Cribropeltis citrullina, 238
Criconema, 314
 civellae, 314
 decalineatum, 314
 spinalineatum, 314
Criconemella xenoplax, 313
Criconemoides annulatum,
 314
 citri, 314
 crotaloides, 314
 curvatum, 314
 cylindricum, 314
 komabaensis, 314
 lobatum, 314
 mutabile, 314
 ornatum, 314
 parvum, 314
 rusticum, 314
 similis, 314
 teres, 314
 xenoplax, 314
Crinkle, strawberry, 495
 mild, 495
Crinum, 612
Crisfolatan, 18
Crisfuran, 42
Cristulariella, 261
 depraedans, 261
 pyramidalis, 261

Crocanthemum, 650
Crocus, 612
 corm rot, 382
Crocus, autumn, 544
Cronartium, 426–427
 appalachianum, 427
 cerebrum, 428
 coleosporioides, 427
 comandrae, 427
 comptoniae, 427
 conigenum, 427
 filamentosum, 427
 fusiforme, 427–428
 harknessii, 428
 occidentale, 428
 quercuum, 428
 ribicola, **425,** 427–430
 stalactiforme, 430
 strobilinum, 430
Crossonema, 314
Crossvine, 555
Crotalaria, 612
Croton, 612
Crotothane, 18
Crowberry, 613
Crowfoot, 791–792
Crownbeard, 857
Crown canker, of dogwood,
 219–220
 of rose, 201–202
Crown gall, 99, 98–100
 Color Plate 4C
Crown-headed lance nema-
 tode, 320
Crown rot, 169–170
 of asparagus, 397
 bacterial, 104
 of corn, 386
 of delphinium, 104, 416
 of snapdragon, 393
Crown vetch, 613
Crucifers, bacterial leaf spot
 of, 110–111
 black leafspot, 249–250
 black leg, 127
 black rot, 117
 club root, 226–227
 downy mildew, 233
 ring spot, 277
 white rust, 501
Cryptantha, 613
Cryptanthus, 613
Cryptochaete polygonia,
 374
Cryptococcaceae, 78
Cryptodiaporthe, 199
 aculeans, 199
 castanea, 199
 salicina, 199
Cryptogramma, 642

Cryptomeria, 613
 needleblight, 173
Cryptomyces maximum, 199
Cryptomycina pteridis, 262
Cryptonol, 18, 41
Cryptospora, longispora, 151
Cryptosporella, 199
 umbrina, 199–201, **200**
 viticola, 201
Cryptosporium, 201
 minimum, 201
 pinicola, 201
Cryptostictis, 151
 arbuti, 262
Cucumber, 613–614
 angular leaf spot, 113
 bacterial wilt, 108–109
 blight, 136
 cyst nematode, 318
 mosaic, 484–485
 scab, 452
 wilt, 508
Cucumber mosaic virus,
 484–485
 Color Plate 4E, 5E
Cucumis melo, 725–726
 sativus, 613–614
Cucurbita, 826–827
Cucurbitariaceae, 71
Cucurbits, angular leaf spot
 of, 113
 Alternaria blight of, 136
 bacterial wilt of, 108–109
 Fusarium root rot, 383
 gummy stem blight, 165
 leaf spot, 256, 291, 294
 powdery mildew, 352–354
Cudweed, 659
Cufram Z, 19
Cufraneb, 19
Culvers-root, 614
Cuman, 19
Cumene, 19
Cumminsiella, 430
 mirabilissima, 430
 texana, 430
Cunila, 625
Cunninghamia, 614
Cuphea, 615
Cupressus, 617
Curamil, 19
Curitan, 19
Curaterr, 42
Curly top, beet, 477–478
Currant, 615
 anthracnose, 97–98
 cane blight, 139–140
 cane-knot canker, 224
 dieback, 218
 leaf spot, 255, 277, 279

Currant *(cont.)*
  mosaic, 485
  Pseudomonas, 112
  root rot, 387
Currant, flowering, 615–616
  leaf spot, 279
  nematode, 311
Curuba, 582
Curvularia, 151
  lunata, 151–152
  trifolii f. sp. gladioli,
    151–152
Curvularia disease of gladiolus,
  151–152
Curzate M, 19
Curzate M8, 19
Cuscuta americana, 230
  arvensis, 230
  californica, 230
  coryli, 230
  epithymum, 230
  exaltata, 230
  gronovii, 230
  indecora, 230
  paradoxa, 230
  pentagona, 230
  planifera, 230
Cushion-pink, 813
Custard-apple, 589
Cyamposis, 669
Cycad, 616
Cycas, 616
Cyclamen, 616
  leaf and bud blight, 159
  wilt, 508
Cycloconium oleaginum, 262
Cycloheximide, 10–11, 19
Cyclomorph, 19
Cydonia, 789
Cylindrocarpon, 374
  cylindroides, 201
  liriodendri, 373
  radicicola, 374
Cylindrocladium, 152
  avesiculatum, 152, 262
  crotalariae, 374
  floridanum, 375
  heptaseptatum, 375
  macrosporium, 262
  pteridis, 262, 375
  scoparium, 152, 201–202,
    375
Cylindrocladium blight, 152
Cylindrosporium, 152, 262
  betulae, 263
  carigenum, 277–278
  chrysanthemi, 263
  clematidinis, 263
  defoliatum, 153
  griseum, 153
  juglandis, 153

  mali, 214
  rubi, 293
  salicinum, 263
Cymbidium mosaic, 489
Cymbopogon, 703
Cynara cardunculus, 579
  scolymus, 539
Cynodon, 665–667
  Rhizoctonia blight, 180
Cynoglossum, 617
Cyperus rotunders, 741
Cyphomandra, 849
Cypress, 617
  Coryneum canker of,
    198–199
  dieback, 212
  root rot, 403–404
Cypress, bald, 546–547
  hinoki, 588
Cypress-vine, 617–618
Cyprex, 19, 21
Cypripedium, leaf spot, 253
  rot, 107
Cyrilla, 618
Cyrtomium, 641
Cyst nematodes, 318–319
Cystopteris, 640
Cystopus, 500
Cytisus, 564
Cytospora, **196,** 202
  abietis, 202
  annularis, 202
  chrysosperma, 202
  kunzei var. piceae, 202–203
  leucostoma, 203
  nivea, 203
  sambucicola, 203
Cytospora canker, of Italian
    prune, 203
  of spruce, 202–203
Cytovirin, 47
Cyttariaceae, 72

Daconil 2787, 19
Dactylis glomerata, 665–667
Daedalea, 375
  confragosa, 375
  quercina, 375
  unicolor, 375
Daffodil, 737–738
Dagger nematode, 331
Dahlia, 618
  Botrytis on, 142
  leaf smut, 460
  mosaic (stunt), 485
  oakleaf, 485
  ring spot, 485
  root nematode, 329
Daisy, oxeye, 619
  shasta, 681
Daldinia concentrica, 375

Dalea, 815–816
Dalibarda, 619
Dames–rocket, 677
Damping off, 227–228,
    249–250, 396–397, 409
Dandelion, 619
Daphne, 619
  leaf spot, 268, 274
Darlingtonia, 574
Dasanit, 42
Dasyscypha, 203
  agassizi, 203
  calycina, 203
  ellisiana, 203
  pseudotsugae, 203
  resinaria, 203
  wilikommii, 203–204
Date calyx-end rot, 367
  fruit rot, 369
Date palm, 750–751
  decline disease, 394
  leaf and stalk rot, 317
Datura, 619
  bacterial canker, 101
Daucus carota, 582
Daucus carota var. sativa,
    581–582
Daylily, 620
  anthracnose, 90
Dazomet, 19, 42
Dazonet Powder, 42
DBCP, 42
DCNA, 19
D-D 92, 42
D-D soil fumigant, 42
DDT injury, 338–339
Dead-arm disease of grapes,
    201
Decline disease of date palms,
    394
  nematodes, 314, 329
Decumaria, 620
Deergrass, 794
Deficiency, calcium, 337–338
  copper, 338
  iron, 340–341
  magnesium, 342
  manganese, 342
  nitrogen, 344
  oxygen, 344
  phosphorus, 344
  potassium, 344
  water, 347
  zinc, 349
Dehydroacetic acid, 19
Deksonal, 19
Delan, 19
Delan-Col, 19
Delphinium, 620–621
  black leg, 106
  black spot, 113

crown rot, 104
ring spot, 485
Delsene, 19
De Man's meadow nematode, 326
Dematiaceae, 78
Dematophora, 413
Demorphotheca, 621
Demosan, 16, 19
Denarin, 19
Dendranthemum grandiflora, 596–597
Dendromecon, 848
Dendrophoma obscurans, 153
Deodar, 585
Dermatea, 204
  acerina, 204
  balsamea, 204
  livida, 204
Dermea pseudotsugae, 204
Desain, 19
Desert Bird of Paradise, 621
Desert-candle, 621
Desert-plume, 621
Desert-thorn, 715
Desert-willow, 622
Desmanthus, 567
Desmella superficialis, 430
Desmodium, 622, 680
Deutzia, 622
Devils-club, 622
Devilwood, 622
Devizeb, 19
Dewberry, 159, 163, 622
Dewdrop, 619
Dexon, 19
Diamidfos, 42
Dianthus, 623
  leaf spot, 291
Dianthus barbatus, 840
  caryophyllus, 580–581
Diaporthe, 153
  arctii, 153
  batatatis, 376
  citri, 376
  cubensis, 204
  decorticans, 218
  eres, 172, 204
  gardeniae, 217
  helianthi, 204
  oncostoma, 204
  phaseolorum, 153–154, 376
    var. caulivora, 154, 204
    var. sojae, 154
  pruni, 204
  prunicola, 204
  vaccinii, 154
  vexans, 154, 172
  vincae, 218
Diatrypaceae, 71
Diazoben, 19

Dibotryon, 124
  morbosum, 124–126, 125
Dibromochloropropane, 42
Dicamate, 19
Dicentra, 629
  spectabilis, 558
Dichlofenthion, 43
Dichlofluamide, 20
Dichlone, 20
Dichloran, 20
Dichlorofenthion, 43
Dichloropropene, 20, 43
Dichondra, 624
Dichotomophthora
  indica, 154
  portulacae, 205, 376
Dichotomophthoropsis
  nymphaerum, 263
Dictyosporae, 77
Didymaria didyma, 263
Didymascella, 154
  thujina, 154, 155
  tsugae, 155
Didymelia, 155
  applanata, 155
  lycopersici, 253
  sepincoliformis, 204
Didymellina, 263
  macrospora, 263–265, 264, 272
  ornithogali, 265
  poecilospora, 265
Didymosphaeria, 155
  populina, 156
Didymosporae, 77
Didymosporina aceris, 274
Didymosporium arbuticola, 265
Dieback, 92, 94, 193, 339
  apple, 213–214
  apricot, 207, 213–214
  ash, 202
  black locust, 204
  Botryosphaeria, 194
  boxwood, 212, 214, 215
  butternut, 212, 213
  camellia, 208
  chestnut, 199
  currant, 139–140, 218
  elm, 193, 206
  Nectria, 213–214
  pines, 147
  poplar, 202
  raspberry, 193
  rhododendron, 219–220
  rose, 205, 209
  sumac, 199
  vinca, 218
Dieffenbachia, 624
  leaf spot, 118, 255
    Color Plate 4A

Diervilla, 624
Diethofencarb, 20
Difolatan, 20
Digitalis, 648–649
Dikar, 20
Dill, 624
  rot (wilt), 383
Dilophospora geranii, 265
Dimanin A, 41
Dimanin C, 41, 47
Dimerium, 122
  juniperi, 128
Dimerosporium, 129
  abietis, 129
  hispidulum, 129
  pulchrum, 129
  robiniae, 129
  tropicale, 129
Dimethirimol, 20
Dimortheca, 578–579
Dinitro compounds, 20
Dinobuton, 20
Dinocap, 20, 26
Dinofen, 20
Dioscorea, 870
Diospyros, 767–768
Dipher, 20
Diplocarpon, 131
  earliana, 246–247
  rosae, 130–135, 132, 133, 275
Diplodia, 156
  camphorae, 205
  coluteae, 156
  gossypina, 156
  infuscans, 205
  juglandis, 205
  longispora, 156
  natalensis, 156, 205, 376
  opuntia, 377
  phoenicum, 377
  pinea, 377
  quercina, 205
  sarmentorum, 156
  sophorae, 205
  sycina, 205
  theobromae, 377
  tubericola, 377
  zeae, 377
Diplodia collar and root rot, 377
Diplodia corn ear rot, 377
Diplodina, 265
  eurhododendri, 265
  persicae, 377
Diplotheca wrightii, 265–266
Dipsacus, 842
Dirca, 703
Direx, 20
Discella carbonacae, 205

Discohainesia oenotherae, 281–282
Discula fraxinea. *See* Gloeosporium aridum quercina
Disease inflorescence, 139
Diseases, bacterial, 98–121
  types of, 518–519
Disinfectant, 7
Disinfestant, 7
Dithane D-14, 20, 29–30
Dithane M-22, 20, 27–28
Dithane M-22 Special, 20, 27–28
Dithane M-45, 20
Dithane Z-78, 21, 39–40
Dithianon, 21
Dithiocarbamates, 21
Ditranil, 21
Di-Trapen, 43
Dittany, 625
Ditylenchus, 315
  destructor, 315
  dipsaci, 315–316
  gallicus, 316
  iridis, 316
Dizygotheca, 625
DMTT, 21, 43
DNC, 21
DNOC, 21
Doassansiaepilobii, 460
Dodder, 228–230, **229**
  bigseed alfalfa, 230
  clover, 230
  common, 230
  field, 230
  hazel, 230
  littleseed alfalfa, 230
Dodder latent mosaic, 485
Dodecatheon, 625
Dodemorfe, 21
Dodemorph, 21
Dodine, 21
Dogbane, 625
Doguadine, 21
Dog-tooth violet, 636
Dogwood, dwarf, 625
Dogwood, flowering, 625–626
  Botrytis on, 142
  crown canker, 219–220
  leaf spot, 253, 256, 291
  Nectria canker, 214–215
  spot anthracnose, 471
  twig blight, 151
Dogwood, gray, 626
  Pacific, 626
  pagoda, 626
Dojyopicrin, 21
Dolichodorus, 316
  heterocephalus, 316
  obtusus, 316

Dolichos, 627
  anthracnose, 89
Dollar spot of turf, 414
Dolochlor, 21
Dorlone, 43, 46
Doronicum, 627
Dorylaimus, 316
Dothichiza, 206
  caroliniana, 266
  populea, 206
Dothideaceae, 69
Dothideales, 69
Dothiora polyspora, 203
Dothiorella, 206
  fraxinicola, 206
  gregaria, 367
  quercina, 206
  ulmi, 206, 506–507
Dothiorella rot of avocado and citrus, 367
Dothiorella wilt of elms, 506–507
Dothistroma pini, 156
Double spot of blueberry, 266
Douglas-fir, 627–628
  bark canker, 196
  canker, 203
  dwarf mistletoe, 299
  needle cast, 302, 306
  needle rust, 437
  Phomopsis canker, 217–218
  root rot, 408
Dove-plum, 603, 809
Dowco 186, 21
Dow-Fume 75, 43
Dowfume 59, 43
Dowfume C, 43
Dowfume EB-5, 43
Dowfume F, 43
Dowfume MC-2, 43
Dowfume MC-33, 21, 43
Dowfume N, 43
Dowfume V, 43
Dowfume W-85, 43
Downy mildew(s), 231–237
  acalypha, 235
  agrimony, 234
  alfalfa, 234
  anemone, 235
  aster, 232
  beet, 234
  Bermuda grass, 237
  blackberry, 234
  carnation, 233
  carrot, 235
  cereals, 237
  crucifers, 233–234
  cucurbits, 236–237
  forget-me-not, 233
  geranium, 235

godetia, 232
gonobolus, 235
grape, 235–236
hackberry, 236
hepatica, 235
lettuce, 232
lima bean, 234–235
lupine, 234
mock strawberry, 234
oats, 237
onion, 232–233
pea, 234
prickly poppy, 232
rhubarb, 234
rose, 234
snapdragon, 232
soybean, 233
spinach, 233
strawberry, 233
tobacco, 234
viburnum, 235
Downy spot of pecan, 277–278
Doxantha, 584
Draba, 628
Dracaena, 628
  leaf spot, 268, 285
    Color Plate 7E
  tip blight, 173
Dracocephalum, 628
Dragonhead, 628
Drawinol, 21
Drazoxolon, 21
Drechslera catenaria, 159
  gigantea, 159
  setariae, 270
Drought, 339
Dry bubble of oyster mushroom, 515
Dryopteris, 643
Dry root rot of bean, 383
Dry rot, of gladiolus, 418
  of pomegranate, 394
DSE, 21
Duchesnea, 730
Du Nema, 43
Duranta, 629
Dusting, 55–56
Dusty miller, 587
Dutch elm disease, 503–506, **505**
Dutch iris, 691
Dutchmans-breeches, 629
Dutchmans-pipe, 629
Duter, 21
Du-Ter Fungicide, 21
Dwarf banana, 548
Dwarf dandelion, 699–700
Dwarf mistletoe, 298–299
  Douglas-fir, 299
    eastern, 299
    hemlock, 299

lodgepole pine, 299
  southwestern, 299
  western, 299
Dwell, 21
Dynone, 21
Dyrene, 21-22
Dyschoriste, 629

Early blight, of carrot, 148
  celery, 147-148
  potato and tomato, 137-138
Ear rot of corn, 377, 384
Earthcide, 22
Easter Cactus, 629
Easter lily, black scale rot of,
  372
Eastern gall rust, 428
Echeveria, 629
Echinacea, 629
Echinocactus, 571
Echinocystis, 730
Echinodontium tinctorium,
  377-378
Echinops, 658
Echium, 860
Ectostroma liriodendri, 266
EDB, 43
EDB-85, 43
E-D-Bee, 43
EDC, 43
ED/CT, 43
Eelworm disease of narcissus,
  315-316
Eelworms, 306
Egg-fruit, 713
Eggplant, 630
  anthracnose, 92
  bacterial canker, 101-102
  leaf spot, 253, 257
  Phomopsis blight, 154, 172
  rust, 448
  tobacco cyst nematode, 319
Eichhornia, 863
EL-273, 22
Elaeagnus, 630-631, 802
Elaphomycetaceae, 73
Elder, 631
  branch canker, 203
Elderberry, 631
  disease, 484
Elecampane, 690-691
Elephants-ear, 632
Elgetol, 22
Elm, 632-633
  black spot, 269
  blight, 172
  canker, 219-220
  dieback, 193, 206
  Dothiorella wilt, 506-507
  Dutch elm disease, 503-506,
    505

leaf blister, 245
leaf spot, 268, 290
mistletoe, 298
mosaic, 485
phloem necrosis, 122
Sphaeropsis canker, 223
twig blight, 172
wetwood, 107-108
wilt, 506-507
yellows, 124
zonate canker, 485-486
Elsinoaceae, 69
Elsinoë, 470
  ampelina, 470-471
  cinnamomi, 471
  corni, 471
  diospyri, 471
  euonymi-japonici, 471
  fawcetti, 471-472
  ilicis, 472
  jasminae, 472
  ledi, 472
  lepagei, 472
  leucosphila, 472
  magnoliae, 472
  mangiferae, 472
  mattirolianum, 472
  parthenocissi, 472
  phaseoli, 472
  piri, 472
  quercicola, 472
  quercus-falcatae, 472
  randii, 473
  rosarum, 473
  solidaginis, 473
  tiliae, 473
  veneta, 473
Elvaron, 22
Elymus, 866
Elytroderma, **302,** 303
  deformans, 303
Emilia, 633
Empetrum, 613
Encelia, 633
Endive, 633-634
Endoconidiophora, **196**
  fagacearum, 503
  fimbriata, 369-370
    var. platani, 195-196
Endocronartium hardnessii,
  430
Endomycetaceae, 68
Endomycetales, 68
Endophyllum, 431
  sempervivi, 431
  tuberculatum, 431
Endosan, 22
Endothia, 156
  gyrosa, 207
  parasitica, 157
Endothia canker, 157

End spot of avocado, 339
Engelmann daisy, 634
Engelmannia, 634
Englerulaceae, 69
English daisy, 634
English holly, 680-681
English iris, 691
English ivy, 693
  bacterial leaf spot, 119
  leaf and stem spot, 251
  leaf spot, 283-284
  scab, 474
English walnut, 861-862
ENT 27164, 43
Entomophthorales, 66
Entomosporium leaf spot, **136,**
  157
Entomosporium maculatum,
  157
  thuemenli, 158
Entyloma, 460
  calendulae, 460
  compositarum, 461
  crastophilum, 461
  dactylidis, 461
  dahliae, 461
  ellisii, 461
  irregulare, 461
  lineatum, 461
  polysporum, 461
Epicoccum, 266
  asterinum, 266
  neglectum, 266
  nigrum, 266
  purpurascens, 207
Epigaea, 634
Epilobium, 634-635
Episcia, 635
Equitdazin, 22
Eradex, 17
Eradicant, 7
Eradication, 5
Eraditon, 22
Eranthemum, 635
Erazidon, 22
Eremochloa, 587
Eremurus, 621
Erianthus, 779
Erica, 674
Erigenia, 671
Erigeron, 635
Eriobotrya, 712-713
Eriodictyon, 871
Eriophyllum, 636
Erodium, 677
  cicutarium, 645
Erostrotheca multiformis, 301
Erwinia, 102-109
  amylovora, **103,** 103-105
  aroideae, 105
  atroseptica, 106

Erwinia (cont.)
  carnegiana, 105
  carotovora, 105–107
  chrysanthemi, 107
  cypripedii, 107
  dissolvens, 107
  herbicola, 107
  lathyri, 107
  nimipressuralis, 107–108
  rhapontica, 108
  stewartii, 108
  tracheiphila, 108–109
Eryngium, 636
Erysimum, 861
Erysiphaceae, 69
Erysiphales, 69
Erysiphe, 351, 352, 353
  aggregata, 352
  cichoracearum, 352–354
  graminis, 354
  heraclei, 354
  lagerstroemiae, 354
  polygoni, 354–355
  taurica, 355
  trina, 355
Erythrina, 636
Erythronium, 636
  black spot, 253
  leaf blight, 149
Escarole, 633–634
Eschscholtzia, 574
ETCMTD, 22
Ethazol, 22
Ethoprop, 44
Ethylene dibromide, 44
Ethylene dichloride, 44
Etridiazole, 22
Etrimin, 22
Eucalyptus, 636–637
  canker, 204
Eucharis, 637
Eugenia, 637
  black spot, 128
Euonymus, 637
  basal canker, 222
  crown gall on, 98–100
  leafspot, 267, 279
  mildew, 356
  mosaic, 486
Euparen, 22
Euparene, 22
Euparen M, 22
Eupatorium, 638
  Botrytis on, 142
Euphorbia, 662
Euphorbia corollata, 825
  cyparissias, 825
  heterophylla, 825
  lathyris, 825
  maculata, 825

marginata, 817
pulcherrima, 779–780
trigona, 638
European dagger nematode,
  329
European larch canker, 203
European Nectria canker,
  214–215
European powdery mildew,
  356
Eurotiaceae, 69
Eurotiales, 68–69
Eustoma, 638
Eutypa armeniaca, 207
Evening-primrose, 743
Everlasting, 638–639
  pearl, 528
Evolvulus, 638
Exacum, 639
Exanthema, 339
Excipulaceae, 78
Exclusion, 5
Exobasidiceae, 75
Exobasidiales, 75
Exobasidium, 239–240
  azaleae, 240
  burtii, 240
  camelliae, 240
  oxycocci, 240
  rhododendri, 240
  symploci, 240
  uvae-ursi, 240
  vaccinii, 240, 241
  vaccinii-uliginosi, 240–241
Exocortis, citrus, 482–483
Exosporium, 266
  concentricum, 267
  liquidambaris, 267
  palmivorum, 267
Exotherm Termil, 22

Fabraea, 157
  maculata, 157–158
  thuemenii, 158
Fagus, 553
Fairy rings, 237–238
Fairy ring spot, 271
Fall-daffodil, 829
Fall dandelion, 671
False-boneset, 700
False bugbane, 848
False-chamomile, 723
False dragonhead, 771
False garlic, 741
False-hellebore, 856
False indigo, 548
False Jerusalem-cherry, 695
False loosestrife, 713
False-mallow, 718
False Mesquite, 575

False root-knot nematode,
  325
False smut of palms, 461
False solomons seal, 814–815
False yeasts, 78
Farkleberry, 639
Fasciation, 101
Favolus, 378
  alveolaris, 378
Feather rot, 408
Feijoa, 639
Felt fungi, 222–223
Fenaminosulf, 22
Fenarimol, 22
Fendlera, 639
Fennel, 639
Fenolovo acetate, 23
Fensulfothion, 44
Fentin acetate, 23
Ferbam, 23
Ferberk, 23
Fermasan, 23
Fermate, 23
Fermid 850, 23
Fern, adders-tongue, 640
  birds-nest, 640
  bladder, 640
  Boston, 640
  bracken, 640
  brake, 641
  chain, 643
  Christmas, 641
  cinnamon, 642
  cliff-brake, 641
  holly, 641
  interrupted, 642
  lady, 641
  leatherleaf, 641
  maidenhair, 642
  osmunda, 642
  ostrich, 642
  polypody, 642
  rock, 642
  rock-brake, 642
  royal, 642
  sensitive, 643
  shield, 643
  silvery spleenwort, 641
  tree, 643
  walking, 643
  wood, 643
  woodwardia, 643
Fern diseases, 640–643
  anthracnose, 95
  leaf blight, 109, 110
  leaf blister, 245
  leaf spot, 250, 262
  nematode, 310
  post harvest decay, 375
  rust, 450

tar spot, 262
tip blight, 173
Ferrous sulfate, 23
Fescues, 665–667
  netblotch, 270
  smut, 467
Festuca, 665–667
Fetterbush, 715
Feverfew, 643
Ficus aurea, 644–645
  carica, 643
  elastica, 801
Fig, 644
  cyst nematode, 318
  dieback, 205
  leaf blight, 168
  leaf spot, 260, 277
  mosaic, 486
  pin nematode, 325
  ripe rot, 380
  rust, 424
  smut, 367
  sooty mold, 469
  spine nematode, 314
  thread blight, 168–169
Fig, Florida strangler, 644–645
Fig-marigold, 727
Figwort, 645
Filaree, 645
  red leaf, 486
Filbert, 645, 673
  blight, 118
Filipendula, 725
Filipin, 23
Fimetariaceae, 70
Fir, 645–647
  black mildew, 129
  canker, 196–197, 203, 204,
    216, 221
  mistletoe, 299
  needle blight, 161, 170,
    179–180
  needle cast, 302, 303, 305,
    306
  Phomopsis canker, 217, 218
  root rot, 403, 408, 421
  rust, 436, 450
  twig dieback, 223
Fire of arbor-vitae, 148
  iris, 263–265
  tulip, 145–146, **146**
Fire blight, **103,** 103–105
Fire-chalice, 872
Firethorn, 647, 788
Fireweed, 634–635
Firmiana simplex, 770
Fisheye rot of apple, 373
Fittonia, 647
Five-leaf aralia, 520
Flannel bush, 649

Flax, flowering, 647
Fleabane, 635
Floating-heart, 741
Floras-paintbrush, 633
Florida yellow-frumpet, 829
Florists' Smilax, 542, 815
Flotation sulfur, 23
Flower blight of camellia,
    170, 181–182, 183
Flowering maple, 519
Flowering tobacco, 740
Flower rot of orchids, 363
Flower smut, 462
Flower spot of azalea, 165–168
Fly speck of fruit, 238–239
FMC 9102, 23
FMC 10242, 44
Foam-flower, 647
Foeniculum, 639
Fog-fruit, 709
Folcid, 20, 23
Foliar nematode, 325
Folicur, 23
Folosan, 23
Folpan, 23
Folpet, 23
Foltaf, 23
Fomes, 378
  annosus, 378
  applanatus, 378
  connatus, 378
  everhartii, 378–379
  fomentarius, 379
  fraxinophilus, 379
  igniarius, 379
  officinalis, 379
  pini, 379
  pinicola, 379
  rimosus, 380
  robustus, 380
  roseus, 380
Fomitopsis officinalis, 379
Fonganil, 23
Fongarid, 23
Fore, 23–24
Forestiera, 647
Forget-me-not, 647–648
  downy mildew, 233
Formaldehyde, 24
Formalin, 24
Forsythia, 648
Fortunella, 700
Forturf, 24
Fouquieria, 648
Four-o'clock, 648
Foxglove, 648–649
  anthracnose, 89
  leaf spot, 288
Fragaria, 831–832
Frangipani, 779

Franklinia, 663
Frasera, 606
Fraxinus, 540–541
Fraxinus holotriocha, 541
Freesia, 649
Fremontia, 649
French-mulberry, 575
Fringe-tree, 649
Fritillaria, 649
Froelichia, 650
Frog-eye disease of soybean,
    258
Frommea obtusa duchesneae,
    431
Frommella duchesnea, 431
Frost injury, 339, **339**
Frost scorch, 468
Frostwort, 650
Frucote, 24
Fruit rot, 149
  Alternaria, 363
  Diplodia, 376
  gray mold, 368
  peach, 377
  pepper, 372, 376
  tomato, 376
Fruit spot, 238–239
Fuberidazol, 24
Fuchsia, 650
Fuklasin, 24
Fumago vagans, 470
Fumazone, 44
Fumigant-1, 44
Fumigants, 7
Fungi, 60–78
  imperfecti, **77,** 77–78
Fungicides, 7, 10
Fungiclor, 24
Funginex, 24
Fungi-Rhap, 24
Fungi-Rhap Cu6, 24
Fungitrol, 24
Fungo 50, 24
Furadan, 44
Furcaspora sp., 158
Furcraea, 724
Fusarex, 24
Fusarium, 380, **380**
  annuum, 507
  avenaceum, 380
  bulbigenum, 382
  buxicola, 214
  culmorum, 380
  graminearum, 384
  heterosporium, 453
  lateritium, 157, 303
  moniliforme, 380
    var. subglutinans, 158,
      207
  nivale, 467–469

Fusarium *(cont.)*
  orthoceras, 380
    var. gladioli, 381
  oxysporum, 380–382,
    507–510
    f. apii, 507
    f. asparagi, 507
    f. barbati, 507
    f. batatas, 381, 507
    f. callistephi, 507
    f. cattleyae, 508
    f. cepae, 381
    f. chrysanthemi, 381, 508
    f. conglutinans, 508
    f. cubense, 508
    f. cucumerinum, 508
    f. cyclaminis, 508
    f. dianthi, 508
    f. gladioli, 381–382, 508
    f. hebae, 509
    f. lilii, 382
    f. lycopersici, 509
    f. melonis, 509
    f. narcissi, 382
    f. niveum, 509
    f. peniciosum, 509–510
    f. pisi, 510
    f. radicis-lycopersici, 382
    f. raphani, 510
    f. spinaciae, 510
  poae, 382
  roseum, 382
    f. cerealis, 382
  solani, 158, 207, 383
    f. cucurbitae, 383
    f. phaseoli, 383
    f. pisi, 383
Fusarium brown rot of
  gladious, 381–382
Fusarium patch of grass,
  467–468
Fusarium root rot of cucurbits,
  383
Fusarium wilts, 507–510,
  passim
Fusicladium, 267
  carpophilum, 452
  dendriticum, 453, 455–457,
    456, 457
  eriobotryae, 453
  photinicola, 453
  pisicola, 267
  pyracanthae, 453
  pyrinum, 457–458
  robiniae, 267
  saliciperdum, 453–454
Fusicoccum, 207
  amygdali, 208
  elaeagni, 208
  putrefaciens, 385

Gaillardia, 650
Galanthus, 817
Galax, 650
Galben, 24
Galinsoga parviflora, 814
Galium, 651
Gall, 239–243
  bacterial, of oleander,
    115–116
  basal, 242
  blueberry bud-proliferating,
    242
  bud, 240
  cane, 98–100
  crown, 98–100, **99**
  leaf, 240, 242
    of azalea, 240, **241**
    of camellia, 240
    of hedge parsley, 242
    red, 242
    of rhododendron, 240
  rust, 435
  stem, 242, 243
Gallex, 41
Galltrol-A, 41
Galtonia, 651
Ganocide, 24
Ganoderma, 383
  applanatum, 378, 383–384
  curtisii, 384
  lucidum, 384
  sulcatum, 384
  zonatum, 384
Garbanzo, 592
Garden balsam, 689
Garden cress, 767
Garden heliotrope, 855
Gardenia, 651–652
  canker, 217–218
  sooty mold, 469
Garden pinks, 623
Garden sorrel, 802
Garden verbena, 856
Garlic, 652
  clove rot, 397
Garlic, false, 741
Garrya, 652, 813
Gas toxicity, 339–340
Gaultheria, 652
  procumbens, 868
  shallon, 804–805
Gaura, 652
Gayfeather, 705–706
Gaylussacia, 687
Gayophytum, 696
Gazania, 653
Gelsemium, 581
Genista, 653
Gentian, 653
Gentiana, 653

Geoglossaceae, 72
Geranium, 653–654
  bacterial leaf spot, 118,
    120
  black leaf speck, 295
  blackleg, 410
  Botrytis on, 142
    Color Plate 8C
  chlorotic spot, 486
  crinkle, 486
  downy mildew, 235
  leaf spot, 256, 265, 281
  mosaic, 486
  petal spot, 270
  rust, 447
Gerbera, 654–655
Germander, 655
German iris, 691–692
Geum, 655
Giant-hyssop, 522
Giant reed, 539–540
Giant sequoia, 811
Gibasis, 841
Gibbago trianthemae, 267
Gibberella, 158
  baccata, 158, 208
  fujikuroi, 380
  zeae, 384
Gibberidea, 126
  heliopsidis, 126
Gilia, 655–656
Gillenia, 656
Ginger, wild, 656
Ginkgo, 656
Ginseng, 656–657
  Alternaria blight, 137
Girdling roots, 340
Gladiolus, 657–658
  bacterial blight, 119
  blue mold, 397
  Botrytis blight, 144
  dry rot, 418
  flower blight, 151–152
  mild mosaic, 477
  mosaic, 486
  Penicillium dry rot, 397
  red leaf spot, 294
  scab, 111
  smut, 464
  topple, 347
  yellows, 381–382
Gleditsia, 682–683
Globe-amaranth, 658
Globe artichoke, 539
Globeflower, 849
Globe-mallow, 658
Globe-thistle, 658
Globe-tulip, 723
Globodera rostochiensis,
  267

Gloeocercospora inconspicua, 267–268
  sorghi, 268, 288
Gloeodes pomigena, 188–189
Gloeophyllum saepiaria, 389
Gloeosporium, 92
  allantosporum, 92
  apocryptum, 92
  aridum, 540
  aridum quercina, 156
  betularum, 268
  cactorum, 97
  foliicolum, 384
  inconspicuum, 268
  limetticolum, 92
  malicorticis, 97
  melongenae, 92
  mezerei, 268
  perennans, 215–216
  piperatum, 93
  quercinum, 93, **93,** 96
  rhododendri, 268
  thuemenii f. tulipae, 93–94
  tiliae, 96
  ulmeum, 269
  ulmicolum, 268
  venetum, 87
Glomerella, 94
  cincta, 268
  cingulata, 94, 159, 208, 268, 384–385
    var. vaccinii, 385
  gossypii, 95
  nephrolepis, 95
  phomoides, 95
Glorybower, 602
Glory-bush, 658, 844
Glory-of-the-snow, 594
Gloxinia, 658–659
Glutinium macrosporium, 209
Glycine max, 819–820
Glyodex, 24
Glyodin, 24–25
Glyoxide, 25
Glyrophene, 25
Gnaphalium, 659
Gnomonia, 95
  caryae, 95
    var. pecanae, 268
  fragariae, 268
  leptostyla, 95
  nerviseda, 269
  platani, 95–96
  quercima, **93,** 96
  rubi, 159, 208
  tiliae, 96
  ulmea, 269
  veneta, 95–96
Gnomoniaceae, 70

Gnomoniella, 269
  coryli, 269
  fimbriata, 269
Goats-beard, 659
Godetia, 659
  downy mildew, 232
Godfrey's meadow nematode, 326
Godronia, 385
  cassandrae, 385
    var. vaccinii, 385
Going-out of grass, 271
Gold-dust tree, 542
Golden aster, 597
Goldenbells, 648
Golden-chain, 659
Golden-club, 660
  Botrytis on, 145
Golden-eye, 660
Golden-glow, 660, 801–802
Golden-larch, 660
Golden nematode, 319
Golden pea, 843
Goldenrain-tree, 660–661
Goldenrod, 661
  black patch, 126
  black spot (scurf), 253
  scab, 473
Goldenseal, 661
Goldentop, 662
Goldentuft, 526
Goldthread, 662
Gomphrena, 658
Gonatobotryum maculicolum, 269
Gonobulus, downy mildew on, 235
Gooseberry, 662–663
  dieback, 218
  leaf spot, 279
  mildew, 359
  scab, 474
Gopher plant, 662
Gordonia, 663
Gouania, 663
Gourd, 663
Grading injuries, 340
Graft canker of rose, 197
Graft incompatibility, 340
Graft mold, 300, 370
Granox PFM, 25
Grape, 663–665
  bitter rot, 390
  black rot, 385–386, **386**
  bunch mold, 367
  dead-arm disease, 201
  downy mildew, 235–236
  fanleaf, 486
  leaf roll, 486
  leaf scorch, 247

leafspot, 261
Pierce's disease, 123–124
powdery mildew, 361
root rot, 412
shoot and twig blight, 183
spiral nematode, 317
spot anthracnose, 470–471
white emperor disease, 486
white rot, 373
yellow mosaic, 486
yellow vein, 486
Grapefruit, 599–600, 665
Grape-hyacinth, 665
Graphiolaceae, 75
Graphiola phoenicis, 461
Graphium, 269
  sorbi, 269
  ulmi, 503–506, **505**
Grass(es), 665–667
  brown blight, 271
  brown leaf blotch, 191
  brown patch, 396, 411
  brown stripe, 290
  char spot, 290
  copper spot, 268, 288
  cyst nematode, 318
  dollar spot, 414
  downy mildew, 237
  fairy ring, 237, 238
  going-out, 271
  helminthosporium diseases, 270–271
  leaf blight, 162, 168, 179
  leaf blotch, 191, 270, 291
  leaf fleck, 274
  leaf mold, 271
  leaf spot, 270, 293
  melting-out, 270
  mosaic, 471
  nematode, 309
  netblotch, 270
  pink patch, 374
  powdery mildew, 354
  purple leaf blotch, 191
  red leaf spot, 270
  red thread, 374
  ring nematode, 313
  rust, 444–446, 447
  sheath nematode, 318
  silver spike, 382
  snowmold, 467–468, 468
    Color Plate 6A, 6B
  speckle, 290
  speckled leaf blotch, 191
  stem rust, 444–446
  tan leaf spot, 280
  tar spot, 283
  white tip blight, 183
  yellow gum disease, 101
  zonate spot, 270

Grass-of-Parnassus, 668
Gray bark of raspberry, 155
Gray blight of pines, 161,
    303
Gray bulb rot of tulips,
    411–412
Gray leaf of palms, 281
Gray leaf spot, 294–295
Gray mold blight, 141–143
Greasy blotch of carnation,
    191
Greek valerian, 780
Greenbrier, 815
Green fruit rot, 416
Green mold, 397, 420
Green scurf, 254
Gremmeniella abietina, 222
Grevillea, 668
Grindelia, 668
Griphosphaeria corticola, 209
Gromwell, 710
Ground-cherry, 668–669
    bacterial canker, 101–102
    purple-flowered, 669
Ground-myrtle, 858–859
Groundsel, 810–811
Ground smoke, 669
Grovesinia pyramidalis, 261
Guanidine, 25
Guar, 669
Guava, 669
    scab, 474
Guayule, 669
Guernsey-lily, 739
Guignardia, 189
    aesculi, 189
    bidwellii, 385–386, 386
        f. parthenocissi, 270
    populi, 290
    vaccinii, 386
Guinea-gold-vine, 678
Gummosis, 340
Gummy stem blight, 165
Gum-tree, 636–637
Gumweed, 668
Gymnoascaceae, 69
Gymnocladus, 698
Gymnoconia interstitialis, 431
    peckiana, 431
Gymnosporangium, 431–432
    aurantiacum, 433
    bermudianum, 432
    bethelii, 432
    biseptatum, 432
    clavariforme, 432
    clavipes, 432–433
    confusum, 433
    corniculans, 433
    cornutum, 433
    cupressi, 433

davisii, 433
effusum, 433
ellisii, 433
exigum, 433
exterum, 433
floriforme, 433
fraternum, 433
globosum, 433
gracile, 434
haraeanum, 434
harknessianum, 434
hyalinum, 434
inconspicuum, 434
japonicum, 434
juniperinum, 436
juniperi-virginianae, 432,
    434–435
kernianum, 435
libocedri, 435
multiporum, 435
nelsoni, 435
nidus-avis, 435
nootkatense, 435
photiniae, 433
speciosum, 435
trachysorum, 435–436
transformans, 433
tremelloides, 436
tubulatum, 436
vauqueliniae, 436
Gypsophila, 670

Hackberry, 670
    downy mildew, 236
    leaf blight, 153
Hackelia, 701
Hadrotrichum globiferum,
    159
Haipen, 25
Hairy mistletoe, 298
Hairy root, 98
Haitin, 25
Halesia, 670
Hamamelis, 869
Hamelia, 671
Harbinger-of-spring, 671
Hardenbergia, 671
Hardhack, native, 823
Hardy orange, 671
Hares-tail, 671
Harven, 25
Hawkbit, 671
Hawksbeard, 671–672
Hawkweed, 672
Hawthorn, 672–673
    leaf blight, 158, 164
    leaf spot, 271
    mistletoe, 298
    rust, 433
Hazel. See Hazelnut

Hazelnut, 673
    canker, 209–210
    leaf blister, 244
    twig blight, 193
Head browning of cauliflower,
    249–250
Head smut of grasses, 464
Heal-all, 787
Heart rot, 340, 372, 375,
    378, 380, 384, 389
    brown mottled, 398
Heat injury, 340
Heath, 674
Heather, 674
Hebe, 674
    Fusarium wilt, 509
Hedera helix, 693
Hedgenettle, 827
Hedge parsley, 674
Hedysarum, 839
Helenium, 674
Helianthemum, 835
Helianthus, 834–835
    tuberosus, 539
Helichrysum, 679, 833
Helicobasidium, 386
    corticioides, 386
    purpureum, 386–387
Helicotylenchus, 317
    dinysteria, 317
    erythrinae, 317
    multicinctus, 317
    nannus, 317
    pseudorobustus, 317
Heliopsis, 674–675
Heliotrope, 674
Heliotropium, 674
Helleborus niger, 596
Helminthosporium, 159, 252
    cactivorum, 387
    carbonum, 159
    catenarium, 159, 270
    cynodontis, 270
    dictyoides, 270
    erythrospilum, 270
    giganteum, 270
    maydis, 159
    papulosum, 238
    rostratum, 270
    sativum, 270
    sesami, 387
    setariae, 270
    siccans, 271
    sorokiniana, 271
    stenacrum, 271
    triseptatum, 271
    tritici-repentis, 271
    turcicum, 160, 387
    vagans, 271
    vignicola, 261

Helotiaceae, 72
Helotiales, **67**, 72
Helvellaceae, 72
Hemerocallis, 620
Hemicriconemoides, 317
  biformis, 317
  chitwoodi, 317
  floridensis, 317
  gaddi, 317
  wessoni, 317
Hemicycliophora, 317
  arenaria, 317
  brevis, 318
  obtusa, 318
  parvana, 318
  similis, 318
Hemisphaeriaceae, 70
Hemlock, 675
  mistletoe, 299
  needle blight, 155, 181
  rust, 437–438, 448–449
  twig canker, 204
Hemp, 676
  leaf curl, 256
  leaf spot, 256
Henbane mosaic, 486
Henbit, 676
Hendersonia, 247
  concentrica, 271
  crataegicola, 271
  opuntiae, 247
  rubi, 193
Hendersonula toruloidea, 209,
  510
Henningsomyces anomala,
  223
Hepatica, 676
  downy mildew, 235
Heracleum, 676
Herb-robert, 654
Hercules-club, 535, 677
Hericium erinaceus, 387
Heronsbill, 676
Herpotrichla, 160
  nigra, 160
Hesperis, 677
Heterodera, 318
  avenae, 318
  cacti, 318
  carotae, 318
  cruciferae, 318
  fici, 318
  glycines, 318
  humuli, 318
  marioni, 318
  mothi, 218
  punctata, 318
  rostochiensis, 319
  schactii, 319
  tabacum, 319

trifolii, 319
zeae, 319
Heterosporium, 160, **252**
  allii, 271
  echinulatum, 271
  esclischoltzlae, 271
  gracile, 272
  gracilis, 263–265, **264**
  iridis, 263–265, **264**, 272
  ornithogaii, 265
  syringae, 160
  variable, 272
Heuchera, 677–678
Hexachlorophene, 25, 41
Hexaferb, 25
Hexa-Nema, 44
Hexasul, 25
Hexathane, 25
Hexathir, 25
Hexazir, 25
Hiba arborvitae, 844
Hibbertia, 678
Hibiscus, 678
  esculentus, 744
  palustris, 800
  sabdariffa, 800
  syriacus, 801
Hibiscus, leaf spot, 255
Hickory, 678–679
  anthracnose, 95
  dieback, 213–214
  heart rot, 378
  leaf spot, 95
  mistletoe, 298
  Nectria canker, 214–215
Hieracium, 672
Higginsia, 257
  hiemalis, 161
  kerriae, 161
Hinoki cypress, 588
Hippeastrum, 526, 679
Hirshioporus abietinus, 406
Hizarocin, 25
Hoarhound, 679
Hoary-thick clover, 680
Hoe 002873, 25
Hoe 017411, 25
Hoe 2784, 25
Hoe 2873, 25
Hoe 2960, 44
Hoe 2989, 25
Hoe 6052, 25
Hoe 6053, 25
Hoe 13764, 25
Hoe 17411, 25
Hollow heart, 340
Hollow pocket, 406
Holly, 680–681
  black spot, 276
  blight, 174

canker, 204, 208
frost injury, 339, **339**
leaf spot, 284
spot anthracnose, 472
tar spot, 282
Hollyhock, 681
  anthracnose, 91
  leaf spot, 252, 255, 283
  mosaic, 486
  rust, 446
Holly-osmanthus, 682
Holodiscus, 682
  witches' broom, 486
Homai, 25
Homalocladium, 796
Homalomena, 682
Honesty, 682
  root rot, 364
Honey locust, 682–683
  canker, 224, 225
  dieback, 213–214
Honey plant, 683
Honeysuckle, 683
  leaf spot, 273, 274
  twig blight, 171
Hop, 683–684
  cyst nematode, 318
  mildew, 359
Hop-hornbeam, 684
Hoplolaimus, 320
  coronatus, 320
  galeatus, 320
  uniformis, 320
Hopperburn, 340
Hop-tree, 684
Hornbeam, 684–685
  bark canker, 217
  leaf curl, 244
  leaf spot, 269
Horse-balm, 605
Horse-chestnut, 565, 685
  bleeding canker, 219–220
  dieback, 213–214
  leaf blotch, 189
  powdery mildew, 361
Horse-gentian, 685–686
Horse-mint, 730–731
Horse purslane, 686
Horse-radish, 686
  leaf spot, 117–118, 252,
    255, 285
Hose-end sprayers, 55
Hosta, 686
  anthracnose, 91
Hostathion, 44
Hot-water treatment, 25
Hounds-tongue, 617
Houseleek, 810
Houstonia, 686–687
Hoya, 687

Huckleberry, 687
Huisache, 687
Humulus, 683–684
Husk-tomato, 668–669
Hyacinth, 687
 Botrytis blight, 144
 ring disease, 315–316
 yellows, 119
Hyacinth-bean, 627
Hyacinthus, 687
Hyalodidymae, **77**
Hyalophragmiae, **77**
Hyalopsora, 436
 aspidiotus, 436
 chelianthus, 436
 polypodii, 436
Hyalosporae, **77**
Hydnaceae, 76
Hydnum abietis, 417
 erinaceus, 387
 septentrionale, 417
Hydrangea, 688
 leaf spot, 284
 phyllody, 487
  Color Plate 3C
 ring spot, 487
Hydrastis, 661
Hydraulic sprayers, 51–52
Hydrophyllum, 688
 leaf spot, 290
Hydroxydiphenyl, 25
Hydroxyisoxazole, 25
Hylocereus, 740
Hymenocallis, 822
Hymenochaete agglutinans,
  209–210
Hymenogastrales, 76
Hymenopappus, 688
Hymexazol, 25
Hypericum, 804
Hyphochytriales, 64
Hypholoma perplexum, 387
Hypocreales, 70
Hypoderma, 161, **302,** 303
 desmazieril, 303
 hedgecockii, 303
 lethale, 161, 303
 robustum, 303
Hypodermella, 161, **302,** 304
 abietis-concolori, 161, 304
 ampla, 304
 concolor, 304
 laricis, 161, 304
 nervata, 304
Hypomyces, 161
 ipomoea, 161
 solani, 383
Hyponectria, 162
 buxi, 162
Hypoxis, 828

Hypoxylon pruinatum, 211
Hypsoperine graminiae, 320
Hyssop, 688
Hyssopus, 688
Hysterlaceae, 70
Hysteriales, 70
Hysterothecia, 303

Iberis, 578
Ice plant, 689
Idriella lunata, 388
Ilex aquifolium, 680–681
 cornuta, 680–681
 crenata, 680–681
 equifolium, 680–681
 glabra, 690
 opaca, 680–681
 verticillata, 868
 vomitoria, 870
Illicium floridanum, 530
Illosporium malifoliorum, 272
Impatiens, 689
Incense-cedar, 689
 mistletoe, 298
India hawthorn, 689
Indian cucumber-root, 690
Indian-cup, 813
Indian grass, 690
Indian Lettuce, 732
Indian mallow, 519
Indian mulberry, 732
Indian paintbrush, 750
Indian paint fungus, 377–378
Indian physic, 656
Indigo, 690
Indigobush, 527, 690
Indigofera, 690
Inflorescence blight of maple,
  149
Injury, alkali, 335
 arsenical, 335
 bordeaux, 13–14, 336
 calcium chloride, 337
 chlorine, 338
 copper, 338
  Color Plate 6C
 DDT, 338–339
 frost, 339, **339**
 grading, 340
 heat, 340
 lightning, 342
 salt, 345
 smog, 345
 soot, 345–346
 sulfur, 346
 weed-killer, 347–349, **348**
 winter, 349
Inkberry, 690
Ink spot of iris, 165–166
 of poplar, 258

Inonotus circinatus, 407
Internal browning, 340
Internal cork, sweetpotato,
  496
Inula, 690–691
Ipomoea, 732–733
 batatas, 838–839
Iprodione, 25
Irene, 129
 arallae, 129
 calastroma, 129
 perseae, 129
Irenina manca, 129
Irenopsis martiniana, 129
Iresine, 691
Iris, bulbous, 691
 basal rot, 381
 leaf spot (ink disease),
  165–166
 leaf spot, 263–265, **264**
 mosaic, 487
 nematode, 316
 rust, 446
Iris, rhizomatous, 691–692
 crown rot, 367–368
 leaf spot, 263–265, **264**
Iron deficiency, 340–341
Ironweed, 692
 yellow, 520
Ironwood, 684
Irpex tulipiferae, 388
Isaria clonostachoides, 388
Isariopsis, 272
 clavispora, 279
 griseola, 272
 laxa, 272
Iscothane, 25
Itchgrass, 692
Itersonilia perplexans, 162
Itersonilia sp., 162
Ivesia, 692
Ivy, Boston, 692–693
 leaf spot, 270
  Color Plate 4D, 4F
Ivy, English, 693
 leaf spot, 119, 251, 283–284
 root rot, 404
 scab, 474
Ivy-arum, 785
Ixia, 693
 mosaic, 487
Ixora, 693

Jacaranda, 693
Jack-bean, 694
Jack-in-the-pulpit, 694
 leaf and stalk blight, 145
Jacobinia, 694
Jacobs-ladder, 780
Jacquemontia, 694

Jacquinia, 694
Jamesia, 694
Japanese andromeda, 771
Japanese lawn grass, 873
Japanese plum-yew, 587
Japanese quince, 790
Japanese-spurge, 749
Japanese walnut, 861–862
Jasmine, 695
    blossom blight, 149
    leaf spot, 260
    scab, 472
Jasminum, 695
Jatropha, 695
Java black rot of sweet potato,
    377
Javanese root-knot nematode,
    324
Jerusalem artichoke, 539
Jerusalem-cherry, 695
    bacterial canker, 101–102
    tobacco cyst nematode, 319
Jetbead, 696
Jobs-tears, 696
Joe-pye weed, 638
Jolt, 44
Jonquil, 737–738
Josephs-coat, 526
Joshua-tree, 871–872
Judas-tree, 794
Juglans, 861–862
Jujube, 696
Juneberry, 527
Juniper, 696–698
    black mildew, 128
    blight, 172
    gall rust, 433
    mistletoe, 298
    root rot, 378
Juniperus, 696–698
Jupiters-beard, 855
Jussiaea, 864–865

Kabatia lonicerae, 273
Kabatina juniperi, 211
Kageneckia, 698
Kalanchoe, 698
    leaf spot, 258
Kale, 569–570
Kalmia, 734–735
    leaf spot, 278, 284–285,
    284
Karabation, 26, 44
Karamate, 26
Karathane, 26
Kasugamycin, 26, 41
Kasumin, 26
Kayafume, 26, 44
K-Cop Liquid Agricultural
    Fungicide, 26

Keithia, 154
Kellermannia, 162
    anomala, 162
    sysyrinchi, 162
    yuccaegena, 162
Kemate, 26
Kentucky bluegrass, 665–667
Kentucky coffee-tree, 698
Kernel spot of pecan, 394
Kerria, 698
    leaf and twig blight, 161,
    172
    leaf spot, 260
Kidney bean, 550–552
Kidney vetch, 699
Kiwi, 699
Kiwi luster, 26
Kluyveromyces marvianus var.
    marvianus, 388
Knapsack sprayers, 54
Kniphofia, 699
Knockmate, 26
Knotroot bristlegrass, 699
Koban, 20
Kobu, 26
Kobutol, 26
Kochia, 699
Kocide, 26
Koelreuteria, 660–661
Kohleria, 699
Kohl-rabi, 569–570
Kolkwitzia, 553
Kop 300, 26
Kop-Fume, 44
Krigia, 699–700
Kroma-Clor, 26
Kromad, 26
K-Tea Algaecide, 26
Kudzu, 700
Kue 13032c, 26
Kue 13183b, 26
Kuehneola, 436
    malvicola, 436
    uredinia, 436
Kuhnia, 700
Kumquat, 700
Kumulan, 26
Kumulus S, 26
Kunkelia nitens, 436–437
Kutilakesa pironii, 241–242
Kypman, 26
Kypzin, 26

Labilite, 26
Laboulbeniales, 71
Labrador-tea, 703
Labrelia aspidistrae, 163
Laburnum, 659
    mosaic, 487
Lachenalia, 578

Lachnellula wilkommii,
    203–204, 211
Lactuca, 584, 704–705
Ladys-fingers, 699
Laestadia, 286
Laetiporus sulphureus, 407,
    407
Lagenaria, 663
Lagenidiales, 65
Lagerstroemia, 611
Lagurus, 671
Lamarckia, 662
Laminum, 676
Lance nematode, 320
Lannate, 44
Lantana, 700
Lappula, 701
Larch, 701
    canker, 203–204
    dieback, 213
    needle and shoot blight,
    161
    rust, 438
Larix, 701
Larkspur, 620–621, 701
    Diaporthe blight, 154
Larrea, 611
Larvacide, 17, 26
Lasiobotrys affinis, 273
Late blight of celery, 183–184
    of potato and tomato,
    175–179, **176, 177,**
    178
Lathyrus, 837–838
Laurel, 702
Laurestinus, 702
Laurus, 702
Lavandula, 702
Lavatera, 702
Lavender, 702
Lawn grasses, 665–667
Lawn-leaf, 624
Lawns, 702
Layia, 702
Lead-plant, 527, 690
Leadtree, 703
Leaf blight, arborvitae,
    154–155
    aspidistra, 163
    boxwood, 162, 187
    calla, 174
    clematis, 171
    fig, 168
    grass, 162, 168, 179
    hackberry, 153
    hawthorn, 158, 164
    iris, 165–166
    lilac, 287
    lupine, 159
    may-apple, 183

Leaf blight *(cont.)*
  mesquite, 181
  mountain-laurel, 172
  osage-orange, 185
  parsley, 187
  parsnip, 162
  salsify, 185
  strawberry, 150
  sweet potato, 173
  viburnum, 164
  yucca, 162
Leaf blister, birch, 244
  buckeye, 244
  chinquapin, 244
  elm, 245
  hazelnut, 244
  hornbeam, 244
  maple, 244, 245
  oak, 244
  poplar, 244, 245
  yellow, 245
Leaf blotch, grasses, 191, 270, 291
  horse-chestnut, 189
  lilac, 188
  pecan, 189–190
  peony, 188
  persimmon, 190
  plum, 190
  pomegranate, 190
  potato, 188
Leaf curl, alder, 245
  cherry, 244
  peach, 244–245
  pelargonium, 486
  peony, 491
  raspberry, 494
Leaf fleck of grass, 274
Leaf gall, anemone, 243
  azalea, 240, **241**
  camellia, 240
  red, 243
  rhododendron, 240
Leaf mold, 300
  of grass, 271
  of tomato, 300
Leaf nematode of chrysanthe-
    mum, **311,** 312
Leaf rust of cereals and grasses,
    444–468, 468
Leaf scorch, 246–248, 341–342
  azalea, 247
  cactus, 247
  grapevines, 247
  narcissus, 248
  palm, 246
  strawberry, 246–247
  trees, 341–342
Leaf smut, 461, 463

Leaf spot, 248–295
  abelia, 255
  angular,
    of bean, 272
    of cucurbits, 113
    of grapes, 277
  apple, 261
  ash, 269, 274, 278
  aspidistra, 252
  aster, 253, 294
  aucuba, 281, 283
  bean, 252, 253
  beet, 255–256
  birch, 263, 268
  black of Crucifers, 249–250
  blackberry, 292–293
  bloodroot, 285
  blueberry, 250, 285
  bluegrass, 271, 292
  Boston ivy, 270
  boxwood, 274
  cabbage, 250, 258
  calathea, 249
  calendula, 255
  calla, 285
  camelia, 281, 283, 293
  catalpa, 250, 283
  cherry, 250, 259–260
  Christmas-rose, 261
  chrysanthemum, 263, 291
  citrus, 291
  clematis, 252–253, 263
  clover, 285
  columbine, 255
  crucifers, 249–250, 255
  cucurbits, 256, 291, 294
  currant, 279
  cypripedium, 253
  daisy, 250
  daphne, 268
  dianthus, 291–292
  dieffenbachia, 118, 255
  dogwood, 253, 256, 291
  dracaena, 285
  eggplant, 257
  elm, 268, 290
  English ivy, 119, 251,
    283–284
  fern, 250
  foxglove, 287
  geranium, 118, 120, 256,
    281
  gladiolus, 151–152, 294
  grass, 270, 293
  gray, 294
  holly, 284
  hollyhock, 252, 255
  horse-radish, 117–118, 252,
    255, 287

  hydrangea, 284
  hydrophyllium, 290
  iris, 263–265, **264**
  ivy, 270, 283–284
  kalanchoë, 258, 294
  kerria, 260
  lettuce, 250, 292
  lupine, 253
  magnolia, 284
  maple, 285, 290
  marigold, 250, 293
  mountain-laurel, 284–285,
    284
  mulberry, 263
  nandina, 257
  nephytis, 255
  oak, 249
  okra, 251, 255
  olive, 262
  palm, 268, 281
  parsnip, 287
  passion flower, 250
  pea, 253, 257
  peanut, 255, 257
  pear, 261
  pecan, 256
  peony, 292
  pepper, 256
  phlox, 292
  pittosporum, 257
  poinsettia, 207
  poplar, 258, 274, 275, 279
  potato, 255
  primrose, 111–112, 287
  radish, 250
  raspberry, 293
  redbud, 278
  rhododendron, 257, 268,
    281, 285
  rose, 257, 257–258
  rose-acacia, 250
  sassafras, 249
  schefflera, 250, 251
  sesame, 112
  smilax, 258
  snapdragon, 283
  soybean, 258, 285, 287
  strawberry, 260
  sycamore, 295
  syngonium, 255
  tobacco, 250
  tomato, 250, 292, 294
  tupelo, 279, 286
  wallflower, 252
  watermelon, 165, 291
  willow, 263
  wisteria, 285
  witch-hazel, 284
Leak, 409

Leatherleaf, 588
Leatherwood, 618, 703
Lebbek, 703
Ledum, 703
  spot anthracnose, 472
Leek, 703
Legumes, powdery mildew
  of, 354–355
Leiophyllum, 806
Lembosia, 129
  cactorum, 130
  coccolobae, 130
  illiciicola, 130
  portoricensis, 130
  rugispora, 130
  tenella, 130
Lemon, 599–600, 703
Lemon grass, 703
Lemon-verbena, 709
Lens, 704
Lentil, 704
Lentinus, 388
  lepideus, 389
  tigrinus, 389
Lenzites, 389
  betulina, 389
  saepiaria, 389
Leonotis, 709
Leontodon, 671
Leonurus, 733
Leopard's-bane, 627
Lepidium, 767
Lepiota morgani, 237
Leptomitales, 64
Leptospermum, 704
Leptosphaeria, 163
  coniothyrium, 163, 197,
    211
  korrae, 163, 389
  lindquistii, 399
  thomasiana, 163
Leptostromataceae, 78
Leptostromella, elastica,
  273
Leptothyrella liquidambaris,
  273
Leptothyrium, 273
  californicum, 274
  dryinum, 274
  periclymeni, 274
  pomi, 239
Lesan, 27
Lesion nematode, 327
Lettuce, 704–705
  anthracnose, 96
  bacterial wilt, 121
  big vein, 487
  bottom rot, 396–397
  downy mildew, 232

drop, 415
leaf spot, 250, 292
marginal blight (Kansas
  disease), 110
mosaic, 487
Rio Grande disease, 121
Leucaena, 703
Leuchtenbergia, 570
Leucojum, 705
Leucothoë, 705
  black spot, 128
  blight, 152
  leaf spot, 281
Leveilla taurica, 355
Lewisia, 705
LFA 910, 27
LH 3012, 27
Liatris, 705–706
Libocedrus, 689
Lichens, 295–296
Lightning injury, 342
Ligustrum, 706, 787
Lilac, 706–707
  blight, 114, 174
  leaf blight, 160
  leaf blotch, 188
  mildew, 355–356
  ring spot, 487
  shoot blight, 174, 183
  witches' broom, 487
Lilium, 707–708
Lily, 707–708
  basal rot, 381
  black scale rot, 372
  Botrytis blight, 142,
    143–144
  color adding, 487
  color removing, 487
  dieback, 310–311
  fleck, 487
  latent mosaic, 487
  ring spot, 487
  rosette, 487
  symptomless virus, 487
  yellow flat, 487
Lily-of-the-valley, 708
Lima bean, 550–552
  downy mildew, 234–235
  pod blight, 154–155
  scab, 472
  yeast spot, 394
Limb blight, 150
Lime, 599–600
  anthracnose, 92
Lime, hydrated, 13–14, 27
Lime-induced chlorosis, 342
Lime sulfur, 27
Limnophila sp., 535
Limonium, 828–829

Linaria, 708
  downy mildew, 233
Linden, 708–709
  anthracnose, 96
  bleeding canker, 219–220
  dieback, 213–214
  leaf blight, 148
  scorch, 96
  spot anthracnose, 473
Lindera, 821
Linnaea, 709
Linospora gleditsiae, 274
Linum, 647
Lions-ear, 709
Lippia, 709
  spot anthracnose, 474
Lipstick vine, 710
Liquidambar, 836
Liriodendron, 851
Lirula macrospora, 304
Lithocarpus, 710
  powdery mildew, 355
Lithophragma, 710
Lithops, 830
Lithospermum, 710
Litsea, 710, 781
Little leaf, 342, 349
  of pine, 402
Lobelia, 710–711
Loblolly-bay, 663
Lobularia, 835
Locust, 711
  canker, 204
  dieback, 213–214
  heart rot, 380
  leaf spot, 259, 267
  witches' broom, 488
Loganberry, 710
  dwarf, 488
Lomatium, 712
Lonacol, 27
London plane, 776–777
  blight, 195
  canker, 206–207, 212
Longidorus, 320
  elongatus, 320
  maximus, 320
  sylphus, 320
Lonicera, 683
  infectious variegation, 488
Loose smut, 462, 464, 467
Loosestrife, 712
  fringed, 712
Lophiostomataceae, 71
Lophodermella, 163
Lophodermium, 274, **302**
  durilabrum, 304
  filiforme, 304
  juniperinum, 304

Lophodermium *(cont.)*
 nitens, 304
 piceae, 305
 pinastri, 305
 rhododendri, 274
 seditiosum, 305
Loquat, 712–713
 scab, 453
Lotus, 713
Louisiana broomrape, 191
Louisiana lettuce disease,
 116
Lousewort, 869
Love-lies-bleeding, 526
Lucuma, 713
Ludwigia, 713
Luffa, 663
Lunaria, 682
Lupine, 713–714
 downy mildew, 234
 leaf blight, 159
 leaf spot, 253
 seedling blight, 180
Lupinus, 713–714
Luzula, 869
Lychee, 714
Lychnis, 714
Lycium, 715, 723
Lycoperdales, 76
Lycopersicon, 845–848
Lycopus, 863
Lycoris, 715
Lyonia, 715
 black mildew, 128
Lysiloma, 715
Lysimachia, 712
Lysol, 27
Lythrum, 715

M 9834, 27
Maackia, 716
Macadamia, 716
Maclura, 748
Macrophoma, 212
 candollei, 212, 274
 cupressi, 212
 phoradendron, 212
 tumefaciens, 212
Macrophomina, 389
 phaseoli, 164, 183, 389–390
 phaseolina, 390
Madrone, 716
 canker, 209
Magnesium deficiency, 342
Magnesium sulfate, 27
Magnetic 70, 27
Magnolia, 716–717
 black mildew, 129, 130
 leaf spot, 250, 266, 276,
 284

Nectria canker, 215
 petal spot, 141–143, **141**
 scab, 472
Mahogany, 717
Mahonia, 717
Maianthemum, 718
Maidenhair-tree, 656
Maize, dwarf mosaic, 488
 stunt, 488
 white line mosaic, 488
Malachite, 27
Malachra, 718
Malacothrix, 718
Maleberry, 715
Mallotus, 718
Mallow, garden, 718
 root rot, 389–390
Malus, 610
 sylvestris, 530–533
Malva, 718
Malvastrum, 718
Malvaviscus, 719
Mammillaria, 570–571
Mancozeb, 27
Maneb, 27–28
Maneba, 28
Manebgan, 28
Manesan, 28
Manex, 28
Manex 80, 28
Manfreda, 719
Manganese deficiency, 342
Mangifera, 719
Mango, 719
 scab, 472
 stem rot, 376
 Verticilium wilt, 511–515
Mangrove, 719
Manihot, 719–720
Manioc, 719–720
Manzanita, 720
 witches' broom, 240–241
Manzate, 28
Manzate 200 Fungicide, 28
Manzeb, 28
Maple, 720–722
 anthracnose, 92
 bacterial leaf spot, 109
 bark patch, 192
 basal canker, 221
 bleeding canker, 219–220
 brown mottled heart rot,
 398
 canker, 204, 223
 dieback, 213
 inflorescence blight, 149
 leaf blight, 92
 leaf blister, 244, 245
 leaf spot, 261, 282, 285,
 290

mistletoe, 298
 tar spot, 288, 289
 twig blight, 171
 wilt, 511–513
Maposol, 28, 44
Maranta, 538, 722
Marasmius oreades, 237
Marbleseed, 746
Marginal browning, 342
Marguerite, 722
Marigold, 722–723
 Botrytis on, 142
 leaf spot, 250, 293
Marigold, pot, 572–573
Mariposa-lily, 723
Marrubium, 679
Marsh marigold, 723
Marssonina, 96
 brunnea, 274
 daphnes, 275
 delastrei, 275
 fraxini, 275
 juglandis, 95, 275
 ochroleuca, 275
 panattoniana, 96
 populi, 275
 rhabdospora, 275
 rosae, 275
 truncatala, 275
Massaria platani, 212
Mastigosporum rubricosum,
 274
Matelea, 723
Matricaria, 723
Matrimony-vine, 723
Matthiola, 830
Maurandya, 724
Mauritius-hemp, 724
May-apple, 724
 leaf blight, 183
Mayflower, 634
MBC, 28
MC 1053, 28
M-Diphar, 28
M-Dipher, 28
Meadow-beauty, 724, 794
Meadow nematodes, 326–327
Meadow parsnip, 873
Meadow-rue, 724
Meadowsweet, 725, 823
Measles, peony, 188
MEB 6447, 28
MeBr, 44
Medeola, 690
Medicago, 725
Medlar, 725
Megachytriaceae, 63
Melampsora, 437, **437**
 abietis-capraearum, 437
 abietis-canadensis, 437

albertensis, 437
artica, 437
bigelowii, 438
farlowii, 437–438
hypericorum, 438
medusae, 438
occidentalis, 438
paradoxa, 438
ribesii-purpureae, 438
Melampsoraceae, 74
Melampsorella caryo-
    phyllacearum, 438
cerastii, 438
Melampsoridium betulinum,
    438
Melancomium, 275
fuligineum, 390
pandani, 275
Melanconiaceae, 78
Melanconiales, 77, 78
Melanconidiaceae, 70
Melanconis juglandis,
    212–213
Melanose of citrus, 376
Melanospora, 301
Melanthium, 567
Melasmia, 275
falcata, 276
menziesii, 276
Melia, 592–593
Melilotus, 725
Meliola, 130
amphitricha, 130
bidentata, 130
camelliae, 130
cookeana, 130
cryptocarpa, 130
lippiae, 130
magnoliae, 130
nidulans, 130
palmicola, 130
tenuis, 130
wrightli, 130
Meliolaceae, 69
Melissa, 547
Melogrammataceae, 71
Meloidodera floridensis,
    320
Meloidoderita, 321
Meloidogyne, 321–323, **321,
    322**
arenaria, 323
arenaria thamesii, 323
chitwoodi, 323
graminicola, 323
hapla, 323–324
incognita, 324
incognita, acrita, 324
incognita incognita, 324
javanica, 324

Melon, 725–726
anthracnose, 90
blight, 156
fruit rot, 409
Melothria, 726
Melprex, 28
Meltatox, 28
Melting-out of grass, 270
Menispermum, 732
Mentha, 729
Mentzelia, 726
Menziesia, 727
Mepronil, 28
Mercuran, 28
Mercury toxicity, 342
Meria laricis, 213
Merpan, 28
Merry-bells, 854
Mertect, 28
Mertensia, 727
Mescalbean, 819
Mesembryanthemum, 727
Mespilus, 725
Mesquite, 727
leaf blight, 181
Metalaxyl, 28
Metam 32.7, 44
Metam 42, 44
Metam-Fluid BASF, 44
Metam-Sodium, 28, 44
Metham, 44
Metham-Sodium, 44
Methanal, 28
Meth-0-Gas, 44
Methomyl, 44–45
Methyl bromide, 28–29, 45
Methyl Isothiocyanate, 45
Methylmetiram, 29
Methyl Thiophanate, 29
Metiram, 29
Metiram-Complex, 29
Mezene, 29
Mezineb, 29
MF-344, 29
Micofume, 29, 45
Micromeria, 871
Micropeltaceae, 70
Micropeltis, 164
alabamensis, 276
viburni, 164
Microsphaera, **351,** 355
alni, 355–356
    var. vaccinii, 356
diffusa, 356
euphorbiae, 356
grossulariae, 356
penicillata, 356
Microstroma juglandis, 276
Microthyriaceae, 70
Microthyriales, 70

Microthyriella, 239
cuticulosa, 276
rubi, 239
Mignonette, 727–728
leaf spot, 257
Mikania, 728
Mil-Col, 29
Milcurb, 29
Mildews, black, 127–130
downy, 231–237
powdery, 349–361, **350**
Mildex, 26, 29
Mildothane, 29
Milesia, 439
fructuosa, 439
laeviuscula, 439
marginalis, 439
pycnograndis, 439
polypodophila, 439
Milk thistle, 728
Milkvine, 859
Milkwort, 728
Miltox, 29
Mimosa, 728
twig canker, 208, 213–214
wilt, 509–510
Mimulus, 728–729
Mint, 729
Mirabilis, 648
Mistletoe, 296–299, **297,** 729
California, 297
canker, 209–210
dwarf types, 298–299, 730
eastern, 298
hairy, 298
incense cedar, 298
juniper, 298
Texas, 298
Mist sprayers, 51
Mitchella, 755
Mitella, 556–557
Mobilawn, 45
Mocap, 45
Mock-cucumber, 730
Mock-orange, 730
Mock-strawberry, 730
downy mildew, 234
Mold(s), 299–301
black, 300, 301
leaf, 300
seed, 287
sooty, 468–470
white, of sweetpea, 301
Mollisiaceae, 72
Molucella, 555
Molybdenum toxicity,
    342–344
Mombin, 823
scab, 475
Momordica, 547

Monarda, 730–731
Monardella, 731
Monceren, 29
Moneses, 731
Moneywort, 712
Monillaceae, 78
Monillales, 78
Monilinia, 164
　azaleae, 164
　fructicola, 164, 390–392,
　　391
　johnsonii, 164
　laxa, 164, 390, 392
　oxycocci, 397
　rhododendri, 164
　urnula, 392
Monilochaetes infuscans, 393,
　458
Monkey-flower, 728–729
Monkey-puzzle tree, 536
Monkshood, 731
Monkshood vine, 731
Monoblepharidales, 64
Monochaetia, 213
　desmazierii, 276
　mali, 213
Monox, 29
Monstera, 732
Montbretia, 849
Montia, 732
Moonflower, 732
Moonseed, 732
Morea, 732
Morenoella, 130
　angustiformis, 130
　ilicis, 130
　orinoides, 130
　quercina, 277
Morestan, 29
Morinda, 732
　scab, 474
Morning-glory, 732–733
　stem canker, 225
Morocide, 12, 29
Morrocid, 29
Morus, 735–736
Mosaic, abutilon, 475
　apple, 476
　bean, 477
　　southern, 477
　　yellow, 477
　canna, 479
　carnation, 479
　cattleya, 489
　cauliflower, 480
　celery, 480
　cherry rugose, 481
　chrysanthemum, 482
　cineraria, 482
　clover, 484
　coleus, 484

corn, 484
cowpea, 484
cucumber, 484–485
currant, 485
cymbidium, 489
dahlia, 485
dodder latent, 485
elm, 485
euonymus, 486
fig, 486
geranium, 486
gladiola, 486
grape yellow, 486
henbane, 486
hollyhock, 486
iris, 487
laburnum, 487
lettuce, 487
mild, of gladiolus, 477
muskmelon, 488
mustard, 488
narcissus, 487
nasturtium, 488
nothoscordum, 488
ornithogalum, 489
pea, 490
pea enation, 489
peach, 490
peach rosette, 491
pelargonium, 486
pepper, 492
pepper vein banding, 492
potato aucuba, 492
potato leaf rolling, 492
potato rugose, 492
primrose, 493
prune constricting, 493
radish, 493
raspberry, 494
raspberry yellow, 494
rose, 494
　Color Plate 2D
rose yellow, 495
soybean, 495
soybean yellow, 495
squash, 495
　Color Plate 5E
stock, 495
streptanthera, 496
teasel, 497
tigridia, 497
tobacco, 497
tomato, 498
tritonia, 499
turnip, 500
watermelon, 500
wisteria, 500
Motherwort, 733
Mottle, carnation, 479
　pea, 490
Mottle leaf, 344

Mountain andromeda, 771
Mountain-ash, 733–734
　fire blight, 103–105
　leaf spot, 269
　mistletoe, 298
Mountain ebony, 549
Mountain-heather, 734
Mountain-holly, 734–735
Mountain-laurel, 734
　leaf blight, 172
　leaf spot, 278, 284–285,
　　284
Mountain-mahogany, 735
Mountain-mint, 735
Mountain-sorrel, 735
Mucilago spongiosa, 459
Mucor, 66, 393
　mucedo, 393
　piriformis, 393
　racemosus, 393
Mucoraceae, 66
Mucorales, 66, **66**
Mulberry, 735–736
　bacterial blight, 113–114
　leaf spot, 263, 278
　popcorn disease, 149
　twig blight, 166
　twig canker, 208, 213–214
Mullein, 736
Mummy berry, 392
Musa nana, 548
Muscari, 665
Mushroom, oxyster, 736
Mushroom root rot, 364–366,
　365, 371–372
Mushrooms, 73–76, **73**
Muskmelon, 725–726
　fruit rot, 409
　mosaic, 488
　wilt, 509
Musk-root, 521, 736
Mustard greens, 736–737
　mosaic, 488
　root rot, 364
Mycodifol, 29
Mycoshield, 41
Mycosphaerella, 96, **252**
　angulata, 277
　arachidicola, 255, 277
　aurea, 277
　berkeleyi, 257, 277
　bolleana, 277
　brassicicola, 277
　caroliniana, 277
　carygena, 277–278
　cerasella, 256, 278
　cercidicola, 278
　citri, 278
　citrullina, 165
　colorata, 278, 284–285, **284**
　cruenta, 278

dendroides, 189–190
diospyri, 190
effigurata, 278
fragariae, 278
fraxiniccla, 278
grossulariae, 279
juglans, 278
laricina, 305
ligulicola, 138
liriodendri, 278
louisianae, 278
lythracearum, 190
melonis, 165
milleri, 257
mori, 278
nigromaculans, 278–279
nyssaecola, 279
opuntiae, 96–97
personata, 279
pinodes, 138–139, 165
pomi, 239
populicola, 279
populorum, 279
psilospora, 279
rabiei, 165
ribis, 279
rosicola, 257–258, 279
rubi, 279
sentina, 279
sequoiae, 165
tecomae, 148
Mycosphaerella blight of pea,
  138–139
Mycosphaerellaceae, 71
Mycosyrinx osmundae, 461
Mylone, 19, 29, 45
Myocentrospora, 279
Myosotis, 647–648
Myrlanglaceae, 69
Myriangiales, 67, 69
Myrica california, 750
carolinensis, 549
cerifera, 865
gale, 836
Myriogenospora atramentosa,
  165
Myriophyllum, 754
Myrothecium roridum,
  279–280, 393
Myrtle, 737
Myrtus, 737
Mystrosporium adustum,
  165–166
Myxosporium, 166
diedickii, 166
everhartii, 166
nitidum, 166

N521, 45
  25EC, 29, 41
Nabam, 29–30

Nabasam, 30
Naccobus, 324
batatiformis, 325
dorsalis, 325
Nacobbodera chitwoodi, 325
Naemacyclus niveus, 305
Nailhead spot of tomato,
  250–251
Nandina, 737
leaf spot, 257
Naramycin, 30
Narcissus, 737–738
basal rot, 382
bulb nematode, 315–316
chocolate spot, 487
fire, 182–183
flower streak, 487
green mold rot, 420
leaf scorch, 247
mosaic, 487
silver leaf, 487
smoulder, 414–415
white mold, 287–288
white streak, 487
yellow stripe, 487
Nasturtium, 738
bacterial spot, 112–113
mosaic, 488
Nasturtium officinale, 862
Natal-Plum, 580
Natriphene, 30
Neck rot, 368
Nectarine, 738–739
dieback, 225
leafspot, 261
Nectria, 196
cinnabarina, 213–214
coccinea var. faginata, 214
desmazierii, 214
ditissima, 214
galligena, 214–215
haematococca, 383
magnoliae, 215
Nectria canker, of beech, 214
of boxwood, 187
Nectriaceae, 70
Nectrioidaceae, 78
Needle blight, 155, 156, 161
fir, 161, 170, 179–180, 305
hemlock, 155, 181
larch, 161, 305
pine, 156
redwood, 165
yew, 185
Needle cast, 301–306
Balsam fir, 302, 304
Douglas-fir, 302, 306
fir, 302, 303, 305, 306
larch, 304, 305
pine, 303, 304, 305
spruce, 304, 306

Needle rust, of pine, 424–425
spruce, 424–425
Nellite, 45
Nelumbo, 713
Nemacur, 30, 45
Nemaspor, 30
Nemafene, 45
Nemafume, 45
Nemagon, 45
Nemanex, 45
Nemaset, 45
Nemasol, 45
Nematanthus, 739
Nematocides, 7, 41–46
Nematodes, 306–331
American dagger, 331
awl, 316
begonia leaf blight, 312
boxwood spiral, 328
bud and leaf, 312
burrowing, 327–328
cabbage cyst, 318
cactus cyst, 318
California dagger, 329
California sessile, 325
Carnation pin, 325
Caroline spiral, 329
carrot cyst, 318
celery pin, 325
Chambers' dagger, 331
Christie's spiral, 329
Christie's stubby root, 329
Chrysanthemum foliar, **311,**
  312
Color Plate 6E
citrus, 329
citrus ring, 314
clover cyst, 319
Cobb's meadow, 326
Cobb's ring, 314
Cobb's spiral, 317
Cobb's stubby root, 329
Columbia root-knot, 323
corn cyst, 319
corn meadow, 327
cotton root-knot, 324
crown-headed lance, 320
currant, 311
cyst, 318
dagger, 331
decline, 314, 329
De Man's meadow, 326
European dagger, 329
false root-knot, 325
fern, 310
fig cyst, 318
fig pin, 325
fig spine, 314
foliar, 325
Godfrey's meadow, 326
golden, 319

Nematodes *(cont.)*
  grass, 309
  grass cyst, 318
  grass sheath, 318
  hop cyst, 318
  Javanese, root-knot, 324
  lance, 320
  lesion, 327
  northern root-knot, 323–324
  oak sheathoid, 317
  oat cyst, 318
  Pacific dagger, 329
  pine sheathoid, 317
  potato rot, 315
  reniform, 328
  rice root-knot, 323
  ring, 313
  root-knot, 323, 325
  Scribner's meadow, 326
  Seinhorst stubby root, 329
  smooth-headed meadow,
    326
  southern root-knot, 324
  soybean cyst, 318
  spring dwarf, 310–311
  Steiner's spiral, 317
  stem and bulb, 315–316
  sting, 313
  sugar-beet, 319
  sugar-cane stylet, 330
  summer dwarf, 310
  Tarjan's sheath, 318
  Tesselate stylet, 330
  Thames' root-knot, 323
  Thorne's meadow, 326
  Thorne's needle, 320
  tobacco cyst, 319
  walnut meadow, 327
  West African spiral, 328
  wheat, 309
  yam, 329
  Zimmerman's spiral, 317
  Zoysia spine, 314
Nematospora, 393
  coryli, 394
  phaseoli, 394
Nemopanthus, 734
Nemophila, 739
Neofabraea, 97
  malicorticis, 97
  perennans, 215–216
Neopeckia coulteri, 166
Neottiospora yuccifolia, 280
Neovossia iowensis, 462
Nepeta, 584
Nephis, 45
Nephrolepis, 640
Nephthytis, 739
  leaf spot, 255
Nerine, 739
Nerium, 744

Neurospora sitophila, 394
New York apple-tree canker,
  219, 401
New Zealand flax, 739
New Zealand spinach,
  739–740
NIA 9044, 30
NIA 9102, 30
NIA 10242, 45
Niacide, 30
Nicandra, 533
Nicotiana, 740
Nidulariales, 76
Night-blooming cereus, 740
Nightshade, 740
  foliar nematode, 325
Nigrospora oryzae, 394
Nimrod, 30
Ninebark, 740
Nitrador, 30
Nitrogen deficiency, 344
Nitrogen excess, 344
Nocardia vaccinii, 242
Nomersan, 30
Nonparasitic diseases, 333–349
Nordox SD-45, 30
Nordox SD-50, 30
Norfolk-Island-pine, 536
Northern corn leaf blight,
  160
Northern root-knot nematode,
  323–324
Northwestern apple anthrac-
  nose, 97
Nothanguina phyllobia, 325
Nothoscordum, 526, 741
  mosaic, 488
Nudrin, 45
Nummilaria discreta, 215
Nuphar, 863
Nursery blight, 172
Nutsedge, 741
  cyst nematode, 318
Nu-Z, 30
Nymphaea, 863
Nymphoides, 741
Nyssa, 852
Nyssopsora clavellosa, 439

Oak, 741–743
  anthracnose, **93**, 93, 96
  bacterial wetwood, 107–108
  bark patch, 192
  black mildew, 276
  canker, 194, 204, 219–220
  Dothiorella canker, 206
  heart rot, 398
  leaf blister, 244
  leaf spot, 249, 272, 274,
    279
  powdery mildew, 355, 359

rust, fusiform, 427–428
  sheathoid nematode, 317
  Sphaeropsis canker, 218–219
  strumella canker, 224
  twig blight, 151, 156
  twig canker, 221
  wilt, 503
Oats, cyst nematode, 318
  downy mildew on, 237
  crown rust, 444
Ocean spray, 682
Ocimum, 549
Oconee-bells, 812
Ocotillo, 648, 743
Odontoglossum ring spot,
  489
Oedema, 344
Oenothera, 743
Ofurace, 30
Oidium, 356
  begoniae, 356
  euonymus japonici, 356
  obductum, 356
  pyrinum, 356
  tingitaninum, 356–357
Okra, 744
  leaf spot, 251, 255
  pod spot, 251
Olea, 744–745
Oleander, 744
  bacterial gall, 115–116
  dodder, **229**
  scab, 474
Oleocuivre, 30
Oleo Nordox, 30
Olive, 744–745
  knot, 115
  leaf spot, 262
Olpidiaceae, 63
Olpidium brassicae, 394–395
OM-2424, 30
Omphalia, 395
  pigmentata, 395
  tralucida, 395
Omphalina, 395
OMS 771, 45
Oncidium ring spot, 489
Oncoba, 745
Onion, 745–746
  bacterial bulb rot, 111
  basal rot, 381
  blast, 335
  bloat, 315–316
  downy mildew, 232–233
  neck rot, 368
  pink root, 408
  purple blotch, 187
  smudge, 372
  smut, 463–464
  sour skin rot, 110
  stemphylium blight, 185

white rot, 416
yellow dwarf, 489
Onobrychis, 806
Onoclea, 643
Onosmodium, 746
Onygenaceae, 69
Oomycetes, 64–65, **65**
Oospora, 395
citri-aurantii, 395
lactis, 395
Ophiodothella vaccinii, 280
Ophioglossum, 640
Ophionectria, 216
balsamea, 216
scolecospora, 216
Oplismenus, 746
Oplopanax, 622
Opuntia, 571
black spot, 266
Orange, 599–600, 746
Orange rust of blackberry,
431, 436–437
Orchardgrass, 665–667
Orchids, 747–748
anthracnose, 89
black spot, 268
blossom brown necrotic
streak, 489
brown rot, 107
mosaics, 489
ring spots, 489
Color Plate 1E, 1F
sabralia blight, 268
wilt, 508
Orchid-tree, 549
Oregon-grape, 717
Oriental flowering spirea, 823
Oriental plane, 776–777
Ornalin, 30
Ornithogalum, 828
mosaic, 489
Orobanche ludoviciana, 191
racemosa, 191–192
Orontium, 660
Orthocarpus, 748
Orthocide, 15, 30
Ortho-Phenylphenol, 30
Orthoxenol, 30
Osage-orange, 748
leaf blight, 185
Osier, 748
blister canker, 199
Osmanthus americanus, 622
fragrans, 837
ilicifolius, 682
Osmaronia, 748
Osmorhiza, 839
Osmunda, 642
Osoberry, 748
Ostropaceae, 72
Ostrya, 684

Ovularia, 280
aristolochiae, 280
pulchella, 280
Ovulinia, **136**, 166
azaleae, 166–168
Owls clover, 748
Oxadixyl, 30
Oxalis, 748–749
Oxeye daisy, 619
Oxybaphus, 854
Oxycarboxin, 30
Oxydendron, 749
twig blight, 185
Oxygen deficiency, 344
Oxyquinoline sulfate, 30
Oxyria, 735
Oxythioquinox, 30
Oyster mushroom, 749

Pachistima, 749
Pachyma cocos, 408
Pachysandra, 749
leaf and stem blight, 187
Pacific dagger nematode, 329
Pacific wax myrtle, 750
Paecilonmyces buxi, 395
Paeonia, 764–765
Pagoda tree, 819
dieback, 205
Paintbrush blister rust, 427
Painted cup, 750
Painted-tongue, 805
Pallinal, 31
Palm, areca, 750
coconut, 750
date, 750–751
fishtail, 751
plumy coconut, 751
queen, 751
rhapis, 751
royal, 752
sugar, 752
Washington, 752
Palm diseases, 750–752
black scorch, 246
bud rot, 397, 404
butt rot, 384
decline disease, 395
false smut, 461
greasy spot, 270
leaf and stalk rot, 377
leaf spot, 116, 251, 262,
266, 267, 268, 270,
281, 282
lethal yellowing, 214
Color Plate 1A, 1B
Penicillium disease, 216–217
Palmetto, 752
Paloverde, 753
Pampas grass, 753
Panax, 656–657

Pandanus, 753
Panicum, 840
mosaic, 489
Pansoil, 31
Pansy, 753
Alternaria blight, 138
anthracnose, 92
Botrytis on, 143
scab, 475
Papaver, 782
Papaya, 754
crown rot, 368
Paper-mulberry, 565
Paratylenchus, 325
anceps, 325
dianthus, 325
epacris, 325
hamatus, 325
projectus, 325
Parinol, 31
Parkinsonia, 754
Parnassia, 668
Parnon, 31
Parrotfeather, 754
Parsley, 754–755
leaf blight, 187
mosaic, 121
stem blight, 154
Parsnip, 755
leaf blight, 162
leaf spot, 287
Parthenium, 669
Parthenocissus quinquefolia,
860
tricuspidata, 692–693
Partridge-berry, 755
Parzate, 31
Passiflora, 755–756
Passion-flower, 755–756
brown spot, 250
leaf spot, 260
Pastinaca, 755
Pateliariaceae, 72
Pathogens, classification of,
60–85
Paulownia, 756
Pawpaw, 756
PCNB, 31
Pea, 756–757
anthracnose, 91
Ascochyta blight, 138–139,
165
bacterial blight, 115
black leaf, 267
blight, 138–139, 165
downy mildew, 234
enation mosaic, 489
foot rot, 366
hop cyst nematode, 318
leaf spot, 253, 257
mosaic, 490

Pea *(cont.)*
  mottle, 490
  root rot, 363–364
  rust, 451
  scab, 453
  streak, 490
  wilt, 490
Peach, 757–759
  anthracnose, 94
  asteroid spot, 490
  black mold, 367
  brown rot, 390–392
  calico, 490
  canker, 194
  dwarf, 490
  fruit rot, 377, 384
  golden net, 490
  leaf curl, 244–245, **245**
  leaf and shoot blight, 164
  little peach, 490
  mildew, 359
  mosaic, 490
  mottle, 490
  necrotic leaf spot, 490
  perennial canker, 225
  phony disease, 490
  red suture, 490–491
  ring spot, 491
  root rot, 375
  rosette, 491
  rosette mosaic, 491
  rust, 449–450
  scab, 452
  shoot blight, 150
  stubby twig, 491
  twig canker, 208, 213–214
  wart, 491
  western X-disease, 122–123
  X-disease, 123
  yellow bud, 491
  yellow leaf roll, 123
  yellows, 123
Peacock spot, 262
Peanut, 759–760
  crown rot, 367
  damping-off, 227–228
  early leaf spot, 255
  leaf spot, 257, 277
  mottle, 491
  pod rot, 414
  rootrot, 414, 415
  rust, 442
  stunt, 491
  web blotch, 190
  wilt, 511
Peanut stunt virus, 491
Pear, 761–762
  bitter rot, 384
  black end, 335
  black pox, 238, 239

  canker, 194, 218
  decline, 491
  dieback, 213–214
  fire blight, **103,** 103–105
  leaf blight, 157
  leaf spot, 279
  mistletoe, 298
  Nectria canker, 214–215
  ripe rot, 394
  scab, 457–458
  stony pit, 491
  twig blight, 172
Pearl everlasting, 528
Pea-tree, 762–763
Pecan, 763–764
  brown leaf spot, 256–257
  canker, 203
  downy spot, 277–278
  kernel spot, 394
  leaf blotch, 189–190
  leaf spot (white mold), 276
  liver spot, 268
  root rot, 371–372
  scab, 452–453
  spot anthracnose, 473
  vein spot, 269
Peconazole, 31
Pedicularis, 869
Pelargonium, 653–654
  bacterial leaf spot, 118,
    120
  blossom blight, 142
  leaf curl, 486
  mosaic, 486
  rust, 447
Pellaea, 641
Pellicularia, 168
  filamentosa, 87, 150, 168,
    396–397
    f. sp. microsclerotia, 168
    f. sp. sasakii, 168
    f. sp. timsii, 168
  koleroga, 150, 168–169
  rolfsii, 169–170, 416
Peltrandra, 538
Penconzeb, 31
Penicillium, 216
  digitatum, 397
  expansum, 397
  gladioli, 397
  italicum, 397
  martensii, 397
  roseum, 397
  vermoeseni, 216–217, 397
Peniophora luna, 398
Pennisetum, 764
Penstemon, 764
Pentachloronitrobenzene,
    31
Pentagen, 31

Peony, 764–765
  anthracnose, 94
  Botrytis blight, 142,
    144–145
  Color Plate 8B
  leaf blotch, 188
  leaf curl, 491
  leaf spot, 292
  measles, 188
  ring spot, 491
Peperomia, 765
  ring spot, 491
Peplis diandra, 535
Pepper, 765–767
  anthracnose, 93
  bacterial canker, 101–102
  cyst nematode, 319
  fruit rot, 372, 376
  Fusarium wilt, 507
  leaf spot, 256
  mosaics, 492
  mottle, 492
  Phoma rot, 399
  Phytophthora blight, 174
  ripe rot, 372
  rot, 399
  vein banding mosaic, 492
Pepper-grass, 767
Peppermint *(see* Mint) rot,
    382
Pepper vine, 767
Peraphyllum, 827
Perecot, 31
Perennial canker, 215–216,
    225
Perennial pea, 837–838
Perenox, 31
Peridermium, 439
  appalachianum, 427
  bethelii, 439
  ornamentale, 439
  rugosum, 439
Perisporiales, 69
Periwinkle, 858–859
  canker, 218
Peronospora, **231,** 232
  antirrhini, 232
  arborescens, 232
  arthuri, 232
  destructor, 232–233
  dianthicola, 233
  effusa, 233
  fragariae, 233
  grisea, 233
  lepidii, 233
  leptosperma, 233
  linariae, 233
  lophanthi, 233
  manshurica, 233
  myosotidis, 233

oxybaphi, 233
parasitica, 233–234
pisi, 234
potentiallae, 234
rubi, 234
rumicis, 234
schactii, 234
sparsa, 234
tabacina, 234
trifoliorum, 234
Peronosporaceae, 65
Peronosporales, 65
Persea americana, 544–545
borbonia, 794
Persimmon, 767–768
leaf blotch, 190
mistletoe, 298
tar spot, 276
twig blight, 172
wilt, 502
Pestalotia, 170
aquatica, 280
aucubae, 281
cliftoniae, 281
funerea, 170, 281
guepini, 281
hartigli, 170
leucothoës, 281
longisetula, 398
macrotricha, 281
palmarum, 281
rhododendri, 281
Pestalozziella subsessilis, 281
Pesticides, sources of, 47–50
Pestmaster, 45
Petal blight of azalea, 166–168
of chrysanthemum, 162
Petalostemon, 768
Petasites, 768
PETD, 31
Petroselinum, 754–755
Petunia, 768–769
crown rot, 404
Pezicula, 217
carpinea, 217
coflicola, 217
pruinosa, 217
Pezizaceae, 70
Pezizales, 70
Pezizella oenotherae, 281–282
pH, 334
Phacelia, 573
Phacidiaceae, 71
Phacidiales, 71
Phacidiella coniferarum, 217
Phacidiopycnis pseudotsugae, 217
Phacidium, 170
abietinellum, 170
balsameae, 170

curtisii, 282
infestans, 171
Phaeocryptopus gäumonni, 301–302
Phaeotrichoconis crotolariae, 282
Phakopsora cherimoliae, 439
jatrophicola, 439
pachyrhizi, 439
vitis, 441
zizyphi-vulgaris, 439
Phalaris, 796
Phallales, 76
Phaltan, 23, 31
Phaseolus acutifolius, 552
angularis, 549
aureus, 552
coccineus, 552
limensis, 550–552
vulgaris, 550–552
Phellinus igniarius, 379
Phenamiphos, 31
Phenostat H, 31
Phentinacetate, 31
Phenylphenol, 31
Phialophora, 398
graminicola, 171
malorum, 398
Philadelphus, 730
Philibertia, 769
Philodendron, 769
leaf rot, **Color Plate 4B**
root rot, 409
Phlebia chrysocrea, 398
Phleomycin, 31
Phleospora, 171
aceris, 282
adusta, 171
Phloem necrosis of elm, 122
Phlox, 770
leaf spot, 292
mildew, 352–354, **353**
stem blight, 179
stem and bulb nematode, 315–316
streak, 492
Phoenix, 750–751
Phoenix-tree, 770
Pholiota adiposa, 398
Phoma, **196,** 242
apiicola, 399
arachidicola, 190
betae, 399
conidiogena, 171
destructiva, 399
eupyrena, 305
fumosa, 171
lingam, 127
macdonaldii, 171, 399

mariae, 171
piceina, 171
strobiligena, 171
Phoma fruit spot, 239
rot, 299
stemgall, 242
Phomopsis, **136,** 172
alnea, 217
ambigua, 172
arnoldia, 217
boycei, 217
diospyri, 172
elaeagni, 217
gardeniae, 217–218
japonica, 172
juniperovora, 172
kalmiae, 172
livella, 218
lokoyae, 218
mali, 218, 399
oblonga, 172
occulta, 172
padina, 218
pseudotsugae, 217
vaccinii, 172, 399
vexans, 154, 172
Phomopsis diseases
blight of eggplant, 154
canker, 217, 218
disease of conifers, 217
stem end rot of citrus, 376
stem gall, 242
Phony disease of peach, 490
Phoradendron, 729
californicum, 297
flavescens, 298
juniperinum, 298
libocedri, 298
serotinum, 298
tomentosum, 298
villosum, 298
Phorate TSK, 31
Phormium tenax, 739
Phosphorus deficiency, 344
Photinia, 771
crown rot, 402
Phragmidium, **437,** 440
americanum, 440
disciflorum, 440
fusiforme, 440
montivagum, 440
mucronatum, 440, 441
rosae-acicularis, 440
rosae-arkansanae, 440
rosae-californicae, 440
rosae-pimpinellifoliae, 441
rosicola, 441
rubi-idaei, 441
speciosum, 441

Phragmidium *(cont.)*
  subcorticium, 440
  tuberculatum, 440
Phragmodothella ribesia, 218
Phragmopyxis acuminata, 441
Phygon, 20, 32
Phyllachora, 283
  graminis, 283
  sylvatica, 283
Phyllactinia, **351,** 357–358
  angulata, 358
  corylea, 358
  guttata, 358
Phyllodoce, 734
Phyllostachys, 547
Phyllosticta, 173, **252**
  althaeina, 283
  antirrhini, 283
  aucubae, 283
  batatas, 173
  brassicicola, 277
  camelliae, 283
  camelliaecola, 283
  catalpae, 283
  circumscissa, 283
  citrullina, 165
  concentrica, 283
  congesta, 190
  cookei, 284
  cryptomeriae, 173
  decidua, 284
  hamamelidis, 284
  hydrangeae, 284
  ilicis, 284
  kalmicola, 278, 284–285,
    284
  lagerstroemia, 173
  liriodendrica, 278
  maculicola, 285
  maxima, 285
  minima, 285
  nyssae, 279
  pteridis, 173
  richardiae, 285
  saccardoi, 285
  sanguinariae, 285
  solitaria, 190
  vaccinii, 285
  viridis, 278
  wistariae, 285
Phyllostictaceae, 77
Phyllostictina vaccinii, 285
Phyltaena ficuum, 282
Phyltidiaceae, 63
Phymatotrichum omnivorum,
    399–401, *passim*
Phymatotrichum root rot,
    400–401, *passim*
Physalis, 593–594, 668–669

Physalospora corticis, 173
  corticis, 218
  dracaenae, 173
  glandicola, 218
  gregaria, 173
  ilicis, 284
  miyabeana, 219
  mutila, 401
  obtusa, 173, 219, 401
  rhodina, 219, 401
Physarum cinereum, 459
Physocarpus, 740
Physoderma maydis, 285–286
Physodermataceae, 63
Physopella, 441
  ampelopsidis, 441
  compressa, 441
  fici, 424
Physostegia, 771
Phytomycin, 32, 41
Phyton-27, 32, 41
Phytophthora, 173–174
  cactorum, 174, 219–220,
    401–402, 511
  capsici, 174, 402
  cinnamomi, 174, 221, 402,
    511
  citricola, 402
  citrophthora, 174, 402–403
  cryptogea, 403
    var. richardiae, 403
  drechsleri, 403
  erythroseptica, 174, 403
  fragariae, 403
  ilicis, 174
  infestans, 175–179, **176,**
    177, 178
  lateralis, 403–404
  megasperma, 404
  megasperma f.sp. glycinea,
    404
  nicotianae var. nicotianae,
    404
  nicotianae var. parasitica,
    174, 404
  palmivora, 404
  parasitica, 177, 404
    var. nicotianae, 404
  phaseoli, 234–235
  sojae, 404
  syringae, 177, 221
  terrestris, 404
Phytotoxicity, 8, 13–14
Pic-Clor, 32
Picea, 824–825
Picfume, 32
Pick-a-back, 771
Picramnia, 683
Pierce's grape disease, 476

Pieris, 771
Pigeon pea, 772
Piggotia fraxini, 278
Pilea, 539
Pileolaria, 441
  cotini-coggyriae, 441
  patzcuarensis, 441
Pimenta, 524
Pimpinella anisum, 529
Pine, 772–775
  bleeding canker, 223
  blight, 156
  blister rust of white,
    427–430
  Botrytis on, 142
  brown felt blight, 166
  brown spot needle blight,
    186
  canker, 193–194, 195, 203,
    222, 225
  cone rust, 427, 430
  dieback, 147
  gall rust, 428
  gray blight, 157
  little leaf, 402
  mistletoe, 299
  needle blight, 156
  needle cast, 163, 303, 304,
    305
  needle rust, 424, 425
  pitch canker, 207
  root rot, 404, 406
  sheathoid nematode, 317
  shoot blight, 185
  stem rust, 426, 427
  twig blight, 147
Pineapple, 775
Pink, garden, 623
Pink patch of turf, 374
Pink root of onion, 408
Pink snowmold, 467–468
Pink watery rot of potato,
    403
Pin nematodes, 325
Pinon blister rust, 428
Pinus, 772–775
Piperalin, 32
Pipron, 32
Pipsissewa, 775
Piptoporus betulinus, 406
Piqueria, 829
Piricularia grisea, 179
Pirostoma nyssae, 286
Pistachio, 775
  leaf spot, 292
  shoot blight, 142
Pistacia, 775
Pisum, 756–757
Pitcher-plant, 775–776

Pitcher-sage, 821
Pithecellobium, 776
Pitted sap rot, 406
Pittosporum, 776
  leaf spot, 257
PKhNB, 32
Placosphaeria, 286
  graminis, 286
  haydeni, 286
Plagiostoma, 286
  asarifolia, 286
  prenanthis, 286
Planera, 863
Plane-tree, 776–777
  canker, 212
  canker stain, 195–196
  leaf spot, 261
Plantago, 777
Plantain, common, 777
Plantain-lily, 686
Plant disease
  control, 5–6
  definition of, 2–3
  history of, 3–5
Plantomycin, 32
Plant pathology, 2–5
Plantvax, 32
Plasmodiophora, 226
  brassicae, 226–227
Plasmodiophoraceae, 64,
  226–227
Plasmodiophorales, 64
Plasmopara, 231, 235
  acalyphae, 235
  geranii, 235
  gonolobii, 235
  halstedii, 235, 242
  nivea, 235
  pygmaea, 235
  viburni, 235
  viticola, 235–236
Platanus acerifolia, 776–777
  occidentalis, 776–777
  orientalis, 776–777
  racemosa, 776–777
Platycodon, 777
Plectospira myriandra,
  404–405
Plectranthus australis, 833
Plenodomus, 221, 405
  destruens, 405
  fuscomaculans, 221
Pleosphaerulina, 287
Pleospora, 286
  betae, 399
  herbarum, 287, 294
  lycopersici, 405
Pleuroceras populi, 275
Pleurotus, 736

Pleurotus, 405
  ostreatus, 405, 749
  ulmarius, 405–406
Plum, 777–778
  black knot, 124–126, 125
  brown rot, 390–392, 391
  canker, 204
  leaf blotch, 190
  line pattern, 492
  mistletoe, 298
  pockets, 244
  silver leaf, 417
  white spot, 492
Plumed thistle, 598
Plumegrass, 779
Plumeria, 779
Plumy coconut palm, 751
Plum-yew, 587
Poa compressa, 665–667
  pratensis, 665–667
Pod blight of lima bean,
  153–154
  of soybean, 154
Podocarpus, 779
Podophyllum, 724
Podosphaera, 351, 358
  clandestima var. tridactyla,
  358
  leucotricha, 358
  oxyacanthae, 358
  tridactyla, 358
Pod spot, of bean, 252, 394
  okra, 251
  pepper, 394
Poinciana, 779
Poinsettia, 779–780
  bacterial stem rot, 105–107
  Botrytis on, 142
  canker, leaf spot, 207
  root and stem rot, 411
  scab, 474
Poison hemlock, 780
Poker-plant, 699
Polemonium, 780
Polyanthes tuberosa, 850
Polyclar MZ, 32
Polyclar S, 32
Polygala, 728
Polygonatum, 818
Polynox, 32
Polypodium, 642
Polyporaceae, 73, 76
Polyoxin, 32
Polyoxin AB, 32
Polyoxin B, 32
Polyporus, 406
  abietimus, 408
  anceps, 406
  balsameus, 406

betulinus, 406
dryadeus, 406
gilvus, 406
hispidus, 406
lucidus, 384, 406
pargamenus, 406
schweinitzii, 406
squamosus, 406
sulphureus, 407, 407
tomentosus var. circinatus,
  407
versicolor, 407–408
Polyram, 32
Polyram-Combi, 32
Polyram M, 32
Polyram-Ultra, 32
Polyram Z, 32
Polystichum, 641
Polystomellacea, 70
Polyscias, 536
Pomasol Forte, 32
Pomasol Z Forte, 32
Pome fruit spot anthracnose,
  472
Pomegranate, 780
  dry rot, 394
  leaf blotch, 190
  rot, 367
  spot anthracnose, 474
Poncirus, 671
Ponderosa pine rust, 427
Pond-lily, yellow, 863
Pond-spice, 710, 781
Poplar, 781–782
  black leaf spot, 258
  canker, 196, 202, 203, 206,
    211, 212
  dagger nematode, 331
  leaf blister, 244, 245
  leaf spot, 274, 275, 279
  mistletoe, 298
  rust, 437
  shoot blight, 150
  wetwood, 101
Poppy, 782
  bacterial blight, 120
Poppy-mallow, 782
Populus, 781–782
Poria, 408
  cocos, 408
  laevigata, 408
  luteoalba, 408
  prunicola, 408
  subacida, 408
  weirii, 408
Port Orford white-cedar, 588
Portulaca, 783
  white rust, 501
Potamogeton sp., 535

Potassium deficiency, 344
Potassium permanganate, 32
Potato, 783–785
  acropetal necrosis, 492
  anthracnose, 88
  apical leaf roll, 124
  aucuba mosaic, 492
  bacterial canker, 101–102
  bacterial ring rot, 102
  black dot disease, 88
  blackleg, 105
  black scurf, 396, 411
  bouquet disease, 492
  calico, 492
  crinkle, 492
  early blight, 137–138
  golden nematode, 319
  green dwarf, 492
  late blight, 175–177, **176, 177**
    Color Plate 5C
  leaf blotch, 188
  leaf roll, 492
  leaf-rolling mosaic, 492
  leaf spot, 253, 256
  mottle, 492
  pink watery rot, 403
  powdery mildew, 352–354
  powdery scab, 454
  rhizoctoniose, 396
  rot nematode, 315
  rugose mosaic, 492
  scab, 454–455
  silver scurf of, 458
  spindle tuber, 493
    Color Plate 3E
  vein banding, 493
  viruses A and X, 493
  wart, 243
  watery leak, 409
  wilt, 511–515, **513**
  witches' broom, 493
  yellow dwarf, 493
  yellow spot, 493
Potato virus X, 493
Potato virus Y, 493
PP-588, 33
PP-675, 33
PP-781, 33
Potentilla, 785
Pothos, 785
Pot marigold, 572–573
Powdery mildews, 349–361, **350**
  alder, 352
  apple, 358
  begonia, 356
    Color Plate 5D
  blueberry, 356

  carrot, 354
  cereals, 354
  cherry, 358
  citrus, 356–357
  crape-myrtle, 354
  cucurbits, 352–354
  euonymus, 356
  European, 356
  gooseberry, 360
  grape, 361
  grass, 354
  hop, 359
  horse-chestnut, 361
  legumes, 354–355
  lilac, 355
  live oak, 359
  oak, 355–356
  peach, 360
  phlox, 352–354, **353**
  rose, 360–361
    Color Plate 5A, 5B
  trees, 358
  willow, 361
Powdery scab of potato, 454
Power sprayers, 51–53
Prairie-clover, 768
Prairie coneflower, 793
Prairie gentian, 638
Pratylenchus, 326
  brachyurus, 326
  coffeae, 326
  crenatus, 326
  hexincisus, 326
  leiocephalas, 326
  minyus, 326
  musicola, 326
  penetrans, 326
  pratensis, 326
  safaenis, 326
  scribneri, 326
  subpenetrans, 326
  thornei, 326
  vulnus, 327
  zeae, 327
Prenanthes, 785–786
Previcur, 33
Prezervit, 33, 45
Prickly-ash, 786
  stem canker, 205
Prickly-poppy, 786
  downy mildew, 232
Primrose, 786–787
  black spot, 253
  Botrytis on, 142
  leaf spot, 111–112, 287
  mosaic, 493
Primula, 786–787
Princes-feather, 526
Princess-tree, 756

Pringsheimia sojaecola, 287
Privet, 787
  anthracnose, 94
  powdery mildew, 355–356
  ring spot, 493
Proboscidea, 854
Proboscis-flower, 854
Profume, 46
Prophos, 46
Propineb, 33
Prosopis, 727
Prospodium, 442
  appendiculatum, 442
  lippiae, 442
  plagiopus, 442
  transformans, 442
Protectant, 7
Protection, 6
Prothiocarb, 33
Protomyces macrosporus, 242
Protomycetaceae, 68
Protomycetales, 68
Prune, 777–778
  constricting mosaic, 493
  Cytospora canker, 203
  diamond canker, 493
  dwarf, 493
Prunella, 787
Pruning disease of pine and fir, 147
Prunus species, 589–590
  amygdalus, 524–525
  armeniaca, 534–535
  domestica, 777–778
  laurocerasus, 591
  lyoni, 583
  persica, 757–759
    var. nectarina, 738–739
  serrulata, 590–591
  subhirtella, 591
  triloba, 525
  virginiana, 595–596
Psalilota compestris, 238
Pseudolarix, 660
Pseudomonas, 109
  aceris, 109
  adzukicola, 109
  alboprecipitans, 109
  alcaligenes, 109
  alliicola, 111
  andropogonis, 109
  angulata, 109
  apii, 110
  asplenii, 109
  berberidis, 109
  caryophylli, 109–110
  cattleyae, 110
  cepacia, 110
  cichorii, 110

corrugata, 110
fluorescens, 110
gladioli, 110
jaggeri, 110
maculicola, 110–111
marginalis, 111
marginata, 111
melophthora, 111
primulae, 111–112
pseudoalcaligenes subsp.
    citrulli, 112
ribicola, 112
sesami, 112
solanacearum, 112
stizolobii, 112
syringae, 112
    pv. aptata, 112–113
    pv. coronafaciens, 113
    pv. delphinii, 113
    pv. glycinea, 113
    pv. helianthi, 113
    pv. hibisci, 113
    pv. lachyrmans, 113
    pv. mori, 113–114
    pv. mors-prunorum, 114
    pv. papulans, 14
    pv. phaseolicola, 114–115
    pv. pisi, 115
    pv. savastanoi, 115
    pv. syringa, 115
    pv. tabaci, 115
    pv. tagetes, 116
    pv. tomato, 115
    pv. tonelliana, 115–116
    pv. zizaniae, 116
tabaci, 116
viburni, 116
viridilivida, 116
washingtoniae, 116
woodsii, 116
Pseudonectria pachysandricola,
    221
rouselliana, 221, 226
Pseudoperonospora, 236
celtidis, 236
cubensis, 236–237
Pseudopezicola tetraspora, 247
Pseudopeziza ribis, 97–98
Pseudosaccharomycetaceae,
    78
Pseudotsuga, 627–628
Pseudovalsa longipes, 221
Psidium, 669
Psophocarpus, 867
Ptelea, 684
Pteretis, 642
Pteridium, 640
Pteris, 641
PTF, 33

Puccinia, **437,** 442
allii, 442
amphigena, 442
andropogonis, 442
antirrhini, 442, **443**
arachidis, 442
aristidae, 442
asparagi, 442–443
avocensis, 423
canaliculata, 444
cannae, 448
carduorum, 444
caricina, 444
caricis var. grossulariata,
    444
carthemi, 444
claytoniicola, 444
conoclinii, 444
coronata, 444
crandalli, 444
cynodontis, 444
cypripedii, 444
dioicae, 444
dracunculi, 444
extensicola, 444
flaveriae, 444
glumarum, 448
grammis agrostidis, 444
    avenae, 444
    phlei-pratensis, 445
    poae, 445
    secalis, 445
    tritici, 445
helianthi, 446
heterospora, 446
heucherae, 446
hieracii, 446
iridis, 446
malvacearum, 446
menthae, 446–447
pelargonii-zonalis, 447
peridermiospora, 448
phragmitis, 447
poae-nemoralis, 447
poae-sudeticae, 447
polysora, 447
porri, 442
pringsheimia, 444
psdii, 447
pygmaea, 447
recondita agropyri, 447
    agropyrina, 447
    apocrypta, 447
    impatientis, 447
    secalis, 447
    tritici, 447
ripulae, 448
rubigo-vera, 447
solheimi, 448

sorghi, 448
sparganoides, 448
stenotaphri, 448
striiformis, 448
substriata, 448
taniceti, 444
thaliae, 448
triticina, 447
Pucciniaceae, 74
Pucciniastrum, 448
americanum, 448
epilobii, 448
ericae, 450
goeppertianum, 448
hydrangeae, 448
myrtilli, 448–449
vaccinii, 448–449
Puccoon, 710
Pueraria, 700
Puffballs, 238
Pumpkin, 788, 826–827
bacterial spot, 118
Punctodera punctata, 318
Puncture vine, 788
Punica granatum, 780
Purple blotch, of onion, 187
of oxydendron, 277
Purple cane spot of raspberry,
    155
Purple coneflower, 629
Purple leaf blotch of grass,
    191
Purple leaf spot of strawberry,
    278
Purple loosestrife, 715
Purslane, 783
leaf spot, 267
stem canker and root rot,
    205
Pussy willow, 866–867
Pustule of soybean, 118
Pycnanthemum, 735
Pycnotysanus azaleae,
    146–147
Pyracantha, 788
fire blight, 105
mistletoe, 298
scab, 453
wilt, 507
Pyracantha angustifolia, 105
coccinea, 105
crenulata, 105
Pyracarbolib, 33
Pyrazophos, 33
Pyrenochaeta, 179
lycopersici, 408
phlogis, 179
terrestris, 408
Pyrenophora lolii, 271

Pyrethrum, 788
Pyrola, 788–789
Pyrus, 761–762
Pythiaceae, 65
Pythium, 409
  acanthicum, 409
  aphanidermatum, 409
  aristosporum, 409
  arrhenomanes, 409
  carolinianum, 409
  catenulatum, 409
  debaryanum, 227–228, 409
  dissotocum, 409
  irregulare, 409
  myriotylum, 409
  periplocum, 409
  splendens, 409
  tracheiphilum, 511
  ultimum, 409

Quamoclit, 617–618
Queens delight, 830
Quercus, 741–743
Quick decline of citrus, 483
Quince, 789
  blotch, 239
  Nectria canker, 214–215
  rust, 432–433
Quince, flowering, 790
Quincula, 669
Quinomethionate, 33
Quintox, 33
Quintozene, 33

Rabbitbrush, 790
Radish, 790–791
  black root, 364
  leaf spot, 121, 250
  mosaic, 493
  wilt, 510
Radopholus similis, 327–328
Ragweed, 791
Rain-lily, 791
Ramularia, 252, 287
  armoraciae, 287
  pastinacae, 287
  primula, 287
  vallisumbrosae, 287–288
  variabilis, 288
Ramulispora sorghi, 268, 288
Ranunculus, 791–792
Raphanus, 790–791
Raphiolepis, 689
Raspberry, 792–793
  anthracnose, 92
  black spot, 294
  cane blight, 159, 163
  cane gall, 98–100
  cane spot (dieback), 193, 221

decline, 494
fire blight, 103–105
late leaf rust, 448
leaf curl(s), 494
leaf spot, 293
mosaic, 494
necrosis, 494
rust, 441
spot anthracnose, 473
spur blight, 155
streak, 494
yellow mosaic, 494
yellows, 494
Ratibida, 793
Rattan vine, 794
Rattlesnake-root, 785–786
Ravenelia, 449
  dysocarpae, 449
  humphreyana, 449
  indigoferae, 449
Raxil, 33
Readex, 33
Red-bay, 794
  black mildew, 129
  black spot, 128
Redbud, 794
  canker, 194
  leaf spot, 278
  root rot, 384
  wilt, 511–515
Red-cedar, 696–698
Red leaf gall, 243
Red leaf spot of gladiolus, 294
Red mottle rot, 408
Red osier, 626
Red ray rot, 406
Red ring rot, 379
Red stele disease of strawberry, 403
Red stem, 645
Red thread of grass, 374
Red valerian, 855
Redwood, 811
  bark canker, 204
  needle blight, 165
  seedling blight, 143
Rehmiellopsis balsameae, 179–180
Remasan Chloroble M, 33
Reniform nematode, 328
Reseda, 727–728
Resisan, 33
Resistance, 6
Rhabdocline pseudotsugae, 305–306
Rhabdospora rubi, 221
Rhamnus, 566
  californicus, 604
Rhapis, 751

Rheum, 796
Rhexia, 724, 794
Rhipsalidopsis, 629
Rhizidiaceae, 63
Rhizina inflata, 410
Rhizobiaceae, 98–100
Rhizoctonia, 180, 411
  bataticola, 389–390, 411
  crocorum, 386, 411
  ramicola, 180
  solani, 87, 168, 180, 228, 396–397, 411
  tuliparum, 411–412
Rhizoctoniose of potato, 396
Rhizome rot, 416
Rhizophora, 719
Rhizopus, 412
  arrhizus, 412
  nigricans, 412
  oryzae, 412
  stolonifer, 180, 412
Rhizosphaera kalkoffi, 306
Rhodesgrass, 794
Rhodianebe, 33
Rhododendron, 545–546, 795
  Botrytis on, 143
  bud and twig blight, 146–147
  dieback, 219–220
  heart rot, 378
  leaf gall, 240
  leaf spot, 257, 265, 268, 270, 274, 281, 285
  roseum, 164
  wilt, 511
Rhodotypos, 696
Rhoes, 796
Rhubarb, 796
  anthracnose (stalk rot), 89
  chlorotic ring spot, 494
  crown rot, 103
  downy mildew, 234
  ringspot, 494
Rhus, 834
Rhytisma, 288
  acerinum, 288
  andromedae, 289
  bistorti, 289
  liriodendri, 289
  punctatum, 289
  salicinum, 289
Ribbon-bush, 796
Ribbon-grass, 796
  leafspot, 270
Ribes, 615–616, 662–663
Rice-paper plant, 797
Ricinus communis, 583
Ridomil, 33
Ridomil MZ, 33
Ridomil MZ58, 33

Ridomil MZ72, 33
Ridomil Plus, 33
Ring nematode, 313
Ring rot of tomato, 393
Ring spot, of African-violet,
344
aster, 476
carnation, 480
cherry, 481
chrysanthemum, 482
crucifer, 277
dahlia, 485
delphinium, 485
hydrangea, 487
lilac, 487
lily, 487
orchid, 489
peony, 491
peperomia, 491
privet, 493
rhubarb, 494
tobacco, 498
tomato, 498
walnut, 253
Ripe rot, of fig, 380
of pear, 394
Rivina, 801
Rizolex, 33
Robinia, 711
brooming, 494
Robinia hispida, 800
Rock-cress, 535
Rock-jasmine, 797
Rock-rose, 797
Rock-spirea, 682
Roesleria hypogaea, 412
Rollinia, 797
Romanzoffia, 797
Ronilan, 33
Root and butt rot, 378
Root and stem rot, 396–397
Root-knot nematodes,
323–325
Root rot, 403–404
currant, 387
cypress, 403–404
mushroom, 364–366, **365**
Texas, 399–401
Rop 500F, 33
Rosa, 797–800
Rose, 797–800
black mold of grafts, 300
blackspot, 130–135, **132,
133**
Color Plate 6D
Botrytis on, 143, 194
Color Plate 8A
brand canker, 197–198
brown canker, 199–201,
200

cane blight, 173
canker, 194, 201, 225
Cercospora spot, 257–258
common canker, 197
Coryneum canker, 209, **210**
crown canker, 201–202
dagger nematode, 331
dieback, 205
downy mildew, 234
graft canker, 197
leaf spot, 257, 257–258
mosaic, 494
Color Plate 3D
powdery mildew, 360–361
Color Plate 5A, 5B
rosette, 494–495
rust, 440, 441
Color Plate 2A, 2B
spot anthracnose, 473
streak, 495
yellow mosaic, 495
Rose-acacia, 800
leaf spot, 250
Rose-gentian, 800
Roselle, 800
Rosellinia, 180, 413
herpotrichioides, 181
necatrix, 413–414
Rose-mallow, 800
Rosemary, 800
Rose-of-sharon, 678, 801
Rosette, 345
chrysanthemum, 482
lily, 487
rose, 494–495
Rosmarinus, 800
Rots, 362–421
Alternaria, of citrus,
362–363
avocado root, 402
bacterial, of chicory, 110
of corn, 107
bacterial crown, of
delphinium, 105–107
bacterial ring, of potato,
102
basal, of iris, 381
lily, 382
narcissus, 382
onion, 381
tulip, 381
basal stem, 368
black, of apple, 401
bulbs, 416
carrot, 363
cranberry, 370
crucifers, 117–118
grapes, 385–386, **386**
winter-cress, 116
black crown, 369

black root, 335, 419, 425
black scale, of lily, 372
black stem, 376
blossom-end, 336, **336**
Botrytis crown, of iris, 367
brown, of blueberry, 392
citrus, 404
orchids, 107
stone fruits, 390–392,
391
Color Plate 8E
brown cubical, 373, 375
brown stem, 369
brown stingy, 377–378
calyx end, of date, 367
apple, 363, 416
carnation bud, 382
charcoal, 389–390
cladode, 366, 377
collar, of tomato, 363
corn ear, 377, 384
cranberry bitter, 385
blotch, 362
end, 385
hard, 392
crown, 169–170
cypress root, 403–404
dill root, 383
Diplodia collar and root,
376
citrus, 401
corn ear, 377
Dothiorella rot of avocado
and citrus, 367
dry, of gladiolus, 418
dry root, of bean, 383
early, of cranberry, 386
feather, 408
fisheye fruit, 373
flower, of orchids, 363
foot, 366
fruit
apple, 363, 369
date, 369
peach, 377, 384
pepper, 372
squash, 149
tomato, 362, 368, 376,
405
gray bulb, of tulip, 411–412
gray mold, 368
green fruit, 416
heart, of trees. *Indexed by
pathogens*
Java black, 377
mushroom root, 364–366,
365, 371–372
mycelial neck rot, 368
onion bulb, 111
onion pink root, 408

Rots (cont.)
  pea root, 363–364
  peppermint, 382
  Phoma, of tomato and
    pepper, 399
  Phomopsis stem end, 376
  pink watery, 403
  pitted sap, 406
  poinsettia, 411
  radish black, 364
  red ray, 406
  red ring, 379
  red stele, strawberry, 403
  Rhizopus, 412
  rhubarb crown, 108
  ripe, of apples, 111
    of figs, 380
    of pear, 394
    of pepper, 372
    of tomato, 394
  root, of avocado, 402
    calla, 403
    cypress, 403
    sweet pea, 383
  sapwood. Indexed by
    pathogens
  soft, 105–107
    of caladium, 140
    of calla, 105–106
  sour, 395
  sour skin, of onion, 110
  storage, 369, 398
  sweetpotato, 418
    black, 369–370
    dry, 375
    stem, 381
    storage, 393
  tomato hypocotyl rot, 380
  tomato ripe fruit, 394
  Texas root, 400–401
  tuber, 383
  violet root, 386–387
  white, of grapes, 373
  white butt, 374
  white root, 374, 406,
    413–414
  wood, 375
  wound, of trees. Indexed
    by pathogens
  yello flaky heart, 378
Rotox, 45
Rottobellia, 692
Rotylenchulus reniformis, 328
Rotylenchus, 328
  blaberus, 328
  buxophilus, 328
  uniformis, 328
Rouge-plant, 801
Rovral, 33
Royoc, 732

Roystonea, 752
RPH, 33
Rubber-plant, 801
  leaf spot, 273
Rubigan, 33
Rubus, 557–558, 622–623,
    792–793
Rudbeckia, 801–802
  lacinata, 660
Rue anemone, 803
Ruellia, 802
Rumex, 802
Rumohra, 641
Russian-olive, 630–631, 802
  canker, 208, 217, 225
"Rust," 345
Rusts, 421–451
  ash, 448
  asparagus, 442–443
  aster, 425
  bean, 451
  beet, 450–451
  birch leaf, 438
  blueberry, 448–449
  bluegrass, 447
  cane, 436
  carnation, 451
  cedar-apple, 432, 434–435
    Color Plate 2C, 2D
  chrysanthemum, 444
  comandria blister, 427
  corn, 448
  crown of oats, 444
  Douglas-fir needle, 437
  eastern gall, 428
  fig, 424
  fir-fern, 436, 450
  fir-huckleberry, 448
  fir-willow, 437
  gall, 435
  grass, 444–446, 447
  hawthorn, 433
  hemlock, 437–438, 448–449
  hemlock-poplar, 437
  hollyhock, 446
  iris, 446
  jack-in-the-pulpit
    Color Plate 7A, 7B, 7C,
      7D
  juniper gall, 433
  larch needle, 438
  larch willow, 438
  late leaf, of raspberry, 448
  leaf, of blueberry, 448–449
    of cereals, 447
    of roses, 440
  leaf and cane, of raspberry,
    441
  mint, 446–447
  needle, 424

  needle blister, of pine, 425
  orange, of blackberry, 431,
    436–437
  paintbrush blister, 427
  pea, 451
  peanut, 442
  pelargonium, 447
  pine cone, 427, 430
  pine needle, 426
  pinon blister, 428
  ponderosa pine, 427
  poplar, 438
  quince, 432–433
  raspberry, 441
  raspberry, late leaf, 448
  rose, 440
    Color Plate 2A, 2B
  snapdragon, 442, 443
  southern corn, 447
  southern fusiform, 427–428
  spearmint, 446–447
  spruce needle, 424, 425
  stem, of grains, 444–446
  stone fruits, 449–450
  stripe, of wheat, 448
  sunflower, 426, 446
  sweet fern blister, 427
  western gall, 427, 428, 430
  white pine blister, 428–430
  white, chrysanthemum, 446
  witches' broom, 433, 435,
    436
  yellow, of blackberry, 436
  yellow witches' broom, 438
Rust, white, 65, 500–502
Rutabaga, 803

S 767, 46
S-3349, 33
Sabal, 752
Sabatia, 800
Saccharomycetaceae, 68
Sacchotherium, 287
Safflower, 803
Sage, 803
Sage-brush, 903–904
Sagittaria, 538
Sago-palm, 616
Saguaro, 571
Sainfoin, 806
St. Andrews cross, 804
St. Augustine grass, 665–667,
    804
St. Johns bread, 581
St. Johnswort, 804
Saintpaulia, 521
St. Peterswort, 804
SAI San, 33
Salal, 804–805
Salix, 748, 866–867

Salpiglossis, 805
  bacterial canker, 101–102
Salsan, 33
Salsify, 805
  leaf blight, 185
  white rust, 501
Salsify, black, 805
Salt bush, 805–806
Salt cedar, 842
Salt injury, 345
Salvia, 803
Sambucus, 631
Sanchezia, 806
Sand-myrtle, 806
Sand-verbena, 806
Sandvine, 806
Sandwort, 806–807
Sanguinaria, 559
Sanguisorba, 567, 807
Sansevieria, 807
Sanspor, 33
Sapindus drummondii, 818
  saponaria, 817
Sapium, 594
Sapodilla, 807
Saponaria, 818
Saprol, 34
Saprolegniales, 64
Sapwood rot, 389, 407–408,
  417
Sarolex, 46
Sarracenia, 775–776
Sarsaparilla, 535
Sassafras, 807–808
  leaf spot, 249
Saururus, 808
Sawara cypress-retinospora,
  588
Saxifraga, 808
Saxifrage, 808
Scab, 451–458
  almond, 452
  apple, 455–457, **456, 457**
    Color Plate 6F
  aralia, 474
  avocado, 474
  camellia, 472
  camphor tree, 471–472
  Christmasberry, 453
  citrus, 471
    Color Plate 8D
  cucumber, 452
  English ivy, 474
  gladiolus, 111
  goldenrod, 473
  gooseberry, 474
  guaya, 474
  head, 453
  jasmine, 472
  lima bean, 472

loquat, 453
magnolia, 472
mombin, 475
mango, 472
morinda, 474
oleander, 474
pea, 453
peach, 452
pear, 457–458
pecan, 452–453
poinsettia, 474
potato, common, 454–455
powdery, 454
pyracantha, 453
sour orange, 471–472
violet and pansy, 475
willow, 453–454, 474
Scabiosa, 808
Scald of apple, 345
Scaly cap, 389
Scarborough-lily, 808
Scarlet-bush, 671
Scarlet eggplant, 695
Scarlet sage, 803
Schefflera, 809
  actinophylla, 809
  arboricola, 809
  dwarf, 809
  leaf spot, 250, 251
Schinus, 573–574
Schirrhia, 181
Schizanthus, 568
Schizothyrium
  gaultheriae, 289
  pomi, 238
Schlizonella, 462
Schlizophyllum commune, 414
Schlumbergera, 570
Schrankia, 809
Sciadopitys, 854
Scilla, 809
Scindapsus, 785, 809
Scirrhia acicola, 181, 186
Sclerodermatales, 76
Scleroderris, 222
  abieticola, 222
  lagerbergli, 222
  lateritium, 222
Sclerophthora macrospora, 237
Scleropycnium aureum, 181
Sclerospora, 237
  farlowii, 237
  graminicola, 237
  maerospora, 237
Sclerotinia bifrons, 181, 258
  cameliiae, 181–182
  fructicola, 164, 390–392
  fructigena, 390
  geranii, 416
  gladioli, 418

homeocarpa, 414
intermedia, 414
laxa, 164, 392
libertiana, 415
minor, 182, 414
narcissicola, 414–415
oxycocci, 392
polyblastis, 182–183
sclerotiorum, 183, 222, 415,
  415, 416
seaveri, 164
trifolium, 416
vaccinii-corymbosi, 392
whetzelii, 258
Sclerotiniaceae, 70
Sclerotium, 183
  bataticola, 183, 389–390
  cepivoram, 416
  delphinii, 169, 416
  hydrophilum, 183
  oryzae, 183
  rhizodes, 183, 468
  rolfsii, 169, 183, 416
Scoleconectria, 216
Scolecotrichum graminis,
  289–290
Scopella sapotae, 449
Scorch, cactus, 247
  leaf, 247, 342
  linden, 96
  palm, 246
Scorias spongiosa, 470
Scorzonera, 805
Screw pine, 753
Scribner's Meadow nematode,
  326
Scrophularia, 645
Scurf, black, of potato, 396,
  411
  green, 254
  silver, of potato, 458
  sweetpotato, 458
Scutellaria, 814
Scutellonema, 328
  blaberum, 328
  brachyurum, 329
  bradys, 329
  christiei, 329
Sea-grape, 603, 809
  black mildew, 130
Sea-kale, 809–810
  root rot, 364
Sea-lavender, 828–829
Sea-pink, 537
Seaverinia geranii, 416
Sechium, 588–589
Sedum, 810
Seedling blight, 186
Seedling root rot, 409
Seed mold, 287

Seed protectant, 228
Seed treatment, 288
Seinhorst stubby root nematode, 329
Selenophoma, 290
    donacis, 290
    everhartii, 290
    obtusa, 290
Self-heal, 787
Selinox, 34
Sempervivum, 810
Senecio, 598, 810–811
Senna, 583
Sensitive fern, 643
Septobasidiaceae, 74
Septobasidium, 222
    burtii, 222
    castaneum, 222–223
    curtisii, 223
    pseudopedicellatum, 223
Septocylindrium hydrophyllis, 290
Septogloeum, 290
    acerinum, 290
    oxysporum, 290
    parasiticum, 290
    rhopaloideum, 290
Septoria, 136, 183
    agropyrina, 191, 291
    apiicola (apii and apiigraveolentis), 183–184
    azaleae, 247
    bataticola, 291
    callistephi, 291
    clamagrostidis, 291
    chrysanthemella, 291
    chrysanthemi, 183
    citri, 291
    citrulli, 291
    cornicola, 291
    cucurbitacearum, 291
    cyclaminis, 291
    dianthi, 291–292
    divaricata, 292
    elymi, 191
    gladioli, 292
    glycines, 728
    lactucae, 292
    leucanthemi, 184
    loligena, 292
    lycopersici, 292
    macropoda, 191
    musiva, 279
    obesa, 292
    oudemansii, 292
    paeoniae, 292
    petrosellini, 183
    pistaciarum, 292
    populicola, 279
    pyricola, 279

querceti, 279
ribis, 279
rubi, 279, 292–293
secalis var. stipae, 293
spraguei, 293
tageticola, 293
tenella, 293
tritici var. lilicola, 293
Septotinia, popdophyllina, 184
Sequoia, 811
    Botrytis on, 143
    canker, 194
Seriocarpus, 811
Serviceberry, 527
    witches' broom, 130
Sesame, 811
    bacterial leaf spot, 112
Sesamum, 811
Sesuvium, 812
Setaria, 564, 699
SF-6505, 34
Shallot, 812
Shasta daisy, 812
    leaf spot, 250
Shepherdia, 566–567
Shepherd's purse, 812
Shinleaf, 788–789
Shoot blight, 183
    of lilac, 174
Shoot and leaf gall, 240–241
Shoot hypertrophy, 240
Shooting star, 625
Shortia, 812
Shot berry, 345
Shot hole, cherry, 256, 259, 259–260, 260
    plum, 260
Shrub-althaea, 801
Siberian iris, 691–692
Sicana, 582
Sicarol, 34
Sida, 813
Sidalcea, 589
Silene, 813
Silk-oak, 668
Silk-tassel bush, 652, 813
Silk-tree, 813
Silky sophora, 819
Silky thread blight, 180
Silphium, 813
Silver-bell, 670
Silverberry, 630–631, 802
Silverleaf, 740
Silver scurf of potato, 458
Silybum, 728
Sinningia, 658–659
Sinox, 34
Sirococcus strobilinus, 185
Sisyrinchium, 560
Skimmia, 814

Skullcap, 814
Skunk-cabbage, 814
Skyrocket, 655–656
Slender false-brome, 563
Slenderflower thistle, 814
Slime flux, 107–108
Slime molds, 458–459
Slipperwort, 572
Small flower galinsoga, 814
SMDC, 28, 34, 46
Smelowskia, 814
Smilacina, 814–815
Smilax, 815
    leaf spot, 258
Smithantha, 815
Smog injury, 345
Smoke injury, 345
Smoke tree, 815
    rust, 441
Smooth-headed meadow nematode, 326
Smother, 186
Smuts, 459–467
    anther, of carnation, 467
    bluegrass blister, 460
    calendula, 460
    corn, 465–467, 466
    covered, 462, 464, 465
    dahlia, 460
    dwarf bunt of wheat, 463
    erythronium, 464
    false, of palms, 461
    fescue, 467
    fig, 367
    flag, of wheat, 463, 464
    floral, 464
    flower, 462
    gladiolus, 464
    head, 462, 464
    inflorescence, 461
    leaf, 461, 463
    loose, 462, 464, 467
    nigra, 467
    nuda, 467
    onion, 463–464
    seed, 462
    spinach, 461
    stinking, 463
    stripe, 467
    white, 460
SN 41703, 34
Snake plant, 807
Snapdragon, 816
    anthracnose, 88
    Botrytis, 144
    crown root, 393
    downy mildew, 232
    leaf spot, 283
    rust, 442, 443
Sneezeweed, 674

Snowball spot anthracnose, 475
Snowbell, 833
Snowberry, 817
  spot anthracnose, 475
Snow blight, 171
Snowdrop, 817
  Botrytis on, 143
Snowdrop-tree, 670
Snowflake, 705
Snowmold, 467–468
  Color Plate 6A, 6B
Snow-on-the-mountain, 817
Soapberry, 817
  blight, 153
Soapberry, southern, 817
Soapberry, western, 818
Soap-plant, 595
Soapwort, 818
Society garlic, 818
Sodium dehydroacetate, 34
Sodium hypochlorite, 34
Sodium methyldithiocar-
    bamate, 34, 46
Sofril, 34
Soft rot, 105–107
  of calla, 105–106
  of fruits, 397
  sweetpotato, 412
Soil-testing kit, 334
Solanum capsicastrum, 695
  dulcamara, 695
  elaeagnifolium, 740
  integrifolium, 695
  melongena, 630
  pseudocapsicum, 695
  tuberosum, 783–785
Solasan 500, 46
Solenia, 223
  anomala, 223
  ochracea, 223
Solidago, 661
Solomons-seal, 818
Sometam, 46
Sonchus, 818
  yellow net, 495
Soot injury, 345–346
Sooty-bark canker of aspen, 195
Sooty blotch of fruit, 189
Sooty mold, 468–470
Sophora, 819
Sopranebe, 34
Sorbus, 733–734
Sordariaceae, 70
Sorghastrum, 690
Sorosporium saponariae, 462
Sorrel, garden, 802
Sorrel-tree, 749
Sour gum, 852

Sour orange scab, 471–472
Sour rot, 110, 395
Sourwood, 749
  twig blight, 185
Southern bacterial wilt, 112
Southern blight, 169–170, 183
Southern corn leaf blight, 159
Southern fusiform rust, 427–428
Southern root-knot nematode, 324
Soybean, 819–820
  bacterial blight, 113
  bacterial pustule, 118
  brown spot, 292
  brown stem rot, 369
  canker, 204
  charcoal blight, 164
  cyst nematode, 318
  downy mildew, 233
  foliar blight, 180
  frog-eye, 258
  mosaic, 495
  pod and stem blight, 154
  root and stem rot, 404
  Sclerotinia blight, 182
  target spot, 261
  yeast spot, 394
  yellow mosaic, 495
Spanish bayonet, 871–872
Spanish iris, 691
Spanish moss, 821
  stem blight, 157
Sparaxis, 821
  mosaic, 495
Spartium, 565
Spathiphyllum, 821
Spearmint rust, 446–447
Speckle, 290
Speckled leaf blotch of grass, 191
Speckled tar spot of maple, 289
Specularia, 821
Speedwell, 857
Spergon, 16, 34
Spermophthoraceae, 68
Sphacele, 821
Sphaceloma, 474
  ampelima, 87
  araliae, 474
  cercocarpi, 474
  hederae, 474
  lippiae, 474
  menthae, 474
  morindae, 474
  murrayi, 474
  oleanderi, 474
  perseae, 474

  poinsettiae, 474
  psidii, 474
  punicae, 474
  ribis, 474
  spondiadis, 475
  symphoricarpi, 475
  viburni, 475
  violae, 475
Sphacelotheca, 462
  cruenta, 462
  reiliana, 462
  sorghi, 462
Sphaeralcea, 658
Sphaerellaceae, 71
Sphaeriaceae, 70
Sphaeriales, 67, 70–71
Sphaerioidaceae, 77
Sphaeronema, 329
Sphaerophragmium acaciae, 449
Sphaeropsidaceae, 77
Sphaeropsidales, 77, 77
Sphaeropsis, 223
  ellisii, 223
  malorum, 219
  quercina, 218
  tumefaciens, 223, 242
  ulmicola, 223
Sphaeropsis canker, 218
Sphaerotheca, 351, 359
  castagnei, 359
  fuliginea, 359
  humuli, 359
  lanestris, 359
  macularis, 359
  mors-uvae, 359
  pannosa var. persicae, 359
    var. rosae, 359–360
  phytoptophila, 360
Sphaerulina, 185, 293
  polyspora, 185
  rubi, 293
  taxi, 185
Sphenospora mera, 449
Spice-bush, 821
Spice-lily, 719
Spider-flower, 602
Spider-lily, 822
Spiderling, 822
Spiderwort, 848
Spinacea, 822
Spinach, 822–823
  anthracnose, 91
  black mold, 301
  blight, 495
  downy mildew, 233
  Fusarium wilt, 510
  smut, 461
  white rust, 501
  yellow dwarf, 495

Spindle-tree, 637
Spiraea, 823
Spiral nematode, 317
Spirea, 823
Spondias, 823
Spondylocladium atrovirens, 458
Spongospora subterannea, 454
Spore formation in fungi, 77
Sporobolomycetaceae, 78
Sporocybe rhois, 199
Sporodesmium, 185
    maclurae, 185
    scorzonerae, 185
Sporonema camelliae, 293
Spot, bacterial, of stone fruits, 120–121
    of tomato and pepper, 121
Spot anthracnose, 87, 470–475
    bramble, 473
    Chinese holly, 472
    dogwood, 471
    grape, 470–471
    ledum, 472
    linden, 473
    lippia, 474
    mint, 474
    pecan, 473
    pome fruit, 472
    pomegranate, 474
    rose, 473
    snowball, 475
    snowberry, 475
    virginia creeper, 472
Spotrete, 34
Spotrete-F, 34
Spotrete-WP75, 34
Spotted wilt, tomato, 498–499
Sprayers, 51–55
    aerosol, 53–54
    atomizer, 53–54
    compressed air, 54
    hose-end, 55
    hydraulic, 51–52
    knapsack, 54
    mist, 51
    slide, 54
    wheelbarrow, 54–55
Spraying, 51–55
    versus dusting, 56
Sprays, all-purpose, 58
    pressurized, 54
Spreading decline of citrus, 327–328
Spring-Bak, 34
Spring beauty, 601
Spring dead spot, 163
Spring dwarf nematode of strawberry, 310–311

Spruce, 824–825
    Cytospora canker, 202–203
    needle cast, 304, 306
    needle rust, 424
    root rot, 364–366
    shoot blight, 172
    twig blight, 138, 171
Spurge, cypress, 825
    anthracnose, 474
    flowering, 825
    painted, 825
    spotted, 825
Squash, 826–827
    bacterial spot, 118
    blossom blight, 149
    leaf and stem blight, 174
    mosaics, 495
    Color Plate 5E
Squaw-apple, 827
Squaw-bush, 827
Squill, 809
Squirrel-corn, 629
SR-406, 34
SS-1451, 34
SS-2074, 34
Stachys, 827
Staggerbush, 715
Stagonospora curtisii, 247–248
Stanleya, 621
Staphylea, 827
Starflower, 828
Stargrass, 828
    golden, 828
Star-of-Bethlehem, 828
    leaf spot, 265
Statice, 828–829
    anthracnose, 94
Steccherinum, 417
    abietis, 417
    septentrionale, 417
Steganosporium sp., 223
Steiner's spiral nematode, 317
Steironema, 712
Stellaria, 592
Stem and bulk nematode, 315–316
Stem anthracnose, 91
Stem blight, 138, 186
Stem canker, 153, 197, 225
Stem gall, 242, 243
Stemphylium, 252, 294
    bolicki, 294
    botryosum, 287, 294
    callistepha, 294
    cucurbitacearum, 294
    floridanum, 294
    radicinum, 363
    sarcinaeforme, 287

    solani, 294–295
    vesicarium, 185, 295
Stem rot, 382, 387, 401–402, 414
Stem rust, 444–446
Stem spot, 278–279
Stenanthium, 829
Stenolobium, 829
Stenotaphrum, 665–667, 804
Stephanomeria, 829
Stephanotis, 829
Stereum, 417
    fasciatum, 417
    hirsutum, 417
    ostrea, 417
    purpureum, 417
    sanguinolentum, 417–418
Sternbergia, 829
Stevensea, 265
Stevia, 829
Stewart's disease of corn, 108
Stictidiaceae, 72
Stictochlorella, 253
Stigmatea, 295
Stigmea, 295
    geranii, 295
    rubicola, 295
Stigmella platani-racemosae, 295
Stigmonose, 346
Stilbaceae, 77, 78
Stilbellaceae, 78
Stillingia, 830
Sting nematode, 313
Stippen of apple, 335
Stizolobium, 855–856
Stock, 830
    bacterial blight, 119
    mosaic, 495
Stokes-aster, 830
Stokesia, 830
    powdery mildew, 352–354
Stonecrop, 810
Stone fruits, bacterial canker, 114, 118, 203
    brown rot of, 164
    Color Plate 8E
    Coryneum blight, 150
    rust of, 449–450
Stonemint, 625
Stone plant, 830
Storage rot of apples, 398
    of carrot, 369
    of sweetpotato, 393
Stranvaesia, 831
Strawberry, 831–832
    anthracnose, 89
    bacterial angular leaf spot, 118

bud nematode, 310–311
chlorosis, 496
crinkle, 495
downy mildew, 233
fruit spot, 260
green petal, 495
latent virus, 495
leaf blight, 150
leaf curl, 495
leaf roll, 495
leaf scorch, 245
leaf spot, 268, 278
mild crinkle, 495
mild yellow edge chlorosis,
   495
mottle, 496
multiplier disease, 496
necrotic shock, 496
red stele disease, 403
root rot, 398
severe crinkle, 496
spring dwarf nematode,
   310–311
stunt, 496
summer dwarf nematode,
   309–310
vein banding, 496
witches' broom, 496
yellow edge, 496
yellows, 496
Strawberry-tree, 832
Strawflower, 833
Streak, of pea, 490
   phlox, 492
   raspberry, 494
   rose, 495
   sweet pea, 107
Strelitzia, 556
Streptanthera, 833
   mosaic, 496
Streptocarpus, 833
Streptomyces, 418
   griseolus, 11
   griseus, 10–11
   ipomoea, 418
   scabies, 454–455
Streptomycin, 34
   nitrate, 41
   sulfate, 40
Streptopus, 833
Streptosolen jamesonii, 565
Streptotinia arisaemae, 145
Striga asiatica, 515
String-of-pearls, 468
Stripe rust, 448
Stripe smut, 467
Stromatinia, 418
   gladioli, 418–419
   narcissi, 419

Strumella coryneoidea, 224
Strumella canker of oak, 224
Stubby root nematode, 329
Stunt, chrysanthemum, 482
   Color Plate 3A, 3E
   dahlia, 485
   strawberry, 496
Stunt nematode, 330
Stylet nematode, 330
Stylosanthes, 833
   leaf spot, 260
Styrax, 833
Subdue 2E, 34
Sugar-beet nematode, 319
Sugarberry, 670
Sugar-cane stylet nematode,
   330
Sugar palm, 752
Sul-Cide, 34
Sulfacop, 34
Sulfur, 3, 34–35
Sulfur injury, 346
Sulkol, 35
Sultan, 689
Sultricop, 35
Sumac, 834
   canker, 199
Summer-cypress, 699
Summer dwarf nematode of
   strawberry, 310
Summer hyacinth, 651
Sunflower, 834–835
   bacterial leaf spot, 113
   basal gall, 242
   black knot, 126
   black patch, 126
   Botrytis on, 143
   canker, 204
   charcoal rot, 390
   head rot, 412
   powdery mildew, 352–354
   premature ripening, 171
   rust, 426, 446
   stem lesion and blight, 137
   stem rot, 415–416
Sunrose, 835
Sunscald, 247, 346, **347**
Sunstroke, 346
Super X Macclesfield, 35
Sup'R Flo, 35
Suzu, 35
Suzu H, 35
Swamp-bay, 794
Swamp-lily, 808
Swamp-privet, 647
Swan River daisy, 563
Swedish ivy, 833
Sweet acacia, 687
Sweet alyssum, 835

Sweet bay, 702
Sweet-fern, 836
   blister rust, 427
Sweet-flag, 836
Sweet gale, 836
Sweetgum, 836
   canker, 209–210, 219–220
   leaf spot, 267, 273
   root rot, 375
Sweetleaf, 837
   budgall, 240
Sweet-olive, 837
Sweet pea, 837–838
   anthracnose, 94
   Botrytis on, 142
   bud drop, 337
   leaf spot, 257
   root rot, 383
   white mold, 301
Sweet pepperbush, 602
Sweetpotato, 838–839
   black rot, 369–370
   canker, 207
   dry rot, 376
   feathery mottle, 496
   foot rot, 405
   internal cork, 496
   Java black rot, 377
   leaf blight, 173
   leaf spot, 291
   mosaic, 496
   russet crack, 496
   scurf, 458
   soft rot, 412
   soil rot, 418
   stem rot, 381
   storage rot, 393
Sweet-root, 839
Sweetshrub, 575
Sweet vetch, 839
Sweet william, 840
   Fusarium wilt, 507
Swietenia, 717
Swiss chard, 840
Switchgrass, 840
Sword bean, 694
Sycamore, 776–777
   anthracnose, 95
   canker stain, 195
   leaf spot, 295
Sydowia polyspora, 223
Syllit, 35
Symphoricarpos, 817
   occidentalis, 869
   orbiculatus, 606
Symplocarpus, 814
Symplocos, 837
Syncarpia, 853
Synchytriaceae, 63

Synchytrium, 243
  anemones, 243
  aureum, 243
  endobioticum, 243
  vaccinii, 243
Syngonium, 841
  black cane rot, 369–370
  leaf spot, 255
Synthyris, 841
Syringa, 706–707
Systremma acicola, 186

Tabebuia, 841
Tabernaemontana, 841
Tachigaren, 35
Taenidia, 841
Tagetes, 722–723
Tahitian bridal veil, 841
Taifen, 35
Tairel, 35
Talan, 35
Tamarind, 842
Tamarindus, 842
Tamarisk, 842
Tamarix, 842
Tanacetum, 842
Tanbark oak, 710
Tan leaf spot, 280
Tansy, 842
Taphrina, 244
  aesculi, 244
  amentorum, 245
  aurea, 244
  australis, 244
  bartholomaei, 244
  caerulescens, 244
  carnea, 244
  castanopsis, 244
  cerasi, 244
  communis, 244
  confusa, 244
  coryli, 244
  deformans, 244–245
  farlowi, 244
  faulliana, 245
  filicina, 245
  flava, 245
  flavorubra, 244
  flectans, 244
  japonica (macrophylla), 245
  occidentalis, 245
  populina, 245
  pruni, 244
  prunisubcordata, 244
  robinsoniana, 245
  rugosa, 245
  sacchari, 245
  struthiopteridis, 245
  thomasii, 244
  ulmi, 245

Taphrinaceae, 68
Taphrinales, 68
Taraxacum, 619
Target spot of soybean, 261
Tar spot, of fern and bracken,
    262
  of goldenrod, 286
  of grass, 283
  of holly, 282
  of maple, 288, 289
  of persimmon, 276
  of tulip tree, 266
  of willow, 289
Tarjan's sheath nematode, 318
Tasselflower, 633
Tassel-tree, 652
Taxodium, 546–547
Taxus, 871
TBCS-53, 35, 41
TBZ, 35
Tea, 842
Teaberry, 652
Tear gas, 17, 42
Teasel, 842
  mosaic, 497
Tecomaria, 578
Tecto, 35
Tecto RPH, 35
Tellima, 843
Telone C, 17, 35, 46
Temik, 46
Temik Brand, 46
Termil, 19, 35
Ternstroemia, 843
Terraclor, 35
Terraclor Super X, 35
Terra-Coat, 35
Terracur, 46
Terrazole, 35
Terr-O-Cide, 36, 46
Terr-O-Gas, 36, 46
Tersan, 36
Tersan 75, 36
Tersan 1991, 36
Tersan SP, 36
Tesselate stylet nematode, 330
Tetragonia, 739–740
Tetrapanax, 797
Tetrapom, 36
Tetylenchus joctus, 329
Teucrium, 655
Texas bluebell, 638
Texas mistletoe, 298
Texas root rot, 399–401
Texasweed, 843
Thalia, 843
Thalictrum, 724
  leaf gall, 243
Thames' root-knot nematode,
    323

Thea sinensis, 842
Thecaphora deformans, 462
Thelephora, 186
  spiculosa, 186
  terrestris, 186
Thelephoraceae, 76
Thermopsis, 843
Thiabendazole, 36
Thibenzole, 36
Thielaviopsis basicola, 419
Thimer, 36
Thio 95, 36
Thioknock, 36
Thiolux, 36
Thion 80, 36
Thioneb, 36
Thiophal, 36
Thiophanate, 36
Thiophanate-Methyl, 36
Thioquinox, 36
Thiotex, 36
Thiovit, 36
Thiram, 36–37
Thiram Tech, 37
Thirama D, 37
Thirasan, 37
Thistle, 843
  blessed, 844
  plumbed, 598
Thiuramin, 37
Thorne's meadow nematode,
    326
Thorne's needle nematode,
    320
Thoroughwax, 844
Thread blight, 150, 168
  silky, 180
Thrift, 537
Thuja, 536–537
Thujopsis, 844
Thunbergia, 844
Thylate, 37
Thyme, 844
Thymus, 844
Thynon, 37
Thyronectria, 224
  austro-americana, 224
  berolinensis, 224
Ti, 844
Tiarella, 647
Tiazin, 37
Tiazon, 46
Tibouchina, 658, 844
Tickseed, 606–607
Tidestrominia, 845
Tidy-tips, 702
Tiezene, 37
Tiger-flower, 845
Tigridia, 845
  mosaic, 497

Tilcarex, 37
Tilia, 708–709
Tillandsia, 821
Tilletia, 463, **465**
 buchloëana, 463
 caries, 463
 foetida, 463
 pallida, 463
Tilletiaceae, 75
Timber rot, 389
Tineston, 37
Tinnate, 37
Tip blight, crape myrtle, 173
 dracaena, 173
 fern, 95, 173
 fir, 179–180
Tipburn, 347
Tirampa, 37
Tirpate, 46
TMTDS, 37
TNCS 53, 37, 41
Tobacco, flowering, 740
Tobacco diseases
 blue mold (downy mildew),
  234
 broad ring spot, 417
 brown spot, 250
 cyst nematode, 319
 etch, 497
 mosaic, 497
 necrosis, 498
 rattle, 498
 ring spot, 498
 streak, 498
 wildfire, 115
Tobaz, 37
Tolmiea, 771
Tolyfluanid, 37
Tomato, 845–848
 anthracnose, 95
 aspermy, 498
 bacterial canker, 101–102
 bacterial speck, 115
 bacterial spot, 121
 big bud, 498
 blossom end rot, 336, **336**
 buckeye rot, 404
 cloudy spot, 394
 collar rot, 363
 early blight, 137–138
 enation mosaic, 498
 fernleaf, 498
 fruit rot, 362, 368, 376,
  405
 Fusarium wilt, 509
 Isaria rot, 388
 late blight, 175, 177–179,
  178
 leaf mold, 300
 leaf spot, 250, 292, 294

mosaic, 498
nailhead spot, 250–251
Phoma rot, 399
ring rot, 393
 Color Plate 4E
ring spot, 498
ripe rot, 394
rootlet necrosis, 404–405
root rot, 409
shoestring, 498
sour rot, 395
spotted wilt, 498–499
streak, 499
western yellow blight, 499
yellow net, 499
yellow top, 499
Tomato ringspot virus, 498
Tomato spotted wilt virus,
 498–499
Topas, 37
Topaz, 37
Topaze, 37
Topple of gladiolus, 347
Topsin E, 37
Topsin M, 37
Topsin Turf and Ornamentals,
 37
Topsin Wettable Powder, 37
Torch-lily, 699
Torenia, 848
Torilis, 674
 leaf gall, 242
Torreya, 848
Torula maculans, 301
Toxicity, aluminum, 335
 boron, 337
 gas, 339–340
 mercury, 342
 molybdenum, 342
Toyon, 771
TPTA, 37
TPTH, 37
TPTOH, 37
Trachelospermum, 606
Trachymene, 560–561
Tradescantia, 848
Tragopogon, 805
Trailing arbutus, 634
Trailing four-o'clock, 524
Trametan, 37
Trametes, 419
 pini, 379
 suaveolens, 419–420
 versicolor, 407–408
Transvaal daisy, 654–655
Tranzschelia, 449
 discolor, 449–450
 pruni-spinosae var. discolor,
  449–450
  var. typica, 450

Trautvetteria, 848
Trechispora alnicola, 238
Tree-huckleberry, 639
Treemallow, 702
Tree-of-heaven, 523
Tree-poppy, 848
Tree-tomato, 849
Tree wound dressings, 53, **54**
Trees, powdery mildew of,
 358
Trembling fungi, 75
Tremellales, 75
Triadimiefon, 37
Trianthemum, 686
Triarimol, 37
Triasyn, 37
Triazophos, 46
Tribasic copper sulfate, 37, 41
Tribulus, 788
Tricarbamix, 37
Tricarbasul, 37
Trichoderma, 420
 harzianum, 420
 viride, 420
Trichodorus, 329
 allius, 329
 christiei, 329
 obtusus, 329
 pachydermis, 329
 primitivus, 329
Trichometasphaeria turcica,
 160
Trichopelteae, 70
Trichosanthes, 663
Trichoscyphella hahniana, 203
 willkommii, 203, 211
Trichostema, 560
Trichothyriaceae, 70
Tri-Clor, 37
Tricon, 38
Tricop, 38
Tricotherium roseum, 225
Trientalis, 828
Trifocide, 38
Triforine, 38
Trifrina, 38
Trift, 537
Trifuncit, 38
Trifungol, 38
Trillium, 849
Trimangol, 38
Trimastan, 38
Trimaton, 38, 46
Tri-Miltox, 38
Triofterol, 38
Tri-ogen, 58
Trioneb, 38
Triosteum, 685–686
Tri-PCNB, 38
Triphagmium ulmariae, 450

Triphenyltin acetate, 38
Triphenyltin chloride, 38
Triphenyltin hydroxide, 38
Triphragmium ulmariae, 450
Triple Tin, 38
Tripomol, 38
Triquintam, 38
Triscabol, 38
Tristeza, citrus, 483
Tritisan, 38
Tritoftorol, 38
Tritoma, 699
Tritonia, 849
   mosaic, 499
Tri-VC, 46
Trizone, 28–29, 38, 46
Trollius, 849
Trombone sprayers, 54
Tropaeolum, 738
Trout-Lily, 636
Truban, 38
Trumpet-creeper, 849–850
Trumpet-tree, 841
Trumpetvine, 849–850
   leaf blight, 148
Trunk canker of apple,
      214–215, 221
Tryblidiaceae, 71
Tryblidiella rufula, 186
Tsitrex, 38
Tsuga, 675
Tuads, 38
Tubakia dryina, 249
Tuberaceae, 72
Tuberales, 72–73
Tuberculariaceae, 78
Tubercularia ulmea, 225
Tuberose, 850
Tubothane, 38
Tubotin, 38
Tuburcinia trientalis, 463
Tulbaghia, 818
Tulip, 850
   anthracnose, 93–94
   basal rot, 381
   Botrytis blight (fire),
      145–146, **146**
   breaking, 499–500
      Color Plate 1C, 1D
   gray bulb rot, 411–412
   sooty mold, 469
Tulipa, 850
Tulip-tree, 851
   leaf spot, 266, 268, 278
   Nectria canker, 215
Tung tree, 851–852
   black rot canker, 219
   leaf spot, 261
   thread blight, 168–169
Tupelo, 852
   leaf spot, 279, 286

Tupidanthus, 852
Turf, 665–667
   blight, 163, 165, 171
   brown patch, 396, 411
   copper spot, 268
   fairy ring, 237
   pink patch, 374
   snowmold, 468
Turfcide, 38
Turkeybeard, 870
Turnip, 852–853
   anthracnose, 89–90
   brown heart, 337
   mosaic, 500
   white spot, 255
Turnip mosaic virus, 500
Turpentine tree, 853
Turtle-head, 853
Tussilago, 605
Tuzet, 38
Twig blight, ailanthus, 157
   bladdernut, 151, 161
   blueberry, 154
   cherry, 164
   citrus, 158
   dogwood, 166, 151
   honeysuckle, 171
   kerria, 172
   maple, 171
   mulberry, 166
   pear, 172
   persimmon, 172
   sourwood, 185
   spruce, 171
   yew, 173
Twig canker, of apple, 199
   oak, 221
   peach, 208
   pine, 195
   willow, 199
Twin-flower, 627, 709
Twisted-stalk, 833
Tylenchorhynchus, 330
   capitatus, 330
   claytoni, 330
   dubius, 330
   martini, 330
   maximus, 330
Tylenchulus semipenetrans,
   330
Tylenchus, 330
Tympanis confusa, 225
Typha, 585
Typhula, 468
   idahoensis, 468
   itoana, 468

UC 19786, 38
UC 21149, 46
Udo, 854
Ulmus, 632–633

Umbellularia, 573
Umbrella-pine, 854
Umbrellawort, 854
Uncinula, 351, 361
   circinata, 361
   clintonii, 361
   flexuosa, 361
   macrospora, 361
   necator, 361
   parvula, 361
   polychaeta, 361
   prosopodis, 361
   salicis, 361
Unicorn-plant, 854
Unicrop Maneb, 38
Urbacid, 39
Uredinales, 74
Urediniomycetes, 74
Uredinopsis, 450
   macrosperma, 450
   osmundae, 450
   phegopteridis, 450
   pteridis, 450
   struthiopteridis, 450
Uredo, 450
   coccolobae, 450
   ericae, 450
   phoradendri, 450
   sapotae, 449
Urocystis, 463
   agropyri, 463
   anemones, 463
   carcinodes, 463
   cepulae, 463–464
   colchici, 463–464
   gladiolicola, 464
   hepaticae-trilobae, 463
   kmetiana, 464
   tritici, 464
Uromyces, 437, 450
   betae, 450–451
   caryophyllinus, 451
   ciceris-arietini, 451
   costaricensis, 451
   dianthi, 451
   fabae, 451
   galii-californici, 451
   phaseoli var. typica, 451
   trifolii, 451
Uropyxis eysenhardtiae, 451
Ustilaginaceae, 74
Ustilaginales, 74
Ustilago, 464, **465**
   avenae, 464
   bullata, 464
   heufleri, 464
   hordei, 464
   kolleri, 465
   maydis, 465–467, **466**
   mulfordiana, 467
   nigra, 467

nuda, 467
perennans, 464
striiformis, 467
violacea, 467
zeae, 465–467, **466**
Ustomycetes, 74
Ustulina vulgaris, 420
Uvularia, 854

Vaccinium, 559–560, 610–611, 866
arboreum, 639
Valerian, 855
red, 855
Valeriana, 855
Valerianella, 855
Validacin, 39
Validamycin, 39
Vallota, 808
Valsa, 225
cincta, 203, 225
kunzei, 225
var. kunzei, 203
var. piceae, 202
var. superficialis, 203
leucosomoides, 420
leucostoma, 225
salicina, 225
sordida, 202, 225
Valsaceae, 71
Vancide-TM Flowable, 39
Vancide-TM-95, 39
Vancouveria, 855
Vanilla, 855
Vanilla-leaf, 855
Vapam, 28, 39, 46
Variegation, 347
infectious, of abutilon, 475
of camellia, 479
VC-13 Nemacide, 46
Velvet bean, 855–856
Velvet leaf, 519
leaf spot, 260
Vencedor, 39
Venturia cerasi, 452, 455
chlorospora, 453–454
inaequalis, 455–457, **456, 457**
populina, 156
pyrina, 457–458
tremulae, 156
Venturol, 39
Venus looking-glass, 821
Veratrum, 856
Verbascum, 736
Verbena, 856
downy mildew, 233
garden, 856
Verbena hortensis, 856
Verbesina, 857
Vermicularia ipomoearum, 226

Vernonia, 692
Veronica, 857
downy mildew, 233
Veronicastrum, 614
Verticicladiella, 421
abietina, 421
penicillata, 421
procera, 421
wagenerii, 421
Verticillium, 511
albo-atrum, 511–515
dahliae, 515
fungicola, 515
Verticillium wilt, 511–515, **512, 513, 514**
Vetch, 857
Vetch, sweet, 839
Viburnum, 702, 857–858
bacterial leaf spot, 116
downy mildew, 235
leaf blight, 164
leaf spot, 261
Vi-Cad, 39
Vicia, 857
Viddin D, 46
Vigna, 552
Vigna sesquipedalis, 549–550
sinensis, 609
Viguiera, 660
Vinca, 858–859
canker, 218
Vincetoxicum, 859
Vinclozolin, 39
Viola odorata, 859–860
tricolor, 753
Viola species, Botrytis on, 143
Violet, 859–860
Alternaria blight, 138
anthracnose, 92
root rot, 386–387, 411
scab, 475
Vipers-bugloss, 860
Virginia cowslip, 727
Virginia creeper, 860
leaf spot, 270
spot anthracnose, 472
Virgins-bower, 601–602
Virus diseases, 475–500
Viscum album, 298
Vitavax, 39
Vitex, 860
Vitis, 663–665
Volutella, **136,** 187
buxi, 187, 221, 226
pachysandrae, 187
Volutella blight, 226
Vondcaptan, 39
Vondodine, 39
Vondozeb Plus, 39
Vorlan, 39

Vorlex, 46
Voronlit, 39
VPM, 39, 46

Wake-robin, 849
Waldsteinia, 549
Wallflower, 860
leaf spot, 252
Wallflower, western, 861
Walnut, 861–862
anthracnose, 95–96
blight, 119, 153
branch wilt, 510
brooming disease, 500
canker, 207, 212–213
dieback, 205
leaf spot, 278
meadow nematode, 327
ring spot, 253
Wampi, 601
Wandering jew, 848
Wandflower, 821
Wart, peach, 491
potato, 243
Washingtonia, 752
Washington palm, 752
Water-cress, 862
mottle, 500
Water deficiency, 347
Water dragon, 808
Water elm, 863
Water-horehound, 863
Water-hyacinths, 863
leaf spot, 257
Rhizoctonia blight, 180
Waterleaf, 688
Water-lily, 863
leaf spot, 263
Watermelon, 863–864
anthracnose, 90
fly speck, 238
Fusarium wilt, 509
leaf spot, 165, 291
mosaic, 500
Watermelon mosaic virus, 500
Water-primrose, 864–865
Water-shield, 865
leaf spot, 263
Watery fruit rot of tomato, 395
Watsonia, 865
Waxflower, 868
Wax-myrtle, 865
black mildew, 129
Web blight, 150, 168, 180
Weed-killer injury, 347–349, **348**
Weigela, 865
West African spiral nematode, 328
Western aster yellows, 121

Western gall rust, 427, 428, 430
Western osier, 626
Western soapberry, 818
Western X little cherry, 122–123
Wetwood, of elm, 107–108
of poplar, 101
Wheat, dwarf bunt of, 463
flag smut, 463
nematode, 309
rust, 444–446
snowmold, 468
stinking smut, 463
streak mosaic, 500
stripe, 448
Whetzellinia. See Sclerotinia
Whipplea, 865
White-alder, 602
White butt rot, 374
White-cedar, 588
White heart rot, 387, 420
White line mosaic, 500
White mold of narcissus, 287–288
sweet pea, 301
White mottled rot, 375, 379, 383–384, 406
White pine blister rust, 428–430
White pocket rot, 406
White root rot, 374, 406, 413–414
White rot, of aspen, 374
cherry, 408
grapes, 373
onion, 416
White rusts, 500–502
crucifers, 501
portulaca, 501
salsify, 501
spinach, 501
White sapwood rot, 372, 405, 406
White smut, calendula, 460
White snakeroot, 638
White spongy rot, 378, 379, 406, 417
White spot of crucifers, 255
White-topped aster, 811
White wood rot, 419–420
Whitlow-grass, 628
Whortleberry, 866
Wildfire, tobacco, 115
Wild ginger, 656
Wild rice, 866
Wildrye, 866
Wild tuberose, 719
Willow, 866–867
black canker, 219
canker, 203, 219–220

crown gall, 98–100
gray scab, 474
leaf spot, 263
mistletoe, 298
powdery mildew, 361
scab, 453–454
tar spot, 289
twig and branch canker, 199, 205, 225
Willow-herb, 634–635
Willow, pussy, 866–867
Willow, weeping, 866–867
Wilts, 502–515
aster, 507
bacterial
of carnation, 109–110
of corn, 108
of cucumber, 113
of lettuce, 121
cattleya orchids, 508
celery, 507
cucumber, 508
Dothiorella, of elm, 506–507
Dutch elm disease, 503–506, **505**
Fusarium, of asparagus, 507
carnation, 508
cattleya orchid, 508
chili pepper, 507
chrysanthemum, 508
cyclamen, 508
hebe, 509
muskmelon, 509
spinach, 510
sweet william, 507
tomato, 509
Granville, 112
maple, 511–515
mimosa, 509–510
oak, 502
pea, 510
persimmon, 502
radish, 510
rhododendron, 511
southern, 112
sunflower, 502
Verticillium, 511–515, **512, 513, 514**
walnut branch, 510
watermelon, 509
Wine-flower, 822
Winged bean, 867
Winterberry, 868
Winter cress, 868
black rot, 116
Wintergreen, 868
Winter injury, 349
Winter's peach mosaic, 491
Wire-lettuce, 829
Wistaria, 868–869

Wisteria, 868–869
leaf spot, 285
mosaic, 500
Witches' broom, 124
cherry, 244
holodiscus, 486
lilac, 124, 487
locust, 488
persimmon, 124
pigeon pea, 124
potato, 493
rhododendron, 240–241
serviceberry, 130
statice, 124
strawberry, 496
Witches' broom rust, 433, 436
Witch-hazel, 869
leaf spot, 269
Witchweed, 515
Withertip, 92, 94
Witloof chicory, 633–634
Wolfberry, 869
Wood-betony, 869
Woodland-star, 710
Wood-nymph, 731
Wood rot, 375, 417
Woodrush, 869
Woodsia, 642
Wood-sorrel, 748–749
Woodwardia, 643
Wood-waxen, 653
Wormwood, 538
Wound dressings, 53
Wound rot, 414
Woundwort, 827
Wyethia, 870

Xanthium, 603
Xanthomonas, 116
barbareae, 116
begoniae, 117
campestris, 117–118
carotae, 118
citri, 118
corylina, 118
cucurbitae, 118
dieffenbachiae, 118
fragariae, 118
geranii, 118
glycines, 118
gummisudans, 119
hederae, 119
hyacinthi, 119
incanae, 119
juglandis, 119–120
nigromaculans f. sp. zinniae, 120
oryzae, 120
papavericola, 120
pelargonii, 118, 120
phaseoli, 120

phaseoli var. sojense, 118
pruni, 120–121
stewartii, 108
vesicatoria, 121
vesicatoria var. raphani,
121
vignicola, 121
vitians, 121
Xanthorhiza, 871
Xanthosma, 870
Xanthosoma, 604
Xanthoxylum americanum,
786
clava-herculis, 677
X-disease of peach, 123
Xerophyllum, 870
Xiphinema, 331
americanum, 331
bakeri, 331
chambersi, 331
diversicaudatum, 331
index, 331
radicola, 331
Xylaria, 421
hypoxylon, 421
mali, 421
polymorpha, 421
Xylariaceae, 71
Xylella, 124
Xyloporosis, citrus, 483

Yaltox, 46
Yam, 870
nematode, 329
Yarrow, 870
Yaupon, 870
Yeast spot, of soybean, 394
of lima bean, 394
Yellow-cedar, 588
Yellow cuprocide, 39
Yellow dwarf of onion, 489
potato, 493
spinach, 495

Yellow flaky heart rot, 378–379
Yellow gum disease, 101
Yellow ironweed, 520
Yellow pondlily, 863
Yellow poplar, 851
Yellow-root, 871
Yellow rust of blackberry,
436
Yellow witches' broom rust,
438
Yellows, 349
aster, 121
bayberry, 477
cabbage, 508
carnation, 480, 508
celery, 121–122
cherry, 482
gladiolus, 381–382
hyacinth, 119
peach, 123
strawberry, 496
Yellowtuft, 526
Yellowwood, 871
Yerba buena, 871
Yerba santa, 871
Yew, 871
needle blight, 185
twig blight, 173
Yucca, 871–872
leaf blight, 162
leaf mold, 301
leaf spot, 260–261, 266,
280

Zamia, 872
Zantedeschia, 574–575
Zanthoxylum clava-herculis,
677
Zauschneria, 872
ZC Spray, 39
Zea mays var. saccharata,
607–608
Zebra plant, 530, 872

Zebtox, 39
Zephyranthes, 872
Zephyr-lily, 872
Zerlate, 39
Zidan, 39
Zigadenus, 872
Ziman-Dithane, 39
Zimmerman's spiral nematode,
317
Zinc deficiency, 349
Zincmate, 39
Zinc Metiram, 39
Zinc sulfate, 39, 41
Zineb, 39–40
Zineb 75%, 40
Zineb 75WP, 40
Zinnia, 873
blight, 138
Botrytis on, 143
leaf spot, 120
mildew, 352–354
Zinosan, 40
Ziram, 40
Zirbeck, 40
Ziride, 40
Zitox, 40
Zizania, 866
Zizia, 873
Zizyphus, 696
Zonate leaf spot, of grass, 270
of cowpea, 251
Zoysia, 873
spine nematode, 314
Zucchini yellow mosaic, 500
Zygocactus truncatus, 596
Zygomycetes, 55–66, 66
Zygophiala jamaicensis, 191,
238